Reservoir Characteristics and Evolution Mechanisms of the Shale

Reservoir Characteristics and Evolution Mechanisms of the Shale

Topic Editors

Ruyue Wang
Mengdi Sun
Shang Xu
Songtao Wu
Jianhua Zhao
Yiquan Ma
Jianhua He

MDPI

Basel • Beijing • Wuhan • Barcelona • Belgrade • Novi Sad • Cluj • Manchester

Topic Editors
Ruyue Wang
SINOPEC Petroleum Exploration and Production Research Institute
Beijing
China

Mengdi Sun
Northeast Petroleum University
Daqing
China

Shang Xu
China University of Petroleum
Qingdao
China

Songtao Wu
Research Institute of Petroleum Exploration and Development
Beijing
China

Jianhua Zhao
China University of Petroleum (East China)
Qingdao
China

Yiquan Ma
Chengdu University of Technology
Chengdu
China

Jianhua He
Chengdu University of Technology
Chengdu
China

Editorial Office
MDPI AG
Grosspeteranlage 5
4052 Basel, Switzerland

This is a reprint of the Topic, published open access by the journals *Energies* (ISSN 1996-1073), *Eng* (ISSN 2673-4117), *Journal of Marine Science and Engineering* (ISSN 2077-1312), *Minerals* (ISSN 2075-163X) and *Processes* (ISSN 2227-9717), freely accessible at: https://www.mdpi.com/topics/O3AQT050Z0.

For citation purposes, cite each article independently as indicated on the article page online and as indicated below:

Lastname, Firstname, Firstname Lastname, and Firstname Lastname. Article Title. *Journal Name* **Year**, *Volume Number*, Page Range.

ISBN 978-3-7258-7831-4 (Hbk)
ISBN 978-3-7258-7832-1 (PDF)
https://doi.org/10.3390/books978-3-7258-7832-1

Contents

About the Editors

Ruyue Wang

Ruyue Wang is a Senior Engineer and Min Enze Young Scientific and Technological Talent of Sinopec, and was named among the World's Top 2% Scientists in 2025. He serves as Deputy Secretary-General of the Fluid Geosciences Committee of the Chinese Geophysical Society, a member of the Expert Committee of the Beijing Energy and Environment Society, and adjunct professor and master's supervisor at multiple universities. He is Associate Editor of *Unconventional Resources* and the *Journal of Geo-Energy and Environment*, and a Young Editorial Board Member for several key geoscience and petroleum journals. He earned his PhD in Energy Geology and Engineering from the China University of Geosciences (Beijing) in 2017. He has led or taken part in over 20 national, provincial and ministerial projects, won four provincial and ministerial Science and Technology Progress Awards, seven provincial and ministerial honors, and over 50 bureau-level awards. Since 2012, he has focused on unconventional oil and gas geology, shale oil and gas reservoir geology, and exploration planning. He has published more than 130 academic papers including over 50 SCI/EI papers and co-authored four monographs, with total citations exceeding 6,000. He has 40 first or corresponding author papers, 14 of which are ESI hot/highly cited papers or national top academic papers, with an H-index of 43 and i-10 index of 132 (Google Scholar).

Mengdi Sun

Mengdi Sun (born in April 1990) obtained his PhD in China from China University of Geosciences (Beijing) in 2017. He was funded by the China Scholarship Council and had been trained at the University of Texas for 18 months during his PhD study. Prior to joining Northeast Petroleum University, he was an associate professor at China University of Geoscience (Wuhan) from 2017 to 2021. He was a visiting scholar at CSIRO, Australia in 2019. He is currently a professor and doctoral supervisor at Northeast Petroleum University. In recent years, Dr. Sun has been mainly engaged in the research work on shale oil and gas reservoirs and has published more than 90 academic papers in this area, including 34 high-impact SCI papers as the first/corresponding author in top international refereed journals such as *Gas Science and Engineering*, the *International Journal of Coal Geology*, *Fuel*, and *Marine and Petroleum Geology*, with more than 2700 citations. Dr. Mengdi Sun's research activity has been closely focused on the frontier and hot issues of shale oil and gas. He has actively participated in and organized academic conferences. He was granted the Outstanding Youth Paper award at the 2019 academic annual conference of the China Geological Society and was invited to give an oral report at the 12th International Small Angle Scattering Conference. In 2023, he was celebrated as the outstanding convener of the Sixth National Youth Geological Conference of China. He is a member of the second Youth Working Committee of China Petroleum Society, and a Youth editorial member of *Advances in Geo-Energy Research*, the *Journal of Northeast Petroleum University*, *Natural Gas Industry*, and the *Journal of Geoenergy*. Dr. Sun has presided over two projects of the National Natural Science Foundation of China, one outstanding Youth Fund of Heilongjiang Natural Science Foundation, and participated in two key projects of the National Natural Science Foundation of China.

Shang Xu

Shang Xu is a professor of the School of Geosciences at UPC. His research interests are mainly about oil and gas accumulation mechanism. In the past five years, he has focused on the differential enrichment mechanism of marine shale gas and continental shale oil in China. He has led and participated in more than 20 research initiatives, including the National Natural Science Foundation of China (NSFC), the National Major Science and Technology Programs, the National Key Research and Development (R&D) Programs, provincial-/ministerial-level key projects, and university–industry collaborations. He is the leader of the Youth Innovation Team under the Colleges and Universities of Shandong Province of China. His accolades include the Excellent Young Scholars Fund (2021) from the National Natural Science Foundation of China (NSFC) and the Powering the Nation Youth Scientist Nomination Award (2022). He has published 108 research papers, including 83 as first author and corresponding author (60 SCI-indexed). He has been invited to guest-edit five Special Issues on unconventional oil and gas for SCI-indexed journals, and co-authored three academic monographs, Special Issues, and textbooks. He has been granted 18 national invention patents and eight software copyrights, with participation in formulating two national energy industry standards. He has served on the editorial boards of several important journals in the geology and energy field. He is an Associate Editor for *AAPG Bulletin, Applied Geochemistry* and *Frontiers in Earth Science.* In addition, he is a member of the Youth Working Committee of the Chinese Petroleum Society and a Member of the Gas Geochemistry Committee, Chinese Society for Mineralogy, Petrology and Geochemistry.

Songtao Wu

Songtao Wu, Ph.D., Professorate Senior Engineer, Deputy Director of the Science and Technology Department, PetroChina Research Institute of Petroleum Exploration & Development (RIPED). He is primarily engaged in research on sedimentology and unconventional oil and gas. He serves as a member of the National Technical Committee on Petroleum Geological Exploration Standardization, and is Chairman of the Subcommittee on Petroleum Geology Standardization. He is Associate Editor of *Unconventional Resources* and a youth editorial board member of journals such as *Petroleum Science, AGER, Acta Petrolei Sinica,* and *Acta Sedimentologica Sinica.* He has undertaken 12 national and provincial-/ministerial-level scientific research projects, and has published 57 papers as first/corresponding author (44 indexed by SCI, 7 by EI). From 2023 to 2025, he was consecutively selected for Elsevier's "World's Top 2% Scientists" list. He has co-authored eight monographs, developed five national standards and six industry standards, and been granted 22 invention patents (US and China). He has received 12 provincial-/ministerial-level science and technology awards, including the First Prize of Science and Technology Project of Qinghai Province, the First Prize of Basic Research of CNPC. He also won the First Prize for Outstanding Technical Report at the World Petroleum Congress (WPC) and the International Young Scientist Award at the IGU GIF Awards. He has been recognized as one of CNPC's "Top Ten Outstanding Youth" and "Outstanding Youth Innovation Talent".

Jianhua Zhao

Jianhua Zhao is an Associate Professor and Master's Supervisor at the China University of Petroleum (East China), primarily engaged in research on unconventional oil and gas geology and sedimentary basin analysis. He has presided over multiple scientific research projects, including those funded by the National Natural Science Foundation of China (NSFC) and sub-projects of the National Science and Technology Major Projects. He has published over 50 academic papers, including more than 20 SCI-indexed papers, and holds an h-index of 32. Two of his publications were selected as Frontrunner 5000 (F5000) Top Articles in Outstanding S&T Journals of China, and he was recognized as a Top 1% Highly Cited Scholar by CNKI in 2024 and 2025. His research achievements have earned him a Second Prize of the Shandong Provincial Natural Science Award and a First Prize for Scientific and Technological Progress from the Sinopec Petroleum Exploration and Production Research Institute. Additionally, he serves as a youth editorial board member for academic journals such as *Petroleum Science*, *Acta Petrolei Sinica*, and *Natural Gas Geoscience*.

Yiquan Ma

Yiquan Ma is a professor and doctoral/master's supervisor at Chengdu University of Technology. He is also a researcher at both the State Key Laboratory of Oil and Gas Reservoir Geology and Exploitation, and the Key Laboratory of Deep-time Geography and Environment Reconstruction and Applications of Ministry of Natural Resources. He earned his PhD in Mineral Resource Prospecting and Exploration from the China University of Geosciences (Wuhan) in 2017. From 2017 to 2020, he undertook postdoctoral research at the same university. He has been a faculty member at Chengdu University of Technology since July 2020. His research and teaching focus on sedimentology, unconventional petroleum geology, sedimentary geochemistry, cyclostratigraphy, and astrochronology. He has led or participated in nearly 20 national and provincial-/ministerial-level research projects. His research findings have been published as the first or corresponding author in leading international journals such as *Geology*, *Earth and Planetary Science Letters*, *Earth-Science Reviews*, *Geological Society of America Bulletin*, *Global and Planetary Change*, and the *International Journal of Coal Geology*. He has been granted twelve Chinese and international invention patents, co-authored two monographs and received three first-prize ministerial- and provincial-level awards. He also serves as a youth editorial board member for the SCI-indexed journals *Petroleum Science* and the *Journal of Earth Science*.

Jianhua He

Jianhua He received his Ph.D. from China University of Geosciences (Beijing), and was jointly trained at the University of Queensland during his doctoral studies. He is currently an Associate Professor at the College of Energy, Chengdu University of Technology, and has been selected for the "Dongjiao Huazhang" Talent Program of Chengdu, Sichuan Province. His research focuses on reservoir geomechanics, the formation and evolution mechanisms and distribution prediction of fractures in unconventional reservoirs, as well as theories and evaluation methods of perturbed in situ stress fields. He has led more than ten national- and provincial-level research projects. Dr. He has published over 50 papers as first or corresponding author, including 20 SCI-indexed papers, one Nature Index (NI) paper, and eight EI-indexed papers. He has an H-index of 25, with more than 2,000 citations. He holds 13 authorized invention patents, including two U.S. patents, and five software copyrights. His awards include the Second Prize of the Science and Technology Progress Award from the Ministry of Natural Resources, the First Prize of a national industry association, and the Gold Medal at the International Exhibition of Inventions of Geneva.He is a fixed researcher at the State Key Laboratory of Oil and Gas Reservoir Geology and Exploitation, a member of the Geological Exploration Committee of the Sichuan Petroleum Society, and a committee member of the Petroleum and Natural Gas Expert Group under the Beijing Energy and Environment Society. He also serves as a young editorial board member for several journals, including *Petroleum Science*, the *International Journal of Coal Science & Technology* (*IJCST*), the *Journal of China University of Mining & Technology*, and *Petroleum Exploration and Experiment*.

minerals

MDPI

Editorial

Editorial: Reservoir Characteristics and Evolution Mechanisms of the Shale

Ruyue Wang [1,2,*], Jianhua He [3], Mengdi Sun [4] and Jianhua Zhao [5]

1 Sinopec Petroleum Exploration and Production Research Institute, Beijing 102206, China
2 State Key Laboratory of Shale Oil and Gas Enrichment Mechanisms and Efficient Development, Beijing 102206, China
3 College of Energy (College of Modern Shale Gas Industry), Chengdu University of Technology, Chengdu 610059, China; hejianhua19@cdut.edu.cn
4 Institute of Unconventional Oil & Gas, Northeast Petroleum University, Daqing 163318, China; sunmd@nepu.edu.cn
5 School of Geosciences, China University of Petroleum (East China), Qingdao 266580, China; zhaojh@upc.edu.cn
* Correspondence: wangruyue.syky@sinopec.com

Shale reservoirs have emerged as a pivotal pillar of global unconventional hydrocarbon resources, driving a paradigm shift in the energy industry over the past two decades [1–3]. The widespread application of horizontal drilling and hydraulic stimulation technologies has unlocked substantial oil and gas production from shale formations with diverse depositional backgrounds, particularly in North America and China [3,4]. While significant progress has been achieved in establishing advanced evaluation theories and development techniques through extensive laboratory studies and field practices, shale resource development still faces prominent challenges such as low single-well productivity and rapid production decline [3,5,6]. These issues are inherently linked to unresolved scientific questions, including quantitative reservoir characterization, hydrocarbon occurrence mechanisms, reservoir evolution dynamics, and optimized evaluation methodologies [3,5–7].

Practical exploration and development experiences have demonstrated that lithofacies, burial depth, formation pressure, and other key factors exert significant control over the formation and evolution of shale reservoirs, the types and characteristics of reservoir spaces, hydrocarbon occurrence patterns, and ultimate productivity [1,3,7–10]. To address these critical issues and synthesize the latest research advances, this Special Issue titled "Reservoir Characteristics and Evolution Mechanisms of Shale" was launched. It aims to showcase cutting-edge progress in theories and methodologies related to shale reservoir characteristics and evolution mechanisms, thereby deepening our fundamental understanding and providing theoretical support for achieving high single-well productivity and low-cost sustainable hydrocarbon development.

The response from the scientific community has been strong and insightful. This Special Issue compiles 21 original research articles covering a wide range of geographical regions and geological settings, including continental shales, marine shales, and marine-continental transitional shales in major petroliferous basins across China. The contributed studies employ a comprehensive array of methodological approaches, integrating mineralogy, petrology, geochemistry, geophysics, and reservoir engineering, to address critical scientific questions in shale reservoir research.

The published works can be thematically categorized into three research directions:

 https://doi.org/10.3390/min16010105

1. Reservoir Characteristics and Quantitative Characterization

Studies in this category focus on the detailed characterization of shale reservoir properties, including mineral composition, refined lithofacies classification, pore-fracture system characteristics, and hydrocarbon storage capacity (Contributions 1–6). Based on comprehensive analyses of TOC content, sedimentary structures, and mineral compositions, targeted lithofacies schemes are established (Contributions 1–3). The important controlling role of lithofacies on reservoir quality has been clarified. A quantitative evaluation of pore systems and natural fracture networks is conducted via an integrated multi-technique approach. Various technologies, including Low-Temperature Nitrogen Adsorption (LTNA), High-Pressure Mercury Injection (HPMI), Nuclear Magnetic Resonance (NMR), Large-Field Splicing Scanning Electron Microscopy (LFS-SEM), Modular Automated Processing System (MAPS), Quantitative Evaluation of Minerals by Scanning Electron Microscopy (QEMSCAN), and micro-CT, are employed to provide a comprehensive basis for reservoir quality assessment (Contributions 1–2,4,6).

2. Reservoir Evolution Mechanisms and Controlling Factors

This thematic area explores the dynamic evolution processes of shale reservoirs and their controlling factors, emphasizing the coupled effects of sedimentary environment (Contributions 7–12), diagenetic processes (Contributions 13–15), and tectonic activities (Contributions 16–17). Key findings indicate that depositional environments (e.g., paleoclimate, paleoredox conditions, paleoproductivity, and sedimentary facies) determine the initial material basis of reservoirs: marine shales are dominated by ferruginous to locally euxinic conditions (Contributions 7–10), with biogenic silica as the main material source (Contributions 7,9); continental shales form in semi-arid brackish lakes, where cyanobacterial blooms enhance organic matter accumulation (Contribution 11). Diagenetic processes significantly modify reservoir spaces: compaction reduces primary pores, while cementation (authigenic quartz, ferroan dolomite) and dissolution (feldspar, carbonate) reshape pore-fracture systems (Contributions 12–13); the formation of secondary minerals during the diagenesis–hydrocarbon generation process affects the reservoir quality and brittleness (Contribution 13); organic matter thermal evolution drives acid fluid release, further promoting dissolution of inorganic minerals (Contributions 13–15). Tectonic activities control the development and distribution of natural fractures: tectonic stress intensity, stratum thickness, and brittleness control fracture development (Contributions 16–17); abnormal pressure and volcanic-hydrothermal activities facilitate fracture propagation and modify diagenetic fluid properties (Contributions 10,16–17).

3. Reservoir Evaluation Method and Evaluation Optimization

This category presents innovative evaluation methodologies and optimization for different shale reservoirs (Contributions 18–21). Researchers proposed brittleness models considering mineral composition (e.g., carbonate), microfractures, and rock mechanics, supporting hydraulic fracturing optimization (Contribution 18). Integrated systems combine lithofacies, rock physical properties, geochemical, and engineering parameters, with intelligent methods like the Stacking model aiding favorable exploration and development zone identification (Contributions 19–20).

The collective contributions to this Special Issue represent significant advancements in our understanding of shale reservoir systems, highlighting three key research trends: (1) integration of multi-scale and multi-disciplinary approaches, bridging nanoscale pore dynamics to basin-scale tectonic evolution; (2) shift from static characterization to dynamic process-oriented research, emphasizing the temporal evolution of reservoirs and their response to geological processes; (3) strengthened linkage between basic research

https://doi.org/10.3390/min16010105

and practical applications, focusing on solving engineering challenges in exploration and development.

Looking forward, several frontier directions warrant further attention: (1) refined characterization of reservoir heterogeneity at multiple scales to improve the accuracy of reservoir quality prediction; (2) in-depth investigation of the coupling mechanisms between reservoir evolution and hydrocarbon generation-migration-accumulation, especially in deep to ultra-deep formations and areas with complex structural preservation conditions, to clarify the enrichment regularity of shale oil and gas; (3) development of intelligent evaluation technologies integrating big data and machine learning to enhance the efficiency of reservoir evaluation; (4) systematic research on the long-term stability of reservoirs under exploitation conditions to support sustainable development.

We sincerely thank all authors for their high-quality contributions, which have laid the foundation for this Special Issue. We also express our gratitude to the reviewers for their rigorous and constructive comments, which have significantly improved the scientific quality of the published works. Additionally, we appreciate the professional support and efficient organization provided by the editorial team of *Minerals*. We hope that this Special Issue will serve as a valuable academic platform for researchers worldwide, promoting academic exchanges and innovation in the field of shale reservoir research and contributing to the efficient and sustainable development of global unconventional hydrocarbon resources.

Funding: This work was funded by the National Oil & Gas Major Project of China (No. 2025ZD1404102-04), the National Natural Science Foundation of China (No. 42502135), and the Sinopec Ministry of Science and Technology Project (Nos. P25180, P24181, KLP24017).

Acknowledgments: As Topic Editors, we wish to express our sincere gratitude to all the authors who contributed their high-quality research to this Special Issue. We deeply appreciate the time, expertise, and rigorous efforts of all the reviewers, whose constructive comments and critical evaluations were instrumental in ensuring the scientific quality and integrity of the published work. We are also thankful to the editorial team of Minerals for their unwavering support, professional guidance, and efficient management throughout the process of organizing and publishing this Special Issue.

Conflicts of Interest: The authors declare no conflicts of interest.

List of Contributions:

1. Wang, Y.; Deng, H.; Wang, Z.; Wang, X.; Cao, Q.; Cheng, D.; Zhu, Y.; Li, A. Characteristics and Factors Influencing Pore Structure in Shale Oil Reservoirs of Different Lithologies in the Jurassic Lianggaoshan Formation of the Yingshan Gas Field in Central Sichuan Basin. *Minerals* **2023**, *13*, 958.
2. Wang, X.; Sun, Y.; Wang, T.; Yan, B.; Liu, R. Analysis of Controlling Factors of Pore Structure in Different Lithofacies Types of Continental Shale—Taking the Daqingzi Area in the Southern Songliao Basin as an Example. *Minerals* **2024**, *14*, 1025.
3. Liu, Y.; Li, W.; Yuan, J.; Li, P.; Ge, X.; Ge, X.; Liu, P.; Wu, H.; Yu, X.; Huang, B. Lithofacies Characteristics of the Lower Cretaceous Qing 1 Member in the Heiyupao Depression, Northern Binbei Area of the Songliao Basin. *Minerals* **2025**, *15*, 125.
4. Huang, X.; Li, Z.; Han, X.; Wang, Y.; Guo, Y. Lithofacies-Controlled Pore Characteristics and Mechanisms in Continental Shales: A Case Study from the Qingshankou Formation, Songliao Basin. *Minerals* **2025**, *15*, 1239.
5. Fu, X.; Gong, L.; Su, X.; Liu, B.; Gao, S.; Yang, J.; Qin, X. Characteristics and Controlling Factors of Natural Fractures in Continental Tight-Oil Shale Reservoir. *Minerals* **2022**, *12*, 1616.
6. Wang, Y.; Jia, Y.; Wang, C.; Lin, W.; Zhang, J.; Han, D.; Ma, B.; Wang, H. Characteristics and Geological Significance of Organic Matter Veins in Shale Reservoir: A Case Study of the Silurian Longmaxi Formation in Luzhou Area, Sichuan Basin. *Minerals* **2023**, *13*, 1080.

https://doi.org/10.3390/min16010105

7. Xu, L.; Huang, S.; Sun, M.; Wen, Y.; Chen, W.; Zhang, Y.; Luo, F.; Zhang, H. Palaeoenvironmental Evolution Based on Elemental Geochemistry of the Wufeng-Longmaxi Shales in Western Hubei, Middle Yangtze, China. *Minerals* **2023**, *13*, 502.
8. Chen, X.; Xu, Q.; Qin, Y.; Chen, R.; Wang, Y.; Shi, D.; Bai, J.; Yuan, K.; Cen, W.; Li, F.; et al. Effect of Paleoenvironmental Conditions on the Distribution of Lower Carboniferous Shale in Yaziluo Rift Trough, South China: Insights from Major/Trace Elements and Shale Composition. *Minerals* **2024**, *14*, 659.
9. Ge, X.; Ge, X.; Chen, D.; Liu, Y.; Wang, R.; Li, M. The Source and Significance of Silicon in the Late Permian Dalong Formation, Northeastern Sichuan Basin. *Minerals* **2025**, *15*, 69.
10. Cheng, W.; Wang, R.; He, T.; Sun, C.; Tian, H.; Zhao, J.; Zhao, Y.; He, J.; Zeng, Q.; Liu, J.; et al. Geochemical Evidence of Organic Matter Enrichment and Depositional Dynamics in the Lower Cambrian Yurtus Formation, NW Tarim Basin: Insights into Hydrothermal Influence and Paleoproductivity Mechanisms. *Minerals* **2025**, *15*, 288.
11. Wang, W.; Cui, H.; Tan, J.; Liu, J.; Song, X.; Wang, J.; Chen, L. Permian Cyanobacterial Blooms Resulted in Enrichment of Organic Matter in the Lucaogou Formation in the Junggar Basin, NW China. *Minerals* **2023**, *13*, 537.
12. Zhang, B.; Wen, S.; Yang, K.; Ma, K.; Wang, P.; Xu, C.; Cao, G. Diagenetic Evolution Sequence and Pore Evolution Characteristics: Study on Marine-Continental Transitional Facies Shale in Southeastern Sichuan Basin. *Minerals* **2023**, *13*, 1451.
13. Yan, J.; Chen, S.; Yu, Z.; Zhang, P.; Feng, G. Origin and Reservoir Significance of Authigenic Minerals in Lacustrine Shales: A Case Study from the Paleogene Dongying Sag, Bohai Bay Basin, East China. *Minerals* **2025**, *15*, 493.
14. Wu, W.; Liang, Z.; Xu, L.; Liu, Y.; Li, Y.; Tang, X.; Yin, Y.; Chen, Y. The Effect of Thermal Maturity on the Pore Structure Heterogeneity of Xiamaling Shale by Multifractal Analysis Theory: A Case from Pyrolysis Simulation Experiments. *Minerals* **2023**, *13*, 1340.
15. Wang, Z.; Yang, X.; Guo, S. Evolution of Pore Spaces in Marine Organic-Rich Shale: Insights from Multi-Scale Analysis of a Permian–Pennsylvanian Sample. *Minerals* **2024**, *14*, 392.
16. Bai, X.; Wang, X.; Wang, Z.; Deng, H.; Li, Y.; Li, A.; Cao, H.; Wang, L.; Zhu, Y.; Lu, S.; et al. Characteristics and Evolution of Tectonic Fractures in the Jurassic Lianggaoshan Formation Shale in the Northeast Sichuan Basin. *Minerals* **2023**, *13*, 946.
17. Wang, W.; Xie, X.; Liu, C.; Cao, F.; Zheng, G.; Wang, Z.; Chen, G.; Deng, H.; He, J.; Li, K. Characteristics and Formation Stages of Natural Fractures in the Fengcheng Formation of the Mahu Sag, China: Insights from Stable Carbon and Oxygen Isotope and Fluid Inclusion Analysis. *Minerals* **2023**, *13*, 894.
18. Liu, Y.; Lai, F.; Wang, R.; Wu, Z.; Zhang, X.; Xu, H.; Li, J. Influencing Factors of the Brittleness of Continental Shales Containing Shell Limestone Interlayer. *Minerals* **2023**, *13*, 460.
19. Zhan, Y.; Lin, C.; Ma, C.; Han, W.; Ma, P.; Li, G. Lithofacies Characteristics and Sweet Spot Distribution of Lacustrine Shale Oil Reservoirs: A Case Study of the Second Member of the Kongdian Formation in the Cangdong Sag, Bohai Bay Basin. *Minerals* **2023**, *13*, 1391.
20. Peng, S.; Feng, C.; Qiu, Z.; Zhang, Q.; Liu, W.; Feng, J.; Hu, Z. Prediction of Lithofacies in Heterogeneous Shale Reservoirs Based on a Robust Stacking Machine Learning Model. *Minerals* **2025**, *15*, 240.
21. Zhang, B.; Liu, C.; Yang, Z.; Qin, Y.; Li, M. Shear Wave Velocity Estimation for Shale with Preferred Orientation Clay Minerals. *Minerals* **2025**, *15*, 738.

References

1. Zou, C.N.; Zhu, R.K.; Chen, Z.-Q.; Ogg, J.G.; Wu, S.T.; Dong, D.Z.; Qiu, Z.; Wang, Y.M.; Wang, L.; Lin, S.H.; et al. Organic-matter-rich shales of China. *Earth-Sci. Rev.* **2019**, *189*, 51–78. [CrossRef]
2. Salygin, V.; Guliev, I.; Chernysheva, N.; Sokolova, E.; Toropova, N.; Egorova, L. Global Shale Revolution: Successes, Challenges, and Prospects. *Sustainability* **2019**, *11*, 1627. [CrossRef]
3. Guo, X.S.; Wang, R.Y.; Shen, B.J.; Wang, G.P.; Wan, C.X.; Wang, Q.R. Geological characteristics, resource potential, and development direction of shale gas in China. *Pet. Explor. Dev.* **2025**, *52*, 17–32. [CrossRef]
4. Davies, R.J.; Mathias, S.A.; Moss, J.; Hustoft, S.; Newport, L. Hydraulic fractures: How far can they go? *Mar. Pet. Geol.* **2012**, *37*, 1–6. [CrossRef]

https://doi.org/10.3390/min16010105

Milliken, K.L.; Curtis, M.E. Imaging pores in sedimentary rocks: Foundation of porosity prediction. *Mar. Pet. Geol.* **2016**, *73*, 590–608. [CrossRef]

Josh, M.; Esteban, L.; Delle Piane, C.; Sarout, J.; Dewhurst, D.N.; Clennell, M.B. Laboratory characterisation of shale properties. *J. Pet. Sci. Eng.* **2012**, *88*, 107–124. [CrossRef]

Wang, R.Y.; Hu, Z.Q.; Dong, L.; Gao, B.; Sun, C.X.; Yang, T.; Wang, G.P.; Yin, S. Advancement and trends of shale gas reservoir characterization and evaluation. *Oil Gas Geol.* **2021**, *42*, 54–65. [CrossRef]

Ilgen, A.G.; Heath, J.E.; Akkutlu, I.Y.; Bryndzia, L.T.; Cole, D.R.; Kharaka, Y.K.; Kneafsey, T.J.; Milliken, K.L.; Pyrak-Nolte, L.J.; Suarez-Rivera, R. Shales at all scales: Exploring coupled processes in mudrocks. *Earth-Sci. Rev.* **2017**, *166*, 132–152. [CrossRef]

McMahon, T.P.; Larson, T.E.; Zhang, T.; Shuster, M. Geologic characteristics, exploration and production progress of shale oil and gas in the United States: An overview. *Pet. Explor. Dev.* **2024**, *51*, 925–948. [CrossRef]

0. Ali, N.; Jamil, S.; Zaheer, M.; Hussain, W.; Hussain, H.; Iqbal, S.M.; Ullah, H. Exploration and development of Shale gas in China: A review. *Iran. J. Earth Sci.* **2022**, *14*, 87–103.

 https://doi.org/10.3390/min16010105

minerals

MDPI

Article

Lithofacies-Controlled Pore Characteristics and Mechanisms in Continental Shales: A Case Study from the Qingshankou Formation, Songliao Basin

Xinshu Huang [1,2], Zhiping Li [1,2,*], Xiangxue Han [3], Yongchao Wang [4,5] and Yiyuan Guo [1,2]

1 School of Energy Resources, China University of Geosciences Beijing, Beijing 100083, China
2 Beijing Key Laboratory of Unconventional Natural Gas Geological Evaluation and Development Engineering, Beijing 100083, China
3 School of Land Science and Technology, China University of Geosciences Beijing, Beijing 100083, China
4 Heilongjiang Provincial Key Laboratory of Continental Shale Oil, Daqing 163712, China
5 Exploration and Development Research Institute, Daqing Oilfield Company Ltd., Daqing 163712, China
* Correspondence: 2002011671@cugb.edu.cn

Abstract

Pore systems in continental shales are controlled by lithofacies and show strong heterogeneity, which challenges shale oil development. The Qingshankou Formation in the Songliao Basin is a major shale oil play in China. Previous studies have focused on macroscopic reservoir properties, with limited analysis of pore differences among lithofacies. This study integrates mineralogy, organic geochemistry, and multi-scale pore structure characterization to examine four typical lithofacies: argillaceous, siliceous, calcareous, and mixed shales. Results show that pore evolution in the Qingshankou Formation can be divided into five stages: immature ($Ro < 0.6\%$), low maturity ($0.6\% < Ro \leq 0.8\%$), middle maturity ($0.8\% < Ro \leq 1.0\%$), high maturity ($1.0\% < Ro \leq 1.2\%$), and over maturity ($Ro > 1.2\%$). The overall pattern follows a "three declines and two increases" trend. Due to differences in mineral composition and organic matter (OM), each lithofacies displays dis-tinct pore characteristics, which further influence oil-bearing potential and mobility. Siliceous shale, rich in felsic minerals, exhibits well-preserved pores and a developed micro-fracture network, providing the largest pore volume and average diameter. This facilitates the storage and flow of free oil, making it the preferred exploration target. Argillaceous shale, characterized by abundant clay minerals and OM, supports micropore development and offers the highest specific surface area (SSA). This yields significant adsorbed oil potential, highlighting its value as a secondary exploration target. This study clarifies the lithofacial controls on pore development in continental shales, providing a scientific basis for predicting favorable intervals and optimizing exploration strategies in the Qingshankou Formation and analogous basins.

Keywords: continental shale; lithofacies type; pore characteristic; pore evolution; Songliao Basin; Qingshankou Formation

1. Introduction

Shale oil, as a core component of unconventional oil and gas resources, has become a critical frontier in the global energy sector [1–5]. China possesses vast shale oil resources, representing a key strategic replacement for future oil and gas reserve and production growth, holding significant importance for national energy security [6,7]. Unlike conventional reservoirs, shale oil reservoirs are typically characterized by "self-generation and self-storage", with their enrichment and production efficiency highly dependent on the development of nano- to micro-scale pore systems [8–12]. Therefore, in-depth research on shale pore characteristics and elucidation of their controlling mechanisms are crucial for evaluating shale oil resource potential and optimizing development technologies [13,14].

In recent years, with the continuous advancement of unconventional oil and gas exploration and development, significant progress has been made in shale pore characterization techniques [15–17]. The integrated application of technologies such as low-pressure carbon dioxide adsorption (LPCO_2A), low-temperature nitrogen adsorption (LTNA), high-pressure mercury intrusion (HPMI), nuclear magnetic resonance (NMR), field emission scanning electron microscopy (FE-SEM), focused ion beam scanning electron microscopy (FIB-SEM), small-angle neutron scattering (SANS), and nanoscale computed tomography (nano-CT) has enabled the qualitative and quantitative characterization of pores across full scales and multiple dimensions [18–21]. Based on multi-technique integration strategies, the academic community has gradually established a relatively comprehensive theoretical framework for shale pore classification, structural parameter evaluation, and controlling factor analysis [22–26]. It is widely accepted that shale storage space can be categorized into three types: inorganic pores, organic pores, and microfractures [27–30], whose development is jointly controlled by mineral compositions [31–35], organic matter (OM) properties [36–40], and diagenesis [41–47].

However, research on pore systems in continental shales faces distinct challenges [48–55]. Continental shales develop in dynamic lacustrine basins with strong lithofacies heterogeneity and pore variability due to water-level fluctuations and multiple sediment sources [56–59]. Although a ternary classification based on mineral composition is widely used to categorize continental shales into argillaceous, siliceous, calcareous, and mixed types [60–66], significant uncertainties remain in establishing a predictive "lithofacies–pore" relationship pattern. For instance, Wang et al. [43] and Li et al. [65] reported contrasting dominant favorable lithofacies in different basins, highlighting the complex controls of regional geology and diagenetic history. Therefore, an integrated quantitative analysis across different lithofacies, considering specific sedimentary and diagenetic contexts, is needed to establish a reliable "lithofacies–pore" pattern for better resource evaluation.

The shale of the Qingshankou Formation in the Songliao Basin is a representative sequence for continental shale oil exploration and development in China [11,12,37,62]. Its diverse lithofacies provide ideal conditions for investigating pore characteristics across different lithofacies [67–69]. Addressing the research gaps above-mentioned, this study focused on four typical lithofacies within the Qingshankou Formation, namely argillaceous, siliceous, calcareous, and mixed shales. By integrating techniques such as X-ray diffraction (XRD), total organic carbon (TOC) measurement, rock pyrolysis, vitrinite reflectance (*Ro*) analysis, FE-SEM, LTNA, and HPMI, this research aims to achieve the following objectives: (1) to quantitatively characterize the lithofacies-controlled pore structure differentiation in the Qingshankou Formation shale; (2) reveal the synergistic control mechanisms of mineral composition and OM on pore development; (3) establish a dynamic pore evolution pattern driven by the coupling of OM hydrocarbon generation and diagenesis; and (4) construct a lithofacies-based grading framework for shale oil exploration to identify preferential targets. This multi-scale pore analysis coupled with lithofacies differentiation provides a methodological advance, deepening the understanding of continental shale reservoirs and supporting sweet spot prediction and efficient development in the Songliao Basin and analogous basins.

2. Geological Setting

The Songliao Basin, located in northeastern China, is the country's largest Mesozoic-Cenozoic continental petroliferous basin (Figure 1a). In the regional tectonic framework, the Central Depression represents the primary depocenter of the basin, characterized by long-term stable subsidence, persistent anoxic conditions, and high OM burial efficiency (Figure 1b). It is recognized as the most significant hydrocarbon source rock development zone and a strategic target for oil and gas exploration in the basin as well as the focal area of this study [62,68–70].

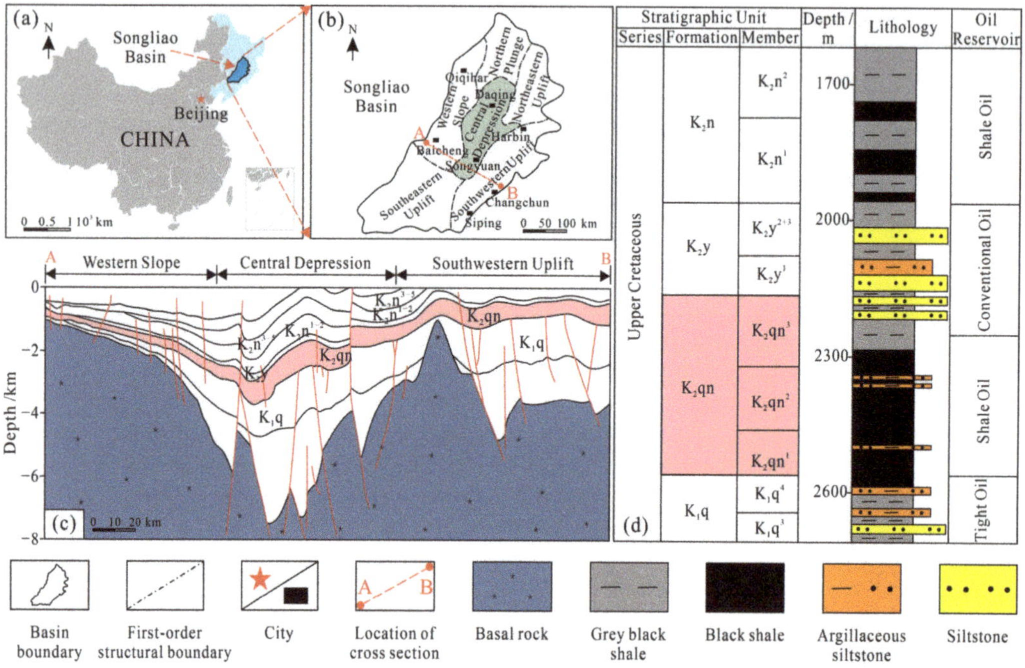

Figure 1. Geological setting of the study area. (**a**) Geographical location of Songliao Basin. (**b**) Tectonic division of Songliao Basin, with the study area of Central Depression specifically highlighted. (**c**) Structural cross-section along the A–B red line in (**b**), with the Qingshankou Formation specifically highlighted. (**d**) Stratigraphic column of the depression period in Songliao Basin, with the Qingshankou Formation specifically highlighted.

From the perspective of geological evolution, the Songliao Basin has undergone three main developmental stages: the rift stage (Jurassic–Early Cretaceous), the depression stage (early Late Cretaceous), and the tectonic inversion stage (late Late Cretaceous). The Qingshankou Formation (K_2qn) was deposited during the peak transgression of the Late Cretaceous lacustrine basin (Figure 1c). Under a warm and humid paleoclimate, it developed high-quality shales with great thickness, wide distribution, and high OM abundance (Figure 1d). These shales are typically buried deeper than 2000 m and have been subjected to complex and intense diagenesis, which significantly influenced their pore systems [68–72]. Recent exploration confirms the formation's substantial shale oil potential [11,12]. However, due to varying depositional environments and diagenetic history, the formation contains diverse lithofacies that create heterogeneous pore systems [68–70]. This heterogeneity directly affects oil storage and fluid flow in the reservoirs [12]. Therefore, studying how lithofacies control pore characteristics is crucial for effective shale oil development in the Qingshankou Formation.

3. Samples and Methods

3.1. Shale Samples

The shale samples were provided by the Daqing Oilfield Core Repository of China National Petroleum Corporation (CNPC). All were collected from the Qingshankou Formation in the Songliao Basin at burial depths of 2200–2500 m. To adequately capture reservoir heterogeneity, 56 samples were obtained with variable spacing: 1–3 m in uniform intervals and 0.1–0.5 m in heterogeneous sections. Mineral composition, as determined by XRD, was used to classify lithofacies based on quantitative thresholds. Samples containing more than 50% clay, felsic, or carbonate minerals were classified as argillaceous, siliceous, or calcareous shale, respectively; the remainder were categorized as mixed shale [60–66]. From these, 16 representative samples (four from each lithofacies) were selected for detailed analysis, with their numbers listed in Table 1. All selected samples were structurally intact, fracture-free, and weighed over 500 g, allowing for preparation into various dimensions to meet the requirements of subsequent experimental procedures (Figure 2).

Table 1. The sample numbers, whole mineral and clay mineral compositions, and geochemical parameters of the shale samples.

Lithofacies	No.	Whole Minerals (%)					Clay Minerals (%)				Geochemical Parameters			
		Clay	Qtz	Fel	Carb	Py	I/S	I	Ch	Kao	TOC (%)	T_{max} (°C)	HI (mg/g)	*Ro* (%)
Argillaceous shale	A-1	61.9	16.6	8.6	7.2	5.7	56.8	22.6	16.9	3.7	2.17	445	682	0.52
	A-2	50.5	17.9	11.3	13.9	6.4	37.2	36.7	24.8	1.3	1.26	437	618	1.33
	A-3	53.8	20.5	10.9	10.6	4.2	46.6	32.2	19.1	2.1	2.09	448	705	0.65
	A-4	55.3	14.7	9.9	17.5	2.6	38.7	35.3	22.4	3.6	1.82	441	678	0.84
	A-Average	55.4	17.4	10.2	12.3	4.7	44.8	31.7	20.8	2.7	1.84	443	671	0.84
Siliceous shale	S-1	34.1	25.1	20.8	13.7	6.3	28.2	41.2	24.1	6.5	1.36	446	615	1.26
	S-2	31.2	31.4	21.7	10.2	5.5	15.4	48.7	30.6	5.3	1.15	435	584	1.45
	S-3	37.3	25.7	23.6	9.1	4.3	31.3	36.4	27.7	4.6	1.62	439	639	0.68
	S-4	30.6	40.1	19.4	6.2	3.7	23.5	33.1	36.3	7.1	1.43	447	663	1.11
	S-Average	33.3	30.6	21.4	9.8	5.0	24.6	39.9	29.7	5.9	1.39	442	625	1.13
Calcareous shale	C-1	11.8	15.5	14.7	54.8	3.2	12.4	52.5	31.5	3.6	1.14	446	528	1.49
	C-2	12.7	19.4	10.1	52.5	5.3	19.3	40.8	38.3	1.6	1.16	434	449	1.58
	C-3	20.5	13.8	12.5	51.3	1.9	20.1	43.1	33.7	3.1	1.33	443	582	0.97
	C-4	16.2	17.2	8.6	55.6	2.4	14.2	35.9	47.2	2.7	1.35	434	488	1.46
	C-Average	15.3	16.5	11.5	53.6	3.2	16.5	43.1	37.7	2.8	1.25	439	512	1.38
Mixed shale	M-1	29.8	21.6	20.4	22.6	5.6	27.5	32.6	36.4	3.5	1.18	436	457	1.37
	M-2	28.2	27.3	14.7	25.3	4.5	23.9	29.4	41.2	5.5	1.49	437	536	1.02
	M-3	42.5	13.5	22.6	19.7	1.7	41.6	45.3	10.8	2.3	1.78	442	488	0.84
	M-4	33.4	19.6	10.7	30.4	5.9	30.1	47.5	14.9	7.5	1.66	445	541	0.59
	M-Average	33.5	20.5	17.1	24.5	4.4	30.8	38.7	25.8	4.7	1.53	440	506	0.96

Note: No.: sample number; Clay: clay mineral; Qtz: quartz; Fel: feldspar; Carb: carbonate mineral; Py: Pyrite; I/S: illite/smectite mixed layer; I: illite; Ch: chlorite; Kao: kaolinite; TOC: total organic carbon; T_{max}: maximum pyrolysis temperature; HI: hydrogen index; *Ro*: vitrinite reflectance.

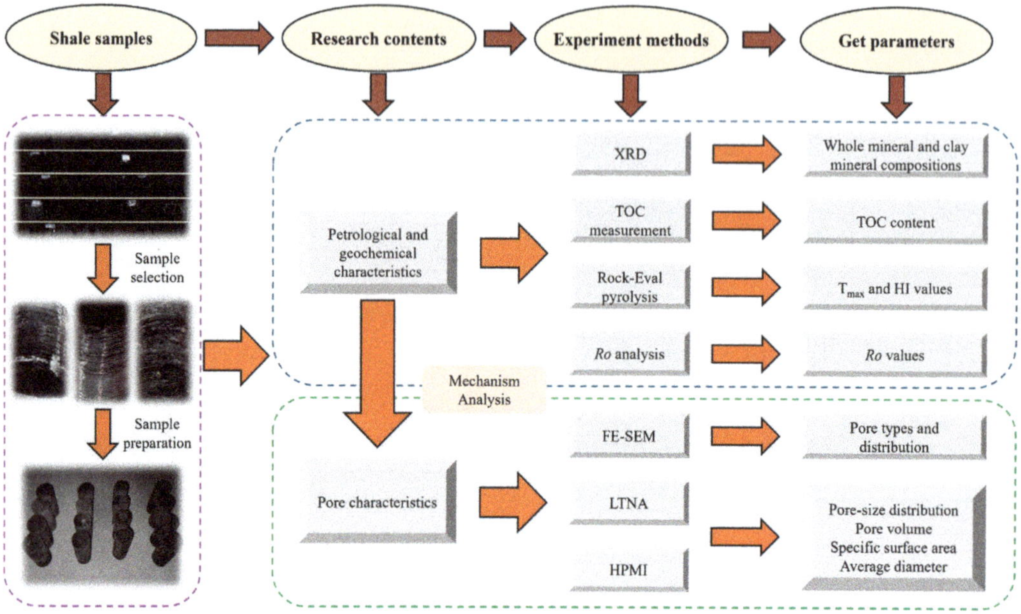

Figure 2. Flowchart illustrating the sample preparation and key experiments for this study.

3.2. Experimental Methods

This study analyzed the petrological, geochemical, and pore structural characteristics of the Qingshankou Formation shale samples using a multi-method workflow [16,17,24–26]. As shown in Figure 2, XRD was used to determine the whole mineral and clay mineral compositions. Geochemical analyses, including TOC measurement, Rock-Eval pyrolysis, and *Ro* analysis, were performed to assess OM abundance, type, and maturity. Pore structure was characterized using FE-SEM for pore type identification and distribution, along with LTNA and HPMI for quantitative pore parameter measurement. All analyses for each sample were performed on subsamples from the same core specimen to ensure data comparability. Integration of these multi-scale datasets clarifies the relationships among mineral composition, OM properties, and pore structure, and elucidates lithofacies-controlled pore characteristics and mechanisms.

XRD analysis was performed using an Olympus BTX III diffractometer (Olympus Corporation, Tokyo, Japan) with Cu-Kα radiation, operating at 40 kV and 40 mA. The measurements were conducted over a 2θ range of 5° to 70° with a step size of 0.02° and a scanning speed of 2°/min. Samples were crushed to below 100-mesh (<150 μm) prior to analysis. For clay mineral identification, glycol-saturated oriented mounts were additionally prepared. Minerals were identified using MDI Jade 9 software, and semi-quantitative analysis was performed using the relative intensity ratio (RIR) method. Analytical reproducibility, monitored by repeated measurements of a standard sample, was within ±5%.

TOC content was determined using a Leco CS-230 carbon/sulfur analyzer (LECO Corporation, St. Joseph, MI, USA). Samples crushed to 100-mesh were treated with dilute hydrochloric acid to remove carbonate minerals and subsequently rinsed with deionized water to neutrality. The TOC content was calculated based on the peak area of CO_2 generated from the combustion of OM. The instrument was calibrated using certified carbon standards, and the analytical uncertainty was less than 2%.

Rock-Eval pyrolysis was conducted using a Rock-Eval®6 instrument (Vinci Technologies, Nanterre, France). Approximately 100 mg of sample (100-mesh) was heated in an inert atmosphere according to the following program: isothermal at 300 °C for 3 min to record the S_1 peak (free hydrocarbons), followed by programmed heating to 650 °C at a rate of 25 °C/min to record the S_2 peak (hydrocarbons from cracking of kerogen). The maximum pyrolysis temperature (T_{max}) was recorded at the peak of the S_2 curve. The hydrogen index (HI) was calculated as $HI = (S_2/TOC) \times 100$.

Ro values was measured using a CRAIC CoalPro micro-photometer (CRAIC Technologies, Inc., San Dimas, CA, USA). Samples were embedded in epoxy resin and polished to a smooth surface. To avoid bias from sample anisotropy, blocks (1 cm × 1 cm × 0.5 cm) were prepared perpendicular to the bedding plane. At least 50 valid measurements were obtained per sample, and the average value was reported as the final *Ro*. The instrument was calibrated with a synthetic reflectance standard prior to analysis, and the standard deviation for replicate measurements was typically below 0.05%.

FE-SEM was conducted using a Zeiss SIGMA 300 microscope (Carl Zeiss AG, Oberkochen, Germany). The instrument was operated at an accelerating voltage of 5–15 kV and magnifications ranging from 100× to 100,000×. Sample blocks (1 cm × 0.5 cm × 0.2 cm) were prepared perpendicular to the bedding plane to obtain representative surface morphological information. After mounting and grinding, the sample surfaces were polished using a Hitachi IM4000 argon ion polisher (Hitachi High-Tech Corporation, Tokyo, Japan) to create a pristine, damage-free surface for high-resolution imaging.

LTNA was performed using a Micromeritics ASAP 2020M analyzer (Micromeritics Instrument Corporation, Norcross, GA, USA). Samples were crushed to 100-mesh, vacuum-degassed at 120 °C for 12 h, and then analyzed at 77.35 K over a relative pressure (P/P_0) range of 0.01 to 0.99. The pore size distribution and pore volume were calculated using the Barrett–Joyner–Halenda (BJH) model from the adsorption branch. The specific surface area (SSA) was determined by the Brunauer–Emmett–Teller (BET) method.

HPMI was conducted using a Micromeritics Autopore 9510 porosimeter (Micromeritics Instrument Corporation, Norcross, GA, USA). Samples were prepared as standard cylinders (2.5 cm in diameter, 5 cm in height) parallel to the bedding plane to facilitate mercury intrusion and extrusion. The test pressure range was 0.01–414 MPa. Pore size was calculated based on the Washburn equation, pore volume was determined from the volume of intruded mercury, and the SSA was calculated based on a cylindrical pore model.

4. Results

4.1. Petrological and Geochemical Characteristics

The mineral composition exerts a primary control on the petrophysical properties of the shale [31–35]. XRD results (Table 1, Figure 3) indicate that the shale mineral composition in the Qingshankou Formation is complex and varies significantly among different lithofacies. Argillaceous shale exhibited the highest clay mineral content (avg. 55.4%), dominated by I/S mixed layer minerals (avg. 44.8%). Siliceous shale had the highest felsic mineral content (avg. 56.9%), with its clay fraction primarily consisting of illite (avg. 39.9%) and chlorite (avg. 29.7%). Calcareous shale contained the highest carbonate mineral content (avg. 53.6%), with a clay mineral composition similar to that of siliceous shale. Mixed shale showed a relatively balanced content of the three main mineral types, with considerable variation in its clay mineral composition and no clearly dominant component.

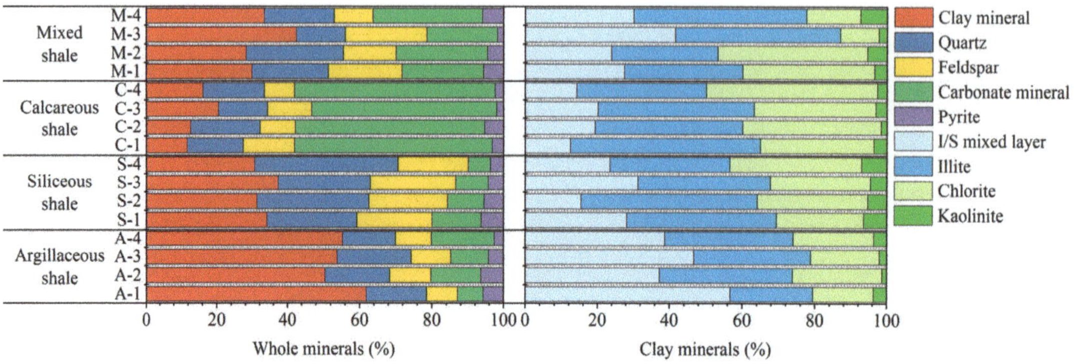

Figure 3. Whole mineral and clay mineral compositions of the shale samples. The left panel presents the whole mineral compositions, and the right panel shows the clay mineral compositions.

Results from the rock pyrolysis and organic carbon analysis (Table 1, Figure 4) showed that the T_{max} values of the shales in the Qingshankou Formation primarily range from 434 to 448 °C. The HI values exhibited a wide variation (449–705 mg/g). Argillaceous shales (avg. 671 mg/g) and siliceous shales (avg. 625 mg/g) displayed high HI values, consistent with Type I kerogen. Mixed shales (avg. 512 mg/g) and calcareous shales (avg. 506 mg/g) showed relatively lower HI values, with some samples indicating characteristics of Type II_1 kerogen. The TOC content distribution revealed a clear lithofacies-dependent pattern. Argillaceous shales possessed the highest TOC content (avg. 1.84%), followed by mixed shales (avg. 1.53%) and siliceous shales (avg. 1.39%), while calcareous shales had the lowest TOC content (avg. 1.25%). *Ro* values varied widely (0.52%–1.58%), indicating that the shales in the Qingshankou Formation have experienced a complete thermal evolution sequence [11,12,68,72]. Calcareous shales exhibited the highest maturity (avg. *Ro* = 1.38%), followed by siliceous shales (avg. *Ro* = 1.13%) and mixed shales (avg. *Ro* = 0.96%), with argillaceous shales showing the lowest maturity (avg. *Ro* = 0.84%).

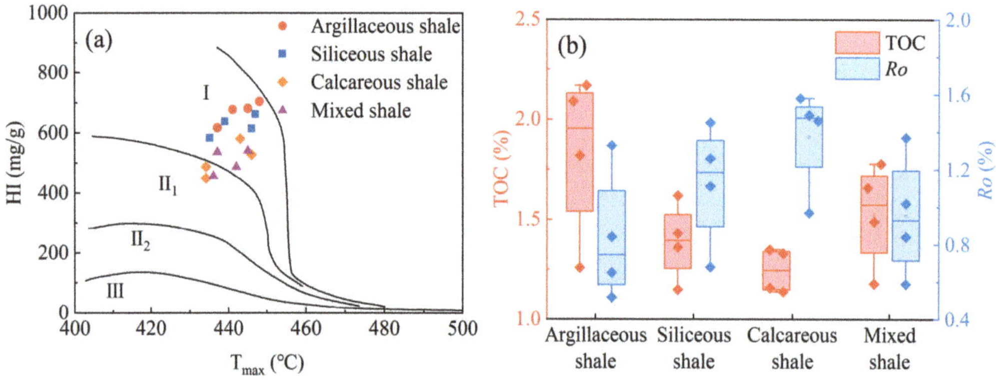

Figure 4. Geochemical characteristics of the shale samples. (**a**) HI vs. T_{max}, indicating kerogen types (denoted by Roman numerals I, II_1, II_2, III) present in different lithofacies; (**b**) TOC content vs. *Ro*, showing OM abundance and maturity across different lithofacies.

4.2. Pore Types and Distribution

FE-SEM revealed three main pore types in the shales of the Qingshankou Formation: inorganic pores, organic pores, and microfractures (Figure 5) [26–29].

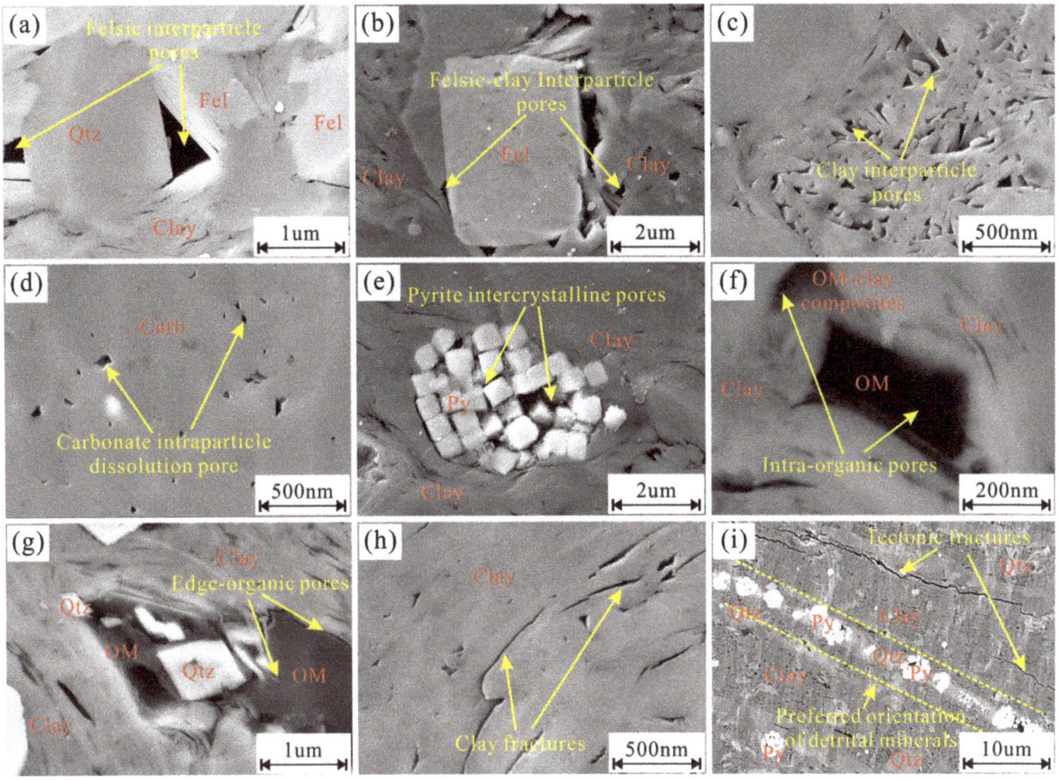

Figure 5. FE-SEM characterization of pore types and their lithofacies associations in the shale samples (Clay: clay mineral; Qtz: quartz; Fel: feldspar; Carb: carbonate mineral; Py: Pyrite; OM: organic matter). (**a**–**c**) Interparticle pores, including (**a**) large felsic interparticle pores common in siliceous shale, (**b**) felsic-clay interparticle pores, and (**c**) fine clay interparticle pores characteristic of argillaceous shale. (**d**) Intraparticle pores, exemplified by dissolution pores within carbonate minerals, predominantly found in calcareous shale. (**e**) Intercrystalline pores within framboidal pyrite aggregates, occurring across all lithofacies. (**f**,**g**) Organic pores, showing (**f**) intra-organic pores and (**g**) edge-organic pores, both most developed in argillaceous shale. (**h**,**i**) Microfractures, comprising (**h**) clay fractures forming a network in argillaceous shale, and (**i**) large-scale, oriented tectonic fractures that enhance connectivity in brittle siliceous shale.

Inorganic pores include interparticle pores, intraparticle pores, and intercrystalline pores [63–71]. Interparticle pores, located at mineral grain contacts, can be subdivided into felsic interparticle pores (Figure 5a), felsic-clay interparticle pores (Figure 5b), and clay interparticle pores (Figure 5c). These pores exhibited diverse morphologies, including triangular, polygonal, and irregular shapes, with pore sizes ranging from several nanometers to hundreds of nanometers [21,31,38]. Among them, felsic interparticle pores were larger (tens to hundreds of nanometers) and common in siliceous shales, whereas clay interparticle pores were smaller (several to tens of nanometers) and enriched in argillaceous shales. Intraparticle pores primarily occurred within feldspar and carbonate minerals (Figure 5d), often appearing circular or elliptical with diameters typically less than 100 nm; these were frequently observed in calcareous shales. Intercrystalline pores mainly developed in framboidal pyrite aggregates (Figure 5e), displaying triangular or polygonal shapes with sizes generally below 100 nm, and were distributed across all lithofacies [43].

Organic pores commonly developed within and at the edges of OM, showing significant spatial heterogeneity [28,29]. Intra-organic pores were mostly circular, elliptical, or sponge-like, with diameters usually less than 100 nm (Figure 5f). Edge-organic pores were primarily slit-like shrinkage pores, reaching lengths up to 1 μm but widths of only a few nanometers (Figure 5g). Organic pores developed in all lithofacies but were unevenly distributed: these were most abundant in argillaceous shales, mainly within the OM and OM-clay composites; less numerous and often isolated in siliceous shales; primarily hosted within bioclastic shells in calcareous shales; and moderately developed in mixed shales, and intermediate between argillaceous and siliceous shales.

Microfractures can be classified by origin into clay fractures and tectonic fractures. Clay fractures developed within the clay-rich matrix (Figure 5h), appearing as curved bands or fracture networks with lengths of several micrometers and widths of tens to hundreds of nanometers; these were most common in argillaceous shales. Tectonic fractures developed in silty bands, aligning with the mineral orientation (Figure 5i), and appeared straight or branched. These were larger in scale, with lengths reaching tens to hundreds of micrometers and widths from hundreds of nanometers to several micrometers; these were frequently observed in siliceous shales [12,62].

4.3. Pore Structure Characterization

LTNA and HPMI are commonly used methods for characterizing the shale pore structure [16–21]. LTNA, based on the physical adsorption behavior of nitrogen molecules (kinetic diameter $\approx$ 0.354 nm) on pore surfaces, offers high sensitivity and accuracy for nanopores but has limited capability for detecting larger pores [18]. HPMI intrudes mercury into pores under high pressure up to 414 MPa, enabling the characterization of pores ranging from several nanometers to hundreds of micrometers. However, the high pressure may alter the original pore structure, leading to data distortion [19,20]. Therefore, this study adopted a combined characterization approach: LTNA for pores < 100 nm and HPMI for pores > 100 nm, achieving continuous and complementary analysis across the full pore size range [19,24,25]. Based on the pore structure of continental shale and the occurrence mode of crude oil, the following pore size classification was used: micropores (<10 nm), mesopores (10–100 nm), and macropores (100–10,000 nm) [37,55]. Pores larger than 10,000 nm are not discussed, as they mainly represent artificial fractures or surface irregularities induced during sample preparation and are considered ineffective storage space. Figure 6 presents the nitrogen adsorption curves, mercury intrusion curves, and combined pore size distributions of different lithofacies. Based on these, the pore volume, SSA, and average diameter were further calculated, as summarized in Table 2 and Figure 7.

Argillaceous shale exhibited an H2–3 type hysteresis loop in the nitrogen adsorption curve, indicating the coexistence of ink-bottle and slit-shaped pores. Significant adsorption at low relative pressure ($P/P_0 < 0.4$) suggests well-developed micropores. The mercury intrusion curve showed a sharp increase after 30 MPa, indicating the presence of microfractures. Combined with the FE-SEM results, these were identified as clay fractures (Figure 5h). Argillaceous shale had the highest cumulative nitrogen adsorption capacity (avg. 20.77 cm^3/g) among the four lithofacies, and the second highest cumulative mercury intrusion volume (avg. 0.0229 mL/g) after siliceous shale. The combined pore size distribution indicates a well-developed pore system, with the average values of pore volume, SSA, and average diameter being 0.06127 cm^3/g, 22.980 m^2/g, and 10.78 nm, respectively.

Siliceous shale showed an H2 type hysteresis loop in the nitrogen adsorption curve, suggesting that slit-shaped pores dominate. A sharp increase in nitrogen adsorption at high relative pressure ($P/P_0 > 0.8$) reflects the developed mesopores. The mercury intrusion curve rose noticeably after 20 MPa, 10 MPa earlier than in argillaceous shale, indicating more large tectonic fractures (Figure 5i). Its cumulative nitrogen adsorption (avg. 20.35 cm^3/g) was similar to that of argillaceous shale, while its cumulative mercury intrusion (avg. 0.0289 mL/g) was the highest among all lithofacies. The combined pore size distribution showed the most developed pore system, with the average values of pore volume, SSA, and average diameter being 0.06419 cm^3/g, 19.703 m^2/g, and 13.15 nm, respectively.

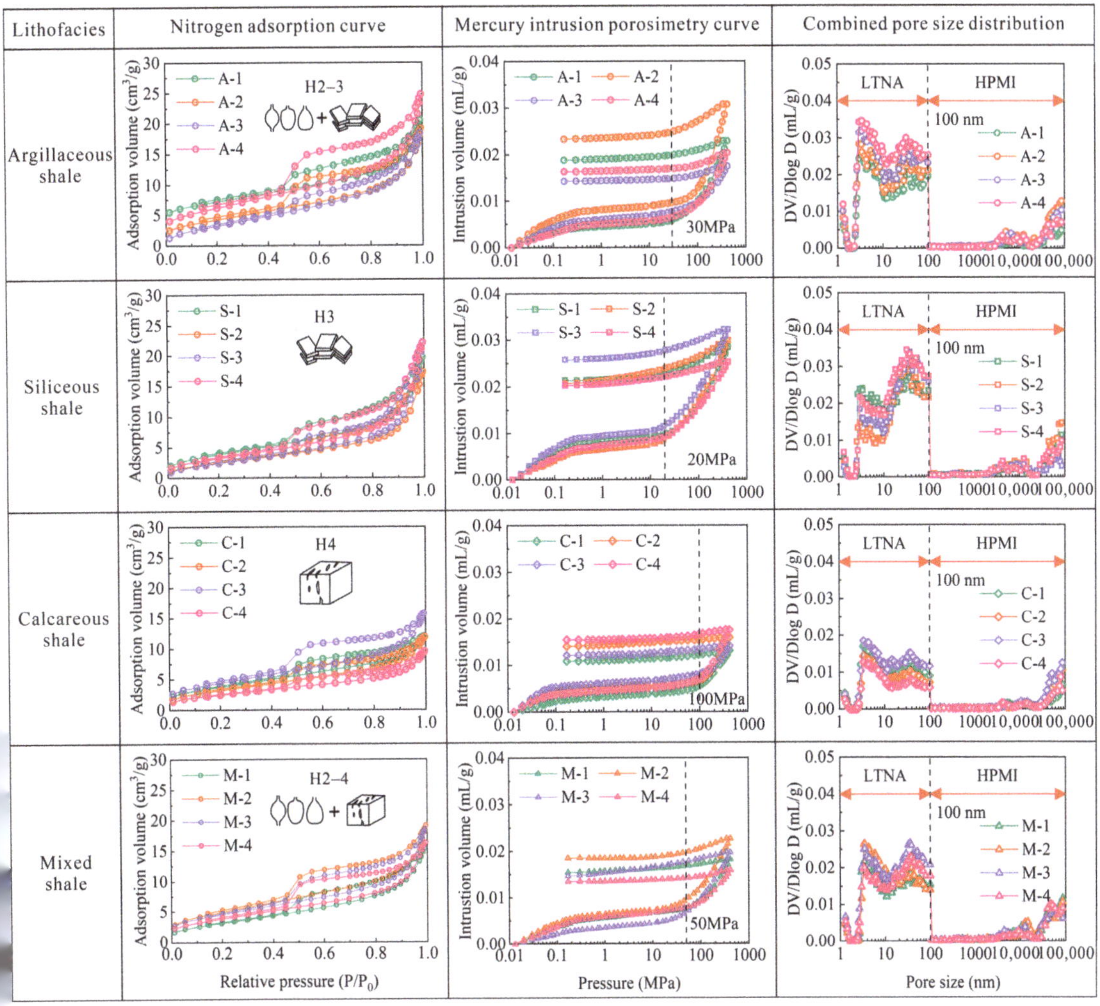

Figure 6. Pore structure characteristics revealed by integrated LTNA-HPMI analysis. The nitrogen adsorption curves reflect pores below 100 nm, while the mercury intrusion curves characterize pores above 100 nm. The combined dataset provides continuous pore-size characterization across the full measurable range.

Table 2. Pore structure parameters of the shale samples.

Lithofacies	No.	Pore Volume (cm^3/g)				SSA (m^2/g)	Average Diameter (nm)
		Micropores	Mesopores	Macropores	Total Pores		
Argillaceous shale	A-1	0.02600	0.02368	0.00372	0.05340	22.270	9.59
	A-2	0.02624	0.02622	0.00632	0.05878	18.471	12.73
	A-3	0.03040	0.02691	0.00554	0.06285	23.633	10.64
	A-4	0.03324	0.03148	0.00534	0.07006	27.546	10.17
	A-Average	0.02897	0.02707	0.00523	0.06127	22.980	10.78
Siliceous shale	S-1	0.02600	0.03220	0.00638	0.06458	22.635	11.41
	S-2	0.01868	0.03131	0.00742	0.05741	18.523	12.40
	S-3	0.02098	0.03641	0.00762	0.06501	17.181	15.14
	S-4	0.02292	0.03896	0.00788	0.06976	20.474	13.63
	S-Average	0.02215	0.03472	0.00733	0.06419	19.703	13.15
Calcareous shale	C-1	0.02390	0.01436	0.00164	0.03990	20.820	7.67
	C-2	0.01991	0.01682	0.00112	0.03785	17.847	8.48
	C-3	0.02596	0.01921	0.00138	0.04655	20.420	9.12
	C-4	0.01568	0.01214	0.00124	0.02906	15.261	7.62
	C-Average	0.02136	0.01563	0.00135	0.03834	18.587	8.22
Mixed shale	M-1	0.02314	0.02288	0.00301	0.04903	17.585	11.15
	M-2	0.02942	0.02382	0.00322	0.05646	23.421	9.64
	M-3	0.02618	0.02774	0.00368	0.05760	24.630	9.35
	M-4	0.02286	0.0249	0.00292	0.05068	19.742	10.27
	M-Average	0.02540	0.02484	0.00321	0.05344	21.345	10.10

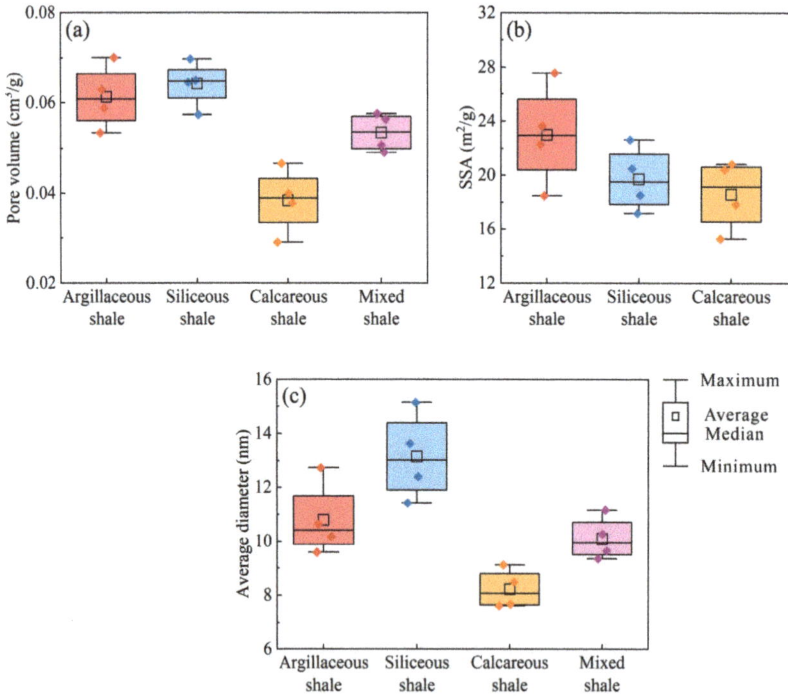

Figure 7. Pore structure parameters in various shale lithofacies. (**a**) Pore volume (PV) and (**b**) SSA, integrating LTNA (<100 nm) and HPMI (>100 nm) data. (**c**) Average diameter, calculated as $4 \times PV/SSA$.

Calcareous shale displayed an H4 type hysteresis loop in the nitrogen adsorption curve, indicating wedge-shaped pores. The mercury intrusion curve rose slowly only after 100 MPa, suggesting poorly developed microfractures. It had the lowest cumulative nitrogen adsorption (avg. 12.34 cm^3/g) and mercury intrusion (avg. 0.0152 mL/g) among the lithofacies. The combined pore size distribution confirms the least developed pore system, with the average values of pore volume, SSA, and average diameter being 0.03834 cm^3/g, 18.587 m^2/g, and 8.22 nm, respectively.

Mixed shale showed an H2–4 transitional hysteresis loop in the nitrogen adsorption curve, reflecting composite development of ink-bottle and wedge-shaped pores. Concentrated mercury intrusion occurred after 50 MPa. Its cumulative nitrogen adsorption averaged 17.39 cm^3/g, and the mercury intrusion averaged 0.0191 mL/g. The combined pore size distribution indicates a moderately developed pore system, slightly better than that of calcareous shale, with the average values of pore volume, SSA, and average diameter being 0.05344 cm^3/g, 21.345 m^2/g, and 10.10 nm, respectively.

5. Discussion

5.1. Controlling Factors on Pore Development

The pore structure of continental shale is co-controlled by multiple geological factors, including the depositional environment, shale composition, and diagenetic evolution [36]. Shale composition serves as the material basis for pore development. The type and content of inorganic minerals and OM are key intrinsic factors controlling the formation and evolution of the pore system, constituting the core focus of this study. Accordingly, this study systematically analyzed the correlations between the contents of clay minerals, felsic minerals, carbonate minerals, TOC, and the volumes of micropores, mesopores, macropores, and total pores in the Qingshankou Formation shale to reveal the underlying control mechanisms. To validate these correlations, published data from the study by Li et al. [65] on the Shahejie Formation are introduced for comparison. It should be noted that due to differences in pore size classification standards among studies, this comparison focused solely on the total pore volume. Furthermore, to minimize the interference of OM abundance on the comparison results, the selected Shahejie Formation samples (n = 9) had a TOC range largely consistent with the Qingshankou Formation samples in this study.

5.1.1. Control of Inorganic Minerals

(1) Clay Minerals

In the Qingshankou Formation, the clay mineral content showed positive correlations with micropore, mesopore, and macropore volumes (Figure 8a; R^2 = 0.37, 0.26, and 0.27, respectively), indicating a positive contribution of clay minerals to pore development across all scales. Previous studies suggest that the layered structure of clay minerals provides favorable conditions for pore formation [34,51,59]. Among them, I/S mixed layer minerals, with their larger interlayer spacing and expansibility, particularly promote micropore development [33,48]. For instance, in this study, argillaceous shale exhibited the highest clay mineral content (average 55.4%, predominantly I/S mixed layer) and the largest micropore volume (average 0.02897 cm^3/g). Figure 8b shows that both the Qingshankou and Shahejie Formations displayed a trend of increasing total pore volume with higher clay mineral content (R^2 = 0.45 and 0.34, respectively), strongly supporting the key role of clay minerals as primary pore builders in continental shales.

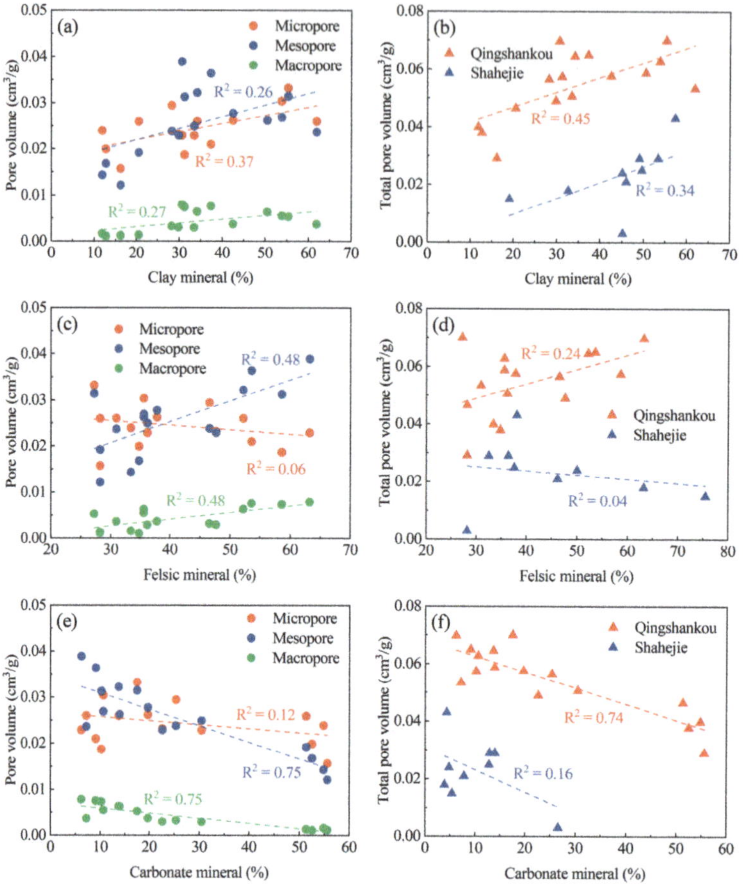

Figure 8. Relationships between mineral composition and multi-scale pore volumes. (**a**) In the Qingshankou Formation, clay mineral content showed positive correlations with micropore, mesopore, and macropore volumes (R^2 = 0.37, 0.26, and 0.27, respectively). (**b**) Positive correlations were observed between clay mineral content and total pore volume in both the Qingshankou and Shahejie Formations (R^2 = 0.45 and 0.34, respectively). (**c**) In the Qingshankou Formation, felsic mineral content exhibited a weak negative correlation with micropore volume but strong positive correlations with mesopore and macropore volumes (R^2 = 0.06, 0.48, and 0.48, respectively). (**d**) A positive correlation between felsic mineral content and total pore volume was observed in the Qingshankou Formation (R^2 = 0.24), while a weak negative correlation was found in the Shahejie Formation (R^2 = 0.04). (**e**) In the Qingshankou Formation, carbonate mineral content showed negative correlations with micropore, mesopore, and macropore volumes (R^2 = 0.12, 0.75, and 0.75, respectively). (**f**) A strong negative correlation between carbonate content and total pore volume was evident in the Qingshankou Formation (R^2 = 0.74), with a weaker negative correlation in the Shahejie Formation (R^2 = 0.16).

(2) Felsic Minerals

In the Qingshankou Formation, felsic mineral content showed a weak negative correlation with micropore volume (Figure 8c; R^2 = 0.06). This is primarily because an increase in felsic minerals leads to a relative decrease in clay minerals [69,72], thereby affecting micropore development. In contrast, mesopore and macropore volumes showed strong positive correlations with felsic mineral content (Figure 8c; R^2 = 0.48 for both). This is mainly attributed to the rigid framework provided by felsic minerals, which effectively resists compaction and helps maintain pore structure integrity [31,34,44,59]. Concurrently, feldspar dissolution and responses to tectonic stress can create secondary pores and microfractures, further expanding the pore space. Overall, felsic mineral content was positively correlated with total pore volume (Figure 8d; R^2 = 0.24), indicating a generally positive impact on pore development. Siliceous shale, with the highest felsic mineral content (average 56.9%) and the largest total pore volume (average 0.0642 cm^3/g), provides strong support for this conclusion. However, in the Shahejie Formation, the felsic mineral content showed a weak negative correlation with total pore volume (Figure 8d; R^2 = 0.04), indicating a suppressive role. This may be related to the greater burial depth of the Shahejie Formation (approximately 4000 m, significantly deeper than the 2200–2500 m range of the samples in this study), where stronger compaction weakens the supportive effect of the felsic mineral framework. Furthermore, Li et al. [65] noted that felsic minerals in the Shahejie Formation are predominantly terrigenous quartz, with low feldspar content, lacking significant secondary pores from dissolution. Additionally, quartz overgrowths further occlude pores, exacerbating pore structure degradation.

(3) Carbonate Minerals

In the Qingshankou Formation, the carbonate mineral content showed negative correlations with micropore, mesopore, and macropore volumes (Figure 8e; R^2 = 0.12, 0.75, and 0.75, respectively), with the most significant negative impact on mesopores and macropores. This is because carbonate minerals in continental shale often exist as calcareous cements. Their precipitation and filling preferentially occur in medium to large pores, leading to substantial loss of these pores, followed by the filling of smaller pores, also causing some damage to them [32–35]. Although existing studies indicate that carbonate dissolution can create some micropores, partially mitigating the reduction in micropores (resulting in the weakest negative correlation) [59], this effect generally cannot offset the negative impact of cementation. The significant negative correlation between total pore volume and carbonate content (Figure 8f; R^2 = 0.74), along with the lowest total pore volume in calcareous shale (average 0.0383 cm^3/g), further confirms the inhibitory effect of carbonate minerals on pore development. This destructive pattern also holds for the Shahejie Formation, but the correlation was weaker (Figure 8f; R^2 = 0.16). This is presumably because carbonate dissolution contributes more significantly to pores in this formation, potentially related to the intrusion of deep acidic hydrothermal fluids [65].

5.1.2. Control of OM

TOC content, a fundamental indicator of OM abundance in shale, is key to understanding the control of OM on pore development [26,35,60]. In the Qingshankou Formation, TOC showed positive correlations with micropore, mesopore, and macropore volumes (Figure 9a; R^2 = 0.54, 0.07, and 0.03, respectively), with the strongest relationship observed for micropores. This is primarily because hydrocarbon generation from OM not only directly creates micropores (e.g., organic pores and dissolution pores), but also helps support the interlayer structures of clay minerals, effectively inhibiting micropore compression during compaction [28,36,48]. The positive correlations with mesopores and macropores are relatively weak, possibly due to the partial pore-occluding effect of some OM [48,60]. Overall, the total pore volume showed a positive correlation with TOC (Figure 9b; R^2 = 0.17), indicating a positive role of OM in pore development. However, the Shahejie Formation exhibited the opposite trend, with a negative correlation between TOC and total pore volume (Figure 9b; R^2 = 0.18). This difference is likely attributed to variations in OM type and maturity between the two formations. This study showed that the OM in the Qingshankou Formation is primarily Type I kerogen, with a small amount of Type II_1 kerogen. The *Ro* ranged from 0.52% to 1.58%, and most values were within the oil window. In contrast, Li et al. reported abundant Type II_2-III kerogen in the Shahejie Formation, with lower *Ro* values of 0.41%–0.75% [65]. This indicates that a substantial portion of the OM is immature and has not reached the oil generation threshold. Consequently, the OM in the Qingshankou Formation possesses strong hydrocarbon generation potential. This promotes both the formation of nanoporous organic pores and the development of secondary pores from mineral dissolution by organic acids generated during thermal maturation. In comparison, the limited hydrocarbon generation potential of the Shahejie Formation means that residual OM is more likely to occlude interparticle pores and microfractures, significantly inhibiting pore development.

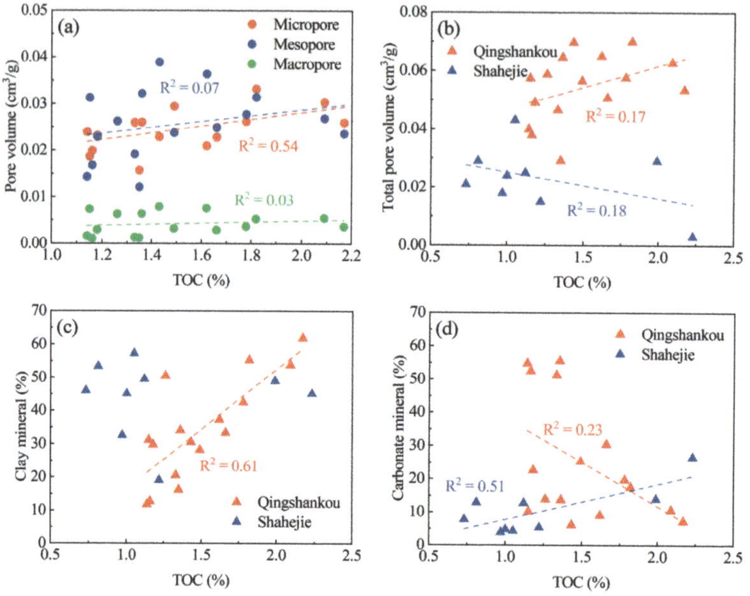

Figure 9. Relationships between TOC content and multi-scale pore volumes as well as clay and carbonate mineral contents. (**a**) In the Qingshankou Formation, TOC content showed positive correlations with micropore, mesopore, and macropore volumes (R^2 = 0.54, 0.07, and 0.03, respectively); (**b**) A positive correlation between TOC and total pore volume was observed in the Qingshankou Formation (R^2 = 0.17), while a negative correlation was found in the Shahejie Formation (R^2 = 0.18); (**c**) TOC was positively correlated with clay mineral content in the Qingshankou Formation (R^2 = 0.61) but showed no significant correlation in the Shahejie Formation; (**d**) TOC was negatively correlated with carbonate content in the Qingshankou Formation (R^2 = 0.23) but positively correlated in the Shahejie Formation (R^2 = 0.51).

Furthermore, the coupling relationship between OM and inorganic minerals is another important reason for this discrepancy [35,39]. In the Qingshankou Formation, TOC showed a significant positive correlation with clay minerals (Figure 9c; R^2 = 0.61), facilitating a favorable pattern where OM and clay minerals synergistically contribute to pore development [11,31,46,73,74]. On the one hand, the layered structure of clay minerals provides ideal hosting spaces for OM, favoring the development of organic pores. On the other hand, OM, through intermolecular forces, effectively supports the interlayer pores of clay minerals. FE-SEM observations revealed well-preserved pores within the OM–clay composites, providing direct evidence for this synergistic mechanism (Figure 5f). Simultaneously, the negative correlation between TOC and carbonate minerals (Figure 9d; R^2 = 0.23) helps avoid strong cementation-related destruction, which is also an important reason why OM effectively promotes pore development in the Qingshankou Formation. Conversely, in the

Shahejie Formation, TOC showed no clear correlation with clay minerals (Figure 9c) but exhibited a significant positive correlation with carbonate minerals (Figure 9d; $R^2 = 0.51$). This suggests that the carbonate minerals in the Shahejie Formation are largely biogenic, sharing a common origin with the OM [65]. Therefore, the OM in the Shahejie Formation is likely hosted within a carbonate matrix prone to cementation. Ions like Ca^{2+}, Mg^{2+}, and Fe^{2+} released during hydrocarbon generation can reprecipitate elsewhere when formation conditions change (e.g., pH, temperature, or pressure) [37]. This process promotes late-stage cement formation and ultimately leads to the negative correlation between TOC and pore volume.

5.2. Pore Evolution Pattern and Lithofacies Differentiation Results

The evolution of shale pores during burial diagenesis is a dynamic process controlled by multiple variables and complex mechanisms [38–43]. *Ro*, as a direct indicator of OM maturity, governs key geological processes such as hydrocarbon generation and diagenetic alteration, making it the most suitable parameter for characterizing pore evolution [38,40,43]. Results from this study show that the *Ro* values of the Qingshankou Formation shale range from 0.52% to 1.58%, reflecting a complete thermal evolution sequence and providing a reliable basis for dividing pore evolution stages. Based on the quantitative relationship between *Ro* and the total pore volume in Figure 10, the pore evolution of the Qingshankou Formation shale can be divided into five stages: immature ($Ro < 0.6\%$), low maturity ($0.6\% < Ro \leq 0.8\%$), middle maturity ($0.8\% < Ro \leq 1.0\%$), high maturity ($1.0\% < Ro \leq 1.2\%$), and over maturity ($Ro > 1.2\%$).

During the immature stage ($Ro < 0.6\%$), the total pore volume decreased. At this stage, OM has not yet entered the hydrocarbon generation threshold, and intense compaction dominated pore evolution, leading to a rapid reduction in primary pores [42]. As shown in Figure 10a, after strong compaction, the clay matrix underwent recrystallization and formed a highly oriented dense schistose structure. Interlayer micropores in clay minerals were largely closed, and the remaining nanopores were mostly slit-shaped and aligned along the clay sheets [12]. Rigid mineral grains were embedded into the plastic clay matrix under compaction, forming a typical "embedded" structure, with only a small number of intergranular pores preserved around them.

In the low maturity stage ($0.6\% < Ro \leq 0.8\%$), the total pore volume began to increase. As OM started to generate hydrocarbons, a critical transition occurred in the pore system [38,40], with the total pore volume gradually rising from 0.051 cm^3/g to approximately 0.073 cm^3/g. Scanning electron microscopy (Figure 10b) revealed the initial development of organic pores during this stage. In addition, organic acids generated during hydrocarbon generation dissolved soluble minerals such as feldspar and carbonates, forming secondary pores (Figure 5d), which collectively promote pore development [36,59].

In the middle maturity stage ($0.8\% < Ro \leq 1.0\%$), the total pore volume decreased again. Although continuous hydrocarbon generation favors the development of organic and dissolution pores, enhanced cementation significantly negatively impacts the pore system [41–45], causing the total pore volume to drop from 0.073 cm^3/g to about 0.048 cm^3/g. As shown in Figure 10c, mineral crystals noticeably filled the intergranular pores, with cements occupying space and blocking throats, directly reducing porosity.

During the high maturity stage ($1.0\% < Ro \leq 1.2\%$), the total pore volume increased for the second time, rising from 0.048 cm^3/g to 0.076 cm^3/g. At this stage, further maturity enhancement strengthens organic acid dissolution [47], and a microfracture network develops [12,39], jointly promoting the formation of secondary pores. Figure 10d shows that the shale in the study area developed distinct microfractures under tectonic stress, significantly increasing pore space and improving flow capacity.

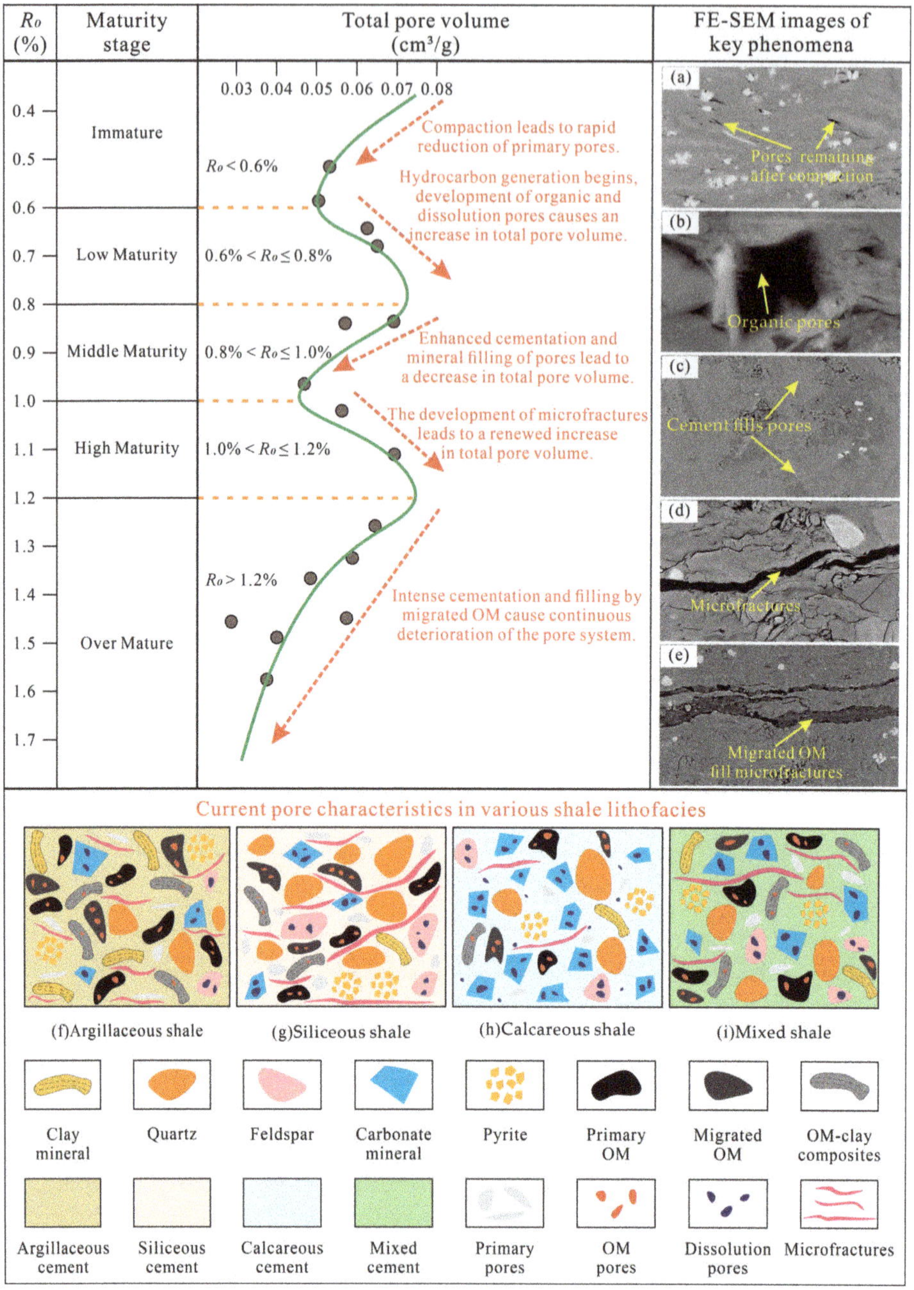

Figure 10. Pore evolution pattern and lithofacies differentiation results of Qingshankou Formation shale. (**a–e**) shows the key phenomena observed during the five stages of pore evolution; (**f–i**) illustrates the current pore characteristics of argillaceous shale, siliceous shale, calcareous shale, and mixed shale, respectively.

In the over maturity stage ($Ro > 1.2\%$), intense cementation and the filling of migrated OM led to renewed deterioration of the pore system, with continuous decline in porosity and effective storage space [48,60]. As shown in Figure 10e, migrated OM filled the pre-existing microfractures, damaging pore structure and reducing reservoir quality.

Overall, the pore evolution of the shale in the study area exhibited a notable "three declines and two increases" pattern. However, due to differences in mineral composition and OM content, different lithofacies showed distinct responses during this multi-stage evolution process, resulting in significant variations in current pore characteristics [41,43]. Argillaceous shale, with its high clay content, was highly sensitive to early compaction, suffering severe loss of primary pores (Figure 10a). However, its high OM abundance promoted the development of numerous organic pores during the low- to high-maturity stages, demonstrating the strongest pore recovery capacity (Figure 10b). Meanwhile, hydrocarbon-generation overpressure facilitates the formation of clay fractures (Figure 5h), improving pore connectivity (Figure 10f). Siliceous shale remained the most stable throughout the evolution process. Its rigid mineral framework effectively resisted early compaction, preserving some primary pores (Figure 5a,b). The high content of brittle minerals made it more prone to forming tectonic fractures under stress (Figure 5i), resulting in the most developed pore network (Figure 10g). Calcareous shale exhibited the poorest pore structure due to intense carbonate cementation (Figure 10c). Although dissolution pores provided limited local porosity, the overall rock remained highly tight, making it the lithofacies with the worst reservoir performance among the four types (Figure 10h). Mixed shale showed transitional evolution characteristics between the other lithofacies. Influenced by mineralogical heterogeneity, it ultimately developed a moderately developed pore system (Figure 10i).

5.3. Implications for Shale Oil Exploration

Continental shale is a typical "self-generation and self-storage" unconventional reservoir, with its commercial development potential primarily determined by two key parameters: oil-bearing potential and mobility. Oil-bearing potential is controlled by OM abundance, type, and maturity, whereas mobility is closely related to mineral composition, pore structure, and fracture development [6–11,45,56,58]. This study systematically evaluated the oil-bearing potential and mobility characteristics of different lithofacies through integrated mineral-petrological, geochemical, and multi-scale pore structure analyses.

Based on the occurrence state of crude oil in shale nano- to micro-scale pore systems, it can be classified into free oil and adsorbed oil (Figure 11) [64,65]. Free oil mainly occurs in fractures and macro-mesopores, serving as the primary contributor to shale oil productivity. In contrast, adsorbed oil is predominantly attached to pore surfaces, especially enriched in micropores, making it difficult to extract effectively. Results of this study show that argillaceous shale had the highest TOC content, with kerogen predominantly Type I and mostly within the oil window, indicating excellent hydrocarbon generation potential. However, its high clay mineral content and large SSA resulted in a pore structure dominated by micropores and mesopores, leading to oil occurring mainly in the adsorbed state. The small average diameter further limits free oil storage. Although the developed clay fracture system enhanced pore connectivity, the proportion of free oil remained low (Figure 11a). In comparison, siliceous shale not only exhibited favorable OM properties but also a high brittle mineral content. This combination facilitated the formation of a pore system dominated by mesopores and macropores. Coupled with a well-developed microfracture network, it demonstrated the best pore volume, diameter, and connectivity, providing ideal conditions for the storage and flow of free oil (Figure 11b). Additionally, mixed shale has a complex mineral composition, and diagenetic differences result in strong pore structure heterogeneity and generally poor pore-throat connectivity. Calcareous shale, affected by strong cementation, suffers a significant loss of primary pores, with an isolated and poorly connected microfracture system, making it difficult to form an effective storage-seepage system. As a result, both mixed and calcareous shales exhibited a significantly lower oil content and mobility compared to argillaceous and siliceous shales (Figure 11c,d).

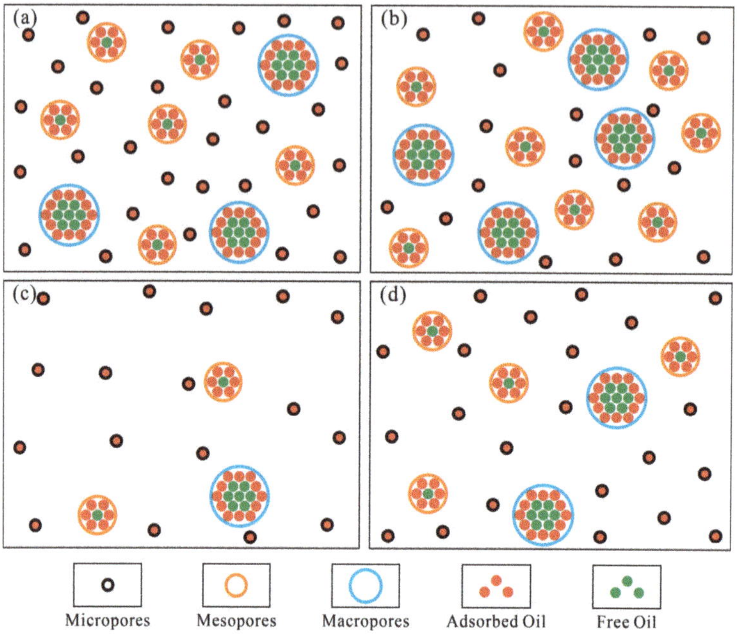

Figure 11. Oil-bearing potential and mobility of Qingshankou Formation shale. (**a**–**d**) illustrate the occurrence states of crude oil in argillaceous, siliceous, calcareous, and mixed shales, respectively.

Comprehensive evaluation confirmed that siliceous shale is the most favorable lithoface in the Qingshankou Formation due to its high free oil content and excellent flow capacity. Although argillaceous shale has poor mobility, its high adsorbed oil content makes it a valuable secondary exploration target. These findings differ from those of Wang et al. [43], who considered argillaceous shale the best target in the Qingshankou Formation, Songliao Basin. The difference may come from sample selection, experimental methods, or regional geology. For example, the limited number of samples in this study may not fully represent the formation's heterogeneity. Also, experimental methods have inherent limitations. For instance, combining LTNA and HPMI can cause errors due to model assumptions and pore size linking issues, which may lead to systematic bias. Given the widespread distribution of both argillaceous and siliceous shales in the Qingshankou Formation, subsequent exploration should prioritize these two lithofacies.

6. Conclusions

This study, through multi-scale pore characterization and mechanism analysis, yielded the following main conclusions:

(1) The pore structure of the Qingshankou Formation shale exhibited significant lithofacial differentiation. Argillaceous shale displayed the most developed micropore system and abundant clay fractures, contributing to the highest SSA. Siliceous shale possessed the most favorable pore structure, characterized by the largest pore volume and average diameter, with tectonic fractures significantly enhancing pore connectiv-

ity. In contrast, calcareous shale had the least developed pore system, while mixed shale showed transitional characteristics.

(2) The close symbiosis between OM and clay minerals established a synergistic pore-enhancement pattern, serving as the primary contributor to micropore development in the Qingshankou Formation shale. Feldspathic minerals formed a rigid framework that effectively supports pores and resists compaction, while also promoting the development of tectonic fractures, which are critical to the formation and preservation of mesopores and macropores. In contrast, carbonate minerals act mainly as cements that fill pore spaces, significantly inhibiting pore development.

(3) The pore evolution of the Qingshankou Formation shale can be divided into five stages, exhibiting a dynamic "three declines and two increases" pattern. During the low- and high-maturity stages, hydrocarbon generation, dissolution, and tectonic stress collectively serve as the primary mechanisms for enhancing secondary porosity, providing the key driving forces for pore expansion and fracture network formation.

(4) Siliceous shale, with its well-developed pore-fracture network, provides excellent storage space and efficient seepage pathways for free oil, making it the preferred exploration target in the Qingshankou Formation. In contrast, argillaceous shale, despite its poor fluid mobility, possesses significant potential as a secondary exploration target due to its abundant adsorbed oil resources. It is proposed that high priority be given to both lithofacies in subsequent exploration.

Limitations and future work: Although this study identified siliceous shale as the most favorable lithofacies, its conclusions remain limited by sample size and experimental methods. To obtain more reliable results, future work should first expand the sample set to better represent the heterogeneity of the Qingshankou Formation. Second, advanced techniques like FIB-SEM should be used for 3D pore imaging to directly characterize pore networks and minimize errors from integrated methods such as LTNA-HPMI. Finally, incorporating multiphase flow simulations and cross-scale data will help clarify the storage and flow mechanisms in shales, offering a more robust basis for identifying favorable lithofacies.

Author Contributions: Conceptualization, X.H. (Xinshu Huang) and Z.L.; methodology, Z.L. and Y.W.; software, X.H. (Xiangxue Han); validation, X.H. (Xinshu Huang), X.H. (Xiangxue Han) and Y.G.; formal analysis, Z.L. and Y.W.; investigation, Y.W.; resources, Z.L. and Y.W.; data curation, Y.G.; writing—original draft preparation, X.H. (Xinshu Huang); writing—review and editing, Z.L.; visualization, X.H. (Xiangxue Han); supervision, Z.L.; project administration, Z.L.; funding acquisition, Z.L. All authors have read and agreed to the published version of the manuscript.

Funding: This research was funded by the China National Petroleum Corporation Major Science and Technology Project (No. 2023ZZ15) and the National Natural Science Foundation of China Joint Fund for Enterprise Innovation Development Project (No. U22B2073). The APC was funded by China University of Geosciences (Beijing).

Data Availability Statement: The data used to support the findings of this study are included within the article.

Acknowledgments: The authors gratefully acknowledge the editors and anonymous reviewers for their constructive comments and valuable suggestions, which have significantly improved the quality of this manuscript.

Conflicts of Interest: Author Yongchao Wang was employed by the Daqing Oilfield Company Ltd. The remaining authors declare that the research was conducted in the absence of any commercial or financial relationships that could be construed as a potential conflict of interest.

References

1. Soeder, D.J. The successful development of gas and oil resources from shales in North America. *J. Pet. Sci. Eng.* **2018**, *163*, 399–420. [CrossRef]
2. Wang, Z.; Fan, Z.; Chen, X.; Li, Y.; Fan, Z.; Wei, Q.; Peng, Y.; Liu, B.; Yue, W.; Wang, X.; et al. Global oil and gas development situation, trends and enlightenment in 2023. *Pet. Explor. Dev.* **2024**, *51*, 1536–1555. [CrossRef]
3. Fauchille, A.L.; Ma, L.; Rutter, E.; Chandler, M.; Lee, P.; Taylor, K. An enhanced understanding of the Basinal Bowland shale in Lancashire (UK), through microtextural and mineralogical observations. *Mar. Pet. Geol.* **2017**, *86*, 1374–1390. [CrossRef]
4. Iqbal, M.A.; Rezaee, R.; Smith, G.; Ekundayo, J.M. Shale lithofacies controls on porosity and pore structure: An example from Ordovician Goldwyer Formation, Canning Basin, Western Australia. *J. Nat. Gas Sci. Eng.* **2021**, *89*, 103888. [CrossRef]
5. Zanoni, G.; Šegvić, B.; Moscariello, A. Clay mineral diagenesis in Cretaceous clastic reservoirs from West African passive margins (the South Gabon Basin) and its impact on regional geology and basin evolution history. *Appl. Clay Sci.* **2016**, *134*, 186–209. [CrossRef]
6. Xia, L.; Zhao, X.; Wu, Y. Status, disadvantages, and prospects of oil shale in China: An economic perspective. *Energy Policy* **2023**, *183*, 113795. [CrossRef]
7. Lei, Q.; Weng, D.; Guan, B.; Shi, J.; Cai, B.; He, C.; Sun, Q.; Huang, R. Shale oil and gas exploitation in China: Technical comparison with US and development suggestions. *Pet. Explor. Dev.* **2023**, *50*, 944–954. [CrossRef]
8. Yang, Z.; Zou, C. "Exploring petroleum inside source kitchen": Connotation and prospects of source rock oil and gas. *Pet. Explor. Dev.* **2019**, *46*, 181–193. [CrossRef]
9. Slatt, R.M.; O'Brien, N.R. Pore types in the Barnett and Woodford gas shales: Contribution to understanding gas storage and migration pathways in fine-grained rocks. *AAPG Bull.* **2011**, *95*, 2017–2030. [CrossRef]
10. Kibria, M.G.; Hu, Q.; Liu, H.; Zhang, Y.; Kang, J. Pore structure, wettability, and spontaneous imbibition of Woodford shale, Permian Basin, West Texas. *Mar. Pet. Geol.* **2018**, *91*, 735–748. [CrossRef]
11. He, W.; Zhu, R.; Cui, B.; Zhang, S.; Meng, Q.; Bai, B.; Feng, Z.; Lei, Z.; Wu, S.; He, K.; et al. The Geoscience Frontier of Gulong Shale Oil: Revealing the Role of Continental Shale from Oil Generation to Production. *Engineering* **2023**, *28*, 79–92. [CrossRef]
12. Wang, X.; Cui, B.; Feng, Z.; Shao, H.; Huo, Q.; Zhang, B.; Gao, B.; Zeng, H. In-situ hydrocarbon formation and accumulation mechanisms of micro- and nano-scale pore-fracture in Gulong shale, Songliao Basin, NE China. *Pet. Explor. Dev.* **2023**, *50*, 1269–1281. [CrossRef]
13. Li, G.; Luo, K.; Shi, D. Key technologies, engineering management and important suggestions of shale oil/gas development: Case study of a Duvernay shale project in Western Canada Sedimentary Basin. *Pet. Explor. Dev.* **2020**, *47*, 791–802. [CrossRef]
14. Zhou, L.; Zhao, X.; Chai, G.; Jiang, W.; Pu, X.; Wang, X.; Han, W.; Guan, Q.; Feng, J.; Liu, X. Key exploration & development technologies and engineering practice of continental shale oil: A case study of Member 2 of Paleogene Kongdian Formation in Cangdong Sag, Bohai Bay Basin, East China. *Pet. Explor. Dev.* **2020**, *47*, 1138–1146. [CrossRef]
15. Josh, M.; Esteban, L.; Delle Piane, C.; Sarout, J.; Dewhurst, D.; Clennell, M. Laboratory characterisation of shale properties. *J. Pet. Sci. Eng.* **2012**, *88*, 107–124. [CrossRef]
16. Qian, Y.; Gao, P.; Fang, X.; Sun, F.; Cai, Y.; Zhou, Y. Microstructure Characterization Techniques for Shale Reservoirs: A Review. *Front. Earth Sci.* **2022**, *10*, 930474. [CrossRef]
17. Wang, Y.; Wang, Z.; Zhang, Z.; Yao, S.; Zhang, H.; Zheng, G.; Luo, F.; Feng, L.; Liu, K.; Jiang, L. Recent techniques on analyses and characterizations of shale gas and oil reservoir. *Energy Rev.* **2024**, *3*, 100067. [CrossRef]
18. Groen, J.C.; Peffer, L.A.; Pérez-Ramírez, J. Pore size determination in modified micro-and mesoporous materials. Pitfalls and limitations in gas adsorption data analysis. *Microporous Mesoporous Mater.* **2003**, *60*, 1–17. [CrossRef]
19. Zhang, J.; Tang, Y.; He, D.; Sun, P.; Zou, X. Full-scale nanopore system and fractal characteristics of clay-rich lacustrine shale combining FE-SEM, nano-CT, gas adsorption and mercury intrusion porosimetry. *Appl. Clay Sci.* **2020**, *196*, 105758. [CrossRef]
20. Wu, Y.; Tahmasebi, P.; Lin, C.; Zahid, M.A.; Dong, C.; Golab, A.N.; Ren, L. A comprehensive study on geometric, topological and fractal characterizations of pore systems in low-permeability reservoirs based on SEM, MICP, NMR, and X-ray CT experiments. *Mar. Pet. Geol.* **2019**, *103*, 12–28. [CrossRef]
21. Zhan, H.; Li, X.; Hu, Z.; Chen, L.; Shen, W.; Guo, W.; He, W.; Zhou, Y. Effect of Particle Size on Pore Structure and Fractal Characteristics of Deep Siliceous Shales in Southern Sichuan, China, Measured Using Small-Angle Neutron Scattering and Low-Pressure Nitrogen Adsorption. *Fractal Fract.* **2025**, *9*, 165. [CrossRef]
22. Pommer, M.; Milliken, K. Pore types and pore-size distributions across thermal maturity, Eagle Ford Formation, southern Texas. *AAPG Bull.* **2015**, *99*, 1713–1744. [CrossRef]
23. Ko, L.T.; Ruppel, S.C.; Loucks, R.G.; Hackley, P.C.; Zhang, T.; Shao, D. Pore-types and pore-network evolution in Upper Devonian-Lower Mississippian Woodford and Mississippian Barnett mudstones: Insights from laboratory thermal maturation and organic petrology. *Int. J. Coal Geol.* **2018**, *190*, 3–28. [CrossRef]
24. Liu, D.; Tian, T.; Liang, R.; Yang, F.; Ye, F. Characterization of shale pore structure by multitechnique combination and multifractal analysis and its significance. *Geofluids* **2020**, *2020*, 8896940. [CrossRef]

25. Iltaf, K.H.; Hu, Q.; Fan, M.; Oware, P.; Wang, Q.; Zhao, C.; Zhang, T.; Awan, R.S.; Khan, D.; Raza, A. Multiscale Pore Characterization of the New Albany Shale: Insights from Complementary Analytical Techniques. *Energy Fuels* **2025**, *39*, 10356–10373. [CrossRef]
26. Chen, S.; Li, X.; Zhu, G.; Lu, Z.; Wang, R.; Huang, Y.; Lan, W.; Jia, F.; Zhu, Y. Classification, controlling factors, and multi-scale characterization techniques in shale reservoir pores: A comprehensive review. *Gas Sci. Eng.* **2025**, *140*, 205662. [CrossRef]
27. Loucks, R.G.; Reed, R.M.; Ruppel, S.C.; Jarvie, D.M. Morphology, Genesis, and Distribution of Nanometer-Scale Pores in Siliceous Mudstones of the Mississippian Barnett Shale. *J. Sediment. Res.* **2009**, *79*, 848–861. [CrossRef]
28. Milliken, K.L.; Rudnicki, M.; Awwiller, D.N.; Zhang, T. Organic matter-hosted pore system, Marcellus Formation (Devonian), Pennsylvania. *AAPG Bull.* **2013**, *97*, 177–200. [CrossRef]
29. Yang, Y.; Yao, J.; Wang, C.; Gao, Y.; Zhang, Q.; An, S.; Song, W. New pore space characterization method of shale matrix formation by considering organic and inorganic pores. *J. Nat. Gas Sci. Eng.* **2015**, *27*, 496–503. [CrossRef]
30. Katz, B.J.; Arango, I. Organic porosity: A geochemist's view of the current state of understanding. *Org. Geochem.* **2018**, *123*, 1–16. [CrossRef]
31. Liu, R.; Zhou, W.; Xu, H.; Zhou, Q.; Jiang, K.; Shang, F.; Gao, W.; Song, W.; Liu, D.; Zhao, H.; et al. Impact of Minerals and Sealing Systems on the Pore Characteristics of the Qiongzhusi Formation Shale in the Southern Sichuan Basin. *ACS Omega* **2022**, *7*, 15821–15840. [CrossRef]
32. Liu, B.; Teng, J.; Li, C.; Li, B.; Bie, S.; Wang, Y. The Control of Shale Composition on the Pore Structure Characteristics of Lacustrine Shales: A Case Study of the Chang 7 Member of the Triassic Yanchang Formation, Ordos Basin, North China. *Energies* **2022**, *15*, 8353. [CrossRef]
33. Kuila, U.; Prasad, M. Specific surface area and pore-size distribution in clays and shales. *Geophys. Prospect.* **2013**, *61*, 341–362. [CrossRef]
34. He, S.; Li, H.; Qin, Q.; Long, S. Influence of Mineral Compositions on Shale Pore Development of Longmaxi Formation in the Dingshan Area, Southeastern Sichuan Basin, China. *Energy Fuels* **2021**, *35*, 10551–10561. [CrossRef]
35. Zhao, R.; Xue, H.; Lu, S.; Li, J.; Tian, S.; Wang, M.; Dong, Z. Multi-scale pore structure characterization of lacustrine shale and its coupling relationship with material composition: An integrated study of multiple experiments. *Mar. Pet. Geol.* **2022**, *140*, 105648. [CrossRef]
36. Xu, J.; Wu, S.; Liu, J.; Yuan, Y.; Cui, J.; Su, L.; Jiang, X.; Wang, J. New Insights into Controlling Factors of Pore Evolution in Organic-rich Shale. *Energy Fuels* **2021**, *35*, 4858–4873. [CrossRef]
37. He, W.; Sun, N.; Zhang, J.; Zhong, J.; Gao, J.; Sheng, P. Genetic mechanism and petroleum geological significance of calcite veins in organic-rich shales of lacustrine basin: A case study of Cretaceous Qingshankou Formation in Songliao Basin, China. *Pet. Explor. Dev.* **2024**, *51*, 1083–1096. [CrossRef]
38. Curtis, M.E.; Cardott, B.J.; Sondergeld, C.H.; Rai, C.S. Development of organic porosity in the Woodford Shale with increasing thermal maturity. *Int. J. Coal Geol.* **2012**, *103*, 26–31. [CrossRef]
39. Yang, G.; Zeng, J.; Qiao, J.; Liu, Y.; Liu, S.; Jia, W.; Cao, W.; Wang, C.; Geng, F.; Wei, W. Role of Organic Matter, Mineral, and Rock Fabric in the Full-Scale Pore and Fracture Network Development in the Mixed Lacustrine Source Rock System. *Energy Fuels* **2022**, *36*, 8161–8179. [CrossRef]
40. Cao, T.; Deng, M.; Cao, Q.; Huang, Y.; Yu, Y.; Cao, X. Pore formation and evolution of organic-rich shale during the entire hydrocarbon generation process: Examination of artificially and naturally matured samples. *J. Nat. Gas Sci. Eng.* **2021**, *93*, 104020. [CrossRef]
41. Wang, Y.; Cheng, H.; Hu, Q.; Liu, L.; Hao, L. Diagenesis and pore evolution for various lithofacies of the Wufeng-Longmaxi shale, southern Sichuan Basin, China. *Mar. Pet. Geol.* **2021**, *133*, 105251. [CrossRef]
42. Fu, Q.; Hu, Z.; Qin, T.; Feng, D.; Yang, B.; Zhu, Z.; Xing, L. Diagenesis and Pore Formation Evolution of Continental Shale in the Da'anzhai Lower Jurassic Section in the Sichuan Basin. *Minerals* **2023**, *13*, 535. [CrossRef]
43. Wang, L.; Liu, B.; Bai, L.; Yu, Z.; Huo, Q.; Gao, Y. Pore evolution modeling in natural lacustrine shale influenced by mineral composition: Implications for shale oil exploration and CO_2 storage. *Adv. Geo-Energy Res.* **2024**, *13*, 218–230. [CrossRef]
44. Taylor, K.; Macquaker, J.H. Diagenetic alterations in a silt-and clay-rich mudstone succession: An example from the Upper Cretaceous Mancos Shale of Utah, USA. *Clay Miner.* **2014**, *49*, 213–227. [CrossRef]
45. Yan, Y.; Wang, M.; Misch, D.; Sachsenhofer, R.F.; Wu, Y.; Li, J. Mineral diagenesis in lacustrine organic-rich shales: Evolution pathways and implications for reservoir characteristics. *J. Asian Earth Sci.* **2024**, *263*, 106026. [CrossRef]
46. Ling, K.; Wang, Z.; Zhang, C.; Dong, L. Coupled Evolution of Clay Minerals and Organic Matter During Diagenesis: Mechanisms of Smectite Illitization in Organic-Rich Shale. *Processes* **2025**, *13*, 2966. [CrossRef]
47. Wang, J.; Zhou, L.; Liu, J.; Zhang, X.; Zhang, F.; Zhang, B. Acid-base alternation diagenesis and its influence on shale reservoirs in the Permian Lucaogou Formation, Jimusar Sag, Junggar Basin, NW China. *Pet. Explor. Dev.* **2020**, *47*, 962–976. [CrossRef]
48. Li, F.; Wang, M.; Liu, S.; Hao, Y. Pore characteristics and influencing factors of different types of shales. *Mar. Pet. Geol.* **2019**, *102*, 391–401. [CrossRef]

49. De Silva, P.N.K.; Simons, S.J.R.; Stevens, P.; Philip, L.M. A comparison of North American shale plays with emerging non-marine shale plays in Australia. *Mar. Pet. Geol.* **2015**, *67*, 16–29. [CrossRef]
50. Yang, C.; Zhang, J.; Tang, X.; Ding, J.; Zhao, Q.; Dang, W.; Chen, H.; Su, Y.; Li, B.; Lu, D. Comparative study on micro-pore structure of marine, terrestrial, and transitional shales in key areas, China. *Int. J. Coal Geol.* **2017**, *171*, 76–92. [CrossRef]
51. Li, H.; He, S.; Radwan, A.E.; Xie, J.; Qin, Q. Quantitative Analysis of Pore Complexity in Lacustrine Organic-Rich Shale and Comparison to Marine Shale: Insights from Experimental Tests and Fractal Theory. *Energy Fuels* **2024**, *38*, 16171–16188. [CrossRef]
52. Jiang, S.; Xu, Z.; Feng, Y.; Zhang, J.; Cai, D.; Chen, L.; Wu, Y.; Zhou, D.; Bao, S.; Long, S. Geologic characteristics of hydrocarbon-bearing marine, transitional and lacustrine shales in China. *J. Asian Earth Sci.* **2016**, *115*, 404–418. [CrossRef]
53. Hackley, P.C.; Dennen, K.O.; Garza, D.; Lohr, C.D.; Valentine, B.J.; Hatcherian, J.J.; Enomoto, C.B.; Dulong, F.T. Oil-source rock correlation studies in the unconventional Upper Cretaceous Tuscaloosa marine shale (TMS) petroleum system, Mississippi and Louisiana, USA. *J. Pet. Sci. Eng.* **2020**, *190*, 107015. [CrossRef]
54. Wang, X.; Yu, Z.; He, K.; Huang, X.; Ye, M.; Guan, M.; Zhang, S. Assessment and exploration prospects of the oldest marine shale oil in North China. *Pet. Explor. Dev.* **2025**, *52*, 630–648. [CrossRef]
55. McMahon, T.P.; Larson, T.E.; Zhang, T.; Shuster, M. Geologic characteristics, exploration and production progress of shale oil and gas in the United States: An overview. *Pet. Explor. Dev.* **2024**, *51*, 925–948. [CrossRef]
56. Hu, S.; Bai, B.; Tao, S.; Bian, C.; Zhang, T.; Chen, Y.; Liang, X.; Wang, L.; Zhu, R.; Jia, J.; et al. Heterogeneous geological conditions and differential enrichment of medium and high maturity continental shale oil in China. *Pet. Explor. Dev.* **2022**, *49*, 257–271. [CrossRef]
57. Lu, H.; Li, Q.; Yue, D.; Wu, Y.; Gao, J.; Wu, S.; Wang, W.; Li, M.; An, K. Quantitative characterization and formation mechanism of the pore system heterogeneity: Examples from organic-rich laminated and organic-poor layered shales of the upper triassic chang 7 member in the southern Ordos Basin, China. *Mar. Pet. Geol.* **2023**, *147*, 105999. [CrossRef]
58. Pang, X.; Li, M.; Li, B.; Wang, T.; Hui, S.; Liu, Y.; Liu, G.; Hu, T.; Xu, T.; Jiang, F.; et al. Main controlling factors and movability evaluation of continental shale oil. *Earth-Sci. Rev.* **2023**, *243*, 104472. [CrossRef]
59. Li, T.; Jiang, Z.; Li, Z.; Wang, P.; Xu, C.; Liu, G.; Su, S.; Ning, C. Continental shale pore structure characteristics and their controlling factors: A case study from the lower third member of the Shahejie Formation, Zhanhua Sag, Eastern China. *J. Nat. Gas Sci. Eng.* **2017**, *45*, 670–692. [CrossRef]
60. Feng, G.; Zhang, L.; Yan, J.; Chen, S.; Pu, X.; Han, W.; Shi, Z.; Zhang, W. Lithofacies-based analysis of pore structure characteristics and controlling factors of shale reservoirs: A case study of the second member of the Kongdian Formation in the Cangdong Sag, Bohai Bay Basin, China. *Mar. Pet. Geol.* **2025**, *182*, 107575. [CrossRef]
61. Fu, Y.; Ruan, Z.; Yu, B.; Hou, G.; Bai, C.; Chang, Q. Reservoir Characterization and Genetic Analysis Based on Shale Lithofacies Classification: A Case Study of the Paleogene Shahejie Formation, East China. *Minerals* **2025**, *15*, 406. [CrossRef]
62. Wang, X.; Sun, Y.; Wang, T.; Yan, B.; Liu, R. Analysis of Controlling Factors of Pore Structure in Different Lithofacies Types of Continental Shale—Taking the Daqingzi Area in the Southern Songliao Basin as an Example. *Minerals* **2024**, *14*, 1025. [CrossRef]
63. Hu, H.; Wang, D.; Li, W.; Zhu, G.; Chen, X. The Heterogeneity Characterization of Lacustrine Shale Pores in the Daanzhai Member of the Ziliujing Formation in the Yuanba Area, Sichuan Basin. *Minerals* **2024**, *15*, 11. [CrossRef]
64. Lu, H.; Xia, D.; Li, Q.; Yue, D.; Wu, S.; Wang, W.; Zhang, X. Quantitative Characterization and Differences of the Pore Structure in Lacustrine Siliceous Shale and Argillaceous Shale: A Case Study of the Upper Triassic Yanchang Formation Shales in the Southern Ordos Basin, China. *Energy Fuels* **2021**, *35*, 15525–15544. [CrossRef]
65. Li, B.; Pang, X.; Dong, Y.; Peng, J.; Gao, P.; Wu, H.; Huang, C.; Shao, X. Lithofacies and pore characterization in an argillaceous-siliceous-calcareous shale system: A case study of the Shahejie Formation in Nanpu Sag, Bohai Bay Basin, China. *J. Pet. Sci. Eng.* **2019**, *173*, 804–819. [CrossRef]
66. Li, P.; Gong, H.; Jiang, Z.; Zhang, F.; Liang, Z.; Wang, Z.; Wu, Y.; Shao, X. Characteristics and Influencing Factors of Multi-Scale Pore Structure Heterogeneity of Lacustrine Shale in the Gaoyou Sag, Eastern China. *Minerals* **2023**, *13*, 359. [CrossRef]
67. Liu, Y.; Li, W.; Yuan, J.; Li, P.; Ge, X.; Ge, X.; Liu, P.; Wu, H.; Yu, X.; Huang, B. Lithofacies Characteristics of the Lower Cretaceous Qing 1 Member in the Heiyupao Depression, Northern Binbei Area of the Songliao Basin. *Minerals* **2025**, *15*, 125. [CrossRef]
68. Liu, B.; Wang, H.; Fu, X.; Bai, Y.; Bai, L.; Jia, M.; He, B. Lithofacies and depositional setting of a highly prospective lacustrine shale oil succession from the Upper Cretaceous Qingshankou Formation in the Gulong sag, northern Songliao Basin, northeast China. *AAPG Bull.* **2019**, *103*, 405–432. [CrossRef]
69. Sun, L.; Zhu, R.; Zhang, T.; Cai, Y.; Feng, Z.; Bai, B.; Jiang, H.; Wang, B. Advances and trends of non-marine shale sedimentology: A case study from Gulong Shale of Daqing Oilfield, Songliao Basin, NE China. *Pet. Explor. Dev.* **2024**, *51*, 1367–1385. [CrossRef]
70. Yang, W.; Li, Y.; Gao, R. Formation and evolution of nonmarine petroleum in Songliao Basin, China. *AAPG Bull.* **1985**, *69*, 1112–1122. [CrossRef]
71. Feng, Z.; Jia, C.; Xie, X.; Zhang, S.; Feng, Z.; Timothy, A.C. Tectonostratigraphic units and stratigraphic sequences of the nonmarine Songliao basin, northeast China. *Basin Res.* **2010**, *22*, 79–95. [CrossRef]

72. Song, Z.; Qin, Y.; George, S.C.; Wang, L.; Guo, J.; Feng, Z. A biomarker study of depositional paleoenvironments and source inputs for the massive formation of Upper Cretaceous lacustrine source rocks in the Songliao Basin, China. *Palaeogeogr. Palaeoclimatol. Palaeoecol.* **2013**, *385*, 137–151. [CrossRef]
73. Kennedy, M.J.; Löhr, S.C.; Fraser, S.A.; Baruch, E.T. Direct evidence for organic carbon preservation as clay-organic nanocomposites in a Devonian black shale; from deposition to diagenesis. *Earth Planet. Sci. Lett.* **2014**, *388*, 59–70. [CrossRef]
74. Rahman, H.M.; Kennedy, M.J.; Löhr, S.C.; Dewhurst, D.N. Clay-organic association as a control on hydrocarbon generation in shale. *Org. Geochem.* **2017**, *105*, 42–55. [CrossRef]

processes

MDPI

Article

Coupled Evolution of Clay Minerals and Organic Matter During Diagenesis: Mechanisms of Smectite Illitization in Organic-Rich Shale

Kun Ling [1,2], Ziyi Wang [1,2], Changhu Zhang [1,2] and Lin Dong [1,2,*]

1 School of Earth and Space Sciences, Peking University, Beijing 100871, China; sqlingkun@pku.edu.cn (K.L.); ziyi-wang@pku.edu.cn (Z.W.); 2201210141@stu.pku.edu.cn (C.Z.)
2 Key Laboratory of Orogenic Belts and Crustal Evolution, MOE, Beijing 100871, China
* Correspondence: lin.dong@pku.edu.cn

Abstract

The transformation of smectite to illite documents multi-scale water–rock–hydrocarbon interaction dynamics. Current studies predominantly emphasize the influence of inorganic systems on this process, while overlooking the dynamic regulation by organic matter and the synergistic effects of multiple controlling factors under actual geological conditions. In this study, we conducted integrated semi-open pyrolysis experiments on natural samples from the Chang-7 Member and hydrothermal experiments using synthetic analogs. The illitization process of smectite was characterized through XRD analysis and SEM observations, while organic geochemical testing was employed to track the corresponding thermal evolution of organic matter. The semi-open pyrolysis results reveal that significant changes in illite–smectite (I/S) mixed layer minerals and illite content/morphology occur above 320 °C, which coincides with the critical threshold for extensive organic matter evolution. Thermal degradation of organic matter generates pore space, thereby enhancing water–rock interactions involving clay minerals. This demonstrates the co-evolution of organic matter and smectite, and indicates that temperature indirectly influences illitization by regulating organic matter thermal evolution. The hydrothermal simulation experiments demonstrate the early-stage characteristics of illitization. Unlike long-term geological evolution, K^+ under experimental conditions primarily originates from the aqueous medium due to kinetic constraints on feldspar dissolution. Notably, organic matter regulates K^+ partitioning dynamics—increased organic matter content hinders K^+ incorporation into smectite interlayers, thereby suppressing the illitization process. Cross-system experimental analysis reveals that organic matter exhibits temporally dependent dual functionality, serving both mediating and modulating roles within inorganic diagenetic systems. This study delineates diagnostic-stage-dependent mechanisms governing smectite illitization through multifactorial synergistic interplay, establishing a predictive framework applicable to organic-rich systems exemplified by the Chang-7 Shale.

Keywords: smectite illitization; organic matter; Chang-7 shale; pyrolysis experiments

1. Introduction

Clay minerals are ubiquitous and essential components of rocks in petroliferous basins, documenting multiscale water–rock–hydrocarbon interactions throughout their formation and evolutionary processes. This renders them of broad scientific significance [1]. Among the various mineralogical transformations, the smectite-to-illite conversion has received particular attention due to its strong links to key geological processes, including thermal maturity reconstruction, hydrocarbon generation and reservoir quality evolution [2–8].

Illitization is a diagenetically-driven transformation in which precursor minerals (particularly smectite) gradually convert to illite, typically initiating under low-temperature conditions during early burial and continuing with increasing temperature and pressure [9]. Regardless of illitization mechanism (dissolution-precipitation vs. solid-state transformation), the rate and extent of illitization are fundamentally affected by time, temperature, and diagenetic fluid chemistry, particularly the concentrations and availability of potassium [10]. Previous studies have conducted extensive kinetic investigations of smectite illitization using both natural samples and laboratory-controlled experiments. These works have established various reaction models based on first-order and second-order kinetic equations, consistently highlighting the dominant role of temperature in the transformation process [11–14]. Furthermore, hydrothermal experiments have systematically examined the influence of other key parameters, including liquid/solid ratio, and reaction duration [15–18].

However, most existing research has focused on illitization in purely inorganic systems, overlooking the co-existence and interactions of clay minerals with organic matter in natural shale systems. In fact, in organic-rich shales (e.g., Chang-7 Shale), clay minerals and organic matter are two principal components, representing the inorganic and organic fractions, respectively. These constituents have been closely associated since deposition, forming widely documented clay-organic nanocomposites that subsequently undergo coupled burial and diagenetic evolution [6,19,20]. During diagenesis, clay minerals and organic matter exert mutual influence, evidenced by the parallel progression of smectite illitization and hydrocarbon generation—demonstrating the intrinsic connection between these two critical geological processes [18,21]. Experimental pyrolysis studies involving smectite and organic matter have further revealed stage-specific correlations between organic decarboxylation/cracking processes and smectite-to-illite transformation [22]. A growing body of evidence indicates that smectite illitization can effectively catalyze hydrocarbon generation [23–25]. Yet, the reverse relationship—how the thermal evolution of organic matter affects the smectite illitization process—remains poorly understood. Even the limited number of studies has typically introduced only a single type of organic compound into experimental setups [26], failing to capture the complex coexisting organic-inorganic conditions characteristic of actual geological settings.

The development of rapid-heating pyrolysis techniques, driven by the temperature-time compensation concept in hydrocarbon kinetics, has enabled simulation of organic matter evolution under controlled thermal conditions [27–29]. This approach allows the reconstruction of geological processes that typically occur over millions of years within a significantly shortened experimental timeframe by employing temperatures significantly exceeding those encountered in natural diagenesis. These pyrolysis systems, classified as open, semi-open, and closed, serve different purposes. Open systems, although widely applied as a classical method for source rock evaluation, exhibit significant discrepancies compared to actual geological conditions [30,31]. Semi-open systems effectively simulate episodic hydrocarbon expulsion, thereby more accurately reproducing the diagenetic processes accompanying hydrocarbon generation [32,33]. Within such systems, it is possible to conduct experimental simulations using intact core samples, which significantly enhances the consistency between thermal simulation experiments and actual geological conditions. By better preserving the natural rock fabric and geochemical characteristics, this approach enables more realistic interactions between organic and inorganic phases, thereby providing a distinct advantage for reconstructing diagenetic evolution processes. In contrast, closed systems prevent material exchange with the external environment. Among the various experimental approaches utilized in such systems, gold-tube pyrolysis is the most extensively applied method [34,35]. A key advantage of closed systems is their ability to facilitate detailed studies of various water–rock reactions in the presence of aqueous media, rendering them particularly suitable for research on the evolution of clay minerals. Salisu et al. successfully reproduced smectite authigenesis and subsequent illitization, replicating a mineral transformation sequence analogous to that observed in natural settings [36]. Moreover, by modulating experimental parameters and performing structural modification of smectite, recent studies have increasingly explored the mechanisms of illitization and identified distinct factors influencing this process [9,37]. With increasingly diverse application scenarios and more sophisticated experimental apparatus, pyrolysis techniques have also become crucial for comprehensively simulating the evolution of organic matter, inorganic minerals, pore structures, and fluids [38–42].

In this study, we aim to bridge the gap between organic and inorganic perspectives by investigating smectite-to-illite transformation under integrated organic-inorganic systems. In contrast to prior studies that examined illitization in isolation, this work is designed to simulate complex geological conditions, with a specific focus on the illitization in the presence of organic matter, addressing the synergistic effects of temperature, potassium ion concentration, K-feldspar and organic components. These factors are inherently interdependent and vary in influence across different diagenetic stages. Through a series of integrated thermal simulation experiments using both natural samples and synthetic powders, we seek to reconstruct the illitization process and unravel the complex, multivariate controls that govern smectite transformation in realistic geological settings.

2. Samples and Methods

2.1. Natural Sample: Semi-Open Pyrolysis Experiments

In this study, natural core samples were used for semi-open pyrolysis experiments. The samples were obtained from the Chang-7 Member of Well Z233, located in the southern Ordos Basin (Figure 1). The shale exhibits high total organic carbon content (TOC = 11.7%) and Type-II kerogen composition (Table 1), with low thermal maturity evidenced by Ro = 0.51%. Six cylindrical samples (25 mm in diameter and 60 mm in height) were wire-cut from the core, with all samples oriented vertically to maintain bedding consistency and ensure experimental reproducibility.

Table 1. Geochemical characterization of the original Z233 sample.

Original Rock	
TOC (%)	11.69
T_{max} (°C)	435
S_1 (mg/g)	2.17
S_2 (mg/g)	38.81
S_3 (mg/g)	0.25
R_O (%)	0.51
Kerogen type	Type-II

The pyrolysis experiments were conducted at the Research Institute of Petroleum Exploration & Development (RIPED) using a semi-open pyrolysis system (Nantong Huaxing Petroleum Devices, Jiangsu, China). The device comprises four subsystems: (1) pyrolysis reaction apparatus, (2) hydrocarbon expulsion system, (3) product separation/collection apparatus, and (4) computerized control apparatus.

The experimental procedure was as follows: first, cylindrical samples were placed in the reaction vessel, with both ends sealed using graphite rings, and the vessel was then mounted in the base of the heating furnace. Subsequently, vertical pressure (35 MPa) was applied via a bidirectional hydraulic apparatus to simulate lithostatic pressure, while a water injection pump introduced fluid to achieve the target formation pressure (20 MPa).

A pre-programmed heating process was initiated to control the pyrolysis temperature. The system's openness was regulated by a backpressure valve and pressure sensor, with the hydrocarbon discharge valve automatically releasing products into a collection container when pressure exceeded the set limit. Once the target pyrolysis temperature was reached, the discharge valve was manually opened to facilitate product collection.

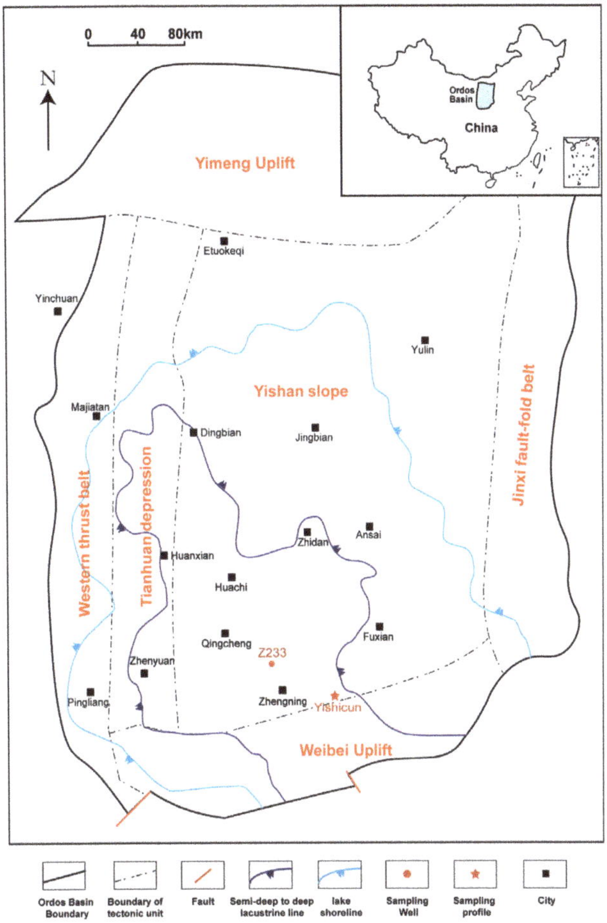

Figure 1. Geological map of the Ordos Basin showing lacustrine facies distribution of the Ch7 Member and locations of the sampling well and section (modified from Wang et al., 2023 [43]). "Reproduced with permission from Guanping Wang, Journal of Asian Earth Sciences; published by Elsevier, 2023".

Six parallel pyrolysis experiments were conducted in this study, employing a sequential heating protocol with final target temperatures ranging from 280 °C to 420 °C (Figure 2). Specifically, all six samples underwent unified pretreatment: initial 1-h preheating from 20 °C to 120 °C with subsequent 1 h of isothermal holding. The system was then heated to 240 °C at a rate of 20 °C/h and maintained for 2 h. During the staged heating phase, all six samples were heated to 280 °C at 20 °C/h and held isothermally, followed by product collection from the 280 °C-targeted specimen. The remaining five replicates were subjected to programmed pyrolysis at 320 °C, maintaining identical thermal parameters, to enable temperature-stage-specific product collection. This sequence was iterated through increasing temperature tiers (350 °C → 380 °C → 400 °C → 420 °C), where each subsequent stage incorporated thermal histories from previous phases. The cumulative thermal progression enabled reconstruction of continuous clay mineral evolution pathways.

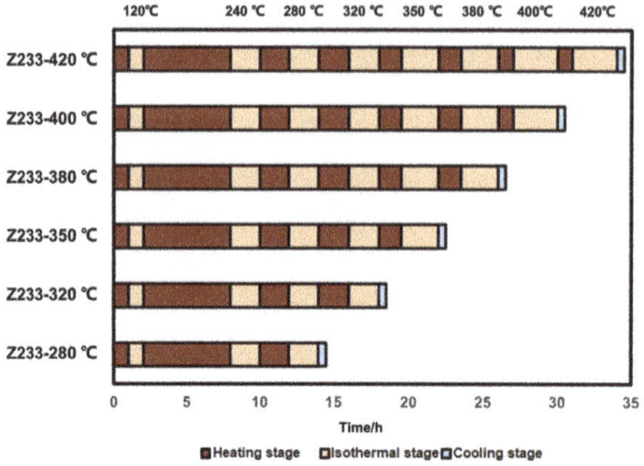

Figure 2. The heating procedure of the semi-open pyrolysis experiments.

Upon completion of the pyrolysis protocol, the reactor underwent rapid cooling. Solid residues were collected for subsequent characterization of clay minerals and organic matter. Clay mineral analysis comprised X-ray diffraction (XRD) and scanning electron microscopy (SEM), while organic geochemical evaluation included vitrinite reflectance analysis, Rock-Eval pyrolysis, and total organic carbon (TOC) quantification.

2.2. Synthetic Samples: Hydrothermal Experiments

The synthetic samples in this study were fabricated through proportioned mechanical blending of pristine mineral components with isolated kerogen concentrates. Mineralogical constituents comprised Na-smectite (Hohhot, Inner Mongolia), chlorite (Chifeng, Inner Mongolia), kaolinite (Chengdu, Sichuan), K-feldspar (Lingshou, Hebei), and AR Grade quartz (Sinopharm Chemical Reagent Co., Ltd., Shanghai, China), with cumulative purity levels ranging from 95% to 99.5%. K-smectite was also produced via cation exchange treatment of Na-smectite using 1 M KCl solution to study the effect of interlayer cations on smectite-to-illite conversion [25]. Fresh shale samples for kerogen isolation were collected from the Yishicun section, exhibiting high organic carbon content (TOC = 16.8%) with negligible thermal maturity (R_o = 0.41%). The kerogen was isolated through a standardized protocol involving: (1) Soxhlet extraction with an azeotropic dichloromethane/methanol mixture (93:7 v/v) of 200-mesh (75 μm) powdered samples for 48 h, followed by (2) sequential acid treatments using 6 M hydrochloric acid for carbonate dissolution and 40% hydrofluoric acid for silicate removal [44].

One reference sample was prepared based on the TOC and primary mineral composition of Chang-7 shale for experimental control (Table 2), along with two supplementary samples configured at identical proportions to investigate aqueous medium effects. Six additional samples were configured through systematic adjustments in kerogen content, K-feldspar proportion, and the smectite type. Hydrothermal experiments were conducted at 250 °C and 300 °C, respectively, therefore duplicate sets of samples were prepared accordingly. To ensure sample homogeneity, all powdered mixtures were thoroughly homogenized through grinding (>30 min) in corundum mortars prior to experimental runs, achieving both consistent particle size distribution and intimate inter-mineral contact.

Table 2. Experimental Conditions of Hydrothermal Experiments and Composition Information of Powder Samples.

Naming Convention	Simulated Temperature (°C)	Solution	Powder Sample Composition Content (wt%)					
			Kerogen	Smectite *	Kaolinite	Chlorite	K-Feldspar	Quartz
250-Na-1 M	250	1 M KCl	10	36 (Na)	3	3	12	36
250-K-1 M		1 M KCl	10	36 (K)	3	3	12	36
250-0.1 M		0.1 M KCl	10	36 (Na)	3	3	12	36
250-0.01 M		0.01 M KCl	10	36 (Na)	3	3	12	36
250-0Kfs		1 M KCl	10	36 (Na)	3	3	0	48
250-6%Kfs		1 M KCl	10	36 (Na)	3	3	6	42
250-15%Kfs		1 M KCl	10	36 (Na)	3	3	15	33
250-5%OM		1 M KCl	5	36 (Na)	3	3	12	41
250-15%OM		1 M KCl	15	36 (Na)	3	3	12	31
300-Na-1 M	300	1 M KCl	10	36 (Na)	3	3	12	36
300-K-1 M		1 M KCl	10	36 (K)	3	3	12	36
300-0.1 M		0.1 M KCl	10	36 (Na)	3	3	12	36
300-0.01 M		0.01 M KCl	10	36 (Na)	3	3	12	36
300-0Kfs		1 M KCl	10	36 (Na)	3	3	0	48
300-6%Kfs		1 M KCl	10	36 (Na)	3	3	6	42
300-15%Kfs		1 M KCl	10	36 (Na)	3	3	15	33
300-5%OM		1 M KCl	5	36 (Na)	3	3	12	41
300-15%OM		1 M KCl	15	36 (Na)	3	3	12	31

* The Na or K in parentheses represents the type of interlayer cations in smectite.

Hydrothermal experiments were conducted at China University of Petroleum (Beijing) and RIPED using flexible gold tubes (8 mm outer diameter, 1mm thickness and 60 mm length) contained within steel pressure vessels (Nantong Huaan Scientific Research Devices Co., Ltd., Jiangsu, China) [45]. The 3 g ground powder sample and 3 mL KCl solution were loaded into a gold tube through six successive additions to ensure thorough wetting of the powder, while maintaining slight fluid mobility within the system. Upon completion of loading, the gold tubes were purged with argon to remove oxygen, sealed, and tested for airtightness.

The hermetically sealed gold tubes were precisely positioned within the high-pressure reactor chamber. Formation pressure conditions (20 MPa) were established through controlled water injection employing a hydraulic pump system. The thermal regime was then initiated with a heating rate of 2 °C/min until reaching predetermined isothermal conditions, which were maintained for 72 h to ensure complete reaction equilibrium. Following simulation experiments, the reactors were allowed to cool naturally to ambient temperature. The pyrolysis products were thoroughly oven-dried at 60 °C under controlled conditions. The dried solids were subsequently reground and homogenized using an agate mortar to achieve particle size uniformity prior to XRD characterization for mineralogical analysis.

2.3. *Organic-Inorganic Composition Analyses*

2.3.1. Mineralogical Characterization

Both bulk rock and clay mineral X-ray diffraction (XRD) analysis were performed using a D8 Advance X-ray diffractometer (Bruker Corporation, Karlsruhe, Germany) with CuKα radiation (λ = 1.5418 Å), voltage at 30 kV, current at 20 mA. Whole-rock samples were ground to <100-mesh size (150 μm), and the <2 μm clay fraction was subsequently isolated from homogenized bulk rock samples following the suspension-centrifugation methodology prescribed by China Petroleum Standard SY/T 5163-2018 [46]. When necessary, pretreatments with H_2O_2 and dilute HCl were applied to remove organic matter and carbonate minerals prior to clay separation. The extracted clay-sized fractions were prepared as oriented mounts. All samples underwent XRD analysis under three standardized treatment conditions: air-dried (N) at room temperature; ethylene glycol solvated (EG) after being saturated at 60 °C for 8 h; and heat-treated state (HT) after heating at 550 °C for 3 h. The XRD patterns were subjected to phase identification and semi-quantitative analysis using MDI Jade 9 software.

Integrated scanning electron microscopy (SEM) analysis was performed on naturally fractured surfaces, obtained by controlled rock cleavage and polished specimens, prepared through mechanical polishing and argon ion milling [47]. Prior to SEM characterization using a FEI Quanta 650 FEG field-emission scanning electron microscope (Field Electron and Ion Company, Hillsboro, OR, USA), samples were sputter-coated with a 5 nm chromium layer. Mineral crystal morphology was characterized through observation of fresh fracture surfaces, while clay mineral and organic matter fabric features were analyzed using secondary electron (SE)/backscattered electron (BSE) imaging combined with energy-dispersive spectroscopy (EDS).

2.3.2. Organic Geochemical Analyses

Total organic carbon (TOC) analyses were performed on a LECO CS230 carbon/sulfur analyzer (LECO Corporation, St. Joseph, MI, USA) [48]. Samples were pulverized to powders finer than 100 mesh and decarbonated by hydrochloric acid treatment. Processed materials were combusted at 550 °C under oxygen-rich conditions for TOC quantification.

Vitrinite reflectance (R_o) analysis was conducted on whole-rock samples through a standardized preparation protocol involving crushing to <10-mesh size (2 mm), homogenization with agar in a 1:1 mass ratio, and consolidation prior to thin-section preparation. The resulting polished sections were analyzed under oil immersion using a Zeiss Scope A1 incident-light microscope equipped with a photomultiplier tube and 546 nm monochromatic illumination (ZEISS Group, Oberkochen, Germany). Between 20 and 55 valid measurement points were systematically acquired per sample, with exclusion of weathered or altered domains to ensure data integrity.

Rock-Eval pyrolysis analysis was conducted on a Rock-Eval 6 instrument (Vinci Technologies, Nanterre, France) employing programmed heating from 300 °C to 650 °C at a rate of 25 °C/min under a nitrogen atmosphere. Approximately 100 mg of homogenized samples with particle sizes between 80 and 100 mesh was subjected to thermal decomposition to quantify free hydrocarbons (S_1), pyrolyzable hydrocarbons (S_2), and maximum pyrolysis temperature (T_{max}). Hydrogen index (HI) was calculated as S_2/TOC × 100. Analytical accuracy was verified through concurrent measurement of certified reference materials.

3. Results

3.1. Mineralogical Characteristics of Semi-Open Pyrolysis Products

XRD analysis reveals that the total clay mineral content in pyrolysis products gradually increased with rising temperature. Specifically, the illite–smectite mixed-layer (I/S) underwent progressive illitization, characterized by a systematic decline in I/S content and a corresponding increase in discrete illite. This mineralogical transformation became geologically significant above the 320 °C threshold. Furthermore, chlorite content follows a trend consistent with bulk clay fraction, suggesting that neoformation of chlorite contributed to the observed compositional changes. These quantitative mineralogical trends are systematically presented in Figure 3 and Table S1.

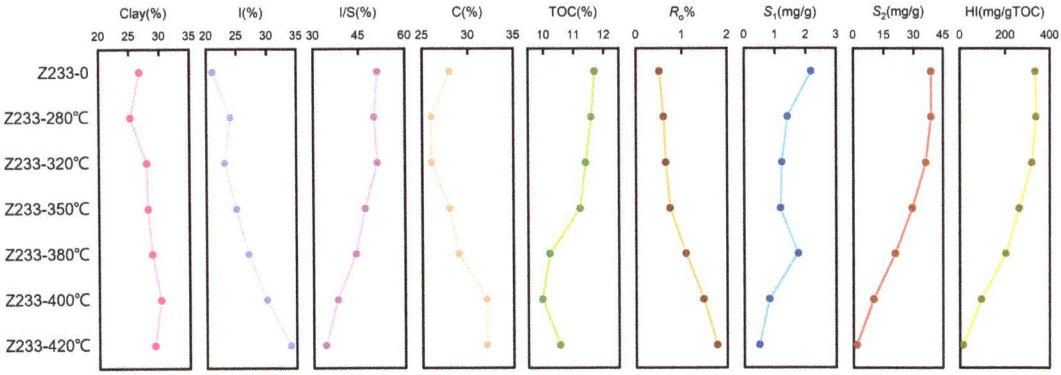

Figure 3. Evolution curves of clay mineral content and organic geochemical parameters in the initial sample (Z233-0) and pyrolysis products. Clay (%) represents the proportion of clay minerals in the total rock, while I (%), I/S (%), and C (%) denote the relative contents of illite, I/S, and chlorite within the clay mineral fraction, respectively.

The SEM analysis of natural fracture surfaces further characterizes the morphological transitions of I/S mixed-layer minerals and illite during thermal evolution. In the sample subjected to 280 °C thermal maturation, clay minerals exhibit matrix-like textural characteristics (Figure 4a). Illite displays a characteristic platy morphology, while I/S mixed-layer minerals appear as thin flakes with slightly curled edges. The absence of the typical honeycomb-like I/S structures suggests an advanced stage of illitization. At this stage, no identifiable chlorite monomers or aggregates were observed. When the pyrolysis temperature was increased to 320 °C, clay minerals developed distinct morphological features (Figure 4b). Illite and I/S mixed-layer minerals maintained their intergrown associations but demonstrated enhanced crystallinity, reflected in more well-defined platy morphologies. Notably, authigenic carbonate minerals were observed precipitating along mineral edges, potentially associated with the co-evolution of clay minerals and organic matter. In addition, sparse acicular chlorite crystals began to appear, although their abundance remained low at this stage (Figure 4c).

At the terminal temperature of 380 °C, clay minerals demonstrated enhanced orientation, reflecting the advancement of illitization and alignment under simulated lithostatic pressure. Authigenic quartz observed within mineral interstices further corroborates active diagenetic fluid-rock interactions (Figure 4d). During this phase, chlorite crystallized rapidly, manifesting two distinct morphological types: acicular microcrystals (~1 μm) and rosette-like aggregates (~10 μm), suggesting temperature-dependent crystallization kinetics (Figure 4e). With continued heating, pyrobitumen derived from organic coking partially filled the pore spaces, progressively occluding pore networks and ultimately obscuring diagnostic crystal morphology along fracture surfaces (Figure 4f). Comparative analysis suggests that the clay mineral transformations observed during pyrolysis experiments resemble those in natural geological samples, particularly in the development of chlorite textures [49].

The argon-ion-polished SEM analysis comprehensively documented the diagenetic fabric evolution of clay minerals and organic matter in semi-open pyrolysis products. The unheated samples exhibited well-preserved interlaminated organic-rich and clay-rich laminae (Figure 5a,a′), containing detrital grains that lack evidence of dissolution. Organic matter was extensively distributed within the interlayers of clay minerals—an association commonly observed in the Chang-7 shale [47]. At 280 °C, the sample retained its microstructural fabric, with only minor organic matter loss (Figure 5b,b′). When the pyrolysis temperature was increased to 350 °C, the organic content decreased significantly, although clay-organic associations persist in certain enriched domains (Figure 5c,c′). By 420 °C, organic laminae are largely replaced by bedding-parallel banded pores, indicating substantial organic degradation and hydrocarbon generation (Figure 5d,d′).

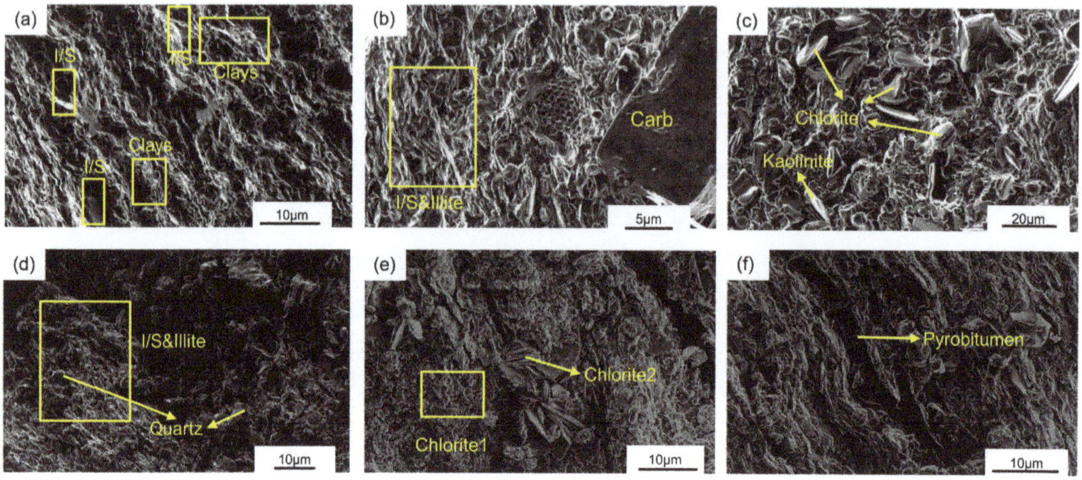

Figure 4. SEM images of natural fracture surfaces in pyrolysis products. (**a**) I/S and matrix-like clays at 280 °C; (**b**) interwoven I/S and illite with carbonate minerals (Carb) indicating clay-organic coevolution at 320 °C; (**c**) acicular chlorite and potential intermediate kaolinite at 320 °C; (**d**) platy illite with higher proportion and authigenic quartz at 380 °C; (**e**) two generations of chlorite: 1 μm-size acicular microcrystals (Chlorite 1) and 10 μm-scale rosette-like aggregates (Chlorite 2) at 380 °C; (**f**) pyrobitumen filling partial pores at 380 °C.

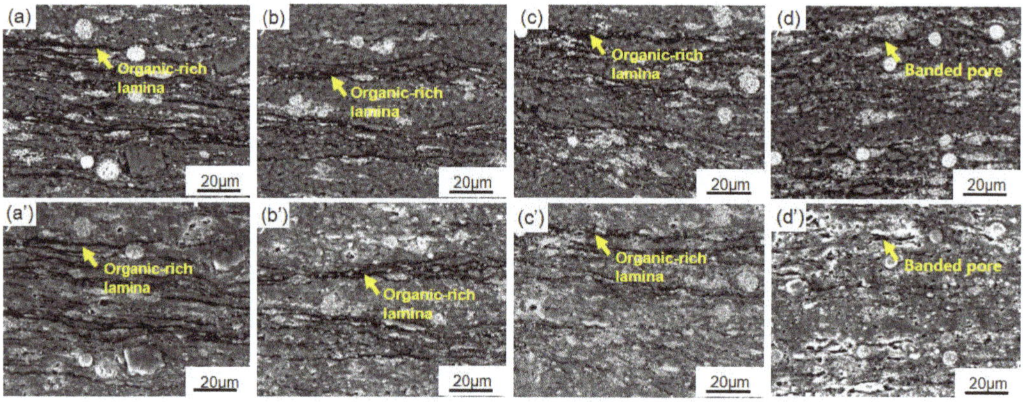

Figure 5. SEM images of argon-ion-polished pyrolysis products. (**a**,**a′**) The initial sample with interlaminated organic-rich and clay-rich laminae; (**b**,**b′**) the essentially unchanged pyrolysis product at 280 °C; (**c**,**c′**) gradually consumed organic-rich laminae at 350 °C; (**d**,**d′**) banded pores replacing organic-rich laminae at 420 °C. (**a**–**d**) are secondary electron images and (**a′**–**d′**) are backscattered electron images.

3.2. Organic Geochemical Analyses of Semi-Open Pyrolysis Products

Organic geochemical indices systematically document the thermal maturation of organic matter throughout the pyrolysis experiments, exhibiting marked evolutionary trends in response to increasing thermal severity (Figure 3 and Table S2). TOC content gradually decreased, with a notably accelerated reduction occurring within the critical temperature window of 350–380 °C. A slight rebound in TOC observed above 400 °C is likely attributed to carbonaceous residue from coking during late-stage pyrolysis, a phenomenon also supported by SEM observation (Figure 4f). Comparative analysis reveals an initial S_1 value reduction in the 280 °C product versus the raw sample, suggesting the expulsion of free hydrocarbons under thermobaric conditions, compounded by limited hydrocarbon generation at this stage. Following this phase, S_1 values remained essentially stable until reaching the 380 °C thermal threshold, where a transient peak in S_1 manifested, indicating significant enhancement in hydrocarbon-generating capacity during this critical stage. Progressive thermal elevation induced expulsion of hydrocarbons formed during earlier maturation stages, leading to concomitant depletion in S_1 values through hydrocarbon migration processes. The S_2 value and Hydrogen Index (HI) remained relatively constant during low-temperature pyrolysis (<320 °C) but underwent rapid attenuation above this threshold. These values approached zero at 420 °C, indicating the effective exhaustion of hydrocarbon generation potential.

3.3. *XRD Spectral Characteristics of Hydrothermal Experiment Products*

3.3.1. Reference Samples

The XRD analysis of hydrothermal experimental products from control samples exhibits diagnostic mineralogical features of the smectite-to-illite transformation. Serving as baseline controls, these samples were also systematically compared with other reaction products to elucidate component-specific controls on illitization dynamics within the system. Figure 6 depicts the XRD patterns of solid residues generated through hydrothermal reactions of Na- and K-smectite with 1 M KCl at 250 °C and 300 °C. Comparative XRD patterns of reaction products relative to pristine Na-smectite reveal significant structural reorganization. A significant decrease in interlayer expandability of clay minerals is demonstrated by upward shifts in the S(001) reflection toward higher °2θ in both oriented air-dried (N) and glycolated (EG) mounts. For instance, when employing Na-smectite as the starting material, the air-dried d(001)-spacing decreased from 12.47 Å for untreated material, to 11.82 Å at 250 °C and 11.31 Å at 300 °C, while d-spacing for EG samples shifted from 16.88 Å to 15.35 Å and 13.92 Å correspondingly.

Given that these peaks are systematically positioned between the characteristic reflections of expandable (smectite) and non-expandable (illite) clay minerals, the ubiquitous presence of I/S mixed-layer phases in the reaction products is conclusively demonstrated. Despite these transformations, discrete illite (d(001)-spacing = 10 Å) remained absent in all products except for minor fluctuations observed in 300 °C K-smectite derivatives.

Notably, the K-smectite reaction products demonstrated greater d-spacing reduction in EG mounts compared with Na-smectite simulation residues, particularly under lower temperature. The d(001)-spacing shifted from 16.88 Å to 15.35 Å at 250 °C and 13.92 Å at 300 °C. A plausible explanation may reside in the incomplete interlayer cation exchange at lower temperatures. In experiments using K-smectite, the higher density of interlayer K^+ ions enhanced attraction between layers, effectively inhibiting ethylene glycol intercalation [9]. With increasing simulated temperature, enhanced cation exchange efficiency and progressive illitization mitigate this initial disparity.

3.3.2. Effects of KCl Solution and K-Feldspar

Considering K^+ ions required for the smectite-to-illite transformation under geological conditions may originate either from formation brines or K-feldspar dissolution [50,51], the K^+ concentration in the aqueous medium and the K-feldspar content were systematically modulated in the experimental setup.

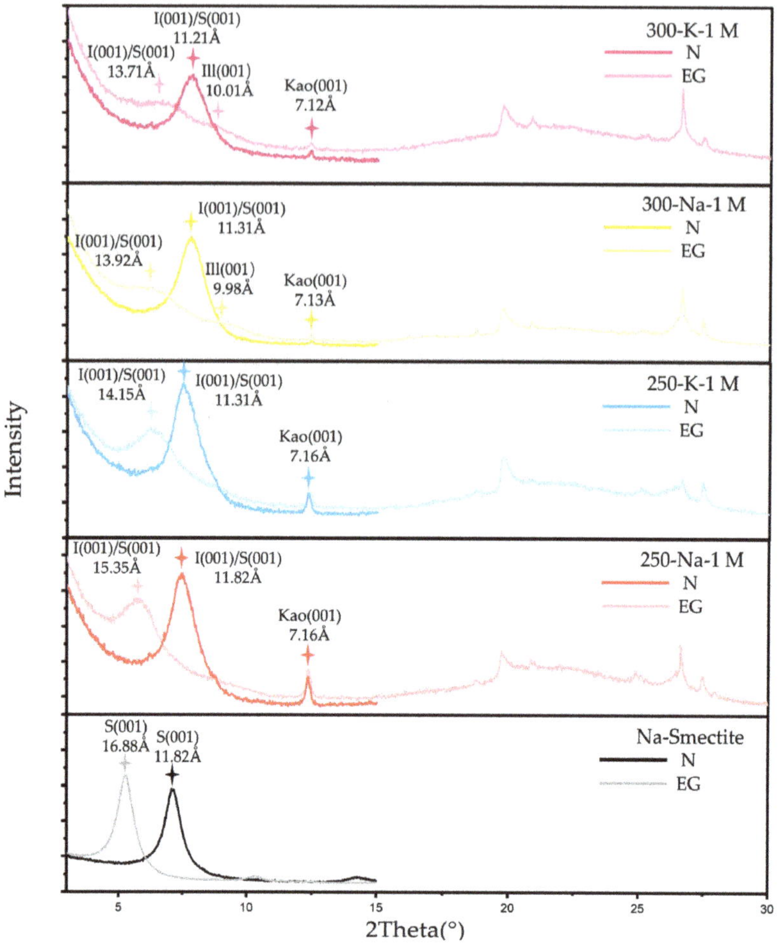

Figure 6. XRD patterns of the untreated Na-smectite and solid residues generated through hydrothermal reactions of Na- and K-smectite with 1 M KCl at 250 °C and 300 °C. N: oriented air-dried mounts; EG: ethylene glycol (EG)-solvated mounts.

In Figure 7, we report the XRD results of simulation products derived from Na-smectite treated with 0.1 M KCl solution at 250 °C and 0.01 M KCl solution at 300 °C, demonstrating distinct structural modifications compared to the reference samples shown in Figure 6. Samples treated at both 250 °C and 300 °C exhibited minimal structural alterations, as evidenced by XRD patterns showing negligible shifts in the EG S(001) reflection and d(001)-spacing consistently around 16.85 Å. Comparative analysis of K-feldspar-bearing systems revealed consistent I-S d-spacing values (11.83–11.88 Å), indicating minimal influence of initial feldspar content on transformation progress. Similarly, no apparent shifts in characteristic peaks were detected in air-dried scans, demonstrating inhibited illitization in K^+-depleted aqueous systems. Figure 8 compiles XRD patterns of solid products generated through experiments with K-feldspar content modulation. In all air-dried mounts, the d(001)-spacing of I/S mixed-layer minerals consistently ranged from 11.83 Å to 11.88 Å, showing negligible inter-sample variations.

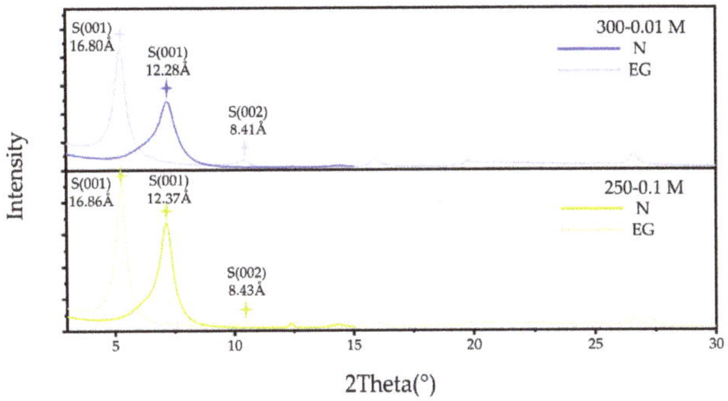

Figure 7. XRD patterns of solid residues generated through hydrothermal reactions of Na-smectite with 0.1 M KCl at 250 °C and 0.01 M KCl at 300 °C. N: oriented air-dried mounts; EG: ethylene glycol (EG)-solvated mounts.

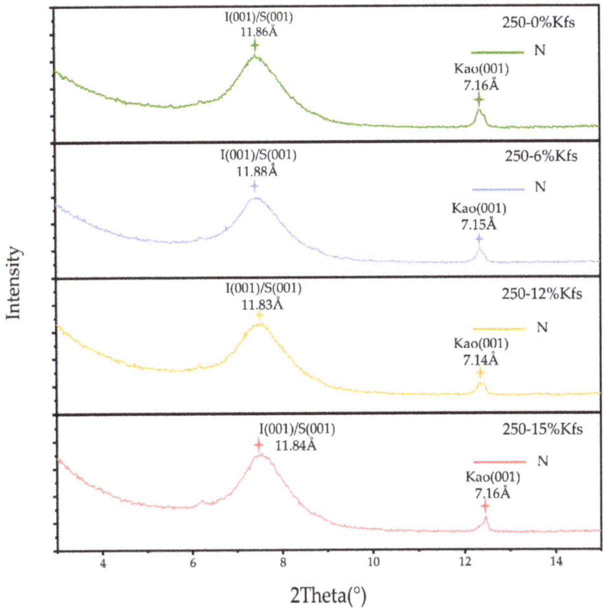

Figure 8. Oriented air-dried XRD patterns of solid residues generated through hydrothermal reactions of Na-smectite with 1 M KCl for various K-feldspar contents at 250 °C. Sample 250-12%Kfs is equivalent to sample 250-Na-1 M.

Under natural geological conditions, when paleosalinity fell below 15‰, such as in the Chang-7 shale, the K^+ required for smectite-to-illite transformation was primarily sourced from the dissolution of potassium-bearing minerals (mainly K-feldspar) [51], with negligible contributions from aqueous media. However, the current experimental results demonstrate that variations in the initial K-feldspar content of samples did not yield significant mineralogical differences. This observation is likely attributable to kinetic constraints on feldspar dissolution under experimental conditions and the masking effect of relatively elevated background K^+ concentrations. Consequently, in low-K^+ concentration experiments, the illitization process was restricted due to insufficient supplementary K^+ supply.

3.3.3. Effects of Organic Matter

The extensive association between organic matter and clay minerals in geological settings indicates their interactive evolution during diagenetic processes. Systematic modulation of kerogen content (5–15 wt%) in precursor assemblages allows effective evaluation of organic controls on smectite illitization. Figure 9 demonstrates corresponding air-dried oriented XRD results under two experimental temperature conditions. Under 250 °C hydrothermal conditions, increasing kerogen content in starting assemblages from 5% to 10% induced expansion of the d(001)-spacing in I/S interstratified minerals from 11.63 Å to 11.85 Å. At 300 °C, the d-spacing expanded from 11.3 Å to 11.5 Å corresponding to kerogen content elevation from 10% to 15%. Notably, increased d-spacing correlates with higher smectite layer proportion in I/S, suggesting that organic enrichment inhibits smectite-to-illite transformation. This finding differs from the organic effects observed in semi-open systems, demonstrating the existence of distinct underlying mechanisms under different geological conditions, such as variations in diagenetic stages. The promotion of organic acid release on K-feldspar dissolution remains unobserved due to restricted thermal evolution degree.

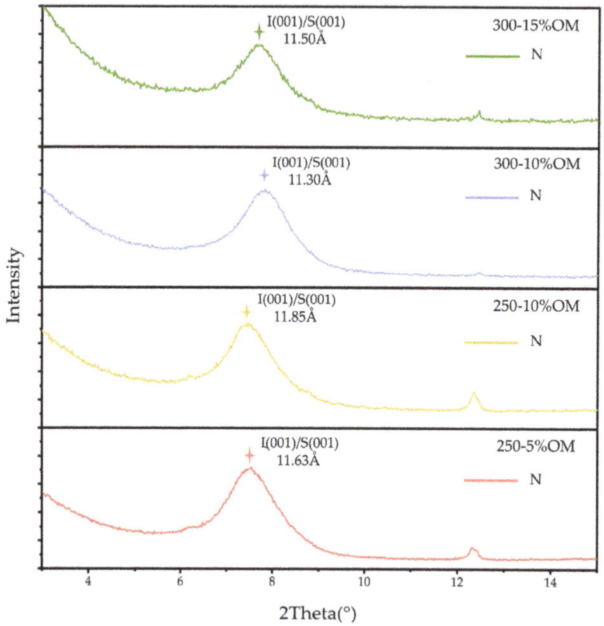

Figure 9. Oriented air-dried XRD patterns of solid residues generated through hydrothermal reactions of Na-smectite with 1 M KCl for various organic matter contents. Sample 250-10%OM is equivalent to sample 250-Na-1 M, the same applies to sample 300-10%OM.

4. Discussion

4.1. Thermally Driven but Organically Mediated Smectite Illitization

Smectite becomes unstable during burial diagenesis due to changes in the chemical environment, and it progressively transforms into illite through mixed-layer intermediates. This transformation is conventionally attributed to thermally activated mechanisms [4,52]. As such, the composition and ordering of illite–smectite minerals serve as critical mineralogical indicators for reconstructing thermal histories and constraining low-grade metamorphism in sedimentary basins. While temperature fundamentally governs reaction kinetics through Arrhenius-type relationships, the transformation of smectite to illite in natural systems is also influenced by complex organic-water–rock interactions. These interactions impose multiple, interrelated controls on the illitization process under geological conditions.

Hydrothermal experiments at 250 °C and 300 °C reveal structural reorganization patterns during the incipient stages of smectite evolution. Elevated temperatures significantly reduce the expandability of clay minerals, as evidenced by corresponding reductions in d(001)-spacing values. However, no fundamental reorganization is observed in the crystallographic ordering of I/S mixed-layer minerals under these thermal regimes. A distinctive feature is that the hydrothermal simulation products obtained at 300 °C exhibit significant peak broadening of I(001)/S(001) in EG XRD scans, revealing pronounced interlayer heterogeneity. In I/S mixed-layer minerals, the rigid illite layers constrain the expansion of adjacent smectite layers during ethylene glycol solvation [53]. Increasing illite layers enhance constraints on the expansion of smectite layers. However, disordered stacking with heterogeneous illite distribution will lead to differential expansion suppression across smectite layers, ultimately resulting in non-uniform swelling of the I/S mixed-layer mineral. This indicates that the progression of illitization does not fundamentally alter the disordered nature of mixed-layer minerals.

Considering the initial I/S mixed-layer minerals with ~80% illite layers, semi-closed pyrolysis experiments reflect an advanced stage of illitization. At 280 °C and 320 °C, the proportions of illite and I/S mixed-layer remain stable. Consistent with the XRD analysis, SEM observations of the corresponding products reveal negligible modifications in both crystal morphologies and textural characteristics throughout this stage. The only notable change is the disappearance of matrix-like clay minerals seen at 280 °C, which were replaced by minor acicular chlorite at higher temperatures. This suggests the neoformation of chlorite through recrystallization. Notably, a similar 'calming stage' persists during smectite illitization in Chang-7 shale samples, where illite-layer proportions remain essentially static despite substantial thermal increase indicated by organic maturation indices [54]. A significant shift occurs at 320 °C, marking a critical turning point, beyond which clear mineralogical and textural transformations are observed. Illitization intensifies, with an increasing number of clay minerals exhibiting a well-defined platy morphology. Under simulated vertical lithostatic pressure, these clay minerals also display a preferred orientation. The concurrent increase in authigenic quartz further supports the occurrence of illitization.

Notably, this stage-wise transformation closely parallels the thermal maturation of organic matter. Beyond 320 °C, S_2 and HI indices demonstrate progressive decrease, ultimately approaching zero, indicative of rapid organic matter consumption under thermal stress. SEM observations confirm that organic laminae become obliterated at 420 °C, accompanied by enhanced bedding-parallel pore networks. This implies that organic degradation actively facilitates illitization by creating porosity and releasing interlayer constraints. The interdependence fundamentally stems from clay-organic associations: under organic-rich conditions, constrained effective specific surface areas (SSA) limit the water–rock interactions [55,56]. Only upon progressive thermal degradation of organic matter are intracrystalline spaces liberated and additional porosity pathways established, enabling sustained illitization.

4.2. Organic Regulation of K^+ Availability and Cation Exchange Dynamics

Clay minerals and organic matter exhibit intimate associations commencing from initial deposition, as evidenced by microscopic observations revealing that organic matter is intercalated with clay minerals as a nanocomposite [19,57]. Within sedimentary basins, the smectite-to-illite transformation demonstrates precise temporal coupling with hydrocarbon generation during progressive burial; this relationship provides critical evidence for coevolutionary feedback mechanisms between clay mineral diagenesis and organic matter maturation pathways [58]. Our semi-open pyrolysis experiments further corroborate this documented geological phenomenon. In particular, the catalytic role of clay minerals in organic matter hydrocarbon generation has been extensively documented [25], particularly regarding the enhanced hydrocarbon generation potential of clay-bound organic fractions attributed to selective adsorption mechanisms.

Conversely, the impact of organic matter on the illitization of smectite is equally noteworthy. In our hydrothermal experiments, variations in organic matter were simulated by adjusting the kerogen content. XRD analysis of I/S mixed-layer minerals demonstrates that increased organic matter content leads to expanded d-spacing, consistent with suppressed illitization. These findings imply that elevated organic loading inhibits smectite illitization under hydrothermal conditions. We propose that this inhibitory effect arises from the dual control of organic matter on K^+ mobility and its accessibility to interlayer sites, through two key aspects.

On one hand, organic matter interacts with clay minerals at the nanometer scale through multiple physicochemical pathways, including ligand interactions, ion exchange, cation bridging, hydrophobic interactions, hydrogen bonding, and van der Waals forces [59,60]. A portion of organic compounds can even intercalate into the interlayers of swelling clay minerals (e.g., smectite), forming tightly bound organic-clay complexes. These adsorbed organics will enhance interlayer stability, reduce cation exchange capacity (CEC), and delay the release of interlayer water via 'water bridges', collectively hindering K^+ exchange [26]. In this study, the majority of organic matter exists as macromolecules that adsorb preferentially at edge sites of clay minerals. This edge-bound configuration acts as a physical barrier, reducing water–rock contact efficiency and further delaying K^+ migration into interlayers.

On the other hand, organic matter competes directly with smectite for K^+ through binding mechanisms [61,62]. In organic-rich systems, functional groups such as carboxyl and phenolic hydroxyl impart negative surface charges that facilitate the electrostatic adsorption of K^+ ions [63]. The oxygen-bearing functional groups also form coordination bonds with K^+, leading to effective K^+ sequestration. Although the binding affinity for K^+ is weaker than that for polyvalent cations, this interaction remains effective in reducing aqueous K^+ concentration given the predominance of K^+ in the system. Consequently, the ion-exchange incorporation of K^+ into smectite interlayers experiences proportional reduction.

Thus, organic matter fundamentally regulates K^+ partitioning in hydrothermal experiments. Elevated organic content imposes a strong constraint on K^+ availability, especially under low-K^+ aqueous conditions, and can effectively stall smectite illitization. The heightened organic matter/K^+ ratio drives tripartite competition for adsorption sites between organic ligands, K^+ ions, and clay minerals. Consequently, the smectite-to-illite transformation is suppressed because of an insufficient supply of K^+. It is important to emphasize that these effects are not static over geological timescales but instead evolve as diagenesis progresses. With increasing burial, thermal degradation of organic matter becomes increasingly important. As thermal maturity rises, decarboxylation, dehydrogenation, and aromatization reactions begin to break down macromolecular organics, liberating previously occupied mineral surfaces and creating secondary porosity. This transformation mitigates the initial inhibitory effect of organic matter and enables renewed K^+ access to interlayers, facilitating illitization. This shift from inhibition to facilitation reflects a dynamic feedback mechanism, in which organic matter first constrains and later promotes clay mineral diagenesis as a function of burial depth and temperature.

In summary, organic matter exerts a temporally evolving influence on smectite illitization, initially impeding and subsequently facilitating the transformation. Recognizing this time-dependent organic–inorganic feedback is essential for accurately reconstructing diagenetic pathways and predicting the evolution of shale reservoir quality under different thermal regimes.

5. Conclusions

This study investigates smectite-to-illite transformation under coupled organic-inorganic conditions through a combined approach of semi-open pyrolysis experiments on natural specimens and hydrothermal simulations using synthesized analogs. The experimental results capture the key geological controls across distinct diagenetic stages, demonstrating strong comparability with naturally evolved systems. The study provides a new perspective on multifactorial, synergistic controls over illitization in lacustrine shale settings and has important implications for evaluating shale reservoir quality and maturity.

1. When aqueous media are deficient in potassium ions, the smectite-to-illite transformation is nearly stagnant. Given that potassium ion (K^+) supply is essential for this process, the extensive illitization observed in low-salinity depositional environments (e.g., the Chang-7 shale) strongly suggests that K-feldspar dissolution plays a critical role in the transformation.
2. Illitization progresses with increasing temperature and is accompanied by the thermal evolution of organic matter. As the primary kinetic regulator, temperature also exerts indirect influences on critical diagenetic processes—including organic matter thermal evolution and K-feldspar dissolution.
3. Organic matter exerts a temporally evolving influence on smectite-to-illite transformation. During early diagenesis, enrichment in organic matter correlates with delayed illitization, likely due to reduced K^+ accessibility. At higher thermal maturity stages, however, organic matter degradation coincides with significantly enhanced transformation rates, facilitated by secondary porosity generation and increased mineral surface exposure. This transition demonstrates a feedback mechanism whereby organic matter functions as both a mediator and a modulator within the diagenetic system.

Supplementary Materials: The following supporting information can be downloaded at: https://www.mdpi.com/article/10.3390/pr13092966/s1, Table S1: Clay mineral content of the initial sample and pyrolysis products; Table S2: Organic geochemical parameters of the initial sample and pyrolysis products.

Author Contributions: Conceptualization, K.L. and L.D.; methodology, K.L. and Z.W.; formal analysis, K.L., Z.W. and L.D.; investigation, K.L. and C.Z.; resources, K.L. and C.Z.; data curation, K.L. and Z.W.; writing—original draft preparation, K.L.; writing—review and editing, K.L. and L.D.; supervision, L.D.; project administration, L.D.; funding acquisition, L.D. All authors have read and agreed to the published version of the manuscript.

Funding: This research was funded by Deep Earth probe and Mineral Resources Exploration-National Science and Technology Major Project, grant number: 2024ZD1001002 and NSFC, grant number: 42090021 and 42373050.

Data Availability Statement: The original contributions presented in the study are included in the article and Supplementary Materials; further inquiries can be directed to the corresponding authors.

Conflicts of Interest: The authors declare no conflicts of interest.

References

1. Cai, L.; Yang, T.; Tian, J.; Yi, J.; Ren, Q. Advances in Studies of Development and Growth Mechanisms of Clay Minerals in Tight Sandstone Reservoirs. *Acta Sedimentol. Sin.* **2023**, *41*, 1859–1889. [CrossRef]
2. Tannenbaum, E.; Huizinga, B.J.; Kaplan, I.R. Role of Minerals in Thermal Alteration of Organic Matter—II: A Material Balance. *AAPG Bull.* **1986**, *70*, 1156–1165. [CrossRef]
3. Pollastro, R.M. Considerations and Applications of the Illite/Smectite Geothermometer in Hydrocarbon-Bearing Rocks of Miocene to Mississippian Age. *Clays Clay Miner.* **1993**, *41*, 119–133. [CrossRef]
4. Ola, P.S.; Aidi, A.K.; Bankole, O.M. Clay Mineral Diagenesis and Source Rock Assessment in the Bornu Basin, Nigeria: Implications for Thermal Maturity and Source Rock Potential. *Mar. Pet. Geol.* **2018**, *89*, 653–664. [CrossRef]
5. Wu, L.; Zhou, C.; Keeling, J.; Tong, D.; Yu, W. Towards an Understanding of the Role of Clay Minerals in Crude Oil Formation, Migration and Accumulation. *Earth-Sci. Rev.* **2012**, *115*, 373–386. [CrossRef]
6. Rahman, H.M.; Kennedy, M.; Löhr, S.; Dewhurst, D.N. Clay-Organic Association as a Control on Hydrocarbon Generation in Shale. *Org. Geochem.* **2017**, *105*, 42–55. [CrossRef]
7. Bjørlykke, K. Clay Mineral Diagenesis in Sedimentary Basins—A Key to the Prediction of Rock Properties. Examples from the North Sea Basin. *Clay Miner.* **1998**, *33*, 15–34. [CrossRef]
8. Zhang, W.; Zhou, S.; Yu, Z.; Liu, X.; Wang, S.; Miao, H.; Liu, D.; Tian, J.; Wang, H. Controls of Clay Mineral Transformation and Organic Matter on Pore Networks of the Paleogene Lacustrine Shale Oil System in the Yitong Basin, NE China. *J. Asian Earth Sci.* **2025**, *280*, 106469. [CrossRef]
9. Mills, M.M.; Sanchez, A.C.; Boisvert, L.; Payne, C.B.; Ho, T.A.; Wang, Y. Understanding Smectite to Illite Transformation at Elevated (>100 °C) Temperature: Effects of Liquid/Solid Ratio, Interlayer Cation, Solution Chemistry and Reaction Time. *Chem. Geol.* **2023**, *615*, 121214. [CrossRef]
10. Zanoni, G.; Egvi, B.; Sweet, D.E. Clay Mineral Diagenesis in Alternating Mudstone-sandstone Beds from Late Palaeozoic Strata of the Anadarko Basin, United States. *Geol. J.* **2023**, *58*, 108–130. [CrossRef]
11. Eberl, D.; Hower, J. Kinetics of Illite Formation. *Geol. Soc. Am. Bull.* **1976**, *87*, 1326–1330. [CrossRef]
12. Roaldset, E.; Wei, H.; Grimstad, S. Smectite to Illite Conversion by Hydrous Pyrolysis. *Clay Miner.* **1998**, *33*, 147–158. [CrossRef]
13. Huang, W.-L.; Longo, J.M.; Pevear, D.R. An Experimentally Derived Kinetic Model for Smectite-to-Illite Conversion and Its Use as a Geothermometer. *Clays Clay Miner.* **1993**, *41*, 162–177. [CrossRef]
14. Elliott, W.C.; Edenfield, A.M.; Wampler, J.M.; Matisoff, G.; Long, P.E. The Kinetics of the Smectite to Illite Transformation in Cretaceous Bentonites, Cerro Negro, New Mexico. *Clays Clay Miner.* **1999**, *47*, 286–296. [CrossRef]
15. Ramseyer, K.; Boles, J.R. Mixed-Layer Illite/Smectite Minerals in Tertiary Sandstones and Shales, San Joaquin Basin, California. *Clays Clay Miner.* **1986**, *34*, 115–124. [CrossRef]
16. Whitney, G. Role of Water in the Smectite-to-Illite Reaction. *Clays Clay Miner.* **1990**, *38*, 343–350. [CrossRef]
17. Bettison-Varga, L.; Mackinnon, I.D. The Role of Randomly Mixed-Layered Chlorite/Smectite in the Transformation of Smectite to Chlorite. *Clays Clay Miner.* **1997**, *45*, 506–516. [CrossRef]
18. Ge, C. Compression Behavior of Smectitic vs. Illitic Mudrocks. Ph.D. Thesis, Massachusetts Institute of Technology, Cambridge, MA, USA, 2019.
19. Kennedy, M.J.; Löhr, S.C.; Fraser, S.A.; Baruch, E.T. Direct Evidence for Organic Carbon Preservation as Clay-Organic Nanocomposites in a Devonian Black Shale; from Deposition to Diagenesis. *Earth Planet. Sci. Lett.* **2014**, *388*, 59–70. [CrossRef]
20. Gu, Y.; Wan, Q.; Li, X.; Han, T.; Yang, S.; Hu, Q. Structure and Evolution of Clay-Organic Nanocomposites in Three Leading Shales in China. *J. Earth Sci.* **2023**, *34*, 824–837. [CrossRef]
21. Wilson, M.J.; Wilson, L.; Shaldybin, M.V. Clay Mineralogy and Unconventional Hydrocarbon Shale Reservoirs in the USA. II. Implications of Predominantly Illitic Clays on the Physico-Chemical Properties of Shales. *Earth-Sci. Rev.* **2016**, *158*, 1–8. [CrossRef]
22. Du, J.; Cai, J.; Lei, T.; Li, Y. Diversified Roles of Mineral Transformation in Controlling Hydrocarbon Generation Process, Mechanism, and Pattern. *Geosci. Front.* **2021**, *12*, 725–736. [CrossRef]
23. Corma, A. Inorganic Solid Acids and Their Use in Acid-Catalyzed Hydrocarbon Reactions. *Chem. Rev.* **1995**, *95*, 559–614. [CrossRef]
24. Gooßen, L.; Döhring, A. Lewis Acids as Highly Efficient Catalysts for the Decarboxylative Esterification of Carboxylic Acids with Dialkyl Dicarbonates. *Adv. Synth. Catal.* **2003**, *345*, 943–947. [CrossRef]
25. Bu, H.; Yuan, P.; Liu, H.; Liu, D.; Liu, J.; He, H.; Zhou, J.; Song, H.; Li, Z. Effects of Complexation between Organic Matter (OM) and Clay Mineral on OM Pyrolysis. *Geochim. Cosmochim. Acta* **2017**, *212*, 1–15. [CrossRef]
26. Cai, J.; Du, J.; Chen, Z.; Lei, T.; Zhu, X. Hydrothermal Experiments Reveal the Influence of Organic Matter on Smectite Illitization. *Clays Clay Miner.* **2018**, *66*, 28–42. [CrossRef]
27. Connan, J. Time-Temperature Relation in Oil Genesis: Geologic Notes. *AAPG Bull.* **1974**, *58*, 2516–2521. [CrossRef]
28. Tissot, B.P.; Pelet, R.; Ungerer, P.H. Thermal History of Sedimentary Basins, Maturation Indices, and Kinetics of Oil and Gas Generation. *AAPG Bull.* **1987**, *71*, 1445–1466. [CrossRef]

29. Waples, D.W. Time and Temperature in Petroleum Formation: Application of Lopatin's Method to Petroleum Exploration. *AAPG Bull.* **1980**, *64*, 916–926. [CrossRef]
30. Ungerer, P. State of the Art of Research in Kinetic Modelling of Oil Formation and Expulsion. *Org. Geochem.* **1990**, *16*, 1–25. [CrossRef]
31. Behar, F.; Kressmann, S.; Rudkiewicz, J.L.; Vandenbroucke, M. Experimental Simulation in a Confined System and Kinetic Modelling of Kerogen and Oil Cracking. *Org. Geochem.* **1992**, *19*, 173–189. [CrossRef]
32. Ma, W.; Hou, L.; Luo, X.; Liu, J.; Tao, S.; Guan, P.; Cai, Y. Generation and Expulsion Process of the Chang 7 Oil Shale in the Ordos Basin Based on Temperature-Based Semi-Open Pyrolysis: Implications for in-Situ Conversion Process. *J. Pet. Sci. Eng.* **2020**, *190*, 107035. [CrossRef]
33. Wang, J.; Li, Z.; Liu, Z.; Dang, H.; Guo, J.; Zhang, C. Semi-Open System Simulation of Organic-Rich Shale to Produce Organic Acids. *ACS Earth Space Chem.* **2024**, *8*, 2420–2427. [CrossRef]
34. Lewan, M.D.; Roy, S. Role of Water in Hydrocarbon Generation from Type-I Kerogen in Mahogany Oil Shale of the Green River Formation. *Org. Geochem.* **2011**, *42*, 31–41. [CrossRef]
35. Huang, Y.; Liao, Y.; Pan, Y.; Hsu, C.S.; Wang, Y.; Peng, P. Improved Light Hydrocarbons Collection Method for the Pyrolysis of Crude Oil in Gold Tube Closed System Experiments. *Org. Geochem.* **2022**, *168*, 104432. [CrossRef]
36. Salisu, A.M.; Bello, A.M.; Amao, A.O.; Lander, R.H.; Al-Ramadan, K. Experimental Study of Smectite Authigenesis and Its Subsequent Illitization: Insights from Modern Estuary Sediments. *J. Sediment. Res.* **2025**, *95*, 562–588. [CrossRef]
37. Bentabol, M.; Lamarca-Irisarri, D.; Van Driessche, A.E.S.; Ryan, P.C.; Huertas, F.J. Illitization of Montmorillonite in Ammonium Solutions under Hydrothermal Conditions. *Appl. Clay Sci.* **2024**, *258*, 107478. [CrossRef]
38. Wu, S.; Zhu, R.; Cui, J.; Cui, J.; Bai, B.; Zhang, X.; Xu, J.; Zhu, D.; You, J.; Li, X. Characteristics of Lacustrine Shale Porosity Evolution, Triassic Chang 7 Member, Ordos Basin, NW China. *Pet. Explor. Dev.* **2015**, *42*, 185–195. [CrossRef]
39. Wang, F.; Guo, S. Influential Factors and Model of Shale Pore Evolution: A Case Study of a Continental Shale from the Ordos Basin. *Mar. Pet. Geol.* **2019**, *102*, 271–282. [CrossRef]
40. Yuan, G.; Cao, Y.; Zan, N.; Schulz, H.-M.; Gluyas, J.; Hao, F.; Jin, Q.; Liu, K.; Wang, Y.; Chen, Z.; et al. Coupled Mineral Alteration and Oil Degradation in Thermal Oil-Water-Feldspar Systems and Implications for Organic-Inorganic Interactions in Hydrocarbon Reservoirs. *Geochim. Cosmochim. Acta* **2019**, *248*, 61–87. [CrossRef]
41. Ma, J.; Ma, Z.; Miao, J.; Zheng, L.; Wang, Q.; He, C. Co-Evolution Simulation Experiment of Source Rock Fluid and Reservoir Rock and Its Geological Implications: A Case Study of Upper Triassic Xujiahe Formation, Western Sichuan Basin. *Pet. Geol. Exp.* **2022**, *44*, 698–704. [CrossRef]
42. Shan, C.; Shi, Y.; Liang, X.; Zhang, L.; Wang, G.; Jiang, L.; Zou, C.; He, F.; Mei, J. Diagenetic Characteristics and Microscopic Pore Evolution of Deep Shale Gas Reservoirs in Longmaxi Formation, Southeastern Sichuan Basin, China. *Unconv. Resour.* **2024**, *4*, 100090. [CrossRef]
43. Wang, G.; Jin, Z.; Zhang, Q.; Zhu, R.; Tang, X.; Liu, K.; Dong, L. Effects of Clay Minerals and Organic Matter on Pore Evolution of the Early Mature Lacustrine Shale in the Ordos Basin, China. *J. Asian Earth Sci.* **2023**, *246*, 105516. [CrossRef]
44. Zhang, D.; Cao, H.; Lei, Y.; Wu, H.; Wang, X.; Guo, X.; Yan, J.; Wang, S.; Shi, T.; Li, H. A Study on Molecular Structural Evolution of Type II Kerogen in a Gold Tube Thermal System: Insights from Solid-State 13C NMR. *Fuel* **2023**, *331*, 125898. [CrossRef]
45. Pan, C.; Jiang, L.; Liu, J.; Zhang, S.; Zhu, G. The Effects of Pyrobitumen on Oil Cracking in Confined Pyrolysis Experiments. *Org. Geochem.* **2012**, *45*, 29–47. [CrossRef]
46. National Energy Administration. *Analysis Method for Clay Minerals and Ordinary Non-Clay Minerals in Sedimentary Rocks by the X-Ray Diffraction: SY/T 5163-2018*; Petroleum Industry Press: Beijing, China, 2018.
47. Wang, Z.; Dong, L.; Jin, Z.; Zou, S.; Fu, J.; Zhu, R. Efforts to Untie the Multicollinearity Knot and Identify Factors Controlling Macropore Structures in Shale Oil Reservoirs. *Adv. Geo-Energy Res.* **2024**, *11*, 194–207. [CrossRef]
48. Wang, R.; Wang, G.; Zhao, G.; Qian, M.; Liu, Y.; He, W.; Li, Z. Geological Characteristics and Resources Potential of Shale Oil in Chang 7 Member of Upper Triassic Yanchang Formation in Fuxian Area, Southern Ordos Basin, Western China. *Unconv. Resour.* **2023**, *3*, 237–247. [CrossRef]
49. Yan, Y.; Wang, M.; Misch, D.; Sachsenhofer, R.F.; Wu, Y.; Li, J. Mineral Diagenesis in Lacustrine Organic-Rich Shales: Evolution Pathways and Implications for Reservoir Characteristics. *J. Asian Earth Sci.* **2024**, *263*, 106026. [CrossRef]
50. Huang, S.; Huang, K.; Feng, W.; Tong, H.; Liu, L.; Zhang, X. Mass Exchanges among Feldspar, Kaolinite and Illite and Their Influences on Secondary Porosity Formation in Clastic Diagenesis—A Case Study on the Upper Paleozoic, Ordos Basin and Xujiahe Formation, Western Sichuan Depression. *Geochimica* **2009**, *38*, 498–506. [CrossRef]
51. Zhao, X.; He, D. *Clay Mineral and Application in Oil and Gas Exploration and Development*; Petroleum Industry Press: Beijing, China, 2016; ISBN 978-7-5183-0360-1.
52. Green, H.; Šegvić, B.; Zanoni, G.; Omodeo-Salé, S.; Adatte, T. Evaluation of Shale Source Rocks and Clay Mineral Diagenesis in the Permian Basin, USA: Inferences on Basin Thermal Maturity and Source Rock Potential. *Geosciences* **2020**, *10*, 381. [CrossRef]

53. Yang, Y.; Adhikari, S.; Xu, G. Molecular Dynamics Simulation in the Interlayer of Mixed-Layer Clays Due to Hydration and Swelling Mechanism. *Crystals* **2021**, *11*, 586. [CrossRef]
54. Ling, K.; Wang, Z.; Cao, Y.; Liu, Y.; Dong, L. Clay Mineral Characteristics and Smectite-to-Illite Transformation in the Chang-7 Shale, Ordos Basin: Processes and Controlling Factors. *Minerals* **2025**, *15*, 951. [CrossRef]
55. Ding, F.; Cai, J.; Song, M.; Yuan, P. The Relationship between Organic Matter and Specific Surface Area in <2 μm Clay Size Fraction of Muddy Source Rock. *Sci. China Earth Sci.* **2013**, *56*, 1343–1349. [CrossRef]
56. Kaiser, K.; Guggenberger, G. Mineral Surfaces and Soil Organic Matter. *Eur. J. Soil Sci.* **2010**, *54*, 219–236. [CrossRef]
57. Löhr, S.C.; Kennedy, M.J. Organomineral Nanocomposite Carbon Burial during Oceanic Anoxic Event 2. *Biogeosciences* **2014**, *11*, 4971–4983. [CrossRef]
58. Chang, J.; Fan, X.; Jiang, Z.; Wang, X.; Chen, L.; Li, J.; Zhu, L.; Wan, C.; Chen, Z. Differential Impact of Clay Minerals and Organic Matter on Pore Structure and Its Fractal Characteristics of Marine and Continental Shales in China. *Appl. Clay Sci.* **2022**, *216*, 106334. [CrossRef]
59. Kleber, M.; Eusterhues, K.; Keiluweit, M.; Mikutta, C.; Mikutta, R.; Nico, P.S. Mineral–Organic Associations: Formation, Properties, and Relevance in Soil Environments. *Adv. Agron.* **2015**, *130*, 1–140. [CrossRef]
60. Yu, W.H.; Li, N.; Tong, D.S.; Zhou, C.H.; Lin, C.X. (Cynthia); Xu, C.Y. Adsorption of Proteins and Nucleic Acids on Clay Minerals and Their Interactions: A Review. *Appl. Clay Sci.* **2013**, *80–81*, 443–452. [CrossRef]
61. Ganor, J.; Reznik, I.; Rosenberg, Y. Organics in Water-Rock Interactions. *Rev. Mineral. Geochem.* **2009**, *70*, 259–369. [CrossRef]
62. Harpreet, K. Forms of Potassium in Soil and Their Relationship with Soil Properties—A Review. *Int. J. Curr. Microbiol. Appl. Sci.* **2019**, *8*, 1580–1586. [CrossRef]
63. Wang, F.L.; Huang, P.M. Effects of Organic Matter on the Rate of Potassium Adsorption by Soils. *Can. J. Soil Sci.* **2001**, *81*, 325–330. [CrossRef]

eng MDPI

Article

Integrated Petrophysical Analysis and Reservoir Characterization of Shaly Sands in the Srikail Gas Field, East Central Bengal Basin, Bangladesh

Shireen Akhter [1,2,*] and Nuno Pimentel [1,2,*]

1 Department of Geology, Lisbon University, 1749-016 Lisboa, Portugal
2 IDL, FCUL, Lisbon University, 1749-016 Lisboa, Portugal
* Correspondence: shiri.geo.ju@gmail.com (S.A.); npimentel@ciencias.ulisboa.pt (N.P.)

Abstract

This study offers a comprehensive petrophysical evaluation and reservoir characterization of the Srikail Gas Field, situated on the Tripura Uplift in the east-central Bengal Basin. Utilizing well log data from four wells (Srikail-1 to Srikail-4), the analysis targets the Bhuban and Bokabil formations of the Surma Group. Standard log suites, including gamma ray, spontaneous potential, caliper, resistivity, neutron, density, and sonic logs, were interpreted using both manual techniques and digital analysis through software. Key petrophysical properties, including shale volume, effective porosity, fluid saturations, permeability, and bulk volume of water, were estimated using a combination of empirical modeling and automated interpretation workflows. Cross-plot methodologies were applied to assist in reservoir evaluation. The study integrated both qualitative and quantitative approaches to characterize each reservoir unit in detail. Results demonstrate significant heterogeneities in reservoir quality across the field. While some intervals exhibit favorable properties suitable for commercial gas production, others are characterized by high carbonate content, poor porosity, and very low permeability (Sand C with 0.05 to 0.08 mD), indicative of tight to semi-conventional reservoirs. The most productive zones, identified as the D sands, are cleaner sands with excellent permeability (102 mD to 355 mD). In contrast, deeper intervals generally exhibit tighter characteristics, with DST-derived permeability values ranging from 0.6 to 0.01 mD. The study recommends integrating core analysis, advanced petrophysical modeling, and 3D seismic interpretation with well log data to enhance reservoir delineation in the Srikail Gas Field. This combined approach would reduce uncertainties, improve input parameter accuracy, and offer a more comprehensive understanding of the Bhuban Formation's heterogeneity, ultimately supporting more effective reservoir evaluation and hydrocarbon recovery planning.

Keywords: petrophysics; reservoir characterization; shaly sands; Bengal Basin

1. Introduction

Wireline logging data can provide valuable information for oil and gas exploration, such as reservoir characteristics, hydrocarbon potential, and formation evaluation. However, interpreting wireline logging data can be challenging, because economic and efficient oil and gas production is highly dependent on understanding key properties of reservoir rock, such as porosity, permeability, and wettability [1].

Previous studies in the Srikail Gas Field have focused on limited sandy zones or single wells, often providing only average petrophysical values. In contrast, this study presents a comprehensive evaluation of all identified sandy intervals across four wells in the field. A combination of manual interpretation and automated analysis using Techlog (2024.2) software was employed to derive key petrophysical parameters. These were calculated using both empirical equations and software-driven methods to ensure consistency and improve interpretation accuracy. This integrated approach allows for a more detailed understanding of formation heterogeneity and overall reservoir quality.

The determination of reservoir quality largely depends on quantitative and qualitative evaluation of petrophysical properties. Petrophysical parameters, i.e., effective porosity (ϕeff), effective water saturation (Sw), formation water resistivity (Rw), hydrocarbon saturation (Shc-eff), and permeability, may all be evaluated using well log data, thus performing a correct reservoir characterization.

This study addresses the formation evaluation and reservoir characterization of shaly sands at the Srikail Gas Field (Bengal Basin), based on a detailed analysis of to all available mud log data and wire line log data. The analysis used the Symandox Method [2,3], with empirical mathematical equations appropriate for shaly sands. Numerous previous studies on the petrophysical analysis and reservoir characterization of Miocene sand [4–8] reservoirs across various gas fields within the Bengal Basin have contributed to the interpretation and analysis presented in this study.

2. Geological Framework

The study area lies in the east-central part of the Bengal Basin (Figure 1) [9,10]. The study was conducted in the Srikail Gas Field, located about 60 km east of Dhaka city and approximately 100 km away from the Surma sub-basin.

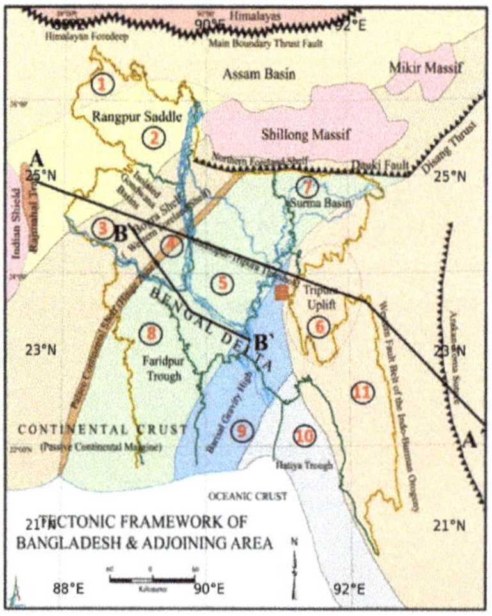

Figure 1. Regional and local tectonic elements of Bengal Basin (adapt. from [9]. Numbers 1 to 11 indicate the different sub-basins, and the Srikail Gas Field is located in the Tripura Uplift (Number 6) of East Central Bengal Basin.

The Bengal Basin formed during the continental extension of the eastern part of Gondwana during the Late Mesozoic and is still ongoing [11]. During the Cenozoic, the Indian plate rifted northwest and then northwards from the combined Antarctica–Australia part of Gondwana, resulting in a major collision between India and Asia in the Miocene. As a result, the Himalayas in the north and the Indo-Burman range in the east were gradually uplifted as the Tethys Sea closed [12]. Tectonically, the Srikail anticline is located on the western part of the folded belt of Bengal Foredeep within the Tripura Uplift of the Bengal Basin (Figure 2) [13].

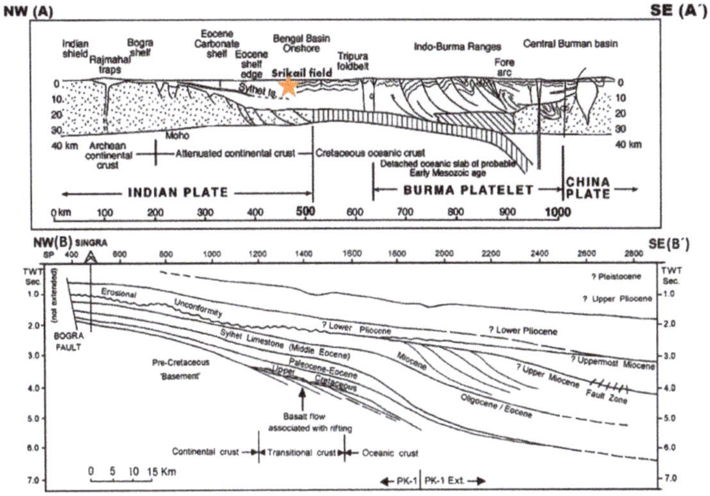

Figure 2. (see Figure 1 for location): (**A**) regional geotectonic cross-section for the Bengal Basin (adapt. from [13]; (**B**) regional cross-section of the Tertiary sedimentary infill of the Bengal basin [13], showing thicker units to the SE, where the Srikail field is located.

The existing stratigraphic system for the Bengal Basin was based exclusively on lithostratigraphic association with the type sections described by Evans [14] in Assam, northeastern India, along the fold belt in the basin's eastern portion. Evans' stratigraphic age estimates for the Assam sequences are questionable since they are based on long-distance correlations between brackish marine macrofauna and vertebrate findings. While some parts of Evans' scheme may be usable in the regional lithostratigraphic or seismic correlation (e.g., the boundary between the Surma and Tipam Groups), other parts of his classification (e.g., the contact between the Bhuban and Bokabil Formations or the internal units of these formations) are difficult to apply to the lithostratigraphic succession throughout the basin. For the Central Basin, which includes the Surma Basin and the Srikail Gas Field, the proposed lithostratigraphy is depicted in Figure 3 [13].

The focus of this study is the Miocene to Early Pliocene Surma Group, which does not outcrop in the study area. The Surma Group sediments were deposited in a large mud-rich prograding delta system, in response to the western encroachment of the Indo-Burma range and rising Himalaya [15]. Most authors have traditionally divided the Surma Group into two units based on Evan's [14] stratigraphic scheme, with the younger Bokabil Formation covering the older Bhubhan Formation [16–18].

The Bhuban Formation has been interpreted by Johnson and Alam [19] as prodelta to delta front deposits of a mud-rich delta system, while the Bokabil sediments correspond to subaerial to brackish sandy deposits [16]. Sultana and Alam [20] interpreted the sediments of this group as shallow marine to tide-dominated coastal deposits within a transgressive–regressive regime based on extensive logging of core samples from the Sylhet Trough. The top of the Surma group is dominated by a shaly unit known as the "Upper Marine Shale" (UMS) [16].

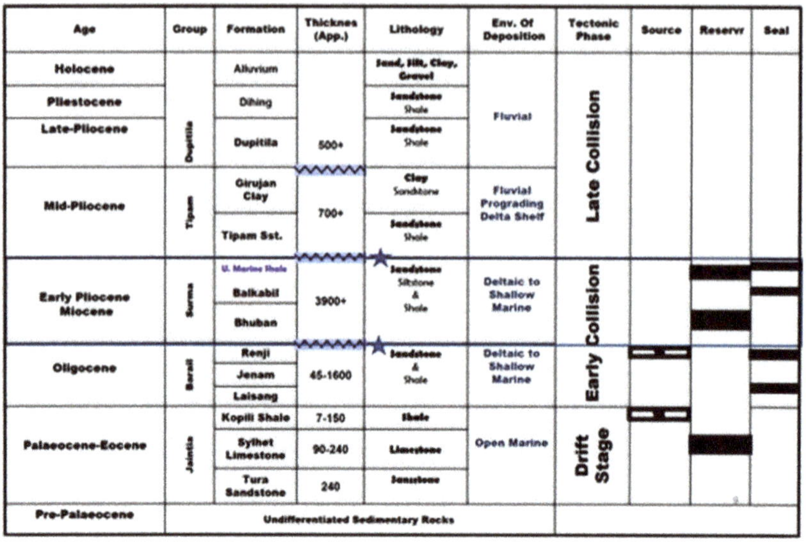

Figure 3. General stratigraphy (thickness in meter) of the study area (modified from [13]), with indication of the main regional unconformities and the studied sequence (blue stars interval).

The thickness of the Surma Group ranges from 2700 m to over 3900 m. Based on seismic data, mud logs, and wireline log responses, multiple prospective sand zones (named A to K, from top to bottom) have been identified and explored by the operating oil and gas companies in the Srikail Gas Field. A detailed description of these sand zones is provided below (Tables 1 and 2).

The absence of sand D and E in Well 1 is due to presence of a thick (over 200 m) canyon fill, which eroded those sands. This erosion is identified in seismic sections and also on the log [21]. Probably this absence is a result of erosion caused by a canyon incision and muddy infill, detected in seismic lines.

Table 1. Total depth of reservoir zones [22,23].

Well Name	Total Depth		Identified Reservoir Zones
	MD	TVDSS	
Srikail Well_1	3583 m	3572 m	A, B, C, F, G, H, I, J, K
Srikail Well_2	3214 m	3198 m	A, B, C, D_{up}, D_{lower}, E
Srikail Well_3	3350 m	3178 m	A, B, C, D_{up}, D_{lower}, E
Srikail Well_4	3512 m	3360 m	A, B, C, D_{up}, D_{lower}, E, F, G

Table 2. Thickness of reservoir zones.

Zone		Well_1	Well_2	Well_3	Well_4
A		11 m	11 m	10 m	08 m
B		31 m	38 m	35 m	33 m
C		24 m	14 m	14 m	10 m
D	D_{upper}	-	60 m	52 m	66 m
	D_{lower}	-	26 m	20 m	27 m
E		-	17 m	26 m	37 m

Table 2. *Cont.*

Zone	Well_1	Well_2	Well_3	Well_4
F	15 m	-	-	24 m
G	39 m	-	-	40 m
H	18 m	-	-	-
I	38 m	-	-	-
J	08 m	-	-	-
K	15 m	-	-	-

3. Materials and Methods

3.1. Materials

This study involves the analysis of four wells (Figure 4), where distinct reservoir zones have been identified in each (Tables 1 and 2).

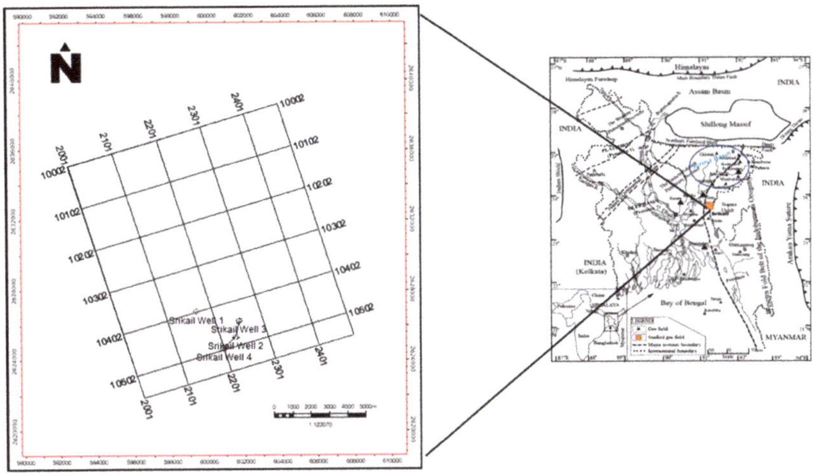

Figure 4. Srikail wells geographical location and position in BAPEX seismic line grid.

Wireline log data have been analyzed to calculate petrophysical parameters and characterize the reservoirs. Srikail wells have been covered by all suites of resistivity and porosity logs, along with GR and SP logs. The open hole composite logs that are used in this study were conducted in different stages and are as follows (Table 3):

Table 3. Open hole composite logs.

Hole Section	Scale	Description
12.25″	1:200/1:500/1:1000 MD	Composite Log
12.25″	1:200/1:500/1:1000 MD	GR/SP/Resistivity
12.25″	1:200/1:500/1:1000 MD	Density/Porosity/Cali/GR
12.25″	1:200 MD	Composite Plot
8.5″	1:200/1:500/1:1000 MD	Composite Log
8.5″	1:200/1:500/1:1000 MD	GR/SP/Resistivity
8.5″	1:200/1:500/1:1000 MD	Density/Porosity/Cali/GR
8.5″	1:200 MD	Full Wave Sonic/GR

All the stages of logging were conducted by China Petroleum Logging Company Ltd. (CPL) (Beijing, China) and covered all sets of logs. The overall quality of logs was found to be satisfactory. Before the petrophysical analysis was started, key quality checks and corrections were applied to ensure data reliability. This included borehole corrections (e.g., for mudcake and hole size), merging the best log passes, removing noisy spikes, and checking repeat log consistency. Intervals with poor data were either corrected or excluded. Environmental corrections for resistivity and porosity logs were also applied during vendor processing to account for mud properties.

Across the four study wells, a total of eight drill stem tests (DSTs) and one production test were conducted. Unfortunately, quantitative permeability results were retrievable for only one well (Well 1). A summary table (Table 4) lists, for each DST, well name, tested depth interval, test type, and data availability (quantitative vs. qualitative) and permeability (when reported).

Table 4. Summary of drill stem test (DST).

Well	Zone	Test Type	Perforation Interval (m)	Result
Well-1	G	Conventional Testing	3334–3341 m and 3343–3350 m	Maximum gas flow 1.2 MMSCFD. Permeability 0.01 mD.
	H	DST-2	3410–3413 m, 3428–3430 m and 3439–3442 m	Highly calcareous. No visual porosity. Gas show 1.49% at 3408 m. Tight. Non-measurable flow. Permeability 0.0149 mD.
	J	DST-1	3530–3534 m	Highly calcareous. No visual porosity. Water and minor gas flow.
Well-2	D upper	DST-2	2963.55–2969 m	Commercial gas flow.
	D lower	DST-1	3026.8–3029.8 m and 3030–3036.5 m	Commercial gas flow.
Well-3	D upper	DST-2	3097–3104 m	Commercial gas flow.
	D lower	DST-1	3185–3190 m	Commercial gas flow.
Well-4	E	DST-1	3220–3231 m	Commercial gas flow.

3.2. Methods

3.2.1. Petrophysical Workflow

Petrophysical analysis was conducted on all the identified reservoirs, based on mud log and wireline log data. Wireline log data were compiled for both manual interpretation [24] and imported into petrophysical software (Techlog). The following flow chart (Figure 5) shows the steps that were undertaken:

In Srikail wells 1, 2, 3, and 4, the calculation of True Vertical Depth (TVD) and True Vertical Depth Subsea (TVDSS) has been determined using measured depth (MD), borehole deviation, and azimuth data.

In this study, digital well-log data in Log ASCII Standard (LAS) format, free from the common limitations of analog records, have been utilized. The dataset includes caliper, Self-Potential (SP), gamma ray, resistivity logs (shallow, medium, and deep), and porosity logs (neutron, density, and sonic), following the guidelines outlined in "Log Interpretation Principles" of Schlumberger [25–28].

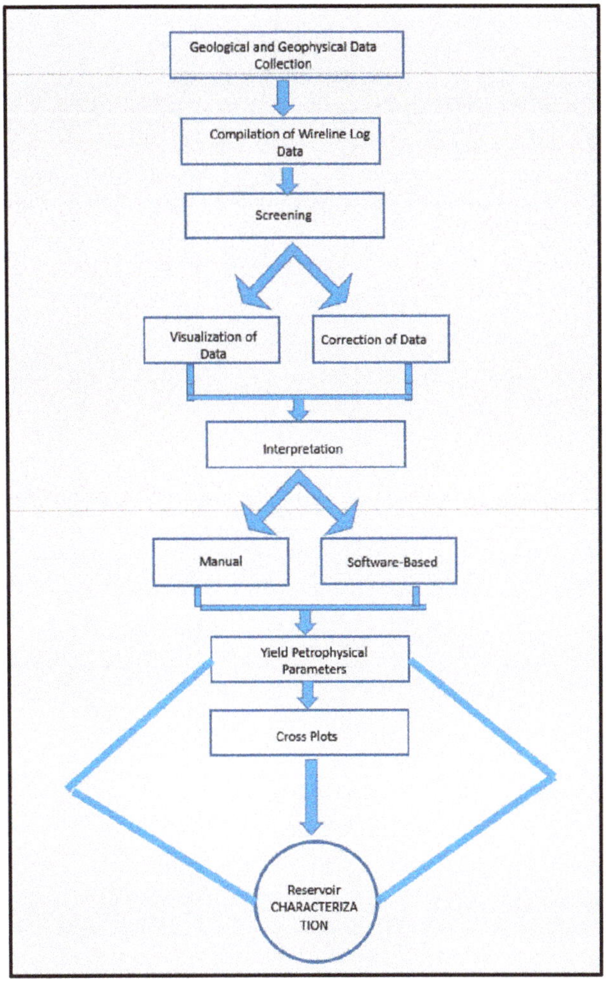

Figure 5. Flow chart of the petrophysical work flow.

Potential hydrocarbon-bearing zones were identified through the integration of gamma ray log interpretation, quick-look analysis of resistivity logs, and the observation of significant negative separation between neutron and density porosity logs. The presence of gas within the reservoirs was corroborated by drill stem test (DST) data and supported by regional geological context. Gas-bearing intervals indicated by neutron–density crossover were further verified using additional wireline log responses to improve interpretation reliability.

Moreover, well-to-well correlation was performed using the available wireline log data. The Gamma Ray log, often referred to as a facies indicator, served as the primary tool for lithological identification and correlation across the studied wells.

The analysis applied the Symandoux model, utilizing empirical equations specifically suited for evaluating shaly sandstone formations [2,3]. Although advanced approaches such as FEM-based resistivity simulation, proposed by Wu et al. [29], provide greater accuracy in laminated shaly systems, the use of the Simandoux model was adopted here due to the absence of high-resolution formation imaging or core-scale shale distribution data.

3.2.2. Petrophysical Parameters

Water Resistivity (R_w)

Two of the most common methods of determining R_w from logs are the inverse-Archie method and the SP method. The inverse-Archie method of determining R_w works under the assumption that water saturation (S_w) is 100%. It is necessary, therefore, that the inverse-Archie method be employed in a zone that is obviously wet. Furthermore, it is desirable to calculate R_w from the inverse-Archie method in a clean formation with relatively high porosity. The following equation (inverse-Archie method) is used here to determine water resistivity.

$$R_{wa} = \frac{\phi^m \times R_t}{a} \quad (1)$$

where

m = cementation exponent = 2;
a = tortuosity factor = 1.

Shale Volume (V_{sh})

Nawab and Islam [30] estimated shale volume in Miocene Bhuban sandstone for selected gas fields of Bangladesh using gamma and porosity logs. In this study gamma is used for shale volume. Because shale is usually more radioactive than sand or carbonate, gamma ray logs can be used to calculate volume of shale in porous reservoirs.

The volume of shale expressed as a decimal fraction or percentage is called V_{shale}.

V_{shale} from Gamma ray:

$$V_{sh} = \frac{GR_{log} - GR_{min}}{GR_{max} - GR_{min}} \quad (2)$$

V_{sh} = volume of shale;
GR_{log} = gamma ray reading of formation;
GR_{min} = minimum gamma ray (clean sand or carbonate);
GR_{max} = maximum gamma ray (shale).

The GR_{min} and GR_{max} values used for V_{shale} calculation were determined directly from well log data, by visually inspecting the gamma ray (GR) curve in each well on a meter-by-meter basis. For each zone, GRmin was selected from the cleanest (lowest-GR) sand intervals, and GRmax from the highest-GR shale intervals, ensuring that the selected points reflected representative lithologies within the respective formation.

Effective Porosity ($\phi_{DN\text{-}eff}$)

Porosity is the fraction of a rock that is occupied by pores. Effective porosity refers to the fraction of the total volume in which fluid flow is effectively taking place and includes catenaries and dead-end pores (as these pores cannot be flushed, but they can cause fluid movement by release of pressure like gas expansion) and excludes closed pores (or non-connected cavities). Porosity is one of the most important rock properties to be determined in petroleum geology, and to determine it, three porosity tools and/or a resistivity tool are used [31].

Effective porosity ϕe = Interconnected pore space/bulk volume

Porosity determination from density and neutron logs.

Porosity from density log (ϕd)

$$\phi d = \frac{(\rho_{ma} - \rho_b)}{(\rho_{ma} - \rho_f)} \quad (3)$$

ϕ_d = porosity from density log, fraction;

ρ_{ma} = density of formation matrix, g/cm^3;
ρ_b = bulk density from log measurement, g/cm^3;
ρ_f = density of fluid in rock pores, g/cm^3.
Effective porosity from neutron and density log:

$$\phi_{\text{n-corrected}} = \phi_n - (V_{cl} \times \phi_{sh})\ \phi_{\text{d-corrected}} = \phi_d - (V_{cl} \times \phi_{sh}) \quad (4)$$

These values of neutron and density porosity corrected for the presence of clays are then used in the equations below to determine the effective porosity (ϕeffective) of the formation of interest.

$$\phi_{\text{effective}} = \left[\frac{(\phi_{\text{n-corrected}})^2 + (\phi_{\text{d-corrected}})^2}{2}\right]^{0.5} \text{for gas} \quad (5)$$

Effective Water Saturation (Swe)

Water saturation is the ratio of water volume to pore volume. It can be expressed as

$$S_w = (\phi w/\phi) \times 100 \quad (6)$$

where
S_w = water saturation of the uninvaded zone;
ϕ_w = conductivity derived or water fill porosity;
ϕ = true porosity from the porosity log.

Effective water saturation (S_{We}): is the ratio of free water volume to effective porosity (PHIe). Since the Archie equation [32] is only applicable to clean sands, it has not been utilized in this study to determine the water saturation in the hydrocarbon-bearing zones. A significant advancement in the 1950s was the realization that shale introduces "excess conductivity," which leads to deviations from the assumptions in Archie's original equations [33]. Therefore, the saturation of water and hydrocarbon has been calculated here using well used formulae for shaly sands provided by Simandoux [2]:

$$S_{we} = \frac{C \times R_w}{\phi_{eff}^2}\left[\frac{5 \times \phi_{eff}^2}{R_w \times R_t} + \left(\frac{V_{sh}}{R_{sh}}\right)^2 - \frac{V_{sh}}{R_{sh}}\right] \quad (7)$$

where
Swe = effective (clay-corrected) water saturation
ϕ_{eff} = effective porosity (corrected for clay/shale content)
R_{sh} = resistivity of adjacent shale
R_w = formation water resistivity (resistivity of the water in the reservoir)
Rt = true resistivity of the uninvaded zone (actual formation resistivity)
V_{sh} = volume of shale (fraction or percentage of shale in the rock)
C = empirical constant: typically, 0.40 for sandstones, 0.45 for carbonates

In this study, a value of C = 0.40 was applied in the Simandoux equation, consistent with standard practice for shaly sandstone formations. This selection is supported by regional analogs, including Rahman & Sarker [34–36], who reported similar values for semi-consolidated deltaic sandstones in the Bengal Basin. However, recent work by Wan Zairani Wan Bakar et al. [37] highlights the benefits of adopting this approach to better account for shale content and heterogeneity in similar lithologies. While localized carbonate presence was noted in certain intervals, it was not laterally extensive or volumetrically dominant. Therefore, the adopted C value was considered appropriate for representing the overall lithological character of the Bhuban Formation.

Effective Hydrocarbon Saturations ($S_{hc\text{-}eff}$)

Hydrocarbon saturation can be determined by the difference between unity and water saturation, and effective hydrocarbon saturation can be determined by the difference between unity and effective water saturation.

$$S_{hc\text{-}eff} = 1 - S_{we} \quad (8)$$

Bulk Water Volume (BWV) and Permeability (K)

Bulk volume water (BWV) is the percentage of the total rock volume that is occupied by water. It compares to the more commonly used water saturation term in that water saturation is the percent of the total pore space occupied by water. It is a critical input to estimating fluid mobility in the reservoir.

$$BWV = S_{we} \times \phi \quad (9)$$

Irreducible water saturation ($\mathbf{S_{wirr}}$) is the ratio of immobile or irreducible water volume to effective porosity.

Permeability is a measurement of fluid mobility, usually expressed in one thousandths of a Darcy or millidarcy. It is a key component of net pay as well; low or no permeability will not produce economic quantities of hydrocarbons.

Timur method [38] is used here to determine the permeability.

$$K = (93 \times \phi 2.2/S_{wirr})^2 \quad (10)$$

Timur's empirical equation relates porosity and irreducible water saturation to permeability. While this method is widely used due to its simplicity and reliance on standard log data, it has known limitations in formations with significant shale content and complex pore structures. Although more advanced models may better account for shale effects and complex pore geometries, their application was limited in this study due to the lack of sufficient core or NMR data for calibration. Nonetheless, the Timur-derived permeability profiles show reasonable trends when compared qualitatively with formation lithology, Vshale content, and the available DST-derived permeability from Well 1.

4. Data Analysis

In this study, data were analyzed in two different ways, for comparison. The first procedure used Techlog software, whereas the second was performed manually, by collecting data at each depth and then calculating the petrophysical parameters using the above-mentioned equations. These calculated parameters were also cross-plotted, from a simple Excel spreadsheet, to analyze their eventual positive or negative correlation in different wells and reservoir zones.

4.1. Computed Analysis (Techlog Software)

Petrophysical analysis (Figures 6–9) was initially conducted using the Techlog software. The gamma ray (GR) log was first corrected for borehole size and subsequently normalized. Since GR readings tend to be elevated in wells with KCl-based mud systems, normalization was performed using data from Srikail-2, the only well drilled without KCl mud. Density log corrections were applied using the Gardner equation. Lithological components and porosity were quantified using the Quanti.Elan module, which performs mineralogical and fluid inversion. Shale volume was estimated from both neutron and density logs.

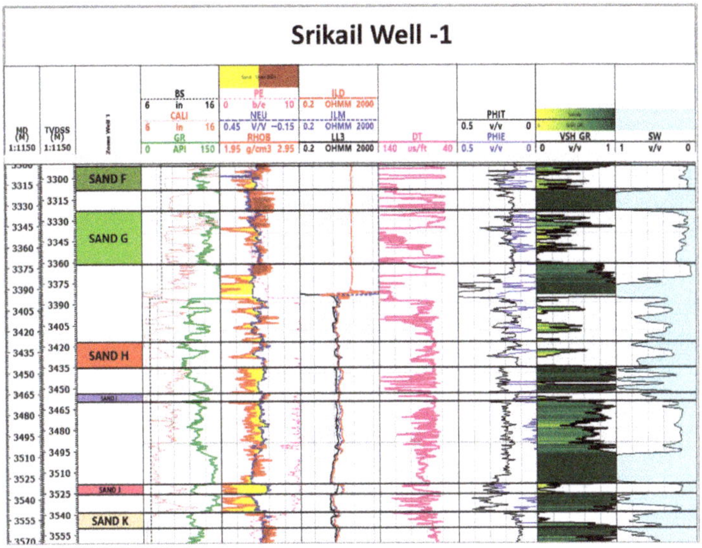

Figure 6. Petrophysical results (Techlog) for F, G, H, I, J, and K sands in Well 1.

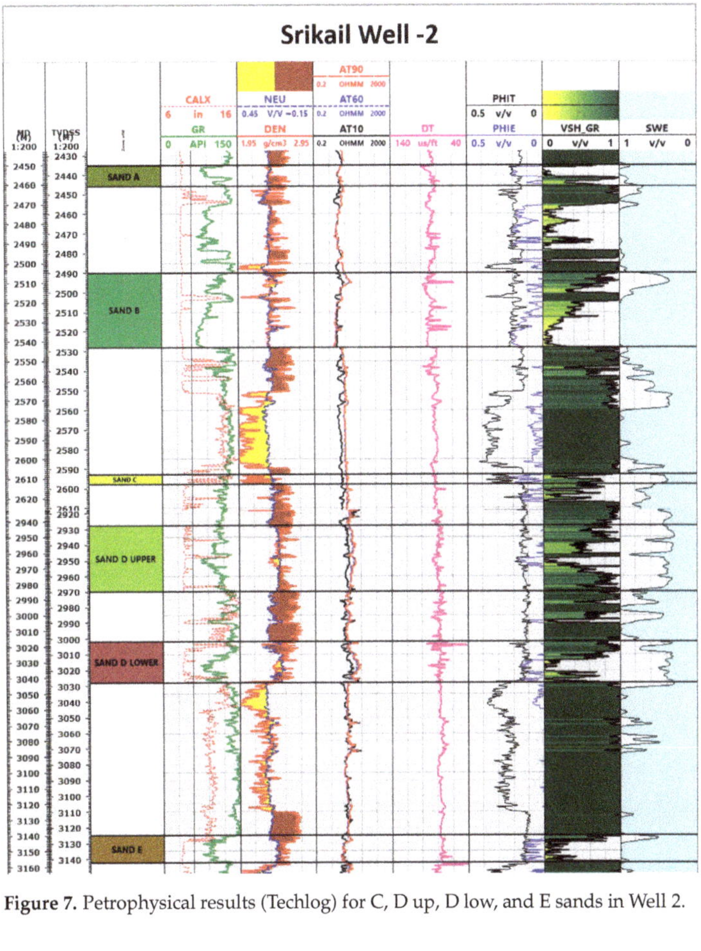

Figure 7. Petrophysical results (Techlog) for C, D up, D low, and E sands in Well 2.

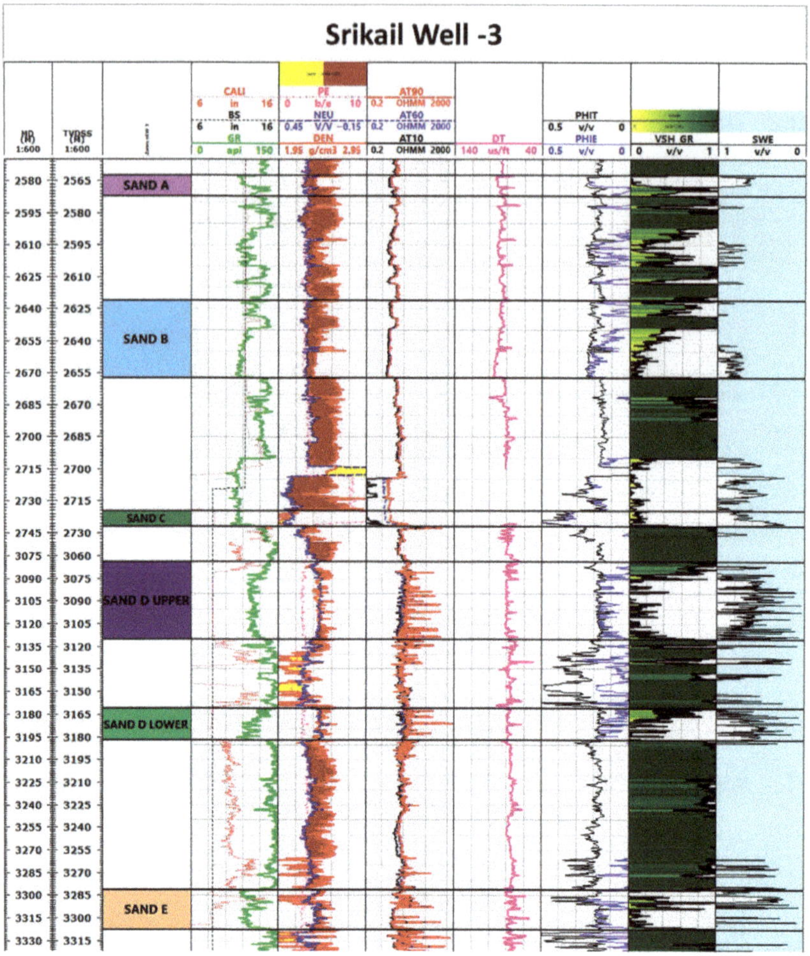

Figure 8. Petrophysical results (Techlog) for C, D up, D low, and E sands in Well 3.

To confirm the presence of gas, shallow (MSFL), medium (LLM), and deep (LLD) resistivity logs were analyzed. Hydrocarbon-bearing zones and gas–water contacts were identified using the quick-look interpretation technique, with particular emphasis on large negative separation between porosity logs, as described by [24] Gas-bearing zones were identified by observing neutron–density crossover behavior.

Total porosity (PHIT_ND) and effective porosity (PHIE_ND) were computed based on neutron–density data. Effective water saturation (SWE_SIM) and effective bulk volume of water (EBV_SIM) were calculated by Techlog software, using the Symandoux equation and method, respectively, and plotted as part of the interpretation results. The parameters applied in this analysis were selected based on previous studies, well reports, and core analysis data.

4.2. Raw Data Analysis

Lithology, shale volume, porosity, water saturation, hydrocarbon saturation, permeability, and hydrocarbon movability were determined using relevant well log data available in both digital and hard copy formats. These parameters were calculated manually using empirical equations in Microsoft Excel to assist in the characterization of the reservoir rocks.

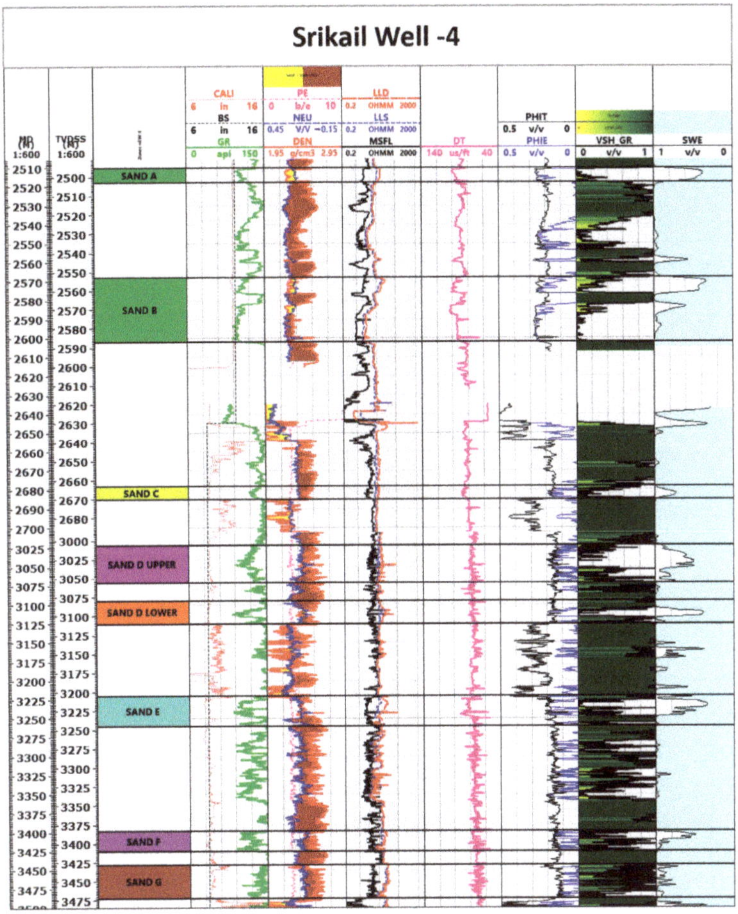

Figure 9. Petrophysical results (Techlog) for C, D up, D low, and E sands in Well 4.

4.2.1. Petrophysical Characteristics

In this paper all the producing and non-producing sands are characterized by different petrophysical parameters. All the parameters are shown in the following tables.

The petrophysical analysis presented in this study is subject to uncertainties due to the complex lithology of the Bhuban Formation and reliance on empirical parameters in models such as Simandoux. In the absence of core or advanced log data, fixed values for key inputs (e.g., m, n, a, C) were used based on regional analogs. Future studies should integrate probabilistic modeling and sensitivity analysis to better constrain parameter ranges and improve confidence in reservoir evaluations.

4.2.2. Cross Plots of Petrophysical Parameters

Petrophysical analysis plays an important role in reservoir characterization. To explain the reservoirs, the petrophysical data was cross-plotted using Excel sheets. Plots of volume of shale vs. depth have been compiled. To further comprehend the reservoirs, porosity vs. permeability was plotted, as well as volume of shale vs. permeability, saturation, and porosity.

Vshale values vary with depth along each individual sand show, with scattered data points suggesting heterogeneous lithologies, characterized by alternating or interbedded sand–shale layers (Figure 10). This heterogeneity results from varying depositional conditions, possibly from fluctuating energy levels in a transitional or marginal marine environment (e.g., deltaic or estuarine system).

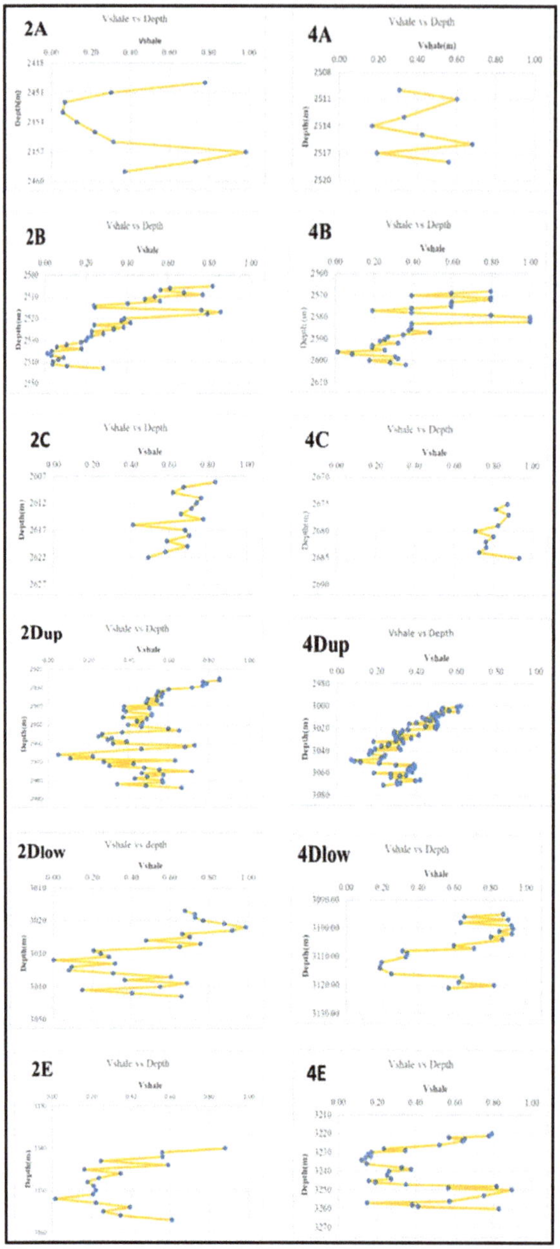

Figure 10. Vsh variation with depth for sands (**A**, **B**, **C**, **D$_{up}$**, **D$_{low}$** and **E**) in Wells 2 and 4.

Looking at individual sands, most of them tend to become more shaly (higher Vshale values) towards the top, probably resulting from a rapid input of clean sandy lobes, followed by an increasing mix with settling clays. Such heterogeneities have direct implications for reservoir quality, as higher Vshale values are typically associated with reduced porosity and permeability, thereby lowering hydrocarbon storage and flow potential.

In Sand A, porosity increases with decreasing Vsh (Figure 11). Both porosity and permeability demonstrate a strong inverse relationship with Vsh, indicating that decreasing shale content increases reservoir quality. A positive linear correlation exists between porosity and permeability.

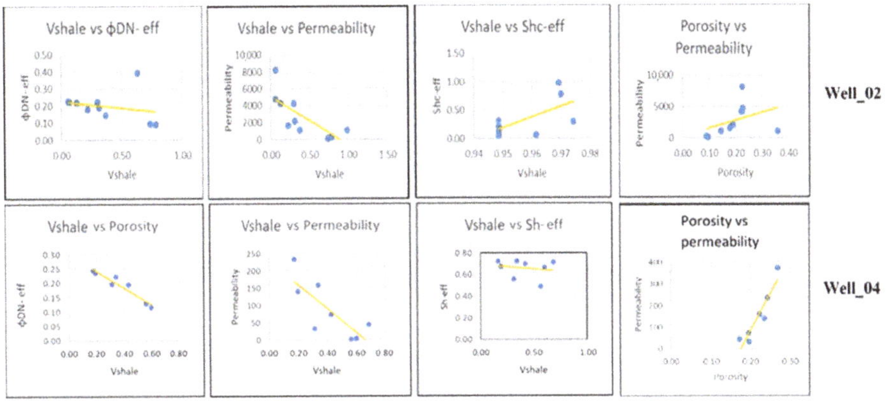

Figure 11. Cross plots of Sand A.

Sand B shows lower porosity in shale-rich zones and higher porosity where Vsh is reduced (Figure 12). Both porosity and permeability are negatively affected by increasing Vsh, and a strong linear relationship persists between porosity and permeability.

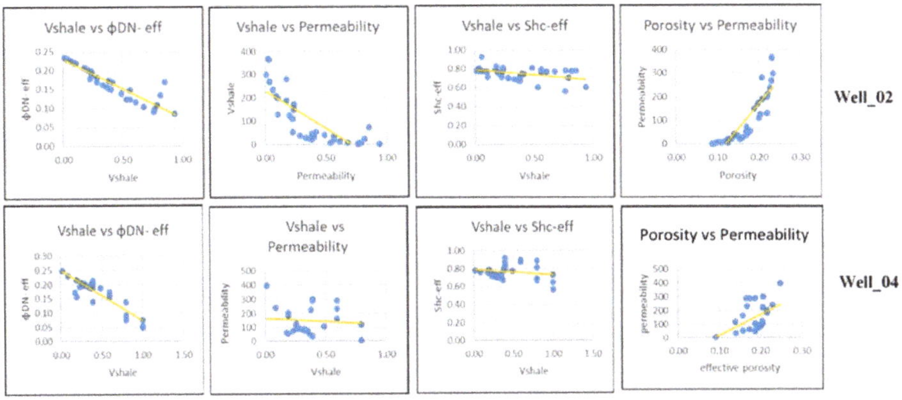

Figure 12. Cross plots of Sand B.

In Sand C porosity and permeability are lower with higher Vsh (Figure 13). Permeability improves with lower shale content and elevated porosity. Hydrocarbon saturation changes gradually throughout the interval. Strong inverse correlations are noted between Vsh and both porosity and permeability. The porosity–permeability relationship is again linear and well-defined.

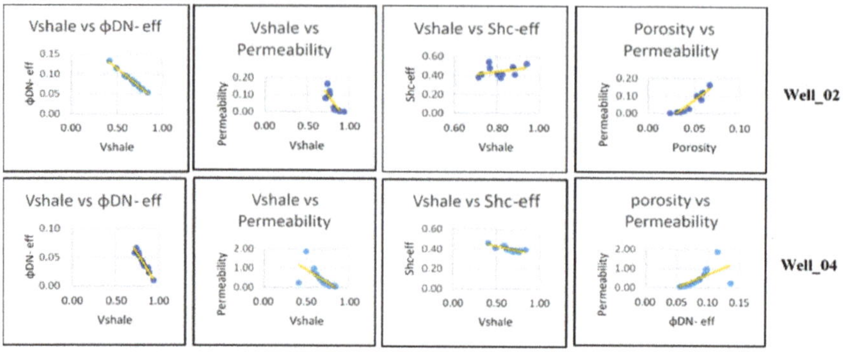

Figure 13. Cross plots of Sand C.

Sand D_{upper} shows higher porosity and permeability in zones with lower Vsh (Figure 14). Permeability increases with decreasing Vsh, and hydrocarbon saturation rises progressively. The data confirm the significant role of Vsh in reducing reservoir quality. A consistent linear relationship exists between porosity and permeability.

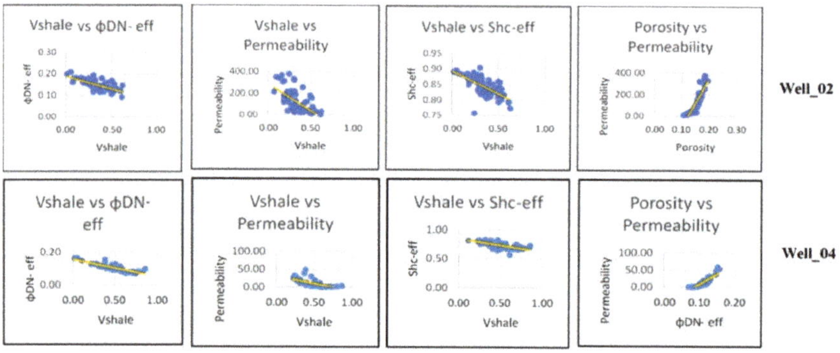

Figure 14. Cross plot of Sand D_{upper}.

In Sand D_{lower}, porosity and permeability improve where Vsh is lower and effective porosity is higher (Figure 15). A clear negative correlation is evident between Vsh and reservoir quality indicators. Vsh remains a critical controlling factor, showing a strong inverse correlation with both porosity and permeability. The linear relationship between porosity and permeability is maintained.

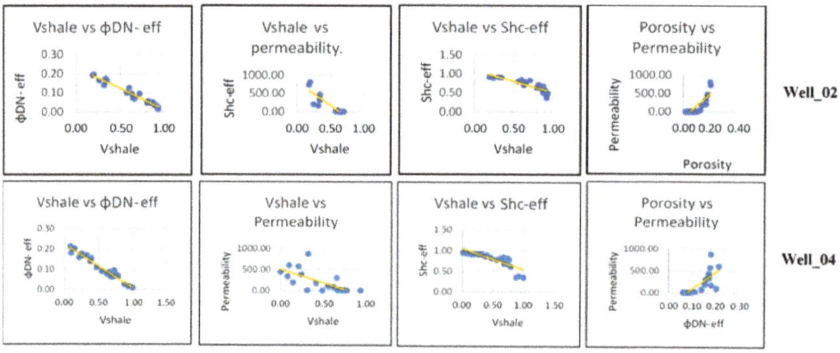

Figure 15. Cross plot of Sand D_{lower}.

In Sand E, porosity is higher in cleaner sands and reduced in zones with increased shale content (Figure 16). Permeability shows a rising trend with decreasing Vsh. Hydrocarbon saturation improves gradually with decreasing Vsh. Shale volume continues to negatively influence reservoir properties. A clear positive linear trend is observed between porosity and permeability.

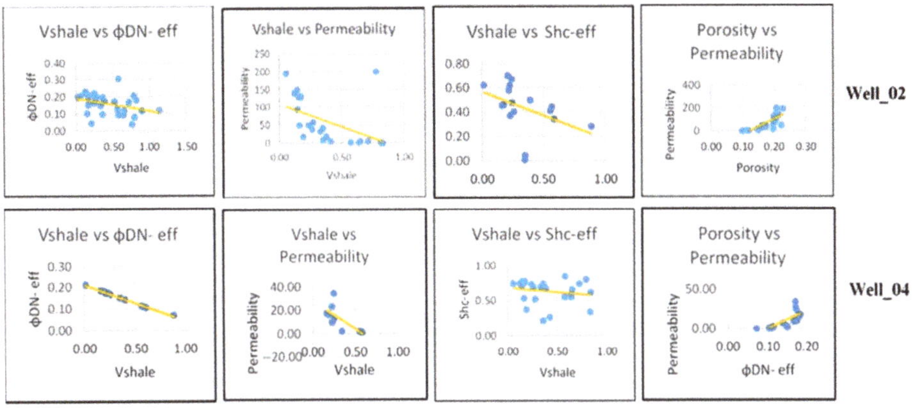

Figure 16. Cross plot of Sand E.

In Sand F porosity increases in sands with lower Vsh, (Figure 17). Hydrocarbon saturation shows a slight increase with improved porosity. Shale content remains a limiting factor for reservoir quality, and the porosity–permeability correlation is positive and linear.

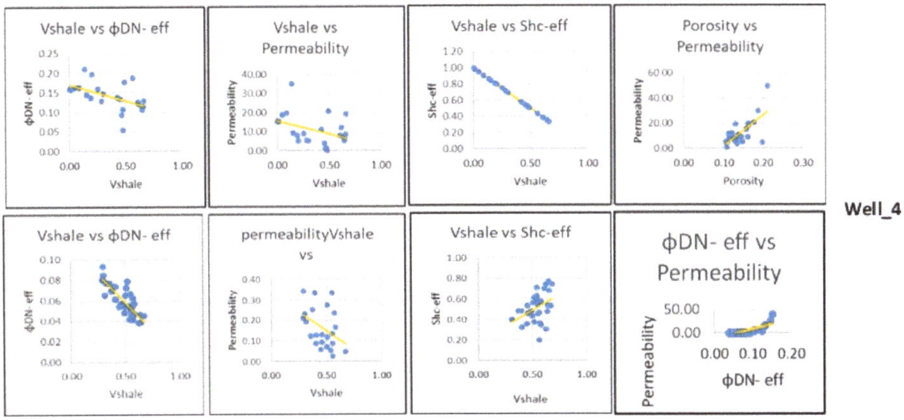

Figure 17. Cross plot of Sands F and G.

In Sand G, porosity increases with shale volume (Vsh) decrease, and permeability also improves (Figure 17), similar to Sand F. The negative influence of Vsh is evident, and a strong linear relationship exists between porosity and permeability.

From the above analysis, it becomes clear that D sands are effectively the best reservoirs [39]. Sand D_{upper} shows good porosity, permeability, and hydrocarbon saturation, all of them reducing in shale-rich zones, probably due also to poorer sorting. Sand D_{lower} presents good porosity and permeability in cleaner intervals in both wells, The consistency in these patterns reinforces the controlling influence of Vsh and depositional sorting on reservoir quality.

5. Results and Discussion

In previous works, Alam et al. [39] characterized the producing quality of three deep sandy zones (D-Upper, D-Lower, and E sand) in three Srikail wells, based in Vsh, porosity, permeability and net-to-gross. Hossain et al. [40] also characterized three sandy units (named Zones 1, 2, and 3) in a single well (Srikail #3), indicating only average values for Vsh, porosity, water and hydrocarbon saturation for each zone. In this work, in-depth petrophysical evaluation and reservoir characterization of all the sandy intervals of the Srikail Gas Field were conducted through a combination of manual analysis and automated interpretation using Techlog software. Petrophysical parameters were derived through both empirical equations and software-based computations, and the results were compared to assess consistency and interpretation accuracy.

From the 11 previously identified reservoir zones (named A to K, from top to bottom), the Srikail-1 well has intersected a total of nine reservoir layers (A to C and F to K), whereas Srikail-4 has intersected the seven reservoir layers (A to G), and both the Srikail-2 and Srikail-3 wells only five (A to E).

The proportion of clay in reservoirs is a crucial parameter influencing its producing quality. The calculated average shale volume (Table 5) varies across different reservoir layers, and among these, Zone C exhibits the highest shale content, whereas Zone B shows the lowest. Within each sandy interval, Vshale variability at specific depths indicates heterogeneous, interbedded sand–shale lithology from fluctuating depositional conditions. An overall increase in Vshale towards the top indicates a transition to finer-grained, lower-energy environments. This heterogeneity impacts reservoir quality by reducing porosity and permeability, affecting hydrocarbon flow.

Table 5. Minimum and maximum gamma ray values (GR_{min}, GR_{max}) and average shale volume (Av. V_{sh}).

Sand	Well_1		Well_2		Well_3		Well_4	
	GR_{min}	GR_{max}	GR_{min}	GR_{max}	GR_{min}	GR_{max}	GR_{min}	GR_{max}
	Av. V_{sh} (%)		Av. V_{sh} (%)		Av. V_{sh} (%)		Av. V_{sh} (%)	
A	73	134	77	142	80	135	96	126
	36%		42%		48%		37%	
B	**94**	**159**	67	146	80	144	96	150
	43%		35%		31%		45%	
C	**97**	**133**	68	146	70	134	65	149
	56%		60%		46%		83%	
D_{upper}	-		80	150	80	150	**99**	186
			48%		49%		34%	
D_{lower}	-		73	148	77	154	90	161
			51%		44%		60%	
E	-		74	155	80	137	90	158
			35%		43%		45%	
F	-		-		-		100	171
							39%	

Table 5. *Cont.*

Sand	Well_1		Well_2		Well_3		Well_4	
	GR_{min}	GR_{max}	GR_{min}	GR_{max}	GR_{min}	GR_{max}	GR_{min}	GR_{max}
	Av. V_{sh} (%)		Av. V_{sh} (%)		Av. V_{sh} (%)		Av. V_{sh} (%)	
G	90	159	-		-		100	171
	52%						50%	
H	74	113	-		-			
	41%							
I	80	124	-		-			
	57%							
J	90	134	-		-			
	42%							
K	85	134	-		-			
	37%							

Porosity estimation is a fundamental step in reservoir evaluation [41], and along with permeability, it dictates the fluid storage capacity and productivity of the reservoir. Effective porosity was derived from a combination of effective density and neutron porosities. The overall effective porosity ranges (Table 6) from 4% to 20%. Sands A, B, D, and E exhibit favorable porosity values and are considered good-quality reservoirs.

Table 6. Average effective porosity ($\phi_{DN\text{-}eff}$) from effective density (ϕ_{D_eff}) and effective neutron porosity (ϕ_{N_eff}).

Sand	Well_1		Well_2		Well_3		Well_4	
	ϕ_{D_eff}	ϕ_{N_eff}	ϕ_{D_eff}	ϕ_{N_eff}	ϕ_{D_eff}	ϕ_{N_eff}	ϕ_{D_eff}	ϕ_{N_eff}
	$\phi_{DN\text{-}eff}$ (%)		$\phi_{DN\text{-}eff}$ (%)		$\phi_{DN\text{-}eff}$ (%)		$\phi_{DN\text{-}eff}$ (%)	
A	15	14	19	20	11	17	17	21
	15%		**21%**		**15%**		**20%**	
B	13	17	15	21	12	24	14	19
	13%		**18%**		**19%**		**17%**	
C	07	08	07	09	9	10	02	05
	08%		**8%**		**7%**		**04%**	
D_{upper}			12	11	12	14	12	20
			12%		**14%**		**15%**	
D_{lower}			10	14	11	15	12	11
			12%		**14%**		**12%**	
E			11	18	12	14	13	17
			15%		**15%**		**16%**	
F							08	19
							14%	

Table 6. *Cont.*

Sand	Well_1		Well_2		Well_3		Well_4	
	ϕ_{D_eff}	ϕ_{N_eff}	ϕ_{D_eff}	ϕ_{N_eff}	ϕ_{D_eff}	ϕ_{N_eff}	ϕ_{D_eff}	ϕ_{N_eff}
	$\phi_{DN\text{-}eff}$ (%)		$\phi_{DN\text{-}eff}$ (%)		$\phi_{DN\text{-}eff}$ (%)		$\phi_{DN\text{-}eff}$ (%)	
G	09	10					04	07
	10%						**7%**	
H	11	09						
	11%							
I	11	05						
	10%							
J	14	09						
	11%							
K	13	11						
	11%							

Hydrocarbon saturation (Shc) values (Table 7) further confirm reservoir potential. Based on the criteria of Asquith and Gibson [42], zones with Shc above 60% are considered hydrocarbon-bearing, making Zones A, B, D, and E the most promising. Zones C, F, G, H, I, J, and K, in contrast, demonstrate higher shale content, lower porosity, and limited hydrocarbon potential.

Table 7. Average effective water (S_w-eff) and hydrocarbon saturation (S_{hc}-eff).

Sand	Well_1		Well_2		Well_3		Well_4	
	S_w-eff (%)	S_{hc}-eff (%)	S_w-eff (%)	S_{hc}-eff (%)	S_w-eff (%)	S_{hc}-eff (%)	S_w-eff (%)	S_{hc}-eff (%)
A	35	65	4	96	40	60	33	67
B	40	60	25	75	31	69	23	77
C	42	58	45	55	40	60	45	55
D_{upper}	-		26	74	20	80	14	86
D_{lower}	-		22	78	17	83	26	74
E	-		40	60	22	78	34	66
F	-		-		-		40	60
G	46	54	-		-		46	54
H	17	83	-		-		-	
I	16	84	-		-		-	
J	19	81	-		-		-	
K	22	78	-		-		-	

Mud log analysis revealed that C and G sands are highly calcareous with poor porosity. In Well-01, the G, H, I, J, and K intervals also exhibited significant calcareous content [43] and sub-optimal porosity. Calcareous tests during mud logging and a lack of gas flow during DSTs (except in Sand I) suggest these sands have low permeability.

Permeability was estimated using the Timur equation for wells 1 through 4 (Table 8). D sands demonstrate the highest permeability and are already in production. Sands A, B, and E also show good permeability and are viable for hydrocarbon extraction. Zone E, in particular, is confirmed as a productive interval in Well-04. C and G sands have very poor permeability. Sands F, H, I, J and K show poor to moderate porosity and permeability, maybe producible with minimal stimulation (e.g., fracturing). Permeability values obtained from DST data in the deeper tested sands range from 0.6 to 0.01 mD [44], indicating a very tight reservoir with limited fluid flow capacity without stimulation.

Table 8. Average permeability (Timur method).

Sand	Well_1	Well_2	Well_3	Well_4
	K (mD)	K (mD)	K (mD)	K (mD)
A	83	206	117	120
B	46	158	161	227
C	0.08	0.11	0.13	0.05
D_{upper}	-	111	102	129
D_{lower}	-	355	117	291
E	-	40	66	58
F	14	-	-	18
G	1.01	-	-	2.6
H	61	-	-	-
I	67	-	-	-
J	64	-	-	-
K	22	-	-	-

Petrophysical parameters cross plots clearly point to an inverse relationship between Vshale and porosity. Higher shale content corresponds with reduced porosity, as shale occupies pore spaces and obstructs pore throats. In this study, Zones C, G, F, H, I, J, and K are identified as low-quality reservoirs due to permeability, although some of them have good hydrocarbon saturation.

The relationship between Vshale, porosity, and permeability was further assessed. Both porosity and permeability decrease as Vshale increases. Cross-plotting porosity versus permeability reveals a strong linear correlation: well-sorted, clean sandstones with low shale content show high porosity and permeability. Conversely, zones with high clay content may display elevated porosity but reduced permeability due to matrix blockage. Considering that in most sands Vsh increases towards the upper layers, it may be concluded that permeability is lower on these more shaly layers. Moreover, the good correlation of porosity decrease with Vhs increase allows us to consider that hydrocarbon saturation also tends to decrease towards the upper layers. Overall, each sand tends to have better reservoir properties at its lower and intermediate layers than at its upper, more shaly layers.

Bulk volume water (BVW) values offer further insight into grain size and reservoir texture. According to Fertl and Vercellino [45], BVW values between 0.035 and 0.07 are indicative of fine to very fine-grained sands. The results (Table 9 suggest that Sands A, B, and C are fine to very fine-grained, D sands are coarse-grained, and Sand E comprises fine-grained and silty material.

The calculated average petrophysical parameters vary across different wells in the reservoir layers. Given the deltaic depositional environment of the Bhuban Formation and the observed variability in petrophysical parameters across wells, future work could benefit from machine learning approaches that integrate seismic attributes. For instance, recent studies [46] have demonstrated the successful application of a probabilistic neural networks (PNN) combined with multi-attribute seismic analysis to predict porosity in complex deltaic reservoirs. Applying similar techniques in the Srikail Gas Field may improve reservoir property prediction, especially where core data are unavailable.

Table 9. Bulk volume of water (BVW, range) for Wells 1, 2, 3 and 4.

Sands	(BVW)
A	0.040–0.060
B	0.030–0.050
C	0.037–0.040
D_{upper}	0.020–0.027
D_{lower}	0.020–0.024
E	0.035–0.080

6. Conclusions

This research presents an integrated petrophysical evaluation and reservoir characterization of the Srikail Gas Field, utilizing both manual techniques and advanced digital interpretation through Techlog software. A combination of digital and hard copy well log data was used to estimate essential petrophysical properties such as shale volume, porosity, fluid saturation, permeability, and bulk volume of water (BVW). The study employed empirical cross plots, analytical models, and automated processing to improve the precision and consistency of interpretations. A quick-look methodology aided in identifying key reservoir intervals and gas-bearing zones.

The analysis reveals that reservoir quality within the field varies from poor to good. Net reservoir thicknesses range from 8 to 66 m, with shale content between 31% and 83%. Effective porosity spans from 4% to 20%, hydrocarbon saturation ranges from 54% to 96%, and effective water saturation falls between 4% and 46%. Permeability estimates vary widely from 0.05 to 355 mD, while BVW values are between 0.02 and 0.08.

Both qualitative and quantitative interpretations were carried out to assess the characteristics of individual reservoir layers. The identified reservoirs predominantly consist of shaly sandstones. Reservoirs A, B, and E show moderate to good reservoir properties, characterized by fine to very fine-grained, silty sandstones with fair to good permeability. Conversely, Reservoirs C and G exhibit poor quality, highly calcareous composition (from mud logging) and extremely low permeability, consistent with tight sandstone classification. Deeper sands such as F, H, I, J, and K display tight to semi-conventional characteristics and are likely to benefit from targeted stimulation and advanced completion strategies.

The most productive zones—Upper D and Lower D—are identified by their cleaner and coarser grain size and superior permeability, contributing significantly to gas production. Meanwhile, DST-derived permeability values from deeper sands (e.g., G, H, J) confirm extremely tight conditions, ranging from 0.6 to 0.01 mD.

Regardless of the different sands and burial depths, there is a clear pattern of upwards decrease in hydrocarbon production properties towards the top layers of each producing sand. This fact indicates the importance of initial sandy inputs, gradually "contaminated" by clay inputs into the depositional environment. In other words, each sand seems to represent a fourth order finning upwards cycle [21].

Integrating the previous analysis, Reservoir Zones A, B, D, and E clearly appear as high-quality reservoirs due their favorable porosity, permeability, hydrocarbon saturation, and low shale content. Among these, D and E sands are already productive, supporting their classification as effective hydrocarbon-bearing reservoirs. In contrast, Zones C and G exhibit poor reservoir characteristics and may be classified as tight sands due to their calcareous nature, poor porosity, and limited permeability. On the other hand, F, H, I, J, and K are tight to semi-conventional and might require partial stimulation and proper completion.

The results of this study contribute to a better understanding of reservoir heterogeneity, fluid flow behavior, and fracture network characterization in unconventional gas systems. Moreover, the outcomes of this research can be applied to other gas fields in and around the Bengal Basin, where similar unconventional sand reservoirs are present.

Future studies should adopt a more integrated and comprehensive approach to enhance reservoir characterization in the Srikail Gas Field. In particular, the use of multiple shaly sand models—following the comparative methodology of Johanna & Kusumah [47]—is recommended to better assess the impact of shale distribution and geometry on key petrophysical parameters. Additionally, incorporating core-based analyses such as porosity and permeability measurements, capillary pressure data, and XRD-derived clay mineralogy would significantly strengthen the calibration and validation of log-derived models. This integration of core data, advanced modeling, and seismic interpretation would reduce uncertainties, refine input parameters, and ultimately provide deeper insights into the reservoir heterogeneity of the Bhuban Formation, thereby supporting more accurate reservoir evaluation and improved hydrocarbon recovery strategies.

Author Contributions: Conceptualization, S.A. and N.P.; methodology, S.A. and N.P.; software, S.A.; validation, N.P.; formal analysis, N.P.; investigation, S.A. and N.P.; resources, N.P.; data curation, S.A.; writing—original draft preparation, S.A.; writing—review and editing, S.A. and N.P.; supervision, N.P.; funding acquisition, N.P. All authors have read and agreed to the published version of the manuscript.

Funding: This research was funded by Portuguese Fundação para a Ciência e a Tecnologia (FCT) I.P./MCTES through national funds (PIDDAC)—UIDB/50019/2020, LA/P/0068/2020 and UID/50019/2025.

Data Availability Statement: The original contributions presented in this study are included in the article. Further inquiries can be directed to the corresponding authors.

Acknowledgments: BAPEX (Bangladesh Petroleum Exploration and Production Company Ltd.) is highly acknowledged by the authors for the access to and use of the various datasets used for this work, particularly the wire-line logs and mud logs of the Srikail Gas Field wells. The Department of Geology and the Instituto Dom Luiz, both at the Faculty of Sciences of Lisbon University, are acknowledged by the authors for their IT and logistical assistance during this work. Schlumberger is acknowledged for the use of an academic license of Techlog, used to analyze all the wireline logs.

Conflicts of Interest: The authors declare no conflict of interests.

References

1. Zeba, J.; Murrell, R. Evaluation of Rock Properties from Logs Affected by Deep Invasion—A Case Study. In Proceedings of the 3rd International Workshop on Rock Physics, Perth, Australia, 13–17 April 2015.
2. Simandoux, P. Mesures Dielectriques en Milieu Poreux, Application à Mesure des Saturations en Eau, Étude du Comportement des Massifs Argileux. *Rev. Inst. Fr. Pétrol.* **1963**, *15* (Suppl. 3).
3. Poupon, A.; Leveaux, J. Evaluation of Water Saturations in Shaly Formations. *Log Anal.* **1971**, *12*, SPWLA-1971-vXIIn4a1.
4. Islam, M.S.; Islam, M.R.; Hossain, M.A.; Rahman, A.M. Petrophysical Analysis of Shaly Sand Gas Reservoir of Titas Gas Field Using Well Logs. *Bangladesh J. Geol.* **2006**, *25*, 106–124.
5. Islam, M.D.; Rahman, M.D.; Woobaidullah, A.S.M. Reservoir Characterization of Srikail Gas Field Using Wireline Log Data. *Dhaka Univ. J. Earth Environ. Sci.* **2020**, *5*, 1–8.
6. Islam, A.R.M.T.; Islam, M.A. Evaluation of Gas Reservoir of the Meghna Gas Field, Bangladesh Using Wireline Log Interpretation. *Univ. J. Geosci.* **2014**, *2*, 62–69. [CrossRef]
7. Islam, A.R.; Islam, T.; Biswas, R.; Jahan, K. Petrophysical Parameter Studies for Characterization of Gas Reservoir of Narsingdi Gas Field, Bangladesh. *Int. J. Adv. Geosci.* **2014**, *2*, 53–58.

8. Mostafizur, R.; Woobaidullah, A.S.M. Evaluation of Reservoir Sands of Habiganj-7 Well by Wireline Log Interpretation. *Dhaka Univ. J. Earth Sci.* **2012**, *2*, 1–8.
9. Reimann, K.U. *Geology of Bangladesh*; Gebrüder Borntraeger: Berlin, Germany, 1993; p. 160.
10. Banglapedia. *National Encyclopedia of Bangladesh*; Asiatic Society of Bangladesh: Dhaka, Bangladesh, 2003.
11. Alam, M. Geology and Depositional History of Cenozoic Sediments of the Bengal Basin of Bangladesh. *Palaeogeogr. Palaeoclimatol. Palaeoecol.* **1989**, *69*, 125–139. [CrossRef]
12. Curray, J.R.; Emmel, F.J.; Moore, D.G.; Raitt, R.W. Structure, Tectonics, and Geological History of the Northeastern Indian Ocean. In *The Ocean Basins and Margins*; Nairn, A.E.M., Stehli, F.G., Eds.; Springer: Boston, MA, USA, 1982; pp. 399–445.
13. Alam, M.; Alam, M.M.; Curray, J.R.; Chowdhury, M.L.R.; Gani, M.R. An Overview of Sedimentary Geology of the Bengal Basin in Relation to the Regional Tectonic Framework and Basin Fill History. *Sediment. Geol.* **2003**, *155*, 179–208. [CrossRef]
14. Evans, P. Tertiary Succession in Assam. *Trans. Mineral. Geol. Inst. India* **1932**, *27*, 155–260.
15. Alam, M.M. *Sedimentology and Depositional Environment of Subsurface Neogene Sediments in the Sylhet Trough, Bengal Basin: Case Study of the Fenchuganj and Beanibazar Structures, Northeastern Bangladesh*; Unpublished Report; Bangladesh Petroleum Institute: Dhaka, Bangladesh, 1993; pp. 1–82.
16. Holtrop, J.F.; Keizer, J. Some Aspects of the Stratigraphy and Correlation of the Surma Basin Wells, East Pakistan. *ECAFE Miner. Resour. Dev. Serv.* **1970**, *36*, 143–154.
17. Hiller, K.; Elahi, M. Structural Growth and Hydrocarbon Entrapment in the Surma Basin, Bangladesh. In *Petroleum Resources of China and Related Subjects*; Wagner, H.C., Wagner, L.C., Wang, F.F.H., Wong, F.L., Eds.; Circum-Pacific Council for Energy and Mineral Resource Earth Science Series; The Circum-Pacific Council for Energy and Mineral Resources (CPCEMR): Menlo Park, CA, USA, 1988; Volume 10, pp. 657–669.
18. Banerjee, M.; Sen, P.; Dastidar, A.G. On the Depositional Condition of the Holocene Sediments of Bengal Basin with Remarks on Environmental Condition of the Soft Grey Clay Deposition in Calcutta. In Proceedings of the Indian Geotechnical Conference 1, Calcutta, India, 21–23 December 1984; Volume Div. I, pp. 63–69.
19. Johnson, S.; Alam, A. Sedimentation and Tectonics of the Sylhet Trough, Bangladesh. *Geol. Soc. Am. Bull.* **1991**, *103*, 1513–1527. [CrossRef]
20. Sultana, D.N.; Alam, M.M. Facies Analysis of the Neogene Surma Group Succession in the Subsurface of the Surma Basin, Bengal Basin, Bangladesh. *Bangladesh Geosci. J.* **2001**, *6*, 53–74.
21. Akhter, S.; Pimentel, N. Sequence Stratigraphy of Miocene Deltaic Sands in the Srikail Gas Field, East-Central Bengal Basin: Insights from Mud Log and Wireline Log Data. *J. Geol. Soc. India*, 2025, *in press*.
22. Bangladesh Petroleum Exploration and Production Company Limited (BAPEX). Well Completion Reports of Srikail #1, #2, #3, and #4. Unpublished report.
23. Petrobangla. *Srikail Geological Study*; Unpublished Report; Petrobangla: Dhaka, Bangladesh, 2009.
24. Rider, M.H. *The Geological Interpretation of Well Logs*; Whittles Publishing: Caithness, UK, 2000.
25. Schlumberger. *Log Interpretation Principles/Applications*; Schlumberger: Houston, TX, USA, 1989.
26. Schlumberger. *Log Interpretation Charts: English Metric*; Schlumberger: Houston, TX, USA, 1979.
27. Schlumberger. *Log Interpretation. Vol. II: Applications*; Schlumberger Well Services Inc.: Houston, TX, USA, 1974.
28. Schlumberger. *Log Interpretation. Vol. I: Principles*; Schlumberger Well Services Inc.: Houston, TX, USA, 1972.
29. Wu, F.; Cong, L.; Ma, W.; Liu, J.; Dai, J.; Nie, X.; Shi, X.; Luo, Y.; Chen, C. Finite Element Method—Based Resistivity Simulation and Water Saturation Calculation of Irregular Laminated Shaly Sandstone. *Geomech. Geophys. Geo-Energy Geo-Resour.* **2023**, *9*, 2. [CrossRef]
30. Nawab, M.M.; Islam, M.A. Estimation of Shale Volume Using Gamma and Porosity Logs: Application to the Selected Gas Fields of Bangladesh. *Bangladesh Geosci. J.* **2005**, *11*, 89–99.
31. Markel, R.H. *Well Log Formation Evaluation*; AAPG Continuing Education Course Note Series: Tulsa, OK, USA, 1979; Volume 14, p. 82.
32. Archie, G.E. The Electrical Resistivity Log as an Aid in Determining Some Reservoir Characteristics. *Petrol. Tech.* **1942**, *5*, 54–62. [CrossRef]
33. Worthington, P.F. The Evolution of Shaly-Sand Concepts in Reservoir Evaluation. *Log Anal.* **1985**, *26*, 23–40.
34. Rahman, M.; Sarker, M. Petrophysical and Reservoir Quality Assessment of Tight Gas Sands in Eastern Bengal Basin. *J. Asian Earth Sci.* **2018**, *45*, 230–245.
35. Haque, M.; Fariduddin, M. Petrophysical Evaluation of Miocene Deltaic Sandstones in the Bengal Basin Using Well Log Interpretation. *J. Pet. Geosci.* **2016**, *22*, 145–158.
36. Chowdhury, S.; Alam, M.; Hossain, Z. Reservoir Characterization of Miocene–Pliocene Gas Sands in the Surma Basin (Bengal Basin) Using Integrated Well Log Analysis. *Mar. Pet. Geol.* **2020**, *120*, 104566.
37. Bakar, W.Z.W.; Saaid, I.M.; Ahmad, M.R.; Amir, Z.; Japperi, N.S.; Fuad, M.F.I.A. Improved Water Saturation Estimation in Shaly Sandstone through Variable Cementation Factor. *J. Pet. Explor. Prod. Technol.* **2022**, *12*, 1329–1339. [CrossRef]

38. Timur, A. An Investigation of Permeability, Porosity and Residual Water Saturation Relationships for Sandstone Reservoirs. *Log Anal.* **1968**, *9*, 8–17.
39. Alam, K.; El-Husseiny, A.; Abdullatif, O.; Babalola, L. Petrophysical Evaluation of Prospective Reservoir Zones in Srikail Gas Field, Bengal Basin Bangladesh. In Proceedings of the 81st EAGE Conference and Exhibition 2019, London, UK, 3–6 June 2019; Volume 2019, pp. 1–5.
40. Hossain, D.; Rahman, M.; Khatu, H.; Haque, R. Petrophysical Properties Assessment Using Wireline Logs Data at Well #3 of Srikail Gas Field, Bangladesh. *China Geol.* **2022**, *5*, 393–401.
41. Ruhovets, N. A Log Analysis Technique for Evaluating Laminated Reservoirs in the Gulf Coast Area. *Log Anal.* **1990**, *31*, 294–303.
42. Asquith, G.B.; Gibson, C. *Basic Well Log Analysis for Geologists*; AAPG: Tulsa, OK, USA, 1982; p. 216.
43. Bangladesh Petroleum Exploration and Production Company Limited (BAPEX). Report on Technical Study for Possibility of Re-Entry/Re-Visit the Suspended, Dry & Abandoned Wells under BAPEX (Case Study: Srikail Gas Field), August 2020. Unpublished report. 2020.
44. Bangladesh Petroleum Exploration and Production Company Limited (BAPEX). Final Technical Report on Srikail Field (3D Project of BAPEX), Schlumberger. Unpublished report.
45. Fertl, W.H. Shaly-Sand Analysis in Development Wells. In Proceedings of the SPWLA 16th Annual Logging Symposium, New Orleans, LA, USA, 4–7 June 1975; Society of Petrophysicists and Well Log Analysts: Houston, TX, USA, 1975; Paper A.
46. Shen, Z.; Zhang, Z.; Liang, Z.; Tao, X.; Wang, X.; Li, X.; Sun, S.; Wang, H.; Mei, L.; Zhou, Y. Optimizing Petrophysical Property Prediction in Fluvial Deltaic Reservoirs: A Multi-Seismic Attribute Transformation and Probabilistic Neural Network Approach. *J. Pet. Explor. Prod. Technol.* 2025, *in press*.
47. Johanna, U.; Kusumah, E.P. Analysis of Petrophysical Parameter on Shaly Sand Reservoir by Comparing Conventional Method and Shaly-Sand Method in Vulcan Subbasin, Northwest Australia. *J. Geosci. Eng. Environ. Technol.* **2023**, *8*, 13880. [CrossRef]

energies

MDPI

Article

Evaluation of Marine Shale Gas Reservoir in Wufeng–Longmaxi Formation, Jiaoshiba Area, Eastern Sichuan Basin

Qiang Yan *, Aiwei Zheng, Li Liu, Jin Wang, Xiaohong Zhan and Zhiheng Shu

Research Institute of Exploration and Development, Sinopec Jianghan Oilfield Company, Wuhan 430233, China; yjyzhaw.jhyt@sinopec.com (A.Z.); liuli.jhyt@sinopec.com (L.L.); yjywangj.jhyt@sinopec.com (J.W.); zhanxh.jhyt@sinopec.com (X.Z.); shuzh.jhyt@sinopec.com (Z.S.)
* Correspondence:qyan2023.jhyt@sinopec.com

Abstract

The Jiaoshiba area, as an important production capacity contribution block for the Fuling shale gas field, is of great significance for its long-term stable production. This study is based on continuous coring, and uses methods such as whole-rock mineral analysis, porosity and permeability analysis, gas content analysis, and organic geochemistry to systematically analyze the influencing factors of reservoir properties and gas content in the studied interval. Combined with the variation law of TOC and other parameters with depth, the target reservoir is comprehensively evaluated, and the evaluation results are verified based on actual production data. The results show that the influence of minerals on permeability is very weak, and cracks can greatly improve permeability, but their contribution to porosity is not significant. Porosity has a certain impact on gas content, but it is not the main controlling factor. The pores related to quartz (organic silicon) are mostly organic pores, which host a large amount of shale gas, while clay minerals are not conducive to the occurrence of shale gas. Organic matter (OM) maturity contributes more to porosity than OM abundance, but OM abundance has a stronger impact on gas content than its maturity. The research intervals can be divided into four categories: Class I (①–③) is the best, followed by Class II (⑦–⑨); Class III (④–⑥) is poor, and Class IV (top, non-gas-bearing layer) is the worst.

Keywords: Jiaoshiba area; Wufeng-Longmaxi formation; marine shale gas; reservoir evaluation; gas bearing capacity

1. Introduction

The Fuling shale gas field (FLSGF), as the first commercial shale gas field in China, has important strategic significance for national energy security. Jiaoshiba, as the main gas-producing area, strongly supports the growth of shale gas production in China. Compared to the United States, the exploration and development of shale gas in China started later, with complex surface and stratigraphic structures, making exploration and development difficult. However, in recent years, China has experienced rapid development. In 2023, China's shale gas production reached 25 billion cubic meters, and national shale gas demonstration zones have been established, such as Fuling, Changning–Weiyuan, and Zhaotong [1–7].

Although there are various methods for evaluating shale gas reservoirs, most of them use the "point to face" method to study. Specifically, limited samples are selected for relevant experimental analysis, including but not limited to thin section analysis, X-ray analysis,

SEM (scanning electron microscopy) analysis, low-temperature nitrogen adsorption, high-pressure mercury intrusion, total organic carbon, chromatography–mass spectrometry, and so on [8–12]. This study selected a typical coring well in the Jiaoshiba area, and conducted continuous coring on the target layer (2330.46–2415.1 m) with intervals mostly less than 1 m. Various experimental methods were used to finely evaluate the Wufeng–Longmaxi Formation, providing reference for shale gas development in the Jiaoshiba area.

2. Geological Background

Most of China's shale gas resources are concentrated in the Paleozoic marine shale reservoirs in the south, which have strong tectonic activity, complex geostress, deep burial, and harsh surface environments [13]. Jiaoshiba is located in Jiaoshi Town, Fuling District, Chongqing, in the eastern part of the Sichuan Basin. Its structural location spans two secondary structural units, with the main area being the Jiaoshiba fault anticline, including four local structures such as the Yanjiang saddle and the Wujiang No. 2 fault anticline. The faults mainly develop in the east and west wings of the structure and its southwest, while the faults in the main structural area are not developed [14–16]. The burial depth gradually increases from northeast to southwest, with a burial depth of less than 3000 m in the central and northern parts and a burial depth of more than 3000 m in some southern areas. The strata of the Wufeng–Longmaxi Formation, the subject of this study, can be vertically divided into sub-layers ①–⑨. Among them, sub-layers ①–③ represent the lower gas layer, sub-layers ④–⑥ represent the middle gas layer, and sub-layers ⑥–⑨ represent the upper gas layer. The three-dimensional development of the three layers has been comprehensively realized (Figure 1).

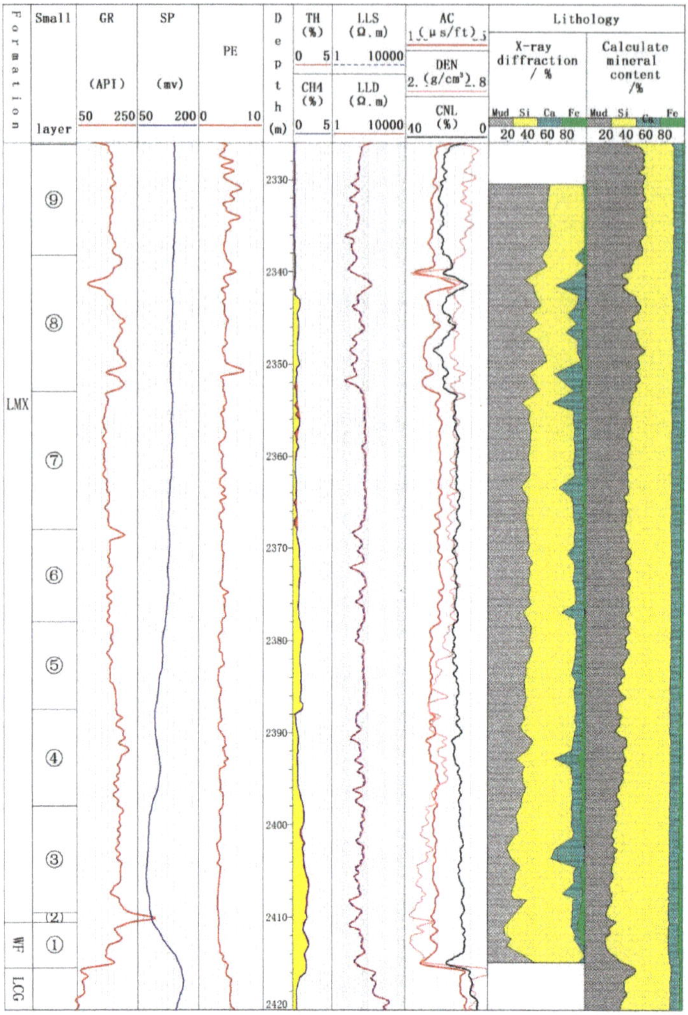

Figure 1. Comprehensive detailed subdivision map of the Wufeng–Longmaxi Formation strata.

3. Samples and Experiments

This study focuses on the core well, and conducts continuous experimental analysis on sub-layers ①–⑨ of the target interval Wufeng–Longmaxi Formation, including total organic carbon analysis, whole-rock mineral analysis, gas content analysis, and physical property analysis.

The TOC analyzer used was the TA-1000 offline total organic carbon analyzer, with the specific experimental procedures as follows: solid cores were ground to 200 mesh, treated with 1 mol/L hydrochloric acid to remove inorganic carbon, washed to neutrality, dried, and then tested.

For whole-rock mineral analysis, a D8 Advance A25 X-ray diffractometer was used. The specific experimental steps were as follows: whole-rock samples are ground to 200 mesh and pressed into tablets; for clay mineral analysis, oriented slices (subjected to ethylene glycol saturation treatment) need to be prepared, followed by scanning and testing.

The porosity testing instrument is a modified helium porosimeter. Cores were processed into cylinders of 25 × 25 × 80 mm, their volume was calculated, then nitrogen was pressurized to equilibrium, and finally the porosity was calculated. The porosity measured in this study was effective porosity, and organic porosity was measured through low-temperature nitrogen adsorption in the experiment.

All experiments were completed at the Experimental Center of the Jianghan Oilfield Exploration and Development Research Institute, Sinopec (Wuhan, China).

4. Results

The TOC test results of 173 shale samples (2330.46–2415.1 m) show that the TOC ranges from 0.55% to 5.89%, mainly concentrated between 1% and 5%. The whole-rock mineral analysis of 87 shale samples shows that they contain quartz, clay minerals, potassium feldspar, plagioclase, calcite, dolomite, pyrite, and hematite, among which quartz and clay minerals have the highest content, ranging from 18.4 to 70.6% and from 16.6 to 62.8%, respectively. Plagioclase ranges from 1.9% to 11.9%, potassium feldspar ranges from 1% to 4.8%, dolomite ranges from 2.3% to 31.5%, calcite ranges from 2% to 11.8%, hematite ranges from 1.4 to 7.5%, and a small amount of samples also detected pyrite, with a content ranging from 1.4% to 7.5%. The physical property (porosity, permeability) tests of 159 samples show that the porosity ranges from 1.17% to 7.22%, and the permeability changes greatly, especially when microcracks and fractures exist, greatly improving the permeability. The overall permeability ranges from 0.0015 mD to 335.81 mD, with an average value of 24.81 mD. The permeability of the samples without cracks ranges from 0.0015 mD to 0.78 mD, with an average value of 0.094 mD; the permeability of samples containing microcracks ranges from 0.40 mD to 9.75 mD, with an average value of 3.56 mD; and the permeability of samples containing fractures ranges from 10.32 mD to 335.21 mD, with an average value of 107.10 mD. A total of 31 shale samples were subjected to on-site gas content experiments, and the results showed that the analytical temperature ranged from 62.3 °C to 64 °C, and the measured gas content ranged from 0.44 to 5.19 m^3/t (Figure 2).

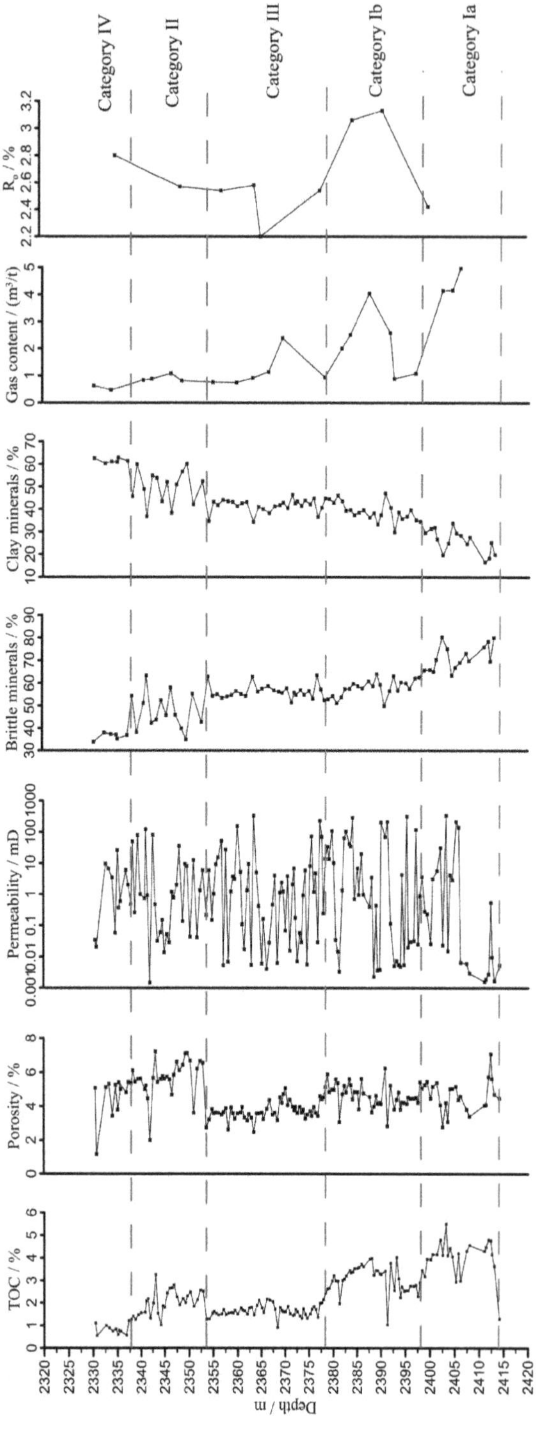

Figure 2. Comprehensive evaluation chart of the reservoir in the studied interval.

5. Discussion

5.1. Factors Affecting Reservoir Physical Properties

Shale reservoir physical properties include porosity and permeability; porosity is the ratio of pore volume to total volume, expressed as a fraction or percentage [16]. Ambrose et al. (2010) obtained reconstructed three-dimensional images using scanning electron microscopy and field emission ion microscopy and found that most of the pores were related to the kerogen network, indicating a close correlation between the pore development degree and the organic matter content [17]. In layers with better sorting (Wufeng Formation and the bottom of the Longmaxi Formation), there are small differences in particle size, uniform distribution of intergranular pores, and a high proportion of effective pore volume (porosity 4~6% [18]). In layers with poor sorting (the middle–upper part of the Longmaxi Formation), there is a mixture of coarse and fine particles, where fine particles fill the gaps between coarse particles, resulting in the blockage of intergranular pores and a decrease in porosity to 2~3% [18]. The permeability refers to the conductivity of the entire pore system (including fractures), especially the development of fractures, which can greatly enhance the permeability [18].

5.1.1. Mineral Composition

According to the properties of the minerals contained in the studied interval, they can be divided into two categories: brittle minerals and ductile minerals. Brittle minerals include quartz, feldspar, calcite, and dolomite, while ductile minerals are mainly clay minerals (with an average content of 37.78%). To eliminate the influence of crack development, all 35 samples selected did not contain cracks. The results show that the permeability ranges from 0.0016 to 0.75 mD, the porosity ranges from 2.63 to 6.43%, and the content of brittle minerals ranges from 33.9 to 80.3%, with an average value of 59.44%. According to Figure 3, it can be seen that neither brittle minerals nor clay minerals have an obvious correlation with each other. Overall, quartz has a relatively weak positive correlation with porosity, while clay minerals have a relatively weak negative correlation with porosity. This can be explained by the fact that brittle minerals such as quartz are more likely to form secondary pores, whereas clay minerals, with strong plasticity, tend to collapse and block pores. Additionally, porosity is more influenced by late diagenesis, thus resulting in such relatively weak correlations.

Feldspar and calcite show a positive correlation with permeability, indicating that feldspar and calcite are prone to forming microfractures, which in turn contribute to permeability but not to porosity (Figure 4). In contrast, clay minerals exhibit a weak positive correlation with permeability, which is a spurious correlation caused by the influence of feldspar and calcite. This phenomenon occurs when there is a strong positive correlation between clay content and feldspar content. This is consistent with previous studies, namely that permeability changes with variations in lithology [19]. Overall, the weak correlations suggest that the influence of mineral content on physical properties (porosity and permeability) is relatively weak.

5.1.2. Development of Cracks

The cracks on the core can be divided into microcracks and small cracks; microcracks are invisible to the naked eye but visible under the microscope, and small cracks are visible to the naked eye. The degree of crack development has little effect on porosity, but greatly increases the permeability. According to Figure 5, the permeability of samples with only developed pores ranges from 0.0015 to 0.7794 mD, and the porosity ranges from 1.17 to 7.22%; the permeability of samples with developed microcracks ranges from 0.3998 to 9.7485 mD, and the porosity ranges from 3.17 to 7.13%; and the permeability of samples with developed small cracks ranges from 10.32 mD to 355.20 mD, and the porosity ranges from 2.49% to 6.24%. Compared to cracks, the contribution of pores to permeability is basically negligible; similarly, compared to pores, the contribution of cracks to porosity can be largely ignored.

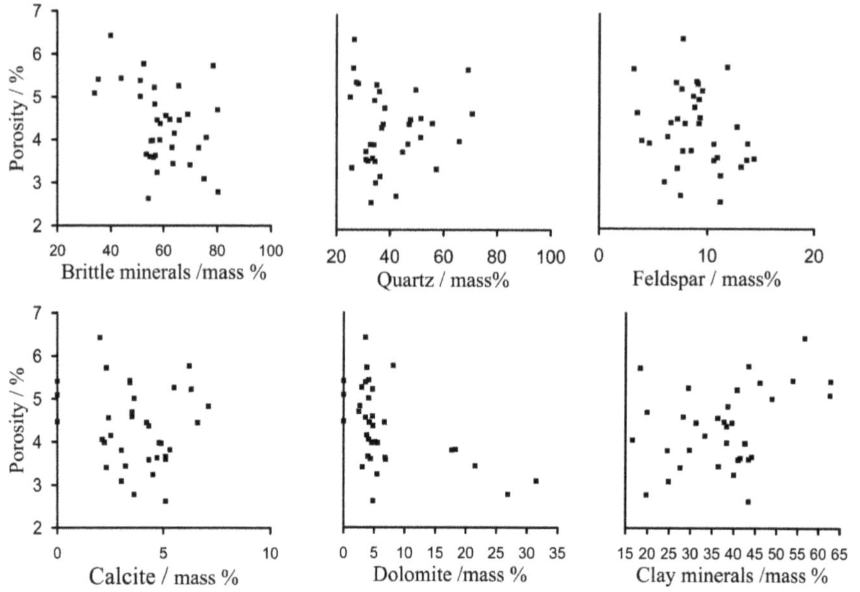

Figure 3. Correlation analysis between porosity and various minerals.

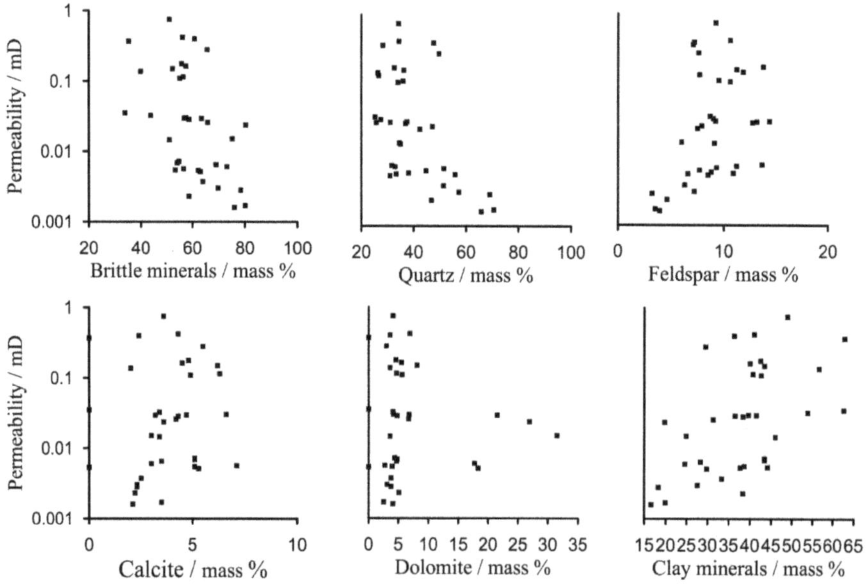

Figure 4. Correlation analysis between permeability and various minerals.

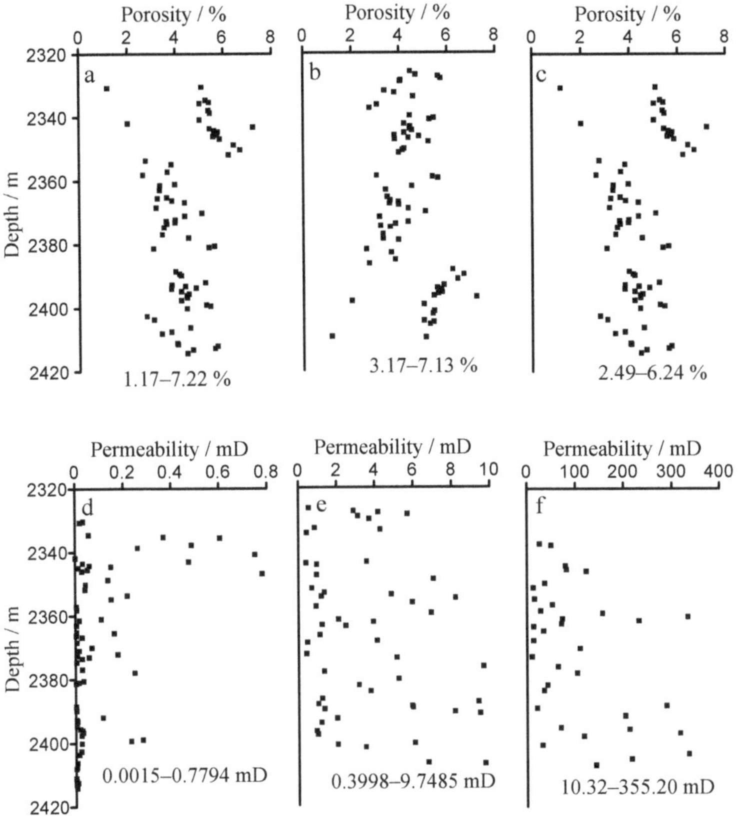

Figure 5. The influence of crack development on porosity and permeability. (**a**,**d**): Only developed pores; (**b**,**e**): developed microcracks; (**c**,**f**): developed cracks.

5.1.3. Organic Matter Abundance and Maturity

OM abundance refers to the OM content in shale, including various organic compounds such as hydrocarbons. This study characterizes the relative OM content based on total organic carbon (TOC), and uses vitrinite reflectance (Ro) to characterize the OM thermal evolution degree. By analyzing the correlation between TOC, Ro, and porosity separately (Figure 6), it can be seen that there is a certain positive correlation (R^2 = 1.95) between TOC and porosity, which fully demonstrates that the OM abundance is beneficial for shale pore development. Shale reservoirs are relatively dense with a large number of nanoscale pores (including organic and inorganic pores) developed. As the content of TOC increases, the development degree of organic pores increases, and their contribution to porosity increases, thus showing a positive correlation. Figure 6b shows a good positive correlation (R^2 = 0.49) between Ro and porosity, with a better correlation than the correlation between TOC and porosity, indicating that OM maturity contributes more to porosity than OM abundance. This can be explained by the fact that as maturity increases, the kerogen degradation not only generates organic pores, but also generates byproducts such as organic acids, which react with shale reservoirs and greatly increase porosity.

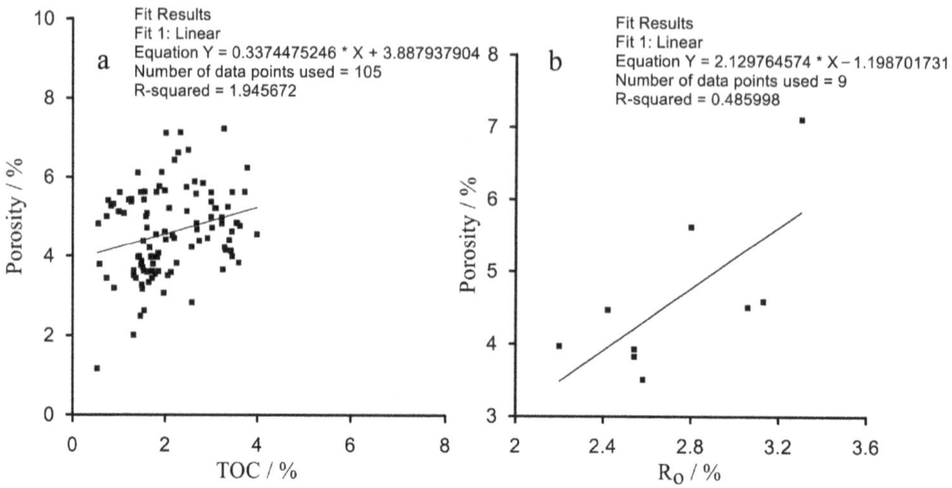

Figure 6. The influence of organic matter characteristics on porosity. (**a**): Correlation analysis between TOC and porosity; (**b**): correlation analysis between R_o and porosity.

5.2. Factors Affecting Gas Content

There are many factors that affect the shale gas content, which can be divided into internal and external factors. The internal factors mainly include porosity, mineral composition, and organic geochemical characteristics (OM abundance and maturity), while the external factors mainly include tectonic movement, preservation conditions, etc. The sample quality for gas content testing ranges from 5.612 to 6.315 Kg, and the test results (including measured gas content, loss gas content, and residual gas content) range from 0.44 to 5.19 m^3/t. This study explores the effects of porosity, mineral composition, and organic geochemical characteristics on its gas content, laying a foundation for the next reservoir evaluation.

5.2.1. Porosity

Pores are the main storage sites for shale gas, and porosity refers to the proportion of all pore volume, which is the basis for the shale gas occurrence and thus affects the gas content. Through the correlation analysis between porosity and gas content (Figure 7), it can be seen that gas content has an upward trend with the increase in porosity, and the correlation is relatively poor, indicating that porosity has a certain impact on gas content, but it is not the main controlling factor.

5.2.2. Mineral Composition

To investigate the influence of minerals on the gas content, this study conducted correlation analysis between various minerals (quartz, plagioclase, potassium feldspar, calcite, dolomite, and clay minerals) and gas content (Figure 8), and then obtained the relationship between each mineral and gas content. Among numerous minerals, only quartz and clay minerals show a good correlation, with quartz showing a significant positive correlation with gas content, while clay minerals show a significant negative correlation with gas content, and other minerals have no significant correlation. The so-called quartz in the Jiaoshiba area is actually organosilicon. In geology, organosilicon refers to naturally occurring organic compounds (or their derivatives) with silicon–carbon (Si-C) covalent bonds in their molecular structure. They are widely distributed in rocks, sediments, soils, oil and gas reservoir fluids, and pore water of geological bodies. A large number of organic pores developed in organosilicon can host a large amount of shale gas. The organosilicon in the shales of the Wufeng–Longmaxi Formation is mainly derived from siliceous biological remains, including siliceous shells of radiolarians, sponge spicules, diatoms, etc. [20,21]. During the sedimentary process, this biogenic silica exists in the form of opal-A (amorphous silica) and is later transformed into quartz through diagenesis [22]. Under the microscope, quartz with biological structures (such as radiolarian skeletal crystals) and cryptocrystalline–microcrystalline quartz aggregates without biological structures can be observed [20,21]. However, the adsorption capacity of clay minerals for shale gas is far inferior to that of organic matter, and it is easy to block the formed pores. Therefore, this correlation is presented, which also proves the widespread development of organic pores.

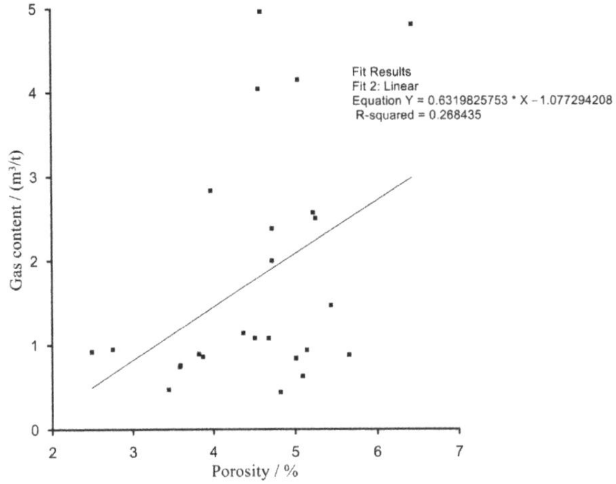

Figure 7. Correlation analysis between porosity and gas content.

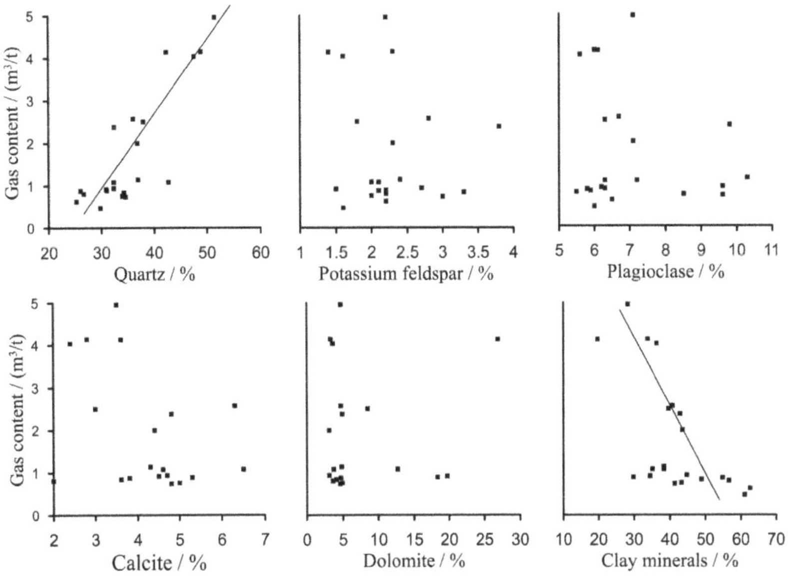

Figure 8. Correlation analysis between various minerals and gas content in the studied interval.

5.2.3. Organic Matter Abundance and Maturity

This study used correlation analysis to determine the impact of OM abundance and maturity on gas content (Figure 9). It can be seen that the OM maturity has a weak positive correlation with gas content, while the OM abundance has a strong positive correlation with gas content, indicating that the influence of OM abundance on gas content is stronger than its maturity. This can be explained by the fact that the studied interval maturity does not change significantly and are all in the mature stage, so their positive correlation is not particularly obvious. However, the OM abundance changes greatly, so there is a more obvious positive correlation.

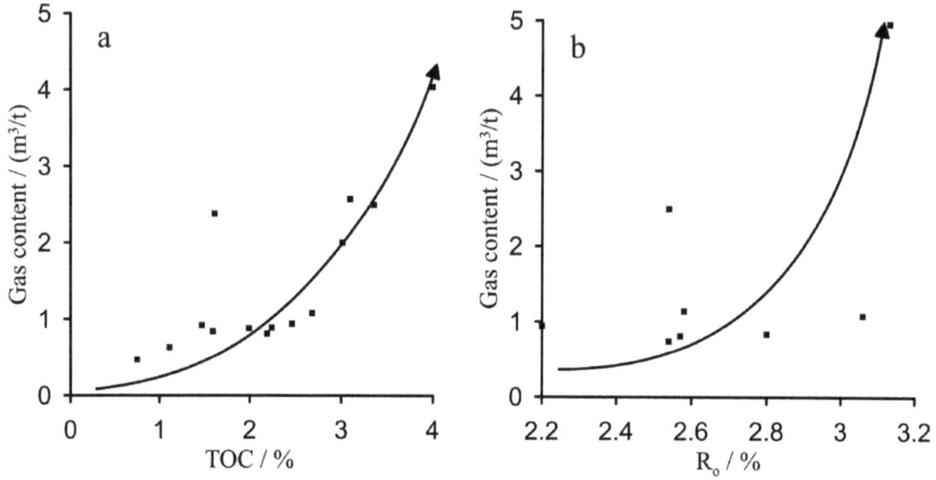

Figure 9. The influence of organic matter characteristics on gas content. (**a**): Correlation analysis between TOC and gas content; (**b**): correlation analysis between R_o and gas content.

5.3. Reservoir Comprehensive Evaluation

This study comprehensively evaluated the studied interval based on various factors such as TOC, porosity, permeability, brittle minerals, clay minerals, and gas content varying with depth (Figure 4), and established reservoir classification standards for the studied interval. The results showed that TOC, porosity, brittle minerals, and gas content showed a stepwise upward trend with depth, while vitrinite reflectance and permeability showed no significant changes. Overall, the lower (①–③) reservoir is superior to the upper (⑦–⑨) reservoir. Based on their changing trends, the studied interval can be divided into four categories: I, II, III, and IV. In Category I, TOC, porosity, and brittle mineral content are at high values, while clay mineral content shows a low value, with the highest gas content. In Category II, TOC and porosity are also at high values, but the brittle mineral content varies back and forth in the low-value range, the clay mineral content varies back and forth in the high-value range, and the gas content shows low values. In Category III, TOC and porosity are at lower values, while the brittle minerals and clay minerals are stable at higher values, and the gas content is at a lower value (only one point has abnormally high gas content). In Category IV, TOC, porosity and brittle mineral are at low values, while clay mineral content is at high values, and gas content is at low values. Category I can be further divided into two sub-classes, Ia and Ib. Ia is the highest-quality reservoir, followed by Ib. Overall, the lower (①–③) gas layer in the studied interval is the best, followed by the upper (⑦–⑨) gas layer, and the middle (④–⑥) gas layer is the worst.

To further verify the rationality of the reservoir classification evaluation criteria proposed in this article, the accumulated gas production from a large number of development wells in the studied area was systematically analyzed. The studied interval can be divided into ①–⑨ small layers from top to bottom, with the small layers ①, ②, and ③ being the lower gas layer, the small layers ④ and ⑤ the middle gas layer, and the small layers ⑦, ⑧, and ⑨ the upper gas layer. This study analyzed a large number of adjustment wells in the studied area, including the upper gas reservoir wells, the middle gas reservoir wells, and in the lower gas reservoir wells, all of which have been in operation for the past 5 years. The results show that the accumulated gas production of the upper gas layer is 0.03–131 million cubic meters, with an average value of 45 million cubic meters; the accumulated gas production of the middle gas layer is 0.15–18 million cubic meters, with an average of 14 million cubic meters; and the accumulated gas production of the lower gas layer is 0.41–171 million cubic meters, with an average of 66 million cubic meters (Table 1).

Table 1. The accumulated gas production analysis.

	Accumulated Gas Production (Million Cubic Meters)		
	Maximum	**Minimum**	**Average**
Upper gas layer (⑦–⑨)	131	0.03	45
Middle gas layer (④–⑥)	86	0.15	14
Lower gas layer (①–③)	171	0.41	66

This result indicates that the lower gas layer is the best, followed by the upper gas layer, and the middle gas layer is the worst, further verifying the reservoir classification and evaluation criteria proposed in this article. This further demonstrates the applicability of the reservoir classification and evaluation method, providing reference for the shale gas development in new areas and even the Sichuan Basin in the next step.

6. Conclusions

The influence of minerals on permeability is very weak, and cracks can greatly increase permeability, but their contribution to porosity is not significant. As the maturity increases, the kerogen degradation produces organic pores, and the byproducts of hydrocarbon generation (organic acids, etc.) that react with shale reservoirs, greatly increasing porosity. OM maturity contributes more to porosity than its abundance. Porosity has a certain influence on gas content, but it is not the main controlling factor. The mineral pores that host shale gas are mostly related to quartz, while clay minerals are not conducive to shale gas occurrence and have a negative impact on the pores formed by minerals such as quartz. OM abundance has a stronger impact on gas content than its maturity.

Overall, the lower reservoir (①–③) is superior to the upper (⑦–⑨) reservoir. Based on their changing trends, the studied interval can be divided into four categories: I, II, III, and IV. In Category I, TOC, porosity, and brittle mineral content are at high values, while clay mineral content is at low values, with the highest gas content. In Category II, TOC and porosity are also at high values, but the brittle mineral content varies back and forth in the low value range, the clay mineral content varies back and forth in the high value range, and the gas content is at low values. In Category III, TOC and porosity are at lower values, while the brittle minerals and clay minerals are stable at higher values, and the gas content is at a lower value (only one point has abnormally high gas content). In Category IV, TOC, porosity, and brittle mineral are at low values, while clay mineral content is at high values, and gas content is at low values. Category I can be further divided into two sub-classes, Ia and Ib. Ia is the highest-quality reservoir, followed by Ib. Overall, the lower (①–③) gas layer in the studied interval is the best, followed by the upper (⑦–⑨) gas layer, and the middle (④–⑥) gas layer is the worst.

Author Contributions: Methodology, Q.Y.; Investigation, Q.Y., X.Z. and Z.S.; Writing—original draft, Q.Y.; Writing—review & editing, A.Z., L.L. and J.W. All authors have read and agreed to the published version of the manuscript.

Funding: This research was funded by the Project of Sinopec Corp (Project Number JHBA250111).

Data Availability Statement: All relevant data are within the paper.

Acknowledgments: In particular, we are very grateful to the reviewers for their constructive suggestions and comments.

Conflicts of Interest: Authors Qiang Yan, Aiwei Zheng, Li Liu, Jin Wang, Xiaohong Zhan and Zhiheng Shu were employed by the Research Institute of Exploration and Development, Sinopec Jianghan Oilfield Company. The authors declare that the research was conducted in the absence of any commercial or financial relationships that could be construed as a potential conflict of interest.

References

1. Juan, L.; Lei, C.; Yun, B.J. Characteristics and main controlling factors of gas content in shallow marine shale: A case study of the Lower Silurian Longmaxi Formation in the Taiyang Block, Zhaotong. *Pet. Geol. Exp.* **2023**, *45*, 296–306.
2. Yang, Z.H.; Tao, G.L.; Bao, Y.J. Discussion on the differential development and retention mechanism of deep marine shale gas reservoir space in the southern region. *Pet. Geol. Exp.* **2022**, *44*, 845–853.
3. Wang, X.; Zhu, Y.; Lash, G.G. Multi-proxy analysis of organic matter accumulation in the upper Ordovician-lower Silurian black shale on the upper Yangtze Platform, South China. *Mar. Pet. Geol.* **2019**, *103*, 473–484. [CrossRef]
4. Zhang, C.; Zhou, T.; Xiao, J.L. Practice and understanding of fracturing technology for infill wells in the Fuling Shale Gas Field. *Fault Block Oil Gas Field* **2022**, *29*, 775–779.
5. Han, S.; Che, M.G.; Su, W. Prediction method and application of single well production of shale gas in Weiyuan Block, Sichuan Basin. *Spec. Oil Gas Reserv.* **2022**, *29*, 141–149.
6. Liu, W.X.; Lu, L.F.; Ye, D.L. Analysis and formation mechanism of abnormal pressure containment boxes of shale gas in the Ordovician Wufeng Formation—Silurian Longmaxi Formation in southeastern Sichuan. *Pet. Geol. Exp.* **2022**, *44*, 804–814.
7. Nie, Z.; Ma, S.; Wu, Q.Z. Development characteristics of natural fractures and their influence on gas content in marine shale in the Changning area. *Fault Block Oil Gas Field* **2022**, *29*, 591–597.
8. Wang, C.; Gao, H.; Fei, E.; Liu, B.; Wang, K.Y. Imbibition law and microscopic oil production characteristics of fracturing fluid in Chang 7 shale reservoir of Ordos Basin. *J. China Univ. Pet.* **2023**, *6*, 95–103.
9. Lin, W.; Guo, W.; Zhang, J.Z.; Li, M.T. Differential pore formation mechanism of organic matter in shale reservoirs with complex lithofacies types of the Longmaxi Formation in southern Sichuan. *Nat. Gas Geosci.* **2024**, *1*, 133–148.
10. Cai, X.W. Study on the Adsorption/Desorption Law and Mechanism of CH_4 in the Deep Marine Shale Matrix in the Jiaoshiba Area, Sichuan Basin. Ph.D. Thesis, Kunming University of Science and Technology, Kunming, China, 2023.
11. Yan, Q.; Zhang, Y.F.; Fu, H.; Jiang, M.L.; Wang, J. Multi-scale characterization of micro-nano pore throat characteristics of tight sandstone reservoirs by high-pressure mercury injection and scanning electron microscopy: A case study of tight sandstone reservoirs in the 4th member of Shahejie Formation in Block Yi 176, Zhanhua Sag, Bohai Bay Basin. *Pet. Geol. Exp.* **2018**, *2*, 280–287.
12. Yan, Q.; Ping, H.W. Evaluation on shale oil mobility for the Eocene Shahejie Formation in Liutun Sag, Dongpu Depression, Bohai Bay Basin. *Energies* **2023**, *16*, 2101. [CrossRef]
13. Guo, T.L.; Zhang, H.R. Formation and enrichment and high-yield model of the Jiaoshiba Shale Gas Field in the Sichuan Basin. *Pet. Explor. Dev.* **2014**, *1*, 28–36.
14. Zhang, B.L.; Zhu, G.; Jiang, D.Z. Numerical simulation and genetic discussion of "Jurassic Mountain-type" folds in eastern Sichuan. *Geol. Rev.* **2009**, *55*, 701–710.
15. Yan, D.P.; Wang, X.W.; Liu, Y.Y. Analysis of fold structural styles and their genetic mechanism in the border area of Sichuan, Hubei and Hunan. *Geoscience* **2000**, *14*, 37–43.
16. Kang, S.M.; Fathi, E.; Ambrose, R.J. Carbon dioxide storage capacity of organic-rich shales. *SPE J.* **2013**, *16*, 842–855. [CrossRef]
17. Sondergeld, C.H.; Ambrose, R.J.; Rai, C.S.; Moncrieff, J. Micro-structural studies of gas shale. In Proceedings of the Society of Petroleum Engineers Unconventional Gas Conference, Pittsburgh, PA, USA, 23–25 February 2010.

18. Liu, C.; Li, Z.; Ma, C.; Meng, Z.; Yaqiu, L. Characterization and source analysis of siliceous materials in the Wufeng-Longmaxi Formation shales in the Jiaoshiba area. *Jianghan Pet. Sci. Technol.* **2022**, *32*, 25–31.
19. Pothana, P.; Ifrene, G.; Ling, K. Integrated Petrophysical Evaluation and Rock Physics Modeling of Broom Creek Deep Saline Aquifer for Geological CO_2 Storage. *Fuels* **2024**, *5*, 53–74. [CrossRef]
20. Mao, X.P.; Chen, X.R.; Li, Z.; Li, S.X.; Li, S.S. Shale Sedimentary Patterns and Organic Matter Enrichment Patterns of the Wufeng Formation—Longmaxi Formation in the Sichuan Basin. *Acta Sedimentol. Sin.* **2025**, *43*, 701–733. [CrossRef]
21. Cai, Q.S.; Chen, X.H.; Zhang, B.M.; Liu, A.; Han, J. Origin of siliceous minerals in the black shale of the Wufeng and Longmaxi Formations in the Yichang area, western Hubei Province:geological significance for shale gas. *Acta Geol. Sin.* **2020**, *94*, 931–946.
22. Guan, Q.S.; Dong, D.Z.; Zhang, H.L.; Sun, S.H.; Zhang, S.R.; Guo, W. Types of biogenic quartz and its coupling storage mechanism in organic-rich shales: A case study of the Upper Ordovician Wufeng Formation to Lower Silurian Longmaxi Formation in the Sichuan Basin, SW China. *Pet. Explor. Dev.* **2021**, *48*, 700–709. [CrossRef]

minerals

MDPI

Article

Shear Wave Velocity Estimation for Shale with Preferred Orientation Clay Minerals

Bing Zhang [1,2,3,*], Cai Liu [4], Zhiqing Yang [3], Yao Qin [3] and Mingxing Li [3]

[1] Key Laboratory of Grain Information Processing and Control (Henan University of Technology), Ministry of Education, Zhengzhou 450001, China
[2] Henan Key Laboratory of Grain Storage Information Intelligent Perception and Decision Making, Henan University of Technology, Zhengzhou 450001, China
[3] College of Information Science and Engineering, Henan University of Technology, Zhengzhou 450001, China
[4] College of Geoexploration Sicence and Technology, Jilin University, Changchun 130026, China
* Correspondence: bzhangc@163.com

Abstract

Accurate shear wave velocity is important for shale reservoir exploration and characterization. However, the effect of the ubiquitous preferred orientation of clay minerals on the velocities of shale has rarely been considered in existing S-wave velocity estimation methods, resulting in limited accuracy of the estimation method. In this study, a S-wave velocity estimation method is proposed for shale while considering the effect of the preferred orientation of clay. First, a compaction model is built by taking the effects of the orientation distribution of clay and the aspect ratio of pores into account. Then, the compaction model is utilized in a workflow to obtain the model parameters by fitting the estimated P-wave velocity with the bedding-normal P-wave velocity from well logging. Finally, the S-wave velocity is estimated using the compaction model and calculated model parameters. The proposed method is verified by laboratory data and successfully applied to a shale gas reservoir. The result shows that the root mean square error almost halves compared with the Xu–White model. Additionally, the correlation coefficient also improves. The improvement in S-wave velocity estimation indicates that the effect of the preferred orientation of clay on the velocities of shale is effectively corrected. The proposed method improves the accuracy of velocity modeling and reservoir characterization for shale.

Keywords: shear wave velocity; preferred orientation of clay; shale; rock physics model; anisotropy

1. Introduction

Shale has emerged as a major research focus because of its unconventional hydrocarbon reserve [1,2]. The shear wave velocity of shale is essential for shale reservoir exploration and characterization [3,4]. The accurate S-wave velocity can be measured by well logging in the reservoir. However, the high cost and absence of logging technology limit the measurement [5]. Therefore, current methods used to obtain the S-wave velocity usually rely on estimation by fitting the modeled and measured P-wave velocity from well logging data. An existing challenge for shale velocity modeling is that the complex microstructures and inherent anisotropy of shale make its P- and S-wave velocities vary with different incident angles [6]. The inherent anisotropy is mainly caused by the preferred orientation of clay minerals in the well logging frequency [7]. Although fractures significantly affect the anisotropy, they only develop in fractured shale. In contrast, the anisotropy caused by the preferred orientation of clay minerals is ubiquitous for shale, which cannot be ignored [8,9].

The rock physics model is an important approach used for S-wave velocity estimation. Many rock physics models describe the microstructures of rock and build the relationship between the properties of microstructures and velocities. Xu and White [10,11] proposed the Xu–White model for shaly sandstone, which was focused on two types of pores with different aspect ratios (the ratio of the minor axis and long axis). The Xu–White model was then extended to carbonate rock, with the type of pores expanding into four categories [12]. Inclusion models introduce ellipsoidal inclusions to describe the minerals and pores of rock. Among them, the self-consistent approximation (SCA) model mixes inclusions together [13], and the differential effective medium (DEM) model gradually adds inclusions into a matrix [14] to model the elastic moduli of rock. Ruiz and Dvorkin [15] found that a narrow aspect ratio range for ellipsoidal inclusions can be applied in DEM for S-wave velocity estimation from the P-wave velocity in nonclastic rock. Vernik et al. [16] proposed a hybrid method to take the effect of the organic content in unconventional shale reservoirs into account, which improved the Greenberg–Castagna S-wave velocity prediction method. Zhang et al. [17] built a statistical rock physics model to estimate the S-wave velocity and provided a confidence interval for the estimation. Then, this model was applied to volcanic rock [18]. Considering the effects of the kerogen distribution and pore shape, a rock physics model for S-wave velocity estimation of organic-rich rock was built [19]. In recent years, rock physics model have combined with neural networks to form new methods for S-wave estimation [20–24]. Overall, rock is generally regarded as isotropic, and the S-wave velocity is estimated based on rock physics models and well logging data by fitting the bedding-normal P-wave velocity.

The accuracy of S-wave velocity estimation for shale is limited by the isotropic rock physics model because shale is generally anisotropic. The main reason for this is that clay minerals are plate-like and have a preferred orientation [25]. The SCA and DEM were extended to an anisotropic medium to model the effect of fully aligned clay minerals on shale [26]. The orientation distribution function (ODF) can be utilized to describe the preferred orientation of clay and model its effect on the elastic tensor of shale [27]. The ODF has also been improved by defining the preferred orientation of clay using a distribution or the reduction in porosity [28,29]. Research studies also regarded kerogen as plate-like and found it to usually be embedded with clay to form the background of shale, in which pores and silt minerals are included [26,30]. The preferred orientation of the mixture also contributes to the inherent anisotropy of shale [31,32]. Figure 1 presents the microstructures of shale, in which, in the preferred orientation of the background, pores with different aspect ratios, spherical-like silt minerals are observed. The inherent anisotropy affects other components, such as pores and fractures, in rock physics modeling. Some models used for the isotropic background are not applicable, such as the Xu–White model, Hudson model [33], and Eshelby–Cheng model [34]. A solution is introducing components as inclusions into the anisotropic background using the anisotropic DEM to model the elastic properties of shale [35,36]. However, the sequence of inclusions added in the background affects the modeling [37]. Also, the differential computation of DEM is computationally intensive [38].

The inherent anisotropy of shale is weak. Research studies have verified the application of the Hudson model and Eshelby–Cheng model for a weak anisotropic background using experiments [39,40]. Typical S-wave velocity estimation methods were evaluated, and the modified Xu–White model was considered as the optimal choice [41]. Additionally, the Xu–White model has been extended in other research studies [12,17,42]. Compared with neural network-related methods, the rock physics model-based approach can obtain model parameters using the measured P-wave velocity and then estimate S-wave velocity, during which the measured S-wave velocity is not entirely necessary. Based on the above knowledge, we propose a new S-wave velocity estimation method considering the effect of preferred orientation of clay minerals on the velocities of shale. Firstly, a compaction model is built to model the orientation distribution of clay and aspect ratio of pores by combining ODF and the Xu–White model. Then, a workflow based on the compaction model is summarized to estimate the velocities of shale using well logging data. The proposed method is verified by laboratory data and applied to a shale gas reservoir, in which the S-wave velocity estimation result is evaluated quantitatively. As prior information or calibration, the improved velocity estimation can benefit seismic inversion and reservoir petrophysical parameter inversion for shale.

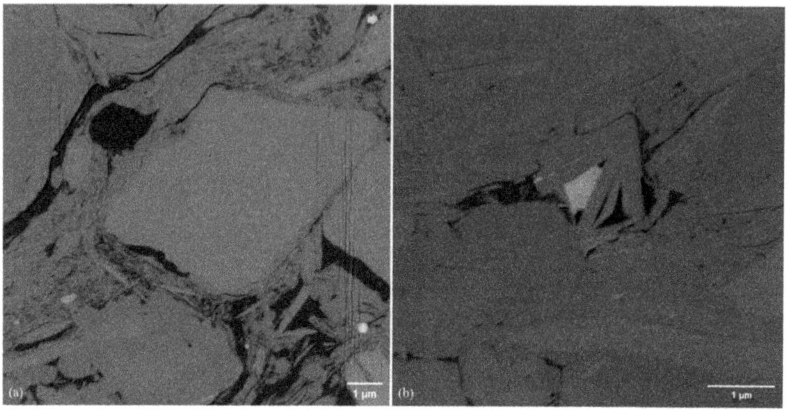

Figure 1. The scanning electron microscope (SEM) image of shale. (**a**) Clay and kerogen mixture and pores; (**b**) silt mineral and pores included in the background. The image shows a preferred orientation plate-like clay and kerogen mixture, pores with different aspect ratios, and spherical-like silt minerals.

2. Theory and Methodology

2.1. The Xu–White Model Used for Velocity Modeling of Rock

The components of rock are minerals, pores, and fluids in the pore space. The geometry and elastic properties of components affect the elastic properties of rock, the relationships between which are built using the rock physics model. P- and S-wave velocities in isotropic rock are determined by its moduli and density:

$$V_P = \sqrt{\frac{K + \frac{4}{3}\mu}{\rho}} \quad (1)$$

$$V_S = \sqrt{\frac{\mu}{\rho}} \quad (2)$$

where K, μ, and ρ indicate the bulk modulus, shear modulus, and density of rock, respectively.

In the modeling of rock velocities, the elastic properties of components are available from rock physics experiments, but the geometry of minerals and pores is more complex, which significantly affects the elastic properties of rock. For rock minerals, clay is plate-like, but silt minerals, such as quartz and calcite, are stiff, with a spheroid-like shape. Meanwhile, pores of rock have different geometries due to the diagenesis and their surroundings. The Xu–White model utilizes two types of pores to describe pores in the clay and sand of rock:

$$\phi_s = f_s \frac{\phi}{1-\phi} \tag{3}$$

$$\phi_c = f_c \frac{\phi}{1-\phi} \tag{4}$$

where ϕ_s and ϕ_c are sand-related porosity and clay-related porosity respectively, whose sum is the porosity of rock (ϕ); f_s and f_c are the volume fraction of sand and clay, respectively.

In the Xu–White model, the matrix of rock consists of sand and clay, whose moduli (K_0 and μ_0) are provided by the time-average [10,11]:

$$T_P^0 = (1-f_c)T_P^s + f_c T_P^c \tag{5}$$

$$T_S^0 = (1-f_c)T_S^s + f_c T_P^c \tag{6}$$

where T_P^0 and T_S^0 are the P- and S-wave transit times of the matrix, T_P^s and T_S^s are P- and S-wave transit times of sand, and T_P^c and T_S^c are P- and S-wave transit times of clay. The transit time of the wave is the reciprocal value of the wave velocity, from which the moduli of the matrix are calculated by combining Equations (1) and (2). The moduli of the matrix can also be calculated using the effective medium theory, such as Hill's average [43].

Then, pores are added into the matrix to obtain the moduli of the dry rock (K_d and μ_d) using the following formulas [38]:

$$K_d = K_0(1-\phi)^p \tag{7}$$

$$\mu_d = \mu_0(1-\phi)^q \tag{8}$$

where p and q are geometry coefficients given by Berryman [13]. For fluid-saturated rock, the moduli (K_s and μ_s) are provided by Gassmann's equations [44]:

$$K_s = K_d + \frac{\left(1-\frac{K_d}{K_0}\right)^2}{\frac{\phi}{K_f} + \frac{(1-\phi)}{K_0} - \frac{K_d}{K_0^2}} \tag{9}$$

$$\mu_s = \mu_d \tag{10}$$

where K_f is the bulk modulus of fluid in pores of rock.

2.2. The Effect of Preferred Orientation Clay Minerals on Velocities of Shale

Shale is usually transversely isotropic with a vertical symmetry axis (VTI) due to the inherent anisotropy. In high-frequency conditions, such as well logging and experiments, the inherent anisotropy is mainly decided by the preferred orientation of inclusions, such as plate-like clay minerals [27]. The orientation distribution function (ODF) describes the orientation property of inclusions and models its effect on the elastic properties of shale, which is decided by a tensor:

$$C_{VTI} = \begin{bmatrix} c_{11} & c_{12} & c_{13} & 0 & 0 & 0 \\ c_{12} & c_{11} & c_{13} & 0 & 0 & 0 \\ c_{13} & c_{13} & c_{33} & 0 & 0 & 0 \\ 0 & 0 & 0 & c_{44} & 0 & 0 \\ 0 & 0 & 0 & 0 & c_{44} & 0 \\ 0 & 0 & 0 & 0 & 0 & c_{66} \end{bmatrix} \tag{11}$$

where c_{ij} is the component of the tensor, and $c_{66} = \frac{1}{2}(c_{11} - c_{12})$. Shale wave velocities are decided by the tensor:

$$c_{11} = \rho V_P^2(90^\circ), c_{12} = c_{11} - 2\rho V_{SH}^2(90^\circ), c_{33} = \rho V_P^2(0^\circ), c_{44} = \rho V_{SH}^2(0^\circ) \tag{12}$$

where the degree indicates the included angle between the wave vector and the symmetry axis of the shale. For the VTI medium, $V_P(0^\circ)$ and $V_{SH}(0^\circ)$ are the measured P- and S-wave velocities from well logging.

Taking the effect of preferred orientation clay into account, the tensor of VTI shale is denoted as c_{ij}^v, and the formulae from Voigt averaging are provided [27]:

$$\begin{aligned}
c_{11}^v &= L + 2M + \frac{4\sqrt{2}\pi^2}{105}\left[2\sqrt{5}a_3W_{200} + 3a_1W_{400}\right] \\
c_{33}^v &= L + 2M - \frac{16\sqrt{2}\pi^2}{105}\left[\sqrt{5}a_3W_{200} - 2a_1W_{400}\right] \\
c_{12}^v &= L - \frac{4\sqrt{2}\pi^2}{315}\left[2\sqrt{5}(7a_2 - a_3)W_{200} - 3a_1W_{400}\right] \\
c_{13}^v &= L + \frac{4\sqrt{2}\pi^2}{315}\left[\sqrt{5}(7a_2 - a_3)W_{200} - 3a_1W_{400}\right] \\
c_{44}^v &= M - \frac{2\sqrt{2}\pi^2}{315}\left[\sqrt{5}(7a_2 + a_3)W_{200} + 24a_1W_{400}\right] \\
c_{66}^v &= \frac{c_{11}^v - c_{12}^v}{2}
\end{aligned} \tag{13}$$

where W_{200} and W_{400} are two Legendre coefficients; $a_1 = c_{11}^a + c_{33}^a - 2c_{13}^a - c_{44}^a$, $a_2 = c_{11}^a - c_{12}^a + 2c_{13}^a - 2c_{44}^a$, $a_3 = 4c_{11}^a - 3c_{33}^a - c_{13}^a - 2c_{44}^a$, $L = \frac{1}{15}\left(c_{11}^a + c_{33}^a + 5c_{12}^a + 8c_{13}^a - 4c_{44}^a\right)$, $M = \frac{1}{30}\left(7c_{11}^a + 2c_{33}^a - 5c_{12}^a - 4c_{13}^a + 12c_{44}^a\right)$; c_{ij}^a is the tensor for the domain, whose components are all fully aligned. For the VTI medium, the following applies:

$$W_{200} = \sqrt{\frac{5}{2}}\int_{-1}^{1} W(\xi)P_2(\xi)d\xi \tag{14}$$

$$W_{400} = \sqrt{\frac{9}{2}}\int_{-1}^{1} W(\xi)P_4(\xi)d\xi \tag{15}$$

where $\xi = \cos\phi$, and ϕ indicates the included angle between the major axis of inclusions and the plane of layers; $W(\xi)$ is the orientation distribution function, which provides the probability density for a particular orientation. $W(\xi)$ can be reconstructed by a Gaussian ODF [28]:

$$W(\xi) = \frac{f_G(\phi) + f_G(\pi - \phi)}{4\pi} \tag{16}$$

$$f_G(\phi) = k_G(\sigma)\exp\left[-\frac{1}{2}\left(\frac{\phi}{\sigma}\right)^2\right] \tag{17}$$

where $k_G(\sigma)$ is a normalization coefficient; σ is the standard deviation of ϕ, named the orientation parameter in this paper. The orientation parameter decides the preferred orientation condition of inclusions.

The tensor for the domain whose components are all fully aligned (c_{ij}^a) can be calculated by the anisotropic SCA:

$$c^a = \sum_{n=1}^{N} v_n c^n\left(I + \hat{G}(c^n - c^a)\right)^{-1}\left\{\sum_{p=1}^{N} v_p\left[I + \hat{G}(c^p - c^a)\right]^{-1}\right\}^{-1} \tag{18}$$

where each n and p indicates an inclusion with the volume fraction v; tensor $\hat{G}$ is decided by the geometry of the inclusion.

2.3. The Compaction Model for Shale Velocity Modeling

Due to the preferred orientation of clay minerals, the velocities of shale are affected by the anisotropy. Thus, a new rock physics model is necessary to obtain accurate velocity estimation for shale. The Xu–White model, a common choice for velocity modeling, has been evaluated as the best V_S predictor [41]. However, this model is only applied to an isotropic medium. In recent years, experiments have validated the application of the isotropic rock physics model to a weak anisotropic medium, including the Hudson model and Eshelby–Cheng model [39,40]. In these research studies, the anisotropic P- and S-wave velocities with a weak anisotropic background are averaged to obtain effective isotropic P- and S-wave velocities, from which the moduli of background are obtained and then used in the isotropic rock physics model. Based on the above research, the rock physics model is built as follows:

(1) Clay and kerogen are first mixed together using the anisotropic SCA (Equation (18)) to form the fully aligned background of shale, whose tensor is indicated as C_b^a. Kerogen and clay often intermingle with each other and form the background of shale. Moreover, similar to clay minerals, kerogen can also be regarded as ellipsoid and has a preferred orientation [8].

(2) The ODF is utilized to model the preferred orientation of a clay and kerogen mixture to form the anisotropic background with partly aligned inclusions. The orientation parameter σ determines the tensor of the mixture, which is indicated as C_b^p. Anisotropic velocities of the background are indicated as $V_P(0°)$, $V_{SH}(0°)$, $V_P(90°)$, and $V_{SH}(90°)$.

(3) It is important to obtain the effective isotropic P- and S-wave velocities (V_P^e and V_S^e) to form the effective isotropic background: $V_P^e = [V_P(0°) + V_P(90°)]/2$, $V_S^e = [V_{SH}(0°) + V_{SH}(90°)]/2$. Then, the bulk and shear moduli of the effective isotropic background (K_0^e and μ_0^e) are obtioned using Equations (1) and (2).

(4) The effective isotropic background and silt minerals of shale are mixed together to form the matrix, whose bulk and shear moduli (K_1 and μ_1) are obtained using Hill's average.

(5) Clay-related pores and sand-related pores are added into the matrix using the Xu–White model to obtain the bulk and shear moduli of the dry shale (K_2 and μ_2).

(6) Gassmann's equation (Equations (9) and (10)) is applied to obtain the bulk and shear moduli of fluid saturated shale (K_3 and μ_3). Then, the P- and S-wave velocities of fluid-saturated shale (V_P and V_S) are calculated from the moduli.

(7) The effect of the effective isotropic background is substituted with the anisotropic background to model the effects of preferred orientation clay and kerogen on the velocities of shale. Although anisotropic velocities of the background are averaged to obtain the effective isotropic velocities for the Xu–White model, measured wave velocities from well logging are directional and parallel to the symmetry axis of shale. So, velocities of the background should be substituted to recover the effects of preferred orientation clay and kerogen on the measured velocities.

The time average is a simple but common choice used to determine the integrity attribute of rock from its components' properties. In the Xu–White model, the moduli of the matrix are provided by the time-average method [10,11]. Similarly, Wyllie's equation provides the P-wave velocity of rock from velocities of matrix and pores using the time-average method [45]. Here, velocities of the fluid-saturated shale can also be provided by the effective isotropic background and the inclusions based on the time average:

$$T_{\mathrm{P}} = \left(1 - f_{\mathrm{c}}^{'}\right) T_{\mathrm{P}}^{\mathrm{i}} + f_{\mathrm{c}}^{'} T_{\mathrm{P}}^{\mathrm{e}} \tag{19}$$

$$T_{\mathrm{S}} = \left(1 - f_{\mathrm{c}}^{'}\right) T_{\mathrm{S}}^{\mathrm{i}} + f_{\mathrm{c}}^{'} T_{\mathrm{S}}^{\mathrm{e}} \tag{20}$$

where T_{P} and T_{S} are the reciprocal value of V_{P} and V_{S} in step 6, respectively; $f_{\mathrm{c}}^{'}$ is the total volume fraction of clay and kerogen; $T_{\mathrm{P}}^{\mathrm{i}}$ and $T_{\mathrm{S}}^{\mathrm{i}}$ are the P- and S-wave transit times of inclusions; and $T_{\mathrm{P}}^{\mathrm{e}}$ and $T_{\mathrm{S}}^{\mathrm{e}}$ are the P- and S-wave transit times of the effective isotropic background in step 3.

For real fluid-saturated anisotropic shale, wave velocities from well logging should be described by the velocities parallel to the symmetry axis of shale and the time average:

$$\widetilde{T}_{\mathrm{P}} = \left(1 - f_{\mathrm{c}}^{'}\right) T_{\mathrm{P}}^{\mathrm{i}} + f_{\mathrm{c}}^{'} T_{\mathrm{P}}(0^{\circ}) \tag{21}$$

$$\widetilde{T}_{\mathrm{S}} = \left(1 - f_{\mathrm{c}}^{'}\right) T_{\mathrm{S}}^{\mathrm{i}} + f_{\mathrm{c}}^{'} T_{\mathrm{SH}}(0^{\circ}) \tag{22}$$

where $\widetilde{T}_{\mathrm{P}}$ and $\widetilde{T}_{\mathrm{S}}$ are vertical incidence P- and S-wave transit times of shale considering the preferred orientation of clay; $T_{\mathrm{P}}(0^{\circ})$ and $T_{\mathrm{SH}}(0^{\circ})$ are the vertical incidence P- and S-wave transit times of the anisotropic background. Here, the P- and S-wave transit times of inclusions are assumed not to be affected by the background. Combining Equations (19)–(22), the formulae for vertical-incidence P- and S-wave velocities ($\widetilde{V}_{\mathrm{P}}$ and $\widetilde{V}_{\mathrm{S}}$) of shale considering the preferred orientation of clay are

$$\widetilde{V}_{\mathrm{P}} = \frac{1}{\frac{1}{V_{\mathrm{P}}} - \frac{f_{\mathrm{c}}^{'}}{V_{\mathrm{P}}^{\mathrm{e}}} + f_{\mathrm{c}}^{'} V_{\mathrm{P}}(0^{\circ})} \tag{23}$$

$$\widetilde{V}_{\mathrm{S}} = \frac{1}{\frac{1}{V_{\mathrm{S}}} - \frac{f_{\mathrm{c}}^{'}}{V_{\mathrm{S}}^{\mathrm{e}}} + f_{\mathrm{c}}^{'} V_{\mathrm{SH}}(0^{\circ})} \tag{24}$$

The velocities $\widetilde{V}_{\mathrm{P}}$ and $\widetilde{V}_{\mathrm{S}}$ are modeled velocities in well logging of shale.

The above processes are used to build a rock physics model for the velocity modeling of shale, which is named the compaction model. The compaction model considers the effect of preferred orientation clay minerals, which can be applied to S-wave velocity estimation for shale.

2.4. S-Wave Velocity Estimation for Shale Based on the Compaction Model

The rock physics model builds the relationship between the wave velocities and microstructures of shale. The properties of microstructures are necessary for velocity estimation based on the rock physics model. Although well logging provides a significant amount of information on the microstructures of shale, some important factors are unavailable from the measurements, such as the aspect ratio of pores and the orientation condition of clay minerals. In the Xu–White model, the aspect ratio of clay-related pores (α_c) significantly affects the velocities of shale, but the effect of the aspect ratio of sand-related pores (α_s) is not significant [17]. The reason for this is that clay minerals always have pores with a small aspect ratio, but sand tends to form pores with a large aspect ratio [10,42].

The P- and S-wave velocities of shale modeled by the compaction model with different aspect ratios of clay-related pores (α_c) and orientation parameters (σ) are presented in Figure 2. In the process, α_s is provided by the following formula [17,46]:

$$\alpha_{\mathrm{s}} = 0.17114 - 0.24477\phi + 0.004314 f_{\mathrm{s}} \tag{25}$$

Figure 2 shows that P- and S-wave velocities increase with α_c and σ. The velocities increase rapidly with α_c when α_c is smaller than 0.05. However, the increase is slower when α_c is bigger than 0.05. This is reflected in the speed variation in the surface elevation (Figure 2a,b) and the positional changes of the differently colored curves along the y-axis (Figure 2c,d). Meanwhile, the velocities increase with σ. Compared with the effect of α_c, the contribution of σ cannot be ignored, especially for bigger α_c. Thus, to model the velocities of shale, both α_c and σ are important and need to be determined.

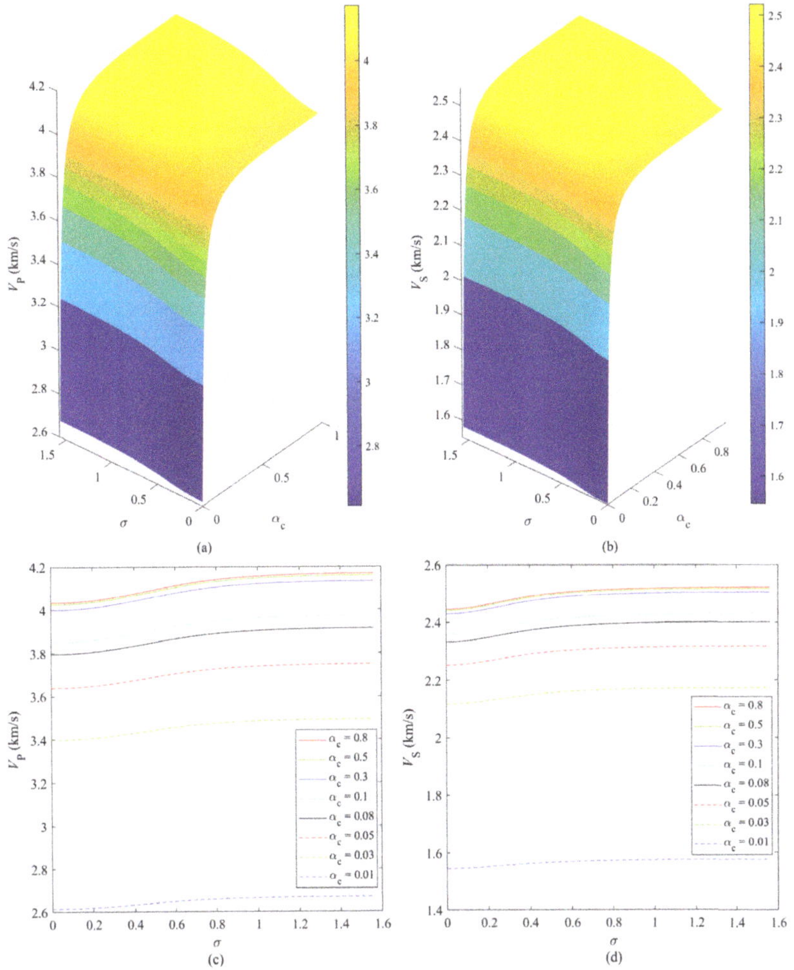

Figure 2. (**a**,**b**) are the modeled P- and S-wave velocities of shale with different aspect ratios of clay-related pores (α_c) and orientation parameters (σ). (**c**,**d**) are the profiles of modeled P- and S-wave velocities of shale with different σ and selected α_c. The shale is made up of clay, kerogen, quartz, and pores, with volume fractions of 0.4, 0.1, 0.45, and 0.05, respectively.

In shale reservoirs, the S-wave velocity is always estimated based on well logging data of the P-wave velocity and other properties of shale. The properties of shale are utilized to initialize the rock physics model. The important factors unavailable from measurements are obtained from the objective function:

$$[\alpha_c, \sigma] = \arg\min\left(\widetilde{V}_P - V_{Pm}\right) \tag{26}$$

where $\widetilde{V}_P$ is the modeled P-wave velocity of shale from the compaction model, and V_{P_m} is the measured P-wave velocity. The particle swarm optimization algorithm improved by simulated annealing (SA-PSO) is utilized in this study to solve the objective function. Zhang et al. [17] provided details of this algorithm in the Appendix. Then, the S-wave velocity is estimated based on the inversion result and compaction model. Figure 3 presents the workflow of the method, in which α_c and σ are valued form $(0, 1)$ and $\left(0, \frac{\pi}{2}\right)$, respectively.

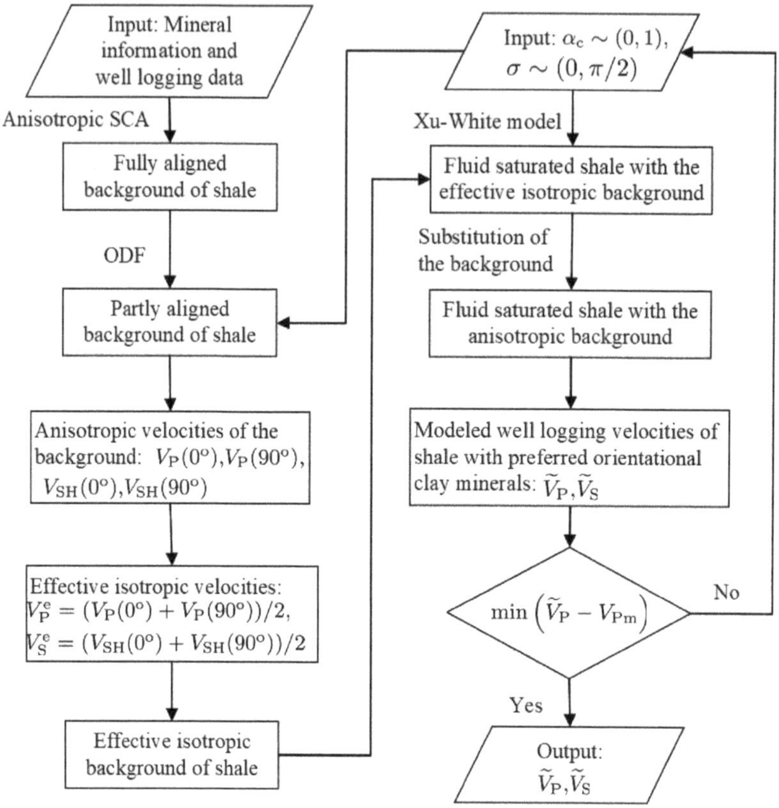

Figure 3. Workflow of S-wave velocity estimation based on the compaction model and well logging data for shale with orientation clay minerals.

3. Result and Discussion

3.1. Model Test Using Laboratory Data

The S-wave velocity estimation method was first verified by laboratory data from a shale gas reservoir. The moduli and densities of rock components are presented in Table 1. The objective function used for the estimation is Equation (26), in which the modeled P-wave velocity approximates the measured P-wave velocity in Figure 4a. The estimated and measured S-wave velocities are shown in Figure 4b. The Xu–White model with the variable α_c is also utilized here as a reference. In Figure 4a, the P-wave velocity calculated from the compaction model is very close to the measured P-wave velocity. In Figure 4b, the S-wave velocity estimated from the compaction model matches well with the measured S-wave velocity, which is better than the Xu–White model.

Table 1. Elastic moduli and density of rock components [5,8,37].

	Bulk Modulus (Gpa)	Shear Modulus (Gpa)	Density ($g \cdot cm^{-3}$)
Quartz	37.9	44.3	2.65
Feldspar	37.5	15	2.62
Calcite	76.8	32	2.71
Dolomite	94.9	45	2.87
Pyrite	147.4	132.5	4.93
Clay	25	9	2.55
Kerogen	5.53	3.2	1.25
Gas	0.18	-	0.26
Brine	2.65	-	0.99

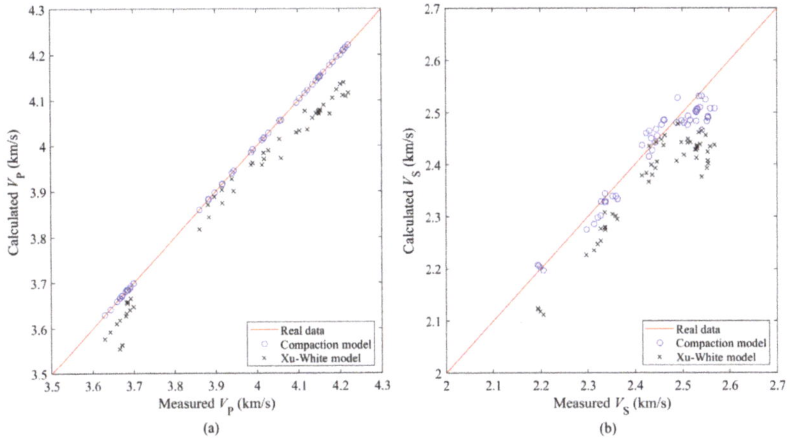

Figure 4. Crossplot of modeled velocities based on the compaction model and Xu–White model using laboratory data. (**a**) P-wave velocities; (**b**) S-wave velocities.

The root mean square error (RMSE) and correlation coefficient (r) are usually applied to quantify errors between the estimation and measurements. Their formulae are as follows:

$$\mathrm{RMSE} = \sqrt{\frac{1}{\mathrm{N}}\sum_{i=1}^{\mathrm{N}}\frac{\left|\widetilde{V}_i - V_i\right|}{V_i}} \tag{27}$$

$$r = \frac{cov\left(\widetilde{V}, V\right)}{\delta\widetilde{V}\delta V} \tag{28}$$

where i indicates sample points, and N indicates the total number of sample points; $\widetilde{V}_i$ and V_i are corresponding estimated and measured velocities, respectively; $cov(\cdot)$ indicates covariance; δ indicates the standard deviation of velocities. For the laboratory data, the RMSE of P-wave velocities from the compaction model and Xu–White model are 0.0011 and 0.0653, respectively. For the S-wave velocities from the compaction model and Xu–White model, the RMSEs are 0.0305 and 0.0828, respectively. The quantification of errors between estimated and measured velocities shows the validity and superiority of the compaction model in S-wave velocity estimation for shale. The modeling and calculation presented in this study were implemented by utilizing MATLAB R2024b software (MathWorks, Inc., Natick, MA, USA) on a PC. Alternative programming languages may also be employed during the implementation phase.

3.2. Application in a Shale Gas Reservoir

The S-wave velocity estimation method based on the compaction model was applied to a shale gas reservoir in the south of China. Most wells in the reservoir adopt conventional well logging, and their silt minerals are simply identified. We used two conventional wells to validate the proposed method for conventional well logging of shale reservoirs. Figures 5 and 6 show the well logging data of well 1 and well 2. The volume fraction of clay mineral is about 0.4–0.6 in the upper section of well 1, while the volume fraction in the lower section is about 0.25–0.4. The volume fractions of kerogen in the upper section and the bottom of well 1 are about 0.05 and 0.1, respectively, which indicates that the bottom is a gas-rich section. The modeled P-wave velocity based on the compaction model is very close to the measured P-wave velocity, except for section 1440–1450, and this result is much better than the Xu–White model. The estimated S-wave velocity based on the compaction model has small errors and good correlation with the measured S-wave velocity. The estimation of S-wave velocity is improved by the proposed method compared with the Xu–White model in the whole well section, especially in section 1400–1450. Relative errors between the estimated S-wave velocity and real data using the two methods, which were calculated using Equation (29), are presented in Figure 5g. Relative errors from the compaction model are about 0.02 for the whole well logging, whereas the errors from the Xu–White model are obviously greater. The volume fraction of clay mineral is normalized by Equation (30) to be presented with errors.

$$Re = \frac{\left|\widetilde{V}_S - V_{Sm}\right|}{V_{Sm}} \tag{29}$$

$$f_{clay}^{n} = \frac{f_{clay}}{\max(Re)} \tag{30}$$

where Re is the relative error of S-wave velocity estimation; $\widetilde{V}_S$ is the estimated S-wave velocity; V_{Sm} is the measured S-wave velocity; f_{clay}^{n} is the normalized volume fraction of clay; f_{clay} is the volume fraction of clay.

It is noticeable that the relative errors from the Xu–White model are related to the volume fraction of clay. Moreover, the relative errors increase with the volume fraction. However, the relative errors from the compaction model are not affected by the volume fraction of clay. The result of well 1 indicates that the compaction model reduces the error deduced by the preferred orientation of clay and improves the accuracy of S-wave velocity estimation. The estimation of S-wave velocity for well 2 is presented in Figure 6, which is similar to well 1. A quantitative evaluation of the estimation result for the two wells is presented in Table 2. The RMSE between the calculated and measured velocities of well 1 and well 2 decreases by half when using the compaction model. Meanwhile, r of well 1 and well 2 slightly increases and is approximately 1. The above results of the two wells indicate the validity and accuracy of the S-wave velocity estimation method for shale based on the compaction model.

The proposed method was also applied in a high-quality well logging from the shale gas reservoir. High-quality well logging uses unconventional logging methods and provides core samples. Therefore, information of shale in these wells is more sufficient and accurate. The estimation result of high-quality well logging, named well 3, is presented in Figure 7. The mineral information of well 3 is more accurate than conventional wells. The compaction model and Xu–White model are utilized in well 3. Calculated P-wave and S-wave velocities, as well as the relative errors, are also presented. The RMSE and r of the estimation are provided in Table 2. The estimation based on the compaction model is significantly better than the Xu–White model. The calculated P-wave velocity is almost identical to the measured P-wave velocity, whose RMSE is 0.0009 and r is 0.9999. S-wave velocity estimation is improved based on the compaction model, whose RMSE decreases by over half, and r increases compared with the Xu–White model. Meanwhile, the improvement in S-wave estimation based on the compaction model is related to the volume fraction of clay. The result of the high-quality well logging is consistent with the two conventional well loggings.

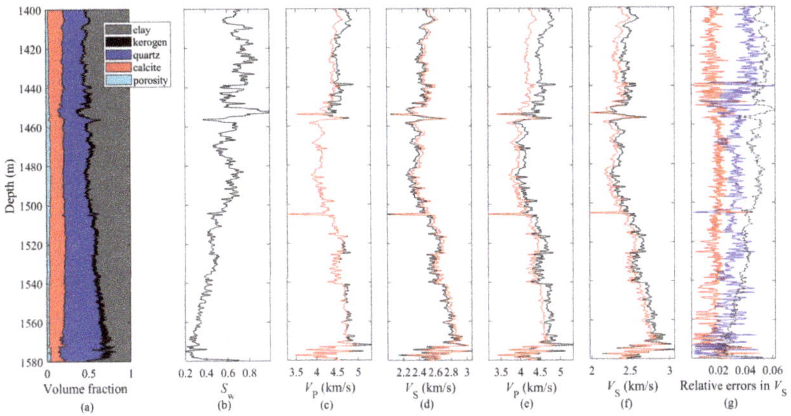

Figure 5. Estimation of S-wave velocity for well 1. From left to right: (**a**) volume fraction of components; (**b**) fluid saturation; (**c**,**d**) estimated velocities based on the compaction model (red) and measured velocities (black); (**e**,**f**) estimated velocities based on the Xu–White model (red) and measured velocities (black); and (**g**) errors from the compaction model (red) and Xu–White model (blue), as well as the normalized volume fraction of clay (dotted black).

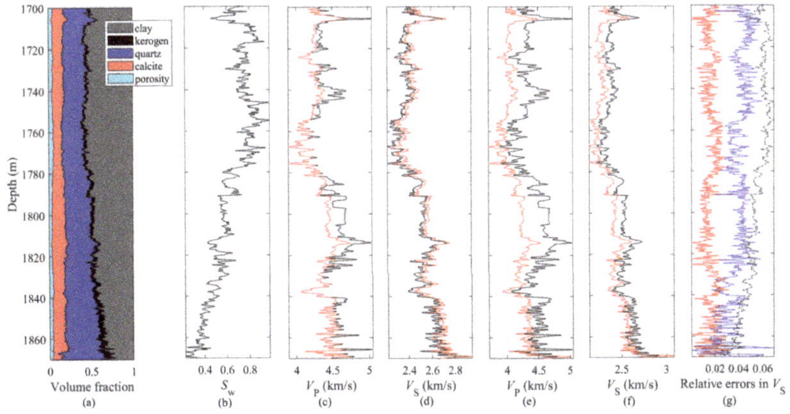

Figure 6. Estimation of S-wave velocity for well 2. From left to right: (**a**) volume fraction of components; (**b**) fluid saturation; (**c**,**d**) estimated velocities based on the compaction model (red) and measured velocities (black); (**e**,**f**) estimated velocities based on the Xu–White model (red) and measured velocities (black); and (**g**) errors from the compaction model (red) and Xu–White model (blue), as well as the normalized volume fraction of clay (dotted black).

Table 2. Errors and correlation coefficients of S-wave velocity estimation for wells using the two models.

	Model	RMSE of V_P	RMSE of V_S	r of V_P	r of V_S
well 1	Xu–White model (variable α_c)	0.2443	0.0855	0.9295	0.9879
	Compaction model	0.1006	0.0455	0.9570	0.9962
well 2	Xu–White model (variable α_c)	0.2855	0.0999	0.8494	0.9724
	Compaction model	0.1150	0.0424	0.9322	0.9932
well 3	Xu–White model (variable α_c)	0.0647	0.0827	0.9879	0.9391
	Compaction model	0.0009	0.0380	0.9999	0.9715

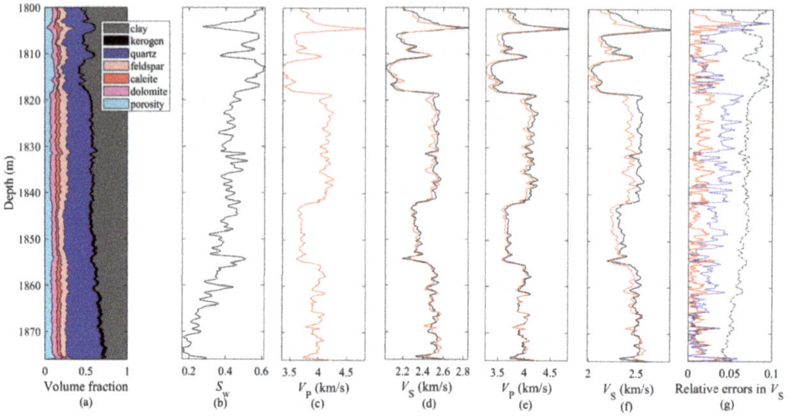

Figure 7. Estimation of S-wave velocity for well 3. From left to right: (**a**) volume fraction of components; (**b**) fluid saturation; (**c**,**d**) estimated velocities based on the compaction model (red) and measured velocities (black); (**e**,**f**) estimated velocities based on the Xu–White model (red); and measured velocities (black) and (**g**) errors from the compaction model (red) and Xu–White model (blue), as well as the normalized volume fraction of clay (dotted black).

4. Conclusions

An S-wave velocity estimation method for shale is proposed in this study. This method takes the effect of the preferred orientation clay minerals on the velocities of shale into account, which improves the accuracy of S-wave velocity estimation for shale. A compaction model is developed based on the ODF and Xu–White model to model the effects of the orientation distribution of clay minerals and aspect ratio of pores on the velocities of shale. It should be noted that the compaction model is only applicable to a weak anisotropic medium. Then, the workflow of S-wave velocity estimation for shale based on the compaction model is summarized. The proposed method is validated by laboratory data and an actual shale gas reservoir, with a result obviously better than the Xu–White model.

Clay is a main component of shale, and its preferred orientation is ubiquitous. This study indicates that the error of S-wave velocity estimation based on the traditional method is related to the volume fraction of clay. But this correlation vanishes using the proposed method; moreover, the error almost halves. The correction of the effect of preferred orientation clay minerals on the velocities of shale will improve velocity modeling, which can provide more accurate information for the following processes of shale reservoir characterization.

Author Contributions: Conceptualization, B.Z.; methodology, B.Z.; software, M.L.; validation, B.Z.; formal analysis, B.Z.; investigation, B.Z.; resources, C.L.; data curation, C.L.; writing—original draft preparation, B.Z.; writing—review and editing, B.Z.; visualization, Z.Y.; supervision, C.L.; project administration, Y.Q.; funding acquisition, B.Z. All authors have read and agreed to the published version of the manuscript.

Funding: This research was funded by the Open Topics of Research Platform from the Grain Information Processing Center of Henan University of Technology (KFJJ2022016), the Key Scientific Research Project of Colleges and Universities in Henan Province (23A170012), and the Research Foundation for Advanced Talents of Henan University of Technology (2021BS037).

Data Availability Statement: The original contributions presented in this study are included in the article. Further inquiries can be directed to the corresponding author.

Conflicts of Interest: The authors declare no conflicts of interest.

References

1. Han, H.; He, C.; Xiao, S.; Chen, J.; Kalmykov, A.; Wu, W.; Wang, Y.; Wang, J. The Formation of Organic Matter Pores in Shales: Implications from Combined Thermal Heating and Scanning Electron Microscopy Imaging. *Minerals* **2025**, *15*, 336.
2. Peng, S.; Feng, C.; Qiu, Z.; Zhang, Q.; Liu, W.; Feng, J.; Hu, Z. Prediction of Lithofacies in Heterogeneous Shale Reservoirs Based on a Robust Stacking Machine Learning Model. *Minerals* **2025**, *15*, 240. [CrossRef]
3. Schuster, G.T.; Li, J.; Lu, K.; Metwally, A.; AlTheyab, A.; Hanafy, S. Opportunities and pitfalls in surface-wave interpretation. *Interpretation* **2017**, *5*, T131–T141. [CrossRef]
4. Guo, Z.; Lv, X.; Liu, C.; Liu, X.; Liu, Y. Prediction of organic richness in shale gas reservoirs using a novel organic-inorganic decoupling seismic inversion method. *Gas Sci. Eng.* **2023**, *110*, 204864. [CrossRef]
5. Tan, M.; Peng, X.; Cao, H.; Wang, S.; Yuan, Y. Estimation of shear wave velocity from wireline logs in gas-bearing shale. *J. Pet. Sci. Eng.* **2015**, *133*, 352–366. [CrossRef]
6. Sethi, C.; Motra, H.B.; Hazra, B.; Ostadhassan, M. Influence of lithological contrast on elastic anisotropy of shales under true-triaxial stress and thermal conditions. *Int. J. Rock Mech. Min. Sci.* **2025**, *190*, 106100. [CrossRef]
7. Vernik, L.; Anantharamu, V. Estimating the elastic properties of mica and clay minerals. *Geophysics* **2020**, *85*, MR83–MR95. [CrossRef]
8. Sayers, C.M. The effect of kerogen on the elastic anisotropy of organic-rich shales. *Geophysics* **2013**, *78*, D65–D74. [CrossRef]
9. Ding, J.; Clark, A.C.; Vanorio, T.; Jew, A.D.; Bargar, J.R. Elastic anisotropy of shales: The roles of crack alignment and compliance ratio. *Geophysics* **2022**, *87*, A13–A17. [CrossRef]
10. Xu, S.; White, R.E. A new velocity model for clay-sand mixtures. *Geophys. Prospect.* **1995**, *43*, 91–118. [CrossRef]
11. Xu, S.; White, R.E. A physical model for shear-wave velocity prediction. *Geophys. Prospect.* **1996**, *44*, 687–717. [CrossRef]
12. Xu, S.; Payne, M.A. Modeling elastic properties in carbonate rocks. *Lead. Edge* **2009**, *28*, 66–74. [CrossRef]
13. Berryman, J.G. Long-wavelength propagation in composite elastic media II. Ellipsoidal inclusions. *J. Acoust. Soc. Am.* **1980**, *68*, 1820–1831. [CrossRef]
14. Norris, A. A differential scheme for the effective moduli of composites. *Mech. Mater.* **1985**, *4*, 1–16. [CrossRef]
15. Ruiz, F.; Dvorkin, J. Predicting elasticity in nonclastic rocks with a differential effective medium model. *Geophysics* **2010**, *75*, E41–E53. [CrossRef]
16. Vernik, L.; Castagna, J.; Omovie, S.J. S-wave velocity prediction in unconventional shale reservoirs. *Geophysics* **2018**, *83*, MR35–MR45. [CrossRef]
17. Zhang, B.; Jin, S.; Liu, C.; Guo, Z.; Liu, X. Prediction of shear wave velocity based on a statistical rock-physics model and Bayesian theory. *J. Pet. Sci. Eng.* **2020**, *195*, 107710. [CrossRef]
18. Qiao, H.; Zhang, B.; Liu, C. Research on Prediction Method of Volcanic Rock Shear Wave Velocity Based on Improved Xu-White Model. *Energies* **2022**, *15*, 3611. [CrossRef]
19. Liu, Z.S.; Lu, H.; Liu, J.Z.; Bao, Q.Z.; Shi, L.; Wang, Z.Y. Estimation of S-wave velocity from wire-line logs for organic-rich rocks. *J. Pet. Sci. Eng.* **2021**, *205*, 108928. [CrossRef]
20. Miah, M.I.; Ahmed, S.; Zendehboudi, S. Model development for shear sonic velocity using geophysical log data: Sensitivity analysis and statistical assessment. *J. Nat. Gas Sci. Eng.* **2021**, *88*, 103778. [CrossRef]
21. Zhang, J.; Liu, Z.; Zhou, Y.; Ai, H.; Han, H. Joint Inversion Method of Rock Physics Based on Hunger Games Search Correction and Bidirectional Long-Short-Term Memory Network. *IEEE Trans. Geosci. Remote Sens.* **2024**, *62*, 5914310. [CrossRef]
22. Zhao, L.; Liu, J.; Xu, M.; Zhu, Z.; Chen, Y.; Geng, J. Rock-physics-guided machine learning for shear sonic log prediction. *Geophysics* **2024**, *89*, D75–D87. [CrossRef]

23. Dang, Y.; Zhang, Y.; Wu, B.; Li, H.; Gao, J. An efficient method of predicting S-wave velocity using sparse Gaussian process regression for a tight sandstone reservoir. *J. Appl. Geophys.* **2024**, *229*, 105480. [CrossRef]
24. Fang, Z.; Ba, J.; Guo, Q.; Xiong, F. Shear-wave velocity prediction of tight reservoirs based on poroelasticity theory: A comparative study of deep neural network and rock physics model. *Geoenergy Sci. Eng.* **2024**, *240*, 213028. [CrossRef]
25. Wang, Z. Seismic anisotropy in sedimentary rocks, part 2: Laboratory data. *Geophysics* **2002**, *67*, 1423–1440. [CrossRef]
26. Hornby, B.E.; Schwartz, L.M.; Hudson, J.A. Anisotropic effective-medium modeling of the elastic properties of shales. *Geophysics* **1994**, *59*, 1570–1583. [CrossRef]
27. Sayers, C. Anisotropic velocity analysis1. *Geophys. Prospect.* **1995**, *43*, 541–568. [CrossRef]
28. Johansen, T.A.; Ruud, B.O.; Jakobsen, M. Effect of grain scale alignment on seismic anisotropy and reflectivity of shales. *Geophys. Prospect.* **2004**, *52*, 133–149. [CrossRef]
29. Bachrach, R. Elastic and resistivity anisotropy of shale during compaction and diagenesis: Joint effective medium modeling and field observations. *Geophysics* **2011**, *76*, E175–E186. [CrossRef]
30. Pan, X.P.; Zhang, G.Z.; Chen, J.J. The construction of shale rock physics model and brittleness prediction for high-porosity shale gas-bearing reservoir. *Pet. Sci.* **2020**, *17*, 658–670. [CrossRef]
31. Bandyopadhyay, K. Seismic Anisotropy: Geological Causes and Its Implications to Reservoir Geophysics. Ph.D. Thesis, Stanford University, Stanford, CA, USA, 2009.
32. Qian, K.; Zhang, F.; Chen, S.; Li, X.; Zhang, H. A rock physics model for analysis of anisotropic parameters in a shale reservoir in Southwest China. *J. Geophys. Eng.* **2016**, *13*, 19–34. [CrossRef]
33. Hudson, J.A. Overall properties of a cracked solid. *Math. Proc. Camb. Philos. Soc.* **1980**, *88*, 371–384. [CrossRef]
34. Cheng, C.H. Crack models for a transversely isotropic medium. *J. Geophys. Res. Solid Earth* **1993**, *98*, 675–684. [CrossRef]
35. Qian, K.R.; He, Z.L.; Chen, Y.Q.; Liu, X.W.; Li, X.Y. Prediction of brittleness based on anisotropic rock physics model for kerogen-rich shale. *Appl. Geophys.* **2017**, *14*, 463–479. [CrossRef]
36. Zhang, F.; Li, X.y.; Qian, K. Estimation of anisotropy parameters for shale based on an improved rock physics model, part 1: Theory. *J. Geophys. Eng.* **2017**, *14*, 143–158. [CrossRef]
37. Mavko, G.; Mukerji, T.; Dovorkin, J. *The Rock Physics Handbook-Tools for Seismic Analysis of Porous Media*, 2nd ed.; Cambridge University Press: Cambridge, UK, 2009.
38. Keys, R.G.; Xu, S. An approximation for the Xu-White velocity model. *Geophysics* **2002**, *67*, 1406–1414. [CrossRef]
39. de Figueiredo, J.J.S.; do Nascimento, M.J.S.; Hartmann, E.; Chiba, B.F.; da Silva, C.B.; de Sousa, M.C.; Silva, C.; Santos, L.K. On the application of the Eshelby-Cheng effective model in a porous cracked medium with background anisotropy: An experimental approach. *Geophysics* **2018**, *83*, C209–C220. [CrossRef]
40. de Figueiredo, J.; Chiba, B.F.; do Nascimento, M.J.; da Silva, C.B.; Santos, L.K. Can Hudson-Crampin effective model be applied in cracked medium in which the background is weakly anisotropic (VTI)? *J. Appl. Geophys.* **2019**, *161*, 255–260. [CrossRef]
41. Sohail, G.M.; Hawkes, C.D. An evaluation of empirical and rock physics models to estimate shear wave velocity in a potential shale gas reservoir using wireline logs. *J. Pet. Sci. Eng.* **2020**, *185*, 106666. [CrossRef]
42. Bai, J.Y.; Yue, C.Q.; Liang, Y.Q.; Song, Z.X.; Ling, S.; Zhang, Y.; Wu, W. Variable aspect ratio method in the Xu-White model for shear-wave velocity estimation. *J. Geophys. Eng.* **2013**, *10*, 035008. [CrossRef]
43. Hill, R. The Elastic Behaviour of a Crystalline Aggregate. *Proc. Phys. Soc. Sect. A* **1952**, *65*, 349–354. [CrossRef]
44. Gassmann, F. Uber die elastizität poröser medien. *Viertel. Naturforsch. Ges. Zürich* **1951**, *96*, 1–23.
45. Wyllie, M.R.J.; Gregory, A.R.; Gardner, G.H.F. An experimental investigation of factors affecting elastic wave velocities in porous media. *Geophysics* **1958**, *23*, 459–493. [CrossRef]
46. Pillar, N.; Yan, J.; Lubbe, R. Variable Aspect Ratio Method in the Xu-White Model for AVO. In Proceedings of the 69th EAGE Conference and Exhibition Incorporating SPE EUROPEC 2007, London, UK, 11–14 June 2007; European Association of Geoscientists and Engineers: Bunnik, The Netherlands, 2007. [CrossRef]

processes

MDPI

Article

Comparative Study on Full-Scale Pore Structure Characterization and Gas Adsorption Capacity of Shale and Coal Reservoirs

Mukun Ouyang [1], Bo Wang [1], Xinan Yu [2], Wei Tang [1], Maonan Yu [1], Chunli You [1], Jianghai Yang [1], Tao Wang [2,3,*] and Ze Deng [3,*]

1 Chongqing Division of Southwest Oil and Gas Field Company, PetroChina, Chongqing 400021, China; youchunli@petrochina.com.cn (C.Y.)
2 Institute of Unconventional Oil and Gas Development, Chongqing University of Science and Technology, Chongqing 401331, China; yxndavid@163.com
3 PetroChina Research Institute of Petroleum Exploration & Development, Beijing 100083, China
* Correspondence: 15090713466@163.com (T.W.); dengze@petrochina.com.cn (Z.D.)

Abstract

Shale and coal in the transitional marine–continental facies of the Ordos Basin serve as unconventional natural gas reservoirs, with their pore structures controlling gas adsorption characteristics and occurrence states. To quantitatively characterize the pore structure features and differences between these two reservoirs, this study takes the Shanxi Formation shale and coal in the Daning–Jixian area on the eastern margin of the Ordos Basin as examples. Field-emission scanning electron microscopy (FE-SEM), high-pressure mercury intrusion, low-temperature N_2 adsorption, and low-pressure CO_2 adsorption experiments were employed to analyze and compare the full-scale pore structures of the shale and coal reservoirs. Combined with methane isothermal adsorption experiments, the gas adsorption capacity and its differences in these reservoirs were investigated. The results indicate that the average total organic carbon (TOC) content of shale is 2.66%, with well-developed organic pores, inorganic pores, and microfractures. Organic pores are the most common, typically occurring densely and in clusters. The average TOC content of coal is 74.22%, with organic gas pores being the dominant pore type, significantly larger in diameter than those in transitional marine–continental facies shale and marine shale. In coal, micropores contribute the most to pore volume, while mesopores and macropores contribute less. In shale, mesopores dominate, followed by micropores, with macropores being underdeveloped. Both coal and shale exhibit a high SSA primarily contributed by micropores, with organic matter serving as the material basis for micropore development. The methane adsorption capacity of coal is 8–29 times higher than that of shale. Coal contains abundant organic micropores, providing a large SSA and numerous adsorption sites for methane, facilitating gas adsorption and storage. This study comprehensively reveals the similarities and differences in pore structures between transitional marine–continental facies shale and coal reservoirs in the Ordos Basin at the microscale, providing a scientific basis for the precise evaluation and development of unconventional oil and gas resources.

Keywords: Daning–Jixian block; deep coalbed methane; pore structure; full-scale pore size distribution; influential factors

1. Introduction

Unconventional oil and gas have the characteristics of wide distribution area and great resource potential. As new fields for China's unconventional natural gas exploration and development, marine–continental transitional shale gas and coalbed methane boast total geological resources of approximately 19.8×10^{12} m^3, with technically recoverable resources reaching 5.1×10^{12} m^3, accounting for 25% of the total shale gas resources [1–3]. In the Upper Paleozoic Shanxi Formation of the Ordos Basin, coal-measure shale and coal seams are widely developed, exhibiting notable features such as large cumulative thickness, extensive planar distribution, and substantial resource potential [4]. In recent years, breakthroughs have been made in both shale gas and coalbed methane (CBM) in the Daning–Jixian block: Five vertical wells obtained industrial gas flow through fracturing and test production in the shale section of the Shanxi Formation, with the maximum unobstructed flow rate reaching 1.0×10^4 m^3/day. The horizontal JP1H well has produced a cumulative gas of 1936×10^4 m^3/day in 18 months and has a stable daily gas production of 3.3×10^4 m^3/day. JS6-7 Well achieved a high-yield industrial gas flow with a daily output of 10.1×10^4 m^3/day in the coal rock section [5]. These achievements highlight the necessity of integrated geological evaluations for transitional shale and coal systems in the Ordos Basin's Upper Paleozoic sequences to optimize exploration strategies.

Shale gas and CBM are important components of unconventional natural gas. Unlike sandstone and carbonate rock reservoirs, shale and coal rock reservoirs have the characteristic of self-storage and self-preservation [6–9]. The occurrence characteristics and production laws of shale gas and CBM are controlled by the complex pore structure of the reservoir. Due to the presence of a large number of organic pores smaller than 2 nm in the organic components of coal reservoirs, CBM mainly exists in the coal rock in an adsorbed state. The pore network of shale reservoirs is mainly composed of interparticle pores, intraparticle pores, and organic pores of micrometer and nanometer scales. Shale gas often exists in the shale reservoir in a free state and an adsorbed state [10]. This storage mechanism is closely associated with the complex nanopore systems in shale and coal reservoirs, which govern gas storage flow behavior and ultimately influence hydrocarbon storage capacity and well productivity. Currently, numerous analytical techniques have been employed to characterize the pore structures of shale and coal, including direct methods such as scanning electron microscopy (SEM) and micro-CT imaging, as well as indirect methods like high-pressure mercury intrusion, low-pressure CO_2 adsorption ($LPCO_2A$), and low-temperature N_2 adsorption (LTN_2A) [5,11–13]. The former enables qualitative observation of pore morphology, size, and distribution, while the latter provides quantitative characterization of pore volume (PV), specific surface area (SSA), and other structural parameters. However, due to the strong heterogeneity and wide pore size distribution of shale and coal, a multi-method approach is required to comprehensively characterize their full-scale pore structures. Previous studies have employed integrated experimental methods to reveal the characteristics of nanopore systems in marine shale reservoirs such as the Wufeng-Longmaxi Formation [4,14–16]. However, similar research on transitional marine–continental facies reservoirs remains limited. Compared to marine shale gas, transitional marine–continental facies shale gas is influenced by complex hydrodynamic conditions and multi-source sediment supply, resulting in frequent sandstone–mudstone interbedding and strong vertical heterogeneity in the distribution of shale and coal layers. These factors significantly constrain the exploration and development of transitional marine–continental facies shale gas [17–20].

This study focuses on the shale and coal samples from the Shanxi Formation in the Daning–Jixian area along the eastern margin of the Ordos Basin. A comprehensive characterization of the full-aperture pore structure characteristics of these shale and coal reservoirs was conducted using an integrated approach combining field emission scanning electron microscopy (FE-SEM), $LPCO_2A$, LTN_2A, and high-pressure mercury intrusion (HPMI) methods. Through comparative analysis, the differences in pore structure between shale and coal samples and their underlying formation mechanisms were systematically investigated. Furthermore, methane isothermal adsorption experiments were performed to examine the natural gas adsorption capacity and its variations within these reservoirs. The research aims to provide theoretical support for the evaluation and development of coal-bearing reservoirs in the Ordos Basin.

2. Geological Setting

The Ordos Basin, located in north-central China, features stable internal structure with limited fault development, exhibiting an overall tectonic framework characterized by north–south uplift and west-thrusting-east-upheaval [21,22]. The basin can be divided into six first-order tectonic units: the Western Thrust Belt, Western Tianhuan Depression, Central Shaanxi Slope, Eastern Shanxi Flexural Fold Belt, Southern Weibei Uplift, and Northern Yimeng Uplift (Figure 1a). As China's largest oil and gas producing basin, the Ordos Basin has accumulated proven natural gas reserves of 6.86×10^{12} m^3. Major gas fields including Sulige, Jingbian, Yulin, Zizhou, and Shenmu have been successively discovered in the Paleozoic strata [23,24].

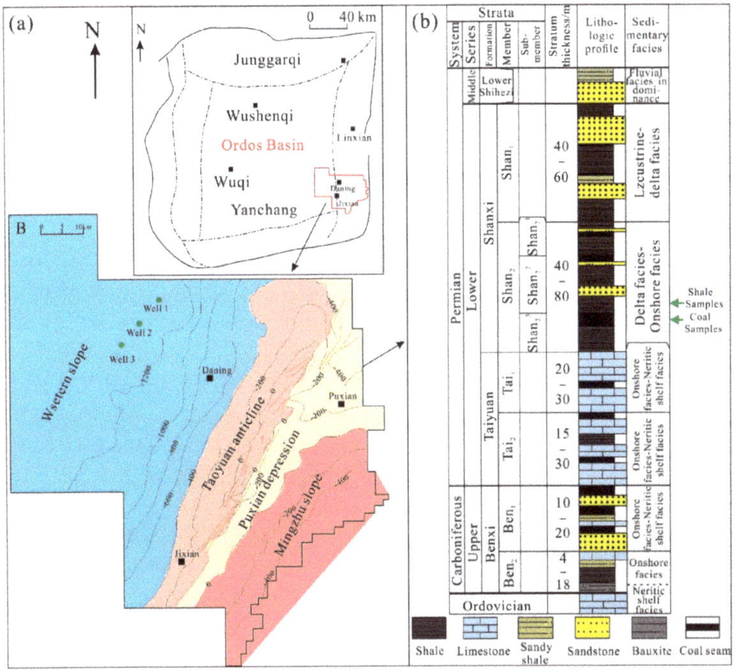

Figure 1. (**a**) Location of the study area; (**b**) Comprehensive stratigraphic column chart.

The Permian system in the Ordos Basin develops, from bottom to top, the Taiyuan Formation, Shanxi Formation, Shihezi Formation, and Shiqianfeng Formation (Figure 1b). Influenced by the Hercynian tectonic movement, the entire North China Platform experienced uplift, causing seawater to gradually retreat from both eastern and western flanks of the basin. This resulted in frequent water body fluctuations in the Ordos Basin, forming multiple sedimentary cycles of shallow marine–delta front–shallow lacustrine facies, which deposited several sets of organic-rich transitional marine–continental shales [23–25]. The study area, located in the southeastern part of the basin, contains well-developed shale and coal seams in the Shan-2^3 submember. The cumulative thickness of shale ranges from 20 to 40 m, while coal seams measure approximately 3–5 m in thickness (Figure 1). These formations currently represent one of the key target intervals for transitional marine–continental shale gas and coalbed methane exploration and development in the Ordos Basin.

3. Materials and Methods

3.1. Experimental Materials

A total of eight fresh samples from the coal-bearing shale and coal seams of the Shanxi Formation in the Ordos Basin were selected for this study, including four fresh coal samples and four fresh shale samples. The collected cylindrical shale and coal specimens were processed through cutting and crushing procedures to prepare subsamples for various analyses: total organic carbon (TOC) testing, Rock-Eval pyrolysis, X-ray diffraction (XRD) analysis for whole-rock and clay mineral composition, maceral identification, FE-SEM observation, HPMI, LPCO_2A and LTN_2A experiments, and CH_4 isothermal adsorption experiments.

3.2. Experimental Methods

The key experiments involved in this study include FE-SEM, HPMI, LPCO_2A, LTN_2A, and CH_4 isothermal adsorption experiments: ① shale and coal rock samples were observed for pore structure through FE-SEM. Before observation, the surface was treated with argon ion polishing (using a Gatan 697 argon ion polishing instrument to polish the sample surface for 6 h) to obtain a smooth observation surface, and 5 nm thick platinum was sprayed to enhance the conductivity of the sample surface [26,27]. FE-SEM imaging is performed at a working distance of 2–4 mm with an acceleration voltage of 1 kV. ② The LPCO_2A and LTN_2A experiments were conducted using a Micromeritics ASAP 2460 SSA, Noklossor, Torrance, CA, USA and pore size analyzer from the United States. The sample particle size was 60–80 mesh. Before the experiment, all samples were dried at 105 °C for 8 h. Then, an appropriate amount of sample was weighed and placed on the machine. The samples were vacuum degassed at 110 °C for 10 h in a degassing station. Finally, high-purity CO_2 and N_2 were subjected to adsorption and desorption experiments at 273 K and 77 K, respectively, to obtain corresponding LPCO_2A and LTN_2A experimental data. The entire experimental process was strictly carried out in accordance with GB/T21650.3-2011 [28] and GB/T21650.2-2008 [29]. ③ The HPMI experiment used the AutoPore9510 fully automatic mercury injection instrument from Micromeritics Instrument in the United States. During the high-pressure testing phase, the instrument injected mercury at a pressure of up to 60,000 psi (approximately 413 MPa), corresponding to a lower limit of 3.75 nm for the test aperture. Before testing, the sample needs to be made into cubic blocks with a size of 1 cm^3, and then the surface of the blocks should be polished to eliminate the hemp skin effect during the testing process. Then, a low-temperature long-term drying scheme (continuous drying at 60 °C for 48 h) should be selected to prevent the original pores in the sample from being damaged by high temperature and effectively remove the internal impurity gases [22]. Before the test starts, the dried sample should be loaded and the instrument should be vacuumed. After starting the test, the data of the mercury injection and removal process should be recorded. The entire experimental process should strictly follow GB/T21650.1-2008 [30]. ④ The high-pressure CH_4 adsorption was carried out using a Gravimetric Isothermal Rig 3 with a large sample weight method. The sample was first crushed to 60–80 mesh, and then the degassed sample was subjected to high-pressure isothermal adsorption experiments using high-purity CH_4 inside the instrument. During the process, 9 testing points were set up at a temperature of 60 degrees Celsius and a maximum pressure of 25 MPa.

4. Results

4.1. Organic Geochemical Characteristics

The TOC measurement and Rock-Eval pyrolysis results reveal that the TOC of the Shanxi Formation shale in the study area ranges from 0.56% to 28.5%, with an average value of 2.66%. Notably, 63% of the samples exceed 2.0%, classifying them as excellent source rocks (Figure 2). Maceral composition analysis demonstrates that the shale contains vitrinite > inertinite > exinite, average contents of 38.3%, 34.5%, and 27.1%, respectively, predominantly indicating Type III organic matter, with a very minor proportion reaching Type II_2. The Shanxi Formation coal samples exhibit TOC values between 66.8% and 86.1%, averaging 74.22%. Their maceral composition is dominated by vitrinite (79.56% average), followed by inertinite (12.58%) and liptinite (7.86%), all classified as Type III organic matter. Compared with shale, the coal samples show significantly lower liptinite but higher vitrinite content (Table 1). Both shale and coal samples from the Shanxi Formation display vitrinite reflectance (R_o) values exceeding 2.0%, average of 2.66%, indicating an over-mature stage favorable for hydrocarbon generation through thermal cracking. The predominance of Type III kerogen in both lithologies, when compared with transitional marine–continental facies Type I–II organic matter, is less conducive to organic pore development. However, the relatively higher inertinite and exinite content in shale macerals compared to coal (which is dominated by vitrinite and inertinite) suggests greater potential for organic pore formation in shale.

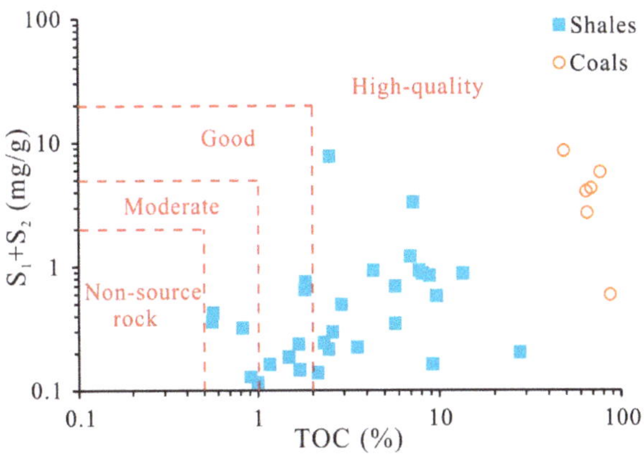

Figure 2. Characteristics of organic matter abundance of coal and shale in the study area.

Table 1. R_o, organic matter abundance, macerals, and minerals of coal and shale in the study area.

Lithology	R_o (%)	Organic Matter Abundance		Macerals (%)			Minerals (%)	
		TOC	$S_1 + S_2$	Vitrinite	Inertinite	Exinite	Brittle Minerals	Clay Minerals
Coal	2.56~2.77 2.65	66.8~86.1 74.22	0.60~8.66 4.35	76.48~82.01 79.56	9.78~15.25 12.58	7.1~13.74 7.86	5~21 9.9	32~56 44
Shale	0.56~28.5 2.66	0.56~6.62 1.86	0.12~7.77 0.77	20~48 38.3	20~57 34.5	15~44 27.1	24~88 61.75	12~76 38.25

The XRD analysis results demonstrate that the Shanxi Formation shale in the study area is primarily composed of brittle minerals and clay minerals (Table 1), with minor amounts of pyrite and carbonate minerals. Among these components: brittle minerals are dominated by quartz, with content ranging from 16% to 65% (average: 46.1%); clay minerals vary between 12% and 76% (average: 38.25%), with kaolinite being the most abundant species. In contrast, coal samples exhibit significantly different mineral compositions: predominantly clay minerals (32–56%, average: 44%); minor quartz and calcite (Table 1); illite/smectite mixed-layer minerals and kaolinite showing the highest concentrations among clay minerals. The substantial variations in mineral composition among shale and coal samples from different depth intervals of the Shanxi Formation indicate strong heterogeneity characteristics of transitional marine–continental facies shales. Furthermore, compared to marine shales, both the transitional facies shales and coals exhibit lower brittle mineral contents [24,29,31].

4.2. Pore Morphology Characteristics

The Shanxi Formation shale and coal samples in the study area exhibit complex micro–nanoscale pore and fracture networks, as revealed by FE-SEM observations (Figure 3). Based on their morphology and genesis, the pores and fractures in shale can be classified into three categories: organic pores, inorganic pores, and microfractures [32]. Organic pores are the predominant pore type in shale, primarily exhibiting individual pore morphologies dominated by circular, elliptical, and irregular polygonal shapes (Figure 3a). Inorganic pores encompass four subtypes: dissolution pores, intergranular residual pores, clay mineral interlayer pores, and pyrite intercrystalline pores [33,34]. Specifically, dissolution pores are typically isolated and predominantly developed within quartz minerals, displaying elliptical or pit-shaped configurations (Figure 3b). Intergranular pores are predominantly hosted between rigid minerals, typically exhibiting fissure-like or irregular polygonal morphologies due to compaction effects (Figure 3c). Clay mineral interlayer pores develop through clay mineral transformation and volume reduction during diagenetic processes; subsequently, these pores either undergo closure under compaction or form elongated, sinuous filamentous pore networks (Figure 3c). In some shale intervals with elevated pyrite content, irregular stacking of multiple crystals creates abundant intercrystalline pores that form reticulated pore networks with favorable connectivity, exhibiting pore diameters ranging from 100 nm to 300 nm (Figure 3c). The Shanxi Formation shales contain observable microfractures with limited distribution density, which are categorized into: those typically propagating along mineral grain boundaries or cross-cutting mineral grains/organic matter (Figure 3a), which effectively interconnect microscopic pore systems, and those predominantly distributed along organic matter margins, formed through hydrocarbon-generation-associated organic matter transformation and progressive volumetric contraction (Figure 3c).

Compared to shales, coal rocks in the study area are primarily composed of organic matter and clay minerals, with notably lower contents of rigid minerals such as quartz and pyrite. Under intense compaction, most pores become occluded, though observable pore types include organic-matter-hosted gas vesicles, residual intergranular pores, cleats, and microfractures (Figure 3d–f). Gas vesicles—a subtype of organic pores—constitute the dominant pore type in the Shanxi Formation coal (Figure 3e), formed during hydrocarbon generation at high-over maturity stages. Scanning electron microscopy reveals these vesicles exhibit elliptical or slit-shaped morphologies with preferential alignment due to compaction (Figure 3e). While residual intergranular pores and microfractures in coal share morphological similarities with their shale counterparts, they occur in significantly lower abundances. Distinctively, localized development of microfractures is observed in coal lithofacies (Figure 3d,e), a feature absent in adjacent shale units.

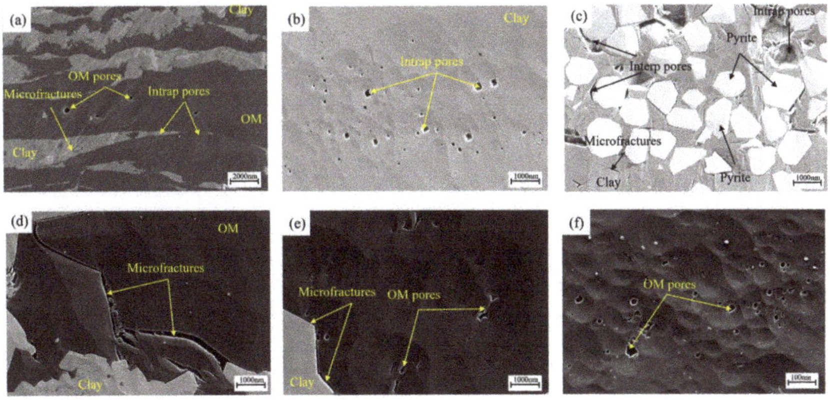

Figure 3. SEM images of coal and shale in the study area. (**a**) DJ-Y-1, Shale; (**b**) DJ-Y-2, Shale; (**c**) DJ-Y-3, Shale; (**d**) DJ-M-1, Coal; (**e**) DJ-M-2, Coal; (**f**) DJ-M-4, Coal.

4.3. Microstructural Pore Characteristics

4.3.1. LPCO_2A

LPCO_2A experiments revealed Type I adsorption isotherms without saturation plateaus for both shale and coal samples from the Shanxi Formation (Figure 4). Comparative analysis of LPCO_2A curves demonstrates distinct adsorption characteristics between coal and shale, primarily manifested in their maximum CO_2 adsorption capacities: coal samples exhibit an average adsorption capacity of 20.18 cm^3/g, significantly exceeding the shale average of 2.13 cm^3/g, indicating more developed micropores in coal. NLDFT model-based calculations of LPCO_2A data yielded pore size distribution curves (0.3–1.5 nm range) and structural parameters (Table 2). Coal samples display micro-PVs and SSAs ranging from 0.038–0.071 cm^3/g and 128.01–242.99 m^2/g, respectively, with mean values of 0.061 cm^3/g and 207.54 m^2/g. In contrast, shale samples show markedly lower values (0.004–0.008 cm^3/g and 14.49–26.99 m^2/g; mean: 0.007 cm^3/g and 21.76 m^2/g), with average pore diameters concentrated at 0.349 nm. The micro-PV and SSA of coal and rock are higher than those of shale, indicating that, compared with shale, coal and rock have more abundant micropores developed. The pore size distribution characteristics based on LPCO_2A P show that the micropore development of shale and coal rock samples generally presents a multi-peak pattern, with peak distributions ranging from 0.3–0.4 nm, 0.4–0.7 nm, and 0.8–0.9 nm (Figure 5). Notably, when the pore diameter is >1.0 nm, the increase rate of micro-PV and SSA slows down significantly, and the contents of PV and surface area corresponding to this pore diameter are both very small. Due to the relatively well-developed micropores in the coal sample and the large SSA of these pores, a large number of adsorption points can be provided for the occurrence of gas.

4.3.2. LTN_2A

LTN_2A isotherms of Shanxi Formation shale and coal samples exhibit characteristic "inverted S shapes" (Figure 6). Following IUPAC classification [33–35], shale samples display Type IV isotherms, indicating the presence of micropores coupled with substantial mesopores and macropores. Significant adsorption capacity differences are observed: coal samples exhibit an average N_2 adsorption capacity of 3.55 cm^3/g, markedly lower than shale's 12.54 cm^3/g, demonstrating superior mesopore development in shale. The isotherms progress through three distinct phases with increasing relative pressure (P/P_0): monolayer adsorption (P/P_0 = 0–0.1): slow convex-curve adsorption with monolayer saturation achieved at $P/P_0 \approx 0.1$, transitioning to multilayer adsorption. Multilayer adsorption

(P/P_0 = 0.3–0.9): gradual adsorption increase following BET theory. Capillary condensation (P/P_0 = 0.9–1): rapid uptake without plateau near saturation pressure (P/P_0 = 1), indicative of macropore/microfracture-facilitated gas condensation. A pronounced hysteresis loop emerges at P/P_0 = 0.45–0.9 (Figure 6), attributable to irreversible capillary condensation in mesopores (2–50 nm). This hysteresis behavior aligns with Type H3 loops in IUPAC nomenclature, suggesting slit-shaped pores formed by plate-like particle aggregates in shale.

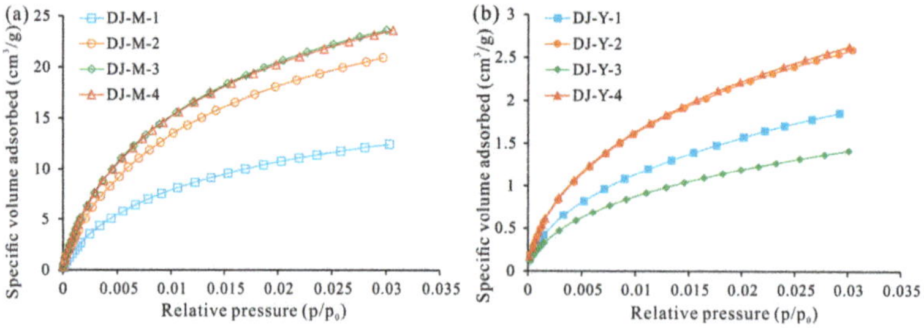

Figure 4. $LPCO_2A$ curves of coals and shales. (**a**) Coal samples, (**b**) shale samples.

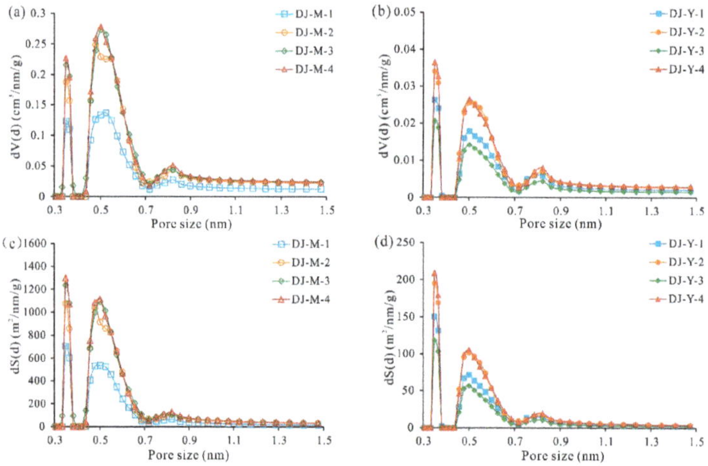

Figure 5. Pore size distribution curves of coal and shale based on $LPCO_2A$. (**a**) PV distribution of the coal samples, (**b**) PV distribution of the shale samples, (**c**) SSA distribution of the coal samples, (**d**) SSA distribution of the shale samples.

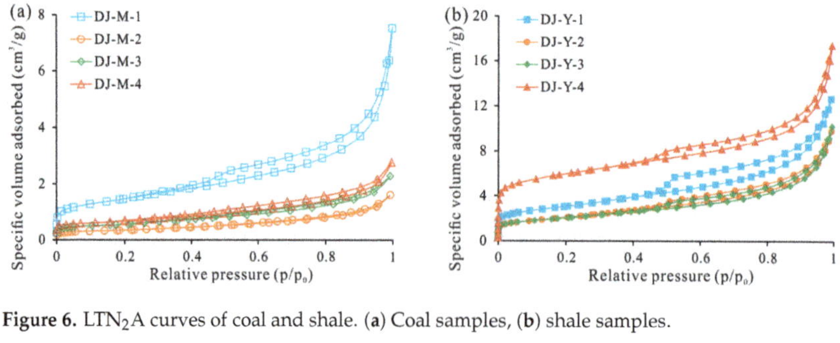

Figure 6. LTN_2A curves of coal and shale. (**a**) Coal samples, (**b**) shale samples.

According to the classification of hysteresis loops by IUPAC [35], the coal and rock conform to the morphology of the H3 type hysteresis loop, which changes slowly at both medium and low relative pressures, and the adsorption capacity increases rapidly at high relative pressures. The vertex of the curve is relatively sharp, indicating that the pore morphology is parallel plate pores with open ends and slit-shaped pores. Furthermore, when the relative pressure is 0.5, the desorption curve has a sharply decreasing inflection point, indicating that a certain number of "ink bottle"-shaped pores have developed in the coal rock. The hysteresis rings of the shale sheet are of type H3, H4, or a mixture of both, indicating that the shale has developed pores in various forms such as wedge, slit, and parallel plate, and the combinations of pore morphology types are diverse.

Non-local density functional theory (NLDFT) analysis of LTN_2A data reveals distinct pore structure characteristics between coal and shale samples within the 1.06–78 nm pore size range (Figure 7; Table 2) [36,37]. Coal samples exhibit DFT PVs and SSAs of 0.002–0.009 cm^3/g and 1.38–5.76 m^2/g, respectively, with mean values of 0.005 cm^3/g and 3.14 m^2/g. The average pore diameter of coal measures 2.18 nm. In contrast, shale samples demonstrate significantly higher pore structure parameters: PVs range from 0.013–0.023 cm^3/g (mean: 0.017 cm^3/g), SSAs from 8.09–22.89 m^2/g (mean: 12.73 m^2/g), and average pore diameter is 1.23 nm. The DFT PV and SSA of shale are higher than those of coal and rock, indicating that, compared with coal and rock, shale has a more abundant mesopore development. The PV and SSA distribution curves of shale and coal rock samples in the study area with pore size are shown in Figure 7. With the increase in pore size, the PV at this stage decreases significantly. When the pore diameter is greater than 10 nm, the change in PV is relatively small, indicating that the proportion of pores at 10 nm is relatively high.

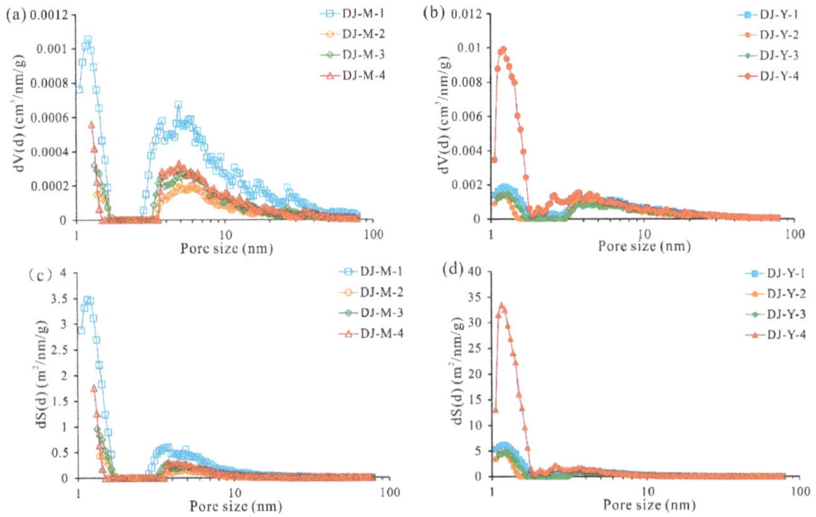

Figure 7. Pore size distribution curves of coal and shale based on LTN_2A. (**a**) PV distribution of the coal samples, (**b**) PV distribution of the shale samples, (**c**) SSA distribution of the coal samples, (**d**) SSA distribution of the shale samples.

4.3.3. HPMI

Figure 8 illustrates mercury intrusion–extrusion curves for shale and coal samples, whose morphological features reflect pore connectivity and throat size distribution [34]. The mercury entry and exit curves of coal and rock have the characteristics of being gentle at the front end and middle and steep at the back end, indicating that there are abundant nanoscale micropores and some mesopores in the coal and rock samples. The mercury intrusion–extrusion curve of shale has the characteristics of being gentle at both ends and steep in the middle, indicating that there are abundant nanoscale mesopores, with some microfractures or large-scale pores in the shale samples. The mercury removal efficiency of high-pressure mercury compression experiments is mainly related to the morphology of pores and throat channels developed in shale. Coal and rock samples have a relatively high organic matter content and usually develop a large number of micro–nanoscale pores and throat channels. The capillary force in the relatively fine and large throat channels is weak, which is conducive to the movement of pore fluids, thereby affecting the higher mercury removal efficiency. Shale samples have a relatively low organic matter content and usually develop a certain amount of nanoscale pores and throat channels. The capillary force in the relatively small throat channels is strong, which is not conducive to the movement of pore fluids, thereby manifested as a lower mercury removal efficiency.

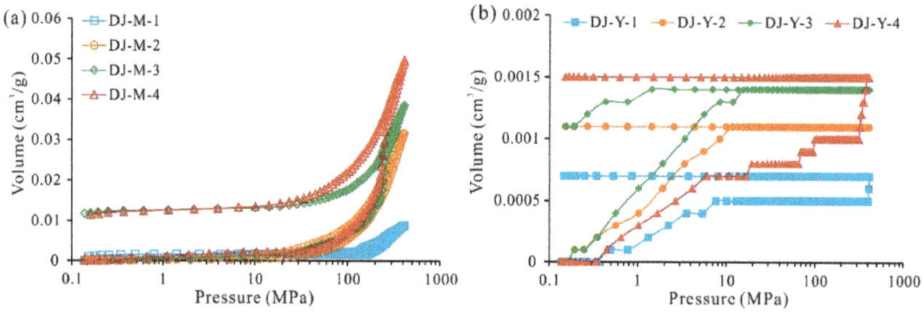

Figure 8. Mercury intrusion–extrusion curves of coal and shale by HPMI. (**a**) Coal samples, (**b**) shale samples.

Based on the Washburn equation [38], the pore size distribution curves and pore structure parameters of shale and coal rock samples in the pore size range of 3 nm~10 μm were calculated using HPMI experimental data (Table 2). The PV and SSA of coal rock samples are 0.009–0.050 cm^3/g and 5.55–30.65 m^2/g, with mean values of 0.032 cm^3/g and 19.72 m^2/g, respectively. The PV and SSA of shale samples are 0.001–0.002 cm^3/g and 0.01–0.56 m^2/g, with mean values of 0.001 cm^3/g and 0.20 m^2/g, respectively. The PV and SSA of shale are lower than those of coal rock, indicating that coal rock has more abundant large pores and fractures compared to shale. The distribution curves of PV and SSA with pore size for coal and shale samples in the research area are shown in Figure 9. The peak of mercury injection occurs within the range of less than 30 nm, and the mercury saturation approaching 80% begins to enter in large quantities only when the pore size is less than 30 nm, indicating that microscale large pores and fractures are mainly developed in coal rocks.

4.4. CH_4 Isothermal Adsorption Characteristics

High-pressure methane isothermal adsorption experiments elucidate the CH_4 adsorption capacity on pore surfaces. The adsorption isotherms (Figure 10) exhibit Type I behavior according to IUPAC classification [35], characteristic of monolayer adsorption dominated by micropore filling. This physisorption mechanism aligns with the Langmuir model assumptions of homogeneous surface coverage and finite adsorption sites [25,26].

Table 2. Pore structure parameters of the coal and shale samples.

Sample ID	Lithology	LPCO$_2$A			LTN$_2$A				HPMI	
		V_{DFT} (cm^3/g)	S_{DFT} (m^2/g)	D_{CO2} (nm)	S_{BET} (m^2/g)	V_{DFT} (cm^3/g)	S_{DFT} (m^2/g)	D_{N2} (nm)	V_{HPMI} (cm^3/g)	S_{HPMI} (m^2/g)
DJ-M-1	Coal	0.038	128.01	0.524	5.08	0.009	5.76	1.22	0.009	5.55
DJ-M-2	Coal	0.064	216.74	0.479	1.20	0.002	1.38	4.887	0.032	19.39
DJ-M-3	Coal	0.071	242.99	0.501	1.96	0.003	2.35	1.326	0.039	23.29
DJ-M-4	Coal	0.071	242.42	0.501	2.23	0.004	3.08	1.273	0.050	30.65
Average		0.061	207.539	0.501	2.618	0.005	3.143	2.177	0.032	19.720
DJ-Y-1	Shale	0.006	19.14	0.349	10.68	0.018	11.73	1.220	0.001	0.25
DJ-Y-2	Shale	0.008	26.43	0.548	7.45	0.013	8.09	1.220	0.001	0.01
DJ-Y-3	Shale	0.004	14.49	0.349	7.13	0.014	8.20	1.273	0.001	0.01
DJ-Y-4	Shale	0.008	26.99	0.349	20.01	0.023	22.89	1.220	0.002	0.56
Average		0.007	21.762	0.399	11.315	0.017	12.729	1.233	0.001	0.205

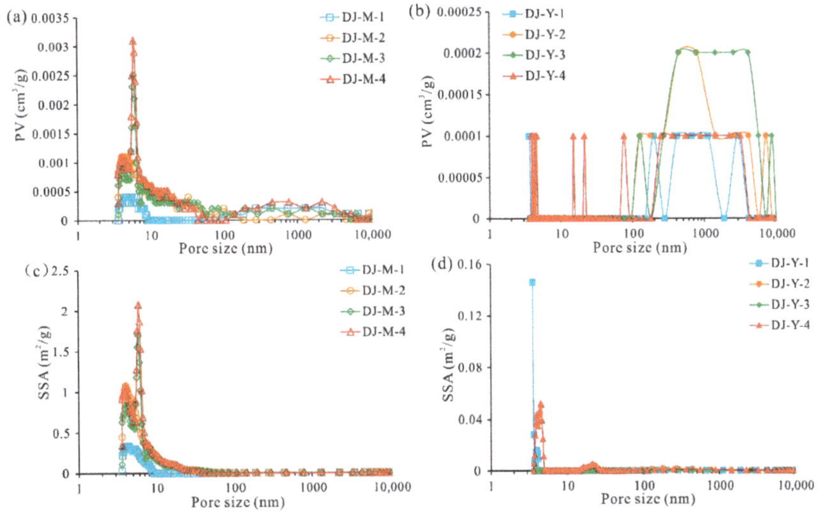

Figure 9. Pore size distribution curves of coal and shale Based on HPMI. (**a**) PV distribution of the coal samples, (**b**) PV distribution of the shale samples, (**c**) SSA distribution of the coal samples, (**d**) SSA distribution of the shale samples.

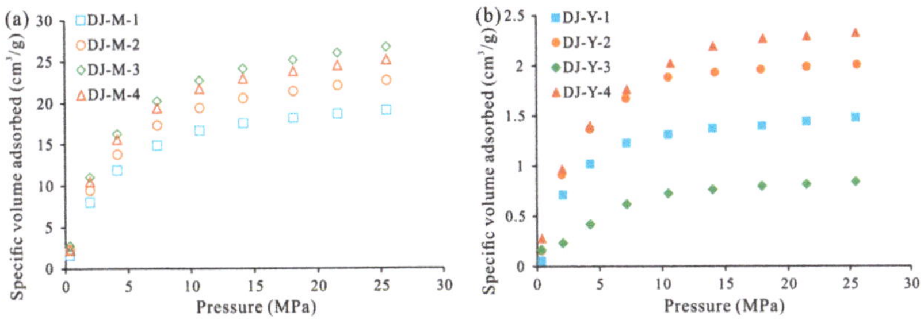

Figure 10. CH_4 isothermal adsorption curves of coal and shale. (**a**) Coal samples, (**b**) shale samples.

The Langmuir equation is expressed as:

$$V = \frac{P}{P + P_L} V_L \quad (1)$$

where P represents equilibrium pressure, MPa, V_L denotes Langmuir volume (monolayer saturation capacity), cm^3/g, and P_L corresponds to Langmuir pressure (half-saturation threshold), MPa.

The fitting results of methane adsorption using the Langmuir model are shown in Table 3. It can be seen that the fitting degree is high, indicating that the Langmuir model can well describe the adsorption behavior of shale and coal rock samples. The V_L of coal rock samples ranges from 21.75 to 30.29 cm^3/g, average of 26.57 cm^3/g. The P_L ranges from 3.47 to 3.59 MPa, average of 3.52 MPa. The V_L of shale samples is 1.04–2.71 cm^3/g, average of 1.92 cm^3/g. The P_L is 2.81–5.54 MPa, with an average of 3.75 MPa. The V_L of shale samples is relatively low, but the P_L is the highest, indicating that their adsorption capacity for gases is the weakest and they struggle to adsorb gases under low pressure conditions. The high V_L of coal rock samples indicates their strongest adsorption ability for gases.

Table 3. Langmuir parameters of coal and shale.

Sample ID	Lithology	V_L (cm^3/g)	P_L (MPa)
DJ-M-1	Coal	21.75	3.47
DJ-M-2	Coal	25.72	3.54
DJ-M-3	Coal	30.29	3.59
DJ-M-4	Coal	28.54	3.48
Average		26.57	3.52
DJ-Y-1	Shale	2.71	3.77
DJ-Y-2	Shale	1.04	5.54
DJ-Y-3	Shale	1.65	2.81
DJ-Y-4	Shale	2.29	2.88
Average		1.92	3.75

5. Discussion

5.1. Quantitative Characterization of Full-Scale Pores in Shale and Coal

According to the joint characterization results of PV and pore size distribution of shale and coal rock samples in the research area (Figure 11), the PV distribution type of coal rock is mainly dominated by single peak microporous dominant type, showing a microporous single peak state. Shale samples exhibit multi-modal characteristics (Figure 11), with micropores, mesopores, and macropores all contributing to the total PV (Figure 11). According to Table 4 and Figure 12a, the total PV of coal rock samples ranges from 0.047 to 0.077 cm^3/g (average 0.067 cm^3/g). Among them, micropores mainly contribute to the PV, followed by mesopores and macropores, with less contribution from both. The micro-PV ranges from 0.038 to 0.071 cm^3/g (average 0.061 cm^3/g), accounting for 80.3% to 95.5% (average 90.6%) of the total PV. Secondly, mesopores have a PV ranging from 0.002 to 0.008 cm^3/g (average 0.004 cm^3/g), accounting for 3.0% to 16.1% (average 6.8%) of the total PV. Macropores are the smallest group, with a volume ranging from 0.001 to 0.002 cm^3/g (average 0.002 cm^3/g), accounting for 1.5% to 3.6% (average 2.6%) of the total PV. The total PV of shale samples ranges from 0.017 to 0.026 cm^3/g (with an average of 0.021 cm^3/g), among which the PV is mainly contributed by mesopores, followed by micropores and macropores, with less contribution from both. The mesoporous volume ranges from 0.011 to 0.016 cm^3/g (average 0.014 cm^3/g), accounting for 56.3% to 69.9% (average 63.4%) of the total PV. Secondly, micropores have PVs ranging from 0.004 to 0.009 cm^3/g (average 0.007 cm^3/g), accounting for 25.8% to 38.4% (average 31.9%) of the total PV. Macropores can be ignored.

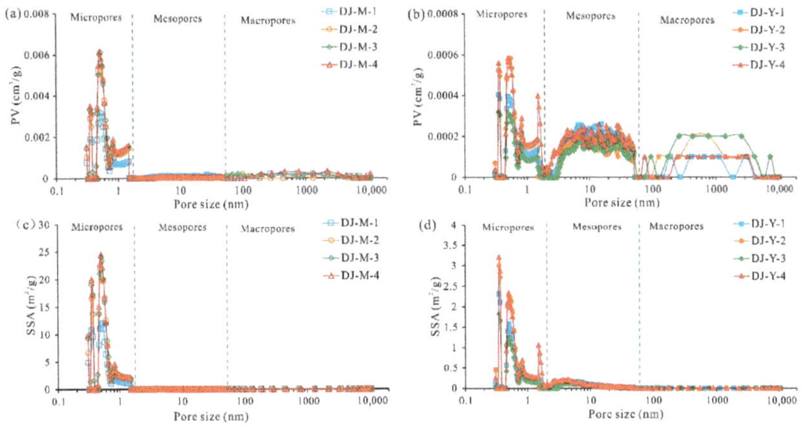

Figure 11. Full-Pore size distribution characteristics of PV and specific surface area of coal and shale. (**a**) PV distribution of the coal samples, (**b**) PV distribution of the shale samples, (**c**) SSA distribution of the coal samples, (**d**) SSA distribution of the shale samples.

Table 4. Statistical results of the PV and SSA of coal and shale.

Sample ID	Lithology	PV (cm^3/g)				SSA (m^2/g)			
		Micropores	Mesopores	Macropores	Total	Micropores	Mesopores	Macropores	Total
DJ-M-1	Coal	0.038	0.008	0.002	0.047	128.18	3.10	0.01	131.30
DJ-M-2	Coal	0.064	0.002	0.001	0.067	216.78	0.82	0.03	217.64
DJ-M-3	Coal	0.071	0.003	0.002	0.076	243.06	1.24	0.04	244.34
DJ-M-4	Coal	0.071	0.003	0.002	0.077	242.42	1.33	0.02	243.76
Average		0.061	0.004	0.002	0.067	207.61	1.62	0.02	209.26
DJ-Y-1	Shale	0.006	0.015	0.000	0.022	19.87	6.57	0.00	26.44
DJ-Y-2	Shale	0.008	0.012	0.001	0.020	26.44	4.91	0.01	31.36
DJ-Y-3	Shale	0.004	0.011	0.001	0.017	14.74	4.36	0.03	19.13
DJ-Y-4	Shale	0.009	0.016	0.001	0.026	30.28	7.86	0.01	38.14
Average		0.007	0.014	0.001	0.021	22.83	5.93	0.01	28.77

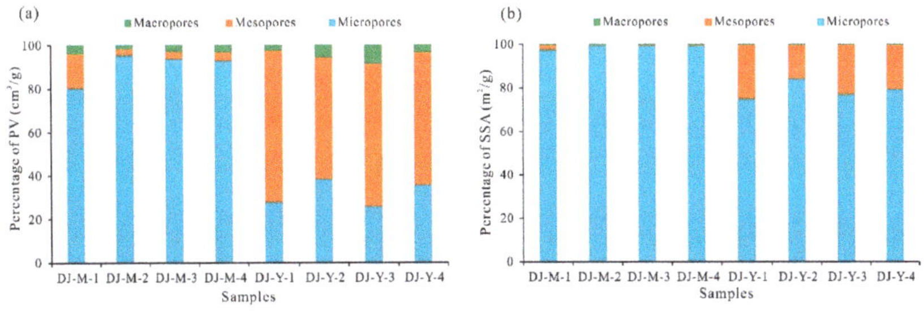

Figure 12. Proportion of PV and SSA of different scales in coal and shale. (**a**) Proportion of PV, (**b**) Proportion of SSA.

According to the joint characterization results of pore SSA and pore size distribution of shale and coal rock samples in the research area (Figure 11), the distribution types of pore SSA in coal rock and shale are mainly dominated by a single peak microporous dominant type, showing a microporous single peak state. According to Table 4 and Figure 12b, the total pore SSA of coal rock samples ranges from 131.30 to 244.34 m^2/g (average 209.26 cm^2/g). Among them, the pore SSA is mainly contributed by micropores, followed by mesopores and macropores, and their contributions are ignored. The SSA of micropores ranges from 128.18 to 243.06 m^2/g (average 207.61 m^2/g), accounting for 97.6%

to 99.6% (average 99.0%) of the total pore SSA. The total pore SSA of shale samples ranges from 19.13 to 38.14 m^2/g (average 28.77 m^2/g), with micropores contributing the most to the pore SSA, followed by mesopores, and the contribution of macropores is ignored. The SSA of micropores ranges from 14.74 to 30.28 m^2/g (average 22.83 m^2/g), accounting for 75.1% to 84.3% (average 79.0%) of the total pore SSA. Next are mesopores, with a pore SSA ranging from 4.36 to 7.86 m^2/g (average 5.93 m^2/g), accounting for 15.7% to 24.9% (average 21.0%) of the total pore SSA. Macropores can be ignored.

In the transitional facies strata of the Shanxi Formation in the research area, the shale pore types are mainly organic pores and clay-mineral-related fractures, with both micropores and mesopores developed. Compared with marine shale, TOC has a greater impact on the development of micropores in transitional shale, because Type III kerogen cracking is more conducive to the formation of smaller diameter organic pores [39–42]. In shale with poor organic matter, clay has a promoting effect on the development of mesopores and macropores, because the mutual transformation between kaolinite, illite, and montmorillonite mixed layers is conducive to the formation of clay mineral pores. For coal rock, the pore characteristics of coal reservoirs are significantly better than shale, with organic matter pores as the dominant pore type and more developed micropores. The pore diameter is smaller, but the development density is higher than that of shale. This is due to the high TOC of coal rock, where organic matter accounts for the vast majority of the total volume, resulting in significantly smaller organic pore diameters than marine shale and transitional shale, with development at the micropore scale. However, it should be noted that, although both coal rock and shale are Type III organic matter, the content of vitreous groups in coal rock is higher, which is more conducive to pore formation. Therefore, the pore development in coal rock is denser than that in shale [26,40].

5.2. Differences in Adsorption Capacity Between Shale and Coal

The pore and fracture structure of coal rock and shale is the most direct factor controlling their adsorption capacity [42]. Previous studies have found a significant positive correlation between V_L and micro-PV, as well as SSA, but a weak correlation with mesopores and macropores [43,44]. The main influencing factor on the maximum adsorption capacity of methane in coal rock and shale samples is the degree of development of micropores, which can provide a large number of adsorption sites for methane and a place for gas adsorption and occurrence. As shown in Figure 13, the adsorption capacity of coal rock for methane in the study area is 5–30 times that of shale reservoirs, with a significant difference between the two. Thanks to the abundant organic matter micropores and large SSA in coal rock, the adsorption capacity of coal reservoirs is high, with a V_L ranging from 21.75 to 30.29 cm^3/g, average of 26.57 cm^3/g, and most exceeding 20 cm^3/g. In shale reservoirs, on the one hand, the PV and SSA are small, and on the other hand, the organic matter content is low. Although shale contains a large amount of clay minerals, these clay minerals are hydrophilic. In the presence of water molecules in the formation conditions, water molecules occupy the majority of the adsorption sites of clay materials, which makes the presence of clay minerals unable to greatly improve the adsorption capacity of the reservoir, resulting in a lower adsorption capacity of shale compared to coal reservoirs. The V_L of shale samples in the research area ranges from 1.04 to 2.71 cm^3/g, average of 1.92 cm^3/g. Comparing the shale and coal rock samples in the study area, the adsorption capacity of coal rock for methane is 8–29 times that of shale reservoirs, and there is a significant difference between the two (Figure 14). Thanks to the abundant organic matter micropores in coal and rock, which have a large SSA, micropores can provide a large number of adsorption sites for methane, providing a place for gas adsorption and occurrence.

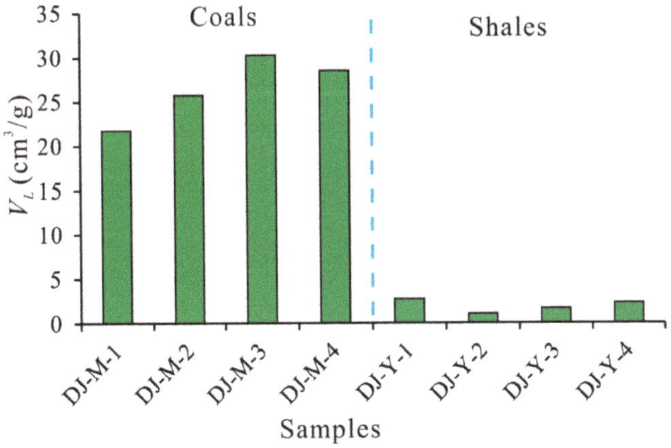

Figure 13. Comparison and analysis of the V_L of coal and shale.

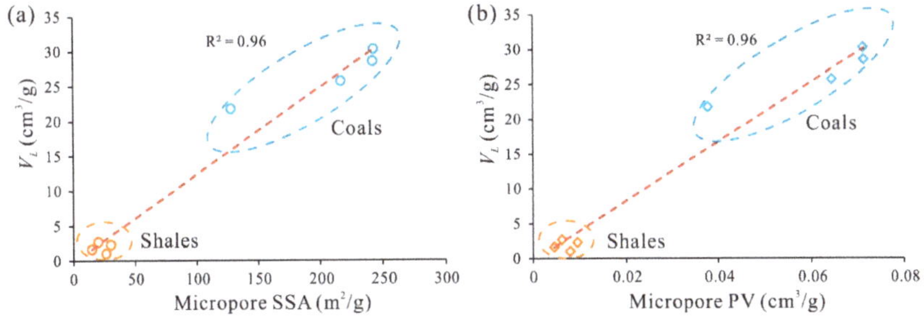

Figure 14. Relationship between the V_L and pore structure parameters of coals and shales. (**a**) V_L vs. micropore SSA, (**b**) V_L vs. micropore PV.

5.3. Comparative Analysis of Coal and Shale Reservoir Characteristics

By comparing the differences in pore structure between shale and coal rock, it can be found that shale has a richer variety of pore types and stronger heterogeneity than coal rock. Coal rock is mainly composed of organic matter components, with less ash content. The organic matter components are mainly composed of vitrinite (Table 1), which contains a large number of micropores below 2 nm. This is the main reason why the micropore content in coal rock is much higher than that in shale (Table 3). At the same time, the fewer inorganic matter components make the pore system more accessible, which is conducive to fluid flow in the pore network. The composition of shale material is more complex compared to coal rock (Table 4), and the internal connectivity of the organic matter pore network is good. However, due to the sealing of peripheral inorganic minerals, the organic matter pores lack pathways to connect with the outside, resulting in a decrease in the overall internal pore connectivity of shale. The research results on closed pores of over-mature marine shale indicate that closed pore systems are mostly related to organic matter pores, mainly formed by the closure of organic matter pore networks by brittle inorganic minerals [45]. Sun et al. (2020) reconstructed the three-dimensional pore network of over-mature marine shale using field emission scanning electron microscopy, which also showed that the organic matter pore network had good pore connectivity [46]. However, the sealing of inorganic minerals around the organic matter reduced the overall pore connectivity of the shale matrix. Mastalerz et al. (2012) also found that the pore connectivity of shale is worse than that of coal rock through small angle neutron scattering experiments after injecting deuterated methane into the sample [47].

The average TOC content of the Shanxi Formation shale reservoir in the research area is 2.66%, and the "honeycomb-like" organic matter pores are developed, which is conducive to methane diffusion and seepage. The shale reservoir has well-developed micropores, mesopores, and macropores, with mesopores being the main type (accounting for an average PV of 63.4%). The average PV of micropores and macropores is 31.9% and 4.7%, respectively. Micropores are the main contributors to its SSA (with an average proportion of 79%), while mesopores account for 21% of its SSA (Table 5). The pore types of coal rock are mainly micropores, followed by mesopores. Macropores are not developed, and the pore structure is worse, which is not conducive to methane permeation. However, the pore SSA of the coal rock reservoir in the research area is more than 10 times that of the Longmaxi Formation shale reservoir (Table 5) and, compared to shale, the micropore structures within coal matrices offer greater availability of sorption sites for methane molecules. Through comparative analysis, it was found that the pore connectivity of the Shanxi Formation shale reservoir in the study area is better than that of the coal rock reservoir in the study area. Within the pore size range of 1–100 nm, the proportion of connected pores in the former is more than twice that of the latter, and the difference between the two is more significant when the pore size is less than 40 nm. There is a natural barrier for seepage in coal rock reservoirs. Comprehensive analysis suggests that the cleavage and fractures of coal rock reservoirs are filled with minerals, and the pores are mainly micropores with poor connectivity, which is the direct cause of their extremely low porosity and permeability. The reservoir properties and pore structure of coal rock reservoirs are poor. Although they have developed cleavage and fractures, they are mostly filled with calcite and other materials. The cleavage fractures form weak planes that are prone to fracture and opening in coal rock. Therefore, the brittleness of coal rock reservoirs is similar to that of shale. Based on the latest understanding of coal rock reservoirs, the concept of ultra-large-volume fracturing should be adopted for deep coal reservoirs. The strength of various fracturing parameters should be higher than that of shale reservoirs to promote more cleavage opening and even cutting of coal matrix to form a complex and ultra-dense fracture network [43,48]. More small-sized proppants should be used to effectively support the multi-level and multi-scale fracture network, establishing an optimized matrix pore cleavage fracture network multi-level secondary microfine-fracture network multi-scale main fracture transport "channel" that conforms to reservoir properties, thereby reducing the difficulty of methane migration from micropores to fractures, increasing the range and depth of fracturing, and achieving a breakthrough in the "natural barrier" of seepage.

Table 5. Comprehensive characteristics of transitional shale and coal reservoirs in the Shanxi Formation.

Sedimentary Environment		**Marine–Continental Transitional Facies**	
Lithology		Coal	Shale
Organic geochemical characteristics	TOC (%)	66.8~86.1/74.22	0.56~6.62/1.86
	Type of organic matter	III	III
	Maceral	Mainly vitrinite group and inertinite Group	Mainly liptinite group and vitrinite group
Mineral composition	Clay content (%)	32~56/44	12~76/38.25
	Brittle minerals (%)	5~21/9.9	24~88/61.75

Table 5. *Cont.*

Sedimentary Environment		Marine–Continental Transitional Facies	
Characteristics of pore structure	Pore type	Organic matter pores	Organic pores and pore fissures in clay minerals
	Micropore structure	The average proportion of PV is 90.6%; The average proportion of SSA is 99.0%	The average proportion of PV is 31.9%; the average proportion of SSA is 79.0%
	Mesopore structure	The average proportion of PV is 6.8%	The average proportion of PV is 63.4%; the average proportion of SSA is 21.0%
	Macropore structure	The average proportion of PV is 2.6%	The average proportion of PV is 4.7%
	Full-scale PV distribution characteristics	The PV is mainly contributed by micropores, followed by mesopores and macropores	The PV is mainly contributed by mesopores, followed by micropores and macropores
	Full-scale SSA distribution characteristics	The SSA is mainly contributed by micropores	The SSA is mainly contributed by micropores, followed by mesopores
Gas-bearing characteristics	V_L (cm^3/g)	21.75~30.29 (26.57)	1.04~2.71 (1.92)
	P_L (MPa)	3.47~3.59 (3.52)	2.81~5.54 (3.75)

6. Conclusions

(1) The average TOC content of shale in the Shanxi Formation of the research area is 2.66%, and the microscopic components are mainly composed of vitrinite and inertinite. The mineral composition is high in quartz and clay minerals, with a small amount of pyrite and carbonate minerals present. The average TOC content of coal rock is 74.22%, and the maceral compositions are mainly composed of vitrinite, with clay minerals and organic matter carbon being the main components.

(2) Organic pores, inorganic pores, and microcracks are all developed in the shale of the Shanxi Formation in the research area. Compared with shale, coal rock mainly develops organic pores, intergranular pores, cleavage, and tectonic fractures, and observations reveal that cleats are mineral-filled. The PV of Shanxi Formation coal rock in the research area is mainly contributed by micropores, with less contribution from mesopores and macropores. The PV of shale is mainly contributed by mesopores, followed by micropores, and macropores are underdeveloped. The SSA of coal rock and shale pores is mainly contributed by micropores.

(3) The adsorption capacity of Shanxi Formation coal rock for methane in the study area is 8–29 times that of shale reservoirs, indicating that its adsorbed gas content is much higher than that of shale. Mainly because coal rock contains a large number of organic matter micropores with a large SSA, micropores can provide a large number of adsorption sites for methane, providing a place for gas adsorption and occurrence.

(4) Based on the understanding of reservoir characteristics and pore structures in the Shanxi Formation coal seams within the study area, a strategic shift in fracturing methodology was implemented. Through the application of targeted, ultra-large-scale volume fracturing technology, significant enhancement in CBM development outcomes has been achieved.

Author Contributions: Conceptualization, M.O. and T.W.; methodology, B.W. and T.W.; formal analysis, T.W. and Z.D.; investigation, W.T. and X.Y.; writing—original draft preparation, M.O. and T.W.; writing—review and editing, M.O. and T.W.; supervision, M.Y. and C.Y.; project administration, J.Y. and B.W.; funding acquisition, X.Y. and Z.D. All authors have read and agreed to the published version of the manuscript.

Funding: This research was funded by the Chongqing Key Project of Technological Innovation and Application Development hnology Project funder grant number CSTB2024TIAD-KPX0096, and the Chongqing Natural Science Foundation Innovation and Development Joint Funfunder grant number CSTB2023NSCQ-LZX0078, the China Petroleum Science and Technology Project funder grant number

2024DJ23, the China Petroleum Science and Technology Project funder grant number 2023ZZ18. And The APC was funded by 2024DJ23 and 2023ZZ18.

Data Availability Statement: The original contributions presented in this study are included in the article. Further inquiries can be directed to the corresponding authors.

Acknowledgments: This study was jointly supported by the China Petroleum Science and Technology Project (2024DJ23), the China Petroleum Science and Technology Project (2023ZZ18), the Chongqing Key Project of Technological Innovation and Application Development (CSTB2024TIAD-KPX0096), and the Chongqing Natural Science Foundation Innovation and Development Joint Fund (CSTB2023NSCQ-LZX0078).

Conflicts of Interest: Authors Mukun Ouyang, Bo Wang, Wei Tang, Maonan Yu, Chunli You, and Jianghai Yang are employed by the Chongqing Division of Southwest Oil and Gas Field Company; Authors Tao Wang and Ze Deng are employed by the PetroChina Research Institute of Petroleum Exploration & Development. The authors declare that the research was conducted in the absence of any commercial or financial relationships that could be construed as a potential conflict of interest.

References

1. Li, Y.; Zhou, D.H.; Wang, W.H.; Jiang, T.X.; Xue, Z.J. Development of Unconventional Gas and Technologies Adopted in China. *Energy Geosci.* **2020**, *1*, 55–68. [CrossRef]
2. Zou, C.; Zhang, G.; Yang, Z.; Tao, S.; Hou, L.; Zhu, R.; Yuan, X.; Ran, Q.; Li, D.; Wang, Z. Concepts, Characteristics, Potential and Technology of Unconventional Hydrocarbons: On Unconventional Petroleum Geology. *Pet. Explor. Dev.* **2013**, *40*, 413–428. [CrossRef]
3. Zou, C.; Yang, Z.; He, D.; Wei, Y.; Li, J.; Jia, A.; Chen, J.; Zhao, Q.; Li, Y.; Li, J.; et al. Theory, Technology and Prospects of Conventional and Unconventional Natural Gas. *Pet. Explor. Dev.* **2018**, *45*, 604–618. [CrossRef]
4. Li, Z.; Liu, D.; Cai, Y.; Ranjith, P.G.; Yao, Y. Multi-Scale Quantitative Characterization of 3-D Pore-Fracture Networks in Bituminous and Anthracite Coals Using FIB-SEM Tomography and X-Ray μ-CT. *Fuel* **2017**, *209*, 43–53. [CrossRef]
5. Zhou, S.; Liu, D.; Cai, Y.; Yao, Y.; Li, Z. 3D Characterization and Quantitative Evaluation of Pore-Fracture Networks of Two Chinese Coals Using FIB-SEM Tomography. *Int. J. Coal Geol.* **2017**, *174*, 41–54. [CrossRef]
6. Wei, Q.; Li, X.; Zhang, J.; Hu, B.; Zhu, W.; Liang, W.; Sun, K. Full-size pore structure characterization of deep-buried coals and its impact on methane adsorption capacity: A case study of the Shihezi Formation coals from the Panji Deep Area in Huainan Coalfield, Southern North China. *J. Pet. Sci. Eng.* **2019**, *173*, 975–989. [CrossRef]
7. Chen, S.; Tao, S.; Tang, D.; Xu, H.; Li, S.; Zhao, J.; Jiang, Q.; Yang, H. Pore Structure Characterization of Different Rank Coals Using N_2 and CO_2 Adsorption and Its Effect on CH4 Adsorption Capacity: A Case in Panguan Syncline, Western Guizhou, China. *Energy Fuels* **2017**, *31*, 6034–6044. [CrossRef]
8. Chalmers, G.R.L.; Bustin, R.M. The Organic Matter Distribution and Methane Capacity of the Lower Cretaceous Strata of Northeastern British Columbia, Canada. *Int. J. Coal Geol.* **2007**, *70*, 223–239. [CrossRef]
9. Wu, J.; Wang, H.; Shi, Z.; Wang, Q.; Zhao, Q.; Dong, D.; Li, S.; Liu, D.; Sun, S.; Qiu, Z. Favorable Lithofacies Types and Genesis of Marine-Continental Transitional Black Shale: A Case Study of Permian Shanxi Formation in the Eastern Margin of Ordos Basin, NW China. *Pet. Explor. Dev.* **2021**, *48*, 1315–1328. [CrossRef]
10. Song, X.; Lu, X.; Shen, Y.; Guo, S.; Guan, Y. A modified supercritical Dubinin–Radushkevich model for the accurate estimation of high-pressure methane adsorption on shales. *Int. J. Coal Geol* **2018**, *193*, 1–15. [CrossRef]
11. Shang, F.; Zhu, Y.; Hu, Q.; Zhu, Y.; Wang, Y.; Du, M.; Liu, R.; Han, Y. Characterization of methane adsorption on shale of a complex tectonic area in Northeast Guizhou, China: Experimental results and geological significance. *J. Nat. Gas Sci. Eng.* **2020**, *84*, 103676. [CrossRef]
12. Pszonka, J.; Schulz, B. SEM Automated Mineralogy applied for the quantification of mineral and textural sorting in submarine sediment gravity flows. *Gospod. Surowcami Miner. Miner. Resour. Manag.* **2022**, *38*, 105–131.
13. Zheng, H.; Yang, F.; Guo, Q.; Pan, S.; Jiang, S.; Wang, H. Multi-scale pore structure, pore network and pore connectivity of tight shale oil reservoir from Triassic Yanchang Formation, Ordos Basin. *J. Nat. Gas Sci. Eng.* **2022**, *212*, 110283. [CrossRef]
14. Yan, J.; Meng, Z.; Zhang, K.; Yao, H.; Hao, H. Pore distribution characteristics of various rank coals matrix and their influences on gas adsorption. *J. Pet. Sci. Eng.* **2020**, *189*, 107401. [CrossRef]
15. He, Q.; Chen, S.; Li, S.; Guo, B.; Lu, J.; Li, Y.; Li, X.; Zhao, L.; Ma, Z. Organic geochemical characteristics and hydrocarbon generation mechanism of marine-continental transitional organic-rich shale: A case study from the Shanxi Formation in the eastern margin of the Ordos Basin. *J. Pet. Sci. Eng.* **2022**, *219*, 11116. [CrossRef]

16. Li, Y.; Tang, D.; Wu, P.; Niu, X.; Wang, K.; Qiao, P.; Wang, Z. Continuous unconventional natural gas accumulations of Carboniferous-Permian coal-bearing strata in the Linxing area, northeastern Ordos Basin, China. *J. Nat. Gas Sci. Eng.* **2016**, *36*, 314–327. [CrossRef]
17. Ahmadi, S.; Khormali, A.; Razmjooie, A. Experimental investigation on separation of water in crude oil emulsions using an oil-soluble demulsifier. *Iran. J. Chem. Chem. Eng.* **2023**, *42*, 2332–2343.
18. Kuang, L.; Dong, D.; He, W.; Wen, S.; Sun, S.; Li, S.; Qiu, Z.; Liao, X.; Li, Y.; Wu, J.; et al. Geological characteristics and development potential of transitional shale gas in the east margin of the Ordos Basin, NW China. *Pet. Explor. Dev.* **2020**, *47*, 435–446. [CrossRef]
19. Niu, X.; Fan, L.; Yan, X.; Zhao, G.; Zhang, H.; Jing, X.; Zhang, M. Enrichment conditions and resource potential of coal-rock gas in Ordos Basin, NW China. *Pet. Explor. Dev.* **2024**, *51*, 972–985. [CrossRef]
20. Wang, Y.; Liu, L.; Zheng, S.; Luo, Z.; Sheng, Y.; Wang, X. Full-scale pore structure and its controlling factors of the Wufeng-Longmaxi shale, southern Sichuan Basin, China: Implications for pore evolution of highly overmature marine shale. *J. Nat. Gas Sci. Eng.* **2019**, *67*, 134–146. [CrossRef]
21. Wu, Z.; He, S.; Li, X.; Liu, X.; Zhai, G.; Huang, Z.; Yang, W. Comparative study on pore structure characteristics of marine and transitional facies shales: A case study of the Upper Permian Longtan Formation and Dalong Formation in the Lower Yangtze area, south China. *J. Pet. Sci. Eng.* **2022**, *215*, 110578. [CrossRef]
22. Wang, J.Y.; Guo, S.B. Study on the Relationship between Hydrocarbon Generation and Pore Evolution in Continental Shale from the Ordos Basin, China. *Pet. Sci.* **2021**, *18*, 1305–1322. [CrossRef]
23. Shan, C.; Zhao, W.; Wang, F.; Zhang, K.; Feng, Z.; Guo, L.; Ma, X.; Liao, T. Nanoscale Pore Structure Heterogeneity and Its Quantitative Characterization in Chang7 Lacustrine Shale of the Southeastern Ordos Basin, China. *J. Pet. Sci. Eng.* **2020**, *187*, 106754. [CrossRef]
24. Liang, L.; Xiong, J.; Liu, X. An Investigation of the Fractal Characteristics of the Upper Ordovician Wufeng Formation Shale Using Nitrogen Adsorption Analysis. *J. Nat. Gas Sci. Eng.* **2015**, *27*, 402–409. [CrossRef]
25. Thommes, M.; Kaneko, K.; Neimark, A.V.; Olivier, J.P.; Rodriguez-Reinoso, F.; Rouquerol, J.; Sing, K.S.W. Physisorption of Gases, with Special Reference to the Evaluation of Surface Area and Pore Size Distribution (IUPAC Technical Report). *Pure Appl. Chem.* **2015**, *87*, 1051–1069. [CrossRef]
26. Loucks, R.G.; Reed, R.M.; Ruppel, S.C.; Hammes, U. Spectrum of Pore Types and Networks in Mudrocks and a Descriptive Classification for Matrix-Related Mudrock Pores. *AAPG Bull.* **2012**, *96*, 1071–1098. [CrossRef]
27. Yang, R.; He, S.; Yi, J.; Hu, Q. Nano-Scale Pore Structure and Fractal Dimension of Organic-Rich Wufeng-Longmaxi Shale from Jiaoshiba Area, Sichuan Basin: Investigations Using FE-SEM, Gas Adsorption and Helium Pycnometry. *Mar. Pet. Geol.* **2016**, *70*, 27–45. [CrossRef]
28. *GB/T 21650.3-2011*; Pore Size Distribution and Porosity of Solid Materials by Mercury Porosimetry and Gas Adsorption-Part 3: Analysis of Micropores by Gas Adsorption. China Quality and Standards Publishing & Media Co., Ltd.: Beijing, China, 2011.
29. *GB/T 21650.2-2008*; Pore Size Distribution and Porosity of Solid Materials by Mercury Porosimetry and Gas Adsorption-Part 2: Analysis of Mesopores and Macropores by Gas Adsorption. China Quality and Standards Publishing & Media Co., Ltd.: Beijing, China, 2008.
30. *GB/T 21650.1-2008*; Pore Size Distribution and Porosity of Solid Materials by Mercury Porosimetry and Gas Adsorption-Part 1: Mercury Porosimetry. China Quality and Standards Publishing & Media Co., Ltd.: Beijing, China, 2008.
31. Zhou, S.; Yan, G.; Xue, H.; Guo, W.; Li, X. 2D and 3D Nanopore Characterization of Gas Shale in Longmaxi Formation Based on FIB-SEM. *Mar. Pet. Geol.* **2016**, *73*, 174–180. [CrossRef]
32. Evans, R.; Marconi, U.M.B.; Tarazona, P. Capillary condensation and adsorption in cylindrical and slit-like pores. *J. Chem. Soc.* **1986**, *82*, 1763–1787. [CrossRef]
33. Sing, K.S.W. Reporting physisorption data for gas/solid systems with special reference to the determination of surface area and porosity (recommendations 1984). *Pure Appl. Chem.* **1985**, *57*, 603–619. [CrossRef]
34. Wang, T.; Tian, F.; Deng, Z.; Hu, H. The Characteristic Development of Micropores in Deep Coal and Its Relationship with Adsorption Capacity on the Eastern Margin of the Ordos Basin, China. *Minerals* **2023**, *13*, 302. [CrossRef]
35. IUPAC (International Union of Pure and Applied Chemistry). Physical chemistry division commission on colloid and surface chemistry, subcommittee on characterization of porous solids: Recommendations for the characterization of porous solids (technical report). *Pure Appl. Chem.* **1994**, *66*, 1739–1758. [CrossRef]
36. Gelb, L.D.; Gubbins, K.E.; Radhakrishnan, R.; Sliwinska-Bartkowiak, M. Phase separation in confined systems. *Rep. Prog. Phys.* **1999**, *62*, 1573–1659. [CrossRef]
37. El-Merraoui, M.; Aoshima, M.; Kaneko, K. Micropore size distribution of activated carbon fiber using the density functional theory and other methods. *Langmuir* **2000**, *16*, 4300–4304. [CrossRef]
38. Washburn, E.W. Note on a method of determining the distribution of pore sizes in a porous material. *Proc. Natl. Acad. Sci. USA* **1921**, *15*, 115–116. [CrossRef]

39. Wang, T.; Deng, Z.; Hu, H.; Wang, H.; Jiang, Z.; Wang, D. Study on pore structure and multifractal characteristics of middle- and high-rank coals based on gas adsorption method: A case study of Benxi Formation in the eastern margin of Ordos Basin. *Energy Fuels* **2024**, *38*, 4102–4121. [CrossRef]
40. Cao, T.; Song, Z.; Wang, S.; Xia, J. Characterization of Pore Structure and Fractal Dimension of Paleozoic Shales from the Northeastern Sichuan Basin, China. *J. Nat. Gas Sci. Eng.* **2016**, *35*, 882–895. [CrossRef]
41. Wang, G.; Zhang, Q.; Zhu, R.; Tang, X.; Liu, K.; Jin, Z. Geological Controls on the Pore System of Lacustrine Unconventional Shale Reservoirs: The Triassic Chang 7 Member in the Ordos Basin, China. *Geoenergy Sci. Eng.* **2023**, *221*, 111139. [CrossRef]
42. He, H.; Liu, P.; Xu, L.; Hao, S.; Qiu, X.; Shan, C.; Zhou, Y. Pore Structure Representations Based on Nitrogen Adsorption Experiments and an FHH Fractal Model: Case Study of the Block Z Shales in the Ordos Basin, China. *J. Pet. Sci. Eng.* **2021**, *203*, 108661. [CrossRef]
43. Wang, T.; Deng, Z.; Hu, H.; Tian, F.; Ding, R.; Zhang, T.; Ma, Z.; Hou, S.; Li, X.; Dai, R.; et al. Pore structure and fractal characteristics of transitional shales with different lithofacies from the eastern margin of the Ordos Basin. *Energy Sci Eng.* **2023**, *11*, 3979–4000. [CrossRef]
44. Liu, J.; Lu, D.; Li, P. Nano-Scale Dual-Pore-Shape Structure and Fractal Characteristics of Transitional Facies Shale Matrix. *J. Nat. Gas Sci. Eng.* **2019**, *68*, 102907. [CrossRef]
45. Deng, Z.; Wang, H.; Jiang, Z.; Sun, M.; Wu, Z.; Xu, Y.; Ke, Y. Pore structure differences between shale and coal and their gas migration mechanisms. *Nat. Gas Ind.* **2022**, *42*, 37–49.
46. Sun, M.; Zhang, L.; Hu, Q.; Pan, Z.; Yu, B.; Sun, L.; Bai, L.; Fu, H.; Zhang, Y.; Zhang, C.; et al. Multiscale connectivity characterization of marine shales in southern China by fluid intrusion, small-angle neutron scattering (SANS), and FIB-SEM. *Mar. Pet. Geol.* **2020**, *112*, 104101. [CrossRef]
47. Mastalerz, M.; He, L.; Melnichenko, Y.; Rupp, J. Porosity of coal and shale: Insights from gas adsorption and SANS/USANS techniques. *Energy Fuels* **2012**, *26*, 5109–5120. [CrossRef]
48. Hu, C.; Tan, J.; Lu, Q.; Zhang, Y. Evolution of organic pores in Permian low maturity shales from the Dalong Formation in the Sichuan Basin: Insights from a thermal simulation experiment. *Gas Sci. Eng.* **2024**, *121*, 205166. [CrossRef]

processes

MDPI

Article

Characteristics of Microorganisms and Origins of Organic Matter in Permian Shale in Northwestern Sichuan Basin, South China

Yuying Zhang [1,2,3,*], Baojian Shen [1,2], Bo Gao [1,2], Dongjun Feng [1,2], Pengwei Wang [1,2], Min Li [1,2], Yifei Li [1,2,3] and Yang Liu [1,2,3]

1 State Key Laboratory of Shale Oil and Gas Enrichment Mechanisms and Efficient Development, Beijing 100083, China; shenbj.syky@sinopec.com (B.S.); gaob.syky@sinopec.com (B.G.); fengdj.syky@sinopec.com (D.F.); wangpw.syky@sinopec.com (P.W.); limin.syky@sinopec.com (M.L.); 2578233032@qq.com (Y.L.); 2578159535@qq.com (Y.L.)
2 Sinopec Key Laboratory of Shale Oil/Gas Exploration and Production Technology, Beijing 100083, China
3 School of Resources and Civil Engineering, Northeastern University, Shenyang 110819, China
* Correspondence: zhangyuying@mail.neu.edu.cn

Abstract

Permian shale gas, a resource-rich energy source, has garnered significant attention in recent years regarding its organic matter enrichment characteristics. This study conducted detailed observations via scanning electron microscopy (SEM) and optical microscopy to clarify the differences in the types and assemblages of hydrocarbon-generating organisms across Permian shale formations in Northwestern Sichuan, as well as to determine the characteristics of organic matter sources. The types and combinations of hydrocarbon-generating organisms in the Gufeng Formation, Wujiaping Formation, and Dalong Formation in Northwestern Sichuan are systematically summarized. Based on this information, the primary sources of organic matter in the Permian shale were analyzed. Hydrocarbon-generating organisms in the Permian shales of the study area are predominantly acritarchs (a type of planktonic algae), followed by higher plants and green algae. In the Gufeng Formation, acritarchs constituted the vast majority of hydrocarbon-generating organisms, with smaller amounts of higher plants and green algae. At the bottom of the Wujiaping Formation, the relative acritarch content decreases significantly, while that of higher plants substantially increases. In the Dalong Formation, acritarchs regain dominance, and higher plants decline, resembling the Gufeng Formation in microorganism composition. The relative content of green algae shows minimal variation across all layers. Overall, the organic matter sources of Permian shale in the study area were mainly acritarchs (derived from planktonic algae), followed by green algae, and terrestrial higher plants. During the Gufeng Formation period, the sea level was relatively high. The Kaijiang–Liangping Trough in Northwestern Sichuan was generally a siliceous deep shelf. The main source of organic matter was aquatic planktonic algae, containing a small amount of terrigenous input. At the bottom of the Wujiaping Formation, the sea level was relatively low, resulting in the overall coastal marsh environment of the Kaijiang–Liangping Trough, which was characterized by mixed organic matter sources, due to an increase in terrigenous organic matter content. The sedimentary environment and organic matter sources of the Dalong Formation were similar to those of the Gufeng Formation. This research can provide a theoretical basis for exploration and development of Permian shale gas.

Processes **2025**, *13*, 2080 https://doi.org/10.3390/pr13072080

Keywords: hydrocarbon-generating organism; organic source; shale; Permian; Northwestern Sichuan Basin

1. Introduction

The Permian shale gas resources in the Northwestern Sichuan Basin exhibit enormous potential as a clean energy source, with significant progress in exploration and development achieved in recent years. The Dalong and Gufeng Formations are core exploration targets [1–4]. Research on the origins of organic matter in shale is critical for shale oil–gas exploration as it clarifies the types and assemblages of hydrocarbon-generating organisms, which could help reveal the hydrocarbon generation potential of source rocks and aid in resource evaluation [5–10]. The organic matter sources of Permian shales were diverse and were mainly influenced by the paleoenvironment, paleoclimate, and biological communities during that time. The planktonic algae that flourished in the ocean (such as cyanobacteria, green algae, etc.) were important sources that were especially well-preserved in oxygen-deficient environments, forming the high organic carbon content in Permian shales [11]. Anaerobic bacteria decompose organic matter in an anoxic environment, contributing to the organic carbon and possibly forming sulfides (such as pyrite). The organic remains of other tiny marine organisms (e.g., foraminifera and radiolarians) could also be components of the organic matter in Permian shales [12]. In addition, the debris of terrestrial higher plants)—especially flourishing gymnosperms and ferns—in the Permian period (particularly the Late Permian) could have been transported into the ocean through rivers, contributing organic matter to marine shale [13]. However, the contribution of higher plants was more significant in the shale close to the continental margin. The spores and pollen from higher plants might have spread very far from land, carried by wind or water, mixing into marine shale and contributing to the organic matter [11]. Researching the types and assemblage characteristics (biological facies) of the hydrocarbon-generating organisms in shale is important for classifying the organic facies of shale and accurately evaluating the quality of potential source rocks for shale gas.

Previous studies indicate that the strata of the Gufeng Formation primarily consist of carbonaceous siliceous mudstone deposited in the basin, with localized development of the outer slope through the deposition of organic-rich argillaceous limestone. The organic matter in the Gufeng Formation mainly originates from primary aquatic organisms, such as plankton and algae, with minor contributions from terrestrial higher plants [14]. The bottom of the Wujiaping Formation consists of tidal-flat argillaceous siltstone interbedded with coal seams, with organic matter derived from marine plankton and algae [12,13]. In the Northwestern Sichuan Basin, the hydrocarbon-generating organisms of the Dalong Formation were predominantly primary aquatic organisms, such as algae and plankton [15,16]. A microscopic analysis of Well Heba-1 revealed that sapropelinite dominated the organic macerals (78.5–85.5%), primarily composed of alginite. Scanning electron microscopy (SEM) also identified the presence of carbonaceous particles—structurally similar to sulfur bacteria—in the Dalong Formation, confirming algae as a key source of organic matter [17]. Most of the algae in Permian shales exist in the form of acritarchs, with the organic matter wall cysts retaining their appearance [11–13].

The Upper Permian shale in the Northwestern Sichuan Basin is characterized by substantial thickness, diverse lithofacies, and multiple sedimentary facies, with Type I-II organic matter [18–20]. These variations imply significant changes in the environment during deposition, leading to heterogeneous organic matter sources and hydrocarbon-generating organisms. However, existing research on the types and assemblages of hydrocarbon-generating organisms of the Upper Permian shale is inadequate, which is unfavorable for the study of organic matter origins. Therefore, it is necessary to analyze the types and distribution of hydrocarbon-generating organisms in the Permian shale of the northwestern Sichuan Basin in detail, to clarify the differences in biogenic assemblages across formations and estimate organic matter enrichment mechanisms.

In this study, hydrocarbon-generating organisms in the Gufeng Formation, at the bottom of the Wujiaping Formation, and in the Dalong Formation have been classified and comprehensively characterized based on SEM and microscopy data. The primary sources of organic matter in these layers are also discussed. This work provides a theoretical foundation for evaluating the potential of Permian shale gas and predicting favorable exploration zones.

2. Geological Setting

The study area was located at the northwestern margin of the Yangtze Plate. During the Permian, this area was influenced by the subduction of the Paleo-Tethys Ocean and the closure of the Qinling Ocean, resulting in alternating extensional and compressional regimes [21,22]. In the Middle–Late Permian, regional compression led to the formation of carbonate platforms and transitional facies, alongside uplift and faulting activities in local areas, which resulted from the impending collision between the Yangtze and North China Plates (Figure 1) [23–25]. The development of Permian marine shale in the north of the Sichuan Basin was controlled by the Kaijiang–Liangping Trough, which was distributed in a banded pattern in the northern part of the Sichuan Basin. The depth of the Kaijiang–Liangping Trough was relatively large at the northern end, ranging from about 100 to 550 m in depth. This trough became shallow at the southern end, eventually disappearing. The morphology of the platform edge could be divided into the faulted steep slope type and the gentle slope type, which played important controlling roles in the formation of large shale gas reservoirs around the trough. The evolutionary history of the Kaijiang–Liangping Trough can be divided into three stages: (a) during the late Middle Permian, the trough was in the embryonic stage, beginning to form under the influence of regional tensile stress, mainly with lateral migration, It presents a northwest–southeast spread pattern, wider in the northwest and narrower in the southeast; (b) during the Late Permian, the trough went into its peak period of the development in the Changxing period, controlled by the expansion of the ocean and basement fault activities, with the internal development of deep-water siliceous limestone and radiolarian siliceous rocks, surrounded by shallow carbonate platform facies to form a complete deep-to-shallow water sedimentary sequence; (3) during the Early Triassic, the trough was in its extinction period—the closure of the ocean led to the gradual filling of the trough with sediments and its eventual extinction [26].

The Chejiaba outcrop is located in the Xibeixiang, about 12 km northwest of Guangyuan City, Sichuan Province. From bottom to top, this crop exposes the Middle Permian Gufeng Formation, the Upper Permian Wujiaping and Dalong Formations, and the Lower Triassic Feixianguan Formation. The strata of the Gufeng Formation are about 18 m thick, in which the lithology is composed of interbedded siliceous rocks and mudstones, with bioclastic limestone as an interlayer, indicating that the deep shelf environment was dominant during this period (Figure 2). The shale in the Gufeng Formation had high organic matter abundance (TOC: 3–10%) dominated by Type II kerogen [27,28]. The bottom of the Wujiaping Formation comprises yellowish or brown mudstone, grey calcareous shale with limestone as an interlayer (~4 m thick) (Figure 2). This member of shale is also known as the Wangpo Shale, with faveolate coal layers in some areas, suggesting a marsh lagoon in a bay environment. The upper member of the Wangpo Shale in the

Wujiaping Formation is mainly grey limestone, indicative of an open carbonate platform environment. The organic matter content of the Wujiaping Formation is relatively low (TOC: 0.5–1.5%) and is primarily Type III kerogen [29]. The Dalong Formation is composed of black siliceous rocks, mudstones, and a thin layer of limestone (Figure 2). According to the types and assemblages of lithologies, this formation could be divided into the first and second members of the Dalong Formation from bottom to top. The first member of the Dalong Formation is about 20 m thick from the bottom, where the rocks are mainly composed of black, thin-layer siliceous rock and mudstone, with marl interbedded and some volcanic ash layers (Figure 2). The whole sedimentary environment was a deep shelf environment near the continental slope. The second member of the Dalong Formation is composed of dark gray argillaceous limestone interbedded with mudstone (Figure 2), meaning that the sedimentary environment during this stage was a slope environment at the edge of the carbonate platform, characterized by relatively deep, weak hydrodynamic and dysoxic water. The shale in the Dalong Formation has high organic matter content (TOC: 0.6–16.6%, avg. 4.58%) and enriched Type I–II kerogen [30–32].

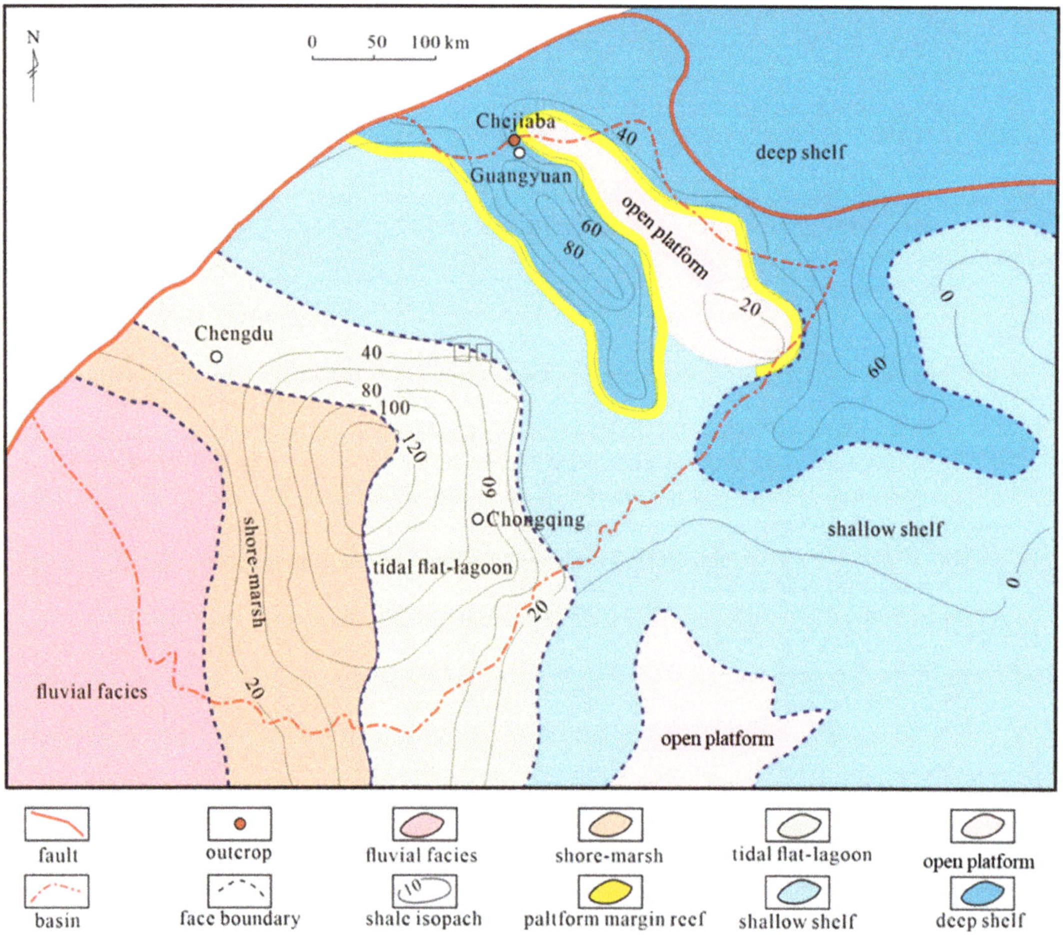

Figure 1. Distribution of sedimentary facies in the Wujiaping Formation in the Sichuan Basin and its surrounding areas [25].

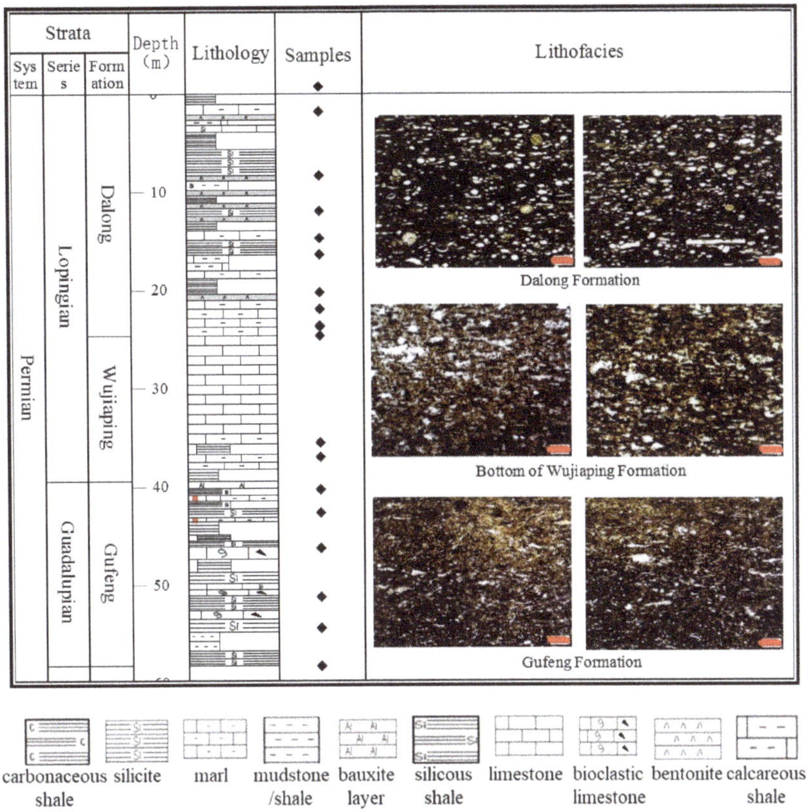

Figure 2. Stratigraphic column and lithofacies distribution in the Chejiaba Outcrop, northwestern Sichuan (the red scale bars in the photographs are 200 μm).

3. Samples and Methods

A total of 20 samples were collected from the Gufeng Formation, the bottom of the Wujiaping Formation (~4 m from the bottom), and the Dalong Formation in the Chejiaba outcrop, Guangyuan area, northwestern Sichuan Basin (the locations of the 20 samples are marked in Figure 2). SEM was conducted at the Shenyang Geological Survey Center, China Geological Survey, using an SEM 3200, to observe microfossil types and quantify their abundance. The samples were systematically prepared for observation using the SEM. First, all samples were cut into equal squares of 10 mm × 10 mm using a precision cutting instrument to ensure consistency in the subsequent observation area. Subsequently, the surface of all samples was embedded with epoxy resin, and then mechanical polishing was carried out. To enhance conductivity, a gold coating with a thickness of 20 nm was sputtered and plated on the surface of all samples. In this study, the observation areas of each sample were the same (within 10 mm × 10 mm), and the quantitative statistical analysis of biological fossils in the SEM images was conducted using the standardized point counting method. The relative content of each microorganic fossil was calculated using the standardized formula $P_i = (N_i/N_t) \times 100\%$, where P_i is the percentage of type i fossil of the total fossils, N_i is the number of type i fossils, and N_t is the number of all observed organisms.

Palynological analysis involved extracting 50 g of shale to isolate sporomorphs (including algae, fungi, and acritarchs), structured organic matter, and amorphous organic matter. These components were identified and quantified under a microscope at the Department of Geology, Northeastern University. All samples were crushed to soybean size, and 50 g of each sample was added to a 1000 mL beaker. Concentrated hydrochloric acid was then slowly poured into each beaker while gently stirring with a glass rod. The mass ratio of hydrochloric acid to the sample was approximately 3:1. The samples were left to react for 24 h, gently stirring every 6 h until no further reaction occurred. Subsequently, the beakers were filled with distilled water and were left for 4–6 h, after which the upper clear liquid was poured out. This step was repeated 3–4 times until the pH value of the liquid in the beaker was approximately 7. The samples were then left to dry. Subsequently, hydrofluoric acid with a 3:1 acid-to-sample mass ratio was slowly poured into the beaker while gently stirring with a glass rod. After 24 h of standing reaction, samples were washed with distilled water 4–5 times. If the sample was not completely dissolved, additional hydrofluoric acid was added, and the sample washing steps were repeated until the samples were completely dissolved. The remaining residue of each sample was poured into a 50 mL centrifuge tube for centrifuging at a speed of 2000 revolutions per second for 10 min, and then the upper clear liquid was dumped. After centrifugation, the centrifuge tubes of each sample were inverted on a table for 24 h. When the liquid in the tubes was drained, the heavy liquid with a density between 2.0 g/cm^3 and 2.2 g/cm^3 was added to the tubes and stirred evenly with a glass rod. The samples in the centrifuge tubes with the heavy liquid were centrifuged for 20 min, after which the upper turbid liquid was poured into a 50 mL glass beaker. Then, 5% acetic acid solution was added to the glass beakers. The mixture in the glass beakers was left to stand for 48 h. When the deposits had completely settled, the upper clear liquid was poured out. The remaining turbid liquid was poured into a 10 mL centrifuge tube, and distilled water was added. The remaining turbid liquid with distilled water was centrifuged for 10 min. The centrifuging step was repeated 3–4 times until all the sediments in the small beaker were centrifuged. Finally, a small amount of the residue from each sample in the centrifuge tubes was taken to make temporary slides. The slides of the samples were observed using a Zeiss Primotech optical microscope at Northeastern University for biological classification and statistical analysis.

4. Results and Discussion

4.1. Types of Hydrocarbon-Generating Organisms

The microorganisms, such as bacteria, algae, fungi, and other types of organisms, lived during historical geological periods and were widely distributed in the oceans and lakes. They underwent complex biochemical and chemical changes during the thermal evolution stage [12]. Putrefaction and humification processes occurred during thermal evolution, converting the microorganisms into organic matter (kerogen), which served as the material basis for the formation of high-quality marine source rocks and could be the main original parent material for the formation of oil and natural gas. These organisms are microscopic and widely distributed in sedimentary rocks.

Alginite constitutes the most prevalent and widely distributed hydrocarbon-generating organism in the Permian shales of the northwestern Sichuan Basin. This organism can be further classified as acritarchs and green algae based on morphological characteristics. Acritarchs are a group of microfossils with multiple sources and various genetic relationships. Previous studies suggest that the main components of the cell walls of acritarchs are cellulose and pectin, similar to those found in the cell walls of planktonic algae [33]. Most of the algae in the Permian shale exist in the form of organic-walled cysts; therefore, the acritarchs may be organs or certain parts of planktonic algae. The types of acritarchs in this study have been classified into seven genera based on morphological dimensions [34]. Common genera such as *Proterocladus, Siphonophycus, Polyedryxium,* and *Veryhachium* are distributed across all stratigraphic units, while others (e.g., *Cymatiosphaera, Lophosphaeridium, Dictyotidium*) occur sporadically. These genera are described as follows:

Proterocladus (Figure 3a–c): the unbranched or occasionally branched filamentous structures could be observed under SEM with distinct septa. Their cell walls are thin and smooth, varying in length (typically longer than they are wide). The apical morphology ranges from simple rounded to capitulum structures.

Siphonophycus (Figure 3d–f): it has a tubular and filamentous form with hemispherical terminations, but without septa or branches. This kind of acritarch often experienced deformation, such as twisting or folding.

Polyedryxium (Figure 3g,h): it is a type of slightly translucent and soft microplankton with polygonal membrane structures. The membrane surface consists of uneven organic layers, forming internal cavities of varying sizes. The single-layered membrane ranges from 10 to 150 μm in diameter and 0.5 to 3.0 μm in thickness, with a smooth or slightly undulating surface.

Veryhachium (Figure 3j): this kind of acritarch is characterized by triangular to quadrangular membrane outlines, with smooth to slightly rough surfaces. The membrane shell is a single-layer structure of the cell wall, in the shape of a cushion without striations or secondary spines.

Cymatiosphaera (Figure 3k–m): these fossils are spherical, 10–150 μm in diameter. They show spherical membranes with polygonal pores (3–6 μm in diameter) and irregular tubercles at pore bases.

Lophosphaeridium (Figure 3n): the membrane shell is spherical (approximately 10 μm in diameter), with small spinose tubercles of uniform thickness distributed in regular rows across the surface. The tubercles are approximately 0.2 μm in height.

Dictyotidium (Figure 3o): they show spherical membranes (10–200 μm in diameter) with polygonal reticulate patterns (mesh diameter: 0.5–3 μm). There are slight differences in appearance among different species.

A large number of spore-like microfossils were observed in Permian shales from the northwestern Sichuan region. These spore-like microfossils were classified as cryptospores and were assigned to the genus *Adinosporus* under the Chlorophyta (green algae) through comparative identification. The SEM images reveal that the cryptospores were well-preserved due to their rigid outer walls (Figure 4). This kind of cryptospore lacks the trilete marks (three radiating fissures) that are characteristic of typical spores [35,36]. These cryptospores typically consist of one or more subspherical spore-like cells, appearing as single spores, tetrads, or flower-like clusters under the microscope. Their spore walls exhibit a single-layered or multi-layered structure and may display pronounced folding (Figure 4).

In addition to alginite, fragments of higher plants were identified in the samples. These fragments retain partial cellular structures, existing as diverse morphologies (commonly arranged in filamentous patterns), which could be easily recognized under a microscope (Figure 5). Abundant plant debris was observed at the bottom of the Wujiaping Formation in the Chejiaba outcrop.

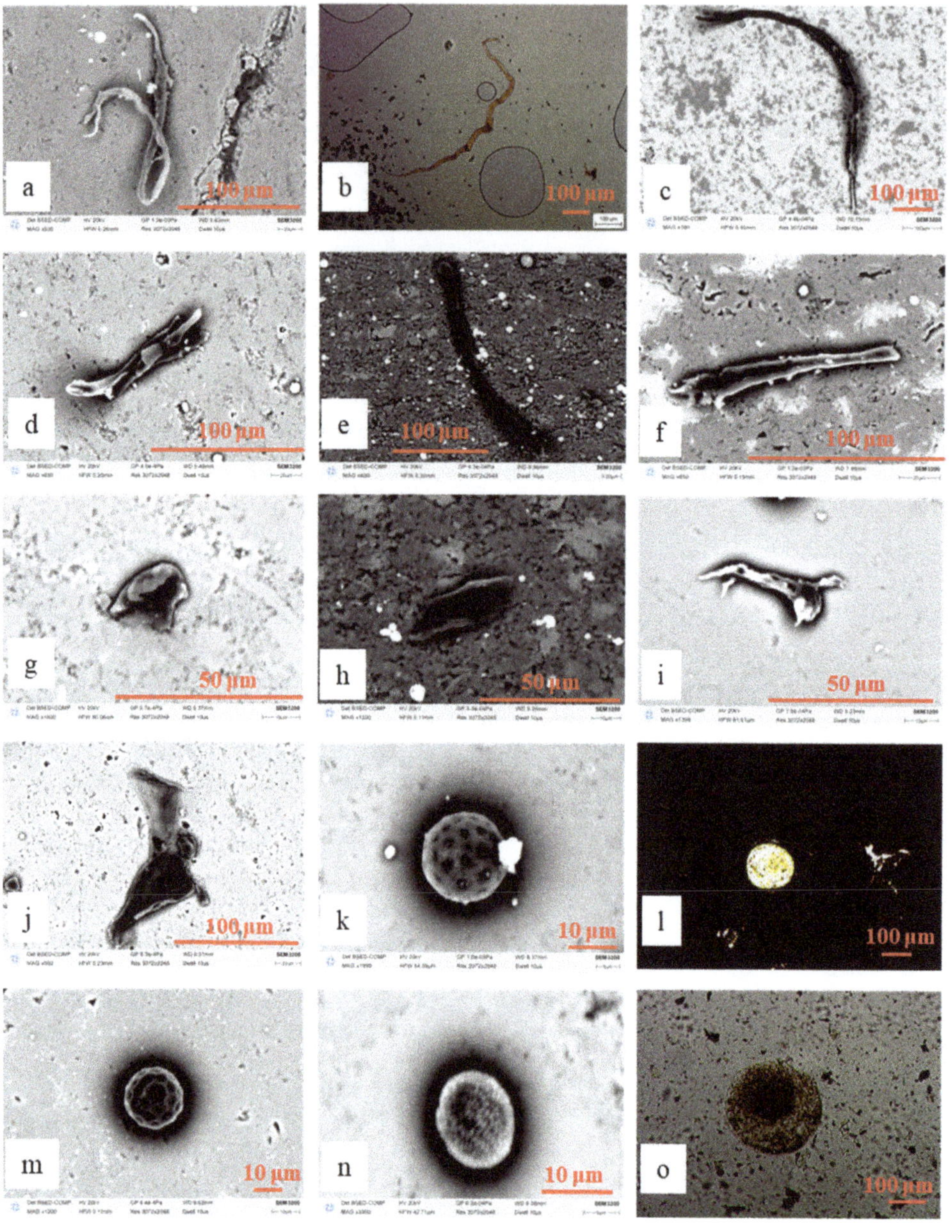

Figure 3. Typical photos of acritarchs in the Permian shale of the Chejiaba outcrop. (**a**) *Proterocladus*, bottom of Gufeng Formation; (**b**) *Proterocladus*, middle of Dalong Formation; (**c**) *Proterocladus*, upper of Dalong Formation; (**d**) *Siphonophycus*, upper of Gufeng Formation; (**e**) *Siphonophycus*, lower of Dalong Formation; (**f**) *Siphonophycus*, upper of Dalong Formation; (**g**) *Polyedryxium*, upper of Gufeng Formation; (**h**) *Polyedryxium*, lower of Dalong Formation; (**i**) *Veryhachium*, middle of Gufeng Formation; (**j**) *Veryhachium*, upper of Gufeng Formation; (**k**) *Cymatiosphaera*, bottom of Gufeng Formation; (**l**) *Cymatiosphaera*, middle of Gufeng Formation; (**m**) *Cymatiosphaera*, upper of Gufeng Formation; (**n**) *Lophosphaeridium*, bottom of Gufeng Formation; (**o**) *Dictyotidium*, lower of Dalong Formation.

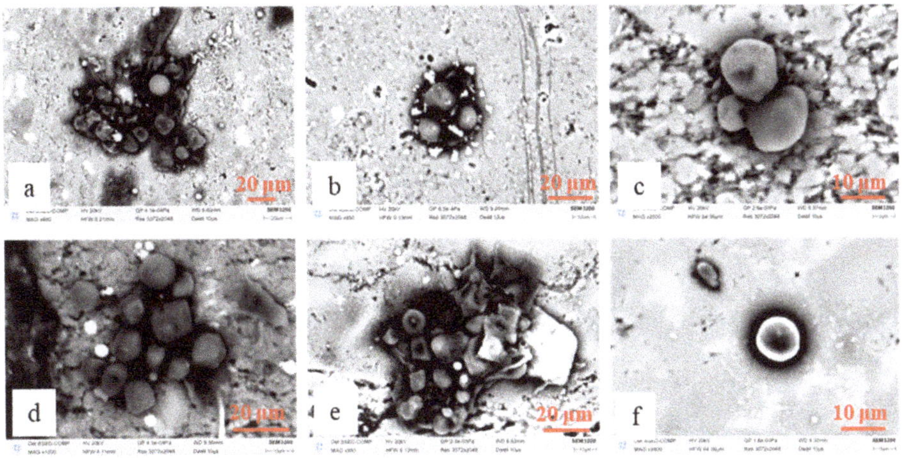

Figure 4. Typical photos of cryptospores in the Permian shale of the Chejiaba outcrop. (**a**) Middle of Gufeng Formation; (**b**) upper of Gufeng Formation; (**c**) middle of Wujiaping Formation; (**d**) lower of Dalong Formation; (**e**) middle of Dalong Formation; (**f**) upper of Dalong Formation.

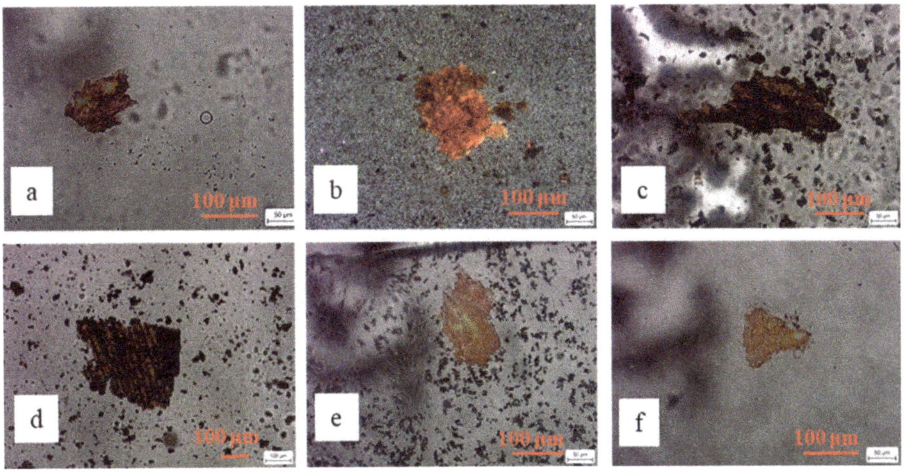

Figure 5. Typical photos of plants in the Permian shale of the Chejiaba outcrop. (**a**) Top of Gufeng Formation; (**b**) bottom of Wujiaping Formation; (**c**) bottom of Wujiaping Formation; (**d**) bottom of Wujiaping Formation; (**e**) lower of Dalong Formation; (**f**) middle of Dalong Formation.

4.2. Distribution of Hydrocarbon-Generating Organisms

The hydrocarbon-generating microorganisms in the Permian shale from the study area could be generally classified into three types: acritarchs, green algae, and fragments of higher plants. In the different Permian shale formations, the relative contents of the three types of hydrocarbon-generating microorganisms show an obvious change from bottom to top (Figures 6 and 7).

The hydrocarbon-generating organisms in the Gufeng Formation of the Chejiaba outcrop are predominantly acritarchs (derived from planktonic algae). The relative acritarch content is very high, averaging 72.5%. Green algae are less abundant, averaging 12.7%, with no obvious changing trend (Figures 6 and 7). Higher plants have an average content of 14.8%, showing an obviously upward-increasing trend in the upper area of the Gufeng Formation, indicating enhanced terrestrial organic input during this period (Figures 6 and 7).

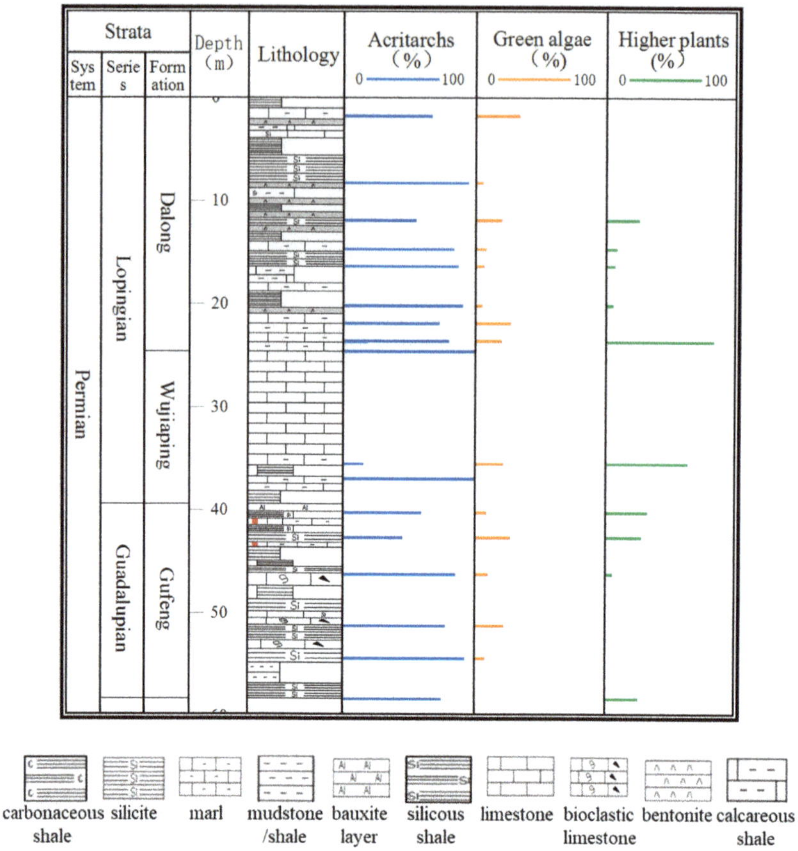

Figure 6. Column of hydrocarbon-forming organisms distribution in the Chejiaba outcrop.

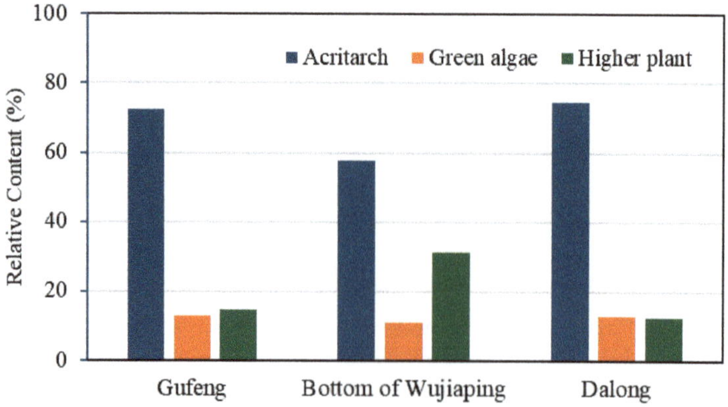

Figure 7. Distribution bar chart of hydrocarbon-generating organisms in different layers of Permian shale in the Chejiaba outcrop.

The hydrocarbon-generating assemblage at the bottom of the Wujiaping Formation consists mainly of acritarchs and higher plants. The average acritarch content decreases to 57.8%, and green algae decline to 10.9%. In contrast, the average content of higher plants increases significantly to 31.3%. The reduced proportion of acritarchs and the marked rise in higher plants reflect increased contributions from terrestrial input compared with the Gufeng Formation (Figures 6 and 7).

The Dalong Formation consists of a similar hydrocarbon-generating assemblage to the Gufeng Formation, dominated by acritarchs (average content: 74.6%). Its higher plant content is only 12.4%, with elevated concentrations primarily observed at the bottom of the Dalong Formation. This suggests that the sea level was low at the beginning of the Dalong Formation, and then transitioned to deeper water later on as higher plants gradually diminished (Figures 6 and 7).

4.3. *Enrichment Model of Multi-Source Organic Matter*

The hydrocarbon-generating organisms of marine source rocks are mainly composed of the debris of plankton (planktonic algae) and benthic organisms (benthic algae, sporangia, and spore walls), bacterial remains, and some plant fragments (linear leaves and higher plants). The organic matter in the high-quality marine shales of South China often has the characteristic of multiple resources. Therefore, the organic matter in marine shale often consists of more than one type of parent material, which can include plankton, benthic organisms, fungi, and higher plants simultaneously [11]. According to the results discussed above, the acritarchs in the Permian shale most likely originated from planktonic algae, especially from the cysts of dinoflagellate, while the macroscopic fragments of planktonic algae are very few. Since the vast majority of planktonic algae mainly exist in the form of acritarchs, the planktonic algae content is represented by acritarchs when determining the source of organic matter in Permian shale.

Previous studies suggest that some kinds of acritarchs, such as *Proterocladus, Siphonophycus, Veryhachium,* and *Dictyotidium,* primarily lived in a shallow marine environment [37,38]. These typical shallow-water acritarchs can be found across all upper Permian layers of the Chejiaba outcrop, implying that the study area was predominantly situated in a shelf environment during the depositional period of Permian shales. Combined with the data of lithology from the Chejiaba outcrop and previously published conclusions, the Permian marine shale in the northwest of the Sichuan Basin was mainly deposited in a deep shelf environment and was controlled by the tectonic evolution of the Kaijiang–Liangping Trough. Therefore, the Permian marine shale was deposited in a stable anoxic reduction environment, which was conducive to the enrichment of organic matter in the shale. The shale in northwestern Sichuan Basin is mainly siliceous shale and silicite, which are characterized by high TOC. The shales of the Gufeng and Dalong Formations are superior among the strata. In particular, the shale in the Dalong Formation formed in a high-productivity environment where distant volcanic activities were superimposed. The volcanic tuff interlayers that frequently occurred in the Dalong Formation could promote increased productivity, as well as the preservation of organic matter. Overall, the sedimentary environment of Permian marine shale in northwestern Sichuan Basin was characterized by deep water, low terrestrial input, high productivity, and anoxic sulfide water, forming a favorable layer for shale gas exploration. Based on these conclusions, the sources and evolution processes of organic matter in the Permian marine shales of the northwestern Sichuan Basin have been elucidated.

During the period of the Gufeng Formation, the deposition was mainly siliceous shale (Figure 2), indicating that the Kaijiang–Liangping Trough in northwestern Sichuan Basin was characterized by a siliceous deep shelf under high sea levels. Organic matter was primarily sourced from planktonic algae, with minor contributions from terrestrial input (Figure 8a). At the bottom of the Wujiaping Formation, the presence of a bauxite layer suggests that the sea level dropped dramatically (Figure 2), leading to a coastal swamp environment in the Kaijiang–Liangping Trough. The coastal swamp setting, caused by lower sea levels, increased terrestrial contributions while reducing planktonic algae dominance (Figure 8b). Afterwards, the study area transitioned to a carbonate platform in the middle–upper Wujiaping Formation (Figure 2), characterized by low organic matter content (Figure 8c). During the deposition stage of the Dalong Formation, the deposits were mainly siliceous shale and marl (Figure 2), which means the Kaijiang–Liangping Trough in the northwestern Sichuan Basin returned to a siliceous deep shelf, leading to a resurgence of planktonic algae as the primary organic matter source. (Figure 8d). On the whole, hydrocarbon-generating organisms in the Permian shales of the study area are dominated by acritarchs (derived from planktonic algae), followed by higher plants and green algae.

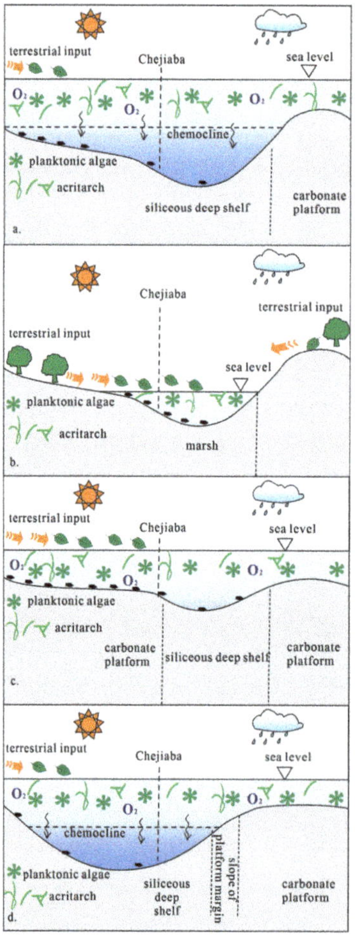

Figure 8. Enrichment model of multi-source organic matter of Upper Permian shale in northwestern Sichuan. (**a**) Gufeng Formation; (**b**) bottom of Wujiaping Formation; (**c**) middle–upper of Wujiaping Formation; (**d**) Dalong Formation.

5. Conclusions

(1) The hydrocarbon-generating organisms in the Permian shales of the Chejiaba outcrop are predominantly acritarchs (derived from planktonic algae), followed by higher plants and green algae. In the Gufeng Formation, the proportion of hydrocarbon-generating organisms that are acritarchs is quite high, followed by higher plants and green algae. At the bottom of the Wujiaping Formation, the proportion of acritarchs gradually decreases, while higher plants increase significantly. In the Dalong Formation, the acritarchs regain dominance, resembling the microorganism composition of the Gufeng Formation. Green algae show minimal variation across all stratigraphic layers.

(2) The organic matter in the Permian shales primarily originates from acritarchs among planktonic algae, supplemented by green algae and terrestrial higher plants. During the stage of the Gufeng Formation, the Kaijiang–Liangping Trough in the study area was characterized by siliceous deep shelf facies with a high sea level. Organic matter was mainly derived from aquatic planktonic algae, with little terrestrial input. At the beginning of the Wujiaping Formation, low sea levels resulted in a coastal marsh environment, where organic matter sources were mixed, composed of planktonic algae and terrestrial input. The Dalong Formation was deposited in siliceous deep-shelf facies, with organic matter primarily sourced from aquatic planktonic algae and minor terrestrial input, similar to the Gufeng Formation.

Author Contributions: Conceptualization, Y.Z. and B.S.; methodology, Y.Z.; software, Y.L. (Yifei Li) and Y.L. (Yang Liu); validation, B.S., B.G. and D.F.; formal analysis, Y.Z.; investigation, M.L.; resources, B.S. and P.W.; data curation, Y.Z.; writing—original draft preparation, Y.Z.; writing—review and editing, D.F.; visualization, P.W.; supervision, B.S.; project administration, B.S.; funding acquisition, Y.Z. All authors have read and agreed to the published version of the manuscript.

Funding: This research was funded by [Sinopec Key Laboratory of Shale Oil/Gas Exploration and Production Technology], titled "Characteristics of Organic Matter Sources in Permian Marine Shale in the Northern Sichuan—Western Hubei Region and Their Impact on Reservoir Development", No. 33550000-24-ZC0613-0045.

Data Availability Statement: The data that support the findings of this study are available from the corresponding author, upon reasonable request.

Conflicts of Interest: The authors declare no conflicts of interest.

References

1. Guo, X.S.; Wang, R.Y.; Shen, B.J.; Wang, G.P.; Wan, C.X.; Wang, Q.R. Geological characteristics, resource potential, and development direction of shale gas in China. *Pet. Explor. Dev.* **2025**, *52*, 15–28. [CrossRef]
2. Liao, F.; Wang, K.; Zhan, J.; Liu, Z.; Du, J.; Gong, S.; Cai, N.; Bai, J.; Zhang, J. Geological Conditions of Shale Gas Accumulation in Coal Measures. *Processes* **2024**, *12*, 1734. [CrossRef]
3. Jin, Z.J.; Liu, G.X.; Wang, P.W.; Nie, H.K.; Li, M.; Wang, G.P. Exploration potential and targets of the Permian shale gas in the Yangtze region, South China. *Oil Gas Geol.* **2025**, *46*, 335–347.
4. Li, R.; Xiong, Z.; Wang, Z.; Xie, W.; Li, W.; Hu, J. Lithofacies Characteristics and Pore Controlling Factors of New Type of Permian Unconventional Reservoir in Sichuan Basin. *Processes* **2023**, *11*, 625. [CrossRef]
5. He, T.H.; Li, W.H.; Lu, S.F.; Yang, E.Q.; Jing, T.T.; Ying, J.F.; Zhu, P.F.; Wang, X.Z.; Pan, W.Q.; Zhang, B.S.; et al. Quantitatively unmixing method for complex mixed oil based on its fractions carbon isotopes: A case from the Tarim Basin, NW China. *Pet. Sci.* **2023**, *20*, 102–113. [CrossRef]
6. Li, W.B.; Wang, J.; Jia, C.Z.; Lu, S.F.; Li, J.Q.; Zhang, P.F.; Wei, Y.B.; Song, Z.J.; Chen, G.H.; Zhou, N.W. Carbon isotope fractionation during methane transport through tight sedimentary rocks: Phenomena, mechanisms, characterization, and implications. *Geosci. Front.* **2024**, *15*, 101912. [CrossRef]
7. Guan, M.D.; Wu, S.T.; Hou, L.H.; Jiang, X.H.; Ba, D.; Hua, G.L. Paleoenvironment and chemostratigraphy heterogenity of the Cretaceous organic-rich shales. *Adv. Geo-Energy Res.* **2021**, *5*, 444–455. [CrossRef]

8. Xia, P.; Hao, F.; Yang, C.; Tian, J.Q.; Fu, Y.; Wang, K. Mechanism of organo-nickel co-enrichment in marine black shale. *Adv. Geo-Energy Res.* **2024**, *13*, 10–21. [CrossRef]
9. Wang, E.Z.; Li, M.W.; Ma, X.X.; Qian, M.H.; Cao, T.T.; Li, Z.M.; Yang, W.W.; Jin, Z.J. Diahopane and diasterane as the proxies for paleoenvironment, hydrocarbon generation condition, and shale oil accumulation. *Chem. Geol.* **2024**, *670*, 122447. [CrossRef]
10. Lu, H.; Pan, L.; Guo, Y.; Xiong, Q.; Dai, F.; Wang, S. Experimental Investigation of the Characteristics and Transformation Mechanism of Jimsar Oil Shale and Derived Shale Oil. *Processes* **2023**, *11*, 411. [CrossRef]
11. Qin, J.Z.; Tao, G.L.; Teng, G.E.; Bian, L.Z.; Xie, X.M.; Fu, X.D. Hydrocarbon-forming organisms in excellent marine source rocks in south China. *Pet. Geol. Exp.* **2010**, *32*, 262–269.
12. Qin, J.Z.; Shen, B.J.; Tao, G.L.; Teng, G.E.; Yang, Y.F.; Zheng, L.J.; Fu, X.D. Hydrocarbon-forming organisms and dynamic evaluation of hydrocarbon generation capacity in excellent source rocks. *Pet. Geol. Exp.* **2014**, *36*, 465–472.
13. Teng, G.E.; Qin, J.Z.; Fu, X.D.; Yang, Y.F.; Xie, X.M. Hydrocarbon source rocks eveluation of the Upper Permian Wujiaping Formation in northeastern Sichuan area. *J. Palaeogeogr.* **2010**, *12*, 334–345.
14. Zhao, Q.F.; Xie, D.Z.; Chen, J.W.; Liang, J.; Zhang, Y.G.; Wang, J.Q.; Dong, H.P.; Yuan, Y. Evaluation and biogenetic derivation and sedimentary environment of source rocks from Permian in the Lower Yangze region. *Geol. Bull. China* **2023**, *42*, 1154–1165.
15. Shen, J.; Lei, Y.; Algeo, T.J.; Feng, Q.L.; Servais, T.; Yu, J.X.; Zhou, L. Volcanic Effects on Microplankton during the Permian-Triassic Transition (Shangsi and Xinmin, South China). *Palaios* **2013**, *28*, 552–567. [CrossRef]
16. Zhang, Y. Forms and Distribution Characteristics of Organic Matter and Controlling Factors on Organic Matter Accumulation in the Latest Permian Dalong Formation, Northern Margin of Upper Yangtze. Ph.D. Thesis, China University of Geosciences, Wuhan, China, 2017.
17. Fu, X.D.; Qin, J.Z.; Teng, G.E.; Wang, X.F. Evaluation on Dalong Formation source rock in the north Sichuan Basin. *Pet. Geol. Exp.* **2010**, *32*, 566–571.
18. Zhang, Z.Y.; Xu, S.; Gou, Q.Y.; Li, Q.Q. Reservoir Characteristics and Resource Potential of Marine Shale in South China: A Review. *Energies* **2022**, *15*, 8696. [CrossRef]
19. Wei, H.Y.; Hu, D.; Qiu, Z.; Zhang, X.; Liu, W.; Kong, W.L.; Ahmed, M. Sedimentological and Geochemical Characteristics of Late Permian Organic-Rich Rocks in North Sichuan and West Hubei Provinces. *Acta Sedimentol. Sin.* **2024**, *42*, 774–798.
20. Yang, X.; Yang, Y.R.; Zhang, J.Y.; Tian, H.; Wang, Q.; Song, F.; Zhang, S.K.; Chen, Y. Characteristics and formation mechanism of Permian marine shale of Kaijiang-Liangping trough in northern Sichuan Basin. *Lithol. Reserv.* **2025**, *37*, 108–119.
21. Kong, W.L. Research of Sedimentary Facies and Paleogeography of Middle and Later Permian in the Northern Margin of Middle and Upper Yangtze Region. Ph.D. Thesis, Hefei University of Technology, Hefei, China, 2011.
22. Li, R.B.; Pei, X.Z.; Liu, Z.Q.; Li, Z.C.; Ding, S.P.; Chen, Y.X.; Liu, Z.G.; Zhang, X.F.; Chen, G.C. Stratigraphic Correlations of Two Different Types of Permian Strata and Their Regional Tectonic Significance in Zhenba-Gaochuan Area, Northern Margin of Yangtze Block. *J. Earth Sci. Environ.* **2011**, *33*, 125–131.
23. Zhang, X.H.; Chen, C.; Huang, J.; Wen, L.; Xie, C.; Xu, S.W.; Gao, Z.L. The discovery of Middle Permian Guangyuan-Bazhong extensional trough in the Sichuan Basin and its petroleum geological significance. *China Pet. Explor.* **2019**, *24*, 466–475.
24. Chen, Y.Z.; Zang, D.G.; Hu, G.; Feng, X.K.; Wang, X.Y.; Xiao, D.; Chen, Y.; Xu, M.; Liang, H.; Wu, Y.L.; et al. Tectonic framework and oil-gas geological significance of late Permian Changxing Period in Sichuan Basin. *Earth Sci. Front.* **2024**, 1–10. [CrossRef]
25. Zhao, P.R.; Gao, B.; Guo, Z.F.; Wei, Z.H. Exploration potential of marine-continental transitional and deep-water shelf shale gas in Upper Permian, Sichuan Basin. *Pet. Geol. Exp.* **2020**, *42*, 335–344.
26. Xu, S.Y.; Xia, M.S.; Zhu, Y.; Li, T.J.; Lin, Y. Evolution stage and tectonic sedimentary characteristics of the Kaijiang-Liangping Trough. *Fault-Block Oil Gas Field* **2023**, *30*, 963–974.
27. Qiu, Z.; Dou, L.R.; Wu, J.F.; Wei, H.Y.; Liu, W.; Kong, W.L.; Zhang, Q.; Cai, G.Y.; Zhang, G.; Wu, W.; et al. Lithofacies Palaeogeographic Evolution of the Middle Permian Sequence Stratigraphy and Its Implications for Shale Gas Exploration in the Northern Sichuan and Western Hubei Provinces. *Earth Sci.* **2024**, *49*, 712–748.
28. Xie, J.P.; Li, Y.X.; Zhang, X.H.; Peng, H.L.; Gao, Z.L.; Li, T.J.; Zhu, J.C.; Hu, G. Evolution and sedimentary conditions of tension trough in Gufeng Member of Maokou Formation, northern Sichuan Basin. *J. Chengdu Univ. Technol. (Sci. Technol. Ed.)* **2025**, *53*, 1–15.
29. Xie, J.R.; Zhao, L.Z.; Shen, P.; Zhang, B.J.; Zhang, X.H.; Chen, Y.G.; Chen, C.; Cen, Y.J.; Hao, Y.; Cao, H.; et al. New exploration discovery of facies controlled porous reservoir in the Wujiaping Formation of the Sichuan Basin and its significance to oil and gas exploration. *Nat. Gas Ind.* **2021**, *41*, 11–19.
30. Guo, J.H.; Hu, G.Y.; He, K.; Mi, J.K.; Tian, L.J.; He, F.; Guo, C.Y.; Lu, M.D. Geochemical characteristics and sedimentary environment of source rocks of Permian Dalong Formation in northern Sichuan Basin. *Lithol. Reserv.* **2023**, *35*, 139–152.
31. Wang, Y.; Bai, L.H.; Zhang, Y.L.; Zhang, X.M.; Yang, B.W.; Duan, K.; Wang, Y.; Xie, T. Reservoir Characteristics and Influencing Factors of Organic-Rich Siliceous Shale of the Upper Permian Dalong Formation in Western Hubei. *Energies* **2023**, *16*, 5130. [CrossRef]

32. Ma, X.G.; Ye, Y.H.; Liu, S.G.; Song, J.M.; Wang, H.; Du, A.Y.; Chen, W.; Li, Z.W.; Dai, X.; Ming, Y.; et al. Sedimentary environment and organic matter enrichment mechanism of highorganic-rich black shale in the Upper Permian Dalong Formation in the eastern Sichuan Basin. *Nat. Gas Geosci.* **2025**, *36*, 621–636.
33. Qin, J.Z.; Pan, A.Y.; Shen, B.J. Insoluble organic matter in source rocks: Derived from organic macromolecules in the skeleton, cell wall and shell of organisms. *Pet. Geol. Exp.* **2020**, *42*, 946–956.
34. Yin, L.M.; Yuan, X.L.; Meng, F.W. *Illustrated Book of Organic Walled Microfossils*; Zhejiang University Press: Hangzhou, China, 2018.
35. Yin, L.M.; Wang, C.J.; Zhao, Y.L.; Ou, Z.Y. Early-Middle Cambrian Palynomorph Microfossils and Related Geochemical Events in South China. *J. Earth Sci.* **2016**, *27*, 180–186. [CrossRef]
36. Song, D. Hydrocarbon-Forming Organism Characteristics and Their Petroleum Geological Significance of High-Quality Source Rocks in the Lower Paleozoic in Southern China. Master's Thesis, Nanjing University, Nanjing, China, 2019.
37. Lei, Y.; Servias, T.; Feng, Q.L.; He, W.H. The spatial (nearshore–offshore) distribution of latest Permian phytoplankton from the Yangtze Block, South China. *Palaeogeogr. Palaeoclimatol. Palaeoecol.* **2012**, *363*, 151–162. [CrossRef]
38. Li, D.D.; Luo, G.M.; Tang, Q.; She, Z.B.; Xiao, S.H. New record of the green algal fossil Proterocladus and coexisting microfossils from the Meso-Neoproterozoic Diaoyutai Formation in southern Liaoning, North China. *Precambrian Res.* **2023**, *393*, 107104. [CrossRef]

Journal of
Marine Science and Engineering

MDPI

Article

Astronomical Forcing of Fine-Grained Sedimentary Rocks and Its Implications for Shale Oil and Gas Exploration: The BONAN Sag, Bohai Bay Basin, China

Jianguo Zhang [1,2,3], Qi Zhong [3,*], Wangpeng Li [1,2,4], Yali Liu [1,2,4], Peng Li [1,2,4], Pinxie Li [3], Shiheng Pang [3] and Xinbiao Yang [5]

1 State Key Laboratory of Shale Oil and Gas Enrichment Mechanisms and Efficient Development, Beijing 102206, China; zhjianguo@cugb.edu.cn (J.Z.)
2 Sinopec Key Laboratory of Shale Oil/Gas Exploration and Production Technology, Beijing 102206, China
3 School of Energy Resources, China University of Geosciences, Beijing 100083, China
4 Petroleum Exploration and Production Research Institute, Sinopec, Beijing 102206, China
5 Exploration and Production Research Institute, Liaohe Oilfield of CNPC, Panjin 124010, China
* Correspondence: 3006230039@email.cugb.edu.cn

Abstract: Fine-grained sedimentary rocks are ideal carriers for astronomical cycle analysis as they can record and preserve significant astronomical cycle signals. Spectral analysis using the Multi-taper Method (MTM) and Evolutionary Harmonic Analysis (EHA) using the Fast Fourier Transform (FFT) were conducted on natural gamma data from key wells in the Es3l sub-member in the Bonan Sag, Bohai Bay Basin, China. Gaussian bandpass filtering was applied using a short eccentricity cycle of 100 ka, and a "floating" astronomical time scale for the Es3l sub-member (Lower 3rd sub-member of Shahejie Formation in Eocene) was established using magnetic stratigraphic ages as boundaries. Stratigraphic divisions were made for single wells in the Es3l of the Bonan Sag, and a stratigraphic framework was established based on correlations between key wells. The research results indicate the following: Firstly, the Es3l of the Bonan Sag records significant astronomical cycle signals, with an optimal sedimentation rate of 8.39 cm/ka identified. Secondly, the cyclical thicknesses corresponding to long eccentricity, short eccentricity, obliquity, and precession cycles are 38.9 m, 9.7 m, 4.6–3.4 m, and 1.96–1.66 m, respectively. Thirdly, the Es3l sub-member stably records 6 long eccentricity cycles and 26 short eccentricity cycles, and the short eccentricity curve is used as a basis for stratigraphic division for high-precision stratigraphic correlations. Fourthly, the quality of sandstone-interbedded mudrock is jointly controlled by the short eccentricity and precession. Eccentricity maximum values result in thicker sandstone interlayers, while minimum precession values promote the thickness of sandstone interlayers. Through astronomical cycle analysis, the depositional evolution mechanism of sandstone-interbedded mudrock is revealed. Combined with the results of high-precision stratigraphic division, this can provide a basis for fine evaluation and "sweet spot" prediction of lacustrine shale oil reservoirs.

Keywords: astronomical cycles; fine-grained sedimentary rocks; high-precision stratigraphic correlation; depositional evolution; Bonan Sag

1. Introduction

Fine-grained sedimentary rocks refer to sedimentary rocks composed of particles with a diameter of less than 0.0625 mm [1,2], which have undergone consolidation and lithification [3], specifically mudrock in this context. Recently, shale oil and gas exploration have gained significant momentum [4–6], and the Bonan Sag of the Bohai Bay Basin in China exhibits substantial potential for shale oil exploration in the Es_3l sub-member. However, a lack of understanding of lithofacies types and their distribution patterns has hindered breakthroughs in shale oil exploration [7], necessitating more in-depth research on mudrock strata. Traditional stratigraphic division methods offer limited precision for fine-grained sedimentary rocks [8]. Due to their continuity in deposition and ability to record significant astronomical cycle signals, fine-grained sedimentary rocks are ideal for astronomical cycle analysis [9].

Milankovitch proposed the theory of Milankovitch cycles in 1941 [10] (referred to as an astronomical cycle in this article), which suggests that variations in Earth's orbital parameters (eccentricity, obliquity, and precession) drive Quaternary glacial–interglacial cycles [11], resulting in sedimentary cyclicity [12–14]. This theory has gained widespread recognition and application in studying sedimentary records throughout geological history. Scholars have identified astronomical cycle signals in various sedimentary materials, including fine-grained sediments [15–17]. Meanwhile, some international studies have focused on marine deposits, arguing that lacustrine deposits are significantly influenced by terrestrial inputs and tectonic activities [18,19]. With the deepening of shale oil and gas exploration, many scholars have identified astronomical cycle signals in lacustrine fine-grained sedimentary rocks, such as the Teruel Basin in Spain [20,21], the Mudurnu-Göynük Basin in Turkey [22], the Upper Cretaceous strata of the Songliao Basin in China [23], and the Paleogene strata of the Bohai Bay Basin [7,9,24,25]. However, the majority of research undertaken by these scholars has concentrated on utilising the astronomical cycle theory to elucidate the palaeoclimatic characteristics, organic matter-enrichment factors, and the stratigraphic division of single wells within lacustrine deposits. These studies remain largely focused on individual "points" and are insufficient to support predictions of lithofacies development types or to ascertain the spatial relationships between lithofacies. Astronomical cycles influence climate change and, subsequently, sedimentary processes by modulating the amount of sunlight received on Earth's surface through variations in its orbital parameters. Research has demonstrated that Earth's orbital parameters can impact climate variations on both local and global scales [24], spanning timescales from tens of thousands to millions of years. Consequently, by identifying astronomical cycle signals within strata, it is feasible to establish an astronomical timescale with a resolution of tens of thousands of years [26]. The crux of identifying astronomical cycle signals lies in comprehending fluctuations in sedimentation rates. The temporal implications inherent in astronomical cycle theory form the cornerstone for high-precision stratigraphic division and correlation. By harnessing the relationship between orbital cycles and sedimentation, the author addresses the challenges of astronomical cycle stratigraphic division and correlation, sandstone-interbedded mudrock to quality prediction, and deposition mechanisms from a sedimentary genesis perspective, offering a novel methodology for precise shale oil and gas exploration. This approach holds considerable significance for guiding the precise exploration of shale oil and gas resources [27].

This study focuses on stratigraphic correlation by selecting representative key wells from the Bonan Sag within the Jiyang Depression of the Bohai Bay Basin. Some scholars have used GR logging curves to identify sedimentary cycles, namely the gamma deviation logging (GDL) method [28]. Since Natural Gamma (GR) data from well logs serve as an ideal palaeoclimatic proxy for cyclostratigraphic analysis [29], GR is adopted as the proxy for astronomical cycle analysis. Spectral analysis, evolutionary harmonic analysis, and correlation coefficient ("COCO") analysis are employed to constrain vertical variations in sedimentation rates, identify astronomical cycle signals, and determine the optimal sedimentation rate for the Es_3l sub-member. Signal validation is conducted, and astronomical cycle signals are extracted from the Es3l sub-member strata. Through filtering and tuning, the cyclical curves are derived, and a "floating" astronomical timescale is established using the 100 ka short eccentricity cycle. This allows for the calculation of sedimentation rates at different depths. The extracted astronomical cycle-filtered curves are then utilized for stratigraphic division and correlation. The 100 ka short eccentricity cycle curve is applied to the stratigraphic division of individual wells, and the results are integrated into well-tie stratigraphic correlations to establish a stratigraphic framework. The well JYC-1, characterized by the most exemplary features of sandstone-interbedded mudrock within the study region, is designated as the primary focus of the research. This well serves as a crucial subject for analyzing astronomical cycles, predicting sandstone-interbedded mudrock quality, and elucidating the mechanisms of depositional evolution. This study applies astronomical cycle theory to cyclostratigraphic analysis, using filtered curves as a basis for high-precision stratigraphic division and correlation, thereby facilitating an intuitive understanding of lithofacies development patterns and predictions of lithofacies types.

2. Regional Geological Overview

The Bonan Sag, situated in the northeastern part of the Jiyang Depression within the Bohai Bay Basin (Figure 1a), is a secondary tectonic unit of the Jiyang Depression, characterized by a structural pattern of "northern faulting and southern overthrusting, with steep slopes in the north and gentle slopes in the south", covering an area of approximately 2800 km^2 [30]. It is bounded by the Chengdong Bulge in the north, the Guxi Fault, the Gudao Bulge to the southeast, the Chenjiazhuang Bulge in the south, and the Yidong Fault in the west, adjacent to the Yihezhuang Bulge [31]. These surrounding uplifts constitute the primary source areas for the Bonan Sag, where the parent rocks are dominated by granite from the Archean–Proterozoic, carbonate rocks from the Lower Paleozoic, and clastic rocks from the Upper Paleozoic [32]. The third member of the Shahejie Formation in the Paleogene can be subdivided into the lower (Es3l), middle (Es3m), and upper (Es3u) sub-members (Figure 1b) [17]. During the deposition of the Es3l sub-member, the climate was humid, promoting a stable and persistent deep-to-semi-deep lake sedimentary environment. This led to the formation of approximately 200–400 m of thick, dark-grey mudstone interbedded with marl deposits. The target interval in the study area is the Es3l submember, comprising primarily endogenic marl and mixed-source mudstone deposits within the deep-to-semi-deep lake environment (Figure 1b).

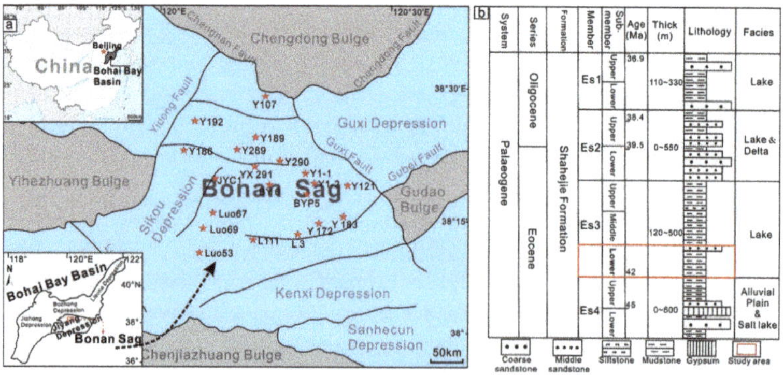

Figure 1. Regional geological background of Bonan Sag [modified from ref. [7]]. (**a**) The Bonan Sag, situated in the northeastern part of the Jiyang Depression within the Bohai Bay Basin, the red pentagram marks the well location; (**b**) the target interval in the study area is the Lower Es3 submember (the red box marked).

3. Data and Methods

3.1. Data Selection

This study selected seven key wells in the steep slope zone and deep sag zone of the Bonan Sag in the Bohai Bay Basin for astronomical cycle analysis, including wells Luo 53, Luo 69, Luo 67, JYC-1, Y289, Y189, and Y107. The main analyzed intervals of these wells are the Es$_3$l of the Shahejie Formation, with thicknesses ranging from 150 to 570 m. Natural Gamma Ray (abbreviated as GR) logging data can serve as a proxy indicator for paleoclimate analysis, with a sampling interval of 0.125 m, meeting the precision requirements for astronomical cycle analysis. The GR values range from 10 to 100 API, where low values correspond to marl, while high values are associated with mudrocks, exhibiting distinct cyclical patterns (Figure 2). For this study, wells dominated by lacustrine deep-water fine-grained sedimentary rocks in the Es$_3$l sub-member in the Bonan Sag were selected for astronomical cycle analysis.

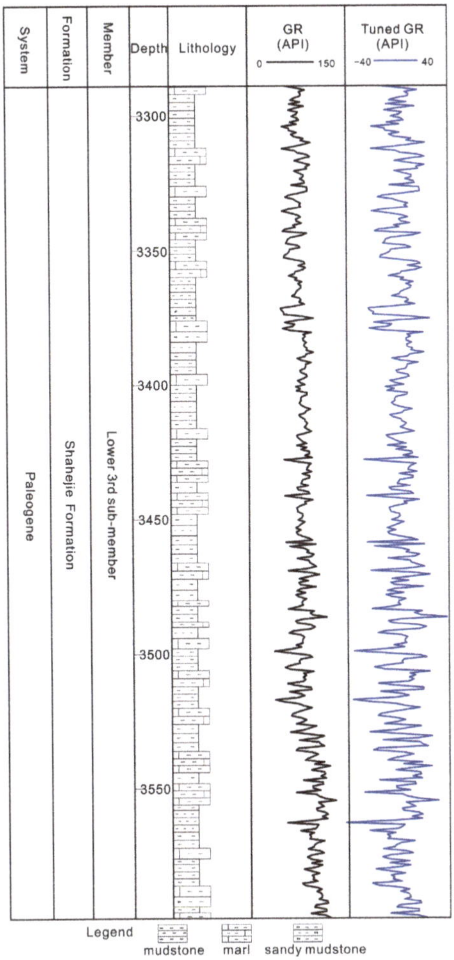

Figure 2. The detrended GR curve for well JYC-1 (presentation of detrended GR data).

The software utilized for data analysis in this research is Acycle v2.2 [33], built on the MATLAB v2019a platform, where all data analyses were conducted. The data analysis procedures encompass Multi-Taper Method (MTM) spectral analysis, Evolutionary Harmonic Analysis based on Fast Fourier Transform (FFT), filtering analysis, and COCO/eCOCO correlation analysis.

Prior to these operations, data preprocessing is essential, which primarily involves: (1) extreme value removal to eliminate outliers from the dataset. (2) Interpolation, where linear interpolation is applied to the GR data to ensure a uniform sampling interval of 0.125 m. (3) Detrending, achieved through a 35% weighted moving average process [34] (Figure 2), to mitigate the influence of trends in the GR data.

3.2. Astronomical Cycle Solutions

Laskar et al. (2011) have refined and improved the initial computational conditions for the La2004 calculation parameters by adopting the new astronomical ephemeris [35], INPOP08 [36], resulting in the La2010 solution that offers a more precise depiction of the eccentricity variation over the past 250 million years (Ma). Notably, La2010 achieves remarkable accuracy in calculating Earth's orbital parameters (eccentricity, obliquity, and precession) since 250 Ma, particularly within the last 50 Ma, thereby providing a theoretical foundation for astronomical cyclicity analysis. Building upon this, Liu et al. (2018) conducted an astronomical cyclicity analysis of the Shahejie Formation in the Bohai Bay Basin [37], establishing a high-resolution astronomical timescale spanning from 22 to 66 Ma. Within this framework, the age of the Es_3l Submember in the Jiyang Depression was anchored at 40.2 to 42.47 Ma, a viewpoint that has been cited by several researchers to conduct related studies [25–38].

This study employs the La2010a astronomical solution [35] (Laskar et al., 2011) within the Acycle v2.4 software to calculate the summer insolation curves for the period 40.2 to 42.47 Ma [33], corresponding to the formation of the Es_3l sub-member in the Jiyang Depression. The sampling interval is set at 1 ka.

4. Results

4.1. Cyclostratigraphic Analysis

4.1.1. Theoretical Astronomical Period

Spectral analysis of these insolation curves reveals theoretical values for eccentricity, obliquity, and precession cycles (Figure 3a). The dominant astronomical cycles identified in the Es_3l sub-member of Bonan Sag (40.2–42.47 Ma) include 405 ka (long eccentricity, E), 125 ka and 100 ka (short eccentricity, e), 51 ka, 40 ka, and 38 ka (obliquity, O), and 23 ka, 22 ka, and 19 ka (precession, P). The ratio of these cycles, 405: 125-100: 51-38: 23-19, approximates 21.3: 5.1: 2: 1. By comparing this ratio with the peak ratios obtained from spectral analysis of GR data, we preliminarily assess whether astronomical cyclicity signals are recorded in the strata. A close correspondence between these ratios suggests the presence of astronomical cyclicity signals.

4.1.2. Deep-Domain Spectrum Analysis

Conducting an MTM spectral analysis on the preprocessed GR data series (Figure 3b–d) of the Es_3l sub-member from selected wells (taking JYC1 as an example) within the Bonan Sag in the depth domain reveals intriguing insights. The results showcase 15 dominant frequency peaks exceeding the 95% confidence curve (Figure 3c), with all peaks validated by the AR1 confidence curve. In this spectral plot, the vertical axis represents energy intensity, while the horizontal axis denotes the frequency corresponding to the depth series. By inverting the frequency values (i.e., 1/frequency), we can derive the cyclic thicknesses associated with the spectral peaks: 38.9, 9.7, 3.7, 3.4, 1.96, and 1.66 m. The overall spectral structure exhibits four prominent frequency bands: 38.9 m, 9.7 m, 3.7–3.4

m, and 1.96–1.66 m. These bands are interpreted as representing the long eccentricity (E), short eccentricity (e), obliquity (O), and precession (P) of Earth's orbit, respectively. The ratio of the spectral peak values within these four bands, taking the maximum within each band, is approximately 20.47:5.1:1.95:1. This closely aligns with the theoretical astronomical cycle ratio of 21.3:5.1:2.1:1. This analysis not only validates the presence of astronomical forcing in shaping the stratigraphic record but also underscores the significance of these orbital parameters in modulating sedimentary processes and patterns in the Bonan Sag.

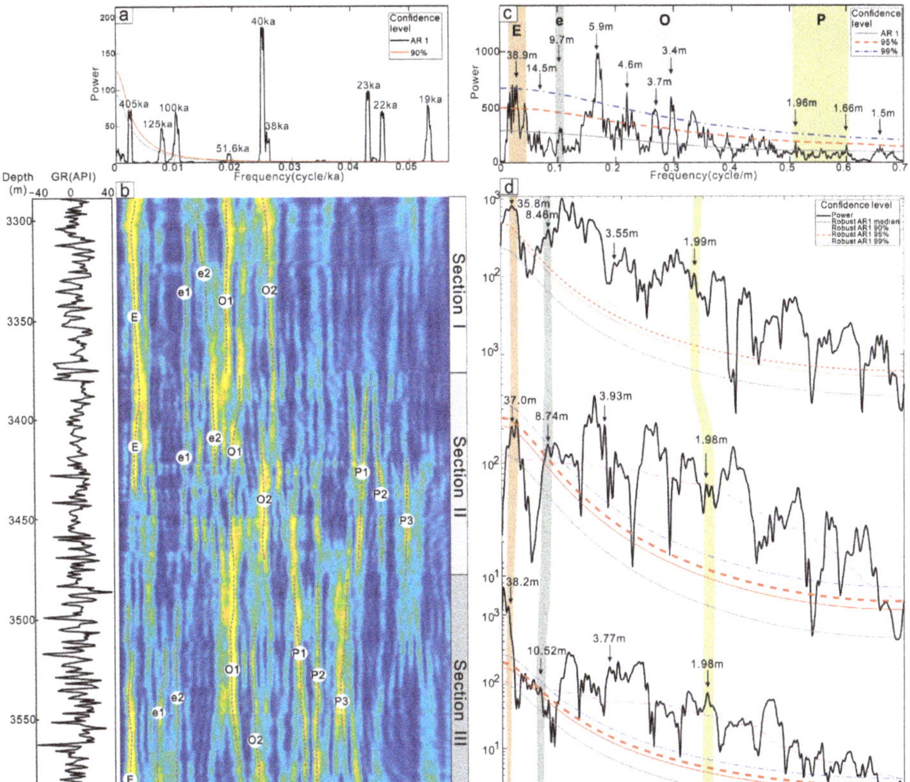

Figure 3. MTM Spectrum Analysis of GR curve in Es$_3$l sub-member of well JYC1. (**a**) MTM spectra analysis of astronomical solutions for 40.2–42.47 Ma. (**b**) Evolutionary Harmonic Analysis in the Es$_3$l sub-member of well JYC1. (**c**) MTM spectra analysis of depth domain in Es$_3$l sub-member of well JYC-1. E = Long Eccentricity (orange band), e = Short Eccentricity (dark green band), O = Obliquity (purple band), P = Precession (light green). (**d**) MTM spectra analysis of segmented GR depth domain in Es$_3$l sub-member of well JYC-1. Orange line: Long Eccentricity; Dark green line: Short Eccentricity; Purple line: Obliquity, Light green: Precession.

To assess the stability of the overall data sequence and analyze the primary frequency characteristics varying with depth in the sedimentary record, an Evolutionary Harmonic Analysis (EHA) of the GR curve is necessary. This method performs Fourier transforms on moving depth segments with different windows, enabling the detection of stability in depth data and the analysis of variations in sedimentation rates across depths. After multiple iterations of tuning, the optimal parameter configuration for this EHA was determined as follows: a sliding window size of 46 m, a window step size of 0.351 m, and a maximum frequency of 1.0, yielding the Evolutionary Spectrum plot (Figure 3b). The peaks in the Evolutionary Spectrum plot align well with those in the MTM spectrum, indicating the preservation of eccentricity, obliquity, and precession signals across various depth segments. Three distinct depth intervals at 3382 m, 3478 m, and 3500 m in the Evolutionary Spectrum plot exhibit notable changes in spectral peak values, suggesting significant variations in sedimentation rates at these locations. The discontinuous peaks observed at 3382 m and 3478 m may be attributed to enhanced terrestrial input at these depths, leading to changes in sedimentation rates and consequently altering the spectral periodicity [39]. For operational convenience and technical feasibility, the JYC1 well was segmented into three intervals: 3289–3382 m, 3382–3478 m, and 3478–3600 m, based on the pronounced fluctuations at 3382 m and 3478 m. The labels E, e, O, and P within the white circles in Figure 3b represent long eccentricity, short eccentricity, obliquity, and precession, respectively. Due to the varying sedimentary environments across different stratigraphic horizons in the study area, separate spectral analyses were conducted on the GR sequences of the aforementioned three depth intervals to mitigate errors arising from differing sedimentation rates in different depositional settings (Figure 3d). In the first interval (3289–3382 m), the four dominant frequency peaks are 35.8, 8.46, 3.55, and 1.99 m, with a ratio close to 405:100:38:19. In the second interval (3382–3478 m), the peaks are 37.0, 8.74, 3.93, and 1.98 m, approximating a ratio of 405:100:40:19. Lastly, in the third interval (3478–3600 m), the peaks are 38.2, 10.52, 3.77, and 1.98 m, with a ratio similar to 405:125:40:19. These findings underscore the influence of varying sedimentary conditions on the spectral signatures and sedimentation rates within the study area.

4.1.3. Optimal Sedimentation Rate Estimation

The selected GR logging sequence is depth-based. Although periodicity in power has been detected above the 95% confidence curve, no estimation has been made regarding the temporal range represented by the depth cycles. Consequently, relying solely on the traditional "ratio method" for astronomical cycle identification yields relatively crude results, necessitating the adoption of additional methodologies to enhance the reliability of cycle recognition. The approach employed in this study involves utilizing Monte Carlo simulation-based COCO/eCOCO (Correlation Coefficient/evolutionary Correlation Coefficient) analysis to test for non-astronomical cycle signals [33]. This is complemented by a significance level test of the null hypothesis (H_0) to validate the results [40]. The significance level in this test represents the probability of falsely rejecting the null hypothesis in the stratigraphic record [41], implying that a lower H_0 value corresponds to a more reliable analysis outcome. For instance, a significance level of 0.001 indicates that 99.9% of the spectral components are detected, with only 0.1% of astronomical cycle signals being erroneously rejected, because the smaller the H_0 in the test, the more reliable the results obtained from the analysis [33]. The sedimentation rate derived from such a rigorous null hypothesis testing is closest to the true sedimentation rate and is designated as the optimal sedimentation rate.

Based on previous estimates of the sedimentation rate for the Es_3l sub-member in the Bonan Sag and the COCO parameter settings [38], the COCO calculation parameters used in this study were configured as follows: minimum frequency of 0, maximum frequency of 4, maximum sedimentation rate of 54 cm/ka, and 2000 Monte Carlo simulations. The COCO analysis yielded an average sedimentation rate of 8.39 cm/ka for the Es_3l submember in Well JYC-1 (Figure 4a), both exceeding the 0.001 significance level (Figure 4b). Moreover, the confidence level for the involvement of all seven astronomical parameters was below 0.05% (Figure 4c), allowing us to reject the null hypothesis of no astronomical cycle signals.

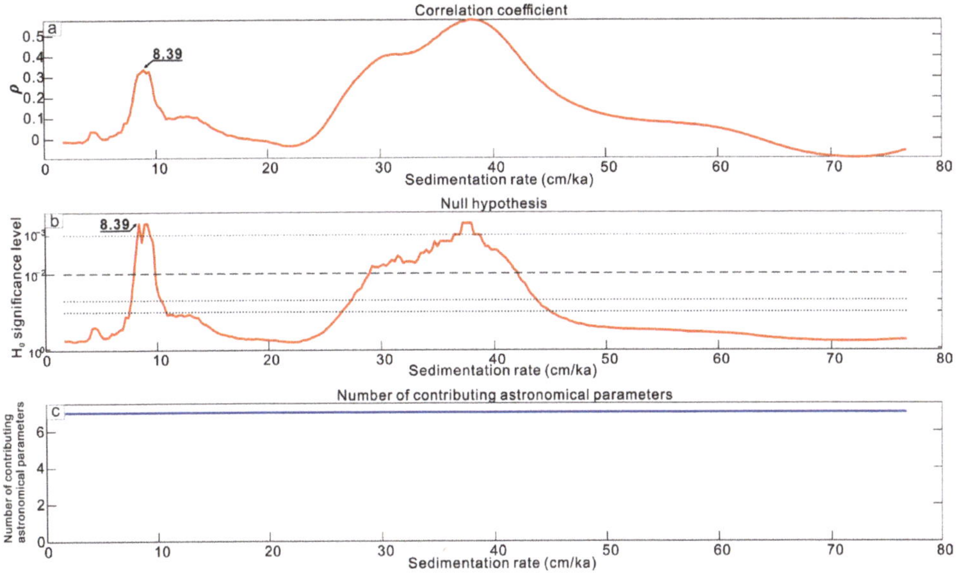

Figure 4. Correlation coefficient analysis results of GR curve "COCO" in the Es_3l of well JYC-1. (**a**) Correlation coefficient. (**b**) H_0 significance level. (**c**) Number of orbital parameters.

4.1.4. Filtering and Tuning

Based on the outcomes of spectral analysis, sedimentary cycles representing astronomical orbital parameters were extracted through Gaussian bandpass filtering. Specifically, for the long eccentricity (E, 405 ka) cycle, the sedimentary cycle (38.9 m) was filtered at a frequency of (0.024 ± 0.008) cycles/m, revealing approximately six long eccentricity cycles recorded in the Es_3l sub-member (Figure 5). For the short eccentricity (e, 100 ka) cycle, the sedimentary cycle (9.7 m) was filtered at a frequency of (0.110 ± 0.01) cycles/m, indicating the presence of roughly 26 short eccentricity cycles in the same Es_3l sub-member (Figure 5). The obliquity (O, 40 ka) cycle, with a sedimentary cycle of 3.7 m, was filtered at a frequency of (0.25 ± 0.01) cycles/m, revealing approximately 56 obliquity cycles in the Es_3l sub-member (Figure 5). Lastly, the precession (P, 22 ka) cycle, manifested as a sedimentary cycle of 1.96 m, was filtered at a frequency of (0.48 ± 0.04) cycles/m, demonstrating the recording of approximately 105 precession cycles within the Es_3l sub-member (Figure 5).

4.2. Astronomical Cycle Stratigraphic Division

4.2.1. GR Curve and Lithological Characteristics

The GR data of the selected key wells in the Bonan Sag, such as Luo53, Luo69, Luo 67, JYC-1, Y289, Y189, and Y107 in the Es3l submember, display significant cyclical characteristics. The GR values range from 5 to 150 API, with low values (<40 API) corresponding to lime–mudstone or sandy interbeds, and high values (>60 API) associated with mudstone. There are notable differences in the average GR values across the wells: the average GR value for JYC-1 is 52.3 API (standard deviation ±18.5); for Luo 53, it is 48.7 API (±16.8); and for Y107, due to its proximity to the source area, the average GR value is higher (63.2 API, ±21.3), reflecting spatial differentiation in the sedimentary environment (Figure 6).

In terms of lithology, the Es_3l submember of the JYC–1 well is mainly composed of dark-grey mudstone in deep-lake facies, interspersed with thin layers of lime–mudstone (single-layer thickness 0.5–3 m), with occasional sandy mudstone (Figure 5). The mudstone layers have a low GR value (30–45 API), reflecting endogenous carbonate deposition, while the mudstone corresponds to higher GR values (60–80 API). The lower part of the Luo53 well is primarily interbedded mud lime–mudstone and mudstone, with the upper part transitioning into mudstone interbedded with thin layers of limestone, showing a stepped increase in the GR curve. The Luo 69 well exhibits relatively uniform lithology, mainly rhythmic layers of mudstone and mud limestone, with the GR curve showing significant periodic fluctuations (±15 API), indicating a stable endogenous sedimentary environment. Additionally, in the Y189 well, located at the edge of the lake basin slope, the Es3l submember has multiple sets of sandstone interbeds (single-layer thickness 0.1–0.3 m), which correspond to a sharp drop in GR values (<40 API), reflecting the periodic transport of terrigenous clastic material (Figure 6).

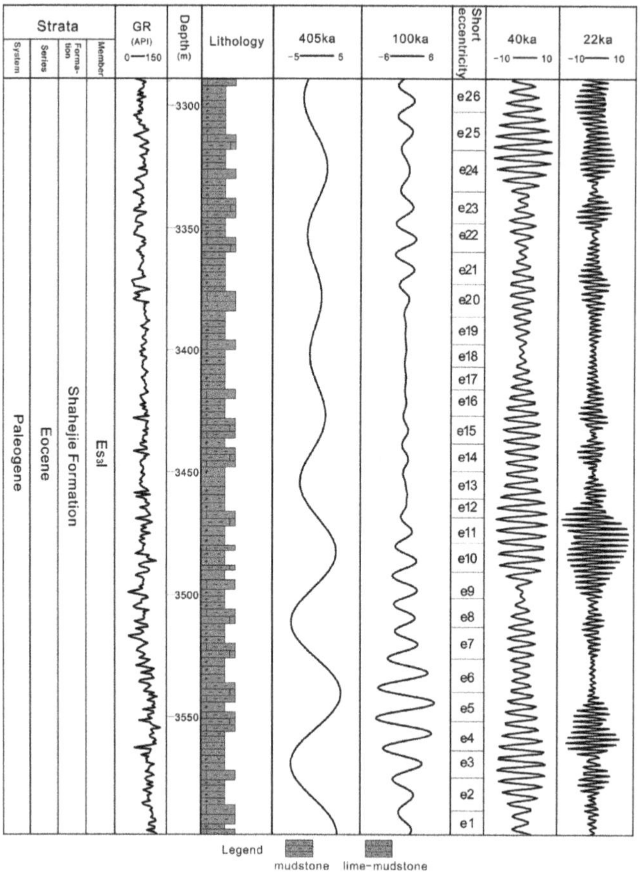

Figure 5. Stratigraphic division in Es_3l sub-member of single Well JYC-1.

4.2.2. Single Well Astronomical Cycle Stratigraphic Division

The periodic variations in Earth's and Sun's orbits allow for fluctuations in the insolation, thereby influencing climate change. Considering the sedimentary environmental implications of GR data indices [29], we utilize the astronomical cycle curves as a datum plane, serving as interfaces for high-resolution stratigraphic subdivision. The filtered astronomical cycle curves from selected single wells are employed for this purpose.

In the stratigraphic subdivision based on astronomical cycles within single wells, the Well L53 Es_3l sub-member comprises six long eccentricity cycles and 25 short eccentricity cycles. Similarly, the Well L69 Es3l sub-member contains six long eccentricity cycles, 25 short eccentricity cycles, 56 obliquity cycles, and 104 precession cycles (Figure 6). Wells L67, JYC1, Y289, Y189, and Y107 each also exhibit six long eccentricity cycles and approximately 25 short eccentricity cycles in their respective Es3l sub-member. This stratigraphic subdivision reveals a relatively stable stratigraphic development across the region, characterized by the consistent presence of roughly six long eccentricity cycles and approximately 25 short eccentricity cycles. Drawing upon the results of astronomical cycle analysis from single wells, we employ the short eccentricity curve as the benchmark for well-to-well stratigraphic correlation, thereby establishing a high-resolution stratigraphic framework for deep-water areas. This framework, based on short eccentricity curves, facilitates detailed stratigraphic subdivision, essential for exploration and development. Additionally, it enables the prediction of lithofacies, as horizontal well fracturing often involves intervals spanning tens of meters. By leveraging the high-resolution stratigraphic framework, fracturing can be targeted within specific sub-cycles, while the extensive horizontal radius of the fracturing (reaching kilometers) necessitates rapid lateral lithofacies changes, which can be anticipated and constrained through the stratigraphic framework.

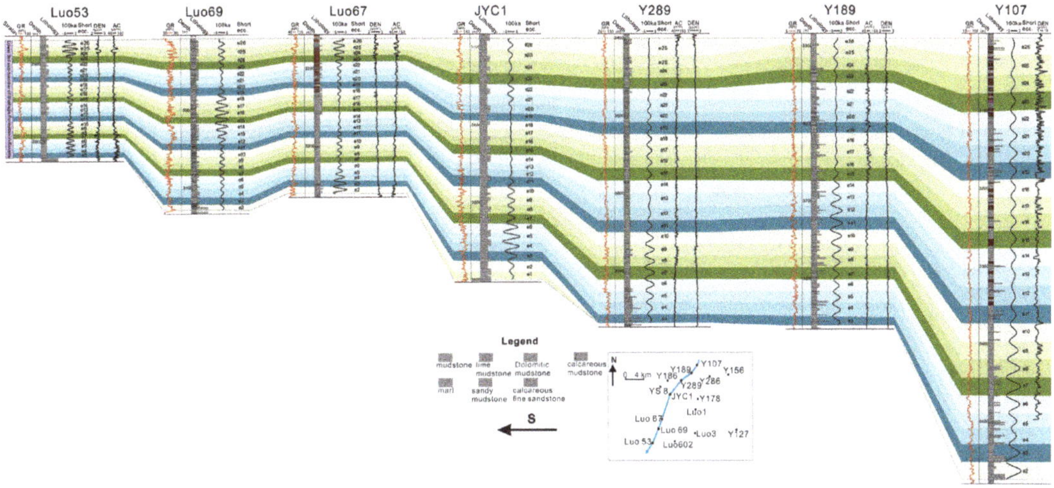

Figure 6. Comparison of well-connected strata with 100 ka short eccentricity cycle.

4.2.3. Stratigraphic Correlation of Connected Wells

By conducting high-precision stratigraphic subdivision based on astronomical cycles from single wells, we anchored the boundaries of the top and bottom of the Es3l sub-member in the Bonan Sag using magnetic stratigraphic ages as reference points [37]. Within this sub-member, the short eccentricity cycles were employed as stratigraphic correlation interfaces. The comparison revealed that the l Es3l sub-member in the Bonna Sag comprises approximately six long eccentricity cycles, numbered sequentially from the base to the top as the 1–6 long eccentricity cycles. Additionally, the number of short eccentricity cycles within this submember remains stable at 25 (Figure 6).

In the well-tie sections (Figure 6), Wells Luo53 Es3l sub-member displays marl and lime–mudstone predominantly in long eccentricity cycles 1–2, indicative of endogenic sedimentation in shallow water environments (dominated by autochthonous algae and carbonate deposition in lake basins) [17]. In contrast, long eccentricity cycles 3–6 are characterized by dark-gray mudstones with intermittent limestone intervals. Well Luo69 exhibits a well-correlated succession of mudstones and marl throughout the Es3l sub-member, dominated by endogenic sedimentation. Well Luo67, the lower half of the Es3l sub-member, comprises mainly mudstones in long eccentricity cycles 1–3, with limited marl development, suggesting mixed-source deposition (a blend of autochthonous and allochthonous components). However, the upper half features abundant marl, particularly in long eccentricity cycle 4, while cycles 5 and 6 alternate between mudstones, likely resulting from mixed-source depositional events [42]. Well JYC1 exhibits a regular succession of mudstones and marl, with a brief interval of sandy mudstone in long eccentricity cycle 5. Well Y289, the lowermost part, long eccentricity cycle 1, is dominated by marl and some dolomite, while cycles 2–6 are dominated by mudstones with scattered marl intervals. The Well Y189 Es3l sub-member primarily comprises mudstones and marl, interspersed with sandstone and siltstone within each long eccentricity cycle, possibly due to its location on the slope margin of the lake basin [7]. Well Y107, being proximal to a provenance area, displays conglomerate at the base and mudstones and shales in the mid-to-upper sections.

There exists an indirect control relationship between long eccentricity and various sedimentary processes (whereby long eccentricity governs precession, which in turn influences the intensity of terrestrial input and ultimately controls lithofacies development). Therefore, by analyzing the lithofacies types within the corresponding sub-intervals of each cycle, we can further leverage the stratigraphic framework to predict lithofacies types.

4.2.4. Depositional Evolution of Sandstone Interbedded Mudrock Reservoirs

Sandy interlayer shale oil represents a significantly important type of shale oil. Its reservoirs are characterized by the development of numerous centimeter- to decimeter-scale thin sandstone layers (Figure 7), which exert a crucial influence on the quality of the "sweet spots". However, vertical lithological distributions and evolutionary features are often indistinguishable in seismic and well logging data. Due to the abundance of sandy interlayer shale reservoirs within the study area, the precise prediction of these thin sandstone layers is imperative to enhance exploration efficiency for this type of shale oil. Therefore, achieving integrated prediction of sandstone and shale quality within interlayered shale oil reservoirs, as well as exploring the depositional and evolutionary patterns of thin sandstone interlayers, holds great significance for the exploration and development of this type of shale oil.

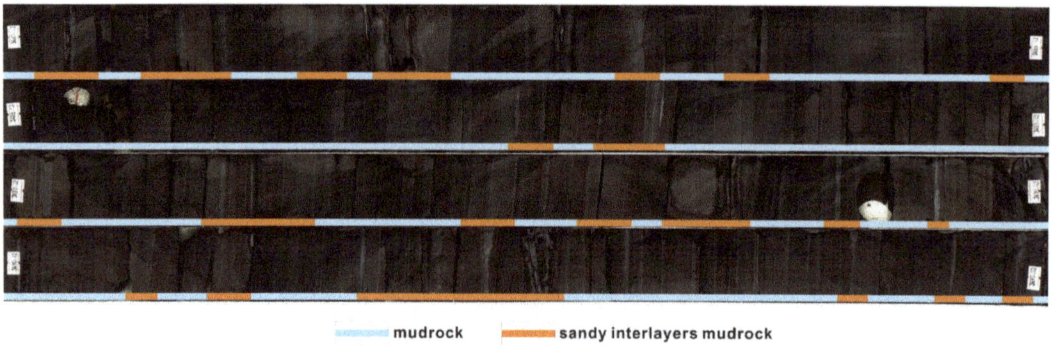

Figure 7. The mudrock core from Well JYC-1, featuring thin sandy interlayers on the centimeter-to-decimeter scale.

The sandstone thicknesses within the three 100 ka cycles (Cycle A, Cycle B, and Cycle C) contained within a single 405 ka cycle in the Es3l of Bonan Sag were individually measured and statistically analyzed. Using well coordinates, points were plotted on a plan view, and contour maps were created to represent the sandstone thickness within each

100 ka cycle (Figure 8). The contour maps revealed that Cycle A had a maximum total thickness of centimeter-to-decimeter-scale sandstone of 8 m, with a common range of 1 to 3 m (Figure 8a). Cycle B showed a maximum total thickness of 18 m, commonly distributed between 8 and 18 m (Figure 8b). For Cycle C, the maximum total thickness was 10 m, with a common range of 1 to 4 m (Figure 8c). Cycle B corresponds to the maximum phase of the 405 ka long eccentricity cycle, while both Cycle B and Cycle C correspond to the minimum phase of the same 405 ka cycle (Figure 9). It is evident that more centimeter-to-decimeter-scale sandstone developed during the maximum phase of the 405 ka long eccentricity cycle, while it was less developed during the minimum phase.

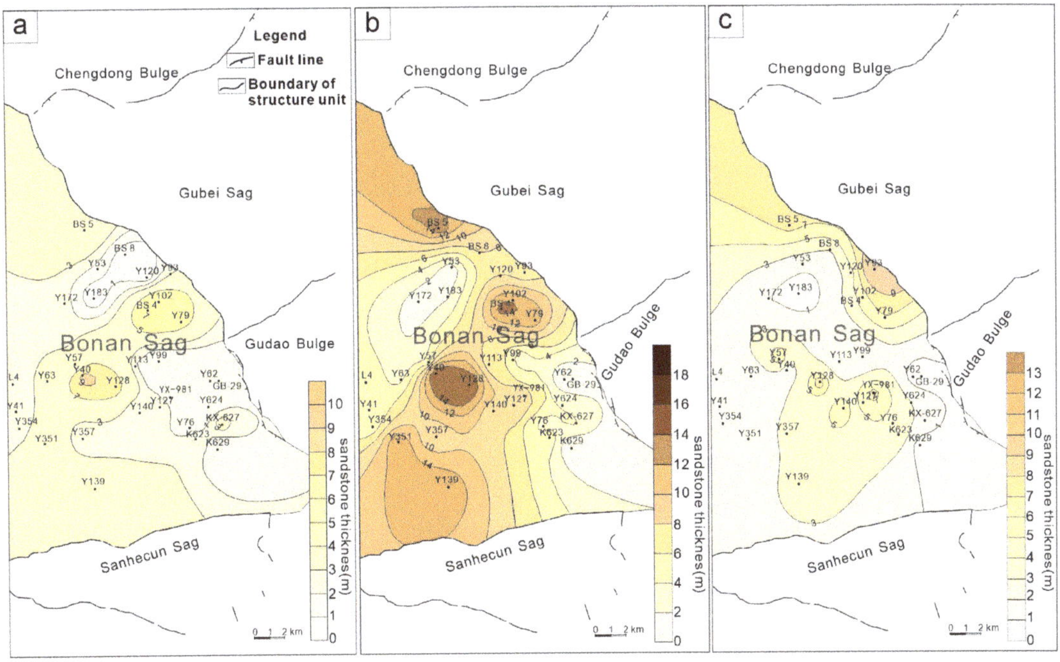

Figure 8. Thickness distribution characteristics of sandstone in the Es_3l sub-member in Bonan Sag with the 100 ka short eccentricity cycle. (**a**) Cycle A: the top 100 ka cycle of a 405 ka cycle. (**b**) Cycle B: the central 100 ka cycle of a 405 ka cycle. (**c**) Cycle C: the bottom 100 ka cycle of a 405 ka cycle.

Within the 405 ka long eccentricity cycle, the short eccentricity cycle with a period of 100 ka and the precession cycle with a period of 22 ka jointly regulate the deposition frequency and thickness of sandstone, as well as the mineral content and organic carbon content variations in mudrock. Specifically, during the high-value phase of the 100 ka short eccentricity cycle, sandstone layers develop frequently and accumulate to a significant thickness, while the felsic mineral content in mudrock is high, and the organic carbon content is low. The 22 ka precession cycle further controls the intricate changes within the depositional internal structure of sandstone-bearing shale. During its low-value phase, the development frequency of sandstone layers is high, with greater thickness, and the felsic mineral content in sandstone-bearing shale is high, while the organic carbon content is low (Figure 9, Table 1).

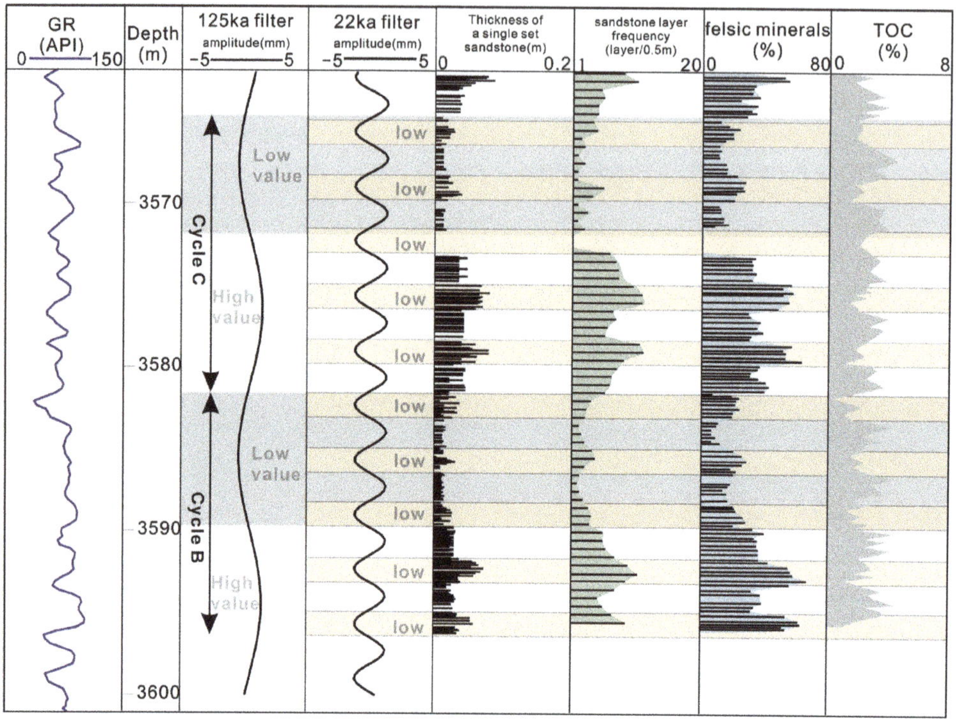

Figure 9. High-frequency evolution characteristics of sandstone–mudrock integration in well JYC-1 of Bonan Sag under the control of the 100 ka short eccentricity cycle and 22 ka precession cycle.

Table 1. Relationship between short eccentricity cycle, precession cycle, and high-frequency evolution of sandstone–mudrock integration in the Es_3l submember of Bonan Sag. The data statistics are based on the data presented in Figure 9.

100 ka Short Eccentricity	22 ka Precession	Thickness of Single-Layer Sandstone	Frequency of Sandstone Development	Felsic Minerals Content	TOC Content
Low value	Low value	Thin	Low	Low	Relatively high
	High value	Extremely thin	Extremely low	Extremely low	High
High value	Low value	Extremely thick	Extremely high	Extremely high	Low
	High value	Thick	High	High	Relatively low

5. Discussion

5.1. Analysis of the Depositional Evolution of Sandstone-Interbedded Mudrock Driven by Astronomical Force

By conducting stratigraphic correlation across wells based on astronomical cycles, it was found that the development of sandstone-interbedded mudrock and its thin sandstone interlayers is driven by astronomical cycles, and the development law of sandstone-interbedded mudrock was identified. The process of deposition constrained by astronomical cycles influences climate change by altering the amount of solar radiation received at the Earth's surface through changes in orbital parameters, which in turn drives deposition [43]. During periods of high value eccentricity, the Earth's orbit approaches an ellipse, resulting in increased solar radiation [44]. Under the influence of the southeast monsoon climate in the Northern Hemisphere, the climate becomes hot and humid with increased rainfall, and the lake basin expands, enhancing the terrigenous input received by the lake basin. Conversely, during periods of low value eccentricity, the Earth's orbit approaches a circle, leading to decreased solar radiation [44]. During these cold and arid times, rainfall decreases, lake levels fall, and the lake basin shrinks, weakening the terrigenous input received by the lake basin. Additionally, variations in the amplitude of precession also affect the strength of terrigenous input, with weakening or enhancement of terrigenous input received by the lake basin at both the maximum and minimum values of precession.

During the variation of the precession's amplitude, when the Northern Hemisphere is at perihelion, it experiences a short and hot summer followed by a long and cold winter at aphelion, indicating enhanced seasonality, whereas at aphelion, it experiences a long and hot summer followed by a short and warm winter, indicating weakened seasonality [14,45]. At the maximum value of the precession curve, the Northern Hemisphere experiences maximum sunshine during summer and minimum sunshine during winter, meaning that the winter solstice occurs at aphelion, while the summer solstice occurs at perihelion, resulting in a brief warm summer and a long, cold winter, during which the lake basin is in a period of contraction, and the deep-water areas of the lake basin receive less terrigenous input (Figure 10a,c). At the minimum value of the precession curve, the precession parameters are opposite to those at the maximum value, meaning that the Northern Hemisphere experiences a long, hot summer followed by a short, warm winter at aphelion (orbital schematic in Figure 10b,d). During this time, the lake basin is in a period of lake level expansion, with enhanced terrigenous input and more sandstone development.

Eccentricity itself cannot directly influence sedimentary evolution. Rather, it requires the combined action of eccentricity and precession to alter the distribution of solar energy [15]. At high eccentricity values, the Earth's orbital path approaches an ellipse, resulting in increased solar radiation (Figure 10a,b), whereas at low eccentricity values, the Earth's orbital path approaches a circle, leading to decreased solar radiation (Figure 10c,d) [44]. Eccentricity influences climate change by modulating the amplitude of precession [46,47], specifically by affecting the amplitude of precession variations [37]. This changes the amount of solar radiation received at the Earth's surface, thereby influencing climate change and terrigenous input and, ultimately, constraining lake basin sedimentary evolution. Therefore, within the 100 ka short eccentricity cycle, sandstone layers in sandstone-interbedded mudrock are thicker during high eccentricity periods and thinner during low eccentricity periods. Moreover, the 100 ka short eccentricity cycle modulates the amplitude of precession, leading to an enhanced climatic constraint on the 100 ka cycle during low precession values, resulting in the thickest sandstone layers in sandstone-interbedded mudrock (Figure 10a). Conversely, during high precession values, the climatic constraint on the 100 ka cycle is weakened, resulting in thinner sandstone layers in sandstone-interbedded shale (Figure 10b).

5.2. The Petroleum Geological Significance of Astronomical Cycle Analysis

The deposition of fine-grained sedimentary rocks is characterized by continuity, which often leads to inconspicuous responses in well logging and seismic data, as well as subtle lithofacies variations. This limitation poses challenges to the application of sequence stratigraphy in the stratigraphic subdivision of fine-grained sedimentary rocks [9], thereby restricting the exploration potential for shale oil in the Es_3l sub-member in the Bonan Sag. Nevertheless, fine-grained sedimentary rocks serve as an ideal medium for astronomical cycle analysis, as the periodic variations in Earth's orbital parameters govern the deposition of lacustrine fine-grained sedimentary rocks. Hence, the temporal significance inherent in astronomical cycle theory can be harnessed for high-precision stratigraphic subdivision and correlation.

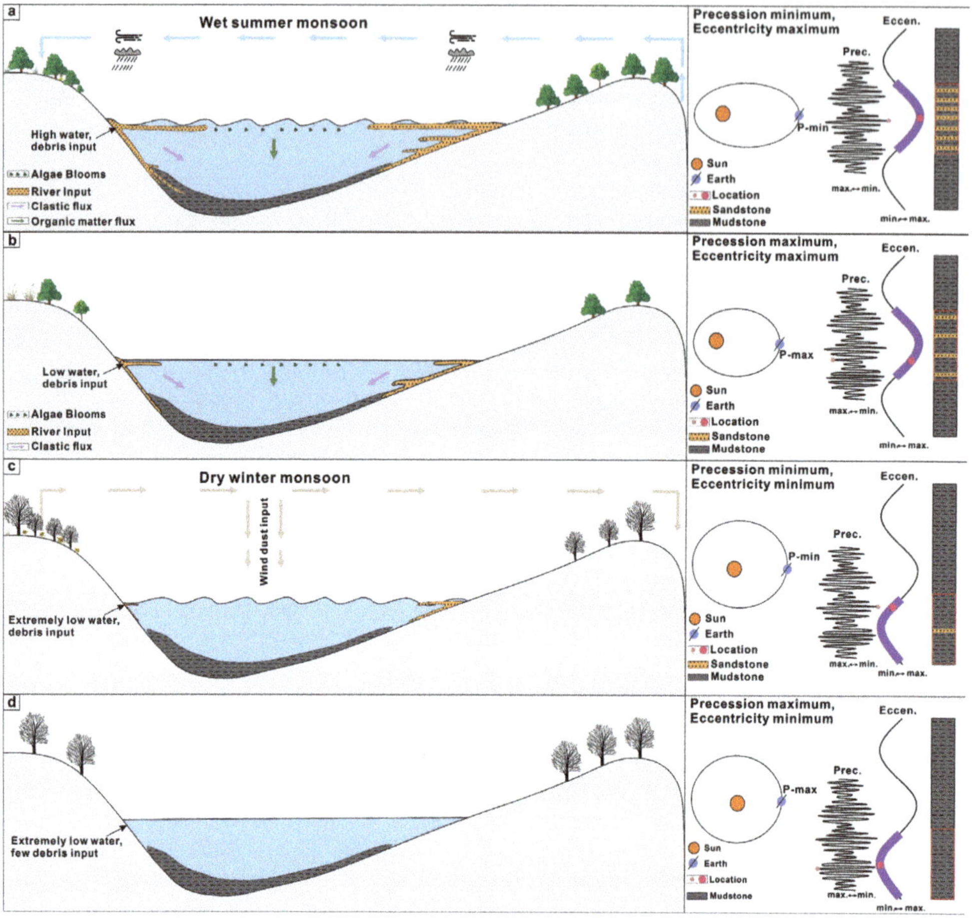

Figure 10. Terrigenous input models driven by astronomical cycle orbits and corresponding Earth orbits. The left side of this figure displays depositional model, while the right side presents the corresponding Earth orbits and astronomical cycle curves. (**a**) When eccentricity reaches its maximum value and precession is at its minimum, terrigenous input is strongest. (**b**) As eccentricity decreases but remains at a relatively high level, and precession is at its maximum, terrigenous input is still strong but weaker than in stage A. (**c**) As eccentricity continues to decrease and precession reaches its minimum, there is some terrigenous input, but it is weaker than in stage B. (**d**) When eccentricity is at its minimum and precession is at its maximum, terrigenous input is minimal. The evolution of terrestrial input is controlled by the synergy of short eccentricity and precession, and the influence of obliquity is almost non-existent. Therefore, only precession and short eccentricity are listed in the Earth orbits and astronomical cycle curves on the right.

Through stratigraphic correlation of connected wells, a more detailed and intuitive understanding of the lithofacies development patterns and spatial distribution relationships within each sub-interval can be achieved. For instance, Zhang et al. (2022) have applied astronomical cycles to perform stratigraphic correlation of connected wells in fine-grained sedimentary rocks of the Dongying Depression, revealing a lithofacies development pattern where the coupling of mudstone–marl is constrained by the precession half cycle under astronomical cycle constraints [17]. This approach also facilitated predictions of lithofacies types in the central part of the lake basin. For the Bonan Sag, the next step involves utilizing astronomical cycle theory to investigate lithofacies development patterns and spatial configuration relationships. Based on the stratigraphic framework established by astronomical cycle curves, and leveraging the temporal characteristics of astronomical cycles [9] and the climate-controlling properties of Earth's orbital periods [18], we aim to genetically interpret and predict lithofacies types. Ultimately, this will enable the prediction of geological "sweet spots" for shale oil and gas in the Bonan Sag, thereby providing valuable insights for the precise exploration of shale oil and gas resources in this region.

6. Conclusions

(1) The sedimentary cycle periods and optimal sedimentation rates were obtained. Utilizing GR as a proxy, spectral analysis has illuminated the intricate sedimentary cycles within the Es_3l sub-member in the Bonan Sag. These cycles encompass periods of 405, 100, 51, 40, 22, and 19 ka. Notably, the stratigraphic thickness modulated by the 405 ka long eccentricity reaches approximately 38.9 m, whereas the 97 ka short eccentricity governs a thickness of roughly 9.7 m. The 51 ka and 40 ka obliquity constrain the thickness to a range of 3.7 to 3.4 m, and the 22 ka and 19 ka precessions further limit it to 1.96 to 1.66 m. Through the application of the "COCO" (Correlation Coefficient) method, the optimal deposition rate for the Es_3l sub-member has been calculated to be 8.39 cm/ka. Leveraging magnetic stratigraphic data, we have anchored the ages of the top and bottom boundaries at 40.20 Ma and 42.47 Ma, respectively, and subsequently constructed a "floating" astronomical timescale for Well JYC1, utilizing the 100 ka long eccentricity filtered curve as a reference.

(2) Different cycles, including long-eccentricity, obliquity, and precession cycles, were delineated, serving as the basis for stratigraphic division. When conducting astronomical cycle stratigraphic division for single wells in the Es_3l of the Bonan Sag, it was found that there were consistently 6 long-eccentricity cycles, 25 short-eccentricity cycles with a slope of approximately 56, and about 104 precession cycles. During well-to-well stratigraphic correlation, within the 405 ka long-eccentricity cycle, stratigraphic division was conducted using a 100 ka short-eccentricity filtered curve, resulting in the identification of approximately 25 short-eccentricity cycles. This approach achieved high-precision stratigraphic division and correlation.

(3) It has been discovered that the synergistic effect of short eccentricity and precession controls the quality of sandstone-interbedded mudrock. Specifically, higher eccentricity values are associated with the development of thicker sandstone interbeds, while minimal precession values tend to enhance the thickness of these sandstone interbeds. Through astronomical cycle analysis, the depositional evolution mechanisms of sandstone-interbedded mudrock have been unveiled. Combining these insights with high-precision stratigraphic classification results provides a basis for detailed evaluation of lacustrine shale oil reservoirs and prediction of "sweet spots".

Author Contributions: Conceptualization, J.Z., Q.Z., W.L., Y.L. and P.L. (Peng Li); Methodology, J.Z., Q.Z., W.L., Y.L. and P.L. (Peng Li); Software, Q.Z., P.L. (Pinxie Li) and S.P.; Validation, J.Z., Q.Z., W.L., Y.L. and P.L. (Peng Li); Formal analysis, Q.Z., P.L. (Peng Li), P.L. (Pinxie Li), S.P. and X.Y.; Investigation, Q.Z., P.L. (Peng Li), P.L. (Pinxie Li) and S.P.; Resources, W.L., Y.L. and P.L. (Peng Li); Data curation, W.L., Y.L. and P.L. (Peng Li); Writing—original draft, J.Z., Q.Z., W.L., Y.L., P.L. (Peng Li), P.L. (Pinxie Li) and S.P.; Writing—review & editing, Q.Z., X.Y.; Visualization, J.Z., Q.Z., W.L., Y.L., P.L. (Peng Li), P.L. (Pinxie Li), S.P. and X.Y.; Supervision, J.Z., Q.Z., W.L., Y.L., P.L. (Peng Li), P.L. (Pinxie Li), S.P. and X.Y.; Project administration, J.Z. and W.L.; Funding acquisition, J.Z., W.L., Y.L. and P.L. (Peng Li). All authors have read and agreed to the published version of the manuscript.

Funding: This article was supported by the Open Fund Project *Fine characterization and reservoir formation mechanism of thick and high-quality massive shale oil reservoirs in the Bonan Sag* [Grant No. 33550000-24-ZC0613-0022] of [Sinopec Key Laboratory of Shale Oil/Gas Exploration and Production Technology]. Funder: Sinopec Key Laboratory of Shale Oil/Gas Exploration and Production Technology. Funder number: 33550000-24-ZC0613-0022. The second funding support item for this project comes from the National Key R&D Program of China. Project name: Interaction Among Different Spheres, Hydrocarbon Resources, and Environmental Effects in the Cenozoic Plate Subduction Zone in Eastern China [Grant No.: 2023YFF0804302].

Data Availability Statement: The original contributions presented in this study are included in the article. Further inquiries can be directed to the corresponding author.

Conflicts of Interest: Author Wangpeng Li, Yali Liu, and Peng Li were employed by the company Sinopec. The remaining authors declare that the research was conducted in the absence of any commercial or financial relationships that could be construed as a potential conflict of interest.

References

1. Aplin, A.C.; Macquaker, J.H.S. Mudstone diversity: Origin and implications for source, seal, and reservoir properties in petroleum systems. *AAPG Bull.* **2011**, *95*, 2031–2059. [CrossRef]
2. Jiang, Z.; Liang, C.; Wu, J.; Zhang, J.G.; Zhang, W.Z.; Wang, Y.S.; Liu, H.M.; Chen, X. Several problems in the study of fine-grained sedimentary rocks containing oil and gas. *Acta Pet. Sin.* **2013**, *34*, 1031–1039. (In Chinese with English abstract)
3. Yan, J.; Yan, Y.; Peng, J.; Li, Z.; Geng, B.; Lai, F. Relationship between astrostratigraphy and cycle stratigraphy, research progress and significance. *Lithol. Reserv.* **2017**, *29*, 147–156. (In Chinese with English abstract)
4. Zou, C.; Zhu, R.; Wu, S.; Yang, Z.; Tao, S.; Yuan, X.; Hou, L.; Yang, H.; Yang, C.; Li, D.; et al. Conventional and unconventional oil and gas accumulation types, characteristics, mechanisms and Prospects—A case study of tight oil and tight gas in China. *Acta Pet. Sin.* **2012**, *33*, 173–187. (In Chinese with English abstract)
5. Yan, J.P.; Liang, Q.; Geng, B.; Feng, C.Z.; Kou, X.P.; Hu, Y. Well logging calculation method and application of geochemical parameters of lacustrine shale—Taking the lower member of Es3 in Bonan sag of Zhanhua sag as an example. *Lithol. Reserv.* **2017**, *29*, 108–116. (In Chinese with English abstract)
6. Zhao, X.Z.; Zhou, L.H.; Pu, X.G.; Jin, M.F.; Shi, Z.N.; Xiao, D.Q.; Han, W.Z.; Jiang, W.Y.; Zhang, W.; Wang, H. Favorable Conditions for Formation and Enrichment of Lacustrine Shale Oil in Faulted Lake Basin: A Case Study of the Second Member of Kongdian Formation in Cangdong Sag, Bohai Bay Basin. *Acta Pet. Sin.* **2019**, *40*, 1013–1029. (In Chinese with English abstract)
7. Zhang, J.G.; Jiang, Z.X.; Liu, L.A.; Yuan, F.; Feng, L.Y.; Li, C.S. Lithofacies and depositional evolution of fine grained sedimentary rocks in the lower submember of the Member 3 of Shahejie Formation in Zhanhua sag, Bohai Bay Basin. *Acta Pet. Sin.* **2021**, *42*, 293–306. (In Chinese with English abstract)
8. Boulila, S.; Galbrun, B.; Huret, E.; Hinnov, L.A.; Rouget, I.; Gardin, S.; Bartolini, A. Astronomical calibration of the Toarcian Stage: Implications for sequence stratigraphy and duration of the early Toarcian OAE. *Earth Planet. Sci. Lett.* **2014**, *386*, 98–111. [CrossRef]
9. Shi, J.; Jin, Z.; Liu, Q.; Huang, Z. Depositional process and astronomical forcing model of lacustrine fine-grained sedimentary rocks: A case study of the early Paleogene in the Dongying Sag, Bohai Bay Basin. *Mar. Pet. Geol.* **2020**, *113*, 103995. [CrossRef]
10. Berger, A. Milankovitch, the father of paleoclimate modeling. *Clim. Past* **2021**, *17*, 1727–1733. [CrossRef]
11. Berger, A. Milankovitch theory and climate. *Rev. Geophys.* **1988**, *26*, 624–657. [CrossRef]
12. Weedon, G.P. *Time-Series Analysis and Cyclostratigraphy*; Cambridge University Press: Cambridge, UK, 2003; pp. 10–14.
13. Strasser, A.H.; Heckel, P.H. Cyclostratigraphy concepts, definitions, and applications. *Newsl. Stratigr.* **2007**, *42*, 75–114. [CrossRef]
14. Huang, C.J. The current status of cyclostratigraphy and astrochronology in the Mesozoic. *Geosci. Front.* **2014**, *21*, 48–66.
15. Wu, H.C.; Zhang, S.H.; Feng, Q.L.; Fang, N.Q.; Yang, T.S.; Li, H.Y. Theoretical Basis, Research Progress and Prospect of Cycle Stratigraphy. *J. Earth Sci. (J. China Univ. Geosci.)* **2011**, *36*, 409–428. (In Chinese with English abstract)
16. Jin, Z.H.; Jiang, Z.X.; Zhang, J.G.; Liu, S.Q.; Chen, J.; Zhao, L.J.; Lei, H.R. Study on cycle stratigraphy of the upper Es4 member in Dongying depression—Taking well Fanye 1 as an example. *Sci. Technol. Eng.* **2017**, *17*, 21–28. (In Chinese with English abstract)
17. Zhang, J.; Jiang, Z.; Liang, C.; Baars, T.F.; Wang, Y.; Abels, H.A. Astronomical forcing of meter-scale organic-rich mudstone–limestone cyclicity in the Eocene Dongying sag, China: Implications for shale reservoir exploration. *AAPG Bull.* **2022**, *106*, 1557–1579. [CrossRef]
18. Kietzmann, D.A.; Palma, R.M.; Iglesia Llanos, M.P. Cyclostratigraphy of an orbitally-driven Tithonian–Valanginian carbonate ramp succession, Southern Mendoza, Argentina: Implications for the Jurassic–Cretaceous boundary in the Neuquén. *Basin. Sediment. Geol.* **2015**, *315*, 29–46. [CrossRef]

19. Fang, Q.; Wu, H.; Hinnov, L.A.; Wang, X.; Yang, T.; Li, H.; Zhang, S. A record of astronomically forced climate change in a late Ordovician (Sandbian) deep marine sequence, Ordos Basin, North China. *Sediment. Geol.* **2016**, *341*, 163–174. [CrossRef]
20. Abels, H.A.; Abdul Aziz, H.; Calvo, J.P.; Tuenter, E. Shallow lacustrine carbonate microfacies document orbitally paced lake-level history in the Miocene Teruel Basin (North-East Spain). *Sedimentology* **2009**, *56*, 399–419. [CrossRef]
21. Abels, H.A.; Aziz, H.A.; Ventra, D.; Hilgen, F.J. Orbital Climate Forcing in Mudflat to Marginal Lacustrine Deposits in the Miocene Teruel Basin (Northeast Spain). *J. Sediment. Res.* **2009**, *79*, 831–847. [CrossRef]
22. Ocakoğlu, F.; Açıkalın, S.; Yılmaz, İ.Ö.; Şafak, Ü.; Gökçeoğlu, C. Evidence of orbital forcing in lake-level fluctuations in the Middle Eocene oil shale-bearing lacustrine successions in the Mudurnu-Göynük Basin, NW Anatolia (Turkey). *J. Asian Earth Sci.* **2012**, *56*, 54–71. [CrossRef]
23. Wu, H.; Zhang, S.; Jiang, G.; Hinnov, L.; Yang, T.; Li, H.; Wan, X.; Wang, C. Astrochronology of the Early Turonian–Early Campanian terrestrial succession in the Songliao Basin, northeastern China and its implication for long-period behavior of the Solar System. *Palaeogeogr. Palaeoclimatol. Palaeoecol.* **2013**, *385*, 55–70. [CrossRef]
24. Yang, Y.; Miall, A.D. Migration and stratigraphic fill of an underfilled foreland basin: Middle–Late Cenomanian Belle Fourche Formation in southern Alberta, Canada. *Sediment. Geol.* **2010**, *227*, 51–64. [CrossRef]
25. Zhao, K.; Du, X.; Lu, Y.; Xiong, S.; Wang, Y. Are light-dark coupled laminae in lacustrine shale seasonally controlled? A case study using astronomical tuning from 42.2 to 45.4 Ma in the Dongying Depression, Bohai Bay Basin, eastern China. *Palaeogeogr. Palaeoclimatol. Palaeoecol.* **2019**, *528*, 35–49. [CrossRef]
26. Zhao, X.X.; Yan, J.P.; Wang, M.; He, X.; Zhong, G.H.; Wang, J.; Geng, B.; Hu, Q.H.; Li, Z.P. Characteristics of lacustrine shale intercalation in Shahejie Formation of Zhanhua Depression and logging identification method. *Lithol. Reserv.* **2022**, *34*, 118–129. (In Chinese with English abstract)
27. Zhou, L.H.; Pu, X.G.; Deng, Y.; Chen, S.Y.; Yan, J.H.; Han, W.Z. Several noteworthy problems in the study of fine-grained sedimentary rocks. *Lithol. Reserv.* **2016**, *28*, 6–15. (In Chinese with English abstract)
28. Tavakoli, V. Application of gamma deviation log (GDL) in sequence stratigraphy of carbonate strata, an example from offshore Persian Gulf, Iran. *J. Pet. Sci. Eng.* **2017**, *156*, 868–876. [CrossRef]
29. Prokoph, A.; Villeneuve, M.; Agterberg, F.P.; Rachold, V. Geochronology and calibration of global Milankovitch cyclicity at the Cenomanian-Turonian boundary. *Geology* **2001**, *29*, 523–526. [CrossRef]
30. Jia, S.; Jiang, Z.X.; Zhang, W.Z. Characteristics and controlling factors of shale oil reservoir in Zhanhua Depression. *Mar. Geol. Front.* **2018**, *34*, 29–38. (In Chinese with English abstract)
31. Liu, H.; Yuan, F.F.; Jiang, Y.L.; Hao, X.F.; Fang, X.Q. Overpressure characteristics and genetic mechanism of Paleogene in Zhanhua Depression. *J. China Univ. Pet. (Ed. Nat. Sci.)* **2021**, *45*, 23–32. (In Chinese with English abstract)
32. Li, C.; Zhu, X.M.; Zhu, S.F.; Geng, M.Y.; Bi, Y.Q.; Shu, Q.L.; Xu, F.G. Characteristics of shale reservoirs in the lower Es3 formation in Luojia area of Zhanhua Depression. *Acta Sedimentol. Sin.* **2015**, *33*, 795–808. (In Chinese with English abstract)
33. Li, M.; Hinnov, L.; Kump, L. Acycle: Time-series analysis software for paleoclimate research and education. *Comput. Geosci.* **2019**, *127*, 12–22. [CrossRef]
34. Cleveland, W.S. Robust locally weighted regression and smoothing scatterplots. *J. Am. Stat. Assoc.* **1979**, *74*, 829–836. [CrossRef]
35. Laskar, J.; Fienga, A.; Gastineau, M.; Manche, H. A new orbital solution for the long-term motion of the Earth. *Astron. Astrophys.* **2011**, *532*, 77–89. [CrossRef]
36. Fienga, A.; Laskar, J.; Morley, T.; Manche, H.; Kuchynka, P.; Poncin-Lafitte, C.L.; Budnik, F.; Gastineau, M.; Somenzi, L. INPOP08, a 4-D planetary ephemeris: From asteroid and time-scale computations to ESA mars express and venus express contributions. *Astron. Astrophys.* **2009**, *507*, 1675–1686. [CrossRef]
37. Liu, Z.; Huang, C.; Algeo, T.J.; Liu, H.; Hao, Y.; Du, X.; Lu, Y.; Chen, P.; Guo, L.; Peng, L. High-resolution astrochronological record for the Paleocene-Oligocene (66–23 Ma) from the rapidly subsiding Bohai Bay Basin, northeastern China. *Palaeogeogr. Palaeoclimatol. Palaeoecol.* **2018**, *510*, 78–92. [CrossRef]
38. Zhang, R.L.; Jin, S.D. Cycle stratigraphy of the lower Es3 member in well Luo 69 in Zhanhua sag, Bohai Bay Basin. *J. Cent. South Univ. (Sci. Technol.)* **2021**, *52*, 1516–1531. (In Chinese with English abstract)
39. Meyers, S.R.; Sageman, B.B.; Hinnov, L.A. Integrated quantitative stratigraphy of the cenomanian-turonian bridge creek limestone member using evolutive harmonic analysis and stratigraphic modeling. *J. Sediment. Res.* **2001**, *71*, 628–644. [CrossRef]
40. Meyers, S.R. Seeing red in cyclic stratigraphy: Spectral noise estimation for astrochronology. *Paleoceanography* **2012**, *27*, PA3228. [CrossRef]
41. Ma, C.; Meyers, S.R.; Sageman, B.B.; Singer, B.S.; Jicha, B.R. Testing the astronomical time scale for oceanic anoxic event 2, and its extension into Cenomanian strata of the Western Interior Basin (USA). *Geol. Soc. Am. Bull.* **2014**, *126*, 974–989. [CrossRef]
42. Zhang, J.G.; Jiang, Z.X.; Liu, P.; Kong, X.X.; Ge, Y.J. Deposition mechanism and geological assessment of continentaultrafine grained shale oil reservoirs. *Acta Pet. Sin.* **2022**, *43*, 234–249. (In Chinese with English abstract)
43. Tuenter, E.; Weber, S.L.; Hilgen, F.J.; Lourens, L.J. The response of the African summer monsoon to remote and local forcing due to precession and obliquity. *Glob. Planet. Change* **2003**, *36*, 219–235. [CrossRef]

44. Abels, H.A.; Kraus, M.J.; Gingerich, P.D.; Sheldon, N. Precession-scale cyclicity in the fluvial lower Eocene Willwood Formation of the Bighorn Basin, Wyoming (USA). *Sedimentology* **2013**, *60*, 1467–1483. [CrossRef]
45. Huang, C.; Hinnov, L.; Fischer, A.G.; Grippo, A.; Herbert, T. Astronomical tuning of the Aptian Stage from Italian reference sections. *Geology* **2010**, *38*, 899–902. [CrossRef]
46. Zachos, J.; Pagani, M.; Sloan, L.; Thomas, E.; Billups, K. Trends, Rhythms, and Aberrations in Global Climate 65 Ma to Present. *Science* **2001**, *292*, 686–693. [CrossRef]
47. Valero, L.; Garcés, M.; Cabrera, L.; Costa, E.; Sáez, A. 20 Myr of eccentricity paced lacustrine cycles in the Cenozoic Ebro Basin. *Earth Planet. Sci. Lett.* **2014**, *408*, 183–193. [CrossRef]

minerals

MDPI

Article

Origin and Reservoir Significance of Authigenic Minerals in Lacustrine Shales: A Case Study from the Paleogene Dongying Sag, Bohai Bay Basin, East China

Jihua Yan [1], Shiyue Chen [1], Zhiyun Yu [2,*], Pengfei Zhang [3] and Guozheng Feng [1]

[1] School of Geoscience, China University of Petroleum, Qingdao 266580, China; 20060003@upc.edu.cn (J.Y.); chenshiyue@upc.edu.cn (S.C.); fgzupc@163.com (G.F.)
[2] College of Geosciences, China University of Petroleum, Beijing 102249, China
[3] Research Institute of Petroleum Exploration and Development, Shengli Oilfield, SINOPEC, Dongying 257022, China; zpf0725@126.com
* Correspondence: 2022310009@student.cup.edu.cn

Abstract: Authigenic minerals in shale are products of the co-evolution of organic and inorganic components, affecting the heterogeneity of shale reservoirs. However, due to their fine granularity and complex rock composition, studies on these minerals in shale are still insufficient. This research focuses on the lacustrine shales from the upper sub-member of the fourth member in the Eocene Shahejie Formation, Dongying Sag, East China. Utilizing core samples, thin sections, scanning electron microscope, X-ray diffraction, elemental geochemistry, and organic geochemistry, we systematically characterized the features and origins of authigenic minerals. The results identified several typical authigenic minerals, including authigenic quartz, framboidal and euhedral pyrite, ferroan dolomite, kaolinite, chlorite, and albite. Authigenic quartz is predominantly diagenetic silica formed through smectite illitization, acidic dissolution of K-feldspar, and alkaline dissolution of detrital quartz. Pyrite is a product of microbial sulfate reduction, with framboidal pyrite forming during an early diagenetic stage under conditions with sufficient solute supply and euhedral pyrite forming during a later stage under conditions with insufficient solute supply. Ferroan dolomite originates from the precipitation of Fe and Mg during smectite illitization, with slight contributions from the acidic dissolution of chlorite and calcite. Kaolinite stems from the acidic dissolution of K-feldspar, while chlorite results from the transformation of kaolinite. Albite primarily arises from the alkaline alteration of anorthite and K-feldspar. Most non-clay authigenic minerals likely enhance reservoir quality by slightly reducing the effects of compaction, whereas authigenic clay minerals typically exert detrimental effects on reservoir properties. This study constrains the genesis of authigenic minerals to assess their influence on reservoir quality in lacustrine shale.

Keywords: Dongying sag; lacustrine shale; authigenic mineral; diagenetic evolution; reservoir quality

1. Introduction

Research on the diagenesis of clastic rocks has evolved over the past few decades from early qualitative observations to quantitative modeling, detailed predictions, and assessments of reservoirs [1–6]. This shift has significantly advanced the exploration of oil basins into more sophisticated domains [7,8]. Compared to clastic rocks and carbonate, organic-rich shales display similar detrital and authigenic minerals, but they are characterized by finer grain sizes, diverse compositions, and coexistence with organic matter [9–12].

Shale diagenesis is governed by both external controls (e.g., tectonic activity, depositional environment, burial history) and internal factors (e.g., lithofacies heterogeneity, organic matter composition and abundance, thermal maturity) [13–17]. The interplay between organic and inorganic diagenetic processes drives the formation of authigenic mineral assemblages, including microquartz, pyrite, sparry calcite, albite, clay minerals, and zeolites, which fundamentally shape reservoir properties [18–20]. Systematic analysis of these minerals provides key insights into microscale diagenetic systems and their control on the development of fine-grained reservoirs [6,14,21–25].

Extensive research has focused on diagenesis in organic-rich shales and its contributions to reservoir quality; however, relatively less attention has been paid to authigenic minerals formed during burial, despite their non-negligible impact on reservoir properties [26–30]. The formation of authigenic minerals is fundamentally controlled by diagenetic fluids. In the relatively closed diagenetic system of shale, the diagenetic fluids are primarily composed of connate sedimentary water, hydrocarbons, and acidic fluids released during the thermal maturation of organic matter, as well as interlayer water released during mineral transformation [9,11,31,32]. Early diagenetic processes in organic-rich shales are largely influenced by inherited primary sedimentary water, which has abundant bicarbonate ions, reduced manganese and iron, and sulfide [33–35]. They can be cemented into early diagenetic minerals such as calcite, pyrite, siderite, and dolomite [36–39]. During shallow burial, hydrocarbons are metabolized by biological processes and redox reactions, such as those involving methanogenic bacteria [40–42]. Key decomposed products of these reactions include phenols, amino acids, and fatty acids [31,43,44]. When thermal maturities greater than Ro~0.5% are reached, hydrocarbon molecules, organic acids, H_2O, CO_2, and N_2 within OM are released and have been described extensively as a mechanism for acidizing diagenetic systems [10,45–48]. Additionally, the dehydration of minerals such as clays and gypsum also affects the properties of diagenetic fluids [49–51]. Therefore, the complex diagenetic environment driven by these varying fluids highlights the inherent difficulties in resolving the complex origins of authigenic minerals in organic-rich shale successions [52,53], and these minerals are considered of significant importance for the reservoir quality of shales in many published studies [17,28,54]. Previous studies have mainly focused on authigenic minerals within marine shale, especially biogenic microquartz crystals, which generate abundant intercrystalline pores and are considered as critical factors enabling efficient shale gas production in both academia and industry [15,17,24,47]. However, related studies reported in lacustrine shales are limited. The complex petrology and fine grain size of lacustrine shales enhance the reactivity of their components, resulting in unique morphologies and distributions of authigenic minerals. These characteristics may lead to implications in reservoirs that are significantly different from those observed in marine shales.

The upper sub-member of the fourth member of the Eocene Shahejie Formation ($Es_4{}^U$) in the Dongying Sag of Bohai Bay Basin, China, is a typical saline lacustrine organic-rich shale succession and an important unconventional target [55,56]. This interval offers an excellent sample set to investigate the origin and diagenetic evolution of authigenic minerals and their impact on reservoir quality in lacustrine shales. This study focuses on detailed observations of core samples, thin sections, and scanning electron microscope (SEM + EDS), as well as quantitative analyses including X-ray diffraction (XRD), major and trace element analysis, total organic carbon (TOC) measurement, rock pyrolysis, and porosity analysis. The aim is to (i) decipher the formation mechanisms of authigenic minerals, (ii) reconstruct their evolutionary sequence, and (iii) assess their impact on lacustrine shale reservoirs. The findings are critical for reservoir quality evaluation and development optimization.

2. Geological Setting

The Dongying Depression is in the southeastern part of the Bohai Bay Basin, located in eastern China (Figure 1A), which is a typical half-graben rift basin (Figure 1B). The Dongying depression can be subdivided into seven secondary structural units, including the Minfeng Subsag, the Lijin Subsag, the Niuzhuang Subsag, the Boxing Subsag, the Southern Gentle Slope Zone, the Central Anticline Zone, and the Northern Steep Slope Zone, with an area of approximately 5700 km^2 (Figure 1C) [57]. The Paleogene strata within the study area are widely distributed and consist, from the bottom up, of the Kongdian Formation, the Shahejie Formation, and the Dongying Formation. The Shahejie Formation can be further divided into the fourth member (Es_4), the third member (Es_3), the second member (Es_2), and the first member (Es_1) [58,59]. The Es_4 is further divided into the Es_4^U and Es_4^L (Figure 1D). The Es_4^L primarily consists of mudstone, gypsum, coarse sandstone, and carbonate rock, while the Es_4^U is mainly composed of calcareous shale and dolomitic shale [58].

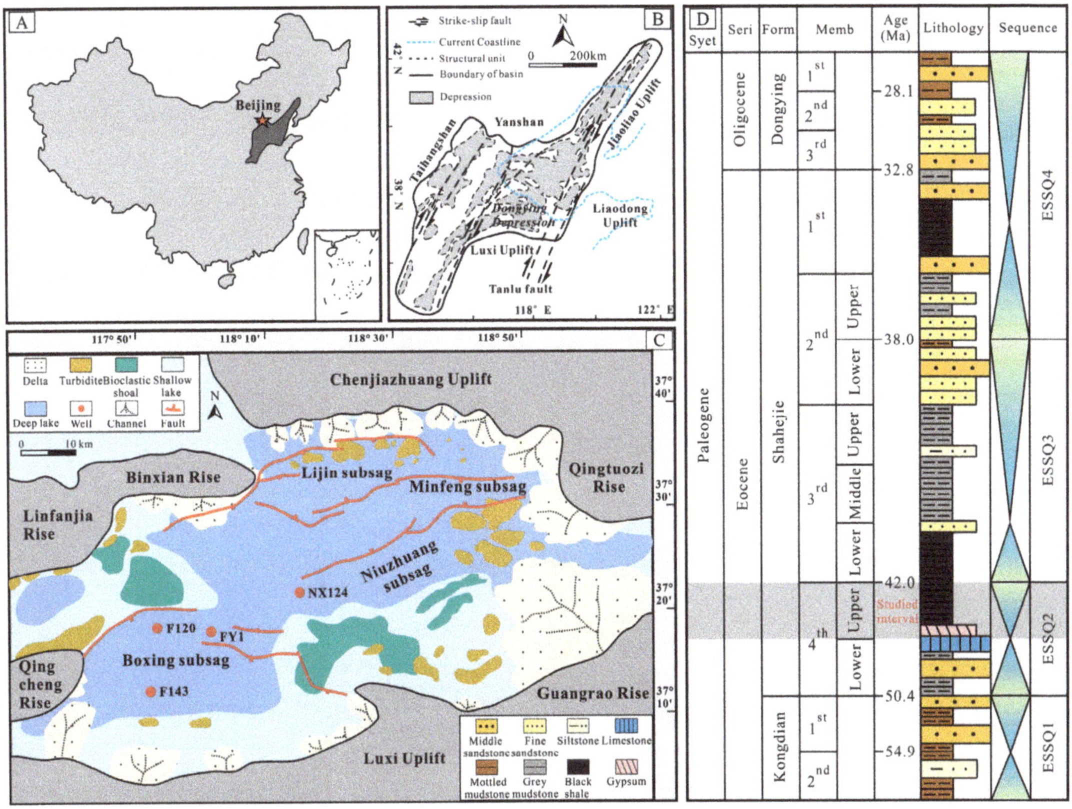

Figure 1. (**A**) Location of the Bohai Bay Basin in China. (**B**) Tectonic divisions of the Bohai Bay Basin. (**C**) The sedimentary environment of the upper sub-member of the 4th member of the Shahejie Formation (Es_4^U) in the Dongying Depression. (**D**) Composite lithostratigraphic charts of the Dongying Depression, Bohai Bay Basin, East China. The study interval is the Es_4^U.

According to previous research, the Dongying Depression quickly subsided with a large-scale lake expansion, reaching a maximum during the Es_4^U. The saline, lacustrine, black organic-rich shale succession formed at the lake bottom (~50 m), which are the primary source rocks for the petroliferous depression [60,61]. The water salinity varied from semi-saline to saline within our target layer [62].

3. Samples and Experimental Methods

In total, 136 samples from the NX124 well (3213 to 3243.15 m), FY1 well (3248 to 3430 m), F120 well (3269.5 to 3289 m), and F143 well (3110 to 3141.08 m) were selected to analyze the genesis of authigenic minerals in organic-rich shale. The methods used included X-ray diffraction (XRD), total organic carbon content (TOC) measurement, rock pyrolysis, vitrinite reflectance (Ro) analysis, major and trace element analysis, sporopollen analysis, scanning electron microscopy (SEM + EDS), and helium porosity testing. These samples were obtained from China's SINOPEC Shengli Oilfield Company and divided into bulk and cylindrical samples. The experiments were conducted in the Reservoir Geology and Basin Analysis Key Laboratory of the China University of Petroleum.

3.1. Petrographic Analysis

We selected 94 bulk samples from the FY1 well and 12 samples from the NX124 well for bulk-rock XRD analysis. First, we ground to powders of less than 300 mesh in size. The X'Pert PRO MPD with CuK α radiation (60 kV, 55 mA) and Ni filtering were used for X-ray diffraction (XRD) analyses to identify the bulk mineralogical composition. X-ray diffraction analysis was performed over a 2θ range of 1–160°, with mineral identification and semi-quantitative analysis based on characteristic peak intensities and positions. Subsequently, 43 samples from the FY1 well were separated for clay XRD. The samples were prepared as oriented mounts and subjected to ethylene glycol vapor treatment under 40–50 °C conditions without employing zincite (or similar) internal standards for quantitative calibration.

Thirty samples from the FY1 well were adhered to a standard glass slide with epoxy resin to analyze mineral and sporopollen characterization using a Leica DM2700P high-precision microscope (Leica Microsystems, Wetzlar, Germany).

3.2. Organic Geochemistry Analysis

The total organic carbon (TOC) contents of 102 bulk samples from the FY1 well and NX124 well were ground to powders of less than 200 mesh in size and treated with 12.5% hydrochloric acid to remove inorganic carbon. After rinsing with deionized water, the samples were filtered to achieve neutrality and dried in an oven at 60–80 °C. Finally, samples were analyzed using a Leco CS-744 carbon-sulfur analyzer with high-temperature oxygen flow combustion.

Rock pyrolysis was performed on 102 bulk samples from the FY1 well and NX124 well. The samples were ground to powders that were 200 mesh in size and placed in the Rock-Eval 6 analyzer. Then, samples were heated isothermally at 300 °C for three minutes to measure the free hydrocarbon content (S_1) released from the rock. Following this, the temperature was increased at a rate of 25 °C/min to 600 °C to obtain the pyrolyzed hydrocarbon content (S_2) generated from the thermal maturity of organic matter and to determine the temperature at which the maximum S_2 yield occurs (Tmax). The hydrogen index (HI) was calculated using the ratio of S_2 to TOC. The mixture was heated on a low-temperature hot plate at 140 °C for 48 h.

In addition, 22 samples from the FY1 well were separated for vitrinite reflectance to determine the diagenetic stage.

3.3. Inorganic Geochemistry Analysis

For major and trace element analysis, 12 samples from the NX124 well were ground to approximately 200 mesh. Then, about a 0.05 g sample was placed into a PTFE crucible with 25 mL of water, followed by the addition of 1 mL hydrochloric acid, 1 mL nitric acid, 2 mL hydrofluoric acid, and 0.5 mL perchloric acid. After evaporation to near dryness, 2 mL of a 1:1 nitric acid solution was added, and the sample was transferred to a 50 mL volumetric flask. The sample was then analyzed using an ELEMENT XR high-resolution inductively coupled plasma mass spectrometer (ICP-MS) (Thermo Fisher Scientific, Waltham, MA, USA).

3.4. SEM and EDS

The scanning electron microscope (SEM) and energy dispersive spectroscopy (EDS) analyses were performed on 12 samples selected from the NX124 well. The samples were first cut into slices with a diameter of approximately 1–3 cm and a thickness of about 0.2–1 cm. Then, 12 samples underwent mechanical and Ar-ion polishing to generate super-flat surfaces suitable for SEM examination. Subsequently, a thin layer of carbon film (approximately a few nanometers thick) was uniformly deposited on the sample surfaces. The Zeiss Crossbeam 550 FIB-SEM analyzer equipped with an energy dispersive spectrometer (EDS) was used as the observation instrument. In addition, 30 SEM samples from the FY1, F120, and F143 wells were analyzed, including Ar-polished images and fresh fracture surfaces images.

3.5. Helium Porosity

The helium porosity of 21 samples from the FY1 well and NX124 well was determined using the KXD-III helium porosity determinator to measure the porosity. Samples were prepared as cylinders with a diameter of 2.5 cm and height of 3 cm. To make helium fully fill the pores in the shale, the working pressure and equilibration time of the helium pycnometry test were 4 MPa for 8 min, respectively. Porosity was obtained from the difference between the bulk density and the skeletal density with a helium expansion method using Boyle's Law helium porosimetry [44].

4. Results

4.1. Petrologic Features and Lithofacies Types

The mineral composition of shale, as shown in Table S1, is diverse in the study area (Figure 2A). The calcite content is the highest, ranging from 1% to 80%, with an average of 39.95%. Quartz content varies from 4% to 55%, averaging 24.52%. Clay minerals range from 3% to 54%, with an average of 16.85%, primarily consisting of illite (average of 86.81%), followed by mixed layers of illite-smectite (average of 12.95%) (Figure 2B). Dolomite content ranges from 1% to 93%, with an average of 11.65%. The contents of plagioclase and pyrite are relatively low, averaging below 5%. The sporopollen identification shows that its color is light yellow to black.

The $Es_4{}^U$ is primarily characterized by calcareous shale, mixed shale, and dolomitic mudstone. Calcareous shale and mixed shale predominantly exhibit a well-lamellar texture, with thin sections revealing three-part laminations of bright carbonate laminae, brown felsic laminae, and dark organo-clay laminae. In contrast, dolomitic mudstone primarily exhibits a massive texture (Figure 2C).

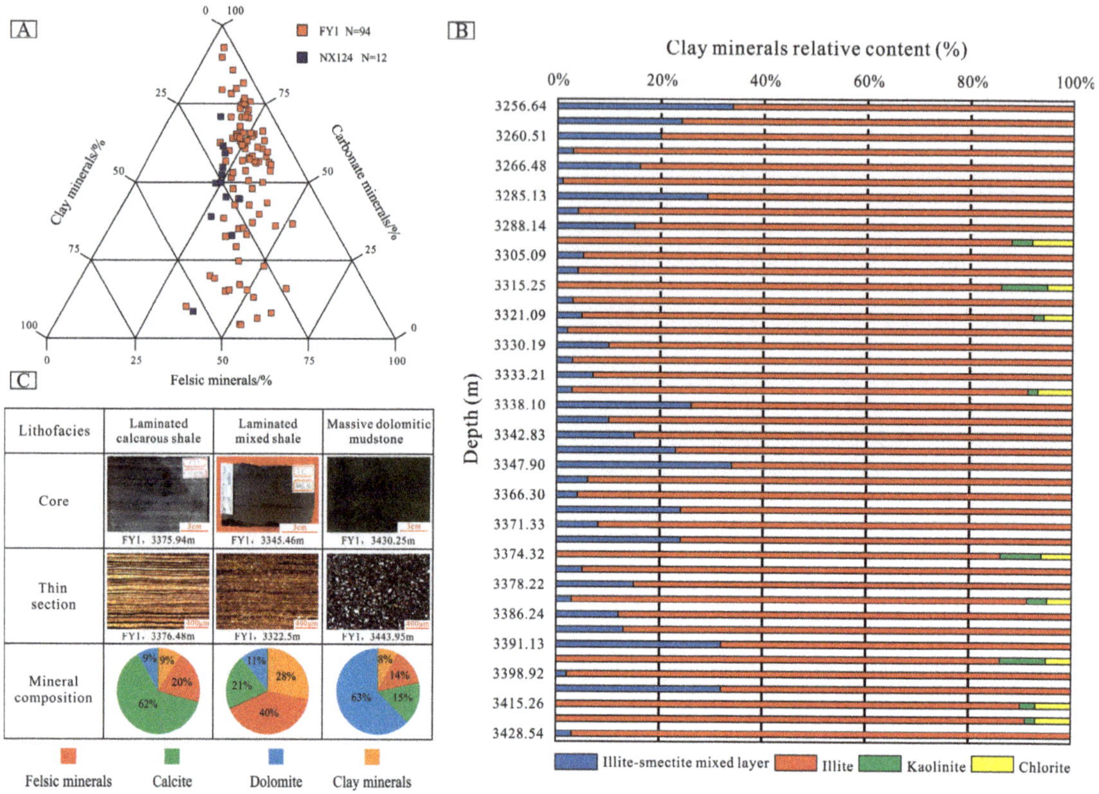

Figure 2. (**A**) Mineralogical composition and classification of shales from Es_4^U. (**B**) Bar graph showing relative content of clay minerals of shales from Es_4^U. (**C**) Core thin sections and XRD data showing lithofacies types and the sedimentary structures of shales from Es_4^U.

4.2. Geochemical Characteristics

4.2.1. Organic Geochemistry

Organic geochemical data indicate that the TOC values of samples from Es_4^U range from 0.55% to 9.05%, with an average of 2.27% (Table S2). Among these, 99.02% of the samples have values greater than 1%, and 51.96% have values greater than 2%. The S_1 values range from 0.34 to 5.8 mg/g, with an average of 2.28 mg/g (Figure 3A), while S_2 values range from 0.93 to 41.08 mg/g, averaging 7.49 mg/g (Figure 3B). The oil saturation index (OSI) can be used to characterize the mobility efficiency of retained hydrocarbons in shale reservoirs [63]. When the OSI exceeds 100, hydrocarbons exceeding the adsorption capacity of the organic matter will exhibit better mobility and oil production potential [19,64]. The OSI values of samples range from 28.82 to 377.52 mg/g, with an average of 107.86 mg/g, indicating that the Es_4^U shale has good oil potential. The Tmax values range from 431 to 461 °C, with an average of 448.04 °C. The HI values range from 129.17 to 544.69 mg/g, averaging 319.66 mg/g. According to the cross plot of Tmax and HI (Figure 3C), the OM is primarily comprised of type I and minor type II. The measured vitrinite reflectance (Ro) values of 22 samples range from 0.52% to 0.93%, indicating a thermal maturity stage spanning from low maturity to the oil window (Table S3).

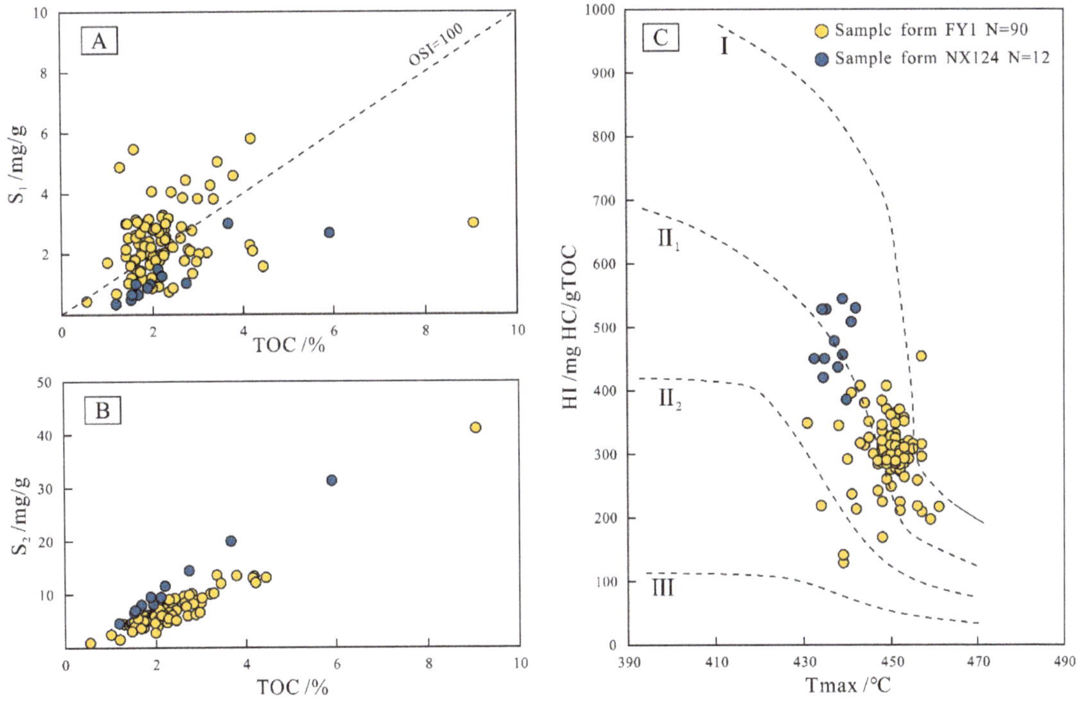

Figure 3. Organic geochemistry characteristics of shales from Es_4^U. (**A**) Relationship between S_1 and TOC, and OSI showing shale oil potential producibility. (**B**) Plot of S_2 versus TOC showing the hydrocarbon potential. (**C**) Plot of the hydrogen index (HI) versus pyrolysis Tmax showing the type and maturity of OM.

4.2.2. Elemental Geochemistry

The major element of samples is shown in Table S4. The most abundant element is SiO_2, with a content ranging from 27.4% to 50.2% and an average of 37.76%. The next most abundant elements are CaO and Al_2O_3, with average contents of 18.26% and 10.43%, respectively. The contents of Fe_2O_3, MgO, and K_2O range from 1% to 5%, while Na_2O is approximately 1%. The contents of TiO_2, P_2O_5, and MnO are generally below 1%.

The trace elements of samples are shown in Table S5. The most abundant trace elements are Sr and Ba, with average concentrations of 1985.58 ppm and 692.25 ppm, respectively. Following these are Li, B, V, Cr, Co, Ni, Cu, Zn, Ga, Rb, Y, and Zr, which have concentrations ranging from 10 to 100 ppm. Be, Sc, As, Nb, Mo, Cd, and Cs have the lowest concentrations, generally falling below 10 ppm.

4.3. Types and Characteristics of Authigenic Minerals

4.3.1. Characteristics of Authigenic Quartz

Authigenic quartz widely appears in forms with short columnar and hexagonal columnar clay-size euhedral microquartz, primarily developing within intergranular pores and adjacent to calcite and illite (Figure 4A).

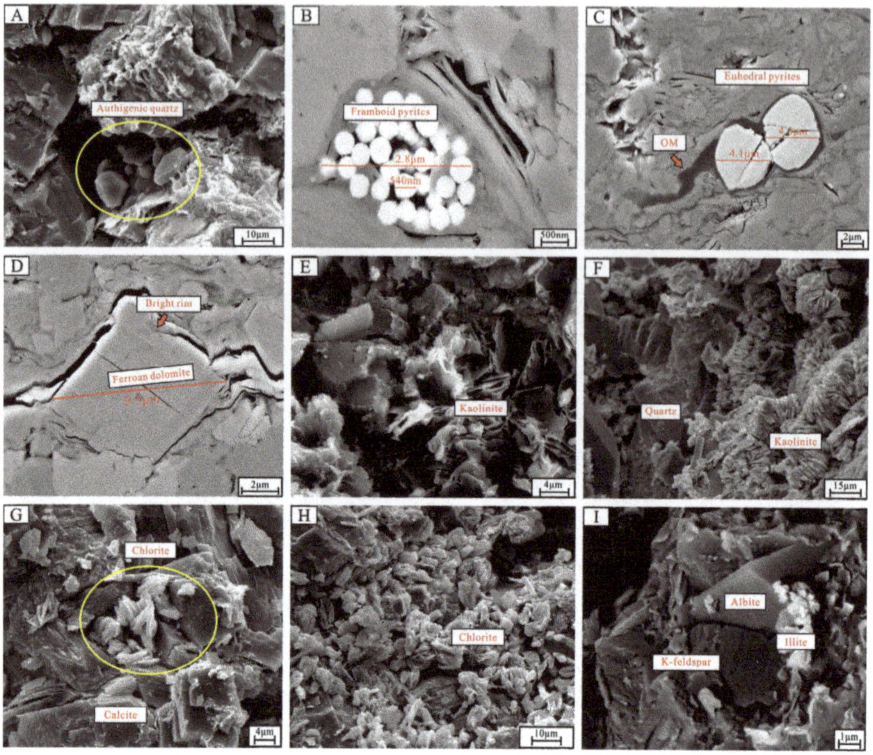

Figure 4. SEM images of typical authigenic minerals. (**A**) Microquartz appears in intergranular pores of calcite and clays, as shown in images of fresh fracture surfaces from the F143 well at a depth of 3116.82 m. (**B**) Framboidal pyrite with nearly equigranular submicron-sized crystals, as shown in Ar-polished images of the NX124 well at a depth of 3216 m. (**C**) Euhedral pyrites with isolated single crystals associated with organic matter, as shown in Ar-polished images of the NX124 well at a depth of 3221 m. (**D**) Ferroan dolomite exhibits an irregular core and annular bright rim with a size of about 9.5 μm, as shown in Ar-polished images of the NX124 well at a depth of 3221 m. (**E**) Kaolinite appears in leaf-like shapes, as shown in images of fresh fracture surfaces from the F143 well at a depth of 3128.39 m. (**F**) Kaolinite appears in book-like shapes with authigenic quartz, as shown in images of fresh fracture surfaces from the F143 well at a depth of 3021 m. (**G**) Chlorite appears in hairy shapes with calcite, as shown in images of fresh fracture surfaces from the F143 well at a depth of 3110 m. (**H**) Chlorite appears in plate-like shapes, as shown in images of fresh fracture surfaces from the FY1 well at a depth of 3112.85 m. (**I**) Euhedral albite appears with dissolved K-feldspar and illite, as shown in images of fresh fracture surfaces from the FY1 well at a depth of 3387.5 m.

4.3.2. Characteristics of Pyrite

Two distinct pyrite morphologies dominate the shale samples. Framboidal aggregates are composed of closely packed, submicron equigranular crystals, forming elliptical clusters (Figure 4B). Euhedral pyrites often appear as discrete crystals with cubic and occasionally elliptical morphologies, with the grain sizes ranging from 1 to 5 μm (Figure 4C). The pyrite concentrations in shales range from 0% to 34%, with an average of 2.84%.

4.3.3. Characteristics of Ferroan Dolomite

The ferroan dolomite is in the form of micron-sized euhedral crystals, which are comprised of an irregular core of dolomite and an annular rim of ankerite. Ferroan dolomite is commonly observed within clays, with grain size ranging from 2 to 20 μm

(Figure 4D). The crystal surfaces are flat and not curved, exhibiting a relatively complete rhombohedral shape.

4.3.4. Characteristics of Authigenic Clay Minerals

The authigenic clays are primarily comprised of kaolinite and chlorite. The kaolinite appears in book-like and leaf-like shapes, often accompanied by microquartz and dissolution of K-feldspar (Figure 4E,F). Chlorite usually appears in hairy and fibrous filaments (Figure 4G,H). The content of kaolinite and chlorite is limited, occurring only at specific depths (Figure 2B).

4.3.5. Characteristics of Albite

The albite is primarily in the form of small granular and columnar euhedral crystals, with grain sizes ranging from 5 to 10 μm. The albite is commonly observed filling the intergranular pore in association with dissolved K-feldspar and illite (Figure 4I). Compared to other feldspar, the albite was characterized by weak dissolution with no evidence of replacement or cementation by other minerals.

4.4. Porosity Characteristics

Twenty-one samples revealed the porosity characteristics of the $Es_4{}^U$ shale, as shown in Table S1. The porosity varies greatly in different wells. Among them, nine samples from the FY1 well showed porosity ranging from 2.65 to 9.43%, with an average of 6.13%. Twelve samples from the NX124 well showed porosity ranging from 7.37 to 15.6%, with an average of 11.25%. Both of them can be classified as low porosity reservoirs.

4.5. Diagenetic Stage

This study used Ro, Tmax, clay mineral types and assemblages, authigenic mineral types, pore types, and sporopollen color as primary indices for classifying the diagenetic stage [65–67]. The Ro ranges from 0.52% to 0.93%, and Tmax ranges from 415 to 451 °C, indicating that the maturity of shales is in the low-mature to mature stage. The clay minerals are composed of illite and illite-smectite mixed layers, with the ratio of the illite-smectite mixed layer being 20%, suggesting an ordered mixed layer. The sparry calcite and ferroan dolomite with bright rims are widely observed, and pore types are mainly classified as dissolved pores and recrystallized intergranular pores in samples. The sporopollen color ranges from light yellow to black. Therefore, we conclude that the $Es_4{}^U$ shale corresponds to the mesodiagenetic A stage (Table 1).

Table 1. Diagenetic division indices of the $Es_4{}^U$ shale.

Dividing Index	Number	Distribution Range (or Main Type)	Diagenetic Stage
Ro/%	22	0.52–0.93	Mesodiagegesis A
Tmax/°C	90	431–461 °C	Mesodiagegesis A
Type and combination of clay minerals	43 43 43 30	(I/S + I + C) The ratio of I/S is 20% Ordered I/S Acicular and flocculent	Mesodiagegesis A Mesodiagegesis A Mesodiagegesis A Mesodiagegesis A
Type and crystallinity of carbonate minerals	30 30	Sparry calcite Ferroan dolomite	Mesodiagegesis A Mesodiagegesis A
Type of pores	30	Dissolved pores and recrystallized intergranular pores	Mesodiagegesis A
Color of sporopollen	39	Light yellow to black	Mesodiagegesis A

5. Discussion

5.1. The Origin of Authigenic Minerals

5.1.1. The Origin of Authigenic Quartz

The extensively exposed continuous deep-water mudstone sequences lack definitive petrological evidence for active hydrothermal or volcanic activity in the study area [60,61]. Hydrothermal authigenic quartz typically occurs as coarse-grained anhedral crystals or vein fillings [68]. However, SEM imaging reveals euhedral, clay-sized microquartz crystals spatially associated with clay minerals (e.g., kaolinite, illite) and dissolved K-feldspar (Figure 5A,B), indicating silica release linked to smectite illitization and acidic dissolution of K-feldspar. Additionally, quartz grains adjacent to pyrite and illite exhibit dissolution features (irregular shapes, diffuse boundaries) (Figure 5C,D) consistent with alkaline-mediated dissolution of detrital quartz precursor [69]. The study area represents a saline lacustrine basin with strongly alkaline conditions (Figure 6), where detrital quartz grains are susceptible to dissolution.

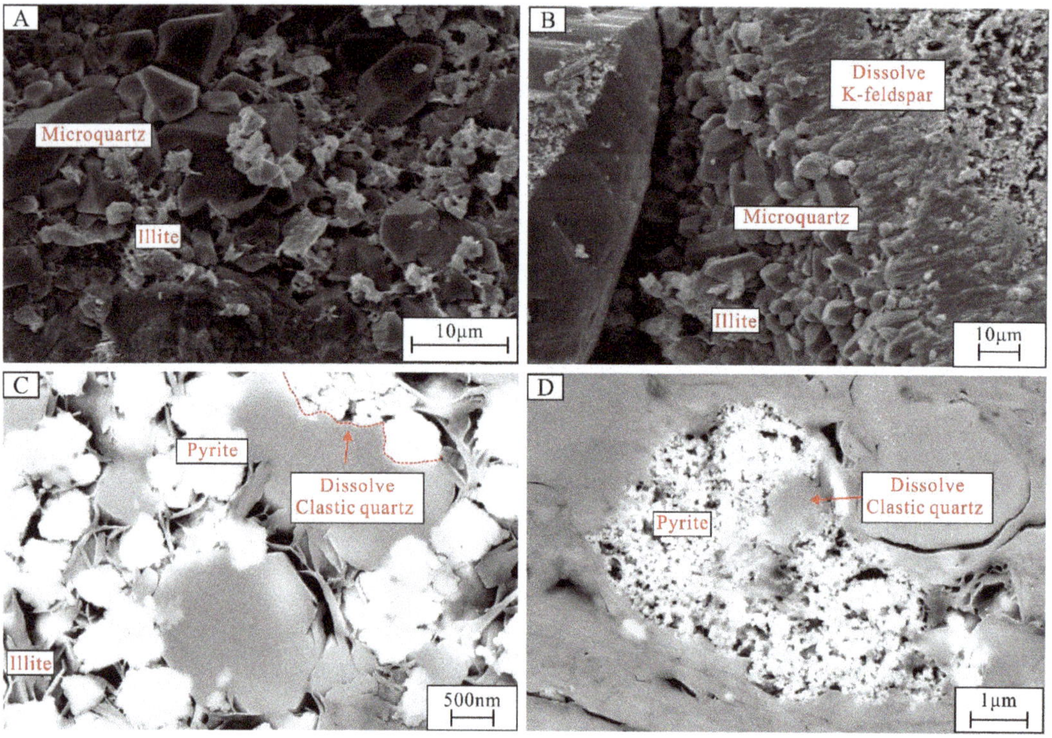

Figure 5. SEM images of typical authigenic minerals. (**A**) Euhedral microquartz crystals distributed with illite, as shown in images of fresh fracture surfaces from the F143 well at a depth of 3019 m. (**B**) Euhedral microquartz crystal growth and dissolved K-feldspar, as shown in images of fresh fracture surfaces from the FY1 well at a depth of 3327 m. (**C**) Partly dissolved detrital quartz in contact with pyrite and illite, as shown in Ar-polished images from the NX124 well at a depth of 3214.15 m. (**D**) Fragmentary detrital quartz and precipitate of pyrite in dissolved pores, as shown in Ar-polished images from the NX124 well at a depth of 3221 m.

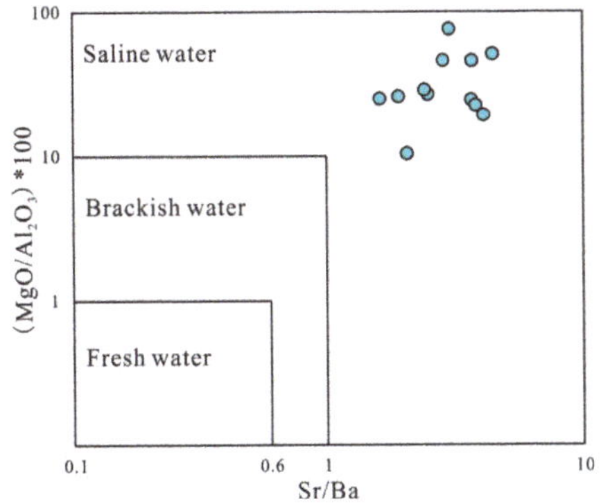

Figure 6. The cross plot of Sr/Ba and n-values ($MgO/Al_2O_3 * 100$) for determining the sedimentary environment.

Following the principle of mass balance, various silicon components can be quantitatively determined using major element analysis combined with XRD data [70,71]. The calculation of detrital silica (Si_{det}) is given by the formula $Si_{det} = Al_{sample} * (Si/Al)_{background}$, where $(Si/Al)_{background}$ is the ratio between Si and Al in the background detrital flux, which is approximately 3.11 [72]. Additionally, a shale containing 787.2 g illite can generate 197.7 g SiO_2 during the smectite illitization [70]. Therefore, the silica released from the smectite illitization (Si_{clay}) can be calculated. Finally, the remaining excess silica (Si_{excess}) can be calculated using the formula $(Si/Al)_{total} - 3.11/(Si/Al)_{total}$ [71]. The results are shown in Table 2. The content of Si_{det} ranges from 21.46% to 50.07%, Si_{clay} ranges from 2.89 to 10.26, and Si_{excess} ranges from 0 to 0.38. The calculated Si_{total} shows a good correlation with measured Si, and their values are relatively close (Figure 7A). The lacustrine shales in our study area show no direct evidence (e.g., sponge spicules or radiolarians) of biogenic silica. However, given documented occurrences of silica fauna in lacustrine shale [73], a biogenic silica component cannot be entirely excluded. The SiO_2-Zr correlation serves as a reliable proxy for silica provenance, where positive correlation reflects terrigenous input and negative correlation implies biogenic sources [73]. Our geochemical data reveal strong Zr correlations with Si_{det} and Si_{clay} (confirming terrigenous derivation), contrasted by a clear negative Zr-Si_{excess} relationship (pointing to potential biogenic contributions) (Figure 7B). Biogenic silica, if present at all, constitutes a volumetrically negligible component (<0.1%) of the total authigenic quartz inventory, with terrigenous sources and diagenetic transformations dominating the silica inventory.

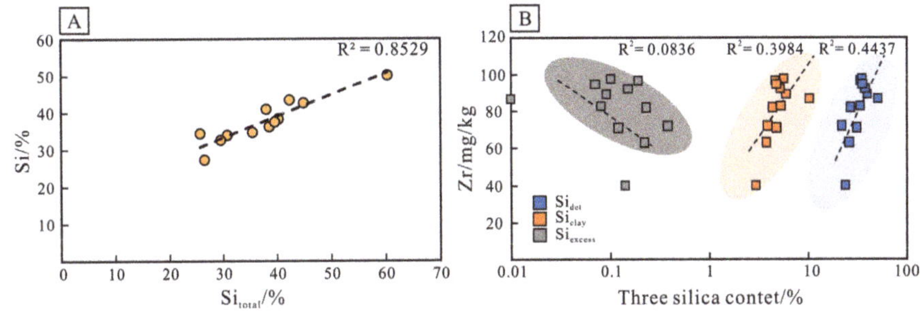

Figure 7. The determination of the origin of different silica types. (**A**) The cross plot of calculated Si_{total} and measured Si. (**B**) The cross plot of three types of silica content and Zr.

Table 2. Calculated percentages of quartz of different origins of shales from $Es_4{}^U$.

Samples	Depth/m	Al/%	Si/%	$(Si/Al)_{total}$	Si_{det}/%	Si_{clay}/%	Si_{excess}/%	Si_{total}/%
NX124-2	3213.00	10.70	41.00	3.83	33.28	4.56	0.19	38.02
NX124-4	3214.15	9.86	34.80	3.53	30.66	4.69	0.12	35.48
NX124-7	3216.00	11.10	38.40	3.46	34.52	5.63	0.10	40.25
NX124-10	3218.00	7.54	27.40	3.63	23.45	2.89	0.14	26.49
NX124-15	3221.00	12.50	42.70	3.42	38.88	5.93	0.09	44.90
NX124-19	3223.00	11.90	43.50	3.66	37.01	5.15	0.15	42.31
NX124-23	3226.25	10.70	36.30	3.39	33.28	5.22	0.08	38.58
NX124-28	3229.10	8.44	34.00	4.03	26.25	4.31	0.23	30.79
NX124-32	3231.60	6.90	34.50	5.00	21.46	3.83	0.38	25.67
NX124-38	3234.15	8.24	32.70	3.97	25.63	3.71	0.22	29.55
NX124-47	3238.70	11.20	37.60	3.36	34.83	4.70	0.07	39.61
NX124-57	3243.15	16.10	50.20	3.12	50.07	10.26	0.00	60.33

5.1.2. The Origin of Pyrite

Pyrites predominantly appear in samples as framboidal and euhedral pyrites, which are often observed in other marine and lacustrine organic-rich shale successions [14,28,72,74]. According to the cross plot of grain size distribution and standard deviation, pyrites in samples have a grain size of less than 5 μm (Figure 8), primarily forming in a sulfurized environment [75]. The supply of S was derived from bacterial sulfate reduction (MSR) with shallow burial and thermochemical sulfate reduction (TSR) with deep burial. These processes were controlled by the openness of the diagenetic system, solute supply, and temperature and pressure conditions [76,77]. Framboidal pyrites primarily form during the syngenetic or early diagenetic stages [78], which have a strong sulfidic environment and a relatively open diagenetic system. Abundant Fe^{2+} and HS^- allow pyrite microparticles to rapidly nucleate, grow, and aggregate, forming an equigranular structure [74,75,79]. Euhedral pyrites form in two stages: (i) directly precipitated from unsaturated solutions of Fe and S during the syngenetic stage [80] and (ii) crystal aggregation and growth of framboidal pyrites during the early stage [38,81]. These origins emphasize that the solute supply in unsaturated fluids is limited, primarily formed by crystal overgrowth [82]. A third formation pathway involves pyrite generation via thermochemical sulfate reduction (TSR) after the mesodiagenetic stage [83]. However, TSR requires higher temperatures and typically occurs after mesodiagenesis [77,83]. The shale of $Es_4{}^U$ is in the mesodiagenetic A stage, which may not exhibit the temperature and pressure conditions required for this reaction.

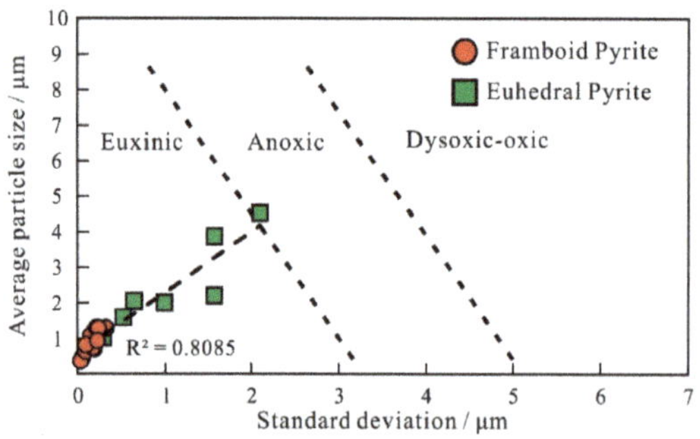

Figure 8. The cross plot of the average particle size versus the standard deviation of the framboid sizes, and the lines at the euxinic–anoxic and the anoxic–dysoxic boundaries are obtained from [75].

5.1.3. The Origin of Ferroan Dolomite

Dolomite is difficult to directly form at normal temperatures. Previous studies indicate that the minimum temperature required for dolomite precipitation in the laboratory is approximately 50–60 °C [84,85]. However, microbial mediation can overcome limitations that dolomite can directly precipitate at room temperature [86,87]. Analogous studies have been reported in the global Precambrian sedimentary rock record [88–91]. Ferroan dolomite is generally considered a product of hydrothermalism and burial dolomitization [92,93]. Hydrothermal ferroan dolomite exhibits coarse grain with a saddle texture and an obvious bright rim, with higher Fe contents usually exceeding 50% [94–96]. However, observations in samples indicate that ferroan dolomite crystals exhibit fine grains with indistinct bright rims and Fe content that is lower than 15% (Figure 9); thus, it is unlikely to be of hydrothermal origin. Therefore, it is more likely that the ferroan dolomite in samples is primarily a product of burial dolomitization. The Mg^{2+} and Fe^{2+} released from the smectite illitization may serve as the major source of ferroan dolomite [97]:

$$4.5K^{+} + 8Al^{3+} + \text{Smectite} \rightarrow \text{Illite} + Na^{+} + 2Ca^{2+} + 2.5Fe^{3+} + 2Mg^{2+} + 3Si^{4+} \quad (1)$$

$$\text{Calcite} + Fe^{2+} + Mg^{2+} \rightarrow \text{Ferroan dolomite} \quad (2)$$

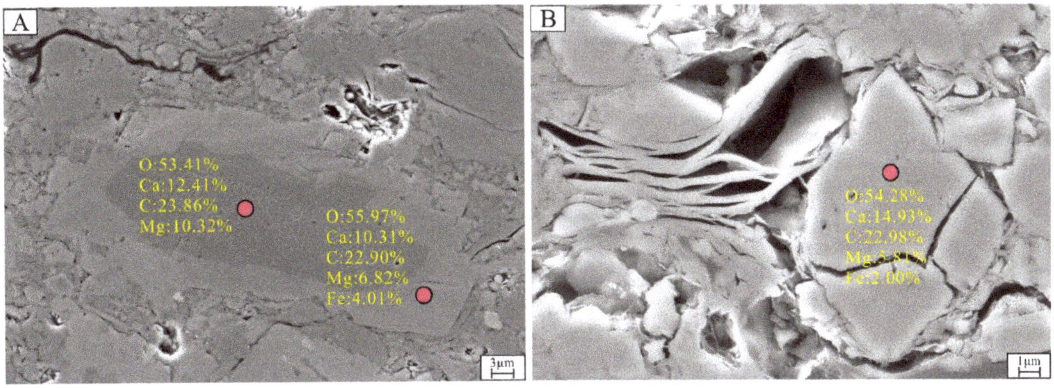

Figure 9. The elemental characteristics and Ar-polished photomicrographs of ferroan dolomite. (**A**) The NX124 well at a depth of 3213 m. (**B**) The NX124 well at a depth of 3223 m.

The cross plot of the dolomite content versus values of Mg + Fe reveals a significant positive correlation (Figure 10A), indicating that dolomite is in the main form of ferroan dolomite. There is also a positive correlation between Mg^{2+} and Fe^{2+} (Figure 10B), suggesting that their sources and enrichment characteristics are likely similar. Additionally, the positive correlation between K and the values of Fe + Mg also indicates a related origin of these ions (Figure 10C). The K-rich diagenetic system is in favor of the smectite illitization that promotes the release of Fe^{2+} and Mg^{2+} as well as the precipitation of ferroan dolomite [98,99]. This is obviously observed in samples where ferroan dolomite is widely distributed within clays.

Additionally, the dissolution of calcite and chlorite driven by CO_2 can induce precipitation of ferroan dolomite:

$$\text{Chlorite} + 5CaCO_3 + 5CO_2 \rightarrow \text{Ferroan dolomite} + \text{Kaolinite} + SiO_2 + 2H_2O \quad (3)$$

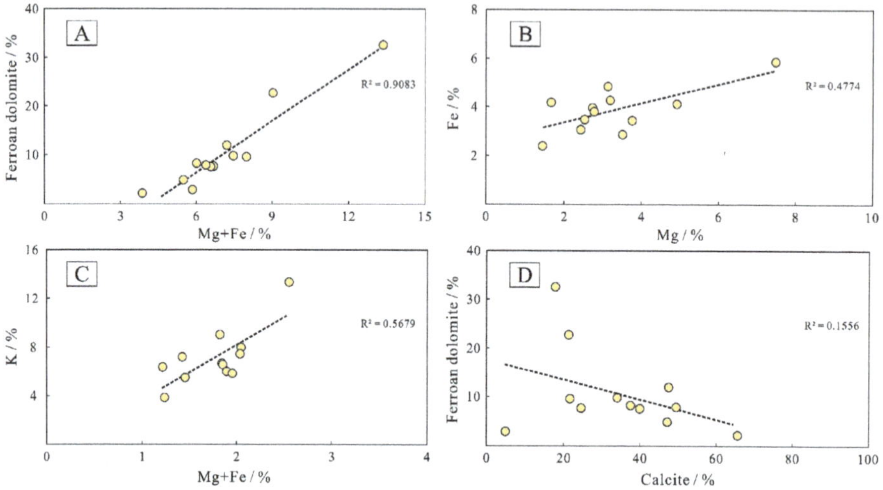

Figure 10. The determination of the origin of ferroan dolomite. (**A**) The cross plot of ferroan dolomite and the value of Mg + Fe. (**B**) The cross plot of Mg and Fe. (**C**) The cross plot of K and the value of Mg + Fe. (**D**) The cross plot of the content of ferroan dolomite and calcite.

In relatively closed diagenetic systems of shale, CO_2 is related to hydrocarbon generation [100]. The dissolution of chlorite releases Mg^{2+} and Fe^{2+}, contributing to the precipitation of ferroan dolomite [101]. Stronger dissolution of chlorite results in increased ferroan dolomite [102], with the reaction being influenced by temperature and pH. The optimal temperature range for this reaction is between 100 °C and 140 °C, and suitable pH conditions are neutral to acidic [103]. Burial temperature can be calculated using vitrinite reflectance (Ro):

$$T = (\mathrm{In}(Ro\%) + 1.26)/0.0081 \quad (4)$$

The values of Ro range from 0.52% to 0.93%, corresponding to calculated temperatures between 120.49 °C and 151.66 °C, which partially fall within the optimal temperature range for this reaction. The cross plot of the calcite and ferroan dolomite reveals an obvious negative correlation (Figure 10D), indicating that the above process probably occurs in diagenetic history. However, the dissolution of chlorite is a slow process [104], and its content is relatively low (Figure 2B). The amount of ferroan dolomite formed through this process may be limited.

5.1.4. The Origin of Clay Minerals

Clay minerals in sedimentary basins primarily originate from terrestrial input and diagenetic alteration [98,105,106]. Illite-smectite mixed layers and illite have extremely high content (Figure 2B), closely related to alkaline connate waters [12,107,108]. Smectite rapidly transforms into illite with burial. Therefore, the precursor of current illite-smectite mixed layers and illite is early smectite. Kaolinite is a product of alteration of K-feldspar grains, which originate from thermodynamic reactions of feldspar dissolution and reprecipitation and are stable in an acidic diagenetic system [98,109].

$$2\text{K-feldspar} + 2H^+ + H_2O \rightarrow \text{Kaolinite} + 4SiO_2 + 2K^+ \quad (5)$$

$$2\text{Albite} + 2H^+ + H_2O \rightarrow \text{Kaolinite} + 4SiO_2 + 2Na^+ \quad (6)$$

$$\text{Anorthite} + 2H^+ + H_2O \rightarrow \text{Kaolinite} + Ca^{2+} \quad (7)$$

Photomicrographs revealed that kaolinite exhibits a booklet shape that grows with microquartz, and obvious dissolved K-feldspar with microquartz was observed in the sample (Figure 5B). The cross plot of kaolinite and feldspar reveals a negative correlation (Figure 11A), further suggesting that the development of kaolinite is primarily related to the acidic dissolution of K-feldspar.

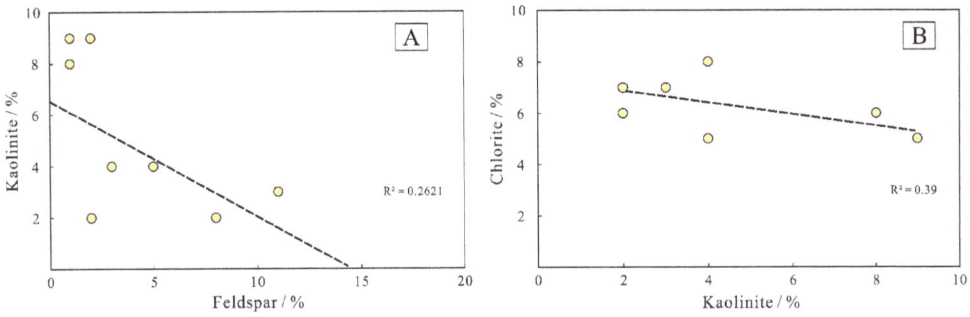

Figure 11. The determination of the origin of clay minerals. (**A**) The cross plot of the content of feldspar and kaolinite. (**B**) The cross plot of the content of kaolinite and chlorite.

Chlorite fibers are primarily products of kaolinite chloritization. Diagenetic fluids change and enrich Mg^{2+} through dissolved dolomite and smectite illitization in mesodiagenesis, triggering the kaolinite chloritization [110]. The cross plot of kaolinite and chlorite indicates a negative correlation (Figure 11B), suggesting the growth of chlorite as dissolution of kaolinite. Photomicrographs reveal that chlorite exhibits hairy and fibrous crystals (Figure 4G,H), indicating its diagenetic origin.

5.1.5. The Origin of Albite

Albite in shales has various origins. The occurrence of volcanic and hydrothermal albite in organic-rich lacustrine shales from the Permian Lucaogou Formation has been well documented [68,111]. However, there is not an obvious volcanic and hydrothermal filling record in Es_4^U shale. SEM observations indicate that albites exhibit well-euhedral crystals with fine-grain size, widely occurring with calcite and illite (Figure 4I). These are interpreted as alkaline alteration products of feldspar [25,69].

$$\text{Anorthite} + 2Na^+ + 4SiO_2 = 2NaAlSi_3O_8 + Ca^{2+} \quad (8)$$

$$2\text{K-feldspar} + 2.5\text{kaolinite} + 2Na^+ + 4SiO_2 = 2\text{albite} + 2\text{illite} + 5H_2O + 2H^+ \quad (9)$$

The alkaline alteration of K-feldspar and anorthite requires a sufficient supply of SiO_2 and Na^+. This process works synergistically with the reaction in Equation (1), where the release of Si^{4+} and Na^+ from smectite illitization promotes the alkaline alteration of K-feldspar and anorthite. The cross plot of Na^+ versus Fe^{2+} and Mg^{2+} indicates a positive correlation (Figure 12A), suggesting that Na, Fe, and Mg may all originate from the smectite illitization. The content of K-feldspar is very low and predominantly features plagioclase in shales (Table S1), which have a strong positive correlation with Na^+ (Figure 12B). SEM observations reveal that albite is the only observable plagioclase phase in samples (Figure 13). However, feldspar from terrestrial input is unlikely to consist entirely of albite.

It is concluded that albite primarily originates from the alkaline alteration of K-feldspar and anorthite.

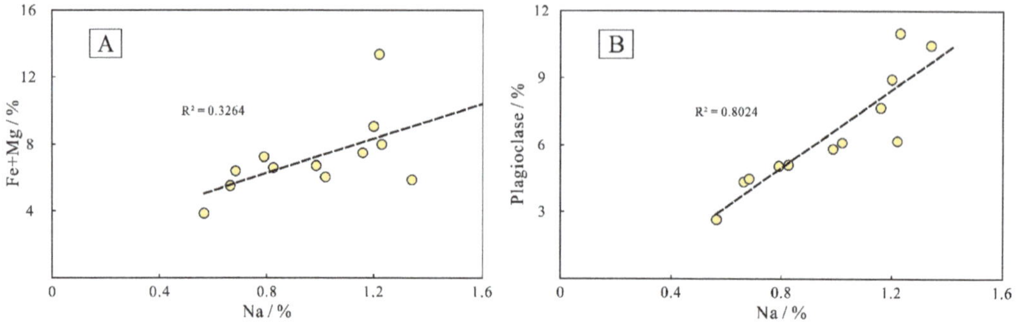

Figure 12. The determination of the origin of albite. (**A**) The cross plot of Na versus the value of Fe + Mg. (**B**) The cross plot of Na and the content of plagioclase.

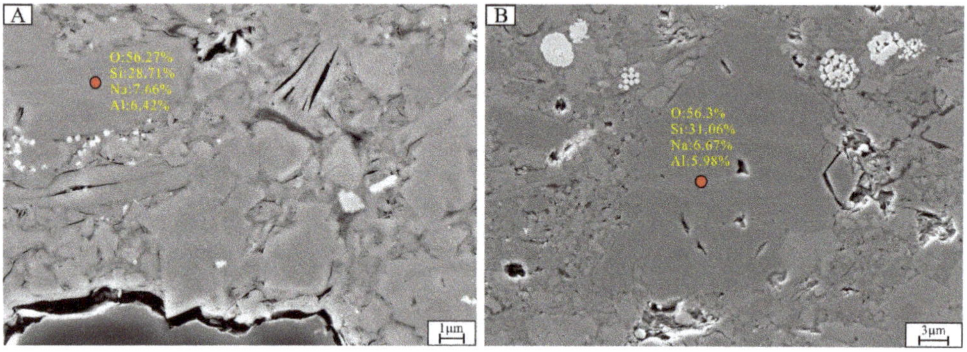

Figure 13. Ar-polished photomicrographs of albite. (**A**) The NX124 well at a depth of 3214.15 m. (**B**) The NX124 well at a depth of 3223 m.

5.2. The Diagenetic Evolution Sequence of Authigenic Minerals

Organic-rich shale has undergone a lengthy and complex process of organic and inorganic diagenesis, which is in the stage of mesodiagenesis A (Table 1). Based on the morphology, mineral assemblage, and diagenetic features of authigenic minerals, their diagenetic evolution sequence has been established (Figure 14).

During the syngenetic diagenesis stage, clay-rich sediments are rich in connate sedimentary water, with primary porosity reaching up to 70% [112], indicating an open diagenetic system. The diagenetic fluid consists of inherited alkaline primary sedimentary water. Due to the abundance of solutes, particularly S and Fe, framboidal pyrite preferentially precipitates [74,79]. The framboidal pyrite crystals were uniform in size and distributed closely, indicating a relatively open diagenetic system.

In the eodiagenetic A stage, the paleotemperature is less than 65 °C, and Ro is below 0.35%. Calcareous sediments rapidly dewater under the compaction, leading to the closure of the diagenetic system. As the abundance of S and Fe solutes decreases, the formation of framboidal pyrite becomes difficult, and precipitation in the form of euhedral pyrite is noted [32]. The euhedral pyrite crystals were of nonuniform sizes, and their distribution was controlled by pores, indicating a relatively closed diagenetic system. Felsic grains dissolve in an alkaline environment, with ions such as Si^{4+}, K^+, Na^+, and Ca^{2+} transferring into the diagenetic fluids.

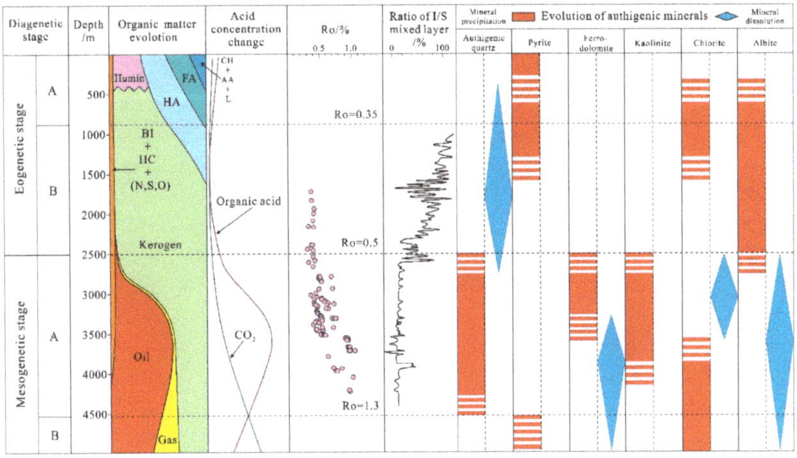

Figure 14. Diagenetic evolution model of authigenic minerals. The thermal evolution characteristics of organic matter and acid concentration changes were modified from [97]. BI + HC + N, S, O-Inherited bitumen + hydrocarbons + nitrogen, sulfur and oxygen compounds; HA-Humic acid; FA-Fulvic acid; CH-Carbohydrates; AA-Amino acid; L-Lipid.

In the eodiagenetic B stage, the paleotemperature ranges from 65 to 85 °C, and Ro ranges from 0.35 to 0.5%. Here, mechanical compaction is the primary factor leading to a closed diagenetic system [4,11]. More ions are released by smectite illitization, such as Si^{4+}, Mg^{2+}, Na^{+}, Ca^{2+}, and Fe^{2+}, into diagenetic fluids [98,113]. The synergistic effect of smectite illitization that continuously consumes K^{+}, as well as alkaline alteration of feldspar that continuously consumes Na^{+}, leads to a significant decrease in K-feldspar and anorthite, with an increasing number of albites. Ca^{2+}, Mg^{2+}, and Fe^{2+} of diagenetic fluids will precipitate as ferroan dolomite. When the temperature was about 70 °C, a few organic acids were released by the hydrocarbon generation of organic matter, which acidized the diagenetic environment [45,114]. Any silica precipitated from diagenetic fluids typically forms as microquartz.

In the mesodiagenetic A stage, the paleotemperature ranges from 85 to 140 °C, and Ro ranges from 0.5 to 1.3%. The organic acids and CO_2 released by organic matter acidize the diagenetic system, which dissolves feldspar and carbonate minerals [115,116]. Dissolved K-feldspar was accompanied by microquartz and kaolinite precipitate in pores, while the release of K^{+} promotes smectite illitization [46,117] and further precipitates as ferroan dolomite [93,99]. Additionally, the Mg^{2+} enriched in diagenetic fluid results from smectite illitization and the dissolution of dolomite, resulting in kaolinite chloritization [103,110]. Small amounts of chlorite and calcite dissolve and precipitate as kaolinite and ferroan dolomite [102,104].

5.3. Implications for Shale Reservoir Quality

Authigenic minerals in shales exhibit complex organic and inorganic interactions that critically influence reservoir quality. The cross plot of diagenetic silica, ferroan dolomite, albite, and pyrite versus porosity shows a relatively positive correlation (Figure 15A–D), suggesting that these minerals have significant implications for shale reservoirs. Although they filled most of the pores, they exhibit important clues for understanding the mechanisms that preserve pores during burial. Beyond overpressure, grain support frameworks constitute the dominant mechanism preserving open intergranular pores throughout shale diagenesis [10,14,118]. Unlike in marine shales where authigenic quartz contributes to pore development [24,38,119], lacustrine authigenic quartz primarily enhances reservoir quality through improved brittleness rather than porosity creation (Figure 4A). While framboidal pyrite develops intercrystalline pores that enhance porosity (Figure 4B), euhedral pyrite promotes hydrocarbon cracking through surface iron ions despite its inability to form pores [100,120]. In some well-documented studies, dolomite, rather than calcite, dominates dissolution within shale reservoirs [118,121]. However, the specific mechanisms are not yet comprehensively understood. A possible explanation was that Mg and Fe incorporation changed the free energy within the crystal and enhanced mineral solubility [122–124]. Thus, the dissolution of dolomite may generate more porosity rather than the dissolution of calcite during burial, although dolomite is less common than calcite. Theoretically, alkaline alteration of K-feldspar and anorthite can generate porosity values of 15.4% and 6.92%, respectively [125]. It looks favorable, but the actual content of feldspar is low. Calcites, dolomites, and microquartz crystals are the dominant non-clay minerals in organic-rich shales [34,63,126]. Feldspar grains are chemically unstable and undergo dissolution, forming dissolved pores during deep burial [127]. Based on the photomicrographs, however, dissolved pores are not well-developed in albite (Figures 4I and 14). This process is associated with an increased concentration of Na^+ within diagenetic fluids, which in turn leads to albite precipitation but not dissolution [128]. Nevertheless, albites still have the potential to enhance the mechanical strength of rock and preserve porosity. As products of diagenetic alteration, authigenic clay minerals degrade pore structure and intensify pore throat heterogeneity [129,130] (Figure 15E,F), but they may contribute to hydrocarbon adsorption and enrichment [48,131].

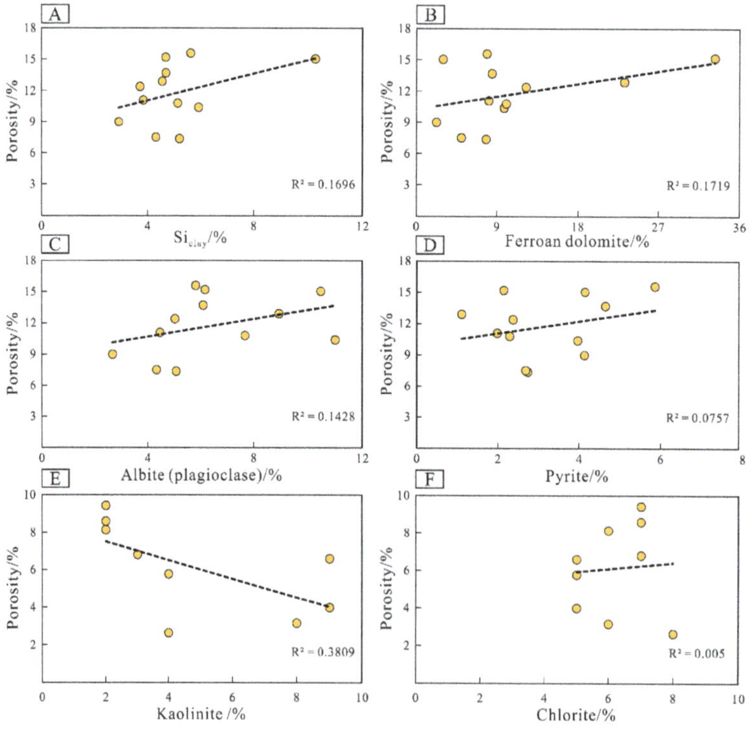

Figure 15. The influence of authigenic minerals on shale reservoir porosity. (**A**) The cross plot of Si_{clay} and porosity. (**B**) The cross plot of ferroan dolomite and porosity. (**C**) The cross plot of albite (albite is almost the only plagioclase phase from SEM) and porosity. (**D**) The cross plot of pyrite and porosity. (**E**) The cross plot of kaolinite and porosity. (**F**) The cross plot of chlorite and porosity.

This study seeks to decode the origin and diagenetic evolution of authigenic minerals in lacustrine shale to evaluate their implications on reservoir quality. However, the number of samples (12) is insufficient to accurately illustrate the complexity and heterogeneity of authigenic minerals in microscale diagenetic systems. Moreover, the methods used are rather conventional. Considerably more sample sets and high-resolution tests are required for a comprehensive understanding of the multiple origins of authigenic minerals. In particular, in-situ high-resolution crystallographic, mineralogical, and element geochemical data are needed to allow for more precise correlations of nucleation, growth, and dissolution of authigenic minerals across the diagenetic history.

6. Conclusions

This study systematically investigates the formation mechanisms of authigenic minerals in lacustrine organic-rich shales and their multiscale impacts on reservoir quality.

1. Authigenic minerals in lacustrine organic-rich shale are dominated by microquartz, framboidal and euhedral pyrites, ferroan dolomite, kaolinite, chlorite, and albite. These authigenic minerals mostly display submicron-sized euhedral crystals and are distributed in large pores.

2. Euxinic and saline primary sedimentary water dominated the type and abundance of early diagenetic authigenic minerals. Framboidal pyrites precipitated directly from solutions with high sulfur saturation. When the sulfur saturation decreased, framboidal pyrites could further grow and transform into euhedral pyrites. Persistent alkaline alteration of feldspar could be an important mechanism for albite growth and preservation.

3. Acid fluids released by the maturity of organic matter are responsible for the formation of an acidified diagenetic system. The dissolution of K-feldspar and carbonate minerals, as well as smectite illitization, provided ions for a variety of mesodiagenetic mineral cements, which resulted for instance in microquartz crystals, ferroan dolomites, kaolinite, and chlorite cemented in pores.

4. Although most rigid authigenic minerals directly precipitate in intergranular pores, they most likely contributed to reservoir quality via a slight reduction in the impact of compaction.

5. This study can be used to investigate authigenic minerals in comparable lacustrine organic-rich shale. However, significantly larger sample sets and in-situ high-resolution tests would be required to conduct case studies and assess their influence on shale reservoirs.

Supplementary Materials: The following supporting information can be downloaded at: https://www.mdpi.com/article/10.3390/min15050493/s1, Table S1: Mineral composition statistics of Es_4^U shale; Table S2: TOC and pyrolysis parameter statistics of Es_4^U shale; Table S3: Vitrinite reflectance parameter statistics of Es_4^U shale in Well FY1; Table S4: Contents of major elements of Es_4^U shale in well NX124; Table S5: Contents of trace elements of Es_4^U shale in well NX124.

Author Contributions: Methodology, J.Y. and Z.Y.; Software, Z.Y. and G.F.; Investigation, G.F.; Resources, J.Y., S.C. and P.Z.; Writing—original draft, J.Y. and Z.Y.; Writing—review & editing, J.Y. and Z.Y.; Supervision, J.Y. and S.C. All authors have read and agreed to the published version of the manuscript.

Funding: This work was supported by the National Natural Science Foundation of China (No. 41572087).

Data Availability Statement: Restrictions apply to the availability of these data. Data were obtained from Shengli Oilfield and are available from the authors with the permission of Shengli Oilfield.

Acknowledgments: We extend our sincere appreciation to three anonymous reviewers for their valuable comments and suggestions. Special thanks are also due to Assistant Editor and Academic Editor for their careful review, constructive comments, and assistance which have significantly improved the manuscript. We express our gratitude to the Shengli Oilfield Branch Company (SINOPEC) for generously providing the samples and permission for publication of this study.

Conflicts of Interest: Author Pengfei Zhang was employed by the company Shengli Oilfield. The remaining authors declare that the research was conducted in the absence of any commercial or financial relationships that could be construed as a potential conflict of interest.

References

1. Giles, M.R.; Marshall, J.D. Constraints on the development of secondary porosity in the subsurface: Re-evaluation of processes. *Mar. Pet. Geol.* **1986**, *3*, 243–255. [CrossRef]
2. Ronald, C.S.; Crossey, L.J.; Hagen, E.S.; Heasler, H.P. Organic-inorganic interactions and sandstone diagenesis. *AAPG Bull.* **1989**, *73*, 1–23.
3. Giles, M.R.; Boer, R.B.d. Secondary porosity: Creation of enhanced porosities in the subsurface from the dissolution of carbonate cements as a result of cooling formation waters. *Mar. Pet. Geol.* **1989**, *6*, 261–269. [CrossRef]
4. Thyne, G. A model for diagenetic mass transfer between adjacent sandstone and shale. *Mar. Pet. Geol.* **2001**, *18*, 743–755. [CrossRef]
5. Taylor, T.R.; Giles, M.R.; Hathon, L.A. Sandstone diagenesis and reservoir quality prediction: Models, myths, and reality. *AAPG Bull.* **2010**, *94*, 1093–1132. [CrossRef]
6. Han, S.; Yu, B.; Ruan, Z.; Bai, C.; Shen, Z.; Löhr, S.C. Diagenesis and fluid evolution in the third member of the Eocene Shahejie Formation, Bonan Sag, Bohai Bay Basin, China. *Mar. Pet. Geol.* **2021**, *128*, 105003. [CrossRef]
7. Dill, H.G.; Luppold, F.W.; Techmer, A.; Chaodumrong, P.; Phoonphun, S. Lithology, micropaleontology and chemical composition of calcareous rocks of Paleozoic through Cenozoic age (Surat Thani Province, central Peninsular Thailand): Implications concerning the environment of deposition and the economic potential of limestones. *J. Asian Earth Sci.* **2004**, *23*, 63–89. [CrossRef]
8. Poursoltani, M.R.; Gibling, M.R.; Pe-Piper, G. Diagenesis, burial history, and hydrocarbon potential of Cambrian sandstone in the northern continental margin of Gondwana: A case study of the Lalun Formation of central Iran. *J. Asian Earth Sci.* **2019**, *172*, 143–169. [CrossRef]
9. Loucks, R.; Reed, R.; Ruppel, S.; Jarvie, D.M. Morphology, Genesis, and Distribution of Nanometer-Scale Pores in Siliceous Mudstones of the Mississippian Barnett Shale. *J. Sediment. Res.* **2009**, *79*, 848–861. [CrossRef]
10. Milliken, K.L.; Rudnicki, M.; Awwiller, D.N.; Zhang, T. Organic matter-hosted pore system, Marcellus Formation (Devonian), Pennsylvania. *AAPG Bull.* **2013**, *97*, 177–200. [CrossRef]
11. Taylor, K.G.; Macquaker, J.H.S. Diagenetic alterations in a silt- and clay-rich mudstone succession: An example from the Upper Cretaceous Mancos Shale of Utah, USA. *Clay Miner.* **2014**, *49*, 213–227. [CrossRef]
12. Chen, S.; Zhang, S.; Wang, Y.; Tan, M. Lithofacies types and reservoirs of Paleogene fine-grained sedimentary rocks Dongying Sag, Bohai Bay Basin, China. *Pet. Explor. Dev.* **2016**, *43*, 218–229. [CrossRef]
13. Aplin, A.C.; Macquaker, J.H.S. Mudstone diversity: Origin and implications for source, seal, and reservoir properties in petroleum systems. *AAPG Bull.* **2011**, *95*, 2031–2059. [CrossRef]
14. Ruppel, S.C.; Reed, R.M.; Loucks, R.G. Spectrum of pore types and networks in mudrocks and a descriptive classification for matrix-related mudrock pores. *AAPG Bull.* **2012**, *96*, 1071–1098.
15. Schicker, A.; Gier, S.; Schieber, J.; Krois, P. Diagenesis of the Malmian Mikulov Formation source rock, Vienna Basin: Focus on matrix and pores. *Mar. Pet. Geol.* **2021**, *129*, 105082. [CrossRef]
16. Gale, J.F.W.; Jon, E.O.; Laubach, S.E. Natural fractures in shale: A review and new observations. *AAPG Bull.* **2014**, *98*, 2165–2216. [CrossRef]
17. Milliken, K.L.; Reed, R.M.; McCarty, D.K.; Bishop, J.; Lipinski, C.J.; Fischer, T.B.; Crousse, L.; Reijenstein, H. Grain assemblages and diagenesis in the Vaca Muerta Formation (Jurassic-Cretaceous), Neuquén Basin, Argentina. *Sediment. Geol.* **2019**, *380*, 45–64. [CrossRef]
18. Cobbold, P.R.; Zanella, A.; Rodrigues, N.; Løseth, H. Bedding-parallel fibrous veins (beef and cone-in-cone): Worldwide occurrence and possible significance in terms of fluid overpressure, hydrocarbon generation and mineralization. *Mar. Pet. Geol.* **2013**, *43*, 1–20. [CrossRef]
19. Sun, B.; Yang, W. Folded calcite cracks in noncalcareous shales: A window into shale diagenesis and hydrothermal influence. *J. Sediment. Res.* **2023**, *93*, 875–894. [CrossRef]
20. Luo, T.; Guo, X.; He, Z.; Dong, T.; Tao, Z.; Yang, R.; Wang, K. Fluid evolution and gas enrichment in the Wufeng-Longmaxi shale reservoirs of the eastern Sichuan Basin, SW China. *J. Asian Earth Sci.* **2024**, *259*, 105905. [CrossRef]

21. O'Brien, N.R. Fabric of Kaolinite and Illite Floccules. *Clays Clay Miner.* **1971**, *19*, 353–359. [CrossRef]
22. Peltonen, C.; Marcussen, Ø.; Bjørlykke, K.; Jahren, J. Clay mineral diagenesis and quartz cementation in mudstones: The effects of smectite to illite reaction on rock properties. *Mar. Pet. Geol.* **2009**, *26*, 887–898. [CrossRef]
23. Rodrigues, N.; Cobbold, P.R.; Loseth, H.; Ruffet, G. Widespread bedding-parallel veins of fibrous calcite ('beef') in a mature source rock (Vaca Muerta Fm, Neuquén Basin, Argentina): Evidence for overpressure and horizontal compression. *J. Geol. Soc.* **2009**, *166*, 695–709. [CrossRef]
24. Peng, J.; Milliken, K.L.; Fu, Q. Quartz types in the Upper Pennsylvanian organic-rich Cline Shale (Wolfcamp D), Midland Basin, Texas: Implications for silica diagenesis, porosity evolution and rock mechanical properties. *Sedimentology* **2020**, *67*, 2040–2064. [CrossRef]
25. Xiong, Z.; Cao, Y.; Liang, C.; Liu, K.; Wang, G.; Zhu, R.; Lei, P.; Wang, Y. Origin and significance of authigenic quartz and albite in lacustrine calcareous fine-grained sedimentary rocks. *Mar. Pet. Geol.* **2022**, *143*, 105799. [CrossRef]
26. Jahren, J.; Bjorlykke, K. Open or closed geochemical systems during diagenesis in sedimentary basins: Constraints on mass transfer during diagenesis and the prediction of porosity in sandstone and carbonate reservoirs. *AAPG Bull.* **2012**, *96*, 2193–2214.
27. Friis, H.; Molenaar, N.; Varming, T. Chlorite meniscus cement–implications for diagenetic mineral growth after oil emplacement. *Terra Nova* **2014**, *26*, 14–21. [CrossRef]
28. Wilson, R.D.; Chitale, J.; Huffman, K.; Montgomery, P.; Prochnow, S.J. Evaluating the depositional environment, lithofacies variation, and diagenetic processes of the Wolfcamp B and lower Spraberry intervals in the Midland Basin: Implications for reservoir quality and distribution. *AAPG Bull.* **2020**, *104*, 1287–1321. [CrossRef]
29. Liang, C.; Wu, J.; Cao, Y.; Liu, K.; Khan, D. Storage space development and hydrocarbon occurrence model controlled by lithofacies in the Eocene Jiyang Sub-basin, East China: Significance for shale oil reservoir formation. *J. Pet. Sci. Eng.* **2022**, *215*, 110631. [CrossRef]
30. Yu, Z.; Chen, S.; Zhang, S.; Liu, X.; Tang, D.; Yan, J. Influence of diagenesis on reservoir performance of shale: A case study of the upper sub-member of Member 4 of Paleogene Shahejie Formation in Dongying sag. *J. Palaeogeogr. (Chin. Ed.)* **2022**, *24*, 771–784.
31. Seewald, J.S. Organic–inorganic interactions in petroleum-producing sedimentary basins. *Nature* **2003**, *426*, 327–333. [CrossRef] [PubMed]
32. Taylor, K.G.; Macquaker, J.H.S.; Keller, M. Compositional controls on early diagenetic pathways in fine-grained sedimentary rocks: Implications for predicting unconventional reservoir attributes of mudstones. *AAPG Bull.* **2014**, *98*, 587–603.
33. Raiswell, R. Evidence for surface reaction-controlled growth of carbonate concretions in shales. *Sedimentology* **1988**, *35*, 571–575. [CrossRef]
34. Loucks, R.G.; Ruppel, S.C. Mississippian Barnett Shale: Lithofacies and depositional setting of a deep-water shale-gas succession in the Fort Worth Basin, Texas. *AAPG Bull.* **2007**, *91*, 579–601. [CrossRef]
35. Wang, M.; Sherwood, N.; Li, Z.; Lu, S.; Wang, W.; Huang, A.; Peng, J.; Lu, K. Shale oil occurring between salt intervals in the Dongpu Depression, Bohai Bay Basin, China. *Int. J. Coal Geol.* **2015**, *152*, 100–112. [CrossRef]
36. Macaulay, C.I.; Haszeldine, R.S.; Fallick, A.E. Distribution, chemistry, isotopic composition and origin of diagenetic carbonates; Magnus Sandstone, North Sea. *J. Sediment. Res.* **1993**, *63*, 33–43.
37. Macquaker, J.H.S.; Taylor, K.G.; Gawthorpe, R.L. High-Resolution Facies Analyses of Mudstones: Implications for Paleoenvironmental and Sequence Stratigraphic Interpretations of Offshore Ancient Mud-Dominated Successions. *J. Sediment. Res.* **2007**, *77*, 324–339. [CrossRef]
38. Soliman, M.F.; El Goresy, A. Framboidal and idiomorphic pyrite in the upper Maastrichtian sedimentary rocks at Gabal Oweina, Nile Valley, Egypt: Formation processes, oxidation products and genetic implications to the origin of framboidal pyrite. *Geochim. Cosmochim. Acta* **2012**, *90*, 195–220. [CrossRef]
39. Dowey, P.J.; Taylor, K.G.; Hendry, J. Diagenetic mineral development within the Upper Jurassic Haynesville-Bossier Shale, USA. *Sedimentology* **2020**, *67*, 47–77. [CrossRef]
40. Canfield, D.E.; Thamdrup, B.; Hansen, J.W. The anaerobic degradation of oraganic matter in Danish coastal sediments: Iron reduction, manganese reduction, and sulfate reduction. *Geochim. Cosmochim. Acta* **1993**, *57*, 3867–3883. [CrossRef]
41. Milkov, A.V. Methanogenic biodegradation of petroleum in the West Siberian Basin (Russia): Significance for formation of giant Cenomanian gas pools. *AAPG Bull.* **2010**, *94*, 1485–1541. [CrossRef]
42. Waliczek, M.; Machowski, G.; Więcław, D.; Konon, A.; Wandycz, P. Properties of solid bitumen and other organic matter from Oligocene shales of the Fore-Magura Unit in Polish Outer Carpathians: Microscopic and geochemical approach. *Int. J. Coal Geol.* **2019**, *210*, 103206. [CrossRef]
43. Schlegel, M.E.; McIntosh, J.C.; Petsch, S.T.; Orem, W.H.; Jones, E.J.P.; Martini, A.M. Extent and limits of biodegradation by in situ methanogenic consortia in shale and formation fluids. *Appl. Geochem.* **2013**, *28*, 172–184. [CrossRef]
44. Mastalerz, M.; Drobniak, A.; Schimmelmann, A. Porosity of Devonian and Mississippian New Albany Shale across a maturation gradient: Insights from organic petrology, gas adsorption, and mercury intrusion. *AAPG Bull.* **2013**, *97*, 1621–1643. [CrossRef]

45. Tissot, B.; Durand, B.; Espitalié, J.; Combaz, A. Influence of Nature and Diagenesis of Organic Matter in Formation of Petroleum1. *AAPG Bull.* **1974**, *58*, 499–506.
46. Curtis, M.E.; Cardott, B.J.; Sondergeld, C.H. Development of organic porosity in the Woodford Shale with increasing thermal maturity. *Int. J. Coal Geol.* **2012**, *103*, 26–31. [CrossRef]
47. Pommer, M.; Milliken, K. Pore types and pore-size distributions across thermal maturity, Eagle Ford Formation, southern Texas. *AAPG Bull.* **2015**, *99*, 1713–1744. [CrossRef]
48. Cai, J.; Du, J.; Song, M.; Lei, T.; Wang, X.; Li, Y. Control of clay mineral properties on hydrocarbon generation of organo-clay complexes: Evidence from high-temperature pyrolysis experiments. *Appl. Clay Sci.* **2022**, *216*, 106368. [CrossRef]
49. Testa, G.; Lugli, S. Gypsum–anhydrite transformations in Messinian evaporites of central Tuscany (Italy). *Sediment. Geol.* **2000**, *130*, 249–268. [CrossRef]
50. Ferrage, E.; Vidal, O.; Mosser-Ruck, R.g.; Cathelineau, M.; Cuadros, J. A reinvestigation of smectite illitization in experimental hydrothermal conditions: Results from X-ray diffraction and transmission electron microscopy. *Am. Mineral.* **2011**, *96*, 207–223. [CrossRef]
51. Wang, G.; Ran, L.; Xu, J.; Wang, Y.; Ma, L.; Zhu, R.; Wei, J.; He, H.; Xi, Y.; Zhu, J. Technical development of characterization methods provides insights into clay mineral-water interactions: A comprehensive review. *Appl. Clay Sci.* **2021**, *206*, 106088. [CrossRef]
52. Li, Y.; Yang, W.; Wang, Q.; Song, Y.; Jiang, Z.; Guo, L.; Zhang, Y.; Wang, J. Influence of the actively migrated diagenetic elements on the hydrocarbon generation potential in tuffaceous shale. *Fuel* **2019**, *256*, 115795. [CrossRef]
53. Du, J.; Cai, J.; Zeng, X.; Zhang, Y.; Mu, X.; Wang, X. Depositional and diagenetic controls on hydrocarbon generation of mudstones through clay–organic matter interactions: A comparative study of the Dongpu Depression, Bohai Bay Basin, China. *J. Asian Earth Sci.* **2024**, *262*, 106000. [CrossRef]
54. Yan, Y.; Wang, M.; Misch, D.; Sachsenhofer, R.F.; Wu, Y.; Li, J. Mineral diagenesis in lacustrine organic-rich shales: Evolution pathways and implications for reservoir characteristics. *J. Asian Earth Sci.* **2024**, *263*, 106026. [CrossRef]
55. Wang, Y.; Hao, X.; Hu, Y. Orderly distribution and differential enrichment of hydrocarbon in oil-rich sags: A case study of Dongying Sag, Jiyang Depression, Bohai Bay Basin, East China. *Pet. Explor. Dev.* **2018**, *45*, 785–794. [CrossRef]
56. Song, M.; Liu, H.; Wang, Y.; Liu, Y. Enrichment rules and exploration practices of Paleogene shale oil in Jiyang Depression, Bohai Bay Basin, China. *Pet. Explor. Dev.* **2020**, *47*, 225–235. [CrossRef]
57. Qiu, N.; Zuo, Y.; Xu, W.; Chang, J. Meso-Cenozoic thermal structure of the lithosphere in the Bohai Bay Basin, eastern North China Craton: Thermal structure in Bohai Bay Basin. *Geol. J.* **2015**, *51*, 794–804. [CrossRef]
58. Fu, C.; Li, S.; Li, S.; Xu, J. Tectonostratigraphic evolution of a rift basin and corresponding source-to-sink systems: Implications for the western Bohai Bay Basin, North China. *Mar. Pet. Geol.* **2022**, *139*, 105587. [CrossRef]
59. Wu, Q.; Xian, B.; Gao, X.; Bai, Q.; Wang, Z.; Liu, J.; Chen, P.; Li, Y.; Rahman, N.U.; Tian, R.; et al. Differences of sedimentary triggers and depositional architecture of lacustrine turbidites from normal regression to forced regression: Eocene Dongying depression, Bohai Bay Basin, East China. *Sediment. Geol.* **2022**, *439*, 106222. [CrossRef]
60. Zhang, J.; Jiang, Z.; Liang, C.; Baars, T.F.; Wang, Y.; Abels, H.A. Astronomical forcing of meter-scale organic-rich mudstone–limestone cyclicity in the Eocene Dongying sag, China: Implications for shale reservoir exploration. *AAPG Bull.* **2022**, *106*, 1557–1579. [CrossRef]
61. Liang, C.; Cao, Y.; Wu, J.; Han, Y.; Liu, K.; Hao, F.; Khan, D.; Mei, J.; Zhang, S.; Wang, Y. Water depth–terrigenous input dynamic equilibrium controls the Eocene lacustrine shale laminae records in Jiyang depression, Bohai Bay Basin, East China. *AAPG Bull.* **2023**, *107*, 1987–2016. [CrossRef]
62. Khan, D.; Zijun, L.; Qiu, L.; Kuiyuan, L.; Yongqiang, Y.; Cong, N.; Bin, L.; Li, X.; Habulashenmu, Y. Mineralogical and geochemical characterization of lacustrine calcareous shale in Dongying Depression, Bohai Bay Basin: Implications for paleosalinity, paleoclimate, and paleoredox conditions. *Geochemistry* **2023**, *83*, 125978. [CrossRef]
63. Zhao, X.; Zhou, L.; Pu, X.; Jin, F.; Shi, Z.; Han, W.; Jiang, W.; Han, G.; Zhang, W.; Wang, H.; et al. Formation conditions and enrichment model of retained petroleum in lacustrine shale: A case study of the Paleogene in Huanghua depression, Bohai Bay Basin, China. *Pet. Explor. Dev.* **2020**, *47*, 856–869. [CrossRef]
64. Wang, M.; Ma, R.; Li, J.; Lu, S.; Li, C.; Guo, Z.; Li, Z. Occurrence mechanism of lacustrine shale oil in the Paleogene Shahejie Formation of Jiyang Depression, Bohai Bay Basin, China. *Pet. Explor. Dev.* **2019**, *46*, 833–846. [CrossRef]
65. Bjorlykke, K. Clay mineral diagenesis in sedimentary basins; a key to the prediction of rock properties; examples from the North Sea Basin. *Clay Miner.* **1998**, *33*, 15–34. [CrossRef]
66. Merriman, R.J. Clay minerals and sedimentary basin history. *Eur. J. Mineral.* **2005**, *17*, 7–20. [CrossRef]
67. Penghui, Z.; Zhiyong, C.H.E.N.; Lu, X.; YanJun, B.; Yan, F. The differential diagenetic evolution and its influencing factors of Lower Cambrian black rock series in the northwestern margin of Tarim Basin. *Acta Petrol. Sin.* **2020**, *36*, 3463–3476. [CrossRef]

68. Li, K.; Xi, K.; Cao, Y.; Wang, Y.; Lin, M. Genesis of granular calcite in lacustrine fine-grained sedimentary rocks and its indication to volcanic-hydrothermal events: A case study of Permian Lucaogou Formation in Jimusar Sag, Junggar Basin, NW China. *Pet. Explor. Dev.* **2023**, *50*, 541–552. [CrossRef]
69. Zhu, H.; Zhong, D.; Yao, J.; Sun, H.; Niu, X.; Liang, X.; You, Y.; Li, X. Alkaline diagenesis and its effects on reservoir porosity: A case study of Upper Triassic Chang 7 tight sandstones in Ordos Basin, NW China. *Pet. Explor. Dev.* **2015**, *42*, 51–59. [CrossRef]
70. van de Kamp, P.C. Smectite-illite-muscovite transformations, quartz dissolution, and silica release in shales. *Clays Clay Miner.* **2008**, *56*, 66–81. [CrossRef]
71. Niu, X.; Yan, D.; Zhuang, X.; Liu, Z.; Li, B.; Wei, X.; Xu, H.; Li, D. Origin of quartz in the lower Cambrian Niutitang Formation in south Hubei Province, upper Yangtze platform. *Mar. Pet. Geol.* **2018**, *96*, 271–287. [CrossRef]
72. Wedepohl, K.H. Environmental influences on the chemical composition of shales and clays. *Phys. Chem. Earth* **1971**, *8*, 307–333. [CrossRef]
73. Li, L.; Bao, Z.; Li, Z.; Chen, L.; Xu, X.; Li, Y.; Zhao, Y.; Song, X. Origin of quartz and its implications for different phase fluids in lacustrine shales from the Qingshankou Formation, southern Songliao Basin, China. *Geoenergy Sci. Eng.* **2024**, *234*, 212673. [CrossRef]
74. Xi, Z.; Tang, S.; Lash, G.G.; Zhang, B.; Lin, D. Geochemical characteristics of organic carbon and pyrite sulfur in Ordovician-Silurian transition shales in the Yangtze Platform, South China: Implications for the depositional environment. *Palaeogeogr. Palaeoclimatol. Palaeoecol.* **2021**, *563*, 110173. [CrossRef]
75. Wilkin, R.T.; Barnes, H.L.; Brantley, S.L. The Size Distribution of Framboidal Pyrite in Modern Sediments—An Indicator of Redox Conditions. *Geochim. Cosmochim. Acta J. Geochem. Soc. Meteorit. Soc.* **1996**, *60*, 3897–3912. [CrossRef]
76. Zuo, Q.; Xu, Y.; Zhang, Y. NanoSIMS sulfur isotope studies of pyrite from the Early Paleozoic marine shale: Implications for the sedimentary environment. *Mar. Pet. Geol.* **2021**, *124*, 104802. [CrossRef]
77. Chao, L.; Shichao, J.; Yingchang, C.; Keyu, L.; Jing, W.; Fang, H. Characteristics, origins, and significance of pyrites in deep-water shales. *Sci. China Earth Sci.* **2024**, *67*, 313–342.
78. Love, L.G.; Amstutz, G.C. Review of microscopic pyrite from the Devonian Chattanooga Shale and Rammelsberg Banderz. *Fortschr Miner.* **1966**, *43*, 277–309.
79. Baronov, A.; Bufkin, K.; Shaw, D.W.; Johnson, B.L.; Patrick, D.L. A simple model of burst nucleation. *Phys. Chem. Chem. Phys. PCCP* **2015**, *17*, 20846–20852. [CrossRef]
80. Steinike, K. A further remark on biogenic sulfides; inorganic pyrite spheres. *Econ. Geol.* **1963**, *58*, 998–1000. [CrossRef]
81. Popa, R.; Kinkle, B.K.; Badescu, A. Pyrite Framboids as Biomarkers for Iron-Sulfur Systems. *Geomicrobiol. J.* **2004**, *21*, 193–206. [CrossRef]
82. Ren, Y.; Yang, X.; Hu, X.; Wei, J.; Tang, C. Mineralogical and geochemical evidence for biogenic uranium mineralization in northern Songliao Basin, NE China. *Ore Geol. Rev.* **2022**, *141*, 104556. [CrossRef]
83. Hao, F.; Zhang, X.; Wang, C.; Li, P.; Guo, T.; Zou, H.; Zhu, Y.; Liu, J.; Cai, Z. The fate of CO_2 derived from thermochemical sulfate reduction (TSR) and effect of TSR on carbonate porosity and permeability, Sichuan Basin, China. *Earth Sci. Rev.* **2015**, *141*, 154–177. [CrossRef]
84. Land, L.S. Failure to Precipitate Dolomite at 25 °C fromDilute Solution Despite 1000-Fold Oversaturation after32 Years. *Aquat. Geochem.* **1998**, *4*, 361–368. [CrossRef]
85. Arvidson, R.S.; Mackenzie, F.T. The dolomite problem: Control of precipitation kinetics by temperature and saturation state. *Am. J. Sci.* **1999**, *299*, 257–288. [CrossRef]
86. Roberts, J.A.; Bennett, P.C.; González, L.A.; Macpherson, G.L.; Milliken, K.L. Microbial precipitation of dolomite in methanogenic groundwater. *Geology* **2004**, *32*, 277–280. [CrossRef]
87. Petrash, D.A.; Bialik, O.M.; Bontognali, T.R.R.; Vasconcelos, C.; Roberts, J.A.; McKenzie, J.A.; Konhauser, K.O. Microbially catalyzed dolomite formation: From near-surface to burial. *Earth Sci. Rev.* **2017**, *171*, 558–582. [CrossRef]
88. Vasconcelos, C.; McKenzie, J.A. Microbial mediation of modern dolomite precipitation and diagenesis under anoxic conditions (Lagoa Vermelha, Rio de Janeiro, Brazil). *J. Sediment. Res.* **1997**, *67*, 378–390.
89. Wright, D.T. The role of sulphate-reducing bacteria and cyanobacteria in dolomite formation in distal ephemeral lakes of the Coorong region, South Australia. *Sediment. Geol.* **1999**, *126*, 147–157. [CrossRef]
90. Vasconcelos, C.; Warthmann, R.; McKenzie, J.A.; Visscher, P.T.; Bittermann, A.G.; Lith, Y.v. Lithifying microbial mats in Lagoa Vermelha, Brazil: Modern Precambrian relics? *Sediment. Geol.* **2006**, *185*, 175–183. [CrossRef]
91. Deng, S.; Dong, H.; Lv, G.; Jiang, H.; Yu, B.; Bishop, M. Microbial dolomite precipitation using sulfate reducing and halophilic bacteria: Results from Qinghai Lake, Tibetan Plateau, NW China. *Chem. Geol.* **2011**, *278*, 151–159. [CrossRef]
92. Hardie, L.A. On the Significance of Evaporites. *Annu. Rev. Earth Planet. Sci.* **1991**, *19*, 131–168. [CrossRef]
93. Bojanowski, M.J. Authigenic dolomites in the Eocene-Oligocene organic carbon-rich shales from the Polish Outer Carpathians: Evidence of past gas production and possible gas hydrate formation in the Silesian basin. *Mar. Pet. Geol.* **2014**, *51*, 117–135. [CrossRef]

94. Li, X.; Zhu, G.; Li, T.; Zhou, L.; Wu, Y.; Shen, B.; Ning, M. Conservative behavior of Mg isotopes in dolomite during diagenesis and hydrothermal alteration: A case study in the Lower Cambrian Qiulitage Formation, Gucheng area, Tarim Basin. *Appl. Geochem.* **2023**, *148*, 105540. [CrossRef]
95. Mitchell, R.H.; Smith, D.L. Geology and mineralogy of the Ashram Zone carbonatite, Eldor Complex, Québec. *Ore Geol. Rev.* **2017**, *86*, 784–806. [CrossRef]
96. Zhao, W.; Shen, A.; Qiao, Z.; Pan, L.; Hu, A.; Zhang, J. Genetic types and distinguished characteristics of dolomite and the origin of dolomite reservoirs. *Pet. Explor. Dev.* **2018**, *45*, 923–935. [CrossRef]
97. Xiong, Z.; Cao, Y.; Liang, C. Characteristics and origin of crystalline dolomite: A case from Paleogene lacustrine fine-grained rocks in Jiyang depression, Bohai Bay Basin, China. *Mar. Pet. Geol.* **2024**, *162*, 106694. [CrossRef]
98. Metwally, Y.M.; Chesnokov, E.M. Clay mineral transformation as a major source for authigenic quartz in thermo-mature gas shale. *Appl. Clay Sci.* **2012**, *55*, 138–150. [CrossRef]
99. Mavromatis, V.; Meister, P.; Oelkers, E.H. Using stable Mg isotopes to distinguish dolomite formation mechanisms: A case study from the Peru Margin. *Chem. Geol.* **2014**, *385*, 84–91. [CrossRef]
100. Chen, Z.; Wang, T.; Liu, Q.; Zhang, S.; Zhang, L. Quantitative evaluation of potential organic-matter porosity and hydrocarbon generation and expulsion from mudstone in continental lake basins: A case study of Dongying sag, eastern China. *Mar. Pet. Geol.* **2015**, *66*, 906–924. [CrossRef]
101. Zhang, S.; Yang, L.; De Paolo, D.J.; Steefel, C.I. Chemical affinity and pH effects on chlorite dissolution kinetics under geological CO_2 sequestration related conditions. *Chem. Geol.* **2015**, *396*, 208–217. [CrossRef]
102. Lowson, R.T.; Brown, P.L.; Comarmond, M.C.J.; Rajaratnam, G. The kinetics of chlorite dissolution. *Geochim. Cosmochim. Acta* **2007**, *71*, 1431–1447. [CrossRef]
103. Liao, R.; Chen, W.; Wang, N.; Zhang, J. Combined effects of temperature, mineral type, and surface roughness on chlorite dissolution kinetics in the acidic pH. *Appl. Clay Sci.* **2021**, *201*, 105931. [CrossRef]
104. Smith, M.; Wolery, T.; Carroll, S. Kinetics of chlorite dissolution at elevated temperatures and CO_2 conditions. *Chem. Geol.* **2013**, *347*, 1–8. [CrossRef]
105. Wooldridge, L.J.; Worden, R.H.; Griffiths, J.; Utley, J.E.P.; Thompson, A. The origin of clay-coated sand grains and sediment heterogeneity in tidal flats. *Sediment. Geol.* **2018**, *373*, 191–209. [CrossRef]
106. Zuo, R.; Cai, J.; Zeng, X.; Cheng, S. Detrital clay mineral input reconstructed based on weathering records and its influence on organic matter enrichment: A case study of the Paleogene Shahejie Formation in the Dongpu Sag, Bohai Bay Basin. *J. Asian Earth Sci.* **2024**, *276*, 106360. [CrossRef]
107. Mosser, C. The use of B, Li and Sn in determining the origin of some sedimentary clays. *Chem. Geol.* **1983**, *38*, 129–139. [CrossRef]
108. D'Ascanio, V.; Greco, D.; Menicagli, E.; Santovito, E.; Catucci, L.; Logrieco, A.F.; Avantaggiato, G. The role of geological origin of smectites and of their physico-chemical properties on aflatoxin adsorption. *Appl. Clay Sci.* **2019**, *181*, 105209. [CrossRef]
109. Xin, C.; Jianhua, Z.; Jing, Y.; Keke, N.; Yuping, Y. Development and formation of Paleogene kaolinite, Bonan subsag. *Pet. Explor. Dev.* **2009**, *36*, 456–462.
110. Beaufort, D.; Rigault, C.; Billon, S.; Billault, V.; Inoue, A.; Inoue, S.; Patrier, P. Chlorite and chloritization processes through mixed-layer mineral series in low-temperature geological systems—A review. *Clay Miner.* **2015**, *50*, 497–523. [CrossRef]
111. Liu, Y.; Ma, K.; Hou, J.; Yan, L.; Chen, F. Diagenetic controls on the quality of the Middle Permian Lucaogou Formation tight reservoir, southeastern Junggar Basin, northwestern China. *J. Asian Earth Sci.* **2019**, *178*, 139–155. [CrossRef]
112. Velde, B. Compaction trends of clay-rich deep sea sediments. *Mar. Geol.* **1996**, *133*, 193–201. [CrossRef]
113. Wang, F.; Feng, Z.; Wang, X.; Zeng, H. Effect of organic matter, thermal maturity and clay minerals on pore formation and evolution in the Gulong Shale, Songliao Basin, China. *Geoenergy Sci. Eng.* **2023**, *223*, 211507. [CrossRef]
114. Milliken, K.L.; Zhang, T.; Chen, J.; Ni, Y. Mineral diagenetic control of expulsion efficiency in organic-rich mudrocks, Bakken Formation (Devonian-Mississippian), Williston Basin, North Dakota, U.S.A. *Mar. Pet. Geol.* **2021**, *127*, 104869. [CrossRef]
115. Liang, C.; Cao, Y.; Liu, K.; Jiang, Z.; Wu, J.; Hao, F. Diagenetic variation at the lamina scale in lacustrine organic-rich shales: Implications for hydrocarbon migration and accumulation. *Geochim. Cosmochim. Acta* **2018**, *229*, 112–128. [CrossRef]
116. Wang, X.; Liu, L.; Wang, Y.; Sheng, Y.; Zheng, S.; Wu, W.; Luo, Z. Comparison of the pore structures of Lower Silurian Longmaxi Formation shales with different lithofacies in the southern Sichuan Basin, China. *J. Nat. Gas Sci. Eng.* **2020**, *81*, 103419. [CrossRef]
117. Fishman, N.S.; Hackley, P.C.; Lowers, H.; Hill, R.J.; Egenhoff, S.O.; Eberl, D.D.; Blum, A.E. The nature of porosity in organic-rich mudstones of the Upper Jurassic Kimmeridge Clay Formation, North Sea, offshore United Kingdom. *Int. J. Coal Geol.* **2012**, *103*, 32–50. [CrossRef]
118. Wilson, R.D.; Schieber, J.; Nie, H.; Sun, C.; Liu, G.; Li, Y.; Schieber, J.; Fan, T.; Li, Z.; Schieber, J. The influence of primary and secondary sedimentary features on reservoir quality: Examples from the geneseo formation of New York, U.S.A. *AAPG Mem.* **2016**, *112*, 167–183.
119. Dowey, P.J.; Taylor, K.G. Extensive authigenic quartz overgrowths in the gas-bearing Haynesville-Bossier Shale, USA. *Sediment. Geol.* **2017**, *356*, 15–25. [CrossRef]

120. Lalonde, K.; Mucci, A.; Ouellet, A.; Gélinas, Y. Preservation of organic matter in sediments promoted by iron. *Nature* **2012**, *483*, 198–200. [CrossRef]
121. Zhang, S.; Liu, H.; Song, G.; Wang, Y.; Chen, S.; Zhang, S. Genesis and control factors of shale oil reserving space in Dongying sag. *Acta Pet. Sin.* **2016**, *37*, 1495–1507,1527.
122. Davis, K.J.; Dove, P.M.; Yoreo, J.J.D. The Role of Mg^{2+} as an Impurity in Calcite Growth. *Science* **2000**, *290*, 1134–1137. [CrossRef] [PubMed]
123. Klasa, J.; Ruiz-Agudo, E.; Wang, L.; Putnis, C.; Valsami-Jones, E.; Menneken, M.; Putnis, A. An atomic force microscopy study of the dissolution of calcite in the presence of phosphate ions. *Geochim. Cosmochim. Acta* **2013**, *117*, 115–128. [CrossRef]
124. Naviaux, J.D.; Subhas, A.V.; Rollins, N.E.; Dong, S.; Berelson, W.M.; Adkins, J.F. Temperature dependence of calcite dissolution kinetics in seawater. *Geochim. Cosmochim. Acta* **2019**, *246*, 363–384. [CrossRef]
125. Stefansson, A. Dissolution of primary minerals of basalt in natural waters I. Calculation of mineral solubilities from 0 °C to 350 °C. *Chem. Geol.* **2001**, *172*, 225–250. [CrossRef]
126. Milliken, K. A Compositional Classification for Grain Assemblages in Fine-Grained Sediments and Sedimentary Rocks. *J. Sediment. Res.* **2014**, *84*, 1185–1199. [CrossRef]
127. Yuan, G.; Cao, Y.; Gluyas, J.; Jia, Z. Reactive transport modeling of coupled feldspar dissolution and secondary mineral precipitation and its implication for diagenetic interaction in sandstones. *Geochim. Cosmochim. Acta* **2017**, *207*, 232–255. [CrossRef]
128. Aagaard, P.; Egeberg, P.K.; Saigal, G.C.; Morad, S.; Bjorlykke, K. Diagenetic albitization of detrital K-feldspars in Jurassic, Lower Cretaceous and Tertiary clastic reservoir rocks from offshore Norway; II, Formation water chemistry and kinetic considerations. *J. Sediment. Res.* **1990**, *60*, 575–581. [CrossRef]
129. Nicholson, P.H.; Esch, W.L.; Ajdukiewicz, J.M. Prediction of deep reservoir quality using early diagenetic process models in the jurassic norphlet formation, gulf of Mexico. *AAPG Bull.* **2010**, *94*, 1189–1227.
130. Porten, K.; Kane, I.; Warchoł, M.; Southern, S.J. A Sedimentological Process-Based Approach to Depositional Reservoir Quality of Deep-Marine Sandstones: An Example from the Springar Formation, Northwestern Vøring Basin, Norwegian Sea. *J. Sediment. Res.* **2016**, *86*, 1269–1286. [CrossRef]
131. Liu, D.; Yuan, P.; Liu, H.; Li, T.; Tan, D.; Yuan, W.; He, H. High-pressure adsorption of methane on montmorillonite, kaolinite and illite. *Appl. Clay Sci.* **2013**, *85*, 25–30. [CrossRef]

minerals

MDPI

Article

Geochemical Evidence of Organic Matter Enrichment and Depositional Dynamics in the Lower Cambrian Yurtus Formation, NW Tarim Basin: Insights into Hydrothermal Influence and Paleoproductivity Mechanisms

Wangming Cheng [1,2,3,4,5,*], Ruyue Wang [6], Taohua He [7,*], Chonghao Sun [1,2,3,4,5], Haonan Tian [1,2,3,4], Jiaqi Zhao [1,2,3,4], Ya Zhao [7], Jiayi He [7], Qianghao Zeng [7], Jiajun Liu [1,8] and Yan Yi [1]

1 Research Institute of Petroleum Exploration and Development, Tarim Oilfield Company, Korla 841000, China; sunch-tlm@petrochina.com.cn (C.S.); tianhn-tlm@petrochina.com.cn (H.T.); zhaojq-tlm@petrochina.com.cn (J.Z.); 18972353901@163.com (J.L.); yiyan-tlm@petrochina.com.cn (Y.Y.)
2 R&D Center for Ultra-Deep Complex Reservoir Exploration and Development, China National Petroleum Corporation, Korla 841000, China
3 Engineering Research Center for Ultra-Deep Complex Reservoir Exploration and Development, Xinjiang Uygur Autonomous Region, Korla 841000, China
4 Xinjiang Key Laboratory of Ultra-deep Oil and Gas, Korla 841000, China
5 Key Laboratory of Gas Reservoir Formation and Development, China National Petroleum Corporation, Korla 841000, China
6 Sinopec Petroleum Exploration and Production Research Institute, Beijing 102206, China; wangruyue.syky@sinopec.com
7 Hubei Key Laboratory of Petroleum Geochemistry and Environment (Yangtze University), Yangtze University, Wuhan 430100, China; 2023710618@yangtzeu.edu.cn (Y.Z.); 2024720583@yangtzeu.edu.cn (J.H.); 2023710623@yangtzeu.edu.cn (Q.Z.)
8 Exploration and Development Research Institute, PetroChina Daqing Oilfield Company Ltd., Daqing 163712, China
* Correspondence: chengwm-tlm@petrochina.com.cn (W.C.); hetaohua@yangtzeu.edu.cn (T.H.)

Abstract: The lower Cambrian Yurtus Formation ($Є_1y$) in the Tarim Basin, characterized by its high organic matter content, serves as a critical source rock for oil and gas exploration in the platform basin. This study presents a high-resolution geochemical analysis of a geological section located near the Aksu Cement Plant in the northwest margin of the Tarim Basin. The focus is on elucidating the sedimentary environment, mechanisms of organic matter enrichment, and the depositional history of the $Є_1y$ source rock. The $Є_1y$ exhibits distinctive geochemical signatures, including elevated concentrations of Mo, Ba, and U, with an average rare earth element (REE) content of 155.75 μg/g. The formation shows significant light REE enrichment (LREE/HREE = 1.74–5.57), a moderate Ce negative anomaly (δCe = 0.4–0.71), and a notable Eu positive anomaly (δEu = 0.94–2.14), indicative of a unique depositional environment influenced by hydrothermal processes. Geochemical evidence suggests that the $Є_1y$ siliceous shales were deposited in a highly reducing, anoxic, and sulfide-rich environment, promoting organic matter preservation and enhancing sedimentary productivity. The presence of hydrothermal trace elements, likely introduced by hydrothermal fluids from volcanic activity along fractures and faults, played a critical role in enriching the sedimentary system, preserving organic matter, and boosting paleoproductivity. The model of organic matter enrichment proposed in this study underscores the dynamic interplay between hydrothermal influences and high primary productivity. These findings provide important insights into the formation of high-quality source rocks and have significant implications for the exploration of deep and ultra-deep oil and gas reserves in the Tarim Basin.

https://doi.org/10.3390/min15030288

Keywords: Tarim Basin; Yurtus Formation; paleoproductivity; sedimentary environment; organic matter enrichment

1. Introduction

The transition from the Precambrian to the Early Cambrian is a critical period in Earth's history, marked by significant tectonic and environmental shifts. This era witnessed extensive marine transgressions, hydrothermal activity, and persistent anoxic conditions, shaping global biogeochemical cycles (Figure 1a) [1–3]. The widespread deposition of Lower Cambrian black shales globally serves as a key record of marine evolution during this time, offering insights into paleoenvironmental conditions [4]. In the Tarim Basin, the Early Cambrian Yurtus Formation (ϵ_1y), predominantly composed of siliceous rocks and siliceous shales (Figure 1b), provides a unique opportunity to explore the paleo-marine environment of this period. Notably, the organic carbon content of the ϵ_1y represents the highest concentrations of marine source rocks in China [5], making it a significant subject of study, especially with the growing focus on deep and ultra-deep strata exploration in the basin [5,6].

The ϵ_1y is overlain by the Xiaoerbulake Formation, which features reef-shoal facies dolomites with numerous fractures and cavities, while the Wusonger Formation consists of thick mudstones and paste rocks. These strata form an important source-reservoir-caprock system, critical for oil and gas exploration in the Tarim Basin [7]. However, the complex and heterogeneous nature of the basal black rock series in the ϵ_1y presents significant challenges in understanding its genesis, depositional processes, and sedimentary background. Despite some progress, research on the mechanisms of organic matter enrichment in this series remains limited and debated [8,9]. The ϵ_1y outcrop in the northwest of the Tarim Basin is thought to have been deposited in environments ranging from restricted shallow shelves to deep-water basins, with the subsurface ϵ_1y forming in semi-restricted or coastal environments [10,11]. These findings have implications for the exploration of Lower Paleozoic organic-rich source rocks in the basin.

Organic matter enrichment in sedimentary rocks is primarily controlled by high primary productivity, which is often enhanced by anoxic conditions that facilitate organic matter preservation [12–14]. The abundance of organic matter is largely governed by anoxic or sulfidic reducing environments, where preservation potential is high [15]. However, neither the "productivity model" nor the "preservation model" alone fully explains the mechanisms of organic matter enrichment. It is likely that these factors interact to produce the observed organic richness. Current models explaining the development of high-quality source rocks include: (1) the "hydrothermal activity-upwelling-anoxic event model," which posits that stratification of the water column allows for oxygenated surface waters with high productivity, while the bottom waters are anoxic and reducing [16]; (2) the combined influence of hydrothermal activity, coastal upwelling, sea-level fluctuations, and organic matter preservation conditions on source rock development [17]; (3) the effect of sea-level changes, topography, hydrothermal events, upwelling, and periodic anoxia on organic matter enrichment [18]; and (4) the interplay between redox conditions and productivity, where the dominant control on organic matter enrichment varies by sedimentary period [19].

This study focuses on a geological section near the Aksu area in the northwest Tarim Basin, where detailed geochemical analyses are employed to reconstruct the depositional environment and sedimentary conditions of the ϵ_1y Formation. The goal is to provide new insights into the mechanisms of organic matter accumulation, preservation, and enrichment, as well as to explore the tectonic evolution of the region.

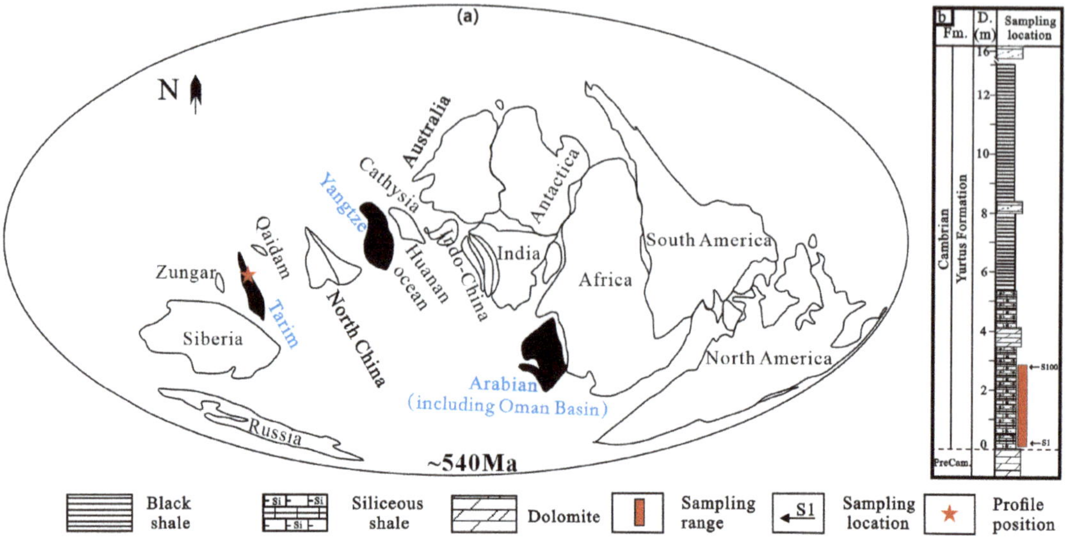

Figure 1. Global palaeogeography during the Early Cambrian (**a**) and locations of the sections and samples (**b**) (Fm., Formation; D., Depth; S1, Sample number; PreCam., Precambrian) [10].

2. Geological Setting

Geologically, the Tarim Basin is divided into three uplifts, four depressions, and seven first-order structural units (Figures 1b and 2). The Aksu area is located on the northwest margin of the Tarim Basin. The Nanhua strata, comprising gray-green and purplish-red feldspar sandstones and siltstones, are approximately 2000 m thick, exhibiting poor sorting and rounding. The Sinian strata consist mainly of sand-mudstone and carbonate rocks of coastal shallow sea-tidal flat facies.

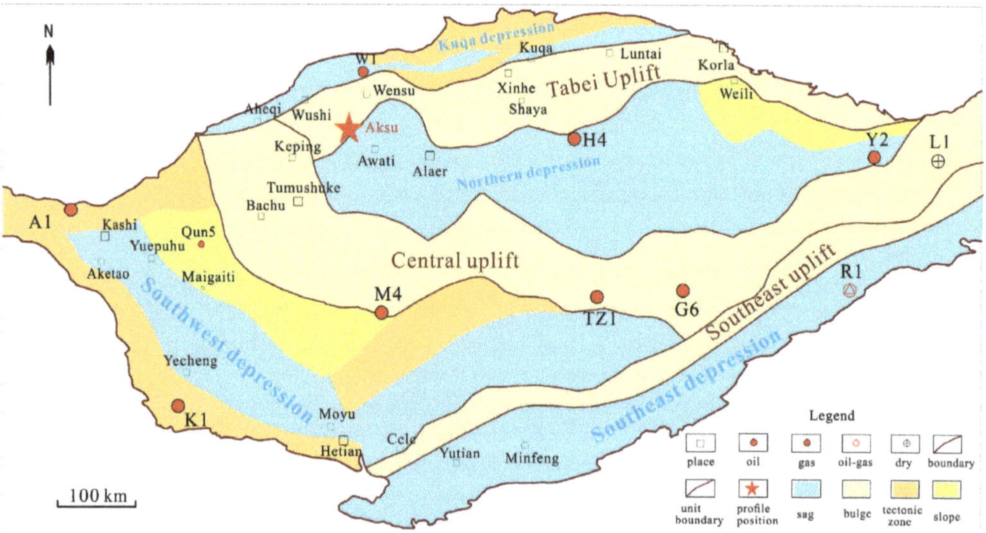

Figure 2. Tectonic unit of Tarim Basin and geological profile location of cement plant.

The Early Cambrian (ϵ_1y) strata were deposited following post-rift subsidence and a large-scale transgression, resulting in the extensive development of black strata across the northwestern region of the Tarim Basin. Recognized as the largest hydrocarbon-bearing basin in China, these strata are primarily composed of black siliceous shales and other dark-colored shales [20], and are situated in the lower section of the $Є_1y$. With a thickness ranging from 10 to 15 m and spanning approximately 26,000 km^2, these strata reflect the unique paleoenvironmental conditions of the Early Cambrian. The extensive development of these black strata across the basin has made them a focal point in petroleum geology, particularly in understanding the genesis and distribution of Lower Paleozoic marine hydrocarbons. Previous studies have highlighted that the organic matter within these shales predominantly consists of Type II kerogen, which is highly effective in generating oil under appropriate thermal maturity conditions. Consequently, these strata are pivotal for current hydrocarbon exploration and production efforts and offer valuable insights into the broader paleoenvironmental and tectonic evolution of the Tarim Basin during the Early Cambrian period.

3. Materials and Methods

This study collected high-resolution samples from Cambrian outcrops in geological profiles near the Aksu Cement Plant, located in the northwest Tarim Basin. Sampling intervals ranged from 0.01 to 0.03 m, yielding a total of 100 samples, labeled Cam_{1y}-1 through Cam_{1y}-100 (Figure 1b), which were subsequently analyzed. To meet space constraints, this paper presents the results for every fifth sample and discusses their geochemical significance. 1-alkyl-2,3,6-trimethylbenzene compounds (ATMBs) from aromatic fractions of typical source rock samples are analyzed with m/z 133/134 to reveal their critical role in the relationship between Cam_{1y} source rocks and major hydrocarbons in the Tarim Basin. Pre-processing and testing procedures for biomarker detection and total organic carbon (TOC) testing can be found in our early work [10].

Major and trace element analyses were performed at the Analytical Laboratory of the Beijing Research Institute of Uranium Geology, China National Nuclear Corporation, following the procedures outlined in previous studies [21]. Major elements were quantified using an AB104L Axioma X-ray fluorescence (XRF) spectrometer, with a relative standard deviation (RSD) of less than 1%. Accuracy was confirmed using shale standard material (GBW 03014) measured alongside the samples [21,22]. Trace elements and REE concentrations were determined following multiple rounds of acid digestion, involving HNO_3, $HClO_4$, and HF. Samples were diluted with 5% HNO_3 to 50 mL before measurement on an ELEMENT XR inductively coupled plasma mass spectrometer (ICP-MS), with an RSD of less than 1.5% for each element. The calculation methods for δCe and δEu are as follows: $\delta Ce = [Ce/(La \times Pr)^{1/2}]_{\text{Post-Archean Australian Shale (PAAS)}}$ and $\delta Eu = [Eu/(Sm \times Gd)^{1/2}]_{PAAS}$.

4. Results

4.1. Major Elements

The experimental results (Table 1) show considerable variation in the silica content across the profile, with SiO_2 mass fractions ranging from 21.61% to 96.42%, averaging 71.56%. This distribution generally follows a consistent fluctuation pattern. Notably, in samples Cam_{1y}-10, Cam_{1y}-35, Cam_{1y}-45, and Cam_{1y}-55, the average SiO_2 content is consistent at 71.56%, while in sample Cam_{1y}-95, the SiO_2 content is exceptionally high, surpassing 90%. The TOC content ranges from 0.2% to 20.28%, with an average of 5.40%. Interestingly, the average TOC content for these five samples is 3.12%, significantly lower than the overall average of 5.40%. Among the other major elements, Al_2O_3 content varies from 0.26% to 10.91%, with an average of 3.85%. In the five samples mentioned, the average Al_2O_3 content is only 0.38%, much lower than the general average. Fe_2O_3 content ranges from 0.15% to 7.46%, with an average of 2.15%. MgO content ranges from 0.064% to

1.9%, with an average of 0.774%, while CaO content varies significantly from 0.11% to 31.92%, with an average of 3.09%. MnO content is generally low and mostly falls below the detection limit. TiO_2 mass fractions range from 0.02% to 0.68%, with a mean value of 0.22%.

Table 1. Major elements (%) of geological profile samples near the Aksu area in the Tarim Basin.

Number	Interval	Horizon	Lithology	SiO_2 /%	Al_2O_3 /%	Fe_2O_3 /%	MgO /%	CaO /%	MnO /%	TiO_2 /%	P_2O_5 /%	LOI /%	FeO /%	TOC /%
Cam$_{1y}$-5	0.21	Є$_1$y	Siliceous shale	93.09	0.539	0.146	0.426	1.21	-	0.018	0.451	3.87	0.11	20.18
Cam$_{1y}$-10	0.33	Є$_1$y	Siliceous shale	95.56	0.387	0.408	0.131	0.177	-	0.016	0.019	3.08	0.35	11.25
Cam$_{1y}$-15	0.43	Є$_1$y	Siliceous rock	54.72	9.19	7.46	1.37	2.06	0.005	0.538	1.03	15.24	0.72	4.87
Cam$_{1y}$-20	0.64	Є$_1$y	Siliceous shale	83.61	2.04	0.758	0.358	0.452	-	0.106	0.208	10.71	0.59	5.96
Cam$_{1y}$-25	0.77	Є$_1$y	Siliceous rock	96.13	0.426	0.618	0.115	0.328	-	0.021	0.173	1.82	0.52	0.2
Cam$_{1y}$-30	0.895	Є$_1$y	Siliceous shale	88.38	0.449	0.568	0.993	2.01	0.006	0.019	0.068	6.6	0.49	2.54
Cam$_{1y}$-35	1.125	Є$_1$y	Siliceous rock	92.72	0.399	0.501	0.069	0.752	-	0.021	0.029	4.76	0.43	1.49
Cam$_{1y}$-40	1.295	Є$_1$y	Siliceous shale	48.41	8.04	6.31	1.6	1.26	-	0.543	1.1	22.25	1.63	4.6
Cam$_{1y}$-45	1.385	Є$_1$y	Siliceous rock	96.32	0.261	0.412	0.066	0.112	-	0.015	0.065	2.04	0.36	0.58
Cam$_{1y}$-50	1.47	Є$_1$y	Siliceous shale	37.01	8.67	4.43	1.71	5.6	0.004	0.584	2.09	25.03	0.95	2.93
Cam$_{1y}$-55	1.575	Є$_1$y	Siliceous rock	96.16	0.462	0.322	0.087	0.126	-	0.023	0.064	2.4	0.28	0.85
Cam$_{1y}$-60	1.632	Є$_1$y	Siliceous shale	51.95	10.91	3.58	1.6	0.662	-	0.676	1.5	18.85	1.21	6.04
Cam$_{1y}$-65	1.747	Є$_1$y	Siliceous rock	96.42	0.375	0.365	0.064	0.193	-	0.019	0.066	2.26	0.3	0.64
Cam$_{1y}$-70	1.867	Є$_1$y	Siliceous shale	47.03	8.07	4.8	1.67	1.68	0.004	0.469	1.94	21.6	0.34	5.92
Cam$_{1y}$-75	1.947	Є$_1$y	Siliceous rock	87.51	0.505	0.514	0.074	2.88	-	0.019	1.71	5.99	0.42	2.05
Cam$_{1y}$-80	2.007	Є$_1$y	Siliceous shale	56.55	5.59	2.78	0.961	3.95	-	0.29	2.64	20.79	0.82	7.88
Cam$_{1y}$-85	2.117	Є$_1$y	Siliceous rock	21.61	2.03	1.58	0.288	31.92	-	0.101	23.44	9.56	1.3	3.6
Cam$_{1y}$-90	2.217	Є$_1$y	Siliceous shale	46.93	9.17	3.92	1.75	2.99	0.007	0.445	1.79	23.43	0.8	10.88
Cam$_{1y}$-95	2.307	Є$_1$y	Siliceous rock	95.01	0.438	0.247	0.248	0.379	-	0.028	0.02	3.31	0.21	1.46
Cam$_{1y}$-100	2.397	Є$_1$y	Siliceous shale	46.04	8.95	3.36	1.9	3.1	0.006	0.41	1.61	26.51	1.58	14.04

Note: "-" indicates that the element content is lower than the detected value; LOI: Loss on ignition; TOC: Total organic carbon.

4.2. Trace Elements

Table 2 presents the trace element compositions of the samples. Notably, the samples show elevated concentrations of Mo, Ba, and U. Mo levels range from 2.29 to 283 μg/g, with an average concentration of 43.85 μg/g. Ba concentrations range from 743 to 3734 μg/g, with an average of 524.71 μg/g, while U content varies from 2.47 to 187 μg/g, with an average of 52.34 μg/g. Moreover, V/(V + Ni) is generally greater than 0.85, indicating a hypoxic sulfurization condition. These elevated trace element levels are indicative of specific geochemical processes influencing the sedimentary environment, and further analysis may reveal insights into the hydrothermal contributions and redox conditions during deposition.

Table 2. The characteristics of trace elements of geological profile samples near Aksu Cement Plant.

Number	Mo /ppm	Ba /ppm	U /ppm	Ba/Sr	V/(V + Ni)	Th/U	U/Th	Zn/Ti	Ni/Co	Rb/K	Th/Sc	Y/Ho	Cd/Mo
Cam_{1y}-5	2.29	300	10.1	3.75	0.88	0.06	15.68	0.55	5.67	30.80	3.32	46.70	1.19
Cam_{1y}-10	7	216	4.11	4.33	0.85	0.05	18.77	0.71	5.96	35.26	0.98	42.60	0.26
Cam_{1y}-15	283	730	113	2.87	0.83	0.06	16.89	0.53	17.39	11.97	0.64	45.97	0.08
Cam_{1y}-20	9.92	362	23	3.38	0.98	0.06	16.43	0.15	25.70	16.76	0.74	43.77	0.37
Cam_{1y}-25	7.98	93.8	6.75	3.57	0.98	0.03	33.92	0.34	13.90	19.93	0.78	43.81	0.11
Cam_{1y}-30	15.9	377	12.8	6.78	0.78	0.01	72.73	1.82	11.55	13.56	0.83	42.50	0.36
Cam_{1y}-35	15.4	441	14.5	7.16	0.93	0.01	73.98	1.55	12.79	14.67	0.51	41.89	0.21
Cam_{1y}-40	176	619	187	2.36	1.00	0.04	25.34	0.07	21.23	20.39	0.74	41.02	0.07
Cam_{1y}-45	14.9	102	9.42	6.46	0.98	0.03	31.82	0.42	11.20	28.35	0.81	41.30	0.10
Cam_{1y}-50	144	308	121	1.49	1.00	0.07	13.55	0.09	17.81	29.48	0.55	41.20	0.13
Cam_{1y}-55	7	74.3	11.2	3.83	0.99	0.03	35.11	0.24	25.32	23.98	0.66	42.29	0.30
Cam_{1y}-60	25.6	519	173	2.39	1.00	0.07	14.42	0.06	21.31	27.22	0.75	45.82	0.51
Cam_{1y}-65	11	233	6.77	7.24	0.97	0.05	21.16	0.78	30.73	36.08	0.62	47.39	0.16
Cam_{1y}-70	54.6	376	80.2	1.57	1.00	0.09	11.44	0.19	32.92	25.71	0.66	51.49	0.82
Cam_{1y}-75	7.37	138	14.6	2.11	0.97	0.03	32.52	0.9	28.27	25.61	0.72	50.50	1.51
Cam_{1y}-80	17.1	306	45.5	1.74	0.99	0.09	11.46	0.27	38.94	22.49	0.61	54.15	3.27
Cam_{1y}-85	8.02	3734	101	4.24	0.92	0.03	38.40	1.7	26.88	19.18	0.71	54.18	7.63
Cam_{1y}-90	50.5	594	64.5	2.93	0.96	0.09	11.50	0.94	31.61	23.77	0.48	39.72	3.29
Cam_{1y}-95	4.57	550	2.47	18.15	0.92	0.07	14.03	1.53	12.76	30.75	0.48	41.40	4.18
Cam_{1y}-100	14.8	421	45.8	1.52	0.97	0.13	7.58	0.62	73.11	22.20	0.64	42.45	7.64
Average	43.85	524.71	52.34	4.39	0.94	0.06	25.84	0.67	23.25	23.91	0.81	45.01	1.61

4.3. REEs

REE concentrations, normalized to PAAS standards [23], are shown in Table 3, with the distribution pattern illustrated in Figure 3. The average ΣREE value is 155.75 μg/g, with the LREE/HREE ratio averaging 3.24, ranging from 1.74 to 5.57. The Y/Ho ratio averages 45.01, and the mean δCe and δEu values are 0.5 (ranging from 0.4 to 0.71) and 1.10 (ranging from 0.94 to 2.14), respectively. These REE characteristics are consistent with a geochemical signature influenced by hydrothermal fluids [24], suggesting that the trace element and REE distributions reflect both the depositional environment and the hydrothermal processes that affected the region during the Early Cambrian.

Table 3. The characteristics of REE of geological profile samples near Aksu Cement Plant.

Number	La /ppm	Ce /ppm	Pr /ppm	Nd /ppm	Sm /ppm	Eu /ppm	Gd /ppm	Tb /ppm	Dy /ppm	Ho /ppm	Er /ppm	Tm /ppm	Yb /ppm	Lu /ppm	Y/Ho
Cam_{1y}-5	2.86	3.62	0.50	1.94	0.37	0.10	0.40	0.08	0.46	0.10	0.30	0.05	0.29	0.04	46.70
Cam_{1y}-10	1.41	1.88	0.26	1.03	0.20	0.06	0.19	0.04	0.22	0.05	0.16	0.03	0.16	0.02	42.60
Cam_{1y}-15	73	85.5	15.4	68.9	14.1	3.04	15.7	3.07	19	4.22	11.7	1.84	10.2	1.39	45.97
Cam_{1y}-20	11.1	10.9	2.26	9.97	1.96	0.45	2.13	0.43	2.87	0.71	2.19	0.40	2.42	0.36	43.77
Cam_{1y}-25	1.86	1.82	0.38	1.8	0.38	0.09	0.43	0.08	0.60	0.14	0.46	0.09	0.50	0.08	43.81
Cam_{1y}-30	2.66	3.09	0.57	2.48	0.51	0.13	0.55	0.10	0.66	0.16	0.44	0.08	0.42	0.06	42.50
Cam_{1y}-35	2.29	2.44	0.58	2.62	0.46	0.14	0.54	0.10	0.65	0.16	0.50	0.09	0.52	0.08	41.89
Cam_{1y}-40	92.1	68.8	17.1	74.7	14.8	3.27	16	2.99	19.4	4.51	13.6	2.38	13.6	1.88	41.02
Cam_{1y}-45	2.43	2.48	0.59	2.77	0.59	0.14	0.68	0.15	0.98	0.25	0.77	0.13	0.79	0.12	41.30
Cam_{1y}-50	90.5	80.2	22.8	114	26.4	6.09	31.9	6.29	40.2	9.03	25	3.83	20	2.69	41.20
Cam_{1y}-55	3.81	3.53	0.99	4.7	0.97	0.23	1.11	0.28	1.43	0.35	1.1	0.19	1.09	0.15	42.29
Cam_{1y}-60	94.1	81.5	19.2	84.5	17.1	3.90	20.8	4.37	30	7.66	24.2	4.28	25	3.72	45.82
Cam_{1y}-65	3.53	3.5	1.00	5.02	1.16	0.29	1.54	0.31	2.19	0.54	1.66	0.30	1.56	0.22	47.39
Cam_{1y}-70	51.8	46.2	9.51	38.6	7.10	1.52	8.07	1.65	11.8	3.03	10.1	1.86	11	1.66	51.49
Cam_{1y}-75	20.7	21.4	5.82	27.3	5.88	1.30	6.89	1.33	8.38	1.81	5.15	0.74	3.66	0.47	50.50
Cam_{1y}-80	37.6	36.7	7.65	34.4	6.63	1.48	7.98	1.56	10.4	2.53	7.55	1.25	6.9	1.01	54.15
Cam_{1y}-85	121	104	24.8	114	23.5	5.51	29.3	5.6	36.6	8.49	25.7	4.02	20.6	2.89	54.18
Cam_{1y}-90	42.4	44.1	9.16	40.7	8.34	1.9	9.54	1.85	12	2.82	8.48	1.5	8.93	1.36	39.72
Cam_{1y}-95	0.76	0.78	0.15	0.64	0.13	0.07	0.16	0.03	0.21	0.05	0.17	0.03	0.25	0.04	41.40
Cam_{1y}-100	43.9	46.9	9.23	40.1	8.12	1.75	8.81	1.68	10.8	2.45	7.14	1.21	7.1	1.04	42.45

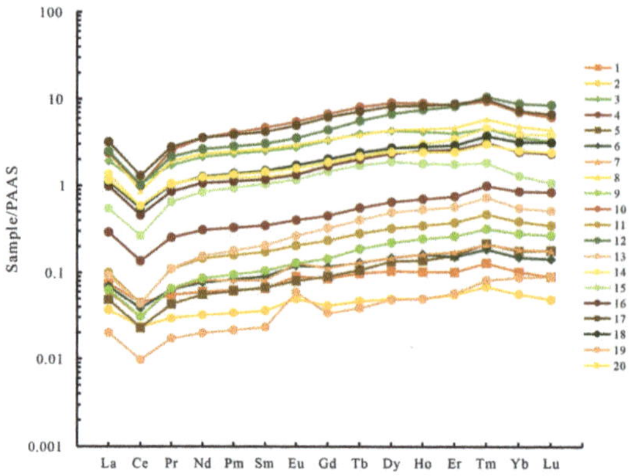

Figure 3. Normalized distribution of REE, normalized value after PAAS. Serial numbers 1–20 in the figure represent samples Cam_{1y}-5, Cam_{1y}-10, . . ., Cam_{1y}-95, Cam_{1y}-100.

4.4. Biomarkers

ATMB represents an extremely unique compound of aromatic biomarkers that are highly resistant to secondary alterations. These biomarkers are particularly significant in geochemical studies because they indicate a distinct source, namely green sulfur bacteria, and are associated with photic zone euxinia (PZE) environments. Such environments are known to enhance the enrichment and preservation of organic matter [10,18]. In this study, these compounds were identified through gas chromatography-mass spectrometry (GC-MS) analysis, where they exhibited characteristic mass spectral patterns (Figure 4). The structural confirmation of these compounds was achieved by comparing them with synthetic standards of 1-alkyl-2,3,6-trimethylbenzenes (C_{13}–C_{22}), as established by pioneering researchers in the field [18,24]. This analysis provides valuable insights into the sedimentary environment and the organic matter preservation processes, reinforcing the significance of hydrothermal and anoxic conditions in the Early Cambrian.

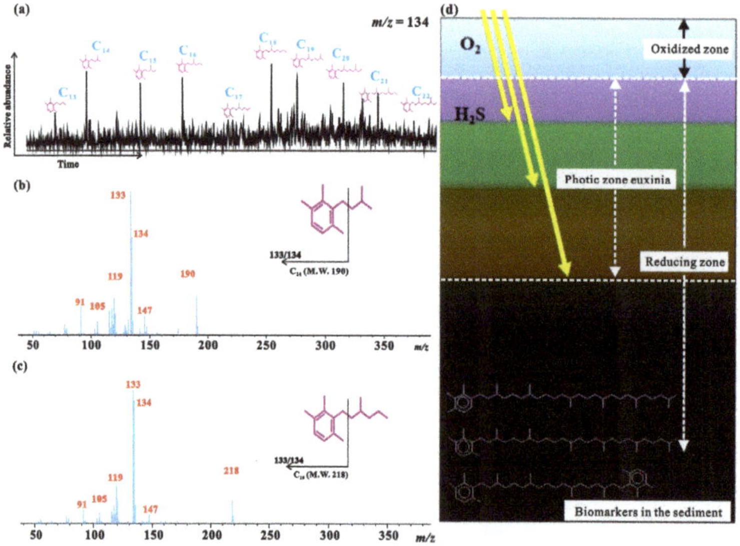

Figure 4. Identification of ATMB (**a**–**c**) and their precursor's habitats (**d**) [10]. (**a**), GC-MS of ATMB; (**b**,**c**), mass spectrogram of ATMB with C_{14} and C_{16}; (**d**), habitats of green sulfur bacteria).

5. Discussion

5.1. Role of Terrigenous Influx

To explore factors such as paleoproductivity and redox conditions, it is necessary to assess whether the samples were influenced by terrigenous clastic components. In this study, thorium (Th) was employed as an index to measure terrigenous clastic input [25]. Additionally, Th, Al, Y/Ho, and Ti were used to assess the contribution of terrigenous detritus to diagenesis [16]. The results reveal a strong correlation between Th and Al, as well as Th and Ti (Figure 5, R^2 = 0.92 and 0.97, respectively), but no correlation with Y/Ho (Figure 5, R^2 = 0.001), indicating that the chemical composition of the samples was influenced by terrigenous detrital input [26]. The correlation between Al and ΣREE (Figure 6, R^2 = 0.84) further corroborates the presence of terrigenous material input [27].

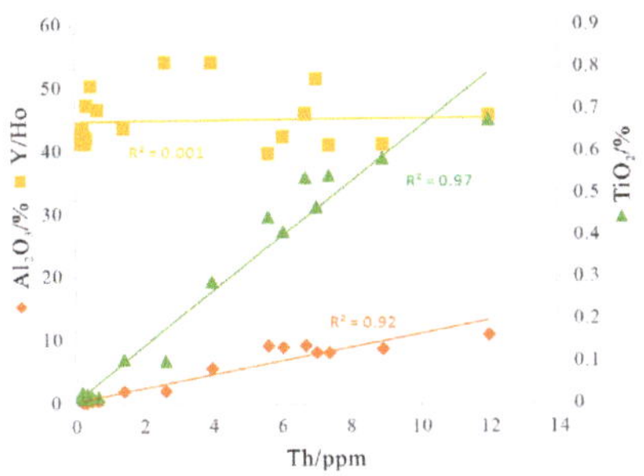

Figure 5. The correlation graph of Th with Al_2O_3, Ti and Y/Ho.

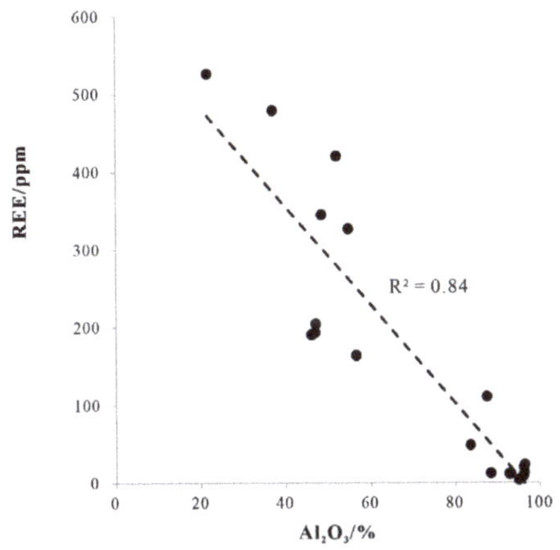

Figure 6. The correlation graph of Al_2O_3 with ΣREE.

The Ti content in sediments is used to calculate the amount of terrigenous detritus with the formula: terrigenous (T, %) = (Ti_{sample}/Ti_{shale}) × 100, where Ti_{shale} = 5995 ppm (PAAS) [28]. The T values range widely, from a minimum of 1.5% to a maximum near 70%, with an average of 21.80%. This indicates that terrigenous input in the target section should be considered, as it may exert a dilution effect on organic matter enrichment, leading to a decrease in TOC.

5.2. Paleoproductivity

Paleoproductivity refers to the amount of organic matter produced per unit area and time during a specific geological period. Geochemical indices serve as reliable tools for quantitatively evaluating paleoproductivity [29], with barium (Ba) being a particularly representative element [7,30]. Ba in marine sediments can originate from terrigenous detrital, biogenic, and hydrothermal sources. However, due to the influence of terrigenous detrital components, Ba cannot directly indicate productivity. Thus, it is necessary to remove the terrigenous Ba component. Typically, titanium (Ti) or aluminum (Al), which have a singular terrigenous origin, are used to eliminate the influence of terrigenous detritus [30]. The formula used to calculate the non-terrestrial Ba content is as follows:

$$w(\mathrm{Ba}_{\text{non-terrestrial}}) = w(\mathrm{Ba}_{\text{total}}) - w(\mathrm{Ti}_{\text{total}}) \times (w(\mathrm{Ba}_{\text{total}})/w(\mathrm{Ti}))_{\mathrm{PAAS}}$$

where $w(\mathrm{Ba}_{\text{non-terrestrial}})$ represents the content of Ba from non-terrestrial sources, $w(\mathrm{Ba}_{\text{total}})$ is the total Ba content in sediments, $w(\mathrm{Ti}_{\text{total}})$ is the total Ti content, and $(w(\mathrm{Ba}_{\text{total}})/w(\mathrm{Ti}))_{\mathrm{PAAS}}$ is the mid-continental Ba/Ti ratio of the PAAS, which is 0.11. The results indicate that the non-terrestrial Ba content in samples Cam_{1y}-5 to Cam_{1y}-45 ranges from 21.7 to 417.9 μg/g, indicating a higher supply of nutrients.

Moreover, the ratio of Zn/Ti is increasingly recognized as a reliable proxy for paleo-productivity, reflecting nutrient availability and primary productivity in ancient oceans [31]. A higher Zn/Ti ratio is often indicative of nutrient-rich environments conducive to enhanced biological activity. For instance, when the Zn/Ti ratio exceeds 0.2, it typically signals abundant nutrients, favoring biological productivity and organic matter accumulation. This is supported by studies suggesting that elevated Zn concentrations, linked with high productivity zones, correlate well with the growth of primary producers in nutrient-rich marine environments. The results indicate that the Zn/Ti ratio of the tested samples is generally greater than 0.2, with an average value of 0.67 and a maximum value of 1.82, which also reflects that the relevant source rocks are developed in water bodies with sufficient nutrient supply.

5.3. Sedimentary Environment

The sedimentary environment plays a critical role in the enrichment of organic matter. Various geochemical indicators have been proposed to identify the sedimentary environment of organic matter (Table 4) [26,32,33]. In this study, the U/Th ratios range from 7.58 to 73.98 (average 25.84), V/Cr ratios from 1.07 to 11.31 (average 4.73), Ni/Co ratios from 5.67 to 76.11 (average 23.25), and V/(V + Ni) ratios from 0.78 to 1.00 (average 0.94), along with high Mo content and a Ce negative anomaly ($\delta Ce < 1$). These characteristics suggest that the deposition environment of the samples was an anoxic sulfide-rich environment. The Rb/K ratio, indicative of sedimentary water depth [32], ranges from 11.97 to 36.08, with an average of 23.91. These values are lower than those of the source rocks from the Heituao Formation in semi-deep sea facies deposits on the northeastern margin of Tarim, where the Rb/K ratio ranges from 37.28 to 50.48 (average 44.98). This suggests that the black rock series of the $Є_1y$ was deposited in shallower waters than those of the Heituao Formation.

Table 4. The trace element ratio criteria of sedimentary environment.

Ratios	Oxidation	Sub-Oxidation	Anoxic
U/Th	<0.75	0.75–1.25	>1.25
V/Cr	<2	2–4.25	>4.25
Ni/Co	<5	5–7	>7
V/(V + Ni)	<0.46	0.46–0.6	0.6–0.85 (>0.85: sulfidic)

Furthermore, evidence for anoxic, sulfide-rich conditions is supported by our earlier organic geochemistry studies, which identified ATMBs indicative of PZE [24,34–38]. ATMBs are derived from isorenieratene, a pigment in green sulfur bacteria (Chlorobiaceae), and are well-preserved in sediments [38]. These compounds serve as reliable biomarkers for PZE in geological records, as green sulfur bacteria perform anoxygenic photosynthesis in environments where light and hydrogen sulfide (H_2S) coexist. The presence of ATMBs indicates H_2S was present in the water column during sediment deposition, with light reaching anoxic layers, thereby confirming PZE conditions. This biomarker distinguishes the source rocks of the $Є_1y$ from other formations [39] and is prevalent in deep to ultra-deep crude oil deposits in the central, northern, and southwestern Tarim Basin (Figure 7). This suggests that the $Є_1y$ source rocks containing these compounds significantly contribute to the deep to ultra-deep oil and gas reserves in the Tarim Basin [10,39–41]. Additionally, it indirectly confirms that the $Є_1y$ source rocks, although relatively thin (10–15 m), have exceptionally high TOC content (exceeding 20% in this study and reaching 29.8% in the well LT1 [42]), which developed under PZE conditions favorable for organic matter enrichment.

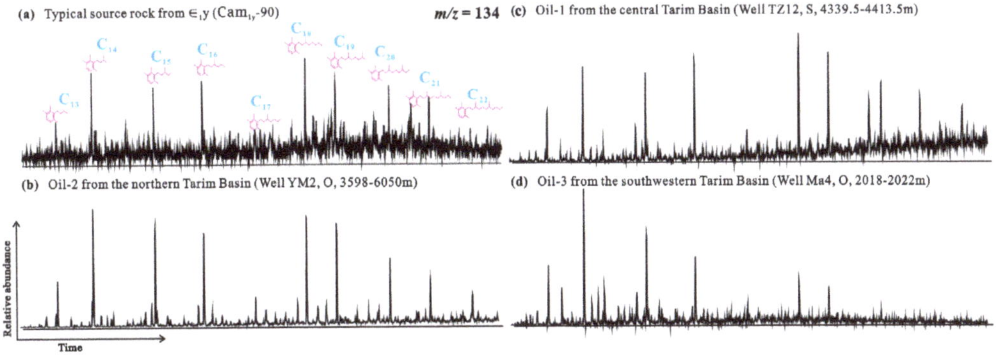

Figure 7. Partial mass chromatograms of ATMBs for typical Cam_{1y} source rocks and crude oil in the Tarim Basin [10]. ((**a**), mass chromatograms of ATMBs for Cam_{1y}-90 source rocks; (**b–d**), mass chromatograms of ATMBs for oils from wells of TZ12, YM2, and Ma4).

5.4. Hydrothermal Activity

Hydrothermal activity plays a crucial role in the enrichment of organic matter, with REE partition patterns and parameters serving as powerful tools for its investigation. A positive Eu anomaly is widely recognized as a significant indicator of hydrothermal fluid activity, though it is essential to account for false positives caused by Ba [23]. Correlation analysis between δEu and Ba or Al reveals no significant correlation, indicating that the positive Eu anomaly in these samples is not due to terrigenous detrital influence.

Additionally, samples influenced by hydrothermal solutions typically exhibit Fe/Ti ratios greater than 15 [10], and Y/Ho ratios close to modern seawater values (~45). In contrast, samples affected by terrigenous debris show lower Y/Ho ratios (Y/Ho = 28). The Fe/Ti ratio of the studied samples ranges from 8.50 to 68.25 (average 32.14), and Y/Ho from 39.7 to 54.2 (average 45.0), indicating the influence of hydrothermal activity. Analysis of samples Cam_{1y}-5 to Cam_{1y}-45 shows that non-continental Ba content varies from 21.7 to 417.9 μg/g, with a significant correlation with δEu (Figure 8a, R^2 = 0.53). Using Al as a control index for terrigenous detritus yields similar results (Figure 8b, R^2 = 0.53). These findings suggest that hydrothermal activity enhanced paleoproductivity. Additionally, correlations between δEu and V/Cr or Ni/Co (Figure 9, R^2 = 0.49 and 0.48, respectively) indicate that hydrothermal fluids influenced the preservation environment of organic matter, though their contribution to paleoproductivity was more significant. This is consistent with the conclusions drawn from previous mercury isotope analyses [43].

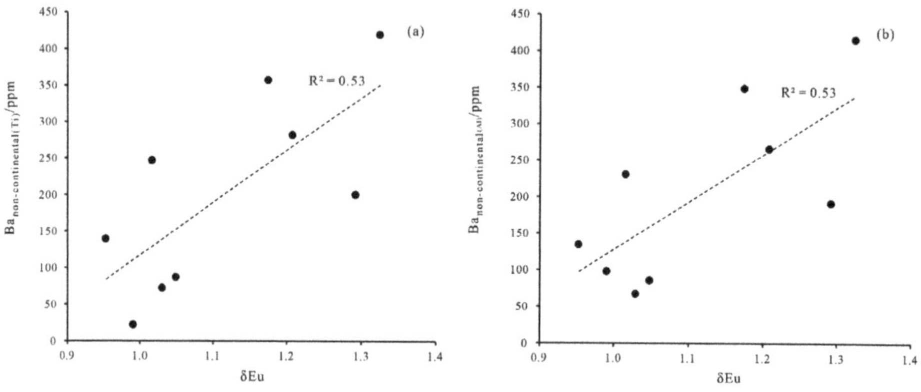

Figure 8. The correlation graph of δEu with $Ba_{non\text{-}continental}$. ((**a**), correlation graph of δEu with $Ba_{non\text{-}continental(Ti)}$; (**b**), correlation graph of δEu with $Ba_{non\text{-}continental(Al)}$).

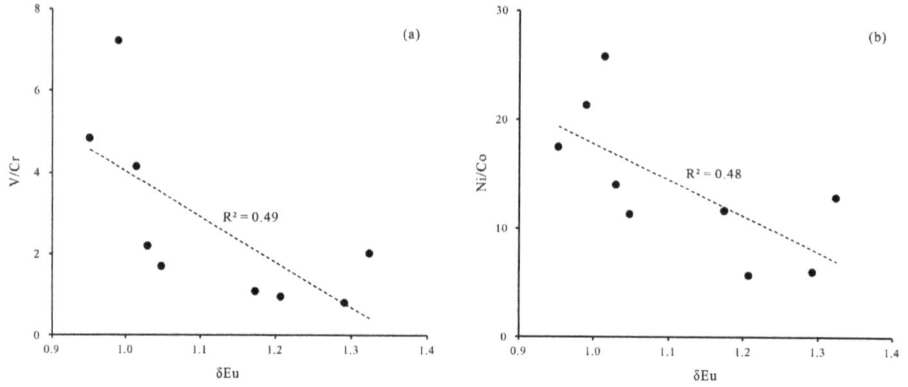

Figure 9. The correlation graph of δEu with V/Cr and Ni/Co. (**a**), the correlation graph of δEu with V/Cr; (**b**), the correlation graph of δEu with and Ni/Co.

5.5. Organic Matter Enrichment Mechanism

Two primary models for source rock development include the preservation model, which emphasizes an anoxic environment, and the productivity model, which highlights high paleoproductivity. The prevailing model is still debated [9]. A pioneering study [44] analyzed and summarized the Cd/Mo ratio across various marine basins with different retention degrees, establishing 0.1 as the threshold between the two models. When Cd/Mo > 0.1, organic matter enrichment is dominated by high productivity; when Cd/Mo < 0.1, an anoxic preservation environment prevails. The Cd/Mo ratio in most samples from this study exceeds 0.1, with only two samples falling below this threshold, indicating that paleoproductivity primarily governs organic matter enrichment. Furthermore, analysis of the relationship between TOC and paleoproductivity indicators Ba and Cu, as well as preservation environment indicators V/Cr and Ni/V, shows a strong correlation with paleoproductivity indicators Ba and Cu (R^2 = 0.50 and 0.98, respectively) (Figure 10) and a weak correlation with preservation environment indicators V/Cr and Ni/V (R^2 = 0.01 and 0.03, respectively). This further supports the conclusion that paleoproductivity, with hydrothermal activity playing a significant role, is the dominant mechanism for organic matter enrichment in the Є$_1$y in the Keping area.

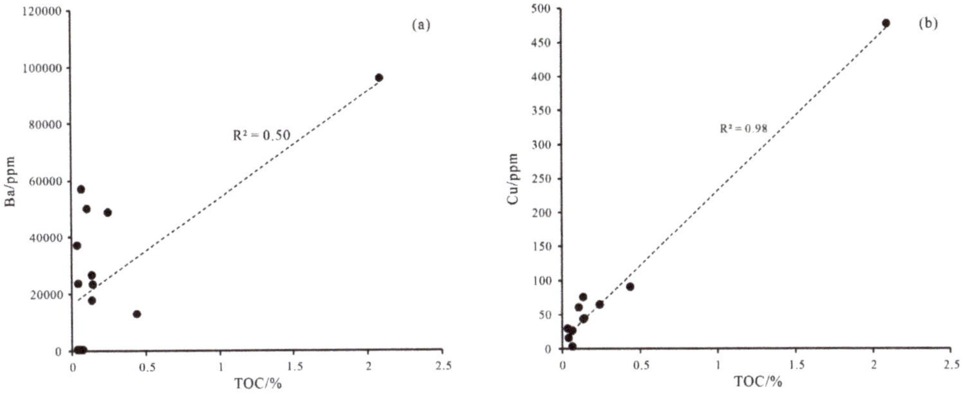

Figure 10. The correlation graph of TOC and paleoproductivity indicators Ba and Cu in the Keping area. ((**a**), the correlation graph of TOC and Ba; (**b**), the correlation graph of TOC and Cu).

The presented composite bar chart (Figure 11) summarizes the high-resolution geochemical data from 100 densely sampled points across a 2.4 m thick siliceous rock layer, which includes both siliceous rocks and siliceous shale interlayers. The chart highlights key geochemical trends that reveal the mechanisms behind organic matter enrichment in the study area. From the bottom to the top of the section, Total Organic Carbon (TOC) displays two prominent features. First, the siliceous rocks are characterized by low organic matter content, with TOC generally less than 2%. In contrast, the siliceous shale interlayers, located in the middle of the siliceous rocks, exhibit much higher TOC values, with some reaching over 20%. The siliceous shale interlayers show two distinct high TOC intervals: one in the lower section (0 to 1.4 m) and another in the upper section (1.4 to 2.4 m). TOC increases at the bottom, decreases in the middle, and then increases again towards the top, which corresponds well with ancient environmental conditions and hydrothermal events.

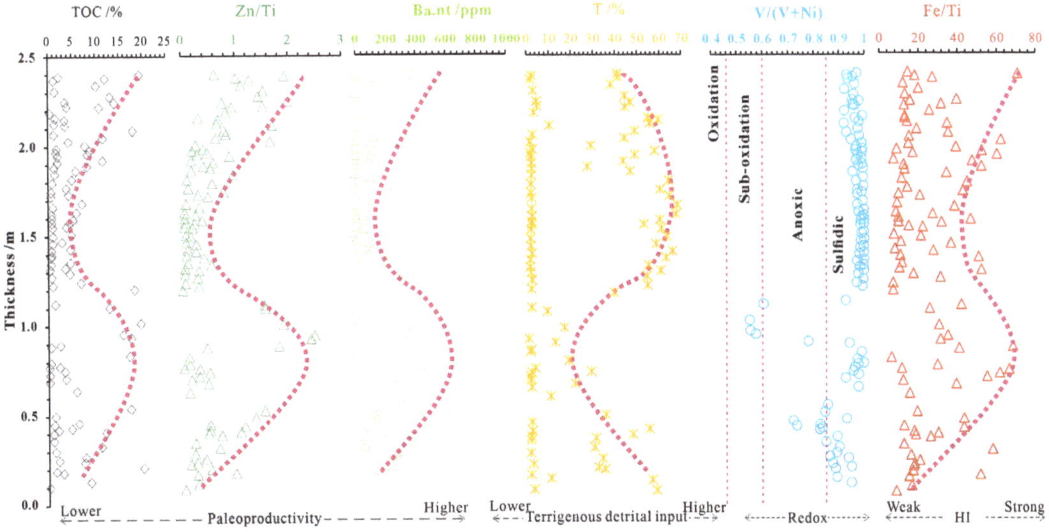

Figure 11. Geochemical composite profile of the black rock series in the Lower Cambrian Yuertusi Formation (Ba-nt.: $w(Ba_{non\text{-}terrestrial})$; T: terrigenous (%) = $(Ti_{sample}/Ti_{shale}) \times 100$ (PAAS with Ti_{shale} = 5995 ppm); HI: Hydrothermal intensity).

As TOC increases, several other key geochemical parameters show parallel trends: (1) nutrient elements (Zn/Ti and $w(Ba_{non\text{-}terrestrial})$): these indicators also exhibit an increasing trend, suggesting that the rise in TOC is significantly influenced by paleoproductivity; terrigenous detrital input; conversely, parameters related to terrigenous input (T) show a decreasing trend, implying that the organic matter enrichment might have been subject to dilution effects from terrigenous sources; (2) V/(V + Ni) ratio: this ratio is consistently greater than 0.85, indicating a sulfur-deficient anoxic environment, which is highly favorable for the preservation of organic matter; (3) hydrothermal activity intensity (Fe/Ti): the Fe/Ti ratio also increases, indicating stronger hydrothermal activity, which further suggests that hydrothermal processes played a significant role in the organic matter enrichment of the siliceous shale.

Thus, the organic matter enrichment mechanism in the siliceous shale can be inferred as follows: Hydrothermal activity induced a sulfur-deficient anoxic environment, which facilitated the preservation of organic matter. The nutrients brought by hydrothermal fluids enhanced ancient productivity, while the increased input of hydrothermal substances relative to terrigenous material led to a reduction in the terrigenous dilution effect. As a result, hydrothermal activity created an optimal environment for organic matter accumulation by promoting a coupling of high ancient productivity, sulfur-deficient anoxia, and low terrigenous dilution. This triple coupling mechanism ultimately contributed to the high organic matter content observed in the siliceous shale, with TOC values exceeding 20%.

5.6. *Depositional Model*

The Th/U ratio (<0.1) of the studied samples is significantly lower than that observed in previous studies, indicating the incorporation of deep-sea materials during the deposition of siliceous rocks [10]. The Cambrian sedimentary environment in Wells Yingdong 2, Tadong 1, and Tadong 2 was characterized by anoxic conditions and high gamma-ray anomalies, with barite nodules observed in the Keping fault uplift [20], suggesting the presence of upwelling currents. Siliceous rocks of hydrothermal origin were deposited not only in the northwest platform, which was characterized by shallow water and gentle slopes, but also in the northeast basin in a deep-water sedimentary environment. This suggests that the Tarim block experienced extensive fracturing during the Early Cambrian, allowing hydrothermal fluids to erupt along these fractures and contemporaneous faults [43]. These hydrothermal fluids introduced reducing gases (H_2, CH_4, H_2S, etc.) into the water, creating anoxic stratification that facilitated the preservation of organic matter. They also transported large quantities of metals and essential elements for life, promoting the growth of thermophilic communities. However, excessive hydrothermal H_2S could lead to the poisoning of deep-water fauna and flora, accelerating the death of marine organisms and enhancing paleoproductivity. The trace elements introduced by these fluids also increased the rate and extent of organic matter decomposition during subsequent evolutionary processes, catalyzing hydrocarbon generation in source rocks [45]. Hydrothermal activity played a significant catalytic role in enhancing paleoproductivity, preserving and generating organic matter, and developing source rocks in the Early Cambrian Tarim block. Thus, combined with the exploration of the organic matter enrichment mechanism in Section 5.5. Organic matter enrichment mechanism, a developmental model of the Early Cambrian source rock was established (Figure 12).

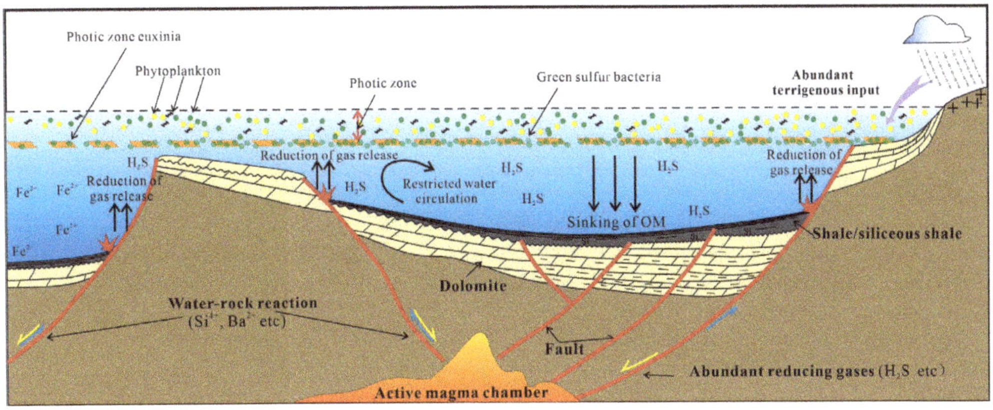

Figure 12. Developmental model of the Early Cambrian source rock, NW Tarim Basin.

6. Conclusions

(1) The Є$_1$y Formation exhibits notable geochemical signatures, including elevated concentrations of Mo, Ba, and U, and a ΣREE of 155.75 μg/g, with a strong enrichment of light rare earth elements (LREE/HREE = 1.74–5.57), moderate Ce negative anomaly (δCe = 0.4–0.71), and a significant Eu positive anomaly (δEu = 0.94–2.14), indicating a unique depositional environment influenced by hydrothermal processes;

(2) Geochemical parameters, including trace element and rare earth element distributions, point to the deposition of the Є$_1$y black shales in a highly reducing, anoxic, and sulfide-rich environment, which was conducive to organic matter preservation and enhanced sedimentary productivity;

(3) The presence of hydrothermal trace elements, introduced via hydrothermal fluids, suggests a critical role in enriching the sedimentary system, preserving organic matter, and boosting the overall paleoproductivity in the shallow marine environment of the Tarim Basin during the Early Cambrian;

(4) Hydrothermal fluids, likely emanating from volcanic activities along fractures and faults during the Early Cambrian, were instrumental in enhancing marine primary productivity, fostering the accumulation of organic matter, and contributing to the formation of high-quality source rocks in the Є$_1$y Formation. The organic matter enrichment model has been established;

(5) The integrated geochemical findings suggest that the Early Cambrian period in the Tarim Basin experienced a dynamic interplay between hydrothermal influences and organic productivity, laying the foundation for significant oil and gas reserves in the deep and ultra-deep strata of the basin.

Author Contributions: Conceptualization and writing—original draft, W.C.; funding acquisition and writing—original draft, T.H.; supervision and resources, C.S.; writing—review and editing, R.W.; methodology and formal analysis, J.H. and Q.Z.; visualization, H.T.; data curation and formal analysis, Y.Z.; software and visualization, J.Z. and Y.Y.; and investigation and software, J.L. All authors have read and agreed to the published version of the manuscript.

Funding: This research was funded by the Major Science and Technology Special Project of China National Petroleum Corporation "Research on Increasing Storage and Production of Marine Carbonate Rock Oil and Gas and Exploration and Development Technology" (No. 2023ZZ16YJ01) and the Open Fund of Key Laboratory of Exploration Technologies for Oil and Gas Resources (Yangtze University), Ministry of Education (NO. K202307).

Data Availability Statement: The data that support the findings of this study are available from the corresponding author upon reasonable request.

Conflicts of Interest: Authors W.C., C.S., H.T., J.Z., J.L. and Y.Y. were employed by the Research Institute of Petroleum Exploration and Development, Tarim Oilfield Company; author W.C., C.S., H.T. and J.Z. was employed by the company R&D Center for Ultra-Deep Complex Reservoir Exploration and Development, China National Petroleum Corporation; author W.C. and C.S. was employed by the company Key Laboratory of Gas Reservoir Formation and Development, China National Petroleum Corporation; author R.W. was employed by the Sinopec Petroleum Exploration and Production Research Institute; and author J.L. was employed by the company Exploration and Development Research Institute, PetroChina Daqing Oilfield Company Ltd. The remaining authors declare that the research was conducted in the absence of any commercial or financial relationships that could be construed as a potential conflict of interest.

References

1. Goldberg, T.; Poulton, S.W.; Strauss, H. Sulphur and oxygen isotope signatures of late Neoproterozoic to early Cambrian sulphate, Yangtze Platform, China: Diagenetic constraints and seawater evolution. *Precambrian Res.* **2005**, *137*, 223–241. [CrossRef]
2. Wang, J.; Chen, D.; Yan, D.; Wei, H.; Xiang, L. Evolution from an anoxic to oxic deep ocean during the Ediacaran–Cambrian transition and implications for bioradiation. *Chem. Geol.* **2012**, *306–307*, 129–138. [CrossRef]
3. Fike, D.A.; Grotzinger, J.P.; Pratt, L.M.; Summons, R.E. Multi-stage Ediacaran ocean oxidation and its impact on evolutionary radiation. *Geochim. Cosmochim. Acta* **2006**, *70*, A173. [CrossRef]
4. Chen, J.; Xu, J.; Wang, J.; Liu, P.; Chen, F.; Li, M. Paleo-environmental variation and its control on organic matter enrichment in the black rock series, Cambrian Yuertusi Formation in the northwestern Tarim Basin. *Earth Sci. Front.* **2023**, *30*, 150–161. (In Chinese with English abstract)
5. Zhu, G.; Chen, F.; Chen, Z.; Zhang, Y.; Xing, X.; Tao, X.; Ma, D. Discovery and basic characteristics of high-quality source rocks found in the Yuertusi Formation of the Cambrian in Tarim Basin, China. *J. Nat. Gas Geosci.* **2016**, *1*, 21–33. [CrossRef]
6. Zhu, G.; Zhang, Z.; Zhou, X.; Li, T.; Han, J.; Sun, C. The complexity, secondary geochemical process, genetic mechanism and distribution prediction of deep marine oil and gas in the Tarim Basin, China. *Earth Sci. Rev.* **2019**, *198*, 102930. [CrossRef]
7. Zhang, C.; Guan, S.; Wu, L.; Ren, R.; Wang, L.; Wu, X. Depositional environments of early Cambrian marine shale, northwestern Tarim Basin, China: Implications for organic matter accumulation. *J. Petrol. Sci. Eng.* **2020**, *194*, 107497. [CrossRef]
8. Li, W.; Wang, J.; Jia, C.; Lu, S.; Li, J.; Zhang, P.; Wei, Y.; Song, Z.; Chen, G.; Zhou, N. Carbon isotope fractionation during methane transport through tight sedimentary rocks: Phenomena, mechanisms, characterization, and implications. *Geosci. Front.* **2024**, *15*, 101912. [CrossRef]
9. Wang, X.; Zhang, S.; He, K.; Wang, H.; Mi, J.; Su, J.; Ye, Y. OMZ and euxinic sulfidic environments dominating hydrocarbon generation poential of organic matter1.4 billion years ago. *Chin. Sci. Bull.* **2021**, *66*, 3005–3017. [CrossRef]
10. He, T.; Li, W.; Lu, S.; Pan, W.; Ying, J.; Zhu, P.; Yang, E.; Wang, X.; Zhang, B.; Sun, D. Mechanism and geological significance of anomalous negative $\delta^{13}C_{kerogen}$ in the Lower Cambrian, NW Tarim Basin, China. *J. Petrol. Sci. Eng.* **2022**, *208*, 109384. [CrossRef]
11. Li, Y.; Xiong, Y.; Liang, Q.; Fang, C.; Chen, Y.; Wang, X.; Liao, Z.; Peng, P.A. The application of diamondoid indices in the Tarim oils. *AAPG Bull.* **2018**, *102*, 267–291. [CrossRef]
12. Tribovillard, N.; Algeo, T.; Lyons, T.; Riboulleau, A. Trace metals as paleoredox and paleoproductivity proxies: An update. *Chem. Geol.* **2006**, *232*, 12–32. [CrossRef]
13. Mansour, A.; Wagreich, M.; Gentzis, T.; Ocubalidet, S.; Tahoun, S.; Elewa, A. Depositional and organic carbon-controlled regimes during the Coniacian-Santonian event: First results from the southern Tethys (Egypt). *Mar. Petrol. Geol.* **2020**, *115*, 104285. [CrossRef]
14. Mansour, A.; Wagreich, M. Earth system changes during the cooling greenhouse phase of the Late Cretaceous: Coniacian-Santonian OAE3 subevents and fundamental variations in organic carbon deposition. *Earth-Sci. Rev.* **2022**, *229*, 104022. [CrossRef]
15. Tyson, R. Sedimentation rate, dilution, preservation and total organic carbon: Some results of a modelling study. *Org. Geochem.* **2001**, *32*, 333–339. [CrossRef]
16. Liu, J.; Liu, Q.; Zhu, D.; Meng, Q.; Liu, W.; Qiu, D.; Huang, Z. The role of deep fluid in the formation of organic-rich source rocks. *J. Nat. Gas Geosci.* **2018**, *3*, 171–180. [CrossRef]
17. Zhang, B.; Zhang, S.; Bian, L.; Jin, Z.; Wang, D. Developmental modes of the Neoproterozoic-Lower Paleozoic marine hydrocarbon source rocks in China. *Chin. Sci. Bull.* **2007**, *52*, 77–91. [CrossRef]
18. Gao, Z.; Shi, J.; Lv, J.; Chang, Z. High-frequency sequences, geochemical characteristics, formations, and distribution predictions of the lower Cambrian Yuertusi Formation in the Tarim Basin. *Mar. Petrol. Geol.* **2022**, *146*, 105966. [CrossRef]

19. Zhou, L.; Algeo, T.J.; Shen, J.; Hu, Z.; Gong, H.; Xie, S.; Huang, J.; Gao, S. Changes in marine productivity and redox conditions during the Late Ordovician Hirnantian glaciation. *Palaeogeogr. Palaeoclimatol. Palaeoecol.* **2015**, *420*, 223–234. [CrossRef]
20. Zhou, X.; Chen, D.; Dong, S.; Zhang, Y.; Guo, Z.; Wei, H.; Yu, H. Diagenetic barite deposits in the Yurtus Formation in Tarim Basin, NW China: Implications for barium and sulfur cycling in the earliest Cambrian. *Precambrian Res.* **2015**, *263*, 79–87. [CrossRef]
21. Zhang, S.; Wang, X.; Hammarlund, E.; Wang, H.; Costa, M.; Bjerrum, C.; Connelly, J.; Zhang, B.; Bian, L.; Canfield, D.E. Orbital forcing of climate 1.4 billion years ago. *Proc. Natl. Acad. Sci. USA* **2015**, *112*, E1406–E1413. [CrossRef] [PubMed]
22. Wang, X.; Zhao, W.; Zhang, S.; Wang, H.; Su, J.; Canfield, D.E.; Hammarlund, E.U. The aerobic diagenesis of Mesoproterozoic organic matter. *Sci. Rep.* **2018**, *8*, 13324. [CrossRef] [PubMed]
23. Taylor, S.R.; McClennan, S.M. *The Continental Crust: Its Composition and Evolution*; Blackwell Scientific Publications: Oxford, UK, 1985; Volume 21, p. 312.
24. He, T.; Lu, S.; Li, W.; Sun, D.; Pan, W.; Zhang, B.; Tan, Z.; Ying, J. Paleoweathering, hydrothermal activity and organic matter enrichment during the formation of earliest Cambrian black strata in the northwest Tarim Basin, China. *J. Petrol. Sci. Eng.* **2020**, *189*, 106987. [CrossRef]
25. Xu, L.; Lehmann, B.; Zhang, X.; Zheng, W.; Meng, Q. Trace element distribution in black shales from the Kunyang phosphorite deposit and its geological significances. *Acta Petrol. Sin.* **2014**, *30*, 1817–1827. (In Chinese with English abstract)
26. Goldberg, E.D.; Arrhenius, G. Chemistry of Pacific pelagic sediments. *Geochim. Cosmochim. Act.* **1958**, *13*, 153–212. [CrossRef]
27. Zhou, X.; Chen, D.; Qing, H.; Qian, Y.; Wang, D. Submarine silica-rich hydrothermal activity during the earliest Cambrian in the Tarim Basin, northwest China. *Int. Geol. Rev.* **2014**, *56*, 1906–1918. [CrossRef]
28. Li, W.; Lu, S.; Tan, Z.; He, T. Lacustrine Source Rock Deposition in Response to Coevolution of the Paleoenvironment and Formation Mechanism of Organic-Rich Shales in the Biyang Depression, Nanxiang Basin. *Energy Fuels* **2017**, *31*, 13519–13527. [CrossRef]
29. Rahman, S.; Henderson, G.M.; Rickaby, R.E.M. Biogenic barium as a productivity proxy in the South Pacific Ocean. *Paleoceanogr. Paleoclimatol.* **2022**, *37*, e2021PA004291.
30. Zhu, G.; Wang, P.; Li, T.; Zhao, K.; Yan, H.; Li, J.; Zhou, L. Nitrogen geochemistry and abnormal mercury enrichment of shales from the lowermost Cambrian Niutitang Formation in South China: Implications for the marine redox conditions and hydrothermal activity. *Global Planet. Change* **2021**, *199*, 103449. [CrossRef]
31. Moghaddam, M.; Hadavi, F. Changes in productivity in the boundary between Abderaz and Abtalkh Formations (East Kopet Dagh, NE Iran) evidenced by calcareous nannofossils and elements (Zn, Cu, Cd, Pb and CaCO3). *Arab. J. Geosci.* **2014**, *7*, 5309–5320. [CrossRef]
32. Algeo, T.J.; Maynard, J.B. Trace-element behavior and redox facies in core shales of Upper Pennsylvanian Kansas-type cyclothems. *Chem. Geol.* **2004**, *206*, 289–318. [CrossRef]
33. Algeo, T.J.; Hinnov, L.; Moser, J.; Maynard, J.B.; Elswick, E.; Kuwahara, K.; Sano, H. Changes in productivity and redox conditions in the Panthalassic Ocean during the latest Permian. *Geology* **2010**, *38*, 187–190. [CrossRef]
34. He, T.; Zeng, Q.; Lu, S.; Li, W.; Li, M.; Wen, Z.; Yang, E.; Jing, T.; Ying, J.; Zhu, P.; et al. Aryl isoprenoids from the Lower Paleozoic in the Tarim Basin, NW China: Insight into deep ancient hydrocarbon exploration. *Geoenergy Sci. Eng.* **2023**, *225*, 211666. [CrossRef]
35. He, T.; Li, W.; Lu, S.; Yang, E.; Jing, T.; Ying, J.; Zhu, P.; Wang, X.; Pan, W.; Chen, Z. Distribution and isotopic signature of 2-alkyl-1,3,4-trimethylbenzenes in the Lower Paleozoic source rocks and oils of Tarim Basin: Implications for the oil-source correlation. *Petrol. Sci.* **2022**, *19*, 2572–2582. [CrossRef]
36. Connock, G.T.; Nguyen, T.X.; Philip, R.P. The development and extent of photic-zone euxinia concomitant with Woodford Shale deposition. *AAPG Bull.* **2018**, *102*, 959–986. [CrossRef]
37. Hays, L.; Beatty, T.; Henderson, C.; Love, G.; Summons, R. Evidence for photic zone euxinia through the end-Permian mass extinction in the Panthalassic Ocean (Peace River Basin, Western Canada). *Palaeoworld* **2007**, *16*, 39–50. [CrossRef]
38. Schwark, L.; Frimmel, A. Chemostratigraphy of the Posidonia Black Shale, SW-Germany: II. Assessment of extent and persistence of photic-zone anoxia using aryl isoprenoid distributions. *Chem. Geol.* **2004**, *206*, 231–248. [CrossRef]
39. Xu, J.; He, T.; Zeng, Q.; Zhao, Y.; Wen, Z. Compound-Specific C/H Isotopic Signature of Ultra-Deep Hydrocarbon Fluids from the Shuntuoguole Area in the Tarim Basin, Northwestern China. *Energies* **2024**, *17*, 1211. [CrossRef]
40. He, T.; Li, W.; Lu, S.; Yang, E.; Jing, T.; Ying, J.; Zhu, P.; Wang, X.; Pan, W.; Zhang, B.; et al. Quantitatively unmixing method for complex mixed oil based on its fractions carbon isotopes: A case from the Tarim Basin, NW China. *Petrol. Sci.* **2023**, *20*, 102–113. [CrossRef]
41. Sun, Y.; Xu, S.; Lu, H.; Cuai, P. Source facies of the Paleozoic petroleum systems in the Tabei uplift, Tarim Basin, NW China: Implications from aryl isoprenoids in crude oils. *Org. Geochem.* **2003**, *34*, 629–634. [CrossRef]
42. Zhu, G.; Milkov, A.V.; Li, J.; Xue, N.; Chen, Y.; Hu, J.; Li, T.; Zhang, Z.; Chen, Z. Deepest oil in Asia: Characteristics of petroleum system in the Tarim basin, China. *J. Petrol. Sci. Eng.* **2021**, *199*, 108246. [CrossRef]

43. Zhao, K.; Zhu, G.; Meng, X.; Yan, H.; Li, T.; Wang, P. Enrichment of mercury in the Lower Cambrian sedimentary successions by submarine hydrothermal venting. *J. Asian Earth Sci.* **2022**, *240*, 105439. [CrossRef]
44. Sweere, T.; van den Boorn, S.; Dickson, A.J.; Reichart, G.-J. Definition of new trace-metal proxies for the controls on organic matter enrichment in marine sediments based on Mn, Co, Mo and Cd concentrations. *Chem. Geol.* **2016**, *441*, 235–245. [CrossRef]
45. Jin, Z.; Yang, L.; Zeng, J.; Zhang, L. Deep fluid activitities and their effects on generation of hydrocarbon in Dongying depression. *Pet. Explor. Dev.* **2002**, *2*, 42–44.

minerals

MDPI

Article

Prediction of Lithofacies in Heterogeneous Shale Reservoirs Based on a Robust Stacking Machine Learning Model

Sizhong Peng [1,2], Congjun Feng [1,2,*], Zhen Qiu [3,4,*], Qin Zhang [3,4], Wen Liu [3,4], Jun Feng [5,6] and Zhi Hu [5]

1 State Key Laboratory of Continental Dynamics, Northwest University, Xi'an 710069, China; 202010337@stumail.nwu.edu.cn
2 Department of Geology, Northwest University, Xi'an 710069, China
3 National Energy Shale Gas R & D (Experiment) Center, Langfang 065007, China; zhangqin2169@petrochina.com.cn (Q.Z.); liuwen0906@126.com (W.L.)
4 PetroChina Research Institute of Petroleum Exploration and Development, Beijing 100083, China
5 School of Information and Technology, Northwest University, Xi'an 710069, China; fengjun@nwu.edu.cn (J.F.); huxiaozhi8848@163.com (Z.H.)
6 Xi'an Key Laboratory of Paleo-Bioinformatics, Northwest University, Xi'an 710069, China
* Correspondence: fengcj@nwu.edu.cn (C.F.); qiuzhen316@163.com (Z.Q.)

Abstract: The lithofacies of a reservoir contain key information such as rock lithology, sedimentary structures, and mineral composition. Accurate prediction of shale reservoir lithofacies is crucial for identifying sweet spots for oil and gas development. However, obtaining shale lithofacies through core sampling during drilling is challenging, and the accuracy of traditional logging curve intersection methods is insufficient. To efficiently and accurately predict shale lithofacies, this study proposes a hybrid model called Stacking, which combines four classifiers: Random Forest, HistGradient Boosting, Extreme Gradient Boosting, and Categorical Boosting. The model employs the Grid Search Method to automatically search for optimal hyperparameters, using the four classifiers as base learners. The predictions from these base learners are then used as new features, and a Logistic Regression model serves as the final meta-classifier for prediction. A total of 3323 data points were collected from six wells to train and test the model, with the final performance evaluated on two blind wells that were not involved in the training process. The results indicate that the stacking model accurately predicts shale lithofacies, achieving an Accuracy, Recall, Precision, and F1 Score of 0.9587, 0.959, 0.9587, and 0.9587, respectively, on the training set. This achievement provides technical support for reservoir evaluation and sweet spot prediction in oil and gas exploration.

Keywords: shale gas; lithofacies prediction; stacking machine learning model; heterogeneous shale reservoir characterization

1. Introduction

The world is currently in the transitional phase from traditional energy to renewable energy, and the unconventional oil and gas industry, including natural gas hydrates, deep coalbed methane, shale gas, and tight oil, still plays an irreplaceable role [1–3]. In China, unconventional oil and gas exploration has gradually shifted towards developing deep shale gas and deep coalbed methane [4–6]. A detailed understanding of deep shale gas and coalbed methane reservoir characteristics and geological conditions is a prerequisite for accurately exploring oil and gas reservoirs [7]. The Ordos Basin is a highly significant oil and gas basin in China, particularly in the eastern part, which contains abundant shale gas resources from the Upper Paleozoic strata. The deep shale gas reservoirs of the

Permian Shanxi Formation, with burial depths greater than 2000 m, feature thin, high-quality shale layers, rapid lithological changes, and highly complex stress states, making it challenging to predict reservoir characteristics with advanced technologies. Lithofacies is a comprehensive assemblage comprising lithology, sedimentary structures, mineral composition, and paleontological features, which collectively provide valuable geological information. The main advantages of lithofacies identification include predicting reservoir properties, identifying geological conditions, optimizing exploit plans, and avoiding risks. Therefore, accurately predicting shale reservoir lithofacies is crucial for determining the sweet spots for oil and gas development [8,9]. Traditional reservoir analysis relies on core sampling during drilling, where geologists subject rock samples to subjective analysis using microscopes, chemical analysis, and other tools, classifying lithofacies based on mineral compositions and proportions obtained from laboratory tests. This method is labor-intensive, requiring extensive core sampling, long laboratory analysis periods, and specialized petrologists' expertise for judgment. Subsequently, geologists utilized well-logging data and traditional intersection plots for lithofacies identification [10]. However, the accuracy of this prediction method is too low (about 70%) [11] to achieve the precision required for identifying the sweet spot zones in oil and gas exploration.

With the rapid advancement of computer technology in recent years, artificial intelligence (AI) has emerged as a key research focus in oil and gas exploration due to its high adaptability, accuracy, and cost-effectiveness with the rapid advancement of computer technology in recent years [12–14]. Scholars worldwide have applied machine learning (ML) algorithms to predict reservoir porosity, permeability, total organic carbon content, elastic modulus, and hydrocarbon volumes, achieving promising results [7,14–22].

Additionally, scholars have explored various machine learning methods for lithology or lithofacies prediction. Wang et al. [23] were pioneers in lithofacies prediction, combining the mineral composition of shale with total organic carbon content. Using pulsed neutron spectroscopy (PNS) log data, they employed an artificial neural network (ANN) model to predict shale lithofacies in the Appalachian Basin, achieving an accuracy of 85.16%. Bhattacharya et al. [24] applied SVM, ANN, SOM, and MRGC models to predict lithofacies in the Bakken and Mahantango-Marcellus Shales, constructing a 3D data-driven lithofacies model. Santos et al. [25] developed a deep recurrent neural network approach for classifying sedimentary facies using well-logging, achieving minimal misclassification. Dong et al. [26] introduced the MKFD method, which utilizes multi-scale kernel functions to achieve optimal nonlinear mapping, enabling more effective extraction of nonlinear lithology features to distinguish lithofacies in carbonate reservoirs. Lu et al. [27] studied the Fengcheng Formation in the Junggar Basin and trained a lithology identification model based on structural graphs using the GraphSAGE method, achieving an accuracy of 90.41%. In a recent study, Zheng et al. [28] proposed an attention-based dual network that combines global (overall microphotographs) and local (differentiating framework particles) features for automatic rock classification. Ali et al. [29] employed machine learning models such as SOM, MRGC, ANN, and KNN to identify lithofacies in subsurface strata. The results demonstrated that MRGC and KNN exhibit superior performance in lithofacies identification, highlighting the critical role of selecting the appropriate machine learning method for accurate lithofacies recognition and modeling in complex geological settings. Table 1 shows some of the latest related studies on lithology and lithofacies prediction in different reservoirs [23–37].

Table 1. Recent studies on lithofacies or lithology prediction using artificial intelligence technology.

References	Reservoirs	Objectives	Employed Models	Optimal Model
Wang et al. (2012) [23]	Marcellus shale	Shale lithofacies with major mineralogy and TOC content	ANN	ANN
Bhattacharya et al. (2016) [30]	Bakken and Mahantango–Marcellus shale	Shale lithofacies with major mineralogy and TOC content	SVM, ANN, SOM, MRGC	SVM
Bhattacharya et al. (2019) [24]	Bakken shale	Shale lithofacies with major mineralogy and TOC content	SVM	SVM
Bressan et al. (2020) [31]	Sandstone and mudstone	The lithology of sandstone and mudstone is grouped for identification	DTr, RF, SVM, MLP	RF
Santos et al. (2022) [25]	Rio Bonito Formation in ParanáBasin	Identification of silt, clay, coal, volcanic sill, and sandstone with different grain sizes	XGBoost, RF, NB, SVM, BiLSTM	BiLSTM
Dong et al. (2022) [26]	Asmari Formation in Zagros Basin	Identification of dolomite, limestone, mudstone, sandstone, and anhydrite	MKFD	MKFD
Zheng et al. (2022) [32]	Xujiahe and Ziliujing Formations	Lithofacies with different depositional environments	SVM, MLP, XGBoost	XGBoost
Lu et al. (2023) [27]	Fengcheng Formation in Junggar Basin	Lithology identification in continental shale reservoirs	SVM, RF, XGBoost, GraphSAGE	GraphSAGE
Ullah et al. (2023) [33]	Lower Goru member in the southern area of Pakistan	Shale and sandstone with different grain sizes	RF, ANN, SVM, XGB, MLP	MLP
Zheng et al. (2024) [28]	Sedimentary rocks	Identification of sedimentary rock types by thin-section photomicrographs	DCNNs	DCNNs
Ali et al. (2024) [29]	Lower Goru Formation	Lithofacies with different porosity and permeability	SOM, MRGC, ANN, KNN	MRGC
Ashraf et al. (2024) [34]	Lower Goru Formation	Lithofacies depositional environment	SVM, KNN, RF, DTr, NB, AB	RF
This study	Shanxi Formation in Ordos Basin	Shale lithofacies classification based on mineral composition and lamination structure	RF, HGB, XGB, Cat, FCNN, Stacking model	Stacking model

This study assembles four base classifiers into an advanced stacking model to perform detailed lithofacies prediction for shale reservoirs in gas fields deposited in the marine–continental transition zone. Core data and corresponding well-logging records were obtained from six wells (Well-W1, -W2, -W3, -W4, -W5, and -W6) and two blind wells (Well-W7 and -W8) in the Permian Shanxi Formation on the eastern margin of the Ordos Basin. In the initial phase, well-log data were processed and matched with lithofacies data to construct the dataset, which was then divided into training and testing sets. On the dataset, we compared the performance of the stacking model with five commonly used models (RF, HGB, XGB, Cat, and FCNN), evaluating prediction accuracy using metrics such as Accuracy, Recall, Precision, and F1 Score, as well as visual tools like confusion matrices and ROC curves. Ablation experiments were conducted to validate the contribution and effectiveness of each base model to the overall ensemble model. Our study demonstrates that the stacking model can reliably identify different shale lithofacies in heterogeneous layers, providing significant assistance for reservoir evaluation and sweet spot prediction in oil and gas exploration. The workflow chart used to accomplish this study is presented in Figure 1.

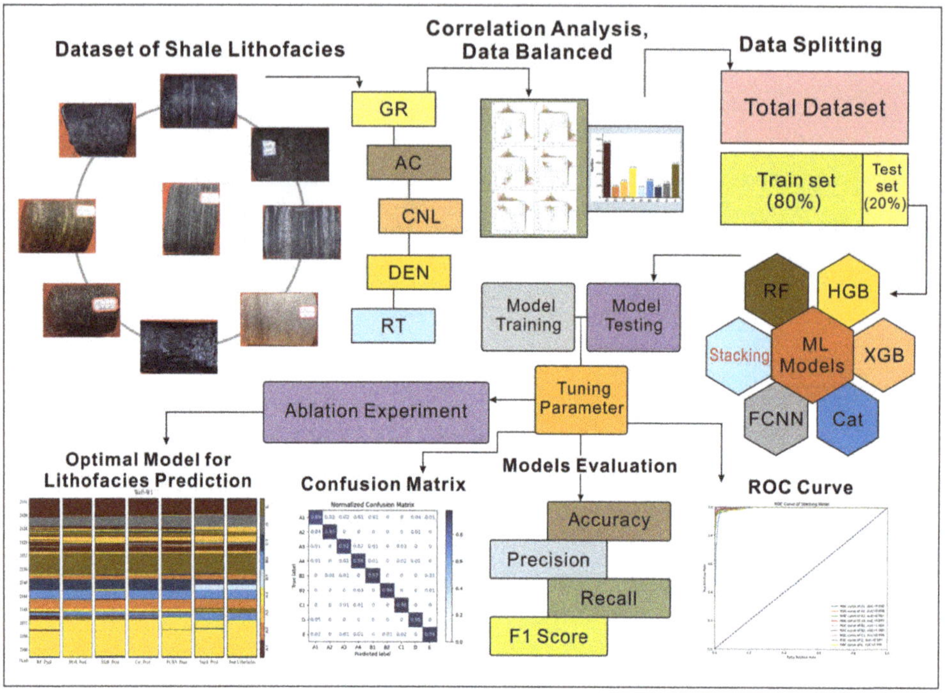

Figure 1. Workflow chart to accomplish this study.

The highlights of this study are as follows:

- Most researchers focus on lithology identification using artificial intelligence, while studies on shale lithofacies identification considering different mineral compositions and laminar structures remain rare;
- By integrating four machine learning models into a stacking model, the accuracy of lithofacies prediction has been significantly improved. Validation on blind wells, which were excluded from the training process, yielded excellent prediction results;
- The effectiveness of the stacking model combination was further validated through ablation experiments, providing a feasible approach for optimizing the model structure.

2. Geological Background and Lithology Classification

Since the Paleozoic era, the Ordos Basin has undergone a palaeogeographical evolution from a shallow marine environment to an inland rift-lake basin [38]. In the Late Carboniferous, the formation of the Central Uplift led to an overall east–west sedimentary structural differentiation in the Ordos Basin. In the eastern part of the basin, barrier-lagoon and shallow marine shelf deposits were predominantly formed [39]. By the Early Permian, the impact of the Hercynian orogeny caused a rapid uplift of the northern margin of the North China platform, resulting in a north-high, south-low topography. Seawater withdrew from the southeast of the basin, and the eastern part of the basin was dominated by a coastal transition zone, where large quantities of organic-rich shale were deposited [40]. As a result, the shale gas resources in the coastal transition zone of the eastern Ordos Basin are abundant, with considerable reserves, making it a key area for increasing natural gas reserves and production in the Ordos Basin.

The study area is located in the eastern Ordos Basin, within the tectonic setting of the Jinxi Fold Belt [41], exhibiting a monocline structure that tilts from east to west (Figure 2). The Shanxi Formation shale reservoirs from the Late Permian are widely distributed, covering over 80% of the effective area in the eastern margin of the Ordos Basin. With an average thickness of approximately 88.6 m and an average total organic carbon content of 4.91%, they have reached the overmature stage of dry gas generation, making them an excellent gas-producing interval [42].

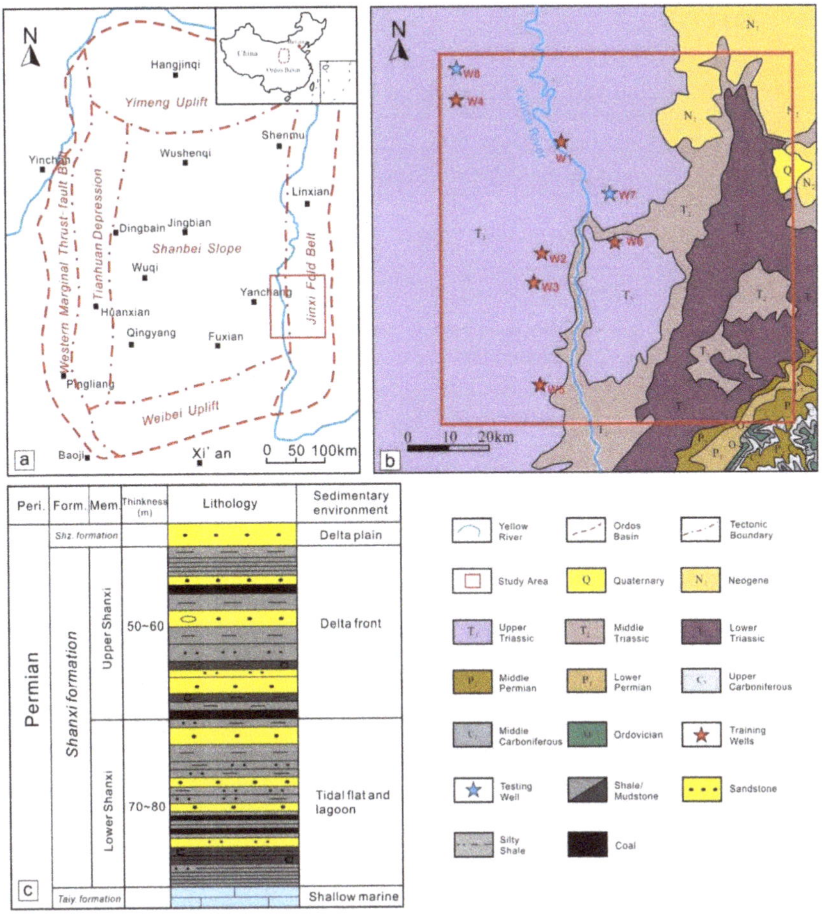

Figure 2. Regional tectonic map (**a**), well location (**b**), and lithological column (**c**) of the study area.

Controlled by the complex and variable sedimentary environment of the coastal transition zone, the lithology of the Shanxi Formation primarily consists of shale, siltstone, sandstone, and coal [43]. To systematically differentiate the characteristics of shale in various depositional environments, this study further classifies the Shanxi Formation into nine major lithofacies based on core and well-logging interpretation. These classifications are based on lithology, mineral composition, sedimentary structures, well-logging signatures, and depositional environment (Figure 3). Mineral composition, rock photographs, and well-logging data were obtained from eight wells in the study area. The detailed definitions of the lithofacies and their corresponding well-logging properties are as follows:

(1) Massive bedding argillaceous shale (A1) with a homogeneous structure. It is primarily composed of clay minerals, with trace amounts of pyrite. It exhibits moderate gamma, acoustic, and neutron responses, with low density and high resistivity. These lithofacies were deposited in lagoonal and proximal deltaic environments;

(2) Laminated bedding argillaceous shale (A2) is characterized by weak, slender, and parallel lamination. It is primarily composed of clay minerals, exhibiting the highest gamma response, a relatively high acoustic response, and a moderate neutron response, as well as moderate density and low resistivity. These lithofacies were primarily deposited in lagoonal environments;

(3) Lenticular bedding argillaceous shale (A3) is characterized by wider parallel lamination, moderate-to-weak lenticular structure, and weak bioturbation. It is primarily composed of clay minerals, with moderate values across all well-logging curves. These lithofacies were primarily deposited in tidal flat environments;

(4) Flaser bedding argillaceous shale (A4) is characterized by wavy or flaser lamination and strong bioturbation. It is primarily composed of clay minerals, with moderate gamma, acoustic, and neutron responses, relatively high density, and the lowest resistivity. These lithofacies were primarily deposited in tidal flat and delta front environments;

(5) Massive bedding siliceous shale (B1), consisting of massive argillaceous siltstone or silty mudstone with strong bioturbation. It is primarily composed of felsic minerals, with low gamma, acoustic, and neutron responses, relatively high density, and low resistivity. These lithofacies were primarily deposited in tidal flat and barrier island environments;

(6) Laminated bedding siliceous shale (B2) is characterized by lamination and cross-lamination. It is primarily composed of felsic minerals, with low gamma, moderate acoustic, and low neutron responses, as well as a relatively high density and low resistivity. These lithofacies were primarily deposited in tidal flat environments;

(7) Massive bedding mixed shale (C1) is characterized by massive bedding. It contains nearly equal proportions of clay minerals and felsic minerals, with moderate gamma, moderate acoustic, and low neutron responses, as well as a relatively high density and low resistivity. These lithofacies were primarily deposited in interchannel and tidal flat environments;

(8) Coal (D) is characterized by massive bedding, possibly containing small amounts of pyrite. It is almost entirely composed of clay minerals, with very low gamma, the highest acoustic, and the highest neutron responses, as well as the lowest density and the highest resistivity. These lithofacies were primarily deposited in shrinking lagoon and swamp environments;

(9) Sandstone (E) is characterized by massive bedding, planar cross-bedding, or trough cross-bedding. It is predominantly composed of felsic minerals, with moderate gamma, moderate acoustic, and low neutron responses, as well as a relatively high density and low resistivity. These lithofacies were primarily deposited in submerged distributary channels and tidal channel environments.

Lithofacies	Pie Plot	Gamma (API)	Acoustic (us/m)	Neutron (%)	Density (g/cm³)	Resistivity (Ω·m)	Photograph
Massive bedding argillaceous shale (A1)	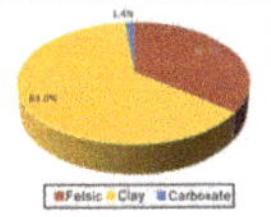	28-208 127	173-422 268	13-72 38	1.3-2.7 2.2	8-413 99	
Laminated bedding argillaceous shale (A2)	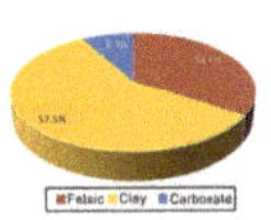	116-267 166	209-393 300	21-48 34	1.9-2.9 2.4	14-189 62	
Lenticular bedding argillaceous shale (A3)	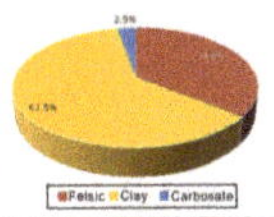	89-203 129	176-375 250	14-68 35	1.6-2.8 2.3	14-229 74	
Flaser bedding argillaceous shale (A4)	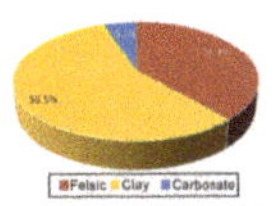	70-156 125	193-389 238	18-54 29	1.7-2.8 2.5	13-143 42	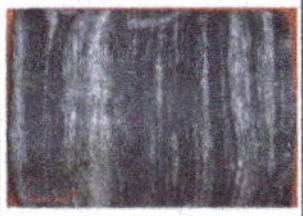
Massive bedding siliceous shale (B1)	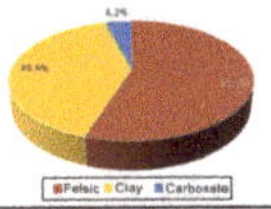	74-150 107	202-368 230	14-61 23	1.7-2.8 2.6	14-167 78	
Laminated bedding siliceous shale (B2)	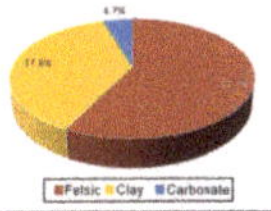	70-154 114	174-306 243	13-47 25	2.2-2.7 2.5	29-153 80	
Massive bedding mixed shale (C1)	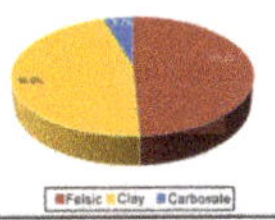	85-146 123	203-287 239	16-31 22	2.2-2.9 2.7	17-208 64	
Coal (D)	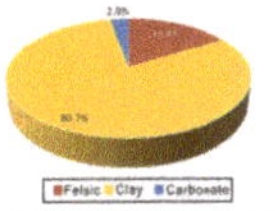	21-183 106	225-433 337	28-58 44	1.3-2.6 1.8	29-5213 1203	
Sandstone (E)	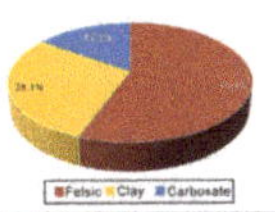	61-150 101	195-300 216	9.3-49 16	1.6-2.8 2.6	22-386 92	

Figure 3. Summary of the mineralogical features, conventional well-log responses, and photograph of the 9 lithofacies defined from core and advanced well logs in Shanxi formation of the Ordos basin.

3. Dataset Preparation

3.1. Data Preprocessing

In the study area located on the eastern margin of the Ordos Basin, lithofacies and well-log data from 8 wells were selected as the focus of this research. The lithofacies for these wells were used as training labels, determined through core observations and X-ray diffraction (XRD) mineral composition analysis. For convenience, the lithofacies data were encoded into numerical labels, with the corresponding lithofacies encoding provided in Figure 3. The well-logging curves corresponding to each lithofacies were then used as features for model training. Six of the eight wells were selected for model training, with 3323 data points collected at a sampling interval of 1 decimeter. These points were split into training and testing sets at an 80% and 20% ratio, respectively, while the remaining two wells served as blind wells to validate the model's performance, ensuring that the model could perform well on unseen data. To ensure the data's accuracy and reliability, the well-log data underwent depth correction, data cleaning, standardization, and normalization. Depth correction was performed based on the interfaces where lithology changes correspond to anomalies in the log curves [30,44]. Standardization was applied to remove inevitable system errors during data collection. The method was achieved by selecting thicker typical shale layers as lateral comparison markers, ensuring that the well-log data from corresponding strata in each well had the same average value or similar frequency distribution. Normalization scaled different data types to a [0–1] range, thus eliminating dimensional differences between the curves. The normalization formula is as follows:

$$Normx_i^n = \frac{x_i^n - xmin^n}{xmax^n - xmin^n} \tag{1}$$

where $Normx_i^n$ represents the normalised value of the i-th data record associated with the n-th variable distribution, x_i^n denotes the actual recorded well-log or attribute value, and $xmax^n$ and $xmin^n$ represent the maximum and minimum recorded values corresponding to the n-th variable, respectively. It is important to note that the resistivity data underwent logarithmic transformation before normalization to enhance the data distribution and the predictive model's performance.

3.2. Feature Selection

Suitable well-log curves were selected as features to improve the efficiency of model training [14]. The gamma-ray log (GR) measures the natural radioactivity of the strata, helping to identify lithology and distinguish different rock types. It is beneficial for identifying shale, as shale typically exhibits higher gamma-ray responses than other rock types. The acoustic log (AC) measures the speed of sound propagation in the strata, providing information about the mechanical properties of the formation and aiding in the identification of lithology, porosity, and fluid content. The compensated neutron log (CNL) measures the hydrogen index of the formation by detecting hydrogen atoms, which can be used to determine porosity and fluid type. The density log (DEN) measures the bulk density of the formation, aiding in the determination of porosity, lithology, and mineral composition. The resistivity log (RT) measures the electrical resistivity of the formation, which is commonly used to evaluate fluid content and lithology. It is susceptible to hydrocarbon content, and since the shale lithofacies in the study area vary in hydrocarbon content, using the resistivity log to distinguish lithofacies is a good choice. Therefore, we selected five conventional well-log parameters closely related to lithology to construct the dataset, including GR, AC, CNL, DEN, and RT.

3.3. Using SMOTE to Deal with Data Imbalance

The lithofacies of the Shanxi Formation are highly unevenly distributed in the study area (Figure 4). For instance, the massive bedding argillaceous shale (A1) constitutes the largest proportion, accounting for up to 28.2% of all lithofacies. The laminated bedding argillaceous shale (A2), massive bedding siliceous shale (B1), and massive bedding mixed shale (C1) account for 5.4%, 4.8%, and 5.4%, respectively. To eliminate prediction bias caused by data imbalance, the data were balanced using SMOTE (Synthetic Minority Over-sampling Technique) [45,46]. SMOTE randomly selects a minority class sample and then identifies the k nearest neighbors (typically k = 5). For each synthetic sample to be generated, a point is randomly chosen along the line segment connecting the minority sample and its neighbor, and a new sample is created. This process is repeated until the dataset is balanced. The changes in lithofacies distribution before and after applying SMOTE are shown in Table 2.

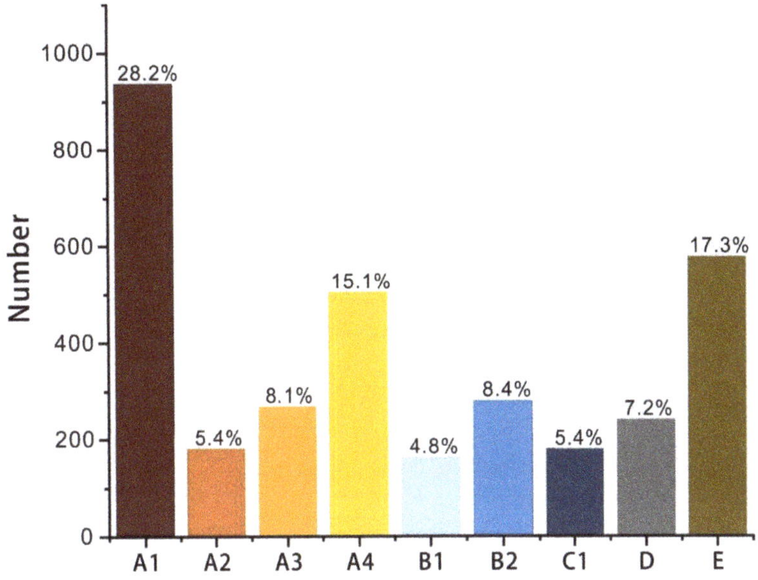

Figure 4. The distributions of lithofacies throughout the datasets.

Table 2. Changes in lithofacies distribution before and after applying SMOTE.

Lithofacies	A1	A2	A3	A4	B1	B2	C1	D	E	Total
Before SMOTE	937	180	268	503	161	279	179	240	576	3323
After SMOTE	969	932	920	931	906	934	923	935	921	8371

4. Methodology

4.1. Machine Learning (ML) Models

4.1.1. Random Forest (RF)

Random Forest uses bootstrap sampling to obtain a data subset for training each decision tree. Specifically, for an original training dataset containing n samples, a subset is created by randomly sampling with replacement, forming the training set for a decision tree [19,47,48]. When constructing the node splits of each decision tree, not all features are used to find the best-split feature and split point. Instead, a subset of features is randomly selected from the total m features (for example, k features are randomly chosen, where k is typically much smaller than m, with common choices being $k = \sqrt{m}$ or $k = log_2 m$, and the best split is determined only within these k features. Then, each decision tree is trained using the conventional decision tree construction method to determine the best feature and split point for each node, allowing the tree to grow until a predefined stopping criterion is reached. In classification problems, each decision tree outputs a predicted class. The final predicted class is the one that appears most frequently among the predictions of all decision trees, thus employing a voting mechanism to determine the final classification result.

4.1.2. HistGradient Boosting Classifier (HGB)

The HistGradient Boosting Classifier is a histogram-based gradient boosting classifier model particularly suited for handling large-scale datasets and sparse data [49,50]. Its core idea is to bin continuous features, dividing feature values into several intervals in a histogram reducing computational complexity during training. Compared to traditional gradient boosting trees (such as GBDT), binning reduces the cost of feature and split point search.

4.1.3. Extreme Gradient Boosting (XGB)

Extreme Gradient Boosting is an efficient and widely used gradient boosting framework [17,51,52]. Gradient boosting is a machine learning method that builds an ensemble model based on decision trees. Its core idea is to train multiple decision trees sequentially, where the training objective is the loss function and regularization term (i.e., the Regularized Learning Objective). The objective function of XGBoost consists of two parts: the loss function (which measures the difference between predicted results and true labels) and the regularization term (which controls the complexity of the model to prevent overfitting). The form of its loss function $\theta b_j(\theta)$ is typically represented as follows:

$$Ob_j(\theta) = \sum_{i=1}^{n} l(y_i, \hat{y}_i) + \sum_k \Omega(f_k) \tag{2}$$

where n is the number of training samples, l is the loss for a single sample, assumed to be a convex function, y_i is the model's predicted value for the training sample, and $\hat{y}$ is the true label of the training sample. The regularization term $\Omega(f_k)$ defines the complexity of the model:

$$\Omega(f) = \gamma T + \frac{1}{2}\lambda\|\omega\|^2 \tag{3}$$

where γ and λ are hyperparameters set manually, ω is the vector of values from all leaf nodes of the decision tree, and T is the number of leaf nodes. When constructing each decision tree, XGBoost uses a second-order Taylor expansion to approximate the objective function, allowing it to find the optimal split points more accurately. During the training phase, a sequence of decision trees is trained one by one according to the predetermined objective function and tree-building method. After each decision tree is trained, it is added to the existing model, and the model's predictions on the training data are updated. The residuals are recalculated, and subsequent decision trees are trained based on the new residuals. This iterative process continues until the convergence condition is met.

4.1.4. Categorical Boosting (Cat)

Categorical Boosting is also a gradient boosting decision tree-based machine learning algorithm, specifically designed to handle categorical feature data [52,53]. It belongs to the class of decision tree-based gradient boosting algorithms (such as GBDT). Its advantage lies in handling categorical features, as it eliminates the need for manual encoding (such as One-Hot Encoding or Label Encoding) and instead uses an efficient method called "target encoding". In CatBoost, the basic predictor is the oblivious tree, where "oblivious" refers to using the same splitting criterion at the same level of the tree, i.e., splitting using the same feature. Such trees are balanced, less prone to overfitting, and accelerate testing speed.

In traditional gradient boosting, the construction of each tree directly depends on the predictions of the previous tree, and this dependency may lead to error accumulation. CatBoost introduces a strategy called Ordered Boosting, which involves selecting a random permutation σ of the samples and then training n supporting models $M_1, M_2, M_3, \ldots, M_n$, where model M_i is taught using the first i samples of permutation σ. In each step, to obtain the residual of the j sample, model $M_{j\text{-}1}$ is used. This algorithm is called Ordered Boosting, which removes bias from the gradient computation, thereby improving the model's stability and robustness.

4.1.5. Framework of Stacking Classifier (Stacking)

The stacking model leverages a hierarchical modeling architecture, where the predictions of multiple base learners serve as input features for the meta-model [54–56]. This forms a two-level decision system, which can be mathematically expressed as follows:

$$\hat{y}_{i_{stack}} = M_{meta}([(M_1(x), (M_2(x), \ldots, (M_n(x)])$$

where m_i represents a base model, and M_{meta} denotes the meta-model. The stacking process begins with K-fold cross-validation, which generates probability predictions from the base models as meta-features, effectively mitigating data leakage. For each base model, a probability prediction vector $P_i \in R^c$ (where C is the number of classes) is obtained from each validation fold. These probability predictions are then concatenated to form the meta-feature matrix:

$$X_{meta} = [P_1, P_2, \ldots, P_n] \in R^{N\times(n\cdot C)}$$

where N is the number of training samples, and n is the number of base models. Next, a meta-model (e.g., a linear model or a shallow neural network) is trained using the complete meta-feature set to produce the final predictions. By leveraging predictions from multiple base models, the stacking framework constructs a decision hyperplane that adapts to high-dimensional classification tasks. The use of cross-validation for generating meta-features prevents the propagation of overfitting from base models trained on the training set. Furthermore, the diversity among base models enhances the adaptability of the final stacking model to new data. After extensive experimentation, we selected four classifiers (RF, HGB, XGB, and CAT) as base models and Logistic Regression as the meta-model, forming the final stacking model. The architecture of the stacking model is presented in Figure 5.

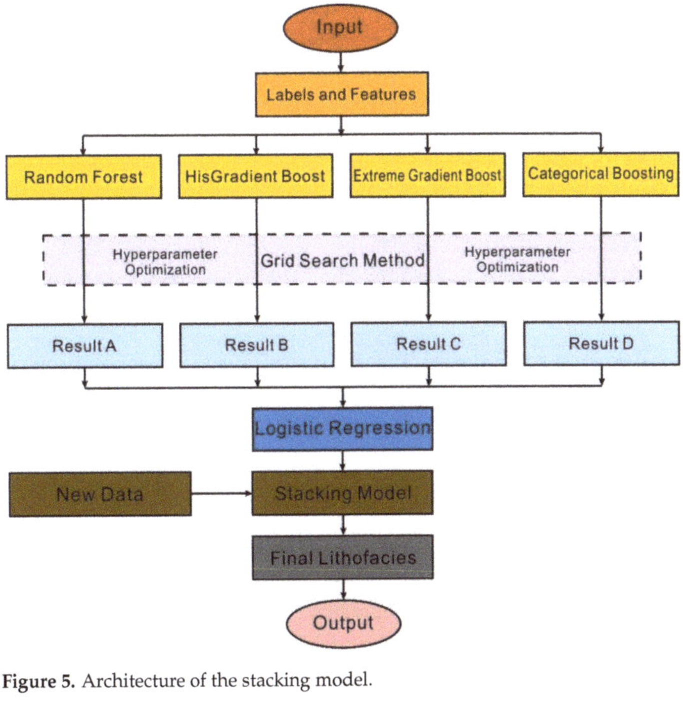

Figure 5. Architecture of the stacking model.

4.2. Evaluation of Machine Learning Models

Using appropriate metrics to evaluate the model is crucial for improving its performance. This study uses Accuracy, Precision, Recall, and F1 Score as evaluation metrics, along with visualizations such as confusion matrices and ROC curves to comprehensively assess the model's efficiency:

$$\text{Accuracy} = \frac{TP + TN}{TP + TN + FP + FN} \tag{4}$$

$$\text{Precision} = \frac{TP}{TP + FP} \tag{5}$$

$$\text{Recall} = \frac{TP}{TP + FN} \tag{6}$$

$$\text{F1Score} = \frac{2TP}{2TP + FP + FN} \tag{7}$$

where *TP*, *TN*, *FP*, and *FN* represent True Positive, True Negative, False Positive, and False Negative, respectively. Accuracy refers to the ratio of correctly detected results to the total number of samples analyzed. Precision is based on our predicted results, representing the proportion of predicted positive samples that are truly positive. The recall is based on the original samples, indicating the proportion of positive instances in the sample that are correctly predicted. The F1 Score is the harmonic mean of Precision and Recall. Accuracy, Precision, Recall, and F1 Score ranges are all between 0 and 1.

5. Result and Discussion

5.1. Correlation Analysis

To analyze the relationship between the response characteristics of different lithofacies well-logging curves, we conducted a cross-plot analysis on 8371 samples from the study area (Figure 6). Additionally, the probability density curves for different lithofacies were displayed above and to the right of the cross plot. The cross-plot and probability density curves help reflect the sensitivity of well-logging curves to different lithologies. The smaller the overlap in the distribution range of the cross plot for different lithofacies, the better the effect of using these two curves to distinguish the lithofacies. Similarly, the greater the separation between the peaks of the probability density curve, the better the ability to distinguish lithofacies using this curve. The figure shows that the GR curve measures the formation's natural radioactivity. Generally, mudstone or shale is more enriched with rare earth elements, so the radioactivity is higher. The GR curve has three peaks on the probability density plot, which can distinguish A2, B2, and E lithofacies. The AC curve reflects the propagation velocity of sound waves in the formation. It shows three peaks on the probability density plot, effectively distinguishing A4, C1, and E lithofacies. The CNL curve uses the interaction between neutrons and atoms in the formation, particularly the slowing effect of hydrogen atoms on neutrons, to assess changes in formation porosity. The CNL curve performs well in classification, clearly dividing into four peaks on the probability density plot, effectively distinguishing A, C1, D, and E lithofacies. The DEN curve reflects the bulk density of the formation based on gamma-ray scattering and absorption. It is divided into two peaks on the probability density curve, distinguishing B2 from other lithofacies. The RT curve measures the resistivity of the formation, which is significantly influenced by the enrichment of organic matter. It is divided into two peaks on the probability density curve, distinguishing B2 from other lithofacies. Therefore, conventional cross-plots can only distinguish sandstone from shale but cannot further distinguish the shale structure and mineral composition.

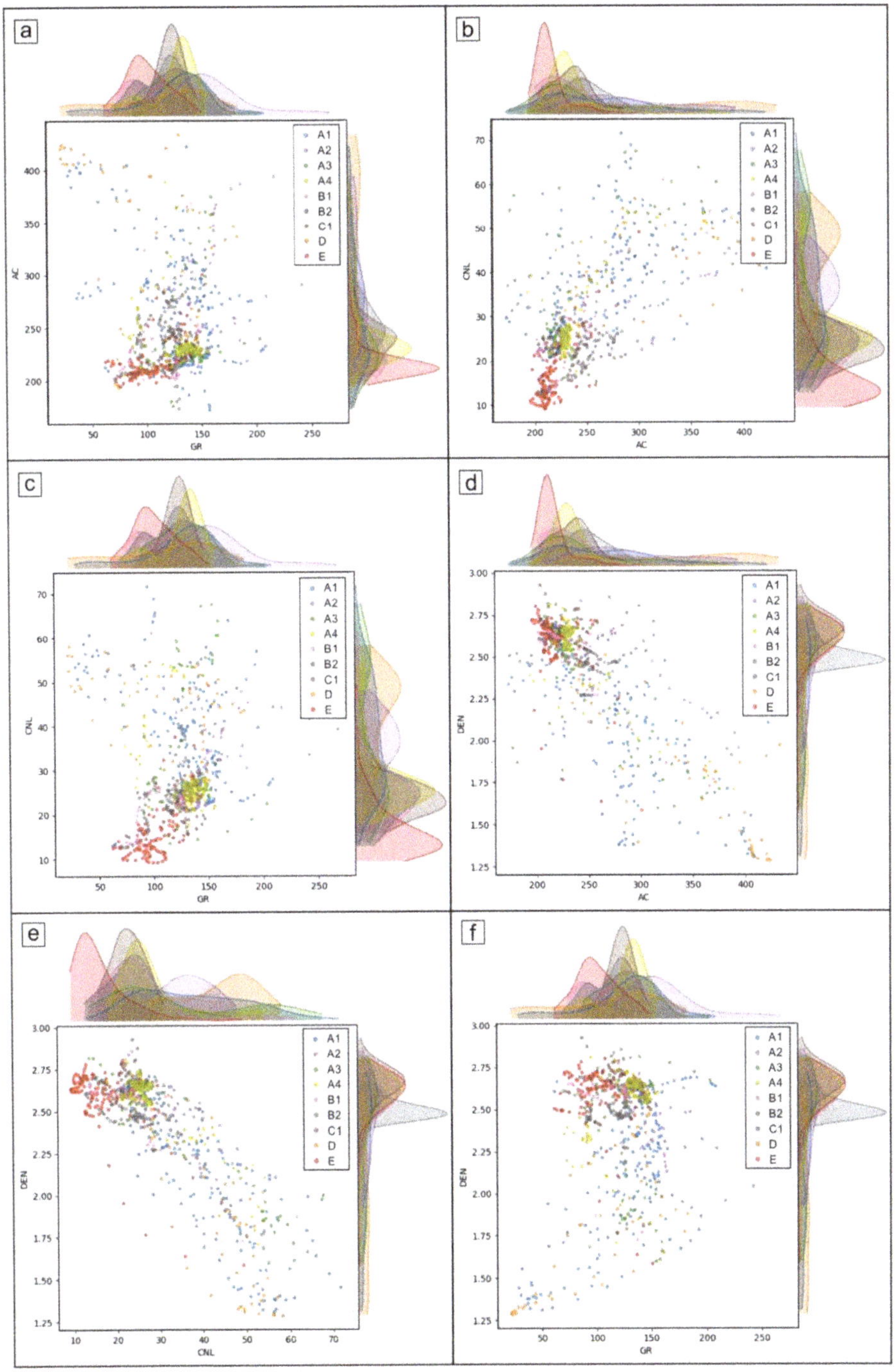

Figure 6. *Cont.*

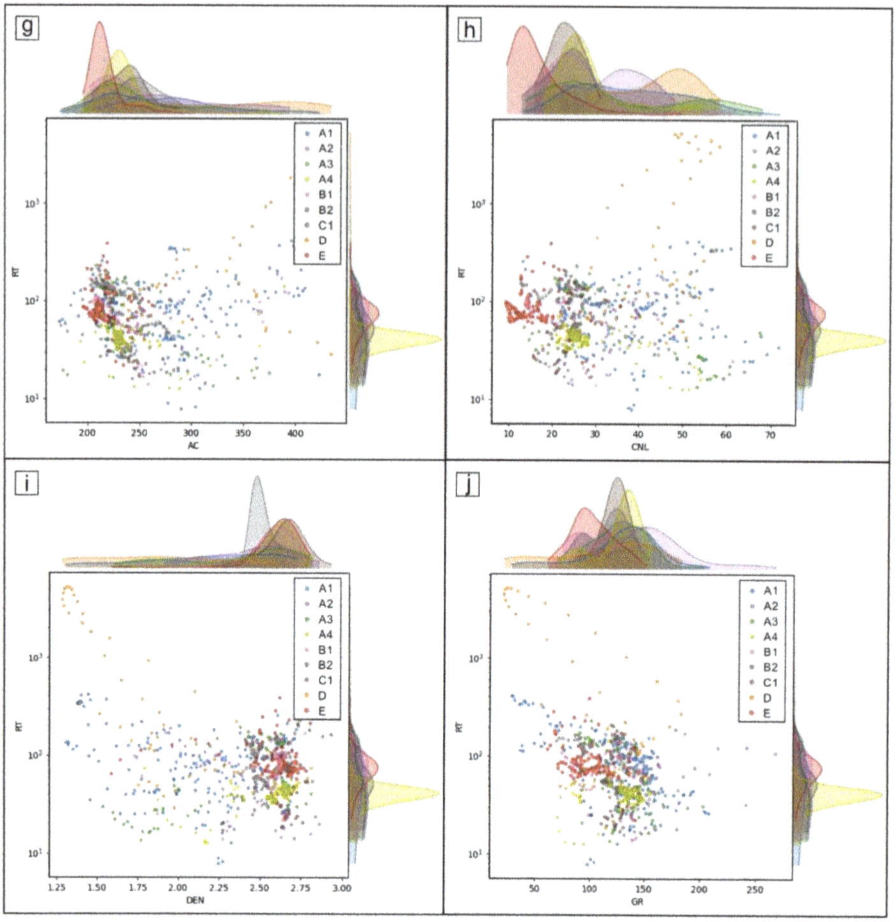

Figure 6. The cross-plot figures showing the correlations among five well logs, whereby nine distinct lithofacies are visually distinguished by using various colors ((**a**), AC-GR; (**b**), CNL-AC; (**c**), CNL-GR; (**d**), DEN-AC; (**e**), DEN-CNL; (**f**), DEN-GR; (**g**), RT-AC; (**h**), RT-CNL; (**i**) RT-DEN; (**j**), RT-GR).

5.2. Optimizing the Hyperparameters of ML Models

Tuning hyperparameters is crucial for improving the performance of machine learning models, as adequately selecting hyperparameters can make the model more accurate and stable and reduce the risk of overfitting. In this study, hyperparameter optimization was determined using the Grid Search Method [57]. The method involves first defining a set of candidate values for the hyperparameters (i.e., the grid) based on the current research and then training and validating the model for each combination. Then, for each set of hyperparameter combinations, cross-validation is performed to evaluate the model's performance, thereby selecting the best parameters. Table 3 shows the types of hyperparameters, candidate values, and the final optimal parameters obtained through the Grid Search Method.

Table 3. Grid parameters and optimal parameters of base models.

Model	Tuning Hyperparameters	Candidates	Optimal Parameters
RF	*n*_estimators	10, 50, 100	100
	max_depth	3, 5	5
	min_samples_split	5, 10	10
	min_samples_leaf	1, 2, 4	2
	bootstrap	True, False	FALSE
Cat	learning_rate	0.01, 0.02, 0.05	0.05
	depth	3, 5, 7	7
	iterations	50, 100	100
HGB	learning_rate	0.01, 0.02, 0.05	0.02
	max_iter	100, 200	100
	max_depth	3, 5, 7	3
	max_bins	200, 255	200
	l2_regularization	0.0, 1.0, 10.0	1
XGB	learning_rate	0.01, 0.02	0.02
	max_depth	2, 3	3
	*n*_estimators	50, 100	100
	subsample	0.85, 0.9	0.85
	colsample_bytree	0.85, 0.9	0.85
	gamma	0, 0.1	0.1
	reg_alpha	0.1, 0.3	0.1
	reg_lambda	1.0, 1.2	1

5.3. Comparison Between Stacking Model and Other Models

To horizontally validate the superiority of the developed stacking model, the stacking model was compared with five individual models to predict lithofacies results. Among them, four classifier models comprise the basic model of the stacking model, and a common neural network model is added. The results (Table 4) show that the Accuracy, Recall, Precision, and F1 Score of all six classification models are more than 0.5. On the training set, the XGB and RF models performed the worst. The RF model's accuracy (0.8272) is the only score exceeding 0.8, with all other scores below 0.8. Similarly, the RF model's accuracy (0.8186) is the only score above 0.8, with other scores below 0.8. The following models are HGB, Cat, and FCNN, all with scores greater than 0.8. The model scores (Accuracy, Recall, Precision, and F1 Score) on the training set, ranked from lowest to highest, are as follows: HGB (0.8423, 0.8605, 0.8423, and 0.8394), Cat (0.8498, 0.8474, 0.8654, and 0.8498), and FCNN (0.8756, 0.8825, 0.8756, and 0.8733). The stacking model performs the best, with all four evaluation metrics exceeding 0.95, and its prediction accuracy far surpasses that of other models. Furthermore, the results for all models on the training set are slightly higher than those on the test set. However, the two are mainly consistent, indicating minimal overfitting during the model training process and strong overall generalization ability.

Table 4. Different models' performances based on Precision, Recall, and F1 Score.

Model	Training Data Set				Testing Data Set			
	Accuracy	Precision	Recall	F1	Accuracy	Precision	Recall	F1
Stacking	0.9587	0.9590	0.9587	0.9587	0.9456	0.9464	0.9456	0.9458
RF	0.7938	0.8272	0.7938	0.7866	0.7564	0.7449	0.7922	0.7564
XGB	0.7919	0.8186	0.7919	0.7837	0.7594	0.7892	0.7594	0.7494
HGB	0.8423	0.8605	0.8423	0.8394	0.8045	0.8247	0.8045	0.8007
Cat	0.8498	0.8474	0.8654	0.8498	0.8331	0.8281	0.8492	0.8331
FCNN	0.8756	0.8825	0.8756	0.8733	0.8638	0.8736	0.8638	0.8602

Figure 7 used a stacked bar chart to visualize the recognition performance of different models for various lithofacies categories. The results show that all models performed well for the sandstone (E), with the Recall, Precision, and F1 Score close to 0.85. The mineral composition of sandstone is significantly different from that of other lithofacies, and sandstone has high porosity. The probability density plot also clearly separates E and other lithofacies (Figure 6). For the recognition of coal (D), the Precision of all models is close to 0.9, but the Recall of XGB, HGB, RF, Cat, and FCNN is poor, close to 0.6, indicating that a few D lithofacies were misclassified. For the recognition of shale lithofacies, unexpectedly, all models performed well in identifying massive bedding argillaceous shale (A1) and flaser bedding argillaceous shale (A4), with the Recall, Precision, and F1 score all close to 0.8. The result may be because, during the transition of clayey shale structures from horizontal lamination to flattened structures, the content of felsic minerals significantly increases, along with its porosity, resulting in a better logging response. XGB, HGB, RF, Cat, and FCNN have considerable misclassifications in predicting lenticular bedding argillaceous shale (A3). Similarly, XGB, RF, and Cat have lower recognition performance for massive bedding siliceous shale (B1) and massive bedding mixed shale (C1). It shows that the mineral compositions of these lithofacies are quite similar, so their logging responses do not differ significantly. In the probability density plot, the peaks of these lithofacies overlap (Figure 6). On the other hand, the stacking model performs excellently in recognizing all lithofacies, with the Recall, Precision, and F1 Score exceeding 0.9 for each lithofacies, with some lithofacies reaching up to 0.99.

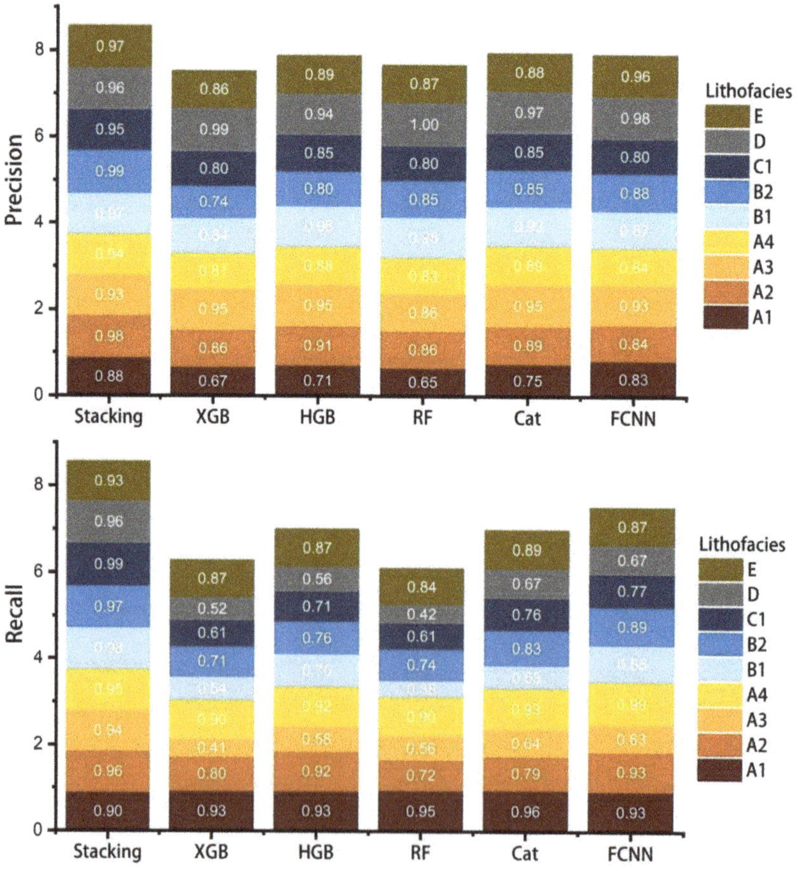

Figure 7. *Cont.*

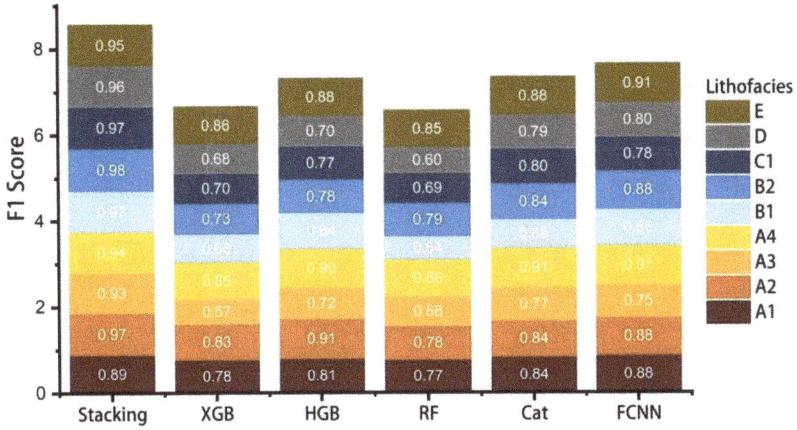

Figure 7. Evaluation performances based on Precision, Recall, and F1 Scores for each lithofacies by all models.

The confusion matrix is commonly used to visualize the model's overall accuracy, and the normalized confusion matrix results for each lithofacies are shown in Figure 8. All models exhibit high accuracy in predicting A1, A4, and E lithofacies, with recognition rates close to 90%, consistent with the results in Figure 5. For shale lithofacies classification, the RF, XGB, and Cat models perform moderately, with RF and XGB performing the worst, as multiple lithofacies have recognition rates below 0.5. The HGB and FCNN models perform better, with recognition rates close to 0.8 for A1, A2, A4, B1, and B2 lithofacies, but the recognition rates for A3 and D lithofacies are lower, around 0.6. Among all models, the stacking model still performs the best, with recognition rates for nearly all lithofacies reaching 0.9, although 13% of other lithofacies are still misclassified as A1 during A1 recognition. Overall, most models are unable to recognize laminated bedding argillaceous shale (A2) and lenticular bedding argillaceous shale (A3), and they also struggle to recognize massive bedding siliceous shale (B1) and massive bedding mixed shale (C1). The main reasons are as follows: first, the distinction between laminated bedding shale and lenticular bedding shale is unclear, and the two often exist in a transitional state. Second, although the felsic mineral content of siliceous shale is more than 50%, it is not much different from mixed shale (Figure 3).

The ROC curve (receiver operating characteristic curve) is also commonly used to evaluate the predictive performance of models, with the horizontal and vertical axes representing the false positive rate (FPR) and true positive rate (TPR), respectively. The AUC (Area Under Curve) is defined as the area under the ROC curve, with values typically ranging from 0.5 to 1, where a larger AUC indicates better classifier performance. The ROC curves for different classification models are shown in Figure 9. All models for sandstone (E) show high predictive classification performance, with AUC values greater than 0.95. For shale lithofacies, the ROC curve characteristics of different models vary. The ROC curves for each lithofacies from the optimal model, stacking, nearly overlap with the top-left corner of the coordinate system, with AUC values of 0.992, 0.998, 0.997, 0.997, 1.000, 1.000, 0.998, 0.997, and 0.996 for A1, A2, A3, A4, B1, B2, C1, D, and E, respectively, all greater than 0.99, indicating excellent recognition performance.

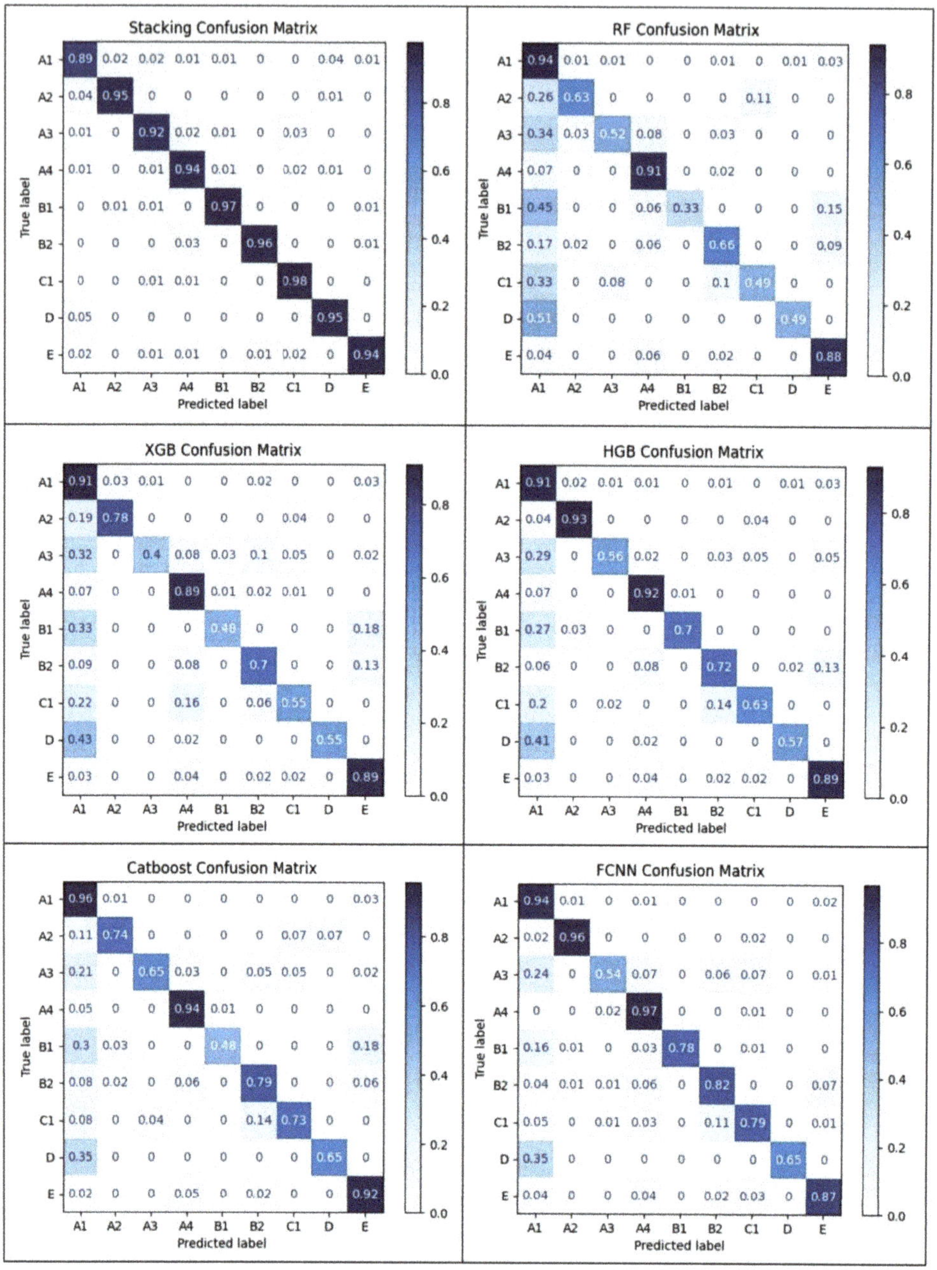

Figure 8. The confusion matrix shows the classification results of all models for each lithofacies.

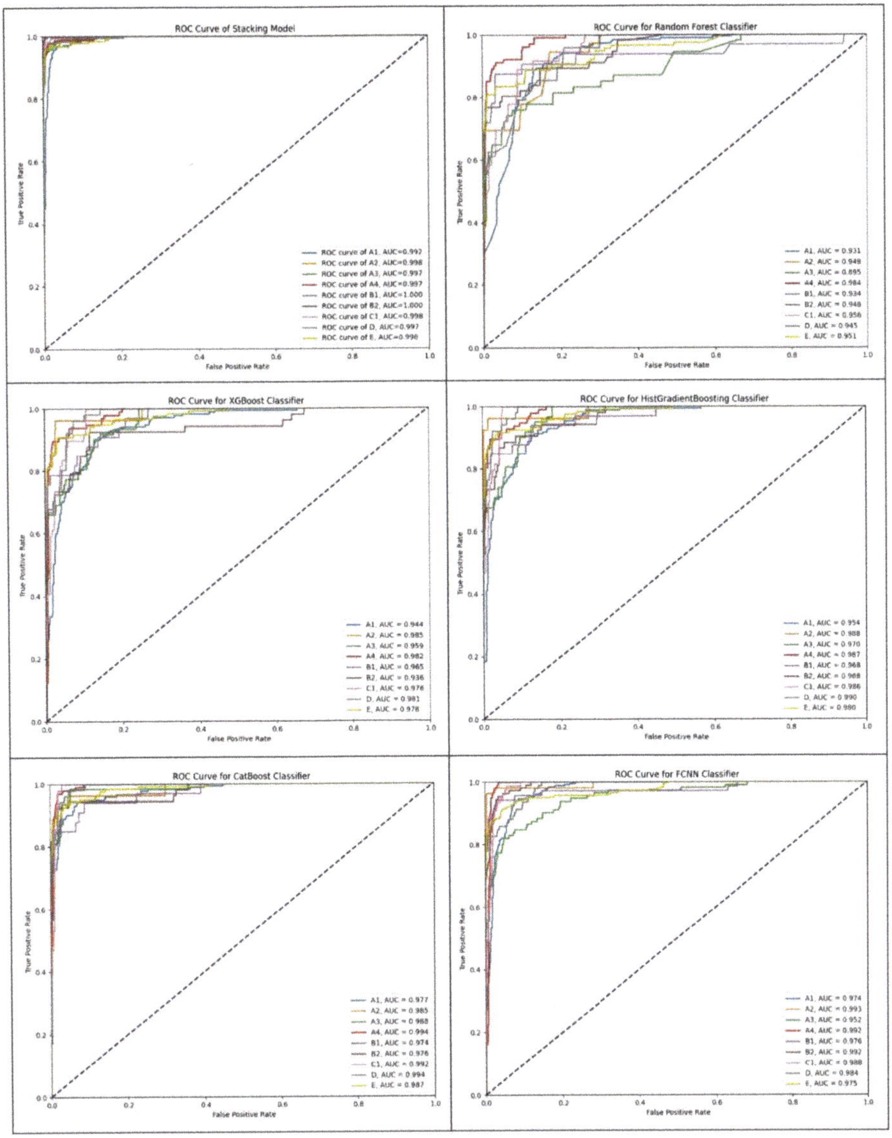

Figure 9. Performance analysis through ROC curves generated for each lithofacies of all ML models.

The prediction results of different models are shown in the lithofacies bar chart in Figure 10. Using W1 and W2 core training wells as examples, we visualized the vertical distribution of lithofacies predicted by different models versus the true observations of lithofacies using bar charts. In Well-W1, at the depth ranges of 2123–2129 m and 2137–2142 m, RF, HGB, XGB, Cat, and FCNN all show significant misclassifications and fail to accurately identify lithofacies A3, B1, and D. Similarly, for the more complex Well-W1, at the depth ranges of 2123–2129 m, 1910–1924 m, and 1934–1940 m, only the stacking model shows the best match in recognition.

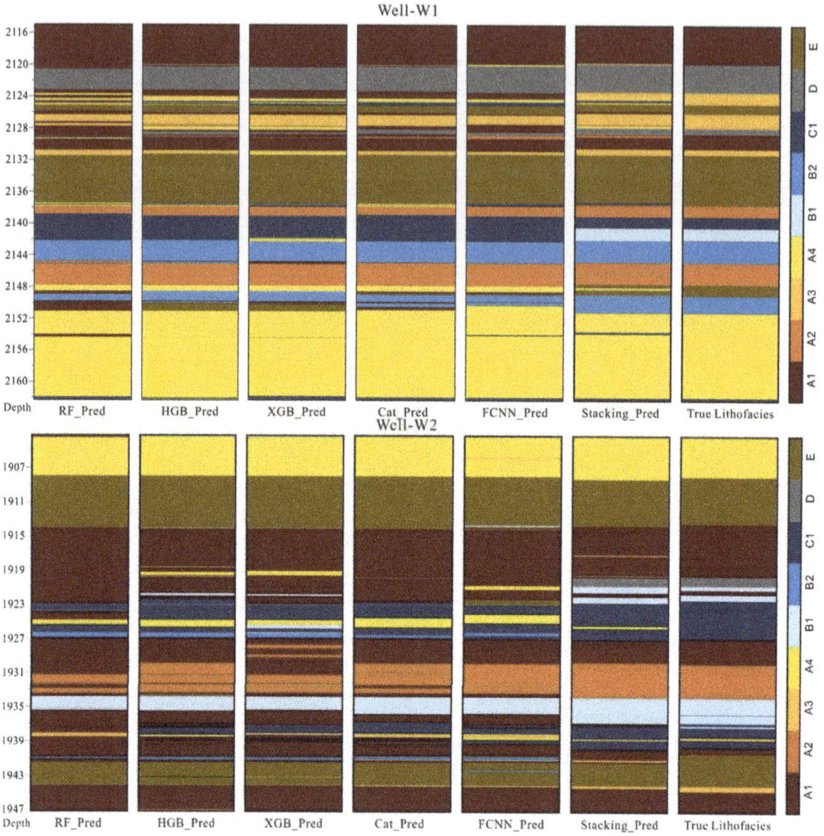

Figure 10. Comparison of predicted lithofacies by RF, HGB, XGB, Cat, FCNN, and stacking models in Well-W1 and Well-W2.

5.4. Ablation Experiments

By constructing combination models of RF, HGB, XGB, and Cat (RF-HGB-Cat, XGB-HGB-Cat, XGB-RF-Cat, and XGB-RF-HGB), the contribution and effectiveness of each base model to the overall model were further validated. Table 5 presents the statistical results of evaluating the ablation experiments by all combination models [58]. The results from the ablation experiments show that, on the training set, compared to the XGB-RF-HGB-Cat model, the RF-HGB-Cat model has reductions of 8.1%, 8.0%, 8.1%, and 8.1% in Accuracy, Precision, Recall, and F1 Score, respectively. The XGB-HGB-Cat model shows reductions of 7.0%, 7.1%, 7.3%, and 7.3%, the XGB-RF-Cat model shows reductions of 3.9%, 4.1%, 4.1%, and 4.1%, and the XGB-RF-HGB model shows reductions of 9.5%, 9.7%, 9.9%, and 10.0%. The results on the test set are generally consistent with those on the training set. The results indicate that the Cat model contributes the most to the overall model, followed by XGB, RF, and HGB. Although the prediction performance of XGB is not ideal as a standalone model, it still contributes significantly to the stacking model. Combining different classifier models results in a higher-performing model, and their relationship is not merely an additive one. The stacking model also performs well in the two blind wells (Well-W7 and Well-W8) that are not involved in training. The prediction results in the lithofacies bar chart are shown in Figure 11. The predicted lithofacies for Well-W7 and Well-W8 are almost identical to the actual lithofacies, with only a tiny portion misclassified. The blind well prediction results indicate that this stacking model demonstrates high accuracy and adaptability, making it applicable to the majority of non-cored wells with similar conventional logging curves.

Table 5. Ablation experimental results of all combination models.

Models	Training Data Set				Testing Data Set			
	Accuracy	Precision	Recall	F1 Score	Accuracy	Precision	Recall	F1 Score
RF-HGB-Cat	0.8812	0.8827	0.8812	0.881	0.864	0.8678	0.864	0.8643
XGB-HGB-Cat	0.8888	0.8912	0.8888	0.889	0.8725	0.8725	0.8725	0.8749
XGB-RF-Cat	0.9196	0.9198	0.9196	0.9193	0.9124	0.9124	0.9124	0.9135
XGB-RF-HGB	0.8635	0.8661	0.8635	0.863	0.8441	0.8466	0.8441	0.8432
XGB-RF-HGB-Cat	0.9587	0.9590	0.9587	0.9587	0.9456	0.9464	0.9456	0.9458

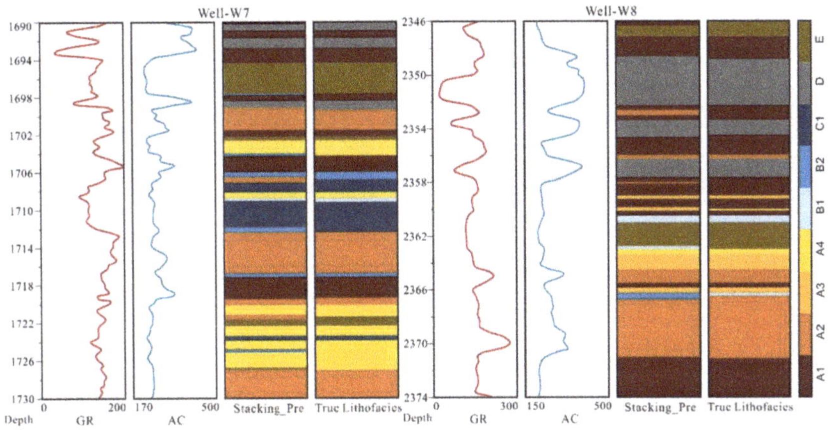

Figure 11. Lithofacies predicted by the stacking model in blind wells (W7 and W8) provided the best matching results.

5.5. Limitation and Significance of ML Models for Future Implications

The stacking model, an advanced machine learning classification method, can provide accurate predictions for complex, heterogeneous shale reservoirs (e.g., the marine-terrestrial transitional Shanxi Formation shale), offering higher precision and lower costs compared to conventional core observation and experimental methods. This is especially true in newly developed well areas where core samples are limited, as the ML model offers unparalleled advantages [59]. Although existing studies have achieved high accuracy in lithofacies recognition, more refined predictive tools are still needed for aspects such as sedimentary structures, fractures, porosity, permeability, and hydrocarbon content [19]. The conventional logging curves used in this study are unlikely to achieve this goal. However, higher-precision logging tools, such as Elemental Capture Spectroscopy (ECS) and Thermal Neutron Imaging Systems (TNI), can significantly enhance the model's predictive accuracy [60]. This study is also of significant importance for oil and gas exploration. Porosity, permeability, and hydrocarbon content vary across different lithofacies. In practical exploration and development, accurately locating sweet spots can be challenging [61,62]. However, the ML model's precise lithofacies prediction offers an effective tool to mitigate the risks of well placement and selection of production intervals. Additionally, detailed lithofacies provide important evidence of the depositional environment, facilitating a deeper understanding of the sedimentary history and reservoir conditions in the study area. In conclusion, the development of accurate ML models is highly beneficial for oil and gas exploration personnel. By accurately identifying reservoir lithofacies, the efficiency of oil and gas exploration can be greatly enhanced.

6. Conclusions

This study proposes a stacking model for predicting shale lithofacies based on four basic ML models in the Ordos Basin. The main conclusions are as follows:

(1) The stacking model performs excellently in predicting shale lithofacies, achieving Accuracy, Recall, Precision, and F1 Scores of 0.9587, 0.959, 0.9587, and 0.9587, respectively, on the training set. These results surpass those of RF, HGB, XGB, Cat, and FCNN models. The results on the test set are consistent with those on the training set. Lithofacies evaluation stack plots, confusion matrices, ROC curves, and lithofacies columnar distribution charts further validate this conclusion;

(2) The ablation experiments validated the contribution and effectiveness of each base model to the overall framework. The results show that the Cat model has the highest contribution to the overall model, followed by XGB, RF, and HGB. The stacking model also demonstrated high predictive accuracy in two blind wells not included in the training process, indicating its strong accuracy and adaptability, making it suitable for application in most uncored wells;

(3) The stacking model, an advanced machine learning classification method, enables the accurate identification of shale lithofacies. Enhancing the quality of logging data can further improve the model's performance. Accurately identifying favorable shale lithofacies is of great importance for oil and gas exploration researchers, as it aids in locating sweet spots and reduces exploration risks.

Author Contributions: Conceptualization, S.P. and C.F.; Methodology, S.P., C.F. and Z.Q.; Validation, Q.Z.; Formal analysis, W.L. and Q.Z.; Data curation, Z.H.; Investigation, W.L. and J.F.; Writing—original draft preparation, S.P. and C.F.; Writing—review and editing, S.P. and C.F.; Visualization, Z.Q.; Supervision, Q.Z.; Project administration, C.F. and Z.Q. All authors have read and agreed to the published version of the manuscript.

Funding: This research was funded by Science Fund for the Natural Science Basic Research Plan in Shaanxi Province of China (Grant nos. 2017JM4013; 2020J0798).

Institutional Review Board Statement: Not applicable.

Informed Consent Statement: Not applicable.

Data Availability Statement: The algorithm that supports the findings of this study are available from the corresponding author upon request. However, due to confidentiality requirements, the data will not be provided.

Conflicts of Interest: The authors declare no conflicts of interest.

Abbreviations

The following abbreviations are used in this manuscript:

ANN	Artificial Neural Network
SVM	Support Vector Machine
SOM	Self-Organizing Map
MRGC	Multiresolution Graph-based Clustering
DTr	Decision Tree
RF	Random Forest
MLP	Multilayer Perceptron
XGBoost	Extreme Gradient Boosting
NB	Naïve Bayes

BiLSTM	Bidirectional Long Short-Term Memory
FCNN	Fully Convolutional Neural Network
MKFD	Multi-Kernel Fisher Discriminant
GraphSAGE	Graph Sample and Aggregated Graph Embedding
DCNNs	Deep Convolutional Neural Networks
AB	Adaptive Boosting
HGB	HistGradient Boosting
Cat	Categorical Boosting

References

1. Zou, C.; Dong, D.; Wang, Y.; Li, X.; Huang, J.; Wang, S.; Guan, Q.; Zhang, C.; Wang, H.; Liu, H.; et al. Shale Gas in China: Characteristics, Challenges and Prospects (I). *Pet. Explor. Dev.* **2015**, *42*, 753–767. [CrossRef]
2. Dong, D.; Zou, C.; Dai, J.; Huang, S.; Zheng, J.; Gong, J.; Wang, Y.; Li, X.; Guan, Q.; Zhang, C.; et al. Suggestions on the Development Strategy of Shale Gas in China. *J. Nat. Gas Geosci.* **2016**, *1*, 413–423. [CrossRef]
3. Li, Y.; Tang, D.; Wu, P.; Niu, X.; Wang, K.; Qiao, P.; Wang, Z. Continuous Unconventional Natural Gas Accumulations of Carboniferous-Permian Coal-Bearing Strata in the Linxing Area, Northeastern Ordos Basin, China. *J. Nat. Gas Sci. Eng.* **2016**, *36*, 314–327. [CrossRef]
4. Kuang, L.; Dong, D.; He, W.; Wen, S.; Sun, S.; Li, S.; Qiu, Z.; Liao, X.; Li, Y.; Wu, J.; et al. Geological Characteristics and Development Potential of Transitional Shale Gas in the East Margin of the Ordos Basin, NW China. *Pet. Explor. Dev.* **2020**, *47*, 471–482. [CrossRef]
5. Zou, C.; Qiu, Z.; Zhang, J.; Li, Z.; Wei, H.; Liu, B.; Zhao, J.; Yang, T.; Zhu, S.; Tao, H.; et al. Unconventional Petroleum Sedimentology: A Key to Understanding Unconventional Hydrocarbon Accumulation. *Engineering* **2022**, *18*, 62–78. [CrossRef]
6. Wang, Z.; Liu, L.; Hu, J.; Wang, F.; Li, D.; Zhang, J.; Zhu, S.; Zhang, R.; Zhao, F.; Zhang, C.; et al. Dispersion of Sandy Sediments during Marine–Continental Transition: An Integrated Study from the Late Paleozoic Western Ordos Basin. *Mar. Pet. Geol.* **2024**, *160*, 106620. [CrossRef]
7. Wang, P.; Peng, S.; He, T. A Novel Approach to Total Organic Carbon Content Prediction in Shale Gas Reservoirs with Well Logs Data, Tonghua Basin, China. *J. Nat. Gas Sci. Eng.* **2018**, *55*, 1–15. [CrossRef]
8. Liu, J.-J.; Liu, J.-C. An Intelligent Approach for Reservoir Quality Evaluation in Tight Sandstone Reservoir Using Gradient Boosting Decision Tree Algorithm—A Case Study of the Yanchang Formation, Mid-Eastern Ordos Basin, China. *Mar. Pet. Geol.* **2021**, *126*, 104939. [CrossRef]
9. Tognoli, F.M.W.; Spaniol, A.F.; De Mello, M.E.; De Souza, L.V. A Machine-Learning Based Approach to Predict Facies Associations and Improve Local and Regional Stratigraphic Correlations. *Mar. Pet. Geol.* **2024**, *160*, 106636. [CrossRef]
10. Verma, S.; Zhao, T.; Marfurt, K.J.; Devegowda, D. Estimation of Total Organic Carbon and Brittleness Volume. *Interpretation* **2016**, *4*, T373–T385. [CrossRef]
11. Ye, T.; A-juan, W.E.I.; Deng, H.; Zeng, J.C.; Gao, K.S.; Sun, Z. Study on volcanic lithology identification methods based on the data of conventional well logging data: A case from Mesozoic volcanic rocks in Bohai bay area. *Prog. Geophys.* **2017**, *32*, 1842–1848. (In Chinese)
12. Mulashani, A.K.; Shen, C.; Nkurlu, B.M.; Mkono, C.N.; Kawamala, M. Enhanced Group Method of Data Handling (GMDH) for Permeability Prediction Based on the Modified Levenberg Marquardt Technique from Well Log Data. *Energy* **2022**, *239*, 121915. [CrossRef]
13. Kadri, M.M.; Ganguli, S.S.; Sen, S.; Hacini, M.; Kumar, P. Characterization and Feature Ranking of Well Log Variables Using Data-Driven Algorithms for Total Organic Carbon Estimation of Organic-Rich Shales. *Energy Fuels* **2023**, *37*, 19575–19589. [CrossRef]
14. Mkono, C.N.; Chuanbo, S.; Mulashani, A.K.; Mwakipunda, G.C. Deep Learning Integrated Approach for Hydrocarbon Source Rock Evaluation and Geochemical Indicators Prediction in the Jurassic—Paleogene of the Mandawa Basin, SE Tanzania. *Energy* **2023**, *284*, 129232. [CrossRef]
15. Mahmoud, A.A.; Elkatatny, S.; Ali, A.Z.; Abouelresh, M.; Abdulraheem, A. Evaluation of the Total Organic Carbon (TOC) Using Different Artificial Intelligence Techniques. *Sustainability* **2019**, *11*, 5643. [CrossRef]
16. Rui, J.; Zhang, H.; Zhang, D.; Han, F.; Guo, Q. Total Organic Carbon Content Prediction Based on Support-Vector-Regression Machine with Particle Swarm Optimization. *J. Pet. Sci. Eng.* **2019**, *180*, 699–706. [CrossRef]
17. Meng, M.; Zhong, R.; Wei, Z. Prediction of Methane Adsorption in Shale: Classical Models and Machine Learning Based Models. *Fuel* **2020**, *278*, 118358. [CrossRef]

18. Shalaby, M.R.; Malik, O.A.; Lai, D.; Jumat, N.; Islam, M.A. Thermal Maturity and TOC Prediction Using Machine Learning Techniques: Case Study from the Cretaceous–Paleocene Source Rock, Taranaki Basin, New Zealand. *J. Pet. Explor. Prod. Technol.* **2020**, *10*, 2175–2193. [CrossRef]
19. Hui, G.; Chen, Z.; Wang, Y.; Zhang, D.; Gu, F. An Integrated Machine Learning-Based Approach to Identifying Controlling Factors of Unconventional Shale Productivity. *Energy* **2023**, *266*, 126512. [CrossRef]
20. Niu, W.; Sun, Y.; Zhang, X.; Lu, J.; Liu, H.; Li, Q.; Mu, Y. An Ensemble Transfer Learning Strategy for Production Prediction of Shale Gas Wells. *Energy* **2023**, *275*, 127443. [CrossRef]
21. Zhang, H.; Wu, W.; Wu, H. TOC Prediction Using a Gradient Boosting Decision Tree Method: A Case Study of Shale Reservoirs in Qinshui Basin. *Geoenergy Sci. Eng.* **2023**, *221*, 111271. [CrossRef]
22. Du, J.; Zhang, C.; Li, S.; Zhang, L.; Zhang, W. Two-Stage Prediction Method for Capacity Aging Trajectories of Lithium-Ion Batteries Based on Siamese-Convolutional Neural Network. *Energy* **2024**, *295*, 130947. [CrossRef]
23. Wang, G.; Carr, T.R. Marcellus Shale Lithofacies Prediction by Multiclass Neural Network Classification in the Appalachian Basin. *Math. Geosci.* **2012**, *44*, 975–1004. [CrossRef]
24. Bhattacharya, S.; Carr, T.R. Integrated Data-Driven 3D Shale Lithofacies Modeling of the Bakken Formation in the Williston Basin, North Dakota, United States. *J. Pet. Sci. Eng.* **2019**, *177*, 1072–1086. [CrossRef]
25. Santos, D.T.D.; Roisenberg, M.; Nascimento, M.D.S. Deep Recurrent Neural Networks Approach to Sedimentary Facies Classification Using Well Logs. *IEEE Geosci. Remote Sens. Lett.* **2022**, *19*, 3001405. [CrossRef]
26. Dong, S.; Zeng, L.; Du, X.; He, J.; Sun, F. Lithofacies Identification in Carbonate Reservoirs by Multiple Kernel Fisher Discriminant Analysis Using Conventional Well Logs: A Case Study in A Oilfield, Zagros Basin, Iraq. *J. Pet. Sci. Eng.* **2022**, *210*, 110081. [CrossRef]
27. Lu, G.; Zeng, L.; Dong, S.; Huang, L.; Liu, G.; Ostadhassan, M.; He, W.; Du, X.; Bao, C. Lithology Identification Using Graph Neural Network in Continental Shale Oil Reservoirs: A Case Study in Mahu Sag, Junggar Basin, Western China. *Mar. Pet. Geol.* **2023**, *150*, 106168. [CrossRef]
28. Zheng, D.; Zhong, H.; Camps-Valls, G.; Cao, Z.; Ma, X.; Mills, B.; Hu, X.; Hou, M.; Ma, C. Explainable Deep Learning for Automatic Rock Classification. *Comput. Geosci.* **2024**, *184*, 105511. [CrossRef]
29. Ali, M.; Zhu, P.; Huolin, M.; Jiang, R.; Zhang, H.; Ashraf, U.; Hussain, W. Data-Driven Machine Learning Approaches for Precise Lithofacies Identification in Complex Geological Environments. *Geo-Spat. Inf. Sci.* **2024**, 1–21. [CrossRef]
30. Bhattacharya, S.; Carr, T.R.; Pal, M. Comparison of Supervised and Unsupervised Approaches for Mudstone Lithofacies Classification: Case Studies from the Bakken and Mahantango-Marcellus Shale, USA. *J. Nat. Gas Sci. Eng.* **2016**, *33*, 1119–1133. [CrossRef]
31. Bressan, T.S.; Kehl De Souza, M.; Girelli, T.J.; Junior, F.C. Evaluation of Machine Learning Methods for Lithology Classification Using Geophysical Data. *Comput. Geosci.* **2020**, *139*, 104475. [CrossRef]
32. Hong, Z.; Yao, J.; Li, K.; Hu, G. Conjunction of Active and Semi-Supervised Learning for Wireline Logs-Based Automatic Lithology Identification. *IEEE Geosci. Remote Sens. Lett.* **2022**, *19*, 3008905. [CrossRef]
33. Koeshidayatullah, A.; Al-Azani, S.; Baraboshkin, E.E.; Alfarraj, M. FaciesViT: Vision Transformer for an Improved Core Lithofacies Prediction. *Front. Earth Sci.* **2022**, *10*, 992442. [CrossRef]
34. Zheng, D.; Hou, M.; Chen, A.; Zhong, H.; Qi, Z.; Ren, Q.; You, J.; Wang, H.; Ma, C. Application of Machine Learning in the Identification of Fluvial-Lacustrine Lithofacies from Well Logs: A Case Study from Sichuan Basin, China. *J. Pet. Sci. Eng.* **2022**, *215*, 110610. [CrossRef]
35. Ullah, J.; Li, H.; Ashraf, U.; Ehsan, M.; Asad, M. A Multidisciplinary Approach to Facies Evaluation at Regional Level Using Well Log Analysis, Machine Learning, and Statistical Methods. *Geomech. Geophys. Geo-Energy Geo-Resour.* **2023**, *9*, 152. [CrossRef]
36. Ashraf, U.; Zhang, H.; Thanh, H.V.; Anees, A.; Ali, M.; Duan, Z.; Mangi, H.N.; Zhang, X. A Robust Strategy of Geophysical Logging for Predicting Payable Lithofacies to Forecast Sweet Spots Using Digital Intelligence Paradigms in a Heterogeneous Gas Field. *Nat. Resour. Res.* **2024**, *33*, 1741–1762. [CrossRef]
37. Hussain, W.; Luo, M.; Ali, M.; Al-Khafaji, H.F.; Hussain, I.; Hussain, M.; Ahmed, S.A.A.; Obaidullah. A Gamma-Ray Spectroscopy Approach to Evaluate Clay Mineral Composition and Depositional Environment: A Case Study from the Lower Goru Formation, Southern Indus Basin, Pakistan. *J. Appl. Geophys.* **2024**, *226*, 105414. [CrossRef]
38. Liu, G.; Ma, S.; Zhang, Z.; Mao, Z.; Han, T.; Shi, Z. U–Pb Ages of Tuff from the Triassic Yanchang Formation in the Ordos Basin: Constraints on Palaeoclimate and Tectonics. *Mar. Pet. Geol.* **2023**, *150*, 106128. [CrossRef]
39. Wang, Z.; Zhou, N.; Lu, S.; Liu, Y.; Lin, L.; Liu, Y.; Song, B. Generation, Accumulation, and Distribution of Upper Paleozoic Tight Sandstone Gas in the Northeastern Margin of the Ordos Basin. *Mar. Pet. Geol.* **2023**, *156*, 106463. [CrossRef]
40. Jiang, F.; Jia, C.; Pang, X.; Jiang, L.; Zhang, C.; Ma, X.; Qi, Z.; Chen, J.; Pang, H.; Hu, T.; et al. Upper Paleozoic Total Petroleum System and Geological Model of Natural Gas Enrichment in Ordos Basin, NW China. *Pet. Explor. Dev.* **2023**, *50*, 281–292. [CrossRef]

41. Chen, R.; Wang, F.; Li, Z.; Evans, N.J.; Chen, H.; Wei, X. Late Paleozoic Provenance Shift in the East-Central Ordos Basin: Implications for the Tectonic Evolution of the North China Craton. *J. Asian Earth Sci.* **2021**, *215*, 104799. [CrossRef]
42. Hou, Z.-S.; Chen, S.-Y.; Liang, Z. Sedimentary Features and Sequence Stratigraphy of the Successions around the Carboniferous–Permian Boundary in the Ordos Basin: Links to Glacial and Volcanic Impacts. *J. Palaeogeogr.* **2023**, *12*, 358–383. [CrossRef]
43. Fu, C.; Yu, X.; Li, S.; Peng, Z.; Shi, S. Carboniferous–Permian Transgression/Regression Mechanisms in the Eastern Ordos Basin and Their Sea-Level Spatiotemporal Variability: Insights from Source-to-Sink Systems. *Mar. Pet. Geol.* **2021**, *123*, 104722. [CrossRef]
44. Ullah, S.; Hanif, M.; Radwan, A.E.; Luo, C.; Rehman, N.U.; Ahmad, S.; Latif, K.; Ali, N.; Thanh, H.V.; Asim, M.; et al. Depositional and Diagenetic Modeling of the Margala Hill Limestone, Hazara Area (Pakistan): Implications for Reservoir Characterization Using Outcrop Analogues. *Geoenergy Sci. Eng.* **2023**, *224*, 211584. [CrossRef]
45. Fonseca, J.; Bacao, F. Geometric SMOTE for Imbalanced Datasets with Nominal and Continuous Features. *Expert Syst. Appl.* **2023**, *234*, 121053. [CrossRef]
46. Chawla, N.V.; Bowyer, K.W.; Hall, L.O.; Kegelmeyer, W.P. SMOTE: Synthetic Minority Over-Sampling Technique. *J. Artif. Intell. Res.* **2002**, *16*, 321–357. [CrossRef]
47. Shi, X.; Wang, J.; Liu, G.; Yang, L.; Ge, X.; Jiang, S. Application of Extreme Learning Machine and Neural Networks in Total Organic Carbon Content Prediction in Organic Shale with Wire Line Logs. *J. Nat. Gas Sci. Eng.* **2016**, *33*, 687–702. [CrossRef]
48. Wood, D.A. Predicting Total Organic Carbon from Few Well Logs Aided by Well-Log Attributes. *Petroleum* **2023**, *9*, 166–182. [CrossRef]
49. Ahmed, A.H.A.; Jin, W.; Ali, M.A.H. Prediction of Compressive Strength of Recycled Concrete Using Gradient Boosting Models. *Ain Shams Eng. J.* **2024**, *15*, 102975. [CrossRef]
50. Alam, M.S.; Akinpelu, A.A.; Nazal, M.K.; Rahman, S.M. Removal of N-Nitrosodiphenylamine from Contaminated Water: A Novel Modeling Framework Using Metaheuristic-Based Ensemble Models. *J. Environ. Manag.* **2024**, *365*, 121503. [CrossRef]
51. Bione, F.R.A.; Venancio, I.M.; Santos, T.P.; Belem, A.L.; Rangel, B.R.; Souza, I.V.A.F.; Spigolon, A.L.D.; Albuquerque, A.L.S. Estimating Total Organic Carbon of Potential Source Rocks in the Espírito Santo Basin, SE Brazil, Using XGBoost. *Mar. Pet. Geol.* **2024**, *162*, 106765. [CrossRef]
52. Ahmed, Y.; Dutta, K.R.; Nepu, S.N.C.; Prima, M.; AlMohamadi, H.; Akhtar, P. Optimizing Photocatalytic Dye Degradation: A Machine Learning and Metaheuristic Approach for Predicting Methylene Blue in Contaminated Water. *Results Eng.* **2025**, *25*, 103538. [CrossRef]
53. Abdi, J.; Hadipoor, M.; Hadavimoghaddam, F.; Hemmati-Sarapardeh, A. Estimation of Tetracycline Antibiotic Photodegradation from Wastewater by Heterogeneous Metal-Organic Frameworks Photocatalysts. *Chemosphere* **2022**, *287*, 132135. [CrossRef] [PubMed]
54. Wang, Y.; Wang, J.; Wang, Q. Enhancing Autonomous Driving Safety: A Robust Stacking Ensemble Model for Traffic Sign Detection and Recognition. *Sustainability* **2024**, *16*, 8597. [CrossRef]
55. Mu, G.; Wei, Q.; Xu, Y.; Zhang, H.; Zhang, J.; Li, Q. Capacity Estimation for Lithium-Ion Batteries Based on Heterogeneous Stacking Model with Feature Fusion. *Energy* **2024**, *313*, 133881. [CrossRef]
56. Usman, U.S.; Salh, Y.H.M.; Yan, B.; Namahoro, J.P.; Zeng, Q.; Sallah, I. Fluoride Contamination in African Groundwater: Predictive Modeling Using Stacking Ensemble Techniques. *Sci. Total Environ.* **2024**, *957*, 177693. [CrossRef]
57. Wang, J.; Li, Y.; Song, W.; Li, A. Research on the Theory and Method of Grid Data Asset Management. *Procedia Comput. Sci.* **2018**, *139*, 440–447. [CrossRef]
58. Liu, Y.; Li, N.; Li, C.; Jiang, J.; Wu, X.; Liang, H.; Zhang, D.; Hu, X. Prediction of Total Organic Carbon Content in Deep Marine Shale Reservoirs Based on a Super Hybrid Machine Learning Model. *Energy Fuels* **2024**, *38*, 17483–17498. [CrossRef]
59. Jia, R.; Lv, Y.; Wang, G.; Carranza, E.; Chen, Y.; Wei, C.; Zhang, Z. A Stacking Methodology of Machine Learning for 3D Geological Modeling with Geological-Geophysical Datasets, Laochang Sn Camp, Gejiu (China). *Comput. Geosci.* **2021**, *151*, 104754. [CrossRef]
60. Polito, C.P. Correlations for Estimating Coefficients for the Prediction of Maximum and Minimum Index Void Ratios for Mixtures of Sand and Non-Plastic Silt. *Geotechnics* **2023**, *3*, 1033–1046. [CrossRef]
61. Johnson, L.M.; Rezaee, R.; Kadkhodaie, A.; Smith, G.; Yu, H. Geochemical Property Modelling of a Potential Shale Reservoir in the Canning Basin (Western Australia), Using Artificial Neural Networks and Geostatistical Tools. *Comput. Geosci.* **2018**, *120*, 73–81. [CrossRef]
62. Chen, Z.; Wu, Q.; Han, S.; Zhang, J.; Yang, P.; Liu, X. A Study on Geological Structure Prediction Based on Random Forest Method. *Artif. Intell. Geosci.* **2022**, *3*, 226–236. [CrossRef]

minerals

MDPI

Article

Lithofacies Characteristics of the Lower Cretaceous Qing 1 Member in the Heiyupao Depression, Northern Binbei Area of the Songliao Basin

Yali Liu [1,2], Wangpeng Li [1,2], Jiapeng Yuan [1,2], Pei Li [1,2], Xun Ge [1,2], Xiaotong Ge [1,2], Pengfei Liu [1,2], Haiguang Wu [1,2], Xuntao Yu [1,2] and Botao Huang [1,2,*]

1 State Key Laboratory of Shale Oil and Gas Enrichment Mechanisms and Efficient Development, Beijing 102206, China
2 Sinopec Key Laboratory of Shale Oil/Gas Exploration & Production Technology, Beijing 102206, China
* Correspondence: 228002010015@stu.nepu.edu.cn

Abstract: Strategic breakthroughs have been made in the exploration and evaluation of Gulong shale oil in the Songliao Basin. However, the Heiyupao Depression, located near the Gulong Depression, hosts a thick section of the Qingshankou Formation shale that has not been extensively studied. This paper presents novel insights into the lithofacies characteristics, depositional environment, and reservoir features of the Qingshankou Formation shale in the Heiyupao Depression, with a specific focus on the origin and maturation of organic-rich shale. Four core wells were drilled, and 152 core samples were analyzed through a variety of techniques, including rock type classification, mineral composition, TOC content, rare earth elements, rock pyrolysis, organic matter type determination, and CT scanning. Results indicate that the Qingshankou shale is dominated by felsic compositions and Type I kerogen, with organic maturity varying across the section. Based on lithology, sedimentary structures, mineralogy, and organic matter abundance, five distinct lithofacies are identified: high-organic mud-rich felsic shale, high-organic sand-rich felsic shale, medium-organic sand-rich felsic shale, medium-organic massive shale, and low-organic sand-rich felsic shale. Notably, the Type A lithofacies (high-organic mud-rich felsic shale) is identified as a primary source rock due to its intergranular and organic matter pores, albeit with low porosity and poor connectivity. In contrast, the Type E lithofacies (low-organic sand-rich felsic shale) have high porosity, well-developed micro- and nano-scale pores, and strong connectivity, marking them as the primary reservoirs. The characteristics of this region differ significantly from those of Gulong shale oil, requiring different extraction strategies. The mineral composition of such shale is predominantly felsic rather than mixed. The findings not only provide theoretical support for the exploration of complex lacustrine shale in the Songliao Basin but also offer valuable insights for the resource development of similar non-marine shale systems worldwide.

Keywords: shale lithofacies; lithofacies characteristics; Songliao Basin; Qingshankou Formation

1. Introduction

As global demand for fossil fuels rises, unconventional oil and gas resources have become key targets for exploration. North American shale formations, predominantly marine and deposited in stable tectonic settings, are widely distributed [1]. In contrast, China's lacustrine shales exhibit greater geological variability, with diverse sedimentary systems and complex lithologies, lacking a standardized lithofacies classification [1]. Inspired

 https://doi.org/10.3390/min15020125

by the North American shale gas revolution, China's shale gas and oil exploration has gained increasing attention [2–8]. Shale oil in China, primarily from continental deposits, is estimated at 800 to 1000 $\times$ 10^8 tons, highlighting its potential for development [9]. Key regions include the Songliao, Junggar, and Ordos Basins, where some lacustrine basins have achieved varying degrees of industrial development.

The Songliao Basin is one of the world's most significant continental oil and gas-rich basins. Its Cretaceous Qingshankou Formation shales serve as the primary hydrocarbon source rocks and are a strategic resource for China. The Binbei area of the Songliao Basin has become a hotspot for exploration, especially in the Heiyupao Depression, where the Qingshankou Formation holds abundant oil and gas resources, yet large-scale fields have not been discovered. The Qing 1 member of the Qingshankou Formation shows widespread mudstone distribution, suggesting a favorable combination of source, reservoir, and cap [10].

Many studies have focused on the Heiyupao Depression. He argues that the Qing 1 member experienced a significant marine transgression, expanding the lake basin and developing a large-scale delta–lake system in the Lindian–Heiyupao region [11]. Jiang used the gas accumulation threshold theory to calculate the oil and gas thresholds, estimating the available oil in the Heiyupao–Keshan system at 5.09 $\times$ 10^8 tons [12].

However, the main constraints to oil and gas development in Binbei are the poor lateral continuity of sand bodies and strong reservoir heterogeneity within the Qingshankou Group. Distinct lithofacies reveal different depositional environments and diagenetic processes, emphasizing the need for further research into the lithofacies in Binbei. Lithofacies classification, based on physical properties like color, granularity, and mineral composition, remains inconsistent. Current research focuses on shale laminations, organic content, and mineral composition. For example, Loucks et al. classified North American Barnett shale into various types [13], while Trabucho-Alexand et al. categorized Jurassic black shales in the Dutch Central Trough [14]. Studies of Gulong shales have also influenced the Binbei area, with Wang et al. identifying organic-rich and poor types [15], while Zhang et al. defined five major lithofacies, including shale, mudstone, and siltstone [16,17].

Research on lithofacies in Binbei remains limited. Given the region's distinct shale characteristics compared to neighboring areas, a specific lithofacies classification for the Heiyupao Depression is necessary to better understand the spatial distribution and formation environments of shale lithofacies.

This study investigates shale core samples from the Qing 1 member in the Heiyupao Depression using microscopic observation, X-ray diffraction (XRD), and organic geochemistry. The goal is to provide a scientific foundation for identifying high-quality reservoirs and hydrocarbon storage.

2. Geological Setting

The Binbei area primarily comprises four first-level tectonic units and 12 s-level tectonic units. As shown in Figure 1b, the Heiyupao Depression is located in the northern part of the central depression area. Figure 1a shows that it is bordered to the south by the Daqing Changyuan and Sanzhao Depressions, to the north by the Qianyuan Tectonic Belt, to the east by the Mingshui Terrace, and to the west by the Qijia-Gulong, Wuyuer Depression, and KeshanYilong Anticline. Based on the distribution and contact relationships of the Cretaceous formations and the tectonic characteristics, the Songliao Basin is divided into three tectonic cycles: the rift, sag, and shrinkage stages. The development characteristics of each stage are uneven, with deep structures resulting from extensional deformation, shallow structures from compressional deformation, and intermediate layers representing transition types. As depicted in Figure 1c, the tectonic movement characteristics of the Binbei area generally correspond to these stages [18,19].

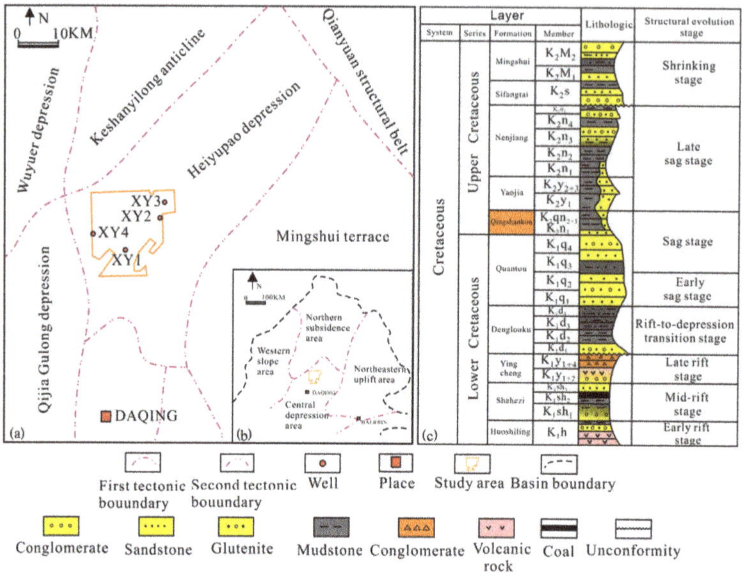

Figure 1. Tectonic unit division and stratigraphic column of Binbei area (modified from [10]). (**a**) Tectonic unit division map of the study area. (**b**) Tectonic unit division map of the Songliao Basin. (**c**) Tectonic unit division and composite stratigraphic column of the research area.

The primary hydrocarbon source rocks in the Songliao Basin were developed during the Upper Cretaceous sedimentary period, consisting sequentially of the Qingshankou Formation (K_2qn), Yaojia Formation (K_2y), Nenjiang Formation (K_2n), Sifangtai Formation (K_2s), and Mingshui Formation (K_2m). During this period, the basin was characterized by extensive sag lake basin deposits, accompanied by several periodic oscillations at the basin's base, which created favorable source–reservoir–seal combinations. Throughout the sagging period of the basin, the Heiyupao Depression not only formed a large continental lake basin but also generated a rich hydrocarbon system centered around the sag lake basin. The Nen 2 (K_2n_2) member possesses a basin-wide mudstone cap layer, while the Yaojia Formation serves as a significant reservoir within the Songliao Basin. The Qing 1 member (K_2qn_1) contains mudstone and shale, and the Qing 2 and 3 members (K_2qn_{2+3}) feature extensive delta complexes [20,21]. Consequently, the Qingshankou Formation is recognized as the principal hydrocarbon source rock sequence within the depression [22–25], establishing the Heiyupao Depression's Qingshankou Formation as a favorable exploration domain.

3. Materials and Methods

The experiment collected a total of 152 core samples from wells XY1, XY2, XY3, and XY4 in the Qingshankou Formation of the Heiyupao Depression in the Songliao Basin. The samples primarily consisted of shale, mudstone, and silty mudstone. In order to understand its specific mineral composition, structural characteristics, organic geochemical features, and reservoir characteristics, twenty-two samples were pulverized for analysis, which included XRD, rock pyrolysis, total organic carbon (TOC) content analysis and rare earth elements analysis. Additionally, 12 samples underwent thin section observation, scanning electron microscopy (SEM), and micro-CT scanning.

Whole-rock mineral analysis of shale was performed using XRD with a Rigaku SmartLabSE X-ray diffractometer (Rigaku Corporation, Tokyo, Japan), operating at 40 kV and 40 mA, in accordance with the "X-ray Diffraction Analysis Method of Clay Minerals and Common Non-clay Minerals in Sedimentary Rocks" (SY/T 5163-2018) [26]. During sample preparation, powdered samples are ground to a maximum of 75 microns, while block and thin film samples are processed according to specific specifications. Clay minerals are prepared as naturally oriented slides (N slides), ethylene glycol-saturated slides (EG slides), and heated slides (T slides). During testing, the samples are placed steadily and accurately, and the measurement program is initiated to record the diffraction patterns. Finally, professional software such as MDI Jade is used for qualitative analysis of the patterns and mineral content is quantified through semi-quantitative calculations using the internal standard method. Rock pyrolysis experiments were executed utilizing a YY3000A type oil rock comprehensive evaluation instrument, adhering to the national standard "Rock Pyrolysis Analysis" (GB/T 18602—2012) [27]. The specific procedure involves crushing the source rock samples to below 200 mesh, drying them at low temperatures, weighing approximately 100 mg for rock pyrolysis analysis, and calculating the total organic carbon (TOC) data. The pyrolysis process, including S0, S1, S2, and Tmax values, was monitored using a hydrogen flame ionization detector, with testing performed according to the temperature programming method specified in the standard. S0, S1, and S2 values represent the hydrocarbon content in a unit mass of source rock at temperatures of 90 °C, 90–300 °C, and 300–600 °C, respectively; Tmax is the temperature corresponding to the peak of the S2 value, indicating the maximum temperature of organic matter thermal cracking.

Rare earth element analysis was performed using an inductively coupled plasma mass spectrometer (ICP-MS) Element II. Samples, stored at low temperature, were placed in an oven at 60 °C for more than 12 h. After drying, the samples were ground to below 200 mesh using a mortar and pestle. The prepared powder samples were subjected to pre-treatment before testing, and the resulting solutions were analyzed for trace and rare earth elements using the inductively coupled plasma mass spectrometer. The analysis error for elements with mass fractions less than 10 ppm is less than 10%, while for elements with mass fractions greater than 10 ppm, the analysis error is less than 5%.

SEM observations were performed using an FEI Quanta200F field emission scanning electron microscope (Thermo Fisher, Waltham, MA, USA) and a Leica EM RES102 argon ion polisher (Leica, Wetzlar, Germany). Following mechanical and argon ion polishing with the Leica EM RES102, the shale sample surfaces were carbon-coated. Observations were made using the FEI Quanta200F at a scanning voltage of approximately 15 kV and an energy spectrum point spacing of 2 μm. Additionally, micro-CT scanning was conducted using a Carl Zeiss three-dimensional stereo X-ray microscope, model Xradia 520 Versa (Carl Zeiss, Oberkochen, Germany). The technical specifications of the micro-CT include a load capacity of 15 kg, with movement ranges of 45, 100, and 50 mm in the X, Y, and Z directions, respectively, and a 360° rotation capability.

4. Results

4.1. *Sedimentary Structure and Geochemical Features of the Qing 1 Member Shale*

4.1.1. Sedimentary Structure

Cross-bedding (Figure 2f), parallel lamination (Figure 2f), scour surfaces (Figure 2h), deformation structures (Figure 2h), and the mudstones display horizontal lamination (Figure 2d), and the shales exhibit notably developed fissility (Figure 2e). Utilizing a single layer thickness of 1 cm and 1 m as boundaries, three structural types can be classified: laminated, bedded, and massive. When the single layer thickness exceeds 1 m, a uniformly massive structure is formed, primarily within silty mudstones. Layers with thicknesses ranging from 1 cm to 1 m predominantly develop bedded structures (Figure 2a), primarily occurring among mudstones, sandstones, and carbonates. Layers with thicknesses of ≤1 cm are characterized by lamination, primarily observed in mudstones and shales. Based on granularity, the types of lamination in mudstone and shale can be further classified into two main categories: sand-grade and mud-grade laminations. Mud-grade laminations consist of finer particles, mostly smaller than 4 μm, and comprise less than 5% of the total composition (Figure 2b). In contrast, sand-grade laminations contain coarser particles with diameters greater than 4 μm, accounting for between 5% and 20% of the total composition (Figure 2c).

Figure 2. Typical sedimentary structural features of the Qing 1 member in the Heiyupao Depression. (**a**) Well XY2, 2050.85 m, fine sandstone with organic-rich mud laminae, locally visible glauconite, under plane-polarized light. (**b**) Well XY1, 1960.32 m, mud-grade laminated felsic shale, under plane-polarized light. (**c**) Well XY2, 2027.75 m, sand-grade laminated felsic shale, under plane-polarized light. (**d**) Well XY1, 1999.41 m, shale with well-developed cleavage, core photo. (**e**) Well XY1, 1994.71 m, mudstone with sandstone bands, core photo. (**f**) Well XY4, 1765.20 m, siltstone with muddy bands, parallel bedding, deformation structures, core photo. (**g**) Well XY2, 2029.95 m, silty mud with calcareous fine sandstone bands, core photo. (**h**) Well XY3, 1935.22 m, dolomite mixed with calcareous fine sandstone including ostracod layers, core photo.

4.1.2. Rock Type

Based on four core-drilling wells within the study area, totaling 234 m, four types of rock were identified: siltstone, sandstone, mudstone (shale), and carbonate. Well XY4 exhibits a total stratigraphic thickness of 72 m, which includes 20 layers of mudstone with a cumulative thickness of 35.11 m, 13.73 m of sandstone, 23.16 m of siltstone, and 0.05 m of carbonate.

The shale is well developed, predominantly black and dark gray, with clearly visible lamination. The mineral composition primarily consists of felsic and clay minerals, with a small amount of carbonates and other minerals (Figure 2d).

Mudstone is commonly observed, appearing in gray-black and dark gray, with some silt-grade particles. It primarily exhibits massive and bedded forms, displaying visible horizontal lamination. Areas containing a significant number of silt particles are classified as siltstone mudstone. The composition is predominantly felsic minerals, followed by clay minerals (Figure 2e).

Sandstone development is moderate, primarily resulting in the formation of siltstone and fine sandstone, with siltstone layers typically measuring between 10 and 30 cm in thickness. Fine sandstone predominantly appears in layered formations, with localized occurrences of calcareous fine sandstone reaching up to 5 m in thickness. The prevailing colors are primarily gray and light gray, with siltstone generally exhibiting a gray hue. Fine sandstone may include sections of siltstone mud laminae and silt laminae, while siltstone is characterized by mud laminae, with additional minerals observable locally (Figure 2f,g). The size and quantity of particles provide further insight into the nature of silty siltstones.

Carbonate development is localized and primarily manifests in layers no thicker than 10 cm as well as in lens-shaped forms. Complete ostracod layers are visible both above and below dolomite layers (Figure 2h), which effervesce vigorously upon exposure to acid.

4.1.3. Mineral Composition

Based on whole-rock and clay mineral XRD analysis, the mineral composition of the Qing 1 member from well XY4 in the Heiyupao Depression of the Binbei area is primarily characterized by clay minerals, quartz, potassium feldspar, and plagioclase, with minor constituents including calcite, dolomite, ankerite, and pyrite (Figure 3a, Table 1). As shown in Figure 3b. The clay minerals are predominantly illite, with the presence of illite-smectite mixed layers and a small amount of glauconite. The ternary diagram of the mineral composition of the Qing 1 member in the Heiyupao Depression (Figure 4) clearly indicates that the mud shales of the Qing 1 member are primarily classified as felsic shales.

Table 1. Composition of whole-rock minerals of the Heiyupao Depression.

Samples	Caly (%)	Quartz (%)	Potassium Feldspar (%)	Plagioclase Feldspar (%)	Calcite (%)	Dolomite (%)	Ferroan Dolomite (%)
XY3-1	37.1	29	5.7	16.5	4.6		
XY3-2	43.8	28.7	7.1	14.3			
XY3-3	37.1	28.6	7.4	19.3	3.5		
XY3-4	37	33.5	5.6	13.1	5.1		
XY3-5	32.8	31	5.2	20	4.3		
XY3-6	26.9	27	4.8	27.7	9.1		
XY3-7	40.3	31.3	4.7	14	4.6		
XY3-8	24.7	29.3	5.2	9.6	21.1		5.6
XY3-9	35.2	29.6	4.6	12.1	7.7	10.9	

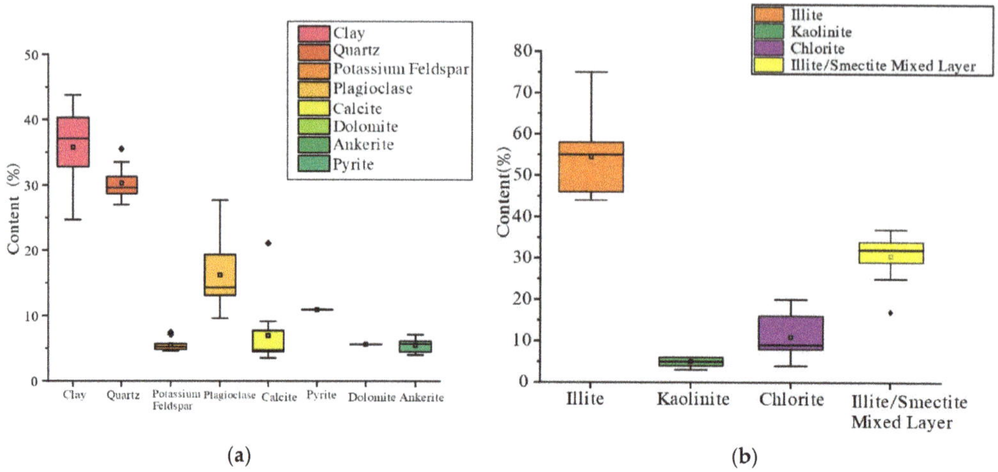

Figure 3. Composition of whole-rock minerals (**a**) and clay minerals (**b**) in the Qing 1 member of the Heiyupao Depression.

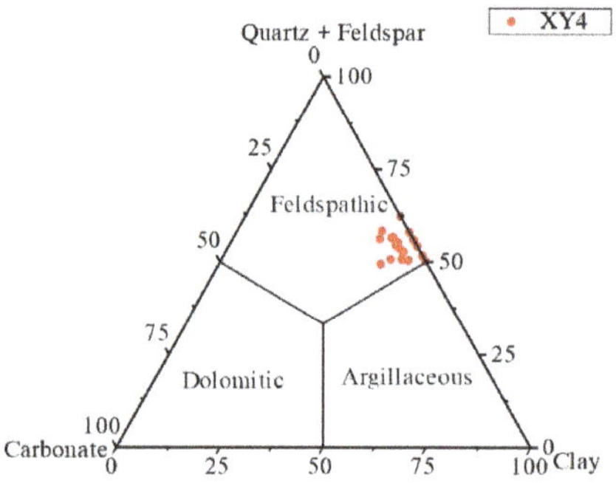

Figure 4. Ternary diagram of mineral composition for the Qing 1 member in the Heiyupao Depression.

Based on the combined characteristics of rock type, sedimentary structure, and whole-rock mineral composition, the Qing 1 member in the Heiyupao Depression is now classified into six lithofacies: layered sandstone, layered silty sandstone, massive silty mudstone, massive felsic mudstone, sand-grade laminated felsic shale, and mud-grade laminated felsic shale (Table 2).

Table 2. Heiyupao Depression Qing 1 formation lithofacies classification.

Rock Type	Sedimentary Structure	Mineral Type	Particle Size
Layered sandstone	Layered or massive	Felsic content > 50%	Silt particles > 75%
Layered silty sandstone	Layered	Felsic content > 50%	50% < Silt particles < 75%
Massive silty mudstone	Massive	Felsic content > 50%	20% < Silt particles < 50%
Massive felsic mudstone	Massive	Felsic content > 50%	Silt particles < 20%
Sand-Grade Laminaefelsic shale	Laminations (thickness < 1 cm)	Felsic content > 50%	5% < Proportion of silt laminations < 20%
Mud-Grade laminated felsic shale		Felsic content > 50%	Proportion of silt laminations < 5%

4.1.4. Rare Earth Element Characteristics

Rare earth elements (REEs) are widely distributed in strata and have stable properties, with their concentration differences and elemental combination parameters effectively reflecting the geochemical information preserved in the source area [26]. As shown in Table 3, the total REE content (∑REE) in the black shale of the Heiyupao Sag ranges from 149.15 to 271.52 μg/g, with an average of 215.73 μg/g. The ratio of light REE to heavy REE (LREE/HREE) ranges from 3.46 to 4.57, with an average of 4.03, indicating some differentiation between light and heavy REE. δCe ranges from 0.56 to 0.66, with an average of 0.90, showing a slight negative cerium anomaly. The (La/Yb) N value ranges from 10.44 to 17.33, with an average of 14.30, indicating significant differentiation between light REE lanthanum and heavy REE ytterbium (Table 3). The higher (La/Yb) N value suggests greater enrichment of light REE relative to heavy REE. The Y/Ho ratio ranges from 20.68 to 29.61, remaining relatively stable. The overall REE distribution curve fluctuates widely, which often indicates mixed source material during deposition. Standardized REE data using PAAS (Post-Archean Australian Shale) indicate that the REE distribution pattern of the black shale of the Qing 1 member in the Qingshankou Formation in the study area is relatively flat (Figure 5), with light REE enrichment and heavy REE depletion. This reflects a similar or identical source material during the deposition period and a relatively consistent tectonic background.

Table 3. Rare earth element characteristics of the Qing 1 member shale in the Qing Shankou formation, Heiyupao Sag.

Samples	∑REE/(μg·g^{-1})	LREE/HREE	δCe	δEu	(La/Yb)N	Y/Ho	Ce/La
XY-4-1	174.17	4.57	0.86	0.66	16.59	27.68	1.89
XY-4-2	169.42	3.93	0.91	0.63	12.34	26.35	2.08
XY-4-4	177.94	3.86	0.88	0.62	12.95	27.94	1.98
XY-4-5	224.1	4.36	0.83	0.64	15.89	26.72	1.8
XY-4-7	149.15	3.99	0.98	0.61	10.44	20.68	2.33
XY-4-8	226.82	4.2	0.92	0.62	15.03	27.84	2.06
XY-4-10	271.52	4.17	0.9	0.63	15.8	28.64	1.95
XY-4-11	215.81	3.46	0.93	0.59	11.8	28.09	2.04
XY-4-12	242.59	3.84	0.89	0.63	14.17	28.29	1.95
XY-4-13	271.13	4.39	0.9	0.65	17.33	29.61	1.93
XY-4-14	250.38	3.93	0.87	0.59	14.97	29.42	1.91
Average	4.06	0.90	0.62	14.30	27.39	1.99	4.06

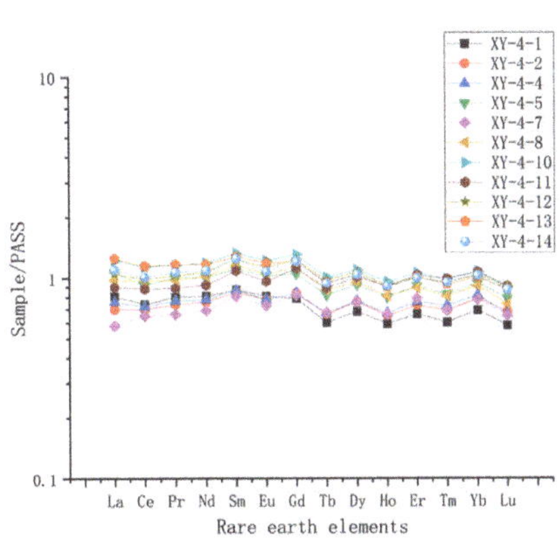

Figure 5. Rare earth elements for the Qing 1 member in the Heiyupao Depression.

4.1.5. Shale TOC Classification

TOC is utilized to indicate the abundance of organic matter, which in turn reflects the characteristics of the organic phase. Based on the relationship diagram of TOC and S1 + S2 in the Qing 1 member shales from well XY4 in the Heiyupao area (Figure 6), and employing the Songliao Basin source rock evaluation methods, organic matter richness is classified as follows: (1) high organic matter (TOC > 2%, S1 + S2 > 11 mg/g), (2) medium organic matter (TOC: 1%–2%, S1 + S2: 6–11 mg/g), and (3) low organic matter (TOC < 1%, S1 + S2 < 6 mg/g) (Table 2). Utilizing 21 core samples from well XY2, a chart illustrating the organic matter richness patterns for different rock types in the study area was developed. The results indicate that within the study area, most sandstones, certain massive mudstones, and massive silty mudstones, along with sand-grade laminated shales that possess a higher sand content, typically exhibit lower organic matter richness. Conversely, the majority of silty mudstones and massive mudstones, along with sand-grade laminated shales, are classified as having medium organic matter, while high organic content is primarily observed in mud-grade laminated shales, with some also found in sand-grade laminated shales.

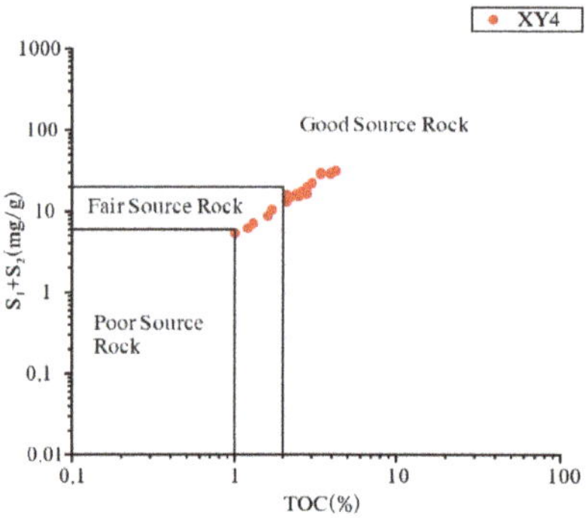

Figure 6. Relationship diagram of TOC and S1 + S2 in fine-grained sedimentary rocks of the Qing 1 member in the Heiyupao Depression.

4.1.6. Shale Organic Matter Types

The hydrocarbon generation potential of source rocks is determined by the hydrogen enrichment level. The intersection diagram of hydrogen index and maximum pyrolysis peak temperature shows that the Qing 1 member source rocks in the Heiyupao Sag exhibit high hydrogen indices (500–700 mg/g), indicating predominantly Type I kerogen with very high hydrocarbon generation potential (Figure 7).

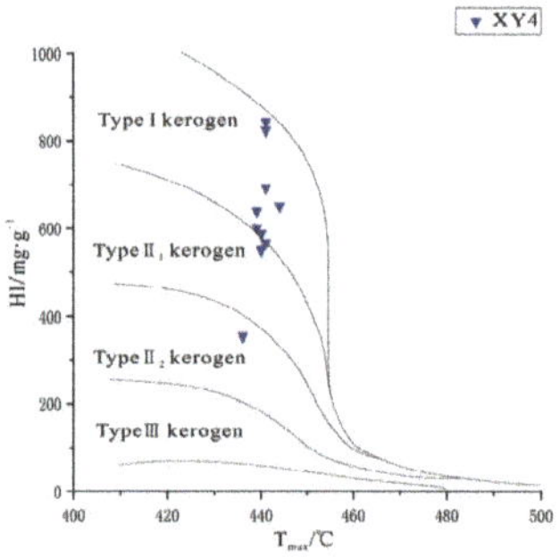

Figure 7. Relationship between hydrogen index and maximum pyrolysis peak temperature in Qing 1 member of the Heiyupao Sag.

4.2. Shale Lithofacies Classification

Based on core observations, thin section identification, and organic geochemical analysis, and taking into account the diversity, frequent changes, developed lamination, and strong heterogeneity of lithofacies types in the Qing 1 member of the study area, a "quaternary" lithofacies classification scheme has been adopted for the shale lithofacies of the Qing 1 member in the Heiyupao Depression of the Binbei area (Figure 8). This classification employs rock types, sedimentary structures, mineral characteristics, and organic matter richness to define five distinct lithofacies: (A) high organic matter mud-grade laminated felsic shale, (B) high organic matter sand-grade laminated felsic shale, (C) medium organic matter sand-grade laminated felsic shale, (D) medium organic matter massive felsic mudstone, and (E) low organic matter sand-grade laminated felsic shale.

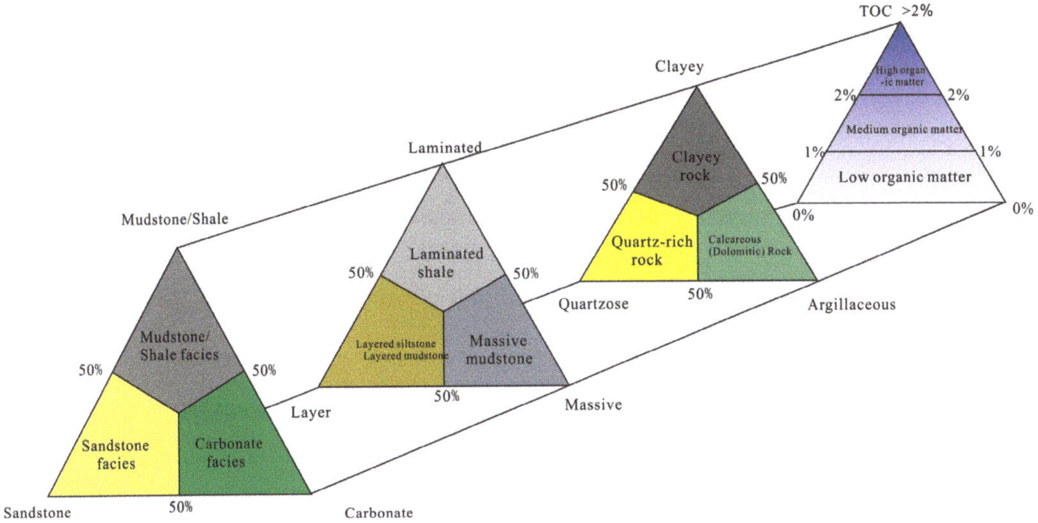

Figure 8. Diagram of the "quaternary" classification scheme for shale lithofacies in the Qing 1 member of the Heiyupao Depression (modified from [28]).

Four-Component Classification Scheme for Shale Lithofacies

Type A lithofacies (high-organic mud-grade laminated felsic shale) exhibit a dark black or dark gray-black coloration, characterized by straight and continuous lamination primarily consisting of mud-grade lamination. The TOC is ≥2%, with certain sections reaching values between 4% and 6%, thereby classifying it as high organic matter (Figure 9a). The lamination features a darker hue with lower sand content and is predominantly composed of felsic, clay, and organic material laminae (Figure 9b). Thick clay and organic laminae overlap with felsic laminae, with isolated quartz grains present but not forming distinct layers. This lithofacies is primarily composed of felsic minerals and contains a large proportion of clay minerals, with a total clay content of 40%, quartz at 31%, potassium feldspar at 5%, and plagioclase at 12%. It also includes minor amounts of pyrite and calcite, each making up 5% (Figure 10a,f).

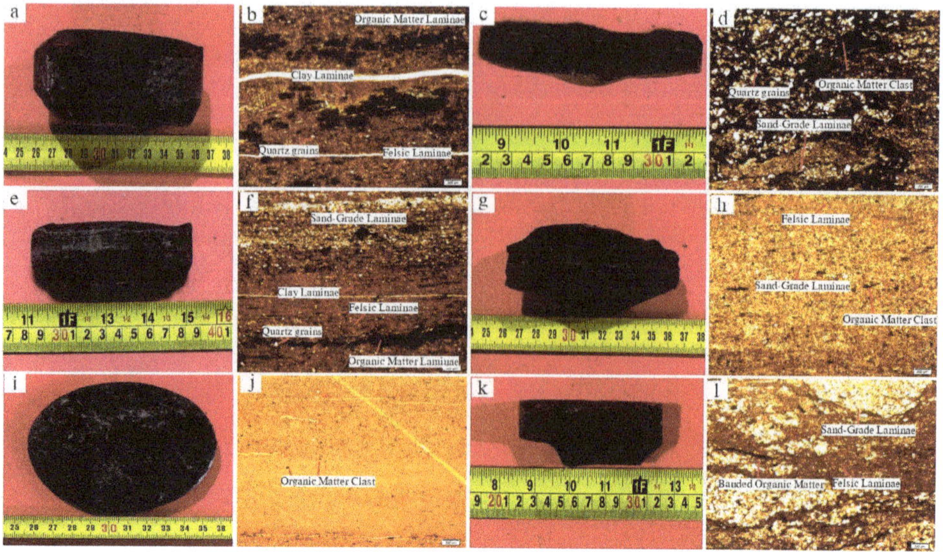

Figure 9. Photographs and microscopic features of shale lithofacies in the Qing 1 member of the Heiyupao Depression. (**a**) XY1 well, 1961.25 m, high-organic mud-grade laminated felsic shale, core photograph; (**b**) XY1 well, 1961.25 m, high-organic mud-grade laminated felsic shale, under plane-polarized light; (**c**) XY3 well, 1933.52 m, high-organic sand-grade laminated felsic shale, core photograph; (**d**) XY3 well, 1933.52 m, high-organic sand-grade laminated felsic shale, under plane-polarized light; (**e**) XY4 well, 1760.68 m, high-organic sand-grade laminated felsic shale, core photograph; (**f**) XY4 well, 1760.68 m, high-organic sand-grade laminated felsic shale, under plane-polarized light; (**g**) XY2 well, 2028.65 m, medium-organic sand-grade laminated felsic shale, core photograph; (**h**) XY2 well, 2028.65 m, medium-organic sand-grade laminated felsic shale, under plane-polarized light; (**i**) XY3 well, 1935.44 m, medium-organic massive felsic mudstone, core photograph; (**j**) XY3 well, 1935.44 m, medium-organic massive felsic mudstone, under plane-polarized light; (**k**) XY2 well, 1760.68 m, low-organic sand-grade laminated felsic shale, core photograph; (**l**) XY2 well, 1760.68 m, low-organic sand-grade laminated felsic shale, under plane-polarized light.

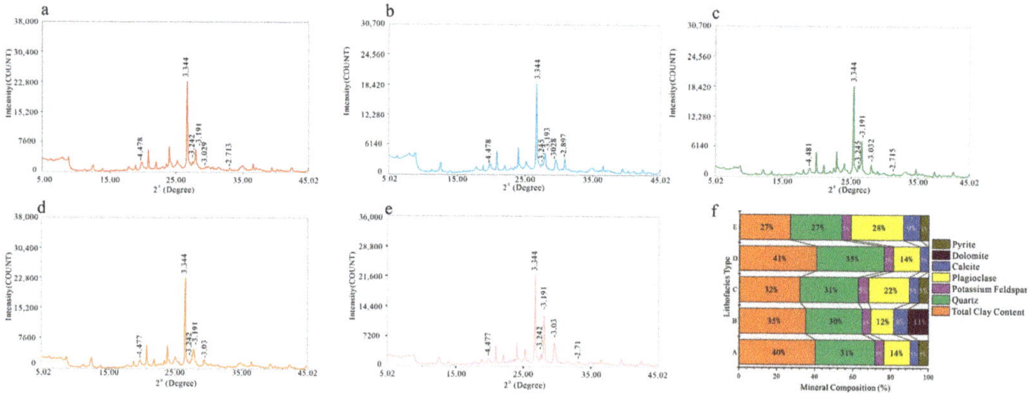

Figure 10. Whole-rock mineral composition across different lithofacies. (**a**) XRD whole-rock mineral spectrum of high-organic mud-grade laminated felsic shale; (**b**) XRD whole-rock mineral spectrum of high-organic sand-grade laminated felsic shale; (**c**) XRD whole-rock mineral spectrum of medium-organic sand-grade laminated felsic shale; (**d**) XRD whole-rock mineral spectrum of medium-organic massive felsic mudstone; (**e**) XRD whole-rock mineral spectrum of low-organic sand-grade laminated felsic shale; (**f**) XRD whole-rock mineral composition content diagram of A-E lithofacies.

Type B lithofacies (high-organic sand-grade laminated felsic shale) are dark gray-black in color (Figure 9c) and display both straight and continuous as well as discontinuous lamination, predominantly consisting of sand-grade laminae (Figure 9d). These sand laminae are interspersed with dark and bright laminae (Figure 9e). The TOC is $\geq$2%, with few values surpassing this threshold; however, it remains classified as high organic matter overall. These lithofacies primarily comprise sand-grade laminae, felsic laminae, clay laminae, and interspersed organic laminae (Figure 9f). Organic material is observed as layers or blocks situated between sand-grade and felsic laminae, with moderate quantities of quartz grains visible, forming sand-grade laminae. This lithofacies is mainly composed of felsic minerals. Compared to the A-type lithofacies, it has a reduced clay mineral content of 35%, with quartz at 30%, potassium feldspar at 5%, and plagioclase at 12%. The contents of calcite and dolomite have increased, with calcite comprising 8% and dolomite 11% (Figure 10b,f).

Type C lithofacies (medium-organic sand-grade laminated felsic shale) are gray-black to light gray-black in color (Figure 9g) and display straight, continuous, and wavy laminae, mainly consisting of sand-grade laminae. The TOC is between 1% and 2%, with most values between 1.5–2%. This is classified as medium organic matter. It predominantly consists of sand-grade and felsic laminae (Figure 9h). Organic matter appears as blocks and smaller organic laminae, with abundant quartz and feldspar particles forming many sand-grade laminae. These lithofacies remain primarily composed of felsic minerals, with a lower clay content compared to the AB-type lithofacies. The clay mineral content is 32%, while the proportion of felsic minerals has increased. Quartz makes up 31%, potassium feldspar 5%, and plagioclase 22%, with pyrite and calcite each comprising 5% (Figure 10c,f).

Type D lithofacies (medium-organic massive felsic mudstone) are light gray-black in color (Figure 9i) and are primarily composed of felsic minerals. The TOC is between 1 and 2%, classifying it as medium organic matter. Laminae are weakly developed, and the overall structure is relatively homogeneous (Figure 9j). Organic matter manifests as bands or blocks, with large quartz particles observable in some sections. The D-type lithofacies have a significant amount of clay minerals, yet their quartz content is the highest among the five lithofacies. The total clay content is 41%, quartz content is 36%, potassium feldspar is 5%, plagioclase is 14%, and calcite is 5% (Figure 10d,f).

Type E lithofacies (low-organic sand-grade laminated felsic shale) are characterized by a color range from light gray-black to dark gray (Figure 9k). The laminae are predominantly straight, continuous, and wavy, primarily consisting of sand-grade laminae. The TOC is below 1%, which classifies it as low organic matter. The silt laminae are more pronounced, while felsic laminae are less developed, exhibiting almost no clay laminae. Organic matter is present in the form of bands or blocks within the sand-grade laminae (Figure 9l). In the E-type lithofacies, the felsic content is the highest, with quartz accounting for 27%, potassium feldspar 5%, plagioclase 28%, calcite 9%, and pyrite 5% (Figure 10e,f).

The overall lithofacies of the Heiyupao Depression in the Binbei area display considerable variation. Based on rock type, structure, mineral composition, and organic matter richness, Type A lithofacies (high-organic mud-grade laminated felsic shale) and Type B lithofacies (high-organic sand-grade laminated felsic shale) are the predominant lithofacies. Their primary characteristics include high organic matter content, with mud-grade laminae being dominant, and fewer sand-grade laminae, where organic matter primarily occurs in laminated and discontinuous band forms.

5. Discussion

5.1. Reservoir Characteristics of Different Shale Lithofacies

5.1.1. Pore Types and Characteristics

After argon ion polishing of representative samples from various shale lithofacies and subsequent SEM analysis, it was determined that the pore types in the Qingshankou Formation of the Heiyupao Depression predominantly include organic matter pores, intergranular pores, intragranular pores, and microfractures. Notable variations in pore types exist among different lithofacies. The high-organic mud-grade laminated felsic shale and high-organic sand-grade laminated felsic shale predominantly feature intergranular pores and organic matter pores, with a limited presence of microfractures (Figure 11a–d). The medium-organic sand-grade laminated felsic shale and medium-organic massive felsic mudstone primarily exhibit intergranular pores, intragranular pores, and microfractures, with a minor occurrence of organic matter pores (Figure 11e–g). The low-organic sand-grade laminated felsic shale mainly displays intergranular pores, intragranular pores, and microfractures, while organic matter pores are poorly developed (Figure 11h,i).

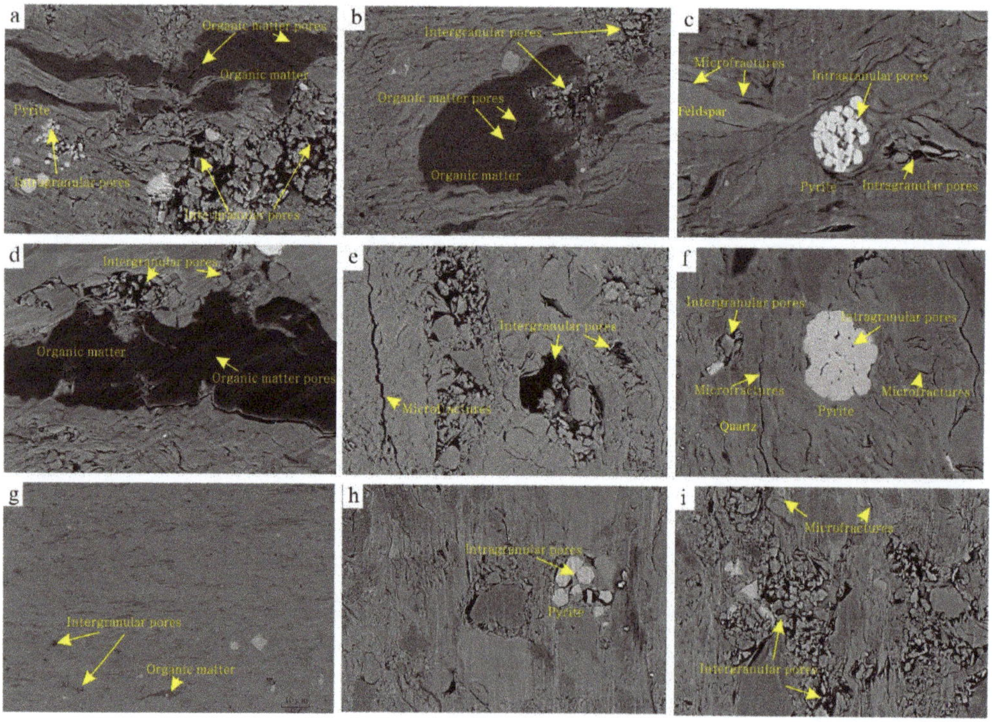

Figure 11. Pore types in the Qing 1 member of the Heiyupao Depression. (**a**) XY1 well, 1961.25 m, high-organic mud-grade laminated felsic shale. (**b**) XY1 well, 1998.98 m, high-organic mud-grade laminated felsic shale. (**c**) XY4 well, 1760.68 m, high-organic sand-grade laminated felsic shale. (**d**) XY4 well, 1762.25 m, high-organic sand-grade laminated felsic shale. (**e**) XY2 well, 2028.65 m, medium-organic sand-grade laminated felsic shale. (**f**) XY2 well, 2038.58 m, medium-organic sand-grade laminated felsic shale. (**g**) XY3 well, 1935.44 m, medium-organic massive felsic mudstone. (**h**) XY2 well, 1760.68 m, low-organic sand-grade laminated felsic shale. (**i**) XY2 well, 1760.68 m, low-organic sand-grade laminated felsic shale.

Organic matter pores serve as the primary storage space for shale gas, predominantly developing within organic matter. These pores are formed through thermal degradation during burial and exhibit elliptical, irregular, and banded shapes (Figure 11a,b). The connectivity of these pores varies to some extent.

Typical intergranular pores predominantly develop in lithofacies with a high content of brittle minerals such as quartz and feldspar. These pores are largely influenced by diagenetic processes. Some primary residual intergranular pores and secondary pores combine to create composite pores, while acidic fluids can completely dissolve grains, leading to the formation of dissolution pores. The pore radii range from tens of nanometers to several micrometers (Figure 11d–f).

Intragranular pores primarily form within the particles of clay mineral flakes or pyrite crystals, along with small amounts of nanoscale intragranular dissolution pores found in feldspar minerals. These pores are generally in an unfilled or semi-filled state. Intragranular pores also encompass those found in framboidal pyrite aggregates or formed through dissolution within grains. The intragranular pores within framboidal pyrite aggregates typically exhibit a polygonal shape, characterized by straight edges and relatively large pore diameters (Figure 11f).

Microfractures primarily develop around brittle minerals such as quartz and feldspar (Figure 11c,e,f), which are susceptible to cracking under stress. Furthermore, hydrocarbons generated during the thermal maturation of organic matter can induce rock expansion, resulting in the formation of fractures. The width of microfractures typically ranges from nanometers to micrometers, while their lengths can vary from a few micrometers to several millimeters.

5.1.2. 3D Pore Structure Characteristics

In contrast, the medium-organic sand-grade laminated felsic shale exhibits a greater dependence on microfracture development. Micron-scale and nanoscale pores (Figure 12c, red, green, and yellow regions) are frequently associated with microfractures (Figure 12d, blue region). The maximum pore throat radius for these facies is 50 nm, while the average pore throat radius ranges from 6 to 50 nm, with an average of approximately 25 nm. The porosity distribution ranges from 3.2% to 7.4%, with an average of about 5.1%.

The low-organic sand-grade laminated felsic shale samples display well-developed micron-scale and nanoscale pores, with larger internal pores visible (Figure 11e, red regions) and strong pore connectivity (Figure 12f, blue regions). The primary reservoir space is formed by interconnected micron-scale and nanoscale intergranular pores in felsic minerals, semi-filled bedding seams, and microfractures (Figure 12e,f). The maximum pore throat radius ranges from 70 to 1765 nm, with an average of 278 nm, while the average pore throat radius ranges from 12 to 117 nm, with an average of 28 nm. The porosity distribution ranges from 4.3% to 12%, with an average of approximately 9%.

In this study, the maximum resolution of the micro-CT used reached 1.4 μm. By integrating the results of three-dimensional pore characterization with two-dimensional electron microscopy, it was observed that in the high-organic sand-grade laminated felsic shale facies, electron microscope images reveal scattered pores, primarily intergranular and intragranular, with poor pore connectivity at the micron scale (Figure 12a,b). The maximum pore throat radius for these facies ranges from 32 to 55 nm, with an average of approximately 46 nm. The average pore throat radius ranges from 6 to 13 nm, with an average of 10 nm. The porosity distribution spans from 1% to 5%, with an average of about 2.5%.

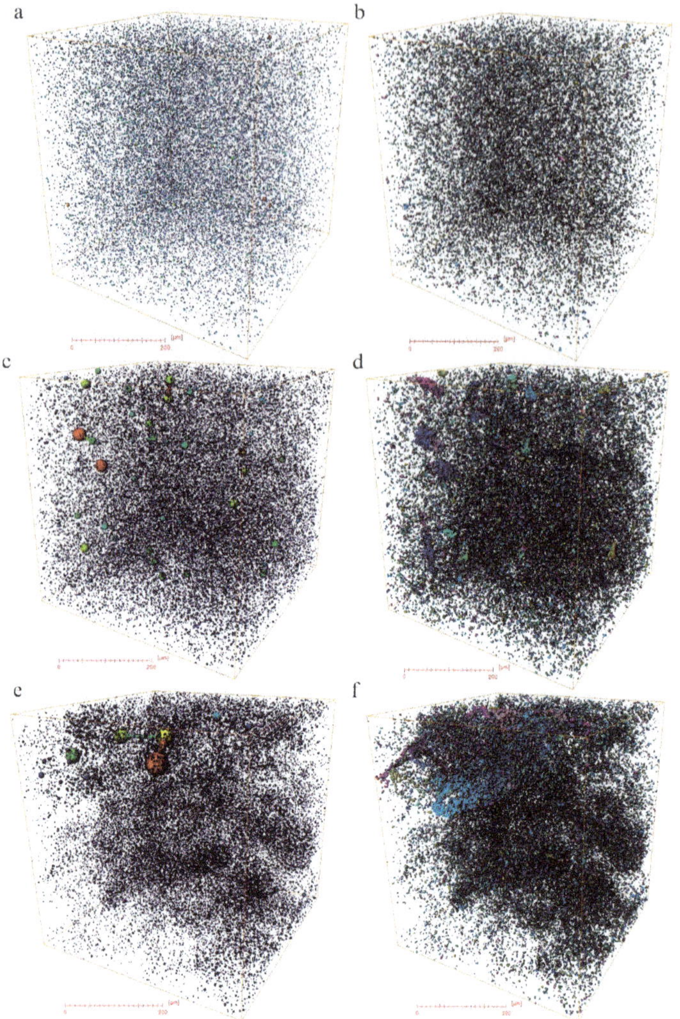

Figure 12. Pore distribution characteristics of different lithofacies (based on micro-CT analysis; regions of the same color represent interconnected pores). (**a**) XY4 well at 1760.68 m, high-organic sand-grade laminated felsic shale facies, pore network model, micro-CT. (**b**) XY4 well at 1760.68 m, high-organic sand-grade laminated felsic, Pore connectivity model, micro-CT. (**c**) XY4 well at 2028.65 m, medium-organic sand-grade laminated felsic, shalepore network model, micro-CT. (**d**) XY2 well at 2028.65 m, medium-organic sand-grade laminated felsic shale, Pore connectivity model, micro-CT. (**e**) XY2 well at 1760.68 m, low-organic sand-grade laminated felsic shale, pore network model, micro-CT. (**f**) XY2 well at 1760.68 m, low-organic sand-grade laminated felsic shale, Pore connectivity model, micro-CT.

This study highlights both similarities and distinct differences from previous research regarding the relationship between lithofacies classification, reservoir performance, and TOC [29]. Both studies establish a strong link between lithofacies types and sedimentary structures, mineral composition, and organic matter abundance, emphasizing their influence on reservoir performance. There is agreement that the relationship between TOC and pore features is complex and nonlinear, with laminar structures and mineral composition having a significant impact on pore development and hydrocarbon flow paths.

This research demonstrates that high-TOC lithofacies (>2%) generally enhance organic pore development, while low-TOC lithofacies (<1%) maintain good fluidity due to the presence of sand-grade intergranular pores. Conversely, previous studies have found that high-TOC lithofacies, despite having abundant micropores, limit fluid mobility, while medium-TOC lithofacies show better fluidity [30]. A notable distinction between the studies is in the sand-grade laminae content of high-TOC lithofacies, which can be attributed to variations in regional geological settings. This study focuses on the geologically complex northern Binbei area of the Heiyupao Depression, where lamination is closely linked to paleoenvironmental conditions [30,31]. Compared to Gulong shale oil [17], the maturity of this region is lower, and the shale lithofacies mineral composition is more homogeneous, predominantly felsic shale. By contrast, the southern Songliao Basin, characterized by a stable sedimentary environment with abundant fine-grained deposits and mature organic source rocks, emphasizes the significance of micro-level analysis [31]. Compared to North America's stable marine shale systems, this study underscores the role of unique sedimentary structures and organic matter distribution in complex continental basins and their influence on hydrocarbon accumulation. Future research should integrate macro and micro analyses to optimize lithofacies classification, investigate the relationship between TOC and pore types in depth, and apply multi-scale analysis to comprehensively reveal reservoir characteristics in continental basins, providing scientific support for precise oil and gas exploration and development.

6. Conclusions

(1) The overall rock types of the Qing 1 member in the Heiyupao Depression can be classified into siltstone, sandstone, mudstone, shale, and carbonate. Notably, mudstone and shale are predominantly composed of felsic minerals. These rocks exhibit medium to high organic matter content and feature well-developed laminae, primarily consisting of mud-grade and sand-grade laminae.

(2) Using a newly refined lithofacies classification system, five distinct lithofacies have been identified: (A) high-organic mud-grade laminated felsic shale, (B) high-organic sand-grade laminated felsic shale, (C) medium-organic sand-grade laminated felsic shale, (D) medium-organic massive felsic mudstone, and (E) low-organic sand-grade laminated felsic shale.

(3) Micro-CT and scanning electron microscope analyses indicate that various lithofacies display distinct pore structure characteristics. High-organic sand-grade laminated felsic shale is characterized by intergranular and organic matter pores, exhibiting low porosity and poor connectivity. Medium-organic sand-grade laminated felsic shale relies on the development of microfractures, demonstrating moderate porosity and good connectivity. In contrast, low-organic sand-grade laminated felsic shale features high porosity, well-developed micron-scale and nanoscale pores, and strong connectivity.

Author Contributions: Conceptualization, H.W.; Data curation, P.L. (Pengfei Liu); Formal analysis, X.G. (Xun Ge) and X.G. (Xiaotong Ge); Funding acquisition, Y.L.; Investigation, P.L. (Pei Li); Methodology, J.Y.; Project administration, Y.L.; Resources, Y.L.; Visualization, B.H.; Writing—original draft, Y.L., W.L. and B.H.; Writing—review and editing, B.H., X.Y. and H.W. All authors have read and agreed to the published version of the manuscript.

Funding: This work was financially supported by Sinopec Science and Technology Department Project (Grant No. P23092).

Data Availability Statement: The author does not have permission to share data due to internal policy.

Conflicts of Interest: The authors declare no conflicts of interest.

References

1. Li, M.W.; Ma, X.X.; Jiang, Q.G.; Li, Z.M.; Pang, X.Q.; Zhang, C.T. Enlightenment from formation conditions and enrichment characteristics of marine shale oil in North America. *Pet. Geol. Recovery Effic.* **2019**, *26*, 13–28.
2. Jia, C.Z.; Jiang, L.; Zhao, W. The shale revolution and basic geological theory problems of shale and tight oil and gas. *Pet. Sci. Bull.* **2023**, *8*, 695–706.
3. Sun, L.D.; Liu, H.; Zhu, R.K.; Cui, B.W.; Lei, Z.D.; Meng, S.W.; Tang, J.Z. Ten noteworthy issues on shale oil revolution in China. *Acta Pet. Sin.* **2023**, *44*, 2007–2019.
4. Shan, J.F.; Han, X.; Mao, J.L. Gas geochemistry and deep gas exploration potential in Eastern Liaohe Depression. *Spec. Oil Gas Reserv.* **2011**, *18*, 7–10.
5. Zou, C.N.; Zhang, G.S.; Yang, Z.; Tao, S.Z.; Hou, L.H.; Zhu, R.K.; Yuan, X.J.; Ran, Q.Q.; Li, D.H.; Wang, Z.P. Geological concepts, characteristics, resource potential and key techniques of unconventional hydrocarbon: On unconventional petroleum geology. *Pet. Explor. Dev.* **2013**, *40*, 385–399, 454. [CrossRef]
6. Yang, Z.; Hou, L.H.; Tao, S.Z.; Cui, J.W.; Wu, S.T.; Lin, S.H.; Pan, S.Q. Formation conditions and "sweetspot" evaluation of tight oil and shale oil. *Pet. Explor. Dev.* **2015**, *42*, 555–565.
7. Zou, C.N.; Yang, Z.; Cui, J.W.; Zhu, R.K.; Hou, L.H.; Tao, S.Z.; Yuan, X.J.; Wu, S.T.; Lin, S.H.; Wang, L.; et al. Formation mechanism, geological characteristics and development strategy of nonmarine shale oil in China. *Pet. Explor. Dev.* **2013**, *40*, 14–26. [CrossRef]
8. Zou, C.N.; Zhu, R.K.; Bai, B.; Yang, Z.; Hou, L.H.; Zha, M.; Fu, J.H.; Shao, Y.; Liu, K.Y.; Cao, H.; et al. Significance, Geologic Characteristics, Resource Potential and Future Challenges of Tight Oil and Shale Oil. *Bull. Mineral. Petrol. Geochem.* **2015**, *34*, 3–17.
9. Zhao, W.Z.; Hu, S.Y.; Hou, L.H.; Yang, T.; Li, X.; Guo, B.C.; Yang, Z. Types and resource potential of continental shale oil in China and its boundary with tight oil. *Pet. Explor. Dev.* **2020**, *47*, 1–11. [CrossRef]
10. Gao, J.B.; Wu, J.F.; Zhang, S.H.; Pang, X.Q.; Chen, D.X. Sedimentary characteristics and petroleum geological conditions analysis of the Qingshankou Formation in the Heiyupao Depression, Northern Binbei area of the Songliao Basin. *West-China Explor. Eng.* **2012**, *24*, 87–90, 93.
11. He, Y.M. *Study on Sedimentary Facies of Low Cretaceous in Binbei, Songliao Basin*; Xi'an Shiyou University: Xi'an, China, 2014.
12. Jiang, F.J.; Jiang, Z.X.; Tian, H.F.; Wang, X.D. Quantitative Study of The Mid-shallow Hydrocarbon Accumulation Systems in Binbei Area, Songliao Basin. *J. Xi'an Shiyou Univ. (Nat. Sci. Ed.)* **2005**, *20*, 6, 13–18.
13. Loucks, R.G.; Ruppel, S.C. Mississippian Barnett Shale: Lithofacies and depositional setting of adeep-water shale-gas succession in the Fort Worth Basin, Texas. *AAPG Bull.* **2007**, *91*, 579–601. [CrossRef]
14. Trabucho-Alexandre, J.; Dirkx, R.; Veld, H.; Klaver, G.; de Boer, P.L. Toarcian black shales in the Dutch Central Graben: Record of energetic, variable depositional conditions during an oceanic anoxic event. *J. Sediment. Res.* **2012**, *82*, 104–120. [CrossRef]
15. Wang, L.; Zeng, W.T.; Xia, X.M.; Zhou, H.Y.; Bi, H.; Shang, F.Z.; Xue, X. Study on lithofacies types and sedimentary environment of black shale of Qingshankou Formation in Qijia-Gulong Depression, Songliao Basin. *Nat. Gas Geosci.* **2019**, *30*, 1125–1133.
16. Zhang, J.Y.; Bai, Y.F.; Lv, J.C.; Zhao, Y.; Liu, X.; Li, D. Types and Characteristics of Shale Oil Lithofacies in Qing 1 Member of Gulong Sag. In Proceedings of the Xi'an International Field Exploration and Development Conference, Xi'an, China, 16–18 October 2019.
17. Lin, T.F.; Fu, X.L.; Ai, X.; Bai, Y.F. Lithofacies types and cycle patterns of shale layers of Qingshankou Formation in Gulong Sag, Songliao Basin. *World Geol.* **2024**, *43*, 378–389.
18. Ying, D.B. *Evaluation on Efficient Source Rock of Qingshankou Formation in Northern Songliao Basin, China*; Chengdu University of Technology: Chengdu, China, 2012.
19. Guo, Y.C.; Su, M.; He, Y.L.; Peng, W. Characteristics of sequence stratigraphy and sedimentary system of Yaojia Formation in Heiyupao Depression, Songliao Basin. *Pet. Geol. Recovery Effic.* **2010**, *17*, 29–32, 113.
20. Yang, Y.B. *Study on Sequence Stratigraphic Framework of Denglouku to the Second Member of Quantou Formation in the Lindian-Hei Yupao Area*; Northeast Petroleum University: Daqing, China, 2012.
21. Ma, Z.Z.; Pang, X.Q. Predicion on Favorable Area for Exploration of Lithologic Reservoirs in Binbei, Songliao Basin. *Nat. Gas Ind.* **2007**, *27*, 44–47, 133.
22. Shi, H. *Assessment on Source Rock of Hydrocarbon in Binbei Area of Songliao Basin*; Northeast Petroleum University: Daqing, China, 2008.
23. Tang, H.F. *Strcutural Styles and Trap's Prediction in Binbei Area of Songliao Basin*; Jilin University: Jilin, China, 2004.
24. Yu, L.M. *Comprehensive Evaluation of Sweet Spots in Interlayer Shale Oil of the First Member of Qingshankou Formation in Daqingzijing Area of the Songliao Basin*; Northeast Petroleum University: Daqing, China, 2023.
25. Xu, M.H.; Wang, F.; Tian, J.C.; Ren, Z.C.; Meng, H.; Yu, W.; Wang, J.; Wu, J.Y.; Xiao, Y.X. Classification of Lacustrine Organic-Rich Mud Shale Petrography and the Depositional Environment: An Example from the Chang 73 Sub-member in the Ordos Basin. *Acta Sedimentol. Sin.* **2023**, 1–24. [CrossRef]
26. *SY/T 5163-2018*; X-Ray Diffraction Analysis Method of Clay Minerals and Common Non-Clay Minerals in Sedimentary Rocks. National Standards of People's Republic of China: Beijing, China, 2018.

27. *GB/T 18602-2012*; Rock Pyrolysis Analysis. National Standards of People's Republic of China: Beijing, China, 2012.
28. Shi, X.; Liu, J.; Zhu, Y.; Xu, L.; Yang, Y.; Luo, C.; Li, Y.; Zhong, K.; Yang, X.; Wu, Q.; et al. Effects of the Sedimentary Environment on Organic-Rich Shale in the Intracratonic Sag of the Sichuan Basin, China. *Appl. Sci.* **2024**, *14*, 8594. [CrossRef]
29. Lin, L.M.; Liu, S.; Xu, Y.; Li, F. Lithofacies and Pore Structure of Fine-grained Sedimentary Rocks of Qing-1 Member of Cretaceous in the Southern Songliao Basin. *Acta Geol. Sin. Engl. Ed.* **2021**, *95*, 570–584. [CrossRef]
30. Cai, Y.; Zhu, R.; Luo, Z.; Wu, S.; Zhang, T.; Liu, C.; Zhang, J.; Wang, Y.; Meng, S.; Wang, H.; et al. Lithofacies and Source Rock Quality of Organic-Rich Shales in the Cretaceous Qingshankou Formation, Songliao Basin, NE China. *Minerals* **2022**, *12*, 465. [CrossRef]
31. Liu, B.; Shi, J.X.; Fu, X.F.; Lyu, Y.F.; Sun, X.D.; Gong, L.; Bai, Y.F. Petrological characteristics and shale oil enrichment of lacustrine fine-grained sedimentary system: A case study of organic-rich shale in first member of Cretaceous Qingshankou Formation in Gulong Sag, Songliao Basin, NE China. *Pet. Explor. Dev.* **2018**, *45*, 884–894. [CrossRef]

minerals

MDPI

Article

The Source and Significance of Silicon in the Late Permian Dalong Formation, Northeastern Sichuan Basin

Xiaotong Ge [1,2,3,*], Xun Ge [1,2,3], Daizhao Chen [4], Yali Liu [1,2,3], Ruyue Wang [1,2,3] and Min Li [1,2,3]

1 State Key Laboratory of Shale Oil and Gas Enrichment Mechanisms and Efficient Development, Beijing 102206, China; wangruyue.syky@sinopec.com (R.W.)
2 Sinopec Key Laboratory of Shale Oil/Gas Exploration and Production Technology, Beijing 102206, China
3 Petroleum Exploration and Production Research Institute, SINOPEC, Beijing 102206, China
4 Key Laboratory of Cenozoic Geology and Environment, Institute of Geology and Geophysics, Chinese Academy of Sciences, Beijing 100029, China
* Correspondence: gexiaotong.syky@sinopec.com

Abstract: The Late Permian was a critical interval in geological history, during which dramatic changes occurred in the Earth's surface system, and a set of black rock series rich in organic matter and silicon, the Dalong Formation, was deposited in the northeastern Sichuan Basin. We conducted a detailed sedimentological and petrological investigation integrated with (major and trace) element contents in the deep-water sequence of the Xibeixiang and Jianfeng sections. It demonstrates the source of silicon, tectonic background, and sedimentary environment of the Dalong Formation, and explores the influence of hydrothermal activities on organic matter enrichment. The results show that the upper part of the Dalong Formation contained more radiolarians in the Xibeixiang section compared to the Jianfeng section. Hydrothermal proxies such as Eu/Eu*, Al-Fe-Mn diagram, Al/(Al + Fe + Mn), and Lu_N/La_N suggest a biotic origin for the chert in the Dalong Formation in the Xibeixiang and Jianfeng sections, while the Xibeixiang section was slightly affected by hydrothermal activities. The La-Th-Sc diagram and the La/Sc and Ti/Zr crossplots point to a continental island arc and active continental margin origins for the Xibeixiang and Jianfeng sections. Combined with previous research, the silicon of the Dalong Formation in the northeastern Sichuan Basin is mainly derived from biological sources. The Xibeixiang section was affected by a small amount of hydrothermal fluid due to its proximity to the Paleo-Tethys Ocean and continental island arcs. Furthermore, the enrichment of organic matter was predominantly driven by high productivity, with minimal impact from hydrothermal activities. These insights hold significant research value and practical implications for shale gas exploration in the Sichuan Basin.

Keywords: northeastern Sichuan Basin; Dalong Formation; source of Silicon; biotic origin; hydrothermal activity

1. Introduction

The Upper Permian deep-water succession in the northeast Sichuan Basin is characterized by organic-rich strata, serving as effective source rocks that supply substantial hydrocarbons to nearby gas fields [1]. This depositional succession captures extensive information about the oceanic chemical and physical changes during the Late Permian, providing valuable insights into the silicon source of the Dalong Formation. The formation is stratified from bottom to top with organic-rich marly limestone, shale, siliceous rocks, and siliceous limestone, spread across troughs. The extensive organic-rich siliceous rocks serve as source materials for shale gas and mineral resources [2]. Understanding the silicon source and sedimentary environment of these rocks is crucial [3], as it governs the formation and accumulation of fossil energy, presenting significant research and economic implications.

Previous studies have reported that the siliceous source of the Dalong Formation in the Yangtze block can be classified into biogenic and hydrothermal origins. In the Lower Yangtze block, the base of the Dalong Formation shows the signature of both origins, with the middle section predominantly biogenic and the top primarily hydrothermal [4]. The bedded chert in the Middle Yangtze block also displays a combination of these origins [5]. However, the silicon source in the Upper Yangtze block remains undetermined. Notably, the Shangsi section and Leba 1 borehole interval are largely characterized by biochemical deposits with some hydrothermal contributions [6,7], while the cherts of the Dalong Formation in the Changjianggou section are believed to have a mixed source of biological origin and volcanic activity [8]. To further elucidate the silicon sources in the northeastern Sichuan region, this study focuses on an organic-rich intrashelf basinal succession of the Upper Permian at the Xibeixiang and Jianfeng sections in Guangyuan, Sichuan Province. Comprehensive analyses including the determination of major and trace elements were conducted. These investigations aimed to clarify the siliceous source, structural background, and sedimentary environment of the organic-rich black rocks within the Dalong Formation. These findings enhance the understanding of the mechanisms behind organic matter enrichment and evaluate the potential of these strata as hydrocarbon source rocks.

2. Geological Background

During the Lopingian (Late Permian), the South China Block, comprising the Yangtze and Cathaysia subblocks, was geographically isolated and migrated to the subtropical latitudes (approximately 23–25° N) of the Equatorial Warm Water Province (EWWP), locating at the eastern margin of the Paleo-Tethys Ocean (Figure 1) [9]. In this period, carbonate platforms were prevalent over the Yangtze subblock, especially in the upper Yangtze region, where several intraplatform or intrashelf basins developed (Figure 1). In terms of ancient geography, the Xibeixiang section was situated close to the opening of the Kaijiang–Liangping intraplatform basin, connected to the Qinling Ocean, also known as the Paleo-Tethys. The Jianfeng section was located at the southwestern edge of the Kaijiang-Liangping trough, approximately 35 km northeast of the Shangsi section, with a similar depositional environment and was once considered as an auxiliary candidate for GSSP (Global Stratotype Section and Point) for the P-T boundary [10].

The Xibeixiang section is classified as a deep-water intrashelf facies, where the Dalong Formation, with a total thickness of 36.74 m, rests unconformably on the Wuchiaping Formation (Figure 2a). The formation is distinguished by abundant deep-water fauna, including ammonoids and radiolarians, affirming its relevance to a deep-water intrashelf basinal environment. Stratigraphically, the Dalong Formation is divided into three main parts. The lower part consists of thin-bedded siliceous mudstone and thin- to medium-bedded lime mudstone/wackestone, interspersed with two volcanic ash layers, indicative of the rapid deepening of water depth. The middle part is predominated by organic-rich marly limestone and intercalated with black shales or organic-rich limestones (Figure 2b), reflecting the trend of progressive transgression. The upper part features bedded radiolarian cherts, intercalated with shales and a limestone lens (Figure 2d), transitioning upwards into a limestone succession. This upper limestone succession is characterized by thin- to medium-bedded lime mudstone (Figure 2e) with nektonic ammonite fossils (Figure 2f) and signs of bioturbation, further interspersed with seven layers of volcanic ash (Figure 2g), illustrating complex depositional dynamics.

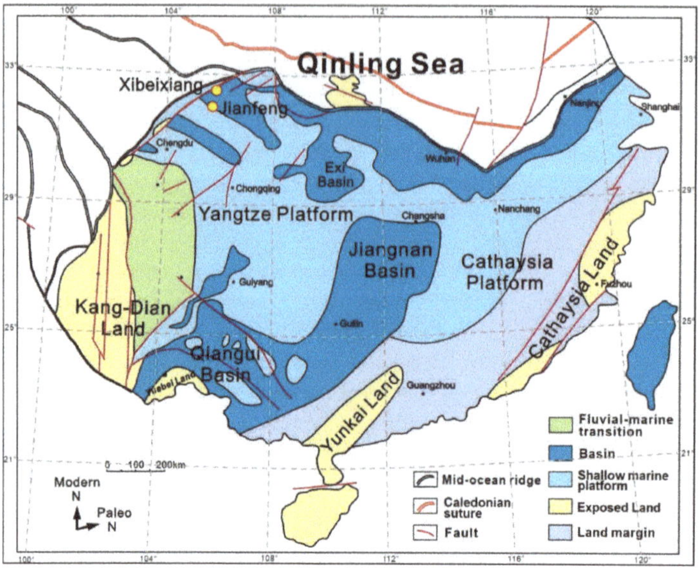

Figure 1. Paleogeographic map of South China during the deposition period of the Dalong Formation, showing the location of the Xibeixiang and Jianfeng sections (modified from [11]).

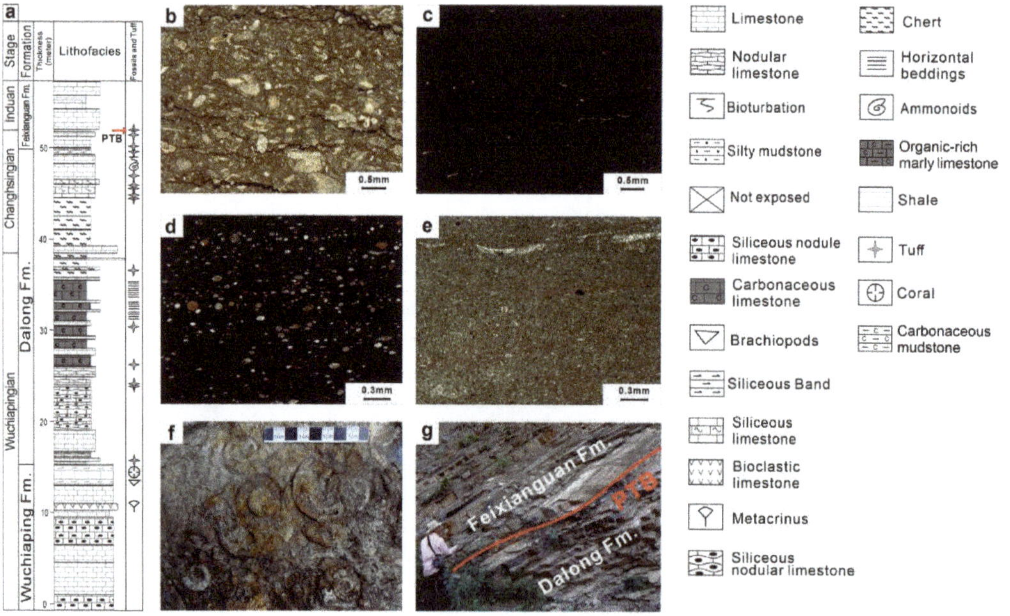

Figure 2. (**a**) The lithological column of the Xibeixiang section. Photomicrographs under plane-polarized light show (**b**) ampelitic limestone in the middle part of the Dalong Formation, (**c**) bedded chert bearing large numbers of radiolarians in the upper part of the Dalong Formation, and (**e**) lime mudstone in the uppermost Dalong Formation. (**d**) The limestone lens of the upper part of the Dalong Formation. (**f**) Ammonite fossils in the upper part of the Dalong Formation. (**g**) The photomacrograph of the Permian-Triassic boundary (PTB). A standing person for scale (1.65 m).

The Jianfeng section exhibits a conformable contact between the Dalong and Wuchiaping formations (Figure 3a). In the lower part of the Dalong Formation, the dominant lithologies are medium-bedded and siliceous limestones. The middle part is characterized by black shale and marly limestone (Figure 3b) interbedded with siliceous rocks and is notable for its abundance of ostracod fossils and an intercalated layer of volcanic ash (Figure 3c), which may have originated from the Emeishan Large Igneous Province [12]. The upper section is characterized by carbonaceous mudstones alternating with radiolarian siliceous rocks (Figure 3d), where the radiolarians are arranged in parallel. This section is capped by two layers of volcanic ash. Above this, the Changhsing Formation is conformably in contact with the Dalong Formation, indicating a continuous depositional sequence.

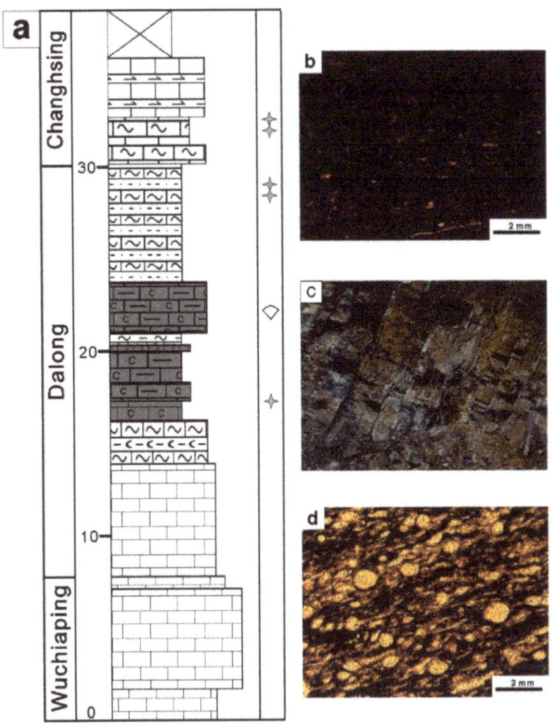

Figure 3. (**a**) Lithological column of the Jianfeng section. (**b**) Photomicrographs under plane-polarized light show carbonaceous limestone in the middle part of the Dalong Formation. (**c**) Volcanic ash in the middle Dalong Formation. (**d**) Photomicrographs under plane-polarized light show bedded chert bearing large numbers of radiolarians in the upper part of the Dalong Formation. See Figure 2 for legends.

3. Materials and Methods

Specimens were extracted from the Xibeixiang and Jianfeng stratigraphic sections, with meticulous attention paid to eliminating any weathered exteriors, as well as any post-depositional veins or voids, and discernible nodules or streaks. These specimens underwent fragmentation using a hammer, followed by a refinement process that employed the Rocklabs BTRM device to achieve a powdery consistency with a granularity of below 200 mesh. The resulting homogenized powders were then collected for the analyses.

For the determination of major elements, the lost-on-ignition (LOI) value was derived from the weight loss of the samples after they were subjected to combustion, a figure that was later utilized to adjust the measurements of elemental concentrations. Each sample was weighed to a precision of 50 milligrams and then dissolved in an alkaline solution. The solution was acidified with nitric acid before being analyzed using an inductively coupled plasma optical emission spectrometer (ICP-OES) at the China University of Geosciences in Beijing. The precision of these analytical results was assured by comparing them against standard reference materials AGV-2, GSR-1, and GSR-5, with an accuracy of over 3%.

For the analysis of trace elements, approximately 100 milligrams of the finely ground samples were subjected to drying at a temperature of 980 °C for a duration of one hour. Subsequently, 40 milligrams of each sample were immersed in a mixture consisting of 1.5 mL of concentrated HNO_3 and 1.5 mL of HF, followed by heating at a temperature of 195 °C for 48 h within sealed Teflon containers. An additional 1 milliliter of HNO_3 was added to each sample, which was then evaporated almost to dryness. The samples were subsequently reconstituted with 2 mL of HNO_3 and 2 mL of Milli-Q water, and heated at 165 °C for 24 h to ensure thorough digestion. The samples were then diluted by a factor of 4000 and analyzed using an inductively coupled plasma-mass spectrometer (ICP-MS) at the China University of Geosciences. The precision of these analytical measurements was commonly greater than 5%, as monitored by the standards AGV-2, BHVO-2, W-2, GSR-1, and GSR-3, with the exception of phosphorus and potassium, which had a precision of less than 15%, and chromium, scandium, copper, zinc, strontium, and tantalum with a precision of less than 10%.

To determine the contribution from terrestrial sources and the authigenic components, the calculation of the trace element excess, denoted as XXS, is essential. This is achieved using the equation: $X_{XS} = X_{sample} - Al_{sample} \times (X/Al)_{PAAS}$, where X_{sample} represents the concentration of the element X in the sample, and Al_{sample} denotes the aluminum concentration within the same sample. The term $(X/Al)_{PAAS}$ refers to the ratio of element X to aluminum in the Post-Archean Australian shale, which serves as a benchmark [13]. In instances where the authigenic components may have been compromised, leading to negative X_{XS} values for some samples in this study, a standardization approach has been implemented. These negative values are adjusted to a uniform value of 0.01 ppm to facilitate the creation of log10 plots that depict the relationship between the trace element excess and depth. This adjustment ensures that the graphical representation remains informative and consistent.

4. Results

The geochemical data from Xibeixiang and Jianfeng sections are provided in the Supporting Information (Tables S1–S8).

The silicon content in the Dalong Formation at the Xibeixiang section varies significantly. The SiO_2 content ranges from 5.82% to 90.56%, with an average of 57.48%. The Fe/Ti ratios span from 1.49 to 225.39, averaging 24.92. Al/(Al + Fe + Mn) varies from 0.08 to 0.87, with an average of 0.57. In terms of trace elements, scandium (Sc) content ranges from 0.45×10^{-6} to 10.04×10^{-6}, averaging at 3.23×10^{-6}. Thorium (Th) content ranges from 0.14×10^{-6} to 5.77×10^{-6}, averaging at 1.85×10^{-6}. Lanthanum (La) content ranges from 0.69×10^{-6} to 42.56×10^{-6}, averaging at 11.28×10^{-6}. The Ti/Zr ratios vary from 5.18 to 36.16, with an average of 17.63, while the La/Sc ratios range from 0.91 to 27.19, averaging 3.92. The europium anomaly (Eu/Eu*) values extend from 0.34 to 1.01, with a mean of 0.79. Standardized lutetium to lanthanum ratios (Lu_N/La_N) range from 0.63 to 2.77, with an average of 1.51.

In the Jianfeng section, the SiO_2 content varies from 5.05% to 93.64% (average value 56.27%). The Fe/Ti ratios span from 3.38 to 32.29, averaging 13.40%. Al/(Al + Fe + Mn) ratios vary from 0.34 to 0.85, with an average of 0.62. Regarding trace elements, Sc content ranges from 0.48×10^{-6} to 22.58×10^{-6}, averaging at 6.17×10^{-6}. The contents of Th range from 0.18×10^{-6} to 47.50×10^{-6}, averaging at 4.83×10^{-6}. La content ranges from 2.35×10^{-6} to 165.02×10^{-6}, averaging at 20.60×10^{-6}. The Ti/Zr ratios range from 1.51 to 38.71, with an average of 18.70, while the La/Sc ratios range from 1.01 to 23.87, averaging at 4.40. The Eu/Eu* values extend from 0.24 to 1.23, with an average of 0.90. Lastly, standardized Lu_N/La_N ranges from 0.63 to 2.01, averaging at 1.30.

5. Discussion

5.1. Origin of Siliceous Rocks

Aluminum (Al) and Titanium (Ti) co-occur in aluminosilicate minerals, serving as reliable indicators of terrestrial input. Conversely, iron (Fe) enrichment in sediments from mid-ocean ridges signifies hydrothermal activities [14]. Utilizing the Al-Fe-Mn triangular chart developed by Adachi et al. [15], which differentiates hydrothermal from non-hydrothermal siliceous rocks based on analyses of modern marine systems, the role of hydrothermal activities in sediment deposition can be assessed. The Al/(Al +Fe + Mn) ratio of marine sediments has also been used to measure the contribution of hydrothermal activities to sediment deposition [14]. It is reported that in the East Pacific Ridge hydrothermal system, this ratio can be as low as 0.01, indicating strong hydrothermal influence, while semi-oceanic siliceous sediments typically show a ratio of 0.60 [16]. The stronger the influence of hydrothermal activities is, the lower the Al/(Al + Fe + Mn) ratio becomes [15]. Similarly, Fe/Ti ratios could rise when sediment undergoes hydrothermal activities.

The Al-Fe-Mn triangle diagrams of the siliceous rocks of the Dalong Formation reveal that chert and siliceous limestone in Xibeixiang section are primarily of non-hydrothermal origin, while carbonate rocks in Xibeixiang section exhibit both hydrothermal and non-hydrothermal origins (Figure 4a). Furthermore, the chert and siliceous limestone in the Jianfeng section are solely of a non-hydrothermal origin, and the carbonate rocks also have no connection with hydrothermal activities (Figure 4b). These conclusions are consistent with the sedimentological characteristics, which show that the bedded chert and siliceous limestone of the Dalong Formation in the Xibeixiang and Jianfeng sections are rich in siliceous radiolarians, indicating that cherts and siliceous limestones have a biogenic origin rather than hydrothermal origin and are therefore biogenic. The mudstone and marly limestone in the Xibeixiang section contain multiple layers of volcanic ashes, which may be partly related to hydrothermal activities.

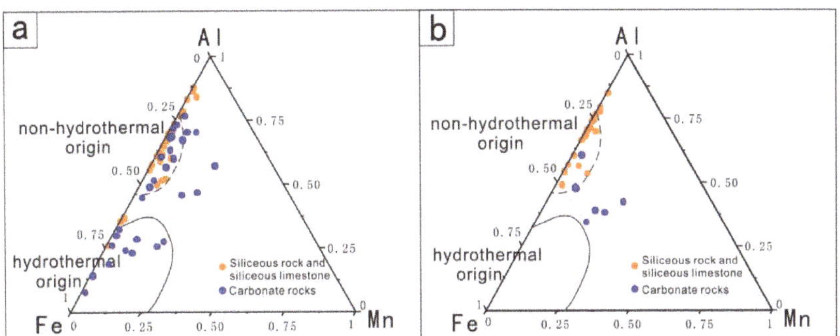

Figure 4. Al-Fe-Mn diagram of the Dalong Formation in the (**a**) Xibeixiang and (**b**) Jianfeng sections (after reference [15]).

The Al/(Al + Fe + Mn) ratios of the siliceous rocks in the Dalong Formation of the Xibeixiang section are predominantly above 0.6, and only a portion of the samples fall below 0.6, appearing in the marly limestone in the middle part of the Dalong Formation

(Figure 5). It suggests the minimal hydrothermal influence in the siliceous rocks in the Xibeixiang section and possible hydrothermal influence in the marly limestone. Similarly, in the Jianfeng section, siliceous rocks and siliceous limestone of the Dalong Formation consistently show Al/(Al + Fe + Mn) ratios greater than 0.6, while micrite, marly limestone, and bioclastic limestone exhibit ratios around 0.6. Fe/Ti ratios in both sections follow a similar pattern, particularly lower in the silicon-rich intervals. These findings align with the Al-Fe-Mn triangle diagram results, confirming that the source of silicon in the silicon-rich layers of the Dalong Formation in the Sichuan Basin is primarily biogenic rather than hydrothermal.

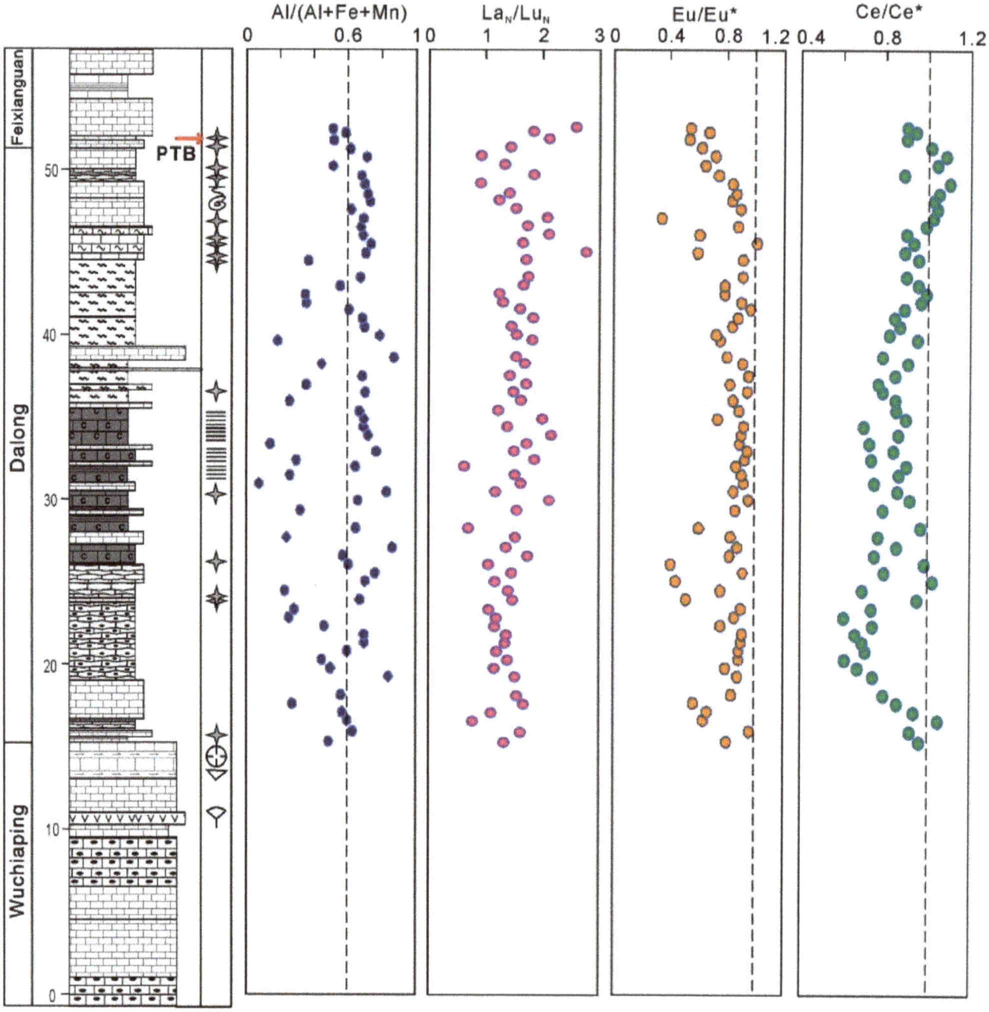

Figure 5. Chemostratigraphic profiles for the hydrothermal proxies of Dalong Formation in the Xibeixiang section. See Figure 2 for legends. The dotted line represents the boundary of the condition, as explained in detail in the text.

The anomaly of the rare earth element Europium (Eu) can indicate hydrothermal activity characteristics. In seawater, where rare earth elements predominantly exist as trivalent ions, Eu is unique as it presents in a divalent state (Eu^{2+}) [17]. This special valence state can lead to abnormal Eu eigenvalues. During magmatic activities, Eu^{3+} is often reduced to Eu^{2+}, which has a larger ionic radius than its neighboring rare earth elements, causing fractionation [18]. This reduction commonly occurs in the hydrothermal systems of mid-ocean ridges. Eu^{3+} can be reduced to Eu^{2+} in severely anoxic environments, and its ionic radius, similar to that of Ba^{2+}, facilitates its adsorption by barite, leading to precipitation [19]. The Eu fractionation often occurs in the hydrothermal system and leads to a positive anomaly of Eu [17], because the reduced Eu^{2+} is mainly absorbed by feldspar or other minerals due to the affinity of crystallization [20], resulting in Eu enrichment. The most pronounced Eu-positive anomalies are observed in seawater above hydrothermal fluids along mid-ocean ridges, with Eu/Eu* values reaching up to 10 [20]. The Eu/Eu* ratio in siliceous rocks typically decreases from 1.35 to 1.02 as the distance from the mid-ocean ridge increases, with Eu positive anomaly signals nearly disappearing 25 km (or 100 km) away from hydrothermal vents, although regional seawater may still be influenced by hydrothermal activities [21].

In the Dalong Formation at the Xibeixiang section, most Eu/Eu* values are below 1.0 (Figure 5), suggesting a predominantly non-hydrothermal origin for the deposits. It is worth noting that Eu/Eu* shows relatively high ratios in the marly limestone of the middle Dalong Formation, demonstrating the same conclusion as before. Furthermore, Eu/Eu* ratios decline near the P-T boundary, probably due to the severe anoxia event in the P-T transition. Similarly, the Dalong Formation at the Jianfeng section shows mostly Eu/Eu* values below 1.0, except for one instance in the marly limestone of the middle Dalong Formation, which exceeds 1.0, suggesting that the cherts of the Dalong Formation in the Jianfeng section are dominated by a non-hydrothermal origin. It is illustrated that the Sichuan Basin might have been close to the hydrothermal vent and affected by hydrothermal fluid to some extent during the deposition of the marly limestone.

In normal seawater, heavy rare earth elements (HREEs) exhibit more stable fluid chemical behavior due to their stronger inorganic complexation compared to light rare earth elements (LREEs) [22,23]. As a result, LREEs are more readily removed from seawater, leading to a seawater distribution pattern characterized by LREE deficiency [17]. Conversely, ocean basin hydrothermal fluids are enriched in LREEs [20], with significant absorption of LREEs from seawater by the hydrothermal fluids of mid-ocean ridges, leading to LREE depletion in surrounding seawater. Siliceous rocks, which primarily adsorb rare earth elements from seawater influenced by hydrothermal fluids, display a distribution pattern indicative of LREE depletion in mid-ocean ridge environments, with Lu_N/La_N ratios as high as 3 [24,25]. This ratio typically decreases with distance from the ridge, from 1.55 to 0.87, and further to 0.37 in normal ocean basins [21]. In open continental marginal seas, where LREEs are also removed from seawater, sediment Lu_N/La_N ratios reach around 1.10 [21].

In the cherts of the Dalong Formation at the Xibeixiang section, the Lu_N/La_N ratios are less than 3 and average at 1.51, suggesting that the Xibeixiang section was weakly influenced by hydrothermal fluids. It is noted that these ratios experience a dramatic increase to approximately 3 near the P-T boundary (Figure 5), which is confirmed by the decrease in Al/(Al + Fe + Mn) ratios. Similarly, Lu_N/La_N peaked in the siliceous limestone of the upper Dalong Formation (Figure 5), which is also supported by the relatively low Al/(Al + Fe + Mn) values and high Eu/Eu* values exceeding 1. Combined with the volcanic ash interlayers, it is illustrated that the volcanic or hydrothermal activities only existed at the end-Permian. Moreover, most rare earth element distribution curves of the Xibeixiang section do not show LREE deficiency, with only a few samples exhibiting mild LREE depletion (Figure 6), suggesting that most of the deposition of Dalong formation was unrelated to hydrothermal events. As for the cherts of the Dalong Formation at the Jianfeng section, the average of Lu_N/La_N ratios (1.30) is less than that at the Xibeixiang section (Figure 7), with no significant loss of LREEs observed (Figure 6). These findings suggest that the siliceous rocks in both the Xibeixiang and Jianfeng sections are predominantly of a non-hydrothermal origin. However, hydrothermal influences might have affected the Xibeixiang section near the P-T transition due to its proximity to the Paleo-Tethys.

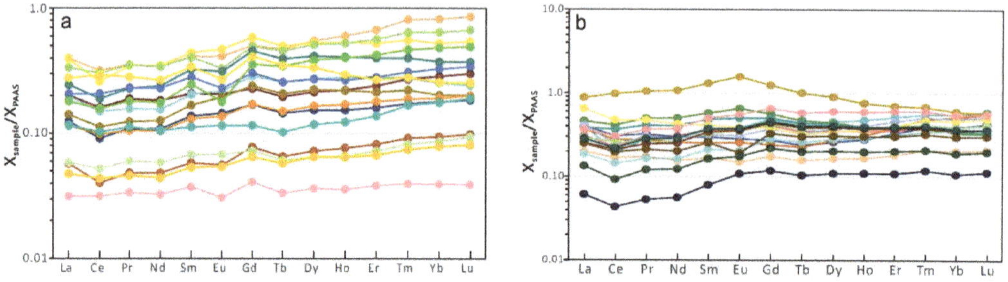

Figure 6. Post-Archean Australian shale (PAAS)-normalized REE pattern of the siliceous rock in the Dalong Formation at (**a**) the Xibeixiang section and (**b**) the Jianfeng section.

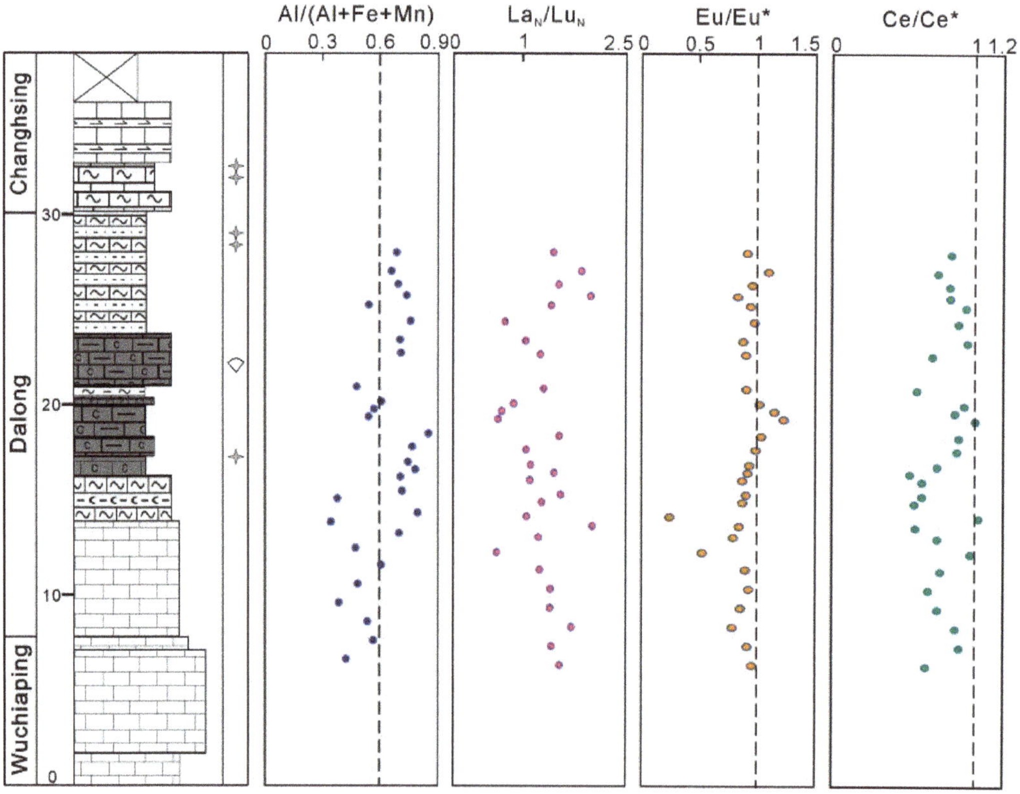

Figure 7. Chemostratigraphic profiles for the hydrothermal proxies of the Dalong Formation in the Jianfeng section. See Figure 2 for legends.

Cerium (Ce) typically exists as Ce^{3+} in marine environments, which can be easily oxidized to the less soluble Ce^{4+} and precipitate under oxidative conditions. Therefore, positive Ce anomalies or Ce/Ce* ratios close to 1 in marine sediments often indicate anoxic conditions [13]. In both the Xibeixiang and Jianfeng sections, the Ce/Ce* values exhibit an upward trend from the bottom to the top of the Dalong Formation, approaching 1 in the middle and upper parts. Combined with former studies [26], it is reported that the depositional environment of the upper and middle Dalong Formation was anoxic, and the organic accumulation was primarily driven by productivity. Data on rare earth element contents and ratios further suggest that the formation of this anoxic environment and the associated enrichment of organic matter were largely independent of hydrothermal activities.

5.2. Sedimentary Tectonic Background of the Dalong Formation

Certain inert trace elements such as La, Zr, Ce, Nd, Y, Hf, Nb, Ti, Sc, and Th are key indicators of the tectonic background and provenance of sedimentary deposits due to their minimal migration during sedimentation and short residence times in seawater [27]. These elements are consistently delivered to clastic sedimentary rocks through weathering and transportation, thereby offering distinct indicators that can be traced back to their original rock formations [27]. The La-Th-Sc triangulation chart effectively differentiates various tectonic settings. Oceanic island arc environments typically show higher Sc contents and lower La/Sc ratios (average 0.55 ± 0.2), continental margins have higher La contents with La/Sc ratios around 4, and continental island arcs exhibit moderate La/SC ratios (average 0.8 ± 0.3) [28]. Although this chart can distinguish between Paleozoic oceanic and continental island arcs, it is difficult to distinguish between active and passive continental margins. However, the intersection chart of La/Sc and Ti/Zr ratios helps in this differentiation [29]. Samples from oceanic island arcs are identified by titanium-to-zirconium (Ti/Zr) ratios that exceed 40 and lanthanum-to-scandium (La/Sc) ratios that surpass 1. In contrast, continental island arcs display Ti/Zr ratios within the range of 10 to 35 and La/Sc ratios that fall between 1 and 3. Active continental margins are distinguished by elevated La/Sc ratios, which range from 3 to 6. Conversely, passive continental margins are indicated by Ti/Zr ratios that are below 10 and La/Sc ratios from 3 to 9 [29].

The siliceous rocks of the Dalong Formation in the Xibeixiang and Jianfeng sections do not display oceanic island arc characteristics. In the Xibeixiang section, the predominant tectonic settings are continental island arcs and active continental margins, with a minor component of passive continental margins (Figures 8a and 9a). On the other hand, the Jianfeng section indicates geologic characteristics of both a continental island arc and an active continental margin, devoid of the attributes typically associated with a passive continental margin, as illustrated in Figures 8b and 9b. Hydrothermal processes are known to be particularly pronounced in regions of active continental margins and oceanic island arcs. The absence of oceanic island arc traits in these sections suggests that the northeast Sichuan basin was distant from major hydrothermal activities. Nonetheless, the proximity of the Xibeixiang section to the Paleo-Tethys may explain its more pronounced hydrothermal signals compared to the Jianfeng section.

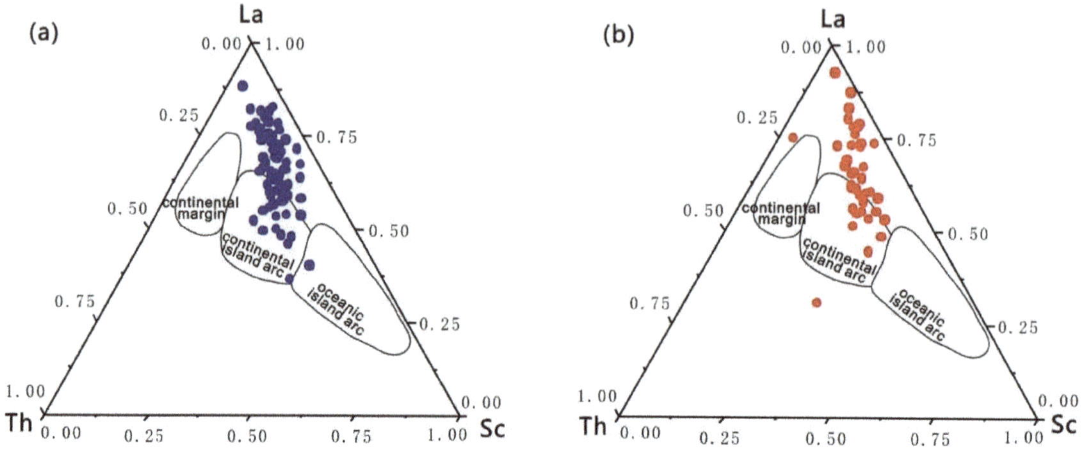

Figure 8. La-Th-Sc diagram of Dalong cherts at (**a**) the Xibeixiang section and (**b**) the Jianfeng section (base map after reference [29]).

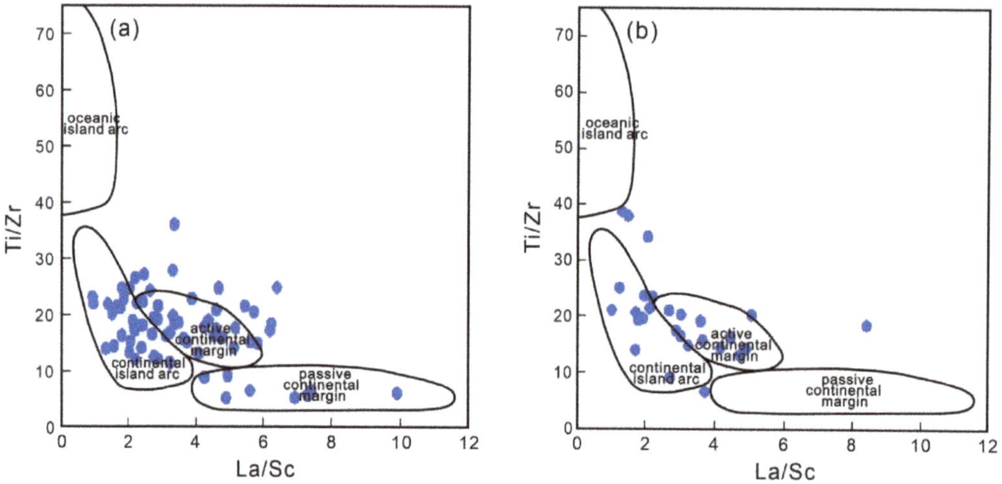

Figure 9. Cross-plot between La/Sc and Ti/Zr of Dalong cherts at (**a**) the Xibeixiang section and (**b**) the Jianfeng section (base map after reference [29]).

5.3. Influence of Hydrothermal Activity on Organic Matter Enrichment

As discussed above, the siliceous rocks in the upper part of the Dalong Formation are more likely to have non-hydrothermal origins. A large number of radiolarians observed in photomicrographs demonstrated that the source of siliceous rocks in both the Xibeixiang and Jianfeng sections leans more toward biological origins. The carbonate rocks of the middle Dalong Formation and P-T boundary in the Xibeixiang section show more hydrothermal characteristics. Previous studies, however, suggested that the organic enrichment and high primary productivity emerged in parallel with the ampelitic limestone, and the anoxic environment appeared during the deposition of marly limestone and chert. The weak correlation between TOC and indicators of hydrothermal activities (Figure 10) indicates that the organic accumulation is less likely to be related to the hydrothermal activities.

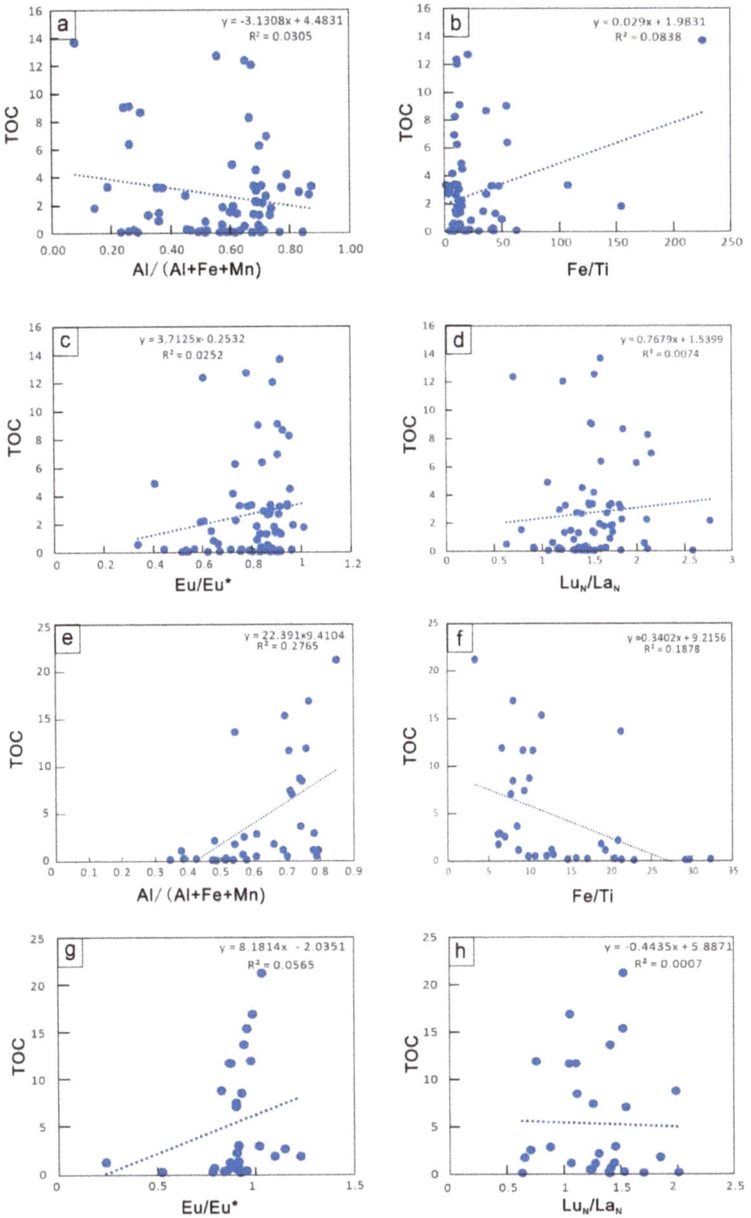

Figure 10. Cross-plot between TOC and indicators of hydrothermal activities in (**a**–**d**) the Xibeixiang section and (**e**–**h**) the Jianfeng section.

6. Conclusions

The silicon source of the cherts and siliceous limestone of the Dalong Formation at the Xibeixiang and Jianfeng sections is predominantly biogenic, as evidenced by the abundance of radiolarians, and shows minimal influence from hydrothermal activities, as documented by the rare earth element proxies. However, the mudstone and marly limestone in the Xibeixiang section could be affected by hydrothermal activities, as suggested by the Eu/Eu* ratios and the presence of several volcanic ash layers observed in the field. Furthermore, the middle part of the marly limestone in the Jianfeng section may retain a weaker hydrothermal signal compared to the Xibeixiang section. The tectono-sedimentary background of the Dalong Formation in the Xibeixiang section primarily consists of continental island arcs and active continental margins, with a minor component of passive continental margins. Conversely, the cherts and siliceous limestone in the Jianfeng section reflect a continental island arc and active continental margin setting. The lack of oceanic island arc characteristics in both sections indicates that the northeast Sichuan basin is distant from significant hydrothermal activity zones. Notably, the active continental margin features of the siliceous rocks in the Xibeixiang section are more marked than those in the Jianfeng section, likely due to its proximity to the open marine environments during the Late Permian, which might have allowed it to retain subtle signs of hydrothermal activities. The weak correlation between TOC and indicators of hydrothermal activities demonstrates that hydrothermal activity had minimal impact on the formation of anoxic environments and the enrichment of organic matter during the deposition of the Dalong Formation.

Supplementary Materials: The following supporting information can be downloaded at: https://www.mdpi.com/article/10.3390/min15010069/s1, Table S1: XBX major elements; Table S2: XBX Eu anomaly; Table S3: XBX Ce anomaly; Table S4: XBX rare elements; Table S5: JF major elements; Table S6: JF Eu anomaly; Table S7: JF Ce anomaly; Table S8: JF rare elements;

Author Contributions: Conceptualization, R.W.; methodology, X.G. (Xiaotong Ge); formal analysis, X.G. (Xiaotong Ge) and X.G. (Xun Ge); investigation, X.G. (Xun Ge); data curation, X.G. (Xiaotong Ge); writing—original draft preparation, X.G. (Xiaotong Ge) and M.L.; writing—review and editing, D.C.; supervision, D.C. and Y.L.; funding acquisition, D.C. and Y.L. All authors have read and agreed to the published version of the manuscript.

Funding: This study was supported by the National Natural Science Foundation of China (91755210) and the Science and Technology Project of SINOPEC (P24207).

Data Availability Statement: The data presented in this study are available in Supplementary Materials.

Acknowledgments: The researchers extend their gratitude to Hongyu Zhang and Zihu Zhang for their contributions to the laboratory work. The valuable feedback and recommendations provided by the editorial team and peer reviewers are also greatly appreciated.

Conflicts of Interest: The authors declare that this study received funding from the National Natural Science Foundation of China (91755210) and the Science and Technology Project of SINOPEC (P24207). The funder had the following involvement with the study: supervision, funding acquisition and writing—review and editing.

References

1. Cao, T.T.; Song, Z.G.; Wang, S.B.; Cao, X.; Li, Y.; Xia, J. Characterizing the pore structure in the Silurian and Permian shales of the Sichuan Basin, China. *Mar. Pet. Geol.* **2015**, *61*, 140–150. [CrossRef]
2. Wei, Z.F.; Wang, Y.L.; Wang, G.; Sun, Z.; Xu, L. Pore characterization of organic-rich Late Permian Da-long Formation shale in the Sichuan Basin, southwestern China. *Fuel* **2018**, *211*, 507–516. [CrossRef]
3. Wang, R.Y.; Hu, Z.Q.; Dong, L.; Gao, B.; Sun, C.; Yang, T.; Wang, G.; Yin, S. Advancement and trends of shale gas reservoir characterization and evaluation. *Oil Gas Geol.* **2021**, *42*, 54–65.
4. Fang, S.Y. Middle-Upper Permian Siliceous Rocks in Lower Yangtze Area Origin and Shale Gas Potential. Master's Thesis, Nanjing University, Nanjing, China, 2021.
5. Xu, Y.T. Genetic geochemistry for the bedded silicalite in the late permian Dalong formation and its sedimehtary setting in Southeastern Hubei. *J. Guilin Inst. Technol.* **1997**, *17*, 204–212. (In Chinese with English Abstract)
6. Li, F.; Liu, Z.J.; Chen, F.R.; Wei, F.B.; Guo, J.C.; Su, Z.X. Origin of siliceous mudstone in Permian Dalong Formation and its influence on compressibility, Northeastern Sichuan Basin. *Nat. Gas Geosci.* **2023**, *34*, 349–358. (In Chinese with English Abstract)
7. Fang, X.; Zhou, Y.Q.; Yao, X.; Liu, M.C.; Liu, J.Z. Geochemical characteristics and petrogenesis of siliceous rocks from Shansi section, Guangyuan, Sichuan Province. *Miner. Pet.* **2017**, *37*, 93–102. (In Chinese with English Abstract)

8. Yang, Y.C. Research on Geochemical Characteristics and Formative Factors of Cherts in the Dalong Formation of the Shangjianggou Area, Northwestern Sichuan Basin. Master's Thesis, Chengdu University of Technology, Chengdu, China, 2015.
9. Mei, S.; Henderson, C.M. Evolution of Permian conodont provincialism and its significance in global correlation and paleoclimate implication. *Palaeogeogr. Palaeoclimatol. Palaeoecol.* **2001**, *170*, 237–260. [CrossRef]
10. Wignall, P.B.; Hallam, A.; Lai, X.; Yang, F. Palaeoenvironmental changes across the Permian/Triassic boundary at Shangsi (N. Sichuan, China). *Hist. Biol.* **1995**, *10*, 175–189. [CrossRef]
11. Luo, J.X.; He, Y.B. Anoxic environments of the Permian of Middle and Upper Yangtze Area. *J. Palaeogeogr.* **2011**, *13*, 11–20. (In Chinese with English Abstract)
12. Shellnutt, J.G.; Denyszyn, S.W.; Mundil, R. Precise age determination of mafic and felsic intrusive rocks from the Permian Emeishan large igneous province (SW China). *Gondwana Res.* **2012**, *22*, 118–126. [CrossRef]
13. Tribovillard, N.; Algeo, T.J.; Lyons, T.; Riboulleau, A. Trace metals as paleoredox and paleoproductivity proxies: An update. *Chem. Geol.* **2006**, *232*, 12–32. [CrossRef]
14. Murray, R.W. Chemical criteria to identify the depositional environment of chert: General principles and applications. *Sediment. Geol.* **1994**, *90*, 213–232. [CrossRef]
15. Adachi, M.; Yamamoto, K.; Sugisaki, R. Hydrothermal chert and associated siliceous rocks from the northern Pacific: Their geological significance as indication of ocean ridge activity. *Sediment. Geol.* **1986**, *47*, 125–148. [CrossRef]
16. Boström, K.; Peterson, M.N.A. The origin of aluminum-poor ferromanganoan sediments in areas of high heat flow on the East Pacific Rise. *Mar. Geol.* **1969**, *7*, 427–447. [CrossRef]
17. De Baar, H.J.W.; Bacon, M.P.; Brewer, P.G.; Bruland, K.W. Rare earth elements in the Pacific and Atlantic Oceans. *Geochim. Cosmochim. Acta* **1985**, *49*, 1943–1959. [CrossRef]
18. Michard, A.; Albarède, F.; Michard, G.; Minster, J.F.; Charlou, J.L.; Albar, F. Rare-earth elements and uranium in high-temperature solutions from East Pacific Rise hydrothermal vent field (13° N). *Nature* **1983**, *303*, 795–797. [CrossRef]
19. Kimata, M. The crystal structure of non-stoichiometric Eu-anorthite: An explanation of the Eu-positive anomaly. *Mineral. Mag.* **1988**, *52*, 257–265. [CrossRef]
20. Michard, A. Rare earth element systematics in hydrothermal fluids. *Geochim. Cosmochim. Acta* **1989**, *53*, 745–750. [CrossRef]
21. Murray, R.W.; Buchholtz Ten Brink, M.R.; Gerlach, D.C.; Russ, G.P., III; Jones, D.L. Rare earth, major, and trace elements in chert from the Franciscan Complex and Monterey Group, California: Assessing REE sources to fine-grained marine sediments. *Geochim. Cosmochim. Acta* **1991**, *55*, 1875–1895. [CrossRef]
22. Goldberg, E.D.; Koide, M.; Schmitt, R.A.; Smith, R.H. Rare-earth distributions in the marine environment. *J. Geophys. Res.* **1963**, *68*, 4209–4217. [CrossRef]
23. Turner, D.R.; Whitfield, M.; Dickson, A.G. The equilibrium speciation of dissolved components in freshwater and sea water at 25 °C and 1 atm pressure. *Geochim. Cosmochim. Acta* **1981**, *45*, 855–881. [CrossRef]
24. Baker, R.A. Trace inorganics in water. In *Neutron Activation Analysis of Lanthanide Elements in Sea Water*; Høgdahl, O.T., Melsom, S., Bowen, V.T., Eds.; Jenne E A: Washington, DC, USA, 1968; pp. 308–325.
25. Klinkhammer, G.; Elderfield, H.; Hudson, A. Rare earth elements in seawater near hydrothermal vents. *Nature* **1983**, *305*, 185–188. [CrossRef]
26. Ge, X.T.; Chen, D.Z.; Zhang, G.J.; Huang, T.; Liu, M.; El-Shafeiy, M. Marine redox evolution and organic accumulation in an intrashelf Basin, NE Sichuan Basin during the Late Permian. *Mar. Pet. Geol.* **2022**, *140*, 105633. [CrossRef]
27. Holland, H.D. *The Chemistry of the Atmosphere and Oceans*; Wiley: New York, NY, USA, 1978.
28. McLennan, S.M.; Taylor, S.R.; Kröner, A. Geochemical evolution of Archean shales from South Africa. I. The Swaziland and Pongola Supergroups. *Precambrian Res.* **1983**, *22*, 93–124. [CrossRef]
29. Bhatia, M.R.; Crook, K.A.W. Trace element characteristics of graywackes and tectonic setting discrimination of sedimentary basins. *Contrib. Mineral. Petrol.* **1986**, *92*, 181–193. [CrossRef]

minerals

MDPI

Article

Analysis of Controlling Factors of Pore Structure in Different Lithofacies Types of Continental Shale—Taking the Daqingzi Area in the Southern Songliao Basin as an Example

Xinrui Wang [1], Yu Sun [1,2,*], Tianxu Wang [3], Baiquan Yan [1,2] and Ruhao Liu [4]

1 School of Earth Science, Northeast Petroleum University, Daqing 163318, China; wangxr_2017@163.com (X.W.)
2 Sanya Offshore Oil & Gas Research Institute, Northeast Petroleum University, Sanya 572024, China
3 Exploration and Development Research Institute of Jilin Oilfield, Songyuan 138000, China
4 Institute of Unconventional Oil and Gas Development, Chongqing University of Science and Technology, Chongqing 401331, China
* Correspondence: sunyu_hc@163.com

Abstract: Due to the influence of terrigenous debris, the internal pore structure of continental shale is highly heterogeneous, and the controlling factors are complex. This paper studies the structure and controlling factors of shale reservoirs in the first member of the Qingshankou Formation in the Southern Songliao Basin using core data and various analytical test data. The results show that the original deposition and subsequent diagenesis comprehensively determine the shale reservoirs' pore structure characteristics and evolution law. According to the severity of terrigenous debris, the shale reservoirs in the study area are divided into four categories and six subcategories of lithofacies. By comparing the characteristics of different shale lithofacies reservoirs, the results show that the lithofacies with a high brittle mineral content have more substantial anti-compaction effects, more primary pores to promote retention and a relatively high proportion of mesopores/macropores. Controlling the organic matter content when forming high-quality reservoirs leads to two possibilities. An excessive organic matter content will fill pores and reduce the pore pressure resistance. A moderate organic matter content will make the inorganic diagenesis and organic hydrocarbon generation processes interact, and the development of organic matter mainly affects the development of dissolution pores. The comprehensive results show that A3 (silty laminated felsic shale) reservoirs underwent the pore evolution process of "two drops and two rises" of compaction, cementation and pore reduction, dissolution and pore increase, and organic matter cracking and pore increase, and they are the most favourable lithofacies of the shale reservoirs in the study area.

Keywords: Songliao Basin; lithofacies type; pore structure; pore evolution

1. Introduction

The United States has experienced a technological revolution in shale oil over the past two decades, achieving the economically viable scale development of hydrocarbon resources in shales such as the Bakken Formation in the Williston Basin and the Eagle Ford Formation in the Western Gulf Basin, thereby transforming the global energy market [1,2]. Distinct from North American marine shales, China's shale oil resources are predominantly terrestrial, characterised by complex tectonics, a low level of maturity, well-developed laminations, frequent sedimentary facies changes, and a pronounced heterogeneity in both macro-distribution and micropore structures [3]. Since shale oil is primarily contained within nanoscale pores [4–6], the pore structure features, including the volume, type and size of storage spaces, significantly impact the oil-bearing capacity and occurrence of shale oil reservoirs [5–7]. Consequently, reservoir pore structure serves as a crucial factor in assessing the quality of fine-grained sedimentary reservoirs, directly determining their storage capacity and exploration potential [8–10]. Thus, studying pore structure characteristics and evolution patterns has remained a focal point in shale reservoir research.

Experimental equipment often limits the characterisation of pore structures. The characterisation techniques of shale microscopic pore structure can be roughly divided into two categories: qualitative description and quantitative characterisation. Among them, experimental methods such as optical microscopy, field emission scanning electron microscopy, focused ion beam scanning electron microscopy (FIB-SEM) and micro- and nano-CT scanning can be used to intuitively and finely observe the pore structure morphology [8,10,11]. However, limited by a high resolution, these instruments can only describe the local pore structure pattern, and the obtained pore information is highly homogeneous and not very representative. The gas (CO_2 or N_2) adsorption method [12,13], constant-speed/high-pressure mercury injection [14,15], nuclear magnetic resonance [16] and gas pulse permeability tests can characterise a wide range of pore areas, and the characteristics of reservoir spaces are easy to quantify. However, each method has unique characteristics, and several methods need to be combined to achieve comprehensive pore-size characterisations [17–19]. Effectively integrating different results, systematically analysing the characteristics of shale reservoir spaces and comprehensively characterising pore types and structural characteristics are key to the quantitative characterisation of reservoir pore characteristics.

The formation and evolution of shale reservoirs are mainly affected by organic–inorganic factors [6,7,20]—for example, organic matter cracking pores [21,22], clay transformation [23,24], and quartz and carbonate cementation [25,26]. However, the current research on reservoir controlling factors pays more attention to one aspect: the study of pore evolution during diagenesis. For example, organic matter produces organic acids, which can dissolve minerals [27,28]. The production of hydrocarbons will cause overpressure and inhibit the reduction in porosity to a certain extent [6,7,29]. It generally involves clay mineral transformation, siliceous cement formation, feldspar dissolution, organic matter evolution and the reservoir's physical properties. In particular, significant progress has been made in the influence of pore evolution characteristics [30–33]. In addition, in the over-mature stage, the cracking of organic matter will produce organic matter pores [21,34]. However, whether an excessive organic matter content will damage the shale reservoir structure or felsic minerals is still uncertain [35–37]. It has a dual effect on shale. Increasing these mineral contents will increase the compressive strength of shale rocks and provide soluble mineral carriers [38]. However, when their content is too high, they will reduce the abundance of organic matter in continental shale [39]. Clay content has good plasticity and is closely related to organic matter content. Organic matter cracking promotes the transformation of clay [34,40].

However, the above results apply more to the pore structure of high-TOC (Total organic content) black lacustrine shale reservoirs in deep lakes. However, most lacustrine shales are often subjected to terrigenous debris and intravaginal biological and chemical processes and contain interlayers of siltstone/fine sandstone and carbonate rocks [33,34,40,41]. The diversity of the original sedimentary types will significantly influence the original pore structure. Therefore, it is urgent to supplement the research on the difference in the diagenetic evolution path and genetic mechanism of reservoirs with different original sedimentary types and fabrics.

This paper analyses the mature lacustrine shale in the first member of the Qingshankou Formation (Qing1) of the Upper Cretaceous in Changling Sag, Southern Songliao Basin. The essential characteristics of lithofacies and organic matter in the study area are analysed through core and laboratory analyses, and the lithofacies types are divided. The primary characteristics and reservoir space types of the study area's first member of the Qingshankou Formation are determined based on scanning electron microscopy, high-pressure mercury injection, nitrogen adsorption and other analytical testing methods. Based on the fractal theory, the classification standard of fine-grained sedimentary pores in the Qingyi member of the study area is established, and the full-aperture characterisation is realized.

The controlling factors affecting the pore structure of the reservoir are analysed from the perspective of sedimentation and organic matter, and the favourable lithofacies types and pore structure evolution mode of the shale reservoir are determined, which provides a theoretical basis for the selection of this kind of shale oil dessert area.

2. Geological Setting

The Songliao Basin is a large-scale Mesozoic-Cenozoic continental petroleum-bearing sedimentary basin with a dual structural feature of an initial fault depression followed by a superimposed depression in the NNE direction in Northeastern China [42–44]. The basin's foundation consists of Carboniferous-Permian metamorphic rocks and granite. The sedimentary cover includes Jurassic, Cretaceous, Paleogene, Neogene and Quaternary deposits (Figure 1c). Based on the properties of the basement rocks and the regional geological characteristics of the caprock, the depression structural layers in the Songliao Basin can be divided into six first-order structural units, which are believed to control the distribution of reservoirs for oil, natural gas, shale gas and oil sands (Figure 1a,c). Among them, the Central Depression has long been the sedimentary and subsidence centre of the Songliao Basin, with almost all of its strata being well developed. The thickness of the Cretaceous stratum here exceeds 6500 m, containing two sets of high-quality hydrocarbon source rocks: the first member of the Qingshankou Formation and the second member of the Nenjiang Formation [6].

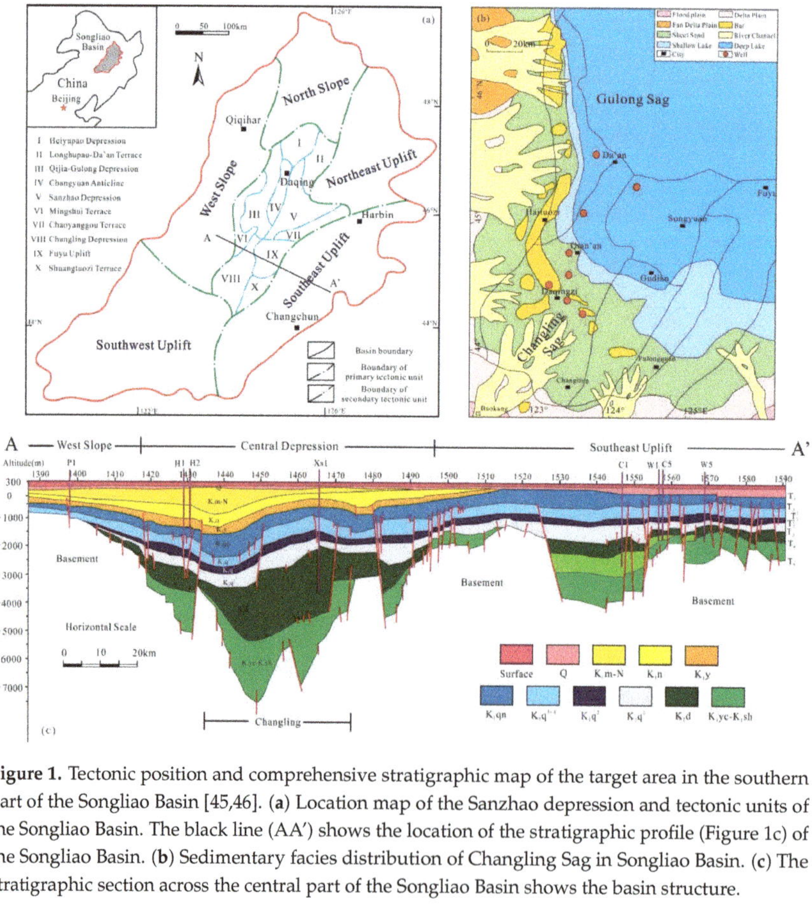

Figure 1. Tectonic position and comprehensive stratigraphic map of the target area in the southern part of the Songliao Basin [45,46]. (**a**) Location map of the Sanzhao depression and tectonic units of the Songliao Basin. The black line (AA') shows the location of the stratigraphic profile (Figure 1c) of the Songliao Basin. (**b**) Sedimentary facies distribution of Changling Sag in Songliao Basin. (**c**) The stratigraphic section across the central part of the Songliao Basin shows the basin structure.

In the Southern Songliao Basin, the sedimentary system's subsidence rate in the lake basin extension was more significant than the sedimentation rate, leading to the expansion of the lake basin and the formation of lake invasions, with only the Tongyu-Baokang sedimentary system remaining relatively intact. Its depositional end is the Southern Songliao Basin shale area [33,34]. Unlike the shale oil-rich Qijia-Gulong Sag in the north of the Songliao Basin, where dark mudstone and shale rich in organic matter are deposited in deep and semi-deep lake facies, the organic-rich shale in the south is located outside the delta front (shallow lake facies) of the Daqingzi area, which is more severely affected by terrigenous debris (Figure 1b).

The Upper Cretaceous is the primary strata in the middle and shallow layers of the Southern Songliao Basin. It is divided into the Quantou Formation (K_1q), Qingshankou Formation (K_2qn), Yaojia Formation (K_2y), Nenjiang Formation (K_2n), Sifangtai Formation (K_2s) and Mingshui Formation (K_2m) from bottom to top (Figure 2a). Among them, the first member of the Qingshankou Formation is the main stratum for shale oil exploration and development in the Songliao Basin [33,45]. The central depression of the Songliao Basin is separated by the Songhua River, which can be divided into Gulong Sag and Changling Sag. In the lower part of the first member of the Qingshankou Formation in Changling Sag, Southern Songliao Basin, large-scale lake floods have occurred, and lithologic assemblages of organic-rich shale, siltstone and carbonate interbeds have been deposited. The upper part of the first member of the Qingshankou Formation is dominated by organic-rich shale (Figure 2b).

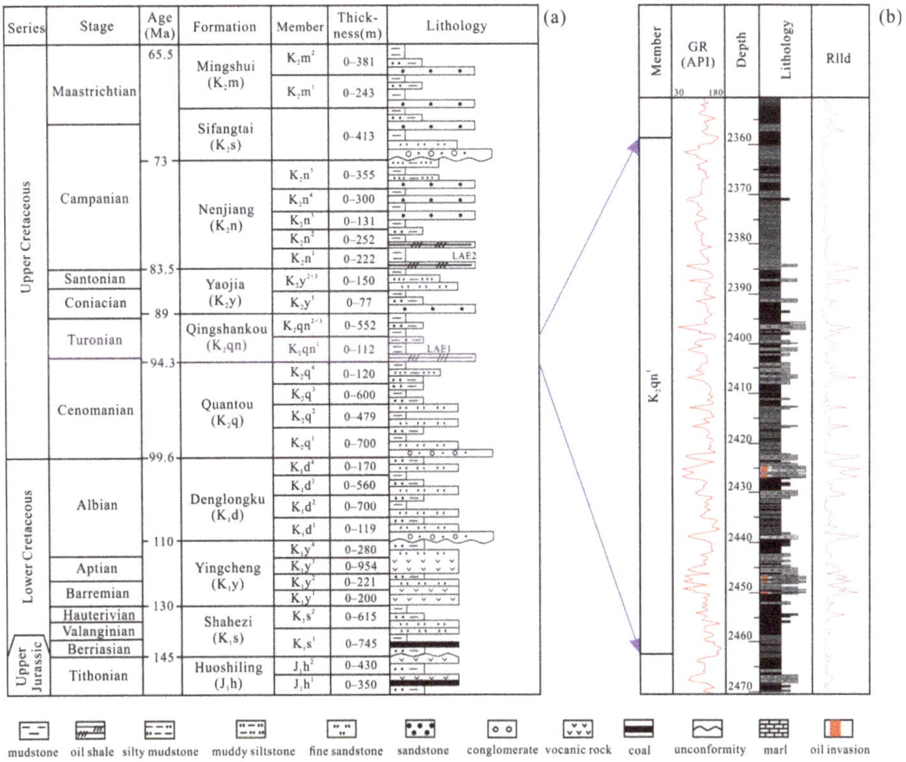

Figure 2. The extent of the target layer and the present study: (**a**) Comprehensive characteristics of the diagenetic system in the basin [45,46]. (**b**) The lithology and logging curve characteristics of typical wells in the study area.

3. Data and Methods

3.1. Samples

The 74 core samples in this study are from 8 critical exploration wells in the first member of the Qingshankou Formation in the Southern Songliao Basin (whose buried depth varies greatly). Figure 1b describes the geographical location of the coring well. Among them, 57 samples (including 45 shale samples) were subjected to TOC tests, X-ray diffraction, vitrinite reflectance, thin-section observation, argon ion polishing field emission scanning electron microscopy, nitrogen adsorption, high-pressure mercury injection, nuclear magnetic resonance and other analytical tests. The maturity of each well is medium (Ro is 1.0%–1.2%). The remaining 17 sandstone samples were analysed for a long time. Only this part of the data is used for the classification of lithofacies types, and this paper does not include the study of pore structure.

3.2. Sample Processing Flow

For the TOC content test, the sample was crushed and pretreated by acidification, washing and drying, and then the TOC content was analysed by a CS-230 carbon-sulphur analyser. The sample was crushed to 200 mesh and prepared on glass slides during XRD whole-rock analysis. An X-ray diffractometer (Model: D8 DISCOVER, UK) was used to test the samples, and the contents of quartz, feldspar, calcite, pyrite and clay minerals were quantitatively identified using the Rietveld full-spectrum simulation method [47].

LTNA and 1-D NMR (Model: D8 GeoSpec2, UK) experiments were used to characterise the shale reservoirs' PSD [48,49]. After washing the sample with oil, grinding it to 60–80 mesh [50] and drying it to remove volatile impurities, an appropriate amount of crushed sample (masses of 1.0 and 2.5 g for shale samples and 4 g for carbonate rock samples) was selected, and the nitrogen adsorption and desorption curves were measured using an specific surface area and porosimetry instrument (Model: ASAP2460-2HD, USA) in the subcritical state (77 K). The specific surface area and PSD were calculated using the Brunauer–Emmett–Teller (BET) and non-local density functional theory (NLDFT) models based on the adsorption curve, respectively. Still, the results can only reflect pores smaller than 200 nm [12,51].

One-dimensional NMR reflects pore size by detecting the relaxation time of the hydrogen nuclei contained in the pores, and the characterised range can vary from a few nanometres to several hundred micrometres. This study used n-dodecane as the saturating medium to prevent clay expansion. The NMR T2 spectra of the samples were measured under dried conditions. The T2 spectrum reflecting the fluid distribution was obtained via basement removal, and the saturated fluid mass and effective porosity were determined using the mass–amplitude conversion relationship.

4. Results

4.1. Shale Lithofacies Types

4.1.1. Basic Petrological Characterisation

Through the meticulous core observation of eight boreholes drilling into the K_2qn^1 Formation in the Daqingzi region, coupled with laboratory thin-section analyses, the lithology of the K_2qn^1 strata in this area was primarily classified into two types: mudstone–shale and siltstone. Based on scanning electron microscopy (SEM) observations and whole-rock mineral X-ray diffraction (XRD) analyses, the mineralogy of the K_2qn^1 mudstone shale is composed primarily of quartz, plagioclase, K-feldspar and clay minerals. Minor amounts of ankerite, calcite, pyrite and siderite are also present. Among the clay minerals, illite is the most predominate, followed by illite/smectite mixed layers, with traces of chlorite (Figure 3a). Examining the ternary diagram of felsic [35–37], clayey and carbonate minerals, it becomes evident that the K_2qn^1 mudstone–shale primarily develops felsic and clayey mudstone–shale, with a minor contribution from dolomitic limestones (Figure 3b). This mineralogical composition reflects the unique sedimentary processes and geochemical environment that shaped the K_2qn^1 Formation in the Daqingzi area.

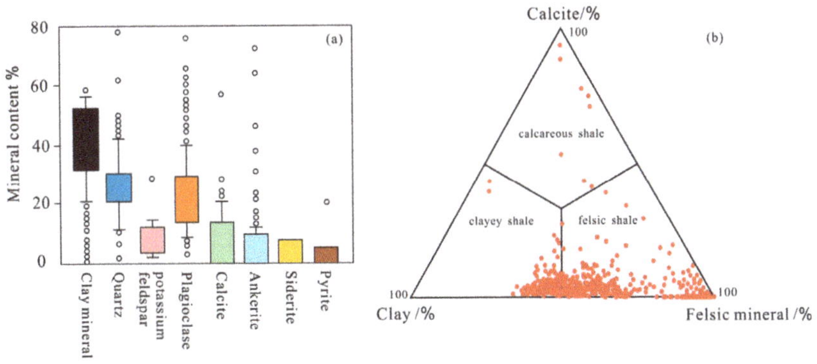

Figure 3. Fundamental petrographic characteristic parameters of the K_2qn^1 reservoir in the Daqingzijing area: (**a**) Box plot of mineral composition, (**b**) triangular diagram of lithological classification.

4.1.2. Sedimentary Structural Characteristics

The fine-grained clastic sediments in the study area are predominantly characterised by laminated, lamellar and massive structures [33]. Transitional lithologies such as siltstone, muddy siltstone and silty mudstone mostly exhibit an enormous structure, accompanied by a considerable amount of deformation and disturbance features. Massive mudstone occurs less frequently within the mudstone shale, and lamination and stratification are more prevalent. Based on grain size, laminae can be further subdivided into silt-grade laminae and mud-grade laminae. Silt-grade laminae possess coarser overall grain sizes, with silt particles (diameter > 4 μm) accounting for over 50% of the content. In contrast, mud-grade laminae display finer grain sizes, with less than 50% of the content consisting of coarser particles (diameter > 4 μm) (Figure 4f,h).

Figure 4. The development characteristics of the sedimentary structure in the first member of Daqingzi. (**a**) Grey white siltstone; wavy bedding; 2129.39 m. (**b**) Gray silty shale; silt lamina and

argillaceous lamina interbedded to form parallel bedding; 2380.02 m. (**c**) Black/grey massive shale; fine silty lamina discontinuous and local silty lenticle; 2101.5 m. (**d**) Black/gray massive mudstone; the main sedimentary structure was destroyed by biological disturbance; 2106.7 m. (**e–g**) macroscopic massive, layered and laminated shale. When the content of silty laminae is high and the interbeds are more frequent, it is easier to rupture and form macroscopic laminae structure; 2402.6 m, 2395.43 m, 2379.41 m. (**h**) Gray argillaceous siltston; sand veins, sandy mass, deformation, with obvious characteristics of traction flow structure and showing signs of burrowing; 2151.7 m. (**i**) Black/gray massive mudstone; grain sequence bedding formed by fine silty lenticles and agglomerates in the middle; 2104.4 m. (**j**) silty shale; 2231.70 m. (**k**) Blocky shale; clay lamina felsic lamina interbedded; 2056.39 m.

Laminae with a single layer thickness of $\leq$1 cm are classified as laminated (bedding planes or shales), characterised by variations in colour, grain size and mineral composition between layers (Figure 4b,c,e,f,k). These primarily correspond to shales dominated by different grain-size laminae, suggesting that shales may also contain relatively high silt contents. When the thickness of a single lamina exceeds 1 cm, it is referred to as stratified (Figure 4h,j) and is typically found in transitional lithologies such as silty mudstone and muddy siltstone, where internal deformation structures are common. Laminae thicker than 1 m are massive, lacking distinct internal structures and exhibiting a uniform colour, grain size and mineral composition (Figure 4a,d,i). These are predominantly present in mudstone and siltstone, with locally observed graded bedding and cross-bedding.

4.1.3. Classification of Lithofacies Types

In the previous lithofacies classification scheme, if there are too many indicators, it may lead to confusion on lithofacies types, which is difficult to predict by logging and seismic activity and not easy to popularize; alternatively, if there are too few indicators, it may result in a lack of obvious distinctions between different petrographic phases in terms of their reservoirs and oil-bearing characteristics. Many scholars often use "TOC + structural type + mineral fraction" to classify shale lithofacies types [37–41].

The K_2qn^1 Formation in the study area exhibits many lithofacies types characterised by frequent variations, pronounced lamination and strong heterogeneity. To effectively describe the unique features of these diverse lithofacies, a ternary classification system based on "Sedimentary Structures—Mineral Composition—Lithology" has been employed [33]. Within this framework, the study area primarily encompasses four significant categories, encompassing a total of six lithofacies (Table 1): clay-laminated claystone (A1), clay-laminated felsic shale (A2), silt-laminated felsic shale (A3), massive mudstone (B), silty mudstone and muddy siltstone (C) and siltstone (D) (Figure 5a–f).

Table 1. Lithofacies classification bases of the K_2qn^1 Formation.

Sample Number Depth (m)		Lithofacies Types	Quartz	Feldspar	Clay Minerals	Mineral Characteristics	Sedimentary Structure Characteristics
A	A1	Clay laminated clayey shale	17.4–33.2 22.55	9.5–18.5 15.47	34.1–46.3 39.12	felsic laminae < clayey laminae	Laminated Single lamina thickness < 1 cm Interbedding of felsic laminae and clayey laminae
	A2	Clay laminated felsic shale	21.7–33.4 28.25	12.9–23.5 18.62	28.4–40.4 36.01	felsic laminae > clayey laminae	
	A3	Silty laminated felsic shale	27.6–35.6 31.78	18.5–26.4 25.3	8.4–39.4 28.29	felsic laminae > 50%	
B	B	Blocky mudstone	25.4–34.1 31.4	15.25–25.7 21.24	20.24–36.1 28.91	Clayey content > 50% Silt-sized particles < 25%	Massive
C	C	Silty mudstone and argillaceous sandstone	31.4–41.1 35.21	24.5–42.7 36.85	10.27–15.1 12.57	Felsic content > 50%	Layered or massive With internal deformation structures present in some parts
D	D	Siltstone	34.7–42.4 38.25	33.2–64.5 51.2	8.4–12.1 11.4	Felsic content > 50% Silt-sized particles > 75%	Layered or massive

Note: 17.4–33.2, 22.55 stands for min. to max. and ave.

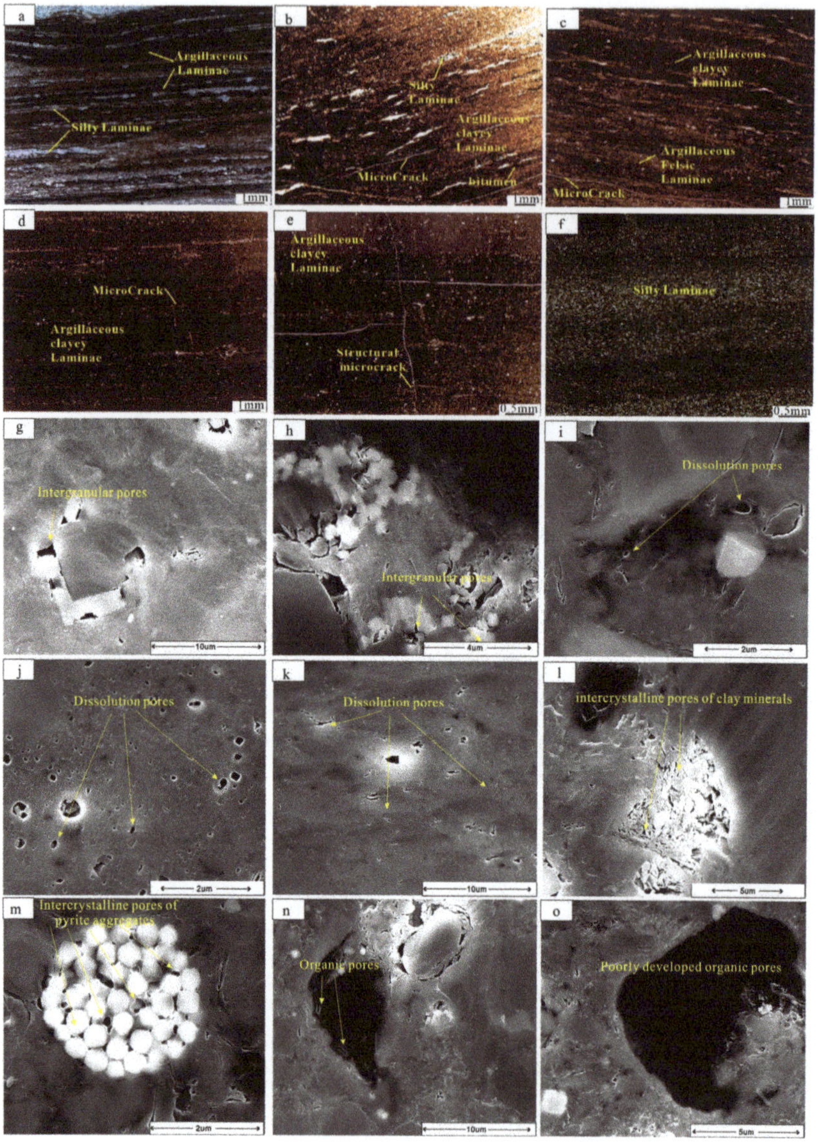

Figure 5. Thin-section characteristics of main shale lithologies in the Qing1 member of the Changling Sag. (**a**) Low-organic-matter-content laminated A3; (**b**) medium-organic-matter-content lamellar A2; (**c**) high-organic-matter-content lamellar A1; (**d**) high-organic-matter-content interlayer microfracture A2; (**e**) high-organic-matter-content interlayer microfracture A2; (**f**) inorganic matter A3; (**g**) interparticle pores in lithofacies type C; (**h**) interparticle pores in lithofacies type A2; (**i**) dissolution pores in lithofacies type A2; (**j**) dissolution pores in lithofacies type A3; (**k**) dissolution pores in lithofacies type A1; (**l**) intercrystalline pores in lithofacies type A1; (**m**) intercrystalline pores within pyrite aggregates in lithofacies type A1; (**n**) organic matter pores in lithofacies type A1; (**o**) poorly developed organic matter pores in lithofacies type A3.

4.2. Differences in the Pore Structures of Different Lithofacies Reservoirs

4.2.1. Differences in Pore Types of Different Types of Lithofacies

The fine-grained sedimentary reservoir within the first section of the Qingshankou Formation exhibits diverse pore types. Drawing upon a comprehensive suite of experiments, including numerous cast thin sections, argon ion polished scanning electron microscopy (SEM), large field-of-view stitched SEM and energy-dispersive X-ray spectroscopy (EDS), this study identified four dominant pore types in the reservoir: interparticle pores, dissolution pores, intercrystalline pores and organic pores. Notably, organic pores are relatively underdeveloped (Figure 5).

The interparticle pores in the study area predominantly exist in the form of micropores or fractures, primarily comprising pore spaces formed by the mutual support between different particles, including brittle particles (quartz, feldspar, carbonate rocks and pyrite) and flexible particles (organic matter and clay minerals). The abundance of interparticle pores is highly susceptible to mechanical compaction and cementation processes. Microscopic observations reveal that these pores often exhibit a polygonal shape, characterised by straight edges and relatively large pore diameters ranging from 200 nm to 1 μm. These pores are developed across various lithofacies types (Figure 5g,h).

The dissolution pores are primarily distributed along the edges of mineral grains, exhibiting a harbour-like or serrated morphology. The pore diameters of interparticle dissolution pores mainly range from 200 nm to 1 μm. These dissolution pores generally exhibit good connectivity, significantly enhancing the reservoir properties of the formation. Intragranular dissolution pores, on the other hand, are primarily developed within feldspar grains, displaying an irregular serrated pattern, with pore diameters distributed mainly from 50 nm to 500 nm (Figure 5i–k).

Intercrystalline pores are predominantly developed between mineral crystals such as pyrite and clay. In the first section of the Qingshankou Formation within the study area, clay mineral-related intercrystalline pores are predominantly present, with pore diameters typically ranging from 100 nm to 500 nm. A minor occurrence of pyrite-related intercrystalline pores is also observed, with pore diameters commonly distributed between 200 nm and 500 nm (Figure 5i,m).

Organic matter pores positively contribute to the adsorptive properties of shale. During the hydrocarbon generation process, organic carbon consumption leads to the formation of organic matter pores within the reservoir. In the first section of the Qingshankou Formation, organic matter pores exhibit complex morphologies within or at the organic matter's edges. The pore diameters typically range from 50 nm to 1 μm, although organic matter pores constitute a relatively small proportion overall (Figure 5n,o).

Among the pore types in various lithofacies of the first section of the Qingshankou Formation in this region, dissolution pores and interparticle pores account for relatively high proportions, averaging approximately 37.1% and 36.9%, respectively. In lithofacies type A3, due to their higher content of brittle minerals, interparticle pores comprise a higher percentage, approximately 39.4%. In lithofacies types C and D, interparticle pores constitute approximately 44.7% of the pore types, while dissolution pores comprise about 33.95%. Intercrystalline pores exhibit similar proportions across all lithofacies, averaging approximately 19.6%. There exist notable differences in the proportion of pore types among various lithofacies, particularly among the mudstone and shale lithofacies types (A1, A2 and A3). Among these, the proportion of intercrystalline pores varies relatively little across lithofacies, whereas interparticle pores, dissolution pores and organic matter pores exhibit regular variations with changes in lithofacies types. As the lithofacies transition from A1 to A3, the content of brittle minerals gradually increases, the TOC content decreases, the clay mineral content rises and the proportion of interparticle pores gradually increases, while the proportions of dissolution pores and organic matter pores gradually decrease (Table 2).

Table 2. Distribution of pore types across diverse lithofacies types.

Lithofacies (Quantity)	Pore (wt.%)	Intercrystalline Pore Percentage (%)	Dissolved Pore Percentage (%)	Intergranular Pore Percentage (%)	Organic Pore Percentage (%)
A1	2.7–4.05 3.21	14.2–22.1 18.6	28.2–48.1 37.3	22.7–31.2 27.1	8.4–12.2 10.1
A2	3.35–5.21 4.21	13.7–23.7 18.1	36.4–51.2 44.2	30.1–39.2 35.2	8.2–10.4 9.4
A3	4.02–6.21 5.27	19.2–24.2 21.1	25.4–43.2 33.1	32.7–49.2 39.4	4.2–8.1 6.4

Note: 2.7–4.05, 3.21 stands for min. to max. and ave.

4.2.2. Differences in Pore Sizes of Different Types of Lithofacies

This study utilizes the self-similarity principle of fractal theory to classify fine-grained sedimentary reservoirs [16,17]. Pores of the same type exhibit identical physical characteristics, known as self-similarity. Based on variations in pore features, pores within sandstone and mudstone reservoirs of fine-grained sediments are systematically categorized. Pores with similar characteristics demonstrate comparable trends, and the inflexion points on the curves align with the pore throat threshold values that signify transitions between pore types. Consequently, the self-similarity evident in the curves, coupled with the relationship diagram between the cumulative porosity and throat radius of key well samples from diverse lithofacies types (Figure 6), reveals that the samples all display four distinct fractal segments. The consistent trends among pores with identical features indicate that the pore systems in the mudstone and shale reservoirs of the study area can be segmented into four types according to the proposed classification scheme. Specifically, the values of lgr = −2.3, lgr = −1.3 and lgr = −0.3 represent the pore throat thresholds for different pore types. Therefore, leveraging the self-similarity observed in the curves, we categorize pores into four classes: macropores (>500 nm), medium pores (50 nm to 500 nm), small pores (5 nm to 50 nm) and micropores (<5 nm).

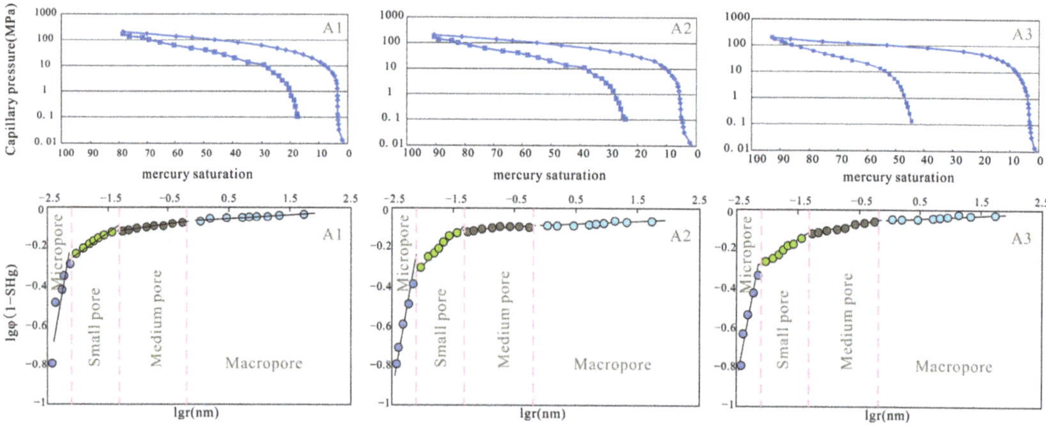

Figure 6. Fractal characteristics of pore throats in fine-grained sediments from the first section of Qingshankou Formation, categorized into four types: macropores (>500 nm), mesopores (50 nm to 500 nm), micropores (5 nm to 50 nm) and nanopores (<5 nm). A1: Clay laminated clayey shale, A2: Clay laminated felsic shale, A3: Silty laminated felsic shale.

Given that the data obtained from the high-pressure mercury injection primarily reflect the pore throat distribution, which deviates from the direct pore distribution, the mercury injection pressures do not often exceed 200 MPa, resulting in the suboptimal characterisation of micro- and nanopores compared to low-temperature nitrogen adsorption (LTNA). On the other hand, nuclear magnetic resonance (NMR) indirectly characterises pore space by detecting hydrogen nuclei within the rock's storage voids, necessitating a calibration factor to convert hydrogen signals into pore radii. This study employs LTNA to calibrate NMR, enabling comprehensive shale pore size characterisation. Firstly, the pore sizes from LTNA are converted into a cumulative curve. Subsequently, the porosity component values corresponding to each T2 point in NMR are divided by the total porosity to obtain the respective pore proportions. These proportions are then obtained from the most minor to the largest, yielding the NMR pore proportion cumulative curve H(T2). Based on the linear relationship between the NMR relaxation time T2 and pore radius r, H(T2) is transformed into a function of the pore radius H(r/k), where k is the linear coefficient. By iteratively adjusting k, the error between H(r/k) and the cumulative curve G(r) from LTNA is minimized, and the corresponding k value is designated as the calibration factor C. Using r = T2 x C, the NMR T2 spectrum is converted into an NMR pore throat distribution curve, quantitatively depicting the distribution of pore sizes across different ranges and generating a comprehensive pore size distribution map for the samples (Figure 7).

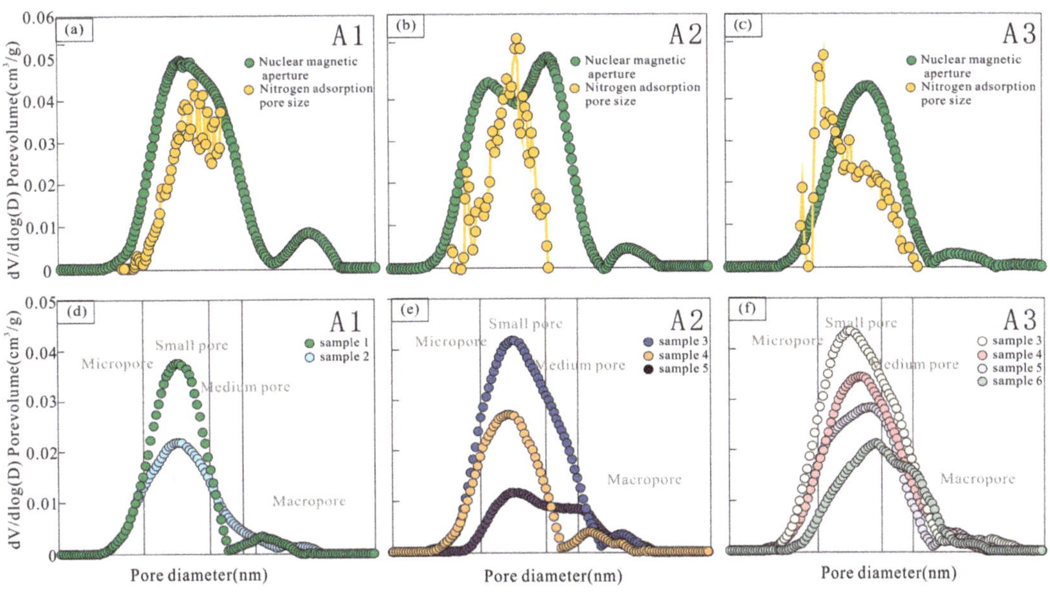

Figure 7. Comparison of pore size distribution between shale from the first section of Qingshankou Formation in the Daqingzijing area. (**a–c**) Low-temperature nitrogen adsorption calibration nuclear magnetic resonance diagram. (**d–f**) Comparison of pore size distribution of full pore size range. A1: Clay laminated clayey shale, A2: Clay laminated felsic shale, A3: Silty laminated felsic shale.

The calibrated pore size distribution of fine-grained sediments from the first section of the Qingshankou Formation primarily spans 5–1000 nm, with a dominant peak frequently located within 5–100 nm. This indicates a prevalence of medium pores, small pores and micropores in this region's fine-grained sediments. Specifically, mudstones and shales exhibit pore sizes mainly distributed between 1 and 100 nm, while siltstones' pore sizes range predominantly from 10 to 1000 nm, with a dominant peak in the 100–1000 nm range.

The difference in pore size between different lithofacies types is more obvious than that of pore type. Combined with the results of the full pore size characterisation (Table 3), it can be seen that in the four shale lithofacies, all kinds of lithofacies are dominated by micropores and small pores. Among them, the proportion of micropores and small pores of the A1 type is about 81.1%, the proportion of micropores and small pores of the A2 type is about 76.6%, the proportion of micropores and small pores of the A3 type is about 68.8% and the proportion of micropores and small pores of the B type is about 82.3%. Regarding mesopores and macropores, type A3 lithofacies have a high proportion of mesopores/macropores, with an average proportion of about 31.2%. The proportion of mesopores/macropores in type A1 lithofacies is the lowest, with an average proportion of about 18.9%. It shows that the classification of lithofacies types can effectively characterise the heterogeneity of reservoir characteristics of terrigenous clastic deposits.

Table 3. TOC and pore structure characteristics of lithofacies.

Lithofacies (Quantity)	TOC (wt.%)	Micropore Percentage (%)	Small Pore Percentage (%)	Medium Pore Percentage (%)	Macropore Percentage (%)
A1	0.45–5.7 2.14	35.2–42.2 40.3	28.2–48.1 40.8	10.5–18.4 14.9	1.4–7.4 4.0
A2	0.42–5.1 1.74	26.2–38.4 34.5	36.4–53.2 44.1	10.2–21.5 17.0	1.6–8.4 4.4
A3	0.35–2.9 1.41	17.4–41.2 24.9	37.4–52.1 43.9	24.2–38.7 21.1	7.2–15.1 10.1

Note: 0.45–5.7, 2.14 stands for min. to max. and ave.

5. Discussion

5.1. Influence of Mineral Fraction and Organic Matter on Pore Space

5.1.1. The Influence of Sedimentary Characteristics on Reservoir Pore Structure

The influence of the original deposition on reservoir pores is mainly reflected in mineral composition. The mineral content also has a certain impact on the reservoir characteristics [6,21,25,38]. Because the B type lithofacies are less developed in this area, this paper does not feature a targeted discussion of the controlling factors. This paper analyses the correlation between the pore size distribution characteristics of the three shale lithofacies types mainly developed in the study area and their different mineral contents (Figure 8).

In terms of feldspar content, with the increase in feldspar content, the proportion of micropores decreases, and the proportion of macropores + mesopores increases. In contrast, in A3 type lithofacies, the positive correlation between the feldspar content and the proportion of macropores + mesopores is stronger. This phenomenon is mainly due to the increase in feldspar content, the increase in the direct intergranular pores of brittle minerals and the fact that a part of the intergranular pores can be retained during diagenesis. At the same time, feldspar minerals are dissolved, which further increases the proportion of dissolved pores, resulting in the promotion of the feldspar mineral content to larger pores.

In terms of quartz mineral content, quartz mineral content has a good positive correlation with the proportion of mesopores and macropores in class A lithofacies, a weak positive correlation with the proportion of small pores and a poor correlation with the proportion of micropores. This phenomenon is mainly because in class A mudstone lithofacies, quartz and feldspar belong to brittle minerals with a similar content. Both of them jointly determine the development ratio of larger pores, while the quartz content is not the main controlling factor in small pores and micropores.

An excessive clay content will reduce the reservoir's compaction resistance. On the other hand, clay minerals contain a large number of intercrystalline pores, which leads to an increase in the proportion of micropores. Overall, the clay mineral content of each type of lithofacies is negatively correlated with the proportion of mesopores and macropores and positively correlated with the proportion of micropores and micropores.

Based on the above factors, it can be judged that for the type A lithofacies, quartz and feldspar have an improved effect on mesopores/macropores, and clay minerals will increase the proportion of micropores and small pores. Therefore, the pore throat structure of the type A3 lithofacies with a higher quartz feldspar content is better.

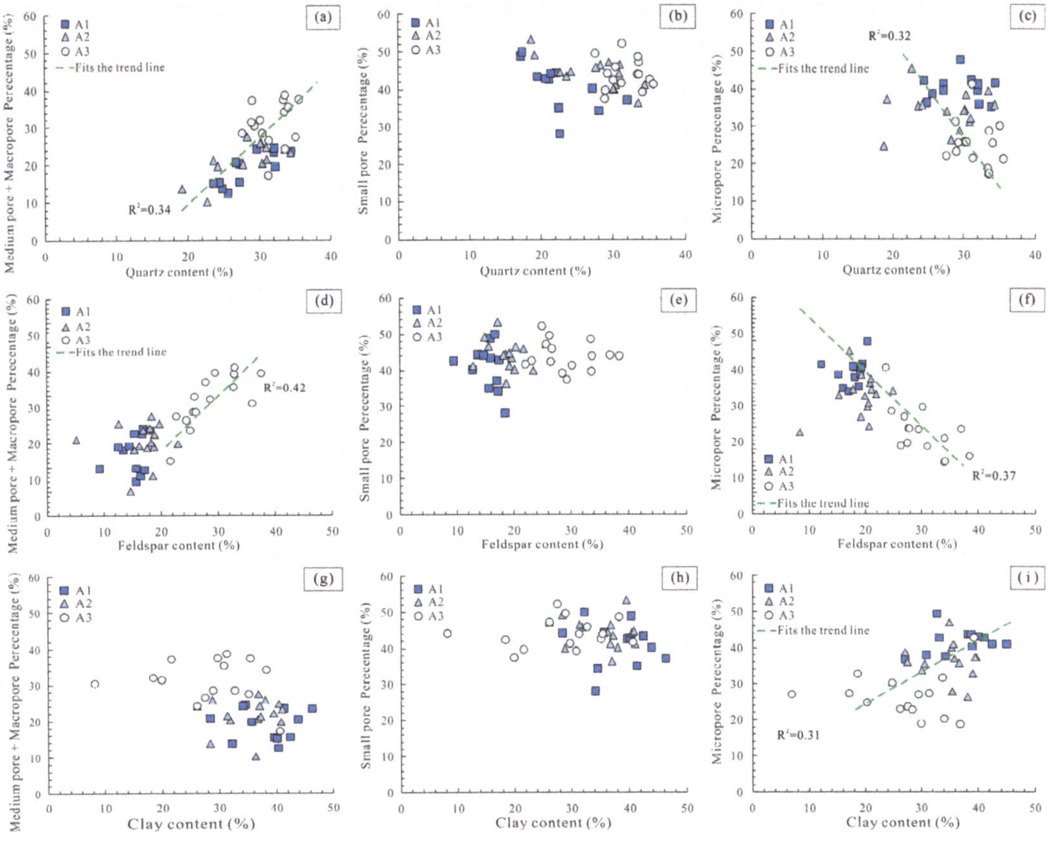

Figure 8. Analysis of the influence of different mineral content on pore structure. (**a**–**c**) The relationship between feldspar content and the proportion of macropores, micropores and micropores in different lithofacies; (**d**–**f**) The relationship between quartz content and the proportion of macropores, micropores and micropores in different lithofacies; (**g**–i) The relationship between clay content and the proportion of macropores, micropores and micropores in different lithofacies.

5.1.2. The Controlling Effect of Organic Matter Characteristics on Reservoir Pore Structure

The content of organic matter in fine-grained sediments also exerts a significant degree of control over the pore structure of reservoirs. By integrating the analysis of organic matter characteristics in the study area, we examine the factors that govern the pore structure characteristics of fine-grained sedimentary reservoirs based on the organic matter content. Tmax is a rapid indicator of organic matter maturity, although its actual results can be influenced by mineral composition and kerogen types. In the first section of the Qingshankou Formation within the study area, the average hydrogen index of fine-grained sediments ranges from 280.6 to 340.9 mg/g. According to the classification standard for organic matter types in source rocks (SY/T5735-2019) [6], the first section of the Qingshankou Formation predominantly comprises Type II kerogen (Figure 9a). Notably, the organic matter in the study area exhibits strong heterogeneity, with a wide distribution span across various lithofacies types (Figure 9b).

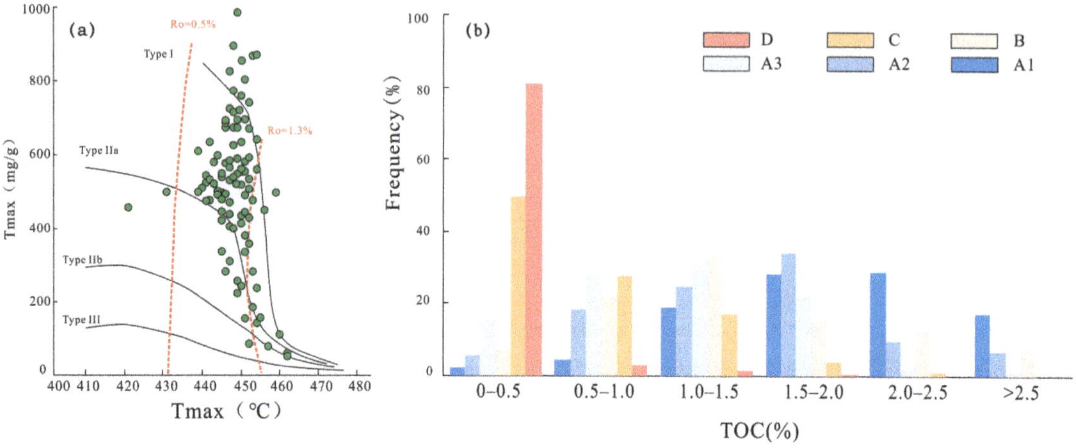

Figure 9. Organic matter characteristics of the first section of Qingshankou Formation. (**a**) Classification of organic matter types in the shale reservoirs of the first section of the Qingshankou Formation was determined based on the relationship between the hydrogen index (HI) and Tmax. (**b**) A statistical analysis is presented of the TOC content distribution characteristics across various lithofacies types within the first section of the Qingshankou Formation in the Daqingzijing area.

The statistical results indicate that the relationship between the total organic carbon (TOC) content and porosity does not always follow a linear pattern [33] (Figure 10a). Instead, an overall trend emerges, initially showing an increase, followed by a deceleration in the rate of increase, and finally, a gradual decline. When the TOC content is small, organic matter plays a positive role in enhancing pore space. Specifically, among various pore size characteristics, the proportion of medium pores and macropores gradually increases with rising TOC content. This is attributed to the fact that within this TOC content range, during hydrocarbon generation, organic acids dissolve minerals such as feldspar, resulting in an overall improvement to the reservoir. The amount of organic matter content is positively correlated with the proportion of dissolution pores (Figure 10d). Additionally, due to the relatively high feldspar content in A3 lithofacies, the increase in medium and macropores is steeper, whereas the rise in A1 and A2 lithofacies is more gradual at this stage.

When the TOC content is high, because the soluble organic matter fills the intergranular pores and the organic matter has flexible characteristics, an excessive organic matter content will reduce the compressive strength of the rock, which is not conducive to pore development. Regarding porosity, with the increase in organic matter content, the porosity and the proportion of mesopores and macropores increase slowly, and even the porosity decreases (Figure 10a,b). In terms of the proportion of micropores, all kinds of lithofacies types increase the proportion of micropores (Figure 10c). That is to say, with the increase in TOC content, some of the macropores and mesopores are filled with organic matter or transformed into micropores or small pores under compaction. At this stage, organic matter plays a destructive role in pore structure. There is a stable positive correlation between the development of organic matter pores and the content of TOC, and the difference between each lithofacies type is small, indicating that the content of organic matter has a significant controlling effect on its pore production (Figure 10e).

In the lithofacies types of categories C and D, the internal organic matter content is extremely low. However, due to the severe interbedding of terrigenous clastic fine-grained sedimentary sandstone and mudstone in this region, the acidic fluids generated by the thermal evolution of organic matter remain to act as significant erosive agents that contribute to pore formation through dissolution under burial conditions. This process leads to an increase in the content of feldspar dissolution pores within the mudstone and adjacent sandstone sections. The graphical representation indicates that as the distance between the sandstone and adjacent mudstone sections increases, the content of medium pores tends to increase. Additionally, a higher TOC content accelerates the dissolution of brittle minerals within the mudstone and subsequent erosion towards adjacent sandstone sections, resulting in a more significant increase in the proportion of medium pores (Figure 10f).

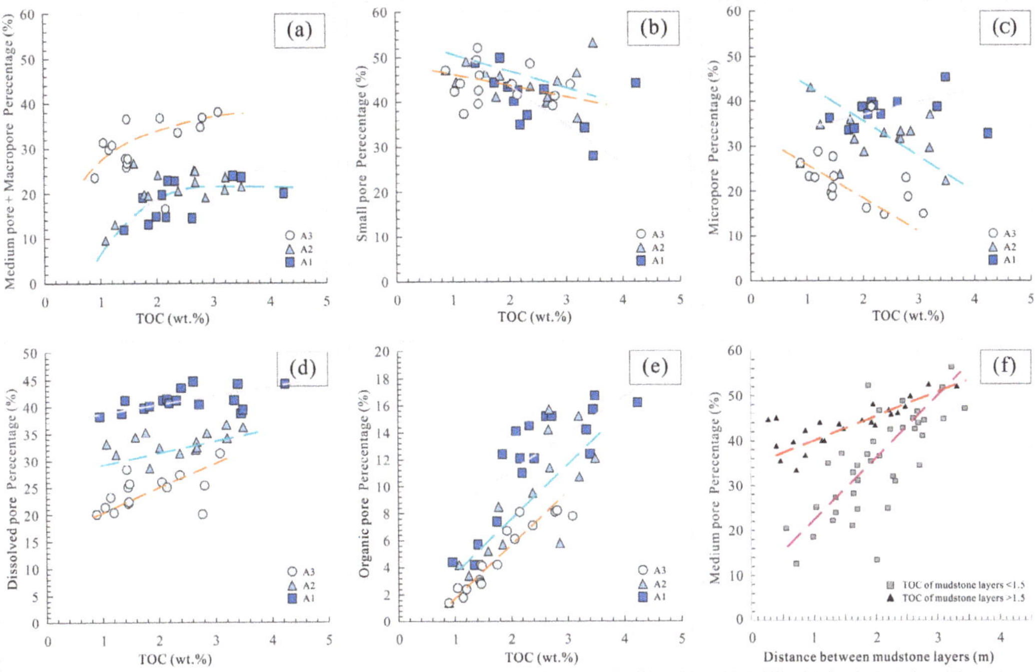

Figure 10. The controlling effect of organic matter content on reservoir characteristics. (**a–c**) The correlation between organic matter content and the proportion of mesopores + macropores, micropores and micropores. (**d**) The correlation between organic matter content and the proportion of dissolution pores. (**e**) The correlation between organic matter content and the proportion of organic matter pores. (**f**) The correlation between the proportion of mesopores in sandstone and the distance from mudstone with different TOC content.

5.2. Impact of Diagenesis on Pore Characteristics

5.2.1. Types of Diageneses

The content of quartz and feldspar in the shale reservoir of the first section of the Qingshankou Formation in the study area is high (Table 1). The soluble components provided by these brittle minerals can improve the pores of the shale reservoir [27,28,51–53]. Dissolution is one of the most important pore-increasing effects in the study area. Its occurrence is related to acidic fluids. Before entering the mature stage of organic matter, kerogen cracking produces a large amount of CO_2 and carboxylic acids, which are dissolved with soluble components to form secondary dissolution pores (Figure 11a–c). However, the intensity of dissolution is mainly controlled by two factors, which can be divided into the number of organic acids and the development of fluid exchange pathways. Both determine the pore-increasing efficiency of dissolution. Combined with the energy spectrum and SEM, the influence of feldspar on dissolution is analysed. Taking type A3 lithofacies as an example (Figure 11d), the thin sections show strong dissolution differences. The porosity of the upper part can reach 13%, and the lower part is only 1.5%. From the plane distribution map of elements (Figure 11e–i), the content of each element is similar, indicating that the distribution of inorganic minerals such as feldspar, quartz and calcite is similar. However, the difference between the upper and lower parts of the C content is obvious, which indicates that the upper part has a higher organic matter content, which can confirm that the TOC content analysed above plays an important role in the occurrence of dissolution [54,55].

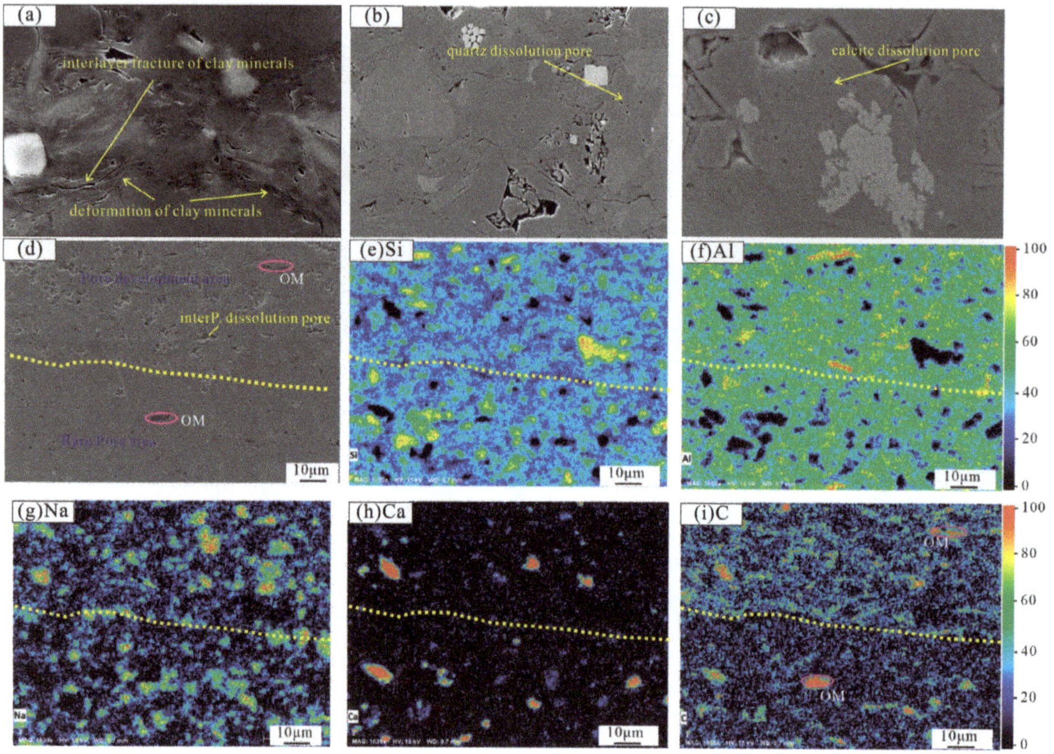

Figure 11. The SEM image and element image of type A3 lithofacies in the study area. (**a**–**c**) Dissolution phenomenon. (**d**) Yellow dotted line represents the boundary between the upper and lower parts, which show different pore development characteristics. (**e**–**i**) represent the spectral distribution of Si, Al, Na, C and C elements, respectively. Except for element C, the difference between the upper and lower elements is relatively small, and the upper part of element C is more developed than the lower part.

5.2.2. Changes in Pore Characteristics during Diagenetic Evolution

Based on the variation patterns of various parameters, the pore evolution can be delineated into four distinct stages (Figure 12). When the Ro maturity value is less than 0.85, compaction dominates diagenesis, leading to a rapid decrease in porosity. During this stage, the proportion of macropores and medium pores decreases while the proportion of small pores increases gradually. The predominant pore types at this stage are intergranular pores and intercrystalline pores.

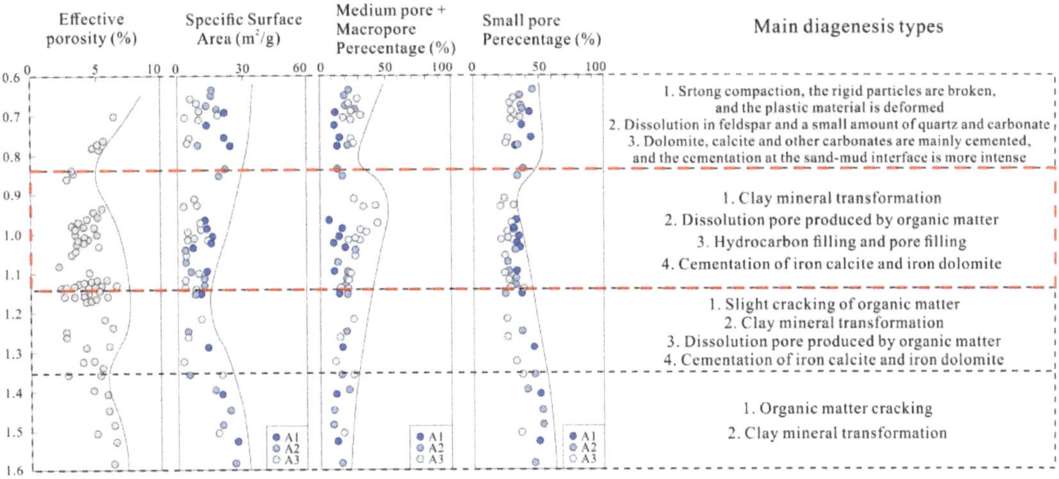

Figure 12. Evolution of shale pore structure in the first section of Qingshankou Formation in the Daqingzijing area.

As Ro ranges between 0.85 and 1.15, dissolution and hydrocarbon generation processes dominate. Affected by dissolution, the porosity exhibits an increasing trend. The proportion of medium pores and macropores increases while the specific surface area decreases, primarily due to the filling of micropores by soluble organic matter. The dominant pore types during this stage are intergranular pores and dissolution pores, with a minor presence of organic pores.

Between Ro values of 1.15 and 1.35, the stages are marked by ferroan dolomite cementation, the enhanced cracking of soluble organic matter and clay transformation. As a result, the porosity and the proportion of macropores and medium pores decrease while the specific surface area increases. This indicates that soluble organic matter or bitumen begins to crack, releasing micropores and small pores. The primary pore types at this stage are intercrystalline pores and a limited number of organic pores.

When Ro exceeds 1.35, the cracking of organic matter and clay transformation intensify further, resulting in the formation of abundant organic pores. Overall, porosity increases while the specific surface area and the proportion of small pores decrease. The proportion of macropores and medium pores continues to decline. The dominant pore types in this final stage are organic pores and intercrystalline pores of clay minerals.

5.2.3. Pore Evolution Model for Dominant Lithofacies Reservoirs

Based on the previous research results, it can be found that among different lithofacies types, A3 lithofacies have a higher porosity and a higher proportion of mesopores/macropores and are often identified as dominant lithofacies reservoirs in the process of production and development. These pore structure characteristics are affected by factors such as mineral content and organic matter content on one hand and organic matter maturity, mechanical compaction and dissolution in the diagenetic process on the other hand. Organic matter is often deposited with clay minerals to form an organic matter–clay mineral complex.

Drawing on the delineation of diagenetic evolution stages and alterations in pore types, a pore evolution model for the premium reservoirs of A3 and D lithofacies is established (Figure 13). At the initial burial stage, when the organic matter maturity is low (Ro < 0.5%), the sedimentary structure is relatively loose. Early compaction leads to a tighter packing of rock grains, resulting in reduced pore sizes and porosity. The dominant pore types are intergranular and intercrystalline pores.

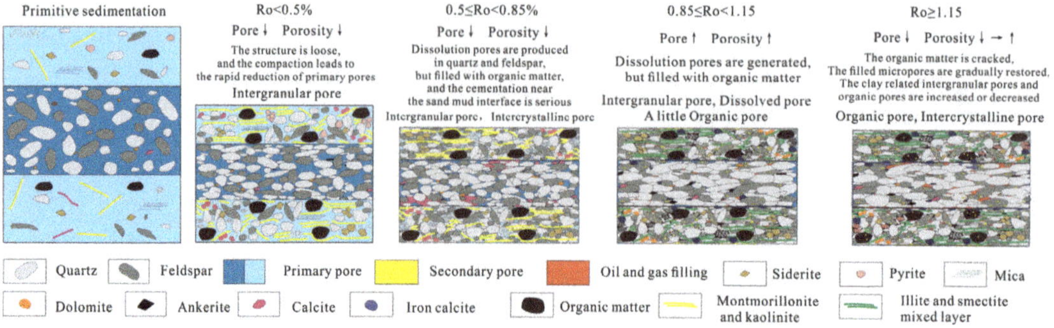

Figure 13. Pore evolution model of shale in the first section of Qingshankou Formation in the Daqingzijing area.

As the burial depth increases, the organic matter maturity ranges between 0.5% and 0.85%, marked by the onset of calcite cementation and dissolution. During this phase, pore sizes and porosity decrease further, with intergranular and intercrystalline pores persisting and some dissolution pores emerging.

In the maturity range of 0.85% to 1.15%, hydrocarbon generation and dissolution by organic matter facilitate the development of additional dissolution pores, while soluble organic matter partially fills micropores. This stage witnesses enlarged pore sizes and an increased porosity, with pore types transitioning from primarily intergranular and intercrystalline to a mix dominated by intergranular and dissolution pores, along with the initial appearance of organic pores.

When Ro > 1.15, soluble organic matter undergoes cracking, and the clay mineral transformation restores previously filled micropores. Clay-mineral-related intercrystalline pores emerge, accompanied by decreased pore sizes and porosity. The pore types gradually shift to being dominated by intercrystalline pores, with a minority of organic pores.

On the whole, the dissolution of organic acid emissions in the process of organic matter hydrocarbon generation can promote porosity and pore connectivity. Although the TOC of type A3 lithofacies is relatively lower than that of A1, the higher content of rigid particles (such as quartz and feldspar) can form a support skeleton and protect the residual primary pores. The A3 type belongs to the reservoir with dominant lithofacies.

6. Conclusions

This paper conducts a detailed investigation into the pore structure characteristics and controlling factors of various lithofacies types within the first section of the Qingshankou Formation in the Daqingzijing area, Southern Songliao Basin. The key findings are summarized as follows:

1. The fine-grained sediments of the first member of the Qingshankou Formation in the Daqingzijing area of the Southern Songliao Basin are mainly composed of clay minerals, quartz and feldspar. The lithofacies types can be divided into four categories and six subcategories. They are mainly A1 (clay laminated clayey shale), A2 (clay laminated felsic shale), A3 (silty laminated felsic shale), B (blocky compacted), C (silty compacted and argillaceous shale) and D (siltstone). The A3 type is due to the high content of silty lamina. It has a high brittle mineral content, and the proportion of pore size in different lithofacies is different.
2. Sedimentation mainly controls the development of reservoir pore structure through the brittle mineral content, organic matter abundance, and degree of thermal evolution. In terms of the mineral content, the lithofacies with a high brittle mineral content have a stronger anti-compaction effect and more retained primary pores. Therefore, the pore throat structure of type A3 lithofacies with a high quartz feldspar content is better. The influence of organic matter on reservoir quality manifests in two ways. An

appropriate organic matter content produces organic acid in the process of thermal evolution to promote the formation of secondary dissolution pores in mudstone and sandstone near the mudstone section and, at the same time, provides a material basis for later organic matter pores. However, an excessive organic matter content will block the pores and reduce the compressive strength of the rock, which is not conducive to the pore preservation of the rock.

3. The A3 (silty laminated felsic shale) shale reservoir has the most favourable lithofacies in the study area and experienced a pore evolution process characterised by "two drops and two rises": compaction, cementation and pore reduction; dissolution and pore increase; and organic matter cracking and pore increase.

Author Contributions: Conceptualization, X.W., Y.S., and T.W.; methodology, B.Y.; validation, R.L. and B.Y.; formal analysis, X.W. and Y.S.; investigation, Y.S.; resources, T.W. and B.Y.; data curation, T.W.; writing—original draft preparation, X.W.; writing—review and editing, X.W. and R.L.; visualization, R.L.; supervision, Y.S. and B.Y.; funding acquisition, T.W. All authors have read and agreed to the published version of the manuscript.

Funding: This research was funded by the Project of Sanya Yazhou Bay Science and Technology City (No. SCKJ-JYRC-2022-09) and Heilongjiang Provincial Natural Science Foundation of China (No. ZD2023D002).

Data Availability Statement: All relevant data are contained within the paper.

Conflicts of Interest: The authors declare no conflicts of interest.

References

1. Zou, C.; Tao, S.; Yang, Z.; Yuan, X.; Zhu, R.; Hou, L.; Jia, J.; Wang, L.; Wu, S.; Bai, B.; et al. New Advance in Unconventional Petroleum Exploration and Research in China. *Bull. Mineral. Petrol. Geochem.* **2012**, *31*, 312–322.
2. Wei, Y.; Ran, Q.; Tong, M.; Wang, Z. A full cycle productivity prediction model of fractured horizontal well in tight oil reservoirs. *J. Southwest Pet. Univ. (Sci. Technol. Ed.)* **2016**, *38*, 99–106.
3. Zhao, W.; Jia, A.; Wei, Y.; Wang, J.; Zhu, H. Progress in shale gas exploration in China and prospects for future development. *China Pet. Explor.* **2020**, *25*, 31–44.
4. Zou, C.; Zhao, Q.; Cong, L.; Wang, H.; Shi, Z.; Wu, J.; Pan, S. Development progress, potential and prospect of shale gas in China. *Nat. Gas Ind.* **2021**, *41*, 1–14.
5. Jin, Z.; Zhu, R.; Liang, X.; Shen, Y. Several issues worthy of attention in current lacustrine shale oil exploration and development. *Pet. Explor. Dev.* **2021**, *48*, 1471–1484. [CrossRef]
6. Liu, B.; Wang, H.; Fu, X.; Bai, Y.; Bai, L.; Jia, M.; He, B. Lithofacies and depositional setting of a highly prospective lacustrine shale oil succession from the Upper Cretaceous Qingshankou Formation in the Gulong sag, northern Songliao Basin, northeast China. *AAPG Bull.* **2019**, *103*, 405–432. [CrossRef]
7. Liu, H.; Zhang, S.; Song, G.; Wang, X.; Teng, J.; Wang, M. Effect of shale diagenesis on pores and storage capacity in the paleogene Shahejie Formation, dongying depression, Bohai Bay Basin, east China. *Mar. Pet. Geol.* **2019**, *103*, 738–752. [CrossRef]
8. Zou, C.; Zhu, R.; Bai, B.; Yang, Z.; Wu, S.; Su, L.; Dong, D.; Li, X. First discovery of nano-pore throat in oil and gas reservoir in China and its scientific value. *Acta Pet. Sin.* **2011**, *27*, 1857–1864.
9. Loucks, R.; Reed, R.; Ruppel, S.; And, D.; Jarvie, D. Morphology, genesis, and distribution of nanometer-scale pores in siliceous mudstones of the mississippian Barnett shale. *J. Sediment. Res.* **2009**, *79*, 848–861. [CrossRef]
10. Kirchofer, A.; Firouzi, M.; Psarras, P.; Wilcox, J. Modeling CO_2 transport and sorption in carbon slit pores. *J. Phys. Chem. C* **2017**, *121*, 21018–21028. [CrossRef]
11. Bernard, S.; Wirth, R.; Schreiber, A.; Schulz, H.; Horsfield, B. Formation of nanoporous pyrobitumen residues during maturation of the Barnett shale (fort worth basin). *Int. J. Coal Geol.* **2012**, *103*, 3–11. [CrossRef]
12. Jie, X.; Yan, Z.; Yang, W.; Chen, S.; Jiang, Z. Structural deformation and its pore-fracture system response of the Wufeng-Longmaxi shale in the Northeast Chongqing area, using FE-SEM, gas adsorption, and SAXS. *J. Pet. Sci. Eng.* **2022**, *209*, 109877.
13. Xie, X.; Deng, H.; Fu, M.; Hu, L.; He, J. Evaluation of pore structure characteristics of four types of continental shales with the aid of low-pressure nitrogen adsorption and an improved FE-SEM technique in Ordos Basin, China. *J. Pet. Sci. Eng.* **2020**, *197*, 108018. [CrossRef]
14. Peng, S.; Zhang, T.; Loucks, R.; James, S. Application of mercury injection capillary pressure to mudrocks: Conformance and compression corrections. *Mar. Pet. Geol.* **2017**, *88*, 30–40. [CrossRef]
15. Han, H.; Guo, C.; Zhong, N.; Pang, P.; Gao, Y. A study on fractal characteristics of lacustrine shales of Qingshankou Formation in the Songliao Basin, northeast China using nitrogen adsorption and mercury injection methods. *J. Pet. Sci. Eng.* **2020**, *193*, 107378. [CrossRef]

16. Zhang, F.; Jiang, Z.; Sun, W.; Li, Y.; Zhang, X.; Zhu, L.; Wen, M. A multiscale comprehensive study on pore structure of tight sandstone reservoir realized by nuclear magnetic resonance, high pressure mercury injection and constant-rate mercury injection penetration test. *Mar. Pet. Geol.* **2019**, *109*, 208–222. [CrossRef]
17. Mastalerz, M.; He, L.; Melnichenko, Y.; Rupp, J. Porosity of coal and shale: Insights from gas adsorption and SANS/USANS techniques. *Energy Fuels* **2012**, *26*, 5109–5120. [CrossRef]
18. Clarkson, C.; Freeman, M.; He, L.; Agamalian, M.; Melnichenko, Y.; Mastalerz, M.; Bustin, R.; Radlinski, A.; Blach, T. Characterization of tight gas reservoir pore structure using USANS/SANS and gas adsorption analysis. *Fuel* **2012**, *95*, 371–385. [CrossRef]
19. Huang, Z.; Chen, J.; Wang, Y.; Deng, C.; Xue, H. Pore Distribution of Source Rocks as Reavealed by Gas Adsorption and Mercury Injection Methods: A Case Study on the First Member of the Cretaceous Qingshankou Formation in the Songliao Basin. *Geol. Rev.* **2013**, *59*, 587–594.
20. Kuang, L.; Hou, L.; Wu, S.; Cui, J.; Tian, H.; Zhang, L. Organic matter occurrence and pore-forming mechanisms in lacustrine shales in China. *Pet. Sci.* **2022**, *19*, 1460–1472. [CrossRef]
21. Curtis, M.; Ambrose, R.; Sondergeld, C.; Rai, C. Transmission and Scanning Electron Microscopy Investigation of Pore Connectivity of Gas Shales on the Nanoscale. In Proceedings of the North American Unconventional Gas Conference and Exhibition, The Woodlands, TX, USA, 14–16 June 2011; Society of Petroleum Engineers: Kuala Lumpur, Malaysia, 2011; Volume 6.
22. Milliken, K.; Rudnicki, M.; Awwiller, D.; Zhang, T. Organic matter–hosted pore system, Marcellus Formation (Devonian), Pennsylvania. *AAPG Bull.* **2013**, *97*, 177–200. [CrossRef]
23. Curtis, M.; Cardott, B.; Sondergeld, C.; Rai, C. Development of organic porosity in the Woodford Shale with increasing thermal maturity. *Int. J. Coal Geol.* **2012**, *103*, 26–31. [CrossRef]
24. He, W.; Zhu, R.; Cui, B.; Zhang, S.; Meng, Q.; Bai, B.; Feng, Z.; Lei, Z.; Wu, S.; He, K.; et al. The geoscience frontier of Gulong shale oil: Revealing the role of continental shale from oil generation to production. *Engineering* **2023**, *28*, 79–92. [CrossRef]
25. Hart, B.; Schieber, J.; Kalinec, J. Clay diagenesis and overpressure development in upper cretaceous and tertiary shales of south Texas. *Mar. Pet. Geol.* **2023**, *147*, 105978. [CrossRef]
26. Liang, C.; Cao, Y.; Liu, K.; Jiang, Z.; Wu, J.; Hao, F. Diagenetic variation at the lamina scale in lacustrine organic-rich shales: Implications for hydrocarbon migration and accumulation. *Geochim. Cosmochim. Acta* **2018**, *229*, 112–128. [CrossRef]
27. Teng, J.; Qiu, L.; Zhang, S.; Ma, C. Origin and diagenetic evolution of dolomites in Paleogene Shahejie Formation lacustrine organic shale of Jiyang Depression, Bohai Bay Basin, East China. *Pet. Explor. Dev.* **2022**, *49*, 1251–1265. [CrossRef]
28. Chen, L.; Zhang, H.; Liang, J.; Zhu, H.; Zheng, Y. Effect of burial processes on the evolution of organic acids and implications for acidic dissolution from a case study of the Nanpu Sag, Bohai Bay Basin, China. *J. Nat. Gas Sci. Eng.* **2017**, *39*, 173–187. [CrossRef]
29. Chen, Q.; Xu, X.; Wang, S.; Li, W.; He, S.; Yang, H. Dissolution pores in shale and their influence on reservoir quality in Damintun Depression, Bohai Bay Basin, East China: Insights from SEM images, N2 adsorption and fluid-rock interaction experiments. *Mar. Pet. Geol.* **2020**, *117*, 104394. [CrossRef]
30. Sun, N.; Chen, T.; Gao, J.; Zhong, J.; Huo, Z.; Qu, J. Lithofacies and reservoir characteristics of saline lacustrine fine-grained sedimentary rocks in the northern Dongpu Sag, Bohai Bay Basin: Implications for shale oil exploration. *Asian Earth Sci.* **2023**, *252*, 105686. [CrossRef]
31. Loucks, R.; Reed, R.; Ruppel, S.; Hammes, U. Spectrum of pore types and networks in mudrocks and a descriptive classification for matrix-related mudrock pores. *AAPG Bull.* **2012**, *96*, 1071–1098. [CrossRef]
32. Zhou, Y.; Ji, Y.; Zhang, S.; Wan, L. Controls on reservoir quality of Lower Cretaceous tight sandstones in the Laiyang Sag, Jiaolai Basin, eastern China: Integrated sedimentologic, diagenetic and microfracturing data. *Mar. Pet. Geol.* **2016**, *76*, 26–50. [CrossRef]
33. Xiao, D.; Zheng, L.; Xing, J.; Wang, M.; Wang, R.; Guan, X.; Guo, X. Coupling control of organic and inorganic rock components on porosity and pore structure of lacustrine shale with medium maturity: A case study of the Qingshankou Formation in the southern Songliao Basin. *Mar. Pet. Geol.* **2024**, *164*, 106844. [CrossRef]
34. Wang, X.; Wang, M.; Li, J.; Shao, H.; Deng, Z.; Wu, Y. Thermal maturity: The controlling factor of wettability, pore structure, and oil content in the lacustrine Qingshankou shale, Songliao Basin. *J. Pet. Sci. Eng.* **2022**, *215*, 110618. [CrossRef]
35. Shen, R.; Xu, R.; Li, Y. Microscopic pore structure and solid-liquid force of Chang 73 shale in Ordos basi. *J. Cent. South Univ. (Sci. Technol.)* **2024**, *55*, 2208–2221.
36. Lin, T.F.; Fu, X.L.; Ai, X. Lithofacies types and cycle patterns of shale layers of Qingshankou Formation in Gulong Sag, Songliao Basin. *World Geol.* **2024**, *43*, 378–389.
37. Zhang, H.; Feng, Y.L.; Yang, Z. Shale oil reservoir characteristics and its formation mechanism within alkline lacustrine basins: A case study of the Fengcheng Formation in Mahu sag, Junggar basin, China. *Acta Geol. Sin.* **2024**, 1–16.
38. Sun, L.; Liu, H.; He, W.; Li, G.; Zhang, S.; Zhu, R. An analysis of major scientific problems and research paths of Gulong shale oil in Daqing Oilfield, NE China. *Pet. Explor. Dev.* **2021**, *48*, 527–540. [CrossRef]
39. Liu, C.; Xu, X.; Liu, K.; Bai, J.; Liu, W.; Chen, S. Pore-scale oil distribution in shales of the Qingshankou Formation in the changling sag, Songliao Basin, NE China. *Mar. Pet. Geol.* **2020**, *120*, 104553. [CrossRef]
40. Bruce, C.H. Smectite dehydration—Its relation to structural development and hydrocarbon accumulation in northern Gulf of Mexico basin. *AAPG Bull.* **1984**, *68*, 673–683.
41. Schmitt, M.; Fernandes, C.; Cunha Neto, J.; Wolf, F.; Dos Santos, V. Characterization of pore systems in seal rocks using nitrogen gas adsorption combined with mercury injection capillary pressure techniques. *Mar. Pet. Geol.* **2013**, *39*, 138–149. [CrossRef]

42. Feng, Z.; Jia, C.; Xie, X.; Zhang, S.; Feng, Z.; Cross, T. Tectonostratigraphic units and stratigraphic sequences of the nonmarine Songliao basin. *Basin Res.* **2010**, *22*, 79–95.
43. Wang, X.; Sun, Y.; Liu, R.; Li, Z. Research progress into fine-grained sedimentary rock characteristics and formation in a continental lake basin. *Acta Sedimentol. Sin.* **2023**, *41*, 349–377.
44. Li, Z.; Chen, J.; Zou, H.; Wang, C.; Meng, Q.; Liu, H.; Wang, S. Mesozoic–Cenozoic tectonic evolution and dynamics of the Songliao Basin, NE Asia: Implications for the closure of the Paleo-Asian Ocean and Mongol-Okhotsk Ocean and subduction of the Paleo-Pacific Ocean. *Earth-Sci. Rev.* **2021**, *218*, 103471. [CrossRef]
45. Liu, R.; Sun, Y.; Wang, X.; Yan, B.; Yu, L.; Li, Z. Production Capacity Variations of Horizontal Wells in Tight Reservoirs Controlled by the Structural Characteristics of Composite Sand Bodies: Fuyu Formation in the Qian'an Area of the Songliao Basin as an Example. *Processes* **2023**, *11*, 1824. [CrossRef]
46. Wu, H.; Zhang, S.; Jiang, G.; Huang, Q. The floating astronomical time scale for the terrestrial Late Cretaceous Qingshankou Formation from the Songliao Basin of Northeast China and its stratigraphic and paleoclimate implications. *Earth Planet. Sci. Lett.* **2009**, *278*, 308–323. [CrossRef]
47. Hazra, B.; Varma, K.; Bandopadhyay, K. FTIR, XRF, XRD and SEM characteristics of Permian shales, India. *J. Nat. Gas Sci. Eng.* **2016**, *32*, 239–255. [CrossRef]
48. Gao, Y.; Wang, M.; Li, Y.; Jiang, Z.; Deng, Y.; Qin, J. Multi-scale pore structure characterization of lacustrine fine-grained mixed sedimentary rocks and its controlling factors: A case study of Lucaogou Formation in Jimusar Sag. *Energy Fuels* **2023**, *37*, 977–992. [CrossRef]
49. Ross, D.J.; Bustin, R.M. The importance of shale composition and pore structure upon gas storage potential of shale gas reservoirs. *Mar. Pet. Geol.* **2009**, *26*, 916–927. [CrossRef]
50. He, H.; Liu, P.; Xu, L.; Hao, S.; Qiu, X.; Shan, C.; Zhou, Y. Pore structure representations based on nitrogen adsorption experiments and an FHH fractal model: Case study of the block Z shales in the Ordos Basin, China. *J. Pet. Sci. Eng.* **2021**, *203*, 108661. [CrossRef]
51. Lastoskie, C.; Gubbins, K.E.; Quirke, N. Pore size distribution analysis of microporous carbons: A density functional theory approach. *J. Phys. Chem.* **1993**, *97*, 4786–4796. [CrossRef]
52. Jin, Z.; Liang, X.; Bai, Z. Exploration breakthrough and its significance of Gulong lacustrine shale oil in the Songliao Basin, Northeastern China. *Energy Geosci.* **2022**, *3*, 120–125. [CrossRef]
53. Loucks, R.; Ruppel, S. Mississippian Barnett shale: Lithofacies and depositional setting of a deep-water shale-gas succession in the Fort Worth Basin, Texas. *AAPG Bull.* **2007**, *91*, 579–601. [CrossRef]
54. Zhang, T.; Ellis, G.; Ruppel, S.; Milliken, K.; Yang, R. Effect of organic-matter type and thermal maturity on methane adsorption in shale-gas systems. *Org. Geochem.* **2012**, *47*, 120–131. [CrossRef]
55. Jiang, Z.; Guo, L.; Liang, C. Lithofacies and sedimentary characteristics of the silurian longmaxi shale in the southeastern Sichuan Basin, China. *J. Palaeogeogr.* **2013**, *2*, 238–251.

minerals

MDPI

Article

Effect of Paleoenvironmental Conditions on the Distribution of Lower Carboniferous Shale in Yaziluo Rift Trough, South China: Insights from Major/Trace Elements and Shale Composition

Xianglin Chen [1,2], Qiuchen Xu [1,2,*], Yinglun Qin [3,*], Rong Chen [1,2], Yufang Wang [1,2], Dishi Shi [1,2], Jing Bai [1,2], Kun Yuan [1,2], Wenpan Cen [4], Fei Li [1,2] and Tuo Lin [1,2]

1 Oil and Gas Survey, China Geological Survey, Beijing 100083, China; chenxianglin@mail.cgs.gov.cn (X.C.); chenrong@mail.cgs.gov.cn (R.C.); wangyufang@mail.cgs.gov.cn (Y.W.); shidishi@mail.cgs.gov.cn (D.S.); baijing@mail.cgs.gov.cn (J.B.); yuankun@mail.cgs.gov.cn (K.Y.); lifeicgs@mail.cgs.gov.cn (F.L.); lintuo@mail.cgs.gov.cn (T.L.)
2 State Key Laboratory of Continental Shale Oil, Beijing 100083, China
3 Guangxi Energy Group Co., Ltd., Nanning 530201, China
4 Geological Survey Institute of Guangxi Zhuang Autonomous Region, Nanning 530023, China; cenwp23@163.com
* Correspondence: xuqiuchen@mail.cgs.gov.cn (Q.X.); qinyinglun1990@163.com (Y.Q.)

Abstract: Paleoenvironmental conditions significantly influence the distribution patterns and organic matter enrichment of shale. This study investigated the vertical variations of major elements, trace elements, and total organic carbon (TOC) in the Lower Carboniferous marine shale from the Yaziluo Rift Trough, South China, to understand the paleoenvironmental conditions, including redox conditions, terrigenous detrital input, paleoproductivity, and paleo-seawater depth. The Lower Carboniferous formation consists of three sedimentary facies: basin facies, lower slope facies, and upper slope facies. From the basin to the lower slope and then to the upper slope facies, TOC, quartz, and pyrite contents gradually decrease, whereas the carbonate mineral content shows an increasing trend. A continuous decline in paleo-seawater depth transformed a deep-water anoxic environment with high paleoproductivity and low detrital input in the basin facies into a semi-deep-water environment with dysoxic-oxic conditions and moderate detrital influx in the lower slope facies, evolving further into a suboxic environment with high detrital flux in the upper slope facies. The geochemistry results suggest that anoxic conditions and high paleoproductivity were the primary controls on organic matter enrichment in the siliceous shale of the basin facies. In contrast, redox conditions significantly influenced organic matter accumulation in the mixed shale of the lower slope facies, attributed to relatively low paleoproductivity in a more restricted marine setting. Additionally, the adsorption of carbon components by clay minerals facilitated the preservation of organic matter in the calcareous shale of the upper slope facies.

Keywords: Lower Carboniferous shales; paleoenvironment; shale distribution; major/trace elements; Yaziluo Rift Trough

1. Introduction

Recently, the Lower Carboniferous marine shales of the Yaziluo Rift Trough have become a primary target for shale gas exploration and exploitation in southern China, which are characterized by medium total organic carbon (TOC) content, considerable thickness, and high levels of carbonate minerals [1–3]. The trough, shaped like a "V" in the cross-section, is a narrow, long trench formed in the continental crust under tectonic tension, bounded by high-angle syngenetic faults [4]. It features a "slope-basin" sedimentary structure that results in an environment with frequent changes in seawater energy and varied material sources during the deposition process [5]. The Yaziluo Rift Trough spans the southwestern Guizhou, southern Guizhou, and northern Guangxi areas in South China, recording the evolution of paleoclimate, paleoenvironment, and paleobios during a global marine transgression in the early Carboniferous period [6–8].

Much geological research has been conducted on the Lower Carboniferous shales, encompassing geological conditions, geochemical characteristics, and paleoenvironments. Yi et al. (2022) [9] suggest that the sedimentary facies and paleoenvironmental conditions of Lower Carboniferous shale on the northwest margin of the rift trough are platform facies and low-oxygen environments, respectively, based on the analysis of core samples. Su et al. (2017) [10] describe the paleoenvironment of Lower Carboniferous shale in the center of the Yaziluo Rift Trough as an oxygen-deficient environment, based on geochemical studies. Chen et al. (2017) [11] identify the paleoenvironment on the eastern side of the Yaziluo Rift Trough as a neritic environment with shallow marine shelf facies. Tao et al. (2022) [12] report that the sedimentary facies in the eastern part of the Yaziluo Rift Trough include basin facies and platform facies. Evidently, there is significant variation in the paleoenvironment, sedimentary facies types, and spatial distribution of lithofacies within the Lower Carboniferous formation of the Yaziluo Rift Trough. Understanding the impact of the paleoenvironment on shale lithofacies distribution patterns and organic matter enrichment is crucial for evaluating reservoir properties and identifying 'sweet spots' for exploration.

Compared to other regions in southern China, such as the Sichuan Basin, western Hubei, and eastern Chongqing, the Yaziluo Rift Trough has received relatively less attention in the early stages of shale gas exploration and exploitation. With increasing exploration and development, several exploratory wells in the Yaziluo Rift Trough have revealed shale gas with content greater than 1.0 m^3/t. The thickness, TOC content, and thermal maturity (R_o value) of the rich organic shale range from 100 to 400 m, 0.5 to 4.5%, and 2.6 to 3.2%, respectively. Despite being regarded as one of the most promising shale gas reservoirs, the depositional distribution patterns and enrichment mechanisms of organic matter in the Lower Carboniferous shales of the Yaziluo Rift Trough have not been systematically studied. Addressing these questions will provide a theoretical basis for the next stage of shale gas exploration, exploitation, and resource evaluation in the Yaziluo Rift Trough.

In this study, we present a suite of geochemical proxy data and sedimentological data from three wells located in the Yaziluo Rift Trough. The primary objectives of this research are as follows:

(i) To analyze the paleoenvironmental conditions of Lower Carboniferous shales in the Yaziluo Rift Trough by examining variations in geochemical proxy data.
(ii) To discuss the various controlling factors on organic matter accumulation in the Lower Carboniferous shale during the deposition process.
(iii) To reconstruct the lithofacies distribution pattern of Lower Carboniferous shale and propose a relevant model for the Yaziluo Rift Trough.

2. Geological Setting

The formation of the Yaziluo Rift Trough was initiated in the early Devonian period due to the expansion of the Paleo-Tethys Ocean and the stretching action along a NE-SW axis [13,14]. It reached its maximum extent during the middle Devonian, with sedimentary successions of shale extending as far as the Zhaotong region in Yunnan Province. In the early Carboniferous period, although the overall range of shale distribution contracted, the sedimentary characteristics largely retained the patterns established during the Devonian period [15,16]. The farthest extension of these formations along the northwestern direction of the Yaziluo Rift Trough reached the Weining area in Guizhou. Subsequent tectonic ruptures and extensional processes further stretched and subsided the crust, creating a pronounced contrast between deep-water and shallow-water depositional environments within the rift trough [13,15]. During the late Carboniferous period, the paleo-seawater depth gradually decreased from both sides of the rift trough along the line from Liupanshui to Hechi, resulting in the division of the Yaziluo Rift Trough into three distinct zones: a depocenter zone, a transition zone, and a margin zone (Figure 1A).

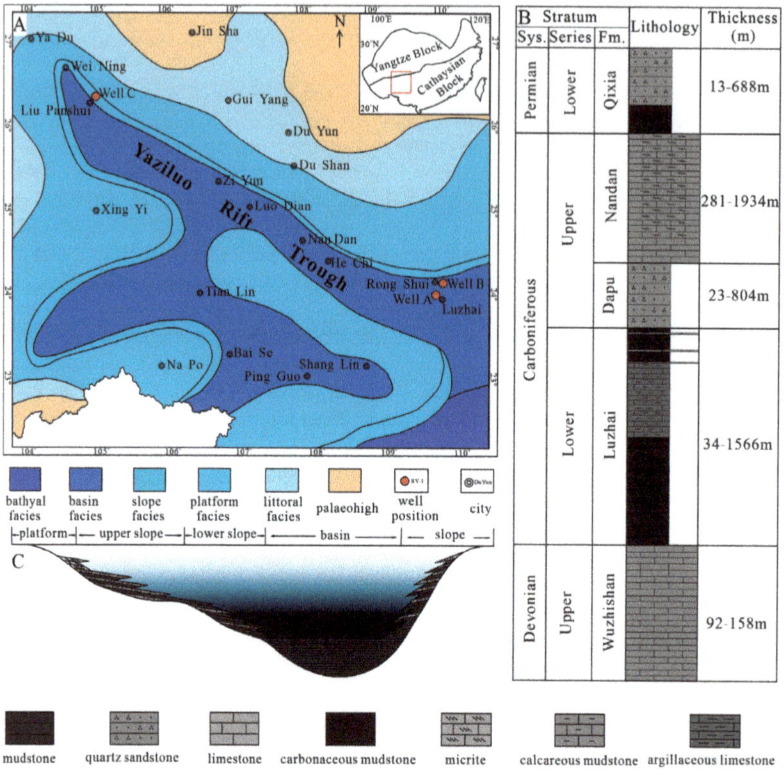

Figure 1. (**A**) Distribution of the Yaziluo Rift Trough during the late Carboniferous period, showing the sampling wells (Well A, Well B and Well C); (**B**) sequence stratigraphic framework according to Geology and Mineral Resources Bureau of Guangxi Zhuang Autonomous Region, 1985 [17].

Specifically, the depocenter zone extends from the Luzhai–Yishan area in the southeast to the Liupanshui–Weining area in the northwest of the Yaziluo Rift Trough, where the sediments are predominantly calcareous mudstone, argillaceous limestone, and carbonaceous mudstone (Figure 1B). In the transition zone between the depocenter and the margin of the Yaziluo Rift Trough, the sediments primarily consist of calcareous mudstone and argillaceous limestone. However, the lithology of the margin zone is composed mainly of micrite and bioclastic limestone. The paleo-seawater depth exhibits a general shallowing trend from southeast to northwest along the Yaziluo Rift Trough, aligning with the direction of paleo-seawater influx during the marine inundation process. Three exploratory wells (Well A, Well B, and Well C) for shale gas development are strategically located in the Luzhai area (depocenter zone), the Rongshui area (margin zone), and the Liupanshui area (margin zone), respectively (Figure 1A).

3. Materials and Methods

3.1. Samples

A total of 107 shale samples were collected from the Lower Carboniferous strata within the Yaziluo Rift Trough in South China, distributed among three wells: 28 samples from Well A, 41 samples from Well B, and 38 samples from Well C. These samples were prepared for the analyses of total organic carbon (TOC), major and trace element concentrations, kerogen types, and mineralogy using X-ray diffraction (XRD) in the China University of Petroleum (East China), Qingdao, China.

3.2. Analytical Methods

Total Organic Carbon (TOC): The TOC contents were measured using the Leco-CS230 Carbon and Sulfur Analyzer, with an accuracy of ±5%, adhering to the Chinese National Standard GB/T 19145-2003 [18] for the determination of TOC in sedimentary rocks. Shale samples were first pulverized to a particle size below 200 mesh (75–90 μm). The powdered samples were then treated with hydrochloric acid at 60 °C to eliminate inorganic carbons. After acid treatment, the samples were rinsed with distilled water and dried in an oven at temperatures between 60 and 80 °C.

Mineralogy: Mineralogical analyses were performed using X-ray diffraction (XRD) on a Rigaku Smart Lab-9, following the Oil and Gas Industry Standard SY/T 5163-2010. All powdered shale samples (finer than 40 μm) were oven-dried at 40 °C for 48 h. The mineral composition was reported as semi-quantitative, based on parameters and formulas detailed by Yang et al. (2018) [19].

Major and Trace Elements: Major elements were analyzed using a Rigaku 100E X-ray Fluorescence Spectrometer (XRF) in the China University of Petroleum (East China), Qingdao, China. Approximately 1 g of powdered shale samples were calcined at 700 °C for over 2 h to remove organic matter and carbonates. Trace elements were quantified using an Agilent 7500A Inductively Coupled Plasma-Mass Spectrometry (ICP-MS). For these analyses, powdered samples were dissolved in a mixture of hydrofluoric acid (HF) and nitric acid (HNO_3), with detailed procedural descriptions provided by Lu et al. (2019) [20].

4. Results

4.1. Major and Trace Elements

Aluminum (Al) and Titanium (Ti) are used as proxies for terrigenous detrital input [21]. In the Lower Carboniferous shales, Al content ranges from 1.55% to 10.59%, with average values of 5.62% in Well A, 5.95% in Well B, and 6.40% in Well C, respectively (Table 1). Ti content varies from 0.02% to 0.60%, with average values of 0.22% in Well A, 0.29% in Well B, and 0.34% in Well C, respectively (Table 1). Phosphorus (P), an indicator of organic matter productivity, ranges from 0.01% to 0.23% (Table 1). Trace elements are also employed to assess paleoenvironmental conditions. For example, trace element ratios in Well A, such as V/Cr (1.49–7.57, avg. 3.54), U/Th (0.17–3.65, avg. 1.72), Co/Ti (34.46–237.17, avg. 100.49), and Sr/Ba (0.16–5.57, avg. 1.60) are generally higher than in the other wells. Furthermore, the ratio of V/Cr (avg. 2.18) in Well B is higher than that in Well C (avg. 0.89), and the ratio of U/Th (avg. 0.60) in Well B is higher than that in Well C (avg. 0.21). However, the average trace element ratio of Sr/Ba in Well A (1.60) is lower than that in Well B (3.15) and Well C (2.45).

Table 1. Sedimentary facies and their average geochemical indexes in the Lower Carboniferous shale.

Well	Geochemical Proxies							
	Redox Conditions		Paleo-Seawater Depth		Detrital Influx		Paleoproductivity	
	V/Cr	U/Th	Co/Ti	Sr/Ba	Al (wt.%)	Ti (wt.%)	P/Ti	Ba/Al (10^{-4})
Well A	3.54	1.72	100.49	1.60	5.62	0.22	0.27	67.99
Well B	2.18	0.60	37.60	3.15	5.95	0.29	0.14	65.13
Well C	0.89	0.21	38.87	2.45	6.40	0.34	0.11	178.77

4.2. Characteristics of Organic Geochemistry

The organic carbon content (TOC) varies significantly across the wells. In Well A, 14 samples exhibited TOC ranging from 1.46% to 5.70%, with an average of 3.63% (Table 2). In Well B, 26 samples showed TOC ranging from 0.86% to 3.51%, with a median of 1.85%

(Table 2). Well C had the lowest TOC, with 20 samples ranging from 0.71% to 1.63% and an average of 1.02% (Table 2). The kerogen types identified in samples from Wells A, B, and C are predominantly Type I, Type II_1, and Type II_2, respectively, reflecting differences in sedimentation influences (Table 2).

Table 2. Lithofacies, kerogen types, and their average mineral compositions in the Lower Carboniferous shale.

Lithofacies	Kerogen Types	TOC (wt.%)	Average Mineral Content (wt.%)					
			Quartz	Feldspar	Calcite	Dolomite	Pyrite	Clay
Siliceous lithofacies	I	3.63	57.56	7.20	8.18	5.15	5.77	19.14
Mixed lithofacies	II_1	1.85	41.85	0.15	26.21	4.20	2.65	24.62
Calcareous lithofacies	II_2	1.02	13.60	0.30	40.20	21.20	2.30	22.40

4.3. Mineralogical Characteristics and Lithofacies

Based on the mineral compositions, four types of lithofacies can be divided in the Lower Carboniferous shale. The clay minerals contain illite, kaolinite, chlorite, and illite/smectite (I/S). The carbonate minerals comprise calcite and dolomite. The siliceous minerals refer to quartz and feldspar. The Lower Carboniferous shale exhibits significant heterogeneity across various lithofacies, reflecting dynamic paleoenvironmental changes (Figure 2). For instance, Well A shows the highest average contents of quartz and pyrite at 57.56 wt.% and 5.77 wt.%, respectively (Table 2). These components significantly decrease in Well C, where the averages are 13.60 wt.% for quartz and 2.30 wt.% for pyrite (Table 2). Conversely, the carbonate content in Well A is the lowest at 13.33 wt.%, but it increases significantly in Well C to an average of 61.40 wt.%, which is approximately four times higher than in Well A (Table 2). The values in Well B are intermediate, with quartz, carbonate, and pyrite contents at 41.83 wt.%, 30.41 wt.%, and 2.65 wt.%, respectively (Table 2). This distribution indicates that the Well A samples are typically siliceous shale lithofacies, characterized by high quartz and pyrite content and low carbonate content (Figure 2). Well C samples are carbonate shale lithofacies with high carbonate and low quartz content (Figure 2), while samples from Well B are classified as mixed shale lithofacies, characterized by medium levels of carbonate, quartz, and pyrite.

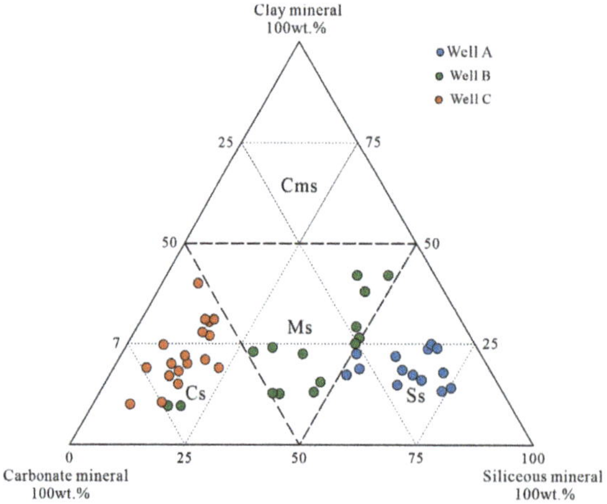

Figure 2. Ternary diagram showing the mineralogy of the three major lithofacies of Lower Carboniferous shale in the Yaziluo Rift Trough (modified from Ma et al. [2], the Cms, Ms, Cs, and Ss represent argillaceous shale lithofacies, mixed shale lithofacies, carboniferous shale lithofacies, and siliceous shale lithofacies, respectively).

5. Discussion

5.1. Reconstruction of Paleoenvironmental Conditions

5.1.1. Redox Proxies

In sedimentary environments, molybdenum (Mo) and uranium (U) are often derived from autogenous enrichment. Under oxidizing conditions, Mo and U exist as molybdate ions ($MoSO_4^{2-}$) and uranyl carbonate complexes, respectively, with limited aggregation. In hypoxic-sulfidic conditions, molybdate can transform into thiomolybdate, which enriches in conjunction with sulfide organic matter and iron sulfide deposits. The limited Mo concentration in the oxic environment represents redox conditions instead of water-mass restriction levels [21,22]. Furthermore, the amount of Mo absorbed by sediments is determined by both the Mo concentrations in seawater and organic matter concentrations [22]. Meanwhile, U precipitates as insoluble uranyl ion (UO_2^{2+}) or as a weakly soluble uranium fluoride complex. Since the reductibility of U is not affected by the H_2S concentrations, U is mainly enriched in sediments compared with Mo enriched in the seawater [21,22]. According to Algeo and Tribovillard (2009) [21], the co-variation of Mo_{EF}-U_{EF} ratios serves as an indicator for assessing redox conditions and seawater residence status, classifying marine environments into non-restricted, weakly restricted, and strongly restricted categories.

In non-restricted marine environments under dysoxic conditions, U enrichment generally precedes that of Mo, with the Mo/U ratio in sediments typically being about 0.3 times that of modern seawater (Mo/U in modern seawater is about 7.5 [22]). As the seawater reduction increases, the Mo enrichment rates surpass U, causing the Mo/U ratio to double compared to modern seawater. In strongly sulfidic and anoxic environments, sediments accumulate higher concentrations of both Mo and U, with free H_2S in the seawater enhancing Mo's accumulation rate, significantly increasing the Mo/U ratio up to three times or more compared to modern seawater [23,24]. In weakly restricted water-mass environments, metal hydride particles such as manganese and iron facilitate rapid Mo enrichment, resulting in a Mo/U ratio generally 3–10 times that of modern seawater. Conversely, in water-mass-restricted environments, the Mo/U ratio is typically lower than in modern seawater, impeding Mo and U replenishment, with U enrichment exceeding that of Mo.

In the depocenter zone, the Mo/U ratio of shale samples is higher than in modern seawater (Figure 3), showing a positive correlation with increasing Mo and U enrichment coefficients. However, the trend does not strictly conform to non-restricted marine environments, as it deviates due to metal hydride adsorption, which usually causes a 3 to 10 times higher ratio than observed in modern seawater. The results indicate that the depocenter zone shale does not align strictly with either weakly or strongly restricted marine environments but suggests a semi-restricted, marine setting with limited connectivity to open seawater. The shale in the transition zone, particularly from Wells B and C, shows low Mo and U enrichment coefficients, with a Mo/U ratio generally 0.3 times that of modern seawater (Figure 3), indicating that the shale originated from a dysoxic to anoxic environment.

Vanadium (V) and Chromium (Cr), sensitive to redox conditions, also play a critical role in identifying environmental oxygen levels. Under oxidizing conditions, V exists as vanadates like HVO_4^{2-} and $H_2VO_4^{-}$ and is not readily enriched in sediments. Conversely, under anoxic conditions, V^{5+} reduces to $V4^{+}$, forming compounds like VO $(OH)_2$ and organo-metal ligands, enriching in sediments [25,26]. Similarly, Cr transitions from soluble chromate (CrO_4^{2-}) in oxidizing conditions to less soluble forms like Cr_2O_3 and Cr $(OH)_3$ under anoxic conditions, enriching at different points of the denitrification interface [27–29]. A V/Cr ratio greater than 4.25 indicates anoxic conditions, a range of 2.0 to 4.25 suggests dysoxic conditions, and less than 2.0 indicates oxic conditions [30].

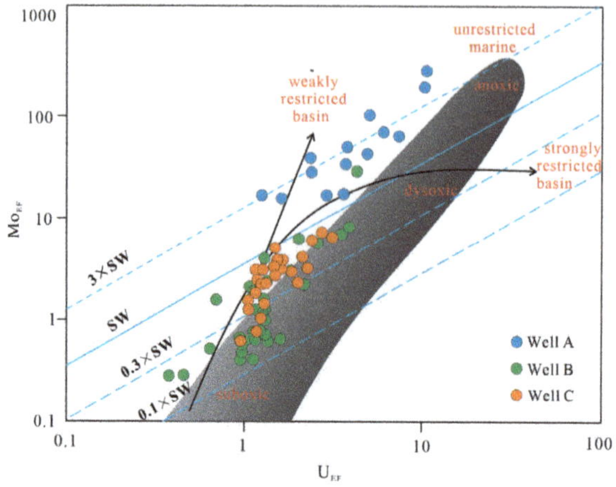

Figure 3. Covariation of Mo_{EF} and U_{EF} from the Lower Carboniferous shale (modified from Algeo and Tribovillard (2009) [21]), exhibiting water-mass restriction and redox conditions.

Implications for Lithofacies Distribution:

In Well A, primarily composed of siliceous lithofacies shale, the progressive reduction in V/Cr and U/Th values from the bottom to top layers suggests a transition from anoxic to dysoxic conditions. The average values indicate an anoxic marine environment at the lower levels, transitioning to a dysoxic state higher up.

In Well B, situated in the transition zone with mixed shale lithofacies, there is a consistent decrease in redox proxy values from bottom to top, indicating a shift from an anoxic-dysoxic to a dysoxic-oxic environment. The lower values in Well B compared to Well A suggest a closer proximity to the edge of the Yaziluo Rift Trough, reflecting different environmental impacts on sedimentation and organic matter preservation.

The carbonate lithofacies shale in Well C, primarily found in the upper slope facies, shows the lowest redox values, indicating higher oxygen content in the paleo-seawater and a partial oxidation environment. This shift is attributed to the southeastern to northwestern paleo-oceanic invasion along the Yaziluo Rift Trough, leading to a shallowing paleo-seawater depth and influencing the overall oxidation state and sedimentary dynamics. The results indicate a higher degree of oxidation and less favorable conditions for organic matter preservation in the upper slope facies, correlating with a lower TOC content in these shales.

The paleoenvironment during the early Carboniferous sedimentary period in the Yaziluo Rift Trough was influenced by seawater intrusion and the development of synsedimentary faults [12,13]. This environment is typified by a long-term, partially oxidized sedimentary environment in the upper slope facies shale [5,10,12]. In contrast, the basin facies and lower slope facies shale in the Yaziluo Rift Trough are characterized by deeper seawater depths and lower oxygen content, resulting in significantly higher redox proxies compared to the upper slope facies. The results indicate that the upper slope facies exhibit a higher degree of oxidation and suboptimal preservation conditions for organic matter, leading to lower total organic carbon (TOC) content in the shale of these facies.

5.1.2. Paleoproductivity Proxies

Barium (Ba), primarily existing in seawater as barite, is characterized by its extended residence time, high preservation rate (exceeding 30%), and exceptional stability [31–33]. Goldberg and Arrhenius (1958) [34] demonstrated a strong correlation between Ba concentrations and paleoproductivity, establishing biogenic Ba as a reliable proxy for assessing productivity levels. To accurately gauge paleoproductivity, it is crucial to eliminate the influence of terrigenous Ba. The Al element serves as an effective indicator of terrigenous input; thus, the Ba/Al ratio is employed in this study to assess paleoproductivity.

Phosphorus (P) plays a vital role as an essential nutrient for marine organisms and significantly influences the paleoproductivity of marine ecosystems [29,32]. Approximately 80% of the P in marine sediments originates from organic matter [31], with a smaller fraction deriving from terrigenous detrital sources [22,35]. Given its critical role in metabolic activities and its enrichment being primarily biologically controlled, P is one of the most reliable indicators for paleoproductivity. Additionally, the Ti element serves as an effective marker of terrigenous input, and the P/Ti ratio is selected to mitigate the influence of terrigenous clastic input on P content.

In Well A, the Ba/Al and P/Ti values of Carboniferous shale show a gradual decrease from the bottom to the top (Figure 4), suggesting relatively stable paleoproductivity. Specifically, the basin facies shale in the lower interval exhibits values ranging from 32.77×10^{-4} to 104.19×10^{-4} for Ba/Al and 0.10 to 0.68 for P/Ti, with averages of 69.67×10^{-4} and 0.30, respectively, indicating relatively high paleoproductivity. The lower slope facies in the upper interval show further declines in paleoproductivity, with values ranging from 51.73×10^{-4} to 85.80×10^{-4} for Ba/Al and 0.08 to 0.24 for P/Ti, and averages of 63.79×10^{-4} and 0.17, respectively.

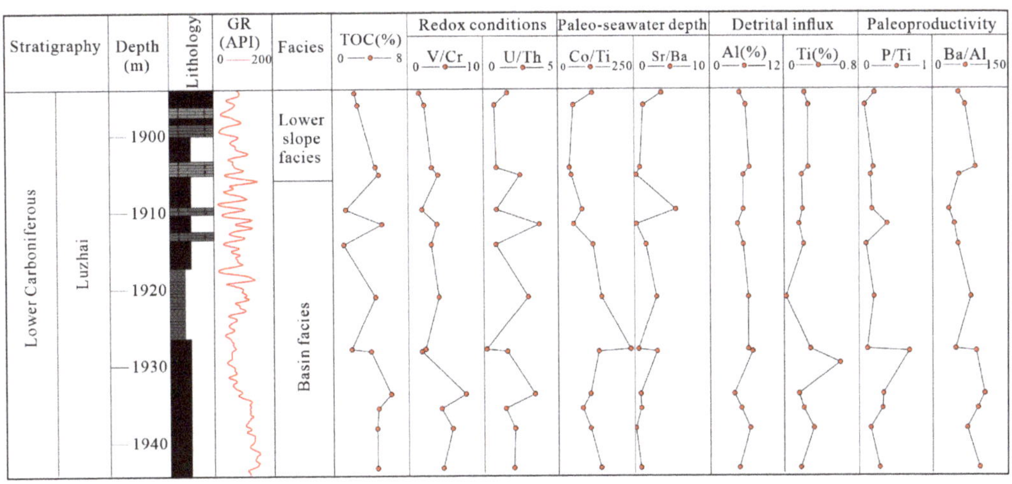

Figure 4. Stratigraphic distributions of redox (V/Cr and U/Th), paleo-seawater depth (Co/Ti and Sr/Ba), detrital influx (Al and Ti), and paleoproductivity (P/Ti and Ba/Al) proxies in Well A, respectively.

Well B displays a slight downward trend in Ba/Al and P/Ti values from bottom to top (Figure 5), maintaining general stability in paleoproductivity. The basin facies shale in the lower interval ranges from 42.84×10^{-4} to 228.04×10^{-4} for Ba/Al and 0.10 to 0.72 for P/Ti, with average values of 77.64 and 0.18, respectively, suggesting relatively high paleoproductivity. The lower slope facies values range from 27.88×10^{-4} to 141.11×10^{-4} for Ba/Al and 0.06 to 0.17 for P/Ti, with averages of 52.63 and 0.10, respectively, indicating a slight decline in paleoproductivity compared to Well A.

In Well C, the ratios of Ba/Al and P/Ti show little fluctuation from bottom to top (Figure 6). The lower slope facies shale exhibits high concentrations, with values ranging from 34.74×10^{-4} to 237.64×10^{-4} for Ba/Al and 0.03 to 0.24 for P/Ti, and average values of 180.36×10^{-4} and 0.12, respectively, indicating elevated paleoproductivity. These results also suggest that terrigenous plant debris entering the lower slope may enhance the seafloor sediment organic content, thereby promoting increased paleoproductivity. Conversely, the upper slope facies exhibit significantly reduced paleoproductivity, with Ba/Al and P/Ti values ranging from 19.41×10^{-4} to 487.16×10^{-4} and 0.04 to 0.21, respectively, with averages of 177.70×10^{-4} and 0.11.

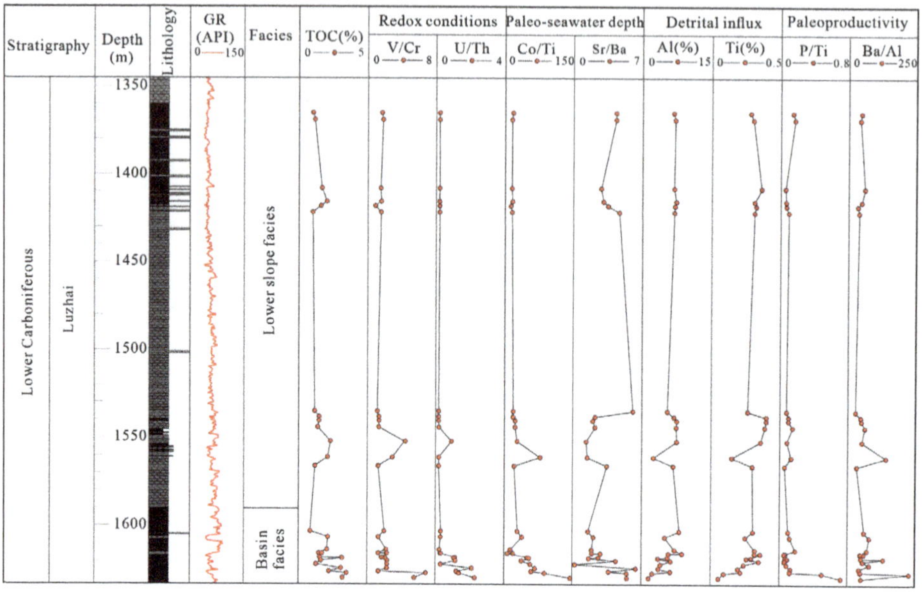

Figure 5. Stratigraphic distributions of redox (V/Cr and U/Th), paleo-seawater depth (Co/Ti and Sr/Ba), detrital influx (Al and Ti), and paleoproductivity (P/Ti and Ba/Al) proxies in Well B, respectively.

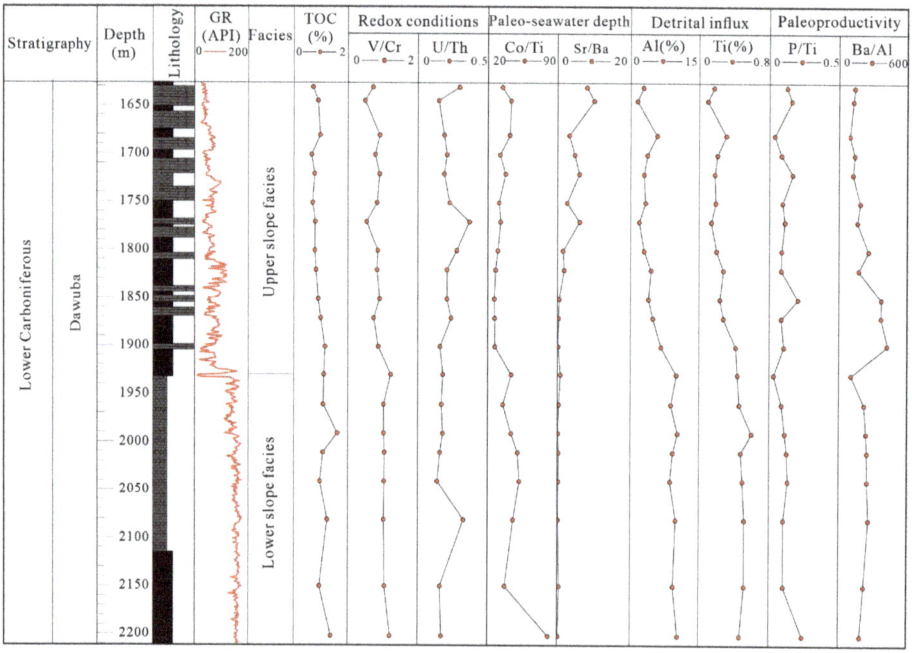

Figure 6. Stratigraphic distributions of redox (V/Cr and U/Th), paleo-seawater depth (Co/Ti and Sr/Ba), detrital influx (Al and Ti), and paleoproductivity (P/Ti and Ba/Al) proxies in Well C, respectively.

5.1.3. Terrigenous Detrital Influx Proxies

Al and Ti are primary components of the continental crust, mainly found in aluminosilicate clay minerals and heavy minerals such as rutile and ilmenite. These elements are notably stable throughout the diagenetic process and are frequently used as indicators of terrigenous detrital influx [26,36]. In Well A, the Al and Ti contents in the Carboniferous shale exhibit a remarkable similarity (Figure 4), suggesting a consistent influx of terrigenous detritus. Specifically, the Al and Ti contents in the basin facies shale of the lower interval range from 4.06 to 6.98 and 0.02 to 0.34, respectively, with average values of 5.58 and 0.21, indicating a relatively low terrigenous detrital influx. In the lower slope facies shale of the upper interval, the contents range from 4.91 to 6.98 for Al and from 0.19 to 0.26 for Ti, with averages of 5.70 and 0.23, suggesting a stable terrigenous detrital influx.

Similarly, in Well B, the variation in Al and Ti contents of the Carboniferous shale is minor, indicating a stable terrigenous detrital influx. Specifically, in the basin facies shale of the lower interval, the contents range from 1.55 to 8.74 for Al and from 0.06 to 0.36 for Ti, with average values of 5.32 and 0.25 (Figure 5). The lower slope facies shale in the upper interval shows Al and Ti contents ranging from 2.53 to 7.50 and from 0.15 to 0.40, respectively, with average values of 6.57 and 0.32 (Figure 5).

However, the Al and Ti contents in the Carboniferous shale of Well C are comparatively high, indicating its proximity to the terrigenous detrital influx source and a heightened susceptibility to further terrigenous detrital deposition. Both elements show a downward trend vertically from bottom to top (Figure 6), signaling a progressive reduction in the influx of terrigenous detritus. Specifically, in the lower slope facies shale of the lower interval, the contents range from 9.10 to 10.59 for Al and from 0.46 to 0.60 for Ti, with averages of 9.77 and 0.51, respectively, denoting a relatively high terrigenous detrital influx. Conversely, in the upper slope facies shale of the upper interval, the Al and Ti contents range from 1.90 to 10.22 and from 0.10 to 0.44, respectively, with averages of 4.58 and 0.24, indicating less terrigenous detrital influx compared to the lower slope facies.

5.1.4. Paleo-Seawater Depth

The geochemical behaviors of strontium (Sr) and barium (Ba), both lithophile elements, differ significantly during the deposition process. Sr tends to precipitate directly in seawater and shows limited affinity for adsorption onto clay material surfaces. In contrast, Ba readily adsorbs onto fine-grained sediments and clay minerals. Thus, the Sr/Ba ratio serves as a commonly used indicator of paleo-seawater depth in marine environments. A low Sr/Ba ratio suggests a relatively deep paleo-seawater depth, characterized by weak hydrodynamic influence and increased adsorption of Ba ions by clay minerals, indicative of a deep- or semi-deep-water environment. Conversely, a high Sr/Ba ratio points to shallow paleo-seawater depth and strong hydrodynamic conditions, typical of shallow marine environments [37,38].

Additionally, paleoenvironmental conditions such as paleoproductivity, redox condition, and terrigenous detrital influx are significantly influenced by sea level fluctuations [39,40]. Rising sea levels can stratify seawater, fostering anoxic conditions at the seabed that hinder benthic organism development and reduce oxidation rates, thus enhancing organic matter preservation [41]. In such anoxic environments, lipid components undergo preferential enrichment and preservation by anaerobic bacteria, increasing organic matter abundance [42,43]. Elements like cobalt (Co) and titanium (Ti) are indicative of aquatic and terrestrial environments, respectively, making the Co/Ti ratio a useful proxy for monitoring sea level changes.

In Well A, the Sr/Ba ratio in Carboniferous shale ranges from 0.16 to 5.57, with an average of 1.60, indicating significant paleo-seawater depth and limited hydrodynamic influence. The Co/Ti values range from 34.46 to 237.17, averaging 100.49 (Figure 4), and exhibit a gradual decline from bottom to top, reflecting a continuous sea level decrease.

In Well B, the Sr/Ba values range from 0.27 to 6.41, with an average of 3.15, suggesting a shallower paleo-seawater depth than in Well A. The Co/Ti values range from 7.42 to

145.83, with an average of 37.60, and also show a downward trend from bottom to top (Figure 5). This indicates that the Rongshui area, closer to the edge of the rift during the early Carboniferous period, likely fostered an anoxic benthic environment conducive to organic matter enrichment. Sea level fluctuations played a significant role in controlling organic matter enrichment.

In Well C, the Sr/Ba values range from 0.14 to 10.31, with an average of 2.45 (Figure 6). Despite the potential for increased Ba content due to clay mineral adsorption, the overall Sr/Ba values remain relatively high, indicating shallower paleo-seawater conditions. The Co/Ti values range from 26.45 to 80.66, with an average of 38.87, showing a gradual vertical decline with intermittent upward fluctuations on a secondary scale. This variability further underscores the complex interactions between sea level changes and terrigenous input in shaping paleo-marine environments.

5.2. Influencing Factors of Organic Matter Enrichment

In the Yaziluo Rift Trough, benthic redox conditions show significant regional variations, primarily influenced by the direction and magnitude of seawater intrusion and the development of synsedimentary faults. Analysis and comparison reveal a positive correlation between both V/Cr and U/Th ratios and TOC content in Well A and Well B, indicating that organic matter enrichment is significantly influenced by redox conditions (Figure 7). The results suggest that anoxic conditions provide a conducive environment for the preservation of organic matter, and an increase in the degree of oxygen depletion enhances this preservation. Conversely, the V/Cr and U/Th ratios in Well C do not show a significant correlation with the TOC content, indicating limited organic matter preservation in environments with relatively high oxidation levels (Figure 7). Clearly, in transition zones with oxic and shallow-water conditions, redox conditions have little influence on organic matter enrichment [44,45].

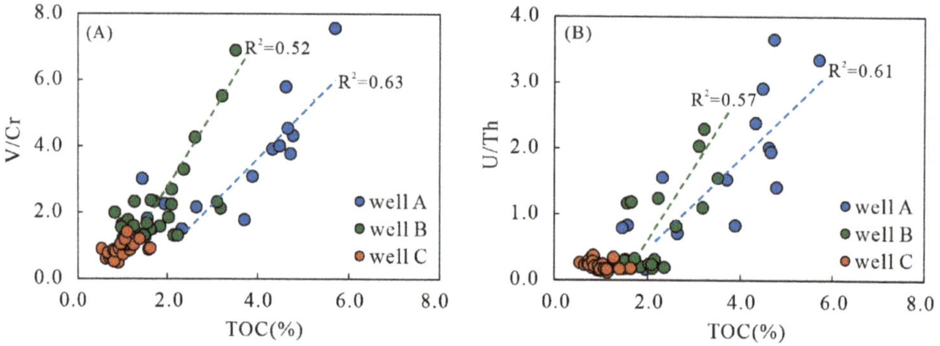

Figure 7. Correlation between redox proxies and TOC in the Lower Carboniferous formation. (**A**) V/Cr and TOC; (**B**) U/Th and TOC.

Regarding paleoproductivity proxies, there is no significant correlation between the Ba/Al ratio and TOC content in Wells A and B. However, a weak positive correlation exists between the P/Ti ratio and TOC content in Well A, suggesting that anoxic-dysoxic conditions may moderately enhance the input of organic matter by marine planktonic organisms; however, the impact is not substantial (Figure 8). Collectively, redox conditions and paleoproductivity synergistically contribute to the enrichment of organic matter. In Well C, correlations between the TOC and both Ba/Al and P/Ti are not statistically significant, indicating that paleoproductivity has limited control over organic matter enrichment in low oxygen benthic conditions (Figure 8). Thus, deep anoxic environments play a fundamental role in preserving organic matter and can partially facilitate paleoproductivity enhancement in the Lower Carboniferous shales of the Yaziluo Rift Trough. Furthermore, paleoproductivity exerts more significant control over organic matter enrichment in anoxic environments.

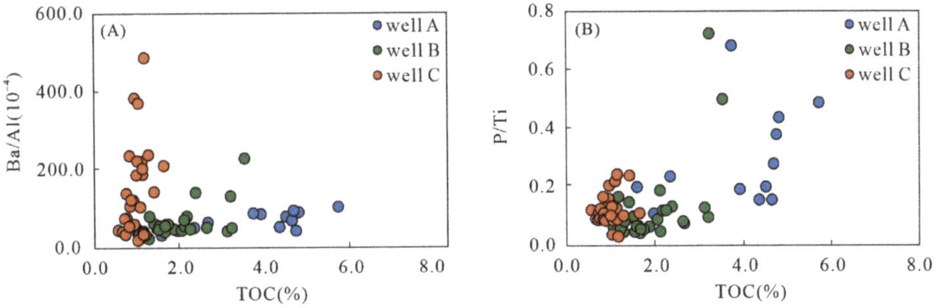

Figure 8. Correlation between paleoproductivity proxies and TOC in the Lower Carboniferous formation. (**A**) Ba/Al and TOC; (**B**) P/Ti and TOC.

When considering terrigenous detrital influx proxies, the concentrations of Al and Ti elements are negatively correlated with the TOC content in Wells A and B, respectively (Figure 9). This indicates that terrigenous detrital input can disrupt water stratification, impair the anoxic environment, and dilute marine primary productivity in anoxic-dysoxic environments, thereby hindering organic matter enrichment. Conversely, in Well C, the concentrations of Al and Ti elements show a positive correlation with the TOC content (Figure 9). This suggests that as paleo-seawater depth becomes shallower, a limited number of higher terrestrial plants provide material sources for the enrichment and supplement of organic matter, facilitating the transition of organic matter from type II_1 to type II_2. Additionally, an appropriate amount of terrigenous detrital input helps form clay-organic complexes through the amalgamation of clay minerals and organic matter. The organic matter adsorbed by clay minerals can rapidly reach the seafloor from surface water, reducing oxidation decomposition and destruction of organic matter, and effectively preserving it in relatively low oxic environments [22,46]. In the transition zone, where shallow seawater depth prevails during the deposition process, conditions are not favorable for the mass propagation of microorganisms, such as algae, which contribute to productivity and organic matter preservation.

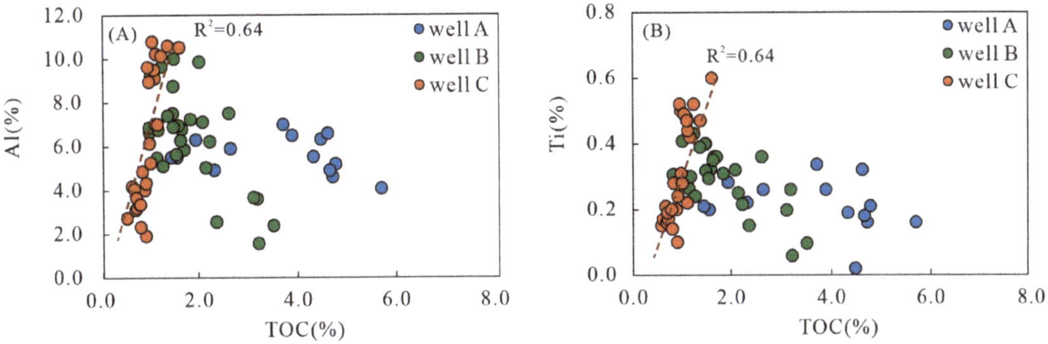

Figure 9. Correlation between detrital proxies and TOC in the Lower Carboniferous formation. (**A**) Al and TOC; (**B**) Ti and TOC.

5.3. Shale Distribution Patterns

The Late Devonian period in the Yaziluo Rift Trough was marked by a significant marine transgression from the southeast to northwest, which facilitated the invasion of the

Paleo-Tethys Ocean. This transgression expanded the distribution area of the rift trough, extending the northwestern terminus as far as Zhaotong, Yunnan Province. Since the early Carboniferous period, the distribution of the rift trough has gradually diminished as the seawater retreated southeastward, continuing the sedimentary pattern established in the Late Devonian [15]. Analysis indicates that the Lower Carboniferous shale has a widespread distribution, suggesting a sedimentary profile characterized by basin facies–slope facies. The sedimentary facies types exhibit a strong correlation with the primary quality of shale, such as mineral composition, TOC content, and kerogen types. In the Yaziluo Rift Trough, carbonate lithofacies shale is predominantly found in the upper slope facies, developed within a paleo-seawater environment characterized by relatively shallow depths, low oxygen levels, and strong water-mass restrictions. This lithofacies is marked by high carbonate mineral content, low pyrite abundance, and low TOC content. The dominant kerogen type in the carbonate lithofacies shale is primarily II_2, indicating a limited organic carbon preservation capacity. The deposition process results in the rapid oxidation and decomposition of organic matter, while a limited number of higher terrestrial plants provide additional sources for enriching organic matter content. Concurrently, paleoproductivity increases with the input of organic matter into seafloor sediments (Figure 10).

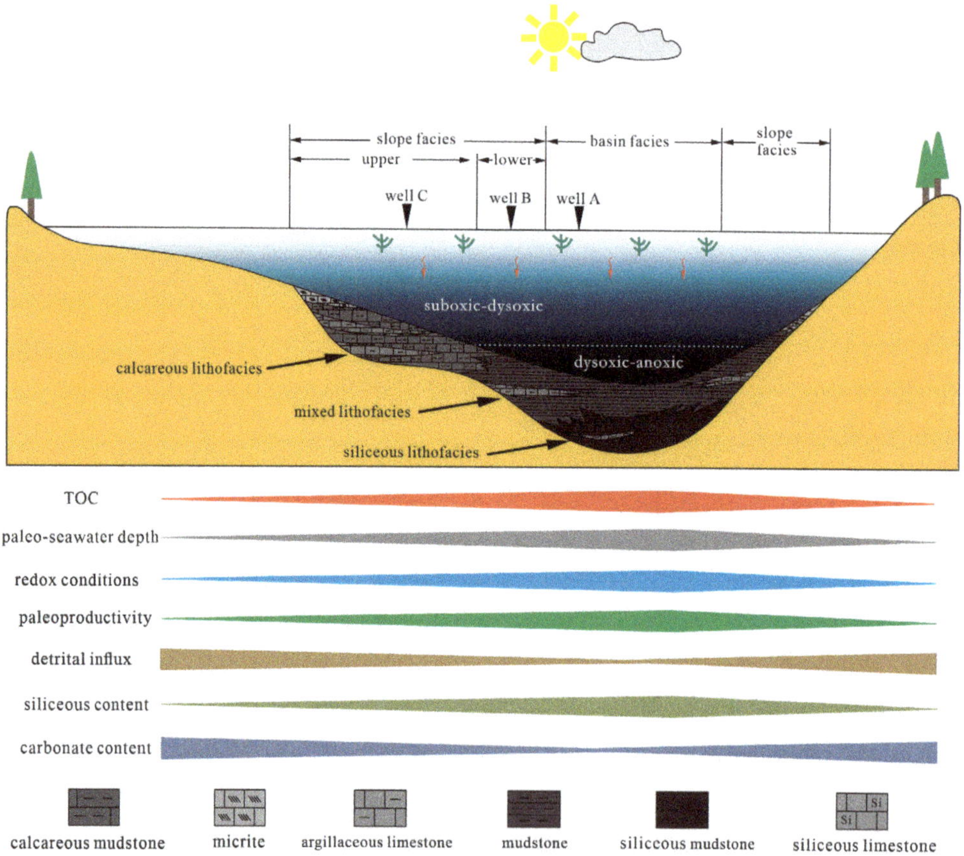

Figure 10. Sedimentary model of the Lower Carboniferous formation. The location of well and variations in paleoenvironment proxies and shale components are schematic.

As the scale of marine transgression and paleo-seawater depth within the rifting trough gradually expanded and deepened from the margin to the depocenter, oxygen deficiency in the paleoenvironment increased. The development of basin facies shale is less influenced by terrigenous input, reducing the dilution effects on organic matter by terrestrial clastics. Consequently, aquatic phytoplankton becomes the main source of organic matter, leading to increased shale organic richness and silica content. This transition facilitates a shift in kerogen type from II_2 to I and II_1. Therefore, siliceous lithofacies shale is predominantly found in the basin facies, developed under anoxic conditions with significant paleo-seawater depths. The TOC and silica mineral content of the siliceous lithofacies shale are relatively high, with enriched pyrite content and developed spicule fossils. The benthic anoxic conditions not only provide a foundation for organic matter preservation but also create a conducive environment for enhancing paleoproductivity. Overall, the degree of organic matter enrichment is primarily regulated by redox conditions and paleoproductivity, with the influx of terrigenous debris disrupting benthic anoxic conditions and diluting the paleoproductivity of seawater, thereby impeding organic matter enrichment (Figure 10).

In contrast, the Yaziluo Rift Trough experienced gradual crustal stretching during the Devonian period, resulting in the formation of depressions and the deposition of slope facies shale along the trough margins. Due to the influence of syndepositional faulting with a NW to SE direction, the deposition of slope facies on both sides of the rift trough became more pronounced since the early Carboniferous period. The Rongshui area, representing lower slope facies, and the Luzhai area, characterized by basin facies, show discernible disparities in redox conditions, terrigenous detrital supply, and paleoproductivity of seawater. In the Rongshui area, the decrease in paleo-seawater depth has led to increased dissolved oxygen content, disrupting benthic anoxic environments and transforming the paleoenvironment into a dysoxic environment unfavorable for organic matter preservation. Additionally, due to its proximity to terrigenous source regions, there has been an increase in terrigenous detrital influx, evidenced by increased concentrations of Al and Ti elements and a heightened abundance of clay minerals, further diluting the organic matter content within shale. Therefore, mixed shale is predominantly found in the lower slope facies, characterized by a dysoxic environment and a relatively shallow paleo-seawater depth. The abundance of organic matter and siliceous minerals in the mixed shale exhibits a decrease, with kerogen type primarily classified as II_1. The enrichment of organic matter is mainly controlled by anoxic preservation conditions (Figure 10).

The Lower Carboniferous shales in the Yaziluo Rift Trough are developed in a paleoenvironment characterized by frequent water changes and diverse material sources. Due to complex control by extensional faulting, the shale exhibits a complex lithological composition with frequent interlayers of limestone, resulting in variations in the depositional position, seawater depth changes, and distance from the terrigenous source area. These factors contribute to differences in the origin and distribution of shales, ultimately controlling their development scale and thickness. This also determines the degree of organic matter enrichment, types of lithology combinations, and mineral composition. Therefore, the establishment of a correlation between sedimentary facies, lithofacies, and paleoenvironments has been achieved. The paleo-seawater depth is relatively shallow for the upper slope facies, exhibiting a suboxic and shallow-water environment. The lithofacies predominantly comprise carbonate lithofacies shale with low organic matter abundance and high carbonate mineral content. Additionally, the preservation of organic matter is influenced by clay derived from terrigenous detrital material, which exerts an impact on its enrichment. The basin facies exhibit a deep paleo-seawater depth with an anoxic environment, low terrigenous detrital influx, high silica mineral content, high organic matter abundance, and predominantly siliceous lithofacies shale type. The enrichment of organic matter is jointly controlled by redox conditions and paleoproductivity. Compared to the basin facies located in the depocenter of the Yaziluo Rift Trough, the lower slope facies exhibit a semi-deep paleo-seawater depth with a dysoxic environment and moderate terrigenous detrital influx. The lower slope facies shale predominantly consists of mixed shale, exhibiting a slight decrease in silica content and organic matter abundance. The enrichment of organic matter is also influenced by the benthic redox conditions.

6. Conclusions

(1) This study investigated the Lower Carboniferous shale in the Yaziluo Rift Trough to understand the impact of paleoenvironmental factors on lithofacies characteristics, shale distribution patterns, and organic matter enrichment. Three primary lithofacies were identified: siliceous lithofacies, mixed lithofacies, and calcareous lithofacies. The siliceous lithofacies shale, predominantly composed of siliceous mudstone and siliceous limestone, was deposited in an anoxic environment characterized by high paleoproductivity and minimal terrigenous detrital input. In contrast, both the mixed and calcareous lithofacies were formed in transition zones. However, the calcareous lithofacies, composed mainly of calcareous mudstone, argillaceous limestone, and interlayers of both, was deposited in relatively shallow waters with suboxic conditions, differing from the deeper paleo-seawater conditions of the mixed lithofacies shale, which is dominated by mudstone and calcareous mudstone with argillaceous limestone layers.

(2) In the basin facies, the accumulation of organic matter was primarily governed by anoxic benthic conditions and high paleoproductivity. However, in suboxic-dysoxic benthic conditions, paleoproductivity had limited influence on organic matter enrichment. Terrigenous detrital input disrupted water stratification and diluted marine primary productivity in anoxic-dysoxic environments. Nevertheless, the adsorption of organic components by clay minerals played a crucial role in promoting the enrichment and preservation of organic matter in environments with suboxic benthic conditions.

(3) The sedimentary profile of the Lower Carboniferous shale in the Yaziluo Rift Trough comprises basin facies, lower slope facies, and upper slope facies. These facies indicate the primary shale quality and are closely associated with specific sedimentary facies types. The lithofacies succession transitions from siliceous lithofacies in the anoxic deep-water basin facies, to mixed lithofacies in the dysoxic semi-deep water lower slope facies, and finally to calcareous lithofacies in the suboxic shallow-water upper slope facies. This sequence reflects a paleoenvironmental transition that coincides with a sea level transition within the rift trough, progressing from deep basin facies to shallower upper slope facies.

Author Contributions: X.C.: Writing and Methodology, Data analyses. Q.X.: Writing and Methodology. Y.Q.: Method, Resources. R.C. and D.S.: Project administration. J.B.: Review and editing. Y.W., K.Y. and W.C.: Resources. F.L. and T.L.: Formal analysis. All authors have read and agreed to the published version of the manuscript.

Funding: This research received funding from the National Natural Science Foundation of China (Grant No. 42002173), the Guangxi Government Procurement Project of "Geological conditions and resource evaluation of Carboniferous shale gas in northern Guizhong Depression" ((2021)3421No.(001-012)), and the China Geological Survey Project of "Investigation and evaluation of shale gas resources in Ninglang Basin" (DD20242219).

Data Availability Statement: The data for this study are available in this manuscript.

Conflicts of Interest: Yinglun Qin are employees of Guangxi Energy Group Co., Ltd. The paper reflects the views of the scientists and not the company.

References

1. Wang, G.C.; Ju, Y.W.; Huang, C.; Long, S.X.; Peng, Y.M. Longmaxi-Wufeng shale lithofacies identification and 3-D modelling in the northern Fuling gas field, Sichuan basin. *J. Nat. Gas Sci. Eng.* **2017**, *47*, 59–72. [CrossRef]
2. Ma, Y.Q.; Lu, Y.C.; Liu, X.F.; Zhai, G.Y.; Wang, Y.F.; Zhang, C. Depositional environment and organic matter enrichment of the lower Cambrian Niutitang shale in western Hubei Province, South China. *Mar. Petrol. Geol.* **2019**, *109*, 381–393. [CrossRef]
3. Wang, N.; Li, M.; Tian, X.; Hong, H.T.; Wen, L.; Wang, W.Z. Climate-ocean control on the depositional watermass conditions and organic matter enrichment in lower Cambrian black shale in the upper Yangtze Platform. *Mar. Pet. Geol.* **2020**, *120*, 104570. [CrossRef]
4. Ma, Y.S.; Mou, C.L.; Tan, Q.; Yu, Q. A discussion on Kaijiang-Liangping ocean trough. *Oil Gas Geol.* **2006**, *27*, 326–331. (In Chinese)

5. Ding, J.H.; Zhang, J.C.; Li, X.Q.; Lang, Y.; Zheng, Y.Y.; Xu, L.F. Characteristics and controlling factors of organic matter enrichment of Lower Carboniferous black rock series deposited in inter-platform region, Southern Guizhou Depression. *Lithol. Reserv.* **2019**, *31*, 83–95. (In Chinese)
6. Awan, R.S.; Liu, C.L.; Gong, H.W.; Chao, D.; Chao, T.; Lawali, G.C. Paleo-sedimentary environment in relation to enrichment of organic matter of Early Cambrian black rocks of Niutitang Formation from Xiangxi area China. *Mar. Pet. Geol.* **2020**, *112*, 104057. [CrossRef]
7. Wu, C.J.; Zhang, L.F.; Zhang, T.W.; Tuo, J.C.; Song, D.J.; Liu, Y.; Zhang, M.F.; Xing, L.T. Reconstruction of paleoceanic redox conditions of the lower Cambrian Niutitang shales in northern Guizhou, Upper Yangtze region. *Palaeogeogr. Palaeoclimatol. Palaeoecol.* **2020**, *538*, 109457. [CrossRef]
8. Zhou, L.; Kang, Z.; Wang, Z.; Peng, Y.Y.; Xiao, H.F. Sedimentary geochemical investigation for paleoenvironment of the Lower Cambrian Niutitang Formation shales in the Yangtze Platform. *J. Petrol. Sci. Eng.* **2017**, *159*, 376–386. [CrossRef]
9. Yi, T.; Chen, J. Occurrence characteristics and exploration potential of Carboniferous shale gas in western Guizhou. *Pet. Reserv. Eval. Dev.* **2022**, *12*, 82–94. (In Chinese)
10. Su, H.M.; Yang, R.D.; Gao, J.B.; Xu, S.L.; Zhang, Z. REE Geochemical Signatures and Sedimentary Environments of the Early Carboniferous Dawuba Formation Black Rock Series in Huishui, Guizhou. *J. Chin. Soc. Rare Earths* **2017**, *35*, 620–631. (In Chinese)
11. Chen, Y.; Huang, W.F.; Liang, Y.P.; Deng, B. Analysis on black shale feature and depositional environment of first member of Luzhai Formation, Luzhai area of Guangxi. *Miner. Resour. Geol.* **2017**, *31*, 605–612. (In Chinese)
12. Tao, J.Y.; Hu, Z.Q.; Shen, B.J.; Pan, A.Y.; Li, C.X.; Wang, R.H.; Zhang, M.L. Sedimentary characteristics and model of platform-trough shale in the Lower Carboniferous, Guizhong Depression, Dianqiangui Basin. *Oil Gas Geol.* **2022**, *43*, 365–377. (In Chinese)
13. Wang, X.W.; Guo, T.L.; Wo, Y.J.; Zhou, Y.; Wu, L.Z.; Zhang, R.Q.; Li, S.J. Characteristics of deep structure segmentation and transformation of Yaziluo fault zone. *Oil Gas Geol.* **2013**, *34*, 220–228. (In Chinese)
14. Han, S.; Zhang, Y.; Huang, J.; Rui, Y.; Tang, Z. Elemental Geochemical Characterization of Sedimentary Conditions and Organic Matter Enrichment for Lower Cambrian Shale Formations in Northern Guizhou, South China. *Minerals* **2020**, *10*, 793. [CrossRef]
15. Wang, S.Y.; Zhang, H.; Wang, T.H.; Wang, C.H.; Peng, C.L.; Hu, R.F.; Chen, M.H.; Shi, L. Filling and evolution the Late Paleozoic Shuicheng-Ziyun aulacogen in western Guizhou, China. *Geol. Bull. China* **2006**, *25*, 402–407. (In Chinese)
16. Tian, S.F.; Yang, R.D. Lithofacies and paleogeography evolution and characteristics of shale gas accumulation in Lower Carboniferous, Guizhou, China. *J. Chengdu Univ. Technol. (Sci. Technol. Ed.)* **2016**, *43*, 291–299. (In Chinese)
17. Geology and Mineral Resources Bureau of Guangxi Zhuang Autonomous Region. *Regional Geology of Guangxi Zhuang Autonomous Region*; Geological Publishing House: Beijing, China, 1985; pp. 853–854. (In Chinese)
18. *GB/T 19145-2003[S]*; General Administration of Quality Supervision, Inspection and Quarantine of the People's Republic of China, Determination of Total Organic Carbon in Sedimentary Rock. Standards Press of China: Beijing, China, 2003.
19. Yang, W.; Zuo, R.S.; Jiang, Z.X. Effect of lithofacies on pore structure and new insights into pore-preserving mechanisms of the over-mature Qiongzhusi marine shales in Lower Cambrian of the southern Sichuan Basin, China. *Mar. Orig. Pet. Geol.* **2018**, *98*, 746–762. (In Chinese) [CrossRef]
20. Lu, B.; Jiang, S.; Lu, Y.C.; Xu, S.; Shu, Y.; Wang, Y.X. Productivity or preservation? The factors controlling the organic matter accumulation in the late Katian through Hirnantian Wufeng organic-rich shale, South China. *Mar. Petrol. Geol.* **2019**, *109*, 22–35. [CrossRef]
21. Algeo, T.J.; Tribovillard, N. Environmental analysis of paleoceanographic systems based on molybdenum-uranium covariation. *Chem. Geol.* **2009**, *268*, 211–225. [CrossRef]
22. Zhang, L.C.; Xiao, D.S.; Lu, S.F.; Jiang, S.; Lu, S.D. Effect of sedimentary environment on the formation of organic-rich marine shale: Insights from major/trace elements and shale composition. *Int. J. Coal Geol.* **2019**, *204*, 34–50. [CrossRef]
23. Tribovillard, N.; Algeo, T.J.; Baudin, F.; Riboulleau, A. Analysis of marine environmental conditions based on molybdenum-uranium covariation-Applications to Mesozoic paleoceanography. *Chem. Geol.* **2012**, *324*, 46–58. [CrossRef]
24. Qiu, Z.; Wei, H.Y.; Liu, H.; Shao, N.; Wang, Y.M.; Zhang, L.F.; Zhang, Q. Accumulation of sediments with extraordinary high organic matter content: Insight gained through geochemical characterization of indicative elements. *Oil Gas Geol.* **2021**, *42*, 931–948. (In Chinese)
25. Jones, B.; Manning, D.A. Comparison of geochemical indices used for the interpretation of palaeoredox conditions in ancient mudstones. *Chem. Geol.* **1994**, *111*, 111–129. [CrossRef]
26. Tribovillard, N.; Algeo, T.J.; Lyons, T.; Riboulleau, A. Trace metals as paleoredox and paleoproductivity proxies: An update. *Chem. Geol.* **2006**, *232*, 12–32. [CrossRef]
27. Sadiq, M. Thermodynamic solubility relationships of inorganic vanadium in the marine environment. *Mar. Chem.* **1988**, *23*, 87–96. [CrossRef]
28. Cranston, R.E.; Murray, J.W. The determination of chromium species in natural waters. *Anal. Chim. Acta* **1978**, *99*, 275–282. [CrossRef]
29. Rimmer, S.M. Geochemical paleoredox indicators in Devonian-Mississippian black shales, central Appalachian Basin (USA). *Chem. Geol.* **2004**, *206*, 373–391. [CrossRef]

30. Wang, S.F.; Dong, D.Z.; Wang, Y.M.; Huang, J.L.; Pu, B.L. Geochemistry Evaluation Index of Redox-sensitive Elements for Depositional Environments of Silurian Longmaxi Organic-rich Shale in the South of Sichuan Basin. *Mar. Orig. Pet. Geol.* **2014**, *19*, 27–34. (In Chinese)
31. Ingall, E.D.; Bustin, R.M.; Van, C.P. Influence of water column anoxia on the burial and preservation of carbon and phosphorus in marine shales. *Geochem. Cosmochim. Acta* **1993**, *57*, 303–316. [CrossRef]
32. Schoepfer, S.D.; Shen, J.; Wei, H.Y.; Tyson, R.V.; Ingall, E.; Algeo, T. Total organic carbon, organic phosphorus, and biogenic barium fluxes as proxies for paleomarine productivity. *Earth Sci. Rev.* **2015**, *149*, 23–52. [CrossRef]
33. Zhang, J.G.; Lv, D.Y.; Chen, H.Y.; Yu, C.; Zhao, K.H.; Liu, X.X.; Liu, Y.M.; Zhang, H.; Liu, B.; Qiang, X.K.; et al. Evolution of sedimentary environment in the Eastern Henan Basin since the Late Pliocene. *Palaeogeogr. Palaeoclimatol. Palaeoecol.* **2024**, *633*, 111896. [CrossRef]
34. Goldberg, E.D.; Arrhenius, G.O.S. Chemistry of Pacific pelagic sediments. *Geo-Chim. Cosmochim. Acta* **1958**, *13*, 153–212. [CrossRef]
35. Liu, K.; Cheng, P.; Fan, C.W.; Song, P.; Huang, Q.T. Evolutions of sedimentary facies and palaeoenvironment and their controls on the development of source rocks in continental margin basins: A case study from the Qiongdongnan Basin, South China Sea. *Petrol. Sci.* **2023**, *20*, 2648–2663. [CrossRef]
36. Xu, Q.L.; Liu, B.; Song, X.M.; Wang, Q.P.; Chen, X.D.; Li, Y.; Zhang, Y. Hydrocarbon generation and organic matter enrichment of limestone in a lacustrine mixed sedimentary environment: A case study of the Jurassic Da'anzhai member in the central Sichuan Basin, SW China. *Petrol. Sci.* **2023**, *20*, 670–688. [CrossRef]
37. Fan, D.; Zhang, T.; Ye, J. Anoxic environment and formation of superlarge ore deposits. *Sci. China Ser. D Ear. Sci.* **1998**, *41* (Suppl. S1), 37–46. [CrossRef]
38. Sun, Z.Q.; Chen, Z.H.; Liu, L.L.; Li, Y.; Zhang, J.L.; Shen, W.L. How to distinguish between marine and lacustrine sedimentary environments?—A case study of Lishui Sag, East China Sea Shelf Basin. *Geoenergy Sci. Eng.* **2023**, *228*, 212032. [CrossRef]
39. Middelburg, J.J. Organic carbon, sulphur, and iron in recent semi-euxinic sediments of Kau Bay, Indonesia. *Geochim. Cosmochim. Acta* **1991**, *55*, 815–828. [CrossRef]
40. Gou, M.X.; Lu, G.; Deng, B.; Wang, C.C.; Li, Z.W.; Yu, Y.; Yang, R.J.; Jin, X. Tectonic–paleogeographic evolution of the Late Triassic in the Sichuan basin, SW China: Constraints from sedimentary facies and provenance nalysis of the Xujiahe Formation. *Mar. Pet. Geol.* **2024**, *160*, 106649. [CrossRef]
41. Feng, Y.; Xiao, X.M.; Gao, P.; Wang, E.Z.; Hu, D.F.; Liu, R.B.; Li, G.; Lu, C.G. Restoration of sedimentary environment and geochemical features of deep marine Longmaxi shale and its significance for shale gas: A case study of the Dingshan area in the Sichuan Basin, South China. *Mar. Pet. Geol.* **2023**, *151*, 106186. [CrossRef]
42. Li, T.; He, S.; Yang, Z. The Marine Source Rock Formation Conditions and Control Factors. *Geol. Sci. Technol. Inf.* **2008**, *27*, 63–70. (In Chinese)
43. He, D.S.; Fang, H.; Zhang, P.H.; Pei, F.G.; Ming, C.D.; He, M.X.; Zhang, X.B. Geochemical characteristics of the Middle–Late Permian sedimentary rocks in the southern Great Xing'an Range, NE China, and their constraints on the closure time of the Paleo Asian Ocean (Eastern Segment). *Sediment. Geol.* **2023**, *450*, 106375. [CrossRef]
44. Kontakiotis, G.; Karakitsios, V.; Maravelis, A.G.; Zarkogiannis, S.D.; Agiadi, K.; Antonarakou, A.; Pasadakis, N.; Zelilidis, A. Integrated isotopic and organic geochemical constraints on the depositional controls and source rock quality of the Neogene Kalamaki sedimentary successions (Zakynthos Island, Ionian Sea). *Med. Geosci. Rev.* **2021**, *3*, 193–217. [CrossRef]
45. Kontakiotis, G.; Karakitsios, V.; Cornée, J.; Moissette, P.; Zarkogiannis, S.D.; Pasadakis, N.; Koskeridou, E.; Manoutsoglou, E.; Drinia, H.; Antonarakou, A. Preliminary results based on geochemical sedimentary constraints on the hydrocarbon potential and depositional environment of a Messinian sub-salt mixed siliciclastic-carbonate succession onshore Crete (Plouti section, eastern Mediterranean). *Med. Geosc. Rev.* **2020**, *2*, 247–265. [CrossRef]
46. Yu, W.; Tian, J.C.; Wang, F.; Liang, Q.S.; Yang, T.; Kneller, B.; Liang, X.W. Sedimentary environment and organic matter enrichment of black mudstones from the upper Triassic Chang-7 member in the Ordos Basin, Northern China. *J. Asian Earth Sci.* **2022**, *224*, 105009. [CrossRef]

energies

MDPI

Article

Paleoenvironment and Hydrocarbon Potential of Salinized Lacustrine Shale with High Terrigenous Input in the Paleogene Biyang Depression (East China): Evidence from Organic Petrography and Geochemistry

Yu Song [1,2,3,*], Paerzhana Paerhati [3], Shilin Xu [1,2,4], Shu Jiang [3,*], Bo Gao [1,2], Shuifu Li [3], Qiang Cao [3], Zhonghui Li [3], Li Wan [3] and Chuang Li [3]

1 State Key Laboratory of Shale Oil and Gas Enrichment Mechanisms and Efficient Development, Beijing 100083, China
2 Sinopec Key Laboratory of Shale Oil/Gas Exploration and Production Technology, Beijing 100083, China
3 Key Laboratory of Tectonics and Petroleum Resources, China University of Geosciences, Ministry of Education, Wuhan 430074, China; lc271531@163.com (C.L.)
4 Sinopec Petroleum Exploration and Production Research Institute, Beijing 100083, China
* Correspondence: songyu@cug.edu.cn or songyu1009@126.com (Y.S.); jiangsu@cug.edu.cn (S.J.)

Abstract: Salinized lacustrine shale (SLS) represents a frontier in the global quest for unconventional hydrocarbon resources. The impact of terrigenous input, which includes terrigenous organic matter (OM) and detrital matter, on the deposition and hydrocarbon potential of SLS is still controversial. Here, we examine this issue using the newly discovered SLS within the Paleogene Biyang Depression, employing a combination of organic petrographic and geochemical analyses. A high influx of terrigenous input (terrigenous OM and detrital matter) promotes the formation of SLS. On the one hand, terrigenous higher plants emerge as the primary source of OM in the SLS, as indicated by the dominance of terrigenous macerals (e.g., terrigenous liptinite) and the abundance of plant-derived biomarkers (e.g., tricyclic terpanes). Additionally, a portion of the OM may originate from bacteria. On the other hand, the rapid input of detrital matter improves the preservation of OM, resulting in the deposition of SLS with high total organic carbon (TOC) contents and low hydrogen index (HI) values. The findings of this study contribute to a deeper understanding of SLS deposition and provide guidance for regional hydrocarbon exploration.

Keywords: lacustrine organic-rich shale; paleo-saline lake; biomarkers; elemental geochemistry; deep resource exploration

1. Introduction

With the growing demand for energy and the dwindling conventional oil reserves, both the exploration and exploitation of unconventional oil reservoirs become imperative. Among these, salinized lacustrine shale systems stand out for their distinct characteristics, presenting both challenges and opportunities for oil exploration and exploitation [1,2]. Salinity exerts a profound influence on sedimentation, diagenesis, and organic matter preservation, ultimately shaping the composition and distribution of hydrocarbon resources [3]. As such, unravelling the complex relationships between organic matter, depositional environments, and hydrocarbon potential in salinized lacustrine shale is essential for resource assessment and exploration strategies [4]. Note that the term 'shale' in this study refers to fine-grained sediments include clay particles (<4 μm), with varying proportions of silt-sized floating grains (up to 62.5 μm) of biogenic and detrital origin [5,6].

Salinized lacustrine shale was extensively deposited during the Paleogene period worldwide, with examples including the Green River Formation of Wyoming in the United States [1]; the Salt IV Formation of the Mulhouse Basin [7] and the Bituminous Limestone of the Camargue Basin in France [8]; the Brown Shale Formation of the Central Sumatra Basin in Indonesia [9]; the Xiaganchaigou Formation of the Qaidam Basin in Northwestern China [10]; and the Kongdian and Shahejie formations of the Bohai Bay Basin [11,12], the Qianjiang Formation of the Jianghan Basin [13], the Funing Formation of the Subei Basin (Eastern China) [14], and the Hetaoyuan Formation of the Nanxiang Basin in Eastern China [15]. Many of these salinized lacustrine shales are renowned for their potential as prolific hydrocarbon source rocks, as well as sources and reservoirs for shale oil resources [4]. Saline lacustrine basins harbor diverse biological organisms and demonstrate significant biological activities. This activity plays a role in enriching the sedimentary organic matter within the basins. The enrichment process includes biological bloom, primary productivity, water stratification, and organic matter preservation. Water stratification in saline lacustrine basins facilitates the preservation of sedimentary organic matter by inducing bottom water anoxia [16].

The Biyang Depression within the Nanxiang Basin represents the first shale oil breakthrough in Eastern China. The Well BYHF1, situated at the center of the depression, achieved a peak daily oil production of 24.2 tons [17]. However, oil production from the well declined rapidly, dropping to less than 5 tons per day after only 100 days [18]. Aside from potential hydraulic fracturing difficulties, the heterogeneity of salinized lacustrine shale is a major contributor to the rapid decline in production [19]. Consequently, in recent years, alongside shale oil exploration and exploitation in the Eh_3^3 and Eh_3^2 strata, which contain salinized lacustrine shale with the highest quality and the largest thickness, preliminary exploration of deeper shale strata (e.g., Eh_3^6 and Eh_3^5) in the depression is also planned [19]. However, there is still a lack of studies on organic matter source, depositional environment, and hydrocarbon potential for these deeper shale strata, significantly impending further shale oil exploration efforts in the Biyang Depression.

This article presents the results of a comprehensive organic petrographic and geochemical study conducted on the newly explored, deeper salinized lacustrine shale strata in the Paleogene Biyang Depression. Through integrated analyses, our objective was to elucidate the origin of organic matter, characterize the depositional environment of the shale strata, and assess their implications for hydrocarbon prospectivity. The findings of this study not only contribute to the understanding of salinized lacustrine shale systems in East China but also provide valuable insights into hydrocarbon exploration and resource assessment in similar geological settings worldwide.

2. Geological Setting

The Nanxiang Basin, situated in East China, is a sedimentary rift basin covering approximately 1.7×10^4 km^2 (Figure 1A). It includes the Xiangzao, Nanyang, and Biyang depressions, as well as the Shigang, Xinye, Tanghe, and Sheqi uplifts (Figure 1B). The hydrocarbon-rich Biyang Depression spans nearly 1000 km^2 and boasts conventional oil reserves of 306 million tons and shale oil reserves of 285 million tons [16,20]. The depression takes the form of a half-graben rift, further divided into the northern slope belt, central deep sag, and southern steep slope belt (Figure 1C,D).

There are four distinct episodes that comprise the structural evolution of a depression. (Episodes I–IV; Figure 2) [21–23]. Episode I signifies the early initial rifting stage, extending from the Paleocene to the very Early Eocene. The Yuhuangding Formation (Ey) was deposited during this phase, characterized by the prevalence of yellow-to-red coarse-grained sandstone interbedded with mudstone, suggesting fluvial facies. Episode II represents the late initial rifting stage and aligns with the deposition of the Early Eocene Dacangfang Formation (Ed). The Ed is marked by an abundance of purple-red coarse- and medium-grained sandstone and grey mudstone, with the occurrence of gypsum at the top, deposited in semi-saline lacustrine environments. The peak of the rifting stage, Episode III, spans the Middle-to-Late Eocene and is denoted by the third and second members of the Hetaoyuan Formation (Eh_3 and Eh_2). The Eh_3 mainly comprises grey-to-dark-grey shale interbedded with carbonate rock, deposited in deep-to-semi-deep saline lacustrine environments. The Eh_2 consists of greyish-green-to-light-grey mudstone, carbonate rock, and fine-grained sandstone, indicating shallow lacustrine environments. Trona is occasionally found interbedded with dolostone in both the upper Eh_3 and Eh_2. The late rifting stage, Episode IV during the Oligocene, coincides with the deposition of the first member of the Hetaoyuan Formation (Eh_1) and the Liaozhuang Formation (El). The Eh_1 exhibits greyish-green-to-light-grey mudstone interbedded with medium and fine sandstone, suggesting shoreline shallow lake environment. The El is characterized by brown-red fine-grained sandstone, mudstone, and gypsum, signifying a transition from a riverine environment to a saline-lake environment.

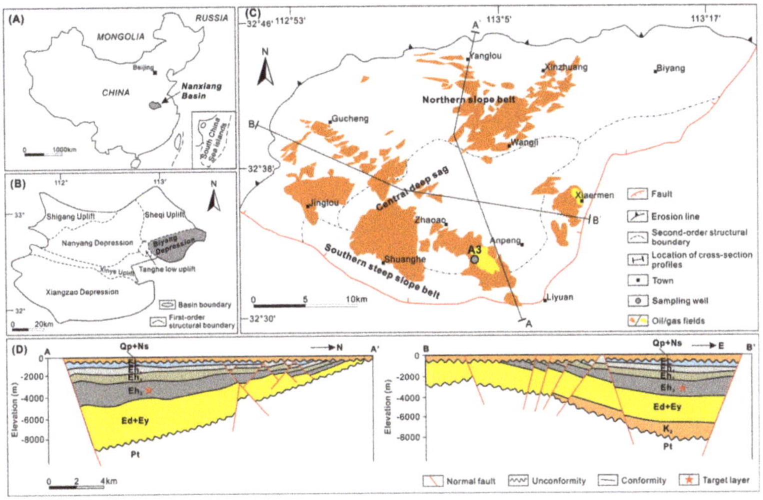

Figure 1. (**A**) Location of the Nanxiang Basin in East China. (**B**) Division of structural units in the Nanxiang Basin and location of the Biyang Depression. (**C**) Division of structural units in the Biyang Depression with locations of discovered oil/gas fields and the sampling well A3. (**D**) Profiles illustrating the vertical structures of the Biyang Depression. Locations for the profiles are given in (**C**). El = the Liaozhuang Formation; $Eh_{1,2,3}$ = first, second, and third member of the Hetaoyuan Formation; Ey = the Yuhuangding Formation; K_2 = Lower Cretaceous; P_t = Permian–Triassic.

The Eh_3 interval is further subdivided into eight units, denoted as Unit VIII to Unit I from bottom to top ($Eh_3{}^8$ to $Eh_3{}^1$; Figure 2). This study focuses on the Units VI and V ($Eh_3{}^6$ and $Eh_3{}^5$), deposited in a deep/semi-deep lake, which comprises salinized lacustrine shale, and they have been prominent targets in recent hydrocarbon exploration efforts [20].

3. Materials and Methods

In the exploration well A3, a total of twenty fine-grained sediments were collected from depths ranging from 3195.2 to 3383.6 m, which is located in the southern part of the central deep sag (Figure 1C). All samples were analyzed for total organic carbon (TOC; wt. %) and Rock-Eval pyrolysis. A Leco CS-230 instrument (LECO Corporation, St. Joseph, MI, USA) was employed to determine the TOC contents after the acidification of samples to eliminate carbonate. A Rock-Eval 6 instrument (Vinci Technologies, Nanterre, France) was used for characterizing free hydrocarbons (S_1; mg HC/g rock), hydrocarbons cracked from kerogen (S_2; mg HC/g rock), and the maximum temperature of hydrocarbon generation (T_{max}; °C).

Ten samples were selected for the organic petrographic analysis. Polished sections were observed using a Nikon microscope (Nikon Corporation, Tokyo, Japan) equipped with reflected and fluorescent light, as well as 20× and 50× oil-immersion lenses. Mean random vitrinite reflectance of six samples was determined in accordance with an established procedure [24].

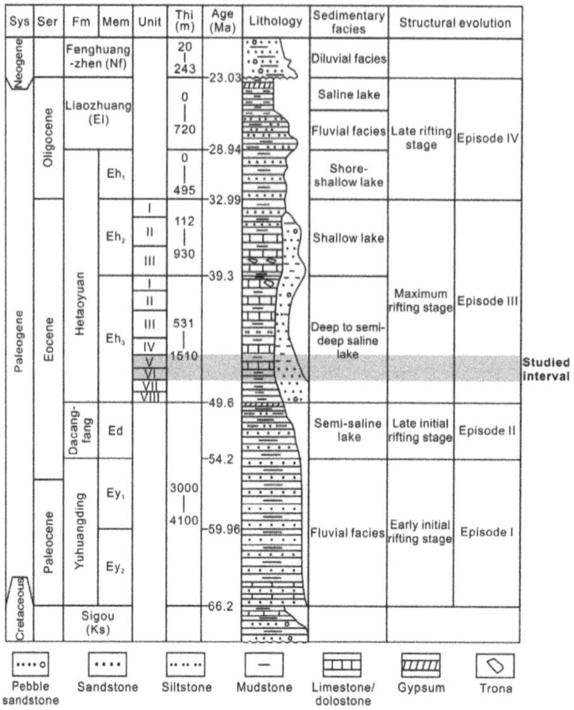

Figure 2. Stratigraphic section, sedimentary facies, and structural evolution of the Biyang Depression. Sys = system, Ser = series, Fm = formation, Mem = member, and Thi = thickness.

Major and trace elemental compositions were analyzed in 19 samples. The determination of major-element oxide contents was conducted using a Shimadzu XRF-1800 instrument (Shimadzu, Kyoto, Japan), while a Thermo Fisher Scientific Element XR ICP-MS (Thermo Fisher Scientific, Waltham, MA, USA) was used to determine the concentration of trace elements. To ensure analytical reliability, both standard materials and duplicate determination of samples were employed.

A total of 11 samples underwent biomarker analysis in this study. Initially, powdered samples were subjected to 72-h extraction with chloroform. Asphaltenes were then precipitated from an 80:1 (*v*:*v*) solution of hexane and dichloromethane (DCM) and subsequently separated via centrifugation. Soluble components of hexane were further divided into saturates, aromatics, and NSO compounds by silica gel and alumina column chromatography (Al_2O_3). This separation process involved successive elution with hexane, a DCM–hexane solution (2:1, *v*:*v*), DCM, and methanol (MeOH). Subsequently, the saturated and aromatic fractions were subjected to analysis using a gas chromatograph equipped with a 30 m DB-5MS fused silica column (inner diameter 0.25 mm; film thickness 0.25 μm) coupled to an Agilent 7890-5977B gas chromatography–mass spectrometry (GC-MS) system (Agilent, America). The oven temperature was programmed to increase from 70 °C to 300 °C at a rate of 3 °C/min, followed by an isothermal period lasting 30 min. The mass spectrometer operated in the electron ionization mode, covering a scan range from mass to charge (*m/z*) of 50 to *m/z* 650, with a total scan time of 0.7 s.

4. Results

4.1. Bulk Geochemistry

The Eh_3^5 shale contains TOC contents ranging from 0.3 wt. % to 1.8 wt. %, which is lower than that of the Eh_3^6 shale (0.8–2.2 wt. %; Table 1). The S_1+S_2 values span 1.3–2.5 mg

HC/g rock and 0.5–4.1 mg HC/g rock in the Eh_3^6 and Eh_3^5 shale, respectively. The hydrogen index (HI = 100 × S_2/TOC) exhibits low values in the Eh_3^6 shale (71–119 mg HC/g TOC) and high values in the Eh_3^5 shale (100–166 mg HC/g TOC). A comparable range of T_{max} values are observed in the Eh_3^6 and Eh_3^5 shale, ranging from 439 to 462 °C and from 431 to 456 °C, respectively. Type II_2 kerogen predominates in the Eh_3^5 shale, while type III kerogen dominates the Eh_3^6 shale, according to cross-plots of HI versus T_{max} and S_2 versus TOC (Figure 3).

Table 1. Bulk geochemical parameters, maceral compositions, vitrinite reflectance, and elemental ratios for the Eh36 and Eh35 shale from the Biyang Depression.

Strata	Depth (m)	TOC (wt. %)	$S_1 + S_2$ (mg/g)	HI (mg/g TOC)	Tmax (°C)	PI	OSI (mg/g TOC)	Maceral Compositions (vol. %)					10^2 × P/Al	Ba_{bio}	B/Ga	Sr/Ba	U_{EF}	M_{oEF}	T_iO_2	Al_2O_3
								Telalg	Ter Lip	Lipto	Vitr	Iner								
Eh_3^5	3195.2	1.0	1.8	149	456	0.18	33	0.3	0.7	0.1	0.9	0.6								
	3196.2	0.8	1.6	162	448	0.19	37						2.0	241	1.0	2.7	4.2	8.6	0.65	15.3
	3205.9	0.6	1.2	130	452	0.29	52						0.7	181	0.5	2.5	1.7	2.1	0.73	17.7
	3206.5	0.8	1.6	148	442	0.31	65	0.1	0.2	0.2	0.4	0.1	0.8	238	0.9	2.6	1.6	4.3	0.62	14.4
	3207	0.4	0.7	123	455	0.37	71						1.0	285	2.6	3.5	3.2	1.5	0.28	6.7
	3208.5	0.9	1.7	142	451	0.24	44	0.3	0.2	0.1	0.5	0.4	0.7	353	1.5	2.8	2.8	4.4	0.46	10.5
	3209.7	1.3	2.8	166	455	0.22	46	0.4	0.0	0.1	1.1	0.6	1.4	82	0.9	3.8	3.7	8.9	0.59	16.1
	3210.2	1.8	4.1	162	444	0.28	62	0.8	0.8	0.2	1.9	0.8	2.4	298	0.8	2.1	4.3	18.4	0.65	15.7
	3212	0.6	1.0	126	447	0.18	27						1.9	367	2.8	2.5	4.9	7.7	0.27	7.3
	3213	1.0	1.9	154	454	0.22	43						1.3	125	0.5	2.6	3.5	4.9	0.76	18.4
	3214	0.3	0.5	100	431	0.39	63						1.0	172	0.7	1.9	1.4	15.5	0.72	15.9
	3214.5	1.1	2.2	151	442	0.23	46							191	1.8	1.7	7.0	12.5	0.56	13.8
Eh_3^6	3342.6	2.2	2.9	81	443	0.40	54						0.8	57	0.8	3.1	3.8	19.8	0.60	14.8
	3342.7	2.0	2.3	71	442	0.37	42	0.3	0.6	0.1	1.6	0.7	1.1	46	0.6	4.6	3.6	28.6	0.66	15.7
	3349.2	1.5	2.0	93	445	0.29	39	0.2	0.2	0.2	1.2	0.6	0.9	87	1.7	5.4	5.2	34.0	0.53	13.4
	3349.6	1.6	2.5	98	459	0.35	54	0.3	0.5	0.2	2.0	0.5	0.8	277	1.0	3.2	2.7	12.2	0.66	12.5
	3357.2	1.2	1.8	111	462	0.28	43	0.1	0.2	0.2	0.7	0.5		232	0.8	3.2	9.8	27.4	0.52	12.8
	3357.6	0.8	1.3	119	452	0.27	44						1.1	15	1.1	2.8	4.4	24.3	0.65	15.7
	3382.8	1.4	2.1	91	459	0.38	55						1.7	97	1.5	3.8	3.7	14.9	0.50	13.2
	3383.6	1.6	2.2	81	439	0.42	58	0.3	0.5	0.1	2.5	0.4	0.7	76	1.0	2.9	2.8	19.3	0.56	15.3

Tel = telalginite. Ter Lip = terrigenous liptinite. Lipto = liptodetrinite. Vitr = vitrinite. Iner = inertinite.

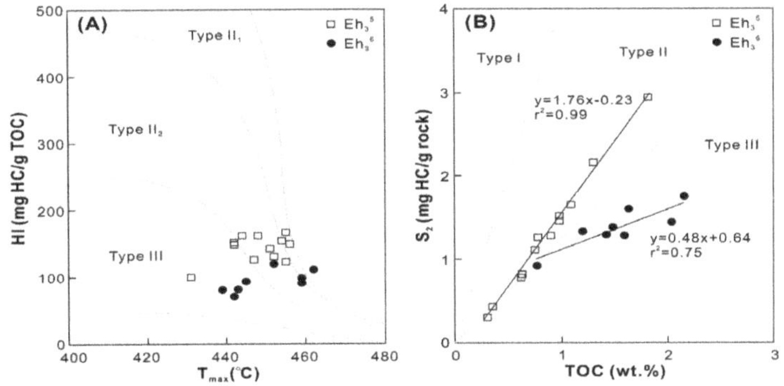

Figure 3. Cross-plots of (**A**) T_{max} versus HI and (**B**) TOC versus S_2.

4.2. *Organic Petrography and Vitrinite Reflectance*

The volume percentages (vol. %) of maceral in the Eh_3^6 and Eh_3^5 shale are less than 5.0 vol. % (Figure 4; Table 1). The Eh_3^5 shale contains a larger concentration of telalginite (including degraded telalginite) than the Eh_3^6 shale, resulting in higher HI values (Figure 3A). Telalginite appears as a unicellular spherical or almost-round shape with yellow-to-orange fluorescence (Figure 5A), whereas degraded telalginite has indistinct boundaries and diminished fluorescence (Figure 5B). Terrigenous liptinite, which is dominated by sporinite and cutinite, ranges from 0.2 to 0.6 vol. % and less than 0.1 to 0.8 vol. % in the Eh_3^6 and Eh_3^5 shale, respectively. Sporinite is distinguished by its elongated thread-like or fusiform shape with yellow-orange fluorescence, whereas cutinite has a serrated edge and poor fluorescent properties (Figure 5C). Both intervals contain trace amounts of liptodetrinite (<0.2 vol. %), which has a small and irregular shape with yellow fluorescence (Figure 5A,C). Among macerals, vitrinite is most prevalent, with values in the Eh_3^6 shale ranging from 0.7 to 2.5 vol. %, and in the Eh_3^5 shale from 0.4 to 1.9 vol. %. The spherical shape of vitrinite suggests that it may transport across long distances, probably suggesting deep-to-semi-deep lacustrine environments (Figure 5D). Inertinite is found in needle-shaped grains and is most likely broken fusinite or semifusinite (Figure 5E). Furthermore, a minor amount of framboidal and semi-cube pyrite is observed, indicating that there were possibly different stages of pyrite formation (Figure 5F). The average values of random vitrinite reflectance (%Rr) vary between 0.9 and 1.0% (Figure 4; Table 1).

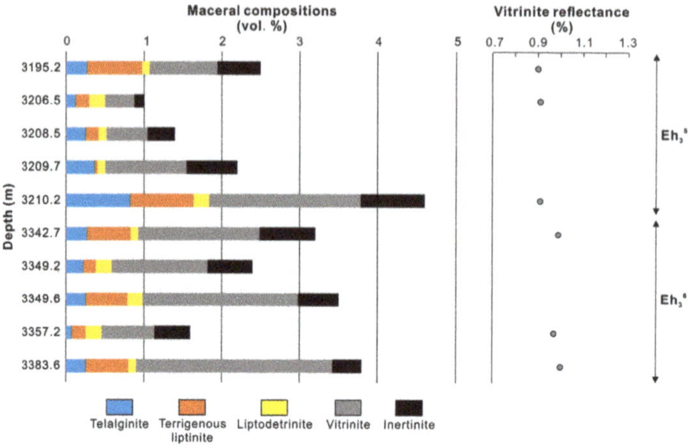

Figure 4. Vertical variations in maceral compositions and vitrinite reflectance.

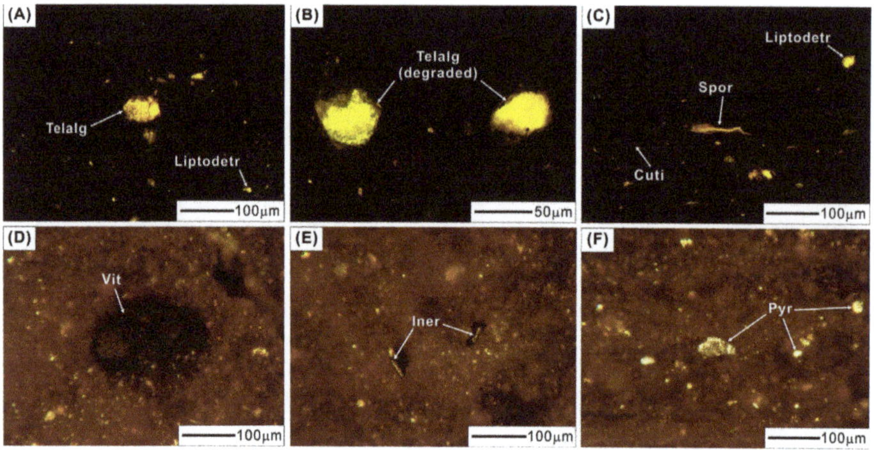

Figure 5. Representative photos of maceral. (**A**) Telalginite (Telalg) and liptodetrinite (Liptodetr), 3342.7 m, Eh_3^6. (**B**) Degraded telalginite, 3349.2 m, Eh_3^6. (**C**) Sporinite (Spor), cutinite (Cuti), and liptodetrinite, 3196.2 m, Eh_3^5. (**D**) Vitrinite (Vit), 3196.2 m, Eh_3^5. (**E**) Inertinite (Iner), 3196.2 m, Eh_3^5. (**F**) Pyrite (Pyr), 3357.2 m, Eh_3^6. (**A–C**) Under fluorescent light. (**D–F**) Under white incident light.

4.3. Elemental Compositions

The elemental metrics used as paleoenvironmental indicators are listed in Table 1, including the P/Al ratio; biogenic Ba value ($Ba_{bio} = Ba_{tot} - Al \times (Ba/Al)_{detrital}$, where Ba_{tot} and Al are the total amounts of Ba and Al, and $(Ba/Al)_{detrital}$ is estimated based on the cross-plot of Ba versus Al) [25]; Sr/Ba and B/Ga ratios; enrichment factors of U and Mo ($TE_{EF} = (TE/Al)_{sample}/(TE/Al)_{UCC}$ [26], where TE stands for the U or Mo concentrations, and UCC refers to elemental composition of the Upper Continental Crust) [27]; and TiO_2 and Al_2O_3 contents.

The 100 × P/Al ratios in the Eh_3^6 and Eh_3^5 shale range from 0.7 to 1.7 and from 0.7 to 2.4, respectively. The Eh_3^6 shale has Ba_{bio} levels ranging from 15 to 277 ppm, which is less than the Eh_3^5 shale's range, which is from 82 to 367 ppm. The Eh_3^6 and Eh_3^5 shale have Sr/Ba ratios that range from 0.6 to 1.7 and from 0.5 to 2.8, respectively. The B/Ga ratios of the Eh_3^6 shale (2.8–5.4) are greater than those of the Eh_3^5 shale (1.7–3.8). The U_{EF} and Mo_{EF} values exhibit the same pattern, with higher values in the Eh_3^6 shale (U_{EF}: 2.7–9.8; Mo_{EF}: 12.2–34.0) and lower values in the Eh_3^5 shale (U_{EF}: 1.4–7.0; Mo_{EF}: 1.5–18.4). Regarding the TiO_2 and Al_2O_3 contents, the Eh_3^6 shale values range from 0.50 to 0.66 wt. % and from 12.5 to 15.7 wt. %, respectively, while in the Eh_3^5 shale, they range from 0.27 to 0.76 wt. % and from 6.7 to 18.4 wt. %, respectively (Figure 6).

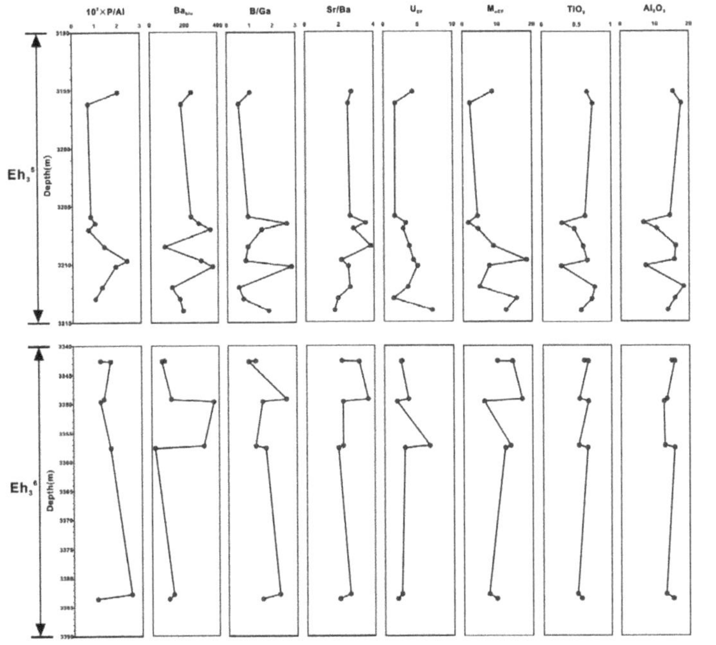

Figure 6. Vertical variations in elemental geochemical parameters of the Eh_3^6 and Eh_3^5 shale.

4.4. Biomarkers

The Eh_3^6 and Eh_3^5 shale contain varying levels of chloroform bitumen 'A', ranging from 0.10 to 0.17 wt. % and from 0.08 to 0.16 wt. %, respectively (Table 2). The Eh_3^6 shale (87.1–94.4 wt. %) contains higher proportions of hydrocarbon fractions compared to the Eh_3^5 shale (82.0–88.2 wt. %). Both intervals exhibit a prevalence of saturated over aromatic fractions, with saturated/aromatic hydrocarbon ratios varying between 2.8 and 4.5.

The Eh_3^6 and Eh_3^5 shale are dominated by short-chain n-alkanes (n-$C_{17\text{-}23}$) with a predominance of odd over even (Figure 7A). The acyclic isoprenoid pristane (Pr) exhibits a higher abundance compared to phytane (Ph), with Pr/Ph ratios ranging from 0.87 to 1.01 in the Eh_3^6 shale and from 0.98 to 1.37 in the Eh_3^5 shale (Table 2). The Eh_3^6 shale has ratios of Pr/n-C_{17} and Ph/n-C_{18} that fall between 0.12 and 0.31 and between 0.17 and 0.32, respectively. On the other hand, the Eh_3^5 shale exhibits two ratios that vary from 0.36 to 0.51 (Pr/n-C_{17}) and from 0.29 to 0.38 (Ph/n-C_{18}), respectively. Drimanes, labdanes, methylheptadecanes, and b-carotane are also assigned in the TIC chromatogram (Figure 7A).

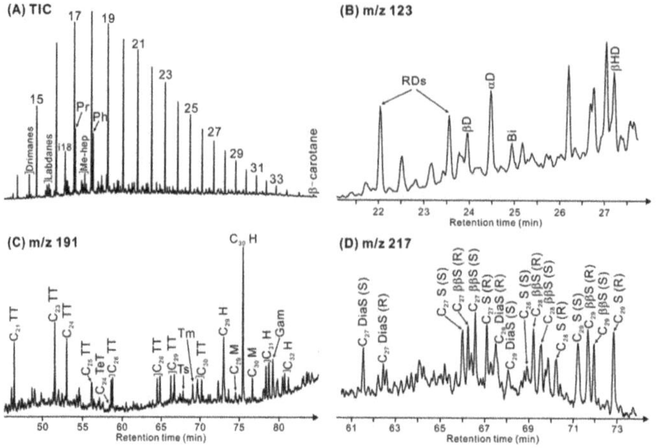

Figure 7. (**A**) Total ion current (TIC) chromatogram of saturated hydrocarbon fraction of a representative sample (3357.2 m, Eh_3^6). (**B**) Mass chromatogram (*m/z* 123) of the corresponding drimanes and bisabolane. (**C**) Mass chromatogram (*m/z* 191) of the corresponding terpenoids. (**D**) Mass chromatogram of the corresponding steranes. A key to labelled peaks in the chromatograms is provided in Table 3.

The mass chromatograms of drimanes (*m/z* 123; Figure 7B) reveal relatively high abundance of 8a(H)-drimane, followed by rearranged drimanes, 8β(H)-homodrimane, 8β(H)-drimane, and bisabolane. The Eh_3^6 shale has ratios ranging from 1.13 to 1.80 for rearranged drimanes/drimanes and from 1.66 to 2.28 for 8β(H)-homodrimane/8β(H)-drimane. In contrast, two ratios in the Eh_3^5 shale fall between 0.70 and 1.11 and between 1.43 and 1.71, respectively (Table 2).

The mass chromatograms of terpenoids (*m/z* 191; Figure 7C) illustrate that tricyclic terpanes are assigned within the range from C_{21} to C_{30} (excluding C_{22} and C_{27}), along with C_{24} tetracyclic terpane. Short-chain tricyclic terpanes (<C_{24}) are more abundant compared to their longer-chain homologues. The ratios of C_{24} tetracyclic/C_{26} tricyclic terpanes vary from 0.38 to 0.66 in the Eh_3^6 shale and from 0.21 to 0.38 in the Eh_3^5 shale (Table 2). The non-aromatic cyclic triterpenoids are primarily composed of hopanoids. The most abundant hopanoid is 17α, 21β C_{30} hopane (C_{30}H), followed by 17a, 21β C_{29} norhopane, 17α, 21β $C_{31\text{-}32}$ homohopane (22S+22R), 17β, 21α $C_{29\text{-}30}$ moretane, and 18α(H)-trisnorhopane (Ts) and 17α(H)-trisnorhopane (Tm). The ratios of Ts/(Ts+Tm) and C_{31} homohopane 22S/(22S+22R) isomers range from 0.67 to 0.76 and from 0.58 to 0.61 in the Eh_3^6 shale, while in the Eh_3^5 shale, these ratios vary between 0.57 and 0.73 and between 0.57 and 0.61, respectively. Gammacerane (Ga) is also assigned in *m/z* 191 mass chromatograms, with Ga/C_{30}H ratios of 0.59–0.82 in the Eh_3^6 shale and 0.50–0.71 in the Eh_3^5 shale.

The mass chromatograms of terpenoids (*m/z* 191; Figure 7C) illustrate that tricyclic terpanes are assigned within the range from C_{21} to C_{30} (excluding C_{22} and C_{27}), along with C_{24} tetracyclic terpane. Short-chain tricyclic terpanes (<C_{24}) are more abundant compared to their longer-chain homologues. The ratios of C_{24} tetracyclic/C_{26} tricyclic terpanes vary from 0.38 to 0.66 in the Eh_3^6 shale and from 0.21 to 0.38 in the Eh_3^5 shale (Table 2). The non-aromatic cyclic triterpenoids are primarily composed of hopanoids. The most abundant hopanoid is 17α, 21β C_{30} hopane (C_{30}H), followed by 17a, 21β C_{29} norhopane, 17α, 21β $C_{31\text{-}32}$ homohopane (22S+22R), 17β, 21α $C_{29\text{-}30}$ moretane, and 18α(H)-trisnorhopane (Ts) and 17α(H)-trisnorhopane (Tm). The ratios of Ts/(Ts+Tm) and C_{31} homohopane 22S/(22S+22R) isomers range from 0.67 to 0.76 and from 0.58 to 0.61 in the Eh_3^6 shale, while in the Eh_3^5 shale, these ratios vary between 0.57 and 0.73 and between 0.57 and 0.61, respectively. Gammacerane (Ga) is also assigned in *m/z* 191 mass chromatograms, with Ga/C_{30}H ratios of 0.59–0.82 in the Eh_3^6 shale and 0.50–0.71 in the Eh_3^5 shale.

Table 2. Chloroform bitumen 'A', group composition, and biomarker ratios for the Eh36 and Eh35 shale from the Biyang Depression.

Strata	Depth (m)	A (wt. %)	B (wt. %)	C	D	E	F	G	H	I	J	K	L	M	N	O	P	Q	R (%)	S (%)	T (%)
Eh_3^5	3195.2	0.10	85.8	3.4	1.31	0.44	0.36	0.88	1.48	0.24	0.60	0.60	0.46	1.66	0.52	0.67	0.88	0.84	32	46	22
	3196.2	0.09	82.0	4.4	1.12	0.39	0.33	0.70	1.43	0.37	0.64	0.59	0.71	2.21	0.52	0.66	0.87	0.96	35	42	23
	3206.5	0.08	85.2	2.9	1.05	0.36	0.38	1.11	1.71	0.37	0.62	0.59	0.51	1.64	0.51	0.67	0.91	0.86	41	29	30
	3208.5	0.11	82.7	3.9	1.12	0.38	0.34	1.09	1.47	0.38	0.63	0.58	0.64	2.32	0.51	0.66	0.96	0.93	31	41	28
	3209.7	0.14	85.0	3.4	0.98	0.42	0.29	0.96	1.66	0.21	0.73	0.61	0.5	1.50	0.54	0.66	1.01	0.91	36	36	28
	3210.2	0.16	88.2	3.0	1.37	0.51	0.33	0.88	1.54	0.32	0.57	0.57	0.51	1.85	0.51	0.63	0.86	0.89	28	51	21
Eh_3^6	3342.7	0.13	88.5	3.2	1.01	0.28	0.28	1.80	2.11	0.47	0.76	0.59	0.64	1.89	0.57	0.70	0.95	0.95	34	30	36
	3349.2	0.17	94.4	4.5	0.99	0.12	0.13	1.63	2.28	0.51	0.68	0.60	0.59	2.64	0.54	0.71	0.97	0.98	28	41	31
	3349.6	0.11	89.7	2.8	0.87	0.24	0.17	1.33	1.66	0.46	0.67	0.61	0.71	2.81	0.58	0.68	1.07	1.02	32	30	38
	3357.2	0.10	87.1	3.8	0.89	0.31	0.32	1.13	1.87	0.38	0.75	0.58	0.82	2.65	0.53	0.67	1.01	0.95	24	28	48
	3383.6	0.10	87.1	3.1	0.91	0.2	0.22	1.21	1.97	0.66	0.72	0.6	0.67	2.43	0.55	0.7	0.89	1.00	22	27	51

A = chloroform bitumen 'A'. B = sum of saturated and aromatic hydrocarbon fractions. C = saturated/aromatic hydrocarbons. D = Pr/Ph. E = Pr/n-C_{17}. F = Ph/n-C_{18}. G = rearranged drimanes/drimanes. H = 8β(H)-homodrimane/8β(H)-drimane. I = C_{24} tetracyclic/C_{26} tricyclic terpanes. J = Ts/(Ts+Tm). K = C_{31} homohopane 22S/(22S+22R). L = gammacerane/ab C_{30} hopane. M = ααα C_{29}/C_{27} sterane (20R). N = ααα C_{29} steranes 20S/(20S+20R). O = ββ/(ββ+αα) C_{29} steranes. P = MPI-1. Q = TNR-2. R = fluorene. S = dibenzofuran. T = dibenzothiophene. The sum of fluorene, dibenzofuran, and dibenzothiophene is 100%.

Table 3. Abbreviations for labelled biomarker compounds in Figures 6 and 7.

Figure	m/z	Abbreviation	Compound	Figure	m/z	Abbreviation	Compound
Figure 7A	TIC	15-33	C_{15}-C_{33} *n*-alkanes	Figure 7D	217	C_{27} ββS (S)	5α,14β,17β-cholestane (20S)
		i18	C_{18} *iso*-alkane			C_{27} S (R)	5α,14α,17α-cholestane (20R)
		Pr	Pristane			C_{28} S (S)	24-methyl-5α,14α,17α-cholestane (20S)
		Me-hep	Methylheptadecanes			C_{28} ββS (R)	24-methyl-5α,14β,17β-cholestane (20R)
		Ph	Phytane			C_{28} ββS (S)	24-methyl-5α,14β,17β-cholestane (20S)
Figure 7B	123	RDs	Rearranged drimanes			C_{28} S (R)	24-methyl-5α,14α,17α-cholestane (20R)
		βD	8β-drimane			C_{29} S (S)	24-ethyl-5α,14α,17α-cholestane (20S)
		αD	8α-drimane			C_{29} ββS (R)	24-ethyl-5α,14β,17β-cholestane (20R)
		Bi	Bisabolane			C_{29} ββS (S)	24-ethyl-5α,14β,17β-cholestane (20S)
		βHD	8β-homodrimane			C_{29} S (R)	24-ethyl-5α,14α,17α-cholestane (20R)
Figure 7C	191	C_{21}-C_{30} TT	C_{21}-C_{30} tricyclic terpanes	Figure 8A	TIC	TeMNs	Tetramethylnaphthalenes
		C_{24} TeT	C_{24} tetracyclic terpane			Phe	Phenanthrene
		Ts	18α-trisnorhopane			C_2 Fls	C_2 fluorenes
		Tm	17α-trisnorhopane			C_4 Fls	C_4 fluorenes
		C_{29} H	17α,21β-norhopane			MPs	Methylphenanthrenes
		C_{29} M	17β,21α-norhopane			DMPs	Dimethylphenanthrenes
		C_{30} H	17α,21β-hopane			TMPs	Trimethylphenanthrenes
		C_{30} M	17β,21α-hopane			Chr	Chrysene
		C_{31} H	17α,21β-homohopane (22S+22R)			MChrs	Methylchrysenes
		Gam	Gammacerane			B[e]P	Benzo[e]pyrene
		C_{32} H	17α,21β-bishomohopane (22S+22R)	Figure 8C	168	3-MB	3-methylbiphenyl
Figure 7D	217	C_{27} DiaS (S/R)	13α,17β-diacholestane (20S/20R)			4-MB	4-methylbiphenyl
		C_{27} S (S)	5α,14α,17α-cholestane (20S)	Figure 8D	170	TMN	Trimethylnaphthalene
		C_{27} ββS (R)	5α,14β,17β-cholestane (20R)				

The mass chromatograms of steroids (*m/z* 217; Figure 7D) indicate the highest abundance of C_{29} regular steranes, followed by C_{27} and C_{28} regular steranes. The ratios of ααα C_{29}/C_{27} sterane (20R) range from 1.89 to 2.64 in the Eh_3^6 shale and from 1.50 to 2.32 in the Eh_3^5 shale (Table 2). Within the range of C_{27-29} regular steranes, the 5α,14β,17β(H) steranes (20S+20R) dominate over the 5α,14α,17α(H) steranes (20S+20R). The ratios of aaa C_{29} steranes 20S/(20S+20R) and ββ/(ββ+αα) C_{29} steranes fluctuate between 0.53 and 0.58 and between 0.68 and 0.71 in the Eh_3^6 shale, whereas the Eh_3^5 shale exhibits two ratios of 0.51–0.54 and 0.63–0.67, respectively. Additionally, minor proportions of 13α,17β C_{27} diasteranes (20S+20R) are also assigned.

Phenanthrene (P) and its mono-, di-, and tri-methyl-substituted homologues (MP, DMP, and TMP) are the predominant aromatic compounds (Figure 8), followed by chrysene and its methyl-substituted homologues (MChr), benzo[e]pyrene (B[e]P), fluorene (F), methyl biphenyl (MB), dibenzofuran (DBF), tri- and tetra-methyl naphthalene (TMN, TeMN), and dibenzothiophene (DBT). The Methylphenanthrene Index 1 (MPI-1 = 1.89×(3-MP+2-MP)/[P+1.26×(9-MP+1-MP)]) [28] values in the Eh_3^6 and Eh_3^5 shale ranges from 0.89 to 1.07 and from 0.86 to 1.01, respectively (Table 2). The trimethylnaphthalene ratio-2 (TNR-2 = (2,3,6-TMN+1,3,7-TMN)/(1,3,5-TMN+1,3,6-TMN+ 1,4,6-TMN)) [29] values vary between 0.95 and 1.02 for the Eh_3^6 shale and between 0.84 and 0.93 for the Eh_3^5 shale. Considering F, DBF, and DBT as 100%, the Eh_3^6 shale has a larger percentage of DBT, but the Eh_3^5 shale has a higher concentration of DBF.

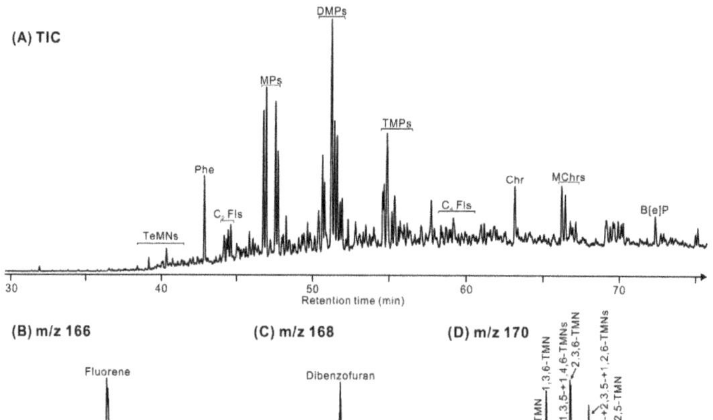

Figure 8. (**A**) Total ion current (TIC) chromatogram of aromatic hydrocarbon fraction of a representative sample (3196.2 m, Eh_3^5). (**B**) Mass chromatogram (*m/z* 166) of the corresponding fluorene. (**C**) Mass chromatogram (*m/z* 168) of the corresponding dibenzofuran and 3- and 4-methylbiphenyl. (**D**) Mass chromatogram (*m/z* 170) of the corresponding trimethylnaphthalene. (**E**) Mass chromatogram (*m/z* 184) of the corresponding tetramethylnaphthalene and dibenzothiophene. (**F**) Mass chromatogram (*m/z* 192) of the corresponding methylphenanthrene. A key to labelled peaks in the chromatograms is provided in Table 3.

5. Discussion

5.1. Maturity

A peak oil window maturity is indicated by the mean values of random vitrinite reflectance (0.9–1.0%) for the Eh_3^6 and Eh_3^5 shale (Figure 4; Table 1). The conclusion is confirmed by the T_{max} values, which are mainly distributed between 440 °C and 460 °C (Figure 3). Maturity indicators derived from terpenoids and steroids, including αβ C_{31} homohopane 22S/(22S+22R), Ts/(Ts+Tm), ααα C_{29} sterane 20S/(20S+20R), and ββ/(ββ+αα) C_{29} steranes ratios, are close to the corresponding equilibrium values (Figure 9A,B), all of which suggest a maturity characteristic of the peak oil window [30]. Furthermore, aromatic maturity indicators like MPI-1 and TNR-2 confirm a maturity stage at the peak oil window (Figure 9C). Overall, the Eh_3^6 shale has greater MPI-1 and TNR-2 values than the Eh_3^5 shale, revealing a comparatively higher maturity. The peak oil window maturity is also evidenced by the presence of degraded telalginite (Figure 5B). Considering that the influences of thermal maturity on both intervals are comparable, paleoenvironmental biomarker indicators (e.g., Pr/Ph) are applied to contrast the paleoenvironmental differences between two intervals.

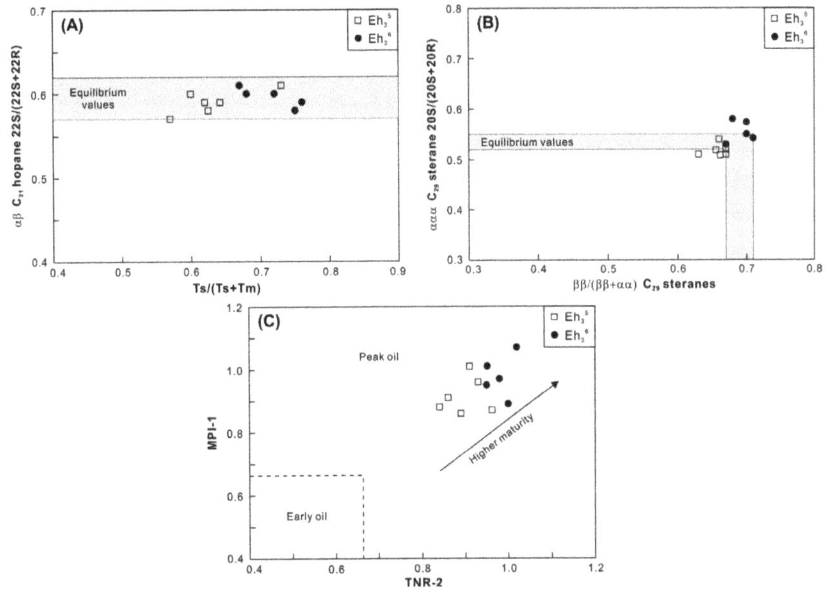

Figure 9. Cross-plots of (**A**) Ts/(Ts+Tm) versus αβ C_{31} hopane 22S/(22S+22R), (**B**) ββ/(αα+ββ) C_{29} steranes versus ααα C_{29} sterane 20S/(20S+20R), and (**C**) TNR-2 versus MPI-1.

5.2. Organic Matter (OM) Source

In contrast to the Eh_3^3 and Eh_3^2 shales, which exhibit a majority of type I-II_1 kerogen [15], the Eh_3^6 and Eh_3^5 shales display dominance by type II_2-III kerogen, suggesting that the organic matter is primarily derived from terrigenous higher plants (Figure 3). This inference is supported by the prevalence of terrigenous macerals (e.g., terrigenous liptinite, vitrinite, and inertinite) in both shale intervals, accounting for 82–91 vol.% (mmf = mineral matter free) and 68–85 vol.% (mmf) in the Eh_3^6 and Eh_3^5 shale, respectively. A negative correlation exists between the HI values and the contents of terrigenous macerals, indicating that the decreased HI values are associated with the increased contribution of higher plants (Table 1).

High ratios of C_{24} tetracyclic/C_{26} tricyclic terpanes (C_{24} Tet/C_{26} TT) are indicative of significant input of terrigenous OM [31]. The Eh_3^6 shale exhibits higher C_{24} Tet/C_{26} TT ratios compared to the Eh_3^5 shale, suggesting a substantial influx of terrigenous OM during the shale's deposition (Figure 10A). It is commonly known that the main source of C_{27} sterols is algae, whereas higher plants are usually linked to C_{29} sterols. [32]. While some biomarker studies have revealed elevated levels of 24-ethylesterols in some microalgae species [33], the likelihood of microalgae-derived C_{29} steranes can be ruled out in this case, given that OM is predominantly sourced from higher plants. Therefore, the elevated ratios of C_{29}/C_{27} $\alpha\alpha\alpha$ steranes (20R) in the Eh_3^6 shale are consistent with the increased terrigenous OM input, which corroborates the findings of C_{24} Tet/C_{26} TT ratios (Figure 10A). Labdanes, which are typical higher plant-derived biomarkers [34], have also been assigned (Figure 7A), further confirming the contribution of terrigenous OM to the shale.

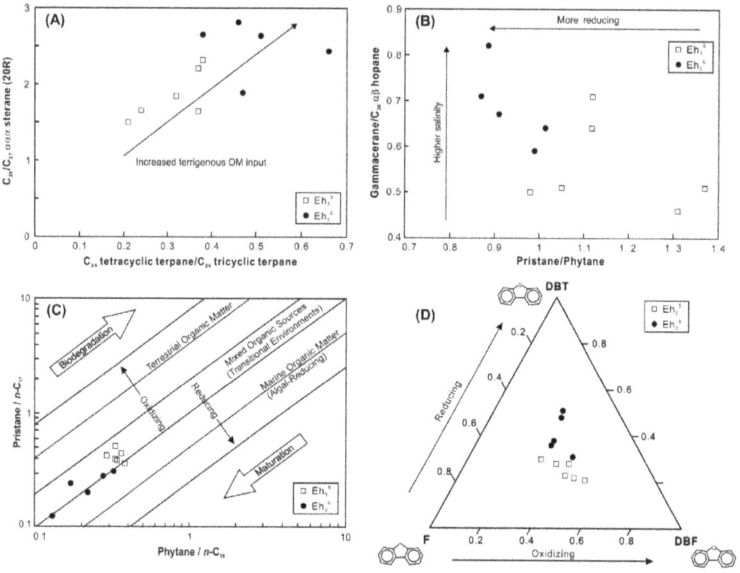

Figure 10. (**A**) Cross-plots of C_{24} tetracyclic terpane/C_{26} tricyclic terpane versus C_{29}/C_{27} $\alpha\alpha\alpha$ sterane (20R). (**B**) Cross-plots of pristane/phytane versus gammacerane/C_{30} $\alpha\beta$ hopane. (**C**) Cross-plots of phytane/n-C_{18} versus pristane/n-C_{17}. (**D**) Ternary plot of dibenzothiophene (DBT), fluorene (F), and dibenzofuran (DBF).

Furthermore, a portion of OM in shale may originate from bacteria, as indicated by the elevated concentrations of hopanoids (particularly $\alpha\beta$ C_{30} hopane) and the assignment of methylheptadecanes (Figure 7A,C) [35]. Hopanoids are mainly produced by prokaryotic organisms such as bacteria [36], whereas 7- and 8-methylheptadecanes are primarily generated by cyanobacteria and are found in significant concentrations in pure cultures of cyanobacteria [37].

5.3. Paleoenvironment Reconstruction

In the middle of the Biyang Depression during the Eh_3^6 and Eh_3^5 periods, deep-to-semi-deep lacustrine environments were generally developed [20,22]. The integration of elemental compositions and biomarkers of shale reflects a variety of depositional environments, including paleosalinity, redox conditions, paleoproductivity, and detrital input. The B/Ga and Sr/Ba ratios, as well as the gammacerane index (GI = Ga/C_{30}H ratios), are utilized to infer paleosalinity levels. The B/Ga ratios range from 2.8 to 5.4 in the Eh_3^6 shale and from 1.7 to 3.8 in the Eh_3^5 shale, indicative of predominantly brackish and freshwater conditions, respectively (Figure 11A; based on the threshold proposed by [38]). This interpretation is corroborated by the higher GI values observed in the Eh_3^6 shale (0.59–0.82) compared to the Eh_3^5 shale (0.50–0.71; Figure 10B). However, the results derived from Sr/Ba ratios differ from those of the preceding parameters (Figure 11A). The impact of carbonate-hosted Sr in the shale is thought to be the cause of this discrepancy, as indicated by a robust positive correlation between the CaO contents and Sr concentrations (the inset figure in Figure 11A).

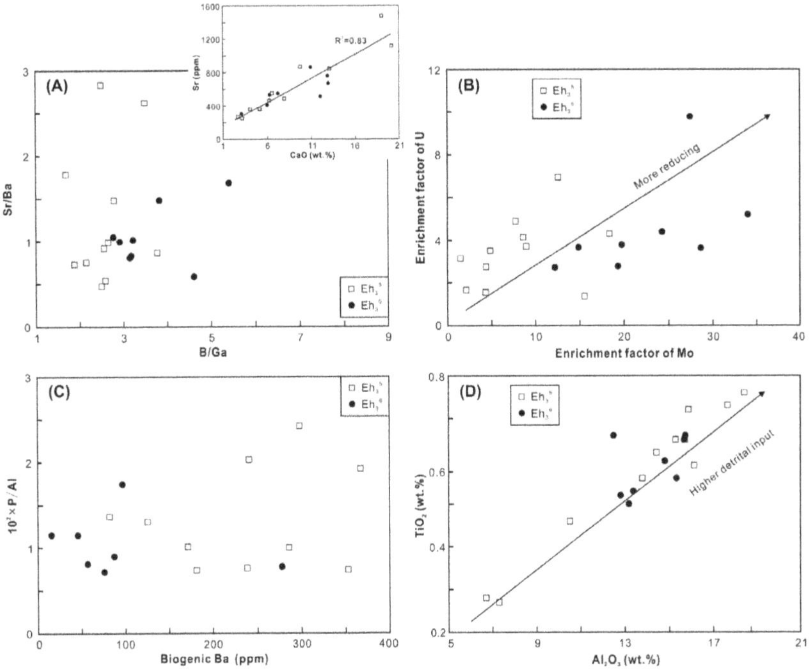

Figure 11. Cross-plots of (**A**) B/Ga versus Sr/Ba, with the inset figure showing CaO versus Sr; (**B**) enrichment factor of Mo versus enrichment factor of U; (**C**) biogenic Ba versus $10^2 \times$ P/Al; and (**D**) Al_2O_3 versus TiO_2.

The Pr/Ph ratios exhibit variability, ranging from 0.87 to 1.0 in the Eh_3^6 shale and 0.98 to 1.37 in the Eh_3^5 shale, suggesting the prevalence of more reducing conditions, probably anoxic (in accordance with the threshold proposed by [39]), during the deposition of Eh_3^6 shale (Figure 10B). Similarly, the cross-plot of Pr/n-C_{17} versus Ph/n-C_{18} indicates that the Eh_3^6 shale plots closer to the region representing reducing conditions (Figure 10C). Fluorene (F) tends to convert into dibenzofuran (DBF) under sufficient oxygen conditions, whereas it tends to generate dibenzothiophene (DBT) in oxygen-depleted conditions [40]. Thus, the positioning of shale in the ternary plot of DBT-F-DBF suggests that the Eh_3^6 shale was formed under comparatively more reducing conditions than the Eh_3^5 shale (Figure 10D). Furthermore, the enrichment factors of U and Mo, serving as redox proxies of sediments [41], also indicate the prevalence of more reducing conditions during the deposition of Eh_3^6 shale (Figure 11B).

The P/Al ratio and Ba_{bio} value are applied to assess paleoproductivity. Overall, the Eh_3^5 shale exhibits high P/Al ratios and Ba_{bio} values, whereas the Eh_3^6 shale displays comparatively low ratios and values (Figure 11C). These findings possibly suggest that the elevated paleoproductivity prevailed during the deposition of Eh_3^5 shale. However, this conclusion may raise questions, as an essential premise for utilizing P and Ba as productivity indicators is the predominance of aquatic-origin OM [42], which is evidently inconsistent with the studied shale. As discussed in Section 5.2, the primary source of OM is terrigenous higher plants, implying potentially low productivity of aquatic organisms. Therefore, it is postulated that aquatic productivity was low during the deposition of both intervals, potentially exerting limited influence on OM enrichment and shale formation.

Aluminum primarily originates from clay minerals, while titanium mainly occurs in clay and heavy minerals such as rutile. Clay minerals are abundant in both the Eh_3^6 and Eh_3^5 shale, whereas rutile is commonly found in the Eh_3 [19,22]. Therefore, the contents of Al and Ti are utilized to reveal detrital influx. The Eh_3^6 shale has a slightly higher average Al_2O_3 and TiO_2 concentration than the Eh_3^5 shale, suggesting an increased detrital supply to the paleo-lake during the deposition of Eh_3^6 shale.

The cartoons illustrating the depositional models of Eh_3^6 and Eh_3^5 shale are depicted in Figure 12. During the deposition of Eh_3^6 shale, there was a notable influx of detrital matter and terrigenous OM into the paleo-lake. Using 2000 Monte Carlo simulations of potential sedimentation rates, the earlier study performed an enhanced correlation coefficient (COCO) analysis for every member of Eh_3, resulting in high sediment accumulation rates (SARs), at 17.8 cm/kyr [23]. These high SARs facilitated the rapid burial of terrigenous OM, thus favoring its preservation. Moreover, because of the occurrence of gypsum at the top of the Ed and the relatively dry paleoenvironment, surface runoff may have transported salt ions released through the weathering of parent rock [43], leading to increased salinity in the paleo-lake and potentially promoting water stratification. Accordingly, the deposition of Eh36 shale, which has a high TOC and low HI value, was facilitated by the combination of a large terrigenous OM input and advantageous preservation circumstances (Figure 12A). In contrast, the deposition of Eh_3^5 shale was characterized by comparatively lower detrital matter and terrigenous OM, corresponding to lower SARs [23]. This lower terrigenous OM input, in conjunction with freshwater and suboxic conditions, resulted in the deposition of Eh_3^5 shale with low TOC and relatively higher HI values (Figure 12B).

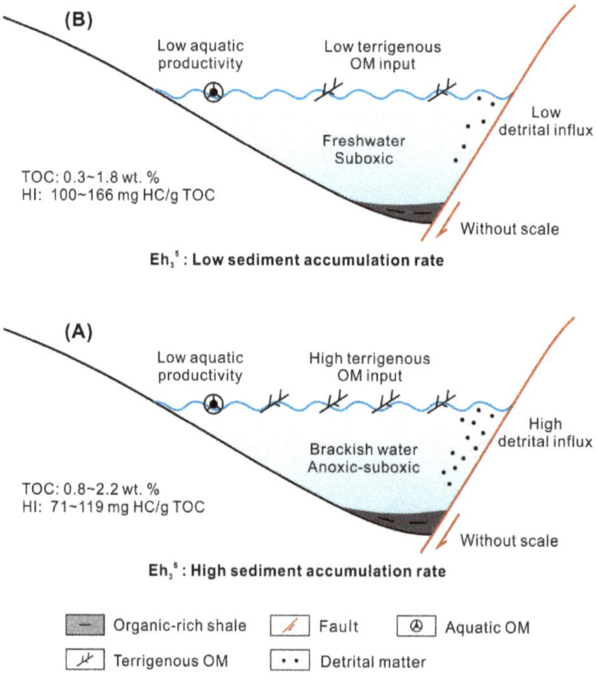

Figure 12. Depositional models of the (**A**) Eh_3^6 shale and (**B**) Eh_3^5 shale.

5.4. Hydrocarbon Potential and Implications for Regional Hydrocarbon Exploration

The hydrocarbon generation potential of the Eh_3^6 and Eh_3^5 shale was evaluated using the cross-plot of TOC versus S_1+S_2 [44]. The Eh_3^6 shale falls within the region characterized by fair potential, whereas the Eh_3^5 shale predominantly exhibits poor-to-fair potential

(Figure 13A). This observation is further corroborated by the cross-plot of TOC versus HI, where the Eh_3^6 shale demonstrates fair oil potential, while the Eh_3^5 shale exhibits little-to-fair oil potential (Figure 13B). The hydrocarbon retention potential of the Eh_3^6 and Eh_3^5 shale is assessed through the cross-plots of TOC versus S_1 and production index ($PI = S_1/(S_1+S_2)$) versus oil saturation index ($OSI = 100 \times S_1/TOC$). In these figures, comparatively higher contents of free (or migrated) hydrocarbons are evident in the Eh_3^6 shale compared to the Eh_3^5 shale (Figure 13C,D).

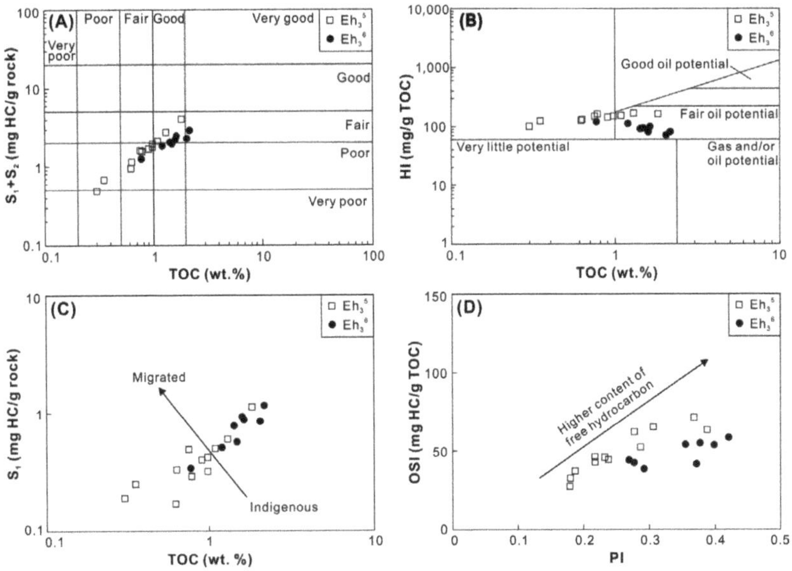

Figure 13. Cross-plots of (**A**) TOC versus S_1+S_2, (**B**) TOC versus HI, (**C**) TOC versus S_1, and (**D**) PI versus OSI.

Overall, the Eh_3^6 and Eh_3^5 shale studied herein exhibit only poor-to-fair oil potential, suggesting their limited contribution to oil reservoirs in the Biyang Depression. However, given the predominance of type II_2-III kerogen, it is hypothesized that these shales may possess significant gas potential in regions where wet/dry gas window maturity is attained. Notably, several gas fields have already been discovered near the Well A3 (Figure 1C). Considering the minimum OSI threshold (~100 mg HC/g TOC) for prospective shale oil successions in the lacustrine basins of China, [4] the studied Eh_3^6 and Eh_3^5 shale exhibit only minimal potential for shale oil exploration. Nevertheless, it is speculated that, in the central region of the paleo-lake, distant from the source area, there may be deposition of salinized lacustrine shale characterized by type I-II_1 kerogen and higher free oil contents. Therefore, the Eh_3^6 and Eh_3^5 shale around the Anpeng area, which was the sedimentary depocenter during the period, could represent a promising target for shale oil exploration.

6. Conclusions

The impact of terrigenous input on the deposition of salinized lacustrine shale was investigated using an example from the newly explored strata (Eh_3^6 and Eh_3^5) of the deep regions in the Biyang Depression, East China. The TOC contents of Eh_3^6 and Eh_3^5 shale range from 0.3 to 2.2 wt. %, with a predominance of type II_2-III kerogen. Both shale intervals exhibit a peak oil window maturity, as evidenced by vitrinite reflectance, T_{max}, and biomarker-derived maturity parameters. The primary source of OM is considered to be terrigenous higher plants for both shale intervals, supported by the dominance of terrigenous macerals (e.g., terrigenous liptinite, vitrinite, and inertinite), low HI values, and biomarker indicators. Additionally, a portion of the OM may originate from bacteria, as indicated by the high abundance of hopanoids and the assignment of 7- and 8-methylheptadecanes. During the deposition of Eh_3^6 shale, a significant influx of detrital matter and terrigenous OM occurred, corresponding to high sediment accumulation rates (SARs). These high SARs facilitated the rapid burial of OM and increased paleosalinity by transporting salt ions from the provenance regions to the paleo-lake, thereby favoring OM preservation. Consequently, the Eh_3^6 shale, characterized by high TOC and low HI values, was deposited under the influence of high terrigenous OM input and favorable preservation conditions. In contrast, the deposition of Eh_3^5 shale involved a comparatively lower flux of detrital matter and terrigenous OM, resulting in shale with lower TOC and relatively higher HI values. Overall, the studied shale exhibits only poor-to-fair potential for hydrocarbon generation and retention. However, areas with higher maturity and a better kerogen type may hold significant potential for conventional gas and shale oil exploration in the Biyang Depression, respectively.

Author Contributions: Conceptualization and writing—original draft preparation, Y.S.; writing—review and editing, P.P.; funding acquisition, S.X.; project administration, S.J.; funding acquisition, B.G.; funding acquisition and methodology, S.L.; investigation, Q.C.; investigation, Z.L.; investigation, L.W.; investigation, C.L. All authors have read and agreed to the published version of the manuscript.

Funding: This research was funded by the general program of National Natural Science Foundation of China (Mechanism and prediction of fracture-induced by in situ heating of shale oil and oil shale reservoirs with different lithofacies), Open funding project of Sinopec Key Laboratory of Shale Oil/Gas Exploration and Production Technology (Reconstruction of biological community structure of organic-rich shale in saline lake basins and its quantitative indication of paleoproductivity), and National Natural Science Foundation of China (Grant Nos. 42072174, 42073067, and 41902139).

Data Availability Statement: The original contributions presented in the study are included in the article; further inquiries can be directed to the corresponding authors.

Acknowledgments: Henan Oilfield Company (SINOPEC) is gratefully appreciated for the permission of sampling.

Conflicts of Interest: The authors declare no conflicts of interest.

References

1. Ruble, T.E.; Lewan, M.D.; Philp, R.P. New insights on the Green River petroleum system in the Uinta Basin from hydrous pyrolysis experiments. *AAPG Bull.* **2001**, *85*, 1333–1371. [CrossRef]
2. Zou, C.; Zhu, R.; Chen, Z.Q.; Ogg, J.G.; Wu, S.; Dong, D.; Qiu, Z.; Wang, Y.; Wang, L.; Lin, S.; et al. Organic-Matter-Rich shales of China. *Earth Sci. Rev.* **2019**, *189*, 51–78. [CrossRef]
3. Warren, J.K. Evaporites through time: Tectonic, climatic and eustatic controls in marine and nonmarine deposits. *Earth Sci. Rev.* **2010**, *98*, 217–268. [CrossRef]
4. Wang, M.; Guo, Z.; Jiao, C.; Lu, S.; Li, J.; Xue, H.; Li, J.; Li, J.; Chen, G. Exploration progress and geochemical features of lacustrine shale oils in China. *J. Pet. Sci. Eng.* **2019**, *178*, 975–986. [CrossRef]
5. Liu, B.; Song, Y.; Zhu, K.; Su, P.; Ye, X.; Zhao, W. Mineralogy and Element Geochemistry of Salinized Lacustrine Organic-Rich Shale in the Middle Permian Santanghu Basin: Implications for Paleoenvironment, Provenance, Tectonic Setting and Shale Oil Potential. *Mar. Pet. Geol.* **2020**, *120*, 104569. [CrossRef]
6. Liu, B.; Wang, H.; Fu, X.; Bai, Y.; Bai, L.; Jia, M.; He, B. Lithofacies and Depositional Setting of a Highly Prospective Lacustrine Shale Oil Succession from the Upper Cretaceous Qingshankou Formation in the Gulong Sag, Northern Songliao Basin, Northeast China. *AAPG Bull.* **2019**, *103*, 405–432. [CrossRef]
7. Hofmann, P.; Huc, A.Y.; Carpentier, B.; Schaeffer, P.; Albrecht, P.; Kelly, B.; Maxwell, J.R.; Damsté, J.S.S.; de Leeuw, J.W.; Leythaeuser, D. Organic matter of the Mulhouse Basin, France: A synthesis. *Org. Geochem.* **1993**, *20*, 1105–1123. [CrossRef]
8. Mascle, A.; Vially, R. The petroleum systems of the southeast basin and Gulf of Lion (France). *Geol. Soc.* **1999**, *156*, 121–140. [CrossRef]
9. Rodriguez, N.D.; Philp, R.P. Source Rock Facies Distribution predicted from oil geochemistry in the Central Sumatra Basin, Indonesia. *AAPG Bull.* **2015**, *99*, 2005–2022. [CrossRef]
10. Guo, P.; Liu, C.; Gibert, L.; Huang, L.; Zhang, D.; Dai, J. How to find high-quality petroleum source rocks in saline lacustrine basins: A case study from the Cenozoic Qaidam Basin, NW China. *Mar. Pet. Geol.* **2020**, *111*, 603–623. [CrossRef]
11. Xin, B.; Hao, F.; Han, W.; Xu, Q.; Zhang, B.; Tian, J. Paleoenvironment evolution of the lacustrine organic-rich shales in the second member of Kongdian Formation of Cangdong Sag, Bohai Bay Basin, China: Implications for organic matter accumulation. *Mar. Pet. Geol.* **2021**, *133*, 105244. [CrossRef]

12. Liang, C.; Jiang, Z.; Cao, Y.; Wu, J.; Wang, Y.; Hao, F. Sedimentary characteristics and origin of lacustrine organic-rich shales in the salinized Eocene Dongying Depression. *GSA Bull.* **2018**, *130*, 154–174. [CrossRef]
13. Kong, X.; Jiang, Z.; Zheng, Y.; Xiao, M.; Chen, C.; Yuan, H.; Chen, F.; Wu, S.; Zhang, J.; Han, C.; et al. Organic geochemical characteristics and organic matter enrichment of mudstones in an Eocene saline lake, Qianjiang Depression, Hubei Province, China. *Mar. Pet. Geol.* **2020**, *114*, 104194. [CrossRef]
14. Guan, M.; Liu, X.; Jin, Z.; Zhao, W.; Liu, W.; Bian, L.; Dong, J.; Zeng, X.; Zeng, B.; Sun, B.; et al. The formation of the Paleocene lacustrine organic-rich shale in the Subei Basin, East China Associated with the Early Late Paleocene event and marine incursions. *Mar. Pet. Geol.* **2024**, *162*, 106730. [CrossRef]
15. Song, Y.; Li, S.; Hu, S. Warm-humid paleoclimate control of salinized lacustrine organic-rich shale deposition in the Oligocene Hetaoyuan Formation of the Biyang Depression, East China. *Int. J. Coal Geol.* **2019**, *202*, 69–84. [CrossRef]
16. Xu, S.; Wang, Y.; Bai, N.; Wu, S.; Liu, B. Organic matter enrichment mechanism in saline lacustrine basins: A review. *Geol. J.* **2024**, *59*, 155–168. [CrossRef]
17. Li, S.; Hu, S.; Xie, X.; Lv, Q.; Huang, X.; Ye, J. Assessment of shale oil potential using a new free hydrocarbon index. *Int. J. Coal Geol.* **2016**, *156*, 74–85. [CrossRef]
18. Li, J.; Liu, Z.; Zhang, X.; Feng, G.; Liu, J.; Ma, Y.; Lu, S.; Li, W.; Zhou, N. Effects of paleoenvironment on continental shale oil enrichment and producibility in the Biyang Depression. *AAPG Bull.* **2022**, *106*, 2043–2071. [CrossRef]
19. Song, Y.; Cao, Q.; Li, S.; Hu, S.; Zhu, K.; Ye, X.; Wan, L. Salinized lacustrine organic-rich shale influenced by marine incursions: Algal-microbial community, paleoenvironment and shale oil potential in the Paleogene Biyang Depression, East China. *Palaeogeogr. Palaeoclimatol. Palaeoecol.* **2021**, *580*, 110621. [CrossRef]
20. Song, Y.; Wan, L.; Xu, S.; Gao, B.; Li, C.; Li, Z.; Paerzhana, P. Lake-level-fluctuation control on shale oil enrichment of the salinized lacustrine organic-rich shale in the Paleogene Biyang Depression, East China. *Minerals* **2024**, *14*, 94. [CrossRef]
21. Yuan, Y.Z.; Luo, J.Q.; Zhu, Y.; Liu, G.; Li, L.; Yu, M. Petroleum exploration history and enlightenment of Biyang Sag and Nanyang Sag in Nanxiang Basin. *Xinjiang Petro. Geol.* **2021**, *42*, 364–373, (In Chinese with English abstract).
22. Dong, Y.; Zhu, X.; Xian, B.; Hu, T.; Geng, X.; Liao, J.; Luo, Q. Seismic Geomorphology Study of the Paleogene Hetaoyuan Formation, Central-South Biyang Sag, Nanxiang Basin, China. *Mar. Pet. Geol.* **2015**, *64*, 104–124. [CrossRef]
23. Xu, K.; Chen, H.; Huang, C.; Ogg, J.G.; Zhu, J.; Lin, S.; Yang, D.; Zhao, P.; Kong, L. Astronomical Time Scale of the Paleogene Lacustrine Paleoclimate Record from the Nanxiang Basin, Central China. *Palaeogeogr. Palaeoclimatol. Palaeoecol.* **2019**, *532*, 109253. [CrossRef]
24. Taylor, G.H.; Teichmüller, M.; Davis, A.; Diessel, C.F.K.; Littke, R.; Petrology, P.R. RobertOrganic Petrology. In *A New Handbook Incorporating Some Revised Parts of Stach's Textbook of Coal Petrology*; Gebrüder Borntraeger: Berlin, Stuttgart, 1998; Geological Magazine; Volume 136, pp. 697–711. ISBN 3-443-01036-9.
25. Shen, J.; Schoepfer, S.D.; Feng, Q.; Zhou, L.; Yu, J.; Song, H.; Wei, H.; Algeo, T.J. Marine Productivity Changes during the End-Permian Crisis and Early Triassic Recovery. *Earth Sci. Rev.* **2015**, *149*, 136–162. [CrossRef]
26. Tribovillard, N.; Algeo, T.J.; Riboulleau, A. Trace metals as paleoredox and paleoproductivity proxies: An update. *Chem. Geol.* **2006**, *232*, 12–32. [CrossRef]
27. Rudnick, R.L.; Gao, S. Composition of the continental crust. In *Treatise on Geochemistry*, 2nd ed.; Newnes: Boston, UK, 2014; pp. 1–51.
28. Cassani, F.; Gallango, O.; Talukdar, S.; Vallejos, C.; Ehrmann, U. Methylphenanthrene maturity index of marine source rock extracts and crude oils from the Maracaibo Basin. *Org. Geochem.* **1988**, *13*, 73–80. [CrossRef]
29. Radke, M.; Welte, D.H.; Willsch, H. Maturity Parameters Based on Aromatic Hydrocarbons: Influence of the Organic Matter Type. *Org. Geochem.* **1986**, *10*, 51–63. [CrossRef]
30. Peters, K.E.; Walters, C.C.; Moldowan, J.M. *The Biomarker Guide: Column 2, Biomarkers and Isotopes in Petroleum Systems and Earth History*, 2nd ed.; Cambridge University Press: Cambridge, UK, 2005.
31. Hanson, A.D.; Zhang, S.C.; Moldowan, J.M.; Liang, D.G.; Zhang, B.M. Molecular organic geochemistry of the Tarim Basin, northwest China. *AAPG Bull.* **2000**, *84*, 1109–1128.
32. Volkman, J.K.; Allen, D.I.; Stevenson, P.L.; Burton, H.R. Bacterial and algal hydrocarbons from a saline Antarctic lake, Ace Lake. *Org. Geochem.* **1986**, *10*, 671–681. [CrossRef]
33. Volkman, J.K.; Barrett, S.M.; Blackburn, S.I. Eustigmatophyte microalgae are potential sources of C29 sterols, C22–C28 n-alcohols and C28–C32 n-alkyl diols in freshwater environments. *Org. Geochem.* **1999**, *30*, 307–318. [CrossRef]
34. Otto, A.; Wilde, V. Sesqui-, di-, and triterpenoids as chemosystematic markers in extant conifers; a review. *Bot. Rev.* **2001**, *67*, 141–238. [CrossRef]
35. Ourisson, G.; Albrecht, P.; Rohmer, M. The hopanoids: Palaeochemistry and biochemistry of a group of natural products. *Pure Appl. Chem.* **1979**, *51*, 709–729. [CrossRef]
36. Rohmer, M.; Bisseret, P.; Neunlist, S. The hopanoids, prokaryotic triterpenoids and precursors of ubiquitous molecular fossils. In *Biological Markers in Sediments and Petroleum*; Moldowan, J.M., Albrecht, P., Philp, R.P., Eds.; Prentice Hall: Englewood Cliffs, NJ, USA, 1992; pp. 1–17.
37. Shiea, J.; Brassell, S.C.; Ward, D.M. Mid-chain branched mono- and dimethyl alkanes in hot spring cyanobacterial mats: A direct biogenic source for branched alkanes in ancient sediments? *Org. Geochem.* **1990**, *15*, 223–231. [CrossRef]

38. Wei, W.; Algeo, T.J. Elemental proxies for paleosalinity analysis of ancient shales and mudrocks. *Geochim. Cosmochim. Acta* **2020**, *287*, 341–366. [CrossRef]
39. Didyk, B.M.; Simoneit, B.R.T.; Brassell, S.C.; Eglinton, G. Organic geochemical indicators of paleoenvironmental conditions of sedimentation. *Nature* **1978**, *272*, 216–222. [CrossRef]
40. Asif, M.; Alexander, R.; Fazeelat, T.; Pierce, K. Geosynthesis of dibenzothiophene and alkyl dibenzothiophenes in crude oils and sediments by carbon catalysis. *Org. Geochem.* **2009**, *40*, 895–901. [CrossRef]
41. Algeo, T.J.; Tribovillard, N. Environmental analysis of paleoceanographic systems based on molybdenum–uranium covariation. *Chem. Geol.* **2009**, *268*, 211–225. [CrossRef]
42. Schoepfer, S.D.; Shen, J.; Wei, H.; Tyson, R.V.; Ingall, E.; Algeo, T.J. Total organic carbon, organic phosphorus, and biogenic barium fluxes as proxies for paleomarine productivity. *Earth Sci. Rev.* **2015**, *149*, 23–52. [CrossRef]
43. Rosen, M. The importance of groundwater in playas: A review of playa classifications and the sedimentology and hydrology of playas. *GSA Bull.* **1994**, *289*, 1–18.
44. Peters, K.E. Guidelines for petroleum source rock using programmed pyrolysis. *AAPG Bull.* **1986**, *70*, 318–329.

minerals

MDPI

Article

Evolution of Pore Spaces in Marine Organic-Rich Shale: Insights from Multi-Scale Analysis of a Permian–Pennsylvanian Sample

Zilong Wang [1,2], Xiaoguang Yang [3,*] and Shaobin Guo [1,2,*]

1 School of Energy Resources, China University of Geosciences (Beijing), Beijing 100083, China; wristzar@outlook.com
2 Key Laboratory of Marine Reservoir Evolution and Hydrocarbon Enrichment Mechanism, Ministry of Education, Beijing 100083, China
3 Oil and Gas Survey, China Geological Survey, Beijing 100083, China
* Correspondence: yangxiaoguang01@mail.cgs.gov.cn (X.Y.); guosb58@cugb.edu.cn (S.G.)

Abstract: The quantitative evolution pattern of pore space and genetic pore types along the maturation process in organic-rich shale reservoirs is unclear, which affects the assessment of shale storage capacity and petroleum production. A black shale outcrop sample from Kansas that is of Permian–Pennsylvanian age was collected and subjected to thermal simulation experiments at 10 different maturity stages to understand the pore sizes and pore types. Scanning electron microscopy (SEM) and image processing were used to characterize the full-scale pore-size distribution and volume evolution of this shale sample by combining low-temperature gas (CO_2 and N_2) physisorption and mercury intrusion porosimetry (MIP) in order to discuss the effects of hydrocarbon generation and diagenesis (HG&D) on pore development at different pore sizes. The study showed that the original shale sample is dominated by slit-like pores, with mainly organic matter (OM) pores distributed in 0–100 nm, intraparticle pores (Intra-P) of clays distributed in 30–100 nm, and interparticle pores (Inter-P) distributed in 100–1000 nm. With the increase in maturity or Ro, the OM pores increased gradually, and the OM pore-size distribution diverged to the two poles. In the oil generation stage, the OM pores were distributed in the range of 30–100 nm, while in the gas generation stage, the OM-hosted pores were mainly distributed in the range of 10–20 nm and 100–500 nm. Further into the over-maturity stage, the OM pores were mainly distributed in the range of 0–10 nm and >100 nm. The pore volume distribution across the whole pore sizes showed that the pore volume of low-maturity shale samples was mainly provided by 100–1000 nm (macropores), and the pore volumes of 0–2 nm, 30–100 nm and 1000+ nm pores gradually increase with increasing thermal maturity, with the final pore-size distribution having four peaks at 0–2, 30–100, 500–1000 nm, and 10–100 μm. Hydrocarbon generation mainly affects the pore volume in the 0–2 nm and 100–1000 nm intervals, with a positive correlation. The 2–30 nm and 30–100 nm pores were likely controlled by diagenesis, such as mineral transformation, illitization, and cementation during the maturation process.

Keywords: quantitative pore evolution; shale; thermal simulation experiments; pore-size distribution

1. Introduction

Shale oil and gas are unconventional resources generated and stored in the interlayers of organic-rich mud or shale, mainly present in the adsorbed and free states [1–4]. With the increasing success of shale oil and gas development in North America and China [5,6], it has become an important component of future fossil fuels. For shale petroleum systems, shales themselves are source rocks that produce hydrocarbons during the process of continuous deep burial [7]. However, shales are reservoirs as well, where pores generated by the diagenesis of minerals and thermal evolution of organic matter provide sites for hydrocarbon storage [8,9]. Due to the features of source-reservoir integration, the joint effects of hydrocarbon generation and diagenesis (HG&D) on pore evolution have become the hotspot for geological studies of shale oil and gas [10,11].

According to the classification of the International Union of Pure and Applied Chemistry (IUPAC), pore sizes are classified into micropores (<2 nm), mesopores (2–50 nm), and macropores (>50 nm) by diameter [12], but there is a mismatch between the classification and pore types developed during shale maturation, and it is difficult to elaborate the intrinsic connection between HG&D and pore evolution. Loucks, Reed, Ruppel and Jarvie [13] first classified the genesis type of shale pores using Ar-ion milling and SEM imaging. With the development of computing resources and image recognition techniques (IRT), it has been possible to quantify pores in images from ~1 nm up, but the small observational area and the strong heterogeneity of shale make the quantification of pores by SEM images still limited [14–18]. For the quantitative characterization of the whole pore-size distribution of shales, pores of the three sizes (0–2, 2–50 and >50 nm) are mostly characterized by the combined method of mercury intrusion porosimetry (MIP), low-temperature N_2 and CO_2 physisorption, and nuclear magnetic resonance [19–23], but the inability to observe the pore morphology and certain assumptions-based processing make the characterization results less meaningful for geological theories and exploration practice.

Thermal simulation experiments carried out at a succession of consecutively high temperatures are regarded as a feasible way to model the longtime low-temperature geological development process in a relatively short experimental period, according to the time-temperature compensation theory put forward by Connan [24]. Originally, thermal simulation experiments were mainly used to reveal the hydrocarbon potential and diagenesis of source rocks [25,26]. With the growing interest in micro- and nano-pores in shales, thermal simulation experiments have been expected to be used to study the evolution of shale oil and gas reservoirs in recent years [10,11]. However, because shale oil and gas reservoirs were always highly mature and ancient rocks that had undergone long-term burial, the organic matter, minerals and fluids within them had already undergone thermal evolution and exchange, making most shales unavailable for thermal simulation experiments. Scholars have used contemporaneous outcrop samples or shallowly buried samples as substitutes [10,11,27–29], but many of them have undergone the early oil generation stage, resulting in loss of oil and cessation of early diagenesis.

Combining thermal simulation experiments, SEM and IRT, and whole pore-size characterization methods will help to explore multi-scale pore evolution from pore genesis classification and clarify the effects of HG&D on pore evolution. In this study, immature shale samples (Ro = 0.64%) in Permian–Pennsylvanian were collected on gold tube thermal simulation experiments with water-bearing closure conditions, SEM, MIP and low-temperature gas adsorption (LTGA, using N_2 and CO_2) were used to characterize the pore structure and whole pore size (0–1000+ nm in dia.) -pore volume distribution of shales at different maturity stages based. The pore evolution was quantified in the intervals of 0–2 nm, 2–30 nm, 30–100 nm, 100–1000 nm, and 1000+ nm, and the effects of HG&D on multi-scale pore evolution were discussed.

2. Geological Setting

Pennsylvanian black shales, pervasive across the central continental United States, have been extensively examined in outcrops as the principal unit of the classic Midcontinent cyclothem (Figure 1). These shales were formerly regarded as exhibiting a high condensation profile within the stratigraphic sequence geological paradigm [30]. In Kansas, they are often thinly bedded but widely distributed as distinctive lithologic outcrops. The organic-rich black shale is consistently present, characterized by its limestone richness, and deposited under open oceanic conditions. Most were formed offshore in deep water under calm, stagnant conditions [31]. Shale samples exhibit elevated levels of total organic carbon (TOC) and low maturity characteristics due to their deposition in deep offshore waters and shallow burial depths (Table 1). Their similar composition to the Barnett and

Woodford Shales, primarily comprising calcite, quartz, and clay [32–34], renders them suitable for thermal simulation experiments and pore evolution studies of marine shales.

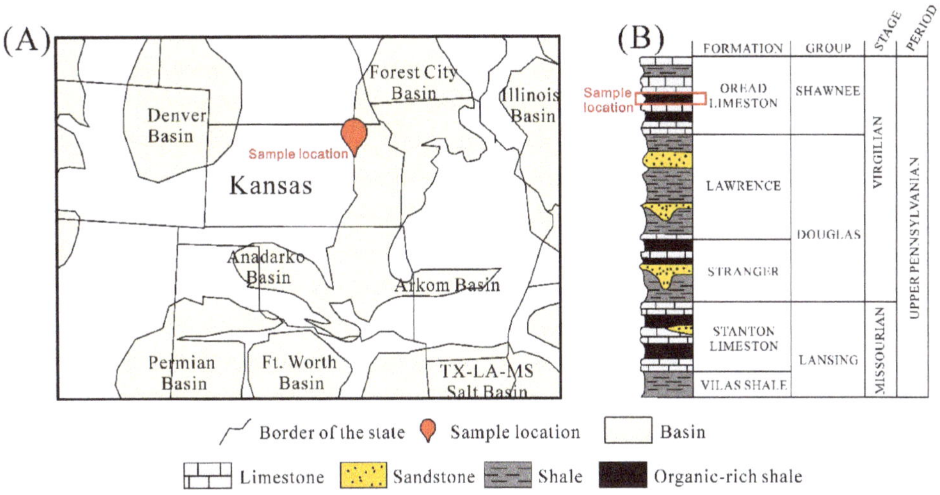

Figure 1. (**A**) Sample location in Kansas; (**B**) stratigraphic position of the Upper Pennsylvanian black shale studied.

Table 1. Basic information of the original shale sample.

Sample ID	Ro (%)	TOC (wt.%)	S2 (mg/g)	Kerogen Type
ye33	0.65	5.52	22.32	II_2

3. Methods

3.1. Thermal Simulation Experiment

Hydrothermolysis in a closed system was used to simulate the formation conditions and to cover the water consumed by OM pyrolysis and water–rock interactions. First, the sample was divided into 11 subsamples of 1 g in weight, each of which was ground to 10–15 mesh. The first sample was the original sample and was tested for TOC and vitrinite reflectance (Ro). The other 10 subsamples were separately sealed into gold tubes under the protection of argon and placed in an autoclave filled with water. To facilitate the kinetic reactions involved in organic matter hydrocarbon generation and mineral transformation [35], a heating rate of 20 °C/min was chosen as the experimental heating rate. The samples were heated to 200, 250, 300, 350, 400, 450, 500, 550, 600, and 650 °C under 20 °C/min heating rate and a 50 MPa confining pressure. After heating, the gold tube was punctured to release the gas into the vacuum system. A GC-6890 gas chromatograph (Purui Instruments, Beijing, China) was used to quantify the gaseous hydrocarbons (C1–C5), and then the liquid light hydrocarbons (C6–C14) were quantified using the internal standard method [28]. The solid residue was filtered, and the organic solvent was evaporated, followed by the weighing of heavy hydrocarbons (C14+). The procedures and details of the thermal simulation experiments can be found in Wang and Guo [28], Yang and Guo [36]. The solid residue was dried at 60 °C and then tested for X-ray diffraction (XRD), low-temperature gas physisorption, MIP, and Ro analyses.

3.2. Vitrinite Reflectance (Ro)

The solid residue adhered to the surface of the plate and was polished. The Ro was measured with a Leica DMRX microscope equipped with a 3Y Post Pro V.2.0.0 microspectrophotometer (Leica Instruments, Wetzlar, Germany). Twenty particles of vitrinite were measured at least in each sample, and the final vitrinite reflectance was obtained according to the average value.

3.3. X-ray Diffraction

The whole-rock minerals were analyzed using a D8 DISCOVER X-ray diffractometer (Brooke Scientific Instruments, Billerica, MA, USA). The shale samples were crushed and ground to particle sizes below 40 μm, made into tablets using the back compression method, and uploaded to the instrument to obtain a diffraction pattern from 3° to 45° at room temperature and 35% humidity [37]. The relative content of minerals was calculated using the area of the main peak corresponding to each mineral and Lorentz polarization correction [38].

3.4. Pore-Size Characterization Methods

The pores of the subsamples after thermal simulation were measured by LTGA (N_2 and CO_2) and MIP, respectively. The subsamples were screened by a 100 mesh sieve before testing.

An Autopore IV9500 automatic mercury injection meter (Huapu General Technology Co., Ltd., Shenzhen, China) was used for MIP tests. The mercury was slowly pressed into the pores of the evacuated sample as the pressure was slowly increased from a vacuum (~0 MPa) up to 400 MPa and then reduced to atmospheric pressure (0.11 MPa), during which the mercury volume change was recorded. The relationship between capillary pressure and area was calculated based on the Young–Dupre equation and Washburn equation, which was then used to characterize the pore volume distribution by the correspondence between pore diameter and pressure [39,40].

A Quadrasorb SI analyzer (Quantachrome Instruments, Boynton Beach, FL, USA) was used for low-temperature N_2 adsorption analyses. The sample was incrementally contacted with N_2 at 77.37 K until the mass remained constant at each relative pressure (P/P_0) from 0.01 to 1.0. The adsorbed N_2 volume is a function of the relative pressure, and the pore volume distribution was calculated by the BJH model [41], which established the relationship between the relative pressure (P/P_0) and the pore size.

A NOVA4200e analyzer (Quantachrome Instruments, Boynton Beach, FL, USA) was used for the low-temperature CO_2 adsorption tests. The maximum balanced pressure was 0.1 MPa at 273 K. Since the saturation vapor pressure (P_0) of CO_2 is 3.48 MPa at this temperature and pressure, the relative pressure variation range was 0~0.03. The pore-size distribution was calculated based on density functional theory (DFT) [7].

3.5. SEM and IRT

FEI Quanta 200F FESEM (FEI company, Hillsboro, OR, USA) was used to observe the subsamples after thermal simulation. The samples were polished with 3000 grit sandpaper, milled with argon ion and sprayed with gold before observations. The SEM images were filtered to remove noise, and the pores were binarized using grayscale thresholding. The number, area and proportion of pores were counted by ImageJ software (version 1.8.0) [42–44].

4. Results

4.1. Maturity and Hydrocarbon Yield

Ro is the classic indicator of maturity, and our discussion on pore development will revolve around different Ro values from 10 thermally matured samples (Table 2). The Ro of the subsamples gradually increased from 0.71%–2.65%, with increasing final thermal simulation temperature (Figure 2). Liquid hydrocarbons (C6+) are generated at the beginning, peaking at Ro = 1.19% and then gradually decreasing until its exhaustion. Gaseous hydrocarbons (C1–C5) begin to be generated in small amounts at Ro = 0.8% and increase rapidly when liquid hydrocarbons begin to decrease. After Ro > 1.98%, the rate of gaseous hydrocarbon generation slows down. Total hydrocarbons were generated rapidly at the beginning up to Ro = 1.19% and then increased slowly.

Table 2. Ro and hydrocarbon yields of the simulated subsamples.

Sample ID	Final Temperature (°C)	Ro (%)	Gaseous Hydrocarbons mg/g	Liquid Hydrocarbons mg/g	Total Hydrocarbons mg/g
ye33-200	200.0	0.71	0.00	2.62	2.62
ye33-250	250.0	0.75	0.00	2.88	2.88
ye33-300	300.0	0.79	0.03	3.47	3.50
ye33-350	350.0	1.1	0.25	9.77	10.02
ye33-400	400.0	1.19	1.68	17.02	18.70
ye33-450	450.0	1.56	6.21	14.11	20.31
ye33-500	500.0	1.78	13.57	7.08	20.65
ye33-550	550.0	1.98	19.13	3.39	22.52
ye33-600	600.0	2.13	20.51	2.21	22.72
ye33-650	650.0	2.65	21.55	1.25	22.80

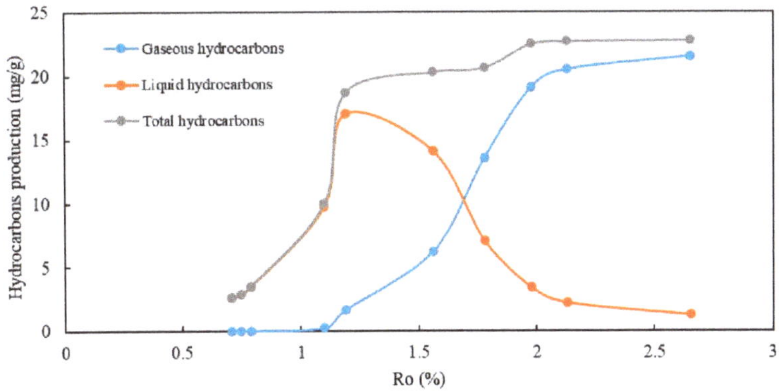

Figure 2. Yields of hydrocarbons at different maturities from the thermal maturation experiments.

4.2. Whole-Rock Mineral Composition

The samples were mainly composed of quartz, clay minerals and calcite, with the total amount of these three components being greater than 95% (Table 3). Calcite was the most dominant mineral at 45.4%–62.3%, followed by clay minerals at 20.6%–30.5% and quartz at 10.1%–18.7%. With the increase in thermal maturity, the calcite content first increased to the highest when Ro = 1.10%, then decreased to the lowest when Ro = 1.78%, and finally gradually increased. The trend of clay minerals content was opposite to that of calcite, and the quartz content is relatively stable (Figure 3).

Table 3. Whole-rock mineral composition by XRD analyses.

Sample ID	Whole-Rock Mineral Content (%)							
	Clay	Quartz	Potassium Feldspar	Plagioclase	Calcite	Dolomite	Pyrite	Siderite
ye33-200	27	12	0.4	1.7	56.9	0.9	0.4	0.7
ye33-250	27.4	14.3	0.8	1.4	54.1	0.6	0.7	0.7
ye33-300	29.9	12	0.5	1.3	53.7	1	0.7	0.9
ye33-350	23.4	10.1	0.4	1.1	62.3	1.1	0.7	0.9
ye33-400	20.6	13.6	0.8	1.6	61.2	1	0.3	0.9
ye33-450	24.4	15	0.5	1.6	55	1.2	0.7	1.6
ye33-500	30.5	18.7	0.9	1.4	45.4	1.8	0.3	1
ye33-550	27.3	13.5	0.8	1.5	54.2	1.4	0.3	1
ye33-600	30.1	13.8	0.5	1	52.8	0.8	0.1	0.9
ye33-650	28.4	13.4	0.5	0.9	54.6	1	0.3	0.9

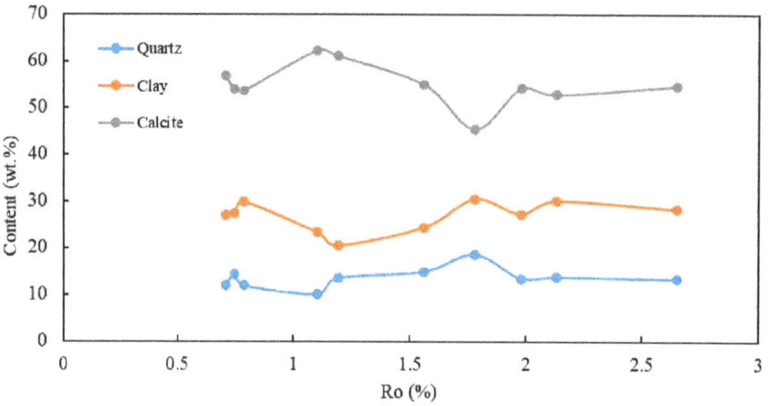

Figure 3. Changes in the compositions of quartz, clays, and calcite at different maturities.

4.3. Pore Volume and Structure Characterization

The CO_2 adsorption capacity of 10 subsamples gradually increased with the relative pressure (P_0/P), and the adsorption quantity was 0.054–0.140 mmol/g at $P/P_0 = 0.03$ (P = 0.1 MPa), with ye33-650 (sample obtained from the artificial maturation experiment at a temperature of 650 °C) having the highest adsorption quantity (Figure 4A), according to the BET equation [12].

$$\frac{P}{Q(P_0 - P)} = \frac{1}{Q_m c} + \frac{c - 1}{Q_m c}\frac{P}{P_0} \quad (1)$$

P is the balance pressure, MPa; P_0 is the saturated vapor pressure of the adsorbent (CO_2), MPa; Q is the actual adsorption quantity of the sample, mmol/g; Q_m is the monolayer saturated adsorption quantity, mmol/g; and c is a constant related to the adsorption capacity.

Therefore, P/P_0 has a linear relationship with $\frac{P}{Q(P_0-P)}$ based on Equation (1), which was a good fit for the 10 subsamples (Figure 4B). The Q_m was 0.086–0.178 mmol/g, which reflected the sample's specific surface area. The term *c* indicates the adsorption capacity of the samples, which was 34.5–124. The *c* of the 10 samples keeps increasing with an increasing maturity (Table 3), which represents a gradually increasing adsorption capacity, which may be the superposition of adsorption potential due to the increasing micropores (<2 nm) and pore surface complexity of the samples in the maturation process. The cumulative pore volumes (Figure 4C) of 0.36–1.06 nm micropores of the samples were calculated based on DFT theory [45] as 0.00131–0.0048 cm^3/g.

According to the IUPAC classification, the N_2 adsorption-desorption curves of the sample belonged to type IV (Figure 5A), representing a porous medium rich in mesopores and micropores. The hysteresis loops and capillary condensation are the most typical features of type IV curves [46], and the hysteresis loop type is type H4, which was dominated by narrow-sized slit pores. Capillary condensation occurred at a P/P_0 of 0.45–0.5, corresponding to a pore size of 3–4 nm. When the relative pressure was less than 0.45, a small quantity of nitrogen remained stagnant in the shales, indicating that the shales have a complex micro- and meso- (<4 nm) pore structure and matrix dissolution [47,48], with the stagnant quantity ranging from 0–0.73 cm^3/g (Table 3) at standard temperature and pressure (STP, 273.15 K and 101.3 kPa). The cumulative pore volumes for 2–200 nm pore sizes were calculated from the BJH equation [23] as 0.033–0.052 cm^3/g (Figure 5B).

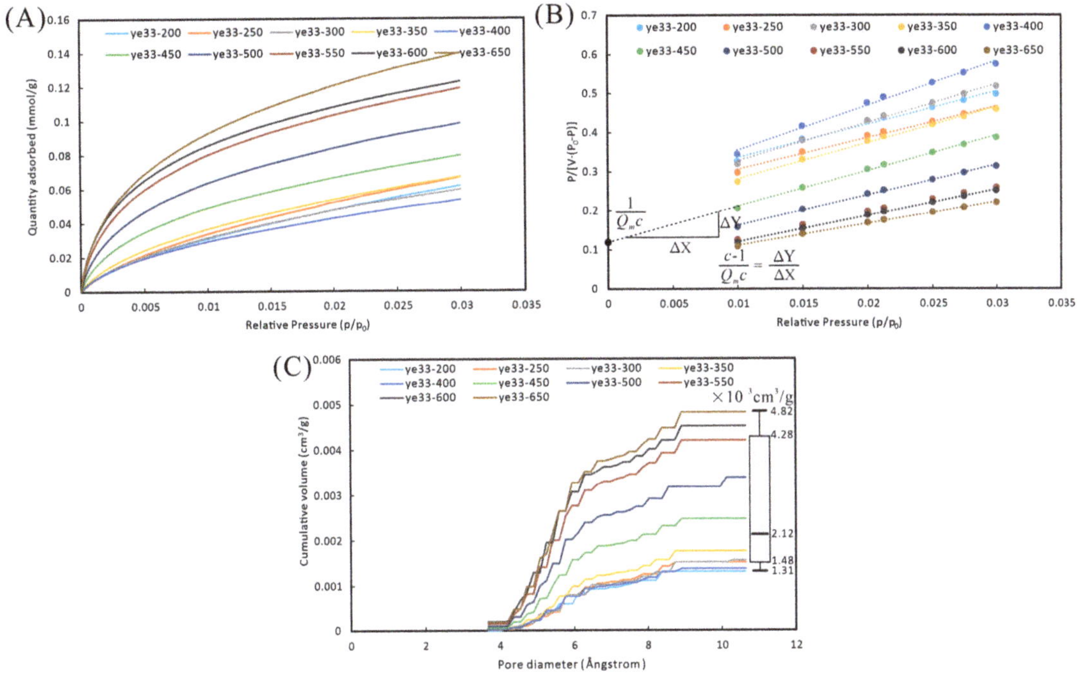

Figure 4. Low-temperature CO_2 adsorption curves (**A**), BET equation fitting results (**B**) and DFT cumulative pore volumes (**C**) for 10 subsamples.

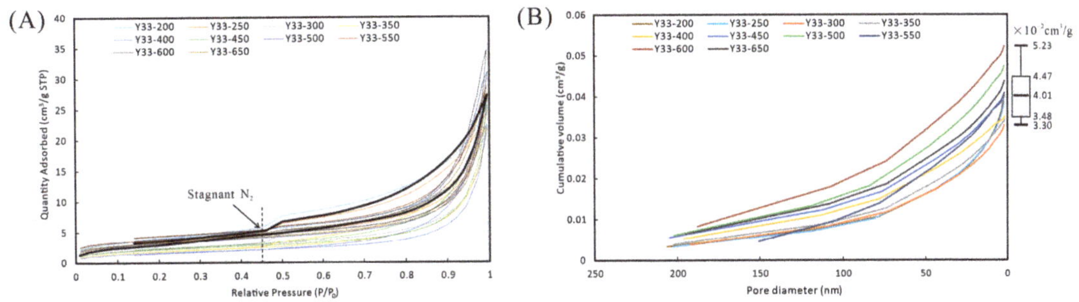

Figure 5. Low-temperature N_2 adsorption-desorption curves (**A**) and BJH cumulative pore volumes (**B**) for 10 subsamples.

MIP has a wide pore-throat size characterization range [22,49]; the mercury intrusion curve increases slowly at pressures less than 1 MPa, the mercury intrusion volume increases rapidly when the pressure is higher than the breakthrough pressure, and then the mercury intrusion rate slows down as the pressure increases (Figure 6A). The maximum mercury intrusion volume of the samples varied widely, from 0.141 to 0.239 cm^3/g. The mercury-extrusion curves were flat at high pressure (>40 MPa), followed by rapid mercury discharge from the sample. After the pressure was reduced to atmospheric pressure (101 kPa), there was still a large amount of mercury that could not be discharged, and the mercury retention efficiency was as large as 55.5%–74.8% (Table 3). The high mercury retention efficiency represents mercury retained in shale samples with poor sample pore connectivity and a large number of single-channel pores. The cumulative pore volumes of 3 nm–300 μm pore-throat sizes were calculated (Figure 6B) based on the Young–Dupre equation, the Washburn equation and the cylindrical pore assumption [39], but the irreversible destruction of pores by high-pressure intrusion may lead to the destruction of tiny pores and unreliable results less than 50 nm.

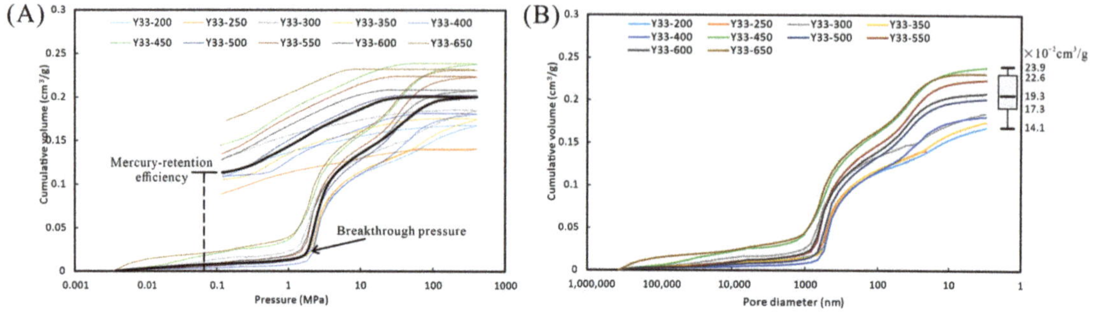

Figure 6. Mercury intrusion-extrusion curves (**A**) and MIP cumulative pore volumes (**B**) for 10 subsamples.

With the increase in Ro, the *c* value and adsorption capacity of the shale increased; the mercury retention efficiency gradually increased, representing that the pore structure became more complex; the breakthrough pressure decreased, indicating that the pore size of macropores grew larger (Table 4).

Table 4. Statistics of key parameters of pore-size characterization methods.

Scheme 2	CO_2			N_2		MIP		
	Q_m (mmol/g)	*c*	Q (mmol/g)	Stagnant N_2 (cm^3/g STP)	Q (cm^3/g STP)	Retention Efficiency (%)	Breakthrough Pressure (MPa)	Injection Volume (cm^3/g)
ye33-200	0.115	34.47	0.062	0.40	25.60	64.30	2.26	0.17
ye33-250	0.121	36.74	0.067	0.29	24.54	62.94	2.26	0.14
ye33-300	0.100	43.91	0.060	0.26	21.09	69.21	1.85	0.18
ye33-350	0.107	48.98	0.067	0.12	21.95	60.27	2.27	0.17
ye33-400	0.086	48.55	0.054	0.00	22.47	60.56	2.26	0.18
ye33-450	0.109	74.80	0.080	0.26	25.52	60.57	1.19	0.24
ye33-500	0.128	91.06	0.099	0.73	30.98	55.52	2.26	0.20
ye33-550	0.149	107.64	0.119	0.70	27.48	60.26	1.85	0.22
ye33-600	0.149	123.77	0.123	0.66	34.55	61.56	1.85	0.21
ye33-650	0.178	100.48	0.140	0.57	28.61	74.82	1.53	0.23

4.4. Pore Characteristics from SEM Imaging

The morphology of minerals, OM, and pores of ye33 were observed at 20 k times and 0.2 kV (Figure 7A). The rigid minerals (quartz and calcite) are off-white and granular in the image, with clean surfaces and a few Intra-P. The clay minerals are narrow lamellar, with slit interlayer pores inside. The organic matter was dark and amorphous, with round and oval pores commonly found inside its particles. The size of Inter-P was larger than that of the other pores. Based on the pore processing (Figure 7B) and statistical results (Figure 7C,D), the original ye33 sample had a plane porosity of 13.02%, representing medium compaction [36]. Less than 30 nm pores were round and elliptical and are mainly distributed in organic matter and clay; the count accounted for 76.33% but only provided 0.88% porosity (Figure 7E). The 30–100 nm pores were mostly in the form of narrow slits, dominated by Intra-P of clay minerals, providing 2.88% porosity (Figure 7D). Pores larger than 100 nm were irregularly shaped, and most of them were Inter-P, providing 9.26% porosity.

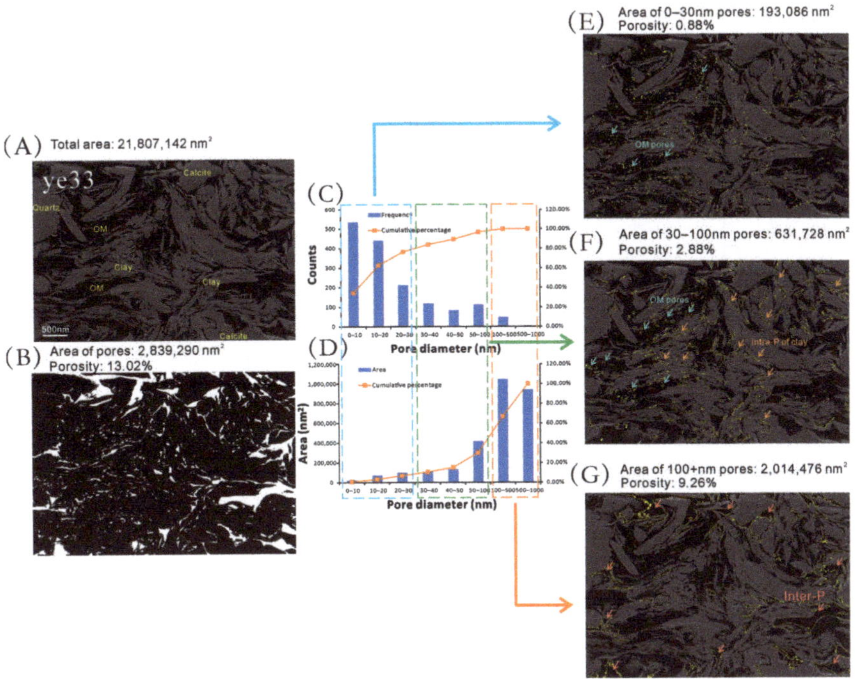

Figure 7. (**A**) SEM image of the ye33 sample; (**B**) Binarized images for pore recognition; (**C**) Distribution of pore counts by statistics from (**B**); (**D**) distribution of pore volume by statistics from (**B**); (**E**) Features of pores in 0–30 nm, (**F**) Features of pores in 30–100 nm and (**G**) Features of pores in 100+ nm.

5. Discussion

5.1. Pore-Size Polarization during Thermal Maturation

5.1.1. Evidence from SEM and IRT

OM pores are formed inside or around OM during thermal maturation due to mass and density reduction and volume contraction in the process of hydrocarbon generation [50,51]. The gas genetic pores inside OM were round and elliptical in the image. Through comparative analyses of OM pore recognition results of samples with different maturities (Figure 8), under the same magnification, it was found that when Ro = 1.1%, the number of intra-OM pores mainly existed in two peaks of 30–40 nm and 50–100 nm in diameters, and the most pore volume was provided by 50–100 nm pores (accounting for 44.62%). A total of 672 pores were recognized, and the plane porosity was 5.67% (Figure 8A). When Ro increased to 1.56%, more tiny OM pores were generated, and the number peaks shifted down to 10–20 nm and 50–100 nm, respectively, while the most pore volume was provided by 100–500 nm pores (accounting for 59.34%). In addition, a total of 910 pores were recognized, and the plane porosity was 9.57% (Figure 8B). When Ro was further increased to 2.65%, some pores continued to expand and connect together, the number peaks changed to 0–10 nm and 50–100 nm, and its main pore volume was provided by 50–500 nm pores (accounting for 72.25%). Furthermore, a total of 514 pores were recognized, and the total plane porosity was 18.21% (Figure 8C). As a whole, the plane porosity of the intra-OM pores increases with an increasing maturity, generating more micropores and mesopores. Adjacent gas genetic pores were interconnected and merged into larger pores while generating more tiny pores in residual organic matter. These two effects cause polarization in the pore-size distribution of organic matter.

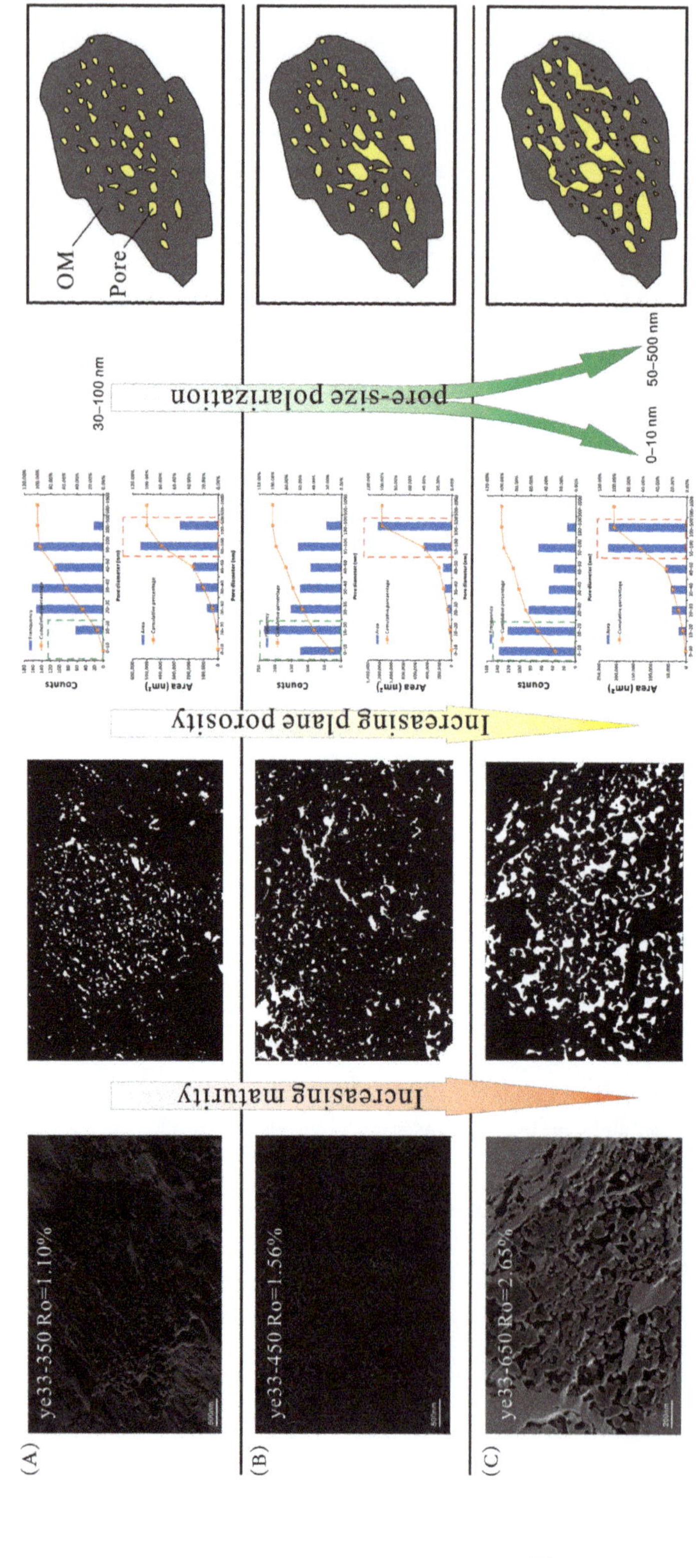

Figure 8. Characterization, recognition, and distribution of pores inside OM of subsamples at the simulated temperatures of 350 °C (**A**), 450 °C (**B**) and 650 °C (**C**).

In low magnification and a large field of view of SEM imaging, the surrounding pores of OM and Intra-P and Inter-P of minerals were observed (Figure 9). The Intra-P and Inter-P of clay minerals are always dominated by slit-shaped pores with an increasing thermal maturation, which was closely related to the lamellar structure of clays [52]; however, the pores are widened, which may be related to hydrocarbon expulsion. Regardless of the maturation stage, the surfaces of rigid minerals (e.g., quartz, calcite, dolomite) are clean, with only a few dissolved pores and fractures, providing negligible pore volume (Figure 9A). The same phenomenon of polarization of pore sizes at 50–100 nm with increasing maturity was found at low magnification, with the formation of pores smaller than 10 nm and more pores larger than 100 nm (Figure 9B,D). At the over-maturation stage, large shrinkage fractures formed at the surrounding edge of the OM. The 100+ nm pores in ye33-650 were dominated by OM shrinkage fractures (Figure 9C), which was different from the original sample (ye33), where the 100+ pores were dominated by Inter-P (Figure 7D,G).

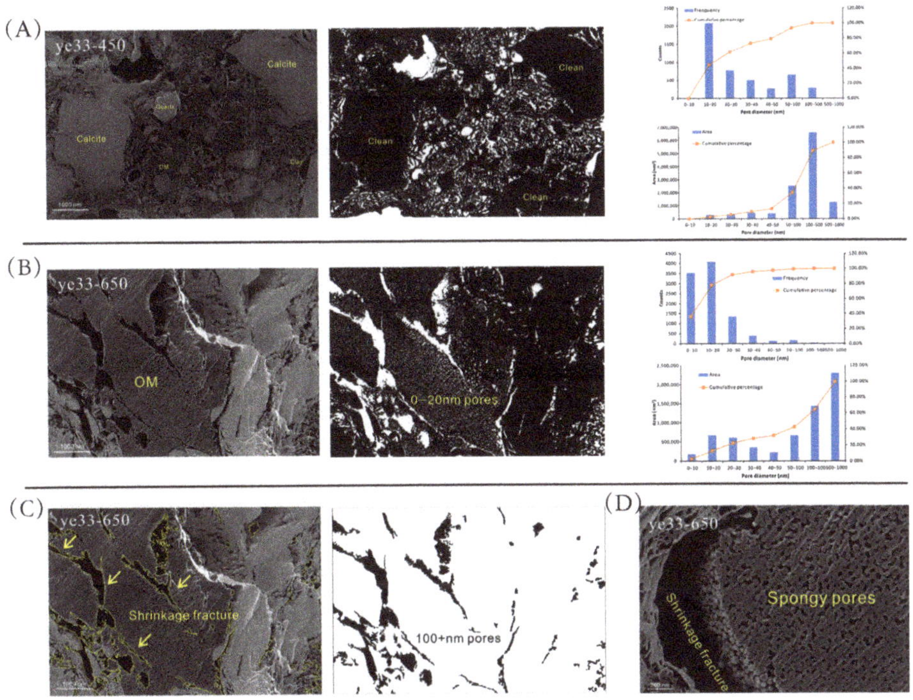

Figure 9. (**A**) Characterization and distribution of the pores of the ye33-450 sample in a large field of view; (**B**) Characterization and distribution of the pores of the ye33-650 sample in a large field of view; (**C**) Recognition results and features of the 100+ nm pores of the ye33-650 sample; (**D**) Spongy OM pores and perimeter shrinkage fracture of the ye33-650 sample.

5.1.2. Evidence from LTGA and MIP

Based on the different methods and associated observational scales [53], the whole pore-size volume distributions of samples with different maturities were characterized (micropores from CO_2 adsorption, mesopores from N_2 adsorption, and macropores from MIP). At a maturity of less than 1.10%, there was only one peak in the pore-size volume distribution, and 100–1000 nm pores dominate (Figure 10), which is also consistent with our SEM observations of the original sample when the pore volume was mainly provided by intergranular pores. When Ro reached 1.19%, a second peak appeared in the 30–100 nm range, proving that the increase in pore space in this range corresponds to the increase in pores inside OM and pores of clay minerals. When Ro further increased to 1.56%, the volume of less than 2 nm pores and greater than 1000 nm pores increased, corresponding to the formation of micropores and surrounding shrinkage fractures of OM. Ultimately, there are four peaks (0–2, 30–100, 500–1000 nm, and 10–100 μm) in the whole pore-size volume distribution (Figure 10).

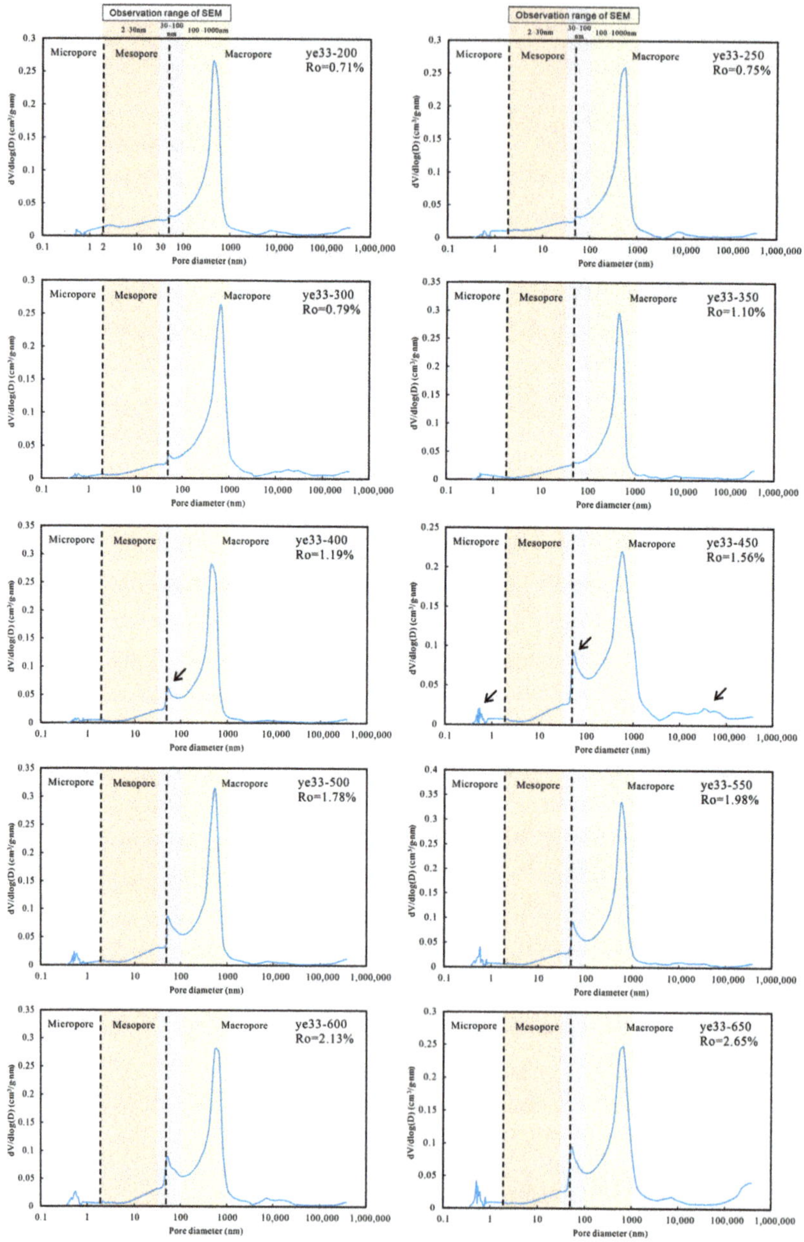

Figure 10. Pore-size volume distribution of subsamples with different maturities. (Arrows represent new peaks generated compared to the previous stage).

Along the increasing temperature and maturation levels, there are different changing patterns of pore development for the following five pore-size intervals (Figure 11): 0–2 nm pores increase continuously with the increase of Ro; 2–30 nm pores decrease first and reach the minimum at Ro = 1.19%, then gradually increase; 30–100 nm pores increase rapidly at Ro = 1.10%~1.56%, then stabilize; 100–1000 nm pores increase with the increasing maturity; 1000+ nm pores fluctuate with Ro, and increase faster in the high over maturity stage (Ro > 1.78%). Therefore, the whole pore-size distribution of the shale also exhibits polarization, with pore volumes of 0–30 nm and 100+ nm gradually increasing after Ro exceeds 1.5%.

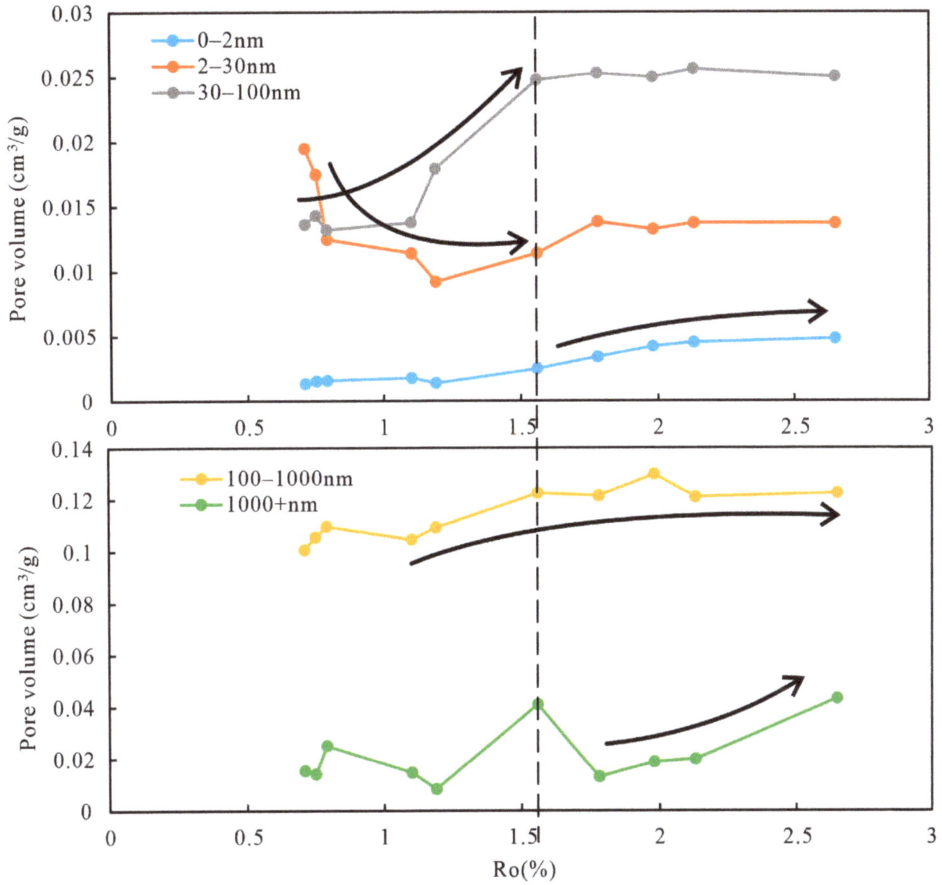

Figure 11. Evolution of pore volume at different pore scales. (Arrows indicate the trend of pore volume variation with maturity at different scales).

5.2. Effects of Hydrocarbon Generation and Diagenesis

The results showed that the yield of hydrocarbons was closely related to the degree of pore development. The gaseous hydrocarbon yield had an extremely positive correlation with the volume of 0–2 nm pores (Figure 12A), representing micropore-level pores in organic matter produced by the generation and expulsion of small molecule hydrocarbons. Due to quality conservation, OM shrank to form macropores and fractures as the total hydrocarbon yield increased and 100–1000 nm pores increased (Figure 12B).

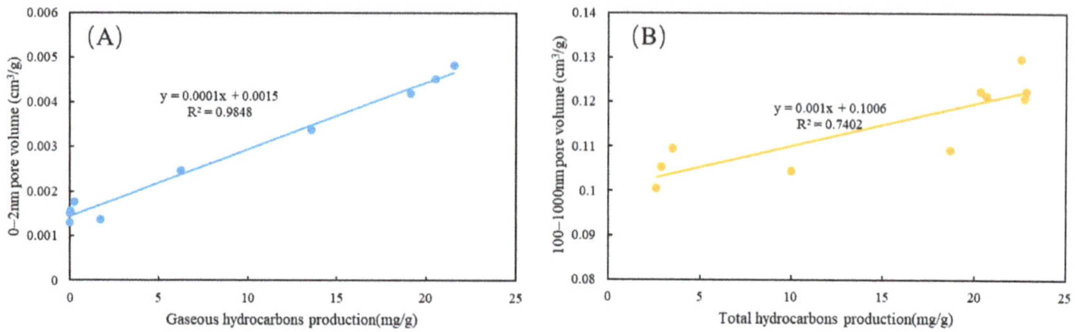

Figure 12. (**A**) Correlation between 0–2 nm pores and the yield of gaseous hydrocarbons; (**B**) correlation between 100–1000 nm pores and the yield of total hydrocarbons.

Based on the characteristics of HG&D and pore evolution, the evolution of organic-rich shale was divided into the following three stages (Figure 13):

(1) Ro at 0.7%–1.2%, in the initial stage, kerogen decarboxylated and formed organic acids [54], calcite was dissolved in the acidic environment, and the Inter-P (100–1000 nm) increased. Subsequently, high-calcium fluids were redeposited in the Inter-P to produce calcium cementation [55], resulting in a decrease in Inter-P (100–1000 nm). The space of loose clay minerals (mainly kaolinite) is compressed, leading to a reduction in the Intra-P of clay minerals (2–30 nm). As the OM matures, the OM begins to generate large amounts of liquid hydrocarbons and 30–100 nm OM inside pores (30–100 nm) [56,57]. However, the generated liquid hydrocarbons and bitumen entered and blocked the Inter-P [58,59], resulting in a large reduction of 100–1000 nm pores. Some microcrystalline authigenic quartz generation also has a negative effect on the pores [60,61].

(2) At 1.2%–2.0%, many gaseous hydrocarbons were generated, bitumen and OM shrank, spongy pores were formed in OM [62,63], and the OM pore size started to polarize, generating a large number of micropores (<2 nm) and a large volume of macropores (100–1000 nm). Illitization and recrystallization [64,65] of clay minerals increased the Inter-P and Intra-P of clay minerals, while gaseous hydrocarbons entered the pores, and high fluid pressure made the pores of clay larger, so the volume of 30–100 nm pores increased considerably.

(3) When Ro was higher than 2.0%, the yield of hydrocarbons gradually decreased, so the change in the 0–100 nm pore volume was small. Most hydrocarbons have been expelled, and the density of the reduced OM resembles a sponge, with many pores developed in it. The OM then continued to shrink, forming shrinkage fractures at the perimeter of the organic matter [29,66], resulting in a continuous increase in the 100–1000 nm pores.

In conclusion, hydrocarbon generation controls organic pore development, diagenesis controls inorganic pore development, and both jointly control the shale pore evolution process.

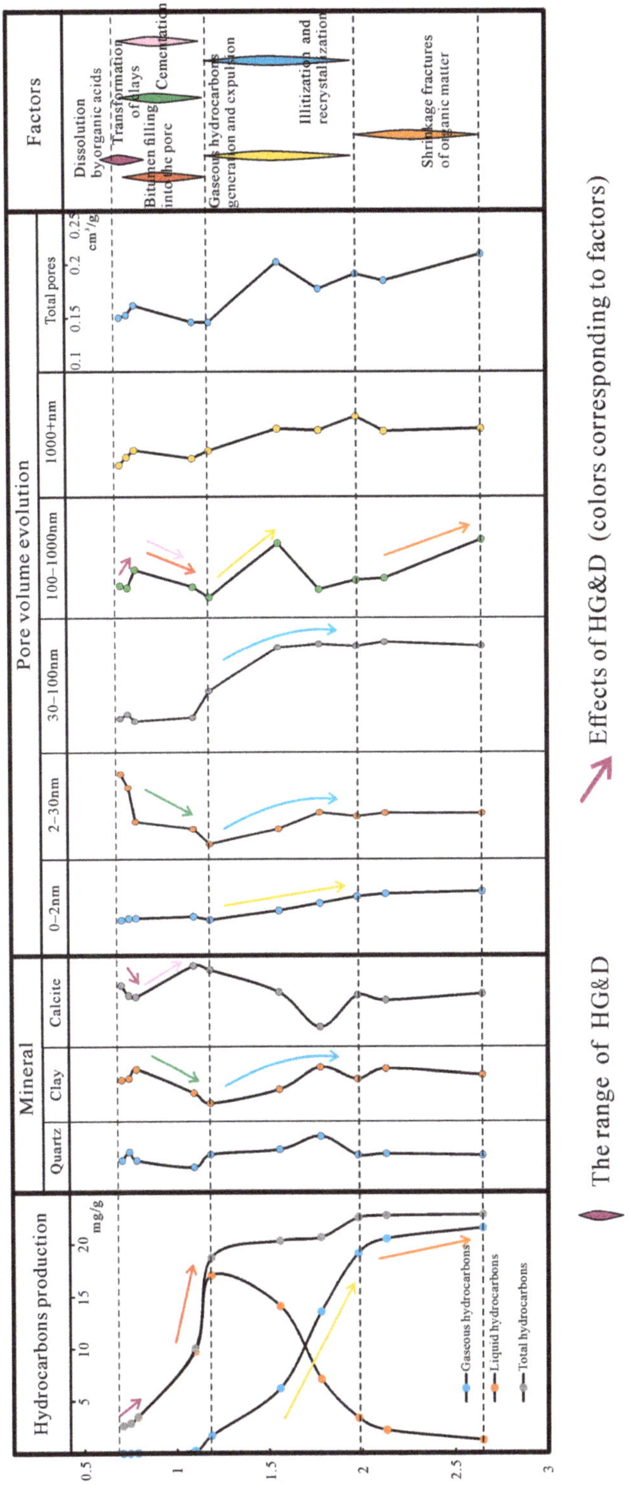

Figure 13. Effects of HG&D on the evolution of pores of different sizes. (Arrows represent the changing trends of different curves).

5.3. Compared to Continental and Transitional Shales

Upon categorizing the evolution of sample pores and analyzing the types of pores formed at various maturity stages, it was observed that as marine shale surpasses the oil generation peak (Ro > 1.2%), there is a noticeable rise in the overall pore volume across various pore sizes. This observation suggests a gradual enhancement in the reservoir's porosity and permeability, thereby fostering favorable conditions for shale gas fracturing [67]. Comparison with published thermal simulation experiments-based pore evolution results [36] shows differences between marine, continental and transitional shales (Figure 14). Marine OM comprised of sapropel-rich organisms such as algae and foraminifera [68,69], which are more capable of oil production and earlier hydrocarbon production than the large amount of plant debris contained in the terrestrial and transitional phases [70–72]. In marine shales, a lot of pre-oil bitumen generates and fills the pores, resulting in a sudden drop in pore volume (e.g., mesopores). Marine shales also cross the oil production peak earlier, which, on the one hand, allows the organic matter pores to reappear [58], and on the other hand, new spongy organic matter [73,74] pore generation due to the cracking and degradation of organic matter (e.g., kerogen and pre-oil bitumen) into gaseous hydrocarbons. The micropores were more abundant in the marine shale after the oil peak, which was favorable to increase the specific surface area and methane adsorption in the future [75,76]. Prior to the oil peak, complex diagenesis (dissolution, cementation, replacement), clay mineral transformations (kaolinization, authigenic chlorite), and compositional differences (quartz origin, calcareous content, clay composition) complicated the macropores volume variation of the three shales [77,78]. Thus, differences in organic matter types and sedimentary mineral components were thought to dominate the differences in pore evolution among the three, which will require more abundant samples for detailed studies in the future. The findings of this study are constrained by the sample size. Moreover, the kerogen type present in the samples is classified as type II_2 rather than type I, leading to notable disparities in the pore evolution process compared to certain marine strata. Additionally, the absence of investigation into various diagenetic processes during experimental design has resulted in unclear identification of influencing factors.

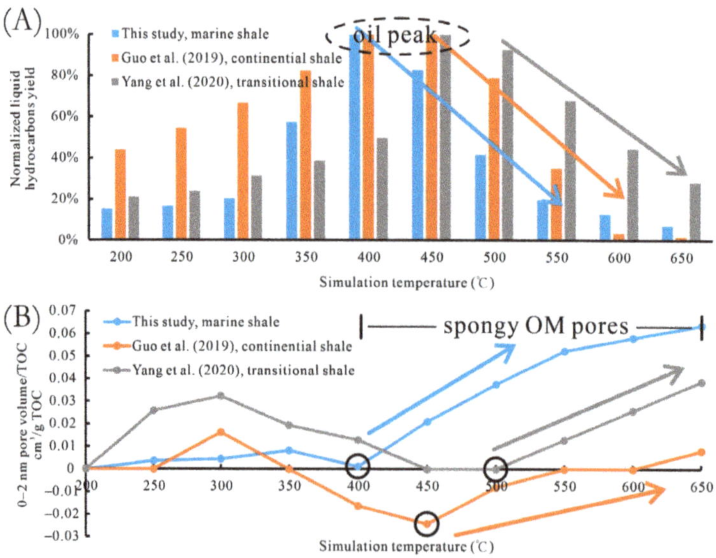

Figure 14. *Cont.*

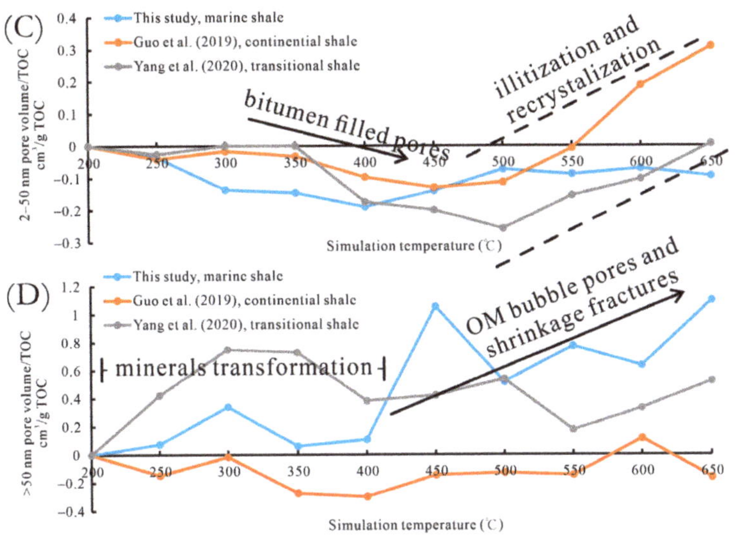

Figure 14. Comparison of thermal simulation results for marine, continental and transitional shales (**A**) variation of normalized liquid hydrocarbon yield with simulation temperature (at oil peak, normalized liquid hydrocarbon yield = 1); (**B**) variation of 0–2 nm pore volume with simulation temperature; (**C**) variation of 2–50 nm pore volume with simulation temperature; (**D**) variation of >50 nm pore volume with simulation temperature [29,36].

6. Conclusions

This study quantified shale pore evolution using IRT and SEM-assisted pore characterization experiments. The original triple classification of pores based on the pore scale was re-classified into six intervals (0–2, 2–30, 30–100, 100–1000, and 1000+ nm) based on the pore genesis. This classification helps to discuss and distinguish the specific effects of HG&D on pores at different scales during the thermal maturation processing of shales. Hydrocarbon generation was strongly correlated with 0–2 nm and 100–1000 nm pores, which was attributed to the polarization characteristics of OM-hosted pores during thermal evolution observed in SEM. Diagenesis (e.g., dissolution, mineral transformation, illitization and cementation) affected 2–30 nm, 30–100 nm, and 100–1000 nm pores at different stages, respectively. In addition, the bitumen filling and formation of shrinkage fractures affected 100–1000 nm or even 1000+ nm pores. Further, we also explored some of the differences and reasons for the thermal simulation experiments-based pore evolution of marine shales from those of continental and transitional shales published in the past. This study serves as a reference for future studies of shale pore evolution and provides a basis for multi-scale studies of genetic pores, especially for the prediction of reservoir space and beneficial stages of shale oil and gas in the underground.

Author Contributions: Conceptualization, S.G.; methodology, S.G.; software, X.Y.; validation, Z.W.; formal analysis, X.Y.; investigation, X.Y.; resources, S.G.; data curation, S.G.; writing—original draft preparation, X.Y.; writing—review and editing, Z.W.; visualization, Z.W.; supervision, S.G.; project administration, S.G.; funding acquisition, S.G. All authors have read and agreed to the published version of the manuscript.

Funding: This study received support from the Ministry of Science and Technology of the People's Republic of China (Grant No. 2016ZX05034) and the China Geological Survey (Grant No. 2023-QN01). Financial support from the China Scholarship Council (CSC) and American Association of Petroleum Geologists (AAPG) Grants-in-Aid are gratefully acknowledged.

Data Availability Statement: Data are contained within the article.

Acknowledgments: We would like to thank anonymous reviewers for their suggestions.

Conflicts of Interest: The authors declare no conflict of interest.

References

1. Bai, L.; Liu, B.; Yang, J.; Tian, S.; Wang, B.; Huang, S. Differences in hydrocarbon composition of shale oils in different phase states from the Qingshankou Formation, Songliao Basin, as determined from fluorescence experiments. *Front. Earth Sci.* **2021**, *15*, 438–456. [CrossRef]
2. Zhang, Y.; Hu, Q.; Long, S.; Zhao, J.; Peng, N.; Wang, H.; Lin, X.; Sun, M. Mineral-controlled nm-μm-scale pore structure of saline lacustrine shale in Qianjiang Depression, Jianghan Basin, China. *Mar. Pet. Geol.* **2019**, *99*, 347–354. [CrossRef]
3. Owusu, E.B.; Tsegab, H.; Sum, C.W.; Padmanabhan, E. Organic geochemical analyses of the Belata black shale, Peninsular Malaysia; implications on their shale gas potential. *J. Nat. Gas Sci. Eng.* **2019**, *69*, 102945. [CrossRef]
4. Tan, J.; Horsfield, B.; Fink, R.; Krooss, B.; Schulz, H.-M.; Rybacki, E.; Zhang, J.; Boreham, C.J.; Van Graas, G.; Tocher, B.A. Shale gas potential of the major marine shale formations in the Upper Yangtze Platform, south China, Part III: Mineralogical, lithofacial, petrophysical, and rock mechanical properties. *Energy Fuels* **2014**, *28*, 2322–2342. [CrossRef]
5. Enomoto, C.B.; Hackley, P.C.; Valentine, B.J.; Rouse, W.A.; Dulong, F.T.; Lohr, C.D.; Hatcherian, J.J. *Geologic Characterization of the Hydrocarbon Resource Potential of the Upper Cretaceous Tuscaloosa Marine Shale in Mississippi and Louisiana, USA*; United States Geological Survey: Reston, VA, USA, 2017.
6. Zou, C.; Zhu, R.; Chen, Z.-Q.; Ogg, J.G.; Wu, S.; Dong, D.; Qiu, Z.; Wang, Y.; Wang, L.; Lin, S. Organic-matter-rich shales of China. *Earth-Sci. Rev.* **2019**, *189*, 51–78. [CrossRef]
7. Jarvie, D.M.; Hill, R.J.; Ruble, T.E.; Pollastro, R.M. Unconventional shale-gas systems: The Mississippian Barnett Shale of north-central Texas as one model for thermogenic shale-gas assessment. *AAPG Bull.* **2007**, *91*, 475–499. [CrossRef]
8. Wang, Y.; Cheng, H.; Hu, Q.; Liu, L.; Hao, L. Diagenesis and pore evolution for various lithofacies of the Wufeng-Longmaxi shale, southern Sichuan Basin, China. *Mar. Pet. Geol.* **2021**, *133*, 105251. [CrossRef]
9. Zhu, W.; Chen, Z.; Liu, K. A new meshless method to solve the two-phase thermo-hydro-mechanical multi-physical field coupling problems in shale reservoirs. *J. Nat. Gas Sci. Eng.* **2022**, *105*, 104683. [CrossRef]
10. Chen, J.; Xiao, X. Evolution of nanoporosity in organic-rich shales during thermal maturation. *Fuel* **2014**, *129*, 173–181. [CrossRef]
11. Sun, L.; Tuo, J.; Zhang, M.; Wu, C.; Wang, Z.; Zheng, Y. Formation and development of the pore structure in Chang 7 member oil-shale from Ordos Basin during organic matter evolution induced by hydrous pyrolysis. *Fuel* **2015**, *158*, 549–557. [CrossRef]
12. Rouquerol, J.; Llewellyn, P.; Rouquerol, F. Is the BET equation applicable to microporous adsorbents. *Stud. Surf. Sci. Catal.* **2007**, *160*, 49–56.
13. Loucks, R.G.; Reed, R.M.; Ruppel, S.C.; Jarvie, D.M. Morphology, genesis, and distribution of nanometer-scale pores in siliceous mudstones of the Mississippian Barnett Shale. *J. Sediment. Res.* **2009**, *79*, 848–861. [CrossRef]
14. Chen, Z.; Liu, X.; Yang, J.; Little, E.; Zhou, Y. Deep learning-based method for SEM image segmentation in mineral characterization, an example from Duvernay Shale samples in Western Canada Sedimentary Basin. *Comput. Geosci.* **2020**, *138*, 104450. [CrossRef]
15. Sun, M.; Zhang, L.; Hu, Q.; Pan, Z.; Yu, B.; Sun, L.; Bai, L.; Fu, H.; Zhang, Y.; Zhang, C. Multiscale connectivity characterization of marine shales in southern China by fluid intrusion, small-angle neutron scattering (SANS), and FIB-SEM. *Mar. Pet. Geol.* **2020**, *112*, 104101. [CrossRef]
16. Tahmasebi, P.; Javadpour, F.; Sahimi, M. Three-dimensional stochastic characterization of shale SEM images. *Transp. Porous Media* **2015**, *110*, 521–531. [CrossRef]
17. Yang, R.; He, S.; Yi, J.; Hu, Q. Nano-scale pore structure and fractal dimension of organic-rich Wufeng-Longmaxi shale from Jiaoshiba area, Sichuan Basin: Investigations using FE-SEM, gas adsorption and helium pycnometry. *Mar. Pet. Geol.* **2016**, *70*, 27–45. [CrossRef]
18. Zhou, S.; Yan, G.; Xue, H.; Guo, W.; Li, X. 2D and 3D nanopore characterization of gas shale in Longmaxi formation based on FIB-SEM. *Mar. Pet. Geol.* **2016**, *73*, 174–180. [CrossRef]
19. Bai, L.-H.; Liu, B.; Du, Y.-J.; Wang, B.-Y.; Tian, S.-S.; Wang, L.; Xue, Z.-Q. Distribution characteristics and oil mobility thresholds in lacustrine shale reservoir: Insights from N2 adsorption experiments on samples prior to and following hydrocarbon extraction. *Pet. Sci.* **2022**, *19*, 486–497. [CrossRef]
20. Chalmers, G.R.; Bustin, R.M.; Power, I.M. Characterization of gas shale pore systems by porosimetry, pycnometry, surface area, and field emission scanning electron microscopy/transmission electron microscopy image analyses: Examples from the Barnett, Woodford, Haynesville, Marcellus, and Doig units. *AAPG Bull.* **2012**, *96*, 1099–1119.
21. Dong, T.; Harris, N.B.; Ayranci, K.; Twemlow, C.E.; Nassichuk, B.R. Porosity characteristics of the Devonian Horn River shale, Canada: Insights from lithofacies classification and shale composition. *Int. J. Coal Geol.* **2015**, *141*, 74–90. [CrossRef]
22. Gao, F.; Song, Y.; Li, Z.; Xiong, F.; Chen, L.; Zhang, X.; Chen, Z.; Moortgat, J. Quantitative characterization of pore connectivity using NMR and MIP: A case study of the Wangyinpu and Guanyintang shales in the Xiuwu basin, Southern China. *Int. J. Coal Geol.* **2018**, *197*, 53–65. [CrossRef]
23. Li, Y.; Schieber, J.; Fan, T.; Wei, X. Pore characterization and shale facies analysis of the Ordovician-Silurian transition of northern Guizhou, South China: The controls of shale facies on pore distribution. *Mar. Pet. Geol.* **2018**, *92*, 697–718. [CrossRef]
24. Connan, J. Time-temperature relation in oil genesis: Geologic notes. *AAPG Bull.* **1974**, *58*, 2516–2521.

25. Waples, D.W. Time and temperature in petroleum formation: Application of Lopatin's method to petroleum exploration: Reply. *AAPG Bull.* **1982**, *66*, 1152.
26. Sweeney, J.J.; Burnham, A.K. Evaluation of a simple model of vitrinite reflectance based on chemical kinetics. *AAPG Bull.* **1990**, *74*, 1559–1570.
27. Songtao, W.; Rukai, Z.; Jinggang, C.; Jingwei, C.; Bin, B.; Zhang, X.; Xu, J.; Desheng, Z.; Jianchang, Y.; Xiaohong, L. Characteristics of lacustrine shale porosity evolution, Triassic Chang 7 member, Ordos Basin, NW China. *Pet. Explor. Dev.* **2015**, *42*, 185–195.
28. Wang, F.; Guo, S. Influential factors and model of shale pore evolution: A case study of a continental shale from the Ordos Basin. *Mar. Pet. Geol.* **2019**, *102*, 271–282. [CrossRef]
29. Guo, S.; Mao, W. Division of diagenesis and pore evolution of a Permian Shanxi shale in the Ordos Basin, China. *J. Pet. Sci. Eng.* **2019**, *182*, 106351. [CrossRef]
30. Doveton, J.; Merriam, D. Borehole petrophysical chemostratigraphy of Pennsylvanian black shales in the Kansas subsurface. *Chem. Geol.* **2004**, *206*, 249–258. [CrossRef]
31. Hakes, W.G. *Trace Fossils from Brackish-Marine Shales, Upper Pennsylvanian of Kansas, USA*; SEPM Society for Sedimentary Geology: Tulsa, OK, USA, 1984.
32. Curtis, M.E.; Cardott, B.J.; Sondergeld, C.H.; Rai, C.S. Development of organic porosity in the Woodford Shale with increasing thermal maturity. *Int. J. Coal Geol.* **2012**, *103*, 26–31. [CrossRef]
33. Dong, T.; Harris, N.B. The effect of thermal maturity on porosity development in the Upper Devonian–Lower Mississippian Woodford Shale, Permian Basin, US: Insights into the role of silica nanospheres and microcrystalline quartz on porosity preservation. *Int. J. Coal Geol.* **2020**, *217*, 103346. [CrossRef]
34. Kibria, M.G.; Hu, Q.; Liu, H.; Zhang, Y.; Kang, J. Pore structure, wettability, and spontaneous imbibition of Woodford shale, Permian Basin, West Texas. *Mar. Pet. Geol.* **2018**, *91*, 735–748. [CrossRef]
35. Jia, B.; Xian, C.; Tsau, J.-S.; Zuo, X.; Jia, W. Status and Outlook of Oil Field Chemistry-Assisted Analysis during the Energy Transition Period. *Energy Fuels* **2022**, *36*, 12917–12945. [CrossRef]
36. Yang, X.G.; Guo, S.B. Porosity model and pore evolution of transitional shales: An example from the Southern North China Basin. *Pet. Sci.* **2020**, *17*, 1512–1526. [CrossRef]
37. Clarkson, C.R.; Solano, N.; Bustin, R.M.; Bustin, A.; Chalmers, G.R.; He, L.; Melnichenko, Y.B.; Radliński, A.; Blach, T.P. Pore structure characterization of North American shale gas reservoirs using USANS/SANS, gas adsorption, and mercury intrusion. *Fuel* **2013**, *103*, 606–616. [CrossRef]
38. Mastalerz, M.; Schimmelmann, A.; Drobniak, A.; Chen, Y. Porosity of Devonian and Mississippian New Albany Shale across a maturation gradient: Insights from organic petrology, gas adsorption, and mercury intrusion. *AAPG Bull.* **2013**, *97*, 1621–1643. [CrossRef]
39. Drummond, C.; Israelachvili, J. Surface forces and wettability. *J. Pet. Sci. Eng.* **2002**, *33*, 123–133. [CrossRef]
40. Washburn, E.W. The dynamics of capillary flow. *Phys. Rev.* **1921**, *17*, 273. [CrossRef]
41. Choma, J.; Jaroniec, M.; Kloske, M. Improved pore-size analysis of carbonaceous adsorbents. *Adsorpt. Sci. Technol.* **2002**, *20*, 307–315. [CrossRef]
42. Tian, S.; Bowen, L.; Liu, B.; Zeng, F.; Xue, H.; Erastova, V.; Greenwell, H.C.; Dong, Z.; Zhao, R.; Liu, J. A method for automatic shale porosity quantification using an Edge-Threshold Automatic Processing (ETAP) technique. *Fuel* **2021**, *304*, 121319. [CrossRef]
43. Schneider, C.A.; Rasband, W.S.; Eliceiri, K.W. NIH Image to ImageJ: 25 years of image analysis. *Nat. Methods* **2012**, *9*, 671–675. [CrossRef]
44. Kapur, J.N.; Sahoo, P.K.; Wong, A.K. A new method for gray-level picture thresholding using the entropy of the histogram. *Comput. Vis. Graph. Image Process.* **1985**, *29*, 273–285. [CrossRef]
45. Lastoskie, C.; Gubbins, K.E.; Quirke, N. Pore size heterogeneity and the carbon slit pore: A density functional theory model. *Langmuir* **1993**, *9*, 2693–2702. [CrossRef]
46. Li, Y.; Wang, Z.; Pan, Z.; Niu, X.; Yu, Y.; Meng, S. Pore structure and its fractal dimensions of transitional shale: A cross-section from east margin of the Ordos Basin, China. *Fuel* **2019**, *241*, 417–431. [CrossRef]
47. Gregg, S. *Studies in Surface Science and Catalysis*; Elsevier: Amsterdam, The Netherlands, 1982; Volume 10, pp. 153–164.
48. Yang, F.; Ning, Z.; Liu, H. Fractal characteristics of shales from a shale gas reservoir in the Sichuan Basin, China. *Fuel* **2014**, *115*, 378–384. [CrossRef]
49. Al Hinai, A.; Rezaee, R.; Esteban, L.; Labani, M. Comparisons of pore size distribution: A case from the Western Australian gas shale formations. *J. Unconv. Oil Gas Resour.* **2014**, *8*, 1–13. [CrossRef]
50. Borjigin, T.; Longfei, L.; Lingjie, Y.; Zhang, W.; Anyang, P.; Baojian, S.; Ye, W.; Yunfeng, Y.; Zhiwei, G. Formation, preservation and connectivity control of organic pores in shale. *Pet. Explor. Dev.* **2021**, *48*, 798–812. [CrossRef]
51. Zhang, K.; Peng, J.; Wang, X.; Jiang, Z.; Song, Y.; Jiang, L.; Jiang, S.; Xue, Z.; Wen, M.; Li, X. Effect of organic maturity on shale gas genesis and pores development: A case study on marine shale in the upper Yangtze region, South China. *Open Geosci.* **2020**, *12*, 1617–1629. [CrossRef]
52. Chen, S.; Han, Y.; Fu, C.; Zhu, Y.; Zuo, Z. Micro and nano-size pores of clay minerals in shale reservoirs: Implication for the accumulation of shale gas. *Sediment. Geol.* **2016**, *342*, 180–190. [CrossRef]
53. Yang, X.; Guo, S. Comparative analysis of shale pore size characterization methods. *Pet. Sci. Technol.* **2020**, *38*, 793–799. [CrossRef]

54. Nie, H.; Sun, C.; Liu, G.; Du, W.; He, Z. Dissolution pore types of the Wufeng Formation and the Longmaxi Formation in the Sichuan Basin, south China: Implications for shale gas enrichment. *Mar. Pet. Geol.* **2019**, *101*, 243–251. [CrossRef]
55. Liu, H.; Zhang, S.; Song, G.; Xuejun, W.; Teng, J.; Wang, M.; Bao, Y.; Yao, S.; Wang, W.; Zhang, S. Effect of shale diagenesis on pores and storage capacity in the Paleogene Shahejie Formation, Dongying Depression, Bohai Bay Basin, east China. *Mar. Pet. Geol.* **2019**, *103*, 738–752. [CrossRef]
56. Bernard, S.; Horsfield, B.; Schulz, H.-M.; Wirth, R.; Schreiber, A.; Sherwood, N. Geochemical evolution of organic-rich shales with increasing maturity: A STXM and TEM study of the Posidonia Shale (Lower Toarcian, northern Germany). *Mar. Pet. Geol.* **2012**, *31*, 70–89. [CrossRef]
57. Loucks, R.G.; Reed, R.M.; Ruppel, S.C.; Hammes, U. Spectrum of pore types and networks in mudrocks and a descriptive classification for matrix-related mudrock pores. *AAPG Bull.* **2012**, *96*, 1071–1098. [CrossRef]
58. Löhr, S.; Baruch, E.; Hall, P.; Kennedy, M. Is organic pore development in gas shales influenced by the primary porosity and structure of thermally immature organic matter? *Org. Geochem.* **2015**, *87*, 119–132. [CrossRef]
59. Wei, L.; Mastalerz, M.; Schimmelmann, A.; Chen, Y. Influence of Soxhlet-extractable bitumen and oil on porosity in thermally maturing organic-rich shales. *Int. J. Coal Geol.* **2014**, *132*, 38–50. [CrossRef]
60. Emmings, J.F.; Dowey, P.J.; Taylor, K.G.; Davies, S.J.; Vane, C.H.; Moss-Hayes, V.; Rushton, J.C. Origin and implications of early diagenetic quartz in the Mississippian Bowland Shale Formation, Craven Basin, UK. *Mar. Pet. Geol.* **2020**, *120*, 104567. [CrossRef]
61. van de Kamp, P.C. Smectite-illite-muscovite transformations, quartz dissolution, and silica release in shales. *Clay Clay Miner.* **2008**, *56*, 66–81. [CrossRef]
62. Huo, Z.; Peng, J.; Zhang, J.; Tang, X.; Li, P.; Ding, J.; Li, Z.; Liu, Z.; Dong, Z.; Lei, Y. Factors influencing the development of diagenetic shrinkage macro-fractures in shale. *J. Pet. Sci. Eng.* **2019**, *183*, 106365. [CrossRef]
63. Meng, Q.; Hao, F.; Tian, J. Origins of non-tectonic fractures in shale. *Earth-Sci. Rev.* **2021**, *222*, 103825. [CrossRef]
64. Berger, G.; Lacharpagne, J.-C.; Velde, B.; Beaufort, D.; Lanson, B. Kinetic constraints on illitization reactions and the effects of organic diagenesis in sandstone/shale sequences. *Appl. Geochem.* **1997**, *12*, 23–35. [CrossRef]
65. Cuadros, J. Modeling of smectite illitization in burial diagenesis environments. *Geochim. Cosmochim. Acta* **2006**, *70*, 4181–4195. [CrossRef]
66. Guo, K.; Song, H.; Chen, X.; Du, X.; Zhong, L. Graphene oxide as an anti-shrinkage additive for resorcinol–formaldehyde composite aerogels. *Phys. Chem. Chem. Phys.* **2014**, *16*, 11603–11608. [CrossRef] [PubMed]
67. Jia, B.; Xian, C.-G. Permeability measurement of the fracture-matrix system with 3D embedded discrete fracture model. *Pet. Sci.* **2022**, *19*, 1757–1765. [CrossRef]
68. Murphy, K.R.; Stedmon, C.A.; Waite, T.D.; Ruiz, G.M. Distinguishing between terrestrial and autochthonous organic matter sources in marine environments using fluorescence spectroscopy. *Mar. Chem.* **2008**, *108*, 40–58. [CrossRef]
69. Meilijson, A.; Finkelman-Torgeman, E.; Bialik, O.M.; Boudinot, F.G.; Steinberg, J.; Waldmann, N.D.; Benjamini, C.; Vinegar, H.; Makovsky, Y. Significance to hydrocarbon exploration of terrestrial organic matter introduced into deep marine systems: Insights from the lower Cretaceous in the Levant Basin. *Mar. Pet. Geol.* **2020**, *122*, 104671. [CrossRef]
70. Agrawal, V.; Sharma, S. Molecular characterization of kerogen and its implications for determining hydrocarbon potential, organic matter sources and thermal maturity in Marcellus Shale. *Fuel* **2018**, *228*, 429–437. [CrossRef]
71. Wang, G.-C.; Sun, M.-Z.; Gao, S.-F.; Tang, L. The origin, type and hydrocarbon generation potential of organic matter in a marine-continental transitional facies shale succession (Qaidam Basin, China). *Sci. Rep.* **2018**, *8*, 6568. [CrossRef]
72. Xia, L.-W.; Cao, J.; Wang, M.; Mi, J.-L.; Wang, T.-T. A review of carbonates as hydrocarbon source rocks: Basic geochemistry and oil–gas generation. *Pet. Sci.* **2019**, *16*, 713–728. [CrossRef]
73. Zhu, X.; Cai, J.; Wang, X.; Zhang, J.; Xu, J. Effects of organic components on the relationships between specific surface areas and organic matter in mudrocks. *Int. J. Coal Geol.* **2014**, *133*, 24–34. [CrossRef]
74. Wu, J.; Liang, C.; Hu, Z.; Yang, R.; Xie, J.; Wang, R.; Zhao, J. Sedimentation mechanisms and enrichment of organic matter in the Ordovician Wufeng Formation-Silurian Longmaxi Formation in the Sichuan Basin. *Mar. Pet. Geol.* **2019**, *101*, 556–565. [CrossRef]
75. Xiong, J.; Liu, X.; Liang, L.; Zeng, Q. Methane adsorption on carbon models of the organic matter of organic-rich shales. *Energy Fuels* **2017**, *31*, 1489–1501. [CrossRef]
76. Zhang, T.; Ellis, G.S.; Ruppel, S.C.; Milliken, K.; Yang, R. Effect of organic-matter type and thermal maturity on methane adsorption in shale-gas systems. *Org. Geochem.* **2012**, *47*, 120–131. [CrossRef]
77. Milliken, K.L.; Olson, T. Silica diagenesis, porosity evolution, and mechanical behavior in siliceous mudstones, Mowry Shale (Cretaceous), Rocky Mountains, USA. *J. Sediment. Res.* **2017**, *87*, 366–387. [CrossRef]
78. Liang, C.; Cao, Y.; Liu, K.; Jiang, Z.; Wu, J.; Hao, F. Diagenetic variation at the lamina scale in lacustrine organic-rich shales: Implications for hydrocarbon migration and accumulation. *Geochim. Cosmochim. Acta* **2018**, *229*, 112–128. [CrossRef]

energies

MDPI

Article

Molecular Simulation of Nitrogen Adsorption in Multidimensional Nanopores and New Insights into the Inversion of Pore Size Distribution for Gas Shale

Zhuo Chen [1,2], Mian Lin [1,2,*], Wenbin Jiang [1,2,*] and Gaohui Cao [1]

1 Institute of Mechanics, Chinese Academy of Sciences, Beijing 100190, China; chenzhuo@imech.ac.cn (Z.C.); caogaohui@imech.ac.cn (G.C.)
2 School of Engineering Science, University of Chinese Academy of Sciences, Beijing 100049, China
* Correspondence: linmian@imech.ac.cn (M.L.); jiangwenbin@imech.ac.cn (W.J.)

Abstract: Low-temperature nitrogen adsorption is a widely used method for the research and evaluation of gas shale's pore structure. The existing interpretation method, utilizing gas adsorption isotherms to obtain pore size distribution (PSD), is always based on the one-dimensional geometry model, while the void space of gas shale has strong multi-dimensional characteristics. It is necessary to investigate the nitrogen condensation and evaporation behavior in multidimensional structures. In this study, a series of two-dimensional and three-dimensional models based on ink-bottle pores were constructed. A hybrid molecular simulation approach combining grand canonical Monte Carlo (GCMC) and molecular dynamics (MD) is proposed to simulate the low-temperature nitrogen adsorption isotherms. Three aspects have been analyzed in detail. Compared with the conventional understanding that the threshold of cavitation in the ink-bottle pore only relates to throat diameter, this study discloses a wider and more comprehensive range of conditions of cavitation that considers both the throat length and diameter. As pore spaces of shale samples consist of many complex interconnected pores, the multi-stage ink-bottle pore model is more suitable than the single ink-bottle pore model to similarly reproduce the wider cavitation pressure range. A more universal parameter is proposed that quantitatively unifies the influence of cavity diameter and length on condensation pressure and has good applicability in cavities with different shapes. This work quantitatively studies the nitrogen adsorption isotherms of three-dimensional complex nanopore structures using molecular simulation and provides a reasonable explanation for the low-temperature nitrogen adsorption isotherms of gas shale.

Keywords: shale; pore size distribution; molecular simulation; low-temperature nitrogen adsorption

1. Introduction

Nanopores are developed in shale gas reservoirs, which are the storage space and migration path for shale gas. To study the pores inside shale, researchers have developed various methods for characterizing micro and nanostructures. The currently mature characterization methods include scanning and transmission electron microscopies (SEM and TEM, respectively), gas physisorption, X-ray diffraction (XRD), mercury porosimetry, nuclear magnetic resonance (NMR), small-angle X-ray (SAXS) and small-angle neutron scattering (SANS). Among these characterization methods, gas physisorption is widely used and is well established. It is relatively inexpensive and is capable of characterizing nanopores with diameters from 0.35 nm to at least 100 nm [1]. Low-temperature nitrogen adsorption is a conventional method by which to obtain the pore size distribution of shale. The nitrogen molecules are used as probes to measure the surface area and volume of the void space of the materials. Briefly, nitrogen condensation occurs in larger pores in which relative pressure increases, and evaporation occurs in smaller pores in which relative pressure decreases. This feature provides a connection between the relative pressure and the size of the void space, and theoretically, the pore size distribution of the void space can be determined with the adsorption and/or desorption isotherms [2–5].

We have carried out low-temperature nitrogen adsorption experiments on more than 1000 shale samples from Wufeng–Longmaxi formations and different wells from southwest China (See Appendix A for details). Typical adsorption isotherms of shale samples obtained from low-temperature nitrogen adsorption experiments at 77.3 K are depicted in Figure 1a. Considering the adsorption and desorption isotherms, from a morphology perspective, a hysteresis loop can be observed between these two isotherms in a large number of samples, which leads to the problem of branch selection for analysis [6]. Furthermore, another common feature of different isotherms is that there is a sudden change in the desorption branch of the isotherm at a relative pressure of 0.5, while the hysteresis loop is closed at a relative pressure of 0.42.

There are two main mechanisms that cause hysteresis. One exists in the one-dimensional cylinder pore or slit geometry with both ends open. The meniscus is cylindrical during adsorption, while it becomes hemispherical on desorption, as depicted by Sarkisov and Monson [7]. The curvature of the spherical meniscus is half that of the cylindrical meniscus, so the condensation pressure is higher than the evaporation pressure [8–10]. Previous experiments on low-temperature nitrogen adsorption on molecular sieves represented by MCM-41 have found that the hysteresis loops of columnar nanopores are mostly of the H_1 type [11]. Another mechanism exists in ink-bottle pores where the wider cavity is controlled by a narrower pore throat, evaporation in the cavity will be delayed until the evaporation in the pore throat or the cavitation happens [12–19]. According to the classification of the hysteresis loops from IUPAC, hysteresis loops of ink-bottle pores are mostly H_2 or H_3 types [20]. It can be seen that most of the experimental adsorption isotherms have H_2-type characteristics, which means that ink-bottle pores are the main pore type of nanopores in shale reservoirs. Meanwhile, through characterization of shale nanopores using techniques such as focused ion beam-scanning electron microscope (FIB-SEM) and micro-computed tomography (CT), it has been found that various types of pores develop in shale, with complex pore shapes, significant differences in pore sizes, and different connectivity modes between pores, and which have strong heterogeneity and anisotropy (Figure 2) [21–24].

In recent years, researchers have continuously investigated the gas condensation and evaporation behavior in multi-dimensional pore structures with molecular simulations. As the basic and most simple unit, the ink-bottle pore is mostly considered. The typical adsorption isotherm obtained by previous simulations of ink-bottle pore is shown in Figure 1b. Via utilization of the molecular simulation of two-dimensional ink-bottle pores, studies have disclosed that the desorption branch in complex structures mainly reflects the size of the controlling pore throat and the volume of the controlled pore cavity, not the size of the pore cavity. The effects of cavity size, throat width, and throat length on the evaporation pressure have been separately discussed [19]. However, the boundary between pore blocking and cavitation type evaporation, considering jointly both throat width and throat length, has not been studied. Furthermore, when cavitation occurs, the predicted declines in desorption branches of single ink-bottle pores are very steep around the same relative pressure, while the window of relative pressure for cavitation for shale samples is wider, as shown in Figure 1. There is a lack of explanation for this phenomenon. Whether it is due to a more complex pore structure is not clear.

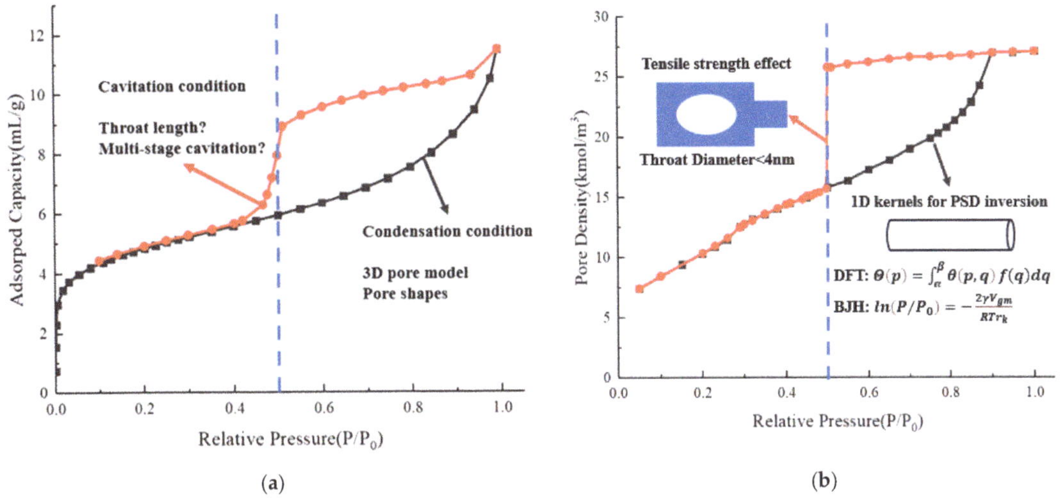

Figure 1. Typical adsorption isotherms. (**a**) Low-temperature nitrogen adsorption isotherm of shale. (**b**) Simulated ink-bottle pore adsorption isotherm from previous works.

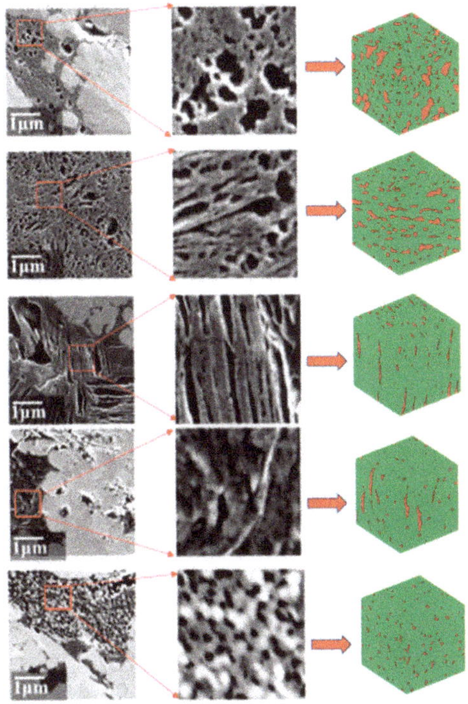

Figure 2. SEM of organic pores of shale samples from southwest China [24].

As the condensation process is less influenced by the pore structure compared with the evaporation process, it is suggested to analyze the adsorption branch to obtain pore size distribution for gas shale according to standards such as ISO 15901-2:2006 [25], IDT. Clearly, the void space of gas shale has strong multi-dimensional characteristics. However, the interpretation method utilizing gas adsorption isotherm to obtain pore size distribution

(PSD) is always based on one-dimensional geometry models, including BJH, NLDFT, QSDFT, etc. [2,26–28]. The underlying assumption when applying these models on gas shale is that the condensation or evaporation mechanism in multidimensional pore space should be the same as that in one-dimensional geometry. Quantitatively, one-dimensional pore models with both ends open are applied directly, without a clear understanding of the potential issues with this assumption. The influence of pore length on condensation pressure has been rarely discussed.

In this study, we aim to employ molecular simulation of two-dimensional and three-dimensional models to further investigate the abovementioned problem. The two-dimensional single ink-bottle pores with different throat diameters and lengths are examined to obtain the boundary between pore blocking and cavitation. The three-dimensional parallel and serial-connected ink-bottle pores are constructed to provide an example of gentle decline in the desorption branch. The two-dimensional single ink-bottle pores with different cavity diameters and lengths are simulated to determine a parameter that correlates with the condensation pressure. The paper is organized as follows: the methodology is described in Section 2, while results and discussions are presented in Section 3, and Section 4 concludes the paper.

2. Methodology

2.1. Simulation Cell Construction and Parameters

Graphite is used to represent the wall of organic nanopores in the molecular model [29,30], and nitrogen molecules are randomly distributed in the pore cavity in the initial state. In previous studies, both the united atomic model and the full atomic model were able to accurately obtain the adsorption–desorption process of nitrogen molecules in nanopores. However, compared with the experimental results of MCM-41, the simulation results of the united atomic model were closer to the experimental results [31,32], and the computational efficiency of using the united atomic model was higher, so the united atomic model is used for calculations. The 12-6 Lennard-Jones potential function is used to describe the van der Waals force between molecules, as follows:

$$V(r) = 4\varepsilon\left[\left(\frac{\sigma}{r}\right)^{12} - \left(\frac{\sigma}{r}\right)^{6}\right] \quad (1)$$

where V is the intermolecular potential between the two atoms, ε is the well depth and reflects how strongly the two particles attract each other, σ is the distance at which the intermolecular potential is zero, and r is the distance between the two atoms. Due to the fact that both the united atomic model and the graphite wall are electrically neutral, this model does not consider the effect of charge force. Table 1 lists the force field parameters of the simulations. The Lorentz–Berthelot mixing rules are used to determine force field parameters between different types of molecules.

Table 1. Force field parameters of the simulation.

Molecular	ε (Kcal/mol)	σ (nm)
C-C	0.0556	0.34
N_2-N_2	0.1892	0.375

The ink-bottle pore model is constructed and taken as a basic unit to construct pores with different connectivity. The basic ink-bottle pore unit contains a cylindrical pore throat and a cylindrical pore cavity. The model of the basic ink-bottle pore unit is shown in Figure 3. The structure is cut from the middle to show its internal wall.

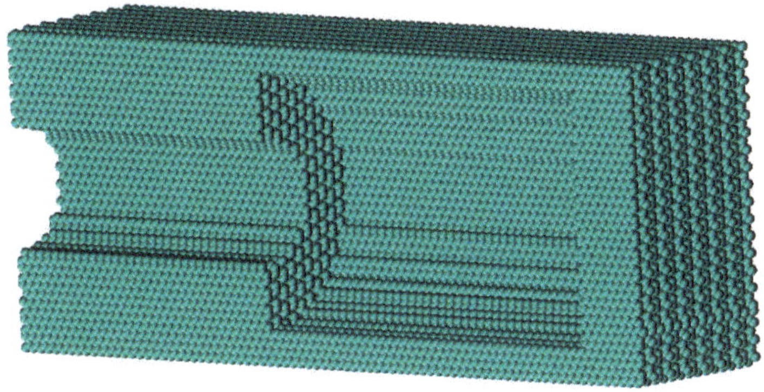

Figure 3. Schematic diagram of single ink-bottle pore unit.

2.2. Simulation Method and Procedure

In previous studies, the GCMC method was often used to simulate adsorption and desorption in nanopores because of the flexible control of the total number of moving molecules in the simulation cell. However, the computational efficiency of GCMC simulation is fairly low and not easy to run parallelly, and it is difficult to simulate the complex and connected nanopore structures. In this work, a hybrid simulation method combining GCMC and MD is used to simulate the adsorption–desorption process of nitrogen molecules. MD simulation can simulate the movement of nitrogen molecules in the pore space based on potential parameters and Newton's equation of motion, while GCMC simulation can adjust the total number of nitrogen molecules in the simulation cell to ensure that the chemical potential is consistent with the target bulk value. The computational accuracy of the GCMC–MD method is shown to be acceptable and the efficiency is significantly improved compared with the conventional GCMC method (See Appendix B for details). The use of the hybrid method provides a basis for the calculation of complex connected nanopore structures.

GCMC–MD simulation is implemented by using GCMC and NVT commands in LAMMPS. A complete simulation procedure contains the following steps: the temperature of the simulation system is set to 77.3 K, and the relative pressure point range of the simulation is arranged between 0.05 and 1. The time step is set to 2 fs, and the motion of nitrogen molecules is determined by the MD simulation with the NVT ensemble at each time step. The GCMC cycle is performed every 10 steps, and 1000 random exchanges (insertions or deletions) of nitrogen molecules are attempted in every GCMC cycle to adjust the total number of nitrogen molecules. A total of 1,000,000 steps of MD simulations are performed, and the last 200,000 steps of simulation data are taken for analysis for every relative pressure point.

To investigate the microscopic behavior of nitrogen molecules in nanopores, we analyze the density of fluid molecules in the pores, which is defined as the amount of adsorbate per unit accessible volume of the pores:

$$\rho_{excess} = \frac{N_{excess}}{V_{acc}} \tag{2}$$

where V_{acc} is the accessible pore volume and N_{excess} is the number of excess nitrogen molecules in the pores, defined as the difference between the total number of fluid molecules in the simulation cell and the total number of free-state molecules filling the accessible pore volume under the same conditions:

$$N_{excess} = N_{box} - \rho_G V_{acc} \tag{3}$$

where ρ_G is the free-state density of the nitrogen gas under pressure.

3. Results and Discussion

3.1. Nitrogen Evaporation Behaviors in a Single Ink-Bottle Pore

Previous studies have disclosed that, for the ink-bottle pore, there are two kinds of mechanisms that will delay gas evaporation in the pore cavity and lead to an obvious adsorption–desorption hysteresis, including pore blocking and cavitation [12–16]. An obvious difference between these two types is the closure point of the hysteresis loop. For the pore-blocking type, the closure point varies with the diameter of the throat, while for the cavitation type, the closure point is fixed when the diameter of the throat is below a threshold value [33]. In this study, we further quantitatively investigate the boundary between these two mechanisms considering the variation of both the diameter and length of the pore throat. Previous studies with a tubular model have disclosed that the diameter of the throat where cavitation occurs was 3–4 nm. Therefore, we established a collection of ink-bottle pores with throats whose diameters are near this range and whose lengths are variant in a certain range. Firstly, the adsorption–desorption curves of ink-bottle pores with different diameters are simulated with a fixed pore diameter of 6 nm, a fixed pore length of 10 nm, and a fixed throat length of 8 nm. The throat diameter varies from 2 nm to 5.5 nm. The hysteresis loops obtained from the simulations are shown in Figure 4.

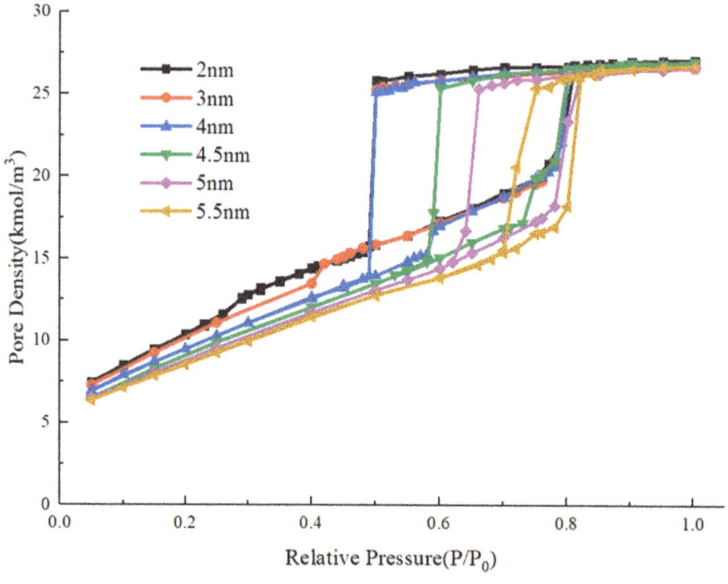

Figure 4. Adsorption isotherms for nitrogen at 77.3 K of ink-bottle pores with different throat diameters.

As the size of the pore cavity is unchanged, the relative pressure of the adsorption branch for condensation is close. Observing the desorption branch, there are two types of variations of evaporation pressure of the pores with different throat diameters. When the throat diameter is larger than 4 nm, the evaporation pressure decreases with the decrease of throat diameter. When the throat diameter is less than 4 nm, the evaporation pressure is always around 0.5. The molecular snapshots of the system with throat diameters of 4 nm before and 4.5 nm after the evaporation are provided in Figure 5 to further examine the inner differences. When the throat diameter is 4 nm, evaporation occurs firstly in the pore cavity while the throat is still condensed, the pore cavity of the system undergoes evaporation through cavitation. When the throat diameter is 4.5 nm, evaporation takes place first in the throat, followed by the cavity, with the pore cavity evaporating through the gas–liquid two-phase meniscus, also known as pore blocking. There are two aspects in which this study is consistent with previous research, indicating the effectiveness of the simulation method in this study. One is the transition from pore blocking to cavitation as the throat diameters decrease. Another is that the critical throat diameter for cavitation is about 4 nm. Furthermore, the cavitation pressure found here, which is around 0.5, is closer to that observed in the isotherms of gas shale, compared with previous studies whose cavitation pressures are always around 0.2 and 0.3 [13–15].

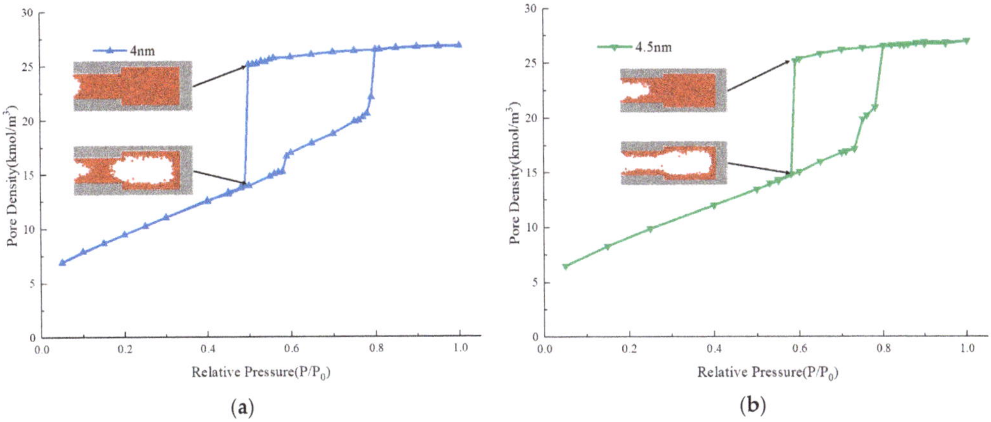

Figure 5. Adsorption isotherms and molecular snapshots of ink-bottle pores. (**a**) Throat diameter of 4 nm. (**b**) Throat diameter of 4.5 nm.

According to Figure 5, the molecular snapshots between and after evaporation can be used to distinguish the underlying mechanism. We further examined the influence of throat length. The simulated adsorption isotherms of ink-bottle pores with different throat lengths and throat diameters of 4 nm and 4.5 nm are depicted in Figure 6. As the length of the throat increases, the evaporation pressure gradually decreases until it approaches a relative pressure of 0.5. For the ink-bottle pores with a throat diameter of 4.5 nm, when the throat length extends from 8 nm to 50 nm, the evaporation pressure changes from 0.57 to 0.5. For the ink-bottle pores with a throat diameter of 4 nm, when the throat length shortens from 8 nm to 2 nm, the evaporation pressure changes from 0.5 to 0.6. Examining the molecular snapshots, it is found that for ink-bottle pores with a throat diameter of 4 nm, evaporation due to cavitation occurs when the throat length is 8 nm and 6 nm, while pore blocking evaporation occurs when the throat length is 4 nm. For the ink-bottle pores with a throat diameter of 4.5 nm, evaporation due to cavitation occurs when the throat length is 50 nm, while pore-blocking evaporation occurs when the throat length is below 30 nm. Based on the above two examples, it can be seen that the transition condition from pore blocking to cavitation is affected by both throat diameter and throat length.

To further obtain the boundary between these two mechanisms quantitatively, the evaporation processes of ink-bottle pores with pore diameters of 2.5–4.5 nm and different pore lengths are simulated. In Figure 7, each point denotes a simulation model with its x-axis value representing throat diameter and the y-axis value throat length. For the black points, the corresponding simulation model experiences cavitation evaporation, while for the blue points, the corresponding simulation model experiences pore blocking evaporation. The blue line indicates the edge of these two mechanisms, which are identified with the molecular snapshots. There is a boundary, depicted with the blue line in Figure 7, that divides the figure into two zones. In the upper zone, cavitation occurs, while in the lower zone, pore blocking dominates.

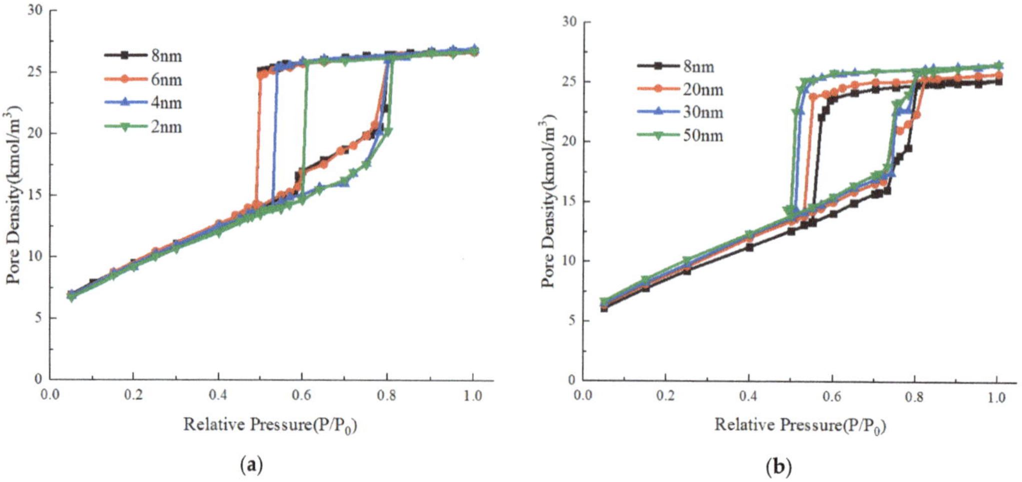

Figure 6. Adsorption isotherms of ink-bottle pores with different throat lengths (**a**) Throat diameter of 4 nm. (**b**) Throat diameter of 4.5 nm.

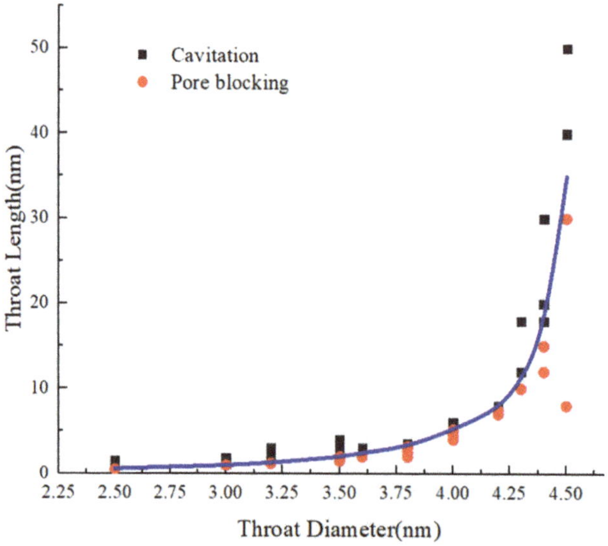

Figure 7. The evaporation mechanism and boundary of the evaporation mechanism transition for ink-bottle pores with different throat lengths and diameters. The blue line is the exponential function fitting relationship between the critical throat length and throat diameter.

The exponential function is applied to fit the boundary of these two zones, and the fitted formula is:

$$y = 1.379 \times e^{4.296(x-3.796)} + 2.082 \tag{4}$$

In a one-dimensional computational model that only considers the diameter of the throat, the throat diameter for which the evaporation mechanism transition occurs is around 4 nm [34]. In the simulation results that comprehensively consider the length and diameter of the throat, based on the characteristics of the exponential function and in the throat with a pore diameter below 3.8 nm, the throat length and diameter comprehensively control the evaporation mechanism. In the throat with a pore diameter above 3.8 nm, the critical throat length for cavitation increases rapidly with the diameter, and the critical condition is mainly controlled by the throat diameter. It must be pointed out that this conclusion is only applicable to systems with throat diameters of 2.5 nm–4.5 nm. In throat diameters less than 2 nm, condensation occurs in the throat at very low pressure. For larger diameter ink-bottle pores, the length of the throat where cavitation occurs exceeds the computational ability of the molecular simulation. Compared with the conventional tubular model the threshold of cavitation only relates to throat diameter, this study discloses a wider range of conditions of cavitation and provides a more comprehensive understanding.

3.2. Nitrogen Evaporation Behaviors in Multi-Stage Ink-Bottle Pores

For a single ink-bottle pore, the decline in the desorption branch is always abrupt in this study and in previous studies within a narrow relative pressure range. However, for gas shale samples, the decline trend is gentle, and the starting point of decline is around 0.5, while the closure point is around 0.42. This indicates that the cavitation process in the nanopores of shale may have a small range of relative pressure, slightly lower than 0.5, rather than being fully completed at the relative pressure of 0.5. From this perspective, simplifying the complex pore space of gas shale so that it is represented by bundles of ink-bottle pores will lose this feature. Therefore, we further consider a series of connected ink-bottle pores.

Firstly, we select three basic ink-bottle pore units with different throat diameters of 3 nm, 4.5 nm, and 5 nm and with throat lengths of the units are all 3 nm. We can obtain the evaporation mechanism of each unit from Section 3.1, wherein the units with throat diameters of 4.5 nm and 5 nm evaporate through the gas–liquid two-phase meniscus, while the units with throat diameters of 3 nm evaporate through cavitation. These three units are connected in different connection sequences. Two systems with these three units connected in series are constructed and named model S1 and S2, while one system with units connected in parallel is constructed and named model P. The common feature of the structure is that the internal structure is connected to the external bulk space through a throat that belongs to the cavitation zone according to Figure 7. The connected structures are shown in Figure 8. The adsorption isotherms of the connected ink-bottle pores are shown in Figure 9.

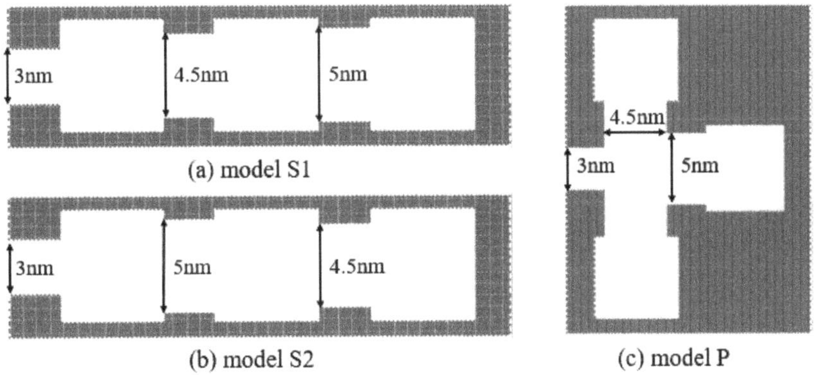

Figure 8. The connected ink-bottle pores with different throat lengths and diameters: (**a**) model S1, comprising three units connected in series with throat diameters of 3 nm, 4.5 nm and 5 nm from left to right, respectively. (**b**) model S2, comprising three units connected in series with throat diameters of 3 nm, 5 nm and 4.5 nm from left to right, respectively. (**c**) model P, comprising three units connected in parallel.

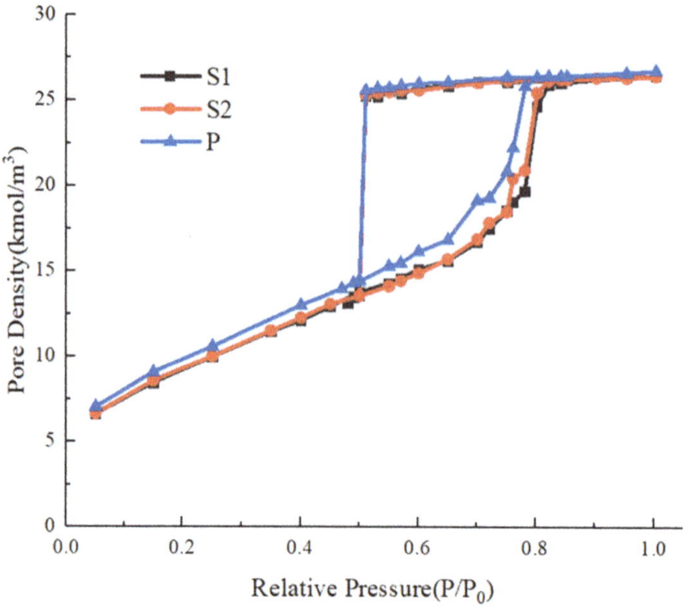

Figure 9. Nitrogen adsorption isotherms of connected ink-bottle pores.

From the simulated isotherms, we can see that there is only one evaporation process in the simulated system with a main throat diameter of 3 nm, and the relative pressure of evaporation is around 0.5. We obtained snapshots of the molecules in the system under this relative pressure, as shown in Figure 10, indicating that cavitation occurs in all three pore cavities simultaneously.

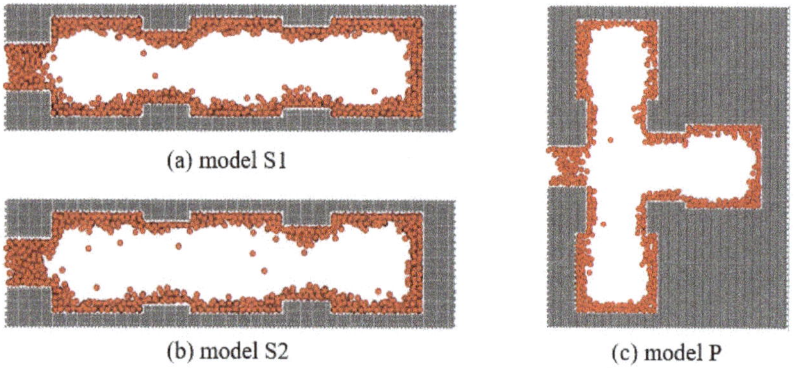

Figure 10. Nitrogen snapshots of connected ink-bottle pores after cavitation. (**a**) model S1, comprising three units connected in series with throat diameters of 3 nm, 4.5 nm and 5 nm from left to right, respectively. (**b**) model S2, comprising three units connected in series with throat diameters of 3 nm, 5 nm and 4.5 nm from left to right, respectively. (**c**) model P, comprising three units connected in parallel.

The pore structures of the models have similar characteristics that are connected to the external bulk phase through a throat with a diameter and length of 3 nm, so the evaporation process is limited by the throat. Although the connection mode and sequence of ink-bottle pore units are different, evaporation occurs through cavitation in all three systems. When the narrow throat that leads to cavitation exists in the connected pore system, it will control the evaporation process of its internal pore structure that would evaporate through the gas–liquid meniscus at a relative pressure higher than 0.5 without the outer throat. Evaporation through cavitation corresponds with the rapid decline of the desorption branch around the relative pressure of 0.5, and the cavitation pressure does not change with the serially- or parallel-connected pore structure and pore size.

Secondly, we constructed another system in which a portion of the internal throats belong to the cavitation zone. The adsorption isotherm and molecular snapshots of the two-stage cavitation are shown in Figure 11.

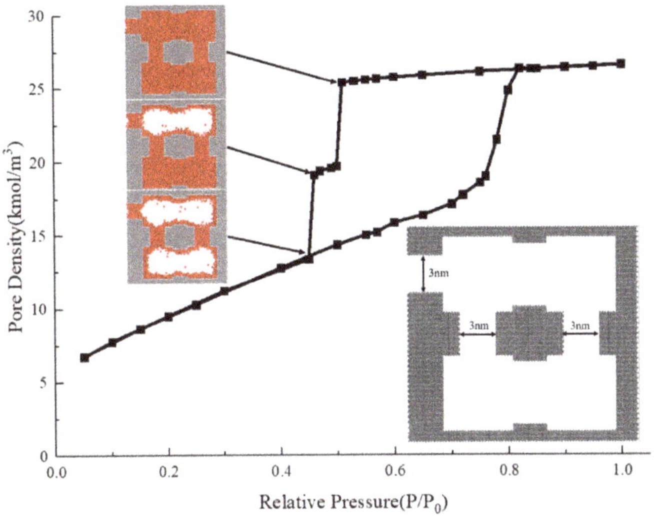

Figure 11. Nitrogen adsorption isotherms and molecular snapshots at cavitation of a two-layer ink-bottle pore.

The hysteresis loop obtained from the simulation shows that there are two rapid declines in the desorption branch of the system, corresponding with the relative pressures of 0.5 and 0.45. The two-stage evaporation process of the desorption branch occurs both through cavitation, which is controlled by the outermost throat, and the vertical throats connected to the internal structure. The upper cavity evaporates through cavitation when the relative pressure is 0.5, while the lower cavity is controlled by multiple throats that belong to the cavitation zone, and cavitation occurs at the relative pressure of 0.45. In such a structure, a multi-stage cavitation process and the evaporation of some internal structures at a cavitation pressure below 0.5 are observed, which is different from the cavitation pressure in a single-throat ink-bottle pore, which is always 0.5. To further investigate the multi-stage cavitation process, multiple ink-bottle pore units with throat diameters and lengths of 3 nm and cavity diameters and lengths of 6 nm are connected head to tail, representing a multi-stage cavitation structure controlled by multiple narrow throats that can lead to cavitation. For these multilayer structures, we calculated the desorption branches of systems with different unit numbers and compared the results with the hysteresis loop of a single ink-bottle pore unit with a throat of 3 nm. The results are shown in Figure 12.

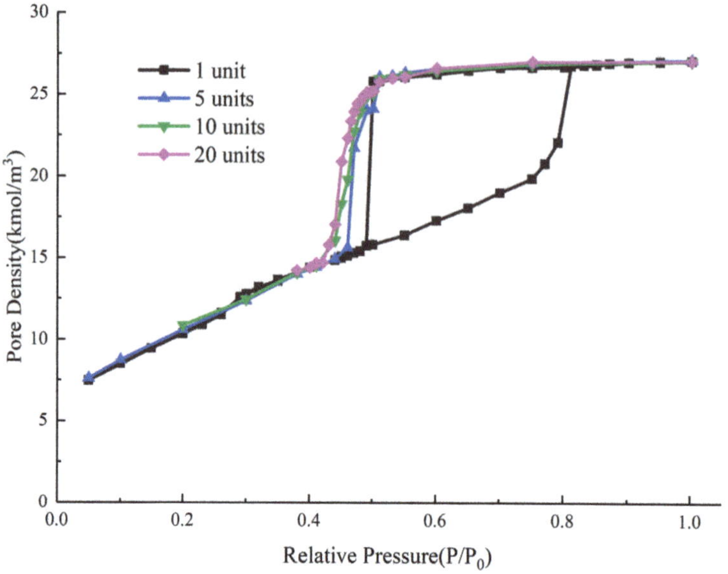

Figure 12. Nitrogen adsorption isotherms of multilayer ink-bottle pore structure with different unit numbers.

The simulation results of the system show that the narrow throats have a controlling effect which will cause the internal structure to evaporate through cavitation at a lower relative pressure. As the number of units increases, the separated dropping steps appearing in the two-level cavitation structure become smoother, and the hysteresis phenomenon of the cavitation process becomes more obvious, but always within a small relative pressure range. Among them, the multi-stage cavitation structures of 10 units and 20 units both overlap with the adsorption branch around the relative pressure of 0.42. This indicates that the hysteresis phenomenon of multi-level cavitation does not infinitely reduce the relative pressure of cavitation, but rather completely ends around 0.42.

From two-dimensional single ink-bottle pores to three-dimensional connected ink-bottle pores, this study found a type of configuration of ink-bottle pore whose desorption branch gently declines between the relative pressure of 0.42 and 0.5. This decline characteristic is like that of gas shale. For the cavitation in the nanopore which is controlled by the single-hierarchy pore throat, the evaporation process occurs through cavitation at the relative pressure around 0.5, which is close to the abrupt drop of the desorption branch of the shale. In the system with multiple throats to cavitate, the multi-stage cavitation process ends around the relative pressure of 0.42, which is close to the closure position of the hysteresis loop of shale. Considering that pore spaces of shale samples consist of many complex interconnected pores, the multi-stage ink-bottle pores are more suitable than the tubular model to equivalently represent the wider cavitation pressure range.

3.3. Nitrogen Condensation Behavior in the Ink-Bottle Pore

In the adsorption isotherms of ink-bottle pores with different throat diameters, although the evaporation mechanism and pressure are different, the condensation pressure does not change due to the unchanged pore size. The condensation pressure of the ink-bottle pore depends on the size of the cavity and is not affected by the throat and connection sequence. In one-dimensional tubular pore models, the condensation pressure corresponds to the size of pores, and the pore size distribution can be determined. For two-dimensional models, it has been found that condensation pressure is not only related to cavity diameters but also to the length [19]. However, there is a lack of quantitative predictive rules. We designed a series of ink-bottle pore systems with different pore cavity sizes but the same throat diameter and length. The diameter and length of the throat are 3 nm and 8 nm, respectively. Firstly, the diameter of the pore cavity is fixed at 6 nm, and the length changes from 5 nm to 30 nm. The simulated adsorption isotherms are plotted in Figure 13. Secondly, the length of the pore cavity is fixed at 10 nm, and the diameter changes from 6 nm to 8 nm. The simulated isotherms are plotted in Figure 14.

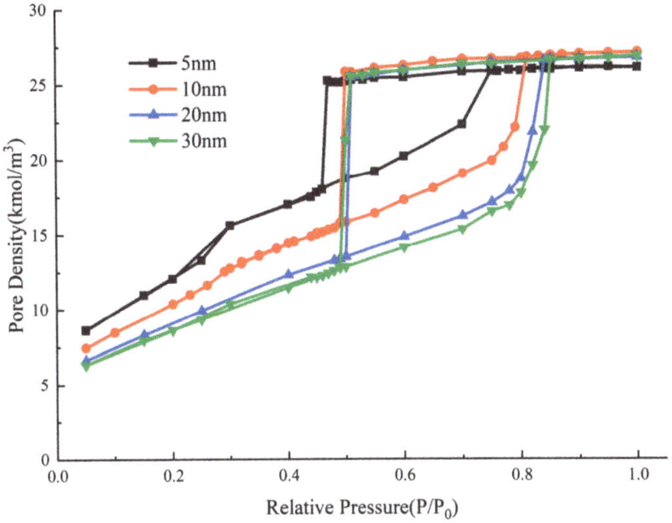

Figure 13. Nitrogen adsorption isotherms with different cavity lengths.

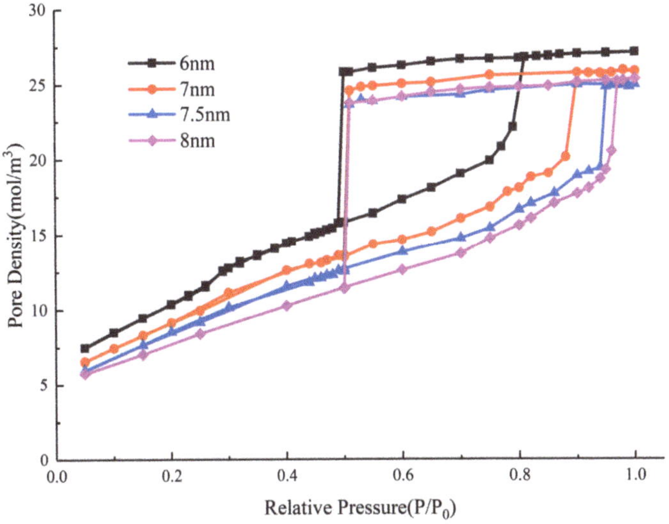

Figure 14. Nitrogen adsorption isotherms with different cavity diameters.

From the hysteresis loop obtained from the simulation, the variation of condensation pressure with the length and diameter of the cavity in the ink-bottle pore has commonalities, i.e., that, as the length and diameter increase, the condensation pressure increases. The larger the space filled in the cavity during the condensation process, the more difficult the condensation process is, and the higher the condensation pressure. The condensation process is an extension of the multi-layer adsorption layer adsorbed on the wall, so the larger the wall area, the more space where the condensation process occurs, and the faster the condensation progresses. On this basis, we defined R_C as related to pore size and investigated the relationship between this parameter and the condensation pressure of the system:

$$R_C = \frac{V - V_A}{V} = 1 - \frac{S \times h_A}{V} \tag{5}$$

where, V is the total pore volume of the pore cavity, V_A is the pore volume occupied by nitrogen molecules in multi-layer adsorption states, S is the surface area of the pore cavity, and h_A is the thickness of the multi-layer adsorption layer.

The key is to determine h_A. Considering that the occurrence of a hysteresis loop indicates that the state of nitrogen molecules is in an incomplete equilibrium state, and that the transition point from the multi-layer adsorption layer corresponds with the lower endpoint of the hysteresis loop. Through the simulation in Section 3.1, the cavitation pressure of the ink-bottle pore in which cavitation occurs is about 0.5, which is the transition point. Under saturated vapor pressure, the number of adsorption layers n of adsorbates in micro/nanostructures is limited. Halsey (1969) proposed the equation for describing the adsorption layer of nitrogen at a liquid nitrogen temperature, also known as the Halsey equation [35]:

$$n = -\left(\frac{5}{In(P/P_0)}\right)^{1/3} \tag{6}$$

When the relative pressure is about 0.5, and when substituting the above equation, the thickness of the adsorption layer of nitrogen molecules is about 0.715 nm. Once the size of the cavity is determined, R_C can be directly obtained through geometric calculations. R_C for the parameters and condensation pressures of different cavity sizes—calculated in Figures 12 and 13, respectively—are summarized in Table 2.

Table 2. Condensation pressure and R_C of ink-bottle pore.

Pore Diameter (nm)	Pore Length (nm)	R_C	Condensation Pressure
6	5	0.72	0.414
6	10	0.8	0.497
6	20	0.86	0.539
6	30	0.9	0.552
6	50	0.92	0.564
7	10	0.9	0.543
7.5	10	0.93	0.561
8	10	0.94	0.578
8.5	10	0.97	0.592
9	10	0.99	0.606

The linear correlation analysis between R_C and condensation pressure is shown in Figure 15.

From the results of linear fitting, the fitted equation is:

$$R_C = 0.6667 \times P_C - 0.0507 \tag{7}$$

The adjusted R^2 of this fitting is 0.97, indicating a strong linear correlation. This indicates that the defined pore parameters, R_C, well integrate the effects of cylindrical pore length and diameter on condensation pressure. Compared with a finite cylindrical cavity with the same diameter, the ratio of S and V is smaller for the one-dimensional tubular pore model with two ends open, hence the parameters R_C and condensation pressure are higher. For a certain low-temperature nitrogen adsorption isotherm, the pore size distribution obtained using the one-dimensional tubular model will classify the pore volume into smaller pore size ranges.

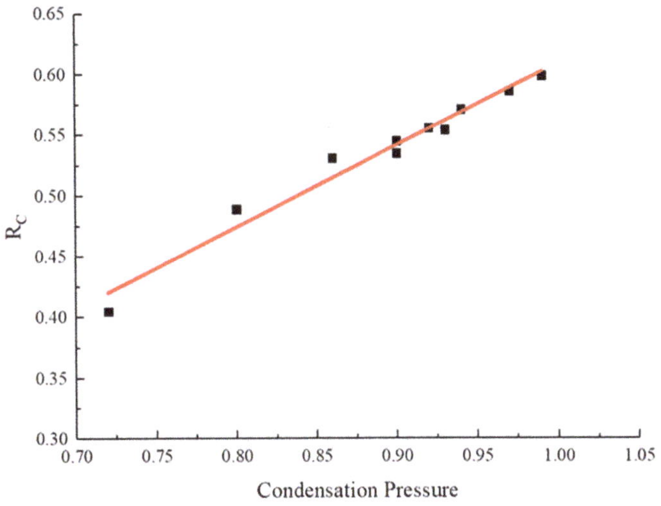

Figure 15. Linear fitting results of R_C and condensation pressure.

In addition, among various pore size distribution calculation methods based on the DFT theory, there are different calculation kernels based on slit, cylindrical, and spherical pore shapes, with corresponding pore sizes as follows: $D_{slit} < D_{\text{cylindrical}} < D_{spherical}$. The selection of kernels with different pore shapes will result in different pore size distributions. We simulated the condensation process of ink-bottle pores in which the cavity shapes are slit and spherical, predicted the condensation pressure using the linear fitting equation of R_C and the condensation pressure obtained through the cylindrical pore and compared these to obtain Table 3.

Table 3. Simulated and predicted condensation pressure and R_C of nanopores with different shapes.

Pore Type	Diameter (nm)	Length (nm)	R_C	Condensation Pressure	Predicted Condensation Pressure	Prediction Error
Slit	4	30	0.586	0.95	0.961	0.011
Slit	4	20	0.562	0.92	0.929	0.009
Slit	5	10	0.562	0.93	0.929	−0.001
Slit	6	10	0.608	0.99	0.997	0.007
Spherical	6	-	0.433	0.72	0.731	0.011
Spherical	7	-	0.496	0.84	0.837	−0.003
Spherical	8	-	0.546	0.91	0.906	−0.004
Spherical	9	-	0.588	0.97	0.968	−0.002

From the predicted results, the relationship between R_C and the condensation pressure obtained through the cylindrical pore cavity is also effective in predicting the condensation pressure of spherical and slit pores, with a maximum prediction error of only 0.011. This further indicates that the condensation process and pressure are directly related to pore space. Compared with pore size, R_C is the more universal parameter that relates to condensation pressure.

Beyond that, this relationship provides an approach by which to determine the surface area of a certain pore without assuming its cross-sectional shape. Conventionally, the shape of the pore needs to be selected first, then the size and volume of the pore are determined along with the relationship between them and the relative pressure and condensation gas amount. Finally, the surface area is calculated based on these two quantities according to the predefined pore shape. Here, the parameter R_C relates the condensation pressure, volume, and surface area according to Equations (5) and (6). The relative pressure pi corresponds with R_{Ci} according to the fitting equation. The increment of nitrogen adsorption is Δv_i, and the surface area increment at this relative pressure is:

$$\Delta s_i = \frac{\Delta v_i(1 - R_{Ci}(P_{Ci}))}{h_A} \tag{8}$$

By accumulating the surface areas of each part, we can obtain the surface area of the pore wall. With this approach, it is possible to break free from the constraints of pore shape assumption and obtain more general calculation results.

The understandings obtained from the former two subsections suggest that for gas shale, there should exist many complex multi-stage ink-bottle-like pore structures. To interpret the pore size distribution, the multi-dimensional model should be applied. However, there is still much effort needed to fulfill this target that is out of the scope of this work and will be addressed in future work. Currently, only the one-dimensional tubular model is mature and optimal selection of the tubular model is more practical. In the ink-bottle-like pore structures, condensation occurs first during the condensation process and could cause the internal structure to condense through the gas–liquid two-phase meniscus rather than the formation of a liquid bridge. Therefore, for the selection of the DFT calculation kernels, it is preferable to use the equilibrium kernel rather than the adsorption kernel to calculate the pore size distribution.

In summary, this study works on statistical adsorption mechanisms, and focuses on the relative pressure of condensation and evaporation feature. The adsorption dynamic is also important and will be investigated in future work [36–38].

4. Conclusions

In this study, we constructed a series of two-dimensional and three-dimensional models based on ink-bottle pores, and molecular simulations of low-temperature nitrogen adsorption in nanopores were performed using the GCMC–MD method. This was undertaken in order to provide insights on the nitrogen condensation and evaporation behaviors associated with multi-dimensional pore structures and their implication for the pore size distribution analysis of gas shale. Three aspects have been analyzed in detail.

Firstly, after carefully simulating and examining the evaporation mechanisms of ink-bottle pores with different throat diameters and lengths, the boundary between pore blocking and cavitation has been determined. Compared with the conventional tubular model in which the threshold of cavitation only relates to throat diameter, this study discloses a wider range of conditions of cavitation and provides a more comprehensive understanding.

Secondly, multi-stage ink-bottle pore models have been constructed and it has been found that the cavitation pressure of the inner pore space controlled by multiple-hierarchal cavitation-zone throats is relatively smaller than that of the outer pore space. Considering that pore spaces of shale samples consist of many complex interconnected pores, the multi-stage ink-bottle pores are more suitable than the tubular model to equivalently represent the wider cavitation pressure range.

Thirdly, ink-bottle pore models with different pore cavity diameters and lengths have been constructed and the condensation process differences investigated. A more universal parameter R_C is proposed. There is a strong linear relationship between R_C and condensation pressure, which not only quantitatively unifies the influence of cavity diameter and length on condensation pressure, but also has good applicability in cavities with different shapes. This breaks free from the constraints of pore shape assumption and obtains more general surface area calculation results. As the condensation process of these pores is carried out through the meniscus formed in the narrow throats, for practical use at this stage, it is more appropriate to use the equilibrium kernels for the calculation of pore size distribution.

Author Contributions: Methodology, Z.C. and W.J.; conceptualization and programming, W.J.; experiment, Z.C.; simulation design, Z.C.; writing, Z.C.; review and editing, M.L. and W.J.; supervision, M.L. and G.C.; funding acquisition, M.L. All authors have read and agreed to the published version of the manuscript.

Funding: This research is funded by the National Natural Science Foundation of China (Grant No. 42030808), the Strategic Priority Research Program of the Chinese Academy of Sciences (Grant No. XDA14010304), the National Natural Science Foundation of China (41690132 and 41872163), and the Technical Innovation Project of Instrument and Equipment Function Development of Chinese Academy of Sciences (2020BA19).

Data Availability Statement: Data are contained within the article.

Acknowledgments: The experimental equipment is supported by Beishide Instrument, Beijing, China.

Conflicts of Interest: We declare no conflict of interest. The funders had no role in the design of the study; in the collection, analyses, or interpretation of data; in the writing of the manuscript; or in the decision to publish the results.

Appendix A

The shale samples for the nitrogen adsorption experiment are selected from the Wufeng–Longmaxi Formation of different blocks in the southern Sichuan Basin. These shale samples are crushed by a compact jaw crusher and sieved between 20 mesh and 40 mesh screens to produce test samples with a particle size of 0.42–0.84 mm. Over 1 g of the sieved sample is loaded into the apparatus for the low-temperature nitrogen adsorption experiment. Before the adsorption procedure, the samples are dried and vacuumed at 110 °C for at least 4 h to fully remove the impurity gas and water in the nanopores. The experimental isotherm is measured at 77.3 K using a surface area and porosity analyzer (3H-2000PS1, Beishide Instrument, Beijing, China). Typical adsorption–desorption curves of shale samples from the Wufeng–Longmaxi Formation are shown in Figure A1.

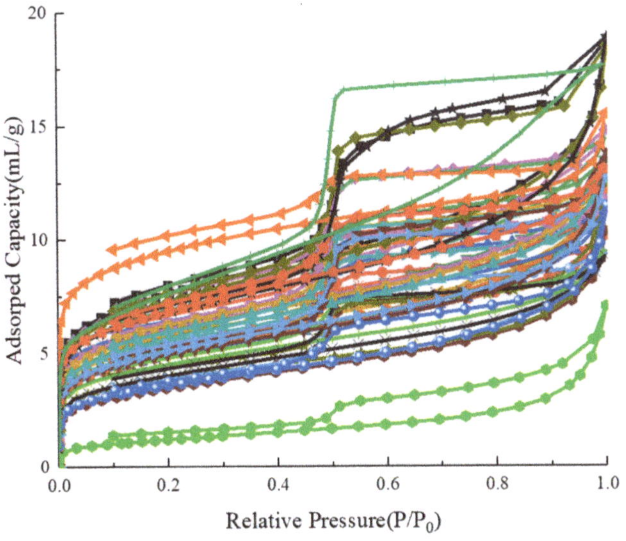

Figure A1. Low-temperature nitrogen adsorption isotherms of shale samples.

Appendix B

To verify the rationality of the GCMC–MD method, we construct a cylindrical nanopore with a diameter of 5.32 nm and periodic boundary conditions are applied at two ends of the pore to simulate the infinite cylinder pores. The GCMC–MD method and GCMC method are both used to simulate the hysteresis loop of adsorption–desorption with the same molecule model and the same parameters, respectively. Each GCMC cycle performs 1000 changes, which include insertion and deletion only but with equal probability. The cutoff radius is chosen as half of the length of the simulation box. In the GCMC–MD mixed simulation case, the GCMC cycle is performed every 10 steps, and nitrogen molecules are added and deleted with equal probability to adjust the total number of nitrogen molecules. A total of 1,000,000 steps of MD simulations are performed for each relative pressure point, and the last 200,000 timesteps of simulation data are taken for analysis at each relative pressure point. The number of GCMC cycles performed by the two methods is the same, but the GCMC method does not rotate and move the molecules in the mixed simulation. The adsorption–desorption curves are shown in Figure A2.

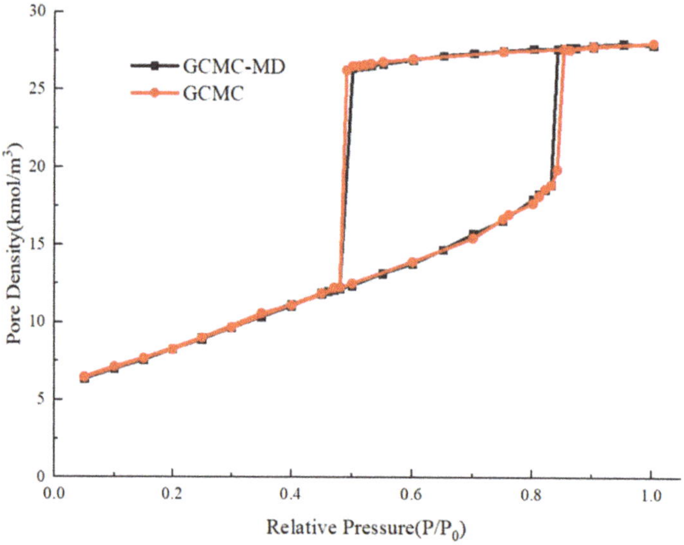

Figure A2. Nitrogen adsorption–desorption curves of cylinder nanopore with GCMC and GCMC–MD methods.

According to the results obtained by the two methods, both the adsorption and desorption isotherms and the relative pressure point of condensation and evaporation are highly coincident, which indicates that the GCMC–MD simulation method has high reliability. The consumption times of these simulations are listed in Table A1.

Table A1. Comparison of calculation time of the GCMC and GCMC–MD simulation methods.

Methods	Calculation Time (h)
GCMC	971
GCMC–MD	124

It can be seen from the calculation time that the calculation efficiency of the GCMC–MD method is much higher than that of the GCMC method under the same conditions. Therefore, the GCMC–MD approach provides a possible means by which to simulate the adsorption–desorption curves of complex nanopore structures.

References

1. Zelenka, T.; Horikawa, T.; Do, D.D. Artifacts and misinterpretations in gas physisorption measurements and characterization of porous solids. *Adv. Colloid Interface Sci.* **2023**, *311*, 102831. [CrossRef] [PubMed]
2. Barrett, E.P.; Joyner, L.G.; Halenda, P.P. The determination of pore volume and area distributions in porous substances, I. Computations from nitrogen isotherms. *J. Am. Chem. Soc.* **1951**, *73*, 373–380. [CrossRef]
3. Wang, M.; Li, Z.; Liang, Z.; Jiang, Z.; Wu, W. Method Selection for Analyzing the Mesopore Structure of Shale—Using a Combination of Multifractal Theory and Low-Pressure Gas Adsorption. *Energies* **2023**, *16*, 2464. [CrossRef]
4. Pang, P.; Han, H.; Hu, L.; Guo, C.; Gao, Y.; Xie, Y. The calculations of pore structure parameters from gas adsorption experiments of shales: Which models are better? *J. Nat. Gas Sci. Eng.* **2021**, *94*, 104060. [CrossRef]
5. Medina-Rodriguez, B.X.; Alvarado, V. Use of Gas Adsorption and Inversion Methods for Shale Pore Structure Characterization. *Energies* **2021**, *14*, 2880. [CrossRef]
6. Sing, K.S.W.; Everett, D.H. Reporting physisorption data for gas/solid systems with special reference to the determination of surface area and porosity (Recommendations 1984). *Pure Appl. Chem.* **1985**, *57*, 603–619. [CrossRef]
7. Sarkisov, L.; Monson, P.A. Modeling of Adsorption and Desorption in Pores of Simple Geometry Using Molecular Dynamics. *Langmuir* **2001**, *17*, 7600–7604. [CrossRef]
8. Hattori, Y.; Kaneko, K.; Ohba, T. Adsorption Properties. In *Comprehensive Inorganic Chemistry II*; Elsevier: Amsterdam, The Netherlands, 2013; Volume 5, pp. 25–44.
9. Pellenq, R.J.M.; Coasne, B.; Denoyel, R.O.; Coussy, O. Simple Phenomenological Model for Phase Transitions in Confined Geometry. 2. Capillary Condensation/Evaporation in Cylindrical Mesopores. *Langmuir* **2009**, *25*, 1393–1402. [CrossRef]
10. Groen, J.C.; Peffer, L.A.A.; Perez-Ramirez, J. Pore size determination in modified micro- and mesoporous materials. Pitfalls and limitations in gas adsorption data analysis. *Microporous Mesoporous Mater.* **2003**, *60*, 1–17. [CrossRef]
11. Barsotti, E.; Piri, M. Effect of Pore Size Distribution on Capillary Condensation in Nanoporous Media. *Langmuir* **2021**, *37*, 2276–2288. [CrossRef]
12. Broekhoff, J.C.; Deboer, J.H. Studies on Pore Systems in Catalysts. 11. Pore Distribution Calculations from Adsorption Branch of a Nitrogen Adsorption Isotherm in Case of Ink-Bottle Type Pores. *J. Catal.* **1968**, *10*, 153–165. [CrossRef]
13. Nguyen, P.T.M.; Do, D.D.; Nicholson, D. Pore connectivity and hysteresis in gas adsorption: A simple three-pore model. *Colloids Surf. A—Physicochem. Eng. Asp.* **2013**, *437*, 56–68. [CrossRef]
14. Klomkliang, N.; Do, D.D.; Nicholson, D. Effects of temperature, pore dimensions and adsorbate on the transition from pore blocking to cavitation in an ink-bottle pore. *Chem. Eng. J.* **2014**, *239*, 274–283. [CrossRef]
15. Phuong, T.M.; Fan, C.Y.; Do, D.D. On the Cavitation-like Pore Blocking in Ink-Bottle Pore Evolution of Hysteresis Loop with Throat Size. *J. Phys. Chem. C* **2013**, *117*, 5475–5484.
16. Nguyen, P.T.M.; Nicholson, D.; Do, D.D. On the Cavitation and Pore Blocking in Cylindrical Pores with Simple Connectivity. *J. Phys. Chem. B* **2011**, *115*, 12160–12172. [CrossRef] [PubMed]
17. Rasmussen, C.J.; Vishnyakov, A.; Thommes, M.; Smarsly, B.M.; Kleitz, F.; Neimark, A. Cavitation in Metastable Liquid Nitrogen Confined to Nanoscale Pores. *Langmuir* **2010**, *26*, 10147–10157. [CrossRef]
18. Loi, Q.K.; Tan, S.J.; Do, D.D.; Nicholson, D. Lower Closure Point for Nitrogen or Argon Adsorption in Mesoporous Solids: Window-Induced Evaporation or Surface-Induced Cavitation? *Ind. Eng. Chem. Res.* **2021**, *60*, 15343–15351. [CrossRef]
19. Zeng, Y.H.; Fan, C.Y.; Do, D.D.; Nicolson, D. Evaporation from an Ink-Bottle Pore: Mechanisms of Adsorption and Desorption. *Ind. Eng. Chem. Res.* **2014**, *53*, 15467–15474. [CrossRef]
20. Thommes, M.; Kaneko, K.; Neimark, A.V. Physisorption of gases, with special reference to the evaluation of surface area and pore size distribution (IUPAC Technical Report). *Pure Appl. Chem.* **2015**, *87*, 1051–1069. [CrossRef]
21. Ji, L.; Lin, M.; Jiang, W.; Cao, G.; Zhou, J. Investigation into the apparent permeability and gas-bearing property in typical organic pores in shale rocks. *Mar. Pet. Geol.* **2019**, *110*, 871–885. [CrossRef]
22. Cao, G.; Lin, M.; Ji, L.; Jiang, W.; Yang, M. Characterization of pore structures and gas transport characteristics of Longmaxi shale. *Fuel* **2019**, *258*, 116146. [CrossRef]
23. Wu, J.; Yuan, Y.; Niu, S.; Wei, X.; Yang, J. Multiscale characterization of pore structure and connectivity of Wufeng-Longmaxi shale in Sichuan Basin, China. *Mar. Pet. Geol.* **2020**, *120*, 104514. [CrossRef]
24. Sun, X.; Luo, C.; Cao, G.H. Differences of Main Enrichment Factors of $S11^1{}_{1\text{-}1}$ Sublayer Shale Gas in Southern Sichuan Basin. *Energies* **2021**, *14*, 5472–5493.
25. *ISO 15901-2:2006;* Pore Size Distribution and Porosity of Solid Materials by Mercury Porosimetry and Gas Adsorption—Part 2: Analysis of Mesopores and Macropores by Gas Adsorption. International Organization for Standardization: Geneva, Switzerland, 2006.
26. Neimark, A.V.; Lin, Y.; Ravikovitch, P.I. Quenched solid density functional theory and pore size analysis of micro-mesoporous carbons. *Carbon* **2009**, *47*, 1617–1628. [CrossRef]
27. Jagiello, J.; Jaroniec, M. 2D-NLDFT adsorption models for porous oxides with corrugated cylindrical pores. *J. Colloid Interface Sci.* **2018**, *532*, 588–597. [CrossRef] [PubMed]
28. Thommes, M.; Cychosz, K.A. Physical adsorption characterization of nanoporous materials: Progress and challenges. *Adsorption* **2014**, *20*, 233–250. [CrossRef]

29. Eugene, A.U. Effect of crystallization and surface potential on the nitrogen adsorption isotherm on graphite A refined Monte Carlo simulation. *Carbon* **2016**, *100*, 52–63.
30. Ambrose, R.J.; Hartman, R.C.; Diaz-Campos, M.; Akkutlu, I.Y.; Sondergeld, C.H. Shale gas-in-place calculations part I: New pore-scale considerations. *SPE J.* **2012**, *17*, 219–229. [CrossRef]
31. Coasne, B.; Galarneau, A.; Di Renzo, F. Molecular Simulation of Nitrogen Adsorption in Nanoporous Silica. *Langmuir* **2010**, *26*, 10872–10881. [CrossRef]
32. Schreiber, A.; Reinhardt, S.; Findenegg, G.H. The lower closure point of the adsorption hysteresis loop of fluids in mesoporous silica materials. *Stud. Surf. Sci. Catal.* **2002**, *144*, 177–184.
33. Nguyen, P.T.M.; Do, D.D.; Nicholson, D. On the Hysteresis Loop of Argon Adsorption in Cylindrical Pores. *J. Phys. Chem. B* **2011**, *115*, 4706–4720. [CrossRef]
34. Lai, W.; Yang, S.; Jiang, Y.; Zhao, F.; Li, Z.; Zaman, B.; Fayaz, M. Artefact peaks of pore size distributions caused by unclosed sorption isotherm and tensile strength effect. *Adsorption* **2020**, *26*, 633–644. [CrossRef]
35. Halsey, G.D. Physical Adsorption of Gases. *J. Am. Chem. Soc.* **1969**, *235*, 1214–1225. [CrossRef]
36. Suzuki, Y.; Ishida, M.; Hata, K.; Ohba, T. Sequential N_2 Adsorption by the Nanopore Entrance Filling Scheme in Nanopores of Carbon Nanotubes. *Langmuir* **2023**, *44*, 15431–15440. [CrossRef] [PubMed]
37. Desouza, A.; Monson, P.A. Modeling fluids confined in three-dimensionally ordered mesoporous carbons. *Adsorption* **2021**, *27*, 253–264. [CrossRef]
38. Schneider, D.; Valiullin, R.; Monson, P.A. Filling Dynamics of Closed End Nanocapillaries. *Langmuir* **2014**, *30*, 1290–1294. [CrossRef]

minerals

MDPI

Article

Diagenetic Evolution Sequence and Pore Evolution Characteristics: Study on Marine-Continental Transitional Facies Shale in Southeastern Sichuan Basin

Bing Zhang [1,2,3], Siyu Wen [1,2,*], Kai Yang [3], Kai Ma [4], Pengwan Wang [5], Chuan Xu [3] and Gaoquan Cao [3]

1 State Key Laboratory of Oil and Gas Reservoir Geology and Exploitation, Chengdu 610059, China; zhangb@cdut.edu.cn
2 Institute of Sedimentary Geology, Chengdu University of Technology, Chengdu 610059, China
3 College of Geophysics, Chengdu University of Technology, Chengdu 610059, China; kaiyang1992@126.com (K.Y.); alex199208@163.com (C.X.); gaoquan@stu.cdut.edu.cn (G.C.)
4 College of Energy, China University of Geosciences (Beijing), Beijing 100080, China; mk211101@163.com
5 PetroChina Hangzhou Research Institute of Geology, Hangzhou 310023, China; wangpw_hz@petrochina.com.cn
* Correspondence: wensiyu18747023@163.com

Abstract: Diagenesis and pore structure are essential factors for reservoir evaluation. marine-continental transitional facies shale is a new shale gas reservoir of concern in the Sichuan Basin. The research on its diagenesis pore evolution model has important guiding significance in its later exploration and development. However, the current research on pore structure changes, diagenesis, and the evolution of marine-continental transitional facies shale is not sufficient and systematic. In order to reveal the internal relationship between pore structure changes and diagenesis, the evolution of marine-continental transitional facies shale was tested by X-ray diffraction, field emission scanning electron microscopy, low-pressure gas adsorption, nuclear magnetic resonance, and the diagenetic evolution sequence and nanopore system evolution of Longtan Formation shale was systematically studied. The results show that the Longtan Formation shale underwent short-term shallow after sedimentation, followed by long-term deep burial. The main diagenetic mechanisms of the Longtan Formation shale include compaction, dissolution, cementation, thermal maturation of organic matter, and transformation of clay minerals, which are generally in the middle-late diagenetic stage. The pore structure undergoes significant changes with increasing maturity, with the pore volumes of both micropores and mesopores reaching their minimum values at Ro = 1.43% and subsequently increasing. The change process of a specific surface area is similar to that of pore volumes. Finally, the diagenetic pore evolution model of Longtan Formation MCFS in Southeastern Sichuan was established.

Keywords: Sichuan Basin; Longtan Formation; marine-continental transitional facies shale; quantitative characterization; diagenesis

1. Introduction

The study of shale gas reservoirs is currently more concerned [1–3]. In the Sichuan Basin, several organic-rich marine shale reservoirs have been discovered in recent years [4–6], providing valuable commercial industrial gas. The marine-continental transitional facies shale (MCFS) of the Longtan Formation in the Sichuan Basin has been highly concerned as the next shale gas reservoir because of its similarities in accumulation conditions and model [7–9]. Studies of shale diagenesis show that compaction, cementation, and mineral composition affect shale reservoir pores [10,11]. Shale reservoirs usually form interconnected pores systems under the interaction effects of diagenesis and the thermal evolution of organic matter (OM) [12]. In order to understand the evolution of pores in shale reservoirs, it is crucial to investigate the diagenesis of shale reservoirs [11,13]. Additionally, the study of mineral composition characteristics, diagenesis, diagenetic evolution, and pore systems evolution of shale reservoirs is the basis of MCFS reservoir evaluation research and also the premise for revealing the accumulation model of MCFS gas. Therefore, establishing a diagenesis-pore evolution model can elucidate the impact mechanism of diagenesis on MCFS reservoirs.

Diagenetic stages of clastic rocks can be divided into syndiagenetic stages, early diagenetic stages, middle diagenetic stages, late diagenetic stages, and epidiagenetis stages (PGEPSC, 2003) [14–16]. The pore characteristics of reservoirs are different under the influence of different diagenetic evolution stages. The early diagenesis stage is dominated by compaction, and the primary intergranular pores decrease [17]. During the diagenetic stage, cementation is well developed, quartz overgrowth and iron-bearing carbonate cements are formed, pore space is destroyed by cement filling, and dissolution is also well developed [18]. In the late diagenetic stage, due to deep burial, the rock experienced compaction. Under deep burial conditions, the increase in temperature favors the thermal evolution of organic matter, leading to its thermal cracking. Therefore, strong acidic dissolution may lead to the formation of secondary pore development zones in the reservoir, providing sufficient space for oil and gas storage [19,20]. During the epigenetic stage, due to the uplift of the Earth's crust and its proximity to the surface environment, reservoirs are often subjected to the leaching of atmospheric fresh water, resulting in the generation of secondary pores, making the rocks loose and porous [21–23]. Compaction, pressolution, cementation, dissolution, metasomatism, recrystallization, etc. all belong to the category of diagenesis [24,25]. During the process of diagenetic evolution, different diagenetic processes have varying degrees of changes in the porosity and permeability space of reservoirs, resulting in different physical properties of reservoirs in different geological periods. At present, research on marine shale gas reservoirs in the Sichuan Basin is relatively mature [26–29], and research on the diagenesis and pore structure evolution of shale reservoirs mainly focuses on marine shale [30–33]. The Upper Permian Longtan Formation shale in the Sichuan Basin is a typical representative of MCFS in China, and it has a complex evolutionary process because of the organic-inorganic diagenesis, which has a significant impact on its reservoir space [12,15,34]. For the Longtan Formation shale in the Sichuan Basin, the related research still focuses on aspects of the sedimentary environment, reservoir distribution, reservoir-forming conditions, and hydrocarbon generation potential [35–38]. The research foundation related to pore structure and diagenetic evolution is relatively weak [7,39], and the comprehensive study of diagenesis and pore structure of MCFS in the Longtan Formation of the Sichuan Basin is still blank.

In this paper, X-ray diffraction (XRD), low-pressure gas adsorption (LPGA), nuclear magnetic resonance (NMR), and field emission scanning electron microscopy (FE-SEM) were used to analyze the geochemical and mineral analysis, quantitative characterization of pore structure, and qualitative characterization of pore morphology of MCFS in the Southeastern Sichuan Basin. The diagenetic evolution sequence and nanopore system evolution of MCFS in the southeastern Sichuan Basin were systematically studied, and the influence of diagenesis on MCFS reservoirs was clarified. It is a supplement to the research content of MCFS reservoirs in the Sichuan Basin and has important reference value for the exploration of marine-continental transitional shale gas in other basins in China.

2. Geological Setting

The Sichuan Basin is rich in natural gas resources and has great exploration potential [40]. The Permian Longtan Formation in the Sichuan Basin is a typical MCFS gas series [41–43]. The Dongwu movement in the late Permian made the Sichuan Basin show a very complex structural pattern. Due to the difference in tectonic framework, the sedimentary facies in different areas of the Sichuan Basin are obviously different in the same period, from river delta facies to deep water shelf facies [44,45]. Among them, the swamp-lagoon facies sedimentary environment is widely developed in the southeastern and central parts (Figure 1) [46], and a large area of coal-bearing MCFS has been deposited. The main lithology is black shale, thin siltstone, and coal seam [47]; the OM type is mainly type III [8,42].

The research well YJ1 selected in this study is located in the south-eastern region of the Sichuan Basin. The study area is characterized by a long axis anticline extending NE-SW, mainly developing NE and NW faults [48,49].

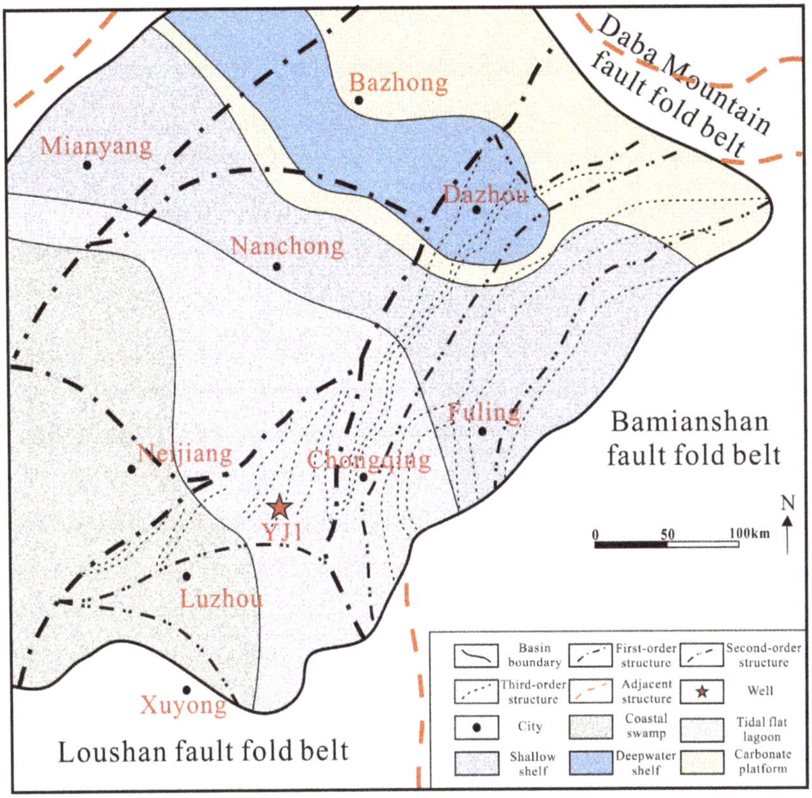

Figure 1. Sedimentary facies distribution in the Southeastern Sichuan Basin.

3. Materials and Methods

The research object of this paper is the MCFS of the third member of the Longtan Formation in the southeastern Sichuan Basin. Fresh shale samples were collected from well YJ1 (Figure 2). In order to ensure the validity of the test data, the samples are mainly taken from the fresh surface of the core sample, avoiding the calcite vein filling and other parts. The test methods adopted are mainly XRD, LPGA, NMR, and FE-SEM.

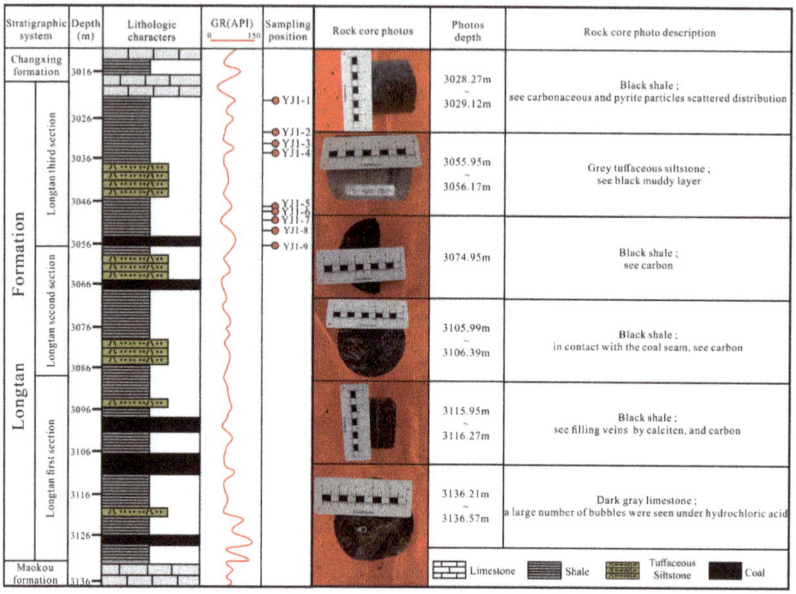

Figure 2. Comprehensive histogram of Longtan Formation MCFS, YJ1 well.

3.1. *Geochemical and Mineralogical Characteristics*

After HCl pretreatment to remove carbonates, the Longtan Formation samples were measured for total organic carbon (TOC) content using a LECO CS-230 analyzer, and the reflectance (Ro) of vitrinite was measured using a microscope photometer. The maximum, minimum, and average reflectance of the vitrinite were recorded. Grind all samples into a 200-mesh fine powder, and then use an AXS X-ray diffractometer (D8ADVANCE) to determine according to industry standard SY/T5163-2010 [50] at 20 °C and 70% humidity. The working voltage and current were 40 kV and 40 mA, respectively. Based on standard powder diffraction analysis data (International Data Center of the Federation), the mineral composition of shale is determined. Mineral content was calculated based on diffraction patterns and the RIR method.

3.2. *Quantitative Characterization of Pore Structure*

3.2.1. Low-Pressure Gas Adsorption (LPGA)

Firstly, grind and screen the shale sample to a size of 40–60 mesh. Then vacuum degassing the sample at 150 °C to remove moisture and other pollutants. Finally, perform LGPA testing on the sample. The adsorption/desorption experiments of low-pressure N_2/CO_2 gas were carried out performed at the Beijing Center for Physical and Chemical Analysis (BCPCA), which were conducted using an ASAP 2460 four-station automatic rapid specific surface area (SSA) and pore analyzer (Micromeritics, Norcross, GA, USA).

3.2.2. Nuclear Magnetic Resonance (NMR)

NMR analysis is an advanced method to analyze the pore structure and fluid distribution characteristics of shale by using NMR phenomena. Samples were saturated with water for 8 h at 25 MPa. Furthermore, NMR measurements were performed on a large sample of saturated water using an NMR analyzer to obtain the pore structure characteristics of the shale.

3.3. *Qualitative Characterization of Pore Morphology*

Field emission scanning electron microscopy (FE-SEM) and MAPS.

With extremely high image resolution, field emission scanning electron microscopy is effective in identifying many types of pores and their shapes in shale. Before scanning, the shale sample was processed into a 1 cm^2 scanning electron microscope thin slice. Quanta250FEG (FEI, Hillsboro, OR, USA) (working conditions: accelerating voltage: 20 kV, magnification: 50 to 300,000 times) was used to scan the samples at 24 °C and 35% humidity. Backscattered two-dimensional (2D) large-area scanning electron microscopy imaging can scan a series of continuous and overlapping high-resolution small images in a selected area for samples that require large-area observation. After the scanning is completed, these small images will be spliced to obtain an ultra-high-resolution, ultra-large-area 2D backscattered electron image.

The above two methods can provide visual evidence for the detailed observation of shale diagenesis and pore types.

4. Results

4.1. Organic Geochemical and Mineralogical Characteristics

4.1.1. Shale Composition

The samples are organic-rich shales, which have a TOC of 1.6%–4.15% (2.5% on average), and the TOC concentration increases with depth. Ro (vitrinite reflectance) content ranges from 0.56% to 3.05%, with an average of 2.16% average. The content of quartz is between 14.3% and 58% (26.5% on average). The carbonate mineral content ranges from 1.8% to 33.8% (15.2% on average), mainly composed of dolomite (0%–30%, 9.9% on average). The content of clay minerals varies between 11% and 59.6% (34.4% on average). There are other minerals (such as siderite) with a content lower than 5% that are not listed in the table (Table 1).

The quantitative characterization of clay minerals shows that the content of illite ranges from 3 to 80% (24.9% on average), the content of smectite is 0%–50% (14.7% on average), and the content of mixed layers of illite and smectite is between 13% and 56% (31.9% on average). The content of chlorite ranges from 5% to 50% (19% on average) (Table 1).

Table 1. Mineral composition and organic geochemical parameters of Longtan Formation MCFS samples, YJ1 well.

Sample ID	Depth (m)	Mineral Content (%)										Clay Mineral Content (%)						
		TOC	Ro	Quartz	Potash Feldspar	Albite	Calcite	Dolomite	Pyrite	Anatase	Manganite	Clay Mineral	S	I/S	I	K	C	I/S (S%)
YJ1-1	3021.5	1.62	0.56	14.3	2.8	5.5	20	0	22	3.4	2.9	29.1	50	28	8	9	5	19
YJ1-2	3030.95	1.6	1.43	16	5	0	7	2	10	3.6	4.4	52	34	40	10	0	16	23
YJ1-3	3031.6	3.04	2.23	21.6	2.7	8.3	2.2	14.8	4.9	4.8	6.4	34.3	19	35	6	11	29	25
YJ1-4	3034.68	2.18	2.74	26.3	0	10.3	0	12.3	6.5	7.7	4.3	32.6	0	35	4	11	50	6
YJ1-5	3050	2.41	2.61	34.2	4.2	9.1	3.8	30	6.2	0	1.5	11	0	37	14	29	20	20
YJ1-6	3050.7	2.37	1.29	22.3	8.9	0	0	1.8	0	4.5	2.9	59.6	29	56	3	4	8	22
YJ1-7	3052.55	1.81	2.51	24.4	4.2	2.6	0	14.7	1.8	8	3.5	40.8	0	13	80	2	5	22
YJ1-8	3054.77	4.15	3.05	58	0	7.6	3	7.6	0	2.2	3.2	18.4	0	20	45	12	23	5
YJ1-9	3056.8	3.32	2.98	21.1	1.7	7.4	11.4	6.6	0	9.9	10.1	31.8	0	23	54	8	15	23

4.1.2. Shale Composition

Based on the XRD test results and the current shale type classification method [51], the types of shale were classified (Figure 3). The results show that the samples taken in this study of YJ1 well in the southeastern Sichuan Basin are basically mixed shale, and there were also two samples that were clay shale.

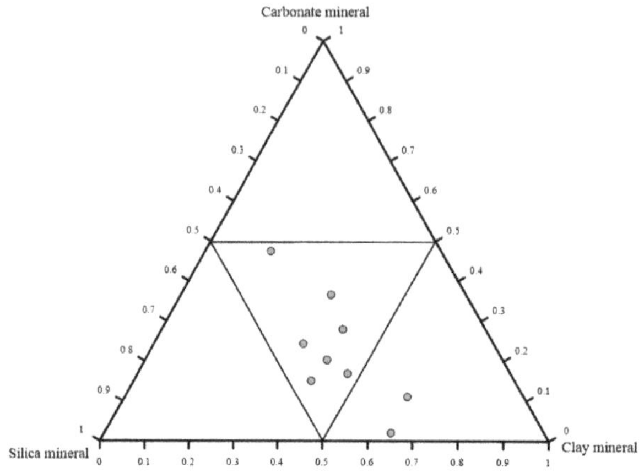

Figure 3. Division of Longtan Formation MCFS lithofacies, YJ1 well [42].

4.2. Qualitative-Quantitative Characterization of Pores

4.2.1. Qualitative Characterization of Pore Morphology

Based on the lithofacies division results of mineral components of MCFS, we use backscattered two-dimensional large-area scanning electron microscopy imaging to finely observe mixed shales and clay shales (Figure 4). Shale samples from two lithofacies contained the following three primary pore types:

1. Framework mineral pores

The framework mineral pores include intergranular pores, intragranular pores, intercrystalline pores, and dissolution pores, which are mainly developed between and inside brittle minerals. This type of pores is mainly common in mixed shale with high brittle mineral content, where dissolution pores are formed by the dissolution of soluble minerals under the action of organic acids. The dissolution pores of dolomite, calcite, and other minerals are common in mixed shale (Figure 4a,c). The intergranular pores and intragranular pores are mainly formed between or within the debris framework particles. Intergranular pores between different types of minerals can be seen in mixed shale (Figure 4c). Intercrystalline pores are typically found between various types of mineral crystals, and the intercrystalline pores of pyrite particles are the most common in mixed shale (Figure 4d).

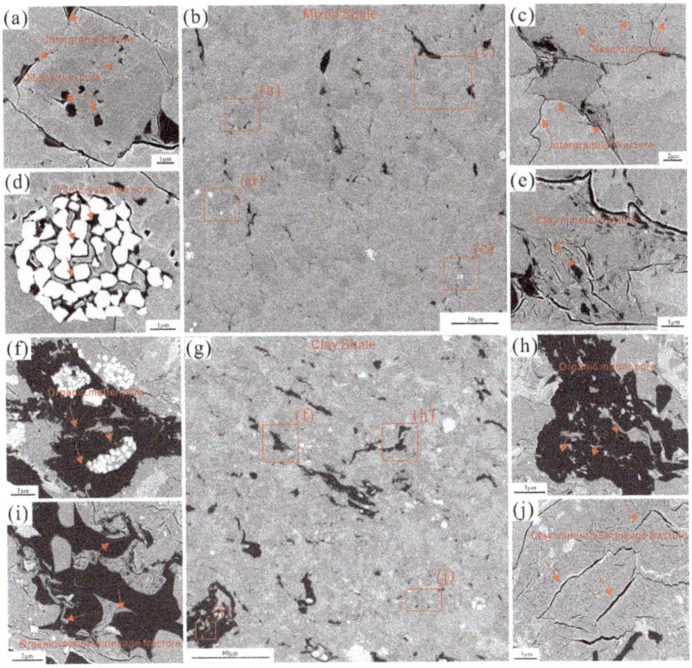

Figure 4. MAPS and FE-SEM images of different types of pores in Longtan Formation MCFS, YJ1 well. (**a**). Interanuler pore and Disslution pore; (**b**). Mixed shale, full view of Figures (a,c,d,e). (**c**). Dissolution pore and Intergranular fracture; (**d**). Intercrystalline pore; (**e**). Clay mineral fracture; (**f**). Oragnic matter pore; (**g**). Clay shale, full view of Figures (f,h,i,j); (**h**). Organic matter pore; (**i**). Organic matter shrinkage fracture; (**j**). Clay minerals Shrinkage fracture.

2. Organic matter (OM) pores

OM pores are formed by hydrocarbon generation within OM and between OM accumulations. The morphology of the pores is diverse, and various forms such as circular and serrated can be seen (Figure 4f,h). The OM pores are mainly distributed in two forms: aggregated and dispersed. The first type of OM pores is mainly developed in the whole piece of OM, which is densely distributed in a honeycomb shape and connected with each other, but with smaller pore diameters. The second type of OM pores are more scattered, have bigger diameters, and are essentially not related to one another.

3. Fracture pores

According to the observation, there are three different types of microfractures in the MCFS. The first type is fractures in skeletal minerals, usually developed within particles or at the edges of debris with good connectivity (Figure 4c). The second type is shrinkage fractures (Figure 4i), usually formed at the edges or inside of OM with a curved shape and good connectivity. The third type is the clay mineral shrinkage joint between clay minerals (Figure 4j).

4.2.2. Pore Structure and Qualitative Characterization

1. Low-pressure gas adsorption

Previous studies on shale suggest that shale pores are mainly mesopores and micropores [52–54]. The method of characterizing mesopores (2–50 nm) in shale by N_2 adsorption and micropores (0–2 nm) by CO_2 adsorption was proposed [55,56]. Adsorption isotherms are divided into five types (types I–VI), hysteresis gyrus is divided into four types (types H1–H4), and pore fractures are divided into micropores less than 2 nm in diameter), mesopores (2 nm to 50 nm), and macropores (more than 50 nm in diameter) by the International Union of Pure and Applied Chemistry (IUPAC) [56].

The results of this study show that the hysteresis loop in the N_2 adsorption-desorption curve of MCFS is H3–H4 type (Figure 5a). The pore size distribution of mesopores is between 2–4 nm and 20–40 nm. The pore size distribution of micropores is near the two peaks of 0.55 nm and 0.85 nm. The pore volume (PV) of micropores in shale ranges from 0.00186 to 0.0296 cm^3/g (0.0023 cm^3/g on average), and that of mesopores varies between 0.0196 and 0.0296 cm^3/g (0.0253 cm^3/g on average).

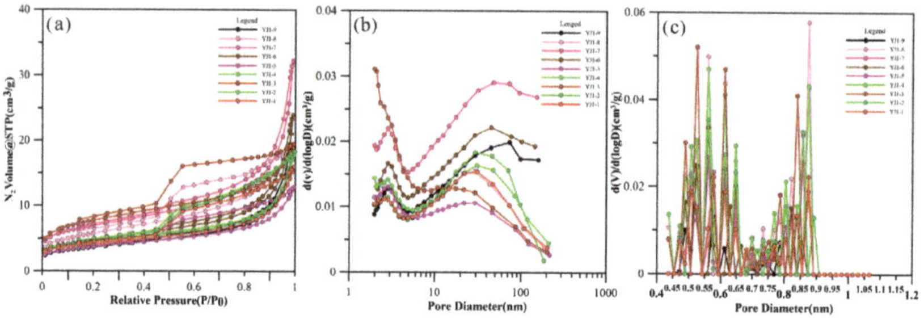

Figure 5. N_2 adsorption-desorption curve and pore size distribution characteristics of mesopores and micropores in Longtan Formation MCFS, YJ1 well. (**a**). N_2 adsorption-desorption curve; (**b**) The pore structure distribution curve of low temperature N_2 adsorption; (**c**) The pore structure distribution curve of low-temperature CO_2 adsorption.

2. Nuclear magnetic resonance

In this paper, porosity and pore structure parameters are analyzed based on the transverse relaxation time distribution of NMR [57–62]. During the testing process, it was found that there was a very high positive correlation between the porosity obtained by weighing and the porosity obtained by nuclear magnetic resonance (Figure 6). Therefore, the porosity obtained by the nuclear magnetic method can represent the porosity of the sample itself. Through nuclear magnetic resonance measurement, the porosity of the selected 9 samples ranged from 1.5% to 5.6% (3.16% on average), and the PV obtained by NMR was between 0.19 cm^3 and 0.57 cm^3 (0.36 cm^3 on average) (Table 2).

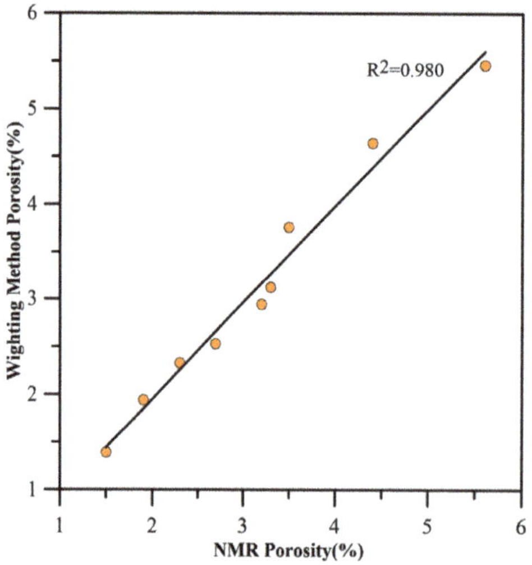

Figure 6. NMR porosity distribution characteristics of Longtan Formation MCFS samples, YJ1 well.

Table 2. Pore structure parameters of Longtan Formation MCFS samples, YJ1 well.

Sample ID	Depth (m)	Lithology	CO_2 Adsorption		N_2 Adsorption		NMR	NMR
			PVDFT (cm^3/g)	SSADFT (m^2/g)	PVBJH (cm^3/g)	SSABET (m^2/g)	PV (cm^3)	Porosity (%)
YJ1-1	3021.5	Mixed shale	0.00287	14.68	0.0226	10.48	0.25	3.2
YJ1-2	3030.95	Mixed shale	0.00186	11.43	0.0196	7.42	0.37	1.5
YJ1-3	3031.6	Mixed shale	0.00225	14.97	0.0258	8.76	0.28	2.7
YJ1-4	3034.48	Mixed shale	0.00173	16.68	0.0235	12.91	0.31	4.4
YJ1-5	3050	Mixed shale	0.0028	15.15	0.0292	12.68	0.39	3.5
YJ1-6	3050.7	Clay shale	0.00207	14.68	0.0234	7.81	0.19	2.3
YJ1-7	3052.55	Clay shale	0.00186	12.43	0.0245	11.79	0.57	1.9
YJ1-8	3054.77	Mixed shale	0.00296	19.459	0.0292	9.55	0.41	5.6
YJ1-9	3056.8	Mixed shale	0.00233	14.73	0.0296	9.31	0.43	3.3

5. Discussion

5.1. Types and Characteristics of Diagenesis

5.1.1. Compaction

Compaction is one of the most important diagenetic processes in the MCFS of the Longtan Formation. The depth of the selected well, YJ1, is greater than 3000 m. Therefore, the primary cause of the shale is the lithostatic pressure generated by the strata above it. The Longtan Formation shale was subjected to long-term subsidence after sedimentation, and the Longtan Formation shale was subjected to strong compaction. Based on the observation of FE-SEM images, it was found that clay minerals were aligned parallel to the bedding (Figure 7a), and the original kaolinite particles were fractured (Figure 7b).

5.1.2. Dissolution

The organic acids formed during the thermal evolution of OM are the main cause of shale dissolution. Organic acids can dissolve soluble minerals like calcite (Figure 7c), feldspar (Figure 7d), and dolomite (Figure 7e) to form irregularly shaped intragranular and intergranular dissolution pores. In the Longtan Formation, dissolution plays a significant role in the formation and growth of shale reservoir spaces [63]. However, the shale in the Longtan Formation is affected by volcanic activity and contains multiple layers of volcanic ash, and an important feature of volcanic ash is that its surface acids, metal salts, and adsorbed gases are highly soluble [64]. Therefore, we speculate that the acid responsible for the dissolution of shale in the Longtan Formation may come from two sources: (1) the thermal evolution of OM; and (2) the input of volcanic material.

5.1.3. Cementation

There are siliceous, carbonate, and pyrite cement filled in the pores of the shale in the Longtan Formation. According to the formation time, it can be divided into two types. The first type is early siliceous cementation, mainly occurring in the early diagenetic stage A. The cement is authigenic microcrystalline quartz formed between pores (<20 μm). The silicon comes from the dissolution of feldspar and likely from the smectite-to-illite conversion (Figure 7f). Early carbonate cementation mainly occurred in the A stage from syndiagenesis to early diagenesis [63], and the cement materials were mainly small microcrystals of calcite and iron dolomite (Figure 7h) filled in the original pores. The second type is late siliceous cementation, mainly occurring in the middle diagenetic stage A, with the cementitious material being amorphous microcrystalline quartz, often coexisting with illite (Figure 7g). The late carbonate cementation mainly occurred in the early diagenetic stage B stage [65], with the cementation mainly composed of calcite and filling some dissolution intergranular pores (Figure 7i).

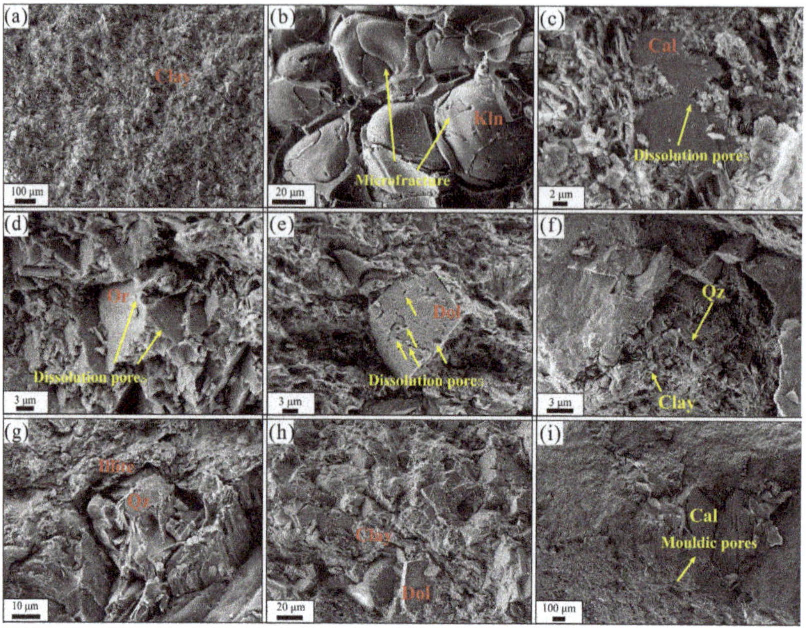

Figure 7. Characteristics of compaction, dissolution, and cementation in Longtan Formation MCFS, YJ1 well. (**a**). Compaction, clay minerals appear to be directionally distributed and have a dense structure, YJ1-1, 3021.5 m; (**b**). Compaction, microfracture appear on the surface of kaolinite particles, YJ1-1, 3021.5 m; (**c**). Dissolution, the development of dissolution pores in calcite, YJ1-6, 3050.7 m; (**d**). Dissolution, with a small amount of dissolution pores developed in granular potassium feldspar, YJ1-1, 3050.7 m; (**e**). Dissolution, the development of dissolution pores in dolomite YJ1-2, 3030.95 m; (**f**). Early siliceous cementation, the formation of authigenic microcrystalline quartz between pores, YJ1-1, 3021.5 m; (**g**). Late siliceous cementation, accompanied by amorphous microcrystalline quartz and illite, YJ1-6, 3050.7 m; (**h**). Early carbonate rock cementation, with the cementitious material being iron dolomite, YJ1-1, 3021.5 m; (**i**). Late carbonate rock cementation, with calcite as the cementitious material, YJ1-2, 3030.95 m. Kaolinite: Kln: Dolomite: Dol; Calcite: Cal; Quartz: Qz; Orthoclase: Or.

5.1.4. Clay Mineral Transformation

The clay minerals in shale transform with an increase in burial temperature. The main transformation mode of clay minerals in shale is from smectite to I/S and illite [66,67]. The organic acids generated during the thermal evolution of OM will accelerate the dissolution of potassium feldspar, promote the transformation of smectite into I/S and illite, and thereby improve the pore connectivity of the formation [68]. In the FE-SEM image, a large number of I/S in the form of flocs, filaments, and flakes can be observed, indicating a higher degree of transformation from montmorillonite to illite and I/S (Figure 8a–e). In addition, there is also kaolinite transformed from smectite in the transitional shale of the Longtan Formation, which is mostly distributed in granular form and has a relatively low content (Figure 8f).

5.1.5. Organic Matter Thermal Maturity

Longtan Formation MCFS has high TOC and Ro ranges from 0.56 to 3.05%, indicating good hydrocarbon generation potential. Under the observation of FE-SEM images, it can be found that there are pit-shaped, slit-shaped, and oval-shaped vents formed by thermal evolution in shale, and honeycomb-shaped, oval-shaped, irregular-shaped, or flaky vent groups can also be seen (Figure 8g–i). During the thermal evolution of OM, organic acids are generated, and the acidified environment promotes dissolution, forming secondary inorganic pores [69]. Therefore, the presence of dissolution pores can also be observed around the OM (Figure 8i).

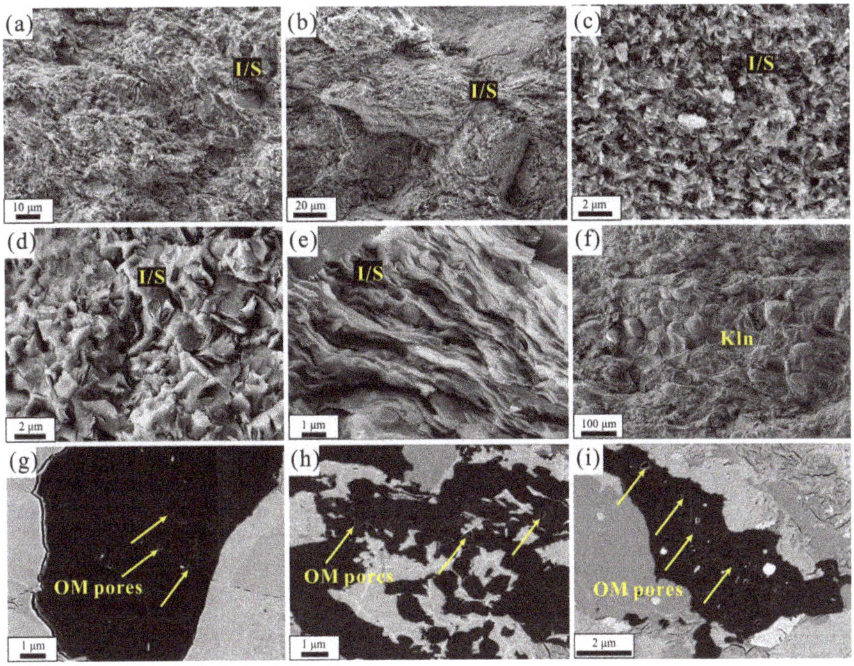

Figure 8. Characteristics of clay mineral transformation and thermal evolution of OM in MCFS, YJ1 well. (**a**,**b**). Different forms of I/S, YJ1-3, 3031.6 m; (**c**–**e**). Different forms of I/S, YJ1-4, 3034.68 m; (**f**). Kaolinite particles, YJ1-3, 3031.6 m; (**g**,**h**). Different forms of OM pores, YJ1-9, 3056.8 m; (**i**). Corrosion pore around the organic matter, YJ1-8, 3054.77 m.

5.2. Diagenetic Evolution Process

5.2.1. Diagenetic Stage

This study divides the diagenetic stages of shale in the Longtan Formation into multiple aspects, such as burial history, thermal maturity, clay mineral morphology, and porosity, relying on the Chinese oil and gas industry standard SY/T5477-2003 [14].

1. The shale porosity of the Longtan Formation varies between 1.5 and 5.6% (3.16% on average). The OM pores, clay mineral lamellar pores, and microfractures are well developed, the residual primary pores are smaller, and a small amount of dissolved pores are observed. The clay mineral lamellar pores and micro-fractures are aligned in an orientation. At low porosity, the secondary pores are mainly diagenetic pores. According to the FE-SEM images, filamentous and flaky illites and leafy chlorites can be identified. Dissolution of carbonate cement and detrital particles (e.g., feldspar) was observed in some samples, but smectite did not exist. It shows that the Longtan Formation shale has basically reached the A stage of the middle diagenetic stage.
2. In clay minerals, the content of smectite ranges between 0 and 50% (14.67% on average), I/S is 13%–56% (31.89% on average), illite ranges between 3 and 80% (24.89% on average), chlorite is 0%–29%, and kaolinite ranges from 5 to 50%. The characteristics of authigenic minerals show that smectite is widely transformed into I/S, illite, chlorite, and kaolinite. And the smectite content in the I/S ranges between 5% and 25% (22.2% on average). It shows that the Longtan Formation shale is in stage A of the middle diagenesis stage. The clay mineral assemblage is composed of mixed layers of illite-smectite + illite + chlorite and illite-smectite mixed layer + illite + chlorite + kaolinite. In addition, through the

analysis of the burial and thermal history of the Longtan Formation in the Sichuan Basin, it is found that the maximum burial depth of the Longtan Formation shale is 4000 m, and the maximum temperature reaches 140 °C–160 °C, indicating that the Longtan Formation shale is in the middle diagenetic stage B [70].

The vitrinite reflectance of the samples from the study area ranges from 0.56 to 3.05%, and the smectite content in the I/S is at least 5%. In summary, the diagenetic evolution of the Longtan Formation shale in the southeastern Sichuan Basin has reached at least the middle-late diagenetic stage (Table 2), which is consistent with previous research results on the diagenetic stage of the Upper Permian Longtan Formation shale in the surrounding areas of the Sichuan Basin [71]. The shale samples located near two layers of tuffaceous siltstone in the study interval exhibit high Ro values compared with other shale samples, which may be related to volcanic activity in the Permian. Volcanic activity can severely roast the surrounding source rocks, generating abnormally high temperatures. This accelerates the thermal evolution of hydrocarbon source rocks, rapidly reaching a high to overly mature stage, thereby improving the generation of hydrocarbons [72–74].

5.2.2. Diagenetic Evolution Sequence

Combining the diagenesis characteristics, diagenetic environment, and burial history (Figure 9), it is considered that the Longtan Formation MCFS has experienced three diagenetic evolution stages. It should be noted that the diagenetic evolution of the Longtan Formation marine continental transitional shale in the southeastern Sichuan region reached at least the middle to late diagenetic stage. However, samples from the early stages of late diagenesis cannot exclude the impact of volcanic activity. Therefore, only samples in middle diagenetic stage B and before correspond to burial histories. The three diagenetic evolution stages are as follows:

1. From the Late Permian to the Middle Triassic, the Longtan Formation MCFS experienced slow subsidence, with a burial depth of <800 m and a low formation temperature (Figure 9). Microcrystalline calcite is formed in Longtan Formation shale, and dissolution pores appear in silicate minerals such as quartz. At that time, the pores in the shale of the Longtan Formation were mainly primary mineral pores [75]. The increase in burial depth, compaction, and pressure dissolution seriously damages the shale reservoir space, leading to a significant reduction in primary pores and a clear and dense pore reduction effect on clay shale [16].
2. From the Middle Triassic to the Late Jurassic, the strata rapidly subsided and were buried deeper, beginning to generate hydrocarbons. The maximum burial depth reaches 3200~4000 m, and the maximum temperature is about 120~140 °C. The shale is rapidly consolidated (Figure 9), and the OM and kerogen begin to produce gas and acid. Numerous OM pores were created during this process. Simultaneously, plenty of organic acids cause soluble minerals (such as calcite, dolomite, feldspar, etc.) to start dissolving heavily, which favors the growth of porosity and the expansion of the matrix PV of mixed shale [76]. It was observed in the FE-SEM images that the primary intergranular pores decreased and the secondary pores increased. Smectite is converted to I/S in large quantities, and an ordered mixed-layer band is produced. At the edge of quartz, the siliceous compounds created during mineral transformation are cemented close by and create tiny secondary pores [16].
3. Since the Late Jurassic, the Longtan Formation shale has undergone a second subsidence, with a burial depth of more than 4000 m and complete consolidation of the rock (Figure 9). The primary intergranular pores essentially faded away, and the OM entered the threshold for generating gas. Clay mineral mixing layers gradually change from disordered to long-range ordered, and smectite basically disappears.

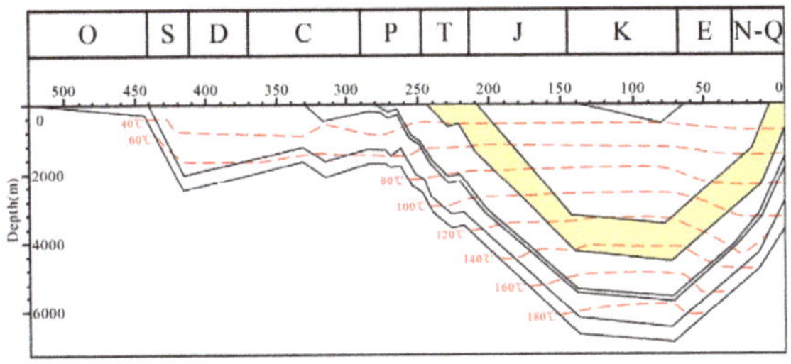

Figure 9. Burial history of Longtan Formation MCFS, YJ1 well (Modified according to [61]).

5.3. Pore Evolution

5.3.1. Relationship between Porosity and TOC, Ro, Clay Minerals

Figure 10a depicts the relationship between porosity and TOC, and there is a positive linear relationship between porosity and TOC ($R^2 = 0.459$). It shows that the enrichment of OM has a certain correlation with the formation of pores. At the same time, with the increase in maturity, the porosity of Longtan Formation shale decreases first and then increases, reaching its minimum value at Ro of 1.5% (Figure 10b). The decrease in porosity at Ro < 1.0% may be due to the blockage of pores and pore throats by hydrocarbons generated by OM, and the mechanical compaction at this time is also an important driving factor for the decrease in porosity. When the maturity is higher than 1.5%, the formation of plenty of organic pores and a rise in porosity in the Longtan Formation shale may be caused by the expelling of produced hydrocarbons from pores and pore throats. The soluble minerals in the Longtan Formation shale are dissolved by the acidic fluid generated at this stage, thus forming more dissolution pores.

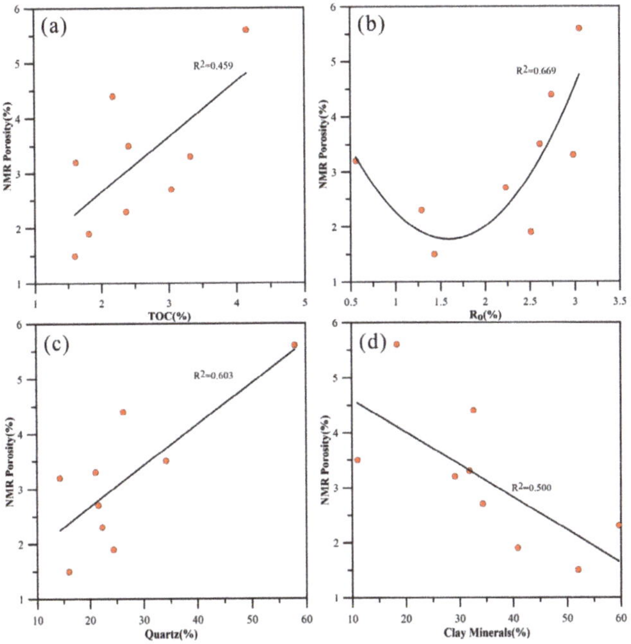

Figure 10. The relationship between porosity and TOC%, Ro and main mineral components of Longtan Formation MCFS, YJ1 well. (**a**). The Relationship between TOC and NMR Porosity; (**b**). The Relationship between Ro and NMR Porosity; (**c**). The Relationship between Quartz content and NMR Porosity; (**d**). The Relationship between Clay Minerals content and NMR Porosity.

Figure 10c,d illustrates the association between porosity and the quartz and clay mineral contents. Porosity and quartz content have a positive linear relationship (Figure 10c, $R^2 = 0.603$). This is because the rock is easier to form microfracture under the action of tectonic action, and the rigid skeleton formed by quartz particles can protect the pore space from compaction damage. The higher the quartz content, the larger the pore space and the stronger the pore compaction resistance, which is helpful in maintaining the original porosity to the greatest extent [77]. Porosity is negatively correlated with clay mineral content (Figure 10d, $R^2 = 0.501$). This is because in the process of mineral transformation, the surface PV, interlayer PV, and aggregate PV of clay mineral particles decrease, and many diagenetic intergranular pores may be formed. Consequently, porosity loss increases as clay mineral content increases [78,79].

5.3.2. Evolution Characteristics of Pore Structure

With the increase in maturity, the micropore volume of Longtan Formation shale decreases first, reaching the minimum at Ro = 1.43%, and then gradually increasing (Figure 11a). The reason for the decrease is that the hydrocarbon generated by OM fills the original pore space, while continuous pyrolysis leads to the continuous release of hydrocarbon gas, and the formation of many organic pores is the reason for the increase in PV. The evolution process of mesopore volume increases with the increase in maturity (Figure 11b). The overall reason may be due to the good connectivity of mesopores. When the generated liquid hydrocarbons are converted into gas, hydrocarbons are more easily released from the pores. Although the mechanical compaction effect is relatively strong, the presence of quartz particles plays a more important supporting role, thus ensuring that the PV of the mesoporous is in an increasing state as a whole.

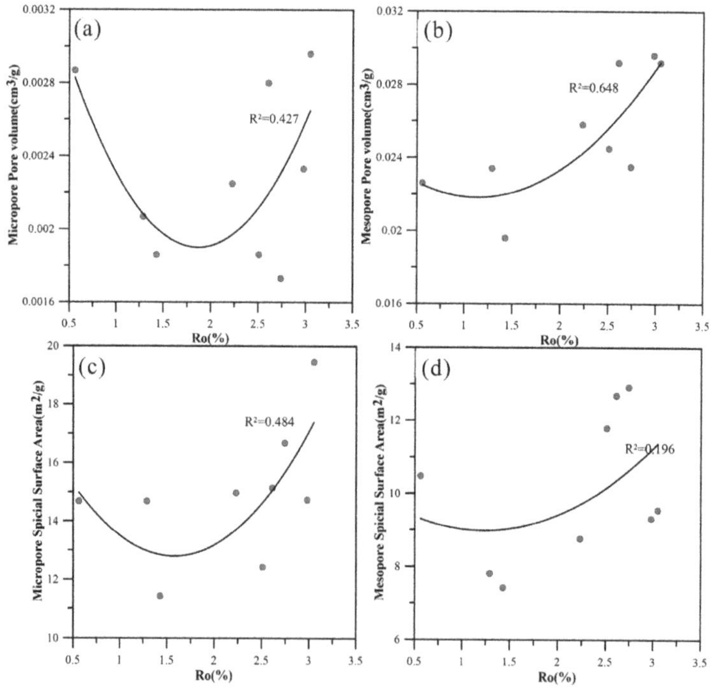

Figure 11. Evolution characteristics of PV and SSA of Longtan Formation MCFS, YJ1 well. (**a**). The Relationship between Ro and PV of Micropore; (**b**). The Relationship between Ro and PV of Mesopore; (**c**). The Relationship between Ro and SSA of Micropore; (**d**). The Relationship between Ro and SSA of Mesopore.

The SSA values of micropores and mesopores in the shale of the study area decrease first and then increase with the increase in maturity. It reaches its minimum value when the Ro ≈ 1.43% (Figure 11c,d). During the maturation process of the shale in the study area, a large amount of liquid hydrocarbons filled the pores, resulting in a decrease in SSA, and then the liquid hydrocarbons gradually converted into gas, thereby increasing the SSA of the shale micropores again. For mesopores, as the dominant mesopores, larger pores are more conducive to the circulation of liquid hydrocarbons. Therefore, the decrease in SSA is small, and on the whole, it is still in an increasing evolution state.

Due to the increase in maturity, the porosity of Longtan Formation shale reaches the minimum value at Ro = 1.43%, and then increases continuously. The volume of micropores decreases first, reaching the minimum value at Ro = 1.43%, and then increases. The volume of mesopores is increasing. The micropore SSA first decreases, reaching a minimum at Ro = 1.43%, and then increases. The mesopores continue to increase (Figure 12).

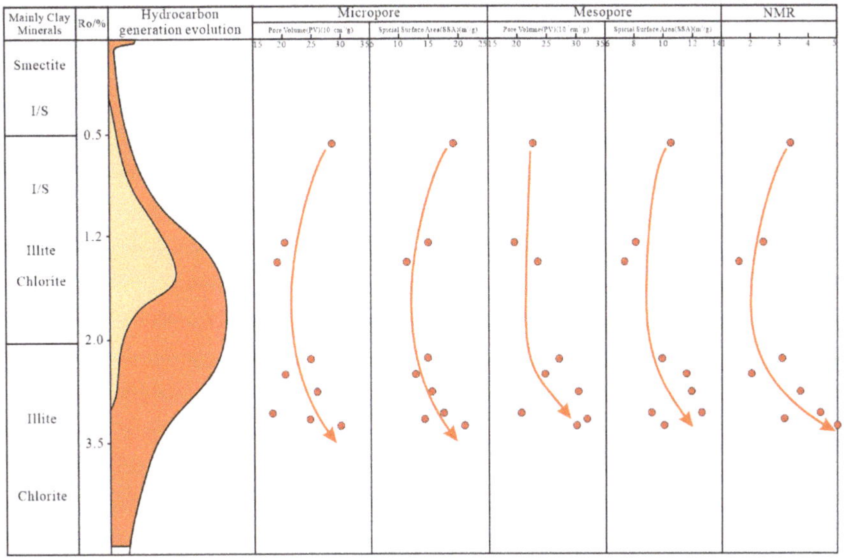

Figure 12. Pore structure evolution of Longtan Formation MCFS under diagenesis, YJ1 well.

5.4. Diagenesis-Pore Evolution Model

The evolution of pore structure in the Longtan Formation MCFS is mainly affected by two main factors: the diagenesis of minerals and the thermal evolution of OM. In different evolution stages, they have different effects on the formation and evolution of the pore system, which lead to different evolution forms (Figure 13).

1. In the immature stage (Ro < 0.5%), the Longtan Formation shale contains a significant smectite content along with a certain amount of I/S and illite. Primary pores like intragranular pores and intergranular pores dominate the pore system.
2. In the low maturity stage (0.5% < Ro < 1.3%), with the increase in maturity, OM began to enter the hydrocarbon generation stage, forming organic acids, dissolving unstable minerals (calcite and feldspar, etc.), and forming secondary matrix dissolution pores. Mixed layer I-S transforms into illite and chlorite during this phase, producing pores in dehydrated clay minerals and enlarging secondary pores [80–84].
3. After the high-over mature stage (Ro > 1.3%), plenty of OM pores are formed in the cracking gas generation stage, and the dissolution is weakened. The clay minerals in shale are mainly stable minerals such as illite and chlorite, and the pore structure of shale is almost stable.

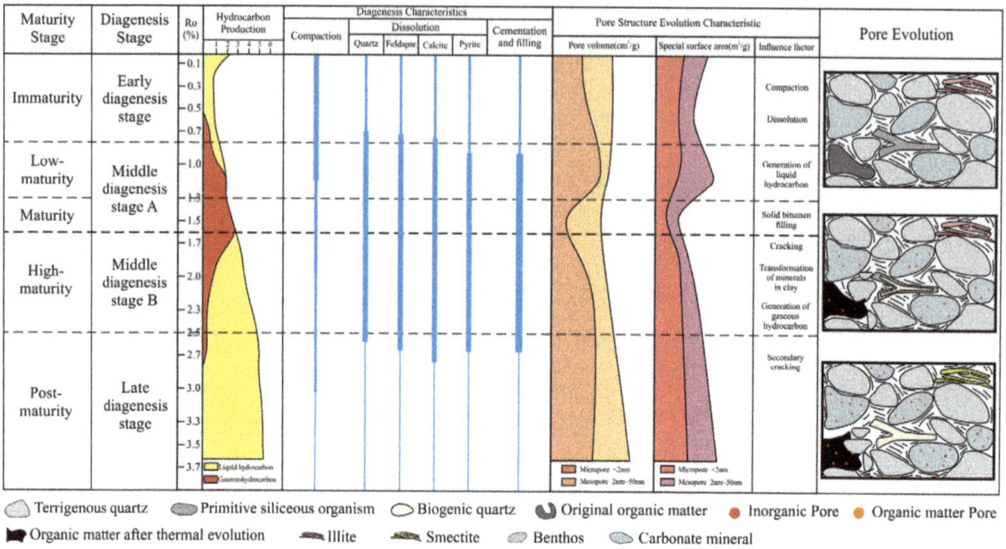

Figure 13. Diagenesis-pore evolution model of Longtan Formation MCFS, YJ1 Well (Modified according to [71]).

6. Conclusions

In this paper, XRD, FE-SEM, LPGA, and NMR were used to subdivide the diagenesis stage according to thermal maturity, clay mineral morphology, and microscopic characteristics. Combined with adsorption characteristics and image analysis, the pore structure characteristics in the process of OM evolution and hydrocarbon generation were studied. The relationship between maturity, mineral composition, PV, and SSA was discussed. The main results are as follows:

(1) The Longtan Formation shale in Southeastern Sichuan is dominated by mixed shale lithologies, and the pore system is dominated by nanoscale pores. The pore morphology of OM is different, and the pore size is small; the mineral pores are mostly irregular, and the pore size is large.

(2) The diagenetic evolution stage of the Longtan Formation MCFS enters the middle-late diagenesis stage, and each stage has corresponding evolution characteristics. In the middle diagenetic stage A, carbonate cement and clastic particles (such as feldspar) were dissolved. In the middle diagenesis stage B, smectite is widely transformed into I/S, illite, chlorite, and kaolinite. In the late diagenesis stage, the maturity varies between 0.56 and 3.05%, and the proportion of smectite in the I/S is at least 5%.

(3) When Ro < 1.43%, the porosity, PV, and SSA of micropores both decrease, while at Ro > 1.43%, the porosity, PV, and SSA of micropores increase. The PV and SSA of mesopores increased with the increase in maturity. There are two kinds of pore-reducing diagenesis: compaction and cementation, and there are two kinds of pore-increasing diagenesis: dissolution and hydrocarbon generation of OM.

(4) Maturity is an important factor controlling the evolution of pore structure. In addition, the composition of clay minerals and brittle minerals also affects the evolution of pore structure. Finally, the diagenetic evolution sequence and nano-scale pore system evolution model of Longtan Formation shale were established.

Author Contributions: Conceptualization, S.W. and B.Z.; methodology, K.M.; software, K.M. and G.C.; validation, S.W., B.Z. and K.Y.; formal analysis, P.W. and G.C.; investigation, S.W.; resources, P.W.; data curation, K.Y.; writing—original draft preparation, S.W. and B.Z.; writing—review and editing, S.W., B.Z. and C.X.; visualization, K.Y. and C.X.; supervision, B.Z.; project administration, B.Z.; funding acquisition, B.Z. All authors have read and agreed to the published version of the manuscript.

Funding: This research was funded by the National Natural Science Foundation of China (Grant No: 42272184).

Data Availability Statement: The data presented in this study are available on request from the corresponding author.

Acknowledgments: We thank the samples provided by Southwest Oil and Gas Field Branch of China National Petroleum Corporation. And I would like to thank Researcher Xu Zhengwei for his suggestions on revising the paper writing.

Conflicts of Interest: The authors declare no conflict of interest.

References

1. Nie, H.; Sun, C.; Li, P.; Jin, Z.; Liu, Q.; Bao, H.; Shen, B.; Dang, W. Carbon Isotope Evidence for Shale Gas Preservation Conditions and Large Methane Release over Million Years: A Case Study of Shale Gas Reservoirs of Wufeng and Longmaxi Formations in the Sichuan Basin. *Geosci. Front.* **2023**, *14*, 101642. [CrossRef]
2. Dong, D.; Liang, F.; Guan, Q.; Jiang, Y.; Zhou, S.; Yu, R.; Gu, Y.; Zhang, S.; Qi, L.; Liu, Y. Development Model and Identification of Evaluation Technology for Wufeng Formation–Longmaxi Formation Quality Shale Gas Reservoirs in the Sichuan Basin. *Nat. Gas Ind. B* **2023**, *10*, 165–182. [CrossRef]
3. Taghavinejad, A.; Sharifi, M.; Heidaryan, E.; Liu, K.; Ostadhassan, M. Flow Modeling in Shale Gas Reservoirs: A Comprehensive Review. *J. Nat. Gas Sci. Eng.* **2020**, *83*, 103535. [CrossRef]
4. Yang, F.; Ning, Z.; Liu, H. Fractal Characteristics of Shales from a Shale Gas Reservoir in the Sichuan Basin, China. *Fuel* **2014**, *115*, 378–384. [CrossRef]
5. MA, X.; XIE, J. The Progress and Prospects of Shale Gas Exploration and Development in Southern Sichuan Basin, SW China. *Pet. Explor. Dev. Online* **2018**, *45*, 172–182. [CrossRef]
6. Nie, H.; Li, P.; Dang, W.; Ding, J.; Sun, C.; Liu, M.; Wang, J.; Du, W.; Zhang, P.; Li, D.; et al. Enrichment Characteristics and Exploration Directions of Deep Shale Gas of Ordovician–Silurian in the Sichuan Basin and Its Surrounding Areas, China. *Pet. Explor. Dev.* **2022**, *49*, 744–757. [CrossRef]
7. Zhang, J.; Li, X.; Wei, Q.; Sun, K.; Zhang, G.; Wang, F. Characterization of Full-Sized Pore Structure and Fractal Characteristics of Marine–Continental Transitional Longtan Formation Shale of Sichuan Basin, South China. *Energy Fuels* **2017**, *31*, 10490–10504. [CrossRef]
8. Luo, Q.; Xiao, Z.; Dong, C.; Ye, X.; Li, H.; Zhang, Y.; Ma, Y.; Ma, L.; Xu, Y. The Geochemical Characteristics and Gas Potential of the Longtan Formation in the Eastern Sichuan Basin, China. *J. Pet. Sci. Eng.* **2019**, *179*, 1102–1113. [CrossRef]
9. Guo, X.; Hu, D.; Liu, R.; Wei, X.; Wei, F. Geological Conditions and Exploration Potential of Permian Marine–Continent Transitional Facies Shale Gas in the Sichuan Basin. *Nat. Gas Ind. B* **2019**, *6*, 198–204. [CrossRef]
10. Wang, X.; Mou, C.L.; Wang, Q.; Ge, X.; Chen, X.; Zhou, K.; Liang, W. Diagenesis of black shale in Longmaxi Formation, southern Sichuan Basin and its periphery. *Acta Petr. Sin.* **2015**, *36*, 1035–1047. (In Chinese with English Abstract)
11. Xie, W.; Wang, M.; Wang, H.; Ma, R.; Duan, H. Diagenesis of Shale and Its Control on Pore Structure, a Case Study from Typical Marine, Transitional and Continental Shales. *Front. Earth Sci.* **2021**, *15*, 378–394. [CrossRef]
12. Ma, K.; Zhang, B.; Wen, S.; Lin, X.; Wang, Y.; Yang, K. Quantitative Characterization and Controlling Factors of Shallow Shale Reservoir in Taiyang Anticline, Zhaotong Area, China. *Minerals* **2022**, *12*, 998. [CrossRef]
13. Zhang, B.; Wen, H.; Qing, H.; Yang, K.; Luo, Y.; Yang, H.; Wang, P.; He, L.; Xiao, W. The Influence of Depositional and Diagenetic Processes on Rock Electrical Properties: A Case Study of the Longmaxi Shale in the Sichuan Basin. *J. Pet. Sci. Eng.* **2022**, *211*, 110119. [CrossRef]
14. *SY/T 5477-2003*; The Division of Diagenetic Stages in Clastic Rocks. Petroleum Geological Exploration Professional Standardization Committee (PGEPSC): Beijing, China, 2003. (In Chinese)
15. Lin, L.; Yu, Y.; Zhai, C.; Li, Y.; Wang, Y.; Liu, G.; Guo, Y.; Gao, J. Paleogeography and shale development characteristics of the Late Permian Longtan Formation in southeastern Sichuan Basin, China. *Mari. Petro. Geo.* **2018**, *95*, 67–81. [CrossRef]
16. Xu, L.; Yang, K.; Wei, H.; Liu, L.; Jiang, Z.; Li, X.; Chen, L.; Xu, T.; Wang, X. Pore Evolution Model and Diagenetic Evolution Sequence of the Mesoproterozoic Xiamaling Shale in Zhangjiakou, Hebei. *J. Pet. Sci. Eng.* **2021**, *207*, 109115. [CrossRef]
17. Luo, J.L.; Liu, X.H.; Lin, T.; Zhang, S.; Li, B. Impact of Diagenesis and Hydrocarbon Emplacement on Sandstone ReservoirOuality of the Yanchang Formation (Upper Triassic) in the Ordos Basin. *Acta Geo. Sin.* **2006**, *5*, 664–673. (In Chinese with English Abstract)
18. Tan, X.F.; Tian, J.C.; Li, Z.B.; Xing, X.L.; Zhang, S.P. Diagenetic character and controling factor of reservoir of Kongdian Formation of Paleogene in Dongying basin. *Coal Geo. Exp.* **2010**, *38*, 27–32. (In Chinese with English Abstract)

19. Lu, S.Q. Reservoir diagenetic seouence and its influence on pore development in the paleogene sandstones in the Dongying sag of the Bohai Bay Basin. *Petro. GeologyExper.* **2008**, *10*, 456–459+466. (In Chinese with English Abstract)
20. You, G.Q.; Pan, J.H.; Liu, S.Q.; Chen, Y.J. Diagenesis and pore evolution of Paleogenes and stone reservoir in Dongying depression. *Acta Petro. ET Mine.* **2006**, *5*, 237–242. (In Chinese with English Abstract)
21. Salman, B.; Franks, S.G. Preservation of Shallow Plagioclase Dissolution Porosity During Burial: Implications for Porosity Prediction and Aluminum Mass Balance. *AAPG Bull.* **1993**, *77*, 1488–1501.
22. Zhang, Z.H.; Hu, W.X.; Zeng, J.H.; Yu, B.S.; Lu, X.C.; Jia, H.Y. Study of Fluid-Rock Interactions in Eogene Formationin Dongying Depression, Bohai Gulf Basin. *Acta Sediment. Sin.* **2000**, *12*, 560–566. (In Chinese with English Abstract)
23. Zhang, W.C.; Li, H.; Li, H.J.; Meng, Y.L.; Yang, F.B. Genesis and distribution of secondary porosity in the deep horizon of Gaoliu area, Nanpu Sag. *Petro. Explo. Dev.* **2008**, *6*, 308–312. (In Chinese with English Abstract) [CrossRef]
24. Yuan, G.; Cao, Y.; Gluyas, J.; Li, X.; Xi, K.; Wang, Y.; Jia, Z.; Sun, P.; Oxtoby, N.H. Feldspar dissolution, authigenic clays, and quartz cements in open and closed sandstone geochemical systems during diagenesis: Typical examples from two sags in Bohai Bay Basin. *East China AAPG Bull.* **2015**, *11*, 2121–2154. [CrossRef]
25. Rahman, M.J.J.; Worden, R.H. Diagenesis and its impact on the reservoir quality of Miocene sandstones (Surma Group) from the Bengal Basin, Bangladesh. *Mar. Pet. Geol.* **2016**, *11*, 898–915. [CrossRef]
26. Guo, X.; Teng, G.; Wei, X.; Yu, L.; Lu, X.; Sun, L.; Wei, F. Occurrence mechanism and exploration potential of deep marine shale gas in Sichuan Basin. *Acta Petro. Sin.* **2022**, *43*, 453–468. (In Chinese with English Abstract)
27. Zou, C.; Dong, D.; Wang, Y.; Li, X.; Huang, J.; Wang, S.; Guan, Q.; Zhang, C.; Wang, H.; Liu, H.; et al. Shale Gas in China: Characteristics, Challenges and Prospects (II). *Pet. Explor. Dev.* **2016**, *43*, 182–196. [CrossRef]
28. Long, S.; Feng, D.; Li, F.; Du, W. Prospect Analysis of the Deep Marine Shale Gas Exploration and Development in the Sichuan Basin, China. *J. Nat. Gas Geosci.* **2018**, *3*, 181–189. [CrossRef]
29. Jiang, S.; Peng, Y.; Gao, B.; Zhang, J.; Cai, D.; Xue, G.; Bao, S.; Xu, Z.; Tang, X.; Dahdah, N. Geology and Shale Gas Resource Potentials in the Sichuan Basin, China. *Energy Explor. Exploit.* **2016**, *34*, 689–710. [CrossRef]
30. Zhang, L.; He, X.; Li, X.; Li, K.; He, J.; Zhang, Z.; Guo, J.; Chen, Y.; Liu, W. Shale Gas Exploration and Development in the Sichuan Basin: Progress, Challenge and Countermeasures. *Nat. Gas Ind. B* **2022**, *9*, 176–186. [CrossRef]
31. Zhai, G.; Wang, Y.; Zhou, Z.; Liu, G.; Yang, Y.; Li, J. "Source-Diagenesis-Accumulation" Enrichment and Accumulation Regularity of Marine Shale Gas in Southern China. *China Geol.* **2018**, *1*, 319–330. [CrossRef]
32. Xiao, H.; Jianfa, W.; Rui, Y.; Shengxian, Z.; Xiaojin, Z.; Dongjun, Z.; Deliang, Z.; Chengxu, Z. Accumulation Conditions and Key Exploration and Development Technologies of Marine Shale Gas Field in Changning-Weiyuan Block, Sichuan Basin. *Acta Pet. Sin.* **2021**, *42*, 259.
33. Wang, Y.; Zhai, G.; Liu, G.; Shi, W.; Lu, Y.; Li, J.; Zhang, Y. Geological Characteristics of Shale Gas in Different Strata of Marine Facies in South China. *J. Earth Sci.* **2021**, *32*, 725–741. [CrossRef]
34. Yang, C.; Xiong, Y.; Zhang, J.; Liu, Y.; Chen, C. Comprehensive Understanding of OM-Hosted Pores in Transitional Shale: A Case Study of Permian Longtan Shale in South China Based on Organic Petrographic Analysis, Gas Adsorption, and X-Ray Diffraction Measurements. *Energy Fuels* **2019**, *33*, 8055–8064. [CrossRef]
35. Zhang, J.; Shi, M.; Wang, D.; Tong, Z.; Hou, X.; Niu, J.; Li, X.; Li, Z.; Zhang, P.; Huang, Y. Fields and Directions for Shale Gas Exploration in China. *Nat. Gas Ind. B* **2022**, *9*, 20–32. [CrossRef]
36. Xiao, W.; Zhang, B.; Yao, Y.; Wang, Y.; Yang, H.; Yang, K. Lithofacies and sedimentary environment of shale of Permian Longtan Formation in eastern Sichuan Basin. *Lith. Reser.* **2022**, *34*, 152–162, (In Chinese with English Abstract).
37. Wen, S.; Zhang, B.; Yao, Y.; Ma, K.; Wang, Y.; Yang, K. Pyrite morphology in shale of Permian Wujiaping Formation in eastern Sichuan Basin and its indicative significance to oceanic anoxic events. *Lith. Reser.* **2023**, *35*, 1–10. (In Chinese with English Abstract)
38. Xiao, W.; Zhang, B.; Yang, K.; Wang, Y.; Wen, S.; Ma, K.; Cao, G. Geochemical Characteristics of the Upper Permian Longtan Formation from Northeastern Sichuan Basin: Implications for the Depositional Environment and Organic Matter Enrichment. *Acta Geol. Sin.* **2023**, *97*, 1196–1213. [CrossRef]
39. Yang, C.; Zhang, J.; Tang, X.; Ding, J.; Zhao, Q.; Dang, W.; Chen, H.; Su, Y.; Li, B.; Lu, D. Comparative Study on Micro-Pore Structure of Marine, Terrestrial, and Transitional Shales in Key Areas, China. *Int. J. Coal Geol.* **2017**, *171*, 76–92. [CrossRef]
40. Dong, D.; Shi, Z.; Guan, Q.; Jiang, S.; Zhang, M.; Zhang, C.; Wang, S.; Sun, S.; Yu, R.; Liu, D.; et al. Progress, challenges and prospects of shale gas exploration in the Wufeng–Longmaxi reservoirs in the Sichuan Basin. *Nat. Gas Ind.* **2018**, *38*, 67–76. (In Chinese with English Abstract) [CrossRef]
41. Cao, Q.; Ye, X.; Liu, Y.; Wang, P.; Jiang, K. Effect of Different Lithological Assemblages on Shale Reservoir Properties in the Permian Longtan Formation, Southeastern Sichuan Basin: Case Study of Well X1. *PLoS ONE* **2022**, *17*, e0271024. [CrossRef]
42. Yang, K.; Zhang, B.; Yao, Y.; Yang, H.; Zhang, H.; Xiao, W.; Wang, Y. Organic Matter Accumulation Mechanism and Characteristics in Marine-Continental Transitional Shale: A Case Study of the Upper Permian Longtan Formation from the Well F5 in Sichuan Basin, China. *J. Pet. Sci. Eng.* **2022**, *208*, 109604. [CrossRef]
43. Zhao, P.R.; Gao, B.; Guo, Z.F.; Wei, Z.H. Exploration Potential of Marine-Continental Transitional and Deep-Water Shelf Shale Gas in the Upper Permian, Sichuan Basin. *Geol. Exp.* **2020**, *42*, 335–344.
44. Chen, Y.; Shi, X.; Xu, H.; Cao, Q.; Pei, X.; Wu, W.; Wang, L.; Yang, X. Study on the Sedimentary Environments and Its Implications of Shale Reservoirs for Permian Longtan Formation in the Southeast Sichuan Basin. *Minerals* **2023**, *13*, 689. [CrossRef]

45. Yang, H.; Lin, L.; Chen, L.; Yu, Y.; Li, D.; Tian, J.; Zhou, W.; He, J. Characteristics of Mineralogy, Lithofacies of Fine-Grained Sediments and Their Relationship with Sedimentary Environment: Example from the Upper Permian Longtan Formation in the Sichuan Basin. *Energies* **2021**, *14*, 3662. [CrossRef]
46. Cao, Q.; Liu, G.; Zhang, C.; Pan, W. Sedimentary environment and its controlling on source rocks during late Permian in Sichuan Basin. Petr. *Geo. Exper.* **2013**, *35*, 36–41. (In Chinese with English Abstract)
47. He, Z.; Nie, H.; Li, S.; Liu, G.; Ding, J.; Bian, R.; Lu, Z. Differential occurence of shale gas in the Permian Longtan Formation of Upper Yangtze region constrained by plate tectonics in the Tethyan domain. *Oil Gas Geo.* **2021**, *42*, 1–15. (In Chinese with English Abstract)
48. Guo, T. Key geological issues and main controls on accumulation and enrichment of Chinese shale gas. *Petr. Expl. Dev.* **2016**, *43*, 317–326. (In Chinese with English Abstract) [CrossRef]
49. Wang, E.; Guo, T.; Liu, B.; Li, M.; Xiong, L.; Dong, X.; Zhang, N.; Wang, T. Lithofacies and Pore Features of Marine-Continental Transitional Shale and Gas Enrichment Conditions of Favorable Lithofacies: A Case Study of Permian Longtan Formation in the Lintanchang Area, Southeast of Sichuan Basin, SW China. *Pet. Explor. Dev.* **2022**, *49*, 1310–1322. [CrossRef]
50. *SY/T 5163-2010*; Analysis Method for Clay Minerals and Ordinary Ono-Clay Minerals in Sedimentary Rocks by the X-ray Diffraction. Petroleum Geological Exploration Professional Standardization Committee (PGEPSC): Beijing, China, 2010. (In Chinese)
51. He, Y.; Tang, X.; Shan, Y.; Liu, G.; Xie, H.; Ma, Z. Lithofacies division and comparison and characteristics of Longtan Formation shale in typical areas of Sichuan Basin and its surrounding. *Natu. Gas Geos.* **2021**, *32*, 174–190. (In Chinese with English Abstract)
52. Zhang, J.; Li, X.; Zhang, G.; Zou, X.; Wang, F.; Tang, Y. Microstructural Investigation of Different Nanopore Types in Marine-Continental Transitional Shales: Examples from the Longtan Formation in Southern Sichuan Basin, South China. *Mar. Pet. Geol.* **2019**, *110*, 912–927. [CrossRef]
53. Wei, Z.; Wang, Y.; Wang, G.; Sun, Z.; Xu, L. Pore Characterization of Organic-Rich Late Permian Da-Long Formation Shale in the Sichuan Basin, Southwestern China. *Fuel* **2018**, *211*, 507–516. [CrossRef]
54. Li, J.; Wang, X.; Zhou, K.; Wang, Y.; Li, N.; Wu, Y.; Wang, M. Characteristics of ultra-deep shale reservoir of marine-continental transitional facies: A case study of lower member of Upper Permian Longtan Formation in well Y4, Puguang Gas Field, northeastern Sichuan Basin. *Petr. Geo. Exper.* **2022**, *44*, 71–84. (In Chinese with English Abstract)
55. Loucks, R.G.; Reed, R.M.; Ruppel, S.C.; Hammes, U. Spectrum of Pore Types and Networks in Mudrocks and a Descriptive Classification for Matrix-Related Mudrock Pores. *AAPG Bull.* **2012**, *96*, 1071–1098. [CrossRef]
56. Thommes, M.; Kaneko, K.; Neimark, A.V.; Olivier, J.P.; Rodriguez-Reinoso, F.; Rouquerol, J.; Sing, K.S.W. Physisorption of Gases, with Special Reference to the Evaluation of Surface Area and Pore Size Distribution (IUPAC Technical Report). *Pure Appl. Chem.* **2015**, *87*, 1051–1069. [CrossRef]
57. Kfeinberg, R.L.; Straley, C.; Kenyon, W.E.; Akkurt, R.; Farooqui, S.A. *Nuclear Magnetic Resonance of Rocks: T1 vs. T2*; OnePetro: Richardson, TX, USA, 1993.
58. Yao, Y.; Liu, J.; Liu, D.; Chen, J.; Pan, Z. A New Application of NMR in Characterization of Multiphase Methane and Adsorption Capacity of Shale. *Int. J. Coal Geol.* **2019**, *201*, 76–85. [CrossRef]
59. Sun, J.; Chen, J.; Yang, Z.; Liu, X.; Liu, Y. Experimental Study of the NMR Characteristics of Shale Reservoir Rock. *Sci. Tech. Rev.* **2012**, *30*, 25–30. (In Chinese with English Abstract)
60. Tan, M.; Mao, K.; Song, X.; Yang, X.; Xu, J. NMR Petrophysical Interpretation Method of Gas Shale Based on Core NMR Experiment. *J. Pet. Sci. Eng.* **2015**, *136*, 100–111. [CrossRef]
61. Xu, H.; Tang, D.; Zhao, J.; Li, S. A Precise Measurement Method for Shale Porosity with Low-Field Nuclear Magnetic Resonance: A Case Study of the Carboniferous–Permian Strata in the Linxing Area, Eastern Ordos Basin, China. *Fuel* **2015**, *143*, 47–54. [CrossRef]
62. Huang, J.; Xu, K.; Guo, S.; Guo, H. Comprehensive Study on Pore Structures of Shale Reservoirs Based on SEM, NMR and X-CT. *Geoscience* **2015**, *29*, 198–205. (In Chinese with English Abstract)
63. Zhao, W.; Yang, L.; Jiang, Z.; He, W.; Huang, L.; Chang, Q.; Tang, X.; Ye, H. Geneses of Multi-Stage Carbonate Minerals and Their Control on Reservoir Physical Properties of Dolomitic Shales. *Mar. Pet. Geol.* **2023**, *153*, 106216. [CrossRef]
64. Zhao, K.; Du, X.; Lu, Y.; Hao, F.; Liu, Z.; Jia, J. Is volcanic ash responsible for the enrichment of organic carbon in shales? Quantitative characterization of organic-rich shale at the Ordovician-Silurian transition. *Bulletin* **2021**, *133*, 837–848. [CrossRef]
65. Zhou, X.; Li, X.; Guo, W.; Zhang, X.; Liang, P.; Yu, J. Characteristics, Formation Mechanism and Influence on Physical Properties of Carbonate Minerals in Shale Reservoirs of Wufeng-Longmaxi Formations, Sichuan Basin, China. *J. Nat. Gas Geosci.* **2022**, *7*, 133–146. [CrossRef]
66. Aldega, L.; Carminati, E.; Scharf, A.; Mattern, F. Thermal maturity of the Hawasina units and origin of the Batinah Mélange (Oman Mountains): Insights from clay minerals. *Mar. Petro. Geo.* **2021**, *133*, 105316. [CrossRef]
67. Aldega, L.; Corrado, S.; Dipaolo, L.; Somma, R.; Maniscalco, R.; Balestrieri, M.L. Shallow burial and exhumation of the Peloritani Mts. (NE Sicily, Italy): Insight from paleo-thermal and structural indicators. *Geo. Soci. of Ame. Bull.* **2011**, *123*, 132–149. [CrossRef]
68. Wang, G.; Jin, Z.; Zhang, Q.; Zhu, R.; Tang, X.; Liu, K.; Dong, L. Effects of Clay Minerals and Organic Matter on Pore Evolution of the Early Mature Lacustrine Shale in the Ordos Basin, China. *J. Asian Earth Sci.* **2023**, *246*, 105516. [CrossRef]
69. Li, J.; Ma, Y.; Huang, K.; Zhang, Y.; Wang, W.; Liu, J.; Li, Z.; Lu, S. Quantitative Characterization of Organic Acid Generation, Decarboxylation, and Dissolution in a Shale Reservoir and the Corresponding Applications—A Case Study of the Bohai Bay Basin. *Fuel* **2018**, *214*, 538–545. [CrossRef]

70. Su, A.; Chen, H.; Feng, Y.; Zhao, J.; Nguyen, A.D. Multistage Fracturing History in the Paleocene Lacustrine Shale Oil Reservoirs of the Subei Basin, Eastern China. *Mar. Pet. Geol.* **2022**, *144*, 105835. [CrossRef]
71. Ma, X.; Guo, S. Study on Pore Evolution and Diagenesis Division of a Permian Longtan Transitional Shale in Southwest Guizhou, China. *Energy Sci. Eng.* **2021**, *9*, 58–79. [CrossRef]
72. Galushkin, Y.I. Thermal effects of igneous intrusions on maturity of organic matter: A possible mechanism of intrusion. *Org. Geochem.* **1997**, *5*, 645–658. [CrossRef]
73. Olsson, I. Regional burial heating vs. *local magmatic heat influence of the Röstånga area, Scania, southern Sweden GFF* **1999**, *121*, 209–214.
74. Stagpoole, V.; Funnell, R. Arc magmatism and hydrocarbon generation in the northern Taranaki Basin, New Zealand. *Pet. Geosci.* **2001**, *7*, 255–267.
75. Wang, R.; Hu, Z.; Bao, H.; Wu, J.; Du, W.; Wang, P.; Peng, Z.; Lu, T. Diagenetic evolution of key minerals and its controls on reservoir quality of Upper Ordovician Wufeng-Lower Silurian Longmaxi shale of Sichuan Basin. *Petr. Geo. Exper.* **2021**, *43*, 996–1005. (In Chinese with English Abstract)
76. Chen, X.; Qu, X.; Xu, S.; Wang, W.; Li, S.; He, H.; Liu, Y. Dissolution Pores in Shale and Their Influence on Reservoir Quality in Damintun Depression, Bohai Bay Basin, East China: Insights from SEM Images, N2 Adsorption and Fluid-Rock Interaction Experiments. *Mar. Pet. Geol.* **2020**, *117*, 104394. [CrossRef]
77. Ismail, I.M.K.; Pfeifer, P. Fractal Analysis and Surface Roughness of Nonporous Carbon Fibers and Carbon Blacks. *Langmuir* **1994**, *10*, 1532–1538. [CrossRef]
78. Ross, D.J.K.; Marc Bustin, R. The Importance of Shale Composition and Pore Structure upon Gas Storage Potential of Shale Gas Reservoirs. *Mar. Pet. Geol.* **2009**, *26*, 916–927. [CrossRef]
79. Ma, Y.; Zhong, N.; Li, D.; Pan, Z.; Cheng, L.; Liu, K. Organic Matter/Clay Mineral Intergranular Pores in the Lower Cambrian Lujiaping Shale in the North-Eastern Part of the Upper Yangtze Area, China: A Possible Microscopic Mechanism for Gas Preservation. *Int. J. Coal Geol.* **2015**, *137*, 38–54. [CrossRef]
80. Liu, B.; Bai, L.; Chi, Y.; Jia, R.; Fu, X.; Yang, L. Geochemical Characterization and Quantitative Evaluation of Shale Oil Reservoir by Two-Dimensional Nuclear Magnetic Resonance and Quantitative Grain Fluorescence on Extract: A Case Study from the Qingshankou Formation in Southern Songliao Basin, Northeast China. *Mar. Pet. Geol.* **2019**, *109*, 561–573.
81. Guo, X.; Qin, Z.; Yang, R.; Dong, T.; He, S.; Hao, F.; Yi, J.; Shu, Z.; Bao, H.; Liu, K. Comparison of Pore Systems of Clay-Rich and Silica-Rich Gas Shales in the Lower Silurian Longmaxi Formation from the Jiaoshiba Area in the Eastern Sichuan Basin, China. *Mar. Pet. Geol.* **2019**, *101*, 265–280. [CrossRef]
82. Liu, D.; Li, Z.; Jiang, Z.; Zhang, C.; Zhang, Z.; Wang, J.; Yang, D.; Song, Y.; Luo, Q. Impact of Laminae on Pore Structures of Lacustrine Shales in the Southern Songliao Basin, NE China. *J. Asian Earth Sci.* **2019**, *182*, 103935. [CrossRef]
83. Xu, H.; Zhou, W.; Hu, Q.; Yi, T.; Ke, J.; Zhao, A.; Lei, Z.; Yu, Y. Quartz Types, Silica Sources and Their Implications for Porosity Evolution and Rock Mechanics in the Paleozoic Longmaxi Formation Shale, Sichuan Basin. *Mar. Pet. Geol.* **2021**, *128*, 105036. [CrossRef]
84. Xu, J.; Wu, S.; Liu, J.; Yuan, Y.; Cui, J.; Su, L.; Jiang, X.; Wang, J. New Insights into Controlling Factors of Pore Evolution in Organic-Rich Shale. *Energy Fuels* **2021**, *35*, 4858–4873. [CrossRef]

minerals

MDPI

Article

Lithofacies Characteristics and Sweet Spot Distribution of Lacustrine Shale Oil Reservoirs: A Case Study of the Second Member of the Kongdian Formation in the Cangdong Sag, Bohai Bay Basin

Yuan Zhan [1,2,*], Chengyan Lin [1,2,*], Cunfei Ma [1], Wenzhong Han [3], Pengjie Ma [4] and Guiang Li [1,2]

1 School of Geosciences, China University of Petroleum (East China), Qingdao 266580, China
2 National Key Laboratory of Deep Oil and Gas, China University of Petroleum (East China), Qingdao 266580, China
3 Research Institute of Exploration and Development, Dagang Oilfield Company, PetroChina, Tianjin 300280, China
4 Sanya Offshore Oil & Gas Research Institute, Northeast Petroleum University, Sanya 572025, China
* Correspondence: b18010020@s.upc.edu.cn (Y.Z.); lincy@upc.edu.cn (C.L.)

Abstract: In contrast to marine shale oil reservoirs, lacustrine shale exhibits rapid lithofacies changes and strong mineral compositional heterogeneity, posing new challenges for the evaluation and distribution prediction of shale oil sweet spots. The oiliness, reservoir properties, oil fluidity, and fracability of different lithofacies were analyzed using emission-scanning electron microscopy (FE-SEM) observation, low-pressure nitrogen physisorption (LNP) analysis, mercury intrusion porosimetry (MIP), nuclear magnetic resonance (NMR), and triaxial compression testing. Based on the mineral composition obtained from X-ray diffraction (XRD) analysis, total organic carbon (TOC) content, and sedimentary structure, four lithofacies were classified, which are organic-rich laminated calcareous shale (LC), organic-rich laminated siliceous shale (LS), organic-rich laminated mixed shale (LM), and organic-poor massive calcareous shale (MC). Considering the factors of oiliness, reservoir properties, oil fluidity, and fracability, the LC and LS lithofacies were determined as being high-quality sweet spots (type I). Within the stratigraphic sequence divided by GR-INPEFA curves, multi-resolution graph-based clustering (MRGC) analysis of sensitive well logs was used to classify the lithofacies, after which the distribution of sweet spots was predicted. The results reveal that the sweet spots exhibit regular changes in their vertical distribution and a ring-like pattern in their planar distribution, influenced by variations in the sedimentary environment. This finding can offer valuable guidance for the future exploitation of shale oil in the Guandong region.

Keywords: lacustrine shale; shale oil; lithofacies; sweet spot; Kongdian Formation; Cangdong Sag

1. Introduction

With the great success of the exploration and exploitation of marine shale reservoirs in North America, shale oil and gas have become an important target for exploitation globally, following conventional reservoirs [1,2]. After exploration and exploitation in recent years, Bohai Bay Basin in the east and the Ordos Basin and Jimsar Basin in the west of China have shown great shale oil and gas potential in the Miocene continental strata [3–5]. Differently from the large set of fine-grained deposits and relatively weak heterogeneity of marine shale reservoirs in North America, such as Barnett and Marcellus shale, the water body of the continental lacustrine basin changes quickly, and the lithofacies transition of shale reservoir is rapid, which both pose great challenges to the evaluation and prediction of shale oil sweet spots [6–8]. Therefore, it is becoming increasingly necessary to understand the heterogeneity of continental shale oil reservoirs and its influence on the comprehensive evaluation of sweet spots.

Due to the extremely poor physical properties of shale reservoirs and their special reservoiring characteristics of self-generation and self-storage, the fracturing engineering, oil generation potential of the source rock, and oil storage space should be considered in sweet spot evaluation [9–13]. Therefore, it is necessary to comprehensively evaluate shale oil reservoirs on the basis of four aspects: oil generation, reservoir physical properties, fluidity, and fracturing properties. Many scholars have conducted research on shale oil reservoir sweet spots based on factors such as mineral composition, organic geochemical characteristics, physical properties, and rock mechanics, predicting the distribution of shale oil sweet spots by integrating these parameters [14–16]. Among these factors, pore structure influences reservoir space and reservoir permeability. However, the study of pore characteristics is affected by the varying resolutions of the different measurement methods. The practical research challenge lies in how to integrate multiple approaches for comprehensive analysis. There is a wide variety of shale oil reservoir sweet spot classification schemes, and their applicability to reservoirs formed under different environments is an important issue. In sweet spot classification methods based on the combination of relevant parameters, reservoir genesis has not been effectively coupled with sweet spots [17]. Lithofacies, as rock composition units influenced by the combined effects of tectonic, sedimentary, and diagenetic processes, form the foundation for sweet spot evaluation and can serve as a guide for predicting sweet spot distribution [18–23]. Using mineral composition, total organic carbon (TOC) content, and sedimentary structures to classify shale lithofacies is a mature method [24–26]. The lithofacies prediction at both vertical and horizontal scales was accomplished by establishing the correlation between core data with well logs and seismic data [27,28]. Moreover, the lithofacies of fine-grained reservoirs change rapidly, and it is difficult for the resolution of traditional geophysical means to meet the requirements for fine prediction of sweet spots. Due to the variety of shale lithofacies and the less distinct electrical signal differences among them compared to conventional reservoirs, there are certain limitations encountered when attempting to accurately predict lithofacies based on well logs. In recent years, the trend of multidisciplinary integration has promoted the application of intelligent methods in the petroleum industry [29,30]. Based on well logs, machine learning techniques leverage the computational power of computers to rapidly and efficiently identify deep-level differences between different lithofacies. The utilization of intelligent methods for high-precision and efficient lithofacies clustering and prediction stands as a pivotal direction for the future development of shale reservoir research [31].

The second member of the Paleocene Kongdian shale formation in the Cangdong Sag of the Bohai Bay Basin exhibits a wide distribution, considerable thickness, and high organic matter content, indicating significant potential for shale oil exploration and exploitation. The initial deployment of horizontal wells in the upper 40–60 m sweet spot objective interval of the Ek_2 formation has yielded promising results. However, the controlling mechanism of lithofacies on sweet spots remains unclear, necessitating the use of lithofacies distribution to predict sweet spots in other intervals. Firstly, in the sequence analyzed according to the GR-INPEFA curve, the lithofacies were classified based on core observation, thin section observation, XRD analysis, scanning electron microscopy (SEM) observation, and other methods. All the core data were sourced from the coring wells G-A, G-B, and G-C. Rock pyrolysis experiments, low-pressure nitrogen physisorption (LNP) analysis, mercury intrusion porosimetry (MIP), nuclear magnetic resonance (NMR), and triaxial compression testing were carried out to study the resource potential, reservoir property potential, and engineering development potential of different lithofacies. By utilizing high-precision lithofacies identification and prediction results at both vertical and horizontal scales, sweet spot prediction was accomplished. The primary objectives of this study are (1) to conduct a comprehensive sweet spot evaluation based on four aspects—oil generation properties, reservoir physical properties, shale oil fluidity, and engineering brittleness—of different lithofacies and (2) to indicate the distribution of sweet spot areas through high-precision lithofacies identification and prediction at both vertical and horizontal scales, thereby providing guidance for subsequent shale oil development.

2. Geological Setting

The Huanghua Depression is located in the central part of the Bohai Bay Basin and is bounded by the Cangxian Uplift in the north and the Chengning Uplift in the south (Figure 1a). The Cangdong Sag is a typical complex fault basin in the Huanghua Depression. It starts from the Kongdian Uplift in the south, ends at the Dongguang Uplift in the north, and is bounded by Cangxian Uplift and Xuhei Uplift in the east and west, with Cangdong Fault and Xuxi Fault, respectively. The total area of the Cangdong Sag exceeds 4000 km^2 [32,33]. It consists of the Nanpi Slope in the south, Shenvsi Faulted Nose Belt in the west, Kongxi Slope in the north, Kongdian Structural Belt, and Kongdong Slope in the east (Figure 1b).

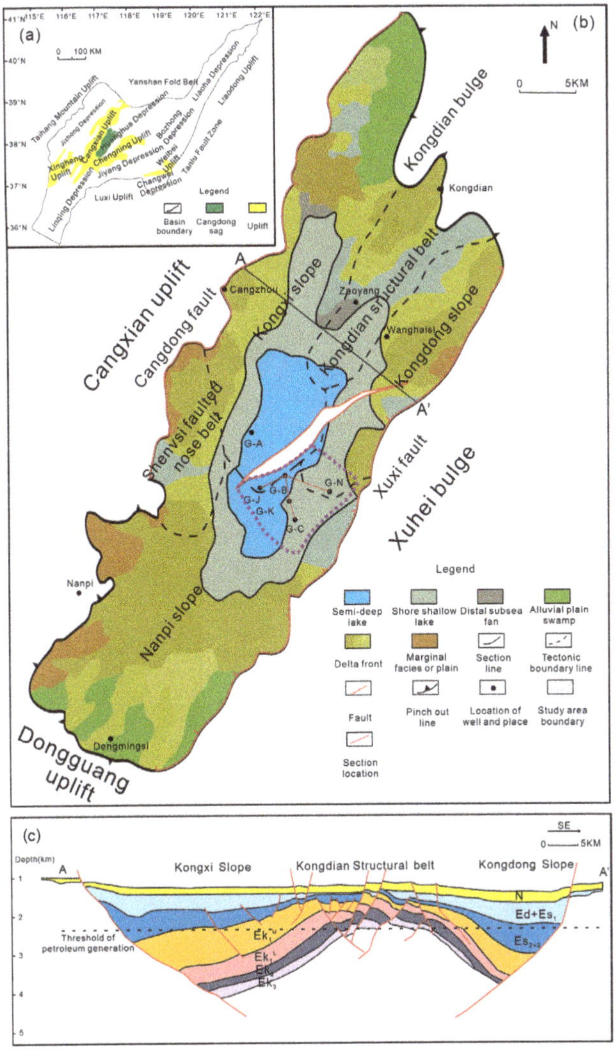

Figure 1. (a) Location of the Cangdong Sag in Bohai Bay Basin (modified after [34]). (b) Structural and sedimentary facies map of the regional setting of the Cangdong Sag (modified after [35]). (c) Typical sedimentary strata section of the Cangdong Sag (modified after [36,37]).

The Cangdong Sag developed in the regional extensional setting. In the Cangdong Sag, the Cenozoic strata include the Kongdian Formation, Shahejie Formation, Dongying Formation, Guantao Formation, and Minghuazhen Formation (Figure 2). In the Paleocene, the second member of the Kongdian Formation (Ek_2) deposited in the initial rift period was a semi-deep to deep lacustrine deposition, including a large and thick fine-grained sediment interval of 400–600 m. The burial depth of the second Kongdian Formation exceeds 2500 m, with the majority of the area situated below the oil generation threshold depth (Figure 1c). Ek_2 is a complete sequence that is divided into four parasequences from bottom to top: Ek_2^4, Ek_2^3, Ek_2^2, and Ek_2^1 [38,39]. The Ek_2^4 member is the fine sandstone and siltstone deposited in the delta sedimentary environment, while Ek_2^3–Ek_2^1 represent the fine-grained sedimentary segment in a semi-deep lake to deep lake environment [35]. This study focuses on the fine-grained sedimentary segment (Ek_2^3–Ek_2^1) located in the Guandong region, which is in the eastern part of the Cangdong Sag. As a pioneering development block for shale oil reservoirs, significant progress has been witnessed in the Guandong region. Since 2017, several horizontal wells developed in the Ek_2^1 fine-grained sedimentary segment, and sustained and stable industrial production has been achieved, which indicates the huge potential for shale oil development in the Cangdong Sag [40].

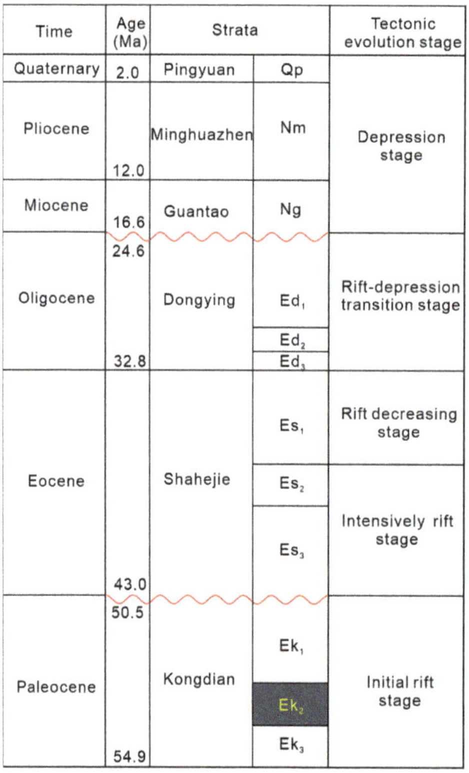

Figure 2. Stratigraphic column of the Cangdong Sag (modified after [41,42]).

3. Materials and Methods

The data for this study were collected from the fine-grained sedimentary interval of the Ek_2 formation in the wells G-A, G-B, and G-C. After a centimeter-level description of the core, more than 70 samples were collected for observation. All samples from different lithofacies were ground and polished into thin sections. In order to achieve clear observations of the organic matter morphology and internal structure, selected samples underwent a 2 h argon ion polishing process using a LEICA EM TIC 3X instrument, with the polishing parameters set at 5 kV and 2 mA. Then, a Zeiss Axio Imager A2m optical microscope was used to take photographs. High-resolution imaging and analysis were performed on typical rock core thin sections using an FEI Quattro S Environmental Scanning electron microscope (ESEM) equipped with four energy dispersive X-ray spectroscopy (EDS, Oxford Ultim Max 40), Carl Zeiss Company, Oxford, United Kingdom. All photo analysis work was completed at the Research Institute of Exploration and Development of the Petrochina Dagang Oilfield Company and the Key Laboratory of Deep Oil and Gas of the China University of Petroleum (East China).

X-ray diffraction (XRD) analysis was carried out at the Key Laboratory of Deep Oil and Gas of the China University of Petroleum (East China). More than 1000 datasets from the wells G-A, G-B, and G-C were obtained using the X′Pert PRO MPD instrument with CuKα radiation. The instrument operated at a voltage of 40 kV and a current of 40 mA. The testing angle 2θ (mineral diffraction angle) range was set from 5° to 60°, and a sampling step size of 0.016° was used. The mass percentage of different minerals can be obtained by the computer of the diffraction pattern.

Tests on the chemical composition of organic matter were conducted at the Key Laboratory of Deep Oil and Gas of the China University of Petroleum (East China), and some data were collected from the Research Institute of Exploration and Development of the Petrochina Dagang Oilfield Company. The Leco CS-744 carbon–sulfur analyzer was used to combust the prepared 100-mesh powdered standard samples. The analysis strictly followed the TOC standard testing technique to obtain the total organic carbon content data for different lithological samples [43]. The free hydrocarbon (S_1; mg_{HC}/g_{rock}) and kerogen-derived hydrocarbon (S_2; mg_{HC}/g_{rock}) contents were analyzed using the Rock-Eval pyrolysis system (Rock-Eval7). The maximum hydrocarbon generation temperature (T_{max}) was recorded, and the hydrogen index ($HI = S_2 \times 100/TOC$; mg_{HC}/g_{TOC}) was calculated.

Low-pressure nitrogen physisorption (LNP) experiments were carried out on the Quantachrome NOVA 4200e analyzer in the Key Laboratory of Deep Oil and Gas of the China University of Petroleum (East China) (Qingdao, China). The sample was first crushed into 50-mesh-sized particles. After deoiling, drying at 110 °C for 10 h, and vacuum degassing at 110 °C for 24 h, the nitrogen adsorption experiment was conducted at −195 °C. The pore volume and pore size distribution were calculated according to the Barrett–Joyner–Halenda (BJH) method [44].

Mercury intrusion porosimetry (MIP) experiments were conducted using a Micromeritics Autopore IV 9500 V1.09 instrument at the Key Laboratory of Deep Oil and Gas of the China University of Petroleum (East China) in strict compliance with experimental operating standards. The pre-treated 1 cm^3 cubic samples were dried and then subjected to a mercury injection process. During the mercury injection process, the pressure was gradually increased up to 275 MPa, and then gradually reduced. The pore radius was calculated by recording the curve of mercury intake and removal. The MIP experiment can measure the pore size distribution in the range of 5.4 nm–200 μm.

Nuclear magnetic resonance (NMR) tests were performed at the Key Laboratory of Deep Oil and Gas of the China University of Petroleum (East China), using the MicroMR02-1 NMR core analyzer. Throughout the NMR experiment workflow, including reagent preparation, sample preparation, selection of measurement parameters, experimental execution, and data processing, there was rigorous adherence to industry standard requirements. The instrument operated under the conditions of 25 °C temperature, magnetic field strength of 0.3 T, and resonance frequency of 12 MHz. The detailed NMR experimental parameters included echo spacing of 0.25 ms, waiting time of 6000 ms, scan number of 128, and echo numbers of 4096. To eliminate the influence of kerogen and crystalline water on pore measurements, the standard core plug samples were subjected to drying and oil extraction. Subsequently, NMR measurements were taken of dried core, saturated water core, and centrifuged water core. The T_2 spectra of the samples in dry, saturation, and centrifugation states were recorded using the Carr–Purcell–Meiboom–Gill (CPMG) sequence. The T_2 spectra were then analyzed to characterize the pore features of different samples.

Triaxial compression tests were conducted to test rock mechanical properties under simulated conditions of underground confining pressure. An RTR-1000 rock triaxial test system was used in the experimental device. The sample was a cylinder with a radius of 12.5 mm and a height of 50 mm. Under a confining pressure of 35 MPa, vertical pressure was applied until the rock was broken. The axial and radial displacements were, respectively, recorded by displacement sensors, and the stress–strain curves were obtained. Young's modulus and Poisson's ratio were calculated based on these curves. The experimental data were collected from the Research Institute of Exploration and Development of the Petrochina Dagang Oilfield Company.

Multi-resolution graph-based clustering (MRGC) is an unsupervised clustering algorithm that uses a multi-resolution clustering algorithm and well logging data as the basis to partition lithofacies [45]. The well logging phase recognition models of different lithofacies are established, and a KNN (K-nearest neighbor) algorithm is used to quickly and accurately classify lithofacies. For shale reservoirs, four sensitive well logs, including gamma ray (GR), acoustic transit time (AC), resistivity (RT), and total organic carbon (TOC) content, are selected for MRGC clustering. After the automatic clustering of lithofacies using the MRGC method, manual adjustments can be applied based on the core data. Ultimately, lithofacies–electrical relationship models are established for different lithofacies, enabling the rapid identification of lacustrine shale lithofacies based on conventional well logging data, and the approach is extended to all wells within the study area.

Since Wheeler introduced the concept of the reference plane, sequence stratigraphy has gradually become an important method in sedimentary research [46]. INPEFA technology can be used to identify the sedimentary cycle interface by integrating the difference between the maximum entropy spectrum analysis value and the conventional curve value and amplifying the subtle changes of typical well logs. Based on cyclic stratigraphy, the variation trend of the INPEFA curve that is calculated based on conventional logging curves can be used toward accurately understanding the formation development trend and ensuring a good correlation with the change in base level. The sequence stratigraphy, established using INPEFA technology, governs the sedimentary patterns of different lithofacies [47]. The GR log is strongly sensitive to mud content and was used to classify the sequence using INPEFA technology.

4. Results

4.1. Geochemical Characteristics, Petrology Characteristics, and Lithofacies Classification

4.1.1. Geochemical Characteristics

Under the control of parasequences reclassified according to the INPEFA curve, the geochemical parameter characteristics of the fine sediment member are obtained. The vitrinite reflectance (Ro) values of samples obtained from the entire fine-grained sedimentary member are consistently distributed within the range of 0.5 to 1, indicating that Ek_2^1–Ek_2^3 have all entered a mature stage of organic matter thermal evolution (Figure 3a). In Ek_2^1, the samples exhibit a wide range of TOC content, with values varying from 0.28 to 12.92 wt.%, with an average of 3.65 wt.%. The free and volatile hydrocarbons (S_1) values are distributed in the range of 0.10 to 8.36 mg_{HC}/g_{rock}, with an average of 0.58 mg_{HC}/g_{rock}. Additionally, the kerogen-derived hydrocarbon (S_2) values span from 0.2 to 72 mg_{HC}/g_{rock}, with an average of 23 mg_{HC}/g_{rock}. Moreover, the hydrogen index (HI) values range from 33 to 1327 mg_{HC}/g_{TOC}, with an average value of 559.8 mg_{HC}/g_{TOC}. The TOC of the Ek_2^2 samples ranges from 0.13 to 12.66 wt.%, with an average of 3.65 wt.%. The S_1 values are distributed in the range of 0.10 to 34 mg_{HC}/g_{rock}, with an average of 0.84 mg_{HC}/g_{rock}. Additionally, the S_2 values span from 0.10 to 69 mg_{HC}/g_{rock}, with an average of 14.04 mg_{HC}/g_{rock}. Moreover, the hydrogen index (HI) values range from 36 to 962 mg_{HC}/g_{TOC}, with an average value of 384.7 mg_{HC}/g_{TOC}. The TOC of the Ek_2^3 samples ranges from 0.15 to 12.47 wt.%, with an average of 4.09 wt.%. The S_1 values are distributed in the range of 0.10 to 13.09 mg_{HC}/g_{rock}, with an average of 1.67 mg_{HC}/g_{rock}. Additionally, the S_2 values span from 0.19 to 67.14 mg_{HC}/g_{rock}, with an average of 21.91 mg_{HC}/g_{rock}. Moreover, the hydrogen index (HI) values range from 79.74 to 797 mg_{HC}/g_{TOC}, with an average value of 484.08 mg_{HC}/g_{TOC}.

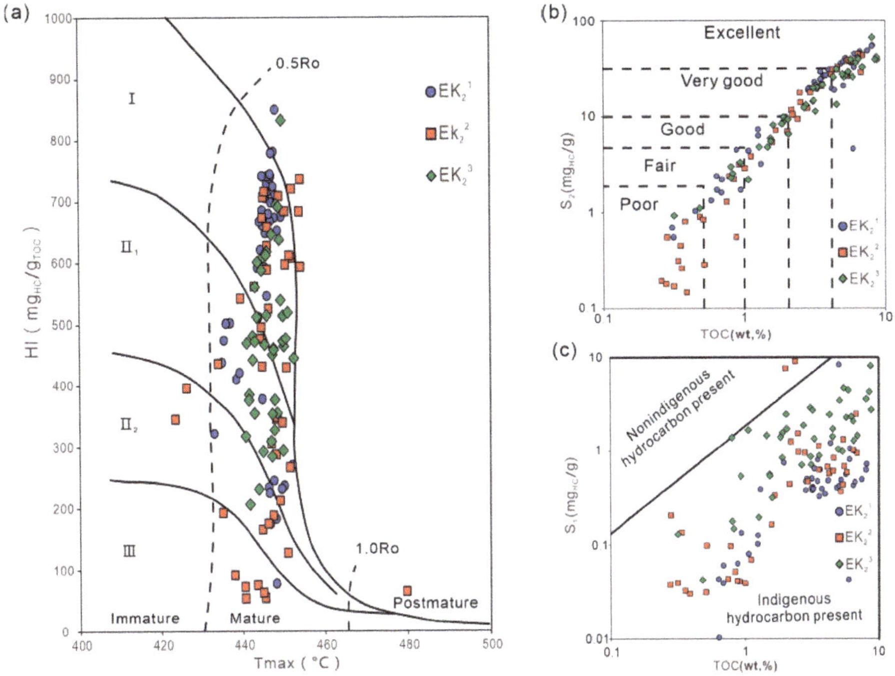

Figure 3. Geochemical parameters map of the Ek_2 shales in the Cangdong Sag. Crossplot of the (**a**) maximum hydrocarbon generation temperature (T_{max}) and hydrogen index (HI); (**b**) kerogen-derived hydrocarbons (S_2) and TOC; and (**c**) free and volatile hydrocarbons (S_1) and TOC.

The crossplot of T_{max} and HI shows that the kerogen of Ek_2^1 and Ek_2^3 is mainly type I and type II_1, and the kerogen of Ek_2^2 is type I, type II, and type III, with strong heterogeneity of kerogen [48]. The crossplot of TOC and S_2 shows that Ek_2^1 and Ek_2^3 samples have fair–excellent hydrocarbon generation potential [49]. The samples of Ek_2^2 show notable heterogeneity, with hydrocarbon generation potential ranging from poor to excellent (Figure 3b). The crossplot of TOC and S_1 (Figure 3c) shows that the samples of Ek_2^1–Ek_2^3 are all indigenous hydrocarbons [50].

4.1.2. Petrology Characteristics

Based on XRD analysis, in the Ek_2 fine-grained sedimentary member, the minerals, according to abundance from highest to lowest, are as follows: dolomite (0.0–95 wt.%, averaging 24.2 wt.%), feldspar (0.0–55 wt.%, averaging 19.4 wt.%), quartz (1.0–46 wt.%, averaging 17.5 wt.%), clay minerals (0.0–41 wt.%, averaging 14.8 wt.%), calcite (0.0–70 wt.%, averaging 14.5 wt.%), and analcime (0.0–59 wt.%, averaging 7.6 wt.%). Additionally, the minor presence of pyrite is observed, with an average proportion of 1.6 wt.%. In Ek_2, the dominant components of the siliceous fraction consist of quartz and feldspar particles derived from terrigenous detritus, with occasional occurrences of authigenic quartz (Figure 4a–c). These siliceous minerals predominantly exhibit a laminated appearance, interlayered with organic-rich clay minerals. Both dolomite and calcite predominantly occur as granular and micritic, often displaying layered distribution patterns (Figure 4d,e). Clay minerals are developed in intergranular pores of calcite and dolomite (Figure 4f).

The laminated structure of the Ek_2 member is constituted by organic-rich clay layers and other mineral layers (Figure 4g). Analcime is observed in various forms, such as clusters or dispersed particles, distributed throughout the entire Ek_2 formation (Figure 4h). Pyrite, identifiable under optical microscopy by its black color, exhibits relatively poor continuity, displaying a patchy distribution pattern (Figure 4i).

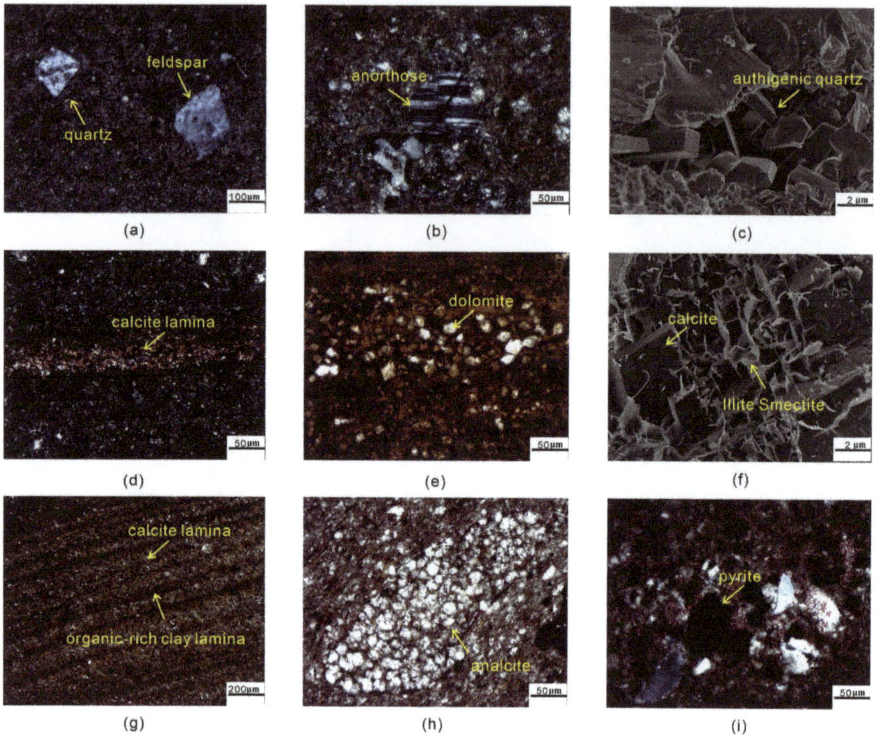

Figure 4. Mineral characteristics of the Ek_2 shale in the Cangdong Sag: (**a**) terrigenous quartz and feldspar particles (3201.14 m, G-A); (**b**) anorthose particles (3196.64 m, G-A); (**c**) authigenic quartz particles image from SEM (2992.79 m, G-A); (**d**) calcite lamina (3199.36 m, G-A); (**e**) dolomite lamina (3191.19 m, G-A); (**f**) SEM image of the fibrous clay minerals in the intergranular pores of calcite cement (3382.41 m, G-A); (**g**) interbedding of calcite laminae and organic-rich clay laminae (3199.36 m, G-A); (**h**) biogenic shell filled with analcite minerals (3194.59 m, G-A); (**i**) porphyritic pyrite in the matrix (3195.70 m, G-A).

From the vertical perspective, mainly calcareous shale and mixed shale developed in the Ek_2^1 formation, and mainly siliceous shale and mixed shale developed in the Ek_2^2 formation. All three lithologies are developed in the Ek_2^3 member, with the strongest heterogeneity observed in this formation.

The sedimentary structure of shale in the Ek_2 member is mostly laminated, and some of it is massive. The lamellar structure is developed in terms of siliceous shale, calcareous shale, and mixed shale, and the thickness of the lamina ranges between 1 μm and 1000 μm. The massive structures are mainly developed in calcareous shale.

4.1.3. Lithofacies Classification

Shale lithofacies can be classified into different types according to mineral composition, organic matter abundance, and sedimentary structure (Figures 5 and 6).

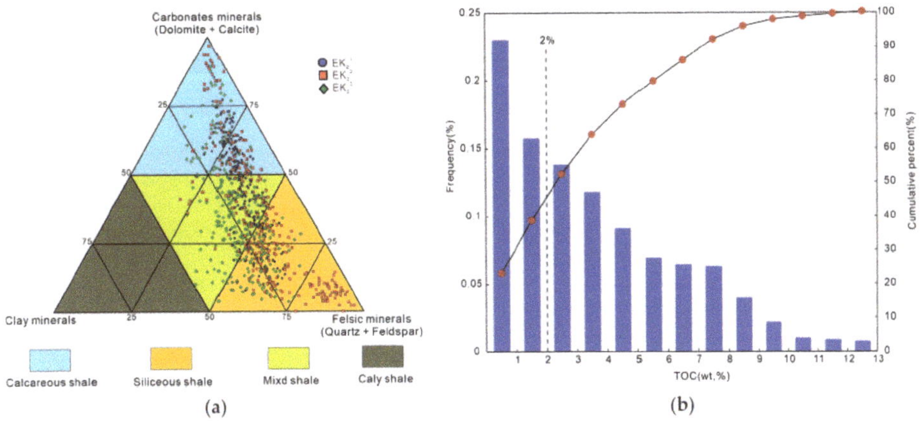

Figure 5. Classification of the Ek_2 shale in the Cangdong Sag: (**a**) ternary diagram of the mineral composition (modified after [51]); (**b**) histogram and cumulative distribution curve of the TOC.

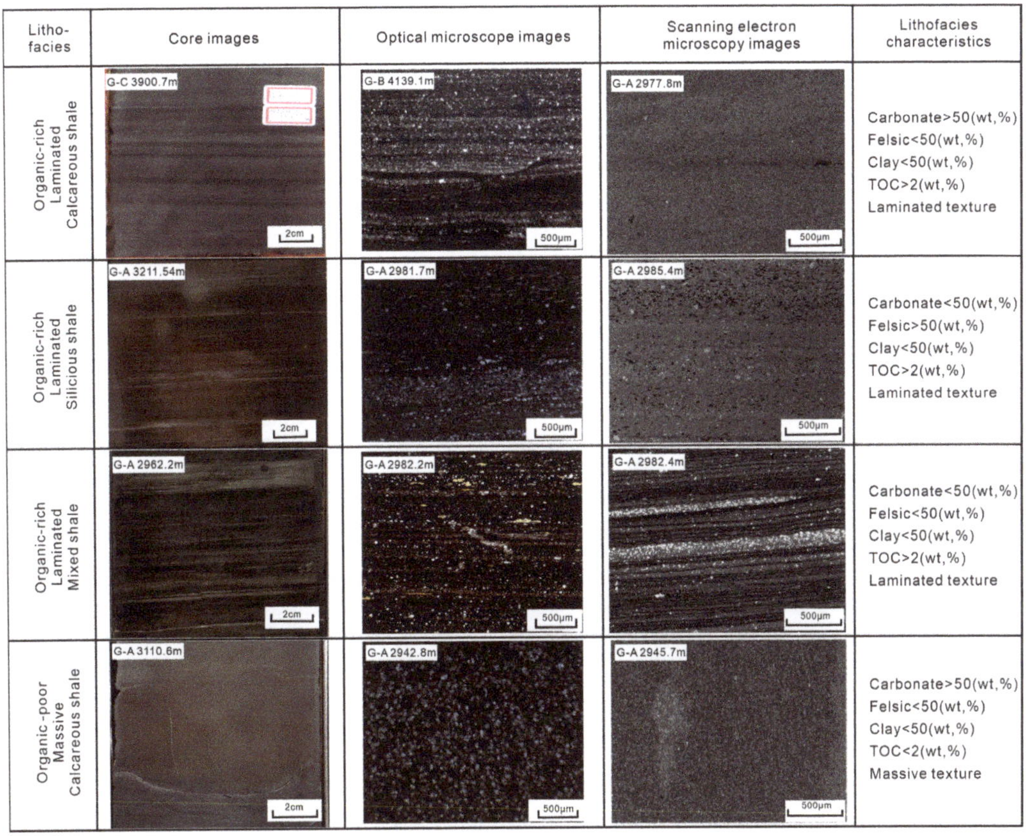

Figure 6. Lithofacies characteristics of the Ek_2 shale in the Cangdong Sag. Examples of each lithofacies shown as core photographs, thin section optical microscope images, scanning electron microscope images, and the mineral composition and structural characteristics.

The lithofacies of the Ek_2 member is classified into four categories based on mineral composition (with 50% as the threshold for determining the dominant lithology), organic matter content (with a TOC value of 2% as the threshold for assessing organic matter enrichment), and mineral compositional structure (laminated and massive features) (Figure 6).

Organic-rich laminated calcareous shale (LC). The organic carbon content exceeds 2 wt.%, the calcareous mineral content is more than 50 wt.%, the content of calcite is higher than that of dolomite, and the developed laminae are highly dense. The shade of color in the calcium-rich and organic-rich laminae varies, with distinct boundaries between the different types of laminae. Microscopic examination reveals oriented mineral arrangements with the development of bedding planes and fractures between the laminae.

Organic-rich laminated siliceous shale (LS). The organic carbon content exceeds 2 wt.%, the siliceous content is more than 50 wt.%, and the dominant laminae consist primarily of siliceous laminae and organic-rich laminae, with a minor occurrence of calcareous laminae. The core color is dark, and the laminae are continuous and straight. Microscopic examination reveals that the mineral arrangement is directional.

Organic-rich laminated mixed shale (LM). The organic carbon content exceeds 2 wt.%, and no dominant minerals are observed, as all mineral contents are less than 50 wt.%. Of these, clay minerals exhibited relatively higher proportions. The content of organic matter is high, and the color is the darkest among the four lithofacies. There is observed development of siliceous, calcareous, and organic-rich laminae, with indistinct boundaries between each type of laminae. Additionally, bedding plane fractures were found to be present between the laminae.

Organic-poor massive calcareous shale (MC). The organic carbon content is mostly less than 2 wt.%, and the calcareous mineral content is more than 50 wt.%. This lithofacies exhibits a massive structure, with dolomite as the predominant mineral composition, often featuring lens-shaped dolomite development. The core has a relatively light color and low organic matter content, and the microscopic observations reveal disordered mineral arrangements.

4.1.4. Vertical Variation in Lithofacies

Although lacustrine shales continuously deposit, their structure and mineral composition are not uniform. Influenced by the depositional environment, lacustrine shales exhibit certain regularities in both vertical and horizontal distributions. Therefore, under the control of sequence stratigraphy theory, it is essential to conduct lithofacies distribution studies within the sequenced framework delineated using the GR-INPEFA method. By utilizing the GR-INPEFA method and core XRD analysis, sequences Ek_2^1–Ek_2^3 were subdivided into seven parasequences from top to bottom, denoted as PS1 to PS7. Upon constructing the vertical lithofacies distribution diagrams for the three cored wells, it becomes evident that the lithofacies distribution characteristics are not same within different parasequences. Taking well G-A as an example (Figure 7a), PS1 is primarily characterized by LC lithofacies, with some occurrence of LS lithofacies. PS2 exhibits a tight sandstone distribution typical of nearshore subaqueous fans sedimentation. As the depth increases, the lithofacies of PS3 to PS5 transition gradually from the MC and LC lithofacies sedimentation to predominantly LS lithofacies sedimentation. In PS6 to PS7, all four lithofacies are present, with a notable abundance of LM lithofacies sedimentation.

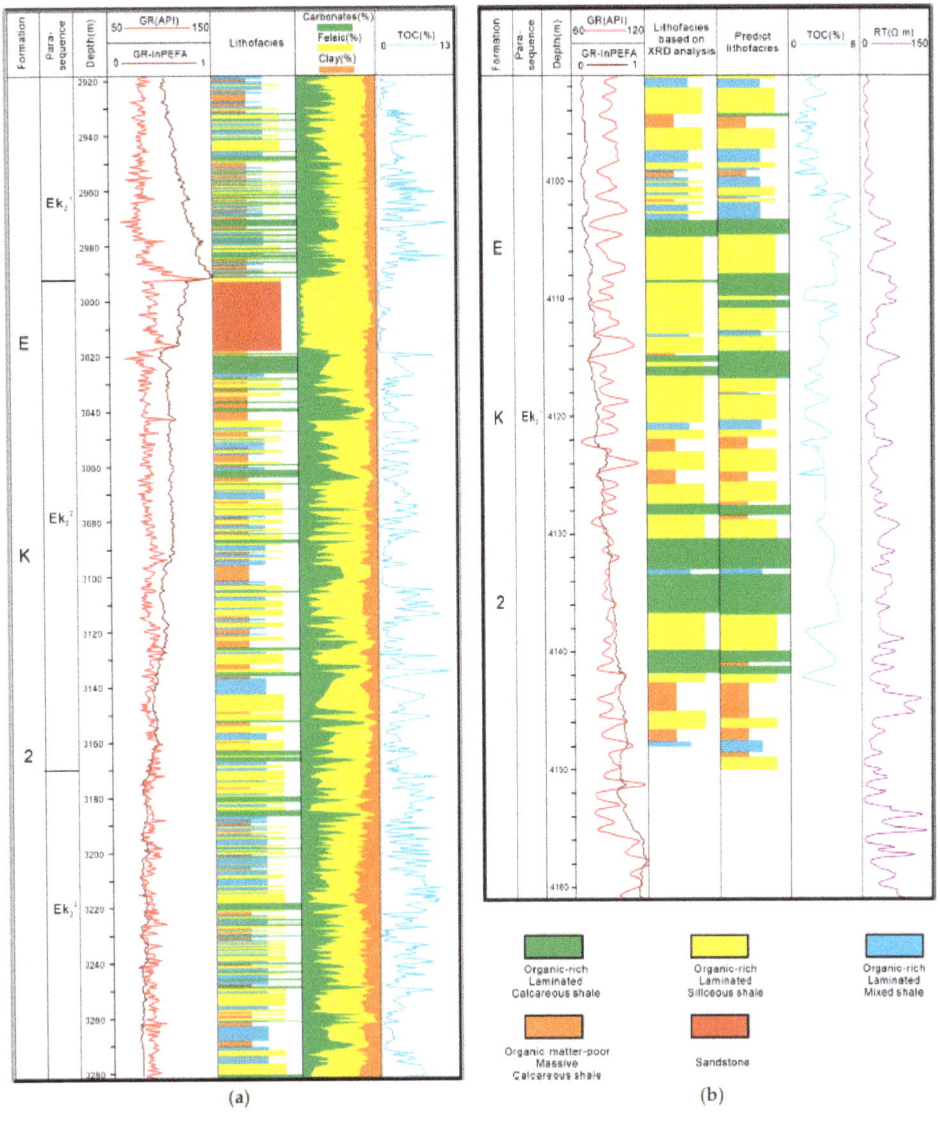

Figure 7. Well section for lithofacies classification and prediction: (**a**) well G-A; (**b**) well G-B.

4.2. Pore Structural Characteristics

4.2.1. Microscopic Pore Types

Given the self-generation and self-storage characteristics of shale oil reservoirs, coupled with the extremely poor physical properties, the development of pores and fractures serves as the primary storage space, becoming a crucial factor influencing shale oil storage and migration. According to the mudstone pore classification scheme, the pore system in the Ek_2 of the Cangdong Sag can be categorized into intergranular pores, intragranular pores, in addition to developed organic pores and microfractures [52]. Upon SEM observation of the samples, various minerals of different origins exhibit distinctive pore type characteristics. The pores within calcareous minerals predominantly encompass intercrystalline pores resulting from calcite recrystallization along with internal and external dissolved pores and fractures, as well as intergranular pores occurring between dolomite grains (Figure 8a,b). Siliceous minerals are mainly intergranular pores contacted by quartz, feldspar, and other brittle minerals, mostly in narrow and long form (Figure 8c). Organic matter is highly abundant in the shale of the Ek_2 member in the study area. Organic matter pores and shrinkage fractures have developed in the process of thermal evolution, and the organic matter pores are mostly circular or irregular polygons (Figure 8d,e). The content of clay minerals is high in mixed shale. The clay minerals mainly develop intergranular pores, and shrinkage fractures form in the process of transformation and dehydration (Figure 8f,g). Pyrite deposition occurred in Ek_2 under an anaerobic reducing environment, with its intercrystalline pores exhibiting ribbon and irregular polygonal shapes (Figure 8h). In the study area, lamina development is pronounced across various lithofacies, and the contact surfaces between laminae composed of different minerals exhibit bedding plane fractures. These bedding plane fractures play a vital role in facilitating oil and gas migration and storage (Figure 8i). By acting as mechanical weak planes, these bedding plane fractures can form a network of fractures during subsequent hydraulic fracturing, thereby enhancing fracturing efficiency.

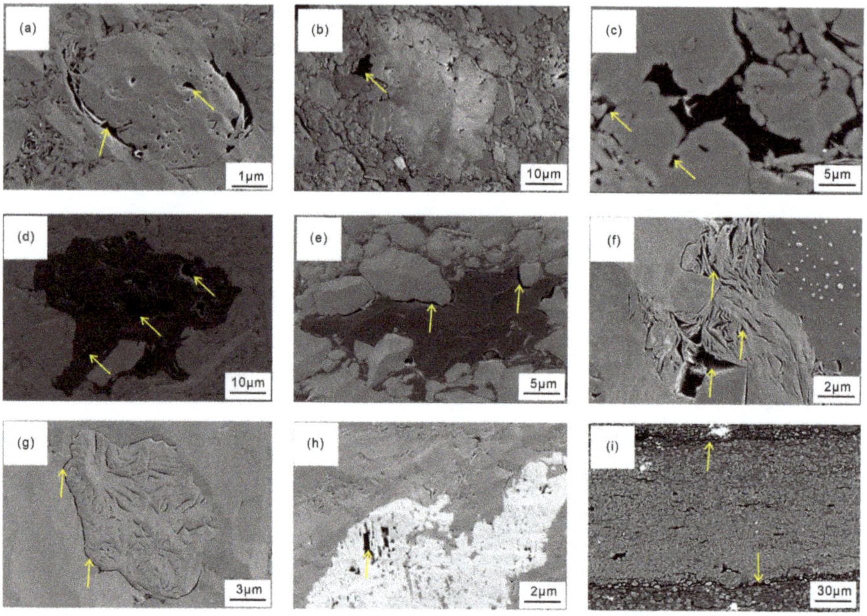

Figure 8. Typical pore characteristics of the Ek_2 shale in the Cangdong Sag: (**a**) calcite intragranular dissolved pores (2984.90 m, G-A); (**b**) dolomite intergranular pores (3085.50 m, G-A); (**c**) quartz intergranular pores (3032.50 m, G-A); (**d**) organic matter internal pores (2985.14 m, G-A); (**e**) organic matter shrinkage fractures (3123.20 m, G-A); (**f**) clay minerals intergranular pores (3116.81 m, G-A); (**g**) clay minerals shrinkage fractures (2987.37 m, G-A); (**h**) pyrite intercrystalline pores (2984.11 m, G-A); (**i**) bedding plane fractures of calcareous lamina (3014.2 m, G-A).

4.2.2. Microscopic Pore Structure Characteristics

LNP technology is a method to effectively characterize the distribution characteristics of micropores and mesopores by using the curves of N_2 adsorption capacity and pressure generated by nitrogen adsorption and desorption processes [53]. Based on the classification results of the LNP hysteresis loop according to IUPAC [54], the hysteresis loops of MC lithofacies samples are H2-type with a small amount of H4-type, indicating that the pore shape is primarily ink-bottle-type with a minor presence of narrow-slit-type (Figure 9a).

The hysteresis loops of the other laminated shale samples are H3-type, indicating that the pores are slit-like in shape. Moreover, the nitrogen adsorption capacity of MC lithofacies is significantly higher than that of other laminated shales, indicating that the massive shale has more developed micropores and mesopores than the laminated shale.

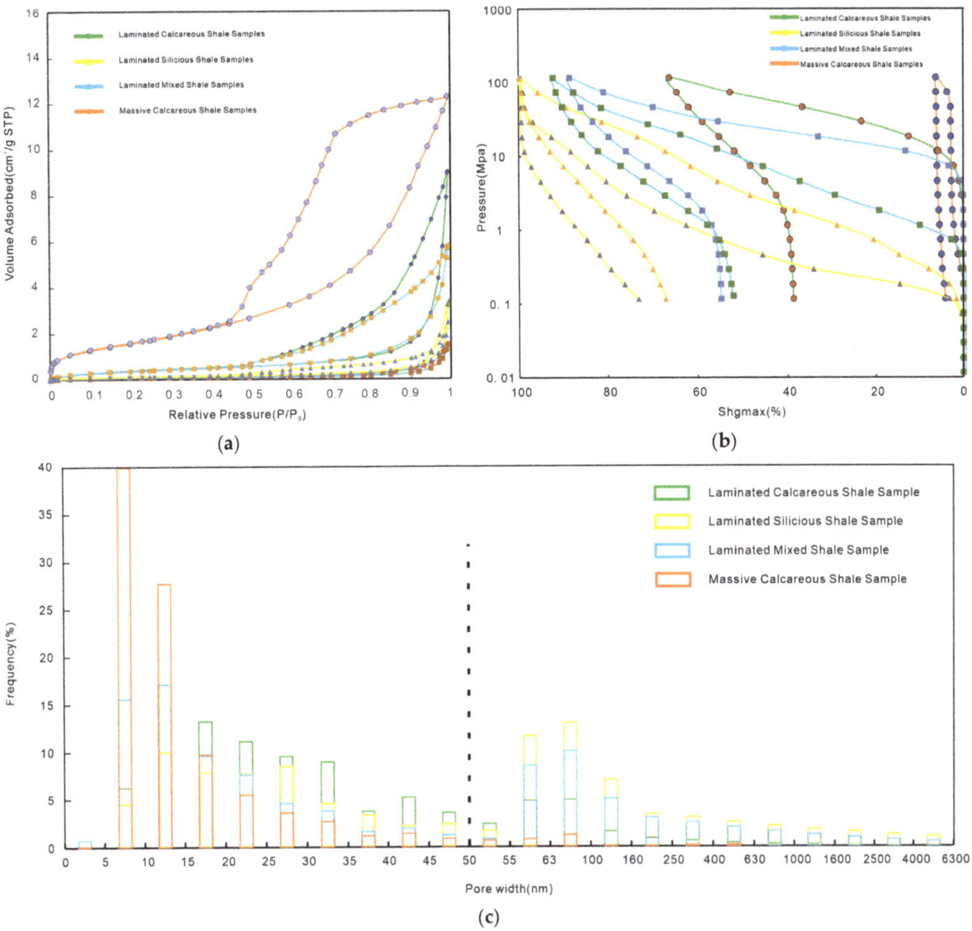

Figure 9. Pore structure characteristics measured by different methods of the Ek_2 shale in the Cangdong Sag: (**a**) low-pressure nitrogen physisorption; (**b**) mercury intrusion porosimetry. (**c**) The full-aperture distribution map.

Mercury intrusion porosimetry (MIP) requires recording the intrusion–extrusion volume of non-wetting phase mercury as a function of pressure increase and decrease to generate a function graph representing the relationship between the cumulative volume change of mercury and pressure. These plots are used to characterize the pore characteristics and connectivity of different lithofacies samples (Figure 9b). The study object of high-pressure mercury injection is nano-microscale pores. According to the characteristics of the mercury saturation–capillary pressure curves of the four lithofacies samples, MC shale has the lowest mercury-extrusion efficiency, and mercury extrusion is concentrated in the high-pressure stage, indicating there is a high proportion of micropores in the lithofacies. The pore characteristics of the three types of laminated shales also vary. LS shale and LM shale have the highest mercury intrusion and extrusion efficiency, and the mercury volume increases at each pressure stage, indicating that both mesopores and macropores are well developed, and the pore connectivity is good. In contrast, LC shale has relatively few macropores, and mesopores are the predominant pores.

NMR can be used to detect information related to pore structure and the fluids within pores. The T_2 relaxation time distribution features can serve as a foundation for the analysis of pore size characteristics in different lithofacies samples (Figure 10). The signal morphology of the T_2 spectra of the dried sample represents solid organic matter, with the main relaxation time peaks of the four lithofacies falling between 0.1 and 1 ms. This pattern of T_2 spectra is consistent with the characteristics observed under saturated water conditions. The LM lithofacies exhibit larger-scale organic matter pores, which are attributed to the abundant occurrence of organic-rich laminae. After eliminating the influence of solid organic matter, analysis of the T_2 spectra under saturated water and centrifugation states was conducted. The T_2 spectral relaxation time of MC lithofacies presents a unimodal distribution, and the main peak is distributed at 0.8 ms, indicating that MC lithofacies mainly develop micropores and mesopores. The T_2 spectral relaxation time of the three laminated shales presents bimodal or trimodal distribution, with the left main peak developing around 0.8 ms and the right main peak developing between 10 ms and 1000 ms, indicating that mesopores and macropores are the predominant pore types in the laminated shales. The T_2 relaxation time of water-saturated shale samples can also be utilized to calculate the NMR porosity [55]. Among the four types of shales, LS lithofacies have the highest porosity (averaging 4.98%), followed by LC lithofacies (averaging 4.08%), and then LM lithofacies (averaging 3.44%), and MC lithofacies (averaging 2.08%) have the lowest porosity. The fluid signals of water saturation and centrifugation processes show that the MC lithofacies has the lowest mobile fluid saturation of the three laminated shales. LS and LM lithofacies have relatively higher mobile fluid saturation, followed by LC lithofacies, which is consistent with the mercury extrusion efficiency results of the MIP experiment.

LNP, MIP, and NMR experiments were used to characterize the pore structure at different scales, and the full-aperture distribution maps for four lithofacies in the fine-grained sedimentary part of Ek_2 were obtained (Figure 9c). Based on the IUPAC classification criteria, the pores can be classified as micropores (<2 nm), mesopores (2–50 nm), and macropores (>50 nm) [56]. It is worth emphasizing that, of the macropores larger than 50 nm, there are also microfractures with pore sizes exceeding 1 μm. Different experiments have varying measurement ranges for pore sizes. For instance, the LNP method typically measures pores in the range of 2 to 100 nm, whereas the MIP method is more accurate for pores in the range of 50 nm to 10 μm. Therefore, in this study, a comprehensive pore size characterization approach was adopted. For micropores and mesopores smaller than 50 nm, the LNP experiment results were used, whereas for larger pores and microfractures larger than 50 nm, the MIP test results were employed. Additionally, the qualitative control of pore size distribution characteristics was achieved through NMR analysis. The full-aperture distribution map reveals that mesopores dominate in different lithofacies of Ek_2, accounting for the largest proportion (54%–97%), followed by macropores and microfractures (up to 43%). The pore size distribution characteristics of different lithofacies exhibit pronounced heterogeneity. The peak pore size of MC lithofacies ranges from 5 nm to 15 nm, and there is little development of macropores. Among the three laminated shales, the peak pore size of LS lithofacies ranges from 55–100 nm, and macropores account for a large proportion. Mesopores of various sizes have developed in the LM lithofacies, and the distribution range of the main peaks is relatively wide, ranging from 5 to 25 nm, with a certain level of macropore development. The main peak of LC lithofacies is not obvious, and there is development of mesopores of various sizes. For all laminated shale lithofacies, there is the presence of a small number of pores with sizes exceeding 1000 nm, indicating the development of microfractures in addition to matrix pores.

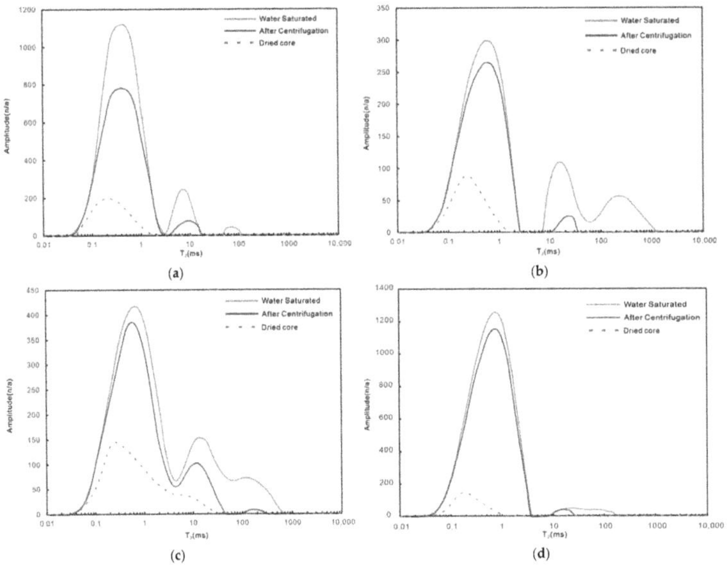

Figure 10. T_2 spectra from NMR experiments for different lithofacies: (**a**) organic-rich laminated calcareous shale; (**b**) organic-rich laminated siliceous shale; (**c**) organic-rich laminated mixed shale; (**d**) organic-poor massive calcareous shale.

4.3. Mechanical Fracturing Tests

Triaxial stress testing is a method to calculate Young's modulus and Poisson's ratio by applying different external forces to a sample under a certain confining pressure, monitoring the strain parameters and obtaining the stress–strain curve. The Young's modulus and Poisson's ratio are then used to further evaluate the brittleness of rock according to the classical mechanical parameter method with Equation (1). The mechanical parameters were obtained by conducting triaxial stress tests on 17 samples of the four shale lithofacies from Ek_2 at a confining pressure of 35 MPa. The accuracy of the mechanical brittleness index can be verified by comparison with the mineral brittleness index calculated by mineral composition content. The mineral brittleness index is calculated according to Equation (2) [57]:

$$BI_{mechanical} = \frac{E_{nor} + \mu_{nor}}{2} \times 100\% \quad (1)$$

$$BI_{minral} = \frac{V_{qua} + 0.3 \times V_{fel} + 0.5 \times V_{cal} + 0.7 \times V_{dol} + 0.5 \times V_{ana}}{V_{qua} + V_{fel} + V_{cal} + V_{dol} + V_{cla} + V_{ana} + V_{toc}} \times 100\% \quad (2)$$

where $BI_{mechanical}$ is the mechanical brittleness index; E_{nor} is the normalized Young's modulus; μ_{nor} is the normalized Poisson's ratio; $BI_{mineral}$ is the mineral brittleness index; V_{qua} is the volume fraction of the normalized quartz; V_{fel} is the volume fraction of the normalized feldspar; V_{cal} is the volume fraction of the normalized calcite; V_{dol} is the volume fraction of the normalized dolomite; V_{cla} is the volume fraction of the normalized clay; V_{ana} is the volume fraction of the normalized analcite; V_{toc} is the volume fraction of the normalized total organic carbon.

The mechanical parameters, along with the mineral composition data (from XRD analysis), are presented in Table 1. The results show that the LC and MC lithofacies have higher average Young's moduli (25.13 MPa and 26 MPa, respectively) and lower average Poisson's ratios (0.24 and 0.19, respectively). The average Young's moduli of the LS and LM lithofacies are relatively low (18.12 MPa and 19.77 MPa, respectively), and the average Poisson's ratios are higher (0.28 and 0.26, respectively). The mechanical brittleness index values of four kinds of lithofacies increase in the following order: MC lithofacies (53.09), LC lithofacies (43.33), LM lithofacies (34.33), and LS lithofacies (31.37). Consequently, it can be inferred that calcareous shale exhibits good brittleness, while siliceous shale and mixed shale demonstrate relatively lower brittleness.

4.4. Lithofacies Prediction

Compared with conventional reservoirs, the lithofacies mineral composition of shale reservoirs is complex and changeable, and the difficulty of lithofacies identification and classification is greatly increased. After preprocessing and standardization of the well logging curves, four conventional well logging curves, natural gamma ray (GR), acoustic transit time (AC), resistivity (RT), and total organic carbon (TOC) content data were selected as input curves. The MRGC method was then employed to establish the well logging facies and lithofacies identification models for the four lithofacies in Ek_2. By comparison with the lithofacies prediction results for $EK_2{}^1$ in well G-B obtained using the MRGC technique with the lithofacies results from the XRD mineral identification, the accuracy rate is over 90%, indicating the favorable applicability of the MRGC technique for lithofacies identification in the Ek_2 (Figure 7b). Based on the lithofacies prediction results from all wells in the study area, two lithofacies well sections were constructed in the east–west and north–south directions (Figure 11). The analysis revealed significant regularities in the vertical and horizontal aspects of lithofacies distribution in the Guandong region.

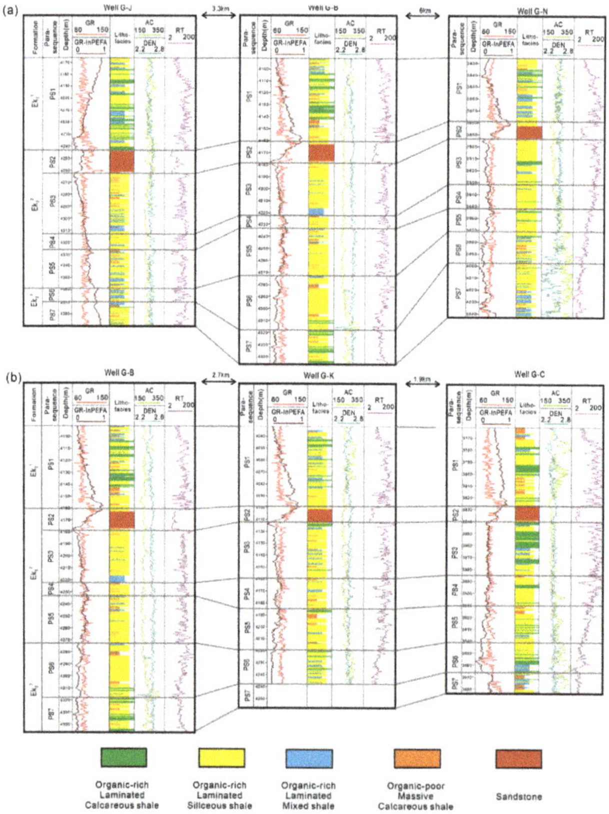

Figure 11. Well sections of predicted lithofacies in Guandong region. The locations of the two well sections are shown in Figure 1b. Well sections in the (**a**) east–west and (**b**) north–south direction.

Table 1. Mineral composition and rock mechanic parameters in different lithofacies in the Ek_2.

Well	Depth (m)	Lithofacies	Quartz (Vol,%)	Feldspar (Vol,%)	Calcite (Vol,%)	Dolomite (Vol,%)	Clay (Vol,%)	Analcite (Vol,%)	TOC (Vol,%)	Mineral Brittleness Index	Elastic Modulus (MPa)	Poisson's Ratio	Mechanical Brittleness Index
G-A	2925.5	MC	18	15	20	11	15	20	1.90	49.91	28.32	0.18	58.17
G-A	2937.6	LM	12	14	14	5	13	22	20.30	35.89	16.14	0.32	19.71
G-A	2955.5	LS	14	18	4	7	13	29	15.68	38.97	18.84	0.35	16.22
G-A	2966.0	LC	13	11	15	33	7	10	11.22	51.92	26.16	0.29	34.34
G-A	2974.0	LC	9	6	4	44	10	15	11.52	51.29	26.40	0.26	39.42
G-A	2999.4	LS	28	53	8	0	4	0	6.66	47.89	16.58	0.21	42.51
G-A	3012.5	LS	20	14	3	0	14	48	0.70	50.12	23.47	0.19	51.91
G-A	3025.1	LS	15	25	11	5	10	25	9.34	43.57	24.73	0.27	37.38
G-A	3032.0	LM	15	14	12	25	12	15	7.61	50.22	27.32	0.21	51.53
G-A	3032.2	LC	13	10	4	38	10	17	7.62	53.75	21.65	0.22	44.91
G-A	3073.8	LM	13	12	2	25	17	17	12.97	43.99	22.31	0.27	33.87
G-A	3187.2	LM	13	22	5	21	18	6	16.26	38.83	15.08	0.32	20.13
G-A	3197.1	LM	22	19	8	18	22	8	2.53	49.12	18.00	0.20	46.43
G-A	3209.0	LS	13	28	5	8	19	0	25.78	28.38	9.97	0.37	6.74
G-A	3281.2	LS	17	20	5	0	26	30	2.69	40.02	15.17	0.25	33.44
G-A	3288.2	MC	16	13	16	20	19	16	0.81	50.18	24.31	0.21	48.00
G-A	3297.8	LC	12	8	8	40	9	8	14.38	50.51	26.33	0.19	54.66

5. Discussion

5.1. Evaluation Parameters of Shale Oil Sweet Spots

5.1.1. Oil-Bearing Capacity

High oil saturation is an important indicator of shale oil enrichment. In shale oil reservoir research, total organic carbon (TOC), chloroform-extractable bitumen "A", and Rock-Eval pyrolysis parameters S_1 and S_2 are commonly utilized as crucial evaluation parameters to distinguish the oil-bearing characteristics of different lithofacies. Due to the abundance of organic-rich laminated layers, three laminated shales (TOC > 2%) in Ek_2 are all considered to be effective source rocks [58]. Based on the crossplot of TOC and S_2 data from well G-A, a strong positive correlation is observed between oil saturation and organic abundance (Figure 12b). The LS and LM lithofacies contain higher organic content, indicating good to excellent hydrocarbon generation potential. The LC lithofacies exhibits a fair to good hydrocarbon generation potential, whereas the MC lithofacies, with fewer organic-rich laminated layers, shows a lower hydrocarbon generation potential, ranging from poor to fair.

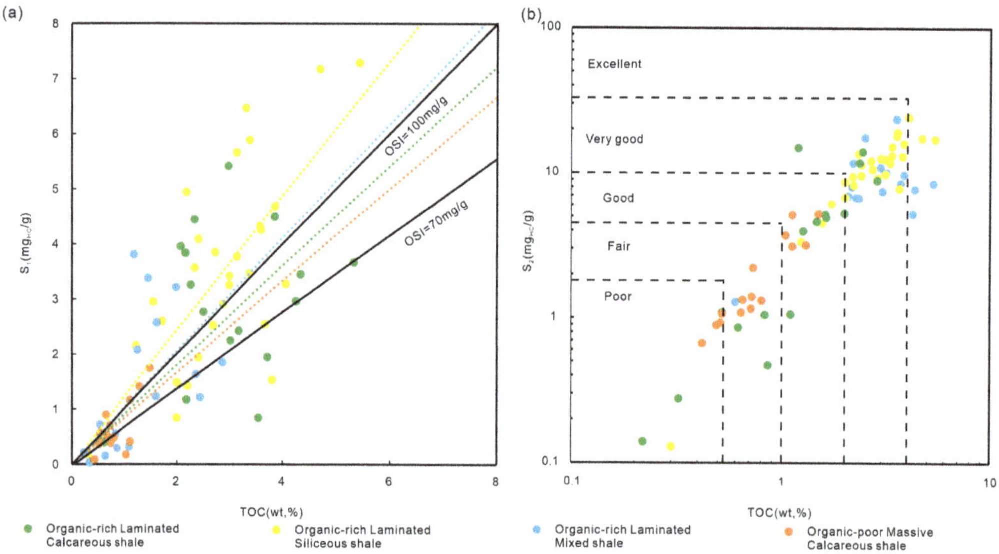

Figure 12. Crossplots of different hydrocarbon generation content and TOC: (**a**) the free hydrocarbon (S_1) content and TOC (with fitting line); (**b**) the kerogen-derived hydrocarbon (S_2) content and TOC.

5.1.2. Reservoir Properties

Pores and fractures, as the main storage space in shale, determine the reservoir's physical properties. Various methods are available to study pore structures, and different measurement techniques offer diverse scales of analysis. Integrating the LNP, MIP, and NMR experiments allows for a comprehensive characterization of the pores across different scales. The comprehensive characterization of pore sizes indicates there is pronounced heterogeneity in the pore distribution characteristics among the different shale lithofacies. The proportion of macropores in the LS lithofacies is higher than 35%, which indicates that the intergranular pores of quartz, feldspar, and other minerals cause siliceous lithofacies to have higher reservoir performance. The macropores in the LC lithofacies are well developed, which may be caused by the dissolution of calcite by acidic fluid generated during the development of intergranular pores and the thermal evolution of dolomite. The complex composition of LM lithofacies, with the deposition of calcareous, siliceous, and clay laminae, contributes to the pronounced heterogeneity in pore size distribution. In addition to the complex mineral composition, the development of micropores, mesopores, and macropores in the LM lithofacies can also be attributed to the clay mineral transformation processes and organic matter thermal evolution. Unlike the abundant microfractures found in laminated shale, the massive shale exhibits limited development of microfractures. The MC lithofacies has a higher abundance of 5 nm–10 nm mesopores and demonstrates the poorest pore connectivity of the four lithofacies. Based on the comprehensive analysis of pore size distribution maps and porosity data measured from nuclear magnetic resonance (NMR) and mercury intrusion porosimetry (MIP), it is evident that the LS lithofacies exhibits the most favorable reservoir properties, followed by the LC lithofacies, and then the LM lithofacies. Conversely, the MC lithofacies exhibits the poorest reservoir properties among the four lithofacies.

5.1.3. Oil Fluidity

Due to the characteristics of fine-grained deposition, a high proportion of clay minerals, and self-generation and self-storage, some hydrocarbons in shale oil reservoirs are adsorbed into source rocks during pyrolysis. Due to the crossover effect, the fluidity of shale oil becomes a critical aspect in sweet spot evaluation. In previous studies, the dynamic oil coefficient OSI (the ratio of S_1 to TOC) has mostly been used to represent the movable efficiency of shale oil [59]. Based on the experience gained from shale oil exploitation in North America, an OSI greater than 100 mg/g indicates high mobility and high exploitation potential. Due to the shallow burial depth of well G-A, the lower Ro values result in relatively low OSI values. Therefore, an OSI value greater than 70 mg/g is considered indicative of the occurrence of a shale oil crossover effect, suggesting that the reservoir has the potential for industrial production. The crossplot of S_1 and TOC for the four lithofacies reveals that the LS and LM lithofacies exhibit OSI values exceeding 100 mg/g, while the LC and MC lithofacies show relatively smaller OSI values, with an average OSI ranging from 70 mg/g to 100 mg/g (Figure 12a). Under the influence of mineral composition and sedimentary structure, the LS and LM lithofacies have the best oil fluidity, while LC lithofacies and MC lithofacies have relatively poor oil fluidity, but all four lithofacies have recoverable potential.

5.1.4. Fracability

The extremely low porosity and permeability of shale oil reservoirs necessitate the use of hydraulic fracturing as a crucial exploitation technique. The integration of shale oil geology and engineering research has become increasingly important in identifying sweet spots for oil exploitation. A comprehensive characterization of reservoir fracability can be achieved by studying the mechanical and mineralogical properties of different lithofacies. In the evaluation of brittleness in Ek_2 shale, the brittle minerals include calcite, dolomite, quartz, feldspar, and analcime, while the ductile minerals consist of clay minerals and organic matter. By normalizing the volume fractions of organic matter and inorganic minerals, a more comprehensive representation of the influence of organic matter on reservoir brittleness can be achieved. By comparing the indices of mechanical brittleness and mineral brittleness for different lithofacies, a clear positive correlation is observed between the two parameters (R^2 = 0.8), indicating that the mechanical properties of shale are influenced by mineral composition and structure (Figure 13). Due to variations in the development density of organic-rich clay laminations, different lithofacies exhibit distinct differences in their brittleness index values. Specifically, the calcareous shale demonstrates higher values for the brittleness index, while the siliceous and mixed lithofacies exhibit relatively lower values. The shale fracability of Ek_2 is in the order of MC lithofacies, LC lithofacies, LM lithofacies, and LS lithofacies.

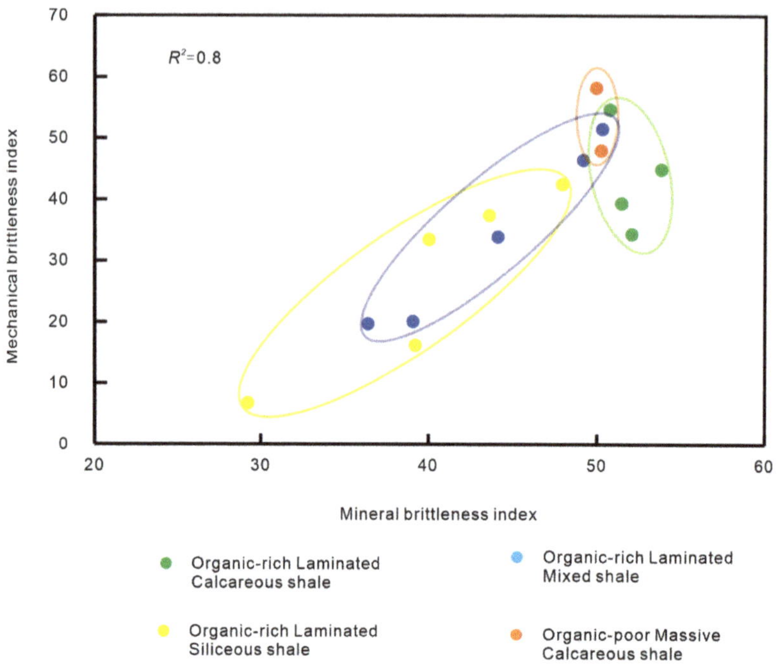

Figure 13. Crossplot of the mechanical brittleness index and the mineral brittleness index.

The sweet spots in shale oil reservoirs are intricately linked with reservoir oil-bearing capacity, reservoir properties, oil fluidity, and fracability. Therefore, in this comprehensive sweet spot evaluation, a multi-parameter equal-weight quantitative evaluation approach was employed. Sweet spot evaluation scores (ES) for the four lithofacies were computed by applying a 25% weighted summation to the normalized TOC, porosity, OSI, and mechanical brittleness index using Equation (3). The evaluation parameters are presented in Table 2. Two levels of sweet spots were identified. Type I includes LC lithofacies (ES of 0.60) and LS lithofacies (ES of 0.75), while type II consists of LM lithofacies (ES of 0.49).

$$ES = \Sigma a_i \cdot P_j, \quad (3)$$

here, ES is the evaluation score; a_i is the weight of the parameter; P_j is the normalized parameter.

Table 2. Evaluation parameters and scores for different lithofacies in the Ek_2.

Lithofacies	Toc (wt%)	OSI (mg/g)	Porosity (%)	Mechanical Brittleness Index	ES
LC	3.2	95.1	4.08	43.33	0.60
LS	3.6	119.4	4.98	31.37	0.75
LM	3.21	102.5	3.44	34.33	0.49
MC	0.82	85.3	2.08	53.09	0.25

5.2. Sweet Spot Distribution and Prediction

In contrast to the stable deposition of thick layers in marine shales, the rapid lithofacies changes in lacustrine basin shales pose challenges for sweet spot prediction. Some researchers have adopted the approach of using a weighted combination of multiple parameters to calculate the sweet spot index and conduct predictive evaluations of sweet spots [40,60]. However, in regions lacking relevant core data and production information, the implementation of this parameter stacking method can be challenging due to the absence of crucial data, leading to relatively low accuracy in the identification of sweet spots. Lithofacies, as a coupled product of tectonics, sedimentation, and diagenesis, can serve as a substitute for various sweet spot parameters in reservoir sweet spot evaluation. Moreover, the sedimentary laws can be used to enhance the predictive accuracy between wells. Different lithofacies exhibit distinct rock physics properties due to variations in mineral composition. Therefore, by relying on the differences in well log responses among different lithofacies, the accurate prediction of shale lithofacies can be achieved based on high-resolution logging data. Compared to traditional manual log interpretation for lithofacies identification, MRGC technology can reveal multi-dimensional lithofacies–electrical relationships, enabling efficient and precise lithofacies prediction. Based on the MRGC method, lithofacies prediction research was conducted on 12 wells in the Guandong region.

Based on the consideration of sedimentation and diagenesis, the vertical and horizontal distribution of the sweet spots are predicted (Figures 14 and 15). The primary type I sweet spot in Ek_2^1 is LC lithofacies. During the PS1 parasequence, the climate was arid, leading to a lowering of the lake level and an increase in water salinity. As a result, a significant amount of carbonate minerals was deposited. The lithofacies deposited during this period exhibit favorable reservoir properties, oil fluidity, oil generation potential, and excellent fracturability.

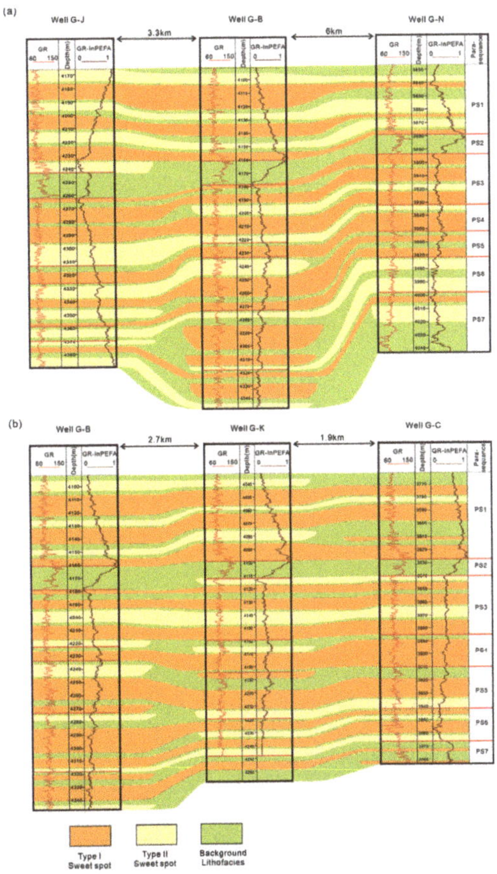

Figure 14. Well sections of sweet spots in the Guandong region. The location of the two well sections is shown in Figure 1b. Well sections in the (**a**) east–west and (**b**) north–south direction.

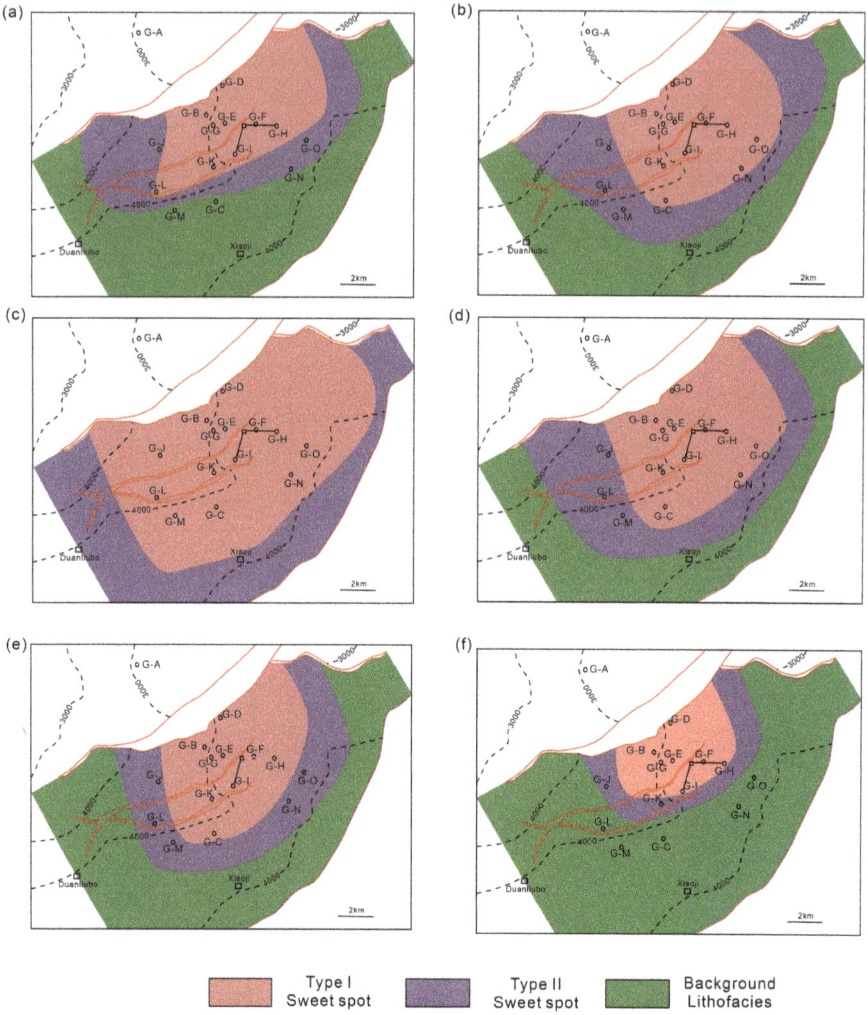

Figure 15. Map of the distribution of sweet spots of different parasequences in the Guandong region. (**a**) PS1; (**b**) PS3; (**c**) PS4; (**d**) PS5; (**e**) PS6; (**f**) PS7.

The type I sweet spot in Ek_2^2 is predominantly composed of LS lithofacies. During the parasequences of PS5 to PS3, the climate tended to be humid, resulting in a significant influx of freshwater into the lake basin, leading to a decrease in water salinity and deepening of the water body (Figure 15c–e). Consequently, a significant amount of organic matter was deposited in the lake basin. The sedimentary environment experienced multiple cycles of fluctuations and an ample supply of detrital materials, which facilitated the extensive development of LS lithofacies. During this period, the lithofacies exhibit excellent reservoir properties, oil fluidity, and oil generation potential, along with moderate fracturability. During the parasequences of PS7 to PS6, as the lake basin deepened and the input of terrigenous debris gradually decreased, the fluctuating climate led to a gradual reduction in the deposition of LS lithofacies, and the sedimentation of LC lithofacies increased. These two lithofacies together constitute the type I sweet spot during the Ek_2^3 period. In the Ek_2 period, the sedimentation of the entire Cangdong Sag was centered in the northwest part of the Guandong region. During the PS7–PS4 period, the lake basin deepened, whereas during the PS4–PS1 period, the lake basin became more shallow. Consequently, throughout these sedimentary processes, the sweet spot areas exhibited a ring-like expansion and contraction pattern, as seen in the planar view.

6. Conclusions

This paper presents the lithofacies-based evaluation and prediction of sweet spots in Ek_2 of the Cangdong Sag. Firstly, based on core analysis, XRD analysis, FE-SEM observations, and the organic geochemical parameters obtained from rock pyrolysis experiments, Ek_2 was categorized into four lithofacies: organic-rich laminated calcareous shale (LC), organic-rich laminated siliceous shale (LS), organic-rich laminated mixed shale (LM), and organic-poor massive calcareous shale (MC).

Subsequently, the organic geochemical parameters obtained from rock pyrolysis experiments were used to investigate the oil-bearing capacity and oil fluidity. The characterization of pore structure and reservoir physical properties was conducted using LNP, MIP, and NMR experiments. The reservoir fracability was evaluated using the brittleness index obtained from triaxial stress tests. Taking into account geological and engineering factors, two types of sweet spots were classified: LS and LC for type I, and LM for type II.

Finally, the distribution of sweet spots was indicated by predicting the distribution of lithofacies. The sweet spots differ slightly for the different sedimentary periods. In Ek_2^1, the optimum lithofacies are LC lithofacies, whereas in Ek_2^2, there is a shift to LS lithofacies. In Ek_2^3, the sweet spots are distributed in both LC and LS lithofacies. The distribution of sweet spots is jointly controlled by climate cyclicity and changes in lacustrine basin water bodies. The dominant distribution characteristics of different lithofacies at vertical intervals and the ring-like distribution patterns in the planar view can provide theoretical guidance for the next stage in shale oil exploration. The findings of this study represent valuable insights for sweet spot research in lacustrine shale oil reservoirs.

The distribution of sweet spots is closely related to the distribution of lithofacies, but the heterogeneity and rapid lithofacies changes in lacustrine shale pose significant challenges for sweet spot evaluation. Therefore, clarifying the sedimentary and diagenetic patterns of lacustrine shale in guiding the prediction of lithofacies and sweet spots in areas without wells represents the future direction of research.

Author Contributions: Conceptualization, C.L.; methodology, P.M.; software, G.L.; validation, C.M.; formal analysis, Y.Z.; investigation, C.M.; resources, W.H.; data curation, W.H.; writing—original draft preparation, Y.Z.; writing—review and editing, C.L.; visualization, Y.Z.; supervision, C.L.; project administration, C.L.; funding acquisition, C.L. All authors have read and agreed to the published version of the manuscript.

Funding: This research was funded by the National Natural Science Foundation of China (NSFC), grant number 41972142 and 42172153.

Data Availability Statement: Not applicable.

Conflicts of Interest: The authors declare no conflict of interest.

References

1. Chalmers, G.R.; Bustin, R.M.; Power, I.M. Characterization of Gas Shale Pore Systems by Porosimetry, Pycnometry, Surface Area, and Field Emission Scanning Electron Microscopy/Transmission Electron Microscopy Image Analyses: Examples from the Barnett, Woodford, Haynesville, Marcellus, and Doig Units. *AAPG Bull.* **2012**, *96*, 1099–1119. [CrossRef]
2. Liang, C.; Cao, Y.; Jiang, Z.; Wu, J.; Guoqi, S.; Wang, Y. Shale Oil Potential of Lacustrine Black Shale in the Eocene Dongying Depression: Implications for Geochemistry and Reservoir Characteristics. *AAPG Bull.* **2017**, *101*, 1835–1858. [CrossRef]
3. Zhang, W.; Xie, L.; Yang, W.; Qin, Y.; Peng, P. Micro Fractures and Pores in Lacustrine Shales of the Upper Triassic Yanchang Chang7 Member, Ordos Basin, China. *J. Pet. Sci. Eng.* **2017**, *156*, 194–201. [CrossRef]
4. Zou, Y.-R.; Sun, J.-N.; Li, Z.; Xu, X.; Li, M.; Peng, P. Evaluating Shale Oil in the Dongying Depression, Bohai Bay Basin, China, Using the Oversaturation Zone Method. *J. Pet. Sci. Eng.* **2018**, *161*, 291–301. [CrossRef]
5. He, W.; Liu, Y.; Wang, D.; Lei, D.; Liu, G.; Gao, G.; Huang, L.; Qi, Y. Geochemical Characteristics and Process of Hydrocarbon Generation Evolution of the Lucaogou Formation Shale, Jimsar Depression, Junggar Basin. *Energies* **2022**, *15*, 2331. [CrossRef]

6. Yang, W.; Zuo, R.; Wang, X.; Song, Y.; Jiang, Z.; Luo, Q.; Zhai, J.; Wang, Q.; Zhang, C.; Zhang, Z. Sensitivity of Lacustrine Stromatolites to Cenozoic Tectonic and Climatic Forcing in the Southern Junggar Basin, NW China: New Insights from Mineralogical, Stable and Clumped Isotope Compositions. *Palaeogeogr. Palaeoclimatol. Palaeoecol.* **2019**, *514*, 109–123. [CrossRef]
7. Lazar, O.R.; Bohacs, K.M.; Macquaker, J.H.S.; Schieber, J.; Demko, T.M. Capturing Key Attributes of Fine-Grained Sedimentary Rocks In Outcrops, Cores, and Thin Sections: Nomenclature and Description Guidelines. *J. Sediment. Res.* **2015**, *85*, 230–246. [CrossRef]
8. Talbot, M.R. The Origins of Lacustrine Oil Source Rocks: Evidence from the Lakes of Tropical Africa. *Geol. Soc. Lond. Spec. Publ.* **1988**, *40*, 29–43. [CrossRef]
9. Ma, P.; Lin, C.; Li, G.; Dong, C.; Jiang, L.; Du, X.; Ren, M.; Liu, W.; Zhao, Z.; Yuan, Y.; et al. Lithofacies Characteristics and Sweet Spot Distribution of Lacustrine Shale Oil: A Case Study from the Dongying Depression, Bohai Bay Basin, China. *Lithosphere* **2022**, *2022*, 3135681. [CrossRef]
10. Zhi, D.; Xiang, B.; Zhou, N.; Li, E.; Zhang, C.; Wang, Y.; Cao, J. Contrasting Shale Oil Accumulation in the Upper and Lower Sweet Spots of the Lacustrine Permian Lucaogou Formation, Junggar Basin, China. *Mar. Pet. Geol.* **2023**, *150*, 106178. [CrossRef]
11. Zou, C.; Yang, Z.; Zhu, R.; Wu, S.; Fu, J.; Lei, D.; Hou, L.; Lin, S.; Pan, S. Geologic Significance and Optimization Technique of Sweet Spots in Unconventional Shale Systems. *J. Asian Earth Sci.* **2019**, *178*, 3–19. [CrossRef]
12. Chen, Z.; Jiang, W.; Zhang, L.; Zha, M. Organic Matter, Mineral Composition, Pore Size, and Gas Sorption Capacity of Lacustrine Mudstones: Implications for the Shale Oil and Gas Exploration in the Dongying Depression, Eastern China. *AAPG Bull.* **2018**, *102*, 1565–1600. [CrossRef]
13. Hill, R.J.; Jarvie, D.M.; Zumberge, J.; Henry, M.; Pollastro, R.M. Oil and Gas Geochemistry and Petroleum Systems of the Fort Worth Basin. *AAPG Bull.* **2007**, *91*, 445–473. [CrossRef]
14. Cai, J.; Cai, C.; Lu, L.; Jiang, Q.; Ma, X.; He, J. Hydrous Pyrolysis of Source Rocks with Different Maturity in Dongying Sag, Bohai Bay Basin, China: Implications for Shale Oil Exploration and Development. *Energies* **2023**, *16*, 6219. [CrossRef]
15. Zhang, S.; Liu, H.; Liu, Y.; Wang, Y.; Wang, M.; Bao, Y.; Hu, Q.; Li, Z.; Zhang, S.; Yao, S.; et al. Main Controls and Geological Sweet Spot Types in Paleogene Shale Oil Rich Areas of the Jiyang Depression, Bohai Bay Basin, China. *Mar. Pet. Geol.* **2020**, *111*, 576–587. [CrossRef]
16. Li, X.; Chen, K.; Li, P.; Li, J.; Geng, H.; Li, B.; Li, X.; Wang, H.; Zang, L.; Wei, Y.; et al. A New Evaluation Method of Shale Oil Sweet Spots in Chinese Lacustrine Basin and Its Application. *Energies* **2021**, *14*, 5519. [CrossRef]
17. Liu, S.; Liu, Y.; Zhang, X.; Guo, W.; Kang, L.; Yu, R.; Sun, Y. Geological and Engineering Integrated Shale Gas Sweet Spots Evaluation Based on Fuzzy Comprehensive Evaluation Method: A Case Study of Z Shale Gas Field HB Block. *Energies* **2022**, *15*, 602. [CrossRef]
18. Tang, X.; Jiang, Z.; Huang, H.; Jiang, S.; Yang, L.; Xiong, F.; Chen, L.; Feng, J. Lithofacies Characteristics and Its Effect on Gas Storage of the Silurian Longmaxi Marine Shale in the Southeast Sichuan Basin, China. *J. Nat. Gas Sci. Eng.* **2016**, *28*, 338–346. [CrossRef]
19. Yang, W.; Zuo, R.; Jiang, Z.; Chen, D.; Song, Y.; Luo, Q.; Wang, Q.; Zhu, H. Effect of Lithofacies on Pore Structure and New Insights into Pore-Preserving Mechanisms of the over-Mature Qiongzhusi Marine Shales in Lower Cambrian of the Southern Sichuan Basin, China. *Mar. Pet. Geol.* **2018**, *98*, 746–762. [CrossRef]
20. Liang, C.; Zhang, J.; Zhu, J.; Wang, W.; Guo, H.; Tan, Q.; Zhao, J.; Luo, H. The Influence of Continental Mixosedimentite Sedimentation on Shale Oil Sweet Spot: A Case Study on the Permian Lucaogou Formation in the Jimsar Sag of the Junggar Basin, NW China. *Front. Earth Sci.* **2023**, *10*, 929768. [CrossRef]
21. Saggaf, M.M.; Nebrija, E.L. A Fuzzy Logic Approach for the Estimation of Facies from Wire-Line Logs. *AAPG Bull.* **2003**, *87*, 1223–1240. [CrossRef]
22. Jungmann, M.; Kopal, M.; Clauser, C.; Berlage, T. Multi-Class Supervised Classification of Electrical Borehole Wall Images Using Texture Features. *Comput. Geosci.* **2011**, *37*, 541–553. [CrossRef]
23. Liu, B.; Wang, H.; Fu, X.; Bai, Y.; Bai, L.; Jia, M.; He, B. Lithofacies and Depositional Setting of a Highly Prospective Lacustrine Shale Oil Succession from the Upper Cretaceous Qingshankou Formation in the Gulong Sag, Northern Songliao Basin, Northeast China. *AAPG Bull.* **2019**, *103*, 405–432. [CrossRef]
24. Li, Y.; Li, Z.-X.; Wang, D.-D.; Song, G.-Z.; Lv, D.-W.; Lusheng-Yin; Liu, H.-Y.; Xu, C.-H.; Kong, F.-F.; Han, Q.-Y. Continental Lacustrine Fine-Grained Lithofacies, Assemblage and Geological Significance of Unconventional Petroleum: A Case Study of the Paleogene of the 4th Member of the Shahejie Formation in the Jiyang Depression. *Energy Explor. Exploit.* **2023**, *41*, 62–83. [CrossRef]
25. Liu, B.; Zhao, X.; Fu, X.; Yuan, B.; Bai, L.; Zhang, Y.; Ostadhassan, M. Petrophysical Characteristics and Log Identification of Lacustrine Shale Lithofacies: A Case Study of the First Member of Qingshankou Formation in the Songliao Basin, Northeast China. *Interpretation* **2020**, *8*, SL45–SL57. [CrossRef]
26. Ou, C.; Li, C.; Rui, Z.; Ma, Q. Lithofacies Distribution and Gas-Controlling Characteristics of the Wufeng–Longmaxi Black Shales in the Southeastern Region of the Sichuan Basin, China. *J. Pet. Sci. Eng.* **2018**, *165*, 269–283. [CrossRef]
27. Chen, S.; Zhang, S.; Wang, Y.; Tan, M. Lithofacies Types and Reservoirs of Paleogene Fine-Grained Sedimentary Rocks Dongying Sag, Bohai Bay Basin, China. *Pet. Explor. Dev.* **2016**, *43*, 218–229. [CrossRef]

28. Raeesi, M.; Moradzadeh, A.; Doulati Ardejani, F.; Rahimi, M. Classification and Identification of Hydrocarbon Reservoir Lithofacies and Their Heterogeneity Using Seismic Attributes, Logs Data and Artificial Neural Networks. *J. Pet. Sci. Eng.* **2012**, *82–83*, 151–165. [CrossRef]
29. Wang, G.; Carr, T.R.; Ju, Y.; Li, C. Identifying Organic-Rich Marcellus Shale Lithofacies by Support Vector Machine Classifier in the Appalachian Basin. *Comput. Geosci.* **2014**, *64*, 52–60. [CrossRef]
30. Zhao, Z.; Su, S.; Shan, X.; Li, X.; Zhang, J.; Jing, C.; Ren, H.; Li, A.; Yang, Q.; Xing, J. Lithofacies Identification of Shale Reservoirs Using a Tree Augmented Bayesian Network: A Case Study of the Lower Silurian Longmaxi Formation in the Changning Block, South Sichuan Basin, China. *Geoenergy Sci. Eng.* **2023**, *221*, 211385. [CrossRef]
31. Kim, J. Lithofacies Classification Integrating Conventional Approaches and Machine Learning Technique. *J. Nat. Gas Sci. Eng.* **2022**, *100*, 104500. [CrossRef]
32. Zhao, X.; Zhou, L.; Pu, X.; Jin, F.; Han, W.; Shi, Z.; Chen, C.; Jiang, W.; Guan, Q.; Xu, J.; et al. Theories, technologies and practices of lacustrine shale oil exploration and development: A case study of Paleogene Kongdian Formation in Cangdong sag, Bohai Bay Basin, China. *Pet. Explor. Dev.* **2022**, *49*, 707–718. [CrossRef]
33. Fang, Z.; Chen, S.; Pu, X.; Yan, J.; Chen, X. Control of Sedimentary Environment on the Lithofacies of Lacustrine Fine-Grained Sedimentary Rocks in the Second Member of the Kongdian Formation in the Cangdong Sag, Bohai Bay Basin, China. *Geol. J.* **2022**, *57*, 2321–2345. [CrossRef]
34. Hao, F.; Zou, H.; Gong, Z.; Deng, Y. Petroleum Migration and Accumulation in the Bozhong Sub-Basin, Bohai Bay Basin, China: Significance of Preferential Petroleum Migration Pathways (PPMP) for the Formation of Large Oilfields in Lacustrine Fault Basins. *Mar. Pet. Geol.* **2007**, *24*, 1–13. [CrossRef]
35. Zhao, X.; Pu, X.; Yan, J.; Jin, F.; Shi, Z.; Chai, G.; Han, W.; Liu, Y.; Jiang, W.; Chen, C.; et al. Cycles of Fine-Grained Sedimentation and Their Influences on Organic Matter Distribution in the Second Member of Paleogene Kongdian Formation in Cangdong Sag, Bohai Bay Basin, East China. *Pet. Explor. Dev.* **2023**, *50*, 534–546. [CrossRef]
36. Deng, Y.; Chen, S.; Pu, X.; Yan, J.; Cui, J. Formation Mechanism and Environmental Evolution of Fine-Grained Sedimentary Rocks from the Second Member of Kongdian Formation in the Cangdong Sag, Bohai Bay Basin. *Oil Gas Geol.* **2020**, *41*, 811–823, 890. [CrossRef]
37. Zhao, X.; Zhou, L.; Pu, X.; Jin, F.; Jiang, W.; Xiao, D.; Han, W.; Shi, Z. Development and exploration practice of the concept of hydrocarbon accumulation in rifted-basin troughs: A case study of Paleogene Kongdian Formation in Cangdong sag, Bohai Bay Basin. *Pet. Explor. Dev.* **2018**, *45*, 1166–1176. [CrossRef]
38. Pu, X.; Zhou, L.; Han, W.; Zhou, J.; Wang, W.; Zhang, W.; Chen, S.; Shi, Z.; Liu, S. Geologic Features of Fine-Grained Facies Sedimentation and Tight Oil Exploration: A Case from the Second Member of Paleogene Kongdian Formation of Cangdong Sag, Bohai Bay Basin. *Pet. Explor. Dev.* **2016**, *43*, 26–35. [CrossRef]
39. Zhao, X.; Pu, X.; Han, W.; Zhou, L.; Shi, Z.; Chen, S.; Xiao, D. A New Method for Lithology Identification of Fine Grained Deposits and Reservoir Sweet Spot Analysis: A Case Study of Kong 2 Member in Cangdong Sag, Bohai Bay Basin, China. *Pet. Explor. Dev.* **2017**, *44*, 524–534. [CrossRef]
40. Zhao, X.; Pu, X.; Jin, F.; Chen, C.; Shi, Z.; Chai, G.; Han, W.; Jiang, W.; Guan, Q.; Zhang, W. Enrichment Law and Favorable Exploration Area of Shale-Type Shale Oil in Huanghua Depression. *ACTA Pet. Sin.* **2023**, *44*, 158–175. [CrossRef]
41. Xin, B.; Hao, F.; Liu, X.; Han, W.; Xu, Q.; Guo, P.; Tian, J. Quantitative Evaluation of Pore Structures within Micron-Scale Laminae of Lacustrine Shales from the Second Member of the Kongdian Formation in Cangdong Sag, Bohai Bay Basin, China. *Mar. Pet. Geol.* **2022**, *144*, 105827. [CrossRef]
42. Liu, Z.; Huang, C.; Algeo, T.J.; Liu, H.; Hao, Y.; Du, X.; Lu, Y.; Chen, P.; Guo, L.; Peng, L. High-Resolution Astrochronological Record for the Paleocene-Oligocene (66–23 Ma) from the Rapidly Subsiding Bohai Bay Basin, Northeastern China. *Palaeogeogr. Palaeoclimatol. Palaeoecol.* **2018**, *510*, 78–92. [CrossRef]
43. Espitalie, J.; Madec, M.; Tissot, B.; Mennig, J.J.; Leplat, P. Source Rock Characterization Method for Petroleum Exploration. In *Offshore Technology Conference*; Offshore Technology Conference: Houston, TX, USA, 1977. [CrossRef]
44. Barrett, E.P.; Joyner, L.G.; Halenda, P.P. The Determination of Pore Volume and Area Distributions in Porous Substances. I. Computations from Nitrogen Isotherms. *J. Am. Chem. Soc.* **1951**, *73*, 373–380. [CrossRef]
45. Bhattacharya, S.; Carr, T.R.; Pal, M. Comparison of Supervised and Unsupervised Approaches for Mudstone Lithofacies Classification: Case Studies from the Bakken and Mahantango-Marcellus Shale, USA. *J. Nat. Gas Sci. Eng.* **2016**, *33*, 1119–1133. [CrossRef]
46. Wheeler, H.E. Baselevel, Lithosphere Surface, and Time-Stratigraphy. *Geol. Soc. Am. Bull.* **1964**, *75*, 599–610. [CrossRef]
47. Yuan, R.; Zhu, R.; Qu, J.; Wu, J.; You, X.; Sun, Y.; Zhou (Nancy), Y. Utilizing Integrated Prediction Error Filter Analysis (INPEFA) to Divide Base-Level Cycle of Fan-Deltas: A Case Study of the Triassic Baikouquan Formation in Mabei Slope Area, Mahu Depression, Junggar Basin, China. *Open Geosci.* **2018**, *10*, 79–86. [CrossRef]
48. Tissot, B.P.; Pelet, R.; Ungerer, P. Thermal History of Sedimentary Basins, Maturation Indices, and Kinetics of Oil and Gas Generation. *AAPG Bull.* **1987**, *71*, 1445–1466. [CrossRef]
49. Applied Source Rock Geochemistry. In *The Petroleum System—From Source to Trap*; American Association of Petroleum Geologists: Tulsa, OK, USA, 1994; pp. 93–120. [CrossRef]
50. Jarvie, D.M. Shale Resource Systems for Oil and GasPart 2—Shale-Oil Resource Systems. In *Shale Reservoirs—Giant Resources for the 21st Century*; American Association of Petroleum Geologists: Tulsa, OK, USA, 2012. [CrossRef]

51. Loucks, R.G.; Ruppel, S.C. Mississippian Barnett Shale: Lithofacies and Depositional Setting of a Deep-Water Shale-Gas Succession in the Fort Worth Basin, Texas. *AAPG Bull.* **2007**, *91*, 579–601. [CrossRef]
52. Loucks, R.G.; Reed, R.M.; Ruppel, S.C.; Hammes, U. Spectrum of Pore Types and Networks in Mudrocks and a Descriptive Classification for Matrix-Related Mudrock Pores. *AAPG Bull.* **2012**, *96*, 1071–1098. [CrossRef]
53. Brunauer, S.; Emmett, P.H.; Teller, E. Adsorption of Gases in Multimolecular Layers. *J. Am. Chem. Soc.* **1938**, *60*, 309–319. [CrossRef]
54. Sing, K.S.W. Reporting Physisorption Data for Gas/Solid Systems with Special Reference to the Determination of Surface Area and Porosity (Recommendations 1984). *Pure Appl. Chem.* **1985**, *57*, 603–619. [CrossRef]
55. Ma, B.; Hu, Q.; Yang, S.; Yin, N.; Qiao, H.; Zhang, T.; Meng, M. Multiple Approaches to Quantifying the Effective Porosity of Lacustrine Shale Oil Reservoirs in Bohai Bay Basin, East China. *Geofluids* **2020**, *2020*, 8856620. [CrossRef]
56. Rouquerol, J.; Avnir, D.; Fairbridge, C.W.; Everett, D.H.; Haynes, J.M.; Pernicone, N.; Ramsay, J.D.F.; Sing, K.S.W.; Unger, K.K. Recommendations for the Characterization of Porous Solids (Technical Report). *Pure Appl. Chem.* **1994**, *66*, 1739–1758. [CrossRef]
57. Han, W.; Dou, Y.; Li, H.; Zhang, W.; Shi, Z.; Dong, J. Optimization and Application of Brittleness Evaluation Methods for Continental Organic-Rich Shale: A Case Study of the 2nd Member of Kongjiadian Formation, Cangdong Sag, Bohai Bay Basin. *Oil Drill. Prod. Technol.* **2022**, *44*, 191–198. [CrossRef]
58. Zou, C.; Dong, D.; Yang, H.; Wang, Y.; Huang, J.; Wang, S.; Fu, C. Conditions of Shale Gas Accumulation and Exploration Practices in China. *Nat. Gas Ind.* **2011**, *31*, 26–39. [CrossRef]
59. Jarvie, D.M.; Hill, R.J.; Ruble, T.E.; Pollastro, R.M. Unconventional Shale-Gas Systems: The Mississippian Barnett Shale of North-Central Texas as One Model for Thermogenic Shale-Gas Assessment. *AAPG Bull.* **2007**, *91*, 475–499. [CrossRef]
60. Han, W.; Zhao, X.; Jin, F.; Pu, X.; Chen, S.; Mu, L.; Zhang, W.; Shi, Z.; Wang, H. Sweet Spot Evaluation and Exploration Practice of Lacustrine Shale Oil of the Second Member of Kongdian Formation in Cangdong Sag, Bohai Bay Basin. *Pet. Explor. Dev.* **2021**, *48*, 900–910. [CrossRef]

minerals MDPI

Article

The Effect of Thermal Maturity on the Pore Structure Heterogeneity of Xiamaling Shale by Multifractal Analysis Theory: A Case from Pyrolysis Simulation Experiments

Wei Wu [1], Zhikai Liang [2,3,*], Liang Xu [1], Yong Liu [4], Yi Li [1], Xianglu Tang [2,3,*], Yingzi Yin [1] and Yao Chen [1]

1 Shale Gas Institute, PetroChina Southwest Oil & Gas Field Company, Chengdu 610056, China; wuweipetrochina@163.com (W.W.); xul_2021@petrochina.com.cn (L.X.); liyi03@petrochina.com.cn (Y.L.); yinyingzi@petrochina.com.cn (Y.Y.); chenyao_ls@petrochina.com.cn (Y.C.)
2 Unconventional Oil and Gas Science and Technology Research Institute, China University of Petroleum, Beijing 102249, China
3 State Key Laboratory of Petroleum Resources and Prospecting, China University of Petroleum, Beijing 102249, China
4 PetroChina Southwest Oil & Gas Field Company, Chengdu 610056, China; liuyongpetrochina@163.com
* Correspondence: liangzhikai2020@163.com (Z.L.); tangxl@cup.edu.cn (X.T.)

Abstract: Shale oil and gas, as source-reservoir-type resources, result from organic matter hydrocarbon generation, diagenesis, and nanoscale pore during the evolution processes, which are essential aspects of shale gas enrichment and reservoir formation. To investigate the impact of diagenetic hydrocarbons on shale pore heterogeneity, a thermal simulation of hydrocarbon formation was conducted on immature shale from the Middle Proterozoic Xiamaling Formation in the Zhangjiakou area, covering stages from mature to overmature. Nuclear magnetic resonance (NMR) instruments analyzed the microstructure of the thermally simulated samples, and the multifractal model quantitatively assessed pore development and heterogeneity in the experimental samples. The results reveal that the quartz and clay mineral contents show alternating trends with increasing temperature. Organic matter dissolution intensifies while unstable mineral content decreases, promoting clay mineral content development. Pyrolysis intensity influences Total Organic Carbon (TOC), which reduces as hydrocarbons are generated and released during simulation. Porosity exhibits a decreasing–increasing–decreasing trend during thermal evolution, peaking at high maturity. At maturity, hydrocarbon generation obstructs pore space, resulting in higher levels of bound fluid porosity than those of movable fluid porosity. Conversely, high maturity leads to many organic matter micropores, elevating movable fluid porosity and facilitating seepage. Shale pore heterogeneity significantly increases before 450 °C due to the dissolution of pores and the generation of liquid and gas hydrocarbons. In the highly overmature stage, pore heterogeneity tends to increase slowly, correlated with the generation of numerous micro- and nano-organic matter pores.

Keywords: shale; pore structure; multifractal; heterogeneity; thermal maturity; nuclear magnetic resonance; OM pore evolution

1. Introduction

With the exponential progress of horizontal drilling technology and extensive hydraulic fracturing techniques, shale oil and gas have achieved remarkable commercial triumphs in the domain of unconventional oil and gas exploration and development. Micro- and nano-pore spaces within shale reservoirs have increasingly captivated scholarly interest as they are considered fundamental in the storage of shale oil and gas [1–4]. In contrast to conventional reservoirs, nanoscale shale reservoir spaces exhibit a diverse range of pore size distribution, intricate pore shapes, and pronounced heterogeneity, imposing significant constraints on the efficient exploitation of shale gas [4–7]. Abundant evidence indicates that the intricate pore structure of shale is frequently synergistically governed by factors such as sedimentation and diagenesis, organic matter hydrocarbon generation, and drainage [8–15].

Thermal maturity exerts a more pronounced influence on the evolution of organic matter porosity than diagenesis on the brittle rock skeleton [12,13,16–19]. Firstly, the thermal maturation of organic matter (OM) governs the development of the organic matter pore network and significantly affects the shale reservoir capacity and microscopic storage mechanisms [20–22]. Secondly, organic matter maturity dictates the stage of shale hydrocarbon evolution, thereby determining the efficiency of shale hydrocarbon generation and release [10,23]. Hence, investigating the hydrocarbon generation and pore system evolution in shales at various stages of development holds immense importance in comprehending the mechanism behind shale gas enrichment and advancing the exploration and development of these formations.

A series of novel experiments have been used to study shale pore characteristics for shale reservoirs. Utilizing cutting-edge techniques such as optical microscopy (OM), scanning electron microscopy (SEM), and focused ion beam scanning electron microscopy (FIB-SEM), these experiments characterize pore morphology, surface porosity, pore connectivity, and other crucial attributes [6,24–28]. Simultaneously, the fluid injection technique was employed to derive the injected fluid volume and rate from various experimental approaches such as gas adsorption, mercury intrusion capillary pressure (MIP), nuclear magnetic resonance (NMR), and small/ultra-small-angle neutron scattering (SAXS/USAXS). These methodologies were utilized to calculate and obtain pertinent data regarding pore volume, pore surface area, average pore diameter, and pore size distribution [4,6,10,25,28–30]. In previous studies, the fractal attributes of the pores were investigated by establishing a power function correlation between the probing scales (relative pressure, transverse relaxation time, detection volume, capillary pressure) and the corresponding probing volumes (adsorption, Hg feed, cumulative saturation) [6,18,31,32]. For shale reservoirs, the fractal dimension is often used to represent the pore surface roughness and the heterogeneity characteristics of the pore structure [7,14,29,31,33]. However, the single fractal dimension proves insufficient in precisely depicting the heterogeneity features, and a multitude of parameters merely partially reflect the pore characteristics and fail to characterize the pore properties adequately. Multifractals, serving as extensions or superpositions of single fractal dimensions, not only enable the definition of system heterogeneity through the continuous dimensional spectrum but also provide adequate characterization for minor local variations [5,7,27,31,34]. In recent years, multifractals have been widely used in the Bakken Formation of the Williston Basin, the Shahezi Formation of the Songliao Basin, the Shahejie Formation of the Dongying Depression, and the Wufeng–Longmaxi Formation of the Sichuan Basin utilizing quantitative image analyses and pore structure data [6,7,27,31].

In recent years, the study of diagenetic processes and evolutionary mechanisms in organic matter-rich shale reservoirs has received increasing attention from geologists [8,9,11,35,36]. Currently, related studies are predominantly conducted through two approaches. The first method involves selecting mature samples and utilizing high-resolution scanning electron microscopy to observe these samples directly. This enables the analysis of vital aspects, such as the evolution of organic–inorganic pores, clay mineral transformations, and kerogen structure evolution, by comparing the pore morphology at various stages of thermal development [15,33,36–38]. However, this method overlooks the variations in the regional geological system and pore space heterogeneity within the geological background. Additionally, it fails to address the influence of tectonic deformation and fracture systems on the pore structure. As a result, direct observations do not unveil the entirety of the evolution of pore characteristics during shale hydrocarbon production. The second approach involves artificial maturation, wherein a low-maturity sample is carefully chosen, and a temperature sequence is applied to induce hydrocarbon production, enabling the quantification of pore evolution at different stages. Prior research has demonstrated that hydrocarbon thermal simulation experiments serve as vital tools in investigating hydrocarbon source rocks and assessing hydrocarbon resources [14,20,22,37,39–41]. They are instrumental in resolving critical hydrocarbon generation and depletion issues. Despite overcoming the challenges posed by sample heterogeneity, this method falls short of providing direct observations of pore morphology. It is impossible to quantitatively analyze the relationship between mineral and pore evolution characteristics within the same field of view and the pore morphology through fluid injection [8,12,18,19,35,42]. A series of results have been obtained regarding pore structure characteristics through thermal simulation experiments, but there is a lack of research on shale pore heterogeneity under conditions of diagenetic processes and hydrocarbon generation. There are limited investigations on the correlation between pore heterogeneity and hydrocarbon production and drainage, particularly concerning the thorough characterization of multifractal parameters during shale's thermal evolution. This inadequacy significantly impedes the comprehensive evaluation of shale reservoirs and hydrocarbon storage.

In this study, Middle Proterozoic Xiamaling shale, selected as a representative low-maturity shale, underwent high-temperature thermal simulation experiments to induce hydrocarbon generation at high maturity and overmaturity. These experiments, coupled with NMR, mineral analysis, and organic geochemistry, were employed to delineate the mineralogical and petrographic attributes as well as the evolution of the pore system. In addition, multifractal theory is utilized to study the pore structure heterogeneity characteristics at different thermal evolution stages, and a pore system evolution model is built. This comprehensive investigation seamlessly integrates the organic matter hydrocarbon generation process and pore heterogeneity, clarifying the favorable reservoir stages in shale evolution and addressing the previously underexplored changes in shale pore system heterogeneity. The findings from this study hold significant relevance for the exploration of highly mature and overmature marine shale gas, not only in China but also in other regions worldwide.

2. Geological Settings

The study area lies within the Zhangjiakou region of Hebei Province, bounded by the Kangbao–Weichang deep fault. It is bordered in the north by the Inner Mongolia–Xing'an plate and in the south by the relatively rigid North China plate (Figure 1). The Yanshan Fold Belt is situated in the central part of Hebei Province and represents one of the most extensively developed intraplate tectonic regions. During the Early Indosinian period, the tectonic direction was predominantly east–west (EW) with a relatively large scale, extensive range, and early formation, thereby influencing the tectonic structure in subsequent stages [12,24,35,43]. The Yanshan Fold Belt, being one of the most developed intraplate tectonic regions, exhibits a combination of NNE- and NS-trending tectonics, along with predominant EW-trending tectonics, which collectively define the primary tectonic features of this area. The deformation in this zone occurred after the Yanshan movement and is characterized by NE- and NNE-striking faults of considerable scale, widely distributed throughout the region [35].

North China is one of the most flourishing and remarkably preserved regions of Middle Neoproterozoic sedimentation in China. Within the Middle Neoproterozoic, the Great Wall System, the Jixian System, and the Qinbaikou System comprise 12 groups. The Jixian Group, in particular, encompasses the Wumishan Formation, the Hongshuizhuang Formation, the Tieling Formation, and the Xiamaling Formation. Measured sections of the Xiamaling Formation at Zhaojiashan, Huailai, reveal that this lower stratum was deposited in a tranquil sedimentary water environment, characterized by dark-colored shales interspersed with thin-slabbed siltstones and argillaceous siltstones as well as sandstone turbidites [35,43,44]. The predominant sediments in the Xiamaling Formation are shallow marine shelf deposits, followed by barrier coastal deposits. From the bottom to top, the sandstone exhibits an increase in coarseness and thickness of grain size. Notably, sea chlorite is observed in the sandstone layer at the bottom and within the middle, indicating the depositional environment as a barrier-free coastal stage [12,35,44].

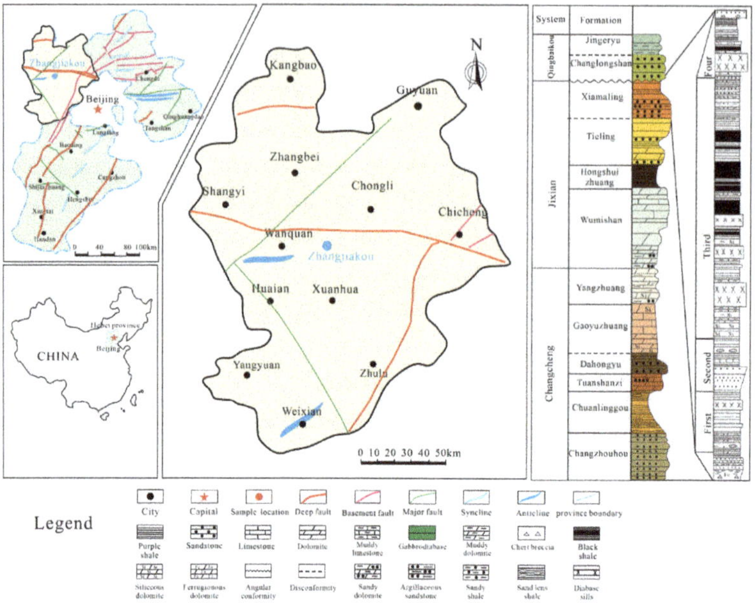

Figure 1. Structure map of Zhangjiakou displaying the tectonic framework and comprehensive lithological column of the Mesoproterozoic Xiamaling shale in Zhangjiakou, Hebei Province (Modified from Xu et al. [12]).

Xiamaling Formation shale hosts the most intricate and exquisitely preserved stratigraphy of Mesopotamia in China [43]. Notably, it represents the earliest example of low-maturity marine shale in China and stands as one of the oldest marine shales discovered globally [12,43]. During the Late Carboniferous period, the Xiamaling Formation in the Zhangjiakou area underwent deposition, embracing a ground temperature gradient of approximately 24 °C per kilometer. It experienced relatively low burial temperatures ($\leq$90 °C) throughout this process and underwent maturation, progressing from immature to early thermal maturity.

3. Samples and Methods

3.1. Samples and Experiments

In this paper, marine shale of the Middle Proterozoic Xiamaling Formation in the Xiahuayuan District of Zhangjiakou, Hebei Province, China, was used as the original sample for the thermal evolution simulation experiment. The TOC of the shales from the Xiamaling Formation ranges from 1.17% to 6.74%, with an average of more than 4.5%; vitrinite reflectance is the lowest at about 0.6%, kerogen type is dominated by type I-II_1, and the mineral composition is dominated by quartz and illite [10,12,35]. For the purpose of this study, a low-maturity, organic matter-rich sample was collected from the Middle Proterozoic Xiamaling Formation in the Xiahuayuan District of Zhangjiakou, Hebei Province, China.

Shale outcrops of the Xiamaling Formation in the field were selected, and 9 core columns were obtained in the horizontal lamination direction in order to avoid the effect of shale pore heterogeneity. At the end of the thermal simulation experiments, organic geochemical and petrological measurements were carried out for each sample to obtain the geochemical parameters and mineralogical compositions of the different degrees of thermal evolution. In order to study the characteristics of shale pore structure, saturation and centrifugal NMR experiments were used to obtain the fluid mobility characteristics of different stages of thermal evolution and combined with the theory of multiple fractals to study the change characteristics of pore heterogeneity. The specific research flow can be seen in the flow chart of Figure 2.

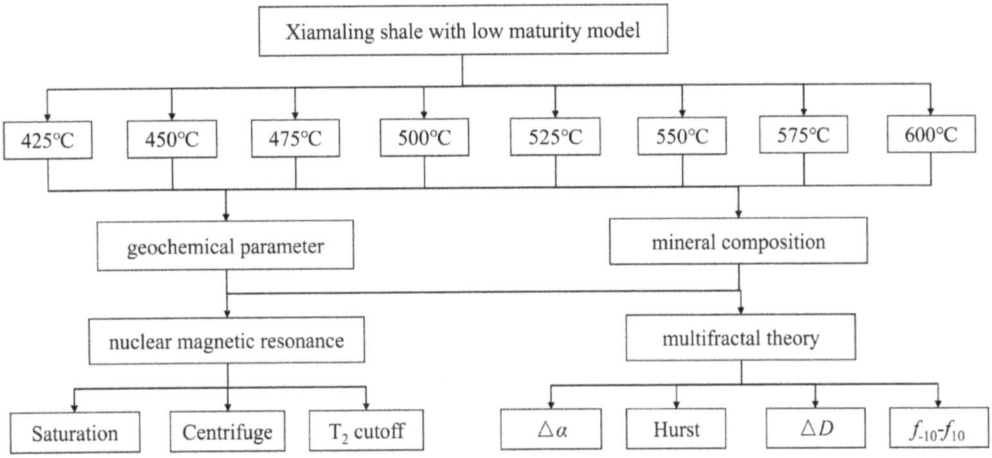

Figure 2. Flowchart of the experiment of this study.

Basic geochemical experiments and mineralogical analysis were performed at the State Key Laboratory of Oil and Gas Resources and Exploration, China University of Petroleum (Beijing). An XRD experiment was conducted to determine the mineral composition of the rock samples using a Bruker D8 Advance X-ray diffractometer with 45 KV Co Ka rays. The TOC test was performed on a Leco CS230 analyzer, and the samples were ground to about 100 μm before the test; the pyrolysis test was performed on an OGE-II analyzer.

Nuclear magnetic resonance (NMR) tests were performed on a RecCore-2500 instrument with a magnetic field at a resonance frequency of 2.38 MHz and a field strength of 0.12 T. The parameters of this NMR experiment were as follows: return interval of 0.2 ms; waiting time of 6 s; number of scans = 64; number of returns = 1024; experimental temperature of 25 °C; and humidity of 55%. The samples were individually centrifuged at 300 Psi to reach the critical bound saturated state and subsequently tested again by NMR to obtain the T_2 spectra of all samples under bound water conditions.

Typically, the T_2 spectrum of NMR is used to calculate porosity. Theoretically, the water-saturated porosity obtained by weighing is equivalent to the NMR porosity (φ_{NMR}). Generally, NMR porosity is divided into movable water porosity (φ_F) and bound water porosity (φ_B) based on the NMR porosity (saturated porosity) results.

$$\varphi_F = \varphi_{NMR}\frac{FFI}{FFI+BVI} \quad (1)$$

$$\varphi_B = \varphi_{NMR}\frac{BVI}{FFI+BVI} \quad (2)$$

where φ_{NMR} is the nuclear magnetic porosity (full water porosity), %; flexible fluid saturation (*FFI*) represents the region between the saturated water state and the curve after centrifugal treatment (percentage of connected pores in the sample) [45]. Bound fluid saturation (*BVI*) represents the region between the sample test curve after centrifugal treatment and the cross-axis (100-*FFI*); φ_F is the movable fluid porosity. The T_2 spectra in the saturated and centrifuged states were converted to the cumulative porosities in the saturated and centrifuged states, corresponding to *BVI* + *FFI*, respectively. The resulting porosities correspond to *BVI* + *FFI*, total porosity (φ_{NMR}), and *FFI* porosity (φ_F). The *BVI* porosity (φ_F) is the difference between φ_{NMR} and φ_F.

3.2. Hydrous Pyrolysis Experiments

Thermal simulation experiments on shale were conducted at the Institute of Unconventional Oil and Gas Science and Technology, China University of Petroleum (Beijing). The experimental setup, graciously provided by Jiangsu Hongbo Machinery Manufacturing Co., Ltd., Nantong, China, consists of four sets of heating core components, allowing for simultaneous multi-sample and multi-temperature point heating (Figure 3). The chosen thermal simulation system operates in an open setup. Different maturity degrees of shale samples were obtained by subjecting them to varying temperature points, following Sweeney and Burnham Easy Ro% parameter at a fixed rate of 10 °C/h under 5 MPa of hydrostatic pressure [46].

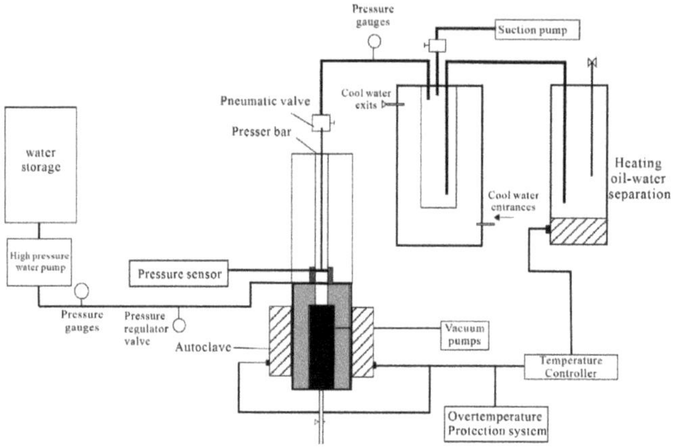

Figure 3. The model to heat cores at high-pressure and temperature.

The specific experimental procedure is as follows:

(1) An original sample of lower-maturity, organic matter-rich marine shale from the Xiamaling Formation was selected and divided into eight parts. Drilling columns in the horizontal laminar direction maximizes overcoming the heterogeneity of the shale;

(2) Before the experiment, the samples were thoroughly washed with water. The initial temperature of heating was 425 °C and the ending temperature was 600 °C. The sample type was a standard core column of 5 cm length and 2.5 cm diameter. Static rock pressure was added during the experiment, but no fluid pressure or water was added. Based on the results of the Easy R_o% equivalent vitrinite reflectance, the samples were heated to different temperatures to create a series of highly mature to overly mature thermal simulation samples;

(3) After the simulation, sample residues were collected and weighed for further geochemical and petrographic tests for NMR analysis. The porosity evolution was controlled by heating time (temperature), assuming constant variability, heating rate, and heating environment.

3.3. Multifractal Theory

Multifractals can be viewed as an extension of fractal theory sued to study the normalized probability distribution of a target over a test range. In contrast to single fractal dimensions, the multifractal approach uses generalized fractal dimension spectra and multifractal singularity spectra to characterize multifractal features [47,48].

Multifractals are described in detail in some works in the literature. First, the pore size distribution curve was log-transformed over the test intervals into a dimensionless interval with 100 equidistant lengths J.

$$\varphi_i = \log(\phi_i/\phi_1),\ i = 1,2,\ldots \tag{3}$$

where ϕ_i is the pore size measured, nm; ϕ_1 is the minimum detection pore size, nm; φ_i is the dimensionless value transformed for the PSD detection range. In interval J, the interior of the PSD interval has $K(\varepsilon) = 2^k$ subintervals of length ε. The interpolation method allows the inclusion of at least 1 measurement in each subinterval. $K_i(\varepsilon)$ is the probability density (percentage) of the pore size distribution inside each subinterval, and a family of partition functions is constructed for $N_i(\varepsilon)$:

$$u_i(q,\varepsilon) = \frac{K_i(\varepsilon)}{\sum_{i-1}^{N(\varepsilon)} K_{i(\varepsilon)}} \tag{4}$$

where $u_i(q,\varepsilon)$ is the q-order probability value of the i-th subinterval in the PSD data and $\sum_{i-1}^{N(\varepsilon)} K_{i(\varepsilon)}$ is the sum of the q-order probabilities of all subinterval values.

Therefore, the multifractal generalized dimension spectrum D_q of PSD is formulated as [49]:

$$Dq = \begin{cases} \frac{1}{q-1}\underset{\varepsilon\to 0}{\mathrm{Lim}}\frac{\log K_{q(\varepsilon)}}{\log \varepsilon}, & q \neq 1 \\ \frac{\underset{\varepsilon\to 0}{\mathrm{Lim}}\sum_{i=1}^{K(\varepsilon)} (P_i(\varepsilon)\ln P_i(\varepsilon))}{\ln \varepsilon}, & q = 1 \end{cases} \tag{5}$$

where q = 0, 1, and 2. The corresponding D_0, D_1, and D_2 represent the capacity, information entropy, and association dimensions, respectively [49]. For partitions of size ε, the probability of quality on a given partition compared to the sum of possibilities on the total partition can be quantified as:

$$u_i(q,\varepsilon) = \frac{P_i(\varepsilon)^q}{K_{q(\varepsilon)}} \tag{6}$$

where the multifractal singularity index $\alpha(q)$ [47] is:

$$\alpha(q) = \underset{\varepsilon\to 0}{\mathrm{Lim}}\frac{\sum_{i=1}^{N(\varepsilon)} u_i(q,\varepsilon)\log P_i(\varepsilon)}{\log \varepsilon} \tag{7}$$

The multifractal spectral function $f(\alpha(q))$ for the PSD particle size distribution of $\alpha(q)$ can be written as [47]:

$$f(\alpha(q)) = \lim_{\varepsilon\to 0}\frac{\sum_{i=1}^{N(\varepsilon)} u_i(q,\varepsilon)\log u_i(q,\varepsilon)}{\log \varepsilon} \tag{8}$$

In summary, the multifractal spectrum $\alpha \sim f(\alpha)$ and the generalized dimension spectrum $q \sim D_q$ are two different sets of parameters that characterize fractals.

4. Results

4.1. Mineral Composition and Organic Geochemistry

In this investigation, a meticulous examination was conducted on the geochemical attributes of samples Z0 to Z8. In particular, the original sample Z0 and the other samples (Z1 to Z8) were used for thermal simulation experiments. In this study, geochemical parameters of shale organic matter were obtained through experiments determining TOC, S_1, S_2, and T_{max}. S_1 represents the number of free hydrocarbons in the rock (mg HC/g rock); if $S_1 > 1$ mg/g, it indicates that there may be oil. S_2 represents the number of rock pyrolysis hydrocarbons (mg HC/g rock); S_2 represents the hydrocarbon generation potential of the source rock during the thermal evolution process. $S_1 + S_2$ is usually used to represent the hydrocarbon generation potential. T_{max} is the temperature at which casein cleavage produces the maximum content of hydrocarbons (i.e., the highest point of the S_2 peak), which is indicative of the thermal maturity of the organic matter. Table 1 shows a substantial Total Organic Carbon (TOC) content of 3.76%, with S_1 registering at 0.98 mg/g and S_2 at 25.43 mg/g of the Z0 sample. These results indicate that the original shale possesses an abundant OM and exhibits promising potential for hydrocarbon production.

T_{max} indicates the temperature corresponding to the maximum content of hydrocarbons produced by kerogen cracking, which can indicate the thermal maturity of OM. Sample Z0 displayed an EqRo of 0.56% and a T_{max} of 431 °C, signifying a state of low maturity. As the temperature increased, the maturity progressed from 1.22% to 4.07%, indicating a transition from maturity to an overmature gas generation stage during the simulation.

Table 1. Organic Geochemical Characteristics with different thermal maturities of the Xiamaling shale in Zhangjiakou, Hebei.

Sample ID	Temperature (°C)	EqRo (%)	TOC (%)	S_1 (mg/g)	S_2 (mg/g)	$S_1 + S_2$ (mg/g)	T_{max} (°C)
Z0	-	0.56	3.76	0.98	25.43	26.41	431
Z1	425	1.22	2.55	2.21	16.53	18.74	515
Z2	450	1.57	2.23	1.53	13.32	14.85	578
Z3	475	1.94	2.37	1.32	11.87	13.19	584
Z4	500	2.35	2.24	1.23	10.23	11.46	596
Z5	525	2.81	2.2	0.86	7.97	8.83	603
Z6	550	3.28	2.21	0.52	5.26	5.78	610
Z7	575	3.73	2.18	0.47	4.41	4.88	623
Z8	600	4.07	2.14	0.39	3.56	3.95	634

At the end of the hydrocarbon thermal simulation experiment, EqRo rose to 4.07% with a T_{max} of 634 °C. A notable reduction in the TOC content and $S_1 + S_2$ content between the simulated samples was observed as the maturity level increased, indicating the generation and expulsion of hydrocarbons during the simulation process. According to previous studies, the main maceral composition in this region is amorphous teleosts and lamellae, which appear flocculent, lamellar, and striated [35]. The maceral composition type is type II, indicating that OM generates hydrocarbons during deposition [43,44].

Table 2 illustrates the mineral compositions of the Xiamaling samples and their corresponding thermal simulations. The initial samples contained significant amounts of quartz (58.1%), less clay minerals (16.1%), and levels of pyrite below 5%. Throughout the simulation process, some minerals changed slightly, with the content of quartz gradually decreasing with maturity and the content of clay minerals slightly increasing. This may be due to the release of K ions from potassium feldspar after being subjected to organic acid dissolution during hydrocarbon production; at the same time, K^+ ions, the illite/smectite mixed layer, and kaolinite can further evolve into illite, resulting in an increase in the content of clay minerals [50].

Table 2. Mineralogical composition characteristics with different thermal maturities of the Xiamaling shale in Zhangjiakou, Hebei.

Sample ID	Quartz (%)	K-Feldspar (%)	Plagioclase (%)	Dolomite (%)	Pyrite (%)	Clay (%)
Z-0	58.1	2.5	2.1	18.1	3.1	16.1
Z-1	52.7	2.4	2	20.9	2.8	22
Z-2	52.7	2.4	1.9	18.3	2.4	22.7
Z-3	49.1	2.9	2.4	25.4	1.9	20.2
Z-4	52.4	2.5	2.2	25.1	1.6	16.2
Z-5	49.3	2.6	2.1	22.9	1.5	21.3
Z-6	45.7	2.5	2	23.1	1.4	24.3
Z-7	48.4	2.6	2.1	20.3	1.5	23.8
Z-8	45.4	2.6	2	22.3	1.4	27.7

4.2. NMR Parameters

In this experiment, the T_2 spectra measured in the saturated water state are presented (Figure 4). Shale exhibits two main types of T_2 spectra: single-peak and bimodal. Previous studies have revealed that most shale samples possess single-peak T_2 spectra, primarily comprising adsorption pores [51]. These spectra display isolated peaks with short relaxation times, typically ranging from 0.01 to 10 ms. Both single-peak and bimodal T_2 spectra indicate that shale porosity is dominated by micropores, while bimodal shales show better-developed connecting pores compared to single-peak ones. In this study, the T_2 spectra were predominantly bimodal, with the right peak (10–100 ms) significantly smaller than the left peak (0.01–10 ms). This indicates the presence of well-developed small pores, some mesopores, and macropores in the shale.

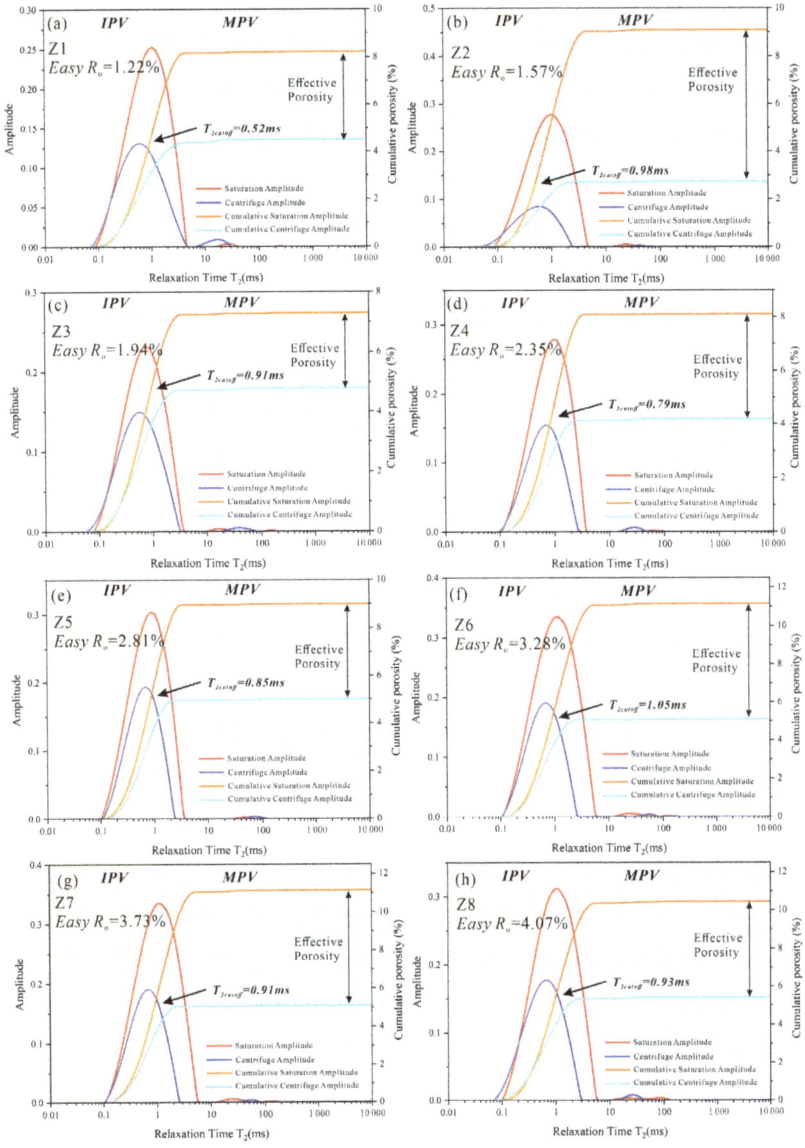

Figure 4. Typical NMR T_2 spectra and effective porosity under fully saturated and centrifuged conditions of different-maturity samples. (**a**) Sample: Z1; (**b**) Sample: Z2; (**c**) Sample: Z3; (**d**) Sample: Z4; (**e**) Sample: Z5; (**f**) Sample: Z6; (**g**) Sample: Z7; (**h**) Sample: Z8.

Bimodal T_2 spectra can be categorized into continuous and discontinuous types based on the presence or absence of interrupted curves. Continuous bimodal T_2 spectra suggest that shale reservoir micropores are well connected to macropores or even microfractures. On the other hand, discontinuous bimodal T_2 spectra indicate relatively closed pores with weak connections between different scales of pores. The NMR T_2 distributions of the natural shale samples from Xiamaling (Figure 4) show that all samples are characterized by a single peak, with T_2 values predominantly ranging from 0.1 to 10 ms. This observation suggests that the shale porosity is predominantly micro–mesoporous, with fewer macropores, during thermally simulated evolution. The pore spaces in the Xiamaling shale are mainly distributed in the range of 0.1–10 ms, with discontinuous curves in the bimodal T_2 spectra, indicating weak connectivity between pores of different scales, which may hinder fluid diffusion.

4.3. Results of the LF-NMR Measurements

The T_2 spectrum of the NMR experiment enables a quantitative description of core pore structure characteristics, allowing the calculation of key physical parameters, such as total porosity, permeability, movable fluid porosity, and bound fluid porosity (Table 3). Porosity is obtained by comparing fluid signals between dry and saturated samples [52,53]. Shale samples from the Xiamaling Formation exhibit a porosity ranging from 7.3% to 11.2%, with a mean value of 9.12%. The relationship between porosity and maturity calculated using NMR shows a gently increasing trend overall. Sample Z7 (Ro = 3.2%) has the highest porosity. T_2 cutoff values, derived from the difference between saturated and centrifuged samples, range from 0.52 to 1.05 ms, with an average of 0.86 ms (Table 3). The difference in T_2 cutoff among samples with different maturity levels is marginal, less than 1.1 ms (Table 3).

Table 3. Results of shale movable and bound fluid saturation, T_2 cutoff, and porosity by NMR.

Samples ID	T_{2c} (mc)	FFI (%)	BVI (%)	Φ_{NMR} (%)	ΦF (%)	ΦB (%)
Z1	0.52	69.73	30.27	9.1	6.35	2.75
Z2	0.98	45.1	54.9	8.23	3.71	4.52
Z3	0.91	34.05	65.95	7.3	2.49	4.81
Z4	0.79	48.17	51.83	8.11	3.91	4.2
Z5	0.85	44.35	55.65	9	3.99	5.01
Z6	1.05	45.31	54.69	9.57	4.34	5.23
Z7	0.91	54.23	45.77	11.16	6.05	5.11
Z8	0.91	47.94	52.06	10.45	5.01	5.44

The T_2 cutoff is larger when the shale contains more bound water, indicating higher bound fluid saturation. Shale reservoir adsorption capacity corresponds to a larger T_2 cutoff in the reservoir pores. FFI values range from 34.05% to 69.73%, with an average of 48.61%. Movable fluid porosities in the thermally simulated samples are all above 2%, ranging from 2.49% to 6.35%, with an average of 2.49%, showing an overall decreasing-to-increasing trend. BVI values range from 30.27% to 65.95%, with an average of 51.39%, and the distribution of bound fluid porosity in the thermally simulated samples ranges from 4.52% to 5.44%, with an average of 2.75%. The FFI parameters showed a U-shaped trend of decreasing and then increasing. A minimum value was observed in the Z3 sample which then increased with maturity.

4.4. Multifractal Characteristics Based on NMR Experiments

Figure 5 displays the enthralling double logarithmic correlation between the partition function $\eta(q, r)$ and the length scale r for the NMR data obtained from samples exhibiting distinct degrees of thermal evolution, with q ranging between −10 and 10. Upon observing Figure 4, it becomes evident that, at the same order, the pore structure of shale remarkably adheres to a segmented linear evolution law concerning $\log(\delta)$ to $\log(x)$, demonstrating an overall first-order linear exponential tendency. This trend is particularly pronounced in the higher-order segment ($q > 0$), which impeccably exhibits strict linear distribution characteristics. Consequently, for a fixed value of q, the curve maintains a consistent slope, while varying q values correspond to distinct slopes. The fitted linear trend lines manifest negative slopes for $q < 0$ and positive slopes for $q > 0$. According to the scale-free and self-similarity of the fractal theory and combined with the $\lg(\delta)$~$\lg(x)$ curve, it can be seen that the pore structure precisely reflects the multifractal characteristic.

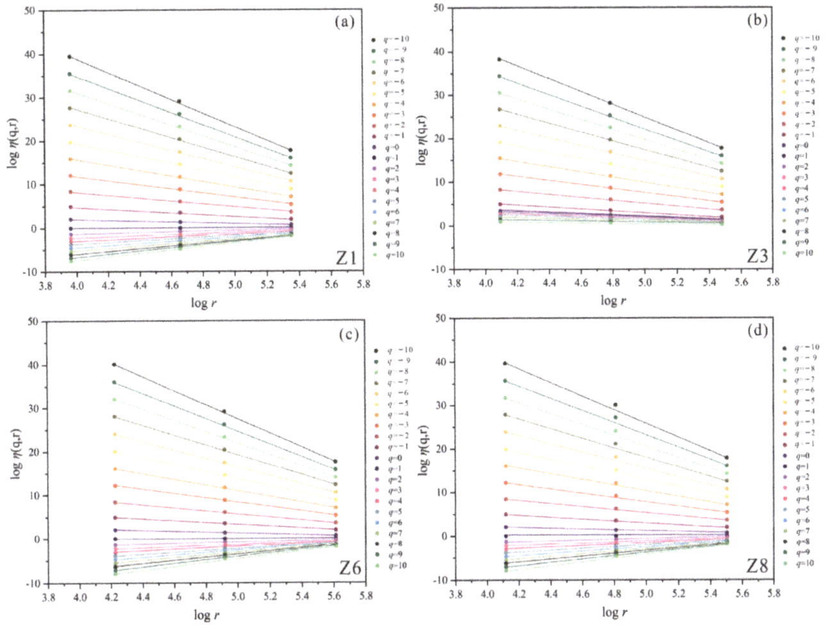

Figure 5. Log–log plot of partition function $\eta(q, r)$ versus length scale r of the different-maturity samples. (**a**) Sample: Z1; (**b**) Sample: Z3; (**c**) Sample: Z6; (**d**) Sample: Z8.

Figure 6 shows that, as thermal evolution varies, the generalized fractal dimension D_q follows an inverted "S" decreasing curve, where D_q decreases as q increases. The curve tends to bend in the opposite direction at $q = 0$, with generalized fractal dimensions at $q < 0$ surpassing those at $q > 0$. Specifically, D_{-10} ranges from 1.1577 to 1.5284, while D_{10} ranges from 0.4593 to 0.5379 (Table 4). Upon comparing different generalized fractal dimension curves, for the section with $q < 0$, D_q increases with higher stages of thermal evolution at the same q value. On the other hand, for the section with $q > 0$, D_q decreases with higher stages of thermal evolution at the same q value. This observation suggests that the pore structure of the shale reservoir becomes more intricate during advanced development, exhibiting a diversified trend. The D_0 values remain similar across all samples, indicating fractal measures with non-empty boxes and density probabilities at varying scales [6,18,31]. D_1 represents the information entropy dimension, often employed to characterize the concentration of shale pore distribution [31]. A larger D_1 implies a more concentrated pore distribution. The results demonstrate a decline in D_1 with increasing maturity, indicating a higher degree of localized pore concentration at high levels of maturity. D_2, the correlation dimension, indicates pore spacing homogeneity, with greater D_2 values indicating more uniform pore spacing [5,31]. The findings display a gradual decrease in D_2 with increasing maturity, signifying a more heterogeneous pore interval.

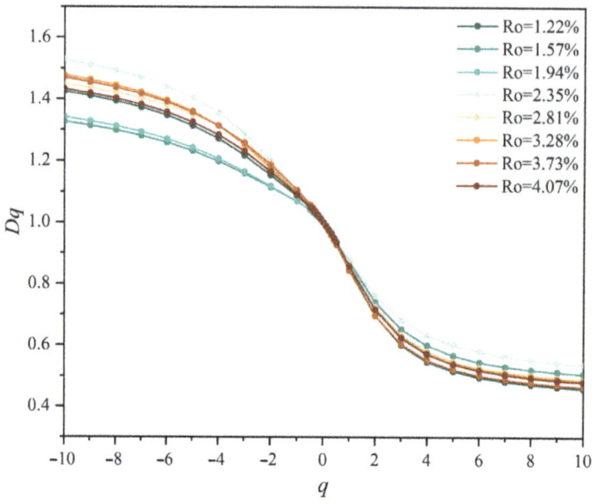

Figure 6. Generalized fractal dimension spectrum of the different-maturity samples.

Table 4. Characteristics of the generalized dimension from all the samples from NMR.

Samples ID	Temperature	D_0	D_1	D_2	$D_0 - D_1$	D_{-10}	D_{+10}	$D_{-10} - D_{+10}$	Hurst
Z0	-	1	0.8997	0.7682	0.1003	1.1577	0.5104	0.6472	0.8841
Z1	425	1	0.8498	0.6969	0.1502	1.4250	0.4593	0.9657	0.8484
Z2	450	1	0.8722	0.7404	0.1278	1.3268	0.5067	0.8201	0.8702
Z3	475	1	0.8661	0.7226	0.1339	1.3414	0.4818	0.8596	0.8613
Z4	500	1	0.8787	0.7631	0.1213	1.5284	0.5379	0.9904	0.8816
Z5	525	1	0.8610	0.7253	0.1390	1.4503	0.4891	0.9612	0.8626
Z6	550	1	0.8568	0.7165	0.1432	1.4786	0.4846	0.9940	0.8583
Z7	575	1	0.8435	0.6960	0.1565	1.4705	0.4646	1.0059	0.8480
Z8	600	1	0.8580	0.7158	0.1420	1.4330	0.4804	0.9527	0.8579

Multifractal spectra serve to quantitatively delineate the intricacy and irregularity of pore distribution in rock samples [5,31,34]. As depicted in Figure 7, the multifractal spectral curves exhibit an upward trend of convexity, all reaching maximum values. The span of the multifractal spectral curves exhibits diverse degrees of variability. $\Delta\alpha$, representing the spectral width of the multifractal spectra, depicts local characteristics of pores concerning their fractal structure and signifies the distribution of pore structures at different size scales [5,7,27,31]. In hydrocarbon thermal simulation conditions, $\Delta\alpha$ gradually increases with thermal maturity, indicating a complex distribution within the pores at high-maturity stage (Table 5).

Table 5. Characteristics of the multifractal singularity spectra from NMR.

Samples ID	Temperature	α_0	α_1	α_2	α_{-10}	α_{10}	$\alpha_{-10} - \alpha_0$	$\alpha_0 - \alpha_{+10}$	$\Delta\alpha$
Z0	-	1.0700	0.8997	0.6457	1.2356	0.4594	0.1655	0.6106	0.7761
Z1	425	1.1153	0.8498	0.5718	1.5662	0.4134	0.4509	0.7019	1.1528
Z2	450	1.0976	0.8722	0.6297	1.4540	0.4561	0.3564	0.6415	0.9979
Z3	475	1.0988	0.8661	0.6019	1.4726	0.4336	0.3737	0.6652	1.0389
Z4	500	1.1089	0.8787	0.6647	1.6807	0.4844	0.5718	0.6245	1.1963
Z5	525	1.1206	0.8610	0.6108	1.5822	0.4403	0.4616	0.6803	1.1419
Z6	550	1.1174	0.8568	0.6012	1.6259	0.4362	0.5085	0.6811	1.1897
Z7	575	1.1314	0.8435	0.5766	1.6156	0.4182	0.4842	0.7132	1.1974
Z8	600	1.1144	0.8580	0.5979	1.5727	0.4324	0.4582	0.6821	1.1403

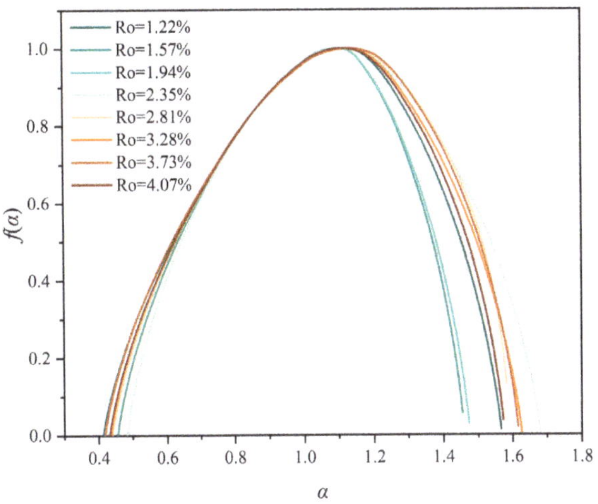

Figure 7. Multifractal singular spectrum of the different-maturity samples.

5. Discussion

5.1. Organic Geochemical Variation during Thermal Maturation

As the temperature increases, various processes, such as organic matter transformation, hydrocarbon generation, retention, and exhaustion, as well as organic–inorganic interactions in shale, contribute to changes in pore structure heterogeneity. The thermal maturity of the shale increased from 1.22% to 4.07%, and T_{max} rose from 515 °C to nearly 634 °C, signifying the peak of hydrocarbon production and the attainment of the overmature stage with increasing temperature (Figure 8). S_1 represents the number of free hydrocarbons in the rock (mg HC/g rock); if S1 > 1 mg/g, it indicates that there may be oil. S_2 represents the number of rock pyrolysis hydrocarbons (mg HC/g rock); S_2 represents the hydrocarbon generation potential of the source rock during the burial/thermal evolution process. $S_1 + S_2$ is usually used to represent the hydrocarbon generation potential. Throughout this process, the TOC content showed a slight decrease from 2.55% to 2.14%, and the $S_1 + S_2$ index decreased from 18.74 mg/g to 3.95 mg/g, indicating significant hydrocarbon generation during thermal evolution. The S_1 and S_2 content demonstrated a slight decrease with increasing temperature. The $S_1 + S_2$ and T_{max} gradually decreased, indicating secondary cracking of generated liquid hydrocarbons and the generation of numerous gas hydrocarbons (Figure 8).

During thermal evolution, notable changes occurred in the content of quartz, feldspar, and clay minerals. The quartz content decreased from 58% to 45% with increasing stages of thermal evolution (Figure 9). As thermal evolution enters the overmature stage and compaction increases with depth of burial, secondary quartz formed early in the diagenetic process experiences a partial dissolution process, resulting in a decrease in quartz content. Additionally, the heterogeneity of shale is believed to contribute to this phenomenon [54,55]. The overall change in feldspar content is not significant with increasing temperature. The organic acid produced by the hydrocarbon production process makes the feldspar content decrease. However, at the high evolutionary stage, the dissolving effect of organic acids in the samples is not strong, while recrystallization due to reverse dissolution results in small changes in feldspar content [56]. The clay content increases with maturity, primarily due to the reduction in montmorillonite content and an increase in illite content as thermal evolution progresses. Furthermore, feldspar and kaolinite are converted to illite when sufficient cations are present in the solution [54,55]. The transformation of smectite of the I/S to illite occurs extensively at 400–450 °C. However, due to the weak water–rock response and insufficient supply of K^+ in the reservoir, the feldspar dissolution remains incomplete [35]. Similarly, smectite/illite is not completely converted to illite due to the closed nature of the reservoir and formation system.

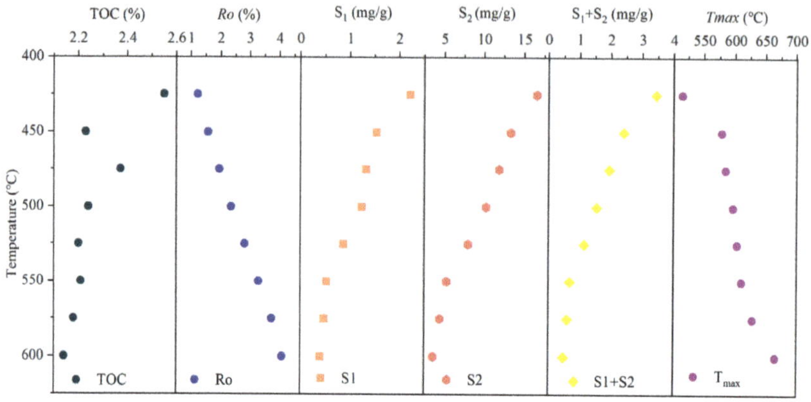

Figure 8. Geochemical indices' variation in the original shale sample and artificially thermal simulated samples.

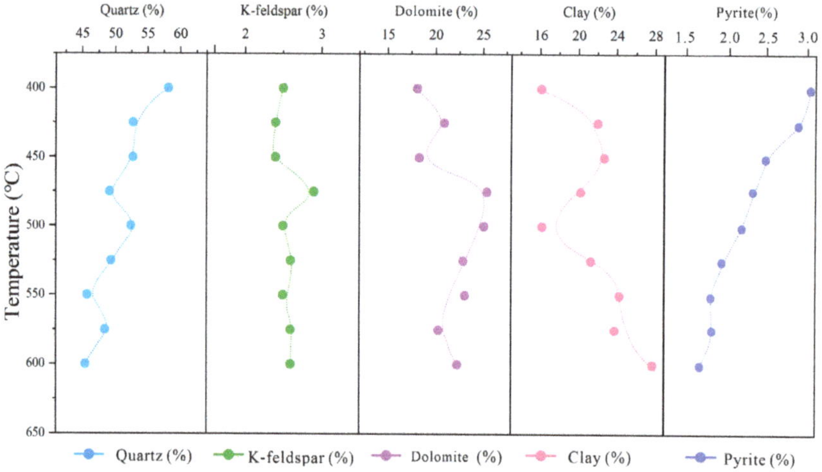

Figure 9. Mineral composition and content variation of the original shale sample and artificially thermal simulated samples.

5.2. The Variation of Pore Structure Heterogeneity during Thermal Maturation

Based on multifractal theory, this research selects various parameters to elucidate pore heterogeneity, connectivity, pore distribution concentration, uniformity of pore spacing, and localized variability at distinct evolutionary stages [5–7,14,27,31]. The thermal evolution findings demonstrate that the shale's multifractal parameters $\Delta\alpha$ and ΔD exhibit relatively parallel evolutionary trends, while Hurst, D_1, and D_2 manifest similar trends (Figure 9). For the analysis, this study considers $\Delta\alpha$ and Hurst as representative parameters, indicating a notable enhancement of pore heterogeneity due to thermal evolution. At 400 °C, the $\Delta\alpha$ value is low, indicating a less heterogeneous pore structure in the original surface sample. As temperature increases, $\Delta\alpha$ rapidly rises, possibly attributable to an increase in large pores and a decrease in mesopores during this stage. Subsequently, $\Delta\alpha$ values experience some reduction, potentially due to the blockage of numerous micropores and mesopores by crude oil and bitumen, leading to a partial decline in pore heterogeneity [11,35]. In the late stage of oil production, the retained oil diminishes, expelled oil increases, certain pore volumes rise, and pore heterogeneity gradually escalates [18]. At temperatures up to 550 °C, extensive solid bitumen cracking generates gas, yielding complex cellular pores and augmenting the specific surface area, resulting in the highest $\Delta\alpha$ value. At 600 °C, a degree of pore heterogeneity reduction occurs, likely attributed to robust compaction, leading to a decrease in the volume of meso- and macropores.

Regarding pore connectivity, the initial stage of hydrocarbon thermal simulation exhibits a rapid decline in the Hurst value, indicating the accumulation of retained hydrocarbons in the pore space during the oil production stage, which leads to the blockage of meso–macropores and a decrease in pore connectivity (Figure 9). Reduced pore connectivity is linked to the diminished pore volume between various pore types associated with this stage [20,57]. As the evolution progresses, the generation of bitumen yields numerous OM pores, resulting in a gradual increase in pore connectivity. The Hurst index reaches its peak value at 500 °C and gradually decreases with rising temperature. Although kerogen pores primarily form network pores, connecting a portion of the inorganic pores and microcracks, the overall effect of increasing temperature leads to elevated pressure, resulting in the closure of some mesopores and macropores, causing partial destruction of the brittle mineral skeleton and, thus, contributing to lower pore connectivity [14,18,22].

5.3. The Effect of Thermal Maturity (Ro) on Pore Structure

Organic matter maturity evolved from low to high during the thermal simulation experiments, and the relationships between NMR porosity, T_2 cutoff, FVI, BVI, and maturity are shown in Figure 10. As the maturity increases, the OM remains in the the hydrocarbon generation stage, and the OM pores develop in large quantities, resulting in a trend of decreasing and then increasing NMR porosity. Milliken et al. noted that shale porosity varies with thermal maturity due to OM pores [23]. From the stage of low maturity to maturity, the OM generates a large number of liquid hydrocarbons that block the pores, decreasing porosity. From the evolutionary stage of maturity to overmaturity, organic matter generates many pores, forming a network of pores, resulting in increasing porosity.

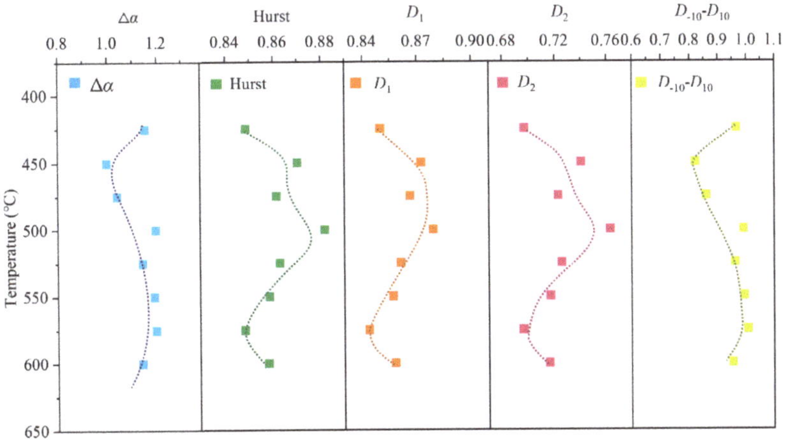

Figure 10. Multifractal parameters of pore structure evolution from the artificially thermal simulated samples.

The T_2 cutoff is commonly employed as a threshold to distinguish between movable and immovable fluids, serving as an indicator of fluid mobility within the shale (Figure 11). Higher T_2 cutoff values signify a larger portion of bound fluid, indicating a higher degree of bound fluid saturation. With increasing thermal evolution, the T_2 cutoff exhibits a rapid rise during maturation, reaches a peak, slightly decreases, and then begins to slowly increase at a Eq of 2.4%. Similarly, bound water porosity follows a similar trend as the T_2 cutoff. During the maturation stage, some hydrocarbons obstruct the inorganic–organic pore space, resulting in pore throat narrowing and higher bound fluid saturation [20,22,40,58]. After reaching 2.0% maturity, the heavy hydrocarbons present in the organic matter persist in undergoing cracking, leading to a continuous augmentation of organic matter pores. Consequently, both the number and volume of mesopores and macropores, along with shale matrix hydrocarbon production, observe a steady ascent [1]. Concurrently, diverse pore types establish connections, such as the formation of reticulated pores within the organic matter, facilitating the linkage of inorganic–organic pores, ultimately resulting in a reduction in bound fluid saturation. Upon reaching a maturity level of 2.5%, the overall BVI (Bulk Volume Irreducible) and T_2 parameters of the samples show marginal alterations, signifying the stabilization phase of kerogen gas and liquid hydrocarbon cracking gas [20,37,41]. In previous studies, it has been suggested that maturity levels greater than 3.5% result in more pronounced carbonization of OM, a decrease in the hydrocarbon production capacity of OM, and a reduction in the organic matter pore space [59,60]. In the present study, with a high level of maturity over 3.5%, gradually reduces at the rate of hydrocarbon evolution and porosity, causing the bound fluid saturation to also increase significantly, which is detrimental to the enrichment of hydrocarbons [59].

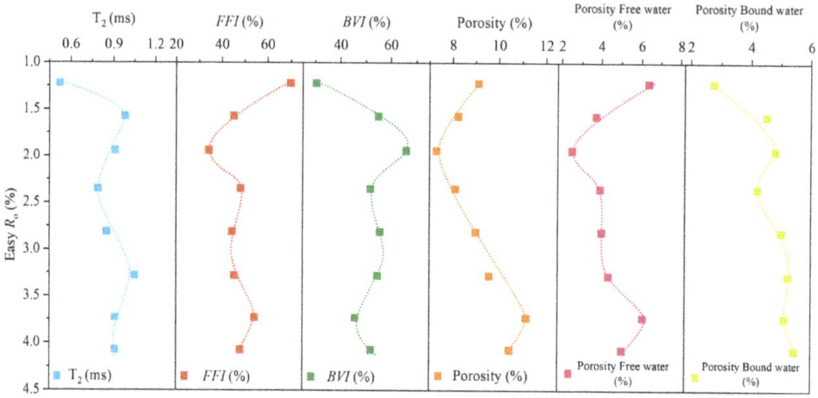

Figure 11. Petrophysical properties and pore structure parameter evolution from the original shale sample and artificially thermal simulated samples.

5.4. Pore Structure and Heterogeneity Evolution Model

A large number of studies have revealed the complex evolution process of the pore system during the hydrocarbon generation and formation stages of shale [6,27,34]. The pore structure characteristics of shale are mainly affected by the thermal development degree of the reservoir and are closely related to sedimentation, structure, and hydrocarbon generation (Figure 12). Previous studies have extensively studied the impact of thermal maturity on shale reservoirs, and various thermal simulation experiments have demonstrated the evolution of shale porosity [11,35,41].

This study focuses on the shale pore development and heterogeneity evolution process under mature-to-overmature conditions. At this stage (1.3% $\leq$ Ro < 2.0%), as the thermal evolution proceeds, some residual liquid hydrocarbons block the intra- and intergranular pores of inorganic mineral particles, causing a significant decrease in porosity. This may be a result of the decline in inorganic porosity of shale reservoirs due to cementation and compaction. The bound fluid saturation content is high and can reach a value higher than 60%. A large number of hydrocarbons results in smoother pore surfaces and reduced roughness, resulting in a reduction in $\Delta\alpha$ [20–22,29,57,58]. As the maturity level continues to increase, it is possible that less oil is retained in the shale reservoir and more oil is expelled, forming a large number of internal connecting pores [61]. Part of the bitumen fills the intergranular pores of brittle minerals and the intragranular pores of clay minerals. As the thermal maturity progresses, micropores begin to form, which enhances the heterogeneity of the pores. In the dry gas generation stage (Ro > 2.00%), OM reaches the overmature dry gas stage, kerogen and asphalt are cracked for a second time, accompanied by the appearance of a large number of connected organic pores, and the total porosity of the shale increases. At the same time, the bound fluid saturation also gradually decreases. As the organic matter pores gradually increase and the pore heterogeneity is enhanced, shale reaches its highest value of $\Delta\alpha$ [1,18]. As the maturity exceeds a level of 3.0%, the hydrocarbon generation ability of organic matter decreases, the total porosity of shale increases, and the formation of pore networks enhances pore connectivity. The reasons for this phenomenon could be that organic matter pores are compressed and closed by the gradual attenuation of compaction and hydrocarbon generation capacity. This may be contributed to pore volume's decrease; the bound fluid saturation increases slightly and the $\Delta\alpha$ value decreases slightly [41,62].

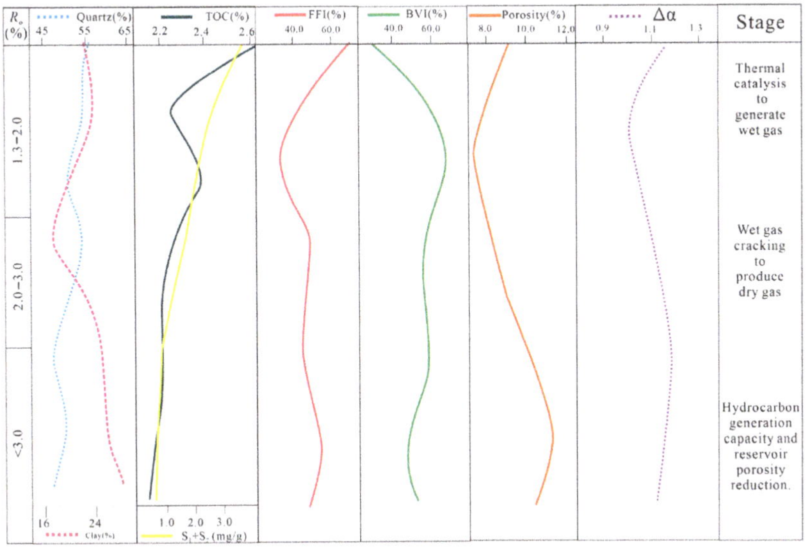

Figure 12. Conceptual evolution model of the diagenetic evolution sequence and multifractal parameters of pore structure from the artificially thermal simulated samples.

6. Conclusions

In this study, multifractal theory was combined with NMR experiments to investigate the pore structure and heterogeneity evolution of shale from natural immature shale to mature shale via artificial thermal simulation. The control effects of organic hydrocarbon generation and removal and diagenetic processes on the pore structure and heterogeneity of OM are systematically analyzed.

As the temperature increases from 425 to 600 °C, the quartz content and clay mineral content show alternating trends, the unstable mineral content is reduced, and the clay mineral content rises. With an increase in pyrolysis intensity, the TOC, S_1, and S_2 content show a trend of decreasing, indicating that OM produces and discharges hydrocarbons during the simulation. With the advancement of thermal maturity, the porosity at the thermal evolution stage shows a pattern of decreasing first, then increasing, and finally decreasing again. At the maturity stage, hydrocarbon generation obstructs the pore space, leading to a higher bound fluid porosity as well as lower total porosity. With the further enhancement of maturity, a large number of liquid hydrocarbons undergo cracking to generate wet gas, and the generation of a large number of OM micropores enhances the moveable fluid porosity within the shale reservoir, increasing the overall porosity of the shale, which is conducive to the seepage process. When maturity reaches the overmature stage, the hydrocarbon generation capacity of OM decreases, while some OM pores are subject to compaction collapse and deformation, resulting in a decrease in porosity and pore structure heterogeneity.

Author Contributions: W.W.: Writing and Methodology, Data analyses. Z.L.: Writing and Methodology. X.T.: Resources, Supervision, Project administration. L.X. and Y.L. (Yi Li): Resources and Supervision. Y.Y.: Method and Writing. Y.C. and Y.L. (Yong Liu): Method, Resources. All authors have read and agreed to the published version of the manuscript.

Funding: This research received funding from the National Natural Science Foundation of China (No. 41872135, No. 42072151, No.42272137), the National Energy Shale Gas R & D (Experiment) Center Grant (No. 2022-KFKT-15), and the Strategic Cooperation Technology Projects of CNPC and CUPB (ZLZX2020-01).

Data Availability Statement: The data presented in this study are available upon request from the corresponding author. The data are not publicly available because some of the basic research involves confidentiality.

Acknowledgments: Many thanks to the State Key Laboratory of Petroleum Resources and Prospecting and the Unconventional Oil and Gas Science and Technology Research Institute. Furthermore, we sincerely appreciate all workers who worked hard with us. Finally, thanks to the reviewers and the editor for their critical input.

Conflicts of Interest: The authors declare that they have no known competing financial interests or personal relationships that could have appeared to influence the work reported in this paper.

References

1. Curtis, M.E.; Cardott, B.J.; Sondergeld, C.H.; Rai, C.S. Development of organic porosity in the Woodford Shale with increasing thermal maturity. *Int. J. Coal Geol.* **2012**, *103*, 26–31. [CrossRef]
2. Loucks, R.G.; Reed, R.M.; Ruppel, S.C.; Hammes, U. Spectrum of pore types and networks in mudrocks and a descriptive classification for matrix-related mudrock pores. *AAPG Bull.* **2012**, *96*, 1071–1098. [CrossRef]
3. Yang, R.; He, S.; Yi, J.; Hu, Q. Nano-scale pore structure and fractal dimension of organic-rich Wufeng-Longmaxi shale from Jiaoshiba area, Sichuan Basin: Investigations using FE-SEM, gas adsorption and helium pycnometry. *Mar. Pet. Geol.* **2016**, *70*, 27–45. [CrossRef]
4. Tang, X.; Jiang, Z.; Jiang, S.; Li, Z. Heterogeneous nanoporosity of the Silurian Longmaxi Formation shale gas reservoir in the Sichuan Basin using the QEMSCAN, FIB-SEM, and nano-CT methods. *Mar. Pet. Geol.* **2016**, *78*, 99–109. [CrossRef]
5. Hou, X.; Zhu, Y.; Chen, S.; Wang, Y.; Liu, Y. Investigation on pore structure and multifractal of tight sandstone reservoirs in coal bearing strata using LF-NMR measurements. *J. Pet. Sci. Eng.* **2020**, *187*, 106757. [CrossRef]
6. Liang, Z.; Jiang, Z.; Li, Z.; Song, Y.; Gao, F.; Liu, X.; Xiang, S. Nanopores Structure and Multifractal Characterization of Bulk Shale and Isolated Kerogen—An Application in Songliao Basin, China. *Energy Fuels* **2021**, *35*, 5818–5842. [CrossRef]
7. Wang, Y.; Cheng, H.; Hu, Q.; Liu, L.; Jia, L.; Gao, S.; Wang, Y. Pore structure heterogeneity of Wufeng-Longmaxi shale, Sichuan Basin, China: Evidence from gas physisorption and multifractal geometries. *J. Pet. Sci. Eng.* **2022**, *208*, 109313. [CrossRef]
8. Chen, K.; Yang, R.; Bao, H.; Dong, T.; Jia, A.; Hu, Q.; Guo, X.; He, S. Depositional-diagenetic process and their implications for pore development of Wufeng-Longmaxi shales in the Jiangdong block, Fuling shale gas field, SW China. *Mar. Pet. Geol.* **2023**, *151*, 106177. [CrossRef]
9. Li, X.; Jiang, Z.; Jiang, S.; Wang, S.; Miao, Y.; Wu, F.; Cao, X. Synergetic effects of matrix components and diagenetic processes on pore properties in the Lower Cambrian shale in Sichuan Basin, South China. *J. Nat. Gas Sci. Eng.* **2021**, *94*, 104072. [CrossRef]
10. Wang, Y.; Liu, L.; Hu, Q.; Hao, L.; Wang, X.; Sheng, Y. Nanoscale Pore Network Evolution of Xiamaling Marine Shale during Organic Matter Maturation by Hydrous Pyrolysis. *Energy Fuels* **2020**, *34*, 1548–1563. [CrossRef]
11. Xu, L.; Yang, K.; Wei, H.; Liu, L.; Jiang, Z.; Li, X.; Chen, L.; Xu, T.; Wang, X. Pore evolution model and diagenetic evolution sequence of the Mesoproterozoic Xiamaling shale in Zhangjiakou, Hebei. *J. Pet. Sci. Eng.* **2021**, *207*, 109115. [CrossRef]
12. Xu, L.; Yang, K.; Wei, H.; Liu, L.; Li, X.; Chen, L.; Xu, T.; Wang, X. Full-Scale Pore Structure Characteristics and the Main Controlling Factors of Mesoproterozoic Xiamaling Shale in Zhangjiakou, Hebei, China. *Nanomaterials* **2021**, *11*, 527. [CrossRef] [PubMed]
13. Liu, H.; Zhang, S.; Song, G.; Xuejun, W.; Teng, J.; Wang, M.; Bao, Y.; Yao, S.; Wang, W.; Zhang, S.; et al. Effect of shale diagenesis on pores and storage capacity in the Paleogene Shahejie Formation, Dongying Depression, Bohai Bay Basin, east China. *Mar. Pet. Geol.* **2019**, *103*, 738–752. [CrossRef]

14. Chen, L.; Jiang, Z.; Jiang, S.; Liu, K.; Yang, W.; Tan, J.; Gao, F. Nanopore Structure and Fractal Characteristics of Lacustrine Shale: Implications for Shale Gas Storage and Production Potential. *Nanomaterials* **2019**, *9*, 390. [CrossRef] [PubMed]
15. Liang, C.; Cao, Y.; Liu, K.; Jiang, Z.; Wu, J.; Hao, F. Diagenetic variation at the lamina scale in lacustrine organic-rich shales: Implications for hydrocarbon migration and accumulation. *Geochim. Et Cosmochim. Acta* **2018**, *229*, 112–128. [CrossRef]
16. Wang, X.P.; Mou, C.L.; Wang, Q.Y.; Ge, X.Y.; Chen, X.W.; Zhou, K.K.; Liang, W. Diagenesis of black shale in longmaxi formation, southern sichuan Basin and its periphery. *Acta Pet. Sin.* **2015**, *36*, 1035–1047.
17. Zhang, S.; Liu, H.M.; Song, G.Q.; Wang, R.S.; Chen, S.Y.; Zhang, S.P. Genesis and control factors of shale oil reserving space in Dongying sag. *Acta Pet. Sin.* **2016**, *37*, 1495–1507. [CrossRef]
18. Guan, M.; Liu, X.; Jin, Z.; Lai, J.; Sun, B.; Zhang, P.; Chen, K. The evolution of pore structure heterogeneity during thermal maturation in lacustrine shale pyrolysis. *J. Anal. Appl. Pyrolysis* **2022**, *163*, 105501. [CrossRef]
19. Gao, Z.; Xuan, Q.; Hu, Q.; Jiang, Z.; Liu, X. Pore structure evolution characteristics of continental shale in China as indicated from thermal simulation experiments. *AAPG Bull.* **2021**, *105*, 2159–2180. [CrossRef]
20. Löhr, S.C.; Baruch, E.T.; Hall, P.A.; Kennedy, M.J. Is organic pore development in gas shales influenced by the primary porosity and structure of thermally immature organic matter? *Org. Geochem.* **2015**, *87*, 119–132. [CrossRef]
21. Wang, X.; Jiang, Z.; Jiang, S.; Chang, J.; Li, X.; Wang, X.; Zhu, L. Pore Evolution and Formation Mechanism of Organic-Rich Shales in the Whole Process of Hydrocarbon Generation: Study of Artificial and Natural Shale Samples. *Energy Fuels* **2019**, *34*, 332–347. [CrossRef]
22. Zhang, Y.; Yu, B.; Pan, Z.; Hou, C.; Zuo, Q.; Sun, M. Effect of thermal maturity on shale pore structure: A combined study using extracted organic matter and bulk shale from Sichuan Basin, China. *J. Nat. Gas Sci. Eng.* **2020**, *74*, 103089. [CrossRef]
23. Milliken, K.L.; Rudnicki, M.; Awwiller, D.N.; Zhang, T. Organic matter-hosted pore system, Marcellus Formation (Devonian), Pennsylvania. *AAPG Bull.* **2013**, *97*, 177–200. [CrossRef]
24. Liu, J.; Zhou, Z.; Liu, X.H.; Ren, S.M.; Liu, Y.M. Geological conditions of the Mesoproterozoic shale gas accumulation in Yanshan area, North China. *Acta Petrol. Sin* **2019**, *40*, 268–278. [CrossRef]
25. Wang, P.; Jiang, Z.; Ji, W.; Zhang, C.; Yuan, Y.; Chen, L.; Yin, L. Heterogeneity of intergranular, intraparticle and organic pores in Longmaxi shale in Sichuan Basin, South China: Evidence from SEM digital images and fractal and multifractal geometries. *Mar. Pet. Geol.* **2016**, *72*, 122–138. [CrossRef]
26. Yang, F.; Ning, Z.; Wang, Q.; Liu, H. Pore structure of Cambrian shales from the Sichuan Basin in China and implications to gas storage. *Mar. Pet. Geol.* **2016**, *70*, 14–26. [CrossRef]
27. Zhang, P.; Yin, Y.; Lu, S.; Li, J.; Chang, X.; Zhang, J.; Pang, Y.; Chen, G.; Liu, Y.; Li, Z. Insights into Pore Structures and Multifractal Characteristics of Shale Oil Reservoirs: A Case Study from Dongying Sag, Bohai Bay Basin, China. *Energy Fuels* **2022**, *36*, 8224–8237. [CrossRef]
28. Zhang, S.; Yan, J.; Hu, Q.; Wang, J.; Tian, T.; Chao, J.; Wang, M. Integrated NMR and FE-SEM methods for pore structure characterization of Shahejie shale from the Dongying Depression, Bohai Bay Basin. *Mar. Pet. Geol.* **2019**, *100*, 85–94. [CrossRef]
29. Ji, W.; Song, Y.; Jiang, Z.; Meng, M.; Liu, Q.; Chen, L.; Wang, P.; Gao, F.; Huang, H. Fractal characteristics of nano-pores in the Lower Silurian Longmaxi shales from the Upper Yangtze Platform, south China. *Mar. Pet. Geol.* **2016**, *78*, 88–98. [CrossRef]
30. Huang, H.; Li, R.; Jiang, Z.; Li, J.; Chen, L. Investigation of variation in shale gas adsorption capacity with burial depth: Insights from the adsorption potential theory. *J. Nat. Gas Sci. Eng.* **2020**, *73*, 103043. [CrossRef]
31. Liu, K.; Ostadhassan, M.; Zou, J.; Gentzis, T.; Rezaee, R.; Bubach, B.; Carvajal-Ortiz, H. Multifractal analysis of gas adsorption isotherms for pore structure characterization of the Bakken Shale. *Fuel* **2018**, *219*, 296–311. [CrossRef]
32. Ji, W.; Song, Y.; Rui, Z.; Meng, M.; Huang, H. Pore characterization of isolated organic matter from high matured gas shale reservoir. *Int. J. Coal Geol.* **2017**, *174*, 31–40. [CrossRef]
33. Li, Z.; Shen, X.; Qi, Z.; Hu, R. Study on the pore structure and fractal characteristics of marine and continental shale based on mercury porosimetry, N2 adsorption and NMR methods. *J. Nat. Gas Sci. Eng.* **2018**, *53*, 12–21. [CrossRef]
34. Zheng, S.; Yao, Y.; Liu, D.; Cai, Y.; Liu, Y.; Li, X. Nuclear magnetic resonance T2 cutoffs of coals: A novel method by multifractal analysis theory. *Fuel* **2019**, *241*, 715–724. [CrossRef]
35. Xu, L.; Yang, K.; Wei, H.; Liu, L.; Li, X.; Chen, L.; Xu, T.; Wang, X. Diagenetic evolution sequence and pore evolution model of Mesoproterozoic Xiamaling organic-rich shale in Zhangjiakou, Hebei, based on pyrolysis simulation experiments. *Mar. Pet. Geol.* **2021**, *132*, 715–724. [CrossRef]
36. Zhao, J.; Jin, Z.; Jin, Z.; Hu, Q.; Hu, Z.; Du, W.; Yan, C.; Geng, Y. Mineral types and organic matters of the Ordovician-Silurian Wufeng and Longmaxi Shale in the Sichuan Basin, China: Implications for pore systems, diagenetic pathways, and reservoir quality in fine-grained sedimentary rocks. *Mar. Pet. Geol.* **2017**, *86*, 655–674. [CrossRef]
37. Chen, J.; Xiao, X. Evolution of nanoporosity in organic-rich shales during thermal maturation. *Fuel* **2014**, *129*, 173–181. [CrossRef]
38. Li, X.; Wang, J.; Li, Z.; Zhang, K.; Feng, X.; Wang, Z. Impact of Brittle Deformation on Pore Structure Evolution in Shale: Samples Collected from Different Fault Positions. *Energy Fuels* **2022**, *36*, 1441–1456. [CrossRef]
39. Hu, G.; Pang, Q.; Jiao, K.; Hu, C.; Liao, Z. Development of organic pores in the Longmaxi Formation overmature shales: Combined effects of thermal maturity and organic matter composition. *Mar. Pet. Geol.* **2020**, *116*, 104314. [CrossRef]
40. Wang, Y.; Qiu, N.; Borjigin, T.; Shen, B.; Xie, X.; Ma, Z.; Lu, C.; Yang, Y.; Yang, L.; Cheng, L.; et al. Integrated assessment of thermal maturity of the Upper Ordovician–Lower Silurian Wufeng–Longmaxi shale in Sichuan Basin, China. *Mar. Pet. Geol.* **2019**, *100*, 447–465. [CrossRef]

41. Hu, H. Porosity evolution of the organic-rich shale with thermal maturity increasing. *Acta Pet. Sin.* **2013**, *34*, 820–825. [CrossRef]
42. Ma, Z.; Zheng, L.; Xu, X.; Bao, F.; Yu, X. Thermal simulation experiment of organic matter-rich shale and implication for organic pore formation and evolution. *Pet. Res.* **2017**, *2*, 347–354. [CrossRef]
43. Fan, W.B. Geological features and research progress of the Mesoproterozoic Xiamaling Formation in the North China Craton: A review after nearly one hundred years of study. *Geol. Rev.* **2015**, *61*, 1383–1406. [CrossRef]
44. Luo, S.S.; Lv, Q.Q.; Shang, F.; Li, L.J. Characteristics of sedimentrary facies of Neoproterozoic Xiamaling formation reservoir in Xuanlong depression. *FaultBlock Oil Gas Field* **2011**, *18*, 26–29.
45. Liu, Z.; Liu, D.; Cai, Y.; Yao, Y.; Pan, Z.; Zhou, Y. Application of nuclear magnetic resonance (NMR) in coalbed methane and shale reservoirs: A review. *Int. J. Coal Geol.* **2020**, *218*, 103261. [CrossRef]
46. Jerry, J.S.; Alan, K.B. Evaluation of a Simple Model of Vitrinite Reflectance Based on Chemical Kinetics (1). *AAPG Bull.* **1990**, *74*, 1559–1570. [CrossRef]
47. Chhabra, A.B.; Meneveau, C.; Jensen, R.V.; Sreenivasan, K.R. Direct determination of the f(alpha) singularity spectrum and its application to fully developed turbulence. *Phys. Rev. A Gen. Phys.* **1989**, *40*, 5284–5294. [CrossRef]
48. Caniego, F.J.; Martí, M.A.; San José, F. Rényi dimensions of soil pore size distribution. *Geoderma* **2003**, *112*, 205–216. [CrossRef]
49. Grassberger, P. Generalized dimensions of strange attractors. *Phys. Lett. A* **1983**, *97*, 227–230. [CrossRef]
50. Wu, S.; Zhu, R.; Cui, J.; Cui, J.; Bai, B.; Zhang, X.; Jin, X.; Zhu, D.; You, J.; Li, X. Characteristics of lacustrine shale porosity evolution, Triassic Chang 7 Member, Ordos Basin, NW China. *Pet. Explor. Dev.* **2015**, *42*, 185–195. [CrossRef]
51. Li, A.; Ding, W.; Wang, R.; He, J.; Wang, X.; Sun, Y.; Gu, Y.; Jiao, N. Petrophysical characterization of shale reservoir based on nuclear magnetic resonance (NMR) experiment: A case study of Lower Cambrian Qiongzhusi Formation in eastern Yunnan Province, South China. *J. Nat. Gas Sci. Eng.* **2017**, *37*, 29–38. [CrossRef]
52. Zhang, Z.; Qin, Y.; Zhuang, X.; Li, G.; Wang, X. Poroperm characteristics of high-rank coals from Southern Qinshui Basin by mercury intrusion, SEM-EDS, nuclear magnetic resonance and relative permeability analysis. *J. Nat. Gas Sci. Eng.* **2018**, *51*, 116–128. [CrossRef]
53. Martínez, G.A.; Davis, L.A. Petrophysical Measurements on Shales Using NMR. In Proceedings of the SPE/AAPG Western Regional Meeting, Long Beach, CA, USA, 19–22 June 2000.
54. Huang, S.J.; Huang, K.K.; Feng, W.L.; Tong, H.P.; Liu, L.H.; Zhang, X.H. Mass exchanges among feldspar, kaolinite and illite and their influences on secondary porosity formation in clastic diagenesis: A case study on the Upper Paleozoic, Ordos Basin and Xujiahe formation, Western Sichuan depression. *Geochimica* **2009**, *38*, 498–506.
55. Peltonen, C.; Marcussen, Ø.; Bjørlykke, K.; Jahren, J. Clay mineral diagenesis and quartz cementation in mudstones: The effects of smectite to illite reaction on rock properties. *Mar. Pet. Geol.* **2009**, *26*, 887–898. [CrossRef]
56. Fan, M.; He, Z.L.; Li, Z.M.; Yu, L.J.; Zhang, W.T.; Liu, W.X.; Jiang, X.Q. Dissolution window of carbonate rocks and its geological significance. *Oil Gas Geol.* **2011**, *32*, 499–500.
57. Zargari, S.; Canter, K.L.; Prasad, M. Porosity evolution in oil-prone source rocks. *Fuel* **2015**, *153*, 110–117. [CrossRef]
58. Han, Y.; Horsfield, B.; Wirth, R.; Mahlstedt, N.; Bernard, S. Oil retention and porosity evolution in organic-rich shales. *AAPG Bull.* **2017**, *101*, 807–827. [CrossRef]
59. Hou, Y.; Zhang, K.; Wang, F.; He, S.; Dong, T.; Wang, C.; Qin, W.; Xiao, Y.; Tang, B.; Yu, R.; et al. Structural evolution of organic matter and implications for graphitization in over-mature marine shales, south China. *Mar. Pet. Geol.* **2019**, *109*, 304–316. [CrossRef]
60. Xue, Z.; Jiang, Z.; Wang, X.; Gao, Z.; Chang, J.; Nie, Z.; Li, H.; Wu, W.; Qiu, H.; Wang, Q.; et al. Genetic mechanism of low resistivity in high-mature marine shale: Insights from the study on pore structure and organic matter graphitization. *Mar. Pet. Geol.* **2022**, *144*, 105825. [CrossRef]
61. Huang, C.; Ju, Y.; Zhu, H.; Lash, G.G.; Qi, Y.; Yu, K.; Feng, H.; Ju, L.; Qiao, P. Investigation of formation and evolution of organic matter pores in marine shale by helium ion microscope: An example from the Lower Silurian Longmaxi Shale, South China. *Mar. Pet. Geol.* **2020**, *120*, 104550. [CrossRef]
62. Borjigin, T.; Lu, L.; Yu, L.; Zhang, W.; Pan, A.; Shen, B.; Wang, Y.; Yang, Y.; Gao, Z. Formation, preservation and connectivity control of organic pores in shale. *Pet. Explor. Dev.* **2021**, *48*, 798–812. [CrossRef]

energies

MDPI

Article

Sand-Carrying Law and Influencing Factors in Complex Fractures of Nano-Clean Fracturing Fluid

Hongkai Han [1,2,3], Zhaoqi Shi [1], Weiqin Zuo [1,2,3,*], Shengjie Wu [1], Yanwei Liu [1,2,3], Kunrong Xie [1], Liqun Long [1] and Hani Mitri [4]

1 School of Safety Science and Engineering, Henan Polytechnic University, Jiaozuo 454000, China
2 Henan International Joint Laboratory of Coal Mine Safety and Occupational Hazards Prevention and Control, Henan Polytechnic University, Jiaozuo 454000, China
3 State Key Laboratory Cultivation Base for Gas Geology and Gas Control, Henan Polytechnic University, Jiaozuo 454000, China
4 Department of Mining and Materials Engineering, McGill University, Montreal, QC H3A0E8, Canada
* Correspondence: zuoweiqin@163.com

Abstract: Nano-clean fracturing fluids have broad application potential in coalbed methane reservoir fracturing owing to their high stability, good temperature resistance, low filtration loss, and strong frictional resistance reduction. However, the sand-carrying regularity of nano-clean fracturing fluids in coalbed methane reservoirs is unclear, especially for complex fractures with variable directions. This study established a sand transport model that considers proppant collision, wall friction blocking, fracture fluid filtration loss, and the fracture branching angle to study the sand-carrying law of nano-clean fracturing fluids and its influencing factors in complex fractures. The degrees of influence on equilibrium height and placement rate from high to low were the proppant particle size, proppant density, fracturing fluid properties, sand ratio, and pumping discharge volume, and the correlation degrees obtained by grey correlation analysis are 0.862, 0.861, 0.855, 0.854, and 0.832, respectively. As the complexity of the fractures deepens and the resistance increases, the flow rate of the fracturing fluid is reduced, making it difficult for the proppant to enter the branching joints. The sand-carrying performance of a nano-clean fracturing fluid is better than that of a common clear-water fracturing fluid. The fluid-structure coupling model of a nano-clean fracturing fluid can accurately characterize the sand-carrying law of nano-clean fracturing fluids, providing a research basis for optimizing high efficiency sand-carrying fracturing fluid parameters in coalbed methane reservoir fracturing construction.

Keywords: clean fracturing fluid; nanoparticles; sand-carrying law; complex fractures

1. Introduction

Coalbed methane (CBM) is a new energy with a high calorific value. However, unconventional natural gas differs from conventional natural gas in its mode of occurrence, reservoir formation mechanism, and reservoir characteristics [1–3]. Generally, it has three "low" characteristics: low saturation, permeability, and reservoir pressure. The method of coalbed methane extraction is different from that of traditional natural gas [4]. It requires reservoir modification methods such as hydraulic fracturing, gas injection to increase production, and heat injection to increase production [5–9]. Hydraulic fracturing is the most widely used method to fracture the target formation. It uses the fracturing fluid to carry the proppant into the fracture, ensuring that the fracturing fluid returns to the formation to create a strong conductivity of the sand-filled fracture and achieving an economical and practical extraction. Therefore, the choice of the fracturing fluid directly determines whether the proppant can effectively fill the fracture.

Alotaibi et al. [10] studied the sand-carrying law of slick water fracturing fluid in complex fractures and confirmed that it could carry proppant into multistage fractures and form sand-filled fractures with a specific inflow capacity. Zhang et al. [11] found that the inlet location significantly influences proppant placement, and the proppant transport capacity can be enhanced by increasing the jet velocity and decreasing the proppant density. Shi et al. [12] found that the fracture spacing and proppant concentration of the injected slurry significantly affected the extension path of the subsequently formed fractures. Fredd et al. [13] reported that a foam fracturing liquid system was superior to ordinary fracturing fluids for improving permeability. Tong et al. [14] found that the proppant spreads more uniformly under the carrying action of a foam fracturing fluid, and the faster the foam drains, the more uneven the final proppant spread is. Wang et al. [15] discussed a new thermosensitive in situ proppant based on the liquid-solid conversion of a supramolecular self-propelled fracturing fluid, and the results showed that the supramolecular self-propelled fracturing fluid had good fluidity and low damage to the reservoir. Li et al. [16] developed a clean fracturing fluid with less damage to the reservoir. Yang et al. [17]. found that a clean fracturing fluid can improve the permeability of the reservoir and the conductivity of fractures, thus improving the recovery rate of coalbed methane. Crews et al. [18] developed a modified nano-clean fracturing fluid that could effectively improve the filtration rate of fracturing fluid. Nettesheim et al. [19] studied the influence of nanoparticles on the performance of a clean fracturing fluid and found that adding a small amount of silica nanoparticles effectively improved the viscosity and viscoelasticity of the system. The members of this paper studied the laws of proppant migration and placement carried by clean fracturing fluid and the influence of fracturing fluid on the coal pore structure [9].

In summary, the current proppant transport law research mainly focuses on conventional clear-water fracturing. However, the clear-water fracturing fluid has poor rheology, large filtration loss, and a small fracturing range, which limits the extension of fracturing fractures; the overall fracturing efficiency is low, and the fracturing effect is not ideal. Moreover, the water viscosity is low, the sand-carrying performance is weak, the filtration loss is significant, and a large amount of water entering the coal seam easily produces a water-lock effect, which shortens the adequate pre-pumping time of coalbed methane and reduces the recovery rate. The foam fracturing fluid has high viscosity, strong fracture-forming ability, and causes low damage to the reservoir; however, the stability and actual fracturing effect of the foam fracturing fluid are not apparent at present, the fracturing cost is high, and the operation is complex. Clean fracturing fluids exhibit good sand-carrying properties, low filtration loss, and no residue; however, they exhibit low viscosity, poor temperature resistance [20], and an insufficient sand-carrying capacity. The modification of clean fracturing fluids by adding nanoparticles can effectively improve the viscosity and temperature resistance of the system and reduce the friction resistance, thus improving the sand-carrying capacity of the proppant in the fracturing fluids [21,22]. Nano-clean fracturing fluids are a new type of fracturing fluid, and there are few studies on the proppant transport law of a nano-clean fracturing fluid. Its sand-carrying law is unclear, and most existing models do not consider the complex fracture branching angle, wall friction blocking, fracturing fluid filtration loss, or other factors. The influence of fracturing fluid rheology on the sand-carrying process is ignored, which deviates significantly from the actual situation.

Therefore, based on the Eulerian multiphase flow model and particle dynamics model, the author constructed a nano-clean fracturing fluid sand-carrying transport model while considering the above influencing factors and the rheology of the fracturing fluid and studied the sand-carrying laws in complex fractures with multi-angle and multi-size clean fracturing fluids, focusing on the sand-carrying effect under different pumping discharge volumes, sand ratios, proppant particle sizes, proppant densities, and rheological parameters. It can provide a specific research basis for optimizing the fracturing parameters of coalbed methane reservoirs and has great significance for the efficient and environmentally friendly development of coalbed methane.

2. Simulation Methods and Modeling

2.1. Mathematical Models

The flow of a sand-carrying fluid composed of proppant and nano-clean fracturing fluid in the fracture is a solid-liquid two-phase flow problem. The Euler-Euler method can be used to simulate the settlement and transport law of the nano-clean fracturing fluid-carrying proppant in the fracture.

The liquid-phase and solid-phase continuity equation is as follows:

$$\frac{\partial(\alpha_i \rho_i)}{\partial t} + \nabla \cdot (\alpha_i r_i v_i) = 0 \tag{1}$$

where α is the volume fraction; ρ is the density, kg/m^3; ∇ is the Hamiltonian operator; t is the time, s; v is the velocity, m/s; the subscript i denotes the phase; and s and l represent the solid phase and liquid phase, respectively.

The liquid-phase and solid-phase momentum equation is as follows:

$$\frac{\partial(\alpha_i \rho_i v_i)}{\partial t} + \nabla \cdot (\alpha_i \rho_i v_i v_i) = -\alpha_i \nabla p_i + \nabla \cdot \tau_i + \alpha_i \rho_i g - M_D + F_S \tag{2}$$

$$\tau_l = \alpha_l \mu_l \left(\nabla \cdot \vec{v}_l + \nabla \cdot \vec{v}_l^{\,T} \right) + \alpha_l (\lambda_l - \frac{2}{3}\mu_l) \nabla \cdot \vec{v}_l \bar{\bar{I}} \tag{3}$$

$$\tau_s = \alpha_s \mu_s \left(\nabla \cdot \vec{v}_s + \nabla \cdot \vec{v}_s^{\,T} \right) + \alpha_{sl} (\lambda_s - \frac{2}{3}\mu_s) \nabla \cdot \vec{v}_s \bar{\bar{I}} \tag{4}$$

where p is the partial pressure, Pa; τ is the shear stress tensor, Pa; M_D is the interphase momentum exchange coefficient, kg/(m^3·s); μ is the particle phase shear viscosity, P$_a$·s; λ is the particle phase volume viscosity, P$_a$·s; F_s is the collision force between solid-phase particles; and I is the unit tensor, N/m^2.

The F_s expression of the collision force between solid particles is as follows:

$$F_s = 2(1+e)\nabla(\alpha_s \rho_l (c_p^2)) \tag{5}$$

where e is the impact recovery coefficient of solid particles, and the value is 0.9; α_s is the particle volume fraction, %; and c_p is the pulsation coefficient of solid particles, m/s.

Considering the relatively high sand content in an actual fracturing operation, the collision and friction between particles may change the momentum and energy exchange between the solid and liquid phases. The particle dynamics theory that considers the interactions between particles is expressed as follows:

$$\frac{3}{2}\left[\frac{\partial(\alpha_s \rho_s \theta_s)}{\partial t} + \nabla \cdot \left(\alpha_s \rho_s \vec{v_s} \theta_s \right) \right] = \left(-\rho_s \bar{\bar{I}} + \bar{\bar{\tau_s}} \right) \nabla \vec{v_s} - \nabla(k_\theta \nabla \theta_s) - \gamma + \Phi \tag{6}$$

where θ_s is the particle temperature, which does not represent the temperature of hot and cold but rather a measure of particle pulsation speed, m^2·s^{-2}; $(-\rho_s \bar{\bar{I}} + \bar{\bar{\tau_s}})\nabla \vec{v_s}$ is the particle pulsation energy due to the shear stress of the particle phase; $(k_\theta \nabla \theta_s)$ is the dissipation of particle pulsation energy along the particle temperature gradient; k_θ is the particle temperature diffusion coefficient, kg·m^{-1}d·s^{-1}; γ is the energy dissipation rate of collisions between particles, which can be calculated according to the formula obtained by Equation (7) [23], kg·m^{-1}·s^{-3}; Φ is the interphase kinetic energy transfer due to particle velocity change, kg·m^{-1}·s^{-3}; e is the impact recovery coefficient of solid particles, g_0 is the radial distribution function; and d_s is the particle size.

$$\gamma = \frac{12(1-e_s^2)g_0}{d_s} \rho_s \alpha_s \theta_s^{3/2} \tag{7}$$

Euler's multiphase flow model considers solid particles as a pseudo-fluid phase and inevitably gives them fluid parameters, such as shear viscosity, which reflects the behavior of collision and friction between particles [24]. The shear viscosity of a solid, μ_s, consists of three components: collisional viscosity, $\mu_{s,col}$; dynamic viscosity, $\mu_{s,kin}$; and frictional viscosity, $\mu_{s,fr}$.

$$\mu_s = \mu_{s,col} + \mu_{s,kin} + \mu_{s,fr} \tag{8}$$

The solid viscosity is described using the particle dynamics and Schaffer models, and the solid viscosity model is expressed as follows:

$$\mu_s = \frac{4}{5}\varepsilon_s\rho_s d_p g_0(1+e)\sqrt{\frac{\theta_s}{\pi}} + \frac{10\rho_s d_p\sqrt{\theta_s\pi}}{96 g_0(1+e)}\left[1+\frac{4}{5}\varepsilon_s g_0(1+e)\right]^2 + \frac{P_{s,f}\sin\phi}{2\sqrt{I_{2D}}} \tag{9}$$

Among them, we have the following:

$$P_{s,f} = 0.05\left[(\varepsilon_s - \varepsilon_{s,\min})^2/(\varepsilon_{s,\max} - \varepsilon_s)^5\right] \tag{10}$$

$$g_0 = \left[1 - \left(\frac{\varepsilon_s}{\varepsilon_{s,\max}}\right)^{\frac{1}{3}}\right]^{-1} \tag{11}$$

where the frictional stress, $P_{s,f}$, and the radial distribution function, g_0, are two important parameters for calculating the particle shear viscosity; $\varepsilon_{s,max}$ is the maximum solid-phase volume fraction, set to 0.63; $\varepsilon_{s,min}$ is the frictional accumulation limit; d_p is the particle diameter, m; and I_{2D} is the second invariant of the partial stress tensor.

For liquid-solid interactions, the liquid phase transfers many forces to the solid phase in the liquid-solid two-phase flow model, including the drag, pressure gradient, and buoyancy forces. In this study, we mainly considered the buoyancy, F_L, and drag, F_D, of the fracturing fluid and the virtual mass force, F_{VM}, of the fluid phase, owing to the accelerated motion of the particle phase.

The interphase momentum transfer phase is calculated as follows:

$$M_D = F_D + F_{VM} + F_L \tag{12}$$

$$F_L = C_L\alpha_s\rho_s(v_l - v_s) \times (\nabla \times v_l) \tag{13}$$

$$F_D = \frac{1}{6}\pi d^3 \frac{\beta|v_l - v_s|}{1-\alpha_l} \tag{14}$$

$$F_{VM} = C_{VM}\alpha_s\rho_l\left(\frac{dv_l}{dt} - \frac{dv_s}{dt}\right) \tag{15}$$

where C_L is the lift coefficient, which is considered to be 0.25; C_{VM} is the virtual mass force coefficient, which is 0.5; d is the particle phase diameter, m; α_s and α_l are the volume fractions of the solid phase and liquid phase, respectively; and v_l and v_s are the liquid-phase and solid-phase velocities, m/s, respectively.

On the basis of the conservation equations for the solid-liquid two-phase flow, the mathematical models that reflect the special properties of the fluid also need to be added. The flow of the sand-carrying fluid in the fracture is a typical solid-liquid two-phase turbulent flow; therefore, the turbulence theory should be used to characterize its flow characteristics. This study simulated the solid-liquid two-phase flow of nano-clean fracturing fluid into the fracturing fracture. The turbulence model is chosen as the k-ε standard model with the turbulent kinetic energy equation and the dissipation rate equation, respectively, as

$$\frac{\partial}{\partial t}(\alpha_l\rho_l k) + \nabla\cdot(\alpha_l\rho_l v_l k) = \nabla\cdot\left(\alpha_l\frac{\mu_t}{\sigma_k}\nabla k\right) + G_{k,l} + \prod\nolimits_k - \alpha_l\rho_l\varepsilon \tag{16}$$

$$\frac{\partial}{\partial t}(\alpha_l\rho_l\varepsilon) + \nabla\cdot(\alpha_l\rho_l v_l\varepsilon) = \nabla\cdot\left(\alpha_l\frac{\mu_t}{\sigma_\varepsilon}\nabla\varepsilon\right) + \alpha_l\frac{\varepsilon}{k}(C_1G_{k,l} - C_2\rho_l\varepsilon) + \prod\nolimits_\varepsilon \tag{17}$$

where k is the turbulent kinetic energy of the continuous phase, m^2/s^2; l represents the liquid phase; ε is the dissipation rate of turbulent kinetic energy, m^2/s^2; μ_t is the viscosity coefficient of the continuous phase, Pa·s; $\prod_k, \prod_\varepsilon$ is the exchange coefficient between solid and liquid phases, kg/(m·s^3); σ_k is the turbulent Prandtl number of energy k, uncaused, and takes the value of 1.0; σ_ε is the turbulent Prandtl number of ε, uncaused, and takes the value of 1.3; $G_{k,l}$ is the turbulent kinetic energy; and C_1 and C_2 are empirical constants, taking the values of 1.44 and 1.92, respectively.

A nano-clean fracturing fluid is a non-Newtonian fluid, and the relationship between its viscosity and shear rate is not linear. Non-Newtonian fluids exhibit shear-dilution properties, in which the viscosity decreases as the shear rate increases, thereby accelerating the proppant settlement rate. Therefore, for non-Newtonian fluids, a significant error occurs when calculating the settling velocity of a proppant using traditional methods. Using the consistency coefficient, k, and rheological index, n, of a power-law fluid to characterize the particle settling velocity in a non-Newtonian fluid, the Reynolds number of a non-Newtonian fluid can be expressed as follows:

$$N_{Rep} = \frac{\rho_f d_p^n V_p^{2-n}}{k} \tag{18}$$

where ρ_f is the fracturing fluid density, kg/m^3; d_p is the particle diameter, m; and V_p is the particle settling velocity, m/s.

Because the model must consider the effect of fracturing fluid elasticity, using the equation of motion of the power rate model to simulate the settling process of proppant particles within a highly viscoelastic solution, such as a nano-clean fracturing fluid, can lead to a significant difference between the calculated and actual results. Based on this, Zhang et al. [25] proposed a new model for calculating the proppant settling velocity through extensive physical modeling experiments that considered the elasticity of the fluid, the flow pattern of the fracturing fluid, and the nature of the proppant. Based on this, after the proppant particles are carried into the fracture by the fracturing fluid, the settlement movement of the proppant is different from the free settlement in the fluid because its speed is affected by the interaction between the particles, the retarding effect of the rough wall surface, and the filtration effect of the fracturing fluid. We propose using Equation (19) to describe the settling process of the proppant within the nano-clean fracturing fluid.

$$V_p = f_c f_w f_f V_F \left[\frac{g\left(\rho_p - \rho_f\right) d_s^{n+1}}{18k} \right]^{\frac{1}{3n-1}} \tag{19}$$

V_F is the settling velocity factor, as shown in Equation (20):

$$V_F = 0.0011^{\frac{2n}{3n-1}} \left[\frac{9\lambda\left(2n^2+n+1\right)}{k} \right]^{\frac{1}{3n-1}} \tag{20}$$

where g is the acceleration of gravity, m/s^2; ρ_p is the particle density, kg/m^3; d_s is the proppant sieve particle size, m; f_c is the correction coefficient of the sand ratio; f_f is the positive coefficient of filtration disrepair; and f_w is the wall correction coefficient.

The settlement of proppant particles in fractures is affected by the interaction between the particles which is a complex dynamic motion caused by collisions or friction. This interference becomes more significant as the proppant to sand ratio increases. In addition, the proppant particles themselves have a specific viscosity, and an increase in the sand ratio increases the overall viscosity of the sand-carrying fluid, thereby increasing the resistance of the particles to sinking. The effect of the sand ratio on proppant interference settlement can be corrected using the sand-concentration correction factor:

$$f_c = (1-c)^m \tag{21}$$

where c is the sand ratio in the carrying fluid, and m is an empirical constant related to the Reynolds number of the sand carrier fluid.

Most fractures formed by onsite fracturing are very narrow, with a width of approximately a few millimeters, and there are many small-scale and discontinuous distributions of rough surfaces on the fracture surface. Therefore, when propagating particles settle freely in the fracture, the wall squeezes the proppant particles, increasing the collision between particles and the friction between the particles and the flow channel. Thus, the kinetic energy loss of the particles increased, and the settlement of the particles was hindered. The wall-effect correction coefficient, f_w, can be used to express the influence of the wall surface of conventional cracks on the settlement rate as follows:

$$\frac{f_w - f_\infty}{f_0 - f_\infty} = \left[1 + \left(\frac{\lambda}{A}\right)^2\right]^B \tag{22}$$

where λ is the ratio of proppant particle diameter to fracture width (ball diameter ratio) without dimensionality; A is the fitting parameter, which is dimensionless and generally equals 0.64; B is the fitting parameter that has no dimensionality and is generally -0.98; and f_0 is the approximate value of the wall factor when the spherical diameter ratio tends to zero, and its general value is 0.001.

When the fracturing fluid flows in the fracture, the pressure difference between the inside and outside of the fracture causes the fracturing fluid to filter out to the formation, which not only causes the fracturing effect to deteriorate but also increases the fracturing construction cost and affects the proppant settlement rate. The positive coefficient, f_f, of filtration disrepair is calculated as follows:

$$f_f = \frac{1}{1 - \frac{C \cdot \sqrt{t_1}}{W}} \tag{23}$$

where C is the comprehensive filtration coefficient; t is time, s; and W is the seam width, m.

2.2. Boundary Conditions

The model was based on a pressure solver with a first-order windward format to discretize the control equations, and the SIMPLE algorithm was used to solve the algebraic system of equations. Setting the left inlet as the velocity inlet and the right outlet as the pressure outlet, and given the corresponding turbulence intensity and hydraulic diameter, the specific values were calculated according to the following equation:

$$I = 0.16(\mathrm{Re})^{-\frac{1}{8}} \tag{24}$$

The hydraulic radius calculation formula is

$$R_h = \frac{\omega H_0}{2(\omega + H_0)} \tag{25}$$

where ω is the inlet width, m; and H_0 is the inlet height, m.

The wall was chosen to have a no-slip boundary, and the tangential velocity and temperature generated by the proppant and wall were calculated using the Johnson-Jackson model with the following equations:

$$\frac{v_{sl}}{|v_{sl}|} \cdot \left(\tau_k + \tau_f\right) \cdot \omega + \frac{\phi \pi \rho_s g_0 \sqrt{\theta_s}}{2\sqrt{3}\alpha_s^{\max}} v_{sl} + \left(\omega \cdot \tau_f \cdot \omega\right) \tan \delta_w = 0 \tag{26}$$

$$\kappa_s \frac{\partial \theta_s}{\partial x} = \frac{\phi \pi |v_s| \rho_s \alpha_s g_0 \sqrt{\theta_s}}{2\sqrt{3}\alpha_s^{\max}} - \frac{\sqrt{3}\pi \rho_s \alpha_s g_0 \left(1 - e_w^2\right) \sqrt{\theta_s}}{4\alpha_s^{\max}} \theta_s \tag{27}$$

where v_{sl} is the slip velocity of the proppant particles at the wall, m/s; τ_k is the shear tensor generated by the proppant particles colliding with the wall, Pa; τ_f is the shear tensor generated by the proppant particles rubbing against the wall, Pa; w refers to the normal vector to the inner side of the wall; ϕ is the reflection coefficient, no factor, taking the value of 0.001; and δ_w is the friction angle between the proppant and the wall, rad, taking the value of $\pi/10$.

3. Model Verification and Simulation Experiment Scheme

3.1. Model Validation

Based on the similarity criterion, a fracture model with a size of 3000 mm × 400 mm × 10 mm was established, and two rectangular injection ports with a size of 20 mm × 10 mm were set on the left side of the fracture. The inlets were located at one-quarter of the fracture height and two-quarters of the fracture height. This inlet arrangement is conducive to increasing the proppant placement at the inlet end of the fracture and improving the connectivity between the borehole and the fracture [26]. In addition, a 20 mm × 10 mm rectangular outlet was set at the center of the right side of the fracture, and the model is shown in Figure 1.

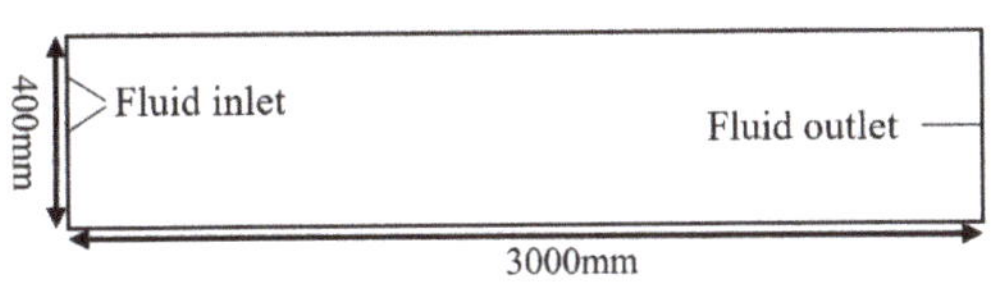

Figure 1. Numerical modeling of a fracture.

To ensure the quality of grid division and accuracy of calculation, a structured grid and positive hexahedral cells were used to grid the entire calculation area, and the grid was encrypted at the entrance; 171,864 grid cells were divided, and the final grid division is shown in Figure 2. According to the mathematical model in Section 2.1 and the boundary conditions in Section 2.2, the sand-carrying law of the clean fracturing fluid under the geometric model in Figure 1 was simulated and studied.

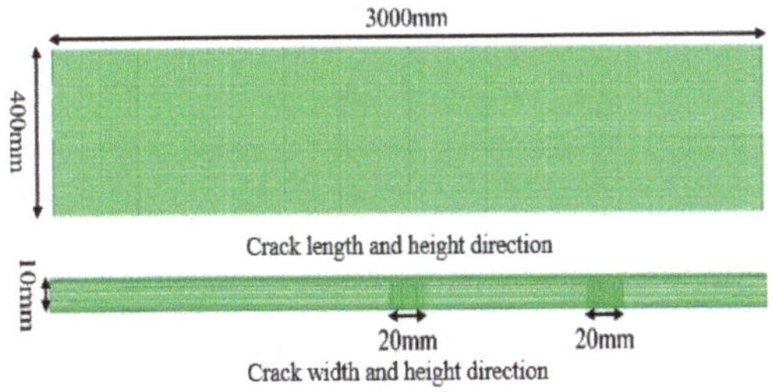

Figure 2. Schematic diagram of grid division.

The accuracy of the proposed model was verified through a comparison with previous experiments, using a large visualized flat-plate single-slit physical model [27]. A rectangular crack with a length, height, and width of 3000, 500, and 6 mm, respectively, was used for the experiment, with an inlet and outlet of the same size on the left and right sides of the crack with a width and height of 6 and 100 mm, respectively. A 30/50 mesh proppant particle (average particle size of 0.45 mm), proppant particles with a density of 2770 kg/m^3, sand ratio of 4%, and pump in discharge conversion speed of 2.31 m/s were selected for the experiment, and the numerical simulation parameters were set. A sand-laying profile was obtained when both parameters were balanced, consistent with the experimental parameter settings, as shown in Figure 3. The physical and numerical simulation results showed that the sand-laying profile distribution and equilibrium time were identical. This indicates that the established model describes the sand-carrying laws of fracturing fluid in a fracture.

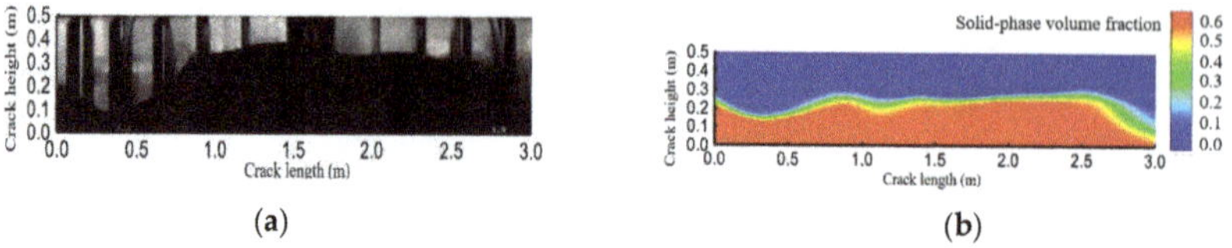

Figure 3. Comparison of physical model and numerical model of sand-laying profiles. (**a**) Physical model of sand-laying profile. (**b**) Numerical model of sand-laying profile.

3.2. Simulation Experiment Scheme of Sand-Carrying Rule of Nano-Clean Fracturing Fluid under Different Factors

The controlled variable method was used to develop an experimental protocol from five aspects, namely the discharge volume, proppant density, proppant particle size, sand ratio, and fracturing fluid parameters, to discuss the changes in sand dike morphology when each factor was changed using the sand dike equilibrium height, equilibrium time, and laydown rate as parameters to characterize the sand dike morphology in the fracture. The experimental protocols are listed in Table 1.

Table 1. Experimental scheme.

Experimental Number	Discharge Volume/(m^3/min)	Proppant Particle Size/(mm)	Density /(kg/m^3)	Sand Ratio/(%)	Rheological Parameters
1#	0.084	0.45	1600	10	$\dot{\lambda} = 0.1$ s; $n = -0.3$; $\mu_0 = 1.0$
2#	0.062	0.45	1600	10	$\dot{\lambda} = 0.1$ s; $n = -0.3$; $\mu_0 = 1.0$
3#	0.104	0.45	1600	10	$\dot{\lambda} = 0.1$ s; $n = -0.3$; $\mu_0 = 1.0$
4#	0.084	0.67	1600	10	$\dot{\lambda} = 0.1$ s; $n = -0.3$; $\mu_0 = 1.0$
5#	0.084	0.30	1600	10	$\dot{\lambda} = 0.1$ s; $n = -0.3$; $\mu_0 = 1.0$
6#	0.084	0.45	1450	10	$\dot{\lambda} = 0.1$ s; $n = -0.3$; $\mu_0 = 1.0$
7#	0.084	0.45	2000	10	$\dot{\lambda} = 0.1$ s; $n = -0.3$; $\mu_0 = 1.0$
8#	0.084	0.45	1600	5	$\dot{\lambda} = 0.1$ s; $n = -0.3$; $\mu_0 = 1.0$
9#	0.084	0.45	1600	20	$\dot{\lambda} = 0.1$ s; $n = -0.3$; $\mu_0 = 1.0$
10#	0.084	0.45	1600	10	$\dot{\lambda} = 0.3$ s; $n = -0.3$; $\mu_0 = 1.0$
11#	0.084	0.45	1600	10	$\dot{\lambda} = 0.5$ s; $n = -0.3$; $\mu_0 = 1.0$
12#	0.084	0.45	1600	10	$\dot{\lambda} = 0.1$ s; $n = -0.4$; $\mu_0 = 1.0$
13#	0.084	0.45	1600	10	$\dot{\lambda} = 0.1$ s; $n = -0.5$; $\mu_0 = 1.0$
14#	0.084	0.45	1600	10	$\dot{\lambda} = 0.1$ s; $n = -0.3$; $\mu_0 = 2.0$
15#	0.084	0.45	1600	10	$\dot{\lambda} = 0.1$ s; $n = -0.3$; $\mu_0 = 3.0$

3.3. Simulation Experiment Scheme of Sand-Carrying Rule of Nano-Clean Fracturing Fluid in Complex Fracture

Owing to the severe inhomogeneity of coal seams and the development of natural fractures, many fractures are generated during the actual fracturing process, comprising main and branch fractures. During construction, the proppant enters not only the main fractures but also the branch fractures, and the transport law of the proppant in the branch fractures is equally essential. The complex fracture network formed by the interlocking of these fractures can effectively reduce the formation pressure and improve the reservoir conditions and provide sufficient inflow channels for subsequent CBM extraction. Therefore, to observe the placement pattern of the proppant in complex fractures more significantly, by establishing a multi-angle and multi-size complex fracture model and setting a larger proppant density and particle size to amplify the effect of clean fracturing fluid on sand dike morphology, the sand-carrying processes of clean fracturing fluid and conventional fracturing fluid in complex fractures were compared and analyzed, and the sand-carrying law of the clean fracturing fluid in the main fracture and branch fracture, respectively, was studied by using the preferred parameters, which can provide theoretical guidance for field fracturing.

3.3.1. Simulation Experiment Scheme of Sand-Carrying Rule for Y-Type Cracks

Three Y-type cracks were established, and the crack form was intersected by the main crack and a branch crack at 30°, 60°, and 90° angles, with the main crack set to the same size of 3000 mm × 400 mm × 10 mm as the single crack model in the previous section, and the branch crack size was set to 1800 mm × 400 mm × 5 mm, which is distributed in the middle of the main crack. The branch fracture with a 90° angle was simulated with clean-water fracturing fluid and nano-clean fracturing fluid. By contrast, the other branch fractures were only simulated with clean fracturing fluid. The inlet flow rate was set to 0.062 m^3/min; the proppant particle size was 0.63 mm, and the density was 2000 kg/m^3.

3.3.2. Simulation Experiment Scheme of Sand-Carrying Rule for Double-Y-Type Cracks

The double-Y crack pattern is a main crack intersected with branch joints at different distances on both sides, and the branch joints are distributed on both sides of the middle of the main crack, with one section of the branch joints set at the inlet end and the other section at the outlet end, and the dimensions of the main crack and branch joints are the same as those in the previous section.

3.3.3. Simulation Experiment Scheme of Sand-Carrying Rule for Well-Type Fracture

The well-type crack consists of one main crack and two orthogonal secondary branch joints, with two tertiary branch joints orthogonal to the secondary branch joints. The geometry of the main crack is consistent with that of the previous section. The dimensions of both the secondary and tertiary branch joints were set to 1800 mm × 400 mm × 5 mm, with the distance of the secondary branch joints from the entrance being 1.6 m and 1.8 m. The tertiary branch joints were parallel to the main joints and 450 mm away from them.

4. Results and Discussions

4.1. Influence of Different Factors on the Sand-Carrying Regularity of Nano-Clean Fracturing Fluid

4.1.1. Effect of Pumping Discharge Volume

Scenarios 2#, 1#, and 3# in Table 1 were selected to analyze the effect of the discharge volume on the proppant spreading morphology of the nano-clean fracturing fluid, and the sand dike morphology in equilibrium was obtained, as shown in Figure 4. The equilibrium parameters for the different discharges are listed in Table 2.

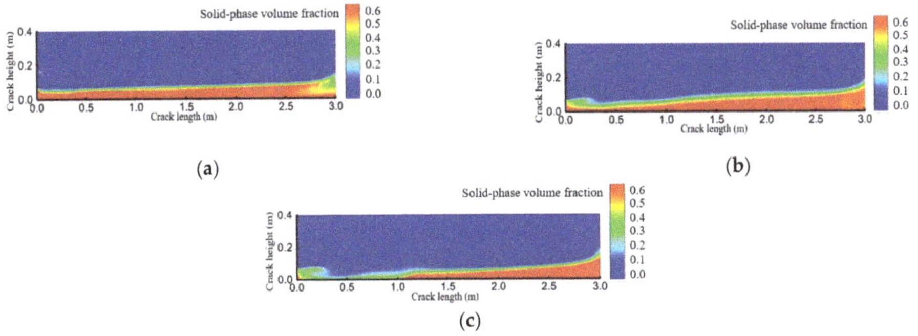

Figure 4. Effect of pumping discharge volume on sand dike morphology. (**a**) Pumping discharge volume of 0.062 m^3/min. (**b**) Pumping discharge volume of 0.084 m^3/min. (**c**) Pumping discharge volume of 0.104 m^3/min.

Table 2. Balance parameters at different pumping discharge volumes.

Pumping Discharge Volume/(m^3/min)	Balance Time/(min)	Balance Height/(cm)	Sand Dike Placement Rate/(%)
0.062	14.5	7.2	23.1
0.084	10.4	6.7	22.4
0.104	5.4	3.6	13.6

As can be seen in Table 2, with an increase in the pumping discharge volume, the equilibrium time decreases from 14.5 min to 5.4 min, a decrease of 62.7%; the equilibrium height decreases from 7.2 cm to 3.6 cm, a decrease of 50%; and the laydown rate decreases from 23.1% to 13.6%, a decrease of 41.1%. The transport power of the proppant particles in the fracture originates from the sand-carrying effect of the fracturing fluid. The variation in the construction discharge volume directly affects the sand-carrying rate of the fracturing fluid, thereby affecting the distribution of the proppant in the fracture. When the pumping discharge volume was 0.062 m^3/min, the fracture tail was less full, and more proppant settled at the entrance. With an increase in the pumping discharge volume, when the pumping discharge volume reaches 0.104 m^3/min, the proppant at 0–1 m of the fracture is in the shape of a "thin film", which indicates that less proppant is placed in the front part of the fracture. By contrast, the proppant content at the back end of the fracture increases significantly.

The experimental results show that the pumping rate is a crucial factor affecting the sand-carrying law of the fracturing fluid and plays a decisive role in the distribution pattern and movement trajectory of the proppant in the fracture. The increase in the pumping discharge volume causes an increase in the turbulence intensity at the inlet end of the main fracture. The vortex formed will involve more proppant at the depth of the fracture, while the settling proppant at the inlet end will be reduced. The unfilled area at the front end of the fracture also increases so that more proppant is transported to the depth of the fracture or even removed from the fracture. The larger the pumping discharge volume, the greater the amount of proppant that was carried out, and the corresponding equilibrium height of the sand dike was reduced. In addition, the larger the pumping discharge volume, the greater the amount of proppant that entered the crack per unit of time, thus decreasing the equilibrium time. Therefore, a low-pumping discharge volume method can be used when using a high-viscosity nano-clean fracturing fluid for fracturing construction.

4.1.2. Effect of Sand Ratio

Scenarios 8#, 1#, and 9# in Table 1 were selected to analyze the effect of the sand ratio on the proppant spreading morphology of the nano-clean fracturing fluid, and the sand dike morphology in the equilibrium state was obtained, as shown in Figure 5. The equilibrium parameters for the different sand ratios are listed in Table 3.

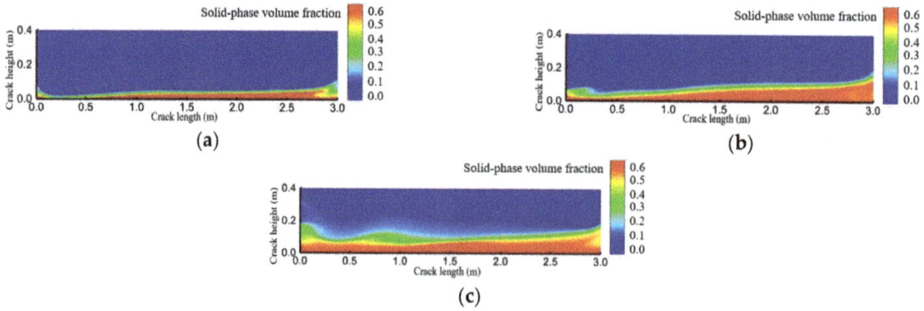

Figure 5. Effect of sand ratio on sand dike morphology. (**a**) Sand ratio of 5%. (**b**) Sand ratio of 10%. (**c**) Sand ratio of 20%.

Table 3. Equilibrium parameters at different sand ratios.

Sand Ratio/(%)	Balance Time/(min)	Balance Height/(cm)	Sand Dike Placement Rate/(%)
5	16.3	2.6	11.7
10	10.4	6.7	22.4
20	6.8	8.1	30.1

As seen in Table 3, the increase in the equilibrium height reached 157.7% when the sand ratio was increased from 5% to 10%, and the increase in equilibrium height was only 17.2% when the sand ratio was increased from 10% to 20%. The equilibrium time gradually decreased from 16.3 min to 6.8 min, with a decrease of 58.2%. Meanwhile, the laydown rate increased from 11.7% to 30.1% with an increase of 157.2% in the sand ratio. When the sand ratio increases from 10% to 20%, the growth rate of the height and placement rate of the sand dike decreases significantly because the settling of the proppant in the process of the fracturing fluid carrying sand disturbs the surrounding fracturing fluid, which will change its flow direction and hinder the settling of the proppant, which is aggravated with an increase in the sand ratio. In addition, the collision between proppant particles consumes a large amount of energy from the sand-carrying fluid, and the increase in the sand ratio aggravates the collision between particles, thus losing more kinetic energy and consuming more energy. Furthermore, the viscosity of the fracturing fluid increased owing to the addition of more solid particles, increasing the viscous drag on the proppant. More proppant was suspended above, with little change in the sand dike morphology.

The experimental results show that the sand ratio has a significant impact on the sand-carrying regularity of the fracturing fluid and the turbulence intensity formed at the entrance of the fracture by the nano-clean fracturing fluid with different sand ratios: the larger the sand ratio, the smaller the vortex and the lower the amount of proppant that is swept away, and the equilibrium height increases gradually as the sand ratio increases. In addition, the larger the sand ratio, the greater the amount of proppant that is involved in the fracture per unit time and the shorter the equilibrium time. When the sand ratio increases, the fracturing fluid carries more proppant into the fracture, the resistance of the fracturing fluid increases, and a large amount of proppant settles at the entrance, resulting in a deep fracture that cannot be effectively laid; in serious cases, the "sand plugging" phenomenon may occur. However, a minimal sand ratio weakens the conductivity of the fractures, and this is not conducive to improving oil recovery. It is essential to select a reasonable proppant-sand ratio; therefore, choosing a low sand ratio in the early stage of fracturing construction and a high sand ratio in the later stage of construction is recommended.

4.1.3. Effect of Proppant Particle Size

Scenarios 5#, 1#, and 4# in Table 1 were selected for analyzing the effect of the proppant particle size on the proppant spreading morphology of the nano-clean fracturing fluid, and the sand dike morphology in the equilibrium state was obtained, as shown in Figure 6. The equilibrium parameters for the different particle sizes are listed in Table 4.

As can be seen in Table 4, with an increase in the particle size, the equilibrium time was reduced from 14.5 min to 5.3 min, a decrease of 63.4%; the equilibrium height increased from 2.3 cm to 9.9 cm, an increase of 330%; and the laydown rate increased from 14.7% to 29.5%, an increase of 100.7%. As seen in Figure 6, when the proppant particle size increased from 0.3 mm to 0.67 mm, the height and placement rate of the sand dike increased significantly because the gravitational effect on the proppant particles strengthened with an increase in particle size. Although the turbulence intensity, buoyancy, and stagnation resistance of the fracturing fluid also increased during this process, their increase was still much lower overall than the increase caused by gravity, thus accelerating the proppant's settling, which increased the equilibrium height of the sand dike.

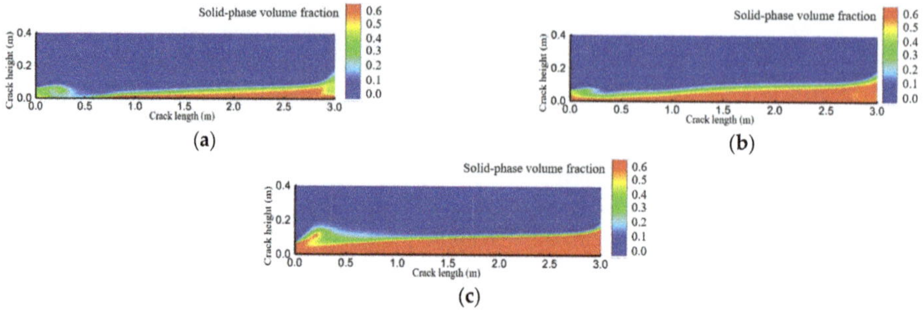

Figure 6. Effect of proppant particle size on sand dike morphology. (**a**) Proppant particle size of 0.3 mm. (**b**) Proppant particle size of 0.45 mm. (**c**) Proppant particle size of 0.67 mm.

Table 4. Equilibrium parameters of different particle sizes.

Grain Size/(m)	Balance Time/(min)	Balance Height/(cm)	Sand Dike Placement Rate/(%)
0.0003	14.5	2.3	14.7
0.00045	10.4	6.7	22.4
0.00067	5.3	9.9	29.5

The experimental results show that the proppant particle size significantly affects the sand-carrying regularity. A proppant with a small particle size is more easily transported to the distal end of the crack; however, the equilibrium height of the formed sand dike is lower, and a blank zone is easily formed at the front end of the crack. As the particle size of the proppant increased, the settling speed of the proppant accelerated, the height of the sand dike increased rapidly, and the flow of the sand-carrying liquid in the crack was blocked, which caused the subsequent proppant to settle more quickly near the entrance of the crack and caused the laying pattern of the sand dike to move toward the crack. Therefore, it is recommended to use a small-particle-size proppant in the early stage of fracturing and a large-particle-size proppant in the later stage to improve the fracturing effect.

4.1.4. Effect of Proppant Density

Scenarios 6#, 1#, and 7# in Table 1 were selected for the study to analyze the effect of the proppant density on the proppant spreading morphology of the nano-clean fracturing fluid, and the sand dike morphology in the equilibrium state was obtained, as shown in Figure 7. The equilibrium parameters at different densities are listed in Table 5.

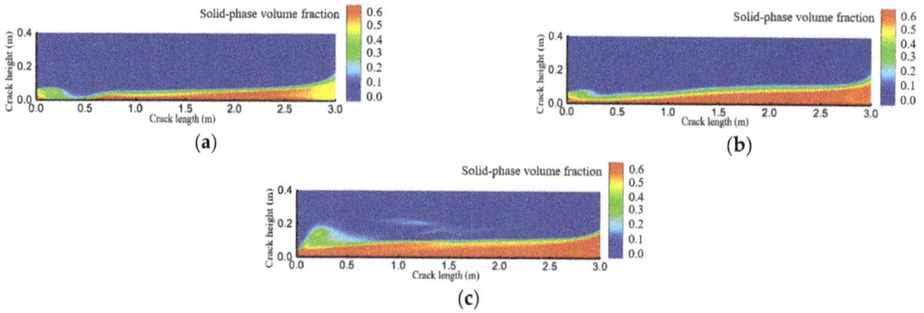

Figure 7. Effect of density on the morphology of sand dikes. (**a**) Proppant with a density of 1450 kg/m^3. (**b**) Proppant with a density of 1600 kg/m^3. (**c**) Proppant with a density of 2000 kg/m^3.

Table 5. Equilibrium parameters for different densities.

Proppant Density/(kg/m^3)	Balance Time/(min)	Balance Height/(cm)	Sand Dike Placement Rate/(%)
1450	13.9	3.0	16.8
1600	10.4	6.7	22.4
2000	6.2	9.7	29.1

As can be seen in Table 5, with an increase in proppant density, the equilibrium time was reduced from 13.9 min to 6.2 min, a decrease of 55.4%; the equilibrium height increased from 3.0 cm to 9.7 cm, an increase of 223.3%; and the laydown rate increased from 16.8% to 29.1%, an increase of 73.2%. Compared to Figure 7, under the same sand volume, the low-density proppant is laid less at both ends of the fracture. The blank zone is easily formed at the front end of the fracture. By contrast, the proppant is more uniformly laid in the fracture than in the medium-density and high-density proppants. This is due to the low gravity of the low-density proppant, as it improves the sand-carrying capacity of the nano-clean fracturing fluid and causes it to be suspended. Furthermore, the medium- and high-density proppants are subject to a greater gravity effect than the low-density proppant. They settle more in the fracture, whereas the low-density proppant is subject to a lower resistance effect and settles less in the fracture. A greater amount of proppant is carried out of the fracture by the fracturing fluid, resulting in the tail of the fracture. The proppant content at the end of the fracture was lower.

The experimental results show that the density of the proppant has a significant effect on the sand-carrying regularity, and the higher-density proppant particles with the same particle size are less likely to be swept to the fracture depth by the vortex formed near the inlet end and tend to settle near the inlet end. By contrast, the low-density proppant can easily carry to the fracture depth owing to its light mass, which enhances the carrying effect of the nano-clean fracturing fluid. Therefore, the equilibrium height gradually increases with an increase in proppant density. An increase in density accelerates the settling of the proppant, leading to the acceleration of sand dike piling and a decrease in the time required to reach the equilibrium height. Therefore, it is recommended to use a low-density proppant in the early stages of fracturing and a high-density proppant in the later stages to improve fracturing results.

4.1.5. Influence of Rheological Parameters

Rheological parameters such as the relaxation time, rheological index, and zero-shear viscosity in the fracturing fluid can affect the viscosity of the fracturing fluid and, thus, reflect the sand-carrying performance of the fluid. Therefore, 1#, 10#, 11#, 12#, 13#, 14#, and 15# in Table 1 were selected to analyze the effects of rheological parameters on the proppant spreading morphology of nano-clean fracturing fluids. The morphologies of the sand dikes at equilibrium are shown in Figures 8–10, and the equilibrium parameters are listed in Tables 6–8.

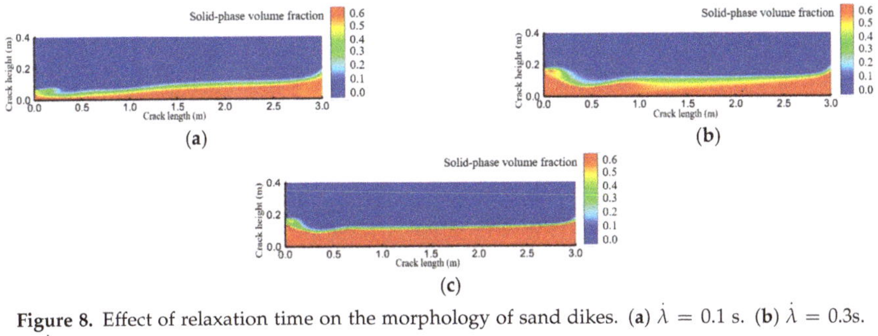

Figure 8. Effect of relaxation time on the morphology of sand dikes. (**a**) $\lambda = 0.1$ s. (**b**) $\lambda = 0.3$s. (**c**) $\lambda = 0.5$ s.

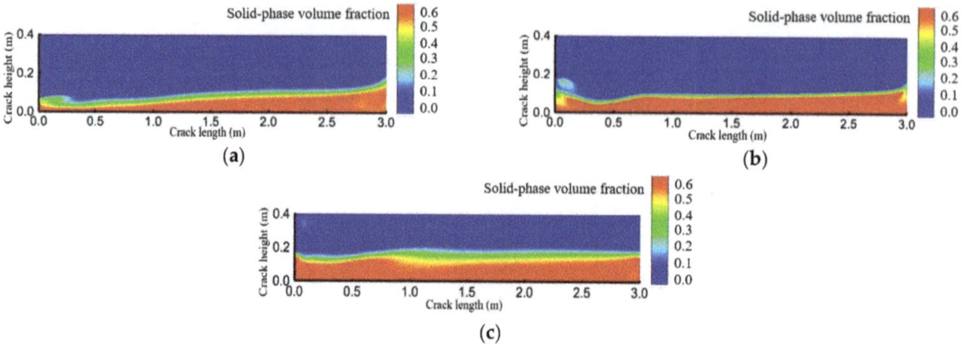

Figure 9. Effect of rheological index on the morphology of sand dikes. (**a**) $n = -0.3$. (**b**) $n = -0.4$. (**c**) $n = -0.5$.

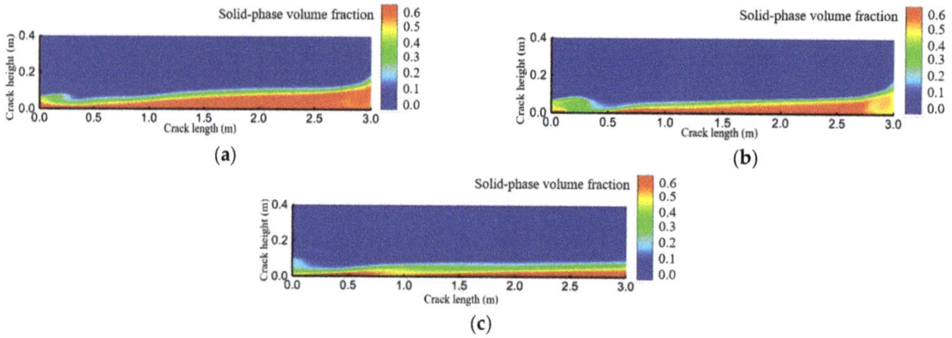

Figure 10. Effect of zero-shear viscosity on sand dike morphology. (**a**) $\mu_0 = 1.0$ Pa·s. (**b**) $\mu_0 = 2.0$ Pa·s. (**c**) $\mu_0 = 3.0$ Pa·s.

Table 6. Equilibrium parameters for different relaxation times.

Relaxation Time/(s)	Balance Time/(min)	Balance Height/(cm)	Sand Dike Placement Rate/(%)
0.1	10.4	6.7	22.4
0.3	8.2	9.6	30.3
0.5	4.3	12.1	32.8

Table 7. Equilibrium parameters for different rheological indices.

Rheology Index	Balance Time/(min)	Balance Height/(cm)	Sand Dike Placement Rate/(%)
−0.3	10.4	6.7	22.4
−0.4	6.8	8.4	24.7
−0.5	5.1	14.7	55.1

Table 8. Equilibrium parameters for different zero-shear viscosities.

Zero-Shear Viscosity μ_0 /(Pa·s)	Balance Time/(min)	Balance Height/(cm)	Sand Dike Placement Rate/(%)
1.0	10.4	6.7	22.4
2.0	11.1	1.9	16.2
3.0	12.6	1.7	15.1

(1) Effect of relaxation time

Scenarios 1#, 10#, and 11# in Table 1 were selected for the study to analyze the effect of the relaxation time on the proppant spreading morphology of the nano-clean fracturing fluid. As shown in Figure 8, with an increase in the relaxation time, the proppant content in the fracture increased, and the equilibrium height of the sand dike increased. The location where the equilibrium height appeared gradually approached the entrance, and many particles settled and gathered at the entrance, resulting in a gradual reduction in the proppant overflow area and easy sand plugging.

As can be seen in Table 6, with an increase in the relaxation time, the equilibrium time was reduced from 10.4 min to 4.3 min, a decrease of 58.6%; the equilibrium height increased from 6.7 cm to 12.1 cm, an increase of 80.6%; and the laydown rate increased from 22.4% to 32.8%, an increase of 31.7%.

(2) Effect of rheological index

Scenarios 1#, 12#, and 13# in Table 1 were selected for the study to analyze the effect of the rheological index on the proppant spreading morphology of the nano-clean fracturing fluid. Figure 9 shows that the height of the sand dike at the entrance gradually increased as the rheological index decreased and that the proppant was collected and settled at the entrance. The height of the sand dike in the horizontal direction increased with the increasing fracture length.

As can be seen in Table 7, with the reduction in the rheological index, the equilibrium time was reduced from 10.4 min to 5.1 min, a decrease of 50.9%; the equilibrium height increased from 6.7 cm to 14.7 cm, an increase of 119.4%); and the laydown rate increased from 22.4% to 55.1% (an increase of 145.9%).

(3) Effects of zero-shear viscosity

Scenarios 1#, 14#, and 15# in Table 1 were selected for the study to analyze the effect of the zero-shear viscosity on the proppant spreading morphology of the nano-clean fracturing fluid. As shown in Figure 10, when changing only the zero-shear viscosity of the nano-clean fracturing fluid, as the zero-shear viscosity increased, the horizontal carrying force and buoyancy of the proppant in the fracture increased. The settling rate of the proppant decreased, resulting in a lower sand dike height at the entrance. Overall, the higher the zero-shear viscosity, the lower the inlet sand dike height, the less prone it is to sand plugging, and the better the sand-carrying performance.

As can be seen in Table 8, with an increase in zero-shear viscosity, the equilibrium time increases from 10.4 min to 12.6 min, an increase of 21.2%; the equilibrium height decreases from 6.7 cm to 1.7 cm, a decrease of 74.6%; and the laydown rate decreases from 22.4% to 15.1%, a decrease of 32.5%.

With all parameters being the same, the smaller the relaxation time, the larger the critical shear rate of fluid shear thinning, the more stable the micelle structure of the fracturing fluid, and the stronger the sand-carrying capacity of the fracturing fluid (the rheological index reflects the characteristics of fluid shear thinning). The smaller the rheological index, the stronger the fluid behaves according to non-Newtonian laws; the larger the zero-shear viscosity, the higher the internal micelle entanglement strength of the fracturing fluid (the orderly arrangement of micelles forms a stable mesh structure) and the stronger the sand-carrying ability of the fracturing fluid. As shown in Figures 8–10, with an increase in zero-shear viscosity and a decrease in relaxation time, the equilibrium height of the sand dike decreased to a certain extent. The proppant moved deeper into the fracture. However, with a decrease in the rheological index, the shear-thinning characteristic of the fluid became more significant, the viscosity of the fracturing fluid inside the fracture decreased rapidly, the sand-carrying capacity weakened, and the proppant accumulated at the entrance.

4.1.6. Sensitivity Analysis of Influencing Factors

To analyze the degree of influence of each influencing factor on the sand-carrying law of the nano-clean fracturing fluid in the fracture, the equilibrium height, sand dike placement rate, and equilibrium time when the sand dike reached dynamic equilibrium were used as reference sequences, and correlation analysis samples were established separately for the grey correlation analysis to determine the degree of influence of each parameter on the morphology of the sand dike and provide a reference for optimizing the fracturing construction. The results of this analysis are shown in Figure 11.

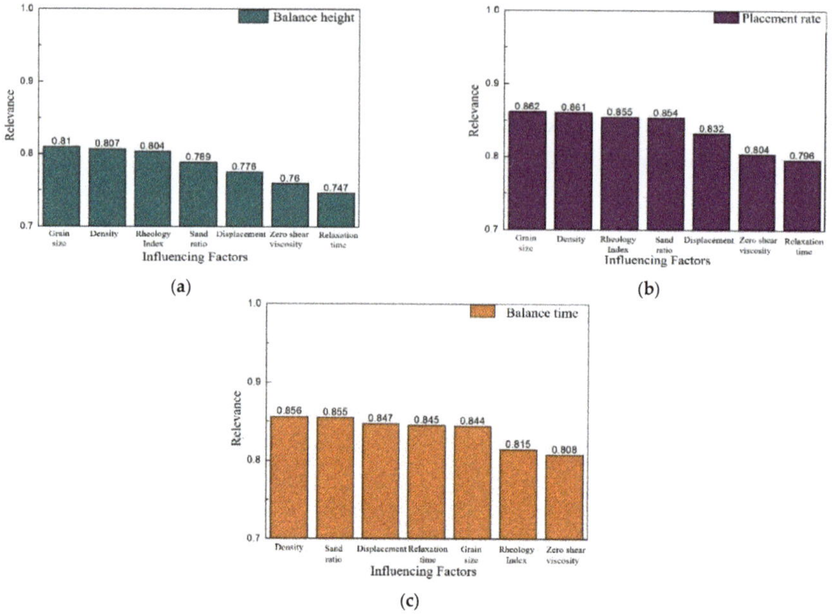

Figure 11. Correlation of each equilibrium parameter. (**a**) Balance height. (**b**) Placement rate. (**c**) Balance time.

As shown in Figure 11, the main factors influencing the equilibrium height and sand dike placement rate are relatively consistent. The degree of influence of each parameter from high to low is as follows: proppant particle size, proppant density, fracturing fluid properties, sand ratio, and discharge volume. The parameters of equilibrium time were the proppant density, sand ratio, discharge volume, fracturing fluid properties, and proppant particle size in descending order. Therefore, a large proppant particle size, high proppant density, and better rheological fracturing fluid significantly affect the equilibrium height and placement rate of the sand embankment and positively improve the fracture conductivity. A low proppant density, small sand ratio, and high displacement can significantly reduce the balance time of the sand bank, which is conducive to the migration of sandy fluid to deep fractures. In the actual construction, reasonable construction parameters can be selected according to the above sequence to ensure the effective placement of the proppant.

4.2. Analysis of Sand-Carrying Laws for Complex Cracks

4.2.1. Analysis of Sand-Carrying Laws of Y-Type Cracks

The Y-type fracture simulation scheme was simulated, and the results are shown in Figure 12. Figure 13 shows the volume distribution of the proppant in the main fracture and branch fracture with the branch fracture angle of 90°. When the fracturing fluid carries sand to the branch joints, the flow rate of the fracturing fluid in the main joints is reduced because part of the fracturing fluid flows into the branch joints, resulting in a poor sand-carrying performance for the fracturing fluid in the main joints and rapid settlement of the proppant. Comparing the sand-carrying performance of the nano-clean fracturing fluid and clean-water fracturing fluid in the main and branch joints, we see that the sand dikes formed in the main joints were more uniform. They had a higher placement rate than those of the clean-water fracturing fluid, and the proppant placement in the branch joints was also higher, indicating that the nano-clean fracturing fluid has a better transport effect in the branch joints, and this is an advantage of the nano-clean fracturing fluid compared with the conventional clean-water fracturing fluid.

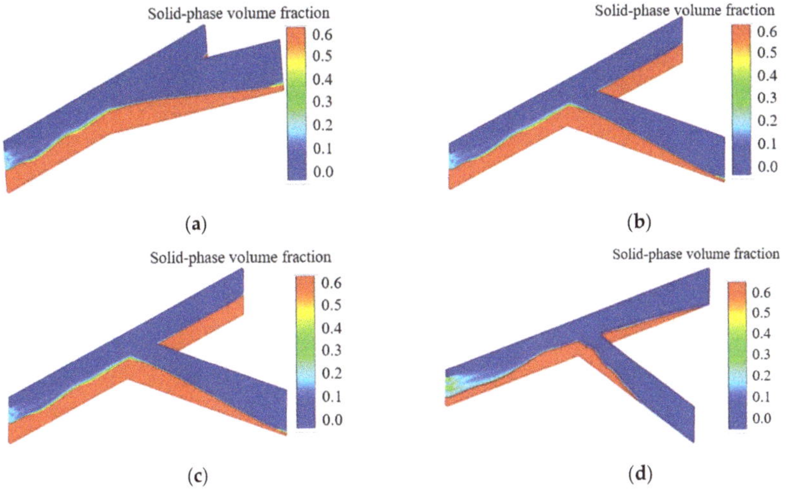

Figure 12. Y-type crack overall sand dike morphology: (**a**) 30° branch fracture nano-clean fracturing fluid, (**b**) 60° branch fracture nano-clean fracturing fluid, (**c**) 90° branch fracture nano-clean fracturing fluid, and (**d**) 90° branch fracture clear-water fracturing fluid.

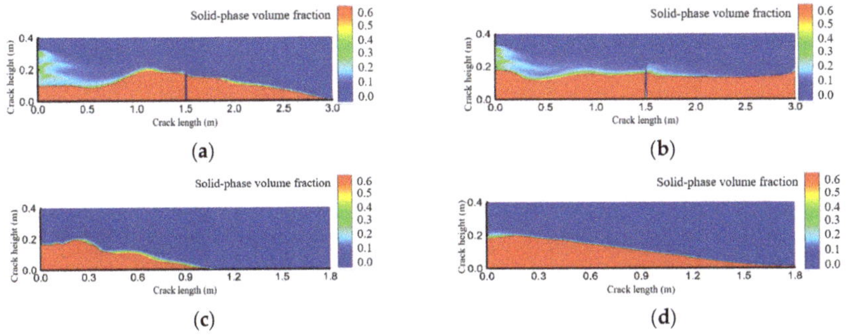

Figure 13. Morphology of sand dike inside the fracture when the angle of the branching joints is 90°. (**a**) Clean-water fracturing fluid main seam. (**b**) Nano-clean fracturing fluid main seam. (**c**) Branch joints of clean-water fracturing fluid. (**d**) Branch joints of nano-clean fracturing fluid.

Figure 14 shows the volume distribution of the proppant within the main and branch joints for branch joints at angles of 30° and 60°. Comparing the sand dike morphology formed at three different angles (30°, 60°, and 90°) of the fractures, we see that the proppant volume distribution in the main joints does not change at different angles. However, when comparing the sand placement in the branch joints at angles of 30° and 90°, the proppant placement can be observed in the tail of the branch joints at an angle of 30°. By contrast, almost no proppant is present in the tail of the branch joints at an angle of 90°. This indicates that the larger the angle between the branch joints and fractures, the greater the obstruction of the sand-carrying movement of the nano-clean fracturing fluid and the more difficult it is for the proppant to enter the branch joints.

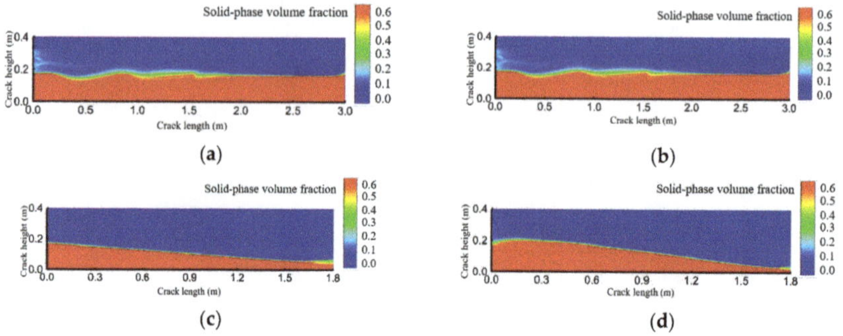

Figure 14. Morphology of sand dike inside the fracture when the angle of branching joints is 30° and 60°. (**a**) Main seam-laying pattern (30°). (**b**) Main seam-laying pattern (60°). (**c**) Branch seam-laying pattern (30°). (**d**) Branch seam-laying pattern (60°).

4.2.2. Analysis of Sand-Carrying Laws of Double-Y-Type Cracks

The double-Y-type crack simulation scheme was simulated, and the overall sand dike shape of the crack is shown in Figure 15. Figure 15 shows the morphological distribution of the overall sand dike for different fracturing fluids. It can be observed that the fracturing fluid carries the proppant particles to settle in the main fracture first. When the proppant is transported to the branch fracture and starts diverting, some proppant exists in both fractures. The sand-carrying fluid drove the transport of the proppant particles in the fracture, and the sand-carrying ability of the nano-clean fracturing fluid was better than that of the conventional fracturing fluid. Therefore, as shown in Figure 16, when the nano-clean fracturing fluid carries sand, the height of the sand dike at the entrance is lower, and sand plugging is less likely to occur. With an increase in time, the accumulation thickness of the sand dike in the horizontal direction increases slowly with the extension of the fracture. The equilibrium height position of the sand dike moves toward the distal end of the fracture. The sand dike formed by the sand-carrying action of the clean fracturing fluid is long and gentle. By contrast, the sand dike formed by the sand-carrying action of the clean-water fracturing fluid has a higher equilibrium height and a larger volume of sand dike in the entire fracture. However, the proppant placement at the end of the fracture is less and more nonuniform, and the equilibrium height and placement rate of the proppant in the branch fracture are lower than those of the clean fracturing fluid.

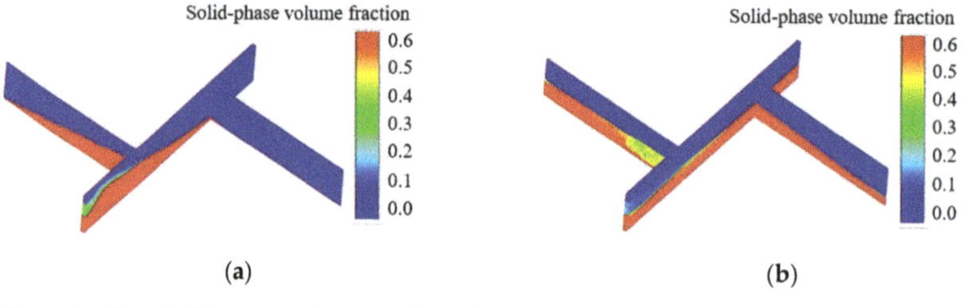

Figure 15. Double-Y-type cracks' overall sand dike morphology. (**a**) Clear-water fracturing fluid. (**b**) Nano-clean fracturing fluid.

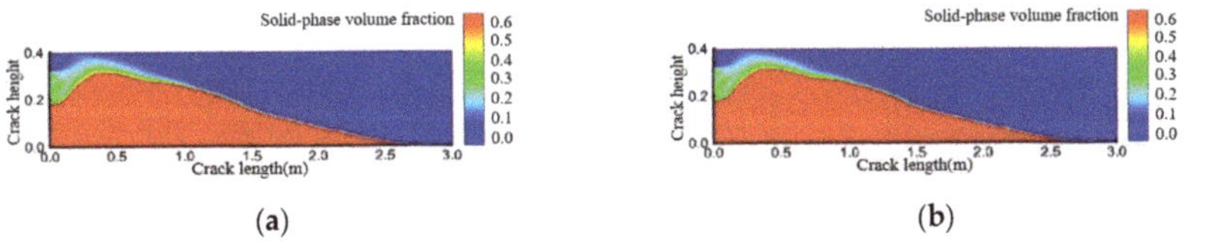

Figure 16. Double-Y-type crack within the main seam sand dike morphology. (**a**) Clear-water fracturing fluid. (**b**) Nano-clean fracturing fluid.

Figure 17 shows that the proppant concentration in the distal branch fracture is much higher than that in the clear-water fracturing fluid. This phenomenon again proves that the sand-carrying performance of the nano-clean fracturing fluid is higher than that of the clear-water fracturing fluid. This is because the shear rate near the proximal branch is more significant than that in the distal branch, and the clean fracturing fluid has good rheology; thus, the apparent viscosity of the fracturing fluid in the proximal branch was low, and the proppant settled more. By contrast, the apparent viscosity was high in the distal branch, and the proppant settled less. When the clear-water fracturing fluid carried the fracturing fluid because the Newtonian fluid was not affected by the shear rate, the sand-carrying fluid first diverted into the proximal branch fracture, and the fluid in the main fracture continued to move in the fracture expansion direction. Moreover, when the sand-carrying fluid moves to the distal branch fracture again, the flow velocity in the fracture decreases because diversion has already occurred, which decreases the proppant concentration in the distal branch fracture.

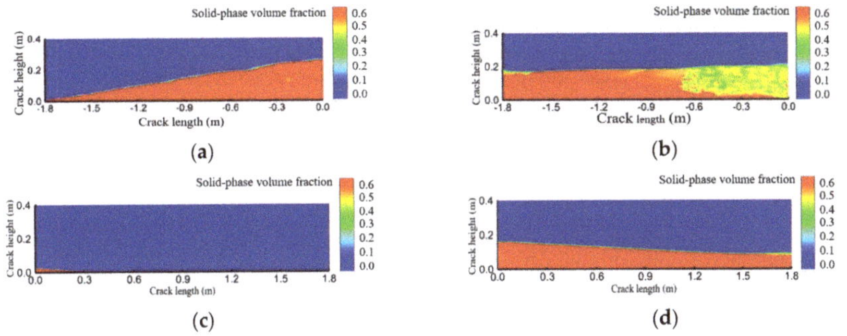

Figure 17. Morphology of sand dike with branching joints within double-Y-type cracks. (**a**) Proximal branch joints of clean-water fracturing fluid. (**b**) Proximal branch joints of nano-clean fracturing fluid. (**c**) Distal branch joints of clean-water fracturing fluid. (**d**) Distal branch joints of nano-clean fracturing fluid.

4.2.3. Analysis of Sand-Carrying Laws of Well-Type Cracks

The overall sand dike morphologies of the cracks are shown in Figure 18. Comparing with the volume distribution of proppant in the main fracture in Figure 19, it can be seen that the sand dike formed under the carrying of clean water still shows a larger nonuniformity compared with the nano-clean fracturing fluid, but the nonuniformity is reduced compared with the double-Y fracture. The equilibrium time of the sand dike increases, and the equilibrium height decreases with an increase in fracture complexity, making it easier to form the sand dike accumulation at the end of the main fracture.

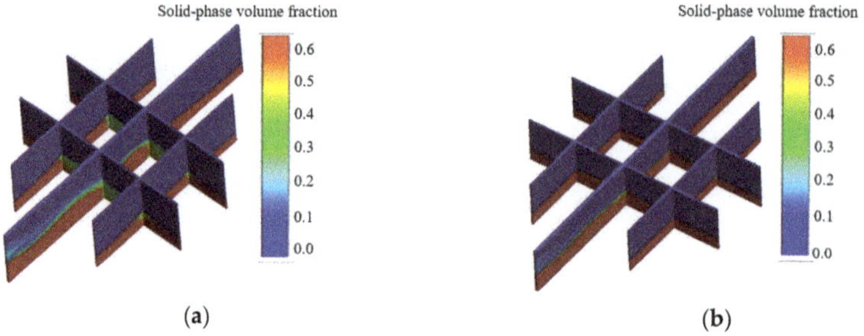

Figure 18. Well-type crack overall sand dike morphology. (**a**) Clear-water fracturing fluid. (**b**) Nano-clean fracturing fluid.

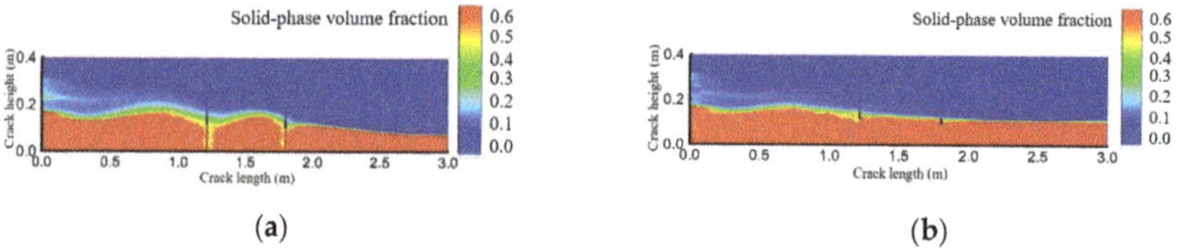

Figure 19. Morphology of sand dike in the main seam of well-type crack. (**a**) Clear-water fracturing fluid. (**b**) Nano-clean fracturing fluid.

As can be seen in Figure 20, the proppant concentration in the secondary branch fracture was significantly higher than that in the clear-water fracture fluid. The placement rate in the proximal tertiary branch fracture was also slightly higher than that in the clear-water fracture fluid; however, the sand-carrying situation of both in the distal tertiary branch fracture was almost the same, which is because, under the sand-carrying situation of the clear-water fracture fluid, the sand dikes accumulated in the main fracture, which reduced the flow channels inside the complex fracture, and the fluid entered the branch fracture. This is because the accumulation of sand dikes in the main fracture reduces the flow channel inside the complex fracture and increases the fluid velocity into the branch fracture, partially suspending the proppant and moving it further into the fracture. In addition, both types of fracturing fluids formed sand dikes at a lower height in the branch joints away from the inlet than in the branch joints at the inlet. This can be mainly attributed to the presence of multiple fractures in the complex fracture for diversion, which reduces the fluid flow rate in the main fracture and causes the fracturing fluid to fail to meet the sand-carrying requirements, thereby accelerating the settlement of the proppant and allowing more fracturing fluid and proppant to enter the branch fracture near the entrance.

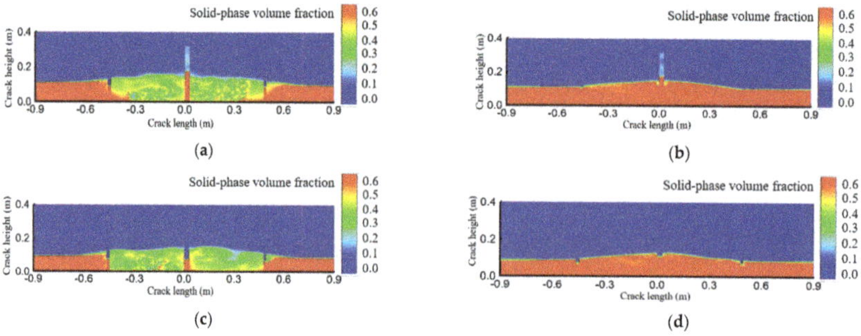

Figure 20. *Cont.*

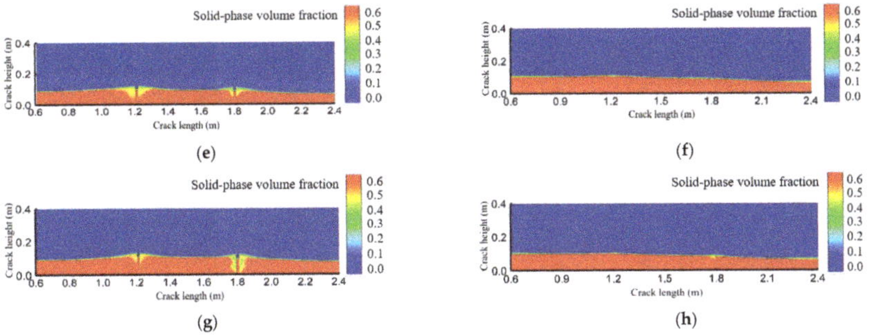

Figure 20. Morphology of sand dike in branching joints within well-type cracks. (**a**) Clear-water proximal secondary branch seam. (**b**) Nano-clean proximal secondary branch seam. (**c**) Clear-water distal secondary branch seam. (**d**) Nano-clean distal secondary branch seam. (**e**) Clear-water proximal tertiary branch seam. (**f**) Nano-clean proximal tertiary branch seam. (**g**) Clear-water distal tertiary branch seam. (**h**) Nano-clean distal tertiary branch seam.

5. Conclusions

(1) Nano-clean fracturing fluids are a new type of fracturing fluid that has a sand-carrying law that is unclear, especially for complex fractures with variable directions. The influence of fracturing fluid rheology, the complex fracture branching angle, or other factors on the sand-carrying process is ignored in the existing models, thus deviating significantly from the actual situation. A new nano-clean fracturing fluid sand-carrying transport model considering factors such as a mutual collision of the proppant, wall friction blocking, fracturing fluid filtration loss, and fracture branching angle was established which can accurately characterize the sand-carrying law of nano-clean fracturing fluid and provide a research basis for the optimization of fracturing construction parameters in coalbed methane reservoirs.

(2) Numerical simulations of the factors influencing the proppant placement pattern under the carrying action of nano-clean fracturing fluid were carried out using the controlled variable method. The grey correlation analysis was used to determine the ranking of the influence of each factor on the sand dike. The results showed that a large proppant particle size, high proppant density, and better rheological fracturing fluid significantly affect the equilibrium height and placement rate of the sand embankment and positively improve the fracture conductivity. A low proppant density, small sand ratio, and high displacement can significantly reduce the balance time of the sand bank, which is conducive to the migration of sandy fluid to deep fractures. In the actual construction, reasonable construction parameters can be selected according to the above sequence to ensure the effective placement of the proppant.

(3) A multi-angle and multi-size geometric model of complex fractures was constructed. A numerical simulation experiment of clean fracturing fluid in complex fractures was conducted to compare and analyze the experimental results of clean-water fracturing fluid and clean fracturing fluid, showing that an increase in the branching angle in complex fractures was not conducive to the sand-carrying of the fracturing fluid. The flow rate of the fracturing fluid would be reduced as the complexity of fractures increased, making it difficult for the proppant to enter the branching joints. However, the overall sand-carrying performance of the nano-clean fracturing fluid in the established complex fractures was better than that of the typical clear-water fracturing fluid, which promoted the application of the nano-clean fracturing fluid in CBM extraction. In this paper, the sand-carrying law of nano-clean fracturing fluid was studied under the assumption fracture, but the fractures generated by in situ fracturing have different trends and shapes, and the formation temperature also makes a difference. Therefore, it is necessary to carry out further research while considering the above influencing factors.

Author Contributions: All authors contributed to the conception and design of the study. W.Z. helped with the analysis through constructive discussions, which led to the idea of this study. Material preparation, data collection, and analysis were carried out by K.X. and L.L. H.H. proposed revisions through constructive discussion, and the first draft of the manuscript was written by H.H. and S.W. K.X. and Z.S. conducted the data analysis and visualized the experimental results to optimize the manuscript. Y.L. played a supervisory and leading role in the essay-writing process, and the accuracy and standardization of the paper's language was modified by H.M. All authors commented on previous manuscripts. All authors have read and agreed to the published version of the manuscript.

Funding: This paper was jointly funded by the National Natural Science Foundation of China, (grants No. 52104190, 52274188, and 52174173); the Outstanding Youth Foundation of Henan Polytechnic University (grant No. J2021-5); Strategic Research and Consulting Project of the Chinese Academy of Engineering (grant No. 2023HENZDB05); the Focus R&D and Promotion for Scientific and Technological Project of Henan Province (grant No. 222102320012); the Project Funded by China Postdoctoral Science Foundation (grants No. 2020M682291 and 2022T150194); and the State Key Laboratory Cultivation Base for Gas Geology and Gas Control (Henan Polytechnic University) (grant No. WS2021A05).

Data Availability Statement: Not applicable.

Conflicts of Interest: The authors declare that there are no conflict of interest.

References

1. Hou, J.; Wang, C.; Liu, P. How to improve the competiveness of natural gas in China with Energy Internet and "The Belt and Road Initiative". *Int. J. Energy Res.* **2018**, *42*, 4562–4583. [CrossRef]
2. Zhou, Y.; Lin, M. System simulation of air-cooling single effect LiBr absorption refrigerating system driven by solar heat source. *Environ. Earth Sci.* **2021**, *766*, 23–25. [CrossRef]
3. Zheng, C.; Jiang, B.; Xue, S.; Chen, Z.; Li, H. Coalbed methane emissions and drainage methods in underground mining for mining safety and environmental benefits: A review. *Process Saf. Environ. Prot.* **2019**, *127*, 103–124. [CrossRef]
4. Lu, Y.Y.; Zhang, H.D.; Zhou, Z.; Ge, Z.L.; Chen, C.J.; Hou, Y.D.; Ye, M.L. Current Status and Effective Suggestions for Efficient Exploitation of Coalbed Methane in China: A Review. *Energy Fuels* **2021**, *35*, 9102–9123. [CrossRef]
5. Wang, Y.; Lu, Y.; Li, Y.; Wang, X.; Yan, X.; Zang, Z. Progress and application of unconventional reservoir fracturing technology. *Acta Pet. Sin.* **2012**, *33*, 149–158.
6. Huang, Z.; Li, G.; Yang, R.; Li, G. Review and development trends of coalbed methane exploitation technology in China. *J. China Coal Soc.* **2022**, *47*, 3212–3238.
7. Chen, H.; Fan, H.; Guo, J.; He, X. Analysis and prospect of hydraulic fracturing fluids for coalbed methane Wells. *Coal Geol. Explor.* **2017**, *45*, 33–40.
8. Men, X.; Han, Z.; Gong, H.; Wang, Y. Challenges and Opportunities of CBM exploration and development in China under the new situation. *Nat. Gas Ind.* **2018**, *38*, 10–16.
9. Zuo, W.; Zhang, W.; Liu, Y.; Han, H.; Huang, C.; Jiang, W.; Mitri, H. Pore Structure Characteristics and Adsorption and Desorption Capacity of Coal Rock after Exposure to Clean Fracturing Fluid. *ACS Omega* **2022**, *7*, 21407–21417. [CrossRef]
10. Alotaibi, M.A.; Jennifer, L. Slickwater Proppant Transport in Hydraulic Fractures: New Experimental Findings and Scalable Correlation. *Miskimins Spe Prod. Oper.* **2018**, *3*, 164–178. [CrossRef]
11. Zhang, T.; Guo, J.; Liu, W. CFD simulation of proppant transport settlement behavior in clear water fracturing. *J. Southwest Pet. Univ.* **2014**, *36*, 74–82.
12. Shi, F.; Wang, X.; Liu, C.; Liu, H.; Wu, H. A coupled extended finite element approach for modeling hydraulic fracturing in consideration of proppant. *J. Nat. Gas Sci. Eng.* **2016**, *33*, 885–897. [CrossRef]
13. Fredd, C.N.; Olsen, T.N.; Brenize, G. Polymer-Free Fracturing Fluid Exhibits Improved Cleanup for Unconventional Natural Gas Well Applications. In Proceedings of the SPE Eastern Regional Meeting, Charleston, WV, USA, 15–17 September 2004.
14. Tong, S.; Singh, R.; Mohanty, K.K. A visualization study of proppant transport in foam fracturing fluids. *J. Nat. Gas Sci. Eng.* **2018**, *52*, 235–247. [CrossRef]
15. Luo, Z.; Zhang, N.; Zhao, L.; Pei, Y.; Liu, P.; Li, N. Innovative Encapsulating Acid with Release Dually Controlled by the Concentration of Hydrogen Ions and Temperature. *Energy Fuels* **2019**, *33*, 4976–4985. [CrossRef]
16. Li, T.; Yang, Q.; Feng, W.; Qian, Z.H.N. Indoor research and field application of a new type of clean fracturing fluid for coalbed methane. *Sci. Tech.* **2012**, *12*, 9828–9832.
17. Yang, F.; Ge, Z.; Zheng, J. Viscoelastic surfactant fracturing fluid for underground hydraulic fracturing in soft coal seams. *J. Pet. Sci. Eng.* **2018**, *169*, 646–653.
18. Crews, J.B.; Huang, T.; Wood, W.R. The Future of Fracturing-Fluid Technology and Rates of Hydrocarbon Recovery. In Proceedings of the SPE Annual Technical Conference and Exhibition, Denver, CO, USA, 21–24 September 2008.

19. Nettesheim, F.; Liberatore, M.W.; Hodgdon, T.K.; Wagner, N.J.; Kaler, E.W.; Vethamuthu, M. Influence of Nanoparticle Addition on the Properties of Wormlike Micellar Solutions. *Langmuir* **2008**, *24*, 7718–7726. [CrossRef]
20. Chen, H.; Fan, H.; Guo, J.; Tang, M.; Ni, F. Evaluation and Prediction of Coalbed Gas Fracturing Fluid. *Adv. Mater. Res.* **2014**, *1008*, 257–263. [CrossRef]
21. Dai, C.; Li, Y.; Xie, Q.; Xu, K.; Hu, Y.; Zhou, Z.; Wu, Y. Reduction of clean fracturing fluid filtration loss by viscosity enhancement using nanoparticles: Is it feasible? *Chem. Eng. Res. Des.* **2020**, *156*, 414–424. [CrossRef]
22. Zhao, M.; Gao, Z.; Dai, C.; Sun, X.; Zhang, Y.; Yang, X.; Wu, Y. Effect of Silica Nanoparticles on Wormlike Micelles with Different Entanglement Degrees. *J. Surfactants Deterg.* **2019**, *22*, 587–595. [CrossRef]
23. Lun, C.K.K.; Savage, S.B.; Jeffrey, D.J.; Chepurniy, N. Kinetic theories for granular flow: Inelastic particles in Couette flow and slightly inelastic particles in a general flow field. *J. Fluid. Mech.* **1984**, *140*, 223–256. [CrossRef]
24. Yao, J.; Yang, Y.; Huang, Z.; Sun, J.; Wang, J.; Yang, Y. Influence of particle viscosity model on the flow behavior of ultra-dense particles simulated by Euler multiphase flow model. *CIESC J.* **2020**, *71*, 4945–4956.
25. Zhang, G.; Li, M.; Geng, K.; Han, R.; Xie, M.; Liao, K. New integrated model of the settling velocity of proppants falling in viscoelastic slick-water fracturing fluids. *J. Nat. Gas Sci. Eng.* **2016**, *33*, 518. [CrossRef]
26. Zhan, Z. Study on the Law of Sand Transport of Slippery Water in Fractured Fracture. Master's Thesis, Xi'an Shiyou University, Xi'an, China, 2018.
27. Zhou, D.; Zhang, Z.; Hui, F. Experiment and numerical simulation on transportation laws of proppant in major fracture during slick water fracturing. *Pet. Drill. Tech.* **2017**, *39*, 499–508.

minerals

MDPI

Article

Characteristics and Geological Significance of Organic Matter Veins in Shale Reservoir: A Case Study of the Silurian Longmaxi Formation in Luzhou Area, Sichuan Basin

Yuanlin Wang [1,2], Yunqian Jia [1,2], Chenchen Wang [2,3,*], Wei Lin [4], Jizhen Zhang [2,5], Denglin Han [1,2], Binyu Ma [1,2] and Huachao Wang [6]

1 School of Geosciences, Yangtze University, Wuhan 430100, China; wangyl.st@yangtzeu.edu.cn (Y.W.); jiayq@yangtzeu.edu.cn (Y.J.); handl@yangtzeu.edu.cn (D.H.); mabinyu2011@163.com (B.M.)
2 Laboratory of Reservoir Microstructure Evolution and Digital Characterization, Yangtze University, Wuhan 430100, China; zhangjz1991@126.com
3 Cooperative Innovation Center of Unconventional Oil and Gas, Yangtze University, Wuhan 430100, China
4 Institute of Geology and Paleontology, Linyi University, Linyi 276000, China; ucaslinwei@126.com
5 College of Resources and Environment, Yangtze University, Wuhan 430100, China
6 No. 3 Oil Production Plant, PetroChina Changqing Oilfield Company, Yinchuan 750006, China; 201671258@yangtzeu.edu.cn
* Correspondence: wcc1220@163.com

Abstract: Organic matter serves as the hydrocarbon-generating parent material for shale reservoirs, in which organic pores are also important reservoir spaces. Different types of organic matter have wide differences in hydrocarbon generation and pore-forming ability. Based on the occurrence state of organic matter, in the over-mature Marine shale organic matter mainly includes in situ and migrated organic matter. It has been extensively studied on in situ organic matter and organic matter migrating into inorganic pores, while there are few reports of organic matter migrating into microfractures. In this study, the over-mature Marine shale reservoir in the first sub-member of the Silurian Longmaxi Formation in the Luzhou area of the Sichuan Basin is taken as an example. Core observation, optical microscope, high-precision large-view scanning (MAPS, modular automated processing system) and mineral analysis scanning (QEMSCAN, quantitative evaluation of minerals by scanning electron microscopy) were conducted to observe the morphological characteristics of organic matter veins, and then analyze the genesis and pore-forming characteristics of such organic matter. The results show that: ① Organic matter veins (OM veins) are soluble organic matter with fractures as an effective channel, and OM veins in the study section is easy to form under the condition of micro-fractures in the shale sweet segment after organic matter generating oil and before gas generation ② Organic matter in the OM veins are less efficient in pore-forming, with sparse pores and smaller pore sizes. The occurrence of fractures varies greatly, including horizontal fractures, oblique fractures and high-angle fractures, which are mostly developed in the $Long1_1{}^1$ and $Long1_1{}^2$ layers. ③ The development of OM veins can indicate better reservoir conditions, that is, the layers have strong hydrocarbon generation intensity (strong pore-forming ability of organic matter) and high brittle mineral content (strong reservoir compressibility). The new findings in this paper reveal that OM veins are instructive for the determination of geological–engineering sweet spots in the Longmaxi Formation in the Sichuan Basin, and also provide guidance for future research on occurrence form and geological significance of different types of organic matter.

Keywords: Sichuan Basin; Luzhou area; Longmaxi Formation; scanning electron microscope; organic matter type

1. Introduction

Organic matter is the parent material of hydrocarbon generation, and organic pores serve as significant reservoir spaces [1,2]. Therefore, the pore-forming ability of organic matter is a key parameter in evaluation of unconventional petroleum systems [3,4]. Previous studies have shown that different types of organic matter have great differences in hydrocarbon generation and pore-forming capacity [5–7]. At present, the classification methods of over-mature Marine organic matter types are mostly based on morphology from the perspective of organic matter sources, and it is widely believed that bitumen and kerogen are the main components of organic matter [8]. Loucks et al. determined seven petrological criteria based on the occurrence of organic matter in SEM micrographs, which can be used to distinguish sedimentary organic matter from migrating organic matter according to the occurrence pore, the continuous distribution, dense texture of organic matter and so on [9]. On this basis the migrating organic matter was further divided into morphological and amorphous forms, and algal fragments, bacterial aggregates, graptolites and microsomes were identified from kerogen [10,11]. The hydrocarbon generation capacity and pore evolution process of different organic matter types are also different, the organic matter filled in the intergranular pores of rigid mineral particles has a high degree of development, which is the best for the development of organic matter macropores, and in the mixture of organic matter and clay minerals, the pore size distribution range is wide but the total amount of organic pores is small [12–14].

In conclusion, many studies have shown that a large amount of migrated organic matter in the pores of inorganic minerals exists in the over-mature Marine shale reservoir of Longmaxi Formation, Sichuan Basin, which plays an important role in the formation of shale gas [15]. In contrast to that the organic matter migrated to microfractures is rarely reported. It is worth noting that the organic matter migrating to the fracture is different from the shrinkage fracture of organic matter developed at the edge of the banded organic matter mentioned in previous studies [16]. Although all of them are in the form of organic matter in the fracture, the organic matter referred to by the predecessors is a long strip of in-situ organic matter. The most obvious feature is that the outline is clear and does not interact with clay minerals. In this study, the organic matter which was migrated and filled into microfractures, was called organic matter veins (OM veins). shale fractures and their corresponding filled veins are important features for restore the process of hydrocarbon generation with important implications for studying shale gas enrichment and preservation [17]. The characteristics of temperature, pressure and diagenetic fluid environment recorded by calcite, quartz and other minerals in fracture veins have been extensively reported [18,19]. Therefore, this study combines organic matter type, veins and fracture open-closure times to explore the indicative significance of OM veins to shale reservoir.

In this manuscript, we characterize OM veins at multiple scales and parameters, and analyses the morphological characteristics, occurrence state and pore-forming ability of organic matter in fractures based on core observation, optical microscope, MAPS high-resolution large-field scanning and QEMSCAN mineral analysis scanning under the same field of view. The research also discusses the distribution, source and formation period of OM veins, and then to explore the geological significance of OM veins. in order to systematically understand the formation and evolution of different organic matter in shale reservoirs, and provide more detailed microscopic evaluation for shale gas desserts.

2. Geological Setting

The study area is mainly the Luzhou area in the southern Sichuan Basin, which is structurally located in the low-steep structural belt in southern Sichuan. From north to south, the comb-shaped anticline structure arranged in an echelon is developed. The syncline structure is wide and gentle, and small-scale faults are mainly developed, which have little damage to shale gas reservoirs [20,21]. In terms of sedimentation, since the early Cambrian, the basin has generally been in a shallow sea environment. Due to the influence of rising ocean currents, the area has formed a paleogeographic environment conducive to the deposition of shale and phosphorus [22].

From Ordovician to Early Silurian, the Sichuan Basin was surrounded by the Chuanzhong uplift, the Xuefeng uplift, and the Qianzhong uplift [23]. The Sichuan Basin was surrounded by the east, west, and south sides, forming a restricted retention basin environment favorable for the development of source rocks, resulting in the terrain of "three uplifts and one depression" [24,25] (Figure 1). At the end of Ordovician and the beginning of Silurian, two global transgressions [26,27] and multi-stage large-scale volcanic activities occurred [28], forming the shale of the Wufeng–Longmaxi formation. Since the early deposition of the Longmaxi formation, the Luzhou area as a whole was located in the deep-water shelf sedimentary environment [29] and was in a quiet anoxic environment [30].

The first member of the Longmaxi formation is divided into $Long1_1$ and $Long1_2$ sub-members. $Long1_1$ sub-member is mainly organic-rich shale, and the TOC value is generally higher than 2.0%. According to the lithology, stratigraphic sequence, sedimentary cycle and paleontology of typical wells, $Long1_1$ sub-member is subdivided into $Long1_1{}^1$, $Long1_1{}^2$, $Long1_1{}^3$ and $Long1_1{}^4$ layers from bottom to top [31].

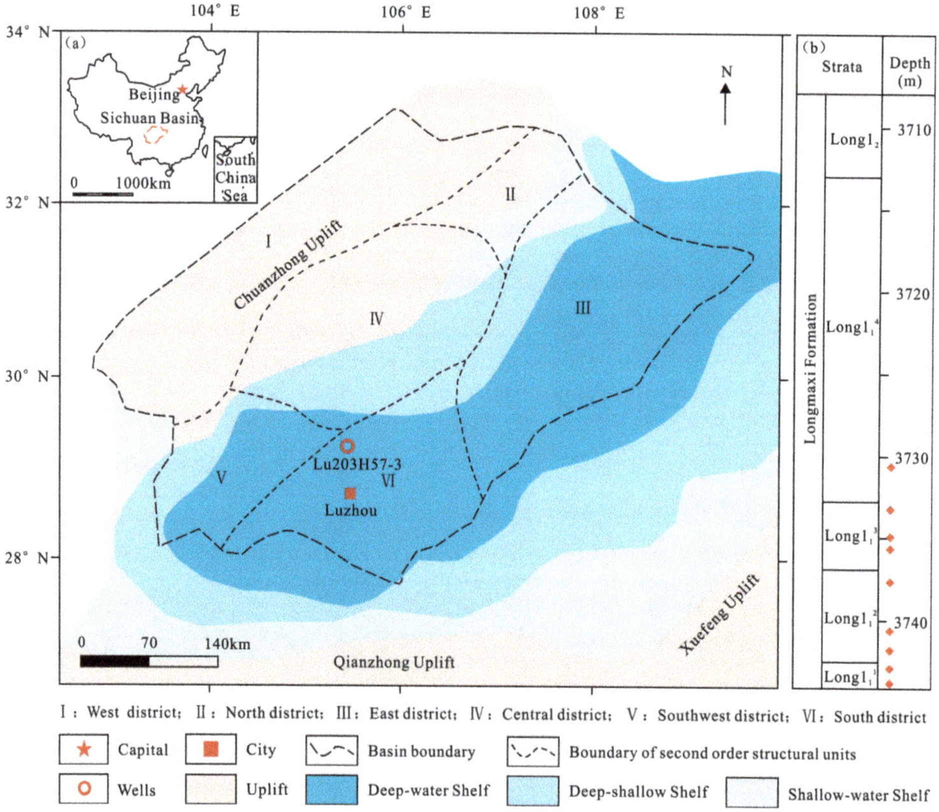

Figure 1. (**a**) Regional geological map of Sichuan Basin (modified after [32–35]). (**b**) Stratigraphic column.

3. Samples and Methods

3.1. Research Samples

A total of 9 representative samples from Silurian $Long1_1$ sub-member of Well Lu 203H57-3 in the Luzhou area were included in this study. The samples covered thin layers of $Long1_1{}^1$–$Long1_1{}^4$ with a depth of 3720~3750 m. After sampling, the fresh samples were packaged and immediately sent to the Laboratory of Reservoir Microstructure Evolution and Digital Characterization of Yangtze University for sample preparation and experiment.

3.2. Experimental Method

In view of the extremely fine-grained sedimentary fabric characteristics and strong heterogeneity of shale, the MAPS scanning technology with high resolution (500 nm) and large field of view (the physical size of the sample is centimeter level) is used to scan the sample as a whole. Scanning technology is used to divide the sample surface area into a series of regular grids, and then scan and create an image of each grid to obtain a series of two-dimensional high-resolution scanning images. Then, all images are spliced to obtain a complete two-dimensional high-resolution large-field scanning image [36]. The test method is based on SY/T 5162-2014.

In view of the fact that the pores of organic matter in shale are at the nanoscale, organic matter in different occurrence states is located and statistically analyzed from the complete MAPS images, and a single image with a higher resolution (4 nm) is obtained. A total of 140 images are captured, including 20 of in situ organic matter, 40 of organic matter in inorganic pores and 80 of OM veins. Using the image-segmentation technology of ImageJ software, the pores in organic matter were extracted from single images of different organic matter types [37], and the pore-forming efficiency was calculated [7]. Porosity measuring standards observe SY/T6103-2019. Specifically, porosity parameters of different samples were collected and the mean of multiple operations by different people was used. On this basis, the statistical errors caused by image factors and artificial factors can be avoided to the maximum extent.

In order to accurately obtain the mineral distribution and content information under MAPS scanning images, QEMSCAN mineral analysis scanning technology was used to scan the same field of view. The scanning technology is based on the energy of X-ray generated by the primary electron in the process of atomic excitation of secondary electrons on the surface of the sample to determine the element type of the object in the scanned point. According to the element distribution information, the actual elements are combined into minerals in the background mineral species database, and then the mineral distribution and content information are obtained. The test method is based on GB/T 17359-2012 and GB/T 20726-2015. The test result is a two-dimensional color image. Different color areas represent different mineral components, and white areas represent non-mineral components such as pores, cracks, and organic matter. Corresponding to the MAPS image in the same field of view, the white area components can be accurately identified.

The instrument used for MAPS scanning is HELIOS NanoLab 650, the voltage is 1~30 KV, the current is 0.78 pA~26 nA, the pixel size of the recognition image is 2~800 nm, and the overlap rate between adjacent spliced small images is 6~8%. The QEMSCAN scanning instrument is QEMSCAN 650F, the voltage is 1~30 KV, the current is 0.78 pA~26 nA, and the pixel size of the recognition image is 0.5~50 μm. The laboratory temperature is 20 ± 5 °C, and the humidity is not more than 60%.

In addition, in order to fully obtain the characteristics of shale reservoirs, LECO carbon and sulfur analyzer CS230 was used to determine the total organic carbon (TOC) content of the samples, and the test method was based on GB/T 19145-2022.

4. Results

4.1. Morphological Characteristics

According to the source of organic matter and its occurrence under SEM, the types of organic matter in the study section are divided into in situ sedimentary organic matter, organic matter migrating to inorganic pores and organic matter migrating to fractures (OM veins). Among them, the in situ organic matter is the most primitive organic matter deposited in the process of sedimentation, so it retains some biological structure characteristics. However, because it comes from different organic components, its morphological performance is not exactly the same (Figure 2a,b). But as a whole, the area of in situ organic matter is large and the boundary between the organic matter and mineral particles is also obvious. The organic matter in the inorganic pores is the most common type of organic matter in the studied section [5], and it is the bitumen generated by in situ organic matter in the process of hydrocarbon generation and expulsion. The area of organic matter in inorganic pores is small and its morphology is controlled by the pores of the mineral matrix before filling, which include intergranular pores, pyrite intercrystalline pores, clay mineral interlayers, and mineral particle cracks (Figure 2c,d). The organic matter in the fracture usually runs through the entire scanning electron microscope field of view with an opening of 5~80 μm (Figure 3).

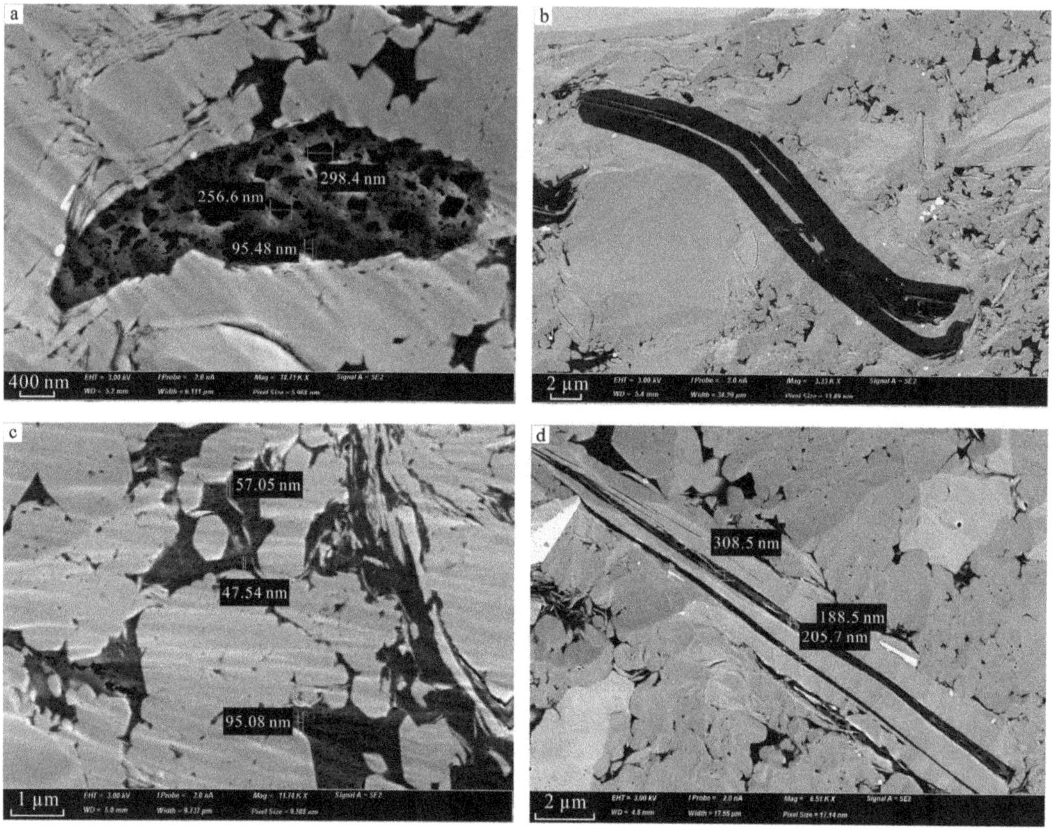

Figure 2. Scanning electron microscope image of Long1_1 sub-member of Lu203H57-3 well. (**a**) In situ organic matter of sapropelinite, in which the pores are alternate in size and dense (Long1_1^3 layer, 3733.2 m). Sapropelinite has neither fixed morphology nor clear contour. It is mainly flocculent or cloudiness and can adapt its morphology continuously to matrix pores. (**b**) In situ organic matter of inertinite, low degree of pore development (Long1_1^3 layer, 3734.9 m). The longitudinal section of inertinite fusinite is fibrous and fusinite is in bedding arrangement. (**c**) Intergranular organic matter, small and dense pores (Long1_1^1 layer, 3743.8 m). (**d**) Organic matter inside clay minerals, small and dense pores (Long1_1^2 layer, 3741.8 m).

OM veins can also be observed in core hand specimens and optical microscope, they obliquely cross bedding and are black and irregular (Figure 4). Ordinary slices show that organic matter is mostly filled into horizontal fractures and clay minerals such as illite are often found around them (Figure 5).

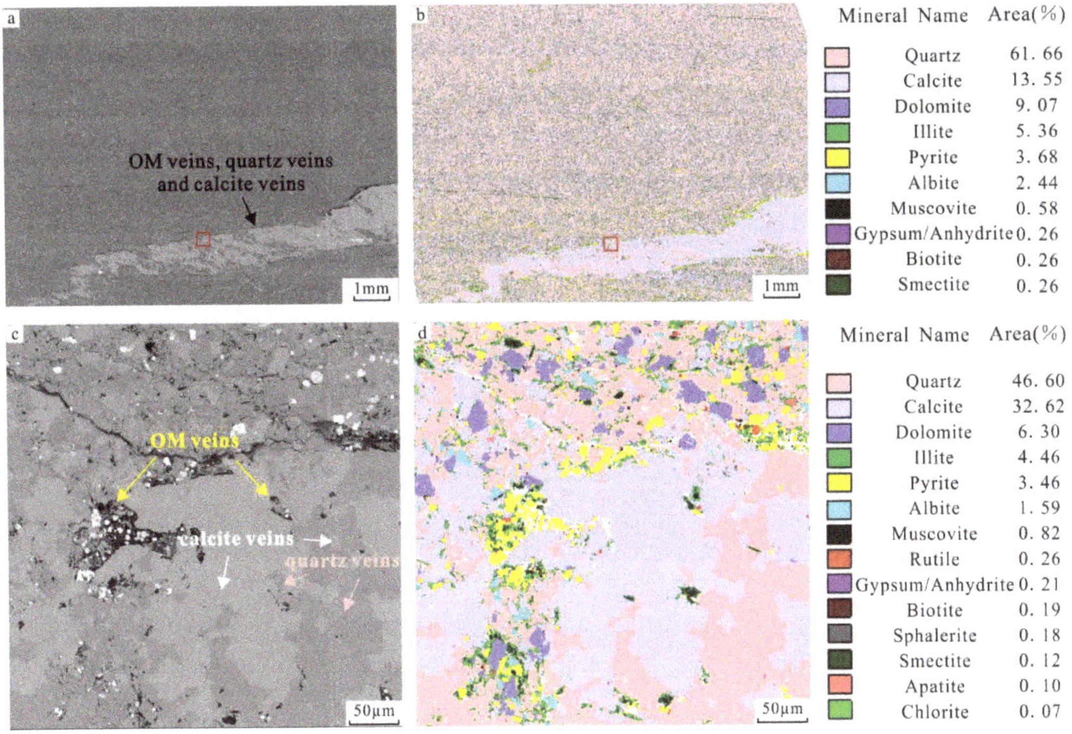

Figure 3. Scanning electron microscope image of Lu203H57-3 well ($Long1_1^1$ layer, 3742.9 m). (**a**) MAPS scan image (500 nm resolution), visible veins. (**b**) QEMSCAN scanning image (15 μm resolution). (**c**) Figure 3a amplified image, MAPS scan image (4 nm resolution), the characteristics of organic matter vein, quartz vein and calcite vein can be seen. (**d**) Figure 3b amplified image, QEMSCAN scanning image (1 μm resolution).

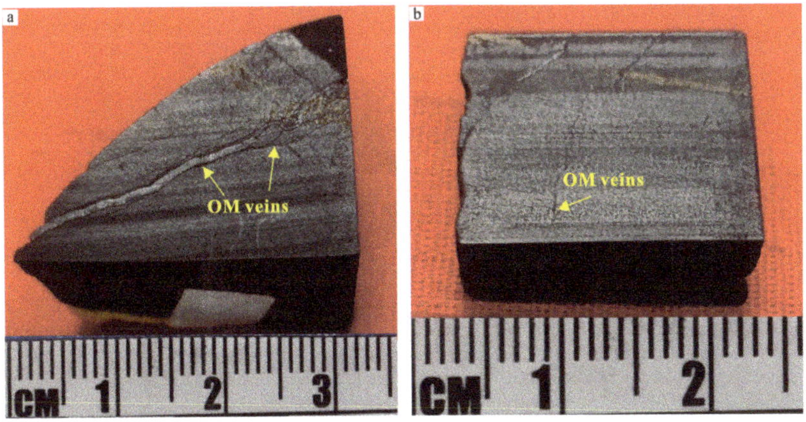

Figure 4. Core photos of $Long1_1$ sub-member of Lu203H57-3 well. (**a**) $Long1_1^2$ layer, 3740.6 m, OM veins can be seen under the light source. (**b**) $Long1_1^1$ layer, 3743.8 m, OM veins can be seen under the light source.

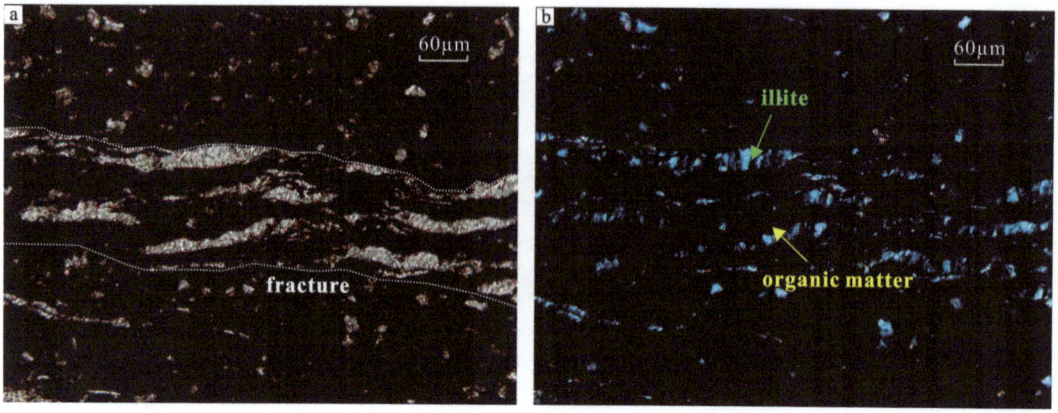

Figure 5. The thin section of Long1_1 sub-member of Lu203H57-3 well. (**a**) Long1_1^2 layer, 3741.8 m, filled with organic matter and illite in horizontal fractures (plane polarized light). (**b**) Long1_1^2 layer, 3741.8 m, filled with organic matter and illite in horizontal fractures (perpendicular polarized light).

At the same time, quartz and calcite mineral veins can be seen around some organic veins (Figure 3). According to the occurrence state between the veins, the OM veins mostly appear on the fracture wall, and some appear inside the quartz and calcite veins (Figure 3c,d). Quartz is mostly isolated growth, discrete distribution, local across the crack wall, visible "quartz bridge" phenomenon [38]. The calcite veins are filled in the main space of the fracture, and the quartz veins are mostly wrapped (Figure 3).

4.2. Reservoir Characteristics

Firstly, the pore-forming ability of different organic matter types was analyzed. The pore size distribution of in situ organic matter pores is wide, mostly in 20~200 nm, and the pore formation efficiency is high, with an average of 28.5% (Figure 2a,b, Table 1). As a contrast, the pores of organic matter migrating to inorganic pores are small and dense, mostly in the range of 20~100 nm, and the average pore formation efficiency is 10.7% (Figure 2c,d, Table 1). OM veins have the worst pore-forming ability, their average pore-forming efficiency is 2.4%. The pores are sparse, and the size of them is mostly 20~70 nm (Figures 6 and 7, Table 1).

Table 1. Comparison of pore-forming efficiency of different organic matter based on MAPS two-dimensional image.

Organic Matter Type	Pore-Forming Efficiency	Pore Characteristics
In situ organic matter	0%~60% (avg. 28.5%)	The pores are alternate in size and dense, mostly 20~300 nm
Organic matter in inorganic pores	3%~22% (avg. 10.7%)	The pores are small and dense, mostly 20~100 nm
Organic matter veins	0.3%~2.9% (avg. 2.4%)	The pores are sparse, mostly 20~70 nm

This study also further identifies the quality of the reservoir in the organic veins. Data display that the brittle mineral content of the samples with OM veins is more than 70%, with an average of 78.98%, while the brittle mineral content of the samples without OM veins is lower, ranging from 50% to 75%, with an average of 61.09%. The average organic porosity of the samples with OM veins is 2.18%, while the average organic porosity of the samples without OM veins is 1.72%. The samples with OM veins have higher brittle mineral content and organic porosity and are mostly developed in Long1_1^1 and Long1_1^2

layers. For TOC content, the average TOC of samples with organic matter veins is 3.69%, while the average TOC of samples without organic matter veins is 3.4%. The TOC content of samples with OM veins development is slightly higher, and the overall difference is not significant (Table 2).

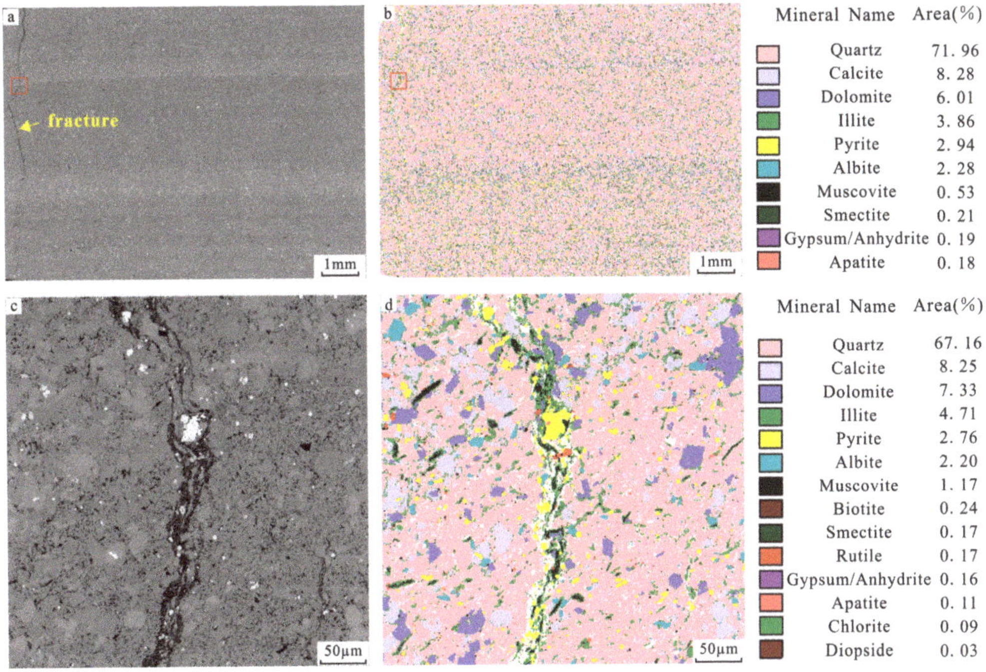

Figure 6. Scanning electron microscope image of Lu203H57-3 well (Long$1_1{}^1$ layer, 3743.8 m). (**a**) MAPS scan image (500 nm resolution), visible fractures. (**b**) QEMSCAN scanning image (15 μm resolution). (**c**) Figure 6a amplified image in red squares, MAPS scan image (4 nm resolution), visible organic matter vein characteristics. (**d**) Figure 6b amplified image in red squares, QEMSCAN scanning image (1 μm resolution), visible mineral composition around the OM veins.

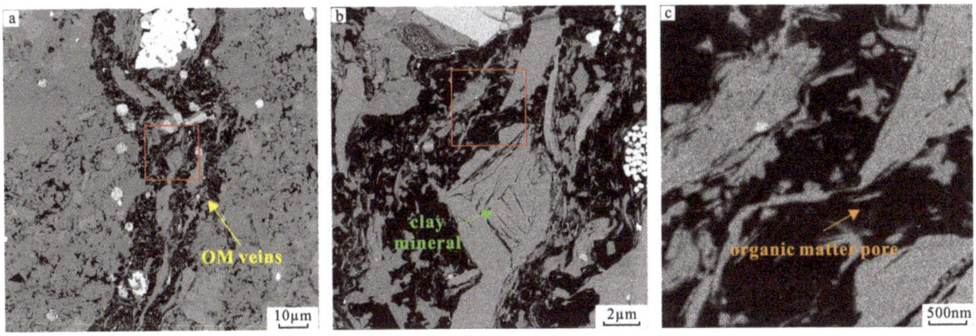

Figure 7. MAPS scanning electron microscopy (4 nm resolution) of Lu203H57-3 well (Long$1_1{}^1$ layer, 3743.8 m). (**a**) OM veins. (**b**) Figure 7a amplified image in red squares. (**c**) Figure 7b amplified image in red squares, visible organic pores.

Table 2. Shale reservoir parameters of $Long1_1$ sub-member in Lu203H57-3 well.

Depth/m	Layer	Quartz Content/%	Calcite Content/%	Brittle Minerals (Quartz + Calcite) Content/%	Organic Porosity/%	Vein Development	Vein Types	TOC
3730.7	$Long1_1^4$	48.04	6.87	54.91	1.0	-	-	2.57
3733.2	$Long1_1^3$	64.28	0.93	71.85	1.8	-	-	2.82
3734.9	$Long1_1^3$	64.70	3.33	65.63	1.5	-	-	3.71
3735.6	$Long1_1^3$	49.99	0.92	50.91	1.8	+	Quartz veins and calcite veins	3.81
3737.6	$Long1_1^2$	58.85	3.28	62.13	2.5	-	-	4.10
3740.6	$Long1_1^2$	59.58	10.71	70.29	1.8	++	OM veins, quartz veins and calcite veins	3.42
3741.8	$Long1_1^2$	76.44	4.46	80.90	1.9	+	OM veins	3.15
3742.9	$Long1_1^1$	69.66	13.55	83.21	2.7	++	OM veins, quartz veins and calcite veins	3.52
3743.8	$Long1_1^1$	73.52	7.99	81.51	2.3	+	OM veins	4.68

Note: - means not seen; + means rare; ++ means common.

5. Discussion

5.1. Formation Mechanism of OM Veins

The occurrence of fractures filled with OM veins varies greatly, including horizontal, oblique and high-angle fractures, the opening is mostly 5–80 μm. According to the above occurrence and the characteristics of filling materials, the formation of this fracture is related to abnormal fluid high pressure [39,40].

Microscopic observation shows that the organic matter filled in the fracture is associated with clay minerals. Previous studies have shown that the mixture of the original deposited structural organic matter and clay minerals is quite common in Marine shales, this is due to the crystal structure and huge specific surface area of clay minerals, which have a strong ability to adsorb organic matter [41]. The mixture of the organic matter and clay minerals in the study section usually occurs in the following two forms. One is acicular clay minerals interwoven in organic matter (Figure 8a,c), and the other is organic matter in clay minerals (Figure 8b,d); organic pores are developed in both of them, which have the potential for hydrocarbon generation. However, the occurrence of clay minerals near OM veins is different from the above; the contact interface between organic matter and clay minerals is sharply defined (Figure 7), there is no obvious tendency for them to contact, react and promote each other, and the lack of pore development of OM veins also indicates that clay minerals here do not promote the hydrocarbon generation process of OM veins. Therefore, we infer that the charging period of OM veins is later based on their morphology and low porosity efficiency.

Therefore, according to the above microscopic characteristics, the formation and evolution process of OM veins is restored. With the continuous burial depth and temperature increase, the maturity of organic matter in the study section gradually increased ($R_o > 0.5\%$) and entered the oil window; kerogen-generated liquid hydrocarbons with low density [42] were transported and filled into inorganic pores in a dispersed form [43]. With the increase in temperature, hydrocarbon generation of organic matter, dehydration of clay minerals, etc., the pore fluid pressure rises [44], and the pore fluid pressure cannot be released, exceeding the shale fracture limit and producing microfractures. The local expansion of micro-fractures forms an effective reservoir space, which releases stress. At the same time, driven by the pressure difference (negative pressure), the soluble organic matter fluid in the source rock of the study section fills into the fractures along the effective channels (pore throats, micro-fractures, kerogen networks, etc.) to form OM veins [45].

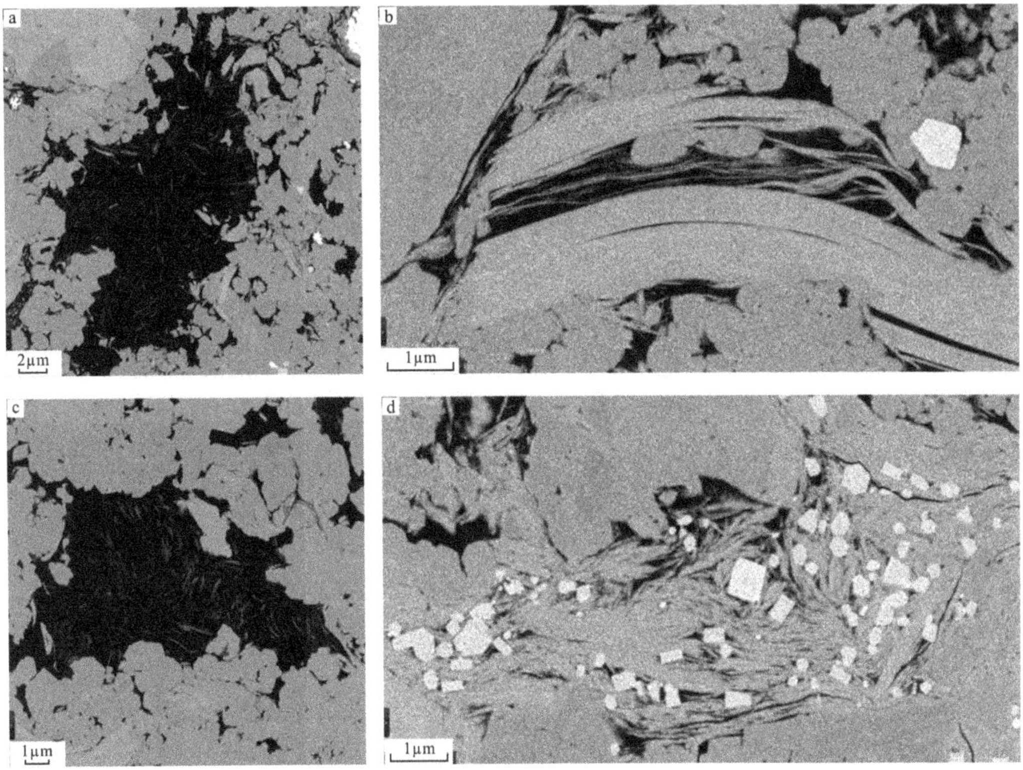

Figure 8. MAPS scanning electron microscopy (4 nm resolution) of the organic matter and clay minerals mixture. (**a**) Acicular clay minerals are interwoven in organic matter, in which a large number of organic pores are developed ($Longl_1^1$ layer, 3743.8 m). (**b**) Organic matter in clay minerals developed well-rounded organic pores ($Longl_1^2$ layer, 3737.6 m). (**c**) Acicular clay minerals are interwoven in organic matter, in which a large number of organic pores are developed ($Longl_1^2$ layer, 3737.6 m). (**d**) Idiomorphic pyrite supports the interlayer fractures of clay minerals, which are filled with organic matter ($Longl_1^3$ layer, 3733.2 m).

5.2. Formation Period of OM Veins

The OM veins and organic matter in inorganic pores are both from the more active organic matter in the in situ organic matter, which migrates in the form of liquid hydrocarbon after the oil generation window and should have good hydrocarbon-generation potential. However, the organic pores in OM veins are poorly developed, even in the presence of a large number of clay minerals as hydrocarbon-generating catalysts (Figure 7). At the same time, the pore-forming efficiency of OM veins is lower than that of in situ organic matter and organic matter in inorganic pores (Table 1), which indicates that organic matter vein formed in a later period, so it experienced a shorter hydrocarbon generation time and a lower degree of pore development. This also makes sense in principle, when organic matter continues to generate hydrocarbons, the intergranular pores are gradually occupied by migrating organic matter, the effect of increasing hydrocarbon-generation pressure will become more and more obvious, and eventually lead to shale reaching the rupture limit and generating abnormally high-pressure-related fractures, then organic matter veins can be formed.

According to the occurrence state of shale fracture composite veins in the study section, organic matter appears inside quartz and calcite veins (Figure 3). Quartz is mostly isolated growth, discrete distribution, local across the crack wall, visible "quartz bridge" phenomenon [38]; the calcite veins are filled in the main space of the fracture, and the quartz veins are mostly wrapped (Figure 9). Based on the above contact relationship, OM veins are formed first, then quartz veins, and then calcite veins. At the same time, according to the previous test methods such as cathodoluminescence and inclusions to restore paleotemperature and pressure of the mineral veins in the study interval, it is shown that the quartz and calcite veins were formed after the kerogen in the interval became gas [46,47], which further indicates that the OM veins were formed before the filling of the mineral veins.

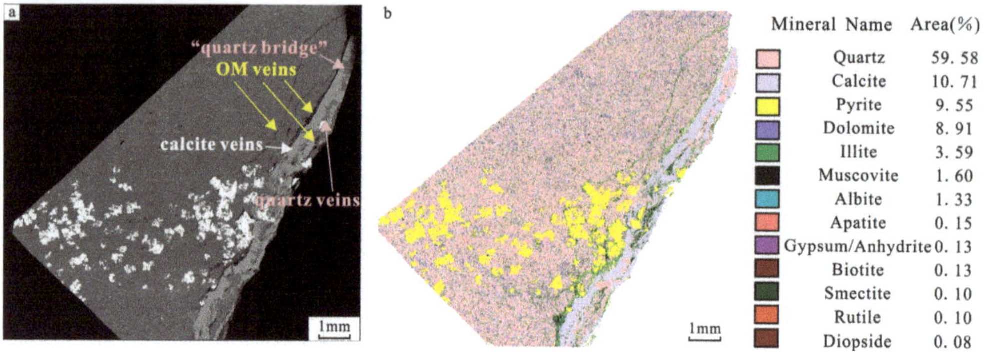

Figure 9. Scanning electron microscope image of Lu203H57-3 well ($Long1_1^2$ layer, 3740.6 m). (**a**) MAPS scan image (500 nm resolution), OM veins, quartz veins and calcite veins can be seen. (**b**) QEMSCAN scanning image (15 μm resolution).

In summary, the OM veins were formed after shale reached the fracture limit, in the late stage of organic matter migration to inorganic pores, and before gas generation and mineral veins are filled.

5.3. Geological Significance of OM Veins

According to the formation mechanism of OM veins, the formation of OM veins requires the existence of overpressure gas reservoirs in the source rocks at that time to reach the fracture limit of shale [48]. Under the same tectonic stress, the higher the brittleness of shale, the smaller the strain before fracture. Therefore, the development degree of fractures in shale is usually positively correlated with the content of quartz and calcite, and negatively correlated with clay minerals [49,50]. The data show that the content of brittle minerals is higher in the interval of OM veins developed in $Long_1$ sub-member of the Luzhou area (Figure 10a), which indicates that the high-pressure-related fractures provided storage space for the occurrence of OM veins.

Simultaneously, the overpressure of the study section is mainly caused by the expansion of fluid volume [44], in which the hydrocarbon generation pressurization of organic matter is the most important influencing factor [40]. In order to compare the difference of hydrocarbon generation intensity between the interval of developed and undeveloped OM veins, TOC and organic porosity of samples were, respectively, used for analysis in this manuscript. The TOC of the samples with OM veins was higher, but the difference was not significant, the obvious difference is that the organic porosity of them is higher (Figure 10b,c), which means the porosity of organic pores is obviously higher in the interval with OM veins. Previous studies have shown that organic matter can break through the large-scale development of organic pores on the surface of organic matter only when the expansion force of gas generation is strong enough [51], so high hydrocarbon generation intensity will lead to high porosity of organic pores. Therefore, the OM veins are mainly due to the high hydrocarbon generation intensity, which makes the organic matter continue to migrate to the fracture after filling the inorganic pores. And the low pore-forming efficiency of OM veins mentioned earlier also suggests that it was formed later than that of most organic matter migrating to inorganic pores.

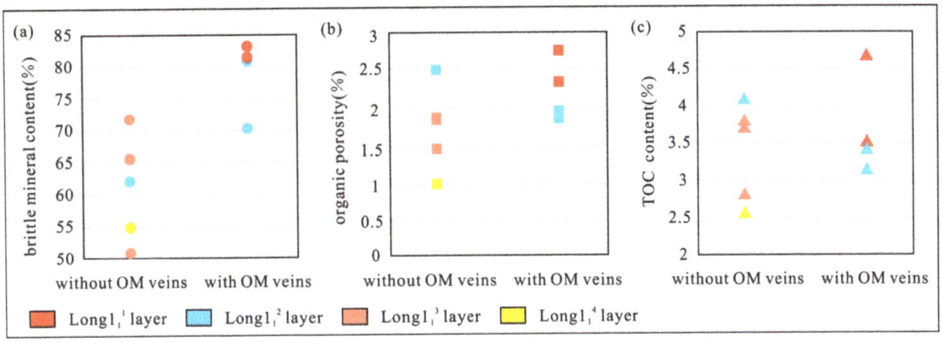

Figure 10. Correlation diagram of OM veins and reservoir parameters. (**a**) Correlation diagram of OM veins and brittle mineral content. (**b**) Correlation diagram of OM veins and organic porosity. (**c**) Correlation diagram of OM veins and TOC.

The OM veins are mostly developed in the $Long1_1^1$ and $Long1_1^2$ layers. The content of brittle minerals in this layer is high, which is conducive to the formation of cracks in the layer before gas generation, so that the basic space conditions for filling the OM veins are available. The high porosity of the organic pores in this section indicates that the organic matter has high hydrocarbon generation intensity and high hydrocarbon generation potential, so the soluble organic matter fills the inorganic pores and continues to migrate to the fractures to form OM veins (Figure 11). Therefore, the quality of $Long1_1^1$ and $Long1_1^2$ shale reservoirs with OM veins is better.

At the same time, combined with the source of OM veins and the factors affecting the formation of OM veins, it can be seen that the conditions that are not conducive to the development of OM veins in each layer of $Long_1^1$ sub-member: some layers have no cracks before gas generation, and soluble organic matter cannot be filled; there are cracks in the interval before gas generation, but the organic matter content is low or the hydrocarbon generation intensity is weak, and the organic matter is not filled into the cracks.

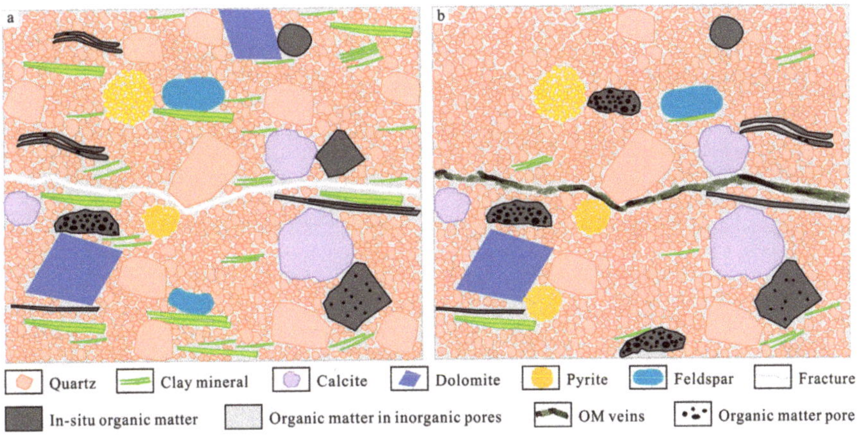

Figure 11. Pattern diagram of OM veins. (**a**) The section without OM veins has higher clay mineral content. (**b**) In the section with OM veins, the content of brittle minerals is higher, and the hydrocarbon generation potential of organic matter is higher.

In summary, the high hydrocarbon generation intensity and high brittle mineral content of shale reservoirs make the formation of OM veins, and the closed diagenesis system also provides good reservoir space for deep shale gas reservoirs [52]. Therefore, the existence of OM veins can indicate better reservoir conditions. The representative horizon has strong hydrocarbon generation intensity and high brittle mineral content, indicating that the horizon has shale gas enrichment conditions and good compressibility. It is a good geological–engineering dessert.

6. Conclusions

(1) The OM veins mainly appear on the fracture wall, and some appear inside the quartz and calcite veins, most of them developed in the $Long1_1{}^1$ and $Long1_1{}^2$ layers.
(2) The pore-forming efficiency of OM veins is lower than that of in-situ organic matter and organic matter in inorganic pores, the pores in the OM veins are sparse and the pore size is small.
(3) The OM veins were formed after shale reached the fracture limit, in the late stage of organic matter migration to inorganic pores, and before mineral veins are filled.
(4) The existence of OM veins can indicate high-quality reservoir conditions. The representative horizons have strong hydrocarbon generation strength (strong pore-forming ability of organic matter) and high brittle mineral content (strong reservoir compressibility), which makes the horizons conducive to shale gas accumulation and fracturing.

Author Contributions: Conceptualization, Y.W. and Y.J.; methodology, D.H.; software, C.W.; validation, W.L., D.H. and J.Z.; formal analysis, Y.J.; investigation, Y.W.; resources, W.L.; data curation, C.W.; writing—original draft preparation, Y.J.; writing—review and editing, Y.W.; visualization, B.M.; supervision, H.W.; project administration, C.W.; funding acquisition, D.H. All authors have read and agreed to the published version of the manuscript.

Funding: This work is funded by the National Natural Science Foundation of China (42072121) and the Natural Science Foundation of Hubei Province (2021CFB182).

Data Availability Statement: The data that support the findings of this study are available from the corresponding author upon reasonable request.

Conflicts of Interest: The authors declare that they have no conflict of interest.

References

1. Qiu, Z.; Zou, C.N. Controlling factors on the formation and distribution of "sweet-spot areas" of marine gas shales in South China and a preliminary discussion on unconventional petroleum sedimentology. *J. Asian Earth Sci.* **2020**, *194*, 103989.
2. Jarvie, D.M.; Hill, R.J.; Ruble, T.E.; Pollastro, R.M. Unconventional shale-gas systems: The Mississippian Barnett Shale of north-central Texas as one model for thermogenic shale-gas assessment. *AAPG Bull.* **2007**, *91*, 475–499.
3. Wang, P.W.; Nie, H.K.; Liu, Z.B.; Sun, C.X.; Cao, Z.; Wang, R.Y.; Li, P. Differences in Pore Type and Pore Structure between Silurian Longmaxi Marine Shale and Jurassic Dongyuemiao Lacustrine Shale and Their Influence on Shale-Gas Enrichment. *Minerals* **2023**, *13*, 190.
4. Loucks, R.G.; Reed, R.M.; Ruppel, S.C.; Hammes, U. Spectrum of pore types and networks in mudrocks and a descriptive classification for matrixrelated mudrock pores. *AAPG Bulletin.* **2012**, *96*, 1071–1098. [CrossRef]
5. Jia, Y.Q.; Han, D.L.; Zhang, J.Z.; Wang, C.C.; Lin, W.; Ren, X.H.; Yang, C.Y.; Chang, L.C. Differences in Pore forming Efficiency among Organic Macerals and Its Restriction against Reservoir Quality: A Case Study Based on the Marine Shale Reservoir in the Longmaxi Formation, Southern Sichuan Basin, China. *Lithosphere* **2021**, *97*, 2700912.
6. Wang, R.Y.; Hu, Z.Q.; Long, S.X.; Du, W.; Wu, J.; Wu, Z.H.; Nie, H.K.; Wang, P.W.; Sun, C.X.; Zhao, J.H. Reservoir characteristics and evolution mechanisms of the Upper Ordovician Wufeng-Lower Silurian Longmaxi shale, Sichuan Basin. *Oil Gas Geol.* **2022**, *43*, 353–364.
7. Jia, Y.Q.; Liu, Z.P.; Ren, X.H.; Zhou, Y.B.; Zheng, A.L.; Zhang, J.; Han, D.L. Organic matter type differentiation process and main control mechanism: Case study of the Silurian Longmaxi Formation shale reservoir in Weiyuan area. *Acta Sedimentol. Sin.* **2021**, *39*, 341–352.
8. Hu, Z.Q.; Du, W.; Peng, Y.M.; Zhao, J.H. Microscopic pore characteristics and the source-reservoir relationship of shale—A case study from the Wufeng and Longmaxi Formations in Southeast Sichuan Basin. *Oil Gas Geol.* **2015**, *36*, 1001–1008.

9. Loucks, R.G.; Reed, R.M. Scanning-electron-microscope petrographic evidence for distinguishing organic-matter pores associated with depositional organic matter versus migrated organic matter in mudrocks. *Gcags Trans.* **2014**, *64*, 51–60.
10. Xie, G.L.; Liu, S.G.; Jiao, K.; Deng, B.; Ye, Y.H.; Sun, W.; Li, Z.Q.; Liu, W.P.; Luo, C.; Li, Z.C. Organic pores in deep shale controlled by macerals: Classification and pore characteristics of organic matter components in Wufeng Formation-Longmaxi Formation of the Sichuan Basin. *Nat. Gas Ind.* **2021**, *41*, 23–34.
11. Cardott, B.J.; Landis, C.R.; Curtis, M.E. Post-oil solid bitumen network in the Woodford Shale, USA-A potential primary migration pathway. *Int. J. Geol.* **2015**, *139*, 106–113.
12. Wang, P.F.; Jiang, Z.X.; Lyu, P.; Jin, C.; Li, X.; Huang, P. Organic matter pores and evolution characteristics of shales in the Lower Silurian Longmaxi Formation and Lower Cambrian Niutitang Formation in peripherfy of Chongqing. *Nat. Gas Geosci.* **2018**, *29*, 997–1008.
13. Curtis, M.E.; Cardott, B.J.; Sondergeld, C.H.; Rai, C.S. Development of organic porosity in the Woodford Shale with increasing thermal maturity. *Int. J. Coal Geol.* **2012**, *103*, 26–31.
14. Gao, Z.; Fan, Y.; Xuan, Q.; Zheng, G. A review of shale pore structure evolution characteristics with increasing thermal maturities. *Adv. Geo-Energy Res.* **2020**, *4*, 247–259.
15. Wang, R.Y.; Hu, Z.Q.; Nie, H.K.; Liu, Z.B.; Chen, Q.; Gao, B.; Liu, G.X.; Gong, D.J. Comparative analysis and discussion of shale reservoir characteristics in the Wufeng-Longmaxi and Niutitang formations: A case study of the well JY1 in SE Sichuan Basin and well TX1 in SE Guizhou area. *Pet. Geol. Exp.* **2018**, *40*, 639–649.
16. Han, J.; Chen, B.; Zhao, X.B.; Zheng, C.; Zhang, J.M. Development characteristics and influential factors of organic pores in the Permian shale in the Lower Yangtze Region. *Nat. Gas Ind.* **2017**, *37*, 17–26.
17. Nie, H.K.; He, Z.L.; Wang, R.Y.; Zhang, G.R.; Chen, Q.; Li, D.H.; Lu, Z.Y.; Sun, C.X. Temperature and origin of fluid inclusions in shale veins of Wufeng-Longmaxi Formations, Sichuan Basin, south China: Implications for shale gas preservation and enrichment. *J. Pet. Sci. Eng.* **2020**, *193*, 107329.
18. Ougier-Simonin, A.; Renard, F.; Boehm, C.; Vidal-Gilbert, S. Microfracturing and microporosity in shales. *Earth-Sci. Rev.* **2016**, *162*, 198–226.
19. Tan, R.; Wang, R.Y.; Huang, Y.H.; Yang, R.; Li, H.B.; Lu, K. Mechanism of the Enrichment and Loss Progress of Deep Shale Gas: Evidence from Fracture Veins of the Wufeng-Longmaxi Formations in the Southern Sichuan Basin. *Minerals* **2022**, *12*, 897.
20. Yang, H.Z.; Zhao, S.X.; Liu, Y.; Wu, W.; Xia, Z.Q.; Wu, T.P.; Luo, C.; Fan, T.Y.; Yu, L.Y. Main controlling factors of enrichment and high-yield of deep shale gas in the Luzhou Block, southern Sichuan Basin. *Nat. Gas Ind.* **2019**, *39*, 55–63.
21. Dong, M.; Gou, W.; Zhang, L.Y.; Wu, Z.H.; Ma, L.C.; Dong, H.; Feng, X.Q.; Yang, Y.H. Characteristics of paleotectonic stress field and fractures of Wufeng-Longmaxi Formation in Luzhou area, southern Sichuan Basin. *Lithol. Reserv.* **2022**, *34*, 43–51.
22. Liu, Z.B.; Gao, B.; Zhang, Y.Y.; Du, W.; Feng, D.J.; Nie, H.K. Types and distribution of the shale sedimentary facies of the Lower Cambrian in Upper Yangtze area, South China. *Pet. Explor. Dev.* **2017**, *44*, 21–31.
23. Gou, T.L.; Zhang, H.R. Formation and enrichment mode of Jiaoshiba shale gas field, Sichuan Basin. *Pet. Explor. Dev.* **2014**, *41*, 28–36.
24. Guo, Y.H.; Li, Z.F.; Li, D.H.; Zhang, T.M.; Wang, Z.C.; Yu, J.F.; Xi, Y.T. Lithofacies palaeogeography of the Early Silurian in Sichuan area. *J. Palaeogeogr.* **2004**, *6*, 20–29.
25. Wang, X.M.; Liu, L.F.; Wang, Y.; Sheng, Y.; Zheng, S.S.; Luo, Z.H. Control of lithofacies on pore space of shale from Longmaxi Formation, southern Sichuan Basin. *Acta Pet. Sin.* **2019**, *40*, 1192–1201.
26. Ge, X.Y.; Mou, C.L.; Yu, Q.; Liu, W.; Men, X.; He, J.L. The geochemistry of the sedimentary rocks from the Huadi No. 1 well in the Wufeng-Longmaxi formations (Upper Ordovician-Lower Silurian), South China, with implications for paleoweathering, provenance, tectonic setting and paleoclimate. *Mar. Pet. Geol.* **2019**, *103*, 646–660.
27. Yan, C.N.; Jin, Z.J.; Zhao, J.H.; Du, W.; Liu, Q.Y. Influence of sedimentary environment on organic matter enrichment in shale: A case study of the Wufeng and Longmaxi Formations of the Sichuan Basin, China. *Mar. Pet. Geol.* **2018**, *92*, 880–894.
28. Wu, L.Y.; Lu, Y.C.; Jiang, S.; Liu, X.F.; He, G.S. Effects of volcanic activities in Ordovician Wufeng-Silurian Longmaxi period on organic-rich shale in the Upper Yangtze area, South China. *Pet. Explor. Dev.* **2018**, *45*, 806–816.
29. Pu, B.L.; Dong, D.Z.; Wang, F.Q.; Wang, Y.M.; Huang, J.L. The effect of sedimentary facies on Longmaxi shale gas in southern Sichuan Basin. *Geol. China* **2020**, *47*, 111–120.
30. Wang, S.F.; Dong, D.Z.; Wang, Y.M.; Li, X.J.; Huang, J.L.; Guan, Q.Z. Sedimentary geochemical proxies for paleoenvironment interpretation of organic-rich shale: A case study of the Lower Silurian Longmaxi Formation, southern Sichuan Basin, China. *J. Nat. Gas Sci. Eng.* **2016**, *28*, 691–699.
31. Zhao, S.X.; Yang, Y.M.; Zhang, J.; Wang, L.S.; Wang, X.Z.; Luo, C.; Tian, C. Micro-layers division and fine reservoirs contrast of Lower Silurian Longmaxi Formation shale, Sichuan Basin, SW China. *Nat. Gas Geosci.* **2016**, *27*, 470–487.
32. Gao, Z.Y.; Fan, Y.P.; Hu, Q.H.; Jiang, Z.X.; Huang, Z.L.; Wang, Q.Y.; Cheng, Y. Differential development characteristics of organic matter pores and their impact on reservoir space of Longmaxi Formation shale from the south Sichuan Basin. *Pet. Sci. Bull.* **2020**, *5*, 1–16.
33. Wu, J.G.; Yuan, Y.; Niu, S.Y.; Wei, X.F.; Yang, J.J. Multiscale characterization of pore structure and connectivity of Wufeng-Longmaxi shale in Sichuan Basin, China. *Mar. Pet. Geol.* **2020**, *120*, 104514.
34. Wang, H.; Zhou, S.; Li, S.; Zhao, M.; Zhu, T. Comprehensive characterization and evaluation of deep shales from Wufeng-Longmaxi Formation by LF-NMR technology. *Unconv. Resour.* **2022**, *2*, 1–11.

35. Guo, X.S.; Li, Y.P.; Borjigen, T.; Wang, Q.; Yuan, T.; Shen, B.J.; Ma, Z.L.; Wei, F.B. Hydrocarbon generation and storage mechanisms of deep-water shelf shales of Ordovician Wufeng Formation-Silurian Longmaxi Formation in Sichuan Basin, China. *Pet. Explor. Dev.* **2020**, *47*, 193–201.
36. Saif, T.; Lin, Q.Y.; Butcher, A.R.; Bijeljic, B.; Blunt, M.J. Multi-scale multi-dimensional microstructure imaging of oil shale pyrolysis using X-ray micro-tomography, automated ultra-high resolution SEM, MAPS Mineralogy and FIB-SEM. *Appl. Energy* **2017**, *202*, 628–647.
37. Schneider, C.A.; Rasband, W.S.; Eliceiri, K.W. NIH Image to ImageJ: 25 years of image analysis. *Nat. Methods* **2012**, *9*, 671–675.
38. Wang, Z.; Lyu, X.X.; Zhang, L.X.; Li, F.; Ouyang, S.Q.; Wang, R. Discovery of "quartz bridge" in Kuqa foreland thrust belt and its geological significance. *Earth Sci.* **2023**, *48*, 342–358.
39. Wang, X.X.; Wan, Y.L. Significance of the organic clay polymer on oil and gas generation. *China Offshore Oil Gas* **1993**, *7*, 27–33.
40. Zeng, L.B.; Ma, S.J.; Tian, H.; Xue, M.; Liu, G.P.; Lyu, W.Y. Research progress of natural fractures in organic rich shale. *Earth Sci.* **2022**, *48*, 1–15.
41. Zeng, L.B.; Qi, J.F.; Wang, Y.X. Origin type of tectonic fractures and geological conditions in low-permeability reservoirs. *Acta Pet. Sin.* **2007**, *28*, 52–56.
42. Lu, X.C.; Hu, W.X.; Fu, Q.; Miao, D.Y.; Zhou, G.J.; Hong, Z.H. Study of combination pattern of soluble organic matters and clay minerals in the immature source rocks in Dongying Depression, China. *Sci. Geol. Sin.* **1999**, *34*, 72–80.
43. Zhang, Q.; Liang, F.; Pang, Z.L.; Zhou, S.W.; Lin, W. Quantitative influence of soluble organic matter on pore structure in transitional shale. *Pet. Geol. Exp.* **2018**, *40*, 730–738.
44. Liu, D.D.; Guo, J.; Pan, Z.K.; Du, W.; Zhao, F.P.; Chen, Y.; Shi, F.L.; Song, Y.; Jiang, Z.X. Overpressure evolution process in shale gas reservoir: Evidence from the fluid inclusions in the fractures of Wufeng Formation-Longmaxi Formation in the southern Sichuan Basin. *Nat. Gas Ind.* **2021**, *41*, 12–22.
45. Zhi, F.Q.; Li, Q.; Fan, D.H.; Zhang, C.M. Study on migration and accumulation of fractured mudstone reservoirs in Zhanhua Sag. *Pet. Geol. Recovery Effic.* **2004**, *11*, 27–29.
46. Deng, M.; Duan, X.G.; Zhai, C.B.; Long, S.X.; Yang, Z.H.; Zheng, L.J.; Li, Z.C.; Cao, T.T. Variation in liquid hydrocarbon content during thermal simulation and its influence on physical property of shale. *Oil Gas Geol.* **2020**, *41*, 1310–1320.
47. Pan, Z.K.; Liu, D.D.; Huang, Z.X.; Jiang, Z.X.; Song, Y.; Guo, J.; Li, C.X. Paleotemperature and paleopressure of methane inclusions in fracture cements from the Wufeng-Longmaxi shales in the Luzhou area, southern Sichuan Basin. *Pet. Sci. Bull.* **2019**, *4*, 242–253.
48. Osborne, M.J.; Swarbrick, R.E. Mechanisms for generating overpressure in sedimentary basins: A reevaluation. *AAPG Bull.* **1997**, *81*, 1023–1041.
49. Tian, H.; Zeng, L.B.; Xu, X.; Shu, Z.G.; Peng, Y.M.; Mao, Z.; Luo, B. Characteristics of natural fractures in marine shale in Fuling area, Sichuan Basin, and their influence on shale gas. *Oil Gas Geol.* **2020**, *41*, 474–483.
50. Wu, Q.Y.; Han, D.L.; Zhang, J.Z.; Wang, C.C.; Ren, X.H.; Lin, Z.Z.; Su, M.M.; Zhu, Y.L.; Zhang, J. Influence of particle size effect on microfracture development in shale reservoir: A case study of the Silurian Longmaxi Formation shale reservoir in Weiyuan area. *J. Cent. South Univ. (Sci. Technol.)* **2022**, *53*, 3603–3614.
51. Ding, J.H.; Zhang, J.C.; Yang, C.; Huo, Z.P.; Lang, Y. Formation Evolution and Influencing Factors of Organic Pores in Shale. *J. Southwest Pet. Univ. (Sci. Technol. Ed.)* **2019**, *41*, 33–44.
52. Yao, C.P.; Fu, H.J.; Ma, Y.Z.; Yan, D.T.; Wang, H.; Li, Y.G.; Wang, J.W. Development characteristics of deep shale fractured veins and vein forming fluid activities in Luzhou block. *Earth Sci.* **2022**, *47*, 1684–1693.

minerals

MDPI

Article

Characteristics and Factors Influencing Pore Structure in Shale Oil Reservoirs of Different Lithologies in the Jurassic Lianggaoshan Formation of the Yingshan Gas Field in Central Sichuan Basin

Youzhi Wang [1], Hucheng Deng [2,3,*], Zhiguo Wang [1], Xiandong Wang [1], Qian Cao [3], Dean Cheng [1], Yanping Zhu [1] and An Li [3,*]

[1] Exploration and Development Research Institute, Daqing Oilfield Limited Company, Daqing 163712, China
[2] State Key Laboratory of Oil and Gas Reservoir Geology and Development Engineering, Chengdu University of Technology, Chengdu 610059, China
[3] College of Energy, Chengdu University of Technology, Chengdu 610059, China
[*] Correspondence: denghucheng@cdut.cn (H.D.); lian@stu.cdut.edu.cn (A.L.); Tel.: +86-198-8115-0998 (H.D.); +86-181-2579-9804 (A.L.)

Abstract: Shale in the Jurassic Lianggaoshan Formation in central Sichuan exhibits strong heterogeneity. The study of the pore structure characteristics of different lithologies is crucial to the selection of the target interval. Shale samples of the Lianggaoshan Formation from well YS5 in the central part of the Sichuan Basin were analyzed using scanning electron microscopy, low-temperature nitrogen adsorption, high-pressure mercury injection (HPMI), and large -field splicing method -based scanning electron microscopy (LFS-SEM) to elucidate the pore structure characteristics of shale and their influencing factors. The mineral composition of the reservoir in the study area was diverse, primarily consisting of clay minerals, followed by quartz and calcite. The reservoir space comprised intergranular, granular, and organic matter pores, and oil was observed to fill the reservoir space. Reservoir characteristics varied with the lithological properties. In clayey shale, intergranular pores located in clay mineral particles and pores between pyrite and natural fractures were mainly observed, with a bimodal distribution of pore size and peak distribution of 10–50 nm and >100 nm. The storage space of ash-bearing shale mainly consisted of intragranular pores and intergranular (crystalline) micropores, with pore sizes primarily concentrated in the 10–50 nm range. The storage space in silty shale mainly developed in clastic mineral particles such as quartz, followed by clay mineral intergranular pores with a relatively wide distribution of sizes. Pores were mainly inkbottle-shaped and slit-type/plate-type pores, with an average specific surface area of approximately 6.9046 $m^2 \cdot g^{-1}$ and an average pore volume of approximately 0.0150 $cm^3 \cdot g^{-1}$. The full-pore capillary pressure curve was established using a combination of gas adsorption–desorption tests and HPMI. The fractal dimension of the sample pore structure was calculated, and a significant linear correlation was found between clay mineral content and the fractal dimension. Thus, the pore structure characteristics were mainly controlled by the content and distribution of clay minerals.

Keywords: central Sichuan Basin; Lianggaoshan Formation; shale; pore structure characteristics; influencing factors

1. Introduction

In recent years, there has been a continuous increase in the exploration and exploitation of shale oil in China, specifically in the Junggar, Ordos, Bohai Bay, Songliao, and Sichuan basins, and shale oil demonstration bases have been established in the Xinjiang, Changqing, Dagang, and Daqing oilfields [1,2]. According to the Ministry of Natural Resources of the People's Republic of China, the geological potential of shale oil resources in China is estimated to be 397.46×10^8 t, with recoverable resources estimated at 34.98

$\times 10^8$ t. This emphasizes the potential of shale oil as a critical factor for future growth in China's oil reserves [3–6]. The Sichuan Basin has a diverse range of natural gas resources, with the Jurassic strata containing multiple sets of organic black shale that serve as the main interval for oil generation and storage. In 2020, significant breakthroughs were made in the extraction of shale oil [7]. The oil production of the Ping'an 1, Taiye 1, Yuanba 9, and Yuanye 3 wells was 112.8 m^3/d, 9.8 m^3/d, 16.6 m^3/d, and 15.6 m^3/d, respectively, and their corresponding gas production amounted to 11.45 $\times 10^4$ m^3/d, 7.5 $\times 10^4$ m^3/d, 1.2 $\times 10^4$ m^3/d, and 1.18 $\times 10^4$ m^3/d, respectively [8]. After nearly 10 years of development, the technical theories related to shale oil reservoir characteristics have gradually made breakthroughs and improvements. Analysis has evolved from early optical microscopic analysis to current field-emission scanning electron microscopic (FESEM) analysis, low-temperature liquid nitrogen adsorption technology, and the high-pressure mercury intrusion method, among others. It has been upgraded from a single experimental perspective to comprehensive characterization from multiple dimensions. In previous studies, advanced techniques such as scanning electron microscopy backscattering (SEM-BS), SEM, nano-CT scanning, HPMI, and low-temperature CO_2 and N_2 adsorption experiments have been employed to study the pore structure characteristics of shale reservoirs [9–14]. The results of these studies hold considerable significance for analysis of the pore structure characteristics of shale reservoirs in Jurassic shales of the Sichuan Basin.

However, the abovementioned methods for analyzing the structure of shale oil reservoirs have certain limitations in describing the pore types, morphology, connectivity, and porosity of shale reservoirs. They cannot fully reflect the pore structure characteristics of shale oil reservoirs and cannot achieve quantitative analysis. Continental shale differs from marine shale in that it has strong heterogeneity, low organic matter abundance, low thermal evolution degree, and low hydrocarbon symbiosis and exhibits interbedded distribution of shale, sandstone, and limestone [15,16]. These characteristics make the evaluation criteria of continental shale oil reservoirs challenging to determine, including the selection and prediction of sweet-spot areas [17]. This study utilizes the large-field splicing method based on scanning electron microscopy (LFS-SEM), low-temperature nitrogen adsorption (LTNA), and HPMI to analyze the micropore characteristics of shale reservoirs in the Lianggaoshan Formation of the of the Yingshan gas field. Moreover, using ImageJ software (Fiji, National Institutes of Health, Bethesda, MD, USA) to analyze shale pores, this new image analysis method is more comprehensive and intuitive than identifying pores in a single scanning electron microscopic field of view. It can avoid errors caused by selecting observation positions and can identify various types of pores on a single image, achieving quantitative analysis of pore structure. This study lays the foundation for further research into the pore structure of shale of the Lianggaoshan Formation and its impact on shale oil and gas enrichment [18–21].

2. Geological Overview

The Yingshan gas field is located in the central portion of the Sichuan Basin region, spanning three counties: Yingshan, Yilong, and Nanbu. Regionally, the Yingshan gas field belongs to the Yilong tectonic group, which is a low–even belt in the Paleo-Middle Depression of North Sichuan, adjacent to the Paleo-High Middle Oblique Gentle [22,23]. During the early stage of the Himalayan tectonic movement, the central part of the Sichuan Basin underwent north–south extrusion, leading to the development of east–west tectonics in the basin. In the latter stage, the northern part of the study area was subjected to north–south tectonic stress, forming east–west compression structures, as well as the northeast, northwest, and northern knob structural belt tectonics in the central portion of the Sichuan Basin [24].

The Jurassic strata in the study area consist of a series of river–delta–lake formations from the bottom to the top, including the Lianggaoshan Formation and the ZhenzhuChong, Dongyuemiao, Maanshan, and D'Anzhai members of the Zilioujing Formation [25]. The Lianggaoshan Formation represents a typical delta sedimentary system, with receding–strongly advancing sedimentary cycles [26]. The entire Lianggaoshan Formation consists of mixed-colored shales interspersed with gray siltstones and black shales. The Lower Lianggaoshan Formation, consisting mainly of purple, mixed-colored, and gray–green shales, primarily developed in a floodplain–lakeshore sedimentary environment. The lake basin of the Upper Lianggaoshan Formation underwent three stages: a lake intrusion period, a maximum lake intrusion period, and a lake retreat period [27]. The overall thickness of the upper section is approximately 55–75 m, with the main lithologies being black shale, light gray fine sandstone, siltstone, and gray–black medium shell shale (20–55 m). This is a set of sand and shale rich in organic matter, and it is the primary source and reservoir layer in the study area [28,29]. The upper section of the Lianggaoshan Formation in the study area is primarily composed of shale, sandstone, or tuff with varying thicknesses, with the thicknesses of shale and interbed being approximately 6.36 m and 6.07 m, respectively.

3. Materials and Methods

3.1. Sample Selection

Eight test samples were selected from the Jurassic Lianggaoshan Formation of well YS5 in the Yingshan area of the Sichuan Basin (Figure 1). The organic matter content of these samples was generally greater than 1.2% and reached up to 3.62%, and the vitrinite reflectance was greater than 0.5%, indicating that they were in the early mature stage for the formation of oil and gas [30–33].

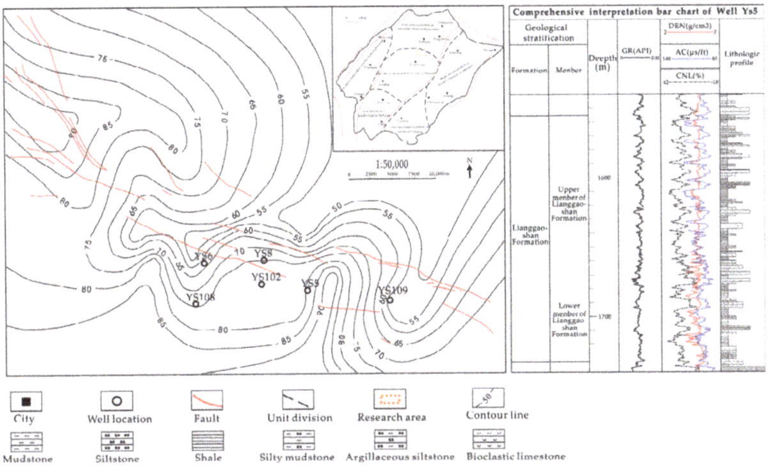

Figure 1. Location of the study area and sampling points.

3.2. Experimental Methods

The pore structure characteristics of the test samples were analyzed using various methods, including LFS-SEM, gas adsorption (CO_2 and N_2 adsorption–desorption tests), and HPMI [34]. The distributions of structural characteristics, such as pore size, specific surface area (SSA), and pore volume (PV), of different reservoir space types were classified and statistically analyzed.

LFS-SEM was used to analyze the mineral composition, fluid distribution, and microscopic pore structure characteristics of the test samples by combining the three imaging modes using the field-emission environment. Based on the observation and analysis of the test samples by field-emission SEM, 4 × 4, 5 × 5, 8 × 8, 10 × 10, or 10 × 12 images were continuously collected from the representative field of view for stitching. Different types of pores were identified by human–computer interaction, and pore sizes and proportions of different storage spaces were classified and counted to achieve qualitative and quantitative analysis (Figure 2).

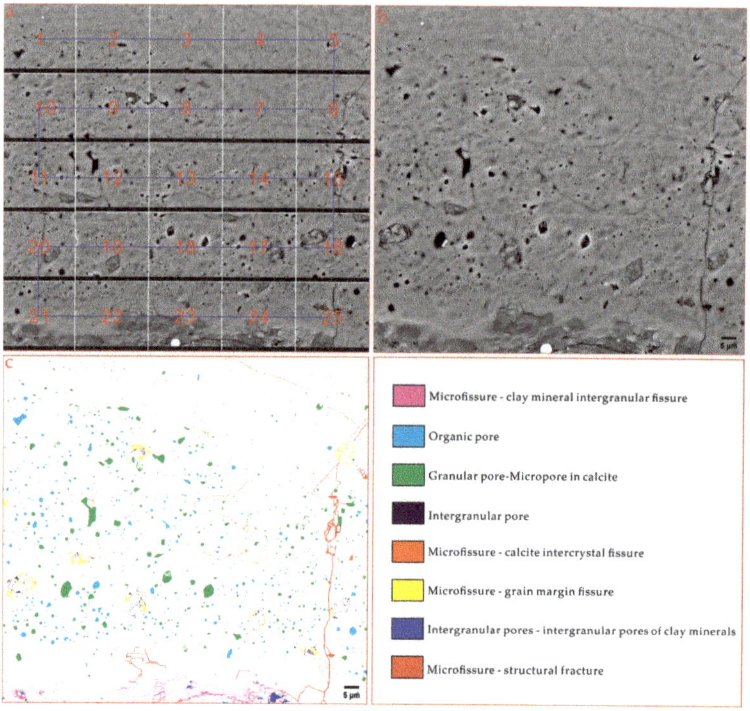

Figure 2. Principle and operation process of the large-field-of-view stitching method. (**a**) Splicing of different images by SEM. (**b**) Images observed using SEM at a magnification of 10,000×. (**c**) Identification of different types of storage spaces.

LTNA as conducted using an ASAP2020 adsorption instrument (Quantachrome, Boynton Beach, FL, USA). Based on the static volume method, the determination process of gas adsorption and desorption was completed using the mass conservation equation and static gas balance and pressure, and the adsorption model was preferred to calculate the specific surface, PV, and pore diameters.

HPMI was conducted using a Poremaster-60 instrument (Quantachrome, Boynton Beach, FL, USA). In HPMI, liquid mercury is gradually injected into the pores of test samples under a certain external pressure. The mercury is a non-wetting phase that is injected to replace the wetted phase, and additional pressure is required to inject this non-wetting-phase liquid into the rock pores. The amount of mercury injected provides the volume of the pores and is controlled by the pore throat sizes. The maximum mercury injection pressure was approximately 418 MPa. The capillary pressure curve and pore distribution curve of the shale sample were obtained by varying the injection pressure.

The X-ray diffraction (XRD) test steps are as follows: the sample is washed to a fluorescence level below 4, then dried at a temperature below 60 °C until the sample is cooled to room temperature, ground until all particles are less than 40μm in size, and finally put into the groove of the sample slide for measurement. After the measurement, the X-ray diffraction pattern is obtained, according to which the relevant data are read, then compared with the standard X-ray diffraction data of minerals to determine the type of minerals, and the contents of various minerals and clays are calculated using the following formula.

$$X_i = \left[\frac{\frac{I_i}{K_i}}{\sum \frac{I_i}{K_i}} \right] * 100$$

where:

X_i is he percentage content of mineral i, expressed as a percentage;
I_i is the diffraction peak intensity of mineral i; and
K_i is the reference strength of mineral i.

4. Characteristics of the Reservoir Space of the Lianggaoshan Formation

4.1. Lithological Characteristics

The reservoir of the Lianggaoshan Formation in the study area is characterized by rocks with diverse mineral compositions, mixed particle sizes, and complex rock structures, as observed through cores, thin sections, and SEM. Clastic particles, mainly composed of quartz, debris, and feldspar, are scattered throughout the rocks, with grain sizes predominantly in the range of 0.03–0.06 mm. Sand rich layers are locally developed, with small amounts of organic matter distributed in the mud as clumps and carbonate particles. such as calcite distributed in the mud as grains, partly filling the intergranular spaces and accounting for the edges of the clastic particles. The lithology of the Lianggaoshan Formation primarily comprises clay shale, sand shale, and laminated blocky gray shale (Figure 3). X-ray diffraction (XRD) reveals that clay minerals dominate the mineral fraction of the shale in well 5 in the study area, followed by quartz, calcite, plagioclase, and pyrite. The clay mineral content ranges from 41.40% to 58.10%, with an average content of 47.28%, whereas quartz content ranges from 28.9% to 39.70%, with an average content of 33.60%. Calcite content varies from 1.70% to 21.90%, with an average content of 9.97%. The predominant clay minerals in the shale are illite, with an average content of approximately 38.00%, followed by kaolinite, chlorite, and illite/smectite (Table 1).

Combined with core observation, thin section identification, scanning electron microscopic observation, and mineral composition determination results, the test samples of the Lianggaoshan Formation mainly include clay shale, silty shale, and lime-bearing shale. The clay shale has a uniform distribution of mud-grade debris and undeveloped laminae. The silty shale lamina is developed, the lamina shape is relatively straight, and the lamina type is mainly a felsic silt thin layer. Mud crystal dolomite and powder crystal dolomite are widely developed in the layered massive limestone-bearing shale. The composition is mainly calcite, with a portion of dolomite. A small amount of clay minerals is distributed between calcite particles. Detritus limestone and debris laminae are developed, and the composition is mainly calcite.

Table 1. Basic information about test samples in the upper section of the study area.

Number	Depth (m)	TOC (%)	Mineral Content (%)						Clay Mineral Content (%)			
			Quartz	Feldspar	Calcite	Dolomite	Pyrite	Clay Mineral	Illite	Kaolinite	Chlorite	Illite/ Montmorillonite Mixed Layers
2	1611.44–1611.55	3.17	32.9	4.5	3.9	0	0.6	58.1	23.2	9.3	17.4	8.1
4	1612.57–1612.71	3.62	28.9	2.4	19.5	1.3	3.6	44.3	14.2	10.2	11.5	8.4
6	1614.69–1614.86	1.35	30.3	2.3	19.8	1	3.1	43.5	17.0	12.2	9.1	5.2
8	1615.64–1615.79	0.64	28.9	2	21.9	2.2	3.6	41.4	14.9	11.2	9.9	5.4
10	1647.04–1647.20	1.64	34.4	7.8	13	0	2.4	42.4	18.7	5.5	12.7	5.5
12	1648.75–1648.96	1.22	35.2	9.2	0	0	2	53.6	24.1	7.0	14.5	8.0
14	1652.54–1652.67	1.46	39.7	9.1	0	0	1.3	49.9	19.0	5.0	9.5	16.5
16	1653.09–1653.27	1.08	38.5	12	1.7	2.8	0	45	13.5	6.3	9.0	16.2

Figure 3. Rock cores, thin sections, and scanning electron microscopic images of the reservoir rocks of the Lianggaoshan Formation. (**a**) Clay shale; 1741.88 m. (**b**) Clay shale with organic matter distributed sporadically in agglomerates; 1658.50 m, 10 × 10(−), scale bae = 200 μm. (**c**) Clay shale with high clay mineral content and intergranular micropores; 1612.57 m, 5000×. (**d**) Siltstone; 1715.34 m. (**e**) Sand shale, quartz, and other mineral grains are dispersed, with visible as streaks; 1714.54 m, 5 × 10(+), scale bae = 500 μm. (**f**) Sand shale showing long, ingrained grains (grain size 10–30 μm); 1652.54 m, 1000×. (**g**) Gray shale showing ash-textured laminae; 1742.58 m. (**h**) Gray shale showing calcite granules distributed in mud; 1717.54 m, 10 × 10(+), scale bae = 200 μm. (**i**) Gray shale showing grainy calcite laminae with intracrystalline pores; 1611.44 m, 4000×.

4.2. Storage Space Type and Distribution Characteristics

The exploration of unconventional oil and gas resources involves the study of reservoir space type, porosity, permeability, and fracture characteristics. These factors are crucial in the process of oil and gas accumulation [35]. Shale oil and gas reservoirs exhibit complex pore structures and diverse space types that serve as both reservoir fluid spaces and flow spaces [36–39]. Several researchers worldwide have proposed various classification schemes for pore structure types and characteristics of shale reservoirs. These schemes classify the pore types into four categories based on factors such as mineral structure, organic matter content, and diagenesis. The pore types include intragranular pores, intergranular pores, organic matter pores, and microfractures. By utilizing SEM and argon ion polishing, the shale storage space of the Lianggaoshan Formation in the study area can be classified into two categories: pores and microfractures. The pores comprise intergranular pores, intragranular pores, organic matter pores, and intergranular microfractures of clay minerals. Moreover, oil stains can be observed filling the reservoir space.

The intergranular pores are mainly distributed between clay or detrital particles, and a large number of intergranular pores were retained during the compaction of clay mineral particles and detrital particles under the action of sedimentation and diagenesis. The intergranular pores are irregularly shaped (e.g., polygons or sheets Figure 4a–d), and they were mainly distributed between different crystals formed by accumulation during the regrowth of different mineral crystals. Samples from the Lianggaoshan Formation are rich in pyrite, calcite, and other carbonate minerals. Calcite exhibited a microcrystalline structure, and pyrite was spherical and showed a cluster-like distribution. Microscopic observation of the intergranular pores in samples from the Lianggaoshan Formation mainly revealed calcite intergranular pores, pyrite intergranular pores, and some clay mineral intergranular pores (Figure 4e–g). These types of pores were irregularly distributed, and the morphology was related to the crystal morphology and accumulation method. The pores were small in size.

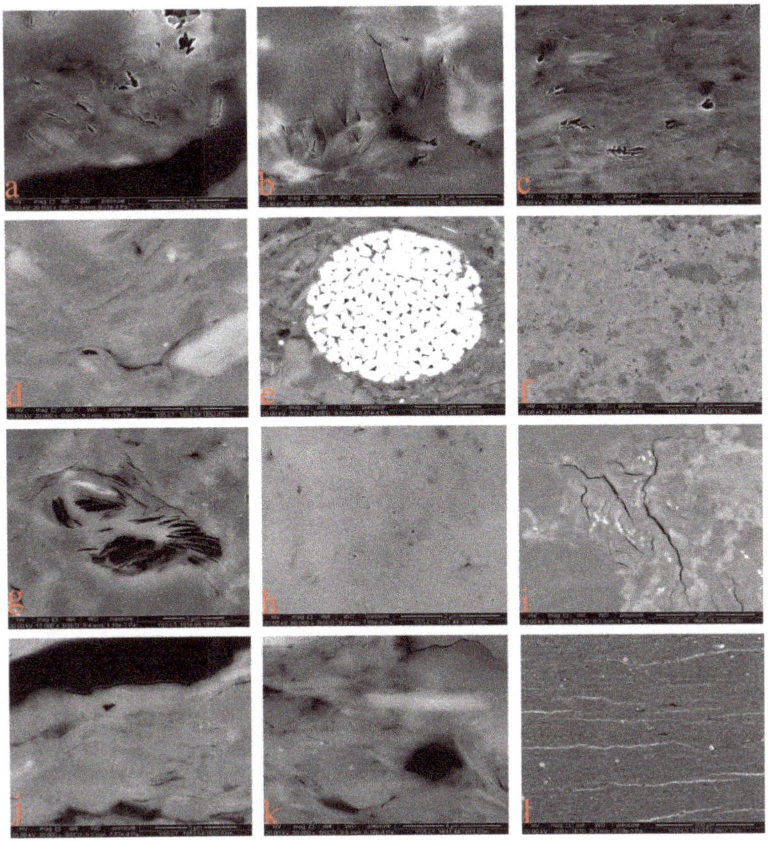

Figure 4. Spatial distribution of different types of shale reservoirs in the study area. (**a**) Intergranular pore development between clay minerals; Yingshan 5-L7, 1652.54 m. (**b**) Intergranular pore development between clay minerals; Yingshan 5-L8, 1653.09 m. (**c**) Intergranular pore development between oil-filled clay minerals; Yingshan 5-L2, 1612.57 m. (**d**) Intergranular pore development between linearly distributed clay minerals; Yingshan 5-L7, 1652.54 m. (**e**) Intercrystalline pores developed between globular pyrite grains; Yingshan 5-L1, 1611.44 m. (**f**) Intercrystalline pores of calcite; Yingshan 5-L1, 1611.44 m. (**g**) Intercrystalline pore development between clay minerals; Yingshan 5-L3, 1614.69 m. (**h**) Intracrystalline solution pores; Yingshan 5-L1, 1611.44 m. (**i**) Intercrystalline microslits of clay minerals; Yingshan 5-L6, 1648.75 m. (**j**) Organic matter with contraction joints developed; Yingshan 5-L7, 1652.54 m. (**k**) Contraction joints in a quartz mineral grain; Yingshan 5-L1, 1611.44 m. (**l**) Laminated joints; Yingshan 5-L2, 1612.57 m.

The intragranular pores were mainly distributed in quartz, clay minerals, and calcite grains. Among these, the content of calcite in the samples from the Lianggaoshan Formation was high, and the pores in the granules were mainly dissolution pores that formed when calcite was dissolved by acidic fluids (Figure 4h). The intragranular pores were independent and irregularly developed, and the pore size was mainly distributed in the orders of nano- and micro-nanometers. The development of microfractures can effectively improve the deep flow capacity of the reservoir. The microfractures in the shales of the Lianggaoshan Formation were mainly laminae, grain edge joints, or organic matter contraction joints, among which the grain edge joints were prominent. These grain edge joints were primarily distributed in a mesh pattern with good interconnectivity. The laminae joints with large pore sizes were mainly distributed among different clay minerals or mineral crystals, such as rhodochrosite (Figure 4i–l). The development of reservoir space types in different lithology samples is quite different.

Observations with a magnification range of 200–20,000× and statistical analysis of all pore types using LFS-SEM revealed variations in the distribution of characteristic parameters such as the number and size of different types of pores. The proportion of different pore types indicated that intergranular pores and microfractures were more prominent, accounting for 49.93% and 25.51% of the total number of pores, respectively. This was followed by organic matter pores and intragranular pores, accounting for 15.62% and 9.44% of the total number of pores, respectively. The sizes of pores ranged from 10 nm to 10 μm, mainly comprising nanoscale pores (≤1 μm). The sizes of intergranular pores varied across a wide range, with the peak pore size mainly distributed in the range of 0.1–1 μm, whereas the sizes of organic matter pores varied across a relatively small range, with the peak pore size being <0.01μm. The sizes of microcracks were observed to range from a few nanometers to a few micrometers (Figure 5).

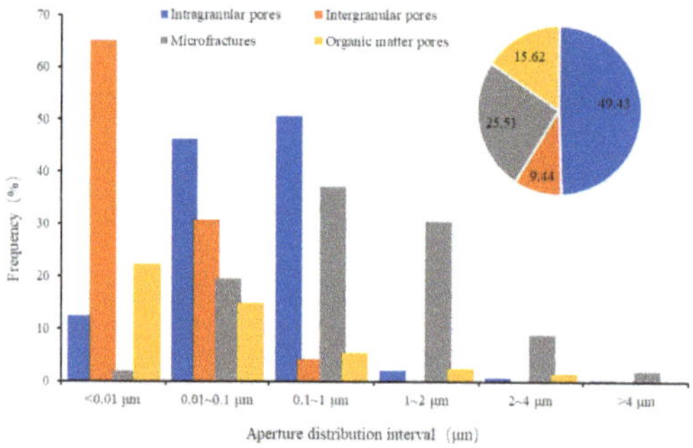

Figure 5. Distribution frequency of different pore types and sizes.

4.3. Pore Structure Distribution Characteristics

To comprehensively characterize the pore structure and the distribution of different pore sizes in the shale of the Lianggaoshan Formation in the central part of the Sichuan Basin, various experimental methods were used in this study. Gas adsorption–desorption data of the samples (CO_2 and N_2) were interpreted using the multipoint BET model and the BJH model, and the samples were characterized in terms of SSA, pore size, and PV. The connectivity of the test samples was evaluated based on the results of HPMI tests. Furthermore, full-pore capillary pressure curves were established by combining LTNA and HPMI tests, and the fractal dimension of the pore structure of different samples was calculated. The fractal dimension of the pore structure was calculated to quantitatively characterize the structural features of the reservoir space distribution across the full pore size range of the samples.

The pore diameter distribution of clay shale and its pore characteristics play a critical role in the evaluation of shale oil reservoirs. There are several pore diameter classification schemes for shale reservoirs, and the International Union of Pure and Applied Chemistry (IUPAC) classification, which defines micropores as pores with diameters between 0 and 2 nm, mesopores as pores with diameters between 2 and 50 nm, and macropores as pores with diameters greater than 50 nm, is currently used. This classification scheme was utilized in the present study. Eight samples from the Lianggaoshan Formation in the study area were subjected to LTNA, and the results indicated that the CO_2 adsorption curves of the samples were similar to the type I isothermal adsorption line, as defined by the IUPAC. When the relative pressure (P/Po) was less than 0.01, the CO_2 adsorption increased as P/Po increased, and when P/Po was greater than 0.01, CO_2 adsorption increased gradually as P/Po increased (Figure 6a), suggesting that the test sample had developed a certain number of micropores and that the filling of these micropores had occurred. The isothermal adsorption–desorption curve displayed an inverse "S" shape as P/Po increased, which is a type II isotherm (Figure 6). The adsorption–desorption curves indicated that both the adsorption and desorption of shale steadily increased when the relative pressure was low, and the desorption curve had a large inflection point when P/Po was in the range of approximately 0.4~0.5, consistent with the inflection point calculated by Harris et al. [40–42]. When P/Po was close to 1, the adsorption volume started to increase rapidly, the adsorption and desorption curves tended to overlap, and the hysteresis loop gradually closed.

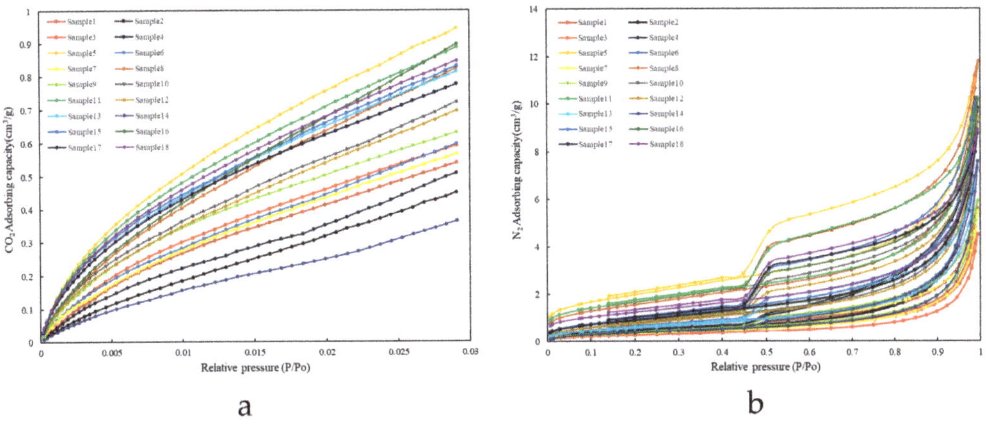

Figure 6. Gas adsorption–desorption curves of shale samples from the Lianggaoshan Formation in the study area. (**a**) CO_2 gas adsorption–desorption curve. (**b**) N_2 gas adsorption–desorption curve.

The pores in the tested samples were categorized into two groups based on their pore structure and whether a hysteresis loop could be generated: class I isothermal adsorption curves were of H2–H3 type, and in the samples at 1614.69 m, 1615.64 m, 1647.04 m, 1648.75 m, and 1653.09 m, the corresponding pores were mainly open pores, including parallel-walled slit-type capillary pores with open edges and cylindrical pores with open ends. Class II isothermal adsorption curves were of H3–H4 type, and in the samples at 1611.44 m, 1612.57 m, and 1652.54 m, the corresponding pores were mainly ink bottle pores with thin necks and wide bodies, which were closed at one end but could produce a hysteresis return line; there was a sharp decreasing inflection point in the desorption curve. The characterization results are consistent with the above SEM observation results.

The gas adsorption and desorption data were analyzed using the multipoint BET model and the BJH model to determine the specific surface area (SSA) and pore volume (PV) of the samples. The SSA values of the test samples ranged from 3.9061 to 11.1376 $m^2 \cdot g^{-1}$, with an average of 6.9046 $m^2 \cdot g^{-1}$, and PV values ranged from 0.0117 to 0.0185 $cm^3 \cdot g^{-1}$, with an average of 0.0150 cm^3 g^{-1}. The SSA and PV values of the samples corresponding to the class I isothermal adsorption curves were large, and micropores were developed. Clay shale has the largest specific surface area, with an average of about 9.13 $m^2 \cdot g^{-1}$, while lime-bearing shale has the smallest specific surface area, mainly distributed at 7.27 $m^2 \cdot g^{-1}$.

The morphology of the mercury pressure curve is indicative of the development of pore space in each pore throat section and the connectivity between pore spaces. Based on

Figure 7, the capillary pressure curve, as a whole, is observed to be skewed towards the upper right, with the central main inlet mercury section being flat and long. To analyze the pore characteristics of the samples, the discharge drive pressure, sorting coefficient, maximum mercury inlet saturation, and mercury withdrawal efficiency were all considered. The pore structures in the study area were broadly categorized into two major types. The first type (Figure 7) included the samples taken at depths of 1611.44 m and 1614.69 m, where the mercury inlet curve exhibited a stepwise increase as the pressure rose. The curve could be divided into two sections, with the displacement pressure being less than 1.0 MPa, corresponding to a maximum pore throat radius of approximately 10 μm. The samples in this category had a relatively large maximum mercury injected saturation (up to 70%) and a mercury withdrawal efficiency of approximately 50%. The morphology of these pores was primarily ink-bottle-like, with micropores that were generally undeveloped, with small pore throats and poor connectivity. For the second type (Figure 7, including the samples taken at depths of 1612.57 m, 1615.64 m, and 1653.09 m), the mercury inlet curve displayed a sudden increase with increasing pressure, and the gentle section of the mercury inlet curve was relatively short. The sorting coefficient was small, and the pore distribution was relatively uniform. The displacement pressure was relatively high, at approximately 4.0 MPa, and the maximum mercury inlet saturation of the samples was small, ranging from 40% to 50%. The PV was relatively small, and the maximum mercury rejection efficiency reached 70%, indicating that the pores in the samples were predominantly slit-type or flat-type micropores with good connectivity.

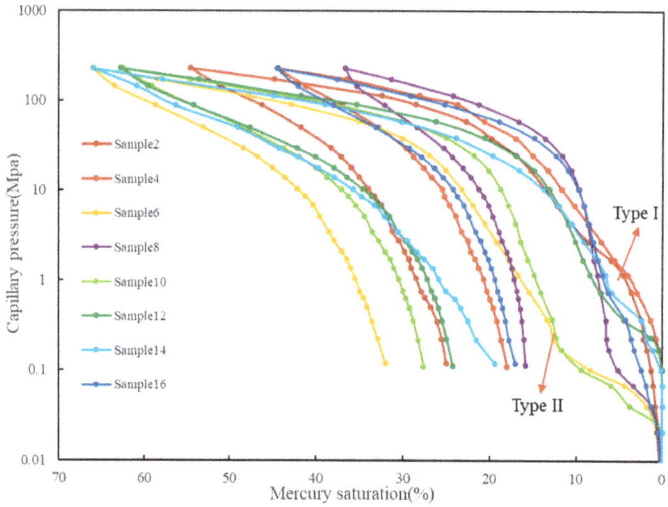

Figure 7. Mercury pressure curves of samples from the Lianggaoshan Formation in well YS5.

The full-scale pore size distributions of various samples were computed based on energy industry standard NB/T 14008-2015 of the People's Republic of China. The results indicated that the full-scale storage capacity of the samples was primarily governed by mesopore (2–50 nm) distributions. The peak values were mainly observed in the range of 20–30 nm. Additionally, some micrometer-scale macropores were observed in the sand shale samples (Figure 8).

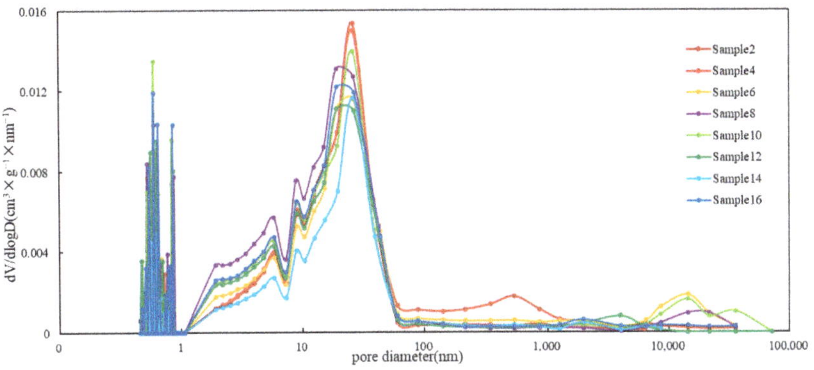

Figure 8. Pore size distribution curve of test samples from the Lianggaoshan Formation in Well YS5 in the central portion of the Sichuan Basin.

5. Factors Influencing the Development of the Pore Structure

5.1. Comparison of Pore Structure Characteristics of Different Lithologies

Using LFS-SEM, the pore size distribution of various lithologies was quantitatively analyzed, revealing notable differences in the pore size amongst different lithologies. As demonstrated in Figure 5, samples 1 and 7 are clay shales that feature a bimodal pore size distribution, with the peaks predominantly distributed in the ranges of 10–50 nm and >100 nm, respectively. Furthermore, samples 6 and 13 are gray shales with a pore size distribution in the range of 10–200 nm, mainly concentrated in the range of 10–50 nm. Samples 2 and 14 correspond to sand shales that exhibit a relatively broad pore size distribution, mainly in the range of 10–2000 nm. The pore size distribution of these samples, particularly that of samples 2 and 14, is relatively broad, with micrometer- and millimeter-sized pores primarily distributed in the range of 10–2000 nm, mainly concentrated in the range of 40–100 nm (Figure 5). Combined with SEM results, it can be seen that a certain number of micropores and microfractures are present in the clay shale samples, primarily consisting of intergranular pores of clay minerals, intergranular pores of pyrite, and natural fractures, indicating good overall pore connectivity. The pore size distribution in the gray shale is quite broad, with micropores mainly consisting of calcite intergrain pores and mineral intragranular pores, and occasionally featuring paleontological intracavity pores.

5.2. Factors Influencing Pore Structure Development

The pore structure characteristics of shale are affected by several factors. The sedimentary environment influences the content of organic carbon, brittle minerals, and clay minerals within shale. The SSA and PV of shale are influenced by the content of organic matter, clay minerals (e.g., illite), and brittle minerals (e.g., authigenic quartz). Not only is the total organic carbon content an important indicator of source rock quality, but it also has a considerable impact on the PV, SSA, and adsorbed gas content of the reservoir, particularly when comparing shale in the high maturity and overmaturity stages. The degree to which intergranular pores, intragranular pores, and microfractures develop is influenced by the content of brittle minerals, which is an important criterion used to determine whether shale oil and gas has commercial exploitation value. The type and content of clay minerals influence the conversion of clay minerals during diagenetic evolution and thereby influence the PV and SSA of micropores within them. The samples examined in this study were primarily fresh samples collected from the same drilling, and the depths at which the samples were obtained did not vary significantly. Thus, organic matter maturity did not have a significant influence on the pore structure, and the effects of organic carbon content, clay mineral content, and brittle mineral content (e.g., quartz and feldspar) on the characteristics of pore structure in different lithologies were mainly studied in this work [43]. The pore fractal dimension, a quantitative measurement of pore structure complexity and surface roughness, exhibited a highly consistent change trend with that of reservoir porosity, SSA, and PV. Based on the results of LTNA and HMPI, a full pore size capillary pressure curve was developed to analyze the full pore size distribution characteristics of shales from the Lianggaoshan Formation in the study area [44]. The fractal dimension of the pore structure within the full pore size range was then calculated using the following equation:

$$lgS = (3 - D)lgR - (3 - D)lgR_{max}$$

where R is the pore size corresponding to the capillary pressure (Pc) in the reservoir (μm); S is the saturation of the wetted phase in the reservoir when the capillary pressure is Pc (%); and D is the fractal dimension, which is a dimensionless quantity.

The fractal dimension values of shale samples were calculated (Table 2), and the correlation between the fractal dimension and organic carbon mass lithology, clay minerals, brittle minerals (e.g., quartz and feldspar), and carbonate mineral content, was analyzed in this study (Figure 9). The correlation between brittle minerals, such as felsic minerals, and the fractal dimension was not very clear. Quartz and feldspar possess a regular crystal structure and form, their intergranular and intragranular pores are not well-developed, their morphology is relatively singular, and their SSA and PV are relatively limited. Thus, these factors had a relatively small effect on the fractal dimension of the pore structure. In contrast, a significant linear correlation was observed between clay mineral content and the fractal dimension, indicating that the pore structure characteristics of the shale in the Lianggaoshan Formation were primarily controlled by clay minerals [45].

Table 2. Distribution of SSA, PV, and the fractal dimension of test samples from the Lianggaoshan Formation.

Number	SSA (m^2/g)	PV (cm^3/g)	Fractal Dimension of Full Aperture Range
1	4.2003	0.008694	
2	5.2561	0.014928	2.7843
3	2.8221	0.006958	
4	5.3439	0.015185	2.7005
5	16.0593	0.017826	
6	7.1863	0.014279	2.7299
7	3.7634	0.008465	
8	14.4304	0.018242	2.7124
9	4.5694	0.009177	
10	9.575	0.015047	2.7568
11	13.7266	0.015825	
12	9.27	0.014558	2.7535
13	6.3728	0.011286	
14	4.1966	0.011727	2.7209
15	10.3870	0.015830	
16	10.3587	0.015705	2.7514
17	9.7045	0.013542	
18	10.6579	0.013794	

The clay minerals in the samples were primarily illite and kaolinite. Illite is a 2:1 type phyllosilicate mineral, while kaolinite is a typical 1:1 type phyllosilicate mineral. Their unique crystal structures give clay minerals porosity and large surface areas. Under SEM, illite often appears as flaky or filamentous, with rectangular, wedge-shaped, or slit interlayer pores often formed between parallel illite plates. As the clay mineral content increases, the development of clay mineral granular (crystalline) pores and microfractures becomes the main contributor to storage space in the samples. Furthermore, organic matter, which can take the form of complexes between clay mineral particles, can also coexist with clay minerals in pores (Figure 3b,e). The interaction between clay minerals and organic matter promotes the thermal evolution of organic matter, leading to the formation of a certain amount of organic matter micropores, thereby increasing the fractal dimension of pore distribution. The heterogeneity of pore distribution is strong. Carbonate minerals, intergranular dissolution pores, and intragranular dissolution pores are relatively developed, and their shapes are mostly circular or elliptical. This type of pore distribution is relatively uniform, and the fractal dimension of pore distribution decreases(Figure 10), indicating that the presence of carbonate minerals such as calcite contributes to the development of pores to a certain extent.

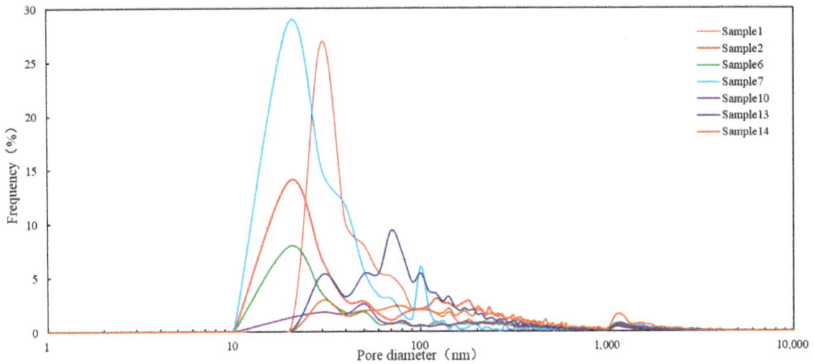

Figure 9. Pore size distribution in samples of different lithologies of the Lianggaoshan Formation in the central part of the Sichuan Basin.

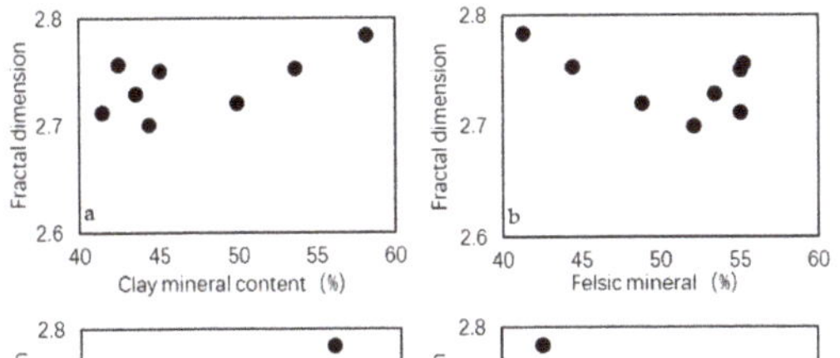

Figure 10. Fractal dimension of TOC and major mineral compositions versus full pore size range. (**a**) Relationship between clay minerals content and fractal dimension. (**b**) Relationship between Felsic mineral and fractal dimension. (**c**) Relationship between TOC and fractal dimension. (**d**) Relationship between Carbonate rock mineral content and fractal dimension.

6. Conclusions

Recommendations for Hydraulic Fracturing Modifications based on In Situ Stress Magnitude Evaluation

(1) The reservoir of the Lianggaoshan Formation in the study area exhibits a diverse rock mineral composition and a complex rock structure. XRD analysis, SEM, and energy spectrum tests reveal that the lithology of the Lianggaoshan Formation primarily consists of clay shale, sand shale, and gray shale. The shale reservoir space of the Lianggaoshan Formation in the study area can be classified into two categories: micropores and microfractures. The pores consist of intergranular pores, intragranular pores, and organic matter pores, and traces of oil filling in the reservoir space are observed. The intergranular pores and microfractures are relatively well-developed (accounting for 49.93% and 25.51% of the total

number of pores, respectively), followed by organic matter pores and intragranular pores. The predominant pore morphologies are ink-bottle-type, slit-type, or plate-type, with the average SSA being approximately 6.9046 $m^2 \cdot g^{-1}$ and the average PV being approximately 0.0150 $mL \cdot g^{-1}$. The storage space is primarily composed of mesopores with a distribution range of 2 to 50 nm, with the peaks mainly located between 20 and 30 nm.

(2) The variations in pore structure characteristics among different lithologies are considerable. In clay shale, a considerable number of micropores and microfractures are observed, which are primarily intergranular pores, interpyrite granular pores, and natural fractures, exhibiting good pore connectivity overall. Gray shale exhibits intragranular pores and intergranular (crystalline) micropores as the primary storage space, with pore sizes ranging from 10 to 50 nm. In sand shale, the reservoir space is primarily developed between clastic mineral particles, such as quartz, with microfractures occurring between clay minerals. The content of clay minerals in various lithologies exhibits a substantial linear correlation with the fractal dimension. As the clay mineral content increases, the quantity of intergranular pores and intergranular microfractures within the samples increases, indicating that clay minerals primarily control the pore structure characteristics.

Author Contributions: Conceptualization, Y.W. and H.D.; writing—original draft preparation, Q.C.; writing—review and editing, A.L.; supervision, Z.W.; resources, X.W.; data curation, D.C.; investigation, Y.Z. All authors have read and agreed to the published version of the manuscript.

Funding: This research was supported by National Natural Science Foundation of China (No. 42072182), and Sichuan Outstanding Young Scientific and Technological Talents Project (No. 2020JDJQ0058).

Institutional Review Board Statement: Not applicable.

Informed Consent Statement: Not applicable.

Data Availability Statement: Not applicable.

Acknowledgments: The authors would like to thank all the reviewers who participated in the review.

Conflicts of Interest: The authors declare no conflict of interest.

References

1. He, W.; Bai, X.; Meng, Q.; Li, J.; Zhang, D.; Wang, Y. Accumulation geological characteristics and major discoveries of lacustrine shale oil in Sichuan Basin. *Acta Pet. Sin.* **2022**, *43*, 885–898.
2. Li, G.; Lei, Z.; Dong, W.; Wang, H.; Zheng, X.; Tan, J. Progress, challenges and prospects of unconventional oil and gas development of CNPC. *China Pet. Explor.* **2022**, *27*, 1–11.
3. Wang, Q.; Tao, S.; Guan, P. Progress in research and exploration & development of shale oil in continental basins in China. *Nat. Gas Geosci.* **2020**, *31*, 417–427.
4. Cao, Q.; Qi, M.; Li, X.; Wang, D.; Deng, Y.; Wu, C. Study on Organic Matter Enrichment and Reservoir Characteristics of Permian Longtan Formation Shale in Southeast Sichuan. *Front. Earth Sci.* **2022**, *10*, 895941. [CrossRef]
5. Cao, Q.; Jiang, K.; Wen, Z.; Wang, X.; Qi, M.; Yin, Z. Characteristics of Organic Matter Pores and the Relationship with Current Pressure System of Lower Silurian Longmaxi Shales in Dingshan Field, Southern Sichuan, China. *Geofluids* **2021**, *2021*, 9967479. [CrossRef]
6. Cao, Q.; Zhou, W.; Deng, H.; Chen, W. Classification and Controlling Factors of Organic Pores in Continental Shale Gas Reservoir based on Laboratory Experimental Results. *J. Nat. Gas Sci. Eng.* **2015**, *27*, 1381–1388. [CrossRef]
7. He, W.; He, H.; Wang, Y.; Cui, B.; Meng, Q.; Guo, X.; Bai, X.; Wang, Y. Major breakthrough and significance of shale oil of the Jurassic Lianggaoshan Formation in Well Ping'an 1 in northeastern Sichuan Basin. *China Pet. Explor.* **2022**, *27*, 40–49.
8. Yang, Y.; Yang, J.; Yang, G.; Tao, S.; Ni, C.; Zhang, B.; He, X.; Lin, J.; Huang, D.; Liu, M.; et al. New research progress of Jurassic tight oil in central Sichuan Basin. *Pet. Explor. Dev.* **2016**, *43*, 873–882. [CrossRef]
9. Clarkson, C.R.; Solano, N.; Bustin, R.M.; Bustin, A.; Chalmers, G.; He, L.; Melnichenko, Y.B.; Radlinski, A.P.; Blach, T.P. Pore structure characterization of North American shale gas reservoirs usingUSANS/SANS, gas adsorption, and mercury intrusion. *Fuel* **2013**, *103*, 606–616. [CrossRef]
10. Labani, M.M.; Rezaee, R.; Saeedi, A.; Hinai, A.A. Evaluation of pore size spectrum of gas shale reservusing low pressure nitrogen adsorption, gas expansion and mercury porosimetry: A case sfrom the Perth and Canning Basins, Western Australia. *J. Pet. Sci. Eng.* **2013**, *112*, 7–16. [CrossRef]
11. Zhou, K.; Li, J.; Xu, Y.; Zhang, Y. Measurement of rock pore structure based on NMR technology. *J. Cent. South Univ. (Sci. Technol.)* **2012**, *43*, 4776–4800.

12. Yang, F.; Ning, Z.; Hu, C.; Wang, B.; Peng, K.; Liu, H. Characterization of microscopic pore structures in shale reservoirs. *Acta Pet. Sin.* **2013**, *34*, 301–311.
13. Guo, X.; Li, Y.; Liu, R.; Wang, Q. Characteristics and controlling factors of micro-pore structures of Longmaxi Shale Play in the Jiaoshiba area, Sichuan Basin. *Nat. Gas Ind.* **2014**, *34*, 9–16.
14. Wang, P.; Nie, H.; Liu, Z.; Sun, C.; Cao, Z.; Wang, R.; Li, P. Differences in Pore Type and Pore Structure between Silurian Longmaxi Marine Shale and Jurassic Dongyuemiao Lacustrine Shale and Their Influence on Shale-Gas Enrichment. *Minerals* **2023**, *13*, 190. [CrossRef]
15. Qing, Y. Formation Mechanism of Tight Oil Reservoirs in Liangshang Member and Shayi Member of Jurassic, Central Sichuan Basin. Ph.D. Thesis, Chengdu University of Technology, Chengdu, China, 2020.
16. Liu, A.; Tian, J.; Zhang, S.; Chen, S.; Yi, J.; Luo, X.; Luo, N.; Cai, L.; Yang, T. Characteristics and controlling factors of tight sandstone reservoirs of the Jurassic Lianggaoshan Formation in the eastern Sichuan Basin. *Mineral. Petrol.* **2023**, *43*, 144–155.
17. Feng, Y. Geological characteristics and exploration direction of continental shale gas in Jurassic Da 'anzhai Member, Sichuan Basin. *Pet. Geol. Exp.* **2022**, *44*, 219–230.
18. Yang, R.; Wang, F.; Yun, N.; Zeng, H.; Han, Y.; Hu, X.; Diao, N. Pore Structure Characteristics and Permeability Stress Sensitivity of Jurassic Continental Shale of Dongyuemiao Member of Ziliujing Formation, Fuxing Area, Eastern Sichuan Basin. *Minerals* **2022**, *12*, 1492.
19. Curtis, M.E.; Cardott, B.J.; Sondergeld, C.H.; Rai, C.S. Development of organic porosity in the Woodford shale with increasing thermal maturity. *Int. J. Coal Geol.* **2012**, *26*, 26–30. [CrossRef]
20. Keller, L.M.; Schuetz, P.; Erni, R.; Rossell, M.D.; Lucas, F.; Gasser, P.; Holzer, L. Characterization of multi-scale microstructural features in Opalinus Clay. *Microporous Mesoporous Mater.* **2013**, *83*, 84–90. [CrossRef]
21. Deng, H.; Hu, X.; Li, H.A.; Luo, B.; Wang, W. Improved pore-structure characterization in shale formations with FESEM technique. *J. Nat. Gas Sci. Eng.* **2016**, *35*, 309–319.
22. Zou, J.; Lei, Y.; Jin, T.; Li, X.; Li, X.; Wei, T. Genesis and Sweet Spot Prediction of Jurassic Tight Sandstone Reservoir in Central Sichuan Basin. *Xinjiang Pet. Geol.* **2018**, *39*, 555–560.
23. Li, N.; Hong, H.; Li, G.; Zhou, H.; Jia, M.; Kang, J. Geological characteristics of shale oil and gas of Lianggaoshan Formation in high-steep structural zone, Sichuan Basin. *Nat. Gas Explor. Dev.* **2022**, *45*, 86–95.
24. Li, J. Reservoir Characteristics of Longmaxi Shale in the Southeast of Chongqing. Master's Thesis, China University of Geosciences (Beijing), Beijing, China, 2013.
25. Liao, Q.; Hu, H.; Lin, J.; Liu, Z.; Mu, X. Petroleum exploration prospect of the Jurassic tight reservoirs in central Sichuan Basin. *Oil Gas Geol.* **2011**, *32*, 815–822+838.
26. Chen, S.; Gao, X.; Wang, L.; Lu, J.; Liu, C.; Tang, H.; Zhang, H.; Huang, M.; Ni, S. Factors controlling oiliness of Jurassic Lianggaoshan tight sands in central Sichuan Basin, SW China. *Pet. Explor. Dev.* **2014**, *41*, 421–427. [CrossRef]
27. Hong, H.; Zhang, S.; Zhang, R.; Guan, X. Establishment of Sequence Stratigraphic Framework and Optimization of Favorable Exploration Areas of Lianggaoshan Formation, Sichuan Basin. J/OL. *Spec. Oil Gas Reserv.* **2023**, *30*, 58–64.
28. Yi, J.; Zhang, S.; Cai, L.; Chen, S.; Luo, X.; Yu, J.; Luo, N.; Yang, T. Strata and Sedimentary Filling Characteristics of the Lower Jurassic Lianggaoshan Formation and Its Hydrocarbon Exploration in Eastern Sichuan Basin. *J. Jilin Univ. (Earth Sci. Ed.)* **2022**, *52*, 795–815.
29. Cheng, D.; Zhang, Z.; Hong, H.; Zhang, S.; Qin, C.; Yuan, X.; Zhang, B.; Zhou, C.; Deng, Q. Sequence structure, sedimentary evolution and their controlling factors of the Jurassic Lianggaoshan Formation in the East Sichuan Basin, SW China. *Pet. Explor. Dev.* **2023**, *50*, 262–272. [CrossRef]
30. Erik, N.Y.; Ay, F. Use of petrological and organic geochemical data in determining hydrocarbon generation potential of coals: Miocene coals of Malatya Basin (Eastern Anatolia-Turkey). *Int. J. Coal Sci. Technol.* **2021**, *8*, 510–533.
31. Corrado, S.; Schito, A.; Romano, C.; Grigo, D.; Poe, B.T.; Aldega, L.; Caricchi, C.; Di Paolo, L.; Zattin, M. An integrated platform for thermal maturity assessment of polyphase, long-lasting sedimentary basins, from classical to brand-new thermal parameters and models: An example from the on-shore Baltic Basin (Poland). *Mar. Pet. Geol.* **2020**, *122*, 104547.
32. Caricchi, C.; Corrado, S.; Di Paolo, L.; Aldega, L.; Grigo, D. Thermal maturity of Silurian deposits in the Baltic Syneclise (on-shore Polish Baltic Basin): Contribution to unconventional resources assessment. *Ital. J. Geosci.* **2016**, *135*, 383–393. [CrossRef]
33. Stach, E. *Stach's Textbook of Coal Petrology*, 3rd ed.; Gerbrüder Barntraeger: Berlin, Germany, 1982.
34. Yu, J.; Liu, H.; Wang, Y.; Liu, H.; Yang, K.; Lin, J. Characteristics and Influencing Factors of Micropore Structures in Jurassic Lianggaoshan Formation in Central Sichuan Basin. *Well Logging Technol.* **2019**, *43*, 612–619.
35. Huang, Y.; Zhang, J.; Zhang, P.; Tang, X.; Su, H. Microscopic Characteristics and Main Controlling Factors of Shale Reservoir Space of Es3 Formation, Northern Dongpu Depression. *J. Shandong Univ. Sci. Technol. (Nat. Sci.)* **2016**, *35*, 8–16.
36. Ge, M.; Ren, S.; Bao, S.; Wang, S.; Guo, T. Continental shale gas-bearing analysis and exploration prospect:a case study of K_1q_1 formation of southern Songliao basin. *China Min. Mag.* **2019**, *28*, 162–168.
37. Zou, G.; She, J.; Peng, S.; Yin, Q.; Liu, H.; Che, Y. Two-dimensional SEM image-based analysis of coal porosity and its pore structure. *Int. J. Coal Sci. Technol.* **2020**, *7*, 350–361. [CrossRef]
38. Fu, D.; Xu, G.; Ma, L.; Yang, F.; He, D.; Duan, Z.; Ma, Y. Gas generation from coal: Taking Jurassic coal in the Minhe Basin as an example. *Int. J. Coal Sci. Technol.* **2020**, *7*, 611–622. [CrossRef]

39. Liu, Y.; Cao, Q.; Ye, X.; Dong, L. Adsorption Characteristics and Pore Structure of Organic-Rich Shale with Different Moisture Contents. *Front. Earth Sci.* **2022**, *10*, 863691. [CrossRef]
40. Yang, F.; Ning, Z.; Kong, D.; Liu, H. Pore Structure of Shales from High Pressure Mercury Injection and Nitrogen Adsorption Method. *Nat. Gas Geosci.* **2013**, *24*, 450–455.
41. Chen, S.; Zhang, C.; Li, X.; Zhang, Y.; Wang, X. Simulation of methane adsorption in diverse organic pores in shale reservoirs with multi-period geological evolution. *Int. J. Coal Sci. Technol.* **2021**, *8*, 844–855. [CrossRef]
42. Liu, A.; Liu, S.; Liu, P.; Wang, K. Water sorption on coal: Effects of oxygen-containing function groups and pore structure. *Int. J. Coal Sci. Technol.* **2021**, *8*, 983–1002.
43. Tan, R.; Wang, R.; Huang, Y.; Yang, R.; Li, H.; Lu, K. Mechanism of the Enrichment and Loss Progress of Deep Shale Gas: Evidence from Fracture Veins of the Wufeng–Longmaxi Formations in the Southern Sichuan Basin. *Minerals* **2022**, *12*, 897. [CrossRef]
44. Cao, Q. Characterization and Techniques of Micropores in Organic-Richshale of Chang 7th of Yanchang Formation, Ordos Basin. Ph.D. Thesis, Chengdu University of Technology, Chengdu, China, 2016.
45. Wang, W.; Shi, W.; Fu, X.; Xie, J.; Ni, K. Oil and gas enrichment regularity of continental shale of Lianggaoshan Formation of Middle Jurassic in Fuling area, Sichuan Basin. *Nat. Gas Geosci.* **2022**, *33*, 764–774. (In Chinese)

minerals

MDPI

Article

Characteristics and Evolution of Tectonic Fractures in the Jurassic Lianggaoshan Formation Shale in the Northeast Sichuan Basin

Xuefeng Bai [1,2], Xiandong Wang [2,*], Zhiguo Wang [2], Hucheng Deng [3,*], Yong Li [3], An Li [3], Hongxiu Cao [3], Li Wang [3], Yanping Zhu [2], Shuangfang Lu [1], Feng Cao [3] and Jianhua He [3]

1 School of Geosciences, China University of Petroleum (East China), Qingdao 266580, China; bxf@petrochina.com.cn (X.B.); lushuangfang@upc.edu.cn (S.L.)
2 Research Institute of Exploration & Development, PetroChina Daqing Oilfeld Company, Daqing 163712, China; wangyouzhi@petrochina.com.cn (Z.W.); zhuyanping@petrochina.com.cn (Y.Z.)
3 College of Energy, Chengdu University of Technology, Chengdu 610059, China; ly15183207470@163.com (Y.L.); lian@stu.cdut.edu.cn (A.L.); caohongxiu@stu.cdut.edu.cn (H.C.); wangli2@stu.cdut.edu.cn (L.W.); cf773992278@163.com (F.C.); hejianhua19@cdut.edu.cn (J.H.)
* Correspondence: wangxd2004@petrochina.com.cn (X.W.); denghucheng@cdut.cn (H.D.); Tel.: +86-13704597931 (X.W.); +86-19881150998 (H.D.)

Abstract: The features and formation stages of natural fractures have significant influences on the fracturing of shale reservoirs and the accumulation of oil and gas. The characteristics and evolution of tectonic fractures in the Lianggaoshan Formation in Northeast Sichuan were investigated based on outcrops, drill cores, geochemical data, and acoustic emission test results. Our results demonstrated that the fracture types of the Lianggaoshan Formation were mainly low-degree bedding-slip fractures, followed by high-degree through-strata shear fractures and vertical tensile fractures. The influences of strike-slip faults on the fractures were stronger than those of thrust faults; fractures in thrust faults were concentrated in the hanging wall. The densities of tensile and shear fractures were inversely proportional to the formation thickness, while the density of interlayer slip fractures was independent of the formation thickness. The density of tectonic fractures was proportional to the quartz content. The fractures of the Lianggaoshan Formation were generated in three stages during uplift: (1) Late Yanshan–Early Himalayan tectonic movement (72~55 Ma), (2) Middle Himalayan tectonic movement (48~32 Ma), (3) Late Himalayan tectonic movement (15 Ma~4 Ma). Fractures greatly improve the oil and gas storage capacity and increase the contents of free and total hydrocarbons. At the same time, they also reduce the breakdown pressure of strata. This study facilitated the prediction of the fracture distribution and oil and gas reservoirs in the Lianggaoshan Formation and provided references for the selection of favourable areas for shale oil and the evaluation of desert sections in the study area.

Keywords: Jurassic; Lianggaoshan formation; shale; fracture characteristics; formation and evolution

1. Introduction

As it is becoming increasingly difficult to stabilize the production of shale gas in the Silurian marine facies of the Sichuan Basin, it is of great significance to actively seek replacement areas for shale oil and gas exploration to ensure an increase in unconventional oil and gas reserves and production in the Sichuan Basin. The Jurassic system is the key strata for oil exploration in the Sichuan Basin, but due to the low porosity and low permeability of tight sandstone and limestone reservoirs, commercial development of conventional oil and gas is difficult, and no breakthrough in industrial production capacity has been achieved thus far [1,2]. In 2020, the daily oil and gas production rates of Well PA1, located in the Jurassic Lianggaoshan Formation in Northeast Sichuan, were 112.8 m^3 and 11.45 × 10^4 m^3, respectively. In 2021, the daily oil and gas production rates of Well TY1, located in the

Jurassic Lianggaoshan Formation in East Sichuan, were 9.8 m^3 and 7.5 × 10^4 m^3, respectively. This series of explorations successfully realized the exploration breakthrough of continental shale oil gas production in the Jurassic Lianggaoshan Formation and confirmed the large reservoir of shale oil and gas in Northeast Sichuan.

Unconventional shale reservoirs are characterized by low porosity and low permeability. As shale gas has an effective storage space and an important channel for seepage, fractures are of great significance to the enrichment and high production of shale gas [3]. The commercial development of shale reservoirs at home and abroad confirmed that the development of natural fractures not only improves the storage conditions of shale reservoirs but also provides favourable conditions for the formation of large-scale volumetric fracture networks during hydraulic fracturing in the development stage [4–9], but the communication of overdeveloped fractures and large-scale fractures is extremely unfavourable to the preservation of oil and gas [10]. The characteristics, controlling factors, formation stage, characterization of natural fracture distribution, distribution prediction, and evolutionary pattern of natural fractures are fundamental issues in the exploration and development of shale reservoirs [11–15]. Among them, tectonic fractures are controlled by regional structural stress or local stress [16]; therefore, the formation stages of tectonic fractures reflect changes in the geomechanical environment. Determining the formation stage of tectonic fractures can not only reflect the evolution stage of the preservation conditions of shale oil and gas but also help to conduct research on the coupling of enrichment and loss of oil and gas.

The key prerequisite for the correct development of shale oil and gas resources is understanding the distribution and causes of fractures [17]. Some studies characterized the distribution of fractures based on fractal theory and micro-CT [13–15]. In addition, numerical simulation also developed into a key tool for assessing the distribution and geometrical features of natural fractures, fluid action in fractures, and the interaction between natural fractures and hydraulic fractures [18–22]. Usually, the fracture cutting relationship of outcrops, cores, and sections is an intuitive and effective way to divide fracturing stages [23,24], but the multistage tectonic movement experienced by the Northeast Sichuan Basin undoubtedly increases the difficulty of analysis, especially for the outcrop fracture zone [12]. There is a specific internal connection between tectonic activity, fracture development and palaeofluid activity, and veins in shale fractures record important information about palaeofluid activity and fracture activity [25]. The captured fluid inclusion records the ancient temperature and pressure information [26], but part of the fluid inclusion phase transition is difficult to observe during temperature measurement; the carbon and oxygen isotopes of the filler can determine the fracture formation stage, but it is affected by the initial isotope composition of the hydrothermal fluid, the temperature of the mineral crystallization and precipitation, and the type of dissolved carbon in the hydrothermal fluid [27]. In addition, fractures usually have the characteristics of multistage genesis, reactivation, and filling. Each method or technology has limitations or multiple solutions [28]. Therefore, they are used to comprehensively assess the formation stage of fractures [12,17,29–32]. Although some research results indicate that it is difficult to use fracture examination as a method, it does provide valuable evidence that reveals previous strata deformation, stress orientations, and tectonic movements [32–35].

However, fractures exhibit characteristics that correspond to different tectonic movement periods, and studies on the development of fractures in multistage tectonic superposition states as well as their effects on oil and gas accumulation and production are still lacking. Here, we investigated the fracture characteristics of the Liangshan Formation in the Northeast Sichuan Basin through outcrops, cores, and thin sections, combined with fluid inclusion tests, carbon and oxygen isotope analysis, and the acoustic emission (AE) Kaiser effect. Afterwards, we investigated the genetic cause of multiperiod fractures and determined the period of fracture formation. Finally, multiphase genesis mechanisms were given, with matching regional burial history and tectonic evolutionary history. We also discussed the impact of fractures on the accumulation and production of oil and gas. This study contributes to a comprehensive understanding of the fracture evolution in the northeast Sichuan Basin and provides a theoretical basis for the exploration and development of shale oil and gas.

2. Geological Setting

Northeast Sichuan experienced the thrust and nappe of Micang Mountain and Daba Mountain towards the basin during the Mesozoic–Cenozoic period. Under this tectonic background, the Micang Mountain–Daba Mountain foreland basin was formed. The main body of the study area is located in the Northeast Sichuan foreland basin, which is part of a piedmont sag controlled by the Daba Mountain nappe belt and the Micang Mountain nappe belt. The western part of the sag is controlled by the northern section of the Longmenshan Fault, the southern part is blocked by the gentle structural belt in the central Sichuan Basin, and the southwestern part is connected with the western Sichuan depression (Figure 1a). Its structural trend is mainly northwest, with compressive and compressive reverse faults and a structural pattern of "two uplifts and one depression" in plan view [2].

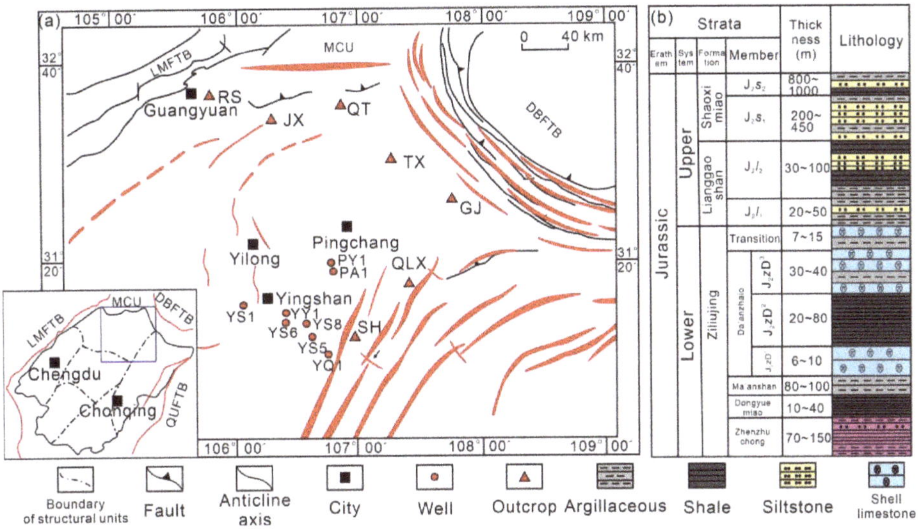

Figure 1. Location and stratigraphic histogram of the Northeast Sichuan Basin. (**a**) Regional location, LMFTB = Longmen Mountain fold-and-thrust belt, DBFTB = Daba Mountain fold-and-thrust belt, QUFTB = Qiyue Mountain fault–fold belt, MCU = Micang Mountain Uplift, NSD = North Sichuan Depression, CSGFB = Central Sichuan Gentle Structural Belt, ESHFB = Eastern Sichuan High and Steep Structural Belt, RS = Rongshan Profile, JX = Jinxi Profile, QT = Qiaoting Profile, TX = Tiexi Profile; GJ = Gujun Profile; QLX = Qilixia Profile; SH = Sanhui Profile; (**b**) Lianggaoshan Formation stratigraphic histogram.

In the Middle–Late Triassic period, with the closure of the Mianlue Ocean and the uplift of the Qinling orogenic belt, the marine sedimentary history ended in the main body of the Sichuan Basin, and the basin began to receive continental clastic deposition in the Late Triassic [36]. Among these deposits, the Jurassic system is widely distributed, and the strata are well developed. The system is a set of typical terrigenous clastic rock deposits dominated by delta-lake phases, with thicknesses of 1500 m–4000 m [37]. In the Northeast Sichuan Basin, the Jurassic system is in parallel unconformity with the underlying Triassic strata, gradually transitions to an angular unconformity towards the basin margin and is in unconformable contact with the overlying Lower Cretaceous strata; the Cretaceous and above strata in the region are severely eroded and less preserved. The Jurassic system contains the Ziliujing Formation (J_1z), the Lianggaoshan Formation (J_1l), the

Shaximiao Formation (J_2s), the Suining Formation (J_3s), and the Penglaizhen Formation (J_3p) sequentially from bottom to top. Herein, the Lianggaoshan Formation is mainly a set of fine-grained deposits with thicknesses of 50–150 m. According to lithological characteristics, it can be divided into an upper segment and a lower segment (Figure 1b). During the sedimentary period of the lower segment, the lake basin was in a wide-basin shallow-water oxidative environment, the depositional centre was located in the central–eastern part, and the lithology was mainly purple-red mudstone interbedded with grey-green siltstone; during the same period of the upper segment, the depositional centre gradually migrated to the northwest from bottom to top, developing multiple sets of organic-rich shale [2].

3. Samples and Experiments

3.1. Sample Source

We observed 645 sets of fractures (including fracture types, fracture strikes, fracture dip angles, fracture lengths, fracture densities, and cutting relationships between fractures) from seven large outcrops in the northeastern Sichuan Basin. The premise of our measurements and observations of outcrop fractures was to use the fresh surface of fractures to avoid natural weathering. For the fracture measurement results to accurately reflect the direction of multistage tectonic stress, the locations of the seven outcrops were scattered around the drillings (Figure 1a). We made similar observations about the fracture characteristics of seven drill cores in the northeastern Sichuan Basin, and the aperture and filling of fractures were also supplemented. The fracture strike parameters on the cores were derived from imaging logging. All experimental test samples were from cores.

3.2. Experimental Analyses

Microfracture characteristics were examined using a Leica polarizing microscope at the State Key Laboratory of Oil and Gas Reservoir Geology and Exploitation in China (SKL) at a temperature of 24 °C and a relative humidity (RH) of 60%.

With a THMS600 heating stage installed on a Leica microscope, fluid inclusions were analysed at the Sichuan Coalfield Geology Bureau, and the uncertainty of temperature detection of the THMS600 heating stage was ± 0.1 °C. During the temperature measurement process, the rate of temperature change in the THMS600 heating stage was controlled to within 0.1~5 °C/min, and the temperature when the inclusions were completely homogenized was observed and recorded. To ensure the reliability of the test results, the homogenization temperatures (Th) of each inclusion were tested three times, with each Th having absolute deviations of no more than 2 °C. The experimental test refers to the China oil and gas industry standards of Micro Temperature Measurement Methods of Fluid Inclusion in Sedimentary Basins (SY/T6010-2011).

The stable carbon and oxygen isotope experimental apparatus was the Delta V isotope ratio mass spectrometer (IRMS) from the Sichuan Coalfield Geology Bureau. The fracture fill sample material (approximately 0.1 mg) was placed into a reaction vessel to react with 100% orthophosphoric acid at 90 ± 1 °C in the Multi-Prep® system. The collected carbon dioxide was purified, and its $\delta^{13}C$ and $\delta^{18}O$ isotopic compositions were then determined by a gas isotope mass spectrometer, with precision exceeding ± 0.2‰. Note that the isotopic analysis followed the Chinese oil and gas industry standards of the Analysis Method for Carbon and Oxygen Isotopes in Organic Matter and Carbonate (SY/T5238—2019). The O and C isotope compositions were represented as standard δ-notation relative to SMOW and PDB, respectively.

X-ray diffraction (XRD) was used to quantitatively analyse the mineral composition at the Sichuan Coalfield Geology Bureau. The samples were all pulverized into a fine powder (200 mesh) and then analysed with an Ultima IV X-ray diffraction with Cu Kα radiation (40 kV, 40 mA) at a scanning speed of 2 °/min and a testing angle range from 0° to 90°. The experimental test referred to the China oil and gas industry standards of Quantitative Analysis of Total Contents of Clay Minerals and Common Nonclay Minerals in Sedimentary Rocks by X-ray Diffraction (SY/T6201-1996).

The acoustic emission experiment adopted an MTS815 rock mechanics tester at the SKL. Selecting experimental samples from nonfractured sections and developed nonbedded sections prevented the influence of nontectonic stress factors on the results. The samples were placed on a rigid testing machine for uniaxial loading, and the loading speed was 15 kN/min. At the same time, the SAEU2S full waveform channel analyser was used to collect and obtain the relevant parameters. The test standards were GB/T 23561.7-2009, GB/T 23561.8-2009, and GB/T 23561.9-2009.

4. Results

4.1. Developmental Properties of Tectonic Fractures

4.1.1. Fracture Types

The results of core drilling and field tests revealed that fractures in the study area are dominantly tectonic fractures. When regional or local structural stress exceeds the rock strength, tectonic fractures are formed [16]. Shear fractures, tensile fractures, and interlayer slip fractures were the most prevalent tectonic fractures in the study area, according to mechanical characteristics.

The shear fractures had a specific direction and regularity (Figures 2a and 3a,b). Meanwhile, the angles of fractures mostly intersected with the rock bedding at high degrees or perpendicular to the bedding, with obvious penetration (Figure 2b). For shear fractures with multiple series and conjugate intersections in the stratigraphic plane (Figure 2b), its acute bisector usually indicates the direction of maximum principal stress.

Figure 2. Characteristics of the tectonic fractures of the Lianggaoshan Formation indicated by outcrops: (**a**) vertical through-strata shear fractures in the Qiaoting Profile of the Lianggaoshan Formation; (**b**) conjugate shear fractures in the Jinxi Profile; (**c**) positive flower-shaped strike-slip faults, which control fracture development, in the Rongshan Profile of the Lianggaoshan Formation (Red solid line: fault; Yellow dotted line: fractures); (**d**) low-degree thrust faults, which control fracture development, in the Gujun Profile of the Lianggaoshan Formation (Red solid line: fault; Yellow dotted line: fractures); (**e**) high-degree thrust faults, which control fracture development, in the Gujun Profile of the Lianggaoshan Formation (Red solid line: fault; Dotted lines of different colors: fractures with different strikes).

Figure 3. Characteristics of tectonic fractures of the Lianggaoshan Formation, as indicated by the core: (**a**) high-degree shear fractures semifilled with calcite, well YS6, 1752.11 m; (**b**) high-degree unfilled shear fractures, well YS6, 1650.41 m; (**c**) tensile fractures semifilled with calcite, well YY1, 2009.76 m; (**d**) tensile fractures semifilled with calcite, well YS1, 2006.41 m; (**e**) interlayer slip fractures, with obvious scratches, well YS6, 1718.18 m; (**f**) interlayer slip fractures with smooth, mirror-like surfaces on the seam surface, well YS6, 1658.31 m.

Affected by tensile stress, the tensile fractures were nearly vertical, and the seam surface was uneven and terminated at the bedding plane, forming intraformational fractures (<100 cm) (Figures 2c and 3c,d). Additionally, low-degree and near-horizontal interlayer slip fractures, which are fractures parallel to layers formed under the action of shearing or sliding during tectonic extrusion, were widely formed in the shale. The surfaces of the fractures had obvious scratches, steps and other characteristics (Figure 3e,f), as well as weak or partial filling.

Tensile and shear fractures associated with larger faults in outcrops were visible near faults (Figure 2c–e). Their strikes were not unified, they intersected the section at a large angle or perpendicularly, and they were mostly filled with calcite or quartz. These fractures next to the fault were directly affected by local stress in the fault within a specific range and had nothing to do with the regional tectonic stress field [38,39].

4.1.2. Fracture Characterization

The measured results of seven fractures in outcrop showed that the shale fractures of the Lianggaoshan Formation had strong penetration ability, the lengths of fractures were mainly 0–200 cm, accounting for 69%, and the length of local fractures exceeded 500 cm (Figure 4a). The angles were mainly high-degree (>45°), accounting for 73% (Figure 4b), and the types of fractures were mostly high-degree shear fractures. The high-degree shear fractures in the study area had many cutting layers and were of large scale, which easily led to the loss of shale oil and gas, which was not conducive to its preservation and high yield.

The observation results of core fractures showed that the tectonic fractures in the study area had obvious developmental differences in the vertical and horizontal directions. Taking well PA1 as an example, the fracture density was vertically uneven, mainly concentrated in the upper segment, and mainly distributed in the thinner rocks (Figure 5). It was mainly related to the rock mineral composition in the vertical direction, and the quartz content generally had a good correspondence with the linear density of fractures.

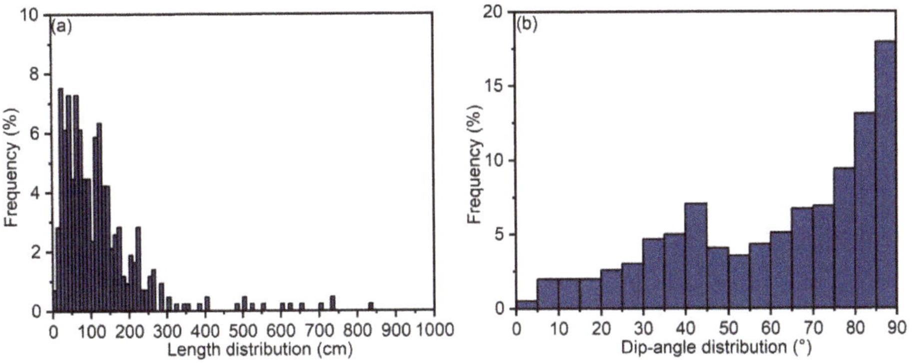

Figure 4. Outcrop-scale fracture parameter distribution.(**a**) Length distribution histogram; (**b**) Dip-angle distribution histogram.

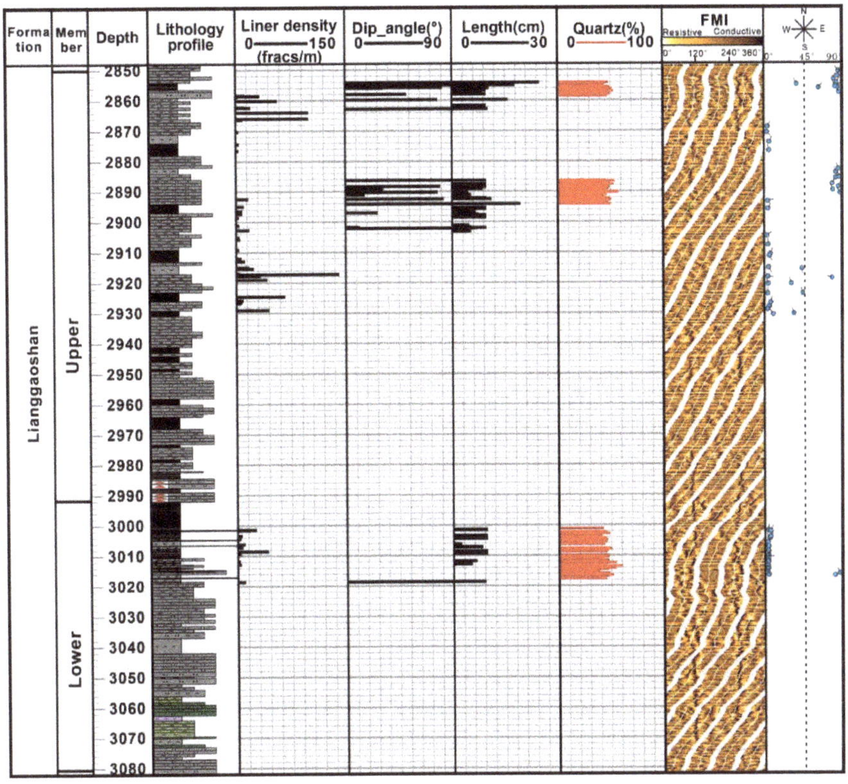

Figure 5. Longitudinal distribution characteristics of fractures (linear density, dip angle, length, and fracture strike) in the PA1 well.

The observation results of core-scale fractures in eight wells showed that the extension lengths of fractures were mainly in the range of 5 cm–15 cm, and there were a small number of fractures with a length of approximately 25 cm (Figure 6a). The angle of fractures was generally low, resulting in mainly interlayer slip fractures, which effectively reduced the penetration of fractures; these fractures were followed by high-degree fractures (Figure 6b). The aperture distribution of fractures was stable, with values mainly in the range of 0~2 mm.

A total of 85.53% of the filler was filled by calcite, asphalt, pyrite, argillite and quartz, and calcite was the main filler (Figure 6d). Fractures that were fully filled with brittle minerals were conducive to the preservation of shale gas. Moreover, in the subsequent fracturing process, the brittle minerals that were filled in the fractures increased the brittleness of the shale, making the fractures easy to reactivate and fracture, thereby improving the fracturing effect [40], and 14.5% of the unfilled fractures were beneficial to the artificial fractures in the later stage of communication.

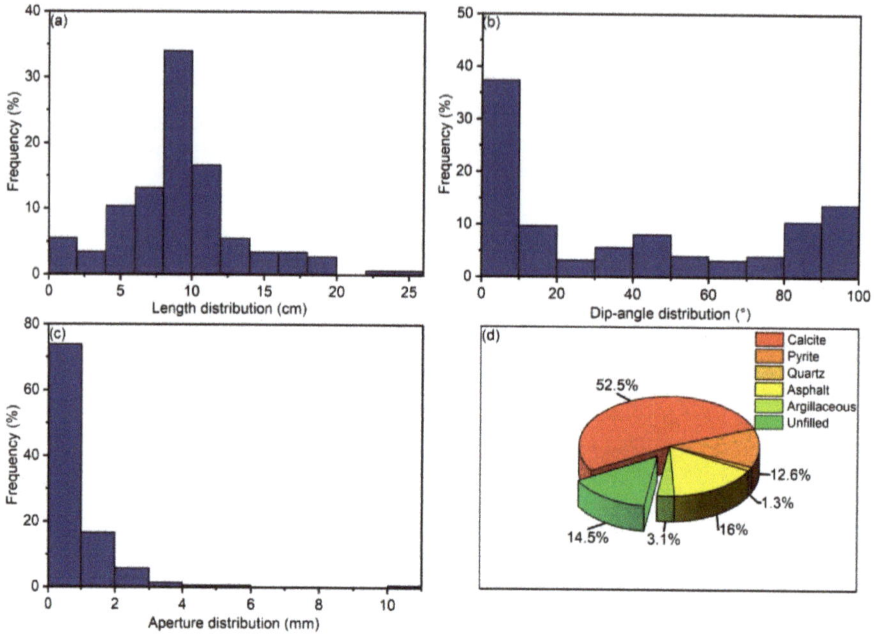

Figure 6. Drill core-scale fracture parameter distribution. (**a**) Length distribution frequency of fractures; (**b**) dip angle distribution of fractures; (**c**) aperture distribution of fractures; (**d**) filling and proportion of fractures.

4.2. Fracture Formation Stages

4.2.1. Fracture Intersection Relationship and Strike Analysis

Fractures are in contact with each other, and their cutting relationship and geometry can be used to determine the formation stage of each other [41]. According to the fracture cutting of the field outcrop observation point and its restricted relationship, at least three phases of fractures were present in the study area (Figure 7), and the existence of X-shaped conjugate shear fractures in outcrops also helps to better evaluate the direction of the ancient stress field [42] (Figure 7c). We calculated statistics on the occurrence of more than 645 sets of tectonic fractures measured in the field, and the results showed that the strikes of the dominant fractures on the rock bedding plane of the outcrops were as follows: (1) WNW- and NNW-trending fractures, (2) ENE- and NNE-trending fractures, and (3) NW-trending fractures (Figure 8a). The fractures in (1) and (2) were all high-degree fractures perpendicular to the rock bedding, and they were X-shaped conjugate shear fractures. Additionally, the outcrop fracture results showed that the NW-trending fractures cut the fractures of other lineages, showing obvious characteristics of late formation. The staging results of fractures and the acute-angle bisector of X-shaped conjugate shear fractures reflected the characteristics of maximum horizontal stress. Therefore, our preliminary analysis suggested that the formation of the tectonic fractures in the study area corresponded to the formation of two sets of conjugate shear fractures under two phases of structural stress and, finally, the formation of NW shear fractures and cutting of the previously formed fractures.

Figure 7. Characteristics of fracture development in Lianggaoshan Formation outcrops: (**a**) Tiexi Profile (The fractures represented by Roman number I were first formed, followed by II, and finally by III); (**b**) Jinxi Profile; (**c**) Qiaoting Profile; (**d**) Gujun Profile.

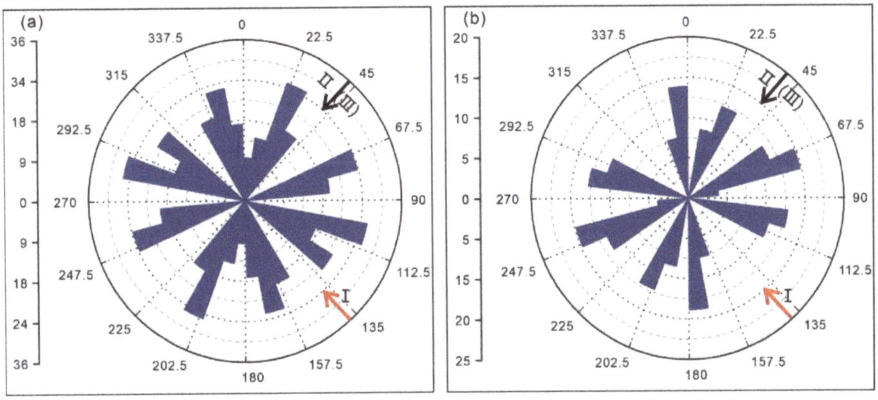

Figure 8. Fractures identified in the field (**a**) and by image logs (**b**) on rosette diagrams. (Roman numerals and black and red arrows represent the compression direction, they are inversed by conjugate shear fractures).

The mutual cutting relationship of different fractures in the core and section can also reveal the formation stage of the corresponding fractures. The core from the study area contained at least three mutually cutting fractures, and the fractures that formed in the early stage were all interrupted in the late stage (Figure 9a). Sliding scratches along the layer in two directions were visible on the adjacent core (Figure 9b). According to the results of fractures interpreted by imaging logs in the study area, Lianggaoshan Formation shale fractures mainly developed in four groups, among which the NNE, ENE, and NW fractures were the most developed, followed by nearly SN fractures (Figure 8b). Compared with those in outcrops, the natural lines present in cores were fewer, but the dominant lines were consistent, and the natural fracture system of the core indicated at least three phases. Additionally, the microfractures of the Lianggaoshan Formation shale were mainly shear fractures, and there were at least two phases of fracture cutting and interlacing. Therefore, based on the comprehensive analysis of core and section fractures by stage, we concluded that the shale of the Lianggaoshan Formation in the Northeast Sichuan Basin should have developed at least three stages of fractures.

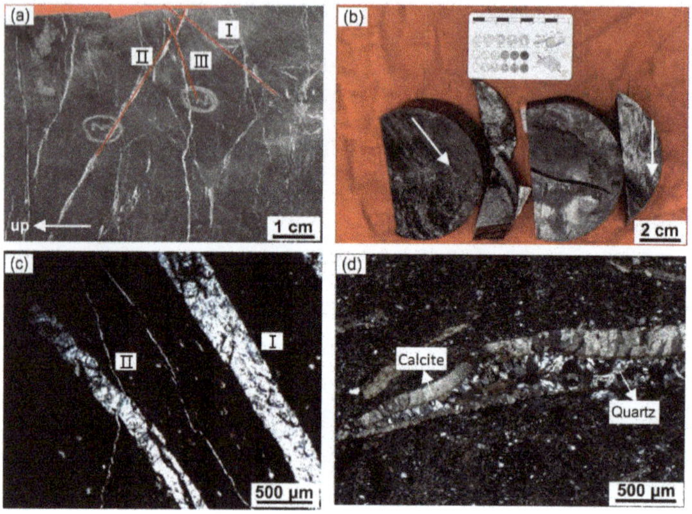

Figure 9. Cutting relationship of core- and section-scale fractures: (**a**) three phases of fractures cut each other, and the late fractures are broken by early fractures, well PY1, 3087.57 m (The fractures represented by Roman number I were first formed, followed by II, and finally by III); (**b**) bedding-slip fractures, with scratches in both directions, well YS6, 1660.57 m; (**c**) two phases of fractures cut each other, and the late cuts the early, well PY1, 3084.12 m (orientation of the thin sections: vertical; the fractures represented by Roman number I were first formed, followed by II); (**d**) two phases of fractures cut each other, and the early fractures are broken by late fractures, well YY1, 2002 m (orientation of the thin sections: vertical; the fractures represented by Roman number I were first formed, followed by II).

4.2.2. Fluid Inclusion Analysis

As the homogenization temperature of fluid inclusions in gas-liquid two-phase brine can usually represent the capture temperature of fluid inclusions, the homogenization temperature of fluid inclusions that formed was also measured, as the fracture filler can reveal the period of fracture formation. The samples of fluid inclusions that filled minerals in this study came from quartz and calcite veins that filled in tectonic fractures. Abundant fluid inclusions are visible in all samples, and the types were mainly hydrocarbon fluid inclusions and gas-liquid two-phase brine fluid inclusions. Hydrocarbon fluid inclusions were abundant in the samples, accounting for approximately 75% of the total fluid inclusions. The sizes of single-phase liquid hydrocarbon fluid inclusions were 4–22 μm; the shapes were ellipse, strip, triangle, etc. The fluid inclusions were mainly distributed in lines or bands along the healed mineral fractures, and they were blue and blue-white under fluorescence (UV) (Figure 10). The gas-liquid two-phase brine fluid inclusions were composed of gaseous hydrocarbons and brine solution, the overall development was less, and they accounted for approximately 25% of the total fluid inclusions. The sizes of fluid inclusions were 3~10 μm; the shapes were square, irregular, etc.; they were produced together with hydrocarbon fluid inclusions (Figure 10). We carried out microthermometry on the gas-liquid two-phase brine fluid inclusions in the quartz and calcite minerals in the veins of 10 sample fractures in the study area. The results showed that the homogenization temperatures of the associated two-phase brine fluid inclusions trapped in the quartz and calcite veins had three distribution intervals (Figure 11), i.e., 55–70 °C, 80–105 °C, and 110–140 °C, indicating that there were three phases of fluid activity and that the fractures were formed in three phases. In addition, the homogenization temperatures of inclusions from quartz cement were 5~15 °C lower than those from calcite.

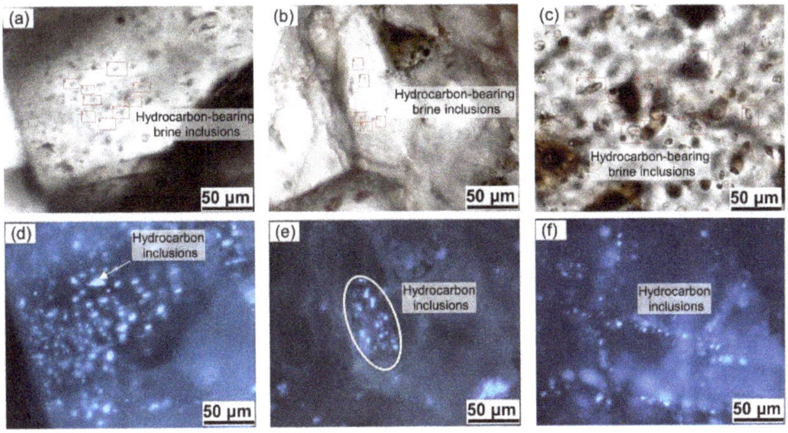

Figure 10. Micrographs of fluid inclusions captured by veins in the Lianggaoshan Formation in the Northeast Sichuan Basin: (**a**) hydrocarbon-bearing brine inclusions distributed in calcite, 1699 m, well YY1, single polarized light; (**b**) hydrocarbon-bearing brine inclusions distributed in calcite, 3077.57 m, well PY1, single polarized light; (**c**) hydrocarbon-containing brine fluid inclusions distributed in quartz, 3089.93 m, well PY1, single polarized light; (**d**) densely distributed hydrocarbon inclusions in calcite, 1699 m, well YY1, fluorescence; (**e**) densely distributed hydrocarbon inclusions in calcite, 3077.57 m, well PY1, fluorescence; (**f**) densely distributed hydrocarbon inclusions in quartz, 3089.93 m, well PY1, fluorescence.

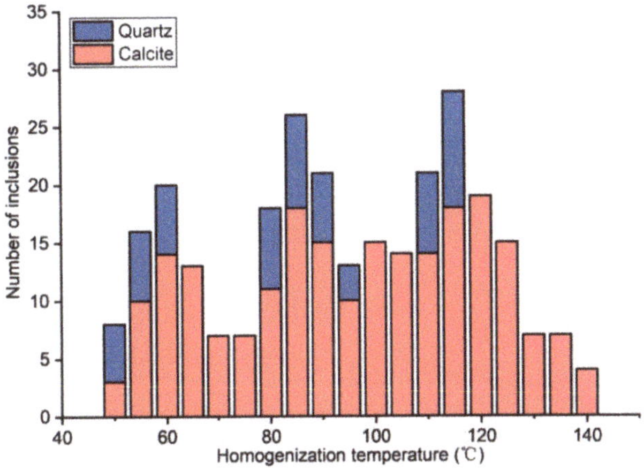

Figure 11. Homogenization temperature distribution of fluid inclusions captured by the tectonic fracture filler in the Lianggaoshan Formation in the Northeast Sichuan Basin.

4.2.3. Stable Carbon and Oxygen Isotope Analysis

Carbon and oxygen isotopes are significantly different in different subsurface environments. Through the determination of carbon and oxygen isotopes in the fracture filler, we calculated the formation stages of fractures [23]. We combined the fracture-filler stable isotope data from different well intervals in the Lianggaoshan Formation in the Northeast Sichuan Basin to analyse the fracture stages in the study area. The results of carbon and oxygen isotope measurements on 18 samples of tectonic fracture filler showed that the $\delta^{18}O$ (PDB‰) values were between −13.3‰ and −6.65‰, the δ13C (PDB‰) values were between −19‰ and 4.53‰, and the $\delta^{18}O$ (PDB‰) and $\delta^{13}C$ (PDB‰) values in the filling minerals were quite different, showing the characteristics of multistage filling (Figure 12). We used the oxygen isotope thermometry equation proposed by Fritz and Smith (1970) to calculate the temperature at which the fractures formed in each sample [43]:

$$T = 31.9 - 5.55(\delta^{18}O - \delta^{18}O_w) + 0.7(\delta^{18}O - \delta^{18}O_w)^2 \tag{1}$$

where T is the temperature when the filling minerals formed, in °C; $\delta^{18}O$ is the oxygen isotope value; $\delta^{18}O_w$ is the oxygen isotope value of the water medium when minerals formed; the Northeast Sichuan Basin was a lacustrine sedimentary system, which is a freshwater environment, and the oxygen isotope value of the water medium is generally −10 ‰. By considering the surface temperature and geothermal gradient, we estimated the ancient burial depth of the fractures and, combined with the stratigraphic burial evolution of the basin, further limited the formation time of the tectonic fractures in each period. We calculated the burial depth of the fractures based on the study area's perennial average surface temperature of 20 °C and a geothermal gradient of 2.2 °C/100 m. There are three partitions in the carbon-oxygen stable isotope distribution of the fracture filler (Figure 12). The $\delta^{18}O$ values of partition I range from −18.37 to −17.5 ‰, and the equivalent depths are 4223 to 4882 m, with an average of 4553 m; the $\delta^{18}O$ values of partition II range from −16.32 to −15.32 ‰, and the equivalent depths are 2784 to 3512 m, with an average of 3184 m; the $\delta^{18}O$ values of partition III range from −14.5 to −13.4 ‰, the equivalent depths are 1584 to 2320 m, and the average is 1883 m.

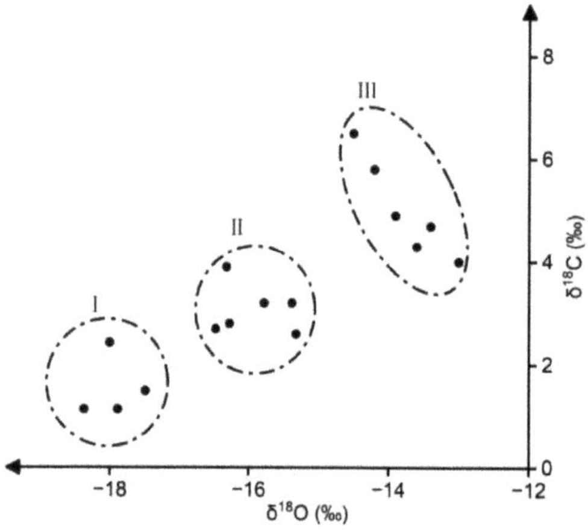

Figure 12. Stable carbon and oxygen isotopes of tectonic fracture filler in the Lianggaoshan Formation in the Northeast Sichuan Basin (I, II, and III represent the partitions of stable carbon and oxygen isotopes).

4.2.4. Acoustic Emission Analysis

The main principle of acoustic emission is that the rock has a memory of the ground stress, and in a specific geological period, when the loading stress exceeds the rock stress, an acoustic emission effect occurs. The inflection point at the cumulative acoustic emission curve suddenly increases to become the Kaiser point. The Kaiser point can show the damage and deformation of the rock and corresponds to the palaeostress stage [43]. We can use the number of Kaiser points on the acoustic emission curve to determine the number of tectonic movements and fracture development stages experienced by rocks [43]. According to the characteristic parameters, such as the number of acoustic emission rings, event rate, and energy intensity received by the acoustic emission detection system, we established the relationship between the loading time and the characteristic parameters of acoustic emission (Figure 13). The acoustic emission curve and calculation test result table show that the sample had four obvious Kaiser effect points, excluding the influence of in situ stress, which shows that the Lianggaoshan Formation experienced three major tectonic movements since its deposition (Figure 13). Therefore, the formation of shale fractures in the Lianggaoshan Formation in Northeast Sichuan mainly underwent three stages.

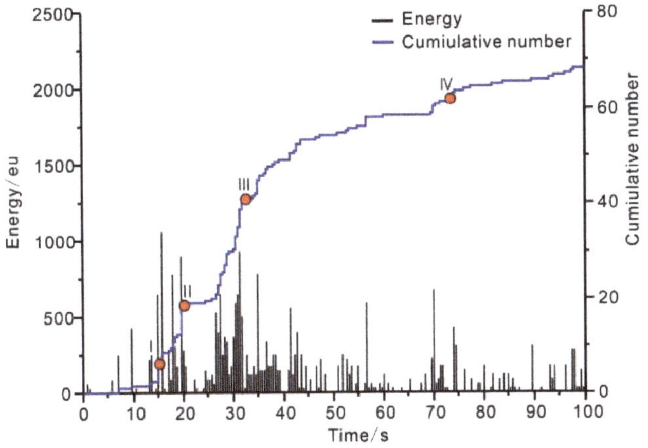

Figure 13. Rock acoustic emission events of the Lianggaoshan Formation in the Northeast Sichuan Basin (I, II, III, IV are the Kaiser points corresponding to the samples).

5. Discussion

5.1. Factors Controlling Fracture Development

5.1.1. Structural Factors

Different structural locations are the key factors leading to the differential development of tectonic fractures, and there are specific differences in the distribution of stress in different parts, which makes the development of tectonic fractures in different parts different. For the parts where the structural stress is concentrated, the scale and development of tectonic fractures are larger. The statistical results of strike-slip faults, inverse faults, and fracture density near the core of the anticline in the study area show that as the distance from the fracture and the core of the anticline increased, the fracture density gradually decreased, and when a specific distance was reached, the fracture density changed steadily (Figure 14). Strike-slip faults have a larger control over fracture size than thrust faults. The former had a more obvious control on the formation of tectonic fractures in Jurassic shale. In addition, the fracture density of the hanging wall of the thrust fault was much higher than that of the footwall; the average fracture density of the hanging wall was 11.7 fracs/m, and the average fracture density of the footwall was 6.4 fracs/m (Figure 14a). In addition, the study of microfractures in the Jurassic Ziliujing Formation showed that the microfractures in the fault–fold zone and the high–steep zone accounted for 40%–90% of the total pore volume, and the microfractures in the underdeveloped areas accounted for approximately 10% of the total pore volume [44].

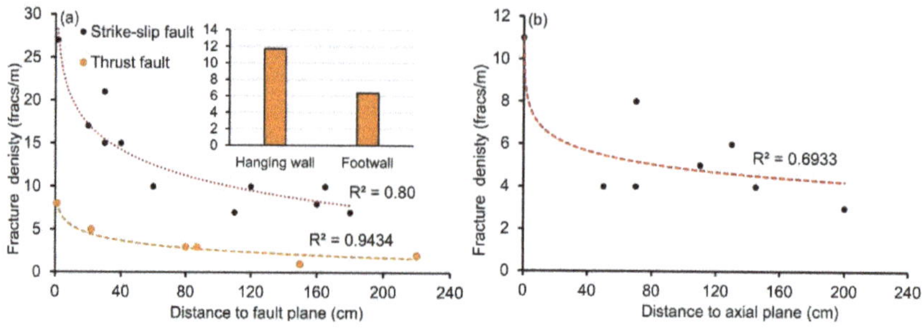

Figure 14. Fracture density controlled by different structural positions. (**a**) The relationship between fracture density and the distance from the fault plane, the histogram in the Figure 14 (**a**) is the comparison of fracture density between the hanging wall and the footwall of the reverse fault; (**b**) the relationship between fracture density and the distance from the axial plane.

5.1.2. Nonstructural Factors

The layered characteristics of shale were obvious because there were specific differences in rock mechanical properties between adjacent layers of layered shale. Tensile fractures are mainly developed in shale formations with relatively high brittleness, and the fractures are perpendicular to the bedding, have good equidistance in the same layer, and terminate on the bedding plane [44,45]. According to the statistical analysis of the densities of tensile fractures, shear fractures and bedding-slip fractures and formation thickness in the study area, it can be concluded that the formation thickness has a significant impact on the development of tensile fractures. Within a specific thickness range, the greater the thickness of the rock mechanical layer is, the lower the density of tensile and shear fractures (Figure 15a). However, bedding-slip fractures had no obvious relationship with formation thickness (Figure 15a). Bedding-slip fractures are caused by horizontal structural stress and formed on the weak plane between layers; therefore, the thickness of the mechanical layer has little effect on these fractures. The optimum shale formation thickness is conducive to the preservation of Jurassic shale oil and gas. However, if the shale formation thickness is too thick and fracture development is low, it is not favourable for the migration and preservation of oil and gas or hydraulic fracturing in the development stages. In contrast, if the shale formation thickness is too thin, fracture development results in unfavourable preservation conditions [46].

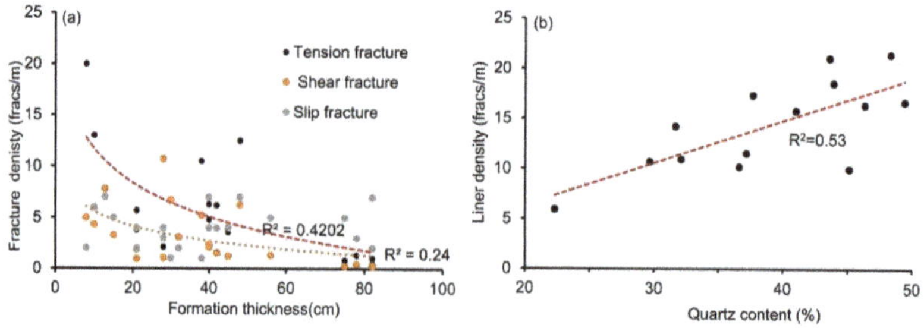

Figure 15. (**a**) Fracture densities under different formation thicknesses; (**b**) Fracture densities under different quartz contents.

The mineral composition of shale is different, and its rock mechanical properties and brittleness are different, which, in turn, affects the development of tectonic fractures in shale. For shale, the higher the content of brittle minerals is, the greater the brittleness [47]. The brittle minerals in the Lianggaoshan Formation shale reservoir were mainly clay and quartz, followed by clay, feldspar, and carbonate minerals (Table 1). Quartz is usually more important than carbonate for shale brittleness, and if the content of carbonate minerals is very low, the mechanical properties of rocks are mainly reflected by the quartz content [48]. Therefore, the effect of minerals on the fractures of the Lianggaoshan Formation mainly depends on the quartz content. The tectonic fracture statistics and mineral composition analysis of the core from the study area showed that the density of tectonic fractures in the Lianggaoshan Formation shale was positively correlated with the quartz content (Figure 15b). Under the same structural stress, shale with a higher degree of brittleness bears less strain before rupture, and so, it is more favourable for shale to rupture and form tectonic fractures [45]. Note that Dong et al. (2017) suggested that not all quartz can affect shale brittleness, and only authigenic quartz has a significant positive correlation with brittleness [49]. We did not effectively distinguished between authigenic quartz and terrigenous clastic quartz, and so, whether there are differences in the effects of different types of quartz on fractures in the Lianggaoshan Formation still needs further study.

Table 1. The mineral composition of the Lianggaoshan Formation in the northeastern Sichuan Basin (Above each line is the range of mineral content, and below is the average value of mineral content).

Well (Number)	Quartz (%)	Feldspar (%)	Calcite (%)	Dolomite (%)	Pyrite (%)	Clay (%)
PA1	22.4–60.1	7.2–28.8	0.4–13.7	1.1–3.2	0.9–2.2	12.2–52.1
(92)	44.3	12.2	4.9	1.8	1.3	34.2
YS5	28.9–40.7	2–12	0–21.9	0–2.8	0–3.6	36.3–58.1
(15)	33.6	6.2	10.2	0.7	2.1	47.2
YS1	28.4–57.7	2.1–21.3	0–8.7	/	0.7–1.8	37.9–62.5
(23)	36.1	9.5	3.5	/	1.2	48.1
YS5	24.6–58.2	0–2.7	0–23.8	0–2.6	/	29.3–48.6
(17)	42.5	1.2	17	0.7	/	35.3
YS8	25.8–62.3	0.2–8.7	2.1–27.6	0–2.7	0–1.3	26.8–52.1
(19)	47.5	4.6	12.3	0.3	0.3	32.2

5.2. Matching the Development Stages of Tectonic Fractures with the Tectonic Evolutionary History

When rocks were fractured due to structural stress, formation water entered the fractures, and crystalline minerals were separated from the fluid and precipitated between the fracture walls. Therefore, the fluid inclusion and isotope features of the fracture filler can be used to efficiently analyse the fracture phase [29]. Through the fracture cutting relationship of outcrops, cores, and sections, combined with fluid inclusions that had a uniform temperature as part of fill minerals, carbon and oxygen isotope analysis of the filler and acoustic emission stage experimental results matching the burial evolutionary history, we can determine the formation period of the fractures of the Lianggaoshan Formation in Northeast Sichuan. The first stage was the Late Yanshan–Early Himalayan tectonic movement (72~55 Ma), and the Late Cretaceous Xuefeng intracontinental orogeny system was part of the NW-trending extension deformation tectonic activity [50]. The study area was affected by far-field compressive stress and SE extrusion and structural stress, from a maximum depth to uplift and denudation; this activity formed the first phase of WNW-NNW-trending conjugate shear fractures, filler-captured fluid inclusions had normalized temperatures within 110~140 °C, and the depths of the Lianggaoshan Formation in this stage were 4223~4882 m (Figure 16). The second stage was Himalayan tectonic movement (48~32 Ma), and the extrusion from the NE that extended to the horizontal structure of the Micang–Daba Mountain thrust nappe zone resulted in local dextral shearing characteristics [51,52]. During this stage, the study area was continuously affected by the compressive stress from the NE to form ENE and NNE conjugate shear fractures, and the filler-captured fluid inclusion normalized temperatures were within 80~105 °C. The depths of the Lianggaoshan Formation in this stage were 2784~3512 m (Figure 16). The third stage was late

Himalayan tectonic movement (15 Ma~4 Ma). The study area continued to be affected by the continuous structural stress from NE of the Dabashan arcuate structural belt [51–53], leading to local fold deformation in the Lianggaoshan Formation and NW shear fractures approximately perpendicular to the principal stress direction at the local structural high, with filler-captured fluid inclusion normalized temperatures within 55~70 °C. The depths of the Lianggaoshan Formation in this stage were 1584~2320 m (Figure 16).

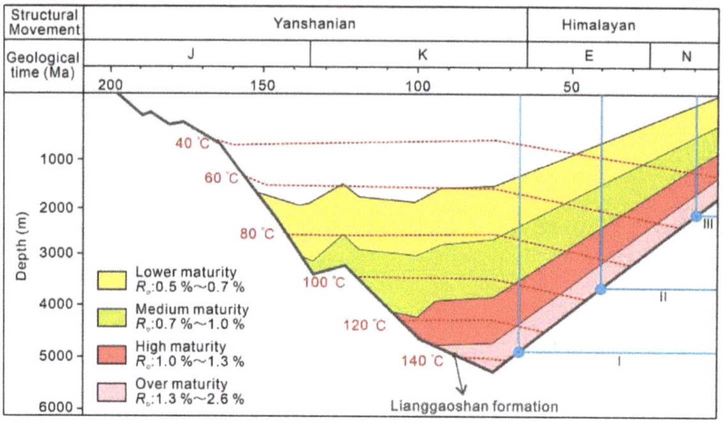

Figure 16. The outcrop burial history of the Lianggaoshan Formation in the Northeast Sichuan Basin (Red-dotted lines: Temperature of the strata; Blue lines: The horizontal and vertical blue lines determine the average formation time and buried depth of the three-stage fractures).

5.3. Effect of Fractures on Hydrocarbon Accumulation or Production

The fractures of the Lianggaoshan Formation have a positive impact on the enrichment and production of shale oil and gas. Figure 17a,b show that if the fracture was in a suitable range, the fracture density had a good positive correlation with the total hydrocarbon and light hydrocarbon contents of gas logging, and their correlations were 0.76 and 0.86, respectively. Obvious pale yellow fluorescence was visible in the areas with fractures, and oil film was also visible in the areas with fractures. These areas had better oil-bearing properties. In addition, Figure 18 shows that there was obvious crude oil enrichment at microfractures, and the oil and gas enrichment was much higher than the enrichment in shale pores. At the same time, they greatly improved the connectivity of shale. In terms of oil and gas production, the existence of fractures reduced the tensile strength of the rock itself [54]. Thus, the shale section was more likely to break during hydraulic fracturing. Therefore, there was a good negative correlation between fracture density and breakdown pressure (Figure 18).

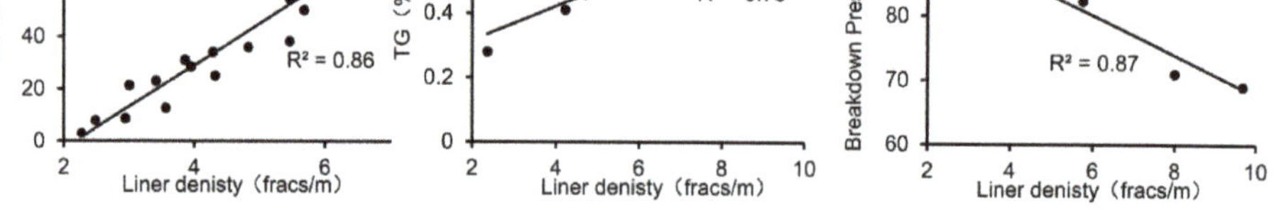

Figure 17. (**a**) Relationship between fracture density and light hydrocarbons; (**b**) relationship between fracture density and total hydrocarbons in gas logging; (**c**) relationship between fracture density and fracture pressure.

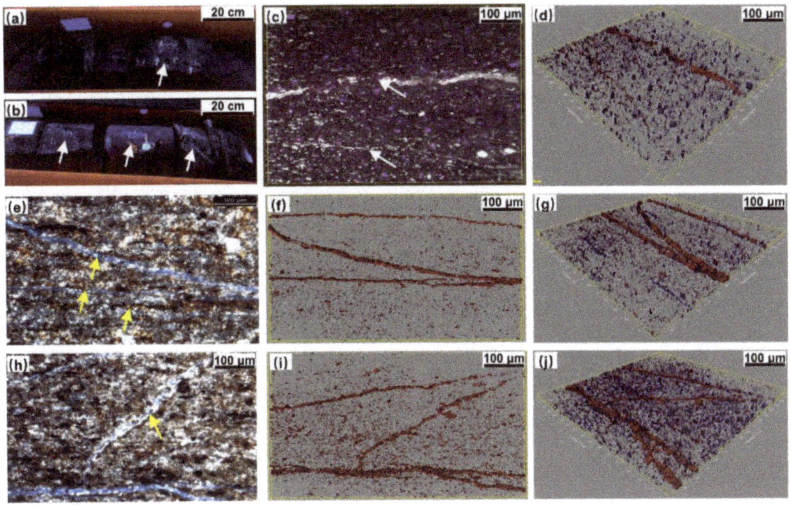

Figure 18. (**a**) Fracture surface after fluorescence irradiation, YQ1 well, 1420.70 m; (**b**) fracture surface after fluorescence irradiation, YQ1 well, 1421.10 m; (**c**) core slices containing fractures, Well PA1, 3002.9 m (orientation of the thin sections: bed-parallel); (**d**) three-dimensional distribution characteristics of oil and gas components in figure panel c; (**e**) thin core with fractures, YQ1 well, 1418.95 m (orientation of the thin sections: vertical); (**f**) distribution characteristics of oil and gas components in the plane of figure panel f; (**g**) three-dimensional oil and gas distribution characteristics of figure panel e; (**h**) thin core with fractures, well YQ1, 1421.65 m (orientation of the thin sections: vertical); (i) distribution characteristics of plane oil and gas components in figure panel h; (j) three-dimensional oil and gas distribution characteristics of figure panel h. (Red in the figures is the light hydrocarbon component, blue is the heavy hydrocarbon component, and grey is the particle skeleton).

5.4. Research Prospects

Our comprehensive geomechanics and geochemical research methods limit the formation period of Jurassic tectonic fractures and preliminarily evaluate the factors influencing tectonic fractures. This study provides theoretical guidance for subsequent fracture prediction and shale oil and gas enrichment process research. However, nontectonic fractures, such as hydrocarbon-generating overpressure fractures and hydraulic fractures, have not been further studied. It is not clear what role these fractures play in oil and gas enrichment and production. U–Pt dating can be used to more accurately determine the formation time of fractures and fluid activities and the relationship between them. Our fracture sampling was obtained from a limited number of drillings and outcrops in a specific area, and it was impossible to directly measure all fractures in the reservoir. Therefore, the distribution characterization of natural fractures was the focus of subsequent research, which is helpful for fracture prediction.

6. Conclusions

1. Three types of tectonic fractures were developed in the Lianggaoshan Formation in the northeastern Sichuan Basin: shear fractures, tensile fractures, and slip fractures. The tectonic fractures were mainly angular interlayer slip fractures. Fractures were filled by calcite, asphalt, pyrite, argillite, and quartz, and only 14.5% were not filled.
2. The fractures of the Lianggaoshan Formation were mainly affected by the structure, thickness, and mineral composition. The density of fracture growth rapidly decreases as the distance between the fault and the fold core increased, while the fracture density changed steadily over a particular distance. In addition, strike-slip faults had

a stronger effect on fractures than thrust faults. With an increase in layer thickness, the density of tensile fractures and shear fractures decreased, while that of interlayer slip fractures was not significantly impacted. Quartz was conducive to the formation of fractures, but the influence of different quartz types on fractures still needs further study.

3. The fractures of the Lianggaoshan Formation were generated in three stages during uplift: (1) Late Yanshan–Early Himalayan tectonic movement (72~55 Ma), wherein WNW-NNW conjugate shear fractures were generated, the normalized temperatures of filler-captured fluid inclusions were 110~140 °C, and the depths of the Lianggaoshan Formation in this stage were 4223~4882 m; (2) Middle Himalayan tectonic movement (48~32 Ma), wherein ENE-NNE conjugate shear fractures were generated, the normalized temperatures of filler-captured fluid inclusions were 80~105 °C, and the depths of the Lianggaoshan Formation in this stage were 2784~3512 m; (3) Late Himalayan tectonic movement (15 Ma~4 Ma), wherein a set of NW shear fractures were generated alone, the normalized temperature of filler-captured fluid inclusions were 55~70 °C, and the depths of the Lianggaoshan Formation in this stage were 1584~2320 m.
4. The existence of fractures greatly improved the oil and gas storage capacity and increased the contents of free and total hydrocarbons. At the same time, they also reduced the rock strength and further reduce the breakdown pressure of the strata.

Author Contributions: Conceptualization, X.B. and H.D.; writing—original draft preparation, Y.L., A.L. and S.L.; writing—review and editing, H.C., L.W. and F.C.; supervision, J.H.; resources, Z.W.; data curation, X.W.; investigation, Y.Z. All authors have read and agreed to the published version of the manuscript.

Funding: This research was supported by the National Natural Science Foundation of China (42072182): study on the mechanical mechanism of particle deformation and displacement alignment during shale page rationalization.

Data Availability Statement: Not applicable.

Conflicts of Interest: The authors declare no conflict of interest.

References

1. Hu, D.; Wei, Z.; Liu, R.; Wei, X.; Liu, Z.; Cheng, F. Major breakthrough of shale oil and gas in Well Taiye 1 in Bashansi Syncline in the Sichuan Basin and its significance. *China Pet. Explor.* **2021**, *26*, 21–32.
2. He, W.; He, H.; Wang, Y.; Cui, B.; Meng, Q.; Guo, X.; Bai, X.; Wang, Y. Major breakthrough and signifcance of shale oil of the Jurassic Lianggaoshan Formation in Well Ping'an 1 in northeastern Sichuan Basin. *China Pet. Explor.* **2022**, *27*, 40–49.
3. Ding, W.; Li, C.; Li, C.; Xu, C.; Jiu, K.; Zeng, W. Dominant factor of fracture development in shale and its relationship to gas accumulation. *Earth Sci. Front.* **2012**, *19*, 212–220.
4. Gale, J.F.W.; Reed, R.M.; Holder, J. Natural Fractures in the Barnett Shale and their Importance for Hydraulic Fracture Treatments. *AAPG Bull.* **2007**, *91*, 603–622. [CrossRef]
5. Bowker, K.A. Barnett Shale Gas Production, Fort Worth Basin: Issues and Discussion. *AAPG Bull.* **2007**, *91*, 523–533. [CrossRef]
6. Engelder, T.; Lash, G.G.; Uzcátegui, R.S. Joint Sets that Enhance Production from Middle and Upper Devonian Gas Shales of the Appalachian Basin. *AAPG Bull.* **2009**, *93*, 857–889. [CrossRef]
7. Gale, J.F.W.; Laubach, S.E.; Olson, J.E.; Eichhuble, P.; Fall, A. Natural Fractures in Shale: A Review and New Observations. *AAPG Bull.* **2014**, *98*, 2165–2216. [CrossRef]
8. Wang, R.; Ding, W.; Zhang, Y.; Wang, Z.; Wang, X.; He, J.; Zeng, W.; Dai, P. Analysis of Developmental Characteristics and Dominant Factors of Fractures in Lower Cambrian Marine Shale Reservoirs: A Case Study of Niutitang Formation in Cen'Gong Block, Southern China. *J. Petrol. Sci. Eng.* **2016**, *138*, 31–49. [CrossRef]
9. Wang, X.; Ding, W.; Cui, L.; Wang, R.; He, J.; Li, A.; Gu, Y.; Liu, J.; Xiao, Z.; Fu, F. The Developmental Characteristics of Natural Fractures and their Significance for Reservoirs in the Cambrian Niutitang Marine Shale of the Sangzhi Block, Southern China. *J. Petrol. Sci. Eng.* **2018**, *165*, 831–841. [CrossRef]
10. Zhang, X.; Wang, R.; Shi, W.; Hu, Q.; Xu, X.; Shu, Z.; Yang, Y.; Feng, Q. Structure- and Lithofacies-Controlled Natural Fracture Developments in Shale: Implications for Shale Gas Accumulation in the Wufeng-Longmaxi Formations, Fuling Field, Sichuan Basin, China. *Geoenergy Sci. Eng.* **2023**, *223*, 211572. [CrossRef]

11. Mcginnis, R.N.; Ferrill, D.A.; Morris, A.P.; Smart, K.J. Mechanical stratigraphic controls on natural fracture spacing and penetration. *J. Struct. Geol.* **2017**, *95*, 160–170. [CrossRef]
12. Zhao, G.; Jin, Z.; Ding, W.; Liu, G.; Yun, J.; Wang, R.; Wang, G. Developmental Characteristics and Formational Stages of Natural Fractures in the Wufeng—Longmaxi Formation in the Sangzhi Block, Hunan Province, China: Insights from Fracture Cements and Fluid Inclusions Studies. *J. Petrol. Sci. Eng.* **2022**, *208*, 109407. [CrossRef]
13. Fu, X.; Gong, L.; Su, X.; Liu, B.; Gao, S.; Yang, J.; Qom, X. Characteristics and Controlling Factors of Natural Fractures in Continental Tight-Oil Shale Reservoir. *Minerals* **2022**, *12*, 1616. [CrossRef]
14. Vega, B.; Kovscek, A.R. Fractal Characterization of Multimodal, Multiscale Images of Shale Rock Fracture Networks. *Energies* **2022**, *15*, 1012. [CrossRef]
15. Wu, W.; Yang, S.; Aguilera, R.; Chen, Z.; Aguilera, R.F. Characterization methods for natural fractures distribution in shale and tight reservoirs. *Int. J. Coal Geol.* **2023**, *273*, 104252. [CrossRef]
16. Ding, W.; Zhu, D.; Cai, J.; Gong, M.; Chen, F. Analysis of the Developmental Characteristics and Major Regulating Factors of Fractures in Marine–Continental Transitional Shale-Gas Reservoirs: A Case Study of the Carboniferous–Permian Strata in the Southeastern Ordos Basin, Central China. *Mar. Petrol. Geol.* **2013**, *45*, 121–133. [CrossRef]
17. Hu, X.; Deng, H.; He, J.; Shen, Z.; Peng, X. Complex genesis of multiperiod fractures in the Middle Triassic Leikoupo Formation in the Pengzhou gas field, western Sichuan Basin. *China Petrol. Sci.* **2023**, *20*, 161–176. [CrossRef]
18. Ju, W.; Wang, J.; Fang, H.; Sun, W. Paleotectonic stress field modeling and prediction of natural fractures in the Lower Silurian Longmaxi shale reservoirs, Nanchuan region, South China. *Mar. Petrol. Geol.* **2019**, *100*, 20–30. [CrossRef]
19. Wang, L.; Chen, W.; Tan, X.; Tan, X.; Yang, J.; Yang, D.; Zhang, X. Numerical investigation on the stability of deforming fractured rocks using discrete fracture networks: A case study of underground excavation. *Bull. Eng. Geol. Environ.* **2020**, *79*, 133–151. [CrossRef]
20. Wang, L.; Golfier, F.; Tinet, A.; Chen, W.; Vuik, C. An efficient adaptive implicit scheme with equivalent continuum approach for two-phase flow in fractured vuggy porous media. *Adv. Water Resour.* **2022**, *163*, 104186. [CrossRef]
21. Kolawole, O.; Ispas, I. How Hydraulic Fractures Interact with Natural Fractures: A Review and New Observations. In Proceedings of the 53rd U.S. Rock Mechanics/Geomechanics Symposium, New York, NY, USA, 23–26 June 2019.
22. Sui, Q.; Chen, W.; Wang, L. Investigation of Nonlinear Flow in Discrete Fracture Networks Using an Improved Hydro-Mechanical Coupling Model. *Appl. Sci.* **2022**, *12*, 3027. [CrossRef]
23. Jiang, L.; Wang, Q.; Wang, X.; Jiang, C.; Zhang, L.; Xue, Z.; Chu, Y. Joint development and paleostress field in Mesozoic strata of the southeastern Ordos Basin. *Acta Petrologica Sin.* **2013**, *29*, 1774–1790.
24. Ghosh, S.; Galvis-Portilla, H.A.; Klockow, C.M.; Slatt, R.M. An Application of Outcrop Analogues to Understanding the Origin and Abundance of Natural Fractures in the Woodford Shale. *J. Petrol. Sci. Eng.* **2018**, *164*, 623–639. [CrossRef]
25. Luo, T.; Guo, X.; Shu, Z.; Bao, H.; He, S.; Qin, Z.; Xiao, Z. Fluid source and formation time of fracture veins of Wufeng Formation and Longmaxi Formation in the south of Jiaoshiba area, Sichuan Basin. *Acta. Pet. Sin.* **2021**, *42*, 611–622.
26. Wang, M.; Chen, Y.; Bain, W.M.; Song, G.; Liu, K.; Zhou, Z.; Steele-Macinnis, M. Direct Evidence for Fluid Overpressure During Hydrocarbon Generation and Expulsion from Organic-Rich Shales. *Geology* **2020**, *48*, 374–378. [CrossRef]
27. Chen, S.; Liu, J. Research progress and prospects of the stages, genesis and fluid evolution of Microfracture veins in petroliferous basins. *Bull. Geol. Sci. Technol.* **2021**, *40*, 81–92.
28. Guo, J.; Ren, J.; Wan, S.; Chen, C.; Lai, J.; Gou, B. Comprehensive study of fracture flow characteristic feasibility of hybrid volume stimulation technique in tight fractured carbonate gas reservoir. *J. Petrol. Sci. Eng.* **2019**, *174*, 362–373. [CrossRef]
29. Kong, X.; Zeng, J.; Tan, X.; Ding, K.; Luo, Q.; Wang, Q.; Wen, M.; Wang, X.; Wang, M. Natural Tectonic Fractures and their Formation Stages in Tight Reservoirs of Permian Lucaogou Formation, Jimsar Sag, Southern Junggar Basin, NW China. *Mar. Petrol. Geol.* **2021**, *133*, 105269. [CrossRef]
30. Fan, C.; Li, H.; Zhao, S.; Qin, Q.; Fan, Y.; Wu, J.; Zhang, J. Formation Stages and Evolution Patterns of Tectonic fractures in Marine Shale: Case Study of the Lower Silurian Longmaxi Formation in the Changning Area of the Southern Sichuan Basin, China. *Energy Fuels* **2020**, *34*, 9524–9539. [CrossRef]
31. Cheng, G.; Jiang, B.; Li, M.; Li, F.; Zhu, M. Structural Evolution of Southern Sichuan Basin (South China) and its Control Effects on Tectonic Fracture Distribution in Longmaxi Shale. *J. Struct. Geol.* **2021**, *153*, 104465. [CrossRef]
32. Li, J.; Li, H.; Xu, J.; Wu, Y.; Gao, Z. Effects of Fracture Formation Stage on Shale Gas Preservation Conditions and Enrichment in Complex Structural Areas in the Southern Sichuan Basin, China. *Front. Earth Sci.* **2022**, *10*, 1–17. [CrossRef]
33. Weil, A.B.; Yonkee, W.A. Layer-parallel shortening across the Sevier fold thrust belt and Laramide foreland of Wyoming: Spatial and temporal evolution of a complex geodynamic system. *Earth Planet Sci. Lett.* **2012**, *357*, 405–420. [CrossRef]
34. Deng, H.; Liu, Y.; Peng, X.; Liu, Y.; Li, A.H. A new index used to characterize the near wellbore fracture network in naturally fractured gas reservoirs. *J. Nat. Gas Sci. Eng.* **2018**, *55*, 52–63. [CrossRef]
35. Aghli, G.; Moussavi-Harami, R.; Tokhmechi, B. Integration of sonic and resistivity conventional logs for identification of fracture parameters in the carbonate reservoirs (A case study, Carbonate Asmari Formation, Zagros Basin, SW Iran). *J. Petrol. Sci. Eng.* **2020**, *186*, 106728. [CrossRef]
36. Cheng, D.; Zhang, Z.; Hong, H.; Zhang, S.; Qin, C.; Yuan, X.; Zhang, B.; Zhou, C.; Deng, Q. Sequence structure, sedimentary evolution and their controlling factors of the Jurassic Lianggaoshan Formation in the East Sichuan Basin, SW China. *Pet. Explor. Dev.* **2023**, *50*, 262–272. [CrossRef]

37. Guo, X.; Hu, D.; Li, Y.; Wei, X.; Liu, R.; Liu, Z.; Yan, J.; Wang, Q. Analyses and thoughts on accumulation mechanisms of marine and lacustrine shale gas:Acase study in shales of Longmaxi Formation and Daanzhai Section of Ziliujing Formation in Sichuan Basin. *Earth Sci. Front.* **2016**, *23*, 18–28.
38. Sun, X.; Hou, G.; Yu, X.; Wei, H.; Mo, T.; Zhou, L.; Xie, Y.; Luo, H. Model for the Fracture Development of the Sand with Low-permeability in the Foreland Faulted Fold Belt. *Geotecton. Metallog.* **2015**, *39*, 808–815.
39. Zhu, X.; Zhang, Q.; Hou, G. The Controlling Factor Study of Development of Carbonatic Rock Fractures in the Kelpin-Bachu Area, Xinjiang and Its Petroleum Exploration significance. *Acta Geol. Sin.* **2017**, *91*, 1181–1191.
40. Zeng, W.; Zhang, J.; Ding, W.; Zhao, S.; Zhang, Y.; Liu, Z.; Jiu, K. Fracture Development in Paleozoic Shale of Chongqing Area (South China). Part One: Fracture Characteristics and Comparative Analysis of Main Controlling Factors. *J. Asian Earth Sci.* **2013**, *75*, 251–266. [CrossRef]
41. Hancock, P.L.; Leitch, E.C.; Shepherd, J. Brittle microtectonics: Principles and practice. *J. Struct. Geol.* **1985**, *7*, 437–457. [CrossRef]
42. Li, Y.; Deng, H.; He, J.; Ma, R. Reconstruction of paleo stress field after the depositional period of Leikoupo Formation in Pengzhoua rea, western Sichuan and its oil and gas geological significance. *Acta. Geo. Sin.* **2022**, *96*, 1–7.
43. Fritz, P.; Smith, D.G.W. The Isotopic Composition of Secondary Dolomites. *Geochim. Cosmochim. Acta* **1970**, *34*, 1161–1173. [CrossRef]
44. Liu, Z.; Qiu, H.; Jiang, Z.; Liu, R.; Wei, X.; Chen, F.; Wei, F.; Wang, D.; Su, Z.; Yang, Z. Types and Quantitative Characterization of Microfractures in the Continental Shale of the Da'anzhai Member of the Ziliujing Formation in Northeast Sichuan, China. *Minerals* **2021**, *11*, 870. [CrossRef]
45. Zeng, L.; Ma, S.; Tian, H.; Xue, M.; Liu, G.; Lv, W. Research progress of natural fractures in organic rich shale. *Earth Sci.* **2022**, *48*, 1–15.
46. Wang, X.; Wang, R.; Ding, W.; Yin, S.; Sun, Y.; Zhou, X.; Li, Q. Development Characteristics and Dominant Factors of Fractures and their Significance for Shale Reservoirs: A Case Study from Є1b2 in the Cen'gong Block, Southern China. *J. Petrol. Sci. Eng.* **2017**, *159*, 988–999. [CrossRef]
47. He, J.; Li, Y.; Deng, H.; Tang, J.; Wang, Y. Quantitative evaluation and influencing factors analysis of the brittleness of deep shale reservoir based on multiple rock mechanics experiments. *Nat. Gas Geosci.* **2022**, *33*, 1103–1116.
48. Yang, S.; Harris, N.; Dong, T.; Wu, W.; Chen, Z. Natural Fractures and Mechanical Properties in a Horn River Shale Core from Well Logs and Hardness Measurements. *SPE Res. Eval. Eng.* **2018**, *21*, 671–682. [CrossRef]
49. Dong, T.; Harris, N.B.; Ayranci, K.; Yang, S. The impact of rock composition on geomechanical properties of a shale formation: Middle and Upper Devonian Horn River Group shale, Northeast British Columbia, Canada. *AAPG Bull.* **2017**, *101*, 177–204. [CrossRef]
50. Ding, Y.; Shao, Z. An Experimental Research into Determination of Highest Paleotectonic Stress State Experienced by Rock Through Geological Ages. *Earth Sci.* **2001**, *26*, 99–104.
51. Mei, L.; Liu, Z.; Tang, J.; Shen, C.; Fan, Y. Mesozoic Intra-Continental Progressive Deformation in Western Hunan-Hubei Eastern Sichuan Provinces of China: Evidence from Apatite Fission Track and Balanced Cross-Section. *Earth Sci.* **2010**, *35*, 161–174.
52. Tang, Y.; Mei, L.; Xiao, A.; Chen, Y.; Luo, K. Transition of tectonic stress field and hydrocarbon significance of the Late Mesozoic-Cenozoicin Xuanhan-Daxian region, northeastern Sichuan Basin. *Acta Pet. Sin.* **2013**, *34*, 59–70.
53. Fan, C.; Qin, Q.; Li, H.; Han, Y.; Xing, J. Formation stages of tectonic fractures of Xujiahe Formation in the fault-fold belt of central Yuanba area, Sichuan Basin. *Acta Pet. Sin.* **2017**, *38*, 1135–1143.
54. Liu, Y.; Lai, F.; Wang, R.; Wu, Z.; Zhang, X.; Xu, H.; Li, J. Influencing Factors of the Brittleness of Continental Shales Containing Shell Limestone Interlayer. *Minerals* **2023**, *13*, 460. [CrossRef]

minerals

MDPI

Article

Characteristics and Formation Stages of Natural Fractures in the Fengcheng Formation of the Mahu Sag, China: Insights from Stable Carbon and Oxygen Isotope and Fluid Inclusion Analysis

Wei Wang [1], Xinhui Xie [2,3,4,*], Caiguang Liu [1], Feng Cao [4], Guoqing Zheng [1], Zhenlin Wang [1], Gang Chen [1], Hucheng Deng [4], Jianhua He [4] and Kesai Li [4]

1 PetroChina Xinjiang Oilfield Company, Karamay 834000, China; wangw_xj@petrochina.com.cn (W.W.); liucaiguang@petrochina.com.cn (C.L.); zgqxjyt@163.com (G.Z.); wzhenl@petrochina.com.cn (Z.W.); chenggang66@petrochina.com.cn (G.C.)

2 Geophysics Post-Doctoral Station, Chengdu University of Technology, Chengdu 610059, China

3 College of Geophysics, Chengdu University of Technology, Chengdu 610059, China

4 College of Energy Resources, Chengdu University of Technology, Chengdu 610059, China; cf773992278@163.com (F.C.); denghucheng@cdut.cn (H.D.); hejianhua19@cdut.edu.cn (J.H.); l.k.s.2016@foxmail.com (K.L.)

* Correspondence: xxh0622@hotmail.com

Abstract: The Fengcheng Formation of the Mahu Sag is an unconventional reservoir that is of paramount importance for exploration and development of hydrocarbon resource. However, current research on natural fractures in the Fengcheng Formation remains limited, posing challenges for exploration of hydrocarbon resource in the region. This study is based on core observations, thin section identification, geochemical testing and the evolution of regional tectonic movements to investigate the characteristics and periods of formation of natural fractures to address this gap. According to the characteristics of natural fractures in the drilling core samples and microsections, the natural fractures in the Fengcheng Formation can be grouped into structural fractures and atectonic fractures. Structural fractures can be further divided into three subtypes: high-angle interlayer shear fractures, along-layer shear fractures, and tensile fractures. Additionally, non-tectonic fractures in this studied area are primarily bedding fractures, hydraulic fractures, and hydrocarbon-generating overpressure fractures. Vertically, fracture development is more prominent at the bottom of Feng #2 Formation and at the top of Feng #3 Formation. Results also indicate that natural fractures primarily formed during three distinct tectonic movement periods. The initial stage of fracture evolution pertains to the Late Permian period (243–266 Ma), filled with fibrous calcite, and exhibiting a uniform temperature of 70–100 °C. The second stage of fracture evolution occurred during the Late Indosinian to Early Yanshanian period (181–208 Ma), mostly filled or semi-filled with calcite, with a uniform temperature of 110–130 °C. The third stage of fracture development happened during the late Yanshanian to early Himalaya period (50–87 Ma), predominantly filled with calcite, and presenting a uniform temperature of 130–150 °C. Among the various types of structural fractures, the density of high-angle interlayer shear fractures demonstrates a positive correlation with daily gas production, indicating their vital role in promoting hydrocarbon resource production and transportation. Furthermore, microfractures generated by hydrocarbon-generating overpressure fractures exhibit small pore sizes and strong connectivity. These microfractures can create an effective permeability system by connecting previously isolated micropores in shale reservoirs, thus establishing interconnected pore spaces in the shale formation.

Keywords: natural fractures; fracture characteristics; formation stages of fractures; Fengcheng Formation; Mahu Sag

1. Introduction

The Junggar Basin in western China is an extensive and intricate sedimentary basin that has captivated the attention of geologists and petroleum exploration companies for numerous years. The Junggar Basin is renowned for its plentiful oil and gas reserves, which are predominantly concentrated in the Permian Fengcheng Formation situated within the Mahu Sag [1–3]. Prior studies propose that the oil and gas reserves in the Fengcheng Formation of the Mahu Sag could potentially reach up to 100 billion tons [4,5]. Natural fractures are crucial in the development of effective unconventional oil and gas reservoirs, as they function as storage spaces and permeable pathways for the flow of oil and gas [6,7]. In shale reservoirs, fractures exert a dual influence on the storage of oil and gas. Fractures, being weak structural planes, facilitate the formation of complex fracture networks during hydraulic fracturing [8]. Furthermore, fractures with high effectiveness can negatively impact storage space, potentially resulting in oil and gas leakage [9]. Therefore, it is essential to investigate the basic characteristics and formation periods of natural fractures for shale oil and gas exploration and development. This study can provide an accurate geological foundation for subsequent oil and gas reservoir deployment, development plan formulation, and optimization [10–12].

The Mahu Sag of the Junggar Basin is a promising exploration area for organic-rich shale formations, particularly the Fengcheng Formation [3,13]. The Fengcheng Formation has experienced multiple tectonic movements, which makes it challenging to determine the periods of fracture formation based solely on outcrop and core observations [4,14–16]. To date, the fluid inclusion test and stable carbon and oxygen isotope analysis of filling materials in the fractures can be regarded as effective and feasible methods for determining the evolution of fracture formation [17–19]. The fractures in the Fengcheng Formation are subjected to various stresses, including regional tectonic compression and shear slip. These stresses result in the development of complex fracture systems. Therefore, it is essential to examine the characteristics and formation period evolution of natural fractures of different origins in the studied area to clarify the favorable exploration area for shale oil and gas [10,15,17].

This study aims to examine the characteristics and formation stages of natural fractures in the Fengcheng Formation of the Mahu Sag using a diverse range of methods, including drill core analysis, thin section petrography, image logging, fluid inclusion testing, and carbon and oxygen isotope analysis. Initially, natural fractures are classified into two main types and six subtypes based on their mechanical origin. Moreover, the evolution of natural fractures throughout the formation period is analyzed by scrutinizing the crosscutting relationships among different fracture groups and interpreting geochemical findings. Subsequently, we also discuss the influence of various fracture types on oil productivity. By elucidating the interrelated dynamics between multistage fractures, this research can provide a geological foundation for shale reservoir exploration in the Mahu Sag.

2. Geological Setting

Figure 1 shows the structural zone and well locations of the Mahu Sag in the Junggar Basin. In this study, shale samples obtained from eleven wells in the Mahu Sag are used to survey the natural fracture characteristics. As one of the six primary hydrocarbon-rich depressions in the Junggar Basin, it occupies an exploration area of approximately 1×10^4 km^2 [20–22]. The Mahu Sag is a foreland basin formed by compression between the Kazakhstan Plate and the West Junggar Ocean. This region is distinguished by a significant depression of hydrocarbon-rich resources.

The northern region of the Mahu Sag has experienced multiple tectonic events, including Hercynian, Indosinian, Yanshanian, and Himalayan movements. These Yanshanian and Himalayan events led to the region's overall elevation, yielding a relatively comprehensive stratigraphy. The stratigraphy of the Mahu Sag, from the top to the bottom, comprises Quaternary, Neogene, Paleogene, Cretaceous, Jurassic, Triassic, Permian, and Carboniferous [23]. The Fengcheng Formation in the Permian is the focus in this study, which constitutes a crucial source rock. The Fengcheng Formation composed of mudstone, marl, siltstone, mudstone sandstone, and mudstone fine sandstone, which can be classified into three sections based on sedimentary and lithofacies characteristics: Fengcheng #3 Formation (P_1f_3), Fengcheng #2 Formation (P_1f_2), and Fengcheng #1 Formation (P_1f_1). On the plane, a northwest-thick and southeast-thin trend is discernible, leading to a thinning trend from northwest to southeast in each section of the Fengcheng Formation.

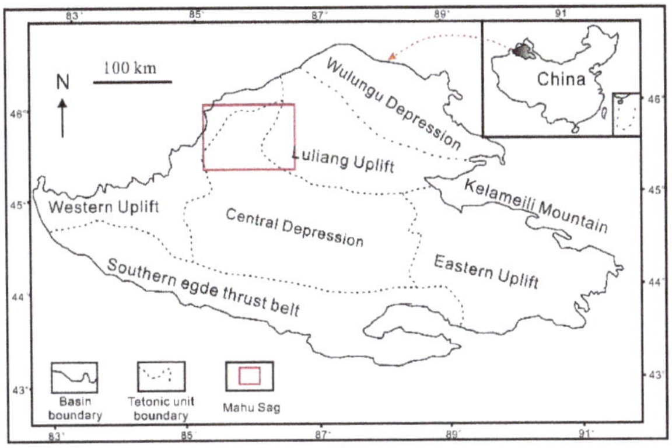

(a)

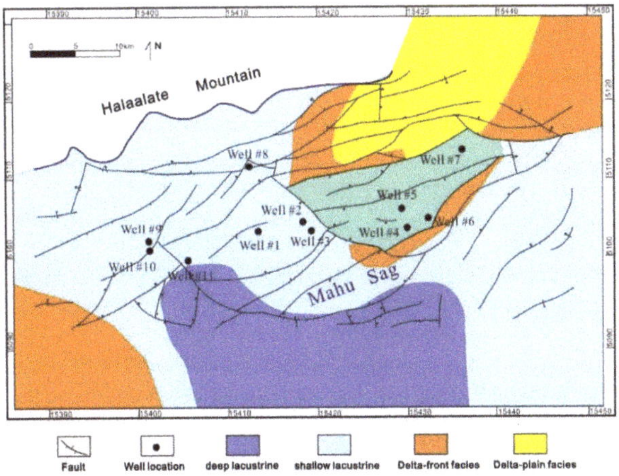

(b)

Figure 1. (**a**) The relative location of the studied areas in Junggar Basin, and (**b**) well locations of shale samples in the Mahu Sag of the Junggar Basin.

3. Methodologies

3.1. Drilling Core and Micro-Section Observation

A total of 35 micro-sections from 10 drilling wells are applied to analyze the microscope of samples. To facilitate pore volume identification, blue dye resin is introduced into the thin sections, while a combination of Alizarin Red-S and potassium ferricyanide is employed to assist in discerning carbonate minerals. Then, a polarizing microscope can be utilized to determine the mineral composition, texture, and validity of pores and fractures.

3.2. Borehole Image Logs

Borehole image logs can provide valuable insights into fracture characteristics, sedimentary structures, geomechanical properties, and depositional trends. Advances in imaging technology have significantly improved the visualization of these features, particularly in non-cored intervals. To evaluate fracture strikes, image log (FMI) data from both cored and non-cored sections are employed, offering a comprehensive understanding of the subsurface fracture patterns. Incorporating FMI data into the analysis allows for a more complete and detailed understanding of the fracture network within the Fengcheng Formation of the Mahu Sag.

3.3. Fluid Inclusion Test

During mineral crystallization, fluid inclusions become entrapped in defects or voids within the mineral lattice. These inclusions are preserved inside the primary mineral. Determining the temperature of primary inclusions that formed simultaneously with the fracture fillings is a commonly accepted method for estimating the period of fracture formation. By assessing the temperature of primary inclusions that formed concurrently with the fracture fillings, the timeframe of fracture formation can be estimated.

3.4. Stable Carbon and Oxygen Isotope Analysis

The sample containing the fracture fill is subjected to reaction with 100% orthophosphoric acid at a constant temperature, resulting in the production of carbon dioxide and water. The carbon dioxide is purified and the isotopic composition of $\delta^{13}C$ and $\delta^{18}O$ is measured using a gas isotope mass spectrometer, with the precision exceeding $\pm 0.2‰$. The experimental apparatus is fDeltaV isotope ratio mass spectrometer.

3.5. Nano-CT Tests

The NanoCT scanner (MicroXRM-L200, ZEISS, Oberkochen, Germany) is used to analyze the distribution characteristics of pores, fractures and grains in a sample. The scanner employs a constant X-ray source voltage of 40 kV with a photon energy of 8 keV. Before testing, the sample should be cut into approximately cylindrical shapes with a diameter of about 100 μm. The scanner has a high resolution in scanning mode, with a large field of view that allows for a visual field of 65 μm and a resolution of 65 nm.

3.6. FE-SEM Experiment

The SEM experiment (Quanta 250 FEG, Thermo Fisher, Waltham, MA, USA) is utilized to determine the microfracture characteristics of shale samples. The experiment is conducted in a highly-vacuumed environment, and images are captured at a magnification ratio ranging from 6 to 1,000,000. Prior to FE-SEM analysis, the samples are coated with a layer of gold and polished using an argon-ion polishing system (Gatan 685, Gatan, Pleasanton, CA, USA).

4. Results and Discussion

4.1. Fracture Characteristics

Natural fractures in the Fengcheng Formation of the Mahu Sag are characterized by the observation of drilling cores and imaging logging images. Specifically, downhole shale samples from five wells and imaging logging images from eight wells are utilized to survey the natural fracture characteristics in the area. Figure 2 depicts the characteristics of the natural fractures in the Fengcheng Formation of the Mahu Sag. Results indicate that the length of the fractures in the Fengcheng Formation falls in the range from 5 cm to 15 cm, with over 85% of the fractures exceeding 5 m in length. The fracture width spans 0.1 mm–2 mm, with the majority falling between 0.2 mm to 1 mm. The inclination angle of the fractures is mostly medium to high, with more than 71% having an angle greater than 60°. Furthermore, approximately 78% of the natural fractures are filled with calcite, quartz, dolomite, and asphalt, with calcite as the principal filling mineral (Figure 2).

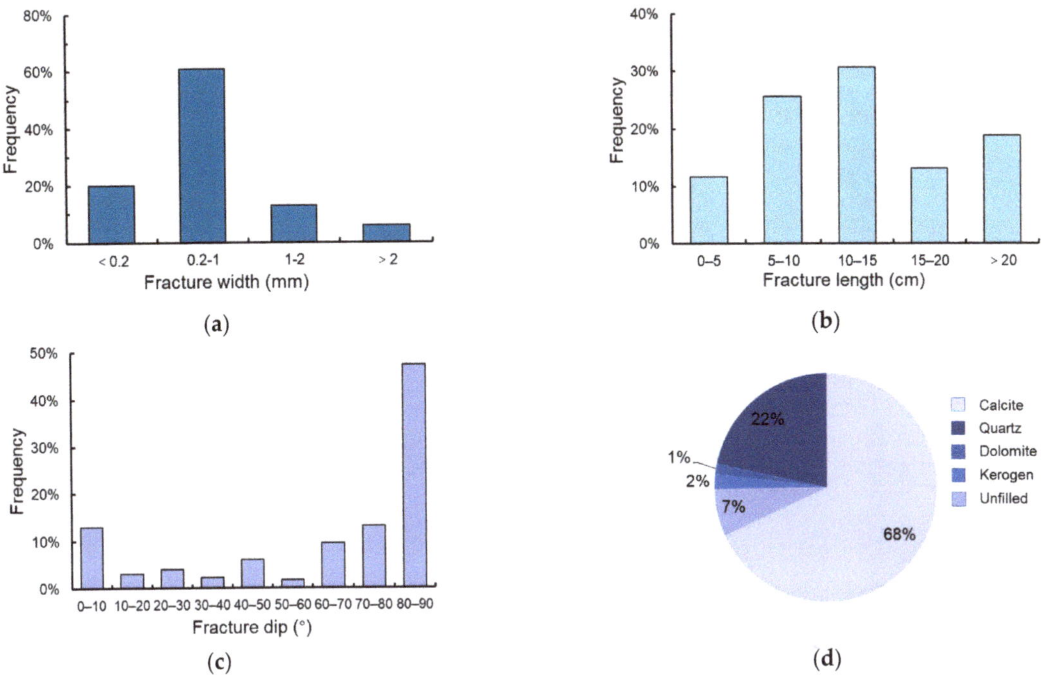

Figure 2. The statistical results of fracture characteristics and fillings of fractures from the drilling cores, microsections, and borehole imaging logging data. (**a**) Frequency distribution of fracture widths in the study area; (**b**) Frequency distribution of fracture lengths in the study area; (**c**) Frequency distribution of fracture dips in the study area; (**d**) Map of the percentage of fracture fills in the study area.

Figure 3 displays a detailed column chart of the fractures profile for Well #3 in the Fengcheng Formation of the Mahu Sag. The results reveal that the distribution of natural fractures in the Fengcheng Formation of the Mahu Sag is relatively uniform. The density of natural fractures decreases progressively with depth. The Feng #3 Formation displays the highest fracture density, spanning from 5 pcs/m to 75 pcs/m with an average value of 14.52 pcs/m. The Feng #2 Formation has a fracture density of 3 pcs/m to 58 pcs/m (average at 25.29 pcs/m), while the Feng #1 Formation exhibits the lowest fracture density, ranging from 3 pcs/m to 10 pcs/m (average at 8.56 pcs/m). The bottom of Feng #2 Formation and the top of Feng #3 Formation demonstrate the most significant degree of fracture development. The degree of fracture development in the Fengcheng Formation of the Mahu Sag appears to be influenced by the mineral composition of the shale samples. Specifically, fracture densities within the shale formation exhibit a positive correlation with increasing brittle mineral content. Shale formations with higher concentrations of brittle minerals (i.e., quartz, feldspar, calcite and dolomite) appear to be more prone to the generation of fractures [22].

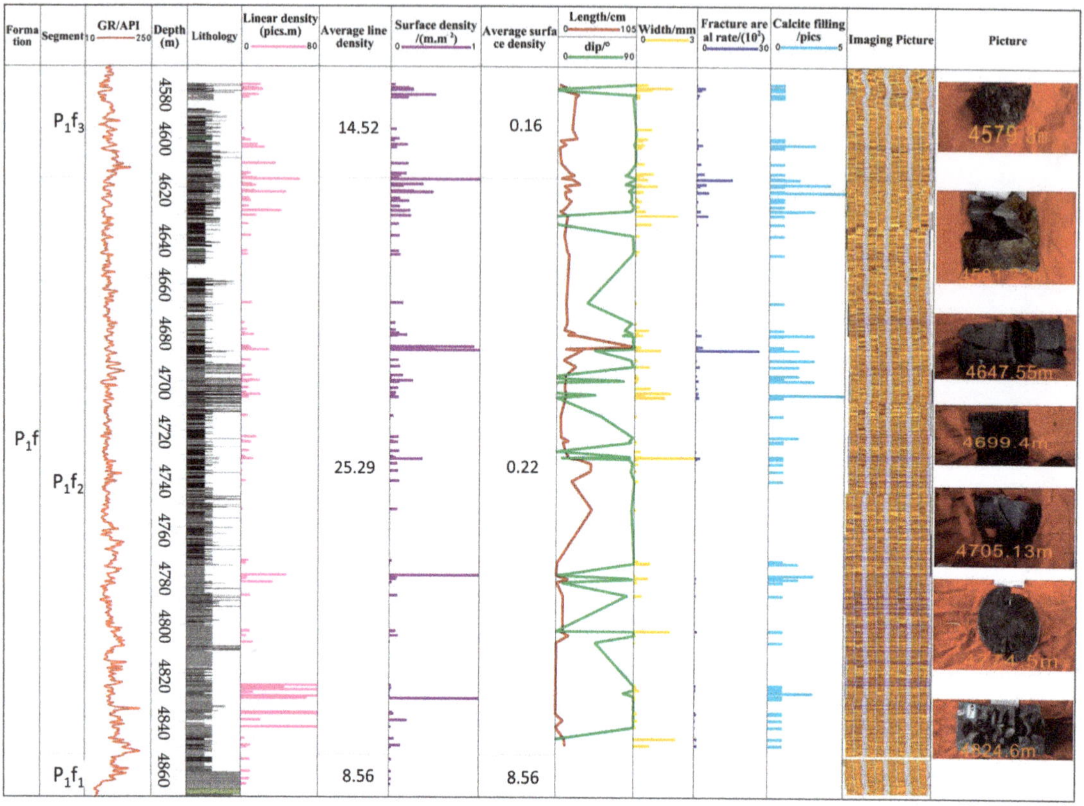

Figure 3. The column chart of the fracture profile for Well #3 in the Fengcheng Formation of the Mahu Sag.

4.2. The Classification of Fractures Based on Mechanical Origin

Based on observations of drilling core samples and microsections, the fractures in the Fengcheng Formation can be divided into two types: structural fractures and atectonic fractures. Structural fractures are primarily controlled by tectonic compression. The Mahu Sag, having experienced multiple episodes of intense tectonic activity, exhibits a complex structural fracture characteristic. Based on mechanical properties, the tectonic fractures in the Fengcheng Formation of the Mahu Sag can be further divided into three subtypes: high-angle interlayer shear fractures, along-layer shear fractures, and tensile fractures. Figure 4 demonstrates core sample photos and microsections of samples that contain tectonic fractures. The high-angle interlayer shear fractures depicted in Figure 4a–c exhibit straight and uniform fracture surfaces. Interactions between distinct sets of interlayer shear fractures involve mutual cutting and constraint. These types of fractures display a conjugated shear distribution filled with calcite and quartz. Generally, interlayer shear fractures are commonly developed in shale formations, which are abundant in brittle minerals. Along-layer slip shear fractures are characterized by distinctive features, including striations, mirror-like surfaces, and step-like structures with low-dip angles (Figure 4d,e). These fractures emerge in the context of tectonic stresses generated by compressional or extensional forces. Figure 4f depicts microsections of shale samples from Well #3, demonstrating the validity of along-layer slip shear fractures within the sample. Tensile fractures, formed in the condition of tensile stresses, are characterized by limited and unstable extension lengths. The widths of tensile fractures are variable, with rough and uneven fracture surfaces. Furthermore, the extension direction of tensile fractures typically aligns parallel to the primary stress direction (Figure 4g,h). Additionally, these types of fractures observed in thin sections are mostly unfilled with mineral (Figure 4i).

Non-tectonic fractures in the Fengcheng Formation of the Mahu Sag are primarily bedding fractures, hydraulic fractures, and hydrocarbon-generating overpressure fractures. Bedding fractures are characterized by near-horizontal or low-angle features, exhibiting rough surfaces (Figure 4j). In the microscope, the bedding fractures appear parallel to each other and display a wavy pattern (Figure 4h,i). Hydraulic fractures, typically filled with fibrous calcite, possess restricted extension lengths. The fracture surface manifests discernible branching configurations, evocative of tree branches (Figure 4m–o). The majority of fractures generated by hydrocarbon-generating overpressure fractures are microscopic fractures that exhibit high effectiveness. Based on the microsection and FE-SEM images, these types of fractures mostly occur at the edges of mineral grains or inside organic matter with irregular shapes (Figure 4p–r).

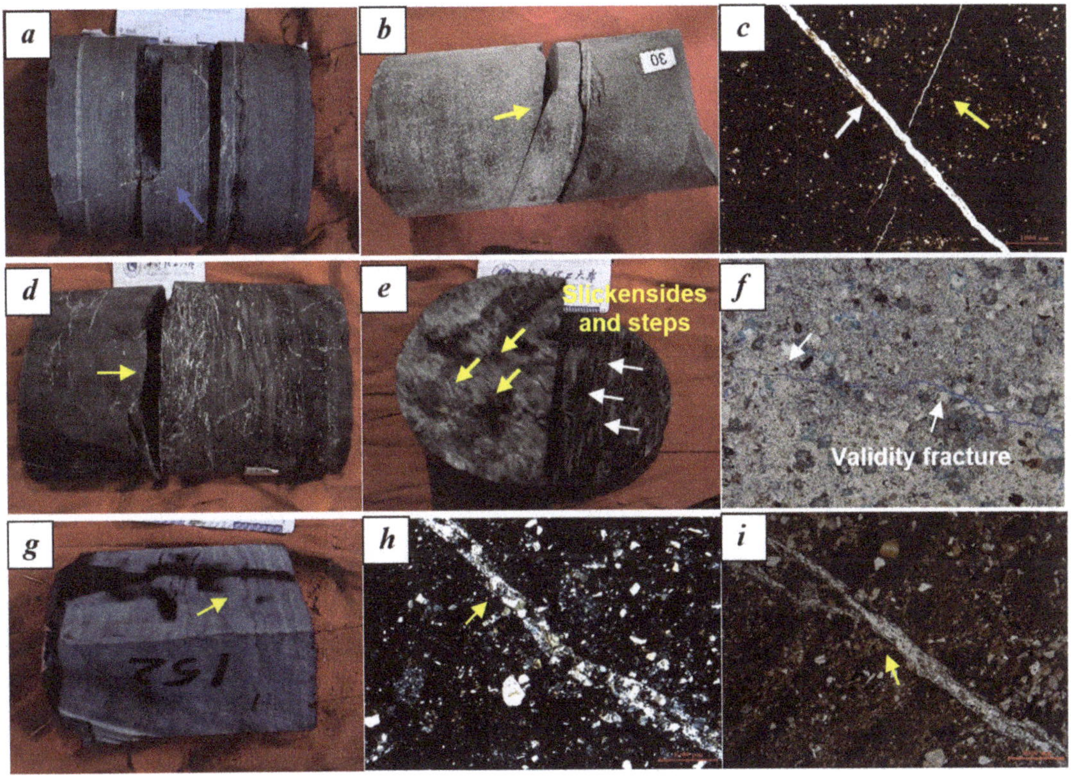

Figure 4. *Cont.*

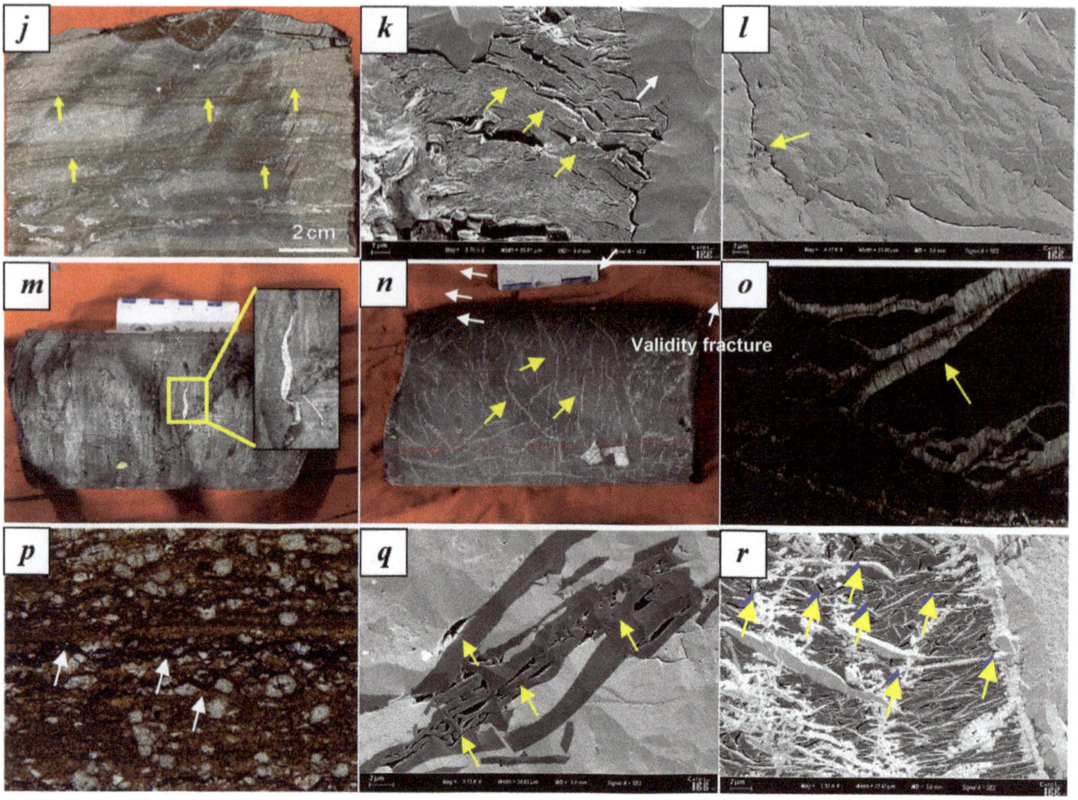

Figure 4. The drilling cores, microsections and FE-SEM images showing the different types of fractures of Fengcheng Formation in the Mahu Sag of the Junggar Basin, China. (**a**) The high-angle interlayer shear fracture with calcite filling, Well #6, 4744.48 m; (**b**) The unfilled high-angle interlayer shear fracture, Well #1, 4744.48 m; (**c**) Straight and uniform high-angle interlayer shear fracture with calcite filling, Well #6, 4647.30 m; (**d**) The unfilled along-layer -shear fracture, Well #6, 4648.10 m; (**e**) Along-layer- shear fracture with step-like structures with low-dip angle, Well #3, 4697.32 m; (**f**) The unfilled along-layer slip shear fracture, Well #3, 4636.14 m; (**g**) Tensile fractures unfilled, Well #3, 4627.53 m; (**h**) Tensile fractures with calcite filling, Well #3, 4614.20 m; (**i**) The unfilled tensile fractures, Well #3, 4616.41 m; (**j**) low-angle bedding fractures, Well #1, 4061.60 m; (**k**) The bedding fractures in the microscope, Well #3, 4740.43 m; (**l**) The unfilled bedding fractures, Well #3, 4802.20 m; (**m**) Hydraulic fractures filled with fibrous calcite, Well #6, 4640.47 m; (**n**) Hydraulic fractures with discernible branching configurations in the fracture surface, Well #6, 4668.90 m; (**o**) Hydraulic fractures in the microscope, Well #3, 4664.85 m; (**p**) Hydrocarbon-generating overpressure fractures with kerogen filling, Well #3, 4744.48 m; (**q**) The unfilled hydrocarbon-generating overpressure fractures, Well #3, 4802.20 m; (**r**) The unfilled Hydrocarbon-generating overpressure fractures, Well #3, 4759.42 m.

4.3. Comprehensive Determination of the Evolution Periods of Fractures

A total of seventeen shale samples were utilized to explore the evolution periods of fractures based on the cutting relationship of natural fractures on the macro- and micro-scales, fluid inclusion test, and stable carbon and oxygen isotope analysis.

4.3.1. The Observation of Drilling Cores and Microsections

The formation periods of natural fractures can be inferred by examining the filling order and intersection relationships among different fractures on core samples. Figure 5a–c display two sets or three sets of shear fractures, which were filled with calcite. In Figure 5a, a late-formed high-angle cross-layer shear fracture is observed cutting through two early-formed cross-layer shear fractures. Figure 5b shows a late-formed high-angle tensile fracture intersecting with an early-formed bedding fracture. Figure 5c illustrates a late-formed shear fracture passing through an unfilled bedding fracture and an earlier-formed calcite-filled bedding fracture. Figure 5d depicts a rose diagram of strikes of tectonic fractures, indicating three formation periods of natural fractures.

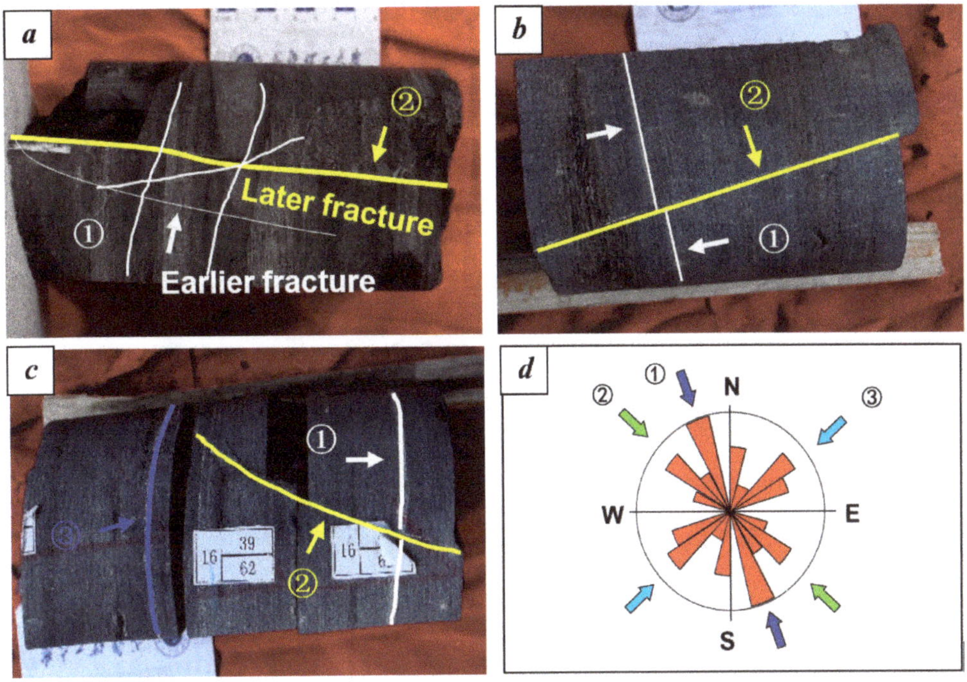

Figure 5. The drilling cores and rose map fracture strike and fracture period division of Fengcheng Formation in the Mahu Sag of the Junggar Basin. (**a**) Two sets or three sets of shear fractures, Well #6, 4744.48 m; (**b**) A late-formed high-angle tensile fracture intersecting with an early-formed bedding fracture, Well #3, 4705.03 m; (**c**) a late-formed shear fracture passing through an unfilled bedding fracture and an earlier-formed calcite-filled bedding fracture, Well #6, 4818.46 m; (**d**) Rose diagram of strikes of tectonic fractures in the Fengcheng Formation of the Mahu Sag. The numbers in the figure stand for the order of fractures generation.

Additionally, imaging logging results show that three sets of tectonic fractures developed in the Fengcheng Formation: NNW-SSE, NW-SE, and NE-SW. Among them, the development of NNW-SSE fractures is the most significant, followed by NW-SE and NE-SW fractures. These results based on imaging logging are highly consistent with the formation periods of fractures observed on the core (Figure 5d). Furthermore, imaging logging results indicate the development of three sets of tectonic fractures in the Fengcheng Formation, oriented NNW-SSE, NW-SE, and NE-SW. Among these, the NNW-SSE fractures exhibit the most significant development, followed by the NW-SE and NE-SW fractures. The results indicate that the formation periods of natural fractures, based on imaging logging, are consistent with those observed in the core samples.

Observation of microsections reveals the presence of microfractures in the Fengcheng Formation exhibiting multiple cutting phenomena. Figure 6a depicts the characteristics of hydraulic fractures at the microscopic scale, with fractures filled by multi-stage calcite.

Due to the recurring opening and closing of the fractures, the calcite crystals exhibit multi-stage growth characteristics. The initial calcite crystal growth is constrained, leading to the formation of tensile blocky-shaped calcite crystals. Subsequently, fractures expand, permitting unrestricted crystal growth. Moreover, an evident mutual cutting relationship exists between the fractures in the microsection shown in Figure 6b. Early calcite veins are intersected by thicker calcite veins that formed later (as illustrated in Figure 6b). Figure 6c illustrates three-stage veins, indicating that the fractures in the Fengcheng Formation have experienced a minimum of three stages. In summary, the microstructure of the Fengcheng Formation is highly intricate and displays characteristics indicative of multiple filling stages.

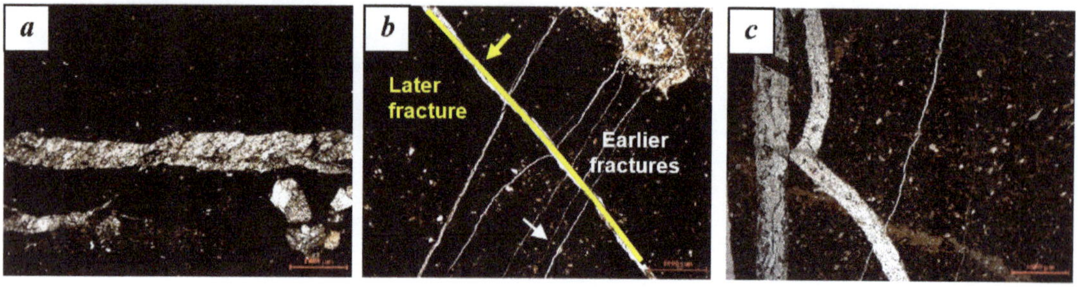

Figure 6. The microsections of Fengcheng Formation in the Mahu Sag of the Junggar Basin, showing the cutting relationship of natural fractures; (**a**) The hydraulic fractures at the microscopic scale with multi-stage calcite filling, Well #6, 4647.30 m; (**b**) Fractures with mutual cutting relationship, Well #3, 4699.40 m; (**c**) Three-stage veins, Well #3, 4609.97 m.

The observations of core and microsections reveal that the Fengcheng Formation shale of the Mahu Sag has developed at least three stages of fractures. The earliest fractures are mostly filled with horizontally fibrous calcite fractures. Mid-stage tectonic fractures, caused by tensional stress, are filled with blocky calcite. Finally, late-stage expansion fractures are primarily filled with blocky calcite and quartz. This indicates that the fractures developed within a relatively open expansion environment, leading to well-defined and regular features of the crystals.

4.3.2. Stable Carbon and Oxygen Isotope Analysis

Analyzing the stable carbon and oxygen isotope compositions of calcite fillings within fractures is a useful method for identifying different stages of fracture development and their associated diagenetic environments. Significant variations in the collected data or calculated formation temperatures could suggest the occurrence of distinct vein mineralization formation stages. In this study, twenty-one samples from the Fengcheng Formation of the Mahu Sag are applied to conduct the stable carbon and oxygen isotope analyses. Results demonstrate that $\delta^{18}O$ and $\delta^{13}C$ values of samples range from −10.17‰ to −3.43‰, −15.76‰ and 4.10‰, respectively. Figure 7 presents a negative correlation between $\delta^{18}O$ and $\delta^{13}C$ isotopes in the samples, suggesting that fracture developed across multiple stages. Based on the oxygen isotope thermometry equation proposed by Fritz and Simith (1970) [24], the paleotemperature of fracture formation can be calculated as follow:

$$T = 31.9 - 5.55\,(\delta^{18}O - \delta^{18}O_w) + 0.7\,(\delta^{18}O - \delta^{18}O_w)^2 \quad (1)$$

where, T is the paleotemperature at which filled calcite formed. $\delta^{18}O$ and $\delta^{18}O_w$ refers to the oxygen isotope value of the fracture filling and aqueous medium, respectively. In this study, the $\delta^{18}O_w$ equals to −0.5‰ according to the previous research [25].

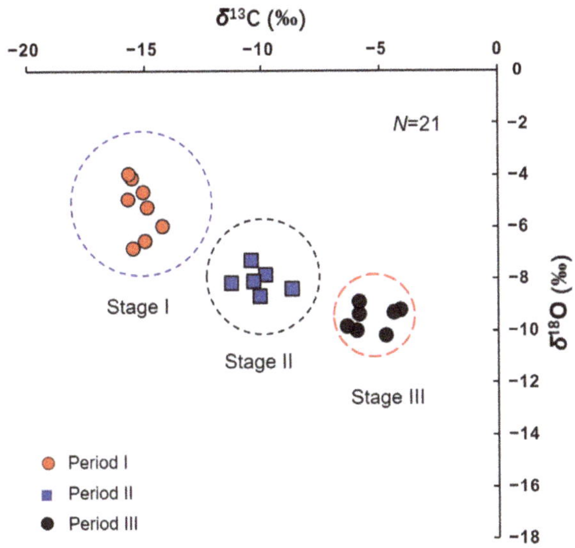

Figure 7. The relationship of $\delta^{18}O$ and $\delta^{13}C$ isotopes of fracture fillings in the Fengcheng Formation of the Mahu Sag.

Given an average ground temperature of 20 °C and a geothermal gradient of 2.65 °C/100 m in the Mahu Sag (Hu et al., 2020) [26], the burial depth at the time of fracture formation can be calculated. The distribution of $\delta^{13}C$ and $\delta^{18}O$ can be divided into three distinct stages. During the first stage of fractures, the $\delta^{18}O$ values of the fracture filling ranged from −3.93 to −6.76‰, with calculated temperatures between 70.22 °C and 109.09 °C, and corresponding burial depths of 1895.24 to 3361.82 m. Fibrous calcite was the primary filling material in these fractures. During the second stage of fractures, the fracture filling material had $\delta^{18}O$ values ranging from −7.32 to −8.02‰. This corresponds to calculated temperatures between 118.11 °C and 130.00 °C, and burial depths of 3702.18 to 4150.92 m. During the third stage of fractures, the $\delta^{18}O$ values of the filling material ranged from −8.28 to −9.07‰, with calculated temperatures between 134.59 °C and 149.12 °C, and corresponding burial depths of 4324.18 to 4872.56 m. The primary filling minerals in these fractures were blocky calcite and granular quartz. Therefore, based on the $\delta^{18}O$ values and temperature calculation results of the filling material in the fractures, we can conclude that the fracture formation can be divided into three stages, corresponding to different depths of burial: 1895.24–3361.82 m, 3702.18–4150.92 m, and 4324.18–4872.56 m.

4.3.3. Fluid Inclusion Tests

The paleotemperature during the time of entrapment can be determined by measuring the homogenization temperature of inclusions in fracture fillings, which assists in identifying the period of fracture formation (Li et al., 2023) [16]. In this study, inclusions were extracted from quartz and calcite that filled the tectonic fractures. The inclusion types primarily consisted of gas-liquid two-phase saline inclusions and hydrocarbon inclusions. Figure 8 indicates the morphologies of the inclusions in fracture fillings of the Fengcheng Formation in the Mahu Sag. The inclusions in quartz, size of 5–10 μm, are mainly irregular polygons with shape of sub-circular or spindle. In addition, the inclusions developed in the calcite fillings are densely distributed in groups or bands, exhibiting predominantly regular quadrilaterals and irregular polygons, with sizes ranging from 3–9 μm. Compared to quartz inclusions, calcite inclusions are more abundant but slightly smaller in size. Results reveal that the homogenization temperature of inclusions in the Fengcheng Formation can be roughly divided into three ranges (Figure 9). Table 1 summarizes the temperature and salinities of the fluid inclusions in the Fengcheng Formation of Mahu Sag. Combined with core observations and microsection identification, it is inferred that the first stage of fractures is primarily filled with fibrous calcite, corresponding to temperature of 70–100 °C; the second stage of fractures are predominantly filled with elongated calcite crystals, corresponding to temperature of 110–130 °C; the third stage of fractures is mainly filled with quartz and calcite, along with some unfilled fractures. These fractures correspond to temperature of 130–150 °C.

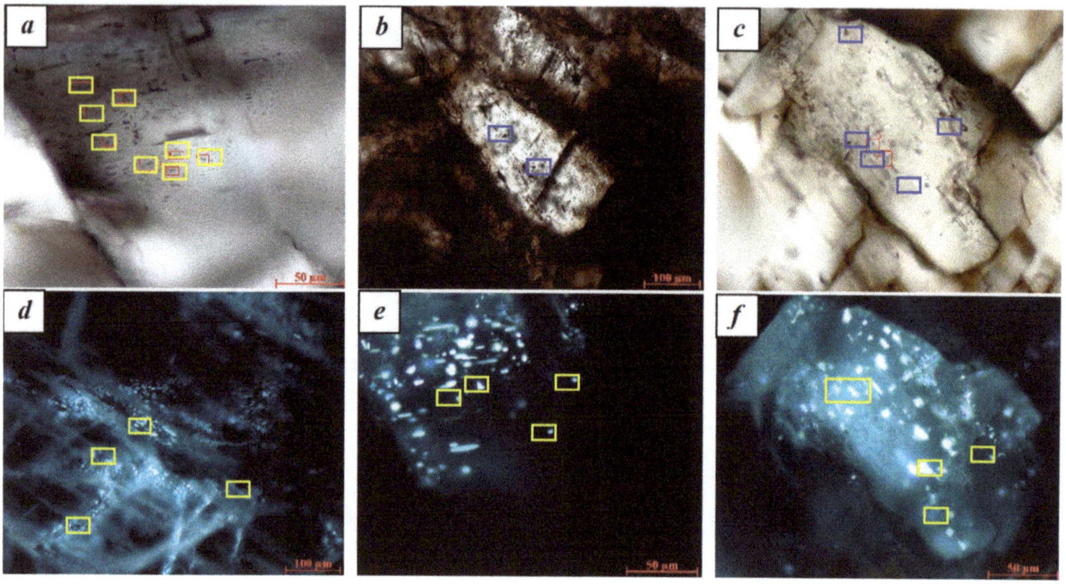

Figure 8. The fluid inclusion photos of Fengcheng Formation in the Mahu Sag of the Junggar Basin, showing the morphologies of the fluid inclusions; (**a**) X203, 4818.61 m, gas-liquid two-phase saline inclusions and hydrocarbon methane inclusions; (**b**) MY1, 4830.21 m, gas-liquid two-phase saline inclusions and hydrocarbon methane inclusions; (**c**) X203, 4812.35 m, gas-liquid two-phase saline inclusions and hydrocarbon methane inclusions; (**d**) X203, 4648.55 m, hydrocarbon methane inclusions; (**e**) X203, 4800.16 m, gas-liquid two-phase saline inclusions and hydrocarbon methane inclusions; (**f**) X203, 4812.35 m, gas-liquid two-phase saline inclusions and hydrocarbon methane inclusions.

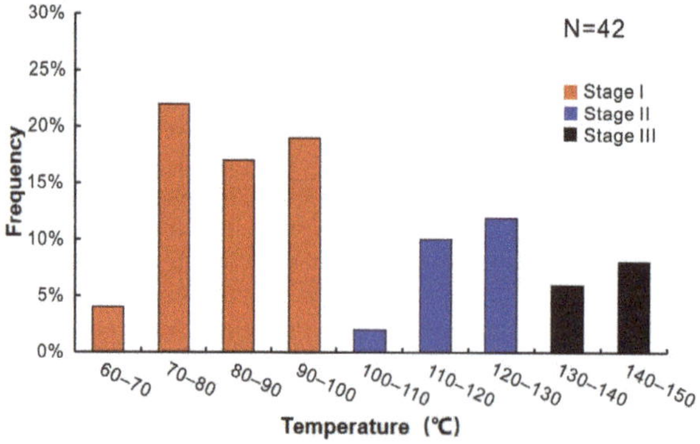

Figure 9. The distribution of the homogenization temperature of fluid inclusions in natural fracture fillings, the sample obtained from the Fengcheng Formation in the Mahu Sag of the Junggar Basin.

Table 1. The fluid inclusion test data of shale in the Fengcheng Formation of the Mahu Sag.

Well	Depth (m)	Inclusion Number	Size (μm)	Phase State	Homogenization Temperature (°C)	Salinities (wt%)
X203	4818.61	1	8	Liquid phase	80	0.18
		2	6	Liquid phase	72	0.35
		3	7	Liquid phase	68	0.18
		4	6	Liquid phase	86	0.53
		5	5	Liquid phase	65	0.18
		6	3	Liquid phase	90	0.7
		7	3	Liquid phase	68	0.18
		8	4	Liquid phase	70	0.35

By examining the cutting relationships of fractures in the drilling cores, and micro-section, we have determined the sequence of tectonic fractures. Results suggest that there are at least three stages of natural fractures in the Fengcheng Formation of the Mahu Sag. These findings have significant implications for understanding the geology of the Mahu Sag and could be valuable for future exploration of natural resources or assessment of geological evolution in the region.

4.4. Genetic Mechanisms of Multiperiod Fractures

Figure 10 presents the burial history and thermal evolution profile for the Mahu Sag. Through a combination of multiple experiments, as well as an analysis of the burial history and evolution, the time frame for tectonic fracture formation can be ascertained. Results suggest that the geological tectonic evolution transpired during three distinct periods—243–266 Ma, 181–208 Ma, and 50–87 Ma—each playing a critical role in fostering the development of tectonic fractures within the Mahu Sag.

The initial stage of fracture evolution pertains to the Late Permian period (243–266 Ma), during which burial depths ranged from 1679 to 3056 m. The Mahu Sag in the Ordos Basin underwent NW-SE compression due to the collision between the Kazakhstan and Junggar plates. Consequently, these compressions led to the development of conjugate NWW-SEE and NNW-SSE shear fractures. The tectonic fractures that emerged during this stage exist in a relatively confined environment, inhibiting the growth of mineral crystals. These fractures are frequently filled with horizontally fibrous calcite. The fluid inclusions within the fracture fillings maintain a consistent temperature of 70–100 °C.

The second stage of fracture evolution occurred during the Late Indosinian to Early Yanshanian period (181–208 Ma), with burial depths ranging from 3396 to 4151 m. Throughout the Mesozoic era, the Chinese mainland experienced compression and collision from the Pacific Plate and the Indian Plate, leading to a gradual uplift of the Mahu Sag. As a result of the westward subduction of the Pacific Plate towards the Eurasian Plate, conjugate NW-SE and NNW-SSE shear fractures can be formed in the Mahu Sag under the NNW-SSE stress. Most of the fractures formed during this period were filled with extensional calcite. These fractures existed in a semi-closed to semi-open environment, which constrained the growth of calcite crystals. Consequently, the calcite fillings within the fractures exhibited elongated and blocky appearances. The fluid inclusions in the calcite maintained a uniform temperature of 110–130 °C.

The third stage of fracture development happened during the late Yanshanian to early Himalaya period (50–87 Ma) at depths ranging from 4528 to 4943 m. The movement and subduction of the Indian plate to the north had a distant effect on the central part of the Central Asian region, causing the Tarim Plate to move northward and resulting in compression in the near-SN direction across the studied area. This compression led to the formation of NNW-SSE and NNE-SSW strike-slip fractures, as well as associated faults near the primary faults. The fractures formed during this stage were mainly filled with blocky calcite and quartz grains, with fluid inclusions exhibiting a uniform temperature of 130–150 °C.

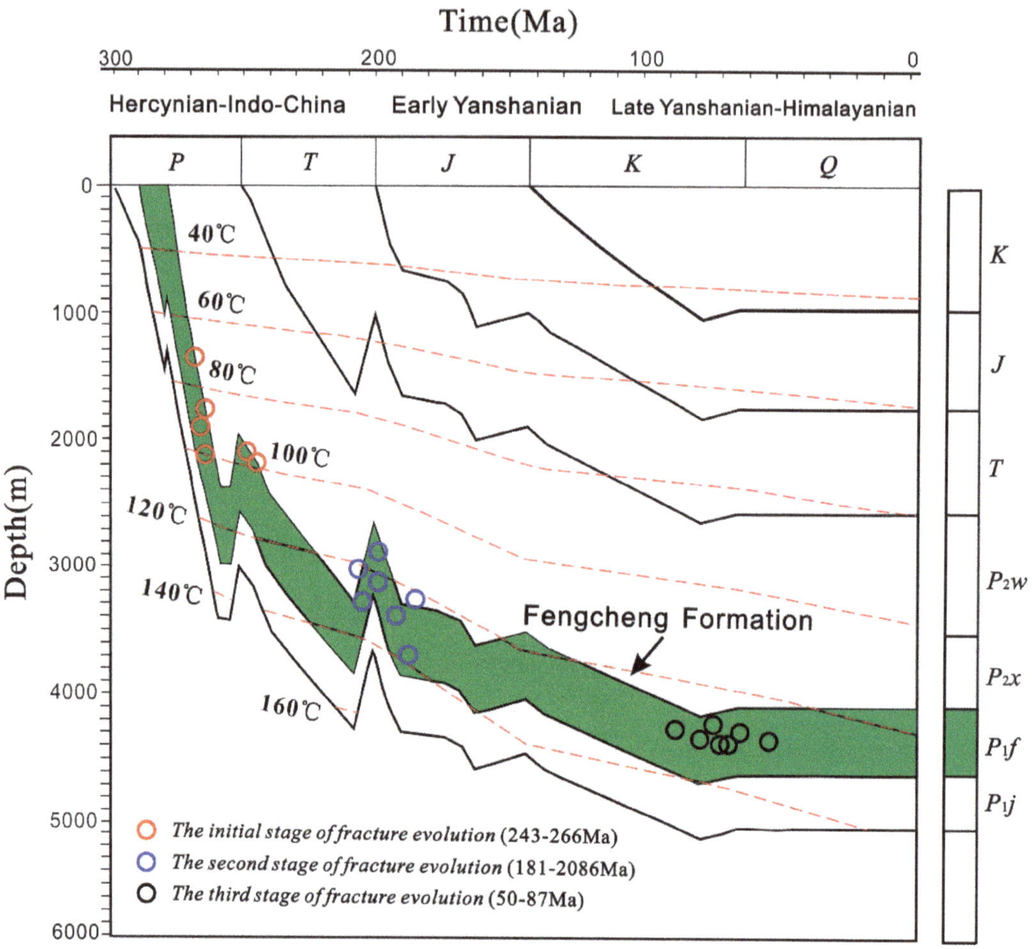

Figure 10. The burial history of Fengnan #4 Well and the estimated time-depth of different stages fractures in the Fengcheng Formation of the Mahu Sag.

4.5. Effect of Fractures on Oil Productivity

4.5.1. Microfractures

Microfractures contribute to the shale reservoir mainly by improving the storage space and enhancing the reservoir properties. In this study, the FESEM technique and nano-CT test were applied to analyze the effect of microfractures on shale reservoir properties.

Figure 11 shows the three-dimensional pore structure of the shale sample, which is mainly composed of dissolved pores, organic pores, diagenetic microfractures, and hydrocarbon overpressure fractures. Figure 11a,b illustrate irregularly shaped inorganic and organic pores with strong heterogeneity, mostly in an isolated state. Figure 11c displays irregularly shaped and unconnected inorganic fractures developed in the shale. Figure 11d exhibits the distribution of organic fractures.

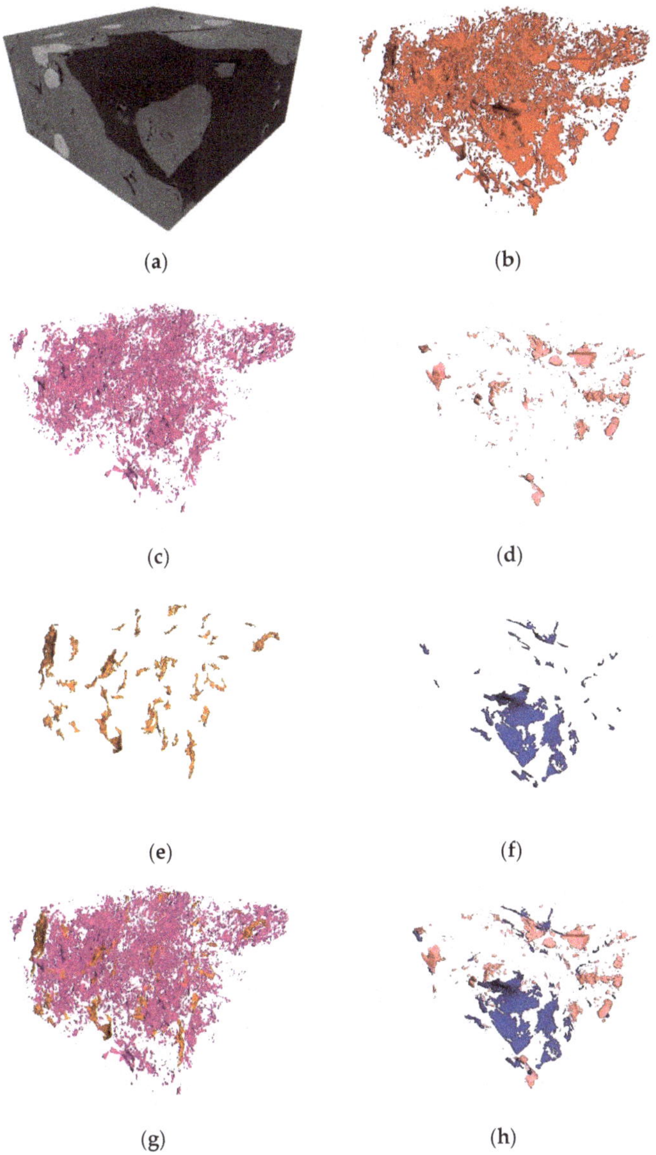

Figure 11. Nano-CT results of the shale sample, obtained from the Maye #1 Well in the Fengcheng Formation of the Mahu Sag; (**a**) 3D image of mineral composition; (**b**) 3D image of the volume of pores and fractures; (**c**) 3D image of the inorganic pores; (**d**) 3D image of the organic pores; (**e**) 3D image of the inorganic fractures; (**f**) 3D image of the organic fractures; (**g**) 3D image of the inorganic pores and fractures; (**h**) 3D image of the organic pores and fractures.

Figure 12a indicates the morphology and shape of the hydrocarbon overpressure microfractures in the Fengcheng Formation of the Mahu Sag. Combined with FE-SEM, it was proven that hydrocarbon overpressure microfractures, with small pore sizes and good connectivity, formed an effective permeability system. Hydrocarbon overpressure microfractures connect the isolated micropores in the shale reservoir, forming a connected pore space.

Figure 12b illustrates the relationship between the densities of hydrocarbon-generating overpressure microfractures and the permeabilities of the shale reservoir. The results reveal a positive correlation ($R^2 = 0.6448$) between the densities of hydrocarbon-generating overpressure microfractures and the permeabilities of the shale reservoir. Figure 12b left shows the SEM image of the shale reservoir in the study area for low permeability and overpressure microfractures density; Figure 12b right shows the SEM image of the shale reservoir in the study area for high permeability and overpressure microfractures density. As the densities of hydrocarbon-generating overpressure microfractures increase, permeabilities of the shale reservoir also increase. Consequently, hydrocarbon-generating overpressure microfractures enhance the effective storage capacity of the reservoir and optimize its pore structure [27]. Furthermore, the microfracture can establish a connection between the organic matter and the external flow channels, ultimately augmenting the reservoir's permeability.

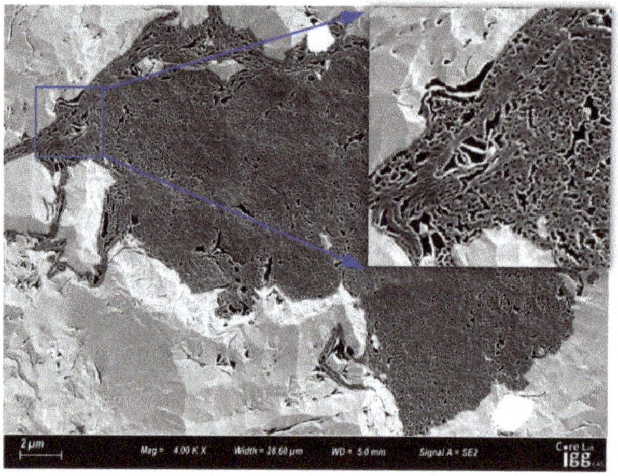

(**a**)

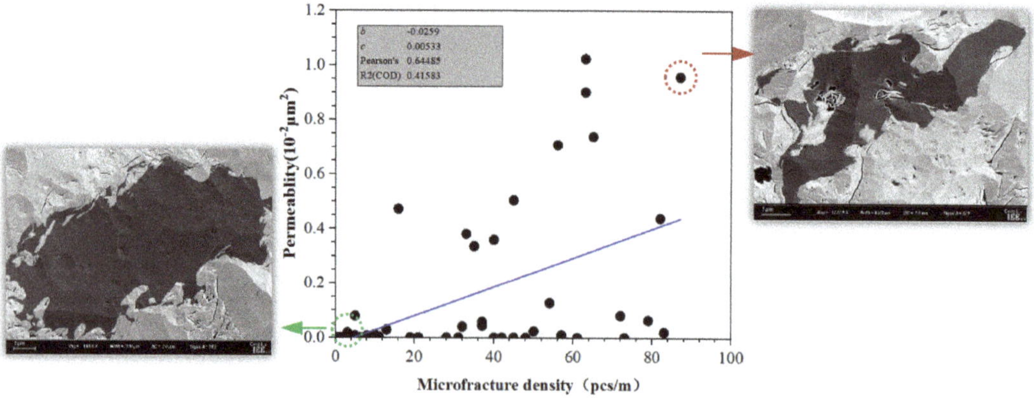

(**b**)

Figure 12. (**a**) The morphology and shape of the hydrocarbon overpressure microfractures; (**b**) The relationship between the microfracture density and shale permeability in the Fengcheng Formation of the Mahu Sag.

4.5.2. Macrofractures

The location of high-yield and industrial gas flow wells predominantly near the fold belt highlights the significant impact of fracture development on natural gas production. In this study, twelve wells of the Fengcheng Formation in the Mahu Sag are investigated to analyze the development features of various fracture types based on imaging logging data. Figure 13a displays the intersection diagram of the fracture development index and the daily gas production of the twelve wells of the Fengcheng Formation in the Mahu Sag. Results indicate that the high-yield gas-bearing layers of the Fengcheng Formation exhibit well-developed fractures. To further examine the effect of the types of tectonic fractures on gas production, the associations between different types of tectonic fractures and daily gas production are plotted separately (Figure 13b–d). Results demonstrate that a positive correlation exists between daily gas production and the density of high-angle interlayer shear fractures, while the correlations with along-layer shear fractures and tensile fractures are not significant.

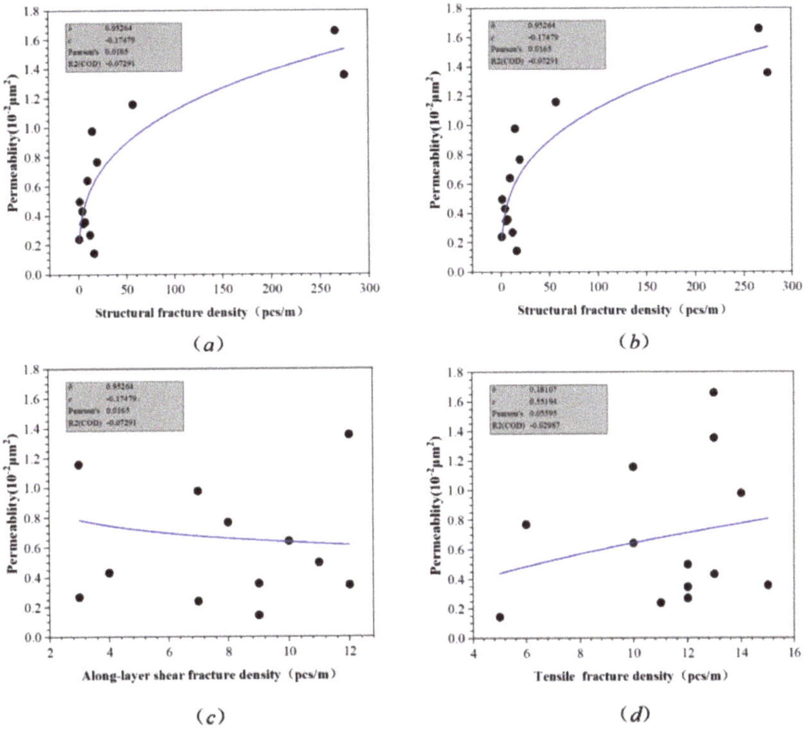

Figure 13. The relationships between the (**a**) tectonic fracture density, (**b**) high-angle interlayer shear fracture density, (**c**) along-layer shear fracture density, (**d**) tensile fracture density, and daily production in the Fengcheng Formation of the Mahu Sag.

The research findings hold considerable importance for natural gas exploration and development in the Mahu Sag. Based on these results, it is recommended that the development of high-angle interlayer shear fractures be prioritized during the exploration and development of high-yield gas-bearing layers. In addition, efforts should be intensified to study along-layer shear fractures and tensile fractures to gain a deeper understanding of their influence on gas production. Moreover, a comprehensive analysis should be undertaken, integrating imaging logging data with seismic data, to determine the characteristics of fractures and assess their impacts on natural gas production more precisely.

5. Conclusions

The fractures of the Fengcheng Formation in the Mahu Sag are analyzed by geological and geochemical methods. The conclusions are as follows:

- The Fengcheng Formation of the Mahu Sag contains two types of natural fractures: tectonic fractures and atectonic fractures. Tectonic fractures consist of high-angle interlayer shear fractures, along-layer shear fractures, and tensile fractures. Atectonic fractures include bedding fractures, hydraulic fractures, and hydrocarbon-generating overpressure fractures. The fractures are typically medium to high angle with filled minerals such as calcite and quartz. Additionally, shale with higher content of brittle minerals (quartz, feldspar, calcite and dolomite) is more susceptible to generating fractures. Vertically, fracture development is more prominent at the bottom of Feng #2 Formation and at the top of Feng #3 Formation.
- Combining the cutting relationships of natural fractures at both macro- and micro-scales, stable carbon and oxygen isotope analysis, and fluid inclusion results with the evolution of regional tectonic movements, the results suggest that the tectonic fractures in the study area can be classified into three stages. The first stage of fracture development occurred during the Late Permian period (243–266 Ma). The second stage of fracture evolution took place during the Late Indosinian to Early Yanshanian period (181–208 Ma). Finally, the third stage of fracture development occurred during the late Yanshanian to early Himalaya period (50–87 Ma).
- Among various types of tectonic fractures, the density of high-angle interlayer shear fractures has been found to exhibit a positive correlation with daily gas production, which suggests that high-angle interlayer shear fracture plays a significant role in promoting the production and transportation of hydrocarbon resources. Furthermore, hydrocarbon-generating overpressure microfractures, characterized by their small pore sizes and good connectivity, serve as an effective permeability system. This system connects previously isolated micropores in the shale reservoir, creating a connected pore space.

Author Contributions: X.X.: Writing—Review and Editing. W.W.: Review and Editing, Experiment, Writing—Review and Editing. C.L.: Data Curation. Z.W. and G.C.: Funding acquisition. F.C.: Review and Editing, Methodology, G.Z. and J.H.: Experiment, Data Curation. H.D. and K.L.: Investigation, Writing—Original Draft, Funding acquisition. All authors have read and agreed to the published version of the manuscript.

Funding: This research was funded by the Sichuan Province Science and Technology Plan Outstanding Youth Talent Project grant number [2020JDJQ0058].

Data Availability Statement: All the research data can be listed in the mabuscript.

Conflicts of Interest: The authors declare no conflict of interest.

References

1. Zhao, W.; Hu, S.; Hou, L.; Yang, T.; Li, X.; Guo, B.; Yang, Z. Types and resource potential of continental shale oil in China and its boundary with tight oil. *Pet. Explor. Dev.* **2020**, *47*, 1–11. [CrossRef]
2. Jin, Z.; Zhu, R.; Liang, X.; Shen, Y. Several issues worthy of attention in current lacustrine shale oil exploration and development. *Pet. Explor. Dev.* **2021**, *48*, 1471–1484. [CrossRef]
3. Lu, G.; Zeng, L.; Dong, S.; Huang, L.; Liu, G.; Ostadhassan, M.; He, W.; Du, X.; Bao, C. Lithology identification using graph neural network in continental shale oil reservoirs: A case study in Mahu Sag, Junggar Basin, Western China. Mar. *Pet. Geol.* **2023**, *150*, 106168. [CrossRef]
4. He, W.J.; Qian, Y.X.; Zhao, Y.; Li, N.; Zhao, X.M.; Liu, G.L.; Miao, G. Exploration Implications of Total Petroleum System in Fengcheng Formation, Mahu Sag, Junggar Basin. *Xinjiang Pet. Geol.* **2021**, *42*, 641–655. (In Chinese) [CrossRef]
5. Tang, Y.; Cao, J.; He, W.-J.; Guo, X.-G.; Zhao, K.-B.; Li, W.-W. Discovery of shale oil in alkaline lacustrine basins: The late paleozoic Fengcheng Formation, Mahu sag, Junggar Basin, China. *Petrol. Sci.* **2021**, *18*, 1281–1293. [CrossRef]
6. Zhang, D.; Ranjith, P.G.; Perera, M.S.A. The brittleness indices used in rock mechanics and their application in shale hydraulic fracturing: A review. *J. Petrol. Sci. Eng.* **2016**, *143*, 158–170. [CrossRef]

7. Huo, J.; Wang, X.H.; Luo, C.; Huang, H.Y.; Chen, J.; Yin, C.; He, J.M. Fracture characteristics of Longmaxi shale in Southern Sichuan. *J. Eng. Geol.* **2021**, *29*, 171–182. [CrossRef]
8. Gale, J.F.W.; Reed, R.M.; Holder, J. Natural fractures in the Barnett Shale and their importance for hydraulic fracture treatments. *AAPG Bull.* **2007**, *91*, 603–622. [CrossRef]
9. Yang, Z.; Wang, X.; Ge, H.; Zhu, J.; Wen, Y. Study on evaluation method of fracture forming ability of shale oil reservoirs in Fengcheng Formation, Mahu sag. *J. Petrol. Sci. Eng.* **2022**, *215*, 110576. [CrossRef]
10. Zeng, L.; Su, H.; Tang, X.; Peng, Y.; Gong, L. Fractured tight sandstone oil and gas reservoirs: A new play type in the Dongpu depression, Bohai Bay Basin, China. *AAPG Bull.* **2013**, *97*, 363–377. [CrossRef]
11. Lyu, W.; Zeng, L.; Zhou, S.; Du, X.; Xia, D.; Liu, G.; Li, J.; Weng, J. Natural fractures in tight-oil sandstones: A case study of the Upper Triassic Yanchang Formation in the southwestern Ordos Basin, China. *AAPG Bull.* **2019**, *103*, 2343–2367. [CrossRef]
12. Liu, J.; Ding, W.; Yang, H.; Liu, Y. Quantitative Multiparameter Prediction of Fractured Tight Sandstone Reservoirs: A Case Study of the Yanchang Formation of the Ordos Basin, Central China. *SPE J.* **2021**, *26*, 3342–3373. [CrossRef]
13. Zhi, D.; Tang, Y.; He, W.; Guo, X.; Zheng, M.; Huang, L. Orderly coexistence and accumulation models of conventional and unconventional hydrocarbons in Lower Permian Fengcheng Formation, Mahu sag, Junggar Basin. *Pet. Explor. Dev.* **2021**, *48*, 43–59. [CrossRef]
14. Zou, C.; Yang, Z.; Cui, J.; Zhu, R.; Hou, L.; Tao, S.; Yuan, X.; Wu, S.; Lin, S.; Wang, L.; et al. Formation mechanism, geological characteristics and development strategy of nonmarine shale oil in China. *Pet. Explor. Dev.* **2013**, *40*, 15–27. [CrossRef]
15. Xie, X.; Deng, H.; Fu, M.; Hu, L.; He, J. Evaluation of pore structure characteristics of four types of continental shales with the aid of low-pressure nitrogen adsorption and an improved FE-SEM technique in Ordos Basin, China. *J. Petro. Sci. Eng.* **2021**, *197*, 108018. [CrossRef]
16. Li, X.Y.; Du, W.; Feng, X.; Shi, F.L.; Chen, Y.; Wang, Y.S.; Jiang, Z.X.; Luo, Q. Pressure Evolution Mechanism of Marine Shale Reservoirs and Shale Gas Accumulation Model: Evidence from Fluid Inclusions in the Wufeng–Longmaxi Formation in the Basin Margin Structural Transition Zone in Northern Guizhou Province, China. *Minerals* **2023**, *13*, 241. [CrossRef]
17. Gonzalez, L.A.; Lohmann, K.C. Carbon and oxygen isotopic composition of Holocene reefal carbonates. *Geology* **1985**, *13*, 811–814. [CrossRef]
18. Frezzotti, M.L.; Tecce, F.; Casagli, A. Raman spectroscopy for fluid inclusion analysis. *J. Geochem. Explor.* **2012**, *112*, 1–20. [CrossRef]
19. Aghli, G.; Moussavi-Harami, R.; Tokhmechi, B. Integration of sonic and resistivity conventional logs for identification of fracture parameters in the carbonate reservoirs: A case study, Carbonate Asmari Formation, Zagros Basin, SW Iran. *J. Petrol. Sci. Eng.* **2020**, *186*, 106728. [CrossRef]
20. Cao, J.; Xia, L.W.; Wang, T.T.; Zhi, D.M.; Tang, Y.; Li, W.W. An alkaline lake in the Late Paleozoic Ice Age (LPIA): A review and new insights into paleoenvironment and petroleum geology. *Earth-Sci. Rev.* **2019**, *202*, 103091. [CrossRef]
21. Song, T.; Huang, F.X.; Wang, S.Y.; Yang, F.; Lv, W.N.; Liu, C.; Jia, P.; Fan, J.J. Characteristics and exploration potential of Jurassic oil and gas reservoirs in Mahu sag of the Junggar Basin. *China Pet. Explor.* **2019**, *24*, 341–350. (In Chinese) [CrossRef]
22. Wang, J.T.; Yang, S.; Zou, Y.; Zhao, Y.; Huang, L.L.; Mou, L.W. Characteristics, genesis and distribution of high-quality reservoir of the first member of the Permian Fengcheng Formation in Mabei area, Junggar Basin. *China Pet. Explor.* **2022**, *27*, 99–109. (In Chinese) [CrossRef]
23. Zou, Y.; Qi, Y.P.; Song, D.; Chen, W.S.; Wei, P.Y. Geological characteristics and sweet spot evaluation of shale oil reservoir in Fengcheng Formation in Well Maye-1. *Xinjiang Pet. Geol.* **2022**, *43*, 714–723. (In Chinese) [CrossRef]
24. Fritz, P.; Smith, D.G.W. The isotopic composition of secondary dolomites. *Geochem. Cosmochim. Acta* **1970**, *34*, 1161–1173. [CrossRef]
25. Xia, Y.; Deng, H.C.; Wang, Y.Y.; He, J.H.; Xie, X.H.; Teng, J. Development characteristics and formation stages of natural fractures of the Triassic Leikoupo Formation in Pengzhou area of Western Sichuan Basin. *Mar. Orig. Pet. Geol.* **2021**, *26*, 81–89. (In Chinese) [CrossRef]
26. Hu, S.J.; Wang, X.J.; Cao, Z.L.; Li, J.Z.; Gong, D.Y.; Xu, Y. Formation conditions andexploration direction oflarge and medium gas reservoirs in the Junggar Basin, NW China. *Pet. Explor. Dev.* **2020**, *47*, 266–279. [CrossRef]
27. Xie, X.; Hu, L.; Deng, H.; Gao, J. Evolution of pore structure and fractal characteristics of marine shale during electromagnetic radiation. *PLoS ONE* **2020**, *15*, e0239662. [CrossRef]

minerals

Article

Permian Cyanobacterial Blooms Resulted in Enrichment of Organic Matter in the Lucaogou Formation in the Junggar Basin, NW China

Wenhui Wang [1,2], Haisu Cui [1], Jingqiang Tan [1,*], Jin Liu [3], Xueqi Song [1], Jian Wang [3] and Lichang Chen [1]

1 Key Laboratory of Metallogenic Prediction of Nonferrous Metals and Geological Environment Monitoring, Ministry of Education, School of Geosciences and Info-Physics, Central South University, Changsha 410083, China
2 State Key Laboratory of Paleobiology and Stratigraphy, Nanjing Institute of Geology and Paleontology, Chinese Academy of Sciences, Nanjing 210008, China
3 Research Institute of Experiment and Testing, Petro-China Xinjiang Oilfield Company, Karamay 834000, China
* Correspondence: tanjingqiang@csu.edu.cn

Abstract: The Permian Lucaogou Formation in the Junggar Basin, NW China is the target layer for shale oil exploration, but its hydrocarbon precursors have remained the focus of debate. In this study, we investigated the Lucaogou source rocks throughout Well J10025 by conducting detailed petrological, paleontological, and geochemical analyses for the purpose of revealing the occurrence of cyanobacterial blooms as specific hydrocarbon events in the upper Lucaogou Formation. The morphological characteristics of the microfossils and the geochemical signatures of the microfossil-bearing layers support a biological affinity with *Microcystis*, a kind of cyanobacteria. *Microcystis* observed as colonial forms embedded in the upper Lucaogou Formation are of great abundance, indicating the presence of cyanobacterial blooms. They were further evidenced by cyanobacteria-derived biomarkers including low terrestrial/aquatic ratio, high 2α-methylhopane index values, and high abundance of 7- and 8-monomethyl heptadecanes. The blooms occurred in a semiarid and brackish paleoenvironment with anoxic to suboxic water conditions and intermittent volcanic eruptions. Permian *Microcystis* blooms contributed to the enrichment of organic matter in the upper Lucaogou Formation in two main ways: by directly promoting the accumulation of algal biomass and by creating an oxygen-depleted environment for better preservation of organic matter. This study adds a new record to the geological occurrences of cyanobacterial blooms in the Permian, and provides unique insight into the hydrocarbon generation of Jimsar shale oil in the Junggar Basin.

Keywords: shale oil; Lucaogou Formation; *Microcystis*; biological precursor; cyanobacterial bloom

1. Introduction

In modern lakes, cyanobacterial blooms (also known as algal blooms) have gained great attention globally due to their triggering factors and harmful impact on aquatic ecosystems [1–3]. However, cyanobacterial blooms might be beneficial for the accumulation of organic matter and generation of oil. In geological history, algal blooms dominated by various cyanobacteria have been widely recorded and might date back to the early Archean [4,5], the Neoarchean [6], or the Proterozoic Eon [7] (see more examples in [8]). In the Phanerozoic Eon, several cyanobacterial bloom peaks are also documented in the Cambrian [9,10], the early Silurian [11,12], and the Late Devonian periods [13,14]. During the transitional Permian–Triassic interval, shallow marine eutrophication resulted in the *Microcystis* bloom that was preserved in spotted and dendroid microbialites [15].

The Lucaogou Formation in the Junggar Basin is an oil-prone source rock containing enormous shale oil and tight oil resources [16–19]. The hydrocarbon precursors preserved in the paleoenvironment have remained the focus of debate [20,21]. Carroll et al. (1992)

inferred that organic matter in the Lucaogou Formation came from methanotrophic bacteria living near the chemocline [22]. Tao et al. (2012) recognized that the dominant maceral group is bituminite, which is the decomposition by-product of algae, animal plankton, bacterial lipids, and other precursors [23]. Xie et al. (2015) observed microbial fossils with spherical structures and suggested microalgal inputs in the Lucaogou oil shales [24]. Su et al. (2019) proposed that the organic materials originated from aquatic organisms and terrestrial plants [25]. Liu et al. (2019) inferred dolomitized, silicified, and carbonized algae and other bacteria as the main contributors of liquid hydrocarbon in the Lucaogou black shales [26]. These works demonstrated microorganisms including bacteria and algae as the material source of the Lucaogou Formation. However, they did not refer to the hydrocarbon potential of cyanobacterial blooms. In fact, previous studies have revealed that the Lucaogou Formation was deposited mainly in a brackish to saline lacustrine system, with relatively stable and anoxic water conditions [27–29]. This kind of paleoenvironment favors microbial growth [8]. Ding et al. (2019) considered that volcanic ash may stimulate cyanobacterial blooming, as the main reason for accumulation of organic matter in sediments [30]. Sun et al. (2021) reported the discovery of methanogens from the Lucaogou Formation and further proposed that methanogenesis had an impact on hydrocarbon generation [31].

In this research, core samples from Well J10025 in the southeastern Jimsar Sag were studied in detail by conducting petrological, paleontological and geochemical analyses to reveal whether any cyanobacterial bloom occurred in the Lucaogou Formation. Based on this work, we hope to provide new insight into research on hydrocarbon generation in the Lucaogou Formation from the aspect of the relationship between cyanobacterial blooms and the enrichment of organic matter.

2. Geological Setting

The Junggar Basin, with an area of 130,000 km^2, is a large superimposed and reformed oil-bearing basin in western China (Figure 1a) [32]. It is underlain by a Precambrian slightly metamorphic crystalline basement. As a consequence of polycyclic tectonic movements, it evolved from a Paleozoic foreland basin to a Mesozoic faulted basin, and finally transformed into a Cenozoic foreland basin [33–35].

The Junggar Basin is divided by basal structures into several structural units (Figure 1b), including three uplifts (the West Uplift, the East Uplift, and the Luliang Uplift), two depressions (the Wulungun Depression and the Central Depression) and the South Margin Piedmont Thrust Zone. The Jimsar Sag, a secondary structure of the East Uplift, is bounded by the Jimsar Fault in the north, the Xidi Fault and Qing 1 Well South No. 1 Fault in the west, the Fukang Fault Zone in the south, and the Guxi Uplift to the east (Figure 1c). The Jimsar Sag is a dustpan depression formed on the lower Carboniferous folded basement [29,36]. According to geological records, the Jimsar Sag underwent several distinct tectonic events in the Hercynian, Indosinian, Yanshan, and Himalayan periods [37,38].

Black shales from the Permian Lucaogou Formation, which are characterized by a large longitudinal span and wide planar distribution, have great exploration and development potential [39–41]. Based on seismic data and detailed core observations, the Lucaogou Formation is divided into two members, the upper member (P_2l_2) and the lower member (P_2l_1), both of which are further subdivided into two beds ($P_2l_2^1$ and $P_2l_2^2$, and $P_2l_1^1$ and $P_2l_1^2$). The upper member is mainly composed of mudstone, dolomitic mudstone, sandstone, muddy dolomite, and sandy dolomite, while siltstone, silty mudstone, and lime mudstone are more common in the lower member (Figure 1d).

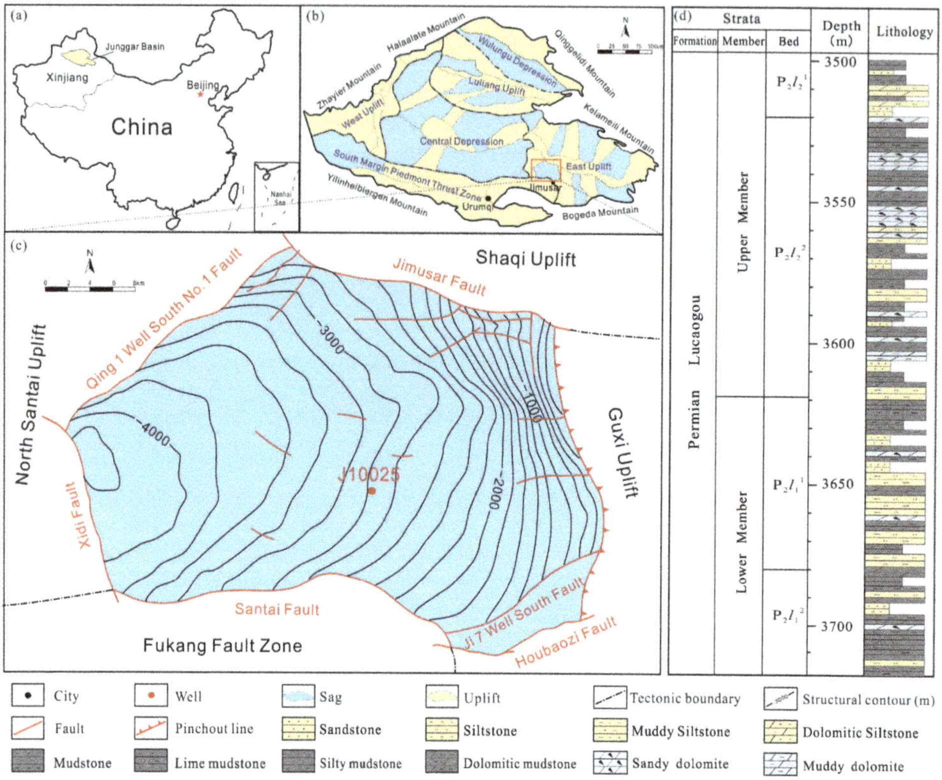

Figure 1. Geological map of the study area and stratigraphic column of the study well (modified from [42]). (**a**) Location of the Junggar Basin in China. (**b**) Structural map of the Junggar Basin. (**c**) Structural framework of the Jimsar Sag and location of the study well. (**d**) Stratigraphic column of Well J10025.

3. Materials and Methods

In this study, 30 samples were collected from Well J10025, including mudstone, siltstone, sandstone, and dolomite. We further selected 26 source rocks exclusive of sandstones for total organic carbon (TOC) and element concentration analyses, and 22 source rocks for biomarker compound analysis (Tables 1 and 2). Individual samples were ground to a particle size of 100 mesh for TOC measurement and to 200 mesh or finer for element and biomarker analysis. For petrological studies, thin sections were cut normal to the bedding.

3.1. *Petrological and Paleontological Analysis*

Thirty samples were systematically analyzed using optical and fluorescence microscopy (Zeiss A1 Scope, Carl Zeiss AG, Oberkochen, Germany) to investigate the basic reservoir characteristics of the Lucaogou Formation. Based on the results, we further selected samples from the microfossil-bearing strata (J-17, 3574.2 m) for scanning electron microscopy (SEM, MIRA3 LMH, Tescan, Brno, The Czech Republic) and in situ element analysis. Samples with freshly broken surfaces were first observed under stereoscope to select the target area. Before SEM analysis, samples were air dried, mounted on stubs using double-sided tape, and coated with gold. Microscope images were taken under an accelerating voltage of 20 kV. In addition, elemental mapping of energy-dispersive X-ray spectrometry (EDS) was carried out on specific fossils. The above tests were conducted at the Geological Lab of Alternative Energy at Central South University (Changsha, China).

Table 1. TOC and element ratios of the Lucaogou Formation.

Sample	Depth (m)	TOC (%)	Sr/Ba	B/Ga	C_{VALUE}	CIA	Mo_{EF}	U_{EF}	P (%)	P/Ti	Ba/Al
J-26	3499.5	5.77	0.99	16.33	1.14	74	11.67	2.56	0.06	0.26	7.02×10^{-3}
J-25	3506.8	1.81	1.40	25.86	0.28	79	0.51	0.56	0.10	0.46	3.06×10^{-3}
J-24	3515.0	7.76	0.76	26.89	0.34	68	6.95	1.95	0.17	0.54	6.78×10^{-3}
J-23	3523.7	6.81	0.62	4.93	0.85	59	23.13	3.46	0.04	0.17	1.68×10^{-2}
J-22	3526.1	5.74	0.49	4.35	0.94	62	15.35	8.51	0.12	0.34	2.60×10^{-2}
J-21	3530.8	14.93	0.85	6.41	1.00	64	29.54	8.22	0.13	0.63	1.33×10^{-2}
J-20	3547.8	5.53	0.26	6.40	0.96	59	10.28	3.98	0.04	0.11	2.02×10^{-2}
J-19	3569.2	3.10	0.48	9.92	0.38	44	14.83	2.44	0.01	0.05	1.91×10^{-2}
J-18	3570.5	6.75	0.60	6.68	0.40	48	11.91	2.72	0.01	0.09	3.15×10^{-2}
J-17	3574.2	19.60	0.71	7.93	0.22	53	6.25	2.41	0.20	1.05	6.79×10^{-3}
J-16	3576.6	16.97	1.04	8.70	0.66	46	19.44	6.17	0.04	0.20	2.17×10^{-2}
J-15	3595.4	15.80	2.16	13.42	0.11	50	2.13	4.42	0.06	0.42	4.36×10^{-3}
J-14	3603.1	7.04	2.11	8.93	0.14	54	3.57	2.81	0.09	0.82	8.60×10^{-3}
J-13	3613.7	6.98	0.63	10.24	0.15	48	6.08	0.79	0.05	0.16	4.84×10^{-3}
J-12	3624.3	3.62	2.10	15.14	0.11	46	9.49	2.11	0.03	0.19	5.59×10^{-3}
J-11	3629.2	2.29	1.15	14.50	0.07	48	3.23	0.85	0.04	0.21	7.32×10^{-3}
J-10	3664.3	4.20	1.73	11.37	0.12	50	1.28	1.66	0.05	0.30	6.21×10^{-3}
J-09	3679.5	3.11	1.57	18.89	0.09	49	1.41	0.93	0.03	0.18	4.82×10^{-3}
J-08	3681.6	2.44	2.24	10.35	0.19	48	3.97	1.31	0.03	0.13	4.66×10^{-3}
J-07	3684.5	3.48	2.84	13.48	0.11	47	1.72	0.74	0.04	0.25	5.93×10^{-3}
J-06	3685.5	5.82	1.26	8.72	0.20	52	12.32	1.18	0.07	0.35	7.76×10^{-3}
J-05	3695.3	1.95	1.81	11.65	0.14	48	15.70	3.98	0.07	0.28	9.83×10^{-3}
J-04	3696.9	2.50	2.28	19.72	0.11	50	2.90	1.49	0.02	0.09	6.09×10^{-3}
J-03	3699.9	5.73	1.30	12.86	0.21	51	4.64	0.80	0.03	0.14	7.56×10^{-3}
J-02	3701.3	7.87	2.54	9.72	0.08	49	63.49	2.42	0.05	0.37	9.71×10^{-3}
J-01	3708.3	4.85	4.38	19.85	0.08	51	4.04	0.77	0.04	0.35	4.72×10^{-3}

Table 2. Biomarker ratios of the Lucaogou Formation.

Sample	Depth (m)	Max-Peak	CPI	OEP	TAR	Pr/n-C_{17}	Ph/n-C_{18}	Pr/Ph	GI	2-MHI	β-Carotane Index	7,8-MMH/C_{max}	Regular Steranes (%) C27	C28	C29
J-26	3499.5	C_{23}	1.08	1.40	0.91	2.14	1.51	1.30	0.10	2.00	18.74	0.04	22.4	24.0	53.7
J-25	3506.8	C_{23}	1.09	1.34	1.26	1.30	1.57	0.74	0.14	3.12	34.69	0.03	42.8	41.6	15.6
J-24	3515.0	C_{23}	1.06	1.23	0.91	1.20	1.09	0.75	0.20	1.41	14.38	0.04	27.7	28.6	43.7
J-23	3523.7	C_{17}	1.08	1.29	0.09	0.86	0.76	1.25	0.21	1.35	3.46	0.12	33.3	26.3	40.4
J-21	3530.8	C_{19}	1.07	1.01	0.31	0.60	0.51	1.01	0.16	1.88	2.86	0.07	25.3	35.0	39.7
J-20	3547.8	C_{25}	1.05	1.06	0.62	10.59	3.67	0.47	0.20	3.64	5.56	0.00	19.6	36.7	43.7
J-19	3569.2	C_{23}	1.05	1.18	0.92	0.91	0.67	0.77	0.24	4.20	8.03	0.04	11.9	39.5	48.5
J-17	3574.2	C_{17}	1.09	1.24	0.11	0.87	0.81	1.33	0.21	4.65	4.69	0.11	23.4	40.2	36.5
J-15	3595.4	C_{19}	1.09	1.14	0.20	0.46	0.30	1.50	0.19	4.69	3.69	0.04	18.0	41.2	40.8
J-14	3603.1	C_{21}	1.08	1.07	0.58	4.37	6.15	0.52	0.24	2.84	6.65	0.01	27.1	20.8	52.1
J-13	3613.7	C_{17}	1.07	1.35	0.13	0.95	0.87	1.31	0.18	4.05	7.34	0.11	17.1	38.4	44.6
J-12	3624.3	C_{21}	1.07	1.05	0.36	1.25	1.02	1.14	0.25	4.01	6.01	0.12	11.0	36.2	52.8
J-11	3664.3	C_{23}	1.05	1.17	0.80	2.93	3.64	0.84	0.25	4.18	31.42	0.24	11.4	34.7	53.9
J-10	3629.2	C_{17}	1.06	1.38	0.29	1.28	1.66	0.82	0.23	4.15	6.77	0.17	10.6	40.2	49.3
J-09	3679.5	C_{23}	1.05	1.14	0.68	1.70	1.95	0.88	0.24	3.33	26.04	0.15	8.3	36.1	55.6
J-08	3681.6	C_{23}	1.05	1.13	1.21	1.69	1.57	0.65	0.22	3.93	66.47	0.05	8.5	36.0	55.5
J-07	3684.5	C_{20}	1.03	1.10	0.79	1.58	1.19	0.20	0.25	3.64	17.15	0.03	9.9	29.3	60.7
J-06	3685.5	C_{22}	1.05	1.08	0.92	0.97	1.20	0.42	0.26	3.48	13.25	0.05	11.7	26.7	61.6
J-04	3696.9	C_{23}	1.06	1.11	2.16	1.79	1.53	0.43	0.20	4.37	28.02	0.04	9.2	35.4	55.5
J-03	3699.9	C_{23}	1.06	1.11	1.52	1.10	1.13	0.49	0.23	2.91	15.66	0.04	9.2	35.2	55.6
J-02	3701.3	C_{17}	1.10	1.43	0.13	0.94	0.95	0.98	0.22	5.26	1.19	0.13	13.3	36.5	50.2
J-01	3708.3	C_{23}	1.05	1.12	1.70	1.23	1.35	0.19	0.22	4.21	28.30	0.03	10.5	32.9	56.6

Note: The carbon preference index (CPI) and the odd–even predominance (OEP); Terrigenous/aquatic ratio (TAR) = $(nC_{27} + nC_{29} + nC_{31})/(nC_{15} + nC_{17} + nC_{19})$; Gammacerane index (GI): gammacerane/17α,18β(H)-C_{30} hopane; 2α-methylhopane index (2MHI) = 2α-methylhopanes/(2α-methylhopanes + 17α,18β(H)-C_{30} hopane); β-carotane index = 100 × (β–carotane/17α,18β(H)-C_{30} hopane); 7-, 8-monomethyl heptadecanes/C_{max} ratio (7,8-MMH/C_{max}).

3.2. Biomarker Compound Analysis

Based on the petrological and paleontological analyses, we carried out biomarker compound analysis on 22 selected samples of the Lucaogou source rocks by QP2020NX gas chromatography–mass spectrometry (GC–MS, Shimadzu Corporation, Kyoto, Japan) at Central South University. To reduce potential contamination, samples were crushed to 100 mesh after the outer weathering surfaces were removed. An Rtx-5MS fused silica column (30 m $\times$ 0.25 mm $\times$ 0.25 μm) was used to separate hydrocarbons, employing helium as a carrier gas. The oven temperature was programmed to start at 80 °C (held for 3 min) and increased to 230 °C at a heating rate of 3 °C/min, and then to 310 °C at 2 °C/min.

3.3. TOC and Element Concentration Analysis

Total organic carbon (TOC) analysis was conducted by a Leco–CS744 instrument (LECO Corporation, St. Joseph, MI, USA) following the procedures of a Chinese national standard GB/T 19145-2003. Twenty-six selected samples of the Lucaogou source rocks were pretreated with dilute hydrochloric acid (5% HCl) to remove inorganic carbon [43,44]. The treated samples were burned in an oven at 1100 °C, adding iron chips and flux. The analytical errors were not more than 0.5%.

Major and trace element analyses were performed on 26 core samples at a petroleum technology corporation in Beijing. A wavelength X-ray fluorescence spectrometer (model: Axios^{mAX}, Malvern Panalytical, Malvern, UK) was used to determine the oxides of major elements, including SiO_2, Al_2O_3, CaO, K_2O, Na_2O, Fe_2O_3, MnO, MgO, TiO_2, and P_2O_5, according to Chinese national standard GB/T 14506.28-2010. Trace elements were analyzed by a NexION300D inductively coupled plasma mass spectrometer (ICP–MS, PerkinElmer, Waltham, MA, USA) in accordance with Chinese national standard GB/T 14506.30-2010. The analytical precision was better than ±5% and detection limits were 1–2 ppm for most trace elements.

The enrichment factor of elements was calculated by the following formula [45,46]. Samples were normalized to the upper continental crust [47].

$$X_{EF} = (X/Al)_{sample}/(X/Al)_{UCC} \tag{1}$$

X and Al represent the weight percentage of elements X and Al, respectively.

The chemical index of alteration (CIA) can be calculated by the equation [48,49]:

$$CIA = [Al_2O_3/(Al_2O_3 + Na_2O + CaO^* + K_2O)] \times 100 \tag{2}$$

CaO^* represents the CaO content in the silicate fraction of the rock, which can be corrected using the formula: $CaO' = CaO - (10/3)\ P_2O_5$. If the content of CaO' is greater than that of Na_2O, then CaO^* is assumed to be equivalent to Na_2O. In contrast, CaO^* is equal to CaO'. All elemental concentrations are stated in percentage units.

The C_{value} is a proxy for paleoclimate, and the calculation is as follows:

$$C_{value} = \Sigma(Fe + Mn + Cr + Ni + V + Co)/\Sigma(Ca + Mg + Sr + Ba + K + Na) \tag{3}$$

The units of all elemental concentrations in formula (3) are in ppm.

4. Results

4.1. Microfossils in the Lucaogou Formation

Detailed core and thin section observations reveal that mudstone, sandy mudstone, and dolomitic mudstone are the main source rocks of the Lucaogou Formation in Well J10025. They are primarily characterized by fine-grained compositions, thin interbedded layers, and frequent changes in lithology (Figures 1d and 2a,b). In the upper member, a majority of well-preserved alginate, especially lamalginite, was visible in several samples, displaying yellowish to green fluorescence under incident blue light illumination

(Figure 2c,f,i). Abundant lamalginite documents the occurrence of algal blooms. These algal laminae are always interbedded with terrigenous clastic laminae and sometimes intermingled with ostracods that were previously identified as *Permiana, Tomiella, Kelameilina*, etc. [50] (Figure 2e,f,h). In Figure 2d,g, dry-cracked fragments and water-escape structures are also present within some cores, indicating aridity and a rapid depositional environment [51,52].

Figure 2. Core images (**left**), thin sections (**middle**), and fluorescent photographs (**right**) of the representative Lucaogou source rocks. (**a**,**b**) Organic matter (OM) laminae intermingled with terrigenous clastic laminae, mudstone, 3530.8 m. (**c**) Lamalginite displaying yellowish to green fluorescence, 3530.8 m. (**d**) OM laminae, terrigenous clastic laminae, and water-escape structure, dolomitic mudstone, 3576.6 m. (**e**) Ostracoda laminae and terrigenous clastic laminae, 3576.6 m. (**f**) Lamalginite and ostracoda laminae, 3576.6 m. (**g**) OM laminae and dry-cracked fragments, dolomitic mudstone, 3595.4 m. (**h**) Ostracoda, and OM laminae, 3595.4 m. (**i**) Lamalginite, yellowish to green fluorescence, 3595.4 m.

Microfossils embedded in the upper Lucaogou Formation were observed using SEM as irregular colonial forms (Figure 3a–d). In Figure 3a, sheet-like algal mats are distributed in a circled region, with an area of approximately 0.12 mm^2. They are characterized and constated by abundant, randomly arranged globular to elliptical shaped individuals, some of them with ornaments on the cell surfaces (Figure 4a–c). Through EDS elemental mapping tests, it was revealed that most of these fossil spheroids were replaced by pyrites, or filled by pyrite framboids inside the spherical chamber, some of which dropped and successively formed shedding molds (Figures 4d and 5a). Furthermore, some cracked spheroids after pyritization are closely attached to silicate minerals (Figure 5b–f).

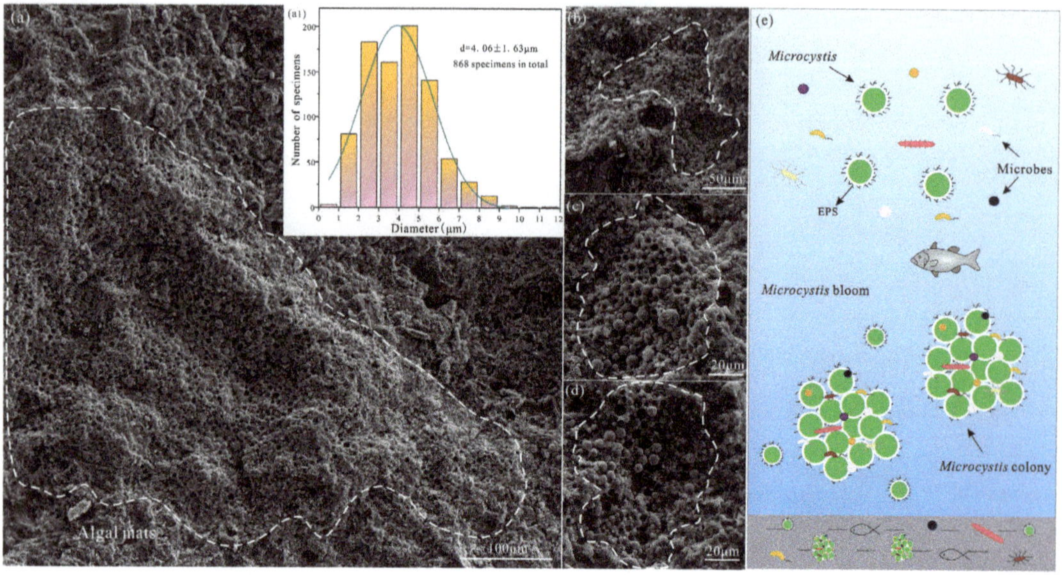

Figure 3. Microscopic photographs of *Microcystis* bloom in the upper Lucaogou Formation (J-17, 3574.2 m). (**a**) Sheet-like algal mats are widely distributed in the circled area. (**a1**) Size distributions and the fitting curve of *Microcystis* spheroids (count: 868, mean: 4.06 μm, standard deviation: 1.63 μm) (**b–d**) *Microcystis* cells are irregularly integrated as colonies. (**e**) Schematic diagram of *Microcystis* colony formation and bloom.

Individual microfossils are of different sizes, ranging from 1 μm to 12 μm in diameter. We counted 868 specimens in total, most of the fossils exhibit a diameter between 1 μm and 7 μm (avg. 4.06 μm) (Figure 3(a1)). They are always encircled by organic matter and attached by clastic minerals. Trilobate and rectangular ornaments are visible in some single individuals (Figure 6). The genesis of these ornaments can be attributed to the metasomatism and crystallization of pyrites. The algal colonies were replaced by pyrites before being squashed and preserved due to their sulfur affinity. During the transformation from initially formed pyrite spheroids to larger particles with higher crystallinity, the specific surface area increased with extra spaces, thus leading to the formation of different ornaments on the cell surfaces. According to EDS testing, the fossils in Figure 6 are mainly composed of C (42.1–52.5 At%), S (24.1–31.2 At%), and Fe (13.1–17.8 At%), with a small amount of O (8.0–9.5 At%) and traces of Si and Ca. The microfossil colonies are interpreted here as cyanobacteria, with affinities to the *Microcystis* colonies (*Chroococcaceae*) based on morphological identification characters.

Microcystis is a kind of greenish planktic *cyanobacterium*, characterized by a coccoid structure, that commonly forms blooms in modern lakes, rivers, and reservoirs [53]. Solitary *Microcystis* cells are commonly spherical or subspherical, with diameters ranging between 1.7 and 7 μm [54,55]. In our study, slightly larger cells with diameters greater than 7 μm can be attributed to mineralization with pyrites. The cells are able to accumulate in large numbers and integrate irregularly with other microbes under suitable environmental conditions (Figure 3e). Therefore, *Microcystis* can exhibit a variety of colonial morphologies, such as ellipsoid, spheroid, and irregular dendric [56,57]. Cyanobacteria have a relatively high probability of becoming incorporated in the fossil record as cellularly intact specimens, especially when they are preserved by permineralization during the early stages of sediment lithification as shown in this study (Figures 4–6). These cyanobacteria colonies are critical contributors to the abundant lamalginite layers and clearly document the occurrence of cyanobacterial blooms.

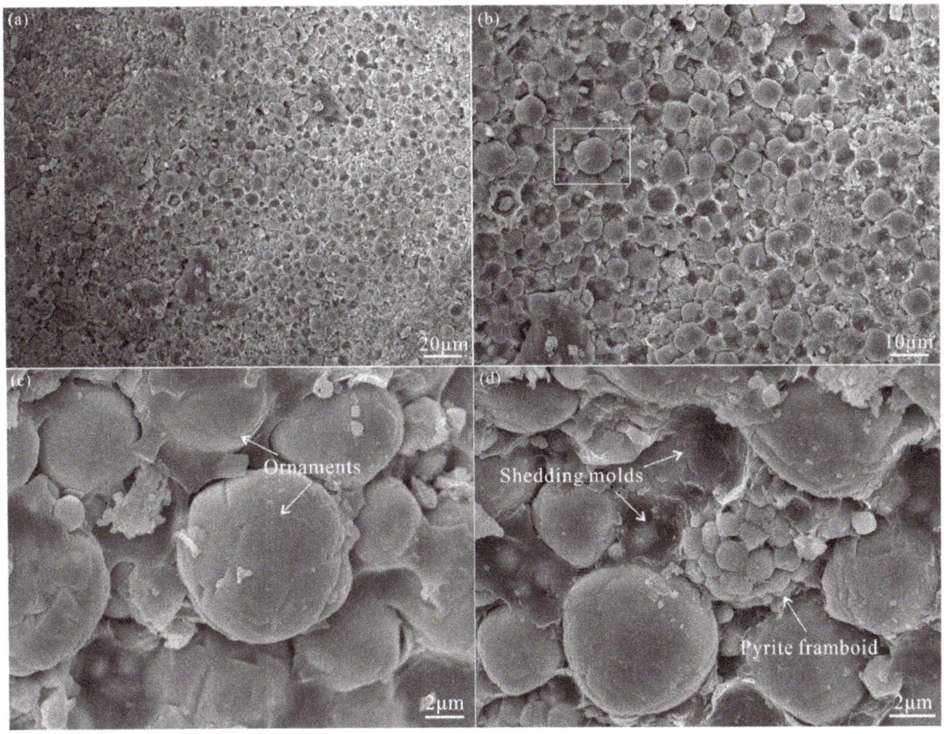

Figure 4. Microscopic photographs of *Microcystis* in the upper Lucaogou Formation (J-17, 3574.2 m). (**a**) *Microcystis* colony. (**b**) Magnified view of (**a**); abundant, randomly arranged globular to elliptical shaped individuals. (**c**) Close-up of the rectangle in (**b**); ornaments are visible on the cell surface. (**d**) *Microcystis* individuals, pyrite framboids, and their shedding molds.

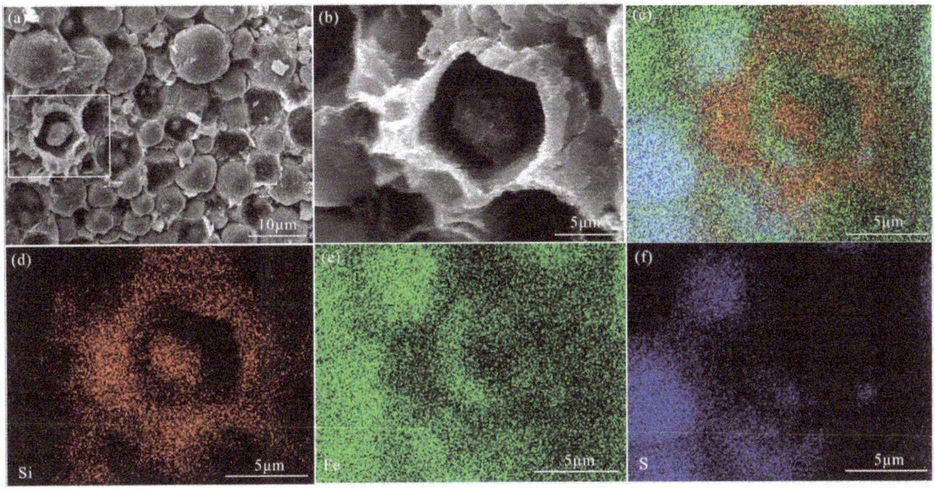

Figure 5. Elemental mapping test results of microfossil *Microcystis*. (**a**,**b**) SEM photographs of a cracked spheroid; (**b**) is the magnified view of the rectangle in (**a**); (**c**–**f**) Elemental mapping test of the cracked spheroid, mainly composed of Si, Fe, and S.

Figure 6. Microscopic photographs of individual *Microcystis* (all bars represent 2 μm). (**a**,**b**) Individuals with trilobate ornaments on the cell surfaces are attached by clastic minerals and encircled by organic matter. (**c**) Microfossil spheroid encircled by organic matter. (**d**) Microfossil spheroid and shedding molds. (**e**,**f**) Microfossil spheroids with rectangular ornaments on the cell surfaces.

4.2. Biomarkers

Biomarkers including normal alkanes, isoprenoids, terpanes, and steranes recorded by mass chromatography are frequently utilized to investigate the origin of organic matter and its depositional environment [18]. Detailed biomarker ratios of the Lucaogou source rocks are listed in Table 2.

The distributions of n-alkanes show unimodal characteristics with carbon peaks located at nC_{17}–nC_{25} (Figure 7a). In addition, the peak numbers of the upper member are smaller than those of the lower member. The terrestrial/aquatic ratio (TAR), which is defined as $(nC_{27} + nC_{29} + nC_{31})/(nC_{15} + nC_{17} + nC_{19})$, varies between 0.09 and 5.97. The carbon preference index (CPI) ranges from 1.05 to 1.10 (avg. 1.07) and the odd–even predominance (OEP) varies from 1.01 to 1.43 (avg. 1.19), indicative of low to moderate maturity of the Lucaogou source rocks.

Abundant isoprenoids such as pristane (Pr) and phytane (Ph) were also detected in the studied samples, with Pr/Ph ratios ranging from 0.19 to 1.50. The ratios of pristine/n-C_{17} are 0.46–10.59 (avg. 2.21) in the upper member and 0.94–2.93 (avg. 1.50) in the lower member. The ratios of phytane/n-C_{17} in the upper and lower members are 0.30–6.15 (avg. 1.63) and 0.95–3.64 (avg. 1.56) respectively. In addition, 7- and 8- monomethyl hepadecanes were identified via a m/z 57 gas chromatogram, representing a predominance over 2–6 monomethyl homologs (Figure 7c). The 7-, 8-monomethyl heptadecanes/C_{max} ratios range between 0.01 and 0.24 (avg. 0.08).

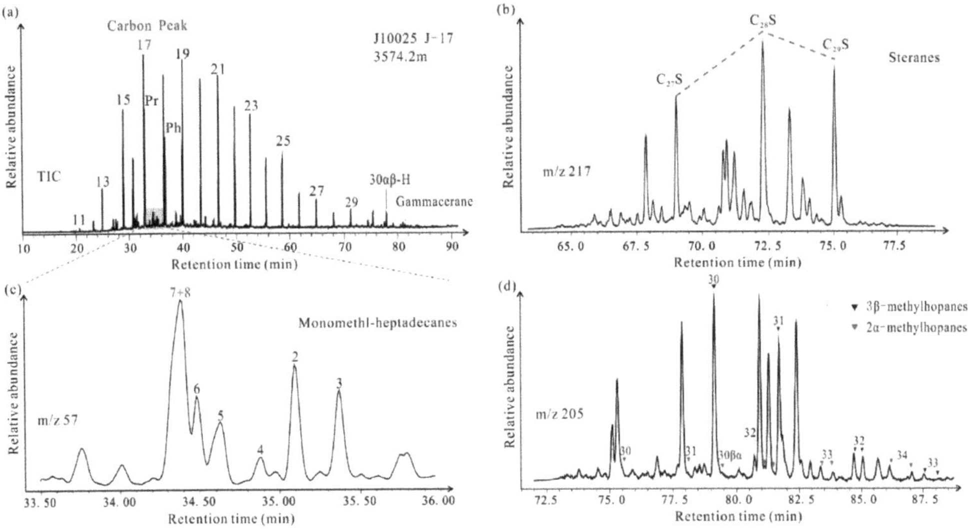

Figure 7. Biomarker chromatograms of cyanobacteria-bearing sample J-17, 3574.2 m. (**a**) Total ion chromatogram of the shale sample. (**b**) Mass chromatograms (m/z = 217) of the steranes. (**c**) Mass chromatograms (m/z = 57) of 2–8 monomethyl heptadecanes. (**d**) Mass chromatograms (m/z = 205) of hopanes.

Differences in the abundance of β-carotane are noticeable between the upper and lower members. The β-carotane index, which is defined as 100 (β-carotane/17α, 18β(H)-C_{30} hopane), is significantly higher in the lower member (1.19–66.47, avg. 21.84) than the upper member (2.86–34.69, avg. 10.01). Moreover, the abundance of gammacerane in samples from the lower member is markedly lower. The gammacerane index (GI: gammacerane/17α, 18β(H)-C_{30} hopane) of the lower member yielded a range of 0.20–0.26 (avg. 0.23), while the values of the upper member were found to be 0.10–0.24 (avg. 0.19).

Regarding regular steranes, samples of the Lucaogou source rocks shows a relatively lower proportion of C_{27} steranes (8.3%–42.8%, avg. 17.4%), a medium concentration of C_{28} steranes (20.8%–41.6%, avg. 34.2%), and a slightly greater abundance of C_{29} steranes (15.6%–61.6%, avg. 48.5%) (Figure 7b). Furthermore, the 2α-methylhopanes of the samples were identified using a representative gas chromatogram of m/z 205 (Figure 7d). The 2α-methylhopane index (2-MHI), defined as 2α-methylhopanes/(2α-methylhopanes + 17α,18β(H)-C_{30} hopane), covers a wide range of 1.35–5.26 (avg. 3.51) (Figure 8).

4.3. TOC and Elemental Geochemical Characteristics

The Lucaogou source rocks in Well J10025 exhibit a wide range of TOC contents that range between 1.81 and 19.60 wt% (Table 1). The upper member has significantly higher values (avg. 9.14 wt%) than the lower member (avg. 4.17 wt%) (Figure 9).

The elemental geochemical data of the Lucaogou source rocks are listed in Table 1. The C-values of the Lucaogou Formation are highly variable (range: 0.07–1.14) and show a significant stratigraphic trend. In general, the C-values increase gradually from the lower member to the upper member. The stratigraphic variation of CIA values exhibits a similar trend to the C-values, varying from low values (46–52) in the lower member to high values (44–79) in the upper member.

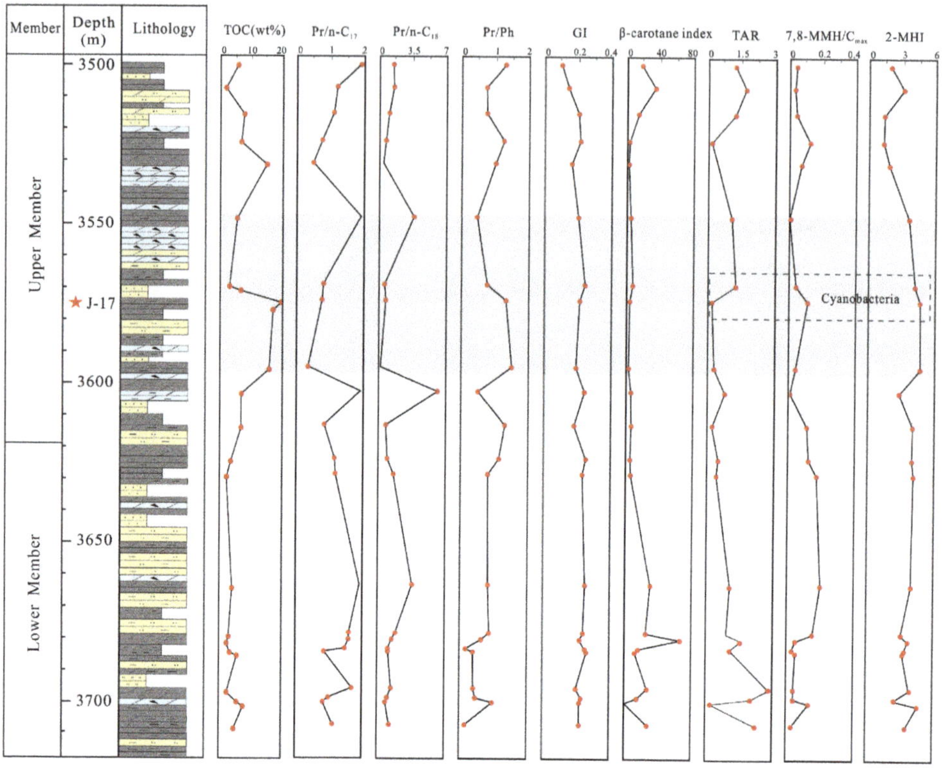

Figure 8. Biomarkers of the Lucaogou Formation (the red star marks the sample location of the microfossil J-17, 3574.2 m).

The Sr/Ba ratios of the lower member are significantly higher (1.15–4.38) than those of the upper member (0.26–2.16). However, it should be noted that some abnormally high values are largely due to the interference of carbonate Sr [58]. The B/Ga ratios also show considerable variations between the upper and lower members. Despite a few high values, most samples in the upper member exhibit lower values (4.35–13.42) than those of the upper member (8.72–19.85). In addition, the enrichment factors of Mo and U (Mo_{EF} and U_{EF}) show similar stratigraphic variation patterns, varying from 0.51 to 63.49 and 0.56 to 8.51, respectively (Figure 9).

Phosphorus (P) is essential to all forms of life on Earth, as it plays a fundamental role in many metabolic processes and acts as a major constituent of the biotic skeleton [46]. In the layers of *Microcystis* fossils, the P concentration increases sharply and reaches its maximum (0.20 wt%) in the sedimentary sequence. The P/Ti and Ba/Al ratios of the upper member (avg. 0.38 and 1.36×10^{-2} respectively) are remarkably higher than those of the lower member (avg. 0.24 and 6.68×10^{-3} respectively).

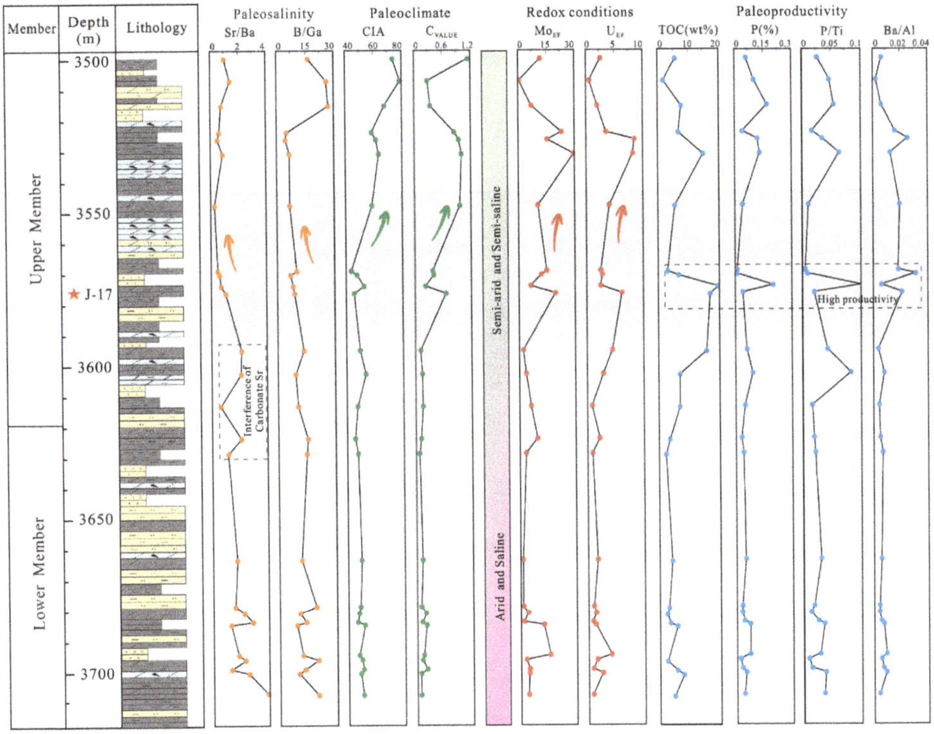

Figure 9. TOC and element geochemical parameters of the Lucaogou Formation (the red star marks the sample location of the microfossil J-17, 3574.2 m).

5. Discussion

5.1. Biological Precursors of the Lucaogou Source Rocks

The distribution of n-alkanes in source rocks can provide information about the origins of organic matter, including algae, bacteria, and higher plants [20,43]. As shown in Table 2 and Figure 7a, the distributions of n-alkanes show unimodal characteristics with carbon peaks located at nC_{17}–nC_{25}, indicating that the source of organic matter is a combination of terrestrial and aquatic organisms. However, most of the samples from the upper member of the Lucaogou Formation exhibit nC_{17}–nC_{21} carbon peak numbers, indicative of a dominance of algal and bacterial input, while the lower member with nC_{23} as the most frequent peak reveals terrestrial higher plant inputs [59,60].

The terrigenous/aquatic ratio (TAR) may reflect the relative abundance of terrigenous plants versus aquatic organisms in sediments [44,61]. The TAR of the Lucaogou Formation exhibits a wide range between 0.09 and 2.16, suggesting a mixed contribution of organic inputs. However, the TAR of the upper member is significantly lower (0.09–1.26, avg. 0.54) than that of the lower member (0.13–2.16, avg. 0.96), especially in the microfossil-bearing layer (0.11), indicating that algae or bacteria are the main biological precursors of the upper member while terrestrial plants contribute more to the lower member.

Regular steranes have been widely used to estimate the relative proportions of organic matter sources [62]. In general, C_{27} regular steranes are primarily derived from algae and zooplankton, C_{28} regular steranes originate from phytoplankton, and C_{29} regular steranes are typically associated with terrestrial plants [63,64]. However, microalgae and cyanobacteria are also considered important sources of C_{29} steranes [30,65]. In Figure 10a, Lucaogou source rocks show a mixed contribution of organic inputs, which in the upper member mainly come from plankton and bacteria while in the lower member they originate from plankton and land plants.

Furthermore, 2α-methylhopanes (2-MHs) are highly diagnostic for identifying cyanobacteria. Thus, the 2α-methylhopane index (2-MHI) is commonly used to evaluate the abundance of 2-MHs in geological sediments [66–68]. High 2-MHI values (1.35–5.26) in Lucaogou source rocks could demonstrate that the biological inputs mainly consisted of cyanobacteria. In addition, monomethyl alkanes including the 7- and 8-carbon substituted methyl heptadecanes are more abundant in modern cyanobacteria [69,70]. The remarkable abundance of 7- and 8-monomethyl heptadecanes compared to their homologs may provide evidence of cyanobacterial blooms [71,72]. Moderate 7-, 8-monomethyl heptadecanes/C_{max} ratios (0.01–0.24) imply that cyanobacteria comprise an important part of the organic matter input of the Lucaogou Formation.

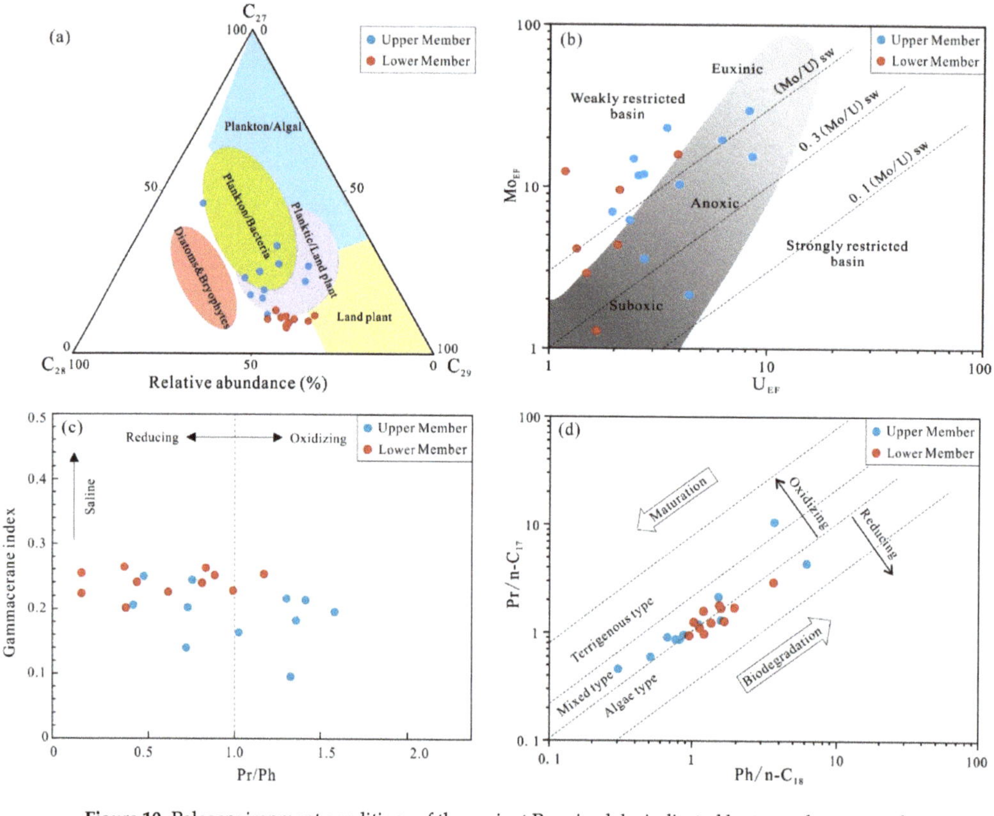

Figure 10. Paleoenvironment conditions of the ancient Permian lake indicated by trace elements and biomarkers. (**a**) Ternary plot of C_{27}–C_{28}–C_{29} regular steranes, illustrating the organic matter input (modified from [73]). (**b**) Plot of U_{EF} versus Mo_{EF}. The diagonal lines represent multiples (0.1, 0.3 and 1) of the Mo:U ratio of present-day seawater (modified from [74]). (**c**) Plot of Pr/Ph versus the Gammacerane index. (**d**) Plot of Ph/nC18 versus Pr/nC17 (following [75]).

In summary, based on a comprehensive analysis of biomarker evidence, cyanobacteria can be reasonably inferred to be the main biological precursors of the upper member of the Lucaogou source rocks deposited in the Permian lake. This inference is in accordance with the results of a paleontological study that identified the microfossil spheroids as *Microcystis*, by using direct SEM observations. In contrast, the hydrocarbon precursors of the lower member were mainly terrestrial higher plants mixed with some planktic algae.

5.2. Paleoenvironmental Conditions of the Permian Microcystis Bloom

As the upper member of the Lucaogou Formation yields better shale oils with cyanobacteria as its main biological precursors, the following discussion focuses mainly on the upper member and tries to decipher the relationship between the Permian paleoenvironment and the *Microcystis* bloom.

5.2.1. Environmental Factors Triggering Modern Cyanobacterial Blooms

Microcystis is one of the predominant taxa involved in cyanobacterial blooms, with others including *Planktothrix, Gloetrichia, Oscillatoriales*, etc. [76]. *Microcystis* prefers to grow in eutrophic waters under strong light, relatively high temperatures (the optimum temperature range is 25–30 °C), and highly alkaline conditions (pH 8.0–9.5) [2,77]. In addition, the growth of *Microcystis* is seasonal and periodic. *Microcystis* is physiologically dormant in the sediment layer during winter but returns to the water column to proliferate rapidly during summer, thus becoming the predominant species [1,78,79]. Furthermore, nutrients such as phosphorus (P), nitrogen (N), iron (Fe), and magnesium (Mg) are also responsible for controlling algal biomass and bloom events in aquatic ecosystems [80,81]. *Microcystis* blooms are closely affected by numerous ecological and environmental factors including climate, sunlight, temperature, pH, nutrients, and hydrodynamic conditions.

5.2.2. Paleoclimate and Paleosalinity

The CIA is a reliable proxy for evaluating the chemical weathering degree of sediments and paleoclimate [49,82]. Low CIA values of the lower member, ranging from 46 to 52, indicate arid conditions under low to moderate degrees of chemical weathering. In contrast, the upper member exhibits relatively higher CIA values (44–79, avg. 58), implying a semiarid climate during the deposition of source rocks. The C-value is another elemental parameter that is utilized for paleoclimate reconstruction [83,84]. The C-values of the lower member (0.07–0.21) reveal an arid environment, while those of the upper member are significantly higher (0.11–1.14), suggesting a transition from an arid to a semiarid climate. In addition, dry-cracked fragments and water-escape structures in core samples are indicative of aridity and a rapid depositional environment (Figure 2d,g).

Sr/Ba and B/Ga ratios are considered effective salinity indicators [85,86]. Wei and Aljeo estimated that the approximate salinity thresholds for Sr/Ba are <0.2 in freshwater, 0.2–0.5 in brackish, and >0.5 in marine facies [58]. For B/Ga, sedimentary ratios of <3, 3–6, and >6 are indicative of freshwater, brackish, and marine facies, respectively. In this study, both ratios exhibit high values (Sr/Ba: avg. 1.47, B/Ga: avg. 12.43), indicative of a saline environment. However, the declining values in the upper member reveal a decreasing salinity trend (Figure 9). While the Sr/Ba proxy is slightly less robust owing to carbonate Sr interference, other proxies can be used to support interpretations of paleosalinity [58,87]. For example, abundant gammacerane is usually interpreted as evidence of stratified water column, which is closely associated with hypersalinity [73,88]. The gammacerane index value (GI: gammacerane/C_{30} hopane) of the upper member ranges from 0.10 to 0.24 (avg. 0.19), which likely indicates a fresh to brackish depositional environment. In contrast, the lower member with values of 0.20–0.26 suggests brackish to saline conditions with unstable stratification of the water column (Table 2, Figure 8).

5.2.3. Redox Conditions

Mo_{EF} and U_{EF} are reliable indicators for paleoredox evaluation [89,90]. The Mo–U covariation model has recently been widely used as a reliable proxy for determining redox and water retention in black shales [74,86]. In the Mo–U covariation model, most samples from the Lucaogou Formation fall into the range above 0.3 times the Mo/U of seawater, yielding greater Mo_{EF} relative to U_{EF}, indicative of an anoxic to euxinic environment (Figure 10b).

The pristine/phytane ratio (Pr/Ph) is normally used as a paleoredox proxy [44,91]. As shown in Table 2 and Figure 8, most of the Pr/Ph ratios are less than 1, suggesting a reducing depositional environment. However, the ratios of the upper member are slightly higher, revealing an anoxic to suboxic environment. Plots of Pr/n-C_{17} versus Ph/n-C_{18} can also be used to infer the redox conditions of source rocks [92], in this case indicating a weakly oxidizing to weakly reducing depositional environment with mixed organic inputs (Figure 10d).

β-carotane generally accumulates in highly reducing and saline lacustrine environments [70,93]. The high β-carotane index (1.19–66.47) of the lower member indicates more reducing and saline conditions. In addition, pyrite framboids in some core samples can also suggest reducing conditions (Figure 4d). Therefore, a suboxic to anoxic depositional environment for the upper member of the Lucaogou Formation was reconstructed by the above-mentioned redox proxies.

5.2.4. Productivity

Productivity plays an important role in organic matter enrichment and source rock potential [46,94]. Various geochemical proxies including C and N isotopes, organic biomarkers, and trace metal abundances (Cu, Ba, Zn, etc.) have been utilized to reconstruct biological productivity [95–98]. In this study, we refer to TOC, P, P/Ti, and Ba/Al as paleoproductivity proxies.

Total organic carbon (TOC) accounts for the largest proportion of organic matter preserved in sediments, acting as a direct indicator of productivity [99–101]. The Lucaogou Formation with TOC values ranging between 1.81 and 19.60 wt% suggest high productivity (Figure 9). In addition, distributions of phosphorous in sediments are closely linked to primary biological productivity and the supply of OM [46,102]. The upper member of the Lucaogou Formation has high concentrations of P (up to 0.20 wt%), reflecting an increasing nutrient input largely due to volcanism. Intermittent volcanic eruptions during the Middle Permian released large amounts of carbon dioxide and volcanic ash, triggering global warming and eutrophication [70,103,104].

Barium (Ba) is always associated with biological metabolism and can be used to represent organic matter input [105,106]. Aluminium (Al) and titanium (Ti) are primarily of detrital origin, therefore P/Ti and Ba/Al are generally considered effective indicators of productivity by eliminating the impact of terrigenous debris [28,107]. The upper member of the Lucaogou source rocks exhibits higher P/Ti and Ba/Al ratios (P/Ti: avg. 0.38, Ba/Al: avg. 1.36×10^{-2}) than the lower member (P/Ti: avg. 0.24, Ba/Al: avg. 6.68×10^{-2}), providing evidence of a nutrient-rich environment with higher productivity in the upper member.

In summary, the sedimentary environment of the Permian Lucaogou Formation transformed from arid and saline conditions in the lower member into semiarid and brackish conditions in the upper member (Figure 9). During the deposition of the upper member, relatively low salinity was favorable for *Microcystis* survival. Furthermore, the warm climate and stratified water body facilitated algal growth, decreased water viscosity, and increased the sedimentation rate of eukaryotic algae, thus strengthening the competitive advantage of *Microcystis* among phytoplankton [108]. In addition, increased nutrient loading from volcanism, especially Fe and P into the lake, was the main cause of eutrophication [30,109]. Fe and P are indispensable elements for cyanobacterial metabolic activities. As water bodies become enriched in nutrients, there is often a shift in the phytoplankton community toward dominance by cyanobacteria [28,110]. Under this combination of favorable conditions, the *Microcystis* bloom occurred in the upper member.

5.3. *Mechanism of Organic Matter Enrichment under Cyanobacterial Blooms*

In terms of organic matter enrichment, several key controlling factors have been proposed, including enhanced preservation, restricted basin conditions, sea level changes, paleoproductivity, and dilution [111–113]. For example, Zhang et al. (2018) regarded the paleoproductivity of surface water and the relatively stable anoxic environment as the main factors controlling the enrichment of organic matter in the Lucaogou Formation [27]. A case study carried out by Zhao et al. (2023) proposed that the relatively high bio-productivity of water and low input of detrital matter were favorable for the preservation of OM [114]. However, these previous studies did not relate the mechanism of organic matter enrichment to cyanobacterial blooms.

The upper member of the Lucaogou Formation was deposited in a brackish lacustrine environment with anoxic to suboxic water conditions (Figure 11). This specific depositional environment in shales is commonly considered an ideal environment for controlling the degradation and preservation of organic matter [115–117]. Favorable paleoenvironmental conditions in the lacustrine system, together with eutrophication, rising CO_2 levels, and a globally warming climate, facilitated the cyanobacteria growth rate, triggered the cyanobacterial bloom, and promoted an accumulation of cyanobacterial biomass [118–121].

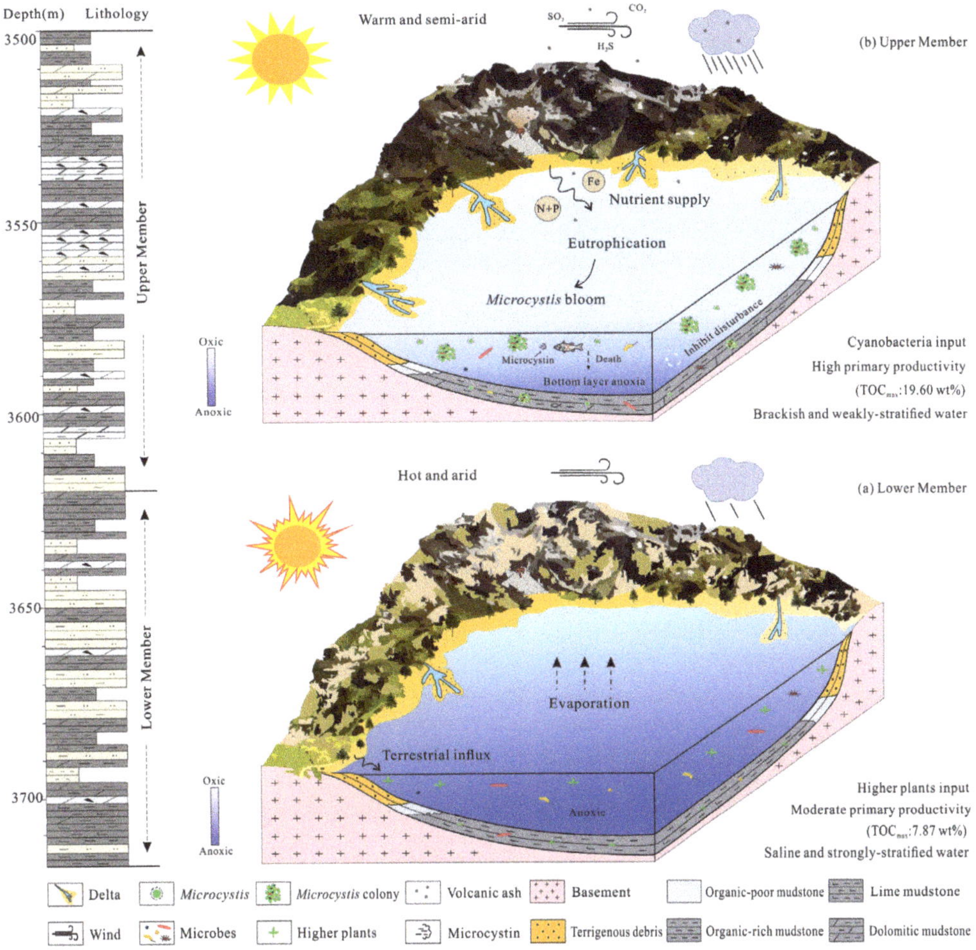

Figure 11. Schematic diagram of a Permian *Microcystis* bloom and its hydrocarbon effect. (**a**) Paleoenvironment of the lower member. (**b**) Paleoenvironment of the upper member.

Abundance of organic matter is usually dependent upon the interplay of a variety of factors including primary productivity, preservation conditions, and sedimentation rate [17,122–124]. The mechanism of organic matter enrichment by cyanobacterial blooms is supported by two main ways. From one perspective, Permian *Microcystis* blooms favored the accumulation of organic matter by directly providing a hydrocarbon source basis for the Lucaogou Formation. The specific mechanism is reflected in modern examples [125]. In modern lakes, seasonal alternation plays a critical role in the formation of *Microcystis* blooms [1,126]. During this process, nutrient-rich bottom water moves upward in spring and summer as the temperature rises, producing favorable conditions for algal blooms. As the temperature decreases in autumn and winter, algal colonies and other organic materials are deposited at the bottom of the lake in the form of sheet-like algal mats and subsequently buried as organic-rich laminae [2,127].

However, previous works have overlooked the other aspect in which cyanobacterial blooms contribute to the enrichment of organic matter. The negative feedback of cyanobacterial blooms to the paleoenvironment may also have facilitated preservation of organic matter. The *Microcystis* bloom that occurred in the upper member of the Lucaogou Formation may have further deteriorated the paleoenvironment by releasing odorous compounds, producing toxins, causing bottom layer anoxia, etc. [57,76]. A large number of *Microcystis* gathered on the lake surface, inhibiting the air circulation between the water and atmosphere, and forming a shield that reduced the oxygen concentrations produced by plant photosynthesis [128]. Moreover, microcystin, a kind of toxin produced by *Microcystis*, is lethal to most metazoans in aquatic ecosystems and can cause mass death of creatures [129,130]. As a result, the decomposition of biological bodies further increases the accumulation of organic matter during cyanobacterial blooms. In addition, the oxygen-depleted sedimentary environment enhanced by the *Microcystis* bloom is favorable for the preservation of organic matter by inhibiting disturbance by benthic organisms and the biodegradation progress of microbes [131,132]. From both aspects, *Microcystis* blooms were of vital importance in the formation of the Lucaogou source rocks, playing a major role and acting as a critical connection point among the numerous hydrocarbon generation events of Jimsar shale oil.

In addition, by their influence on cyanobacteria, volcanism and other geological events could also have had an impact on the accumulation and preservation of organic matter. Permian volcanic eruptions released large amounts of ion-rich volcanic ash, providing abundant nutrients which could stimulate the proliferation of cyanobacteria and advance the productivity-based biogeochemical progress [30,108]. Some key marine events in the Permian, such as photic zone euxinia and faunal mass extinction, were also closely associated with cyanobacterial blooms [133,134]. All these biotic and abiotic factors lead to the formation of Permian Lucaogou source rocks.

6. Conclusions

In this paper, we have presented a comprehensive study of the Lucaogou source rocks including petrological, paleontological, and geochemical analyses. The conclusions are as follows:

(1) Abundantly distributed, globular to elliptical shaped microfossils in the upper member of the Lucaogou Formation are interpreted as *Microcystis*, a kind of cyanobacteria. The great abundance of these fossils referred to the Chroococcaceae indicates the occurrence of Permian cyanobacterial blooms, which are further proven by cyanobacteria-derived biomarkers.

(2) The upper member of the Lucaogou Formation was deposited in a brackish lacustrine environment with anoxic to suboxic water conditions, under a semiarid paleoclimate. Such a paleoenvironment together with abundant nutrients from volcanism ultimately led to cyanobacterial blooms in Permian lakes.

(3) Permian *Microcystis* blooms contributed to organic matter enrichment in two ways. The first was by directly promoting the accumulation of algal biomass, and the second by creating an oxygen-depleted environment for better preservation of organic matter. This study thus provides new insight into shale oil exploration and development in the Junggar Basin, from the perspective of cyanobacterial blooms.

Author Contributions: Formal analysis, H.C., X.S., and L.C.; Funding acquisition, J.T.; Project administration, J.L. and J.W.; Resources, J.L. and L.C.; Software, X.S.; Validation, J.T.; Writing—original draft, W.W. and H.C.; Writing—review & editing, W.W. All authors have read and agreed to the published version of the manuscript.

Funding: This research is supported by the Fundamental Research Funds for the Central Universities of Central South University (No. 2022ZZTS0530), the Department of Natural Resources of Hunan Province, China (No. 2022-5), China Hunan Provincial Science & Technology Department (No. 2022WK2004). This work is also supported by the State Key Laboratory of Paleobiology and Stratigraphy (Nanjing Institute of Geology and Paleontology, CAS) (No. 213120); the Natural Science Foundation of Hunan Province (No. 2023JJ20063 and 2021JJ30816).

Data Availability Statement: The data involved in this paper are all included in the text of the manuscript.

Acknowledgments: We would like to extend our thanks to the Experiment and Testing Research Institute of Xinjiang Oilfield Company, PetroChina, for kindly providing the samples for this study. We are also grateful for numerous conversations with several researchers including Wenquan Xie, Yong Wang, and Xiao Ma (Central South University) on how to strengthen the ideas presented in this paper.

Conflicts of Interest: The authors declare no conflict of interest.

References

1. Latour, D.; Giraudet, H. Factors influencing the spatio-temporal distribution of benthic *Microcystis aeruginosa* colonies (Cyanobacteria) in the hypertrophic Grangent Reservoir (Loire, France). *Comptes Rendus Biol.* **2004**, *327*, 753–761. [CrossRef]
2. Imai, H.; Chang, K.-H.; Kusaba, M.; Nakano, S.-I. Temperature-dependent dominance of *Microcystis* (*Cyanophyceae*) species: *M. aeruginosa* and, *M. wesenbergii*. *J. Plankton Res.* **2009**, *31*, 171–178. [CrossRef]
3. Bullerjahn, G.S.; McKay, R.M.; Davis, T.W.; Baker, D.B.; Boyer, G.L.; Anglada, L.; Doucette, G.J. Global solutions to regional problems: Collecting global expertise to address the problem of harmful cyanobacterial blooms. A Lake Erie case study. *Harmful Algae* **2016**, *54*, 223–238. [CrossRef] [PubMed]
4. Schopf, J.W.; Packer, B.M. Early archean (3.3-billion to 3.5-billion-year-old) microfossils from warrawoona group. *Aust. Sci.* **1987**, *237*, 70–73. [CrossRef]
5. Schopf, J.W.; Kudryavtsev, A.B.; Czaja, A.D.; Tripathi, A.B. Evidence of Archean life: Stromatolites and microfossils. *Precambrian Res.* **2007**, *158*, 141–155. [CrossRef]
6. Kazmierczak, J.; Altermann, W. Paleontology-neoarchean biomineralization by benthic cyanobacteria. *Science* **2002**, *298*, 2351. [CrossRef]
7. Yang, H.; Chen, Z.Q.; Papineau, D. Cyanobacterial spheroids and other biosignatures from microdigitate stromatolites of Mesoproterozoic Wumishan Formation in Jixian, North China. *Precambrian Res.* **2022**, *368*, 106496. [CrossRef]
8. Whitton, B. *Ecology of Cyanobacteria II: Their Diversity in Space and Time*; Springer: Berlin/Heidelberg, Germany, 2012.
9. Riding, R. Temporal variation in calcification in marine cyanobacteria. *J. Geol. Soc. Aust.* **1992**, *149*, 979–989. [CrossRef]
10. Lee, H.; Chen, I.; Chough, S.K. The middle–late Cambrian reef transition and related geological events: A review and new view. *Earth-Sci. Rev.* **2015**, *145*, 66–84. [CrossRef]
11. Kremer, B.; Kamierczak, J. Cyanobacterial Mats from Silurian Black Radiolarian Cherts: Phototrophic Life at the Edge of Darkness? *J. Sediment. Res.* **2005**, *75*, 897–906. [CrossRef]
12. Kremer, B. Mat-forming coccoid cyanobacteria from early Silurian marine deposits of Sudetes, Poland. *Acta Palaeontol. Pol.* **2009**, *51*, 143–154.
13. Wu, Y.B.; Feng, Q.; Gong, Y.M. Blooming of bacteria and algae is a biokiller for mass-extinction of Devonian coral-stromatoporoid reef ecosystems. *Sci. China-Earth Sci.* **2013**, *56*, 1221–1232. [CrossRef]
14. Kazmierczak, J.; Kremer, B.; Racki, G. Late Devonian marine anoxia challenged by benthic cyanobacterial mats. *Geobiology* **2012**, *10*, 371–383. [CrossRef]
15. Wu, Y.S.; Yu, G.L.; Li, R.H.; Song, L.R.; Jiang, H.X.; Riding, R.; Liu, L.J.; Liu, D.Y.; Zhao, R. Cyanobacterial fossils from 252 Ma old microbialites and their environmental significance. *Sci. Rep.* **2014**, *4*, 3820. [CrossRef]
16. Carroll, A.R. Upper Permian lacustrine organic facies evolution, southern Junggar Basin, NW China. *Org. Geochem.* **1998**, *28*, 649–667. [CrossRef]
17. Pan, S.Q.; Horsfield, B.; Zou, C.N.; Yang, Z. Upper Permian Junggar and Upper Triassic Ordos lacustrine source rocks in Northwest and Central China: Organic geochemistry, petroleum potential and predicted organofacies. *Int. J. Coal Geol.* **2016**, *158*, 90–106. [CrossRef]
18. Zhang, S.H.; Liu, C.Y.; Liang, H.; Wang, J.Q.; Yang, M.H.; Liu, G.H.; Huang, H.X.; Guan, Y.Z. Paleoenvironmental conditions, organic matter accumulation, and unconventional hydrocarbon potential for the Permian Lucaogou Formation organic-rich rocks in Santanghu Basin, NW China. *Int. J. Coal Geol.* **2018**, *185*, 44–60. [CrossRef]

19. Cao, Z.; Liu, G.; Kong, Y.; Wang, C.; Niu, Z.; Zhang, J.; Geng, C.; Shan, X.; Wei, Z. Lacustrine tight oil accumulation characteristics: Permian Lucaogou Formation in Jimusaer Sag, Junggar Basin. *Int. J. Coal Geol.* **2016**, *153*, 37–51. [CrossRef]
20. Gao, G.; Zhang, W.; Xiang, B.; Liu, G.; Ren, J. Geochemistry characteristics and hydrocarbon-generating potential of lacustrine source rock in Lucaogou Formation of the Jimusaer Sag, Junggar Basin. *J. Pet. Sci. Eng.* **2016**, *145*, 168–182. [CrossRef]
21. Meng, Z.Y.; Liu, Y.Q.; Jiao, X.; Ma, L.T.; Zhou, D.W.; Li, H.; Cao, Q.; Zhao, M.R.; Yang, Y.Y. Petrological and organic geochemical characteristics of the Permian Lucaogou Formation in the Jimsar Sag, Junggar Basin, NW China: Implications on the relationship between hydrocarbon accumulation and volcanic-hydrothermal activities. *J. Pet. Sci. Eng.* **2022**, *210*, 110078. [CrossRef]
22. Carroll, A.R.; Brassell, S.C.; Graham, S.A. Upper Permian lacustrine oil shales, southern Junggar Basin, Northwest China. *AAPG Bull.* **1992**, *76*, 1874–1902.
23. Tao, S.; Wang, Y.B.; Tang, D.Z.; Wu, D.M.; Xu, H.; He, W. Organic petrology of Fukang Permian Lucaogou Formation oil shales at the northern foot of Bogda Mountain, Junggar Basin, China. *Int. J. Coal Geol.* **2012**, *99*, 27–34. [CrossRef]
24. Xie, X.M.; Borjigin, T.; Zhang, Q.Z.; Zhang, Z.R.; Qin, J.Z.; Bian, L.Z.; Volkman, J.K. Intact microbial fossils in the Permian Lucaogou Formation oil shale, Junggar Basin, NW China. *Int. J. Coal Geol.* **2015**, *146*, 166–178. [CrossRef]
25. Su, Y.; Zha, M.; Ding, X.J.; Qu, J.X.; Cao, C.H.; Jin, J.H.; Iglauer, S. Petrographic, palynologic and geochemical characteristics of source rocks of the Permian Lucaogou formation in Jimsar Sag, Junggar Basin, NW China: Origin of organic matter input and depositional environments. *J. Pet. Sci. Eng.* **2019**, *183*, 106364. [CrossRef]
26. Liu, Y.Q.; Zhou, D.W.; Jiao, X.; Feng, Q.; Zhou, X.H. A preliminary study on the relationship between deep-sourced materials and hydrocarbon generation in lacustrine source rocks: An example from the Permian black rock series in Jimusar sag, Junggar Basin. *J. Palaeogeogr.* **2019**, *21*, 16. [CrossRef]
27. Zhang, X.; Zhuang, X.G.; Tu, Q.J.; Xv, S.Q.; Zhang, Y. Depositional process and mechanism of organic matter accumulation of Lucaogou shale in Southern Junggar Basin, Northwest China. *Earth Sci.* **2018**, *43*, 538–550.
28. Jiang, Y.; Hou, D.; Li, H.; Zhang, Z.; Guo, R. Impact of the Paleoclimate, Paleoenvironment, and Algae Bloom: Organic Matter Accumulation in the Lacustrine Lucaogou Formation of Jimsar Sag, Junggar Basin, NW China. *Energies* **2020**, *13*, 1488. [CrossRef]
29. Kuang, L.C.; Tang, Y.; Lei, D.W.; Chang, Q.S.; Ou, Y.M.; Hou, L.H.; Liu, D.G. Formation conditions and exploration potential of tight oil in the Permian saline lacustrine dolomitic rock, Junggar Basin, NW China. *Pet. Explor. Dev.* **2012**, *039*, 657–667.
30. Ding, X.J.; Qu, J.Q.; Imin, A.; Zha, M.; Jiang, Z.F.; Jiang, H. Organic matter origin and accumulation in tuffaceous shale of the lower Permian Lucaogou Formation, Jimsar Sag. *J. Pet. Sci. Eng.* **2019**, *179*, 696–706. [CrossRef]
31. Sun, F.N.; Hu, W.X.; Wang, X.L.; Cao, J.; Fu, B.; Wu, H.G.; Yang, S.C. Methanogen microfossils and methanogenesis in Permian lake deposits. *Geology* **2021**, *49*, 13–18. [CrossRef]
32. Li, Y.; Xue, Z.; Cheng, Z.; Jiang, H.; Wang, R. Progress and development directions of deep oil and gas exploration and development in China. *China Pet. Explor.* **2020**, *25*, 45–57.
33. Carroll, A.R.; Graham, S.A. Late paleozoic tectonic amalgamation of northwestern China: Sedimentary record of the northern Tarim, northern Turpan, and southern Junggar Basins. *Geol. Soc. Am. Bull.* **1995**, *107*, 571–594. [CrossRef]
34. Kuang, L.H.; Hu, W.X.; Wang, X.L.; Wu, H.G.; L. W.X. Research of the Tight Oil Reservoir in the Lucaogou Formation in Jimusar Sag: Analysis of Lithology and Porosity Characteristics. *Geol. J. China Univ.* **2013**, *19*, 7.
35. Wang, J.; Yuan, B.; Liu, J.; Li, Y.; Li, E. Genesis and pore development characteristics of Permian Lucaogou migmatites, Jimsar Sag, Junggar Basin. *Pet. Geol. Exp.* **2022**, *44*, 413–424.
36. Jiang, Y.Q.; Liu, Y.Q.; Yang, Z.; Nan, Y.; Wang, R.; Zhou, P.; Yang, Y.J.; Kou, J.Y.; Zhou, N.C. Characteristics and origin of tuff-type tight oil in Jimusar Depression, Junggar Basin, NW China. *Pet. Explor. Dev.* **2015**, *42*, 741–749. [CrossRef]
37. Tao, K.Y.; Cao, J.; Chen, X.; Nueraili, Z.; Hu, W.X.; Shi, C.H. Deep hydrocarbons in the northwestern Junggar Basin (NW China): Geochemistry, origin, and implications for the oil vs. gas generation potential of post-mature saline lacustrine source rocks. *Mar. Pet. Geol.* **2019**, *109*, 623–640. [CrossRef]
38. Guo, P.; Wen, H.G.; Gibert, L.; Jin, J.; Wang, J.; Lei, H.Y. Deposition and diagenesis of the Early Permian volcanic-related alkaline playa-lake dolomitic shales, NW Junggar Basin, NW China. *Mar. Pet. Geol.* **2021**, *123*, 104780. [CrossRef]
39. Wang, X.; Yang, Z.; Guo, X.; Xiatian, W.; Feng, Y.; Huang, L. Practices and Prospects of Shale Oil Exploration in Jimsar Sag of Junggar Basin. *Xinjiang Pet. Geol.* **2019**, *40*, 402–413.
40. Zhi, D.; Song, Y.; He, W.; Jia, X.; Yang, Z.; Huang, L. Geological Characteristics, Resource Potential and Exploration Direction of Shale Oil in Middle-Lower Permian, Junggar Basin. *Xinjiang Pet. Geol.* **2019**, *40*, 389–401.
41. Kang, J.L.; Wang, J.H.; Ma, Q.; Li, C.Q.; Chen, X. Fine-grained sublacustrine fan deposits and their significance of shale oil reservoirs in the Lucaogou Formation in Jimsar Sag, Junggar Basin. *Bull. Geol. Sci. Technol.* **2022**, *4*, 1–12.
42. Wu, H.; Hu, W.; Tang, Y.; Cao, J.; Wang, X.; Wang, Y.; Kang, X. The impact of organic fluids on the carbon isotopic compositions of carbonate-rich reservoirs: Case study of the Lucaogou Formation in the Jimusaer Sag, Junggar Basin, NW China. *Mar. Pet. Geol.* **2017**, *85*, 136–150.
43. Tissot, B.; Welte, D. *Petroleum Formation and Occurrence*; Earth Science Reviews; Springer: Berlin/Heidelberg, Germany, 1984.
44. Peters, K.E.; Walters, C.C.; Moldowan, J.M. *The Biomarker Guide: Interpreting Molecular Fossils in Petroleum and Ancient Sediments*; Cambridge University Press: Cambridge, UK, 2005.
45. Calvert, S.E.; Pedersen, T.F. Geochemistry of Recent oxic and anoxic marine sediments: Implications for the geological record. *Mar. Geol.* **1993**, *113*, 67–88. [CrossRef]

46. Tribovillard, N.; Algeo, T.; Lyons, T.; Riboulleau, A. Trace metals as paleoredox and paleoproductivity proxies: An update. *Chem. Geol.* **2006**, *232*, 12–32. [CrossRef]
47. Mclennan; Scott, M. Relationship between the trace element composition of sedimentary rocks and upper continental crust. *Geochem. Geophys. Geosystems* **2001**, *2*, 203–236.
48. Nesbitt, H.W.; Young, G.M. Early Proterozoic climates and plate motions inferred from major element chemistry of lutites. *Nature* **1982**, *299*, 715–717. [CrossRef]
49. Mclennan; Scott, M. 100th Anniversary Symposium: Evolution of the Earth's Surface | | Weathering and Global Denudation. *J. Geol.* **1993**, *101*, 295–303.
50. Liu, J.; Wang, J.; Zhang, X.G.; Luo, Z.J.; Shang, L.; Wang, G.J. Scanning electron microscope study on oil generation characteristics of ostracod fossils from the Lucaogou formation in eastern Junggar Basin. *J. Chin. Electron Microsc. Soc.* **2021**, *40*, 270–278.
51. Wu, H.G.; Hu, W.X.; Cao, J.; Wang, X.L.; Liao, Z.W. A unique lacustrine mixed dolomitic-clastic sequence for tight oil reservoir within the middle Permian Lucaogou Formation of the Junggar Basin, NW China: Reservoir characteristics and origin. *Mar. Pet. Geol.* **2016**, *76*, 115–132. [CrossRef]
52. Wang, R.Y.; Ding, W.L.; Zhang, Y.Q.; Wang, Z.; Wang, X.; He, J.; Zeng, W.; Dai, P. Analysis of developmental characteristics and dominant factors of fractures in Lower Cambrian marine shale reservoirs: A case study of Niutitang formation in Cen'gong block, southern China. *J. Pet. Sci. Eng.* **2016**, *138*, 31–49. [CrossRef]
53. Caracciolo, A.B.; Dejana, C.; Fajardo, P.; Grenni, M. A new fluorescent oligonucleotide probe for in-situ identification of *Microcystis aeruginosa* in freshwater. *Microchem. J.* **2019**, *148*, 503–513.
54. Taylor, T.N.; Krings, M. A colony-forming microorganism with probable affinities to the *Chroococcales* (cyanobacteria) from the Lower Devonian Rhynie chert. *Rev. Palaeobot. Palynol.* **2015**, *219*, 147–156. [CrossRef]
55. Xiao, M.; Li, M.; Reynolds, C.S. Colony formation in the cyanobacterium *Microcystis*. *Biol. Rev.* **2018**, *93*, 1399–1420. [CrossRef]
56. Phelan, R.R.; Downing, T.G. A growth advantage for microcystin production by *Microcystis* PCC7806 under high light. *J. Phycol.* **2011**, *47*, 1241–1246. [CrossRef]
57. Van Le, V.; Srivastava, A.; Ko, S.-R.; Ahn, C.-Y.; Oh, H.-M. *Microcystis* colony formation: Extracellular polymeric substance, associated microorganisms, and its application. *Bioresour. Technol.* **2022**, *360*, 127610.
58. Wei, W.; Algeo, T.J. Elemental proxies for paleosalinity analysis of ancient shales and mudrocks. *Geochim. Et Cosmochim. Acta* **2020**, *287*, 341–366. [CrossRef]
59. Gelpi, E.; Schneider, J.; Mann, J.; Oró, J. Hydrocarbons of geochemical significance in microscopic algae. *Phytochemistry* **1970**, *9*, 603–612. [CrossRef]
60. Reed, J.D.; Illich, H.A.; Horsfield, B.J.O.G. Biochemical evolutionary significance of Ordovician oils and their sources. *Org. Geochem.* **1986**, *10*, 347–358. [CrossRef]
61. Cranwell, P.A.; Eglinton, G.; Robinson, N. Lipids of aquatic organisms as potential contributors to lacustrine sediments-II. *Org. Geochem.* **1987**, *11*, 513–527. [CrossRef]
62. Huang, W.Y.; Meinschein, W.G. Sterols as ecological indicators. *Geochim. Et Cosmochim. Acta* **1979**, *43*, 739–745. [CrossRef]
63. Volkman, J. Sterols in microorganisms. *Appl. Microbiol. Biotechnol.* **2003**, *60*, 495–506. [CrossRef] [PubMed]
64. Grantham, P.J.; Wakefield, L.L. Variations in the sterane carbon number distributions of marine source rock derived crude oils through geological time. *Org. Geochem.* **1988**, *12*, 61–73. [CrossRef]
65. Strobl, S.; Sachsenhofer, R.F.; Bechtel, A.; Gratzer, R.; Gross, D.; Bokhari, S.; Liu, R.; Liu, Z.; Meng, Q.; Sun, P. Depositional environment of oil shale within the Eocene Jijuntun Formation in the Fushun Basin (NE China). *Mar. Pet. Geol.* **2014**, *56*, 166–183. [CrossRef]
66. Summons, R.E.; Jahnke, L.L.; Hope, J.M.; Logan, G.A. 2-Methylhopanoids as biomarkers for cyanobacterial oxygenic photosynthesis. *Nature* **1999**, *400*, 554–557. [CrossRef] [PubMed]
67. Hou, M.; Qu, J.; Zha, M.; Swennen, R.; Ding, X.; Imin, A.; Liu, H.; Bian, B. Significant contribution of haloalkaliphilic cyanobacteria to organic matter in an ancient alkaline lacustrine source rock: A case study from the Permian Fengcheng Formation, Junggar Basin, China. *Mar. Pet. Geol.* **2022**, *138*, 105546. [CrossRef]
68. Doughty, D.M.; Hunter, R.C.; Summons, R.E.; Newman, D.K. 2-Methylhopanoids are maximally produced in akinetes of Nostoc punctiforme: Geobiological implications. *Geobiology* **2009**, *7*, 524–532. [CrossRef] [PubMed]
69. Boudou, J.P.; Trichet, J.; Robinson, N.; Brassell, S.C. Profile of aliphatic hydrocarbons in a recent polynesian microbial mat. *Int. J. Environ. Anal. Chem.* **1986**, *26*, 137–155. [CrossRef]
70. Ding, W.J.; Hou, D.J.; Jiang, L.; Jiang, Y.H.; Wu, P. High abundance of carotanes in the brackish-saline lacustrine sediments: A possible cyanobacteria source? *Int. J. Coal Geol.* **2020**, *219*, 103373. [CrossRef]
71. Dobson, G.; Ward, D.M.; Robinson, N.; Eglinton, G. Biogeochemistry of hot spring environments: Extractable lipids of a cyanobacterial mat. *Chem. Geol.* **1988**, *68*, 155–179. [CrossRef]
72. Luo, G.; Hallmann, C.; Xie, S.; Ruan, X.; Summons, R.E. Comparative microbial diversity and redox environments of black shale and stromatolite facies in the Mesoproterozoic Xiamaling Formation. *Geochim. Et Cosmochim. Acta* **2015**, *151*, 150–167. [CrossRef]
73. Ding, X.J.; Gao, C.H.; Zha, M.; Chen, H.; Su, Y. Depositional environment and factors controlling β-carotane accumulation: A case study from the Jimsar Sag, Junggar Basin, northwestern China. *Palaeogeogr. Palaeoclimatol. Palaeoecol.* **2017**, *485*, 833–842. [CrossRef]

74. Tribovillard, N.; Algeo, T.J.; Baudin, F.; Riboulleau, A.J.C.G. Analysis of marine environmental conditions based onmolybdenum–uranium covariation—Applications to Mesozoic paleoceanography. *Chem. Geol.* **2012**, *324–325*, 46–58. [CrossRef]
75. Shanmugam, G. Significance of coniferous rain forests and related organic matter in generating commercial quantities of oil, Gippsland Basin, Australia. *AAPG (Am. Assoc. Pet. Geol.) Bull.* **1985**, *69*, 1241–1254. [CrossRef]
76. Burford, M.A.; Gobler, C.J.; Hamilton, D.P.; Visser, P.M.; Lurling, M.; Codd, G.A. *Solutions for Managing Cyanobacterial Blooms: A Scientific Summary for Policy Makers*; IOC/UNESCO: Paris, France, 2019.
77. Ji, X.; Verspagen, J.M.H.; Van de Waal, D.B.; Rost, B.; Huisman, J. Phenotypic plasticity of carbon fixation stimulates cyanobacterial blooms at elevated CO_2. *Sci. Adv.* **2020**, *6*, 2926. [CrossRef] [PubMed]
78. Verspagen, J.M.H.; Snelder, E.; Visser, P.M.; Huisman, J.; Ibelings, B.W. Recruitment of benthic *Microcystis* (*Cyanophyceae*) to the water column: Internal buoyancy changes or resuspension? *J. Phycol.* **2004**, *40*, 260–270. [CrossRef]
79. Preston, T.; Stewart, W.D.P.; Reynolds, C.S. Bloomforming cyanobacterium *Microcystis aeruginosa* overwinters on the sediment. *Nature* **1980**, *288*, 365–367. [CrossRef]
80. Schindler, D.W. Evolution of phosphorus limitation in lakes. *Science* **1977**, *195*, 260–262. [CrossRef]
81. Paerl, H.W. Mitigating harmful cyanobacterial blooms in a human-and climatically-impacted world. *Life* **2014**, *4*, 988–1012. [CrossRef]
82. Wang, P.; Du, Y.; Yu, W.; Algeo, T.J.; Pan, W. The chemical index of alteration (CIA) as a proxy for climate change during glacial-interglacial transitions in Earth history. *Earth-Sci. Rev.* **2019**, *201*, 103032. [CrossRef]
83. Moradi, A.V.; Sari, A.; Akkaya, P. Geochemistry of the Miocene oil shale (Hançili Formation) in the Çankırı-Çorum Basin, Central Turkey: Implications for Paleoclimate conditions, source–area weathering, provenance and tectonic setting. *Sediment. Geol.* **2016**, *341*, 289–303. [CrossRef]
84. Xie, W.Q.; Tan, J.Q.; Wang, W.H.; Schulz, H.; Liu, Z.J.; Kang, X.; Wang, Y.; Shahzad, A.; Jan, I.U. Middle Jurassic lacustrine source rocks controlled by an aridification event: A case study in the northern Qaidam Basin (NW China). *Int. J. Coal Geol.* **2021**, *242*, 103779. [CrossRef]
85. Rohling, E.J. Paleosalinity: Confidence limits and future applications. *Mar. Geol.* **2000**, *163*, 1–11. [CrossRef]
86. Jiang, Z.; Ding, X.; Wang, Z.; Zhao, X. Sedimentary paleoenvironment of source rocks of Permian Lucaogou Formation in Jimsar Sag. Lithologic Reservoirs. *Lithol. Reserv.* **2020**, *32*, 11.
87. Salminen, R.; Batista, M.J.; Bidovec, M.; Demetriades, A.; Tarvainen, T. *FOREGS Geochemical Atlas of Europe, Part 1: Background Information, Methodology and Maps*; Electric publication; Geochemical Survey of Finland: Espoo, Finland, 2005.
88. Summons, R.E.; Hope, J.M.; Swart, R.; Walter, M.R. Origin of Nama Basin bitumen seeps: Petroleum derived from a Permian lacustrine source rock traversing southwestern Gondwana. *Org. Geochem.* **2008**, *39*, 589–607. [CrossRef]
89. Algeo, T.J.; Tribovillard, N. Environmental Analysis of Paleoceanographic Systems Based on Molybdenum–Uranium Covariation. *Chem. Geol.* **2009**, *268*, 211–225. [CrossRef]
90. Liu, D.D.; Fan, Q.Q.; Zhang, C.; Gao, Y.; Du, W.; Song, Y. Paleoenvironment evolution of the Permian Lucaogou Formation in the southern Junggar Basin, NW China. *Palaeogeogr. Palaeoclimatol. Palaeoecol.* **2022**, *603*, 111198. [CrossRef]
91. Walters, C. *The Biomarker Guide*; Cambridge University Press: Cambridge, UK, 2005.
92. Peters, K.E.; Fraser, T.H.; Amris, W.; Rustanto, B. Geochemistry of crude oils from eastern Indonesia. *AAPG Bull.* **1999**, *83*, 1927–1942.
93. Jiang, Z.; Fowler, M.G. Carotenoid-derived alkanes in oils from northwestern China. *Org. Geochem.* **1986**, *10*, 831–839. [CrossRef]
94. Wang, R.Y.; Gu, Y.; Ding, W.L.; Gong, D.; Yin, S.; Wang, X.; Zhou, X.; Li, A.; Xiao, Z.; Cui, Z. Characteristics and dominant controlling factors of organic-rich marine shales with high thermal maturity: A case study of the Lower Cambrian Niutitang Formation in the Cen'gong block, southern China. *J. Nat. Gas Sci. Eng.* **2016**, *33*, 81–96. [CrossRef]
95. Calvert, S.; Pedersen, T. Elemental proxies for paleoclimatic and palaeoceanographic variability in marine sediments: Interpretation and application. *Dev. Mar. Geol.* **2007**, *1*, 568–644.
96. Cheng, D.W.; Zhou, C.M.; Zhang, Z.J.; Yuan, X.J.; Liu, Y.H.; Chen, X.Y. Paleo-Environment Reconstruction of the Middle Permian Lucaogou Formation, Southeastern Junggar Basin, NW China: Implications for the Mechanism of Organic Matter Enrichment in Ancient Lake. *J. Earth Sci.* **2022**, *33*, 963–976. [CrossRef]
97. Algeo, T.J.; Kuwahara, K.; Sano, H.; Bates, S.; Lyons, T.; Elswick, E.; Hinnov, L.; Ellwood, B.; Moser, J.; Maynard, J.B. Spatial variation in sediment fluxes, redox conditions, and productivity in the Permian–Triassic Panthalassic Ocean. *Palaeogeogr. Palaeoclimatol. Palaeoecol.* **2011**, *308*, 65–83. [CrossRef]
98. Shen, J.; Schoepfer, S.D.; Feng, Q.; Zhou, L.; Yu, J.; Song, H.; Wei, H.; Algeo, T.J. Marine productivity changes during the end-Permian crisis and Early Triassic recovery. *Earth Sci. Rev.* **2015**, *149*, 136–162. [CrossRef]
99. Schoepfer, S.D.; Shen, J.; Wei, H.; Tyson, R.V.; Ingall, E.; Algeo, T.J.J.E.-S.R. Total organic carbon, organic phosphorus, and biogenic barium fluxes as proxies for paleomarine productivity. *Earth Sci. Rev.* **2015**, *149*, 23–52. [CrossRef]
100. Pedersen, T.E.; Calvert, S.E. Anoxia vs. productivity: What controls the formation of organic-carbon-rich sediments and sedimentary rocks? *Am. Assoc. Pet. Geol. Bull.* **1990**, *74*, 454–466.
101. Guan, M.; Wu, S.; Hou, L.; Jiang, X.; Ba, D.; Hua, G. Paleoenvironment and chemostratigraphy heterogenity of the Cretaceous organic-rich shales. *Adv. Geo-Energy Res.* **2021**, *5*, 444–445. [CrossRef]
102. Goldhammer, T.; Brüchert, V.; Ferdelman, T.G.; Zabel, M. Microbial sequestration of phosphorus in anoxic upwelling sediments. *Nat. Geosci.* **2010**, *3*, 557–561. [CrossRef]

103. Langmann, B. On the Role of Climate Forcing by Volcanic Sulphate and Volcanic Ash. *Adv. Meteorol.* **2014**, *2014*, 340123. [CrossRef]
104. Pan, Y.S.; Huang, Z.L.; Li, T.J.; Guo, X.; Chen, X. Environmental response to volcanic activity and its effect on organic matter enrichment in the Permian Lucaogou Formation of the Malang Sag, Santanghu Basin, Northwest China. *Palaeogeogr. Palaeoclimatol. Palaeoecol.* **2020**, *560*, 110024. [CrossRef]
105. Liao, Z.; Hu, W.; Cao, J.; Wang, X.; Hu, Z.J.M.; Geology, P. Petrologic and geochemical evidence for the formation of organic-rich siliceous rocks of the Late Permian Dalong Formation, Lower Yangtze region, southern China. *Mar. Pet. Geol.* **2019**, *103*, 41–54. [CrossRef]
106. Anderson, L.D.; Delaney, M.L. Use of multiproxy records on the Agulhas Ridge, Southern Ocean (Ocean Drilling Project Leg 177, Site 1090) to investigate sub-Antarctic hydrography from the Oligocene to the early Miocene. *Paleoceanogr. Palaeoclimatol.* **2005**, *20*. [CrossRef]
107. Rimmer, S.M.; Thompson, J.A.; Goodnight, S.A.; Robl, T.L. Multiple controls on the preservation of organic matter in Devonian–Mississippian marine black shales: Geochemical and petrographic evidence. *Palaeogeogr. Palaeoclim. Palaeoecol.* **2004**, *215*, 125–154. [CrossRef]
108. Duggen, S.; Croot, P.; Schacht, U.; Hoffmann, L. Subduction zone volcanic ash can fertilize the surface ocean and stimulate phytoplankton growth: Evidence from biogeochemical experiments and satellite data. *Geophys. Res. Lett.* **2007**, *34*, 95–119. [CrossRef]
109. Wignall, P.B. Large igneous provinces and mass extinctions. *Earth-Sci. Rev.* **2001**, *53*, 1–33. [CrossRef]
110. Qu, C.S.; Qiu, L.W.; Yang, Y.Q.; Yu, K.H.; Tang, L.L.; Wan, M. Environmental response of the Permian volcanism in Lucaogou formation in Jimsar sag, Junggar Basin, Northwest China. *Seismol. Geol.* **2019**, *560*, 110024.
111. Li, Y.F.; Zhang, T.W.; Ellis, G.S. Depositional environment and organic matter accumulation of Upper Ordovician-Lower Silurian marine shale in the Upper Yangtze Platform, South China. *Palaeogeogr. Palaeoclimatol. Palaeoecol* **2017**, *466*, 252–264. [CrossRef]
112. Wang, R.Y.; Hu, Z.Q.; Zhou, T.; Bao, H.Y.; Wu, J.; Du, W.; He, J.H. Characteristics of fractures and their significance for reservoirs in Wufeng-Longmaxi shale, Sichuan Basin and its periphery. *Oil Gas Geol.* **2021**, *42*, 1295–1306.
113. Zhao, D.F.; Guo, Y.H.; Yin, S.; Ren, C.Y.; Wang, Y.J. Prediction of geomechanical sweet spots in a tight gas sandstone reservoir: A case study of lower Permian strata in the southern Qinshui Basin, China. *Interpretation* **2019**, *7*, 207–219. [CrossRef]
114. Zhao, D.F.; Guo, Y.H.; Wang, G.X.; Zhou, X.Q.; Zhou, Y.Y.; Zhang, J.M.; Ren, G.Y. Organic matter enrichment mechanism of Youganwo Formation oil shale in the Maoming Basin. *Heliyon* **2023**, *9*, 13173. [CrossRef]
115. Hedges, J.I.; Keil, R.G. Sedimentary organic matter preservation: An assessment and speculative synthesis. *Mar. Chem.* **1995**, *49*, 81–115. [CrossRef]
116. Lash, G.G.; Blood, D.R. Organic matter accumulation, redox, and diagenetic history of the Marcellus Formation, southwestern Pennsylvania, Appalachian basin. *Mar. Pet. Geol.* **2014**, *57*, 244–263. [CrossRef]
117. Zhao, D.F.; Guo, Y.H.; Zhu, Y.M.; Zhao, S.X.; Chen, Z.H.; Jiao, W.W. Comments on the evaluation system of accurate evaluation and selection of deep marine shale reservoirs. *Unconv. Oil Gas* **2022**, *9*, 1–7.
118. Paerl, H.W.; Huisman, J. Climate change: A catalyst for global expansion of harmful cyanobacterial blooms. *Environ. Microbiol. Rep.* **2009**, *1*, 27–37. [CrossRef]
119. O'Neil, J.M.; Davis, T.W.; Burford, M.A.; Gobler, C.J. The rise of harmful cyanobacteria blooms: The potential roles of eutrophication and climate change. *Harmful Algae* **2012**, *14*, 313–334.
120. Verspagen, J.M.H.; Van, D.; Finke, J.F.; Visser, P.M.; Donk, E.V.; Huisman, J.; Dam, H.G. Rising CO_2 Levels Will Intensify Phytoplankton Blooms in Eutrophic and Hypertrophic Lakes. *PLoS ONE* **2014**, *9*, 104325.
121. Huisman, J.; Codd, G.A.; Paerl, H.; Ibelings, B.W.; Verspagen, J.M.H.; Visser, P.M. Cyanobacterial blooms. *Nat. Rev. Microbiol.* **2018**, *16*, 471–483. [CrossRef]
122. Hackley, P.C.; Fishman, N.; Wu, T.; Baugher, G. Organic petrology and geochemistry of mudrocks from the lacustrine Lucaogou Formation, Santanghu Basin, northwest China: Application to lake basin evolution. *Int. J. Coal Geol.* **2016**, *168*, 20–34. [CrossRef]
123. Zheng, R.; Zeng, W.; Li, Z.; Chen, X.; Man, K.; Zhang, Z.; Wang, G.; Shi, S. Differential enrichment mechanisms of organic matter in the Chang 7 Member mudstone and shale in Ordos Basin, China: Constraints from organic geochemistry and element geochemistry. *Palaeogeogr. Palaeoclimatol. Palaeoecol.* **2022**, *601*, 111126. [CrossRef]
124. Bohacs, K.M.; Grabowski, J.G.; Carroll, A.R.; Mankiewicz, P.J.; Simo, T.J. Production, Destruction, and Dilution—The Many Paths to Source-Rock Development. In *The Deposition of Organic-Carbon-Rich Sediments: Models, Mechanisms, and Consequences*; Society for Sedimentary Geology Special Publication; GeoScienceWorld: McLean, VA, USA, 2005; Volume 82, pp. 61–101.
125. Paerl, H.W.; Tucker, C.S. Ecology of Blue-Green Algae in Aquaculture Ponds. *J. World Aquac. Soc.* **1995**, *26*, 109–131. [CrossRef]
126. Elliott, J.A. The seasonal sensitivity of Cyanobacteria and other phytoplankton to changes in flushing rate and water temperature. *Glob. Change Biol.* **2010**, *16*, 864–876. [CrossRef]
127. Qin, J.Z.; Pan, A.; Shen, B.J. Insoluble organic matter in source rocks: Derived from organic macromolecules in the skeleton, cell wall and shell of organisms. *Pet. Geol. Exp.* **2020**, *42*, 6.
128. Yang, Y.; Zheng, L.; Wu, Y.L.; Yang, F.Y.; Liu, F.; Han, R. Research progress on hydrodynamic mechanisms and simulation of *Microcystis* blooms. *J. China Inst. Water Resour. Hydropower Res.* **2022**, *20*, 449–463.
129. Chorus, I. *Toxic Cyanobacteria in Water: A Guide to Their Public Health Consequences, Monitoring and Management*; E&FN Spon: London, UK, 1999.

130. Song, H.; Xu, J.; Lavoie, M.; Fan, X.; Liu, G.; Sun, L.; Fu, Z.; Qian, H.J.A.M.; Biotechnology. Biological and chemical factors driving the temporal distribution of cyanobacteria and heterotrophic bacteria in a eutrophic lake (West Lake, China). *Appl. Microbiol. Biotechnol.* **2017**, *101*, 1685–1696. [CrossRef]
131. Wu, Y.S.; Yu, G.L.; Jiang, H.X.; Liu, L.J.; Zhao, R.; Wu, Y.S.; Yu, G.L.; Jiang, H.X.; Liu, L.J.; Zhao, R. Role and lifestyle of calcified cyanobacteria (*Stanieria*) in Permian-Triassic boundary microbialites. *Palaeogeogr. Palaeoclimatol. Palaeoecol.* **2015**, *448*, 9. [CrossRef]
132. Mort, H.; Jacquat, O.; Adatte, T.; Steinmann, P.; Föllmi, K.; Matera, V.; Berner, Z.; Stüben, D. The Cenomanian/Turonian anoxic event at the Bonarelli Level in Italy and Spain: Enhanced productivity and/or better preservation? *Cretac. Res.* **2007**, *28*, 597–612. [CrossRef]
133. Xie, S.; Pancost, R.D.; Huang, J.H.; Wignall, P.B.; Yu, J.; Tang, X.; Chen, L.; Huang, X.; Lai, X. Changes in the global carbon cycle occurred as two episodes during the Permian-Triassic crisis. *Geology* **2007**, *35*, 1083–1086. [CrossRef]
134. Xie, S.; Pancost, R.D.; Yin, H.; Wang, H.; Yang, H.; Wignall, P.B.; Luo, G.; Jia, C.; Chen, L. Cyanobacterial blooms tied to volcanism during the 5 m.y. Permo-Triassic biotic crisis. *Geology* **2010**, *38*, 447–450. [CrossRef]

minerals

MDPI

Article

Palaeoenvironmental Evolution Based on Elemental Geochemistry of the Wufeng-Longmaxi Shales in Western Hubei, Middle Yangtze, China

Lulu Xu [1], Saipeng Huang [2,3,*], Mengdi Sun [2], Yaru Wen [1], Wei Chen [1], Yanling Zhang [1], Fan Luo [1] and Hao Zhang [4]

1 Department of Shale Gas Geological Survey, Hubei Geological Survey, Wuhan 430000, China
2 Key Laboratory of Continental Shale Hydrocarbon Accumulation and Efficient Development, Ministry of Education, Northeast Petroleum University, Daqing 163318, China
3 Departament de Mineralogia, Petrologia i Geologia Aplicada, Facultat de Ciències de la Terra, Universitat de Barcelona (UB), c/Martí i Franquès s/n, 08028 Barcelona, Spain
4 Hubei Coal Geological Exploration Institute, Wuhan 430000, China
* Correspondence: huangspcugb@hotmail.com

Abstract: The organic-rich shales found in the Wufeng–Longmaxi Formation are typically deposited in oxygen-deficient reducing environments. One of the primary sources of debate revolves around the question of whether the anoxic bottom water found in these shales is either euxinic or ferruginous, and this matter remains unresolved. Previous studies have mostly focused on the Wufeng–Longmaxi Formation as a whole in order to understand the key factors that control organic matter accumulation (OMA). However, research on OMA for each member, including the Wufeng Formation (WF), the lower Longmaxi Formation (LLM), and the upper Longmaxi Formation (ULM), has been insufficient. This paper aims to investigate the palaeoenvironmental conditions and OMA mechanisms of the Wufeng–Longmaxi shales in western Hubei by integrating data on total organic carbon (TOC) content, mineral compositions, major and trace elements, and iron speciation. The results indicate that the Wufeng–Longmaxi shales were deposited under highly restricted hydrographic conditions, except for relatively open and upwelling conditions in the upper WF. Silica in the upper WF was primarily biogenic origin and not hydrothermal. Ferruginous conditions were the primary redox conditions for organic-rich shales except for minor formations in the lower LLM that were deposited under euxinic conditions. Due to the tectonic uplift caused by the Kwangsian Orogeny in the upper LLM, the palaeoenvironment was characterized by a warmer and wetter climate, high terrigenous influx, oxic conditions, and low productivity as the result of the insufficient nutrients caused by the weak upwelling, leading to the turnover of graptolite biozones from LM5 to LM6. The factors influencing OMA changed vertically. TOC contents have a highly positive correlation with Al content, indicating that terrigenous influx was the main factor affecting OMA in the WF, which significantly differed from patterns found in other regions. This suggests that the sedimentation rate of organic matter was higher than the terrigenous dilution rate during the WF stage. The combination of redox conditions and productivity were the main factors affecting OMA in the LLM, while terrigenous influx was the key factor controlling OMA in the ULM, resulting in the dilution of organic matter. Regions in the eastern Yiling block, which are close to the Qinling Ocean, show better prospects for shale gas exploration. This research will further facilitate the development of shale gas in this area.

Keywords: palaeoredox conditions; palaeoproductivity; organic matter accumulation; Wufeng–Longmaxi shales; Western Hubei

1. Introduction

During the late Ordovician to early Silurian period, black shales in the Wufeng–Longmaxi Formation were widespread on the upper and middle Yangtze Platforms [1–5], and are now considered among the most promising hydrocarbon source rocks in South China [6,7]. In recent years, large-scale commercial developments in shale gas have achieved success in the Jiaoshiba block of the southeastern Sichuan Basin [8–10], and significant shale gas sources have been discovered in the Yiling block of western Hubei [2,3,11]. Consequently, considerable attention has been given to studying the palaeoenvironmental characteristics and organic matter accumulation (OMA) mechanisms of these blocks [5,12–15], which are vital for hydrocarbon generation and play a crucial role in determining the physical properties of shale reservoirs [1–3].

Previous studies have demonstrated that the Wufeng Formation (WF) and lower Longmaxi Formation (LLM) are organic-rich shales [3,13,16] deposited in the deep shelf [15,17] during a large-scale transgression [16,18]. These formations experienced warm and humid climates [5,14], dysoxic to anoxic conditions [13], and high productivity [3,15,16]. Moving upward to the upper Longmaxi Formation (ULM), the palaeoenvironmental became characterized by a warmer and wetter climate, high terrigenous influx [15,16], and the oxic conditions of bottom water due to the fall in sea level caused by the intense tectonic uplift of the Chuanzhong and Xuefeng uplifts [16,19], leading to the formation of organic-lean shales. Remarkably, the fall in sea level during the ULM stage was caused by the Kwangsian Orogeny, which was a tectonic event dominated by horizontal movement with local vertical uplift movement, occurring from the southeast of Guangxi province to the northwest of Hunan province during the late Ordovician to the late Silurian [16,20,21], leading to the change of the environment from anoxic to oxic conditions, and turnover of the graptolite biozone LM5 in the Rhuddanian to LM6 in the Aeronian (Figure 1d) [21]. However, further research is necessary to fully understand the transition from LM5 to LM6.

Recently, iron speciation has been utilized to distinguish ferruginous and euxinic bottom water conditions in the Wufeng–Longmaxi Formation of the Sichuan Basin [22–25]. For instance, formations in the upper WF and lower LLM in well PY1 in the Sichuan Basin and in well Shenci-1 near the Qinling Ocean are apparently deposited under euxinic conditions [23]. However, the anoxic bottom water conditions in the Yichang area of western Hubei may be different from those in the regions near the depocenter of the Sichuan Basin and Qinling Sea in northern Yangtze, due to the different sedimentary facies. Therefore, further studies are required to determine the anoxic-ferruginous or anoxic-euxinic conditions of organic-rich shales in the WF and LLM in western Hubei. Additionally, the upwelling intensity of the bottom water may differ in different regions of the middle and upper Yangtze, owing to the varying sedimentary facies [12].

Generally, the OMA mechanism can be explained by the preservation model and productivity model [2,26], which indicate that redox conditions [5,27] and productivity [3,28] are the two main factors that influence OMA in the Wufeng–Longmaxi Formation. Previous studies have mostly focused on the Wufeng–Longmaxi Formation as a whole in order to understand the key factors that control organic matter accumulation (OMA). However, research on OMA for each member, including the WF, LLM, and ULM, has been insufficient.

Considering the inadequate research on the palaeoenvironmental evolution and OMA mechanism, we collected 24 samples from well YD1 of the Yiling block for geochemical testing to analyze the associated palaeoenvironmental conditions and OMA mechanism (Figure 1d). Moreover, we conducted further studies on the differences in geological conditions between the Yiling block and Jiaoshiba block [6,13]. The geological significance of this study lies in providing a geochemical basis for shale gas exploration in the area.

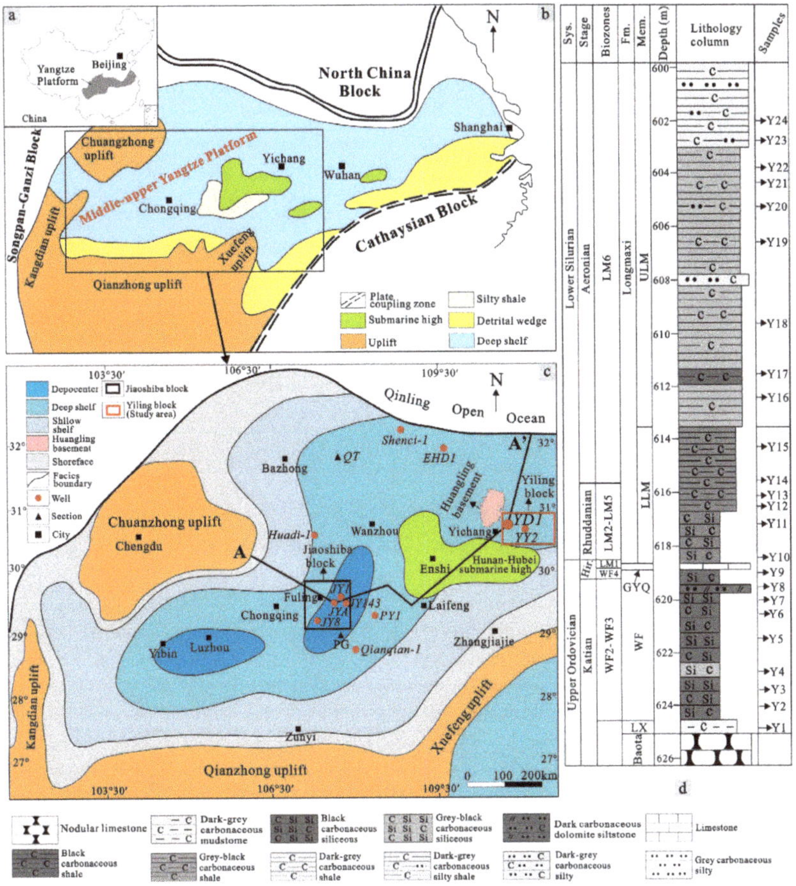

Figure 1. (**a**) Location of the Yangtze Platform in South China. (**b**) Palaeogeography of the Yangtze Platform in South China in the early Silurian (simplified from Zhang et al., 2019a [15]). (**c**) Palaeogeography of the middle-upper Yangtze Platform in the early Silurian showing the study area of the Yiling block in the southeastern Huangling basement (modified from Huang et al., 2020 [16]). (**d**) Lithostratigraphic column of well YD1 in the Yiling block showing black siliceous rock in the WF, black siliceous rock and argillaceous shales in the LLM are organic-rich shales. Note: LX, WF, GYQ, LLM, and ULM are the abbreviations for the Linxiang, Wufeng, Guanyinqiao, and lower and upper Longmaxi formations, respectively.

2. Geological Setting

The Yangtze Platform, located in central China, is divided into the upper, middle, and lower regions based on sedimentary facies (Figure 1a). The middle-upper Yangtze Platform lies in the central and western parts of the region and is bordered by various uplifts and blocks, such as the Kangdian uplift to the southwest, Chuanzhong uplift to the northwest, Qinling Ocean and North China Block to the north, Qianzhong uplift to the south, and Xuefeng uplift and Cathaysia Block to the southeast (Figure 1b) [4,16]. During the late Ordovician to early Silurian, the Cathaysia Block subducted and extruded the middle-upper Yangtze Platform due to the Caledonian movement, forming a tectonic pattern with alternating uplifts and depressions and a semi-closed marine environment [4,15]. The Wufeng–Longmaxi shales were deposited in these depressions, with two main depocenters in the southern and eastern Sichuan Basin, respectively [29]. Additionally, the Hunan–Hubei submarine high was formed between the cities of Chongqing and Yichang (Figure 1b and c) during the WF stage of the late Ordovician due to the collision between the Cathaysia Block and the Yangtze Platform [15,16,30].

The Fuling Jiaoshiba block, located in the southeastern Sichuan Basin and depicted in the black rectangle in Figure 1c, is the oldest commercial shale gas development block in China [6]. Recently, significant shale gas sources have also been discovered in the Yiling block in the study area [2], which is situated southeast of the Huangling basement and indicated by the red rectangle in Figure 1c. Well YD1, which targets the Wufeng–Longmaxi shales, is located in the Yiling block (Figure 1c). Generally, the Wufeng–Longmaxi Formation can be vertically divided into the Katian, Hirnantian, Rhuddanian, and Aeronian stages according to graptolite biozones [21]. According to lithological characteristics, it can be further divided into the LX, WF, GYQ, LLM, and ULM sections [2]. Among these, the LX section is composed of dark grey carbonaceous mudstone, the WF section is primarily composed of black carbonaceous siliceous rock interbedded with minor dolomite siltstone, the lower and upper LLM sections are mainly composed of black carbonaceous siliceous rock and black carbonaceous argillaceous shale, respectively, and the ULM section is mainly composed of dark grey carbonaceous shale (Figure 1d).

3. Methodology

3.1. Samples and Methods

Twenty-four fresh shale samples were collected from well YD1 in the Yiling block, comprising one sample from the LX (Y1), eight samples from the WF (Y2–Y9), six samples from the LLM (YA10–YA15), and nine samples from the ULM (YA16–YA24). After recovery, the samples were immediately placed in paper bags and analyzed by the State Key Laboratory of Biogeology and Environmental Geology in China for their TOC content, mineral compositions, major, trace, and rare earth elements, as well as iron speciation.

Among these, the TOC content was tested with a Multi EA 4000 carbon-sulfur tester, and 200 mesh samples of 0.1~0.3 g were treated with dilute hydrochloric acid to remove inorganic carbon and dried at 60~80 °C before the experiment, using the test methods and procedures following Chinese National Standard GB/T19145-2003. Mineral compositions were tested by a Rigaku SmartLab SE X-ray diffractometer, and approximately 5 g of evenly mixed sample powder was taken out for X-ray diffraction analysis, using test methods and procedures following China Petroleum Industry Standard SY/T 5163-2018. Major elements were analysed by a ZSXPrimus II X-ray fluorescence spectrometer, using test methods and procedures following Chinese National Standard GB/T 14506.28-2010. Trace and rare earth elements were determined with a multireceiver cup inductively coupled plasma source mass spectrometer, using test methods and procedures following Chinese National Standard GB/T14506.30-2010. Total iron content (FeT) was determined by the sulfosalicylic acid spectrophotometric method. Highly reactive iron (FeHR) includes iron in pyrite (Fepy), carbonate (Fecarb), magnetite (Femag), and ferric oxides (Feox). Fepy content was calculated from the concentration of pyrite sulfur (FeS2) extracted using CrCl2 reduction and precipitated as Ag2S in silver nitrate traps [26]. Fecarb, Feox, and Femag fractions were extracted by the sequential procedure described in Poulton and Canfield (2005) [31] and measured by atomic absorption spectroscopy. All of the tests were conducted under an ambient laboratory temperature approximately 24 °C; relative humidity was maintained at 65%; and analytical errors were less than 5%.

3.2. Data Presentation

The enrichment factor (EF) can be applied to evaluate the degree of elements enrichment in sediments using the following equation [32]:

$$EF = (X/Al)_{sample}/(X/Al)_{PAAS} \quad (1)$$

where $(X/Al)_{sample}$ is the ratio of element X to Al measured in the sample, and $(X/Al)_{PAAS}$ is the ratio of element X to Al in post Archean Australian shale (PAAS) [33]. Generally,

EF > 3, 1 < EF < 3, and EF < 1 were interpreted to reflect apparent enrichment, moderate enrichment, and depletion, respectively, relative to the PAAS standard [34].

In palaeomarine environments, the element concentration in sediments includes both non-biogenic and biogenic components. The concentration of elements derived from biogenic sources equals the total concentration in every sample minus its non-biogenic influx, and can therefore be calculated as follows [33,35]:

$$X_{bio} = X_{Sample} - Al_{Sample} \times (X/Al)_{PAAS} \quad (2)$$

where X_{bio} is the biogenic part of element X, and $X_{sample,}$ and Al_{sample} are the contents of elements X and Al measured in the sample, respectively. $(X/Al)_{PAAS}$ is the ratio of X to Al in the PAAS.

The chemical index of alteration (CIA), which was originally proposed by Nesbitt and Young (1982) to determine the chemical weathering intensity of the provenance of clastic rock, was calculated following the formula below [36]:

$$CIA = molar\ (Al_2O_3/[(Al_2O_3 + CaO^* + Na_2O + K_2O)] \times 100) \quad (3)$$

All major element concentrations are calculated in mole fractions. CaO* refers to CaO in silicate. The method of indirect calculation of the value of CaO* as follows [5]: CaO_{Remain} = molar ($CaO\text{-}P_2O_5 \times 10/3$), if molar ($CaO_{Remain}$) < molar ($Na_2O$), then molar ($CaO^*$) = molar ($CaO_{Remain}$); if molar ($CaO_{Remain}$) > molar ($Na_2O$), then molar ($CaO^*$) = molar ($Na_2O$). Generally, CIA values change greatly in different palaeoclimates, and CIA values range from 50 to 70, 70 to 80, and 80 to 100, indicating low, moderate, and high degrees of chemical weathering, respectively [14]. In general, weak weathering is associated with cold and arid climates, whereas intense weathering is associated with hot and wet climates. New K was introduced by K metasomatism during diagenesis, resulting in low CIA values. Therefore, it was necessary to correct it. Generally, the CIA values before K metasomatism (CIA_{corr}) value can be calculated by Al_2O_3–CaO^* + Na_2O–K_2O ternary diagrams [35].

Additionally, La–Th–Sc and Th–Sc–Zr/10 ternary discrimination diagrams are used to infer the tectonic settings [37,38]. The CIA_{corr} value, Sr/Cu, and Rb/Sr ratios are used to reconstruct palaeoclimate conditions [39,40]. The Al/(Al + Fe + Mn) and (Fe+Mn)/Ti ratios are used to evaluate hydrothermal deposition [15,41]. The $Si_{bio}/(Si_{bio} + Fe_{bio} + Al + Ca_{bio})$ ratio is used to distinguish biogenic sources, where a ratio of > 0.8 indicates an absolute biogenic source [17,35]. The Al content is used to study the level of terrigenous influx [15,16]. The V/Cr, Ni/Co, U/Th, and Fe/Al ratios are used to reflect palaeoredox conditions [14,42–45]. A Fe_{py}/Fe_{HR} vs. Fe_{HR}/Fe_T discrimination diagram is used to distinguish euxinic and ferruginous conditions [46,47]. The Mo vs. TOC discrimination diagram and proxy Co × Mn are used to evaluate palaeohydrography [48,49]. The proxies Si_{bio}, $(Ni + Cu + Zn)_{bio}$, P/Ti, and Cu/Al are used to evaluate palaeoproductivity [3,13,15,40]. The La_N/Yb_N ratio is used to reflect the deposition rate [35], where subscripts N represents the chondrite normalization [50].

4. Results

4.1. Lithology and Graptolite Biozones

Core and thin section images are shown in Figure 2. The LX is composed of dark grey carbonaceous mudstone (Figure 2a). The WF is primarily composed of black carbonaceous siliceous rock (Figure 2b) with larger quartz grains and many bioclasts (Figure 2c) interbedded with minor dolomite siltstone (Figure 2d) containing much calcite and dolomite (Figure 2e). Rectograptus of the WF2-3 graptolite biozones was clearly observed in the bedding plane (Figure 2f). The LLM is mainly composed of black carbonaceous argillaceous shale with clear laminated bedding (Figure 2g), the mineral compositions are slightly oriented, and the particle size of quartz is obviously smaller than that in the WF (Figure 2h). Campograptus of graptolite biozone LM6 was clearly observed in the bedding plane (Figure 2i). The ULM is mainly composed of dark grey carbonaceous shale (Figure 2j), and the mineral compositions are apparently oriented with many hydromica minerals (Figure 2k). Demirastrite graptolites of the LM6 biozone are clearly observed in the bedding plane (Figure 2l).

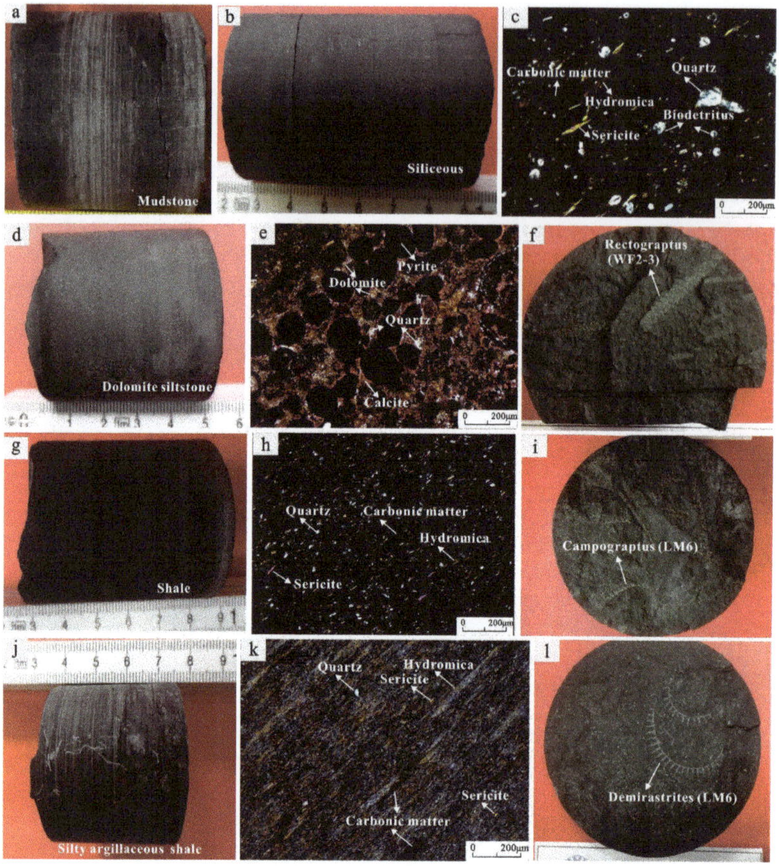

Figure 2. (**a**) Core image of dark grey carbonaceous mudstone in the LX, sample Y1 with a depth of 625 m. (**b**,**c**) Core and thin section images of black carbonaceous siliceous rock in the WF, sample Y9 with a depth of 618.46 m. (**d**,**e**) Core and thin section images of grey-black carbonaceous dolomite siltstone in the WF, sample Y8 with a depth of 619.5 m. (**f**) Rectograptus image of graptolite biozones WF2-3 in the WF, depth 621.26 m. (**g**,**h**) Core and thin section images of black carbonaceous shale in the LLM, sample Y13 with a depth of 616.1 m. (**i**) Campograptus image of graptolite biozone LM6 in the LLM, depth 615.6 m. (**j**,**k**) Core and thin section images of dark grey carbonaceous silty shale in the UMX, sample Y20 with a depth of 605.26 m. (**l**) Demirastrite image of graptolite biozone LM6 in the ULM, depth 610.71 m.

4.2. TOC Content

Generally, organic-rich and organic-lean shales are characterized by TOC contents of greater than 2%, and lower than 2%, respectively [51]. The TOC values are presented in Table 1 and Figure 3a. The TOC content is less than 1% in the LX with a value of only 0.7%, and it ranges from 1.90% to 4.94% (avg. 3.21%), 2.61% to 6.29% (avg. 4.14%), and 0.76% to 2.74% (avg. 1.50%) in the WF, LLM, and ULM, respectively. The results show that the WF and LLM include organic-rich shales, and the LX and ULM include organic-lean shales. Additionally, the results show a significant decrease in TOC content from 6.18% of sample Y13 to 3.05% of sample Y14 in the LLM.

Table 1. TOC and major elements contents of the Wufeng–Longmaxi Formation in well YD1.

Sample No.	TOC (%)	SiO_2 (%)	TiO_2 (%)	Al_2O_3 (%)	Fe_2O_3 (%)	MnO (%)	MgO (%)	CaO (%)	Na_2O (%)	K_2O (%)	P_2O_5 (%)	Fe_T (%)	Fe_{py} (%)	Fe_{HR} (%)	Al (%)	Si_{bio} (%)	$Si_{bio}/(Si_{bio} + Fe_{bio} + Al + Ca_{bio})$	Al/(Al + Fe + Mn)	(Fe + Mn)/Ti	CIAcorr	P/Ti	Fe_{py}/Fe_{HR}	Fe_{HR}/Fe_T	Fe_T/Al
Y24	0.76	61.5	0.69	16.79	6.55	0.06	2.68	0.48	1.21	3.9	0.11	4.59	0.79	1.44	8.89	2.66	0.24	0.66	11.19	76.8	0.12	0.55	0.31	0.52
Y23	0.80	61.6	0.7	17.1	6.38	0.05	2.71	0.43	1.18	4.05	0.11	4.47	0.68	1.2	9.05	2.23	0.21	0.67	10.73	77.4	0.11	0.57	0.27	0.49
Y22	1.74	60.2	0.67	16.41	6.34	0.05	2.59	0.51	1.04	3.92	0.11	4.44	0.97	1.57	8.69	2.64	0.24	0.66	11.14	78.7	0.12	0.62	0.35	0.51
Y21	1.37	60.4	0.67	16.62	6.65	0.05	2.66	0.55	1.04	3.95	0.11	4.65	1.02	1.67	8.80	2.42	0.22	0.65	11.68	79.0	0.12	0.61	0.36	0.53
Y20	1.11	59.3	0.7	16.88	6.44	0.05	2.71	0.62	1.09	4.04	0.1	4.51	0.89	1.62	8.94	1.48	0.15	0.66	10.83	78.2	0.10	0.55	0.36	0.50
Y19	1.73	61	0.69	16.12	6.23	0.05	2.6	0.61	1.04	3.84	0.1	4.36	0.94	1.61	8.53	3.48	0.30	0.66	10.63	78.3	0.11	0.58	0.37	0.51
Y18	1.58	61.3	0.65	15.74	6.23	0.04	2.5	0.42	0.98	3.76	0.11	4.36	1.06	1.6	8.33	4.20	0.34	0.65	11.26	79.0	0.12	0.66	0.37	0.52
Y17	2.74	59.3	0.62	14.24	6.33	0.03	2.28	0.57	1.1	3.32	0.13	4.43	0.85	2.45	7.54	5.57	0.41	0.63	11.97	75.9	0.15	0.35	0.55	0.59
Y16	1.71	62.8	0.6	14.8	6.2	0.04	2.38	0.36	1.01	3.51	0.1	4.34	1.06	2.14	7.84	6.34	0.45	0.64	12.14	77.7	0.12	0.50	0.49	0.55
Y15	2.61	61.1	0.6	13.8	6.61	0.04	2.36	0.94	0.98	3.3	0.1	4.63	1.67	2.71	7.31	7.11	0.46	0.61	12.94	77.0	0.12	0.62	0.59	0.63
Y14	3.05	58.4	0.62	13.83	6.57	0.03	2.08	0.75	1.09	3.27	0.12	4.6	1.23	3.27	7.32	5.80	0.42	0.61	12.43	75.4	0.14	0.38	0.71	0.63
Y13	6.18	70.4	0.48	8.9	3.65	0.02	1.24	0.96	0.98	1.98	0.11	2.56	1.41	2.09	4.71	19.04	0.79	0.65	8.93	69.7	0.17	0.67	0.82	0.54
Y12	6.29	48.9	0.6	11.01	7.55	0.06	2.41	2.5	1.27	2.33	0.2	5.29	3.19	4.41	5.83	5.76	0.38	0.52	14.81	68.8	0.24	0.72	0.83	0.91
Y11	3.45	82	0.26	4.99	1.87	0.02	0.6	1.28	0.61	0.99	0.07	1.31	0.83	1.22	2.64	30.55	0.90	0.67	8.49	67.8	0.20	0.68	0.93	0.50
Y10	3.23	82	0.26	4.55	1.98	0.03	0.71	1.75	0.66	0.88	0.07	1.38	0.78	1.28	2.41	31.19	0.90	0.63	9.03	64.2	0.20	0.61	0.92	0.58
Y9	2.99	78	0.24	6.64	2.7	0.02	0.62	1.45	0.61	1.47	0.06	1.89	1.31	1.93	3.52	26.09	0.86	0.65	13.23	73.0	0.18	0.68	1.02	0.54
Y8	2.58	34.7	0.37	8.34	4.34	0.22	3.62	18.9	0.58	1.85	0.03	3.04	1.9	3.18	4.42	3.28	0.15	0.58	14.45	77.4	0.06	0.60	1.05	0.69
Y7	2.05	82.1	0.2	3.73	1.85	0.04	0.62	2.87	0.46	0.75	0.06	1.29	0.82	1.25	1.97	32.55	0.89	0.60	11.05	67.6	0.22	0.66	0.97	0.66
Y6	3.86	83.9	0.22	4.57	1.66	0.01	0.63	0.7	0.42	1.04	0.05	1.16	0.66	0.98	2.42	32.07	0.92	0.67	8.86	72.9	0.17	0.67	0.84	0.48
Y5	4.49	74.3	0.44	8.07	2.91	0.02	1.28	0.94	0.65	1.99	0.08	2.03	0.85	1.34	4.27	22.17	0.83	0.68	7.77	75.2	0.13	0.63	0.66	0.48
Y4	1.9	89.9	0.13	2.95	0.94	0.01	0.41	0.81	0.26	0.66	0.04	0.66	0.27	0.46	1.56	37.39	0.95	0.70	8.54	73.7	0.22	0.59	0.70	0.42
Y3	2.88	77.2	0.33	7.12	2.81	0.03	1.03	1.41	0.47	1.78	0.07	1.96	0.78	1.25	3.77	24.99	0.85	0.65	10.05	77.9	0.15	0.62	0.64	0.52
Y2	4.94	63.6	0.68	13.98	4.29	0.03	2.17	0.65	0.92	3.71	0.11	3	0.55	1.01	7.40	7.99	0.55	0.71	7.42	78.1	0.12	0.54	0.34	0.41
Y1	0.7	60.8	0.84	17.07	5.52	0.03	2.81	0.95	1.01	4.6	0.09	3.86	0.56	1.12	9.04	1.89	0.19	0.70	7.71	79.5	0.08	0.50	0.29	0.43
PAAS	/	62.8	1	18.9	7.22	0.11	2.2	1.3	1.2	3.7	0.16	/	/	/	/	/	/	/	/	/	/	/	/	/

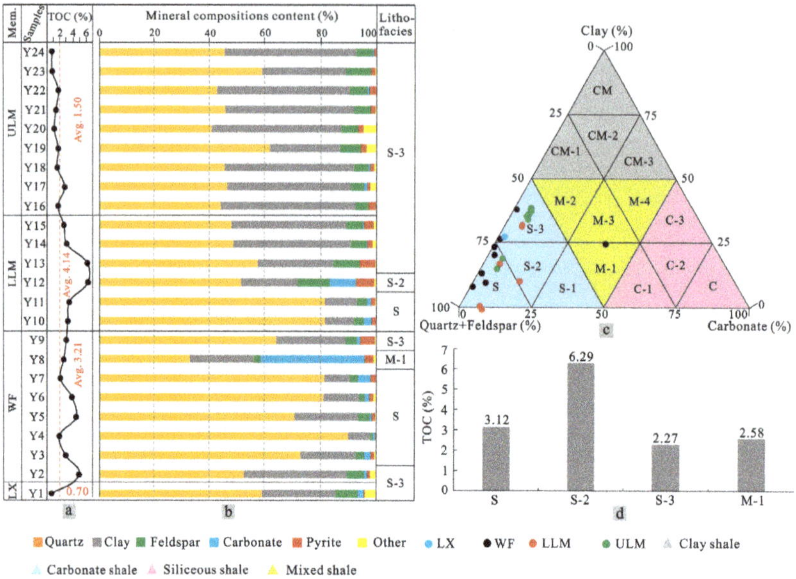

Figure 3. (**a**) Vertical variation in TOC content shows that the WF and LLM include organic-rich shales. (**b**) The percentage of bulk mineral compositions shows that quartz contents decrease and clay contents increase from the bottom to the top. (**c**) The lithofacies classification shows that most samples are potted in the field of siliceous shales. (**d**) TOC contents in different subclass lithofacies show that S-2 and S have higher TOC values. Note: CM, CM-1, CM-2, and CM-3 represent mudstone, siliceous argillaceous shale, argillaceous shale, and calcareous argillaceous shale, respectively. C, C-1, C-2, and C-3 represent limestone, siliceous calcareous shale, calcareous shale, and argillaceous calcareous shale, respectively. S, S-1, S-2, and S-3 represent siliceous rock, calcareous siliceous shale, siliceous shale, and argillaceous siliceous shale, respectively. M-1, M-2, M-3, and M-4 represent calcareous siliceous mixed shale, argillaceous siliceous mixed shale, mixed shale, and argillaceous calcareous mixed shale, respectively.

4.3. Mineral Compositions and Lithofacies

Mineral compositions are mainly composed of quartz and clay in all of the samples, for which quartz contents reach 59.3%, 33.0%–89.9% (avg. 68.3%), 48.1%–81.5% (avg. 61.6%), and 40.8%–61.9% (avg. 48.0%) in the LX, WF, LLM, and ULM, respectively, showing a decreasing trend from the bottom to the top. In contrast, clay contents reach 26.0%, 7.8%–36.6% (avg. 19.6%), 10.4%–41.9% (avg. 25.3%), 24.9%–48.2% (avg. 42.2%), showing an increasing trend from bottom to top. Contents of feldspar, carbonate, and pyrite are low in all of the samples, and the feldspar, carbonate, and pyrite contents are higher in the ULM, WF, and LLM, respectively (Figure 3b).

Shale can be divided into four lithofacies types (siliceous, calcareous, clay, and mixed shales) [52]. These can further be subdivided into sixteen lithofacies types according to a three-end element map of clay, carbonate, and felsic rocks content [53]. The results show that most samples are plotted in the siliceous shale field except for sample Y8 of the WF, which is plotted in the mixed shale field, for which organic-rich shales in the WF and LLM are primarily plotted as subclass lithofacies type S, and organic-lean shales in the LX and ULM are primarily plotted as subclass lithofacies type S-3 (Figure 3c).

4.4. Major and Trace Elements

The results regarding major elements are listed in Table 1. Major oxides mainly include SiO_2 and Al_2O_3 in all of the formations, where organic-rich shales have higher contents of

SiO_2 in the WF (34.74%–89.92%, avg. 72.97%) and LLM (48.92%–82.04%, avg. 67.13%), and organic-lean shales have higher contents of Al_2O_3 in the LX (17.07%) and ULM (14.24%–17.10%, avg. 16.08%). Additionally, organic-rich shales are relatively rich in CaO, while organic-lean shales are relatively rich in MgO, Na_2O, and K_2O. The EF value shows that Si, Ca, Na, P, and Fe contents in the organic-rich shales are higher than those of the organic-lean shales with moderate to apparent enrichment relative to PAAS (Figure 4a).

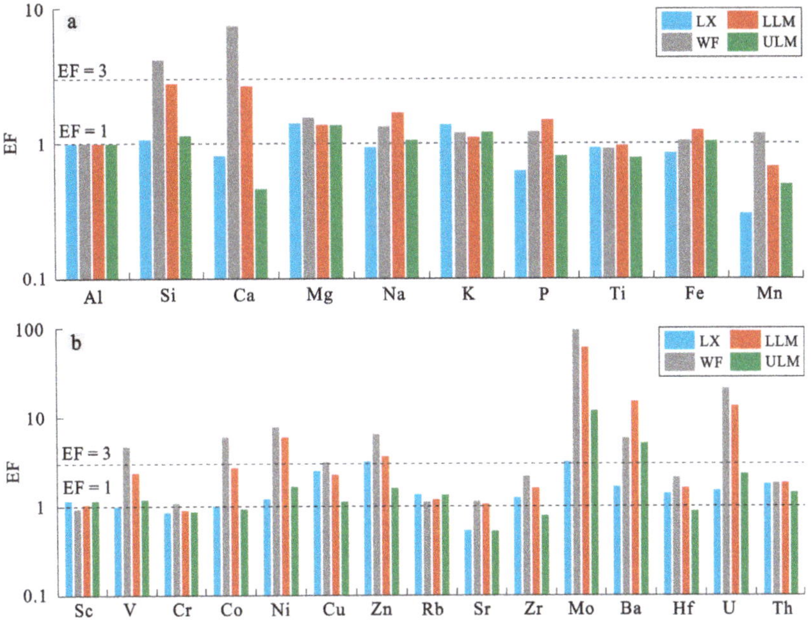

Figure 4. (**a**) Enrichment factors of the major element's values show that Si, Ca, Na, P, and Fe contents in organic-rich shales are higher than those of organic-lean shales, with moderate to apparent enrichment contents compared to those of PAAS. (**b**) Enrichment factors of the trace elements show that V, Co, Ni, Cu, Zn, Mo, Ba, and U contents in organic-rich shales are higher than those of organic-lean shales, with moderate to apparent enrichment relative to those of PAAS. The dotted line with EF = 1 highlights enrichment or depletion, and the dotted line with EF = 3 highlights apparent or moderate enrichment.

The results of the trace elements are listed in Table 2. V, Co, Ni, Mo, and U are rich in organic-rich shales, while Sc, Cr, Rb, Hf, and Th are rich in organic-lean shales. The EF values show that V, Co, Ni, Cu, Zn, Mo, Ba, and U contents in organic-rich shales are higher than those of organic-lean shales with moderate to apparent enrichment relative to PAAS (Figure 4b).

Table 2. Trace elements contents of the Wufeng–Longmaxi Formation in well YD1.

Sample No.	Y24	Y23	Y22	Y21	Y20	Y19	Y18	Y17	Y16	Y15	Y14	Y13	Y12	Y11	Y10	Y9	Y8	Y7	Y6	Y5	Y4	Y3	Y2	Y1	PAAS
Sc (ppm)	16.33	16.85	16.52	16.87	17.54	15.45	15.16	14.15	13.90	13.75	13.40	7.65	10.20	3.88	3.52	4.82	1.57	2.69	3.96	8.00	2.44	6.31	16.02	16.84	16
V (ppm)	139.18	134.34	158.23	175.24	166.84	171.58	166.72	131.83	125.06	134.55	144.00	218.36	228.99	128.61	99.23	123.32	77.90	83.90	279.34	571.00	234.40	96.02	329.21	134.65	150
Cr (ppm)	83.48	86.69	84.90	87.13	87.04	81.64	77.71	74.68	71.64	70.79	71.10	47.48	59.30	28.37	23.18	26.20	4.11	19.80	34.60	64.70	27.00	56.40	115.48	85.18	110
Co (ppm)	18.08	18.26	18.68	20.02	19.15	18.37	18.31	16.00	17.66	18.66	24.40	16.34	29.40	26.89	30.98	16.02	18.20	70.00	22.51	21.60	66.77	24.77	12.00	21.14	23
Ni (ppm)	64.23	64.53	87.47	92.86	83.46	87.25	83.68	64.84	75.25	110.77	144.00	132.13	340.22	93.91	97.50	120.29	195.00	154.00	105.60	150.00	105.29	98.04	83.62	60.83	55
Cu (ppm)	50.48	39.34	46.07	51.68	48.93	48.63	50.48	49.17	47.11	64.75	75.30	58.07	78.58	31.12	26.74	32.53	14.80	19.30	51.26	82.40	32.03	74.03	159.46	113.71	50
Zn (ppm)	124.79	111.17	123.66	122.29	124.99	120.96	107.02	107.72	102.87	116.80	119.00	140.06	355.78	85.89	69.91	110.44	308.00	102.00	135.52	258.00	135.07	172.61	244.17	245.27	85
Rb (ppm)	184.08	194.66	193.04	193.38	193.17	180.57	175.65	164.80	167.13	162.99	153.00	88.32	117.18	42.99	37.66	53.26	54.80	29.40	46.27	87.80	29.80	71.74	172.53	197.96	160
Sr (ppm)	86.66	89.16	107.48	88.65	90.98	86.69	96.85	91.79	74.15	84.31	85.90	65.02	260.65	60.25	53.33	58.84	257.00	61.90	38.54	59.70	33.19	53.26	81.58	97.36	200
Zr (ppm)	146.84	144.92	150.92	146.13	144.68	139.68	137.26	137.50	125.47	130.30	144.00	134.52	213.58	91.31	152.60	146.99	186.00	243.00	56.26	113.00	77.54	98.31	216.38	236.25	210
Mo (ppm)	2.16	3.25	8.17	5.45	6.20	10.52	6.27	31.29	12.75	16.23	34.10	45.00	53.05	13.54	12.68	11.75	33.00	33.50	43.26	85.20	7.51	6.58	25.20	2.84	1
Ba (ppm)	1606.3	2037.3	6096.1	1374.8	1202.8	1662.1	5645.0	4164.0	1127.6	879.7	958.0	873.2	26770.3	896.7	1025.6	1284.2	2732.0	1106.0	893.3	951.0	743.9	995.6	818.6	981.3	650
Hf (ppm)	3.88	3.98	4.03	3.97	3.88	3.63	3.60	3.72	3.34	3.58	3.89	3.61	5.53	2.21	2.89	3.18	7.98	5.15	1.32	2.95	0.87	2.69	4.31	6.37	5
U (ppm)	4.82	4.91	6.78	4.98	6.13	5.90	5.83	8.43	6.38	10.13	14.50	14.97	51.17	8.54	14.53	25.56	87.70	21.50	8.06	13.50	4.31	12.88	7.89	4.22	3.1
Th (ppm)	17.54	17.61	18.57	18.54	18.63	17.69	16.78	17.29	16.12	17.21	17.50	12.65	17.71	6.69	7.05	10.96	11.50	6.15	5.81	11.10	3.24	9.94	18.80	23.42	14.6
La (ppm)	45.95	47.08	48.20	48.57	47.14	45.75	45.31	44.07	42.62	43.59	44.10	32.83	46.47	18.88	23.05	29.09	62.10	25.50	19.57	34.10	11.10	26.54	52.84	60.33	38
Yb (ppm)	2.79	2.93	2.91	2.95	2.80	2.78	2.62	2.73	2.51	2.69	2.84	2.27	3.90	1.46	1.67	2.27	1.36	2.10	1.21	2.19	1.03	1.57	4.43	3.89	2.8
Sr/Cu	1.72	2.27	2.33	1.72	1.86	1.78	1.92	1.87	1.57	1.30	1.14	1.12	3.32	1.94	1.99	1.81	17.36	3.21	0.75	0.72	1.04	0.72	0.51	0.86	/
Rb/Sr	2.12	2.18	1.80	2.18	2.12	2.08	1.81	1.80	2.25	1.93	1.78	1.36	0.45	0.71	0.71	0.91	0.21	0.47	1.20	1.47	0.90	1.35	2.11	2.03	/
Sr/Ba	0.05	0.04	0.02	0.06	0.08	0.05	0.02	0.02	0.07	0.10	0.09	0.07	0.01	0.07	0.05	0.05	0.09	0.06	0.04	0.06	0.04	0.05	0.10	0.10	/
V/Cr	1.67	1.55	1.86	2.01	1.92	2.10	2.15	1.77	1.75	1.90	2.03	4.60	3.86	4.53	4.28	4.71	18.95	4.24	8.07	8.83	8.68	1.70	2.85	1.58	/
U/Th	0.28	0.28	0.37	0.27	0.33	0.33	0.35	0.49	0.40	0.59	0.83	1.18	2.89	1.28	2.06	2.33	7.63	3.50	1.39	1.22	1.33	1.30	0.42	0.18	/
Ni/Co	3.55	3.53	4.68	4.64	4.36	4.75	4.57	4.05	4.26	5.94	5.90	8.08	11.57	3.49	3.15	7.51	10.71	2.20	4.69	6.94	1.58	3.96	6.97	2.88	/
Cu/Al	5.68	4.35	5.30	5.87	5.47	5.70	6.06	6.52	6.01	8.86	10.28	12.32	13.48	11.78	11.10	9.25	3.35	9.77	21.19	19.29	20.51	19.64	21.55	12.58	/
Mo/TOC ($\times 10^{-4}$)	2.85	4.04	4.70	3.98	5.59	6.08	3.97	11.42	7.46	6.22	11.18	7.28	8.43	3.92	3.93	3.93	12.79	16.34	11.21	18.98	3.95	2.28	5.10	4.06	/
Co × Mn ($\times 10^{-8}$)	0.84	0.71	0.72	0.78	0.74	0.71	0.57	0.37	0.55	0.58	0.57	0.25	1.37	0.42	0.72	0.25	3.10	2.17	0.17	0.33	0.52	0.58	0.28	0.49	/
$(Ni + Cu + Zn)_{bio}$ (ppm)	70.72	49.03	92.23	99.75	87.68	94.79	82.95	78.58	76.46	153.60	199.27	240.79	663.90	160.75	148.41	196.50	441.22	237.80	246.44	409.27	242.73	273.10	346.71	248.21	/
La_N/Yb_N	10.96	10.71	11.06	10.96	11.22	10.98	11.54	10.77	11.32	10.78	10.35	9.62	7.94	8.61	9.18	8.55	30.44	8.10	10.76	10.38	7.20	11.27	7.95	10.34	9.05

4.5. Iron Speciation

The iron speciation results are listed in Table 1. Fe_{py} and Fe_{HR} contents are characterized as follows: the LLM (avg. 1.52% and 2.50%, respectively) > the ULM (avg. 0.92% and 1.70%, respectively) > the WF (avg. 0.89% and 1.43%, respectively) > the LX (0.56% and 1.12%, respectively). Average contents of Fe_T are characterized as follows: the ULM (avg. 4.46%) > the LX (3.86%) > the LMX (avg. 3.29%) > the WF (avg. 1.88%).

5. Discussion

5.1. Tectonic Setting

Trace element discrimination diagrams have also been used to distinguish tectonic settings [37,38]. In the diagrams of La-Th-Sc and Th-Sc-Zr/10 (Figure 5a,b) proposed by Bhatia and Crook (1986) [37], most samples mainly plot within or near the field of the active continental margin, and minor samples fall within the field of the continental island arc, which is consistent with that of the Huadi-1 well in the southeastern Sichuan Basin [38]. These results show that from the late Ordovician to the early Silurian, the subduction of the Proto-Tethys oceanic crust towards the South China Plate led to the convergence and compression between the Cathysia Block and Yangtze Platform [16], forming a tectonic setting of mainly active continental margins.

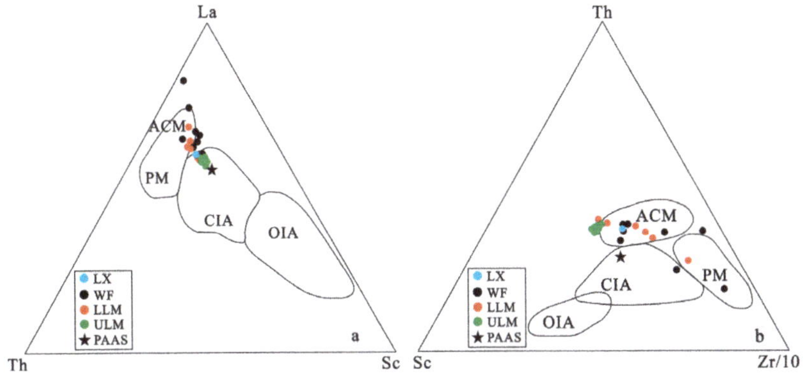

Figure 5. (**a**) La–Th–Sc and (**b**) Th–Sc–Zr/10 ternary diagrams (modified from Ge et al., 2019 [38]) showing mainly active continental margin tectonic settings. Note: OIA, CIA, ACM, and PM are the abbreviations for the oceanic island arc, continental island arc, active continental margin, and passive margin, respectively.

5.2. Palaeoenvironment

5.2.1. Palaeoclimate

The palaeoclimate controls levels of the humidity and weathering, thus affecting terrigenous influx and the sediment supply in sedimentary rocks [41]. The chemical index of alteration (CIA) has been widely used to reconstruct the palaeoclimate, and the formula of CIA is shown in Section 3.2 [5,16,35]. The Sr/Cu ratio is an effective indicator for evaluating palaeoclimate change, with ratios of <5, 5–10, and >10 indicating humid, semihumid-semiarid, and arid palaeoclimates, respectively [40,54]. Additionally, because Rb is relatively stable under weathering, whereas Sr is prone to leaching during weathering, the Rb/Sr ratio is often used to study palaeoclimate in provenance, with high and low ratios indicating a warm and humid climate with a high degree of weathering and an arid climate with a low degree of weathering, respectively [39,55].

The Al_2O_3–CaO^* + Na_2O–K_2O ternary diagrams (Figure 6) show that the predicted weathering trend line (T′) is parallel to the Al_2O_3–CaO^* + Na_2O line and points towards the endmember of Al_2O_3. Starting from the K_2O end member, crossing the sample data point and intersecting with T′, the CIA_{corr} values can be obtained [35].

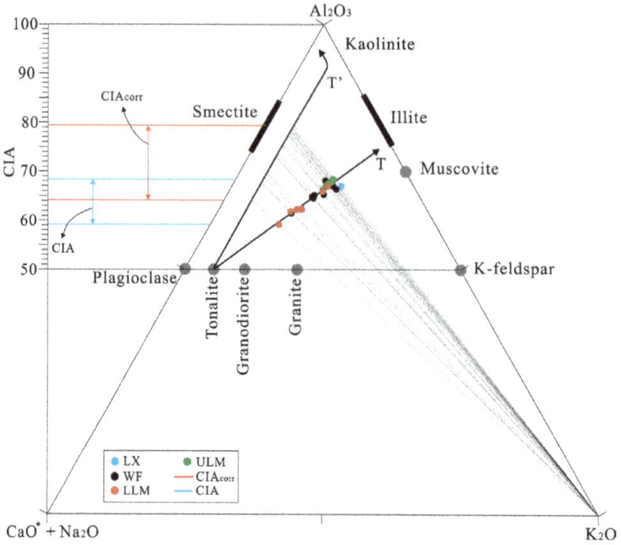

Figure 6. Al_2O_3–CaO^* + Na_2O–K_2O ternary diagrams exhibiting weathering degree (all in mole percentages). The CIA_{corr} values of most samples range from 70 to 80, indicating a moderate degree of chemical weathering and a warm and humid palaeoclimate. Note: T and T′ are the actual and predicted weathering trends, respectively.

The results show that the CIA_{corr} value is highest in the LX with the value of 79.5, and it ranges from 67.6 to 78.1 (avg. 74.5), 64.2 to 77.0 (avg. 70.5), 75.9 to 79.0 (avg. 77.9), and 57.4 to 75.7 (avg. 72.1), respectively, which indicate mainly a moderate degree of chemical weathering and imply a warm and humid climate in the Wufeng–Longmaxi Formation. Comparatively speaking, the CIA_{corr} values of the LX and ULM are higher than that of the WF and LLM, which indicates a much warmer and wetter climate in the organic-lean shales (Table 1; Figures 6 and 7).

As shown in Figure 7, the Sr/Cu ratios of most samples are lower than 5 except for sample Y8 of dolomite siltstone from the WF, which has a Sr/Cu ratio of 17.4, denoting a primarily warm and humid climate in the Wufeng–Longmaxi Formation and a much drier climate in the formation enriched with carbonate minerals.

The profiles of Rb/Sr ratios and CIA_{corr} values are nearly consistent, which shows that from the bottom to the top, the climate was moderately warm and humid in the LX and the lower WF. Then, there was slight warmth and humidity in the upper WF. Further upwards, cold and arid conditions occurred in the GYQ due to Gondwanan continental glaciation at the end of the late Ordovician [16,56]. Subsequently, with the melting of Gondwanan continental glaciers in the early Silurian [16,18], the climate gradually changed to slight warmth and humidity in the lower LLM. Then, the Kwangsian Orogeny intensified in the Yangtze Platform with strong tectonic uplift, leading to the sea level dropping rapidly [20,21] and the palaeoenvironment changing significantly. Remarkably, there is a significant decrease in TOC content from 6.18% in sample Y13 below the red dotted line to 3.05% in sample Y14 above the red dotted line, a significant increase in Al content from 8.9% to 13.83%, and a significant increase in Rb/Sr ratios and CIA_{corr} value, indicating that the climate rapidly changed from slightly warm and humid in the lower LLM to moderately warm and humid in the upper LLM. Accordingly, the significant variation in the palaeoclimate during the LLM stage was one of the key factors that led to the turnover of graptolite biozones from LM5 to LM6. After this, the climate was much hotter and wetter in the ULM with high Rb/Sr ratios, leading to a high degree of chemical weathering [13,16].

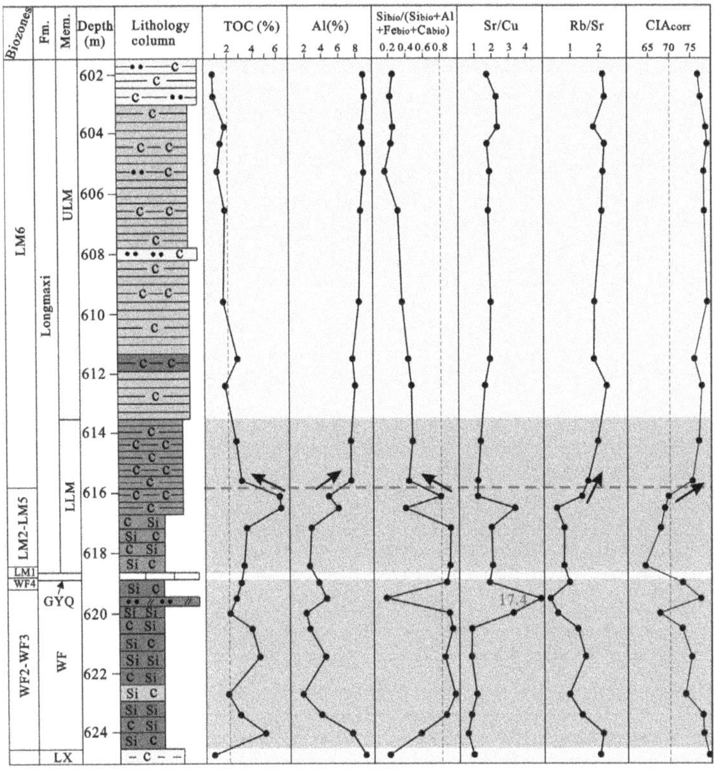

Figure 7. Vertical variation in TOC contents, terrigenous influx proxy Al contents, biogenic source discrimination proxy $Si_{bio}/(Si_{bio} + Fe_{bio} + Al + Ca_{bio})$, palaeoclimate proxies Sr/Cu and Rb/Sr, and CIA_{corr} of the Wufeng–Longmaxi Formation in well YD1, showing that the organic-rich shales in the WF and LLM mainly deposited in the warm and humid climate, and the organic-lean shales in the LX and ULM mainly deposited in a warmer and wetter climate. Remarkably, the significant variation of palaeoenvironment caused by the Kwangsian Orogeny [20,21] directly led to the turnover of graptolite biozones from the LM5 in Rhuddanian to the LM6 in Aeronian. Note: The red dotted line represents the boundary of Rhuddanian and Aeronian.

The variation of the palaeoclimate in well YD1 is consistent with that of well YY2 in the Yiling block [2] and well JY8 in the Jiaoshiba block [16], indicating that the palaeoclimate of the middle Yangtze Platform is similar to that of the upper Yangtze Platform. Additionally, we found that the climate of organic-rich shales in the WF and LLM is neither cold and arid like that of the GYQ nor hot and wet like that of organic-lean shales in the LX and ULM. The climate during the WF and LLM stages can be described as moderately warm and humid.

5.2.2. Origin of Silica and Terrigenous Influx

Previous studies have identified three primary sources of siliceous sediments in marine environments: hydrothermal fluids, siliceous organisms, and silica-rich terrigenous clastics [57]. In sedimentary rocks, Fe and Mn are generally enriched with hydrothermal deposits, and in typical sediments influenced by hydrothermal events, the (Fe + Mn)/Ti and Al/(Al + Fe + Mn) ratios are generally valued at >15 and < 0.4, respectively [41]. The analysis shows that all of the samples plot in the nonhydrothermal field, implying that silica mainly had a nonhydrothermal origin (Figure 8a).

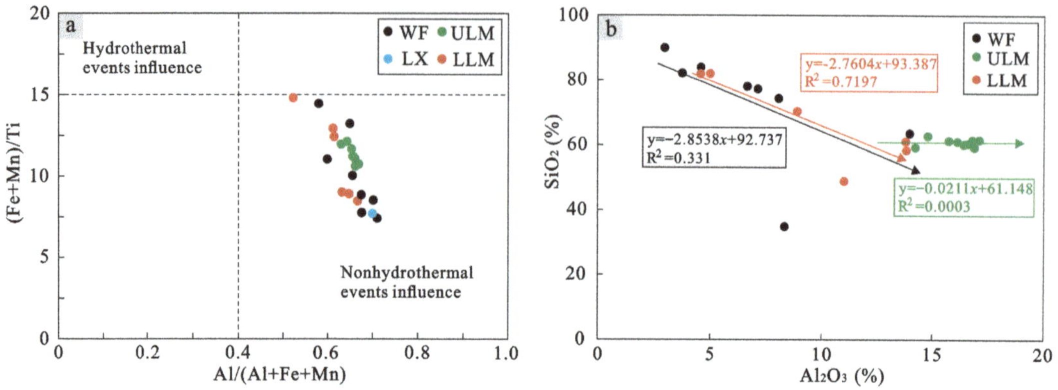

Figure 8. (**a**) Crossplots of (Fe + Mn)/Ti *vs.* Al/(Al + Fe + Mn) showing that the silica mainly had a nonhydrothermal origin. (**b**) Crossplots of SiO_2 vs. Al_2O_3 showing a negative correlation in the WF and LLM, implying a mainly source of biogenic silica and no significant correlation in the ULM, implying that silica probably had a mixed biogenic and terrigenous clastic origin.

Al is generally considered an indicator element for terrigenous clastic material [5,15]; therefore, the crossplots of SiO_2 vs. Al_2O_3 are usually used to distinguish terrestrial and biogenic silica with a positive correlation, no apparent correlation, and a negative correlation reflecting mainly terrestrial silica, mixed origin of terrestrial and biogenic silica, and biogenic silica, respectively [35]. The results show that silica mainly had a biogenic origin in the WF and LLM, and a mixed terrestrial and biogenic origin in the ULM (Figure 8b). The results show that graptolites are the main sources of biogenic silica in the Wufeng–Longmaxi Formation [13,48] (Figure 2), and there is also a small amount of sponge spicules and radiolarians [13,48,58].

Additionally, the $Si_{bio}/(Si_{bio} + Fe_{bio} + Al + Ca_{bio})$ ratios of most samples of organic-rich shales in the WF and the lower LLM are higher than 0.8, indicating an absolute biogenic silica [17,59]. In contrast, the low $Si_{bio}/(Si_{bio} + Fe_{bio} + Al + Ca_{bio})$ ratios of samples from the LX, upper LLM, and ULM suggest that silica partly originated from terrigenous clasts. Remarkably, the $Si_{bio}/(Si_{bio} + Fe_{bio} + Al + Ca_{bio})$ ratios of the LM6 graptolite biozone of the Aeronian above the red dotted line are significantly lower than those of the LM5 graptolite biozone of the Rhuddanian below the red dotted line (Table 1; Figure 7). The reason is that, as a result of the Kwangsian Orogeny [60], the sea level dropped rapidly due to the large-scale uplift in the Yangtze Platform, which led to a decrease in the quantity of graptolites and contents of biogenic silica.

The influx of terrigenous clastic material is a key factor that influences black shale composition and OMA [15,35,42]. The results show that Al contents are higher in organic-lean shales in the LX (9.04%) and ULM (avg. 8.51%) and lower in organic-rich shales in the WF (avg. 3.67%) and LLM (avg. 5.04%) (Table 1; Figure 7). The variation of Al content in the Yiling block is consistent with that in the JY8 well and PG section of the southeastern Sichuan Basin (Figure 1c) [16,34]. Specifically, Al content of the LM6 graptolite biozone in the Aeronian is significantly higher than those of the LM5 graptolite biozone in the Rhuddanian (Table 1; Figure 7), which indicates that as a result of the Kwangsian Orogeny [58], the Chuangzhong and Qianzhong uplifts were significantly uplifted, coupled with a warm and wet climate (Figure 7) [16], resulting in a significant increase in terrigenous influx.

As shown in Figure 9, Al contents show a highly positive correlation with the palaeoclimate proxy Rb/Sr ratio, implying that terrigenous influx is significantly influenced by the palaeoclimate. Generally, warm and humid palaeoclimates lead to a relatively high degree of weathering [14,35,36], which facilitates the weathering and transport of terrigenous clasts from surrounding palaeohighs into the Yangtze Sea. Additionally, Xuefeng uplift was exposed to erosion and became one of the source areas during the early Silurian [16], which is a direct influencing factor leading to a much stronger terrigenous influx in the ULM (Table 1 and Figure 7).

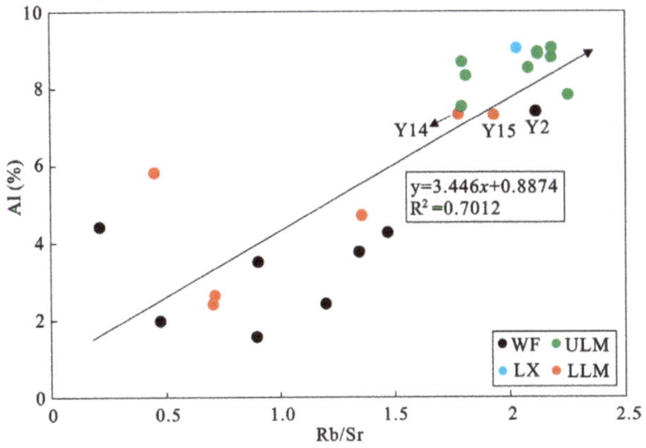

Figure 9. Crossplots of terrigenous influx proxy Al vs. palaeoclimate proxy Rb/Sr show that higher terrigenous influx is closely associated with the warmer and wetter palaeoclimate in the LX, lower WF (sample Y2), upper LLM (samples Y14 and Y15), and ULM.

5.2.3. Palaeoredox Conditions

Some trace elements such as Ni, Co, V, Cr, Th, U, and Mo in sedimentary rocks are redox sensitive. Thus, they are usually used to reflect redox conditions due to their differentiation under different bottom water conditions [32]. The ratios of some trace elements such as V/Cr, Ni/Co, and U/Th are widely used tools in the reconstruction of benthic oceanic redox conditions since these trace metals can coprecipitate with organic matter or metal sulphides under anoxic conditions [61,62].

In general, $V/Cr < 2$, $Ni/Co < 5$, and $U/Th < 0.75$ indicate oxic conditions; $2 < V/Cr < 4.25$, $5 < Ni/Co < 7$, and $0.75 < U/Th < 1.25$ indicate dysoxic conditions; and $V/Cr > 4.25$, $Ni/Co > 7$, and $U/Th > 1.25$ indicate anoxic conditions [14,15,44,45,61,62]. Additionally, iron speciation proxies Fe_{HR}/Fe_T and Fe_{py}/Fe_{HR} are usually used to evaluate the bottom water redox conditions of ancient oceans [46,63]. In general, Fe_{HR}/Fe_T values < 0.38 and >0.38 indicate oxic and anoxic conditions, respectively. For anoxic conditions, Fe_{py}/Fe_{HR} < 0.7−0.8 and >0.7−0.8 indicate ferruginous and euxinic conditions, respectively [47]. Furthermore, the Fe_T/Al ratio is an effective indicator with which to evaluate bottom water redox conditions [43]. Generally, Fe_T/Al ratios are valued at >0.5 in deep-water basins under anoxic conditions. Remarkably, the Fe_T/Al ratios in the sediments under euxinic conditions are 2~3 times those of the sediments under oxic conditions.

The results show that the curve graphs of V/Cr and U/Th are relatively consistent, and they are apparently different from the Ni/Co curve graph (Figure 10). Therefore, the V/Cr and U/Th ratios are used to evaluate redox conditions. From the bottom to the top, the LX and the bottom of the WF are mainly deposited under oxic conditions that rapidly changed to anoxic conditions in the middle-upper WF. At the end of the late Ordicivian, the GYQ was largely deposited under oxic conditions in the middle Yangtze Platform as a result of the significant drop in global sea level [56]. Subsequently, with the melting of the Gondwanan continental glaciers and the influxing of the freshwater in the early Silurian, the sea level rose rapidly, and the seawater formed a stable stratified water body due to density differences, which resulted in clear vertical difference between the surface and bottom layers with the low content of dissolved oxygen in the bottom water [18,64,65], leading to mainly anoxic conditions of bottom water in the lower LLM. During the upper LLM and ULM stages, as the Kwangsian Orogeny with compressive stress continuously transferred from the South China Plate to the Yangtze Platform in the northwest, Xuefeng uplift was elevated rapidly and even exposed to erosion [16,60], leading to the fall of sea level in the long term. Thus, from sample Y14 in the upper LLM, the bottom water started to change from anoxic to mainly oxic, which is not beneficial to the preservation of organic matter such as the species of graptolite, radiolarian, and sponge spicule, leading to a significant drop in TOC content above the red dotted line.

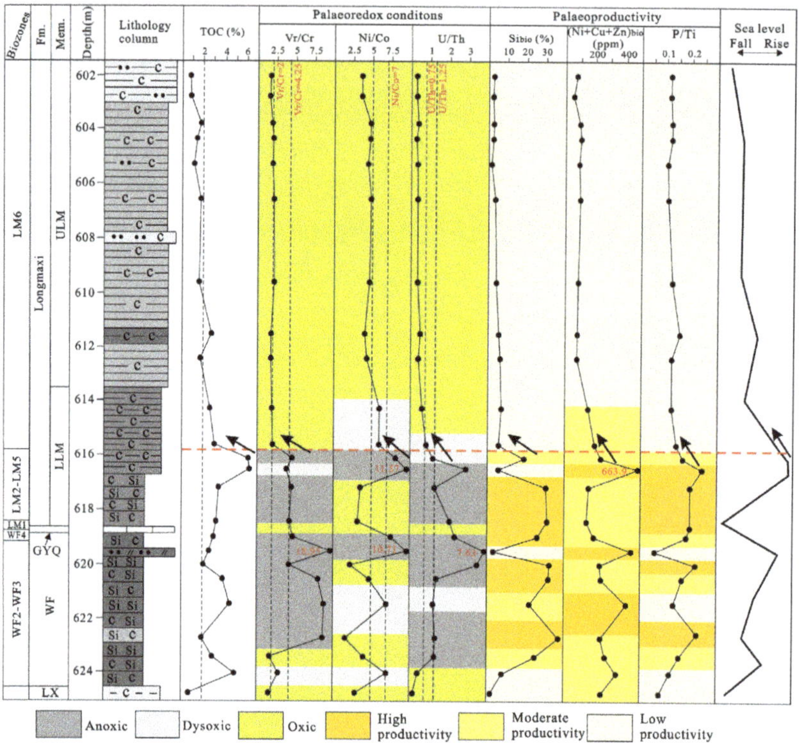

Figure 10. Vertical variations in TOC content; palaeoredox proxies V/Cr, Ni/Co, and U/Th and palaeoproductivity proxies Si_{bio}, $(Ni + Cu + Zn)_{bio}$, and P/Ti of the Wufeng–Longmaxi Formation in well YD1 showing mainly anoxic conditions and high productivity in organic-rich shales of the middle-upper WF and the lower LLM.

As shown in Figure 11, the average Fe_{py}/Fe_{HR} and Fe_T/Al ratios of the LX, WF, LLM, and ULM are 0.50 and 0.43, 0.62 and 0.52, 0.61 and 0.63, and 0.55 and 0.53, respectively. Therefore, the bottom water in the LLM shows the highest level of anoxia, especially in sample Y12, and the bottom water is in a euxinic condition with peak values of Fe_{py}/Fe_{HR} ratios, Fe_T/Al ratios, and TOC content. Subsequently, the Fe_{py}/Fe_{HR} ratios significantly decreased from graptolite biozones LM5 to LM6, which implies that bottom water conditions transformed from ferruginous conditions in the Rhuddanian to oxic conditions in the Aeronian (Figure 11a).

The Fe_{py}/Fe_{HR} vs. Fe_{HR}/Fe_T discrimination diagram shows that the LX and ULM were mainly deposited under oxic conditions and that the WF and the LLM were primarily deposited under ferruginous conditions except for sample Y12 with a Fe_{py}/Fe_{HR} ratio of 0.72, implying euxinic conditions during this period (Figure 11b). There are three main reasons for the euxinic conditions found in sample Y12 from the LLM. with the first reason regards the melting of Gondwanan continent glaciers and the influencing of the salinity stratification, which led to an anoxic reducing environment in bottom water [18,65]. In addition, in the early Silurian, tectonic compression and uplift continuously transferred from Xuefeng uplift in the southeast to the Yangtze Platform in the northwest [16], leading to strong tectonic subsidence and the formation of a deep-water basin during the Longmaxi stage. Furthermore, the seawater in the study area was confined by the Huangling basement (a submarine high), Hunan–Hubei submarine high, and Xuefeng uplift (Figure 1c), leading to a strongly restricted basin, which was beneficial for forming euxinic conditions.

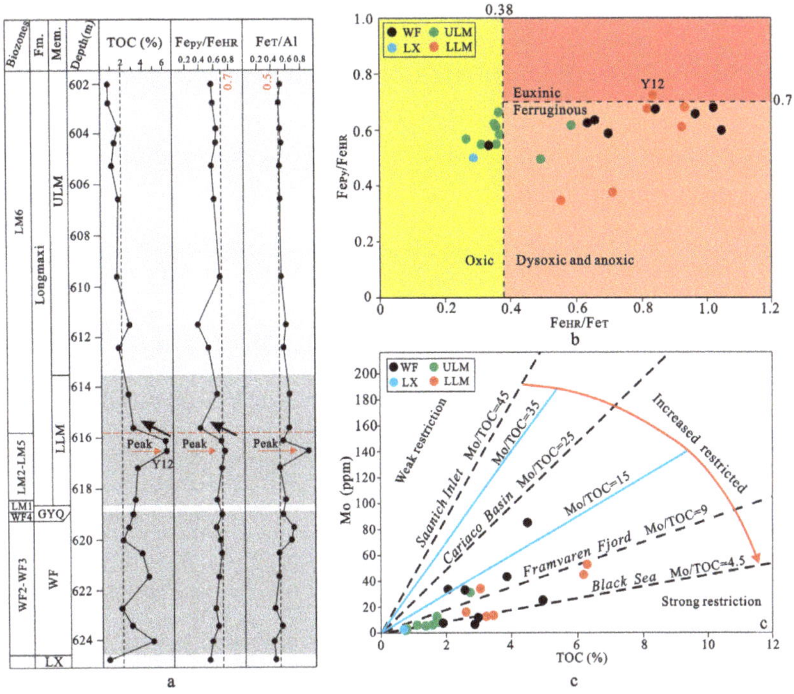

Figure 11. (**a**) Vertical variation in palaeoredox condition proxies Fe_{py}/Fe_{HR} and Fe_T/Al in the Wufeng–Longmaxi Formation in well YD1 showing mainly ferruginous conditions except for one sample in the lower LLM that was deposited under euxinic conditions. (**b**) Crossplots of redox proxies Fe_{py}/Fe_{HR} vs. Fe_{HR}/Fe_T showing that LX and ULM were mainly deposited under oxic conditions and WF and LLM were primarily deposited under ferruginous conditions. (**c**) Crossplots of TOC vs. Mo showing that the Mo/TOC values of most samples are lower than 15×10^{-4} [15], indicating a strong degree of restriction.

The results show that the redox conditions of organic-rich shales in well YD1 of the Yiling block are similar to those in well Qianqian-1 [24] and different from those in well PY1 in the southeastern Sichuan Basin [23] and well Shenci-1 in northern Hubei [22]. The bottom water of well Qianqian-1 is mainly under oxic to dysoxic conditions because it is close to the Qianzhong uplift, and the bottom water is primarily under ferruginous to euxinic conditions in well PY1 in the southeastern Sichuan Basin and well Shenci-1 in northern Hubei close to the Qingling Ocean because of the oxygen-deficient environment in the deep shelf resulting in anoxic conditions. In contrast, the bottom water of well YD1 mainly shows oxic to dysoxic conditions except for minor formations deposited under euxinic conditions (Figure 11a,b), which infer a much shallower palaeoocean in the Yiling block relative to the regions of the southeastern Sichuan Basin and northwestern Qinling Ocean.

Overall, the redox conditions of the Wufeng–Longmaxi Formation in the Yiling block show oxic conditions mainly in the LX, lower WF, upper LLM, and ULM, and anoxic conditions in the middle-upper WF and lower LLM, which is consistent with the Jiaoshiba block [13]. Previous studies have shown that anoxic conditions are directly caused by the rapid transgressive events during the middle-upper WF and lower LLM stages [12,16]. Oxic conditions are mainly attributed to the shallow water environment present during the LX and lower WF stages [12], while oxic conditions are directly caused by the fall of sea level as a result of the continuous elevation of the Chuanzhong and Xuefeng uplifts during the upper LLM and ULM stages [16].

5.2.4. Palaeohydrography

Hydrographic conditions are essential to controlling nutrients obtained from the deep ocean to the shallow shelf, and thus can influence palaeoproductivity and OMA [35]. Usually, Mo is a common redox-sensitive element and is abundant only under sulfurated conditions [66], and the Mo–TOC discrimination diagram is usually used to determine hydrographic conditions. Basins can be classified as presenting three levels of restriction: strongly restricted (Mo/TOC $< 15 \times 10^{-4}$), moderately restricted ($15 \times 10^{-4} <$ Mo/TOC $< 35 \times 10^{-4}$), and weakly restricted (Mo/TOC $> 35 \times 10^{-4}$) [67]. The results show that the average Mo/TOC ratios in the LX, WF, LLM, and ULM are 4.06×10^{-4}, 9.32×10^{-4}, 6.83×10^{-4}, and 5.56×10^{-4}, respectively (Table 2), and the Mo/TOC ratios of most samples are lower than 15×10^{-4} except for two samples in the WF with Mo/TOC ratios of greater than 15×10^{-4}, indicating mainly a strongly restricted basin in the Wufeng-Longmaxi Formation, and minor formations in the middle-upper WF were deposited under a moderately restricted basin (Figure 11c).

Given that the Co × Mn values were observed to show significant differences between sediments from the upwelling and restricted settings, Sweere et al. (2016) [49] proposed that Co × Mn values can be used to evaluate upwelling intensity, with values of <0.4 and >0.4 indicating upwelling and restricted settings, respectively. The results show that most samples plot within the restricted field except for minor samples of organic-rich shales mainly found in the WF that fall within the open and upwelling field, implying that the WF showed much stronger upwelling intensity, followed by the lower LLM, and that the LX, upper LLM and ULM showed weak upwelling intensity (Figure 12a,b), leading to the insufficient nutrients transportation from the deep ocean to the shallow shelf [2,3], which was not beneficial for the survival of graptolites at the stages of the upper LLM and ULM. Thus, the graptolite abundance and quantity rapidly decreased due to the oligotrophic seawater caused by the weak upwelling, which was probably another important factor that led to the turnover of graptolite biozones from LM5 to LM6.

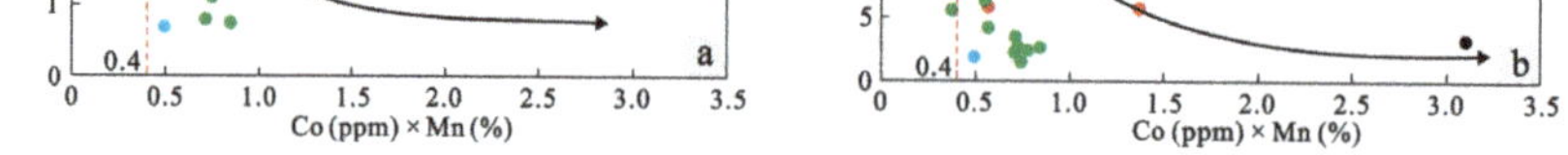

Figure 12. (**a**) Crossplots of TOC content vs. Co × Mn values and (**b**) crossplots of Si_{bio} content vs. Co × Mn values indicating that the WF showed much stronger upwelling and a negative correlation between palaeoproductivity and upwelling intensity, implying that strong upwelling could bring nutrients from the deep ocean to the shallow shelf, leading to high palaeoproductivity.

Overall, both the Mo/TOC and Co × Mn proxies show that the Wufeng–Longmaxi Formation of well YD1 mainly formed in a highly restricted setting except for minor samples that formed in a moderately restricted setting in the WF. In fact, the hydrographic conditions of the Wufeng–Longmaxi Formation are closely related to the sedimentary location [12]. The sedimentary basin in the Yiling block is a highly restricted setting, as it is surrounded by the Xuefeng uplift, the Hunan–Hubei submarine high, and Huangling basement, a submarine high formed in the Sinian (Figure 1c) [68]. The southeastern Sichuan Basin, where wells PY1 and JY143 are located, is a moderately restricted setting because it is a depocentre that has a relatively good connection with the Qinling Ocean [12,68]. The sedimentary basin, where well EHD1 is located, is weakly restricted because it is much closer to the Qinling Ocean, showing the highest level of upwelling intensity [12].

5.2.5. Palaeoproductivity

TOC is the surface productivity of the ocean reflected in sediments, and thus, it is regarded as a most direct primary productivity indicator [69]. Cu, Ni, and Zn are micronutrients for marine microbes that are complex with organic matter under reducing conditions [15,32]. Thus, these elements are typical indicators of marine primary productivity. A large number of radiolarians and graptolites in the shale confirmed the biological contribution of silica in the Wufeng–Longmaxi Formation. Thus, Si_{bio} can be used as a productivity indicator [16,70]. P is involved in most of the metabolic activities of organisms and is usually regarded as one of the most extensively applicable and reliable productivity indicators [71]. To eliminate the terrestrial part of P content, proxy P/Ti is used to estimate primary productivity [72].

The results show that the profiles of TOC, Si_{bio}, $(Ni + Cu + Zn)_{bio}$ and P/Ti generally have similar variation trends with only a small difference in local formation (Figure 10). In the LX and lower WF, palaeoproductivity was low to moderate, apparently increased to high palaeoproductivity in the middle-upper WF, and suddenly decreased in the upper WF. Palaeoproductivity was generally low in the GYQ, which was caused by mass extinction as a result of Hirnantian glaciation at the end of the late Ordovician [14,16]. In the lower LLM in the early Silurian, palaeoproductivity was high and peaked in sample Y12 of graptolite biozone LM5, and then gradually decreased in graptolite biozone LM6 in the upper LLM shown above the red dotted line, maintaining a low level of palaeoproductivity throughout the ULM stage. The decline in palaeoproductivity in the upper LLM and ULM is closely related to the Kwangsian Orogeny, which resulted in the uplift of the Longmaxi Formation and the fall of sea level [20,21,61], and the sedimentary facies gradually changed from deep shelf to shallow shelf, which resulted in a large number of graptolites that cannot be deposited and preserved due to the influence of tides and waves in the shallower water, leading to a decrease in graptolite abundance and quantity [21,73–75], as well as a low palaeoproductivity level with low TOC and Si_{bio} content and $Si_{bio}/(Si_{bio} + Fe_{bio} + Al + Ca_{bio})$ ratios (Figures 7 and 10). Overall, theses analyses show that palaeoproductivity was high in organic-rich shales with peak values occurring in the lower LLM and significantly low values occurring in organic-lean shales in the Yiling block, and the trend of palaeoproductivity variation is consistent with that found in other regions of the middle-upper Yangtze Platform [12,13,15].

Previous studies have shown that palaeoproductivity is strongly related to the upwelling of nutrients [3] because upwelling can usually bring nutrients from the deep ocean to the shallow shelf, which results in plankton flourishing in surface seawater. The scatter plots show that the Co $\times$ Mn values have a perceptible negative correlation with palaeoproductivity proxies TOC and Si_{bio} in well YD1 (Figure 12a,b), which indicates that palaeoproductivity is apparently influenced by upwelling. Remarkably, palaeoproductivity is primarily controlled by upwelling in the WF and the high TOC and Si_{bio} contents of the four samples with Co$\times$ Mn values of less than 0.4. Additionally, in the region close to the Qinling Ocean (well EHD1) with strong upwelling, palaeoproductivity is higher than that far from the Qinling Ocean (well JY143) (Figure 1c) [12].

5.3. OMA Mechanism

5.3.1. Lithofacies Types Influencing OMA

The sedimentary environment controlled the lithofacies types of shale. Previous studies have suggested that siliceous shale in the Wufeng–Longmaxi Formation has relatively high TOC content, which is beneficial for preserving organic matter [51,53]. The results show that siliceous shale has the highest TOC content (Figure 3d), which is consistent with previous research results [52]. TOC levels are higher in subclass lithofacies S and S-2, and relatively lower in subclass lithofacies M-1 and S-3. Lithofacies S and S-2 are mainly distributed in the WF and LLM with anoxic conditions, high palaeoproductivity, and low terrigenous influx, which are conducive to OMA (Figures 3 and 10). In contrast, lithofacies M-1 is distributed in the WF with low palaeoproductivity and moderate terrigenous influx, and the lithofacies S-3 is mainly distributed in the ULM with oxic conditions, low palaeoproductivity, and high terrigenous influx (Figures 3 and 10). Thus, they are generally not beneficial to OMA.

5.3.2. Main Controlling Factor of OMA

In general, OMA occurred for three reasons: organic matter input (palaeoproductivity), organic matter preservation (palaeoredox conditions and deposition rates), and organic matter dilution (terrigenous detrital influx) [35,76], and the correlations between TOC content and the terrigenous influx proxy Al, deposition rate proxy La_N/Yb_N, palaeoredox proxy Ni/Co, and palaeoproductivity proxy $(Ni + Cu + Zn)_{bio}$ are used to study the main controlling factors of OMA in well YD1. The results show that TOC contents have a weak positive correlation with Al content (Figure 13a), have no obvious correlation with the La_N/Yb_N ratio (Figure 13b), and have an apparent correlation with the Ni/Co ratios and $(Ni + Cu + Zn)_{bio}$ concentrations (Figure 13c,d). These analysis results are consistent with previous studies showing redox conditions and productivity to be the key factors of OMA in the Wufeng–Longmaxi Formation [3,5].

However, we further found that the factors that influence OMA changed vertically (Figure 13e–h). In the WF, TOC contents have a highly positive correlation with Al content, indicating that terrigenous influx is the main controlling factor of OMA, which is significantly different from other regions, inferring that the sedimentation rate of organic matter is higher than the terrigenous dilution rate in well YD1. In the LLM, TOC contents have no correlation with Al content, a weak negative correlation with the La_N/Yb_N ratios, and positive correlations with the Ni/Co ratios and $(Ni + Cu + Zn)_{bio}$ concentrations, indicating that combined redox conditions and productivity influence OMA. In the ULM, TOC contents have strongly negative correlations with Al content and no significant correlations with the La_N/Yb_N and the Ni/Co ratios, as well as the $(Ni + Cu + Zn)_{bio}$ concentrations, indicating that terrigenous influx is the main controlling factor of OMA and implying that terrigenous influx led to the dilution of organic matter, which was not conducive to OMA.

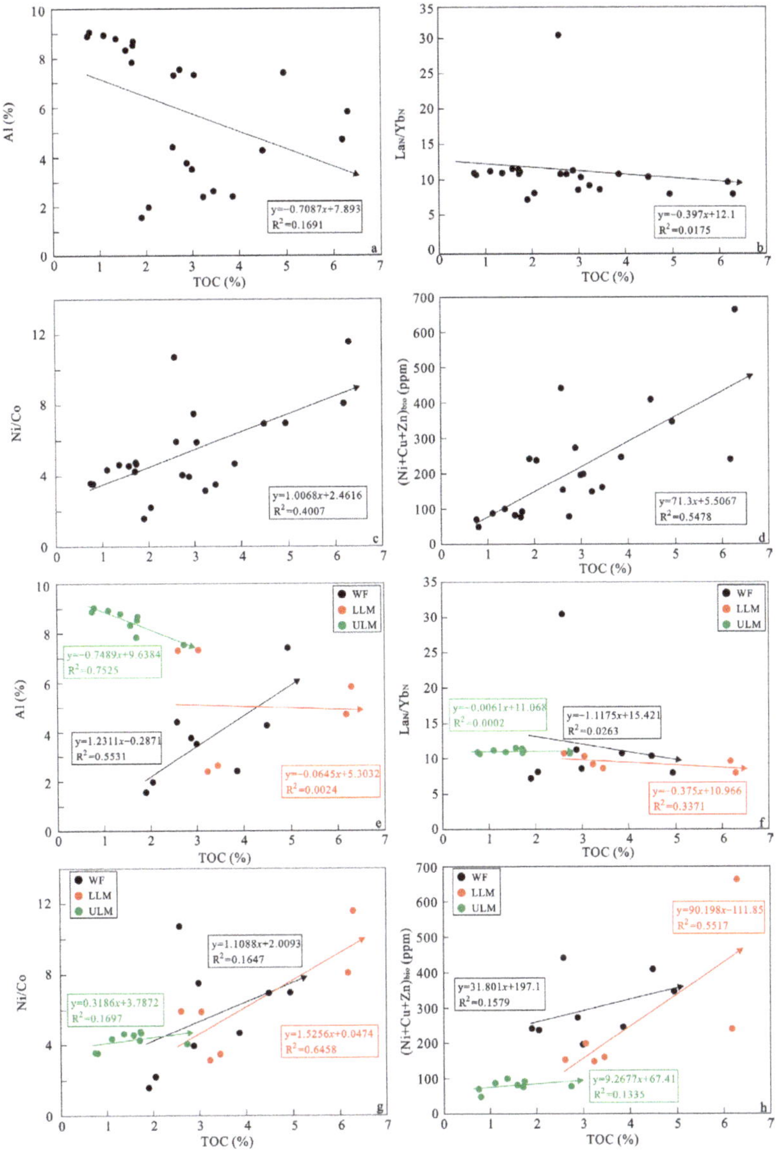

Figure 13. Correlations between TOC content and the factors that influence OMA. (**a–d**) Crossplots of TOC vs. terrigenous influx proxy Al, deposition rates proxy La_N/Yb_N, palaeoredox proxy Ni/Co, and palaeoproductivity proxy $(Ni + Cu + Zn)_{bio}$ of the Wufeng–Longmaxi Formation, respectively. (**e–h**) Crossplots of TOC vs. Al, La_N/Yb_N, Ni/Co, and $(Ni + Cu + Zn)_{bio}$ of the WF, LLM, and ULM, respectively. Overall, OMA is controlled by both redox conditions and productivity. However, the factors that influence OMA change vertically, and OMA is mainly influenced by terrigenous influx in the WF and ULM and controlled by redox conditions and productivity in the LLM.

5.4. Evolution Model

A palaeoenvironmental evolutionary model of the Wufeng–Longmaxi Formation is established for the study area, including the LX, WF, GYQ, LLM, and ULM stages

(Figure 14). The routing of the model starts from the Chuanzhong uplift (point A), passes through the Jiaoshiba block, the Hunan–Hubei submarine high, the Yiling block, and ends in the Qinling Ocean (point A') (Figure 1c).

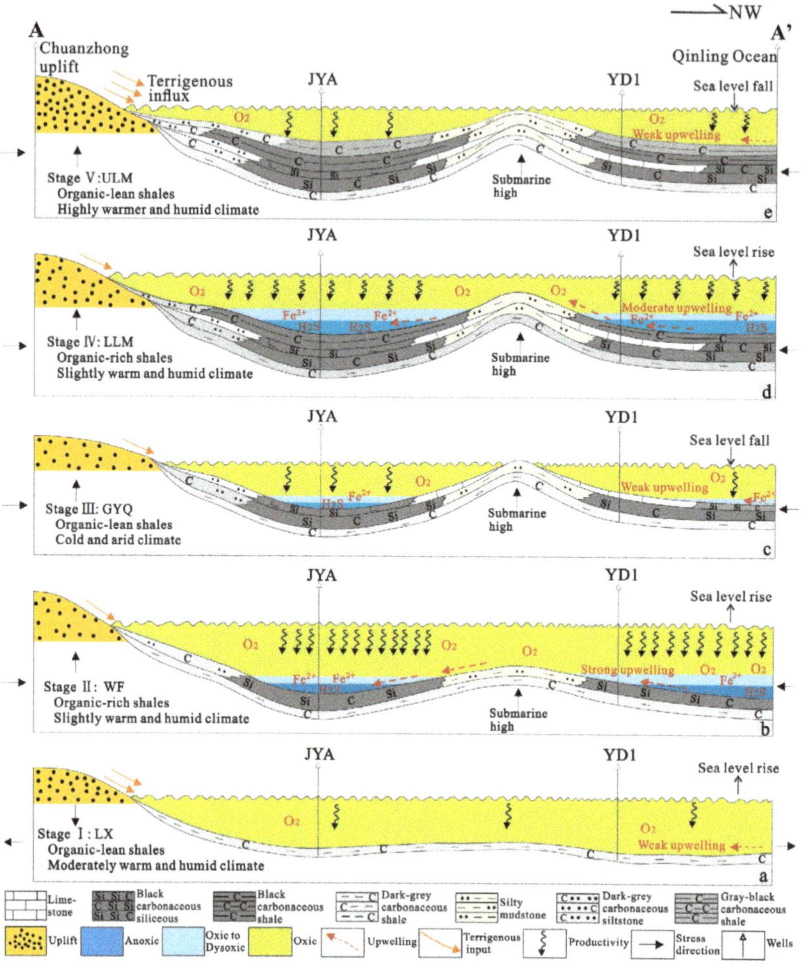

Figure 14. The model of the palaeoenvironmental evolution of the Wufeng–Longmaxi Formation in well YD1. (**a**) The palaeoenvironmental conditions in the LX stage showing moderate terrigenous influx, moderately warm and humid climate, oxic conditions and low productivity. (**b**) The palaeoenvironmental conditions in the WF stage showing low terrigenous influx, slightly warm and humid climate, ferruginous anoxic conditions conditions and high productivity. (**c**) The palaeoenvironmental conditions in the GYQ stage showing low terrigenous influx, cold and arid climate, oxic conditions and low productivity. (**d**) The palaeoenvironmental conditions in the LLM stage showing low terrigenous influx, slightly warm and humid climate, ferruginous to euxinic anoxic conditions conditions and moderate to high productivity. Remarkably, euxinic conditions in the lower LLM are mainly attributed to the rise in sea level caused by the melting of Gondwanan continental glaciers and strong tectonic subsidence caused by continuous compression from the Xuefeng uplift to the Yangtze Platform [16], leading to the formation a deep-water basin. (**e**) The palaeoenvironmental conditions in the ULM stage showing high terrigenous influx caused by the Kwangsian Orogeny, highly warm and humid climate, oxic conditions and low productivity.

During the LX stage, the middle-upper Yangtze Platform was deposited under a tension setting [16] with a broad, flat, and shallow sea environment [30]. The palaeoenvironment was characterized by a moderately warm and humid climate, moderate terrigenous influx, low productivity, and oxic bottom water conditions, which were not beneficial to OMA (Figures 7, 10 and 14a).

During the WF stage, the sea level started to rise, and the middle-upper Yangtze platform was deposited under a compressive tectonic setting caused by the subduction of the Cathaysia Block towards the Yangtze Platform [16], which led to the formation of the Hunan-Hubei submarine high and sedimentary depressions. The palaeoenvironment was characterized by a slightly warm and humid climate, low terrigenous influx, anoxic bottom water conditions, and high productivity with strong upwelling, which was conducive to OMA and the formation of black siliceous rock (Figures 7, 10 and 14b).

During the GYQ stage, the Hirnantian glaciation caused a significant drop in global sea levels [16,56], and the Hunan–Hubei submarine high was gradually exposed at Earth's surface [30]. At this time, the climate was cold and arid [14,77], and weak upwelling led to low productivity in both the middle and upper Yangtze Platforms. The depositional basins were mainly deposited under oxic conditions, except for minor formations in the depocentre of the Sichuan Basin that were still deposited under anoxic conditions [14,78]. The cold glacial climate and poor nutrition in the water body are the important reasons for the global biological extinction event in late Ordovician, with a large number of graptolites disappearing [79]. However, Hirnantia Fauna (Brachiopoda), which benefited from a cold climate and oxygen-rich bottom water, survived (Figures 7, 10 and 14c) [80].

During the LLM, the sea level rose rapidly and seawater formed a stable stratified water body due to the melting of the Gondwanan Glacier [16,64,65]. The palaeoenvironment was characterized by a slightly warm and humid climate, low terrigenous influx, anoxic bottom water and even euxinic conditions (Figure 11), and high palaeoproductivity (Figures 7, 10 and 14d). In addition, the accommodating space of the deep basin significantly increased as a result of continuous tectonic compression [16], leading to the deposition of the thickest organic-rich siliceous and argillaceous shales in the Wufeng–Longmaxi Formation. Remarkably, during the upper LLM stage, the climate became much warmer and wetter, and the sea level gradually dropped due to the strong tectonic uplift as a result of the Kwangsian Orogeny [21,81], which led to dysoxic to oxic bottom water conditions. Meanwhile, the graptolite biozones changed significantly from LM5 in the Rhuddanianto LM6 in the Aeronian (Figures 7 and 10).

During the ULM stage, the palaeoenvironment was characterized by a highly warm and humid climate, similar to that of the upper LLM stage, as well as oxic and low productivity conditions. Remarkably, the Xuefeng uplift continued to rise and was gradually exposed at Earth's surface, finally forming an uplift due to continuous orogenic movement [16], which became a provenance and supplied terrigenous clasts to the middle-upper Yangtze Platform, leading to the highest terrigenous influx in the Wufeng–Longmaxi Formation, which diluted organic matter and was not conductive to OMA (Figures 7, 10 and 14e).

Overall, organic-rich shales in the WF and LLM of the Yiling block are the high-quality shale gas reservoirs, which are characterized by shale thicknesses of 10–20 m, TOC contents of 3%–4%, and total gas contents of 1–3 m^3/t [11]. Although gas contents in the Yiling block are slightly lower than those in the Jiaoshiba block [6,13], prospects for shale gas exploration are still excellent due to high TOC contents, anoxic conditions, and high productivity [2,11]. Remarkably, regions in the eastern Yiling block close to the Qinling Ocean show better prospects for shale gas exploration because of the thicker organic-rich shales caused by deeper water, euxinic bottom water conditions, and higher productivity as a result of the strong upwelling.

6. Conclusions

The Wufeng–Longmaxi shales were primarily deposited in a highly restricted basin, except for the relatively open and upwelling setting of the upper WF. These organic-rich shales were formed in a warm and humid climate with low terrigenous influx, anoxic conditions, and high productivity. The silica in the shales had a primarily nonhydrothermal origin, with an entirely biogenic source only in the upper WF.

In the southeastern Sichuan Basin, the bottom water of organic-rich shales is mainly characterized by anoxic–euxinic conditions, while in the study area, it is primarily dominated by anoxic–ferruginous conditions, suggesting a shallower water depth.

Compared to previous studies, our research provides further evidence that the Kwangsian Orogeny had a rapid and significant impact on the palaeoenvironment of the Wufeng–Longmaxi Formation in western Hubei. This environmental shift is observed in the black shale of the LLM stage, resulting in a warmer and wetter climate, a notable decrease in productivity due to inadequate nutrients caused by weak upwelling, an evident increase in terrigenous input, and a shift in bottom water from anoxic to oxic conditions. These changes led to a decrease in graptolite abundance and the transition of graptolite biozones from LM5 to LM6.

Previous studies have established that redox conditions and productivity were the primary controlling factors for OMA, while our research indicates that the factors influencing OMA change vertically, with terrigenous influx mainly influencing OMA in the WF and ULM, and both redox conditions and productivity affecting OMA in the LLM. Notably, the WF stands out from other regions in that TOC contents have a highly positive correlation with Al content, which indicate that the terrigenous influx is the primary factor influencing OMA, suggesting a higher sedimentation rate of organic matter during this stage.

Despite having slightly lower gas contents than the Jiaoshiba block, the organic-rich shales in the WF and LLM of the Yiling block are still considered high-quality shale gas reservoirs with the organic-rich shale thicknesses exceeding 10 meters. Notably, the eastern regions of the Yiling block that are closer to the Qinling Ocean exhibit even better prospects for shale gas exploration.

Author Contributions: Conceptualization, L.X. and Y.W.; methodology, S.H. and M.S.; software, L.X.; validation, L.X., S.H. and Y.W.; formal analysis, L.X.; investigation, L.X.; resources, L.X.; data curation, W.C.; writing—original draft preparation, L.X.; writing—review and editing, S.H. and M.S.; visualization, F.L. and H.Z.; supervision, L.X. and Y.Z.; project administration, Y.Z.; funding acquisition, L.X. All authors have read and agreed to the published version of the manuscript.

Funding: The authors disclosed receipt of the following financial support for the research, authorship, and/or publication of this article: this work is supported by the scientific research project of Department of Natural Resources of Hubei Province (Grant No. ZRZY2021KJ22). The authors also acknowledge fundings by National Natural Science Foundation of China (Nos. 42202155 and 42272159), China Postdoctoral Science Foundation (No. 2021MD703807), Heilongjiang Provincial Postdoctoral Science Foundation (No. LBH-Z20121), China Scholarship Council (No. 202008230018), and Youth Foundation of the Northeast Petroleum University (No. 2019QNL-21).

Institutional Review Board Statement: Not applicable.

Informed Consent Statement: Not applicable.

Data Availability Statement: Data is unavailable due to privacy restrictions.

Acknowledgments: We would like to thank laboratory staff who helped with the experiments. Careful reviews and constructive suggestions of the manuscript by anonymous reviewers are also greatly appreciated.

Conflicts of Interest: The authors declared no potential conflict of interest with respect to the research, authorship, and/or publication of this article.

References

1. Zhang, L.C.; Xiao, D.S.; Lu, S.F.; Jiang, S.; Chen, L.; Guo, T.L.; Wu, L.Y. Pore development of the Lower Longmaxi shale in the southeastern Sichuan Basin and its adjacent areas: Insights from lithofacies identification and organic matter. *Mar. Pet. Geol.* **2020**, *122*, 104662. [CrossRef]
2. Chen, L.; Jiang, S.; Chen, P.; Chen, X.H.; Zhang, B.M.; Zhang, G.T.; Lin, W.B.; Lu, Y.C. Relative sea-level changes and organic matter enrichment in the Upper Ordovician-Lower Silurian Wufeng-Longmaxi Formations in the Central Yangtze area, China. *Mar. Pet. Geol.* **2021**, *124*, 104809. [CrossRef]
3. Dong, T.; Wang, C.; Liang, X.; Wang, G.C.; Jiang, S. Paleodepositional conditions and organic matter accumulation mechanisms in the Upper Ordovician-lower Silurian Wufeng-Longmaxi shales, Middle Yangtze region, South China. *Mar. Pet. Geol.* **2022**, *143*, 105823. [CrossRef]
4. Huang, H.Y.; He, D.F.; Li, Y.Q.; Li, J.; Zhang, L. Silurian tectonic-sedimentary setting and basin evolution in the Sichuan area, southwest China: Implications for palaeogeographic reconstructions. *Mar. Pet. Geol.* **2018**, *92*, 403–423. [CrossRef]
5. Chen, C.; Mu, C.L.; Zhou, K.K.; Liang, W.; Ge, X.Y.; Wang, X.P.; Wang, Q.Y.; Zheng, B.S. The geochemical characteristics and factors controlling the organic matter accumulation of the Late Ordovician-Early Silurian black shale in the Upper Yangtze Basin, South China. *Mar. Pet. Geol.* **2016**, *76*, 159–175. [CrossRef]
6. Guo, X.S.; Hu, D.F.; Li, Y.P.; Wei, Z.H.; Wei, X.F.; Liu, Z.J. Geological factors controlling shale gas enrichment and high production in Fuling shale gas field. *Petrol. Explor. Dev.* **2017**, *44*, 513–523. [CrossRef]
7. Guo, T.L. Key geological issues and main controls on accumulation and enrichment of Chinese shale gas. *Petrol. Explor. Dev.* **2016**, *43*, 349–359. [CrossRef]
8. Jin, Z.J.; Nie, H.K.; Liu, Q.Y.; Zhao, J.H.; Jiang, T. Source and seal coupling mechanism for shale gas enrichment in upper Ordovician Wufeng Formation—Lower Silurian Longmaxi Formation in Sichuan Basin and its periphery. *Mar. Pet. Geol.* **2018**, *97*, 78–93. [CrossRef]
9. Tang, X.L.; Jiang, S.; Jiang, Z.X.; Li, Z.; He, Z.L.; Long, S.X.; Zhu, D.Y. Heterogeneity of Paleozoic Wufeng-Longmaxi formation shale and its effects on the shale gas accumulation in the Upper Yangtze Region, China. *Fuel* **2019**, *239*, 387–402. [CrossRef]
10. Yi, J.Z.; Bao, H.Y.; Zheng, A.W.; Zhang, B.Q.; Shu, Z.G.; Li, J.Q.; Wang, C. Main factors controlling marine shale gas enrichment and high-yield wells in South China: A case study of the Fuling shale gas field. *Mar. Pet. Geol.* **2019**, *103*, 114–125. [CrossRef]
11. Cai, Q.S.; Chen, X.H.; Zhang, B.M.; Liu, A.; Han, J.; Zhang, G.T.; Li, Y.G. Origin of siliceous minerals in the black shale of the Wufeng and Longmaxi Formations in the Yichang area, western Hubei Province: Geological significance for shale gas. *Acta Geol. Sin.* **2020**, *94*, 931–946.
12. Lu, Y.B.; Jiang, S.; Lu, Y.C.; Xu, S.; Shu, Y.; Wang, Y.X. Productivity or preservation? The factors controlling the organic matter accumulation in the late Katian through Hirnantian Wufeng organic-rich shale, South China. *Mar. Pet. Geol.* **2019**, *109*, 22–35. [CrossRef]
13. Wang, Y.X.; Xu, S.; Hao, F.; Lu, Y.B.; Shu, Z.G.; Yan, D.T.; Lu, Y.C. Geochemical and petrographic characteristics of Wufeng-Longmaxi shales, Jiaoshiba area, southwest China: Implications for organic matter differential accumulation. *Mar. Pet. Geol.* **2019**, *102*, 138–154. [CrossRef]
14. Yan, D.T.; Chen, D.Z.; Wang, Z.Z.; Li, J.; Yang, X.R.; Zhang, B. Climatic and oceanic controlled deposition of Late Ordovician-Early Silurian black shales on the North Yangtze platform, South China. *Mar. Pet. Geol.* **2019**, *110*, 112–121.
15. Zhang, L.C.; Xiao, D.S.; Lu, S.; Jiang, S.; Lu, S.F.; Jiang, S.; Lu, S.D. Effect of sedimentary environment on the formation of organic-rich marine shale: Insights from major/trace elements and shale composition. *Int. J. Coal Geol.* **2019**, *204*, 34–50. [CrossRef]
16. Huang, H.Y.; He, D.F.; Li, D.; Li, Y.Q.; Zhang, W.K.; Chen, J.J. Geochemical characteristics of organic-rich shale, Upper Yangtze Basin: Implications for the Late Ordovician–Early Silurian orogeny in South China. *Paleogeogr. Paleoclimatol. Paleoecol.* **2020**, *554*, 109822. [CrossRef]
17. Lei, Z.H.; Dashtgard, S.E.; Wang, J.; Li, M.; Feng, Q.L.; Yu, Q.; Zhao, A.K.; Du, L.T. Origin of chert in Lower Silurian Longmaxi Formation: Implications for tectonic evolution of Yangtze Block, South China. *Paleogeogr. Paleoclimatol. Paleoecol.* **2019**, *529*, 53–66. [CrossRef]
18. Brenchley, P.J.; Marshall, J.D.; Carden, G.A.F.; Robertson, D.B.R.; Long, D.G.F.; Meidla, T.; Anderson, T.F. Bathymetric and isotopic evidence for a short-lived Late Ordovician glaciation in a greenhouse period. *Geology* **1994**, *22*, 295–298. [CrossRef]
19. Yao, W.H.; Li, Z.X. Tectonostratigraphic history of the Ediacaran-Silurian Nanhua foreland basin in South China. *Tectonophysics* **2016**, *674*, 31–51. [CrossRef]
20. Chen, X.; Zhang, Y.D.; Fan, J.X.; Tang, L.; Sun, H.Q. Onset of the Kwangsian Orogeny as evidenced by biofacies and lithofacies. *Sci. China Earth Sci.* **2012**, *55*, 1592–1600. [CrossRef]
21. Chen, X.; Chen, Q.; Zhen, Y.Y.; Wang, H.Y.; Zhang, L.N.; Zhang, J.P.; Wang, W.H.; Xiao, Z.H. Circumjacent distribution pattern of the Lungmachian graptolitic black shale (early Silurian) on the Yichang Uplift and its peripheral region. *Sci. China Earth Sci.* **2018**, *61*, 1195–1203. [CrossRef]
22. Li, N.; Algeo, T.J.; Cheng, M.; Jin, C.S.; Zhu, G.Y.; Fan, J.X.; Sun, Z.Y. Redox changes in the outer Yangtze Sea (South China) through the Hirnantian Glaciation and their implications for the end-Ordovician biocrisis. *Earth-Sci. Rev.* **2021**, *212*, 103443. [CrossRef]

23. Li, N.; Li, C.; Fan, J.X.; Algeo, T.J.; Yan, D.T.; Zhu, G.Y.; Wu, S.Y.; Tang, S.D.; Cheng, M.; Jin, C.S. Sulfate-controlled marine euxinia in the semi-restricted inner Yangtze Sea (South China) during the Ordovician-Silurian transition. *Paleogeogr. Paleoclimatol. Paleoecol.* **2019**, *534*, 109–281. [CrossRef]
24. Jin, C.S.; Liao, Z.W.; Lash, G.G. High-frequency redox variation across the Ordovician–Silurian transition, South China. *Paleogeogr. Paleoclimatol. Paleoecol.* **2021**, *566*, 110218. [CrossRef]
25. Zhou, X.L.; Liu, Y.; Cao, H.Y.; Zhong, H.T.; Li, Y.C. Responses of oceanic chemistry to climatic perturbations during the Ordovician-Silurian transition: Implications for geochemical proxies and organic accumulations. *Mar. Pet. Geol.* **2021**, *134*, 105341. [CrossRef]
26. Tribovillard, N.; Bialkowski, A.; Tyson, R.V.; Lallier-Verges, E.; Deconinck, J.F. Organic facies variations in the late Kimmeridgian of the Boulonnais area (northernmost France). *Mar. Petrol. Geol.* **2001**, *18*, 371–389. [CrossRef]
27. Rimmer, S.M.; Thompson, J.A.; Goodnight, S.A.; Robl, T.L. Multiple controls on the preservation of organic matter in Devonian–Mississippian marine black shales: Geochemical and petrographic evidence. *Paleogeogr. Paleoclimatol. Paleoecol.* **2004**, *215*, 125–154. [CrossRef]
28. Sageman, B.B.; Murphy, A.E.; Werne, J.P.; Straeten, C.A.V.; Hollander, D.J.; Lyons, T.W. A tale of shales: The relative roles of production, decomposition, and dilution in the accumulation of organic-rich strata, Middle-Upper Devonian, Appalachian basin. *Chem. Geol.* **2003**, *195*, 229–273. [CrossRef]
29. Wang, X.; Tian, J.C.; Lin, X.B.; Chen, W.Z.; Yi, D.X. Sedimentary Environment and Controlling Factors of Organic Matter Accumulation in Wufeng-Longmaxi Formations: A case study of Jielong section in eastern Chongqing. *Acta Sedimentol. Sin.* **2022**. [CrossRef]
30. Rong, J.Y.; Chen, X. Faunal differentiation, biofacies and lithofacies pattern of late Ordovician (Ashgillian) in south China. *Acta Palaeontol. Sin.* **1987**, *26*, 507–535.
31. Poulton, S.W.; Canfield, D.E. Development of a sequential extraction procedure for iron: Implications for iron partitioning in continentally derived particulates. *Chem. Geol.* **2005**, *214*, 209–221. [CrossRef]
32. Tribovillard, N.; Algeo, T.J.; Lyons, T.; Riboulleau, A. Trace metals as paleoredox and paleoproductivity proxies: An update. *Chem. Geol.* **2006**, *232*, 12–32. [CrossRef]
33. Ross, D.J.K.; Bustin, R.M. Investigating the use of sedimentary geochemical proxies for paleoenvironment interpretation of thermally mature organic-rich strata: Examples from the Devonian-Mississippian shales, Western Canadian Sedimentary Basin. *Chem. Geol.* **2009**, *260*, 1–19. [CrossRef]
34. Wang, X.Q.; Zhu, Y.M.; Lash, G.G.; Wang, Y. Multi-proxy analysis of organic matter accumulation in the Upper Ordovician–Lower Silurian black shale in the upper Yangtze Platform, south China. *Mar. Pet. Geol.* **2019**, *103*, 473–484. [CrossRef]
35. Wu, W.; Liu, W.Q.; Mou, C.L.; Liu, H.; Qiao, Y.; Pan, J.N.; Ning, S.Y.; Zhang, X.X.; Yao, J.X.; Liu, J.D. Organic-rich siliceous rocks in the upper Permian Dalong Formation (NW middle Yangtze): Provenance, paleoclimate and paleoenvironment. *Mar. Pet. Geol.* **2021**, *123*, 104728. [CrossRef]
36. Nesbitt, H.W.; Young, G.M. Early proterozoic climates and plate motions inferred from major element chemistry of lutites. *Nature* **1982**, *299*, 715–717. [CrossRef]
37. Bhatia, M.R.; Crook, K.A.W. Trace element characteristics of graywackes and tectonic setting discrimination of sedimentary basins. *Contrib. Mineral. Petrol.* **1986**, *92*, 181–193. [CrossRef]
38. Ge, X.Y.; Mou, C.L.; Yu, Q.; Liu, W.; Men, X.; He, J.L. The geochemistry of the sedimentary rocks from the Huadi No. 1 well in the Wufeng-Longmaxi formations (Upper Ordovician-Lower Silurian), South China, with implications for paleoweathering, provenance, tectonic setting and paleoclimate. *Mar. Pet. Geol.* **2019**, *103*, 646–660. [CrossRef]
39. Fritz, M.; Unkel, I.; Lenz, J.; Gajewski, K.; Frenzel, P.; Paquette, N.; Lantuit, H.; Körte, L.; Wetterich, S. Regional environmental change versus local signal preservation in Holocene thermokarst lake sediments: A case study from Herschel Island, Yukon (Canada). *J. Paleolimnol.* **2018**, *60*, 77–96. [CrossRef]
40. Xu, C.; Shan, X.L.; He, W.T.; Zhang, K.; Rexiti, Y.L.K.; Su, S.Y.; Liang, C.; Zou, X.T. The influence of paleoclimate and a marine transgression event on organic matter accumulation in lacustrine black shales from the Late Cretaceous, southern Songliao Basin, Northeast China. *Int. J. Coal Geol.* **2021**, *246*, 103842. [CrossRef]
41. Shi, J.; Zou, Y.R.; Cai, Y.L.; Zhan, Z.W.; Sun, J.N.; Liang, T.; Peng, P.A. Organic matter enrichment of the Chang 7 member in the Ordos Basin: Insights from chemometrics and element geochemistry. *Mar. Pet. Geol.* **2022**, *135*, 105404. [CrossRef]
42. Rimmer, S.M. Geochemical paleoredox indicators in Devoniane Mississippian black shales, central Appalachian basin (USA). *Chem. Geol.* **2004**, *206*, 373–391. [CrossRef]
43. Lyons, T.W.; Severmann, S. A critical look at iron paleoredox proxies: New insights from modern euxinic marine basins. *Geochim. Cosmochim. Acta* **2006**, *70*, 5698–5722. [CrossRef]
44. Trela, W.; Podhalańska, T.; Smolarek, J.; Marynowski, L. Llandovery green/grey and black mudrock facies of the northern Holy Cross Mountains (Poland) and their relation to early Silurian sea-level changes and benthic oxygen level. *Sediment. Geol.* **2016**, *342*, 66–77. [CrossRef]
45. Smolarek, J.; Marynowski, L.; Trela, W.; Kujawski, P.; Simoneit, B.R.T. Redox conditions and marine microbial community changes during the end-Ordovician mass extinction event. *Global Planet Change* **2017**, *149*, 105–122. [CrossRef]

46. Yuan, Y.Y.; Cai, C.F.; Wang, T.K.; Xiang, L.; Jia, L.Q.; Chen, Y. Redox condition during Ediacaran-Cambrian transition in the Lower Yangtze deep water basin, South China: Constraints from iron speciation and $\delta^{13}C_{org}$ in the Diben section, Zhejiang. *Chin. Sci. Bull.* **2014**, *59*, 3638–3649. [CrossRef]
47. Lei, L.D.; Shen, J.; Li, C.; Algeo, T.J.; Chen, Z.Q.; Feng, Q.L.; Cheng, M.; Jin, C.S.; Huang, J.H. Controls on regional marine redox evolution during Permian-Triassic transition in South China. *Paleogeogr. Paleoclimatol. Paleoecol.* **2017**, *486*, 17–32. [CrossRef]
48. Ma, Y.Q.; Fan, M.J.; Lu, Y.C.; Guo, X.S.; Hu, H.Y.; Chen, L.; Wang, C.; Liu, X.C. Geochemistry and sedimentology of the Lower Silurian Longmaxi mudstone in southwestern China: Implications for depositional controls on organic matter accumulation. *Mar. Petrol. Geol.* **2016**, *75*, 291–309. [CrossRef]
49. Sweere, T.; van den Boorn, S.; Dickson, A.J.; Reichart, G.J. Definition of new trace-metal proxies for the controls on organic matter enrichment in marine sediments based on Mn, Co, Mo and Cd concentrations. *Chem. Geol.* **2016**, *441*, 235–245. [CrossRef]
50. Nancet, W.B.; Taylor, S.R. Rare earth element patterns and crustal evolution–I. Australian post-Archean sedimentary rocks. *Geochim. Cosmochim. Acta* **1976**, *40*, 1539–1551. [CrossRef]
51. Wang, C.; Zhang, B.Q.; Shu, Z.G.; Lu, Y.C.; Lu, Y.Q.; Bao, H.Y.; Li, Z.; Liu, C. Lithofacies types and reservoir characteristics of marine shales of the Wufeng Formation-Longmaxi Formation in Fuling area, the Sichuan Basin. *Oil Gas Geol.* **2018**, *39*, 485–497.
52. Wang, E.Z.; Guo, T.L.; Li, M.W.; Li, C.R.; Dong, X.X.; Zhang, N.X.; Feng, Y. Exploration potential of different lithofacies of deep marine shale gas systems: Insight into organic matter accumulation and pore formation mechanisms. *Mar. Petrol. Geol.* **2022**, *102*, 104563. [CrossRef]
53. Zhai, G.Y.; Wang, Y.F.; Liu, G.H.; Lu, Y.C.; He, S.; Zhou, Z.; Li, J.; Zhang, Y. Accumulation model of the Sinian-Cambrian shale gas in western Hubei Province, China. *J. Geomech.* **2020**, *26*, 696–713.
54. Adegoke, A.K.; Abdullah, W.H.; Hakimi, M.H.; Sarki Yandoka, B.M. Geochemical characterisation of Fika Formation in the Chad (Bornu) Basin, northeastern Nigeria: Implications for depositional environment and tectonic setting. *Appl. Geochem.* **2014**, *43*, 1–12. [CrossRef]
55. Chang, H.; An, Z.S.; Wu, F.; Jin, Z.D.; Liu, W.G.; Song, Y.G. A Rb/Sr record of the weathering response to environmental changes in westerly winds across the Tarim Basin in the late Miocene to the early Pleistocene. *Paleogeogr. Paleoclimatol. Paleoecol.* **2013**, *386*, 364–373. [CrossRef]
56. Haq, B.U.; Schutter, S.R. A chronology of Paleozoic sea-level changes. *Science* **2008**, *322*, 64–68. [CrossRef] [PubMed]
57. Van den Boorn, S.H.J.M.; van Bergen, M.J.; Nijman, W.; Vroon, P.Z. Dual role of seawater and hydrothermal fluids in Early Archean chert formation: Evidence from silicon isotopes. *Geology* **2007**, *35*, 939–942. [CrossRef]
58. Khan, M.Z.; Feng, Q.L.; Zhang, K.; Guo, W. Biogenic silica and organic carbon fluxes provide evidence of enhanced marine productivity in the Upper Ordovician-Lower Silurian of South China. *Paleogeogr. Paleoclimatol. Paleoecol.* **2019**, *534*, 109–278. [CrossRef]
59. Garbán, G.; Martínez, M.; Márquez, G.; Rey, O.; Escobar, M.; Esquinas, N. Geochemical signatures of bedded cherts of the upper La Luna Formation in Táchira State, western Venezuela: Assessing material provenance and paleodepositional setting. *Sediment. Geol.* **2017**, *347*, 130–147. [CrossRef]
60. Chen, X.; Fan, J.X.; Chen, Q.; Tang, L.; Hou, X.D. Toward a stepwise Kwangsian Orogeny. *Sci. China Earth Sci.* **2014**, *57*, 379–387. [CrossRef]
61. Jones, B.; Manning, D.A.C. Comparison of geochemical indices used for the interpretation of palaeoredox conditions in ancient mudstones. *Chem. Geol.* **1994**, *111*, 111–129. [CrossRef]
62. Algeo, T.J.; Maynard, J.B. Trace-element behavior and redox facies in core shales of Upper Pennsylvanian Kansas-type cyclothems. *Chem. Geol.* **2004**, *206*, 289–318. [CrossRef]
63. Poulton, S.W.; Canfield, D.E. Ferruginous conditions: A dominant feature of the ocean through Earth's history. *Elements* **2011**, *7*, 107–112. [CrossRef]
64. Kwak, D.H.; Song, Y.S.; Choi, Y.H.; Kim, K.M.; Jeong, Y.H. Influence of sluice gate operation on salinity stratification and hypoxia development in a brackish estuary dam. *Reg. Stud. Mar. Sci.* **2023**, *57*, 102731. [CrossRef]
65. Van Soelen, E.E.; Twitchett, R.J.; Kürschner, W.M. Salinity changes and anoxia resulting from enhanced run-off during the late Permian global warming and mass extinction event. *Clim. Past* **2018**, *14*, 441–453. [CrossRef]
66. Algeo, T.J.; Tribovillard, N. Environmental analysis of paleoceanographic systems based on molybdenum–uranium covariation. *Chem. Geol.* **2009**, *268*, 211–225. [CrossRef]
67. Algeo, T.J.; Lyons, T.W. Mo-total organic carbon covariation in modern anoxic marine environments: Implications for analysis of paleoredox and paleohydrographic conditions. *Paleoceanography* **2006**, *21*, PA101623. [CrossRef]
68. Zhang, J.F.; Xu, H.; Zhou, Z.; Ren, P.F.; Guo, J.Z.; Wang, Q. Geological characteristics of shale gas reservoir in Yichang area, western Hubei. *Acta Pet. Sin.* **2019**, *40*, 887–899.
69. Canfield, D.E. Factors influencing organic carbon preservation in marine sediments. *Chem. Geol.* **1994**, *114*, 315–329. [CrossRef]
70. Zhang, K.; Li, X.; Wang, Y.; Liu, W.; Yu, Y.; Zhou, L.; Feng, W. Paleo-environments and organic matter enrichment in the shales of the Cambrian Niutitang and Wunitang Formations, south China: Constraints from depositional environments and geochemistry. *Mar. Pet. Geol.* **2021**, *134*, 105329. [CrossRef]
71. Brumsack, H.J. The trace metal content of recent organic carbon-rich sediments: Implications for cretaceous black shale formation. *Palaeogeogr. Palaeoclimatol. Palaeoecol.* **2006**, *232*, 344–361. [CrossRef]

72. Algeo, T.J.; Kuwahara, K.; Sano, H.; Bates, S.; Lyons, T.; Elswick, E.; Maynard, J.B. Spatial variation in sediment fluxes, redox conditions, and productivity in the Permian–Triassic Panthalassic Ocean. *Palaeogeogr. Palaeoclimatol. Palaeoecol.* **2011**, *308*, 65–83. [CrossRef]
73. Cai, Q.S.; Chen, X.H.; Wang, C.S.; Zhang, B.M.; Han, J.; Zhang, G.T.; Liu, A.; Luo, S.Y.; Li, H.; Zhang, M.; et al. Occurrence characteristics and depositional model of Graptolites from the black shale in the Upper Ordovician Wufeng Formation and lower Silurian Longmaxi Formation. *Bull. Geol. Sci. Tech.* **2020**, *39*, 43–53.
74. Wang, H.Y.; Shi, Z.S.; Sun, S.S. Biostratigraphy and reservoir characteristics of the Ordovician Wufeng Formation Silurian Longmaxi Formation shale in the Sichuan Basin and its surrounding areas, China. *Petrol. Explor. Dev.* **2021**, *48*, 1019–1032. [CrossRef]
75. Qiu, Z.; Zou, C.N.; Li, X.Z.; Wang, H.Y.; Dong, D.Z.; Lu, B.; Zhou, S.W.; Shi, Z.S.; Feng, Z.Q.; Zhang, M.Q. Discussion on the contribution of graptolite to organic enrichment and gas shale reservoir: A case study of the Wufeng–Longmaxi shales in South China. *J. Nat. Gas Geosci.* **2018**, *3*, 147–156. [CrossRef]
76. Cirilli, S.; Panfili, G.; Buratti, N.; Frixa, A. Paleoenvironmental reconstruction by means of palynofacies and lithofacies analyses: An example from the Upper Triassic subsurface succession of the Hyblean Plateau Petroleum System (SE Sicily, Italy). *Rev. Paleobot. Palynol.* **2018**, *253*, 70–87. [CrossRef]
77. Armstrong, H.A.; Turner, B.R.; Makhlouf, I.M.; Weedon, G.P.; Williams, M.; Al Smadi, A.; Salah, A.A. Origin, sequence stratigraphy and depositional environment of an upper Ordovician (Hirnantian) deglacial black shale, Jordan. *Palaeogeogr. Palaeoclimatol. Palaeoecol.* **2005**, *220*, 273–289. [CrossRef]
78. Wang, Y.M.; Dong, D.Z.; Huang, J.L.; Li, X.J.; Wang, S.F. Guanyinqiao Member lithofacies of the Upper Ordovician Wufeng Formation around the Sichuan Basin and the significance to shale gas plays, SW China. *Petrol. Explor. Dev.* **2016**, *43*, 42–50. [CrossRef]
79. Finnegan, S.; Heim, N.A.; Peters, S.E.; Fischer, W.W. Climate change and the selective signature of the Late Ordovician mass extinction. *Proc. Natl. Acad. Sci. USA* **2012**, *109*, 6829–6834. [CrossRef]
80. Rong, J.Y.; Chen, X.; Happer, D.A.T. The latest Ordovician Hirnantia Fauna (Brachiopoda) in time and space. *Lethaia* **2002**, *35*, 231–249.
81. Yang, X.R.; Yan, D.T.; Zhang, B.; Zhang, L.W.; Wei, X.S.; Li, T.; He, J.; Shangguan, Y.F.; Zhang, M.X.; She, X.H. The depositional mechanism of organic-rich siliceous shales in Upper Yangtze area: Response to the Kwangsian Orogeny in South China. *J. Pet. Sci. Eng.* **2020**, *192*, 107–310. [CrossRef]

minerals

MDPI

Article

Influencing Factors of the Brittleness of Continental Shales Containing Shell Limestone Interlayer

Yuejiao Liu [1], Fuqiang Lai [1,*], Ruyue Wang [1,2,3,4,*], Zhonghu Wu [5], Xiaoshu Zhang [1], Hao Xu [1] and Jiao Li [1]

[1] Chongqing Key Laboratory of Complex Oilfield Exploration and Development, Chongqing University of Science and Technology, Chongqing 401331, China
[2] State Key Laboratory of Shale Oil and Gas Enrichment Mechanisms and Efficient Development, Beijing 102206, China
[3] Sinopec Key Laboratory of Shale Oil/Gas Exploration and Production Technology, Beijing 102206, China
[4] Sinopec Petroleum Exploration and Production Research Institute, Beijing 102206, China
[5] College of Civil Engineering, Guizhou University, Guiyang 550025, China
* Correspondence: laifq@cqust.edu.cn (F.L.); wangruyue.syky@sinopec.com (R.W.)

Abstract: Brittleness is important in the evaluation of the fracturing ability of shale reservoir and has a significant impact on shale gas exploration and development. This paper discusses the characteristics and controlling factors of brittleness of continental shale in the Da'anzhai Member of the Ziliujing Formation of Lower Jurassic age in the northeast Sichuan Basin. Continental shale lithofacies and their associations were grouped into four main rock types: clayey shale, silty shale, shell calcareous clayey shale, and silty clayey shale, characterized by the high clay content and local enrichment of carbonate minerals as a whole. Compared with the marine shale, the continental shale contained a low content of siliceous minerals, a high content of carbonate minerals, and a large number of shell limestone interlayers. Carbonate minerals play an important role in controlling the brittleness of continental shale. The shale interlayers were mainly shell limestone interlayers with a thickness of several centimeters and a large number of shell laminates with thicknesses of several millimeters were also observed. The shell laminates were mainly filled with calcite. Due to the dissolution process, a large number of bedding joints and corrosion joints were formed in the calcite shell layers. In the interlayers with a high shell content, a large number of microfractures developed. The energy consumption required for maintaining fracture expansion was lower after fracturing; the fractures greatly improved the reservoir's brittleness.

Keywords: continental shale; brittleness; mineral component; fracture; elastic parameter

1. Introduction

Shale oil and gas are widely distributed around the world and the large reservoirs of shale oil and gas deeply affect the international political and economic backgrounds [1–4]. After the United States realized energy independence through the shale oil and gas revolution, China also made great achievements in shale gas exploration and development over the past five years. In 2020, the output of shale gas was nearly 2×10^{10} m^3 and the cumulative proven geological reserves of shale gas exceeded 2 trillion m^3, displaying huge potential [5–7]. In the continental basins widely distributed in China, many sets of shale series with rich organic matter have developed, and continental shale oil and gas is an important exploration field [8–12]. Recently, in the Fuling Area of northeastern Sichuan, an industrial oil and gas flow was achieved, suggesting a key breakthrough in the exploration of Jurassic continental shale oil and gas [13].

Brittleness is important in shale reservoir evaluation because it indicates the fracturing performance of shale reservoirs and largely affects shale gas exploration and development [14–16]. In the comprehensive evaluation of continental shale reservoir, brittleness is an indispensable indicator. Compared with the marine shale in North America and

China [17–21], the continental shale in China is characterized by the low content of siliceous minerals, the high content of clay minerals, complex mineral composition, strong heterogeneity of reservoirs, the thinness of single layers, significant changes of lithology and lithofacies, more developed fractures, and frequent interaction between shale and interlayers (such as sandstone and limestone) [22–26]. The evaluation system of marine shale is not applicable to continental shale [27,28], so it is necessary to explore its complexity and particularity and clarify the brittleness of continental shale. Therefore, the brittleness evaluation method of continental shale gas reservoirs is significant in the exploration of geological characteristics and enrichment mechanisms of continental shale gas as well as in the geological evaluation of continental shale gas and continental strata.

In this study, with the shale series of the Da'anzhai Member of the Lower Jurassic Ziliujing Formation in the Yuanba and Fuling Area in northeastern Sichuan as the research object, lithofacies, mineral components, dynamic and static rock mechanical properties, and diagenetic stages were explored to improve the understanding of the brittleness of continental shale reservoirs and clarify the influences of minerals, fractures, and interlayers on brittleness of continental shale. This study provides the basis for the optimization of desert sections in geological engineering of continental shale reservoirs.

2. Geological Setting

The Jurassic System is widely distributed in the Sichuan Basin and the strata are well developed. In the Early Jurassic period, a set of relatively stable shallow to semi-deep lake facies shale systems was deposited [29]. The diverse interlayers in the shale mainly included shell limestone (argillaceous) and siltstone (argillaceous). The thickness center is located in the Langzhong–Liangping–Wanzhou Area (Figure 1) [30]. The TOC content in shale ranges from 0.5% to 2.0% and the average TOC content is greater than 1%. The type of organic matter is mainly Type II2 and the vitrinite reflectance (*Ro*) gradually increases from southeast to northwest. Ro from eastern Sichuan to northeastern Sichuan is generally greater than 1.3%, indicating the condensate wet gas stage. The Yuanba and Fuling Areas are located in the northeast of the Sichuan Basin. The Yuanba Region is located at the junction between the depression structure in the western Sichuan area and the uplift structure in central Sichuan. The Fuling Area is located in the high and steep fold belt in eastern Sichuan.

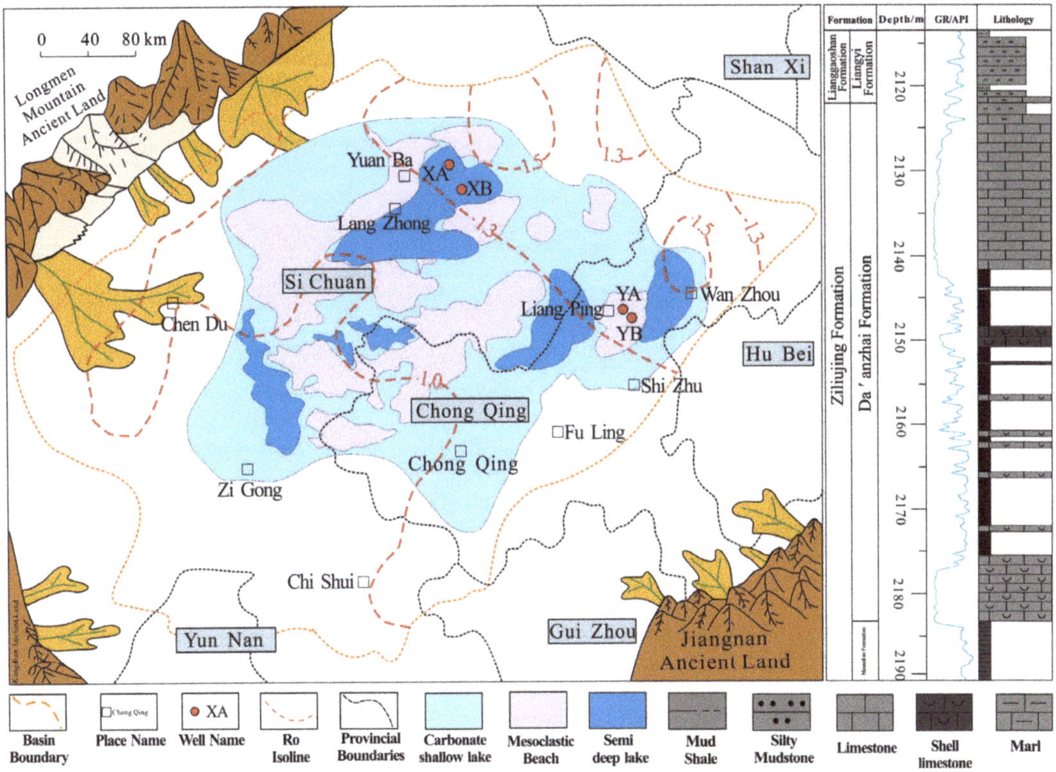

Figure 1. Sedimentary and distribution characteristics of the Jurassic Ziliujing Formation in the Sichuan Basin.

3. Samples and Methods

The core and logging data in this study were collected from the continental shale sequence in the Yuanba and Fuling Areas in the Sichuan Basin, China. Forty-six small core samples with a length of about 50 mm and a width of about 25 mm were drilled from the core. The logging data included conventional logging, array acoustic logging, element logging (ECS), and electrical imaging logging data.

The samples were subjected to the analysis of organic carbon content, identification of organic macerals, determination of helium porosity and permeability, whole-rock mineral analysis, argon ion polishing–scanning electron microscope analysis, pore structure analysis, rock pyrolysis experiments (including field analysis), and a triaxial rock mechanics experiment. The TOC test was conducted in the carbon and sulfur analyzer CS-230 according to GB/T 19145-2003. The reflectance of vitrinite was measured using the MSP200 microphotometer according to the SY/T 5124-2012 standard. The whole rock mineral X-ray diffraction (XRD) analysis was carried out according to the SY/T6210-1996 standard with an X'Pert Pro MPD X-ray diffractometer at Wuxi Institute of Petroleum Geology, Wuxi, China. Porosity and permeability were measured, respectively, with a KX-07F porosity instrument and a GDS-90F permeability instrument according to the GB/T 29172-2012 standard. Argon ion polishing–scanning electron microscope observation was carried out with the Helios 650 scanning electron microscope according to the SY/T 5162-2014 standard at Wuxi Institute of Petroleum Geology, Wuxi, China. The pore structure was determined using the mercury porosimetry adsorption method according to the NB/T 14008-2015 standard. Rock pyrolysis was determined with a Rock-EVAL6 instrument according to the

GB/T 18602-2012 rock pyrolysis analysis method at Wuxi Institute of Petroleum Geology, Wuxi, China. The triaxial compressive strength test of rock mechanics was conducted with a TAW-2000 electro-hydraulic servo rock mechanics tester according to the ASTM D2664-04 standard at Wuxi Institute of Petroleum Geology, Wuxi, China. The interpretation of conventional logging data, array acoustic logging data, FMI resistivity imaging logging data, and element logging data was conducted on the large logging processing and interpretation software platform CIFLOG Version 2.1 developed by the Institute of Logging and Remote Sensing Technology of the China Petroleum Exploration and Development Research Institute.

4. Results and Discussion

4.1. Lithofacies and Mineral Fabric Characteristics

4.1.1. Lithofacies Characteristics

The fine identification and classification of continental shale lithofacies types is an important basis for the evaluation of shale gas exploration and development potential. In the previous studies on this area, the main factors considered in the lithofacies classification scheme of Daanzhai Section were mineral fabric, TOC, and sedimentary structure. According to the report by Liu Zhongbao et al. [24], the three-step lithofacies division scheme (Figure 2) of whole-rock mineral partition, TOC classification, and correction of mineral and sedimentary structures was adopted in the study (I is a muddy (shale) rock area, II is a mixed rock area, III is a sandstone area, and IV is a limestone (dolomite) rock area). Based on the observations and descriptions of the cores of four typical drilling wells in the Yuanba Area (Well XA and Well XB), Fuling Area (Well YA and Well YB), the whole-rock mineral X-ray diffraction analysis, and rock slice identification analysis of 46 samples, 4 shale types (10 shale lithofacies) were determined as follows: clayey shale (high-carbon, medium-carbon, and rich-carbon clayey shale), silty shale (low-carbon, medium-carbon, high-carbon, and rich-carbon silty shale), clayey shell calcareous shale (medium-carbon and high-carbon clayey shell calcareous shale), and silty clay shale (low-carbon and medium-carbon silty clay shale). A large number of shell limestone interlayers in the lithofacies were observed. Among them, high-carbon clayey shale (TOC: 0.59%~3.89%; average: 1.46%) and low-to-medium-carbon silty sand shale were the most common shales (TOC: 0.20%~1.60%, average: 0.89%), followed by high-carbon clayey shell calcareous shale (TOC: 0.45%~3.06%, average: 1.29%) (Figure 3).

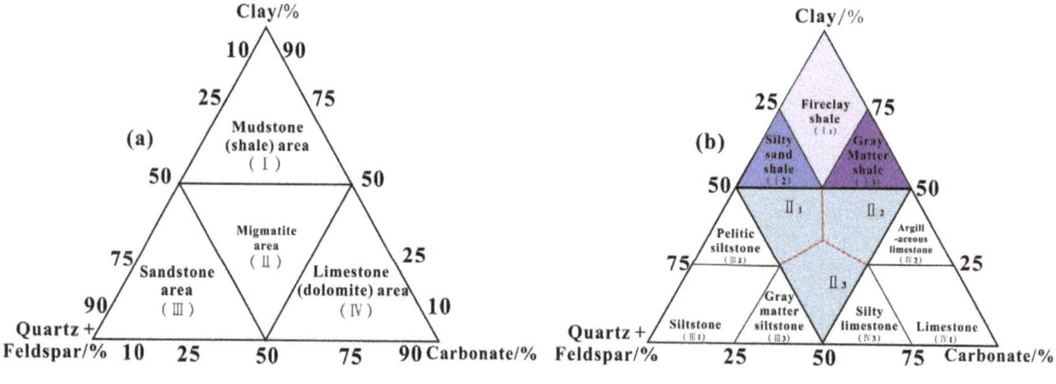

Figure 2. *Cont.*

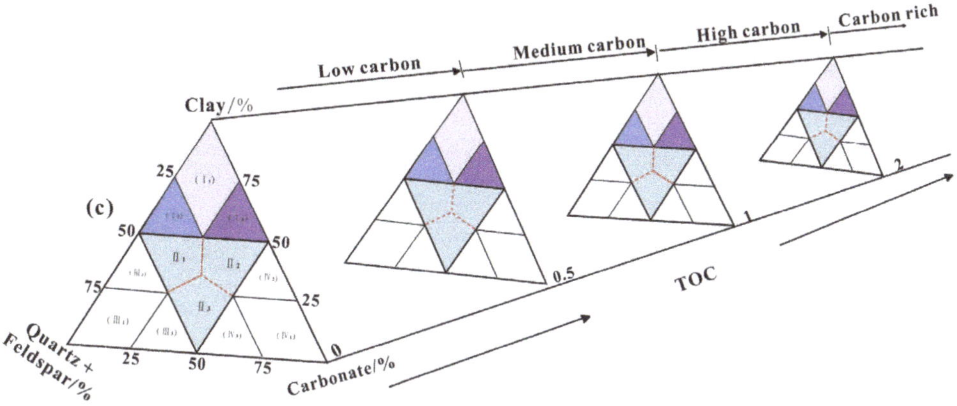

Figure 2. Lithofacies Division Scheme (revised from the method by Liu Zhongbao et al. [24]).

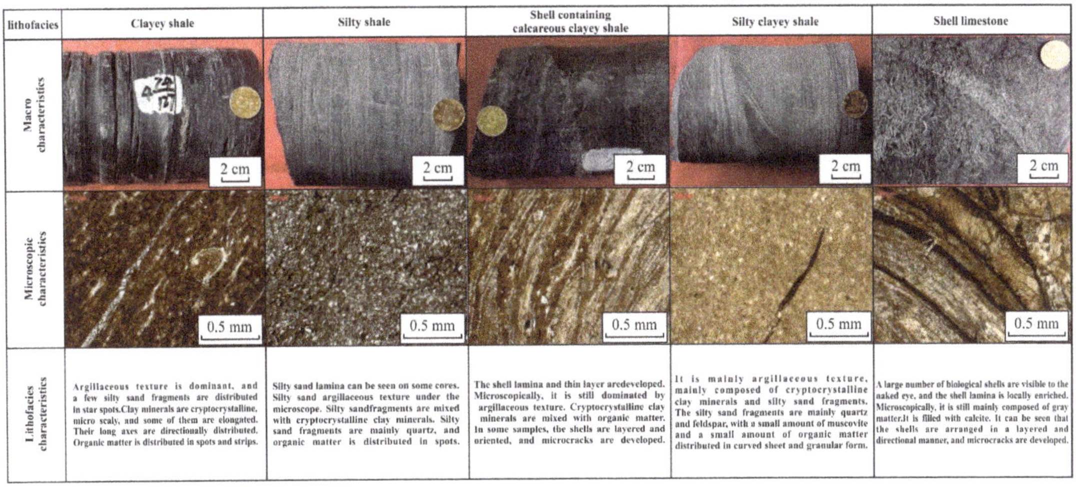

Figure 3. Main lithofacies types of shale in the Da'anzhai Member.

In the Fuling Area, the shale lithofacies were dominated by medium- and high-carbon clayey shale, followed by high-carbon clayey calcareous shale (Figure 4a–c). The shale lithofacies in the Yuanba Area were mainly composed of medium- and high-carbon clayey shale and mixed shale (Figure 4d–f).

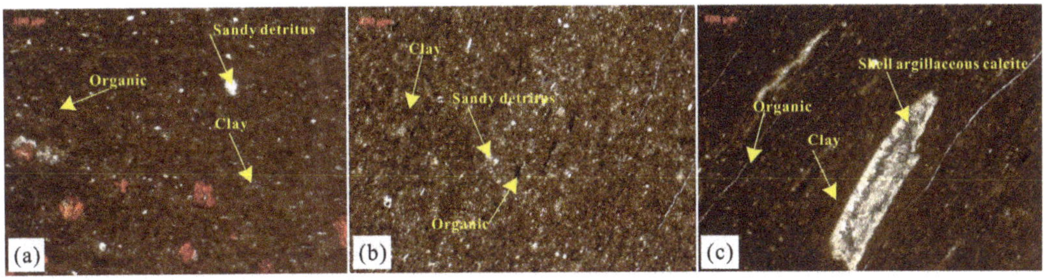

Figure 4. *Cont.*

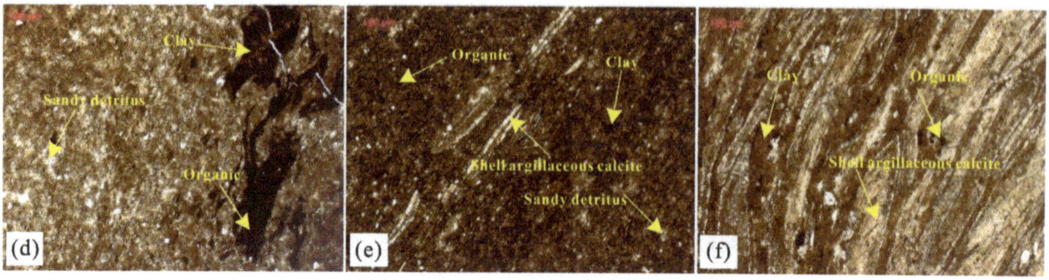

Figure 4. Microscopic photos of continental shale: (**a**) High-carbon clayey shale. (**b**) Medium-carbon calcareous clayey shale. (**c**) Medium-carbon clayey shale. (**d**) Medium-carbon clayey shale. (**e**) Medium-carbon clayey shale. (**f**) Medium-carbon clayey shell calcareous shale.

4.1.2. Mineral Fabric Characteristics

The ECS element logging data (Figure 5a) and the results of the whole-rock mineral X-ray diffraction analysis (Figure 5b) showed that the mineral components of continental shale in the Yuanba Area mainly included clay minerals and quartz, followed by calcite and a small quantity of feldspar and mica. The clay mineral content ($n = 46$) ranged from 21.54% to 68.3% with an average of 48.29%. The total content of quartz and feldspar ranged from 17.1% to 60.3% with an average of 35.69%. The carbonate mineral content ranged from 0 to 48.8% with an average of 12.57%. In general, the contents of clay mineral and quartz were high and carbonate minerals were locally enriched. Continental shale in the Yuanba and Fuling Areas was deposited in shallow and semi-deep lake environments of carbonate rocks (Figure 1). Compared with the marine shale of the Wufeng Formation–Longmaxi Formation (Figure 6a), the continental shale of the Da'anzhai Member had the lower content of siliceous minerals and the highest content of carbonate minerals (Figure 6b). The drilling data and rock slice identification results indicated that the enrichment of carbonate minerals in the continental shale layer was ascribed to the existence of a large number of local shell limestone interlayers. The minerals in the shell limestone mainly included argillaceous calcite, bright calcite, argillaceous, and a large number of developed microfractures (Figure 7).

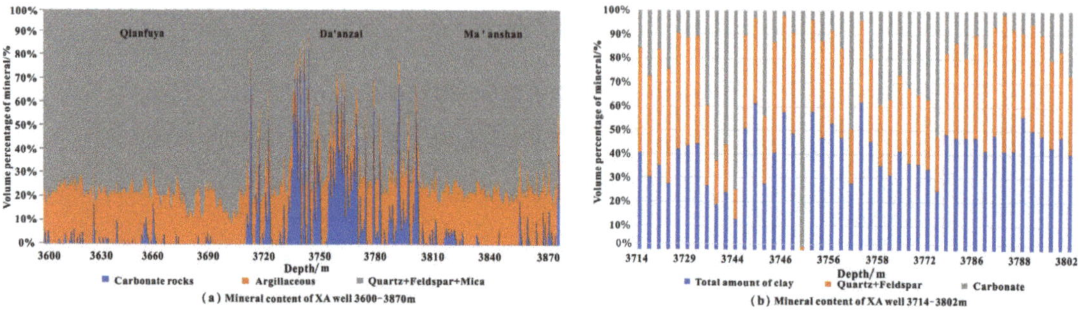

Figure 5. Mineral composition characteristics of the whole rock. (**a**) The ECS element logging data of XA well 3600–3870 m. (**b**) The whole-rock mineral X-ray diffraction analysis of XA well 3714–3802 m.

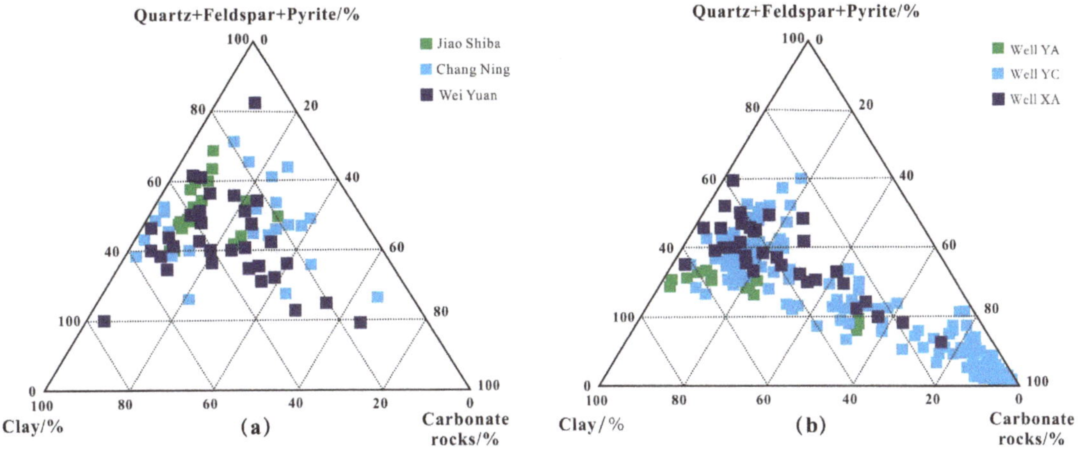

Figure 6. Comparison Diagram of Mineral Composition of Wufeng Formation–Longmaxi Formation Marine Shale and Da'anzhai Member Continental Shale: (**a**) Wufeng Formation–Longmaxi Formation, (**b**) Da'anzhai Member.

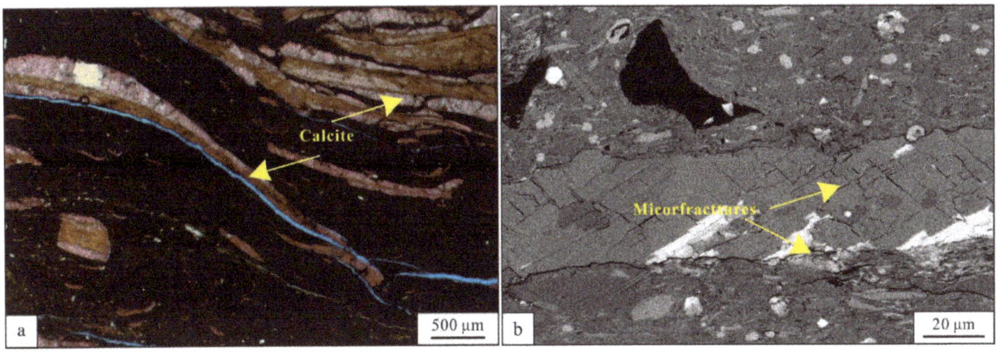

Figure 7. Characteristics of microfractures in the shell limestone interlayer. (**a**) A large amount of calcite in the shell. (**b**) A large number of microfractures develop in the shell.

The continental shale series were deposited in shallow and semi-deep lake environments of carbonate rocks. Compared with the marine shale of the Wufeng Formation–Longmaxi Formation, continental shale is characterized by the lower content of siliceous minerals, higher content of carbonate minerals, frequent interlayers, and strong heterogeneity. Therefore, the evaluation of continental shale brittleness needs to focus on the influences of the contents and distributions of carbonate minerals as well as shell limestone interlayers on brittleness.

4.2. *Characteristics and Influencing Factors of Shale Brittleness*

4.2.1. Mineral Content

The brittleness of continental shale was affected by mineral content and mineral structure. The dynamic and static rock mechanical parameters of continental shale in Well XA and their relationships with mineral content were analyzed with array acoustic wave and ECS logging results. The content of felsic minerals showed a negative correlation with Young's modulus, whereas the content of carbonate minerals had a positive correlation with Young's modulus (Figure 8a,b). The results of the continental shale were the opposite of the positive correlation between quartz and Young's modulus of southern marine shale [31] and similar to that of the marine shale of the first member of the Shahejie Formation in Bonan Sag, Bohai Bay Basin [32]. In addition, the content of felsic and carbonate minerals had a good positive correlation with Young's modulus (Figure 8c), indicating that the brittleness of shale series rocks was mainly determined by the total content of carbonate and felsic minerals. Therefore, the mineral brittleness index method based on mineral composition and the elastic parameter method based on Young's modulus and Poisson's ratio [33,34] were used to calculate the brittleness index Brit1 (felsic and carbonate minerals) and the rock mechanical brittleness index Brit2 (average mechanical parameter) of continental shale and analyze their relationships. The analysis results showed that the mechanical brittleness index was correlated with the brittleness index of felsic and carbonate minerals (Figure 8d).

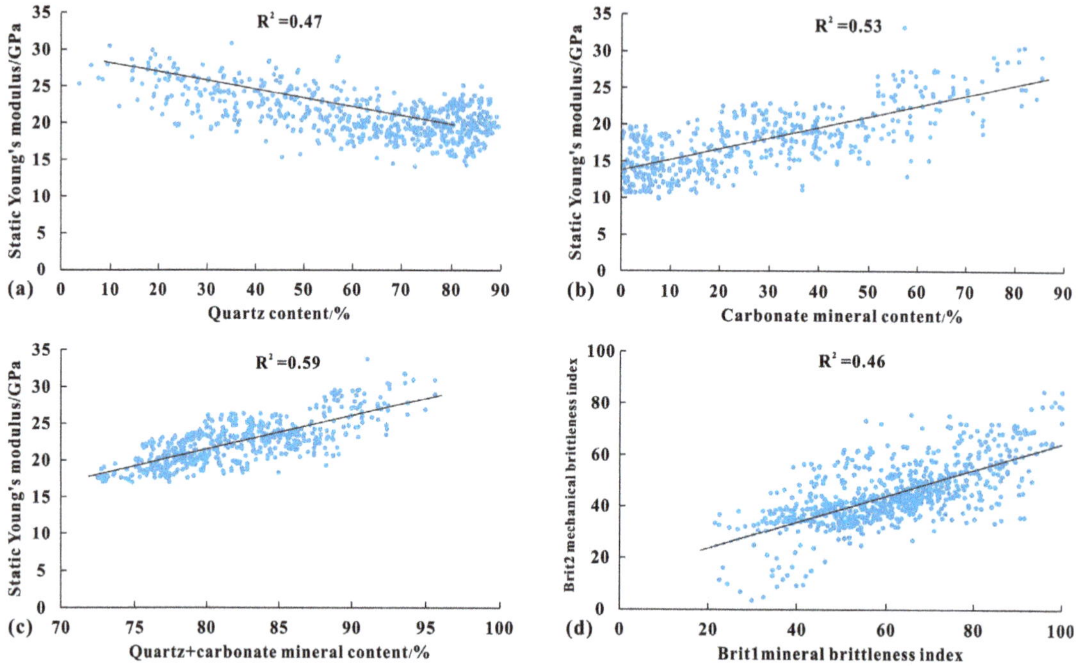

Figure 8. Relationship between rock mineral composition and mechanical parameters. (**a**) Relationship between quartz content and Young's modulus. (**b**) Relationship between carbonate content and Young's modulus. (**c**) Relationship between quartz + carbonate content and Young's modulus. (**d**) Relationship between mineral brittleness index and mechanical brittleness index.

4.2.2. Fractures

According to the previous results [35], the main component of continental shell was calcite (Figure 9). In the early diagenetic stage of continental shale, some aragonite shells were dissolved and precipitated at the edge of aragonite to form the first generation of iron-free calcite cement. In the late diagenesis, rocks were dissolved and the dissolved pores or corrosion fractures were produced in calcite and could be filled with clay minerals, organic matter, or the second generation of iron calcite cement. In addition, the muddy calcite in the shell formed a large number of fine crystals or columnar calcite under recrystallization action.

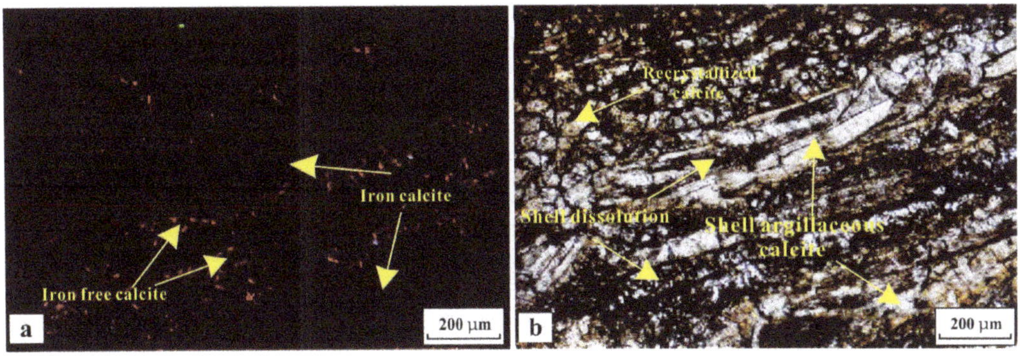

Figure 9. Calcite cement in shell [35]. (**a**) Iron-free calcite cement. (**b**) Dissolution and recrystallization of calcite.

Fracture development affected the size and anisotropy of rock mechanical parameters [36,37] and the development of a large number of calcite bedding fractures. Dissolution fractures in the whole continental shale played an important role in brittleness. The chemical properties of calcite were unstable. During the whole diagenetic stage of the Da'anzhai Member, calcite experienced multiple stages and types of diagenesis actions, including dissolution, compaction, cementation, and recrystallization. Dissolution occurred in various diagenetic stages. In the syngenetic and early diagenetic stages, atmospheric fresh water could selectively dissolve the unstable components in dissolved rocks, such as aragonite, to form dissolved pores. In the late diagenetic stage, calcite was unselectively dissolved to form dissolved pores and fractures. The dissolution of calcite in the whole diagenetic stage was beneficial to fracture development. After dissolution, the calcite-rich shell layer formed a large number of bedding fractures and solution fractures (Figure 10), not only provided a seepage channel for the flow of shale oil and gas, but also greatly improved the brittleness and physical properties of the reservoir.

Figure 10. Coring photos with bedding fractures and dissolution fractures: (**a**) Shell mudstone has developed intermediate shell and rich bedding fractures. (**b**) Shell limestone–mudstone interbedding with developed dissolution fractures and bedding joints.

4.2.3. Shell Interlayers and Lamina

On the basis of the identification results of shale lithofacies, interlayer lithofacies, and the analysis results of lithologic profiles of typical drilling (logging) wells in different regions, continental shale of Well XA was divided into medium-carbon silty shale with argillaceous shell limestone, high-carbon clayey shale and clayey calcareous shale with argillaceous shell limestone, and low-to-medium-carbon silty shale with argillaceous shell limestone and calcareous siltstone from bottom to top. The number of macro-scale shell limestone interlayers (meter scale) reached 14 layers/94 m. Through the fine observations and descriptions of cores in continental shale of the Well XA Member from Well XA and the rock slice identification of shale samples, the large set of shale sections in the above macro-scale lithofacies combination were observed to acquire the fine descriptions of the micro-scale (1 mm to 1 cm) lithofacies combination. A large number of millimeter-to-centimeter-scale shell calcareous laminae or bands were still developed in the single-layer continuous shale in the macro-scale lithofacies combination. In the two coring cycles of continental shale of the artesian well group of Well XA, more calcareous shell laminates were also developed in clayey shale, clayey calcareous shale, and silty shale to varying degrees, and the number of shell limestone interlayers reached 19 layers/5 m in clayey shale (Figure 11). Shell interlayers played an important role in the brittleness of clayey shale.

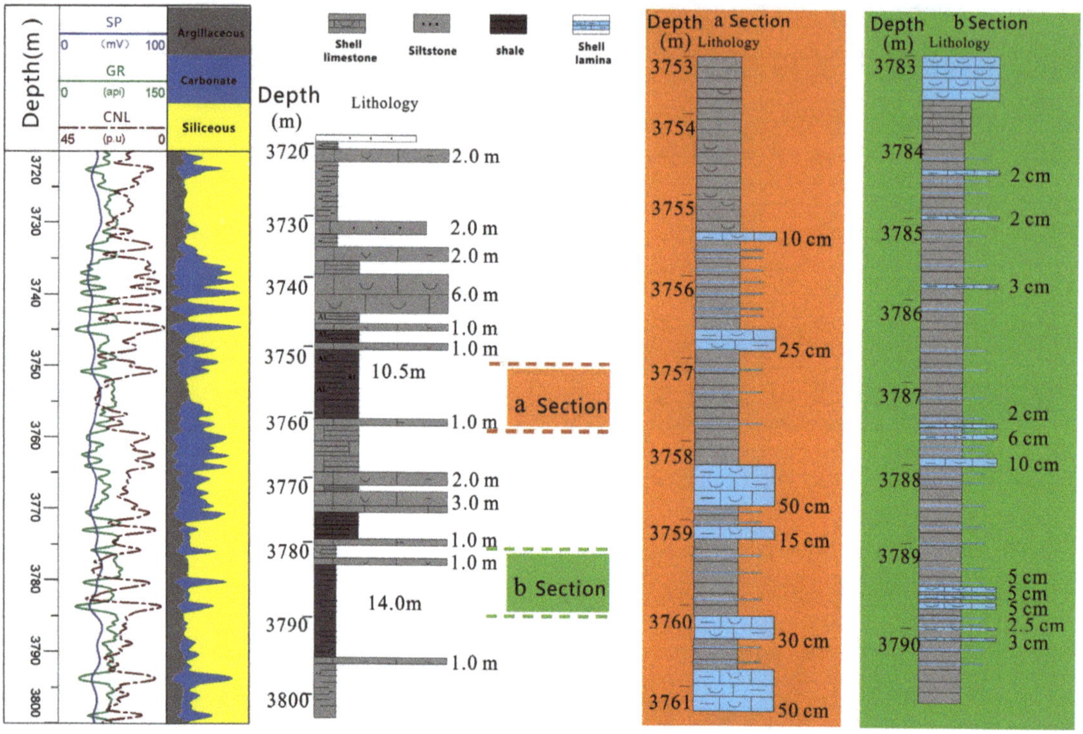

Figure 11. Lithofacies combination characteristics of continental shale in Well XA.

In order to more intuitively display the influences of shell interlayers (lamina) on reservoir brittleness, based on FMI borehole wall micro-resistivity imaging data, conventional logging curves, and logging data, different rock types and interlayers in continental shale were identified. The FMI borehole wall micro-resistivity logging tool was equipped with 192 microelectrodes on 8 plates. Each electrode had a diameter of 0.2 inches and an electrode spacing of 0.1 inches. The vertical and horizontal (around the borehole wall) resolution of the obtained electric imaging image was 0.2 inches (5 mm) [38], which was enough to identify the grain size and shape of fine conglomerates. The strata of continental shale mainly included grayish black mudstone and grayish black shale interbedded with grayish brown shell limestone layer. The corresponding electrical imaging lithological characteristics are summarized below (Figure 12). The conventional curve characteristics of mudstone were similar to those of shale. The electrical imaging results were brownish black to dark black with block characteristics and layered characteristics. Shale had a high natural gamma radioactivity level, high neutron flux value, and low resistivity. Its electrical imaging results were brownish black to dark black with layered structures. The conventional curve of limestone was characterized by the low natural gamma radioactivity level, high density, and extremely high resistivity. The electrical imaging results were bright yellow to white with layered and massive sedimentary structures.

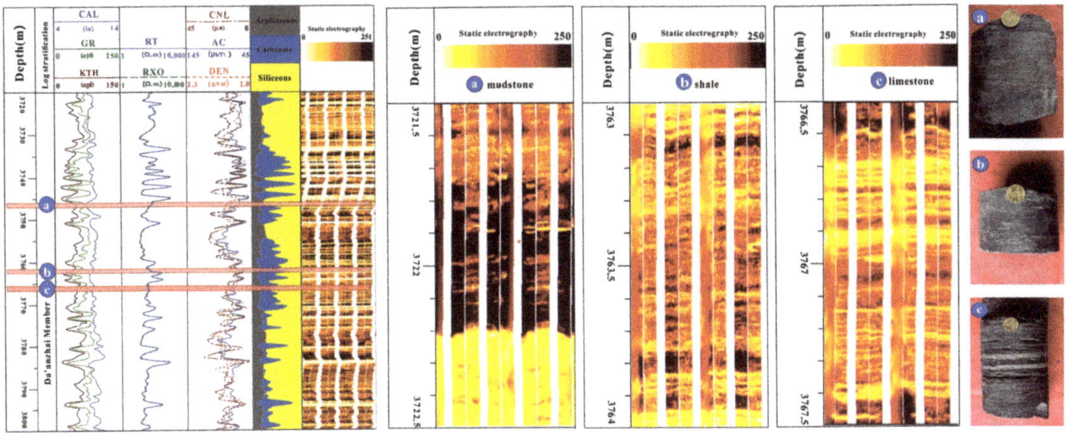

Figure 12. FMI electrical imaging characteristics of Well XA. (**a**) Electrical imaging characteristics of mudstone. (**b**) Electrical imaging characteristics of shale. (**c**) Electrical imaging characteristics of limestone.

Shale samples with low and high shell contents in the study area (Figure 13) were selected for triaxial compression tests. The stress–strain curve showed that the samples with high shell content had multiple peaks and local serration characteristics compared with the samples with low shell content. The ultimate compressive strength and residual strength were lower and the slope of the post-peak elastic modulus was steeper (Figure 14d–f), indicating that the samples with the higher shell content required lower energy to fracture and maintain fracture expansion after fracturing [36]. Therefore, the increase in shell content was more conducive to reservoir reconstruction. Through slice observations, it was found that the curve characteristics might be ascribed to a large number of microfractures developed in the shell interlayers (lamina) (Figure 7). Therefore, compared with the content of felsic minerals, the content of calcareous played the more important role in controlling the brittleness of continental shale.

The above descriptions indicated that a large number of shell interlayers (lamina) were developed in the whole continental shale. Due to the development of a large number of microfractures, the stress–strain curve of the interlayer section with a large number of shell interlayers (lamina) or a high shell content generally showed the characteristics of multiple peaks and local serration. The extreme compressive strength and residual strength were low. The slope of the post-peak elastic modulus was steep and the energy consumption required for maintaining the fracture development was lower. Therefore, compared with the content of felsic minerals, the content of calcareous played the more important role in controlling the brittleness of continental shale.

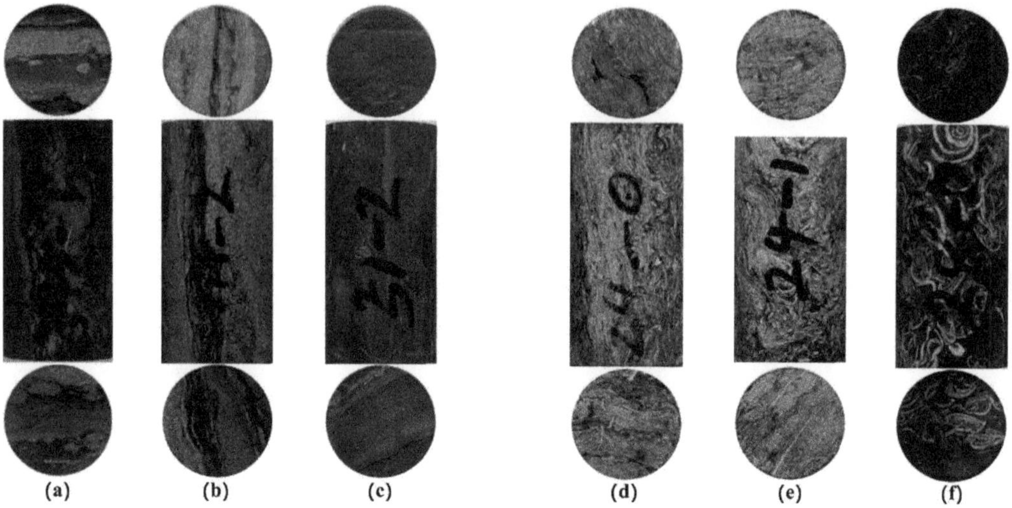

Figure 13. Shale samples with different shell contents. (**a**–**c**) Low shell content. (**d**–**f**) High shell content.

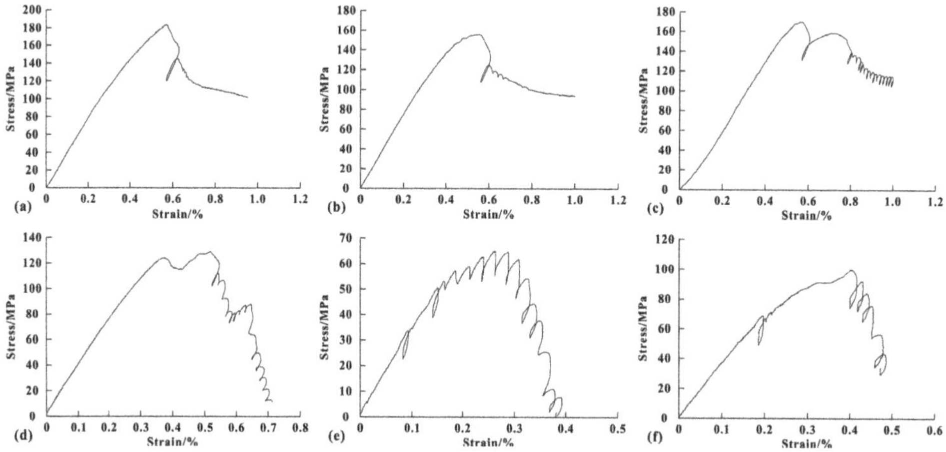

Figure 14. Stress–strain curves with different shell content. (**a**–**c**) Low shell content. (**d**–**f**) High shell content.

4.3. Actual Case Analysis

The mineral brittleness index simulation results and mechanical brittleness index model calculation results for Well XA in this study are shown in Figure 15, including mineral brittleness index Brit1 (calcite, quartz, and dolomite), average brittleness index Brit2 of mechanical parameters, quotient Brit3 of mechanical parameters, and comprehensive brittleness index Brit4 (average of Brit1 and Brit3). Compared with the brittleness index obtained from mechanical experiments, Experimental Brit (1/2 of the sum of mechanical parameters after standardization), Brit1 was relatively small and showed an insignificant brittleness change, so it was unable to accurately indicate the position of the high brittleness development stage. Brit2 was similar to Brit1 and showed the smallest overall brittleness. Brit3 was the largest brittleness index and showed a significant brittleness change. However, due to the complex stress conditions of the reservoir in the study area, the applicability of the results remain to be verified. Brit4 comprehensively considered the influences of mineral and mechanical parameters. Compared with the first three brittleness indexes, Brit4 was highly correlated with Experimental Brit, indicating that this evaluation method was the most reasonable.

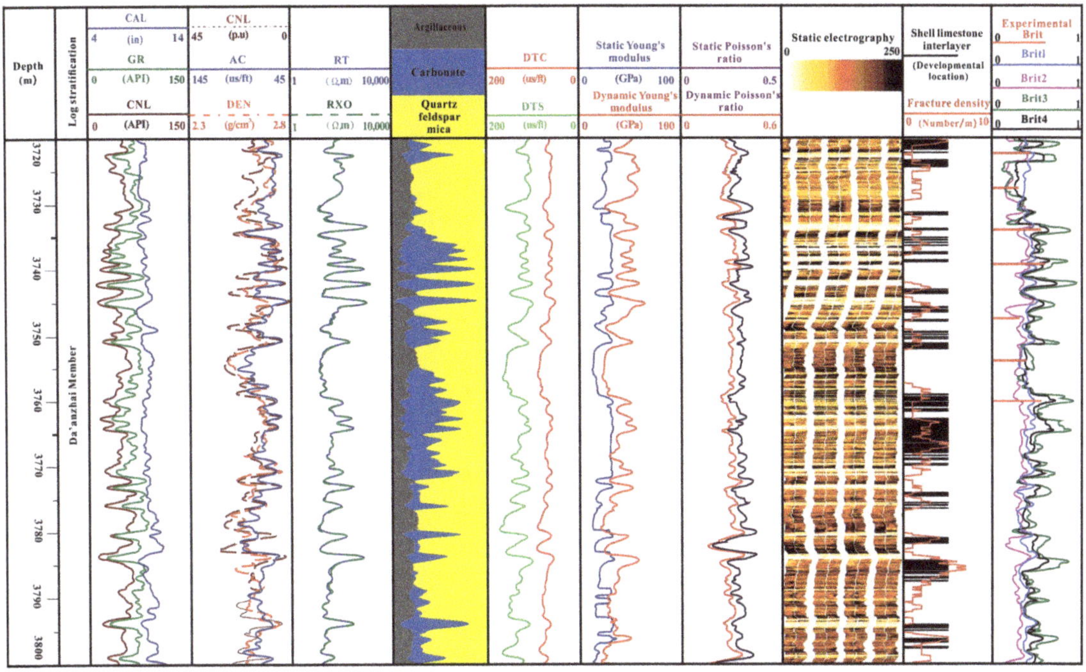

Figure 15. Calculation of brittleness indexes of Well XA.

The location of shell interlayers (lamina) and the density of the fractures of continental shale in the Da'anzhai Member were identified, and the development depth was calculated (the second track from the right in Figure 15). With the increase in the number of developed interlayers (lamina) and the fracture density, the reservoir brittleness increased significantly. Therefore, in the calculation of the brittleness index, the lamina and fractures should be quantified and comprehensively considered.

The above analysis showed that individual brittle minerals or mechanical parameters could not accurately represent the brittleness of a continental shale reservoir. The factors such as shell interlayers (lamina) and fracture development should be comprehensively considered for developing an applicable evaluation system of the brittleness and fracturing of continental shale.

5. Conclusions

(1) The continental shale lithofacies and their association types included four main rock types: clay shale, silty shale, shell calcareous clay shale, and silty clay shale, which were characterized by the high clay content and local enrichment of carbonate minerals.

(2) The brittleness of continental shale was affected by mineral content and mineral structure. The continental shale in the Da'anzhai Member was deposited in shallow and semi-deep lake environments of carbonate rocks. Compared with the marine shale of the Wufeng Formation–Longmaxi Formation, the continental shale had a lower content of siliceous minerals and higher content of carbonate minerals. The

content of felsic minerals showed a weak influence on brittleness. The total content of felsic and carbonate minerals largely determined the brittleness of shale.

(3) The continental shale contained a large number of shell laminates. The shell was mainly filled with calcite. After dissolution, the calcite shell formed a large number of bedding joints and corrosion joints. In the interlayers with a high shell content, a large number of microfractures also developed.

(4) A large number of fractures developed in continental shale, making the energy consumption required for maintaining fracture expansion after rock fracture and fracturing lower, thus greatly improving the brittleness of the shale reservoir.

(5) The main control factors for the brittleness of continental shale and marine shale are not consistent. Therefore, the lithology, interlayer type, shell content, and rock mechanical properties should be fully considered in the establishment of the corresponding brittleness and fracturing evaluation system of continental shale.

Author Contributions: Conceptualization, Y.L. and R.W.; methodology, Y.L.; software, H.X. and Y.L.; validation, J.L. and X.Z.; formal analysis, Y.L. and F.L.; investigation, Z.W. and Y.L.; resources, X.Z.; data curation, R.W.; writing—original draft preparation, Y.L. and R.W.; writing—review and editing, F.L. and Z.W.; visualization, Y.L., F.L. and R.W.; supervision, Y.L. and R.W.; project administration, F.L. and R.W.; funding acquisition, Y.L., F.L. and R.W. All authors have read and agreed to the published version of the manuscript.

Funding: This research was jointly supported by the National Natural Science Foundation of China (41402118519640075210408O), Chongqing Basic Research and Frontier Exploration Projects (cstc2018jcyjAX0503), Chongqing Postgraduate Research and Innovative Projects (CYS22722), and Guizhou Science and Technological Supporting Plan Project (2021-401).

Data Availability Statement: Not applicable.

Acknowledgments: Thank you Sinopec Petroleum Exploration and Production Research Institute, Beijing 102206, China, for the assistance and support, and I would like to express my heartfelt gratitude.

Conflicts of Interest: The authors declare no conflict of interest.

References

1. Zou, C.N.; Yang, Z.; Dong, D.Z.; Zhao, Q.; Chen, Z.H.; Feng, Y.L.; Li, J.R.; Wang, X.N. Formation, Distribution and Prospect of Unconventional Hydrocarbons in Source Rock Strata in China. *J. Earth Sci.* **2022**, *47*, 1517–1533.
2. Adeyilola, A.; Zakharova, N.; Liu, K.Q.; Gentzis, T.; Carvajal-Ortiz, H.; Ocubalidet, S.; Harrison, W.B. Hydrocarbon potential and Organofacies of the Devonian Antrim Shale, Michigan Basin. *Int. J. Coal Geol.* **2022**, *249*, 1039509. [CrossRef]
3. Paul, D.; Malik, S.; Mishra, D.K.; Hiwale, R. Shale gas: An indian market perspective. *Int. J. Energy Econ. Policy* **2021**, *11*, 126–136. [CrossRef]
4. Cudjoe, S.E.; Barati, G.R.; Goldstein, R.H.; Tsau, J.S.; Nicoud, B.; Bradford, K.; Baldwin, A.; Mohrbacher, D. An integrated workflow to characterize lower Eagle Ford shale for hydrocarbon gas huff-n-puff simulation. *Fuel* **2021**, *289*, 119854. [CrossRef]
5. Dou, L.R.; Wen, Z.X.; Wang, J.J.; Wang, Z.M.; He, Z.J.; Liu, X.B.; Zhang, N.N. Analysis of the world oil and gas exploration situation in 2021. *Pet. Explor. Dev.* **2022**, *49*, 1195–1209. [CrossRef]
6. Davis, L.A. The Shale Oil and Gas Revolution. *Engineering* **2018**, *4*, 438–439. [CrossRef]
7. Satyavani, N.; Uma, V.; Pavan, K.P.; Srinivas, K.N.S.S.S.; Mysaiah, D.; Dhanam, K.; Srinivas, G.S.; Satyanarayana, H.V.S.; Vedanti, N. Geophysical Exploration of Unconventional Hydrocarbons. *J. Geol. Soc. India* **2021**, *97*, 1274–1279. [CrossRef]
8. Zhang, S.M.; Yang, Y.M.; Hong, H.T.; Zeng, J.H.; Zhang, R.; Li, Y.C.; He, X.Q. Reservoir characteristics and its petroleum significance of Jurassic Da anzhai shale interval in central Sichuan Basin, SW China. *J. China Univ. Min. Technol.* **2022**, *51*, 718–730.
9. Wang, R.Y.; Hu, Z.Q.; Liu, J.S.; Wang, X.H.; Gong, D.J.; Yang, T. Comparative analysis of characteristics and controlling factors of fractures in marine and continental shales: A case study of the Lower Cambrian in Cengong area, northern Guizhou Province. *Oil Gas Geol.* **2018**, *39*, 631–640.
10. Pu, B.L.; Wang, F.Q.; Dong, D.Z.; Sun, J.B. Challenges of terrestrial shale gas exploration and development from Chang 7 shale in the Ordos Basin. *Arab. J. Geosci.* **2021**, *14*, 644. [CrossRef]
11. Liu, B.; Shi, J.X.; Fu, X.F.; Lv, Y.F.; Sun, X.D.; Gong, L.; Bai, Y.F. Petrological characteristics and shale oil enrichment of lacustrine fine-grained sedimentary system: A case study of organic-rich shale in first member of Cretaceous Qingshankou Formation in Gulong Sag, Songliao Basin, NE China. *Pet. Explor. Dev.* **2018**, *45*, 828–838. [CrossRef]
12. He, J.H.; Ding, W.L.; Li, R.N.; Wang, R.Y.; Zhao, W. Forming condition of the continental shale gas of Shaheji Formation in the central-north Huanghua depression and its resource prospect. *Pet. Geol. Recovery Effic.* **2016**, *23*, 22–30.

13. Liu, H.T.; Li, X.; Wan, Y.Q.; Liu, Z.B.; Zhou, L.; Che, S.Q. Formation conditions and exploration and development potential of continental shale gas: A case of Dongyuemiao Member of the Jurassic in north Fuling area, eastern Sichuan Basin. *Mar. Orig. Pet. Geol.* **2020**, *25*, 148–154.
14. Iyare, U.C.; Blake, O.O.; Ramsook, R. Brittleness evaluation of Naparima Hill mudstones. *J. Pet. Sci. Eng.* **2021**, *196*, 107737. [CrossRef]
15. Li, W.; Liu, J.S.; Zeng, J.; Leong, Y.K.; Elsworth, D.; Tian, J.W.; Li, L. A fully coupled multidomain and multiphysics model for evaluation of shale gas extraction. *Fuel* **2020**, *278*, 118214. [CrossRef]
16. Liu, B.; Yan, M.; Sun, X.D.; Bai, Y.F.; Bai, L.H.; Fu, X.F. Microscopic and Fractal Characterization of Organic Matter within Lacustrine Shale Reservoirs in the First Member of Cretaceous Qingshankou Formation, Songliao Basin, Northeast China. *J. Earth Sci.* **2020**, *31*, 1241–1251. [CrossRef]
17. Iqbal, M.A.; Rezaee, R.L.C.; Pejcic, B.; Smith, G. Integrated sedimentary and high-resolution mineralogical characterisation of Ordovician shale from Canning Basin, Western Australia: Implications for facies heterogeneity evaluation. *J. Pet. Sci. Eng.* **2022**, *208*, 109347. [CrossRef]
18. Liu, Y.J.; Lai, F.Q.; Zhang, H.J.; Tan, Z.J.; Wang, Y.F.; Zhao, X.T.; Tan, X.F. A novel mineral composition inversion method of deep shale gas reservoir in Western Chongqing. *J. Pet. Sci. Eng.* **2021**, *202*, 108528.
19. Wang, R.Y.; Ding, W.L.; Zhang, Y.Q.; Wang, Z.; Wang, X.H.; He, J.H.; Zeng, W.T.; Dai, P. Analysis of developmental characteristics and dominant factors of fractures in Lower Cambrian marine shale reservoirs: A case study of Niutitang formation in Cen'gong block, southern China. *J. Pet. Sci. Eng.* **2016**, *138*, 31–49. [CrossRef]
20. Wang, R.Y.; Hu, Z.Q.; Long, S.X.; Du, W.; Wu, J.; Wu, Z.H.; Nie, H.K.; Wang, P.W.; Sun, C.X.; Zhao, J.H. Reservoir characteritics and evolution mechanisms of the Upper Ordovician Wufeng-Lower Silurian Longmaxi shale, Sichuan Basin. *Oil Gas Geol.* **2022**, *43*, 353–364.
21. Wang, R.Y.; Hu, Z.Q.; Zhou, T.; Bao, H.Y.; Wu, J.; Du, W.; He, J.H.; Wang, P.W.; Chen, Q. Characteristics of fractures and their significance for reservoirs in Wufeng-Longmaxi shale, Sichuan Basin and its periphery. *Oil Gas Geol.* **2021**, *42*, 1295–1306.
22. Li, B.K.; Nie, X.; Cai, J.C.; Zhou, X.Q.; Wang, C.C.; Han, D.L. U-Net model for multi-component digital rock modeling of shales based on CT and QEMSCAN images. *J. Pet. Sci. Eng.* **2022**, *216*, 110734. [CrossRef]
23. Liu, B.; Gao, Y.F.; Liu, K.Q.; Liu, J.Z.; Ostadhassan, M.; Wu, T.; Li, X.L. Pore structure and adsorption hysteresis of the middle Jurassic Xishanyao shale formation in the Southern Junggar Basin, northwest China. *Energy Explor. Exploit.* **2021**, *39*, 761–778. [CrossRef]
24. Liu, Z.B.; Liu, G.X.; Hu, Z.G.; Feng, D.J.; Zhu, T.; Bian, R.K.; Jiang, T.; Jin, Z.G. Lithofacies types and assemblage features of continental shale strata and their significance for shale gas exploration: A case study of the Middle and Lower Jurassic strata in the Sichuan Basin. *Nat. Gas Ind. B* **2019**, *39*, 10–21.
25. Liu, J.S.; Ding, W.L.; Xiao, Z.K.; Dai, J.S. Advances in comprehensive characterization and prediction of reservoir fractures. *Prog. Phys.* **2019**, *34*, 2283–2300.
26. Wei, Y.; Nie, X.; Jin, L.D.; Zhang, C.; Zhang, C.M.; Zhang, Z.S. Investigation of sensitivity of shale elastic properties to rock components based on a digital core technology and finite element method. *Arab. J. Geo-Sci.* **2018**, *11*, 1–14. [CrossRef]
27. Ding, W.L.; Zhu, D.W.; Cai, J.J.; Gong, M.L.; Chen, F.Y. Analysis of the developmental characteristics and major regulating factors of fractures in marine–continental transitional shale-gas reservoirs: A case study of the Carboniferous–Permian strata in the southeastern Ordos Basin, central China. *Mar. Pet. Geol.* **2013**, *45*, 121–133. [CrossRef]
28. Ma, C.F.; Dong, C.M.; Lin, C.Y.; Elsworth, D.; Luan, G.; Sun, X.L.; Liu, X.C. Corrigendum to "Influencing factors and fracability of lacustrine shale oil reservoirs". *Mar. Pet. Geol.* **2021**, *124*, 104861. [CrossRef]
29. Guo, X.S.; Zhao, Y.Q.; Zhang, W.T.; Li, Y.P.; Wei, X.F.; Shen, B.J. Accumulation conditions and controlling factors for the enrichment of shale oil and gas in the Jurassic Qianfoya Formation, Yuanba Area, Sichuan Basin. *Pet. Geol. Exp.* **2021**, *43*, 749–757.
30. Hu, Z.Q.; Wang, R.Y.; Liu, Z.B.; Liu, G.X.; Feng, W.J.; Yang, Z.H.; Wang, P.W. Source-reservoir characteristics and coupling evaluations for the Lower Jurassic lacustrine shale gas reservoir in the Sichuan Basin. *Geosci. Front.* **2021**, *28*, 261–272.
31. Wang, R.Y.; Gong, D.J.; Ding, W.L.; Leng, J.G.; Yin, S.; Wang, X.H.; Sun, Y.X. Brittleness evaluation of the Lower Cambrian Niutitang shale in the Upper Yangtze region: A case study in the Cengong Block, Guizhou Province. *Geosci. Front.* **2016**, *23*, 87–95.
32. Zhang, J.J.; Salad, H.O.; Shi, S.M.; Cao, Y.F.; Sun, J.Y. Indicators for identification of lacustrine sandstones of sandy debris-flow origin: A case study of Es1 Member, Palaeogene Shahejie Formation, Nanpu Sag, Bohai Bay Basin, China. *Geol. J.* **2022**, *57*, 3178–3198. [CrossRef]
33. Cao, D.S.; Zeng, L.B.; Lv, W.Y.; Xu, X.; Tian, H. Progress in brittleness evaluation and prediction methods in unconventional reservoirs. *Pet. Sci. Bull.* **2021**, *6*, 31–45.
34. He, J.H.; Li, Y.; Deng, H.C.; Tang, J.M.; Wang, Y.Y. Quantitative evaluation and influencing factors analysis of the brittleness of deep shale reservoir based on multiple rock mechanics experiments. *Nat. Gas Geosci.* **2022**, *33*, 1102–1116.
35. Wang, X.Y.; Jin, Z.K.; Guo, Q.H.; Wang, J.Y.; Ren, Y.L.; Wang, L.; Wang, Z.F. Genesis of Calcite in Nonmarine Shale of the Lower Jurassic Da'anzhai Member, Northeastern Sichuan. *Acta Sedimentol. Sin.* **2021**, *39*, 704–712.
36. Liu, J.S.; Ding, W.L.; Yang, H.M.; Dai, P.; Wu, Z.H.; Zhang, G.J. Natural Fractures and Rock Mechanical Stratigraphy Evaluation in theHuaqingarea, Ordos Basin: A Quantitative Analysis Based on Numerical Simulation. *J. Earth Sci.* **2022**, 1–19. Available online: http://kns.cnki.net/kcms/detail/42.1874.P.20220711.0942.004.html (accessed on 1 February 2023).

37. Ding, W.L.; Zeng, W.T.; Wang, R.Y.; Jiu, K.; Wang, Z.; Sun, Y.X.; Wang, X.H. Method and application of tectonic stress field simulation and fracture distribution prediction in shale reservoir. *Geosci. Front.* **2016**, *23*, 63–74.
38. Huang, Y.Y.; Wang, G.W.; Han, Z.Y. Micro resistivity imaging logging—The "perspective eye" of petroleum explorers. *Pet. Knowl.* **2022**, *1*, 14–19.

minerals

MDPI

Article

Characteristics and Controlling Factors of Natural Fractures in Continental Tight-Oil Shale Reservoir

Xiaofei Fu [1], Lei Gong [2,*], Xiaocen Su [3], Bo Liu [1], Shuai Gao [2,*], Jianguo Yang [4] and Xinnan Qin [3]

1 Key Laboratory of Continental Shale Hydrocarbon Accumulation and Efficient Development, Ministry of Education, Northeast Petroleum University, Daqing 163318, China
2 Bohai-Rim Energy Research Institute, Northeast Petroleum University, Qinhuangdao 066004, China
3 College of Geosciences, Northeast Petroleum University, Daqing 163318, China
4 Shenyang Center of China Geological Survey, Shenyang 110034, China
* Correspondence: kcgonglei@nepu.edu.cn (L.G.); gaoshuai@nepu.edu.cn (S.G.)

Abstract: Natural fracture growth plays an important role in shale-oil enrichment. Systematically investigating fracture features and their controlling factors in shale-oil reservoirs is essential for accurately predicting fracture distribution. The controlling factors of fracture distribution in the continental shale of the Qingshankou Formation in the Songliao Basin, China, were systematically analyzed based on the quantitative fracture characterization of outcrops and cores. Strata-confined fractures, throughgoing fractures, bedding-parallel fractures, and stylolites can be observed in the Qingshankou shale reservoir in the study area. Fracture distribution is not only controlled by internal factors, e.g., mineral composition, mechanical stratigraphy, and lithofacies, but also by external factors, e.g., faults and abnormally high pressure readings. Mineral composition is the primary factor governing fracture development, and it not only controls fracture abundance, but it also affects fracture filling and effectiveness. Mechanical stratigraphy determines the spatial morphology and developmental pattern of a fracture. Fractures are well-developed in brittle strata, with fracture spacing being proportional to bed thickness. Lithofacies can determine fracture development by controlling the variation of mineral composition, rock structure, bed thickness, etc. Stress concentration is commonly high at fault tips, intersections, and overlaps, where fracture density is high and has good connectivity. The existence of abnormally high pressure reduces effective stress, promoting shear fracture development. Tensile overpressure fractures can also be generated under small levels of differential stress.

Keywords: continental shale; natural fracture; controlling factor; Songliao Basin

1. Introduction

Shale oil is a new bright spot in global unconventional oil exploration with potential resource prospects [1–4]. It can be divided into fractured shale oil, brittle interlayered shale oil, and matrix shale oil [2,5]. Fractures of various scales are regarded as primary storage spaces in fractured shale-oil reservoirs [6–9]. Fractures are also the key factor that determine whether shale oil can be enriched. In the absence of fractures, it is difficult to obtain natural productivity from shale [2]. The initial production capacity of a fractured shale-oil reservoir is generally high, but it then declines rapidly due to the limited storage space in the fracture system, where oil in matrix pores is important for replenishing productivity [10,11]. Therefore, a fractured reservoir is beneficial for shale-oil discovery at the early exploration stage, while the presence of matrix pores in a shale reservoir is the key to long-term, stable production [12–15]. These two types of storage spaces contribute significantly to shale-oil exploration and development.

Shale-oil resources have been discovered in the north of the Songliao Basin, mainly distributed in the Cretaceous Qingshankou Formation [16]. Currently, 45 wells have been tested in the Qingshankou Formation, while commercial oil flow has been obtained from

25 of them. Previous studies estimated that the Qingshankou Formation shale-oil resource contained 180–440 million tons with a favorable exploration area of up to 2×10^4 km^2, exhibiting huge exploration potential. Systematic research and exploration have been deployed since the shale reservoirs were discovered in the 1980s; the productive experimental area of well Y12 was established, and horizontal well G1 was drilled to improve productivity. In August 2021, Daqing Oilfield announced the discovery of a major strategic breakthrough in shale-oil exploration in the Songliao Basin, which confirmed the prediction that the geological reserves of the shale oil equal 1.268 billion tons. However, the reservoir's characteristics are special: natural fractures control the distribution of high-quality reservoirs. Gong et al. [17] characterized and discussed the effectiveness of fractures in these shale reservoirs, but the main controlling factors of fracture distribution need to be further analyzed. A significant challenge in understanding fractures in the subsurface is the incomplete sampling that is inherent in using wellbores (cores and image logs). It is common for subsurface core-based studies to supplement direct observations of the subsurface by using outcrops as analogs and guides to the subsurface, as the present study does, but the use of such analogs must be undertaken with caution because, inevitably, outcrops have experienced loading and thermal histories different from those of the rocks still in the subsurface. This potential limitation of the use of outcrops needs to be acknowledged, and the possible limitations of conclusions drawn from the subsurface should be acknowledged [18]. This paper focuses on fracture characterization in outcrops, and combined with core fracture characterization and experimental analysis, defines the fracture characteristics in this shale-oil reservoir and discusses the main controlling factors of fractures in this shale-oil reservoir.

2. Geological Setting

The Songliao Basin is located in northeast China (Figure 1) and has an area of about 26×10^4 km^2 [19]. It is a large, continental depression basin formed during the Late Cretaceous period where dark shale was widely developed during the Qingshankou period [20]. The Qingshankou Formation was deposited during a large-scale lacustrine transgression under a humid and hot environment. It can be divided, from bottom to top, into the Qing-1, Qing-2, and Qing-3 Members. The Qing-1 Member, the target of study in this paper, is dominated by a set of deep lacustrine organic-rich black shale with a large thickness, a high level of organic carbon content (TOC), and moderate maturity in the center of the lake, which was deposited under a rapid lacustrine transgression [21,22]. It is primarily shale, silty shale, argillaceous siltstone, and limestone. The average thickness of the organic-rich shale in the Qing-1 Member is 70–80 m. The TOC is in the range of 1–5%, with an average value of 2.9%. The vitrinite reflectance (Ro) varies between 0.72% and 1.20%, with an average value of 0.91%. After the deposition of the Qing-1 Member, three tectonic events occurred in the study area, and a series of normal faults was formed at the end of the Qing-3 Member, the Nen-2 Member, and the Nen-3 Member, respectively.

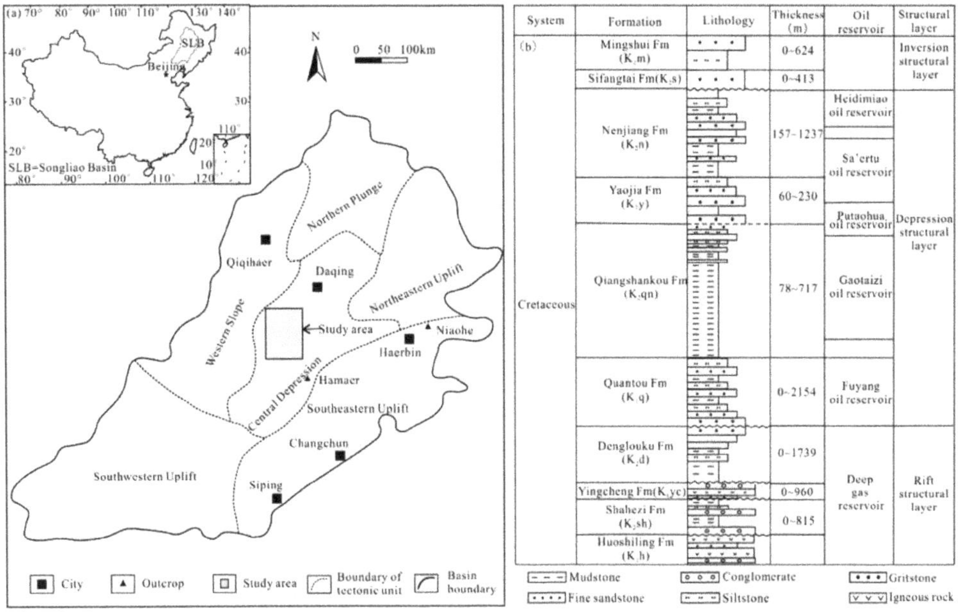

Figure 1. Structural units of the Songliao Basin and the location of the study area.

3. Materials and Methods

Outcrop and core fractures were observed and samples were collected to study fracture development and its controlling factors in Qingshankou shale. Two well-exposed outcrops (Niaohe and Hamaer) at the edge of the Songhua River were selected to describe fracture growth in the shale (Figure 1). In addition, 215-meter cores from 5 wells at Songliao Basin were also observed so as to describe fracture growth. The 1D scanline method was used to accurately describe the fractures [23]. Scanlines were arranged in each mechanical layer and were perpendicular to fracture traces. Information (location, occurrence, height, filling property, etc.) about each fracture intersecting with the scanning line was measured and recorded. Fracture intensity was expressed by fracture spacing and fracture density. Specifically, the fracture spacing is the vertical distance between two adjacent fractures, while the fracture density is the number of fractures per meter, which is inversely related to fracture spacing.

Mineral composition analysis, rock mechanical testing, and lithofacies classification were carried out to study the controlling factors of fracture development. The whole-rock mineral composition of 18 outcrop samples were analyzed using the X-ray diffraction method; corresponding brittleness indexes (BI) were calculated using the following model: BI = quartz content + carbonate mineral content. The rock mechanical parameters (elastic modulus, Poisson's ratio, and compressive strength) of 9 core plugs were tested on a TAW-2000 triaxial rock compression test system at Northeast Petroleum University. The three-step classification criteria proposed by Liu et al. [2] were adopted for the lithofacies classification.

4. Fracture Characterization

4.1. Fracture Types and Characteristics

Four types of fractures were identified in the Qingshankou Formation, namely, strata-confined fractures, throughgoing fractures, bedding-parallel fractures and stylolites.

Strata-confined fractures refer to fractures that develop in a single layer and end at a lithologic interface or a bedding surface [24] whose distribution is controlled by mechanical stratigraphy (Figure 2a). Strata-confined fractures are widely developed, commonly run parallel to each other, and end at the layer interface. They are nearly perpendicular to the bedding surface, and consequently, their heights are generally equal to the bed thickness. These strata-confined fractures appear to be equally spaced on the outcrops. However, data from horizontal wells show that fractures at this depth are highly clustered (not regularly spaced), whereas the same fracture sets in outcrops of the same unit have regular spacing [25]. The reason for the difference is that fractures forming in the deep subsurface are subject to different chemical effects [18,26].

Figure 2. (**a**) Strata-confined fractures (blue lines) and throughgoing fractures (red lines); (**b**) throughgoing fractures (red lines).

Throughgoing fractures can develop in different patterns, e.g., as a single fracture or as high-density and nearly parallel-oriented fracture groups [27–30]. They tend to extend across multiple mechanical units, occurring as multilayered structures (Figure 2b). The throughgoing fractures have a steep dip or are even vertical, while their strikes are consistent with those of the strata-confined fractures. Many fracture segments in throughgoing fracture are single strata-confined, and they are widened or interconnected at the bedding surface to form throughgoing fractures. These throughgoing fractures are also equally spaced, indicating that they are universally penetrative structures rather than isolated or local structures [31,32]. Furthermore, they are commonly formed within the same tectonic stress field.

The bedding-parallel fractures in shale were formed by slow deposition under relatively stable hydrodynamic conditions, which consist of a series of flat and fine bedding-parallel layers. Bedding-parallel fractures are usually non-structural fractures caused by diagenesis, e.g., compaction, cementation, and clay mineral transformation, that commonly occur along the bedding surface (Figure 3a). Shale bedding is interfaced with a weak mechanical property and easily peels off to become a bedding-parallel fracture. These bedding-parallel fractures are well-developed in organic-rich shales and are commonly perpendicular to the overburden pressure, occurring as closed fractures.

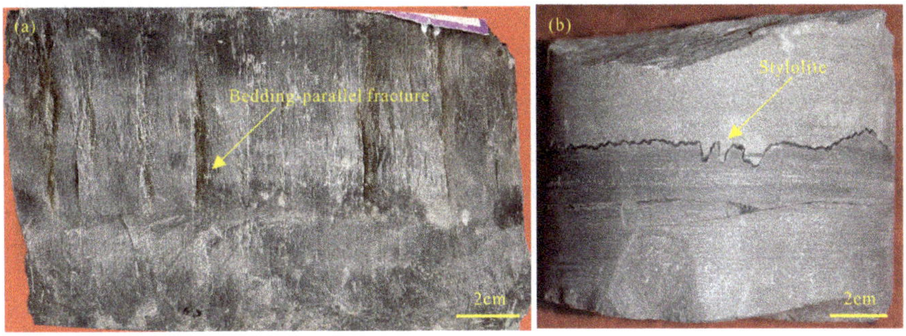

Figure 3. (**a**) Bedding-parallel fractures; (**b**) stylolite.

The stylolites occur as serrated or columnar fractures filled by dark minerals. These serrations or columns are nearly vertical and are perpendicular to the fracture surface (Figure 3b), suggesting that they are diagenetic stylolites. They have an uneven surface that is parallel to the bedding surface, indicating that they are formed by a pressurized solution, where insoluble minerals are retained and soluble minerals are taken away by fluid. Most stylolites are typical seepage barriers since they are filled by insoluble material.

4.2. Quantitative Fracture Characterization

Field measurements show three sets of fractures in the study area, e.g., NNW–SSE trending, NE–SW trending, and NEE–SWW trending (Figure 4a). Fractures with a high dip angle are the most important, e.g., fractures with a dip angle > 70° account for more than 93% of the total (Figure 4b). Fractures in the core are also high-dip, while a few oblique fractures and low-dip fractures are also developed.

Fractures in the Qingshankou Formation are well-developed, especially in the fault zone. Fracture spacing in outcrops is mainly 5–20 cm, while fracture density is 5–12.5 m^{-1}. Fracture density in the core is about 0.3–1.0 m^{-1}, with a maximum value of 1.52 m^{-1}. The ratio of total fracture height to core length is generally less than 0.1, with an average value of 0.05.

The height of the strata-confined fractures in the outcrops ranges from 10 to 40 cm, following a normal distribution pattern. The height of throughgoing fractures is the range of 50–200 cm, which is also normally distributed. However, the height of all of the fractures (strata-confined fractures and throughgoing fractures) obeys a power law distribution, indicating that they are formed under a uniform tectonic stress field. The fracture height in the core is generally less than 20 cm due to mechanical stratigraphy, while the maximum value of a few of the fractures, which penetrate multiple layers, can be up to 2.8 m.

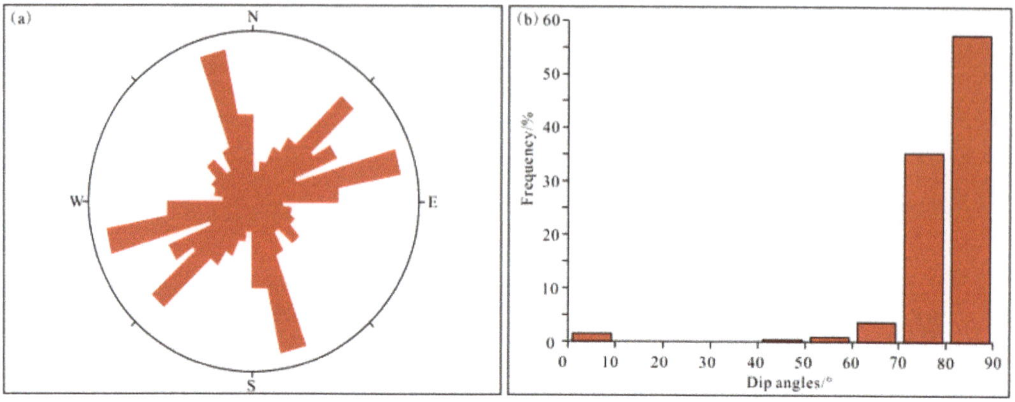

Figure 4. (**a**) Rose diagram of fracture strikes; (**b**) histogram of dip angles in outcrops.

Most fractures are filled by different minerals (calcite, quartz, and bitumen); specifically, fully-filled, half-filled and unfilled fractures account for 59%, 16%, and 25%, respectively. The half-filled and unfilled fractures are effective ones, while the fully-filled fractures have weak surfaces that are conducive to hydrofracturing.

5. Discussion

Fracture development varies greatly in different positions in the shale due to strong heterogeneity. The unevenly developed fractures bring great challenges to the study of tight shale reservoirs and significantly increase oil/gas exploration and development risks in such reservoirs. Although fracturing is complex and heterogeneous in distribution, its occurrence commonly follows certain rules. Geological factors controlling fracture formation and distribution are discussed based on outcrop, core, and thin section observation and experimental analysis, which provides an insight into fracture prediction in shale.

5.1. Mineral Composition and Content

Lithology is the most basic factor influencing fracture development in tight reservoirs [33,34] since mineral composition, texture, and structure, as well as mechanical properties, vary greatly with lithology [35,36]. Therefore, fracture development is different in various forms of lithology under the same tectonic stress [37]. The influence of mineral composition, rock mechanical properties, and brittleness on fracture development was analyzed based on X-ray diffraction analysis and rock mechanical tests.

Fracture line density is positively correlated with quartz content and carbonate mineral content (Figure 5a,b), but it is negatively correlated with clay mineral content (Figure 5c). However, an obscure relationship occurs between feldspar content and fracture development (Figure 4d). These phenomena indicate that, under the same tectonic stress, fractures can be well-developed in brittle rocks with a high content of quartz, calcite, and dolomites, but it can be poorly developed in ductile rocks with high levels of clay minerals (Figure 6). Additionally, fractures in shale with high carbonate mineral content are usually fully filled with calcite [1,38,39], while fractures in shale dominated by quartz and feldspar are weakly filled, commonly unfilled, or half-filled by euhedral quartz (Figures 6 and 7).

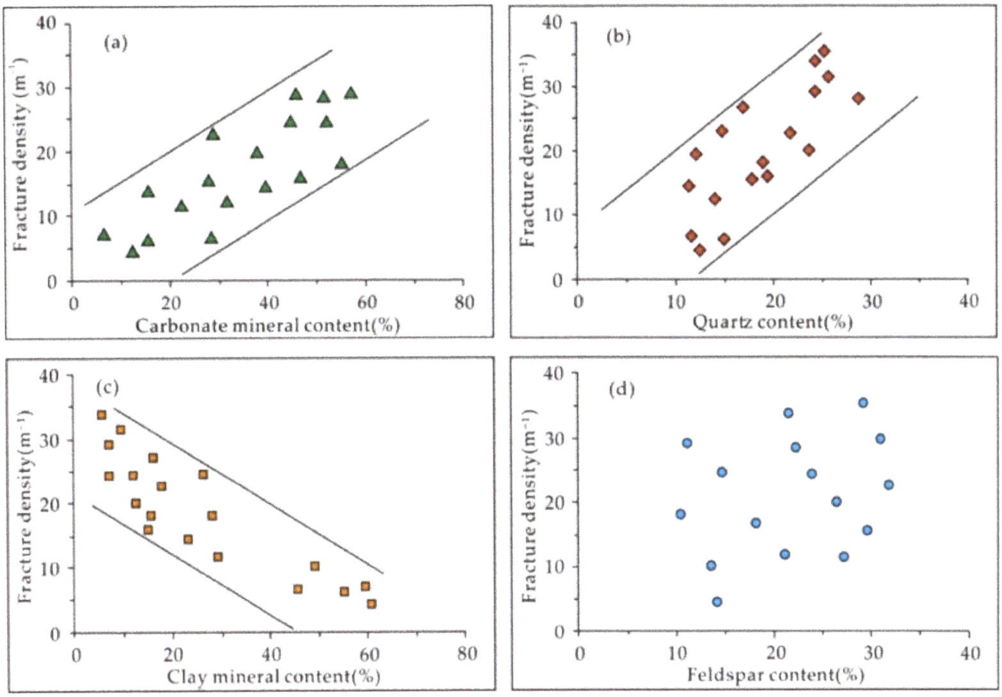

Figure 5. Relationship between different mineral contents and fracture density in Qingshankou shale. (**a**) Carbonate mineral content vs. fracture density; (**b**) quartz content vs. fracture density; (**c**) clay mineral content vs. fracture density; (**d**) feldspar content vs. fracture density. Samples were collected from outcrops.

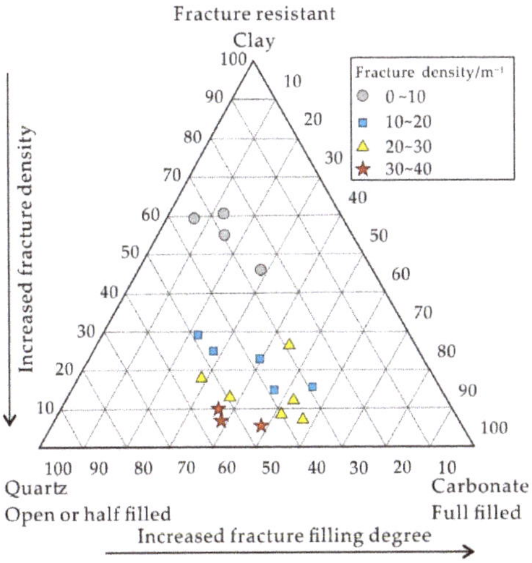

Figure 6. Host rock mineral composition and fracture distribution. Samples were collected from outcrops.

Figure 7. Images showing fractures with different fillings: (**a**) fracture half-filled with euhedral quartz; (**b**) fracture fully filled with calcite. The yellow arrows indicate the location of the fractures.

The brittleness index (BI = quartz content + carbonate mineral content) was calculated to analyze its relationship with fracture density. The results show that fracture development has a better correlation with the BI compared with single-mineral content (Figure 8), suggesting that the brittleness index is a good indicator of fracture growth. Consequently, the rock brittleness index can be used as a constraint when predicting fracture development.

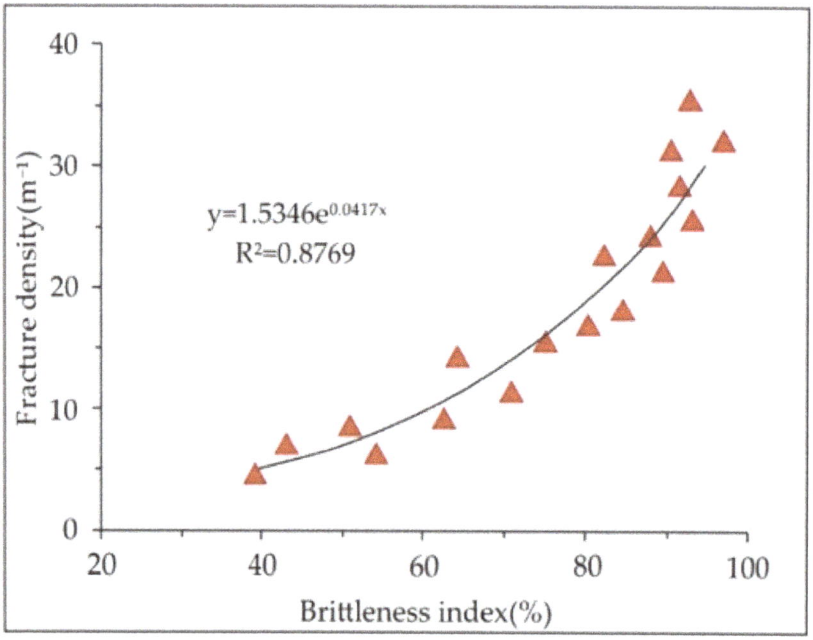

Figure 8. Relationship between brittleness index and fracture density in Qingshankou shale. Samples were collected from outcrops.

Under the same stress field, the fracture development is mainly governed by rock mechanical properties (such as the elasticity modulus, Poisson's ratio, the bulk modulus, compressive strength, etc.). Fracture intensity in different lithologies vary greatly within these parameters. Comparing fracture development and rock mechanical parameters shows that the fracture development is in a mirror-image relationship (inverse ratio) with the compressive strength and Poisson's ratio, and it is positively correlated with the elasticity modulus and the bulk modulus. In other words, lower levels of compressive strength and Poisson's ratio and a larger elasticity modulus are beneficial for fracture development (Figure 9).

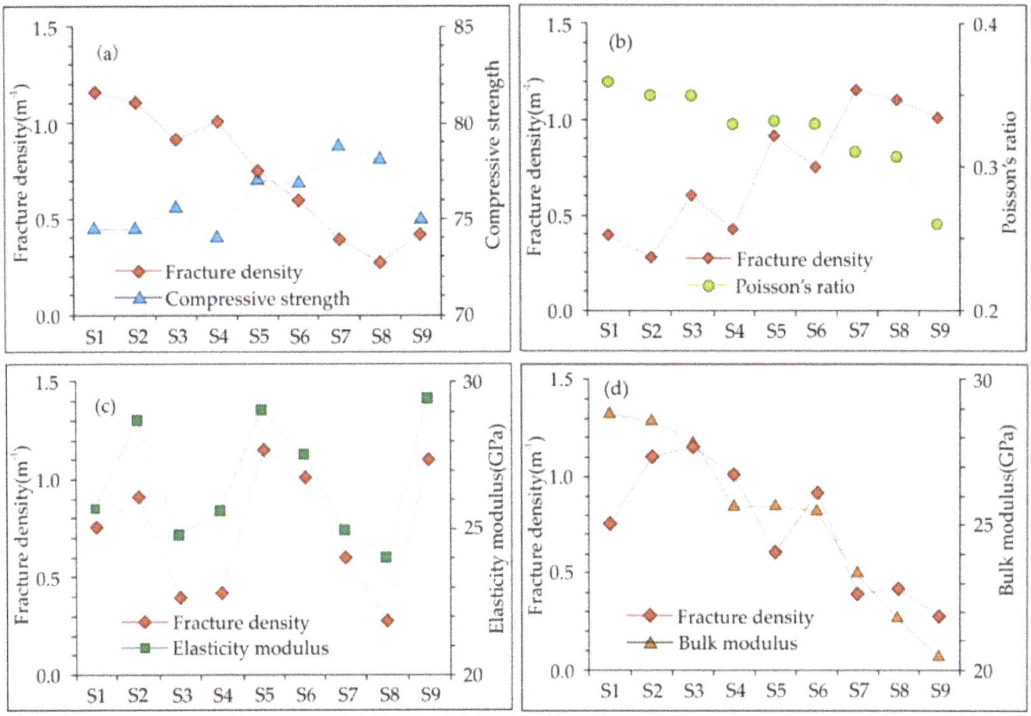

Figure 9. Relationship between rock mechanical parameters and fracture density in Qingshankou shale: (**a**) Compressive strength vs. fracture density; (**b**) elasticity modulus vs. fracture density; (**c**) Poisson's ratio vs. fracture density; (**d**) Bulk modulus vs. fracture density. Samples were collected from cores. The 9 samples have been marked as S1–S9, respectively.

5.2. Mechanical Stratigraphy

Fracture observations on outcrops and cores prove that mechanical stratigraphy is one of the most important factors controlling natural fracture types, nucleation, propagation, geometry, and spatial distribution [7,11,40–42]. Previous studies point out that mechanical stratigraphy controls the formation and distribution of natural fractures [43,44].

On the one hand, mechanical stratigraphy controls fracture morphology and developmental pattern [45,46]. Fractures are usually nucleated at defects (e.g., fossil inclusions, pyrite concretions, pores, and cusps) along mechanical interfaces, then spread into the brittle strata, and eventually terminate at the interface between brittle and ductile strata [47,48]. In a relatively homogeneous lamelleted rock, fractures may also end at the mechanical interface and deflect due to interbedded sliding and/or debonding along the interface [49]. These fractures that are limited by mechanical layers are called strata-confined fractures. In the same mechanical layer, strata-confined fractures have equal spacing that follows a normal distribution. On the other hand, mechanical layer thickness controls fracture abundance [48,50]. The statistics on two observation outcrops in Songliao Basin show that the fracture spacing has a good linear relationship with the mechanical layer thickness; that is, the average fracture spacing is proportional to the mechanical layer thickness (Figure 10). Therefore, it is of great significance to establish a relationship between fracture density and mechanical layer thickness to understand fracture geometry and spatial distribution in different strata units, which is helpful for fracture prediction at subsurface [45,51,52].

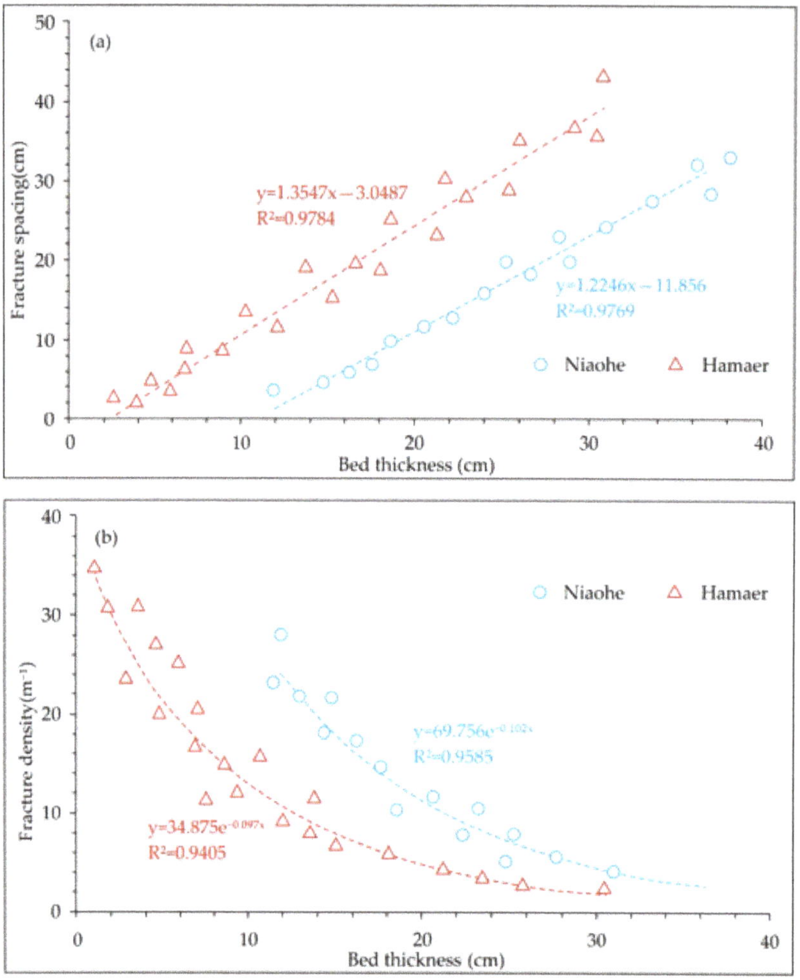

Figure 10. Impacts of mechanical layer thickness on fracture intensity in Qingshankou shale: (**a**) Mechanical layer thickness vs. fracture spacing; (**b**) Mechanical layer thickness vs. fracture density. NHT-1, NHT-2, NHT-3, and YHT-1 represent survey locations, respectively.

Meanwhile, mechanical stratigraphy also controls fault nucleation, the fracturing mode, fault geometry, the displacement gradient and distribution, segment growth, the fault core and damage zone, and the deformation process [48,52]. The mechanical properties of the host rock and mechanical stratigraphy division are important for fault interpretation, internal structure, and seal property characterization [53,54].

5.3. Lithofacies

Lithofacies determines fracture development by controlling TOC, mineral composition, rock structure, and mechanical layer thickness [2]. Seven types of lithofacies were identified in the Qingshankou Formation based on the three-step classification criteria of "total organic carbon abundance—sedimentary structure—mineral composition" [2], i.e., type I: high TOC massive siliceous shale; type II: medium TOC massive siliceous shale; type III: medium TOC laminated siliceous shale; type IV: low TOC laminated siliceous shale; type V: low TOC lamelleted siltstone; type VI: low TOC lamelleted limestone; and type VII: high TOC massive argillaceous shale.

Statistical analysis of the fracture development in different lithofacies shows that fractures are well-developed in medium TOC laminated siliceous shale (type III), low TOC lamelleted siltstone (type V), and low TOC lamelleted limestone (type VI), with an average fracture density > 1.0 m^{-1} (Figure 11). Fractures are moderately developed in low TOC laminated siliceous shale (type IV), with an average fracture density of 0.93 m^{-1}. They are poorly developed in high TOC massive siliceous shale (type I), medium TOC massive siliceous shale (type II), and high TOC massive argillaceous shale (type VII), especially in high TOC massive argillaceous shale (type VII), where the average fracture density is only 0.35 m^{-1}. This indicates that mineral composition is the most important factor controlling fracture development, e.g., quartz and carbonate mineral contents determine fracture development by controlling rock brittleness. Single mechanical layers in laminar shale are lower in thickness, where the fracture density is obviously higher in comparison with massive shale. Under the same conditions, a high TOC can result in a higher fracture density in shale.

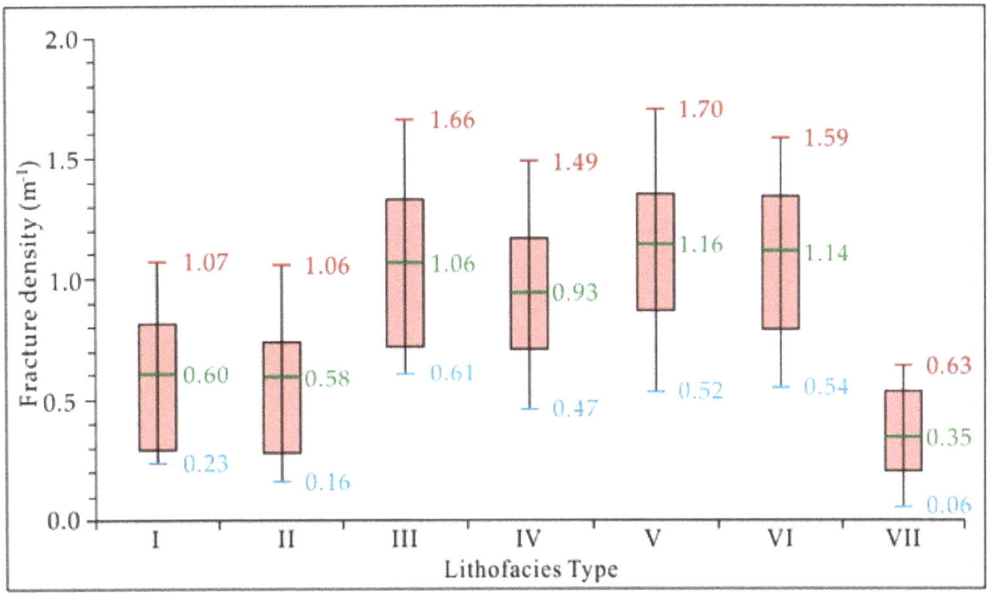

Figure 11. Fracture density in different lithofacies: Type I: high TOC massive siliceous shale; type II: medium TOC massive siliceous shale; type III: medium TOC laminated siliceous shale; type IV: low TOC laminated siliceous shale; type V: low TOC lamelleted siltstone; type VI: low TOC lamelleted limestone; and type VII: high TOC massive argillaceous shale. The short red, green, and blue lines represent the maximum, average, and minimum fracture densities, respectively.

5.4. Fault

It is generally recognized that a fault zone has a dual structure, including a fault core and a surrounding damage zone [55,56]. The fault core is generally composed of the sliding surface, fault gouge, breccia, and tectonic lens, while the damage zone is generally composed of fractures of different scales and secondary faults [57–59]. Fault-controlling stress distribution in different positions is an important external factor for determining fracture development in the Qingshankou Formation. For example, an obvious stress concentration along a fault zone increases the fracture density significantly.

Four small faults can be found at two outcrops. Figure 12a shows two small, normal faults at the Niaohetun outcrop, forming a small graben structure. The first fault (F1) is 20 cm in throw with azimuth and dip angle of 235° and 42°, respectively. The second fault (F2) is 10 cm in throw with azimuth and dip angle of 52° and 42°, respectively. These faults can be divided into fault cores and damage zones, while the fault core is composed of fault breccia (Figure 12b) and structural lens (Figure 12c) with a thickness of 5–10 cm, which is a highly permeable seepage channel. A large number of fractures have developed in the damage zone. Fracture density decreases with the increase in distance from the fault core.

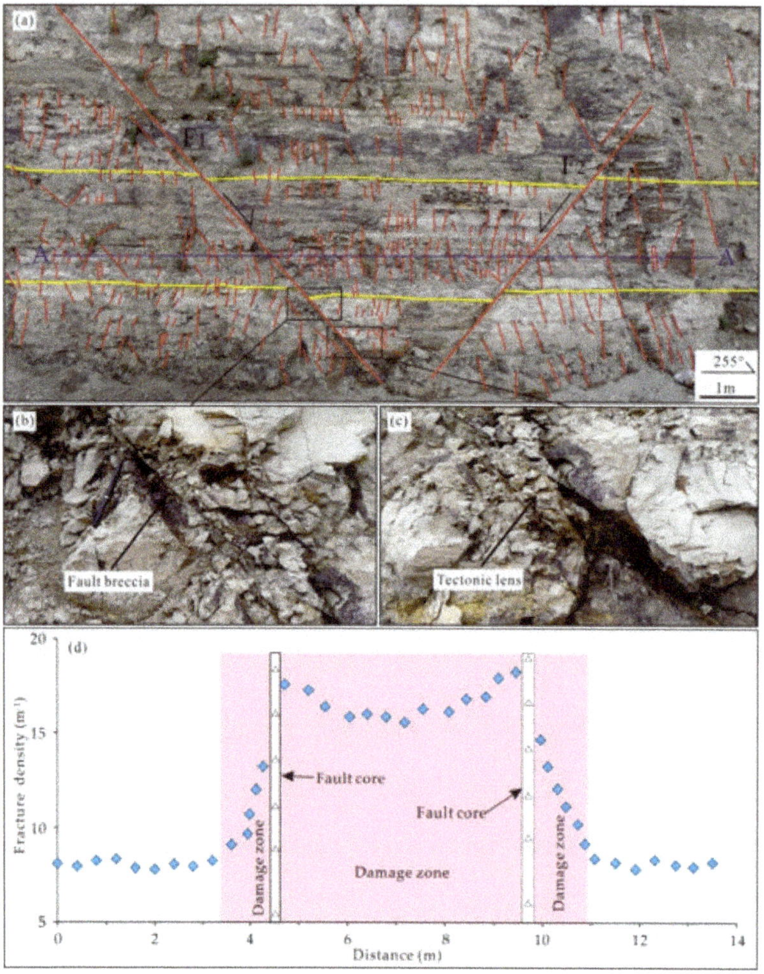

Figure 12. Two small, normal faults at Niaohetun outcrop in Bin County, Haerbin City: (**a**) Photos of these two small, normal faults; (**b**) fault breccia; (**c**) tectonic lens; (**d**) internal structure of fault zone and fracture distribution.

Fracture spacing and density around faults on an outcrop show that fractures have high density and little spacing around fault cores, especially at fault tips, intersections, and overlaps. Fracture density follows an exponentially decreasing trend with an increase in the distance from the fault core, which tends to be stable at a certain distance (Figure 12d). The boundary of the damage zone can be identified at the position where the fracture density is equal to the regional fracture density. The width of the damage zone is proportional to the fault displacement. However, the growth rate of the damage zone's width decreases when the displacement is higher than a certain value (after a few hundred meters) [60,61]. In addition, a difference in fracture density can be found between two walls; specifically, the fracture density of the hanging wall is significantly higher than that of the foot wall [62]. This can be explained by the fact that the hanging wall is commonly the active one, where stress disturbance is more obvious.

5.5. Abnormal High Pressure

Drilling and log data indicate significant abnormally high pressure levels in the Qingshankou Formation, which are caused by under-compaction, hydrocarbon generation, and clay mineral transformation dehydration [2,63]. Abnormally high pressure is an important driving force for fracture development and has an important influence on fracture distribution in shale [64–66].

The existence of abnormal fluid pressure in a reservoir decreases the effective stress when the differential stress is large, moving the stress Mohr circle to the left to intersect with the failure envelope, facilitating the formation of shear fractures (Figure 13, big Mohr circle). Statistical analysis shows a good positive correlation between the fracture density and the formation pressure coefficient, which indicates that the presence of abnormally high pressure contributes to fracture development. This can be explained by the fact that the minimum effective principal stress can vary from compression stress to tensile stress with an increasing pore fluid pressure under small differential stress (Figure 13, small Mohr circle), thus developing tensile fracture.

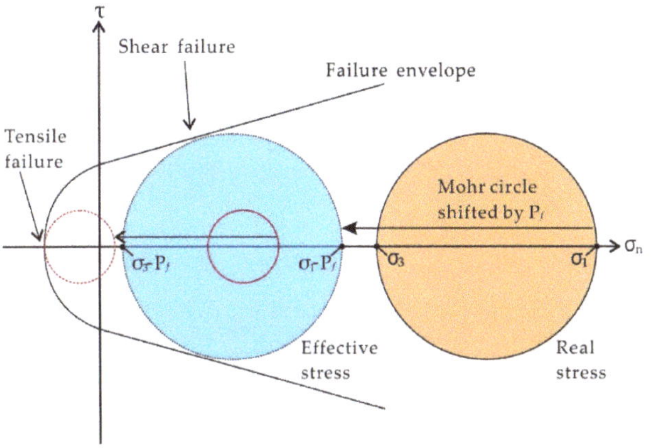

Figure 13. Schematic diagram showing formation mechanism for overpressure fractures (modified from Bons et al. [63] and Gong et al. [64]).

6. Conclusions

Strata-confined fractures, throughgoing fractures, bedding-parallel fractures, and stylolites were identified in Qingshankou shale. Mineral composition, abnormally high pressure, tectonic position, and mechanical stratigraphy are primarily factors controlling fracture development in shale. High levels of quartz and carbonate mineral content can result in high levels of brittleness, which are beneficial for natural fracture development, while clay minerals are unfavorable to natural fracture development. Fracture development is negatively proportional to compressive strength and Poisson's ratio, and it is positively proportional to Young's modulus and the volume modulus. The existence of abnormally high pressure moves the stress Mohr circle to the left with minimal principal stress becoming a negative value, resulting in tensile fractures. Mechanical stratigraphy is another factor controlling fracture formation and distribution, e.g., fracture spacing will increase and fracture density will decrease with increases in the mechanical layer thickness. Also, fracture density will decrease with increases in the distance from fault cores. The boundary of the damage zone can be identified at the position where the fracture density is equal to the regional fracture density, which is generally affected by fault size and activity intensity.

Author Contributions: X.F.: conceptualization and investigation; L.G.: conceptualization and writing—review and editing; X.S.: data curation; B.L.: formal analysis; S.G.: methodology and original draft preparation; J.Y.: investigation; X.Q.: writing—review and editing. All authors have read and agreed to the published version of the manuscript.

Funding: This research was funded by the National Natural Science Foundation of China (grant nos. U20A2093, 42072155, and 41902150); the Natural Science Foundation of Heilongjiang Province (grant no. YQ2022D006); the Young Innovative Talents Training Program for Universities in Heilongjiang Province (grant no. UNPYSCT-2020147); and the Postdoctoral Scientific Research Developmental Fund of Heilongjiang Province (grant no. LBH-Q21001). The APC was funded by the National Natural Science Foundation of China (grant no. 42072155).

Data Availability Statement: The data involved in this paper are all included in the text of the manuscript.

Acknowledgments: The authors thank Jianguo Huang and Jiaqi Yao at Northeast Petroleum University for their constructive help.

Conflicts of Interest: The authors declare that they have no conflict of interest.

References

1. Gale, J.F.W.; Olson, J.E.; Laubach, S.E. Natural fractures in shale: A review and new observations. *AAPG Bull.* **2014**, *98*, 2165–2216. [CrossRef]
2. Liu, B.; Shi, J.X.; Fu, X.F.; Lyu, Y.F.; Sun, X.D.; Gong, L.; Bai, Y.F. Petrological characteristics and shale oil enrichment of lacustrine fine-grained sedimentary system: A case study of organic-rich shale in first member of Cretaceous Qingshankou Formation in Gulong Sag, Songliao Basin, NE China. *Pet. Explor. Dev.* **2018**, *45*, 884–894. [CrossRef]
3. Luo, Q.Y.; Gong, L.; Qu, Y.S.; Zhang, K.H.; Zhang, G.L.; Wang, S.Z. The tight oil potential of the Lucaogou Formation from the southern Junggar Basin, China. *Fuel* **2018**, *234*, 858–871. [CrossRef]
4. Gao, S.; Zeng, L.B.; Ma, S.Z.; He, Y.H.; Gong, L.; Zhao, X.Y.; Xu, W.G.; Tang, X.M. Quantitative prediction of fractures withdifferent directions in tight sandstone reservoirs J. *Nat. Gas Geosci.* **2015**, *26*, 427–434.
5. Luo, Q.R.; Goodarzi, F.; Zhong, N.N.; Wang, Y.; Qiu, N.S.; Skovsted, C.; Such, V.; Schovsbo, N.H.; Morga, R.; Xu, Y.H.; et al. Graptolites as fossil geo-thermometers and source material of hydrocarbons: An overview of four decades of progress. *Earth-Sci. Rev.* **2020**, *200*, 103000. [CrossRef]
6. Li, Y.G.; Gong, L.; Zeng, L.B.; Ma, L.H.; Yang, H.; Zang, B.J.; Zhu, K.W. Characteristics of fractures and their contribution to the deliverability of tight conglomerate reservoirs in the Jiulongshan Structure, Sichuan Basin J. *Nat. Gas Ind.* **2012**, *32*, 22–26.
7. Gottardi, R.; Mason, S.L. Characterization of the natural fracture system of the Eagle Ford Formation (Val Verde County, Texas). *AAPG Bull.* **2018**, *102*, 1963–1984. [CrossRef]
8. Ladevèze, P.; Séjourné, S.; Rivard, C.; Lavoie, D.; Lefebvre, R.; Rouleau, A. Defining the natural fracture network in a shale gas play and its cover succession: The case of the Utica Shale in eastern Canada. *J. Struct. Geol.* **2018**, *108*, 157–170. [CrossRef]
9. Shovkun, I.; Espinoza, D.N. Geomechanical implications of dissolution of mineralized natural fractures in shale formations. *J. Pet. Sci. Eng.* **2018**, *160*, 555–564. [CrossRef]
10. Zeng, L.B.; Lyu, W.Y.; Li, J.; Zhu, L.F.; Weng, J.Q.; Yue, F.; Zu, K.W. Natural fractures and their influence on shale gas enrichment in Sichuan Basin, China. *J. Nat. Gas Sci. Eng.* **2016**, *30*, 1–9. [CrossRef]
11. Smart, K.J.; Ofoegbu, G.I.; Morris, A.P.; McGinnis, R.N.; Ferrill, D.A. Geomechanical modeling of hydraulic fracturing: Why mechanical stratigraphy, stress state, and pre-existing structure matter. *AAPG Bull.* **2014**, *98*, 2237–2261. [CrossRef]
12. Gale, J.F.W.; Gale, R.M.R.; Holder, J. Natural fractures in the Barnett Shale and their importance for hydraulic fracture treatments. *AAPG Bull.* **2007**, *91*, 603–622. [CrossRef]
13. Luo, Q.; Zhong, N.; Dai, N.; Zhang, W. Graptolite-derived organic matter in the Wufeng–Longmaxi Formations (Upper Ordovician–Lower Silurian) of southeastern Chongqing, China: Implications for gas shale evaluation. *Int. J. Coal Geol.* **2016**, *153*, 87–98. [CrossRef]
14. Yin, S.L.; Feng, K.Y.; Nie, X.; Chen, Q.; Liu, Y.; Wang, P.L. Characterization of marine shale in Western Hubei Province based on unmanned aerial vehicle oblique photographic data. *Adv. Geo-Energy Res.* **2022**, *6*, 252–263. [CrossRef]

15. Li, B.K.; Nie, X.; Cai, J.C.; Zhou, X.Q.; Whang, C.C.; Han, D.L. U-Net model for multi-component digital rock modeling of shales based on CT and QEMSCAN images. *J. Pet. Sci. Eng.* **2022**, *216*, 110734. [CrossRef]
16. Liu, B.; Sun, J.H.; Zhang, Y.Q.; He, J.L.; Fu, X.F.; Yang, L.; Xing, J.J.; Zhao, X.Q. Reservoir space and enrichment model of shale oil in the first member of Cretaceous Qingshankou Formation in the Changling sag, southern Songliao Basin, NE China. *Pet. Explor. Dev.* **2021**, *48*, 608–624. [CrossRef]
17. Gong, L.; Wang, J.; Gao, S.; Fu, X.F.; Liu, B.; Miao, F.B.; Zhou, F.B.; Meng, Q.K. Characterization, controlling factors and evolution of fracture effectiveness in shale oil reservoirs. *J. Pet. Sci. Eng.* **2021**, *203*, 108655. [CrossRef]
18. Laubach, S.E.; Lander, R.H.; Criscenti, L.J.; Anovitz, L.M.; Urai, J.L.; Pollyea, R.M.; Hooker, J.N.; Narr, W.; Evans, M.A.; Kerisit, S.N.; et al. The Role of Chemistry in Fracture Pattern Development and Opportunities to Advance Interpretations of Geological Materials. *Rev. Geophys.* **2019**, *57*, 1065–1111. [CrossRef]
19. Gong, L.; Gao, S.; Fu, X.F.; Chen, S.M.; Lyu, B.Y.; Yao, J.Q. Fracture characteristics and their effects on hydrocarbon migration and accumulation in tight volcanic reservoirs: A case study of the Xujiaweizi fault depression, Songliao Basin, China. *Interpretation* **2017**, *5*, 57–70. [CrossRef]
20. Zhu, X.M.; Zeng, H.L.; Li, S.L.; Dong, Y.L.; Zhu, S.F.; Zhao, D.N.; Huang, W. Sedimentary characteristics and seismic geomorphologic responses of a shallow-water delta in the Qingshankou Formation from the Songliao Basin, China. *Mar. Pet. Geol.* **2017**, *79*, 131–148. [CrossRef]
21. Liu, B.; Bai, H.L.; Chi, Y.A.; Jia, R.; Fu, X.F.; Yang, L. Geochemical characterization and quantitative evaluation of shale oil reservoir by two-dimensional nuclear magnetic resonance and quantitative grain fluorescence on extract: A case study from the Qingshankou Formation in Southern Songliao Basin, northeast. *Mar. Pet. Geol.* **2019**, *109*, 561–573. [CrossRef]
22. Huo, Q.; Zeng, H.; Zhang, X.C.; Fang, Q.H.; Wang, Z.Y.; Fu, L. Shale Oil Occurrence and Reservoir Characteristics of the Qijia-Gulong Depression in Songliao Basin. *Acta Geol. Sin.-Engl. Ed.* **2015**, *89*, 151–153. [CrossRef]
23. Marrett, R.; Gale, J.F.W.; Gómez, L.A.; Laubach, S.E. Correlation analysis of fracture arrangement in space. *J. Struct. Geol.* **2018**, *108*, 16–33. [CrossRef]
24. Hooker, J.N.; Laubach, S.E.; Marrett, R. Fracture-aperture size–frequency, spatial distribution, and growth processes in strata-bounded and non-strata-bounded fractures, Cambrian Mesón Group, NW Argentina. *J. Struct. Geol.* **2013**, *54*, 54–71. [CrossRef]
25. Li, J.Z.; Laubach, S.E.; Gale, J.F.W.; Marrett, R. Quantifying opening-mode fracture spatial organization in horizontal wellbore image logs, core and outcrop: Application to Upper Cretaceous Frontier Formation tight gas sandstones, USA. *J. Struct. Geol.* **2018**, *108*, 137–156. [CrossRef]
26. Gale, J.F.W.; Elliott, S.J.; Rysak, B.G.; Ginn, C.L.; Zhang, N.; Myers, R.D.; Laubach, S.E. Fracture description of the HFTS-2 slant core, Delaware Basin, West Texas. In Proceedings of the Unconventional Resources Technology Conference, Houston, TX, USA, 26 July 2021; Volume 6, pp. 26–28.
27. Finn, M.D.; Gross, M.R.; Eyal, Y.; Draper, G. Kinematics of throughgoing fractures in jointed rocks. *Tectonophysics* **2003**, *376*, 151–166. [CrossRef]
28. Gong, L.; Gao, M.Z.; Zeng, L.B.; Fu, X.F.; Gao, Z.Y.; Gao, A.; Zhu, K.W.; Yao, J.Q. Controlling factors on fracture development in the tight sandstone reservoirs: A case study of Jurassic-Neogeneinthe Kuqa foreland basin. *J. Nat. Gas Geosci.* **2017**, *28*, 199–208.
29. Cooke, M.L.; Simo, J.A.; Underwood, C.A.; Rijken, P. Mechanical stratigraphic controls on fracture patterns within carbonates and implications for groundwater flow. *Sediment. Geol.* **2006**, *184*, 225–239. [CrossRef]
30. Gross, M.R.; Eyal, Y. Throughgoing fractures in layered carbonate rocks. *GSA Bull.* **2007**, *119*, 1387–1404. [CrossRef]
31. Zhu, H.Y.; Song, Y.J.; Tang, X.H. Research progress on 4-dimensional stress evolution and complex fracture propagation of infill wells in shale gas reservoirs. *Pet. Sci. Bull.* **2021**, *3*, 396–416.
32. Zeng, L.B.; Zhu, R.K.; Gao, Z.Y.; Gong, L.; Liu, G.P. Structural diagenesis and its petroleum geological significance. *Pet. Sci. Bull.* **2016**, *2*, 191–197.
33. Ding, W.L.; Dai, P.; Zhu, D.W.; Zhang, Y.Q.; He, J.H.; Li, A.; Wang, R.Y. Fractures in continental shale reservoirs: A case study of the Upper Triassic strata in the SE Ordos Basin, Central China. *Geol. Mag.* **2016**, *153*, 663–680. [CrossRef]
34. Wang, W.D.; Zheng, D.; Sheng, G.L.; Zhang, Q.; Su, Y.L. A review of stimulated reservoir volume characterization for multiple fractured horizontal well in unconventional reservoirs. *Adv. Geo-Energy Res.* **2017**, *1*, 54–63. [CrossRef]
35. Lyu, W.Y.; Zeng, L.B.; Zhang, B.J.; Miao, F.B.; Lyu, P.; Dong, S.Q. Influence of natural fractures on gas accumulation in the Upper Triassic tight gas sandstones in the northwestern Sichuan Basin, China. *Mar. Pet. Geol.* **2017**, *83*, 60–72. [CrossRef]
36. Gong, L.; Fu, X.F.; Wang, Z.S.; Gao, S.; Jabbari, H.D.; Yue, W.T.; Liu, B. A new approach for characterization and prediction of natural fracture occurrence in tight-oil sandstones with intense anisotropy. *AAPG Bull.* **2019**, *103*, 1383–1400. [CrossRef]
37. Liu, B.; Song, Y.B.; Zhu, K.; Su, P.; Ye, X.; Zhao, W.C. Mineralogy and element geochemistry of salinized lacustrine organic-rich shale in the middle Permian Santanghu Basin: Implications for paleoenvironment, provenance, tectonic setting and shale oil potential. *Mar. Pet. Geol.* **2020**, *120*, 104569. [CrossRef]
38. Zeng, L.B.; Tang, X.M.; Wang, T.C.; Gong, L. The influence of fracture cements in tight Paleogene saline lacustrine carbonate reservoirs, Western Qaidam Basin, Northwest China. *AAPG Bull.* **2020**, *96*, 2003–2017. [CrossRef]
39. Gale, J.F.W.; Fall, A.; Yurchenko, I.A.; Ali, W.A.; Laubach, S.E.; Eichhubl, P.; Bodnar, R.J. Opening-mode fracturing and cementation during hydrocarbon generation in shale: An example from the Barnett Shale, Delaware Basin, West Texas. *AAPG Bull.* **2022**, *106*, 2103–2141.

40. Aydin, A. Failure modes of shales and their implications for natural and man-made fracture assemblages. *AAPG Bull.* **2014**, *98*, 2391–2409. [CrossRef]
41. Wang, R.Y.; Gu, Y.; Ding, W.L.; Gong, D.J.; Yin, S.; Wang, X.H.; Zhou, X.H.; Li, A.; Xiao, Z.K.; Cui, Z.X. Characteristics and dominant controlling factors of organic-rich marine shales with high thermal maturity: A case study of the Lower Cambrian Niutitang Formation in the Cen'gong block, southern China J. *J. Nat. Gas Sci. Eng.* **2016**, *33*, 81–96. [CrossRef]
42. Laubach, S.E.; Olson, J.E.; Gross, M.R. Mechanical and fracture stratigraphy. *AAPG Bull.* **2009**, *93*, 1413–1426. [CrossRef]
43. Ferrill, D.A.; McGinnis, R.N.; Morris, A.P.; Smart, K.J.; Sickmann, Z.T.; Bentz, M.; Lehrmann, D.; Evans, M.A. Control of mechanical stratigraphy on bed-restricted jointing and normal faulting. *Eagle Form. South-Cent. Tex. AAPG Bull.* **2014**, *98*, 2477–2506.
44. Zeng, L.B.; Su, H.; Tang, X.M.; Peng, Y.M.; Gong, L. Fractured tight sandstone oil and gas reservoirs: A new play type in the Dongpu depression, Bohai Bay Basin, China. *AAPG Bull.* **2013**, *97*, 363–377. [CrossRef]
45. Helgeson, D.E.; Aydin, A. Characteristics of joint propagation across layer interfaces in sedimentary rocks. *J. Struct. Geol.* **1991**, *13*, 897–911. [CrossRef]
46. Lamarche, J.A.; Lavenu, P.C.; Gauthier, B.D.M.; Guglielmi, Y.; Jayet, O. Relationships between fracture patterns, geodynamics and mechanical stratigraphy in Carbonates (South-East Basin, France). *Tectonophysics* **2012**, *581*, 231–245. [CrossRef]
47. Petrie, E.S.; Evans, J.P.; Bauer, S.J. Failure of cap-rock seals as determined from mechanical stratigraphy, stress history, and tensile-failure analysis of exhumed analogs. *AAPG Bull.* **2014**, *98*, 2365–2389. [CrossRef]
48. Ferrill, D.A.; Morris, A.P.; McGinnis, R.N.; Smart, K.J.; Wigginton, S.S.; Hill, N.J. Mechanical stratigraphy and normal faulting. *J. Struct. Geol.* **2017**, *94*, 275–302. [CrossRef]
49. Corradetti, A.; Tavani, S.; Parente, M.; Iannace, A.; Vinci, F.; Pirmez, C.; Torrieri, S.; Giorgioni, M.; Pignalosa, A.; Mazzoli, S. Distribution and arrest of vertical through-going joints in a seismic-scale carbonate platform exposure (Sorrento peninsula, Italy): Insights from integrating field survey and digital outcrop model. *J. Struct. Geol.* **2018**, *108*, 121–136. [CrossRef]
50. Procter, A.; Sanderson, D.J. Spatial and layer-controlled variability in fracture networks. *J. Struct. Geol.* **2018**, *108*, 52–65. [CrossRef]
51. Bai, T.; Pollard, D.D. Fracture spacing in layered rocks: A new explanation based on the stress transition. *J. Struct. Geol.* **2000**, *22*, 43–57. [CrossRef]
52. Wang, R.Y.; Hu, Z.Q.; Liu, J.T.; Wang, X.H.; Gong, D.J.; Yang, T. Comparative analysis of characteristics and controlling factors of fractures inmarine and continental shales:A case study of the Lower Cambrian inCengong area, northern Guizhou Province. *J. Oil Gas Geol.* **2018**, *39*, 631–640.
53. Cao, D.S.; Zeng, L.B.; Lyu, W.Y.; Xu, X.; Tian, H. Progress in brittleness evaluation and prediction methods in unconventional reservoirs. *Pet. Sci. Bull.* **2021**, *01*, 31–45.
54. Agosta, F.; Wilson, C.; Aydin, A. The role of mechanical stratigraphy on normal fault growth across a Cretaceous carbonate multi-layer, central Texas (USA). *Ital. J. Geosci.* **2015**, *134*, 423–441. [CrossRef]
55. Zeng, L.B.; Wang, H.J.; Gong, L.; Li, B.M. Impacts of the tectonic stress field on natural gas migration and accumulation: A case study of the Kuqa Depression in the Tarim Basin, China. *Mar. Pet. Geol.* **2010**, *27*, 1616–1627. [CrossRef]
56. Wang, R.Y.; Hu, Z.Q.; Zhou, T.; Bao, H.Y.; Wu, J.; Du, W.; He, J.H.; Wang, W.P.; Chen, Q. Characteristics of fractures and their significance for reservoirs in Wufeng-Longmaxi shale, Sichuan Basin and its periphery. *J. Oil Gas Geol.* **2021**, *42*, 1295–1306.
57. Choi, J.; Edwards, P.; Ko, K.; Kim, Y. Definition and classification of fault damage zones: A review and a new methodological approach. *Earth-Sci. Rev.* **2016**, *152*, 70–87. [CrossRef]
58. Dimmen, V.; Rotevatn, A.; Peacock, D.C.P.; Nixon, C.W.; Nærland, K. Quantifying structural controls on fluid flow: Insights from carbonate-hosted fault damage zones on the Maltese Islands. *J. Struct. Geol.* **2017**, *101*, 43–57. [CrossRef]
59. Peacock, D.C.; Dimmen, P.V.; Rotevatn, A.; Sanderson, D.J. A broader classification of damage zones. *J. Struct. Geol.* **2017**, *102*, 179–192. [CrossRef]
60. Berg, S.S.; Skar, T. Controls on damage zone asymmetry of a normal fault zone: Outcrop analyses of a segment of the Moab fault, SE Utah. *J. Struct. Geol.* **2005**, *27*, 1803–1822. [CrossRef]
61. Faulkner, D.R.; Jackson, C.A.L.; Lunn, R.J.; Schlische, R.W.; Shipton, Z.K.; Wibberley, C.A.J.; Withjack, M.O. A review of recent developments concerning the structure, mechanics and fluid flow properties of fault zones. *J. Struct. Geol.* **2010**, *32*, 1557–1575. [CrossRef]
62. Torabi, A.; Berg, S.S. Scaling of fault attributes: A review. *Mar. Pet. Geol.* **2011**, *28*, 1444–1460. [CrossRef]
63. Bons, P.D.; Elburg, M.A.; Gomez-Rivas, E. A review of the formation of tectonic veins and their microstructures. *J. Struct. Geol.* **2012**, *43*, 33–62. [CrossRef]
64. Gong, L.; Su, X.C.; Gao, S.; Fu, X.F.; Jabbari, H.; Wang, X.X.; Liu, B.; Yue, W.T.; Wang, Z.S.; Gao, A. Characteristics and formation mechanism of natural fractures in the tight gas sandstones of Jiulongshan Gas Field, China. *J. Pet. Sci. Eng.* **2019**, *175*, 1112–1121. [CrossRef]
65. Gudmundsson, A.; Kusumoto, S.; Simmenes, T.H.; Philipp, S.L.; Larsen, B.; Lotveit, I.F. Effects of overpressure variations on fracture apertures and fluid transport. *Tectonophysics* **2012**, *581*, 220–230. [CrossRef]
66. Zeng, L.B.; Gong, L.; Guan, C.; Zhang, B.J.; Wang, Q.Q.; Zeng, Q.; Lyu, W.Y. Natural fractures and their contribution to tight gas conglomerate reservoirs: A case study in the northwestern Sichuan Basin, China. *J. Pet. Sci. Eng.* **2022**, *210*, 110028. [CrossRef]

MDPI AG
Grosspeteranlage 5
4052 Basel
Switzerland
Tel.: +41 61 683 77 34

MDPI Books Editorial Office
E-mail: books@mdpi.com
www.mdpi.com/books

www.ingramcontent.com/pod-product-compliance
Lightning Source LLC
LaVergne TN
LVHW070503061126
851858LV00011B/284